Nonlinear Optics
of
Organic Molecules
and Polymers

Edited by

Hari Singh Nalwa
Hitachi Research Laboratory
Hitachi Ltd.
Hitachi City, Ibaraki, Japan

Seizo Miyata
Tokyo University of Agriculture and Technology
Kogenei, Tokyo, Japan

CRC Press
Taylor & Francis Group
Boca Raton London New York

CRC Press is an imprint of the
Taylor & Francis Group, an **informa** business

CRC Press
Taylor & Francis Group
6000 Broken Sound Parkway NW, Suite 300
Boca Raton, FL 33487-2742

© 1997 by Taylor & Francis Group, LLC
CRC Press is an imprint of Taylor & Francis Group, an Informa business

First issued in paperback 2019

No claim to original U.S. Government works

ISBN 13: 978-0-367-44837-0 (pbk)
ISBN 13: 978-0-8493-8923-8 (hbk)

Visit the Taylor & Francis Web site at
http://www.taylorandfrancis.com

and the CRC Press Web site at
http://www.crcpress.com

Acquiring Editor: *Tim Pletscher*
Senior Project Editor: *Susan Fox*
Cover Design: *Jason Toemmes*

Library of Congress Card Number 96-20281

Library of Congress Cataloging-in-Publication Data

Nonlinear optics of organic molecules and polymers / edited by Hari Singh Nalwa, Seizo Miyata.
    p.   cm.
  Includes bibliographical references and index.
  ISBN 0-8493-8923-2 (alk. paper)
  1. Optoelectronics—Materials. 2. Nonlinear optics—Materials. 3. Organic compounds—Optical properties. 4. Polymers—Optical properties. I. Nalwa, Hari Singh, 1954- . II. Miyata, Seizo, 1941- .
TA1750.N658 1996
621.36′9—dc20   96-20281
                                                                                              CIP

PREFACE

The field of nonlinear optics emerged three decades ago with the development of the first operating laser and the demonstration of frequency doubling phenomena. These milestone discoveries not only generated much interest in laser science, but also set the stage for future work on nonlinear optics. The invention of lasers led to the discovery of interesting nonlinear optical phenomena in inorganic materials and the development of structure–property relationships. Nonlinear optical effects in organic materials were reported in the early 1970s, and the importance of novel materials was realized through theory, models, and synthesis. The extraordinary growth and development of nonlinear optical materials during the past decade has rendered photonic technologies an indispensable part of our daily life as we enter the 21st century of the "Information Age." Our society will be benefited from advanced computer networks and telecommunication systems that would bring photonics technology from infancy to maturity probably within a decade. Recognizing the importance of photonic technology, this book covers all aspects of nonlinear optics and contains a wealth of information related to computational analysis, techniques, materials, and devices. With the emerging demand for information systems, nonlinear optical materials have been considered the key elements for the future photonic technologies of optical computing, telecommunications, optical interconnects, high density data storage, sensors, image processing, and switching. Considerable effort has been made to develop photonic devices from nonlinear optical organic materials in research laboratories around the world.

The early texts dealt with the problems of setting up the targets of materials and devices through identifying promising nonlinear optical materials for industrial applications, and understanding structure–property relationships. Finding answers to such problems stimulated much research work in novel materials. These problems have been extensively addressed and views on nonlinear optical materials and their based devices are clear now. The rapid growth of this field is evidenced by the several-fold increase in the number of papers and patents published on nonlinear optics in the past few years. Organic materials with sufficiently large optical nonlinearities have been identified to be practical in a variety of photonic devices.

This book, written by leading experts in industry and academia, covers all aspects of nonlinear optics in an in-depth and comprehensive manner; the scope ranges from modern computational analysis of optical nonlinearities, through description of experimental techniques for measuring nonlinear optical coefficients, and provides an in-depth coverage of advances in organic molecules and polymers. This comprehensive text offers an interdisciplinary approach toward correlating molecular engineering with optimization of optical nonlinearities. The field is multidisciplinary, as it needs expertise in computing, chemistry, physics, material science, and optical engineering. During the past decade, the field of nonlinear optics has developed remarkably and, in particular, there has been significant growth in the past 6 to 7 years from the synthetic chemistry viewpoint. A wide range of excellent optical quality single crystals is now available, through there is an increasing trend towards the use of organic polymers as photonic components because of their ease of processing and fabrication; compatibility with metals, ceramics, semiconductors; and glasses; good mechanical strength and flexibility to tailor nonlinear optical properties; and material performance to specific end uses which make them fascinating materials for applications. As a result, much progress on the materials front has been made since the recognition of polymeric materials. This book illustrates in a clear and concise way the structure–property relationship necessary to understand a wider range of organic and organometallic materials with exciting potential for future photonic technologies. It emphasizes how novel organic superstructures can be tailored to optimize nonlinear optical responses by carefully viewing the desired functions originating from the chemical species. Owing to their versatility, organic materials have great potential because multifunctions can be introduced by applying chemical strategies.

The state of the art in the field of nonlinear optics is presented by some of the most renowned scientists in the world. An introduction to nonlinear optical phenomena is provided in Chapter 1 by Garito and co-workers. The in-depth theoretical treatments of second-order optical nonlinearities are presented in Chapter 2 by Morley and Pugh, in order to enable understanding of correlations between molecular structures and second-order optical nonlinearities. In Chapter 3, Watanabe et al. describe various measurement techniques for refractive index and, second-order nonlinear optical susceptibilities measurements with powder methods, the phase-matching method, the Maker-fringe method, the SHEW technique and molecular hyperpolarizabilities from the EFISH method and the HRS method. Nalwa et al. provide a complete coverage of second-order NLO materials in Chapter 4, where many second-order NLO materials, including single crystals, guest–host systems, NLO dye-functionalized polymers, polar polymers, Langmuir-Blodgett films, self-assembled multilayer systems, liquid crystals, and organometallic materials, are summarized. Applications of second-order NLO materials in photonic technologies are also described. Phase-matched second harmonic generation of organic materials with phase-matching techniques such as collinear and noncollinear phase matching in uniaxial and biaxial crystals and in optical waveguides, mode dispersion-type phase matching, collinear-type phase matching, Crenkov-type phase matching, phase matching in periodically poled waveguides, are discussed in Chapter 5 by Watanabe et al.

In Chapter 6, Bosshard and Günter describe various measurement techniques for determining electro-optic coefficients and summarize organic single crystals and polymers for electro-optical applications. Applications of EO polymers in optical modulators are discussed by Teng in Chapter 7. With the current upsurge of interest in electro-optic devices, one can foresee the ultimate applications in coming years. Future high-density optical storage systems that need short wavelengths are expected from organic NLO materials. In Chapter 8, Kippelen, Meerholz, and Peyghambarian describe measurement techniques and organic polymeric materials for photorefractive effects. Photorefractive materials are newly emerging photonic elements with capabilities to store many holograms at a data rate of storage of a hundred times faster than commercially available magnetic hard disks.

Nalwa provides the state-of-the-art reviews on quantum chemical approaches, measurement techniques, and organic materials for third-order nonlinear optics. Theoretical calculations of microscopic third-order optical nonlinearities using empirical, semiempirical, and *ab initio* methods, a comparison between different computational methods and theoretical and experimental results and the effects of nature of π-bonding sequence, delocalization length, donor–acceptor groups, conformation, symmetry, dimensionality and charge-transfer complex formation on optical nonlinearities have been dealt with in detail in Chapter 9. Chapter 10 describes the measurement techniques for third-order nonlinear optical susceptibilities and second hyperpolarizabilities using third-harmonic generation (THG), degenerate four-wave mixing (DFWM), four-wave mixing (FWM), optical Kerr gate (OKG), optical power limiter (OPL), and Z-scan techniques. Third-order nonlinear optical effects are an indispensable molecular property, therefore the studies vary to a great extent. Chapter 11 focuses on third-order NLO organic materials with complete coverage of liquids, molecular solids (dyes, charge-transfer complexes, fullerenes, hellicenes), π-conjugated polymers, NLO-chromophore grafted polymers, organometallic compounds, NLO dyes and semiconductors doped polymers and glass composites, liquid crystals, and biomaterials. Time-resolved spectroscopy to study the ultrafast nonlinear optical response and excited state dynamics are also briefly discussed. Applications of organic NLO materials in third-order nonlinear optics are discussed by Stegeman in Chapter 12. Perry provides an excellent description of optical limiting phenomena in organic materials such as fullerenes, metallo-phthalocyanines, porphyrins, and other dyes in Chapter 13. The applications of optical limiting materials are discussed in Chapter 14 by Van Stryland and co-workers.

Nonlinear optics is now established as one of the best alternatives to electronics for the future photonic technologies. This book presents an overview of the exciting new advances in nonlinear optical materials and their industrial applications. It will provide a useful reference for graduate and advanced undergraduate students as well as researchers, scientists, and engineers in solid-state physics, materials science, chemistry, electrical, optical, fiber and polymer engineering, and computational science.

The editors wish to express gratitude to all contributors for sharing their knowledge and expertise in the field of nonlinear optics by submitting excellent manuscripts which made this book a reality. Their efforts deserve appreciation as these contributors are the pillars of this novel foundation. At Hitachi Research Laboratory, Dr. H. S. Nalwa gratefully acknowledges the support of Dr. Akio Mukoh, a connoisseur of many aspects of nonlinear optics, who has inspired many of us in this field of science and who whole-heartedly encouraged the writing of this monograph. He would also like to thank Drs. Atsushi Kakuta, T. Iwayanagi, Akio Takahashi, M. Isogai, M. Sagawa, T. Nakayama, Y. Imanishi, T. Hamada, S. Ishihara, and R. Inaba at Hitachi Research Laboratory for their continuous support and cooperation. Drs. A. Ticktin, K. H. Hass, and A. Esser of BASF are especially acknowledged for their valuable comments and generosity in providing several original illustrations. Professor M. Poplodous kindly read the computational evaluation of third-order optical nonlinearity and provided useful comments. Dr. Nalwa is also very grateful to Professor Richard T. Keys of California State University at Los Angeles for critically reading some manuscripts and for his valuable comments and to Dr. Toshiyuki

Watanabe of Tokyo University of Agriculture and Technology for his tremendous help in compiling this monograph. Dr. Nalwa would also like to acknowledge Professor Padma Vasudevan, Professor Prem Vrat, and Professor R. P. Dahiya of Indian Institute of Technology in New Delhi and Professor Satya Vir Arya of C. C. R. College at Muzaffarnagar for their continuous support. Thanks are also due to Sri Ram Singh, Jiley Singh, Kadam Singh, Braham Singh, Sardar Singh, Jagmer Singh, Dharam Pal Singh, Ranvir Singh Chaudhary, Yash Pal Singh, Braj Pal Singh, Satyendra Singh, Ashish Kumar, Bhanwar Singh, and Arvind Kumar, and to his friends, Yogesh Malik, Krishi Pal Raghuvanshi, Rakesh Misra, Deepak Singhal, Michael McMurray, Dr. Dharam Pal Singh, Dr. V. B. Reddy, Professor G. K. Surya Prakash, and Dr. Claus Jessen, for their continuous encouragements. Dr. Nalwa would like to acknowledge the cooperation and patience of his wife Dr. Beena Singh Nalwa for her understanding while writing manuscripts during the evenings, weekends and holidays at home and in lessening entropy from my kids, Surya and Ravina. Dr. Nalwa gratefully acknowledges the editiorial support of Betsy Winship of Innodata Corporation. Finally, the assistance of the staff members at CRC Press who helped in this project is gratefully acknowledged.

Hari Singh Nalwa
Seizo Miyata

THE EDITORS

Hari Singh Nalwa is a Research Scientist in the Materials Department at the Hitachi Research Laboratory, Hitachi Ltd., Japan. Dr. Nalwa is also an Honorary Professor at the Indian Institute of Technology in New Delhi. He has edited several books: *Ferroelectric Polymers* (1995), *Organic Electroluminescent Materials and Devices* (1996), and *Handbook of Organic Conductive Molecules and Polymers* (1997), Vols. 1–4. He has authored over 100 scientific publications in leading refereed journals and has 18 Japanese patents either applied for or issued on electronics and photonics materials and devices. He serves on the editorial board of *Applied Organometallic Chemistry, Journal of Macromolecular Science-Physics,* and *Photonics Science News,* and is a member of the American Chemical Society and American Association for the Advancement of Science. He has been awarded a number of prestigious fellowships in India and abroad. He was a guest scientist at Hahn-Meitner Institute in Berlin, Germany (1983), Research Associate at University of Southern California in Los Angeles (1984–1987), and lecturer at Tokyo University of Agriculture and Technology in Tokyo, Japan (1988–1990). He has been working with Hitachi Ltd since 1990. Dr. Nalwa received a B. Sc. degree (1974) in biosciences from the Meerut University, a M. Sc. degree (1977) in organic chemistry from the University of Roorkee, and a Ph. D. degree (1983) in polymer science from the Indian Institute of Technology in New Delhi, India.

Seizo Miyata is the Dean and Professor of the Graduate School of Bio-applications and Systems Engineering at the Tokyo University of Agriculture and Technology. He has more than 200 patents and 150 scientific publications on electronics and photonics materials and devices to his credit. He has authored/edited several books: in Japanese: *Experimental Techniques of Polymeric Materials* (1981), *Chemistry of Polymeric Materials* (1982), *World of Modern Science* (1986), *Intelligent Materials* (1987), and *Photonics* (1995); and in English: *Organic Electroluminescent Materials and Devices* (1996). Professor Miyata serves on the editorial board of *Supramolecular Science, Polymer Journal,* and *International Journal of Nonlinear Optical Physics.* He is the Vice President of the Polymer Society of Japan, and is a member of the Japanese Science Council. He has organized and presided over 15 national and international conferences. He is a leader of the national project in Japan on lightwave manipulation using organic materials and belongs to the Ministry of Education in Japan, and he has received an award from the Polymer Society of Japan on the discovery of novel piezoelectric polymers. He was a guest professor at California Institute of Technology in Pasadena, USA (1982–1983) and a guest scientist at AT&T Bell Laboratories (1984). He has been working with Tokyo University of Agriculture and Technology since 1969. Professor Miyata received a B. Sc. degree (1964) in sciences from the Tokyo Educational University and a Ph. D. degree (1969) in polymer science from the Tokyo Institute of Technology.

CONTRIBUTORS

Christian Bosshard
Nonlinear Optics Laboratory
Institute of Quantum Electronics
Eidgenossische Technische Hochschule (ETH)
CH-8093 Zurich, Switzerland

A. F. Garito
Department of Physics and Astronomy
University of Pennsylvania
Philadelphia, Pennsylvania 19104-6396, USA

Peter Günter
Nonlinear Optics Laboratory
Institute of Quantum Electronics
Eidgenossische Technische Hochschule (ETH)
CH-8093 Zurich, Switzerland

D. J. Hagan
Center for Research and Education in Optics
    and Lasers (CREOL)
University of Central Florida
Orlando, Florida 32826, USA

B. Kippelen
Optical Sciences Center and
Materials Science and Engineering Department
University of Arizona
Tucson, Arizona 85721, USA

K. Meerholz
Optical Sciences Center and
Materials Science and Engineering Department
University of Arizona
Tucson, Arizona 85721, USA

Seizo Miyata
Graduate School of Bio-applications and
    Systems Engineering
Tokyo University of Agriculture and Technology
Koganei, Tokyo 184, Japan

John O. Morley
Chemistry Department
University of Wales, Swansea
Swansea, SA2 8PP, UK

Hari Singh Nalwa
Hitachi Research Laboratory, Hitachi Ltd.,
7-1-1 Ohmika-cho
Hitachi City, Ibaraki 319-12, Japan

Joseph W. Perry
Jet Propulsion Laboratory
California Institute of Technology
4800 Oak Grove Drive
Pasadena, California 91109, USA

N. Peyghambarian
Optical Sciences Center and
Materials Science and Engineering Department
University of Arizona
Tucson, Arizona 85721, USA

David Pugh
Department of Pure and Applied Chemistry
University of Strathclyde
Glasgow, G1 1XL, UK

A. A. Said
Center for Research and Education in Optics
    and Lasers (CREOL)
University of Central Florida
Orlando, Florida 32826, USA

George I. Stegeman
Center for Research and Education in Optics
    and Lasers (CREOL)
University of Central Florida
Orlando, Florida 32826, USA

Chia-Chi Teng
Hoechst Celanese Corporation
Robert L. Mitchell Technical Center
86 Morris Avenue
Summit, New Jersey 07901, USA

Eric W. Van Stryland
Center for Research and Education in Optics
    and Lasers (CREOL)
University of Central Florida
Orlando, Florida 32826, USA

Toshiyuki Watanabe
Department of Applied Chemistry
Tokyo University of Agriculture and Technology
Koganei, Tokyo 184, Japan

K. Y. Wong
Department of Physics
Chinese University of Hong Kong
Shatin, N. T., Hong Kong

T. Xia
Center for Research and Education in Optics
    and Lasers (CREOL)
University of Central Florida
Orlando, Florida 32826, USA

Y. Z. Yu
SEQ Ltd.
181 Cherry Valley Road
Princeton, New Jersey 08540, USA

CONTENTS

Chapter 1

Introduction to Nonlinear Optics

Y. Z. Yu, K. Y. Wong, and A. F. Garito

CONTENTS

Nonlinear optical processes in π-electron organic and polymer systems have attracted considerable interest because their understanding has led not only to compelling technological promise but also to new phenomena, new theoretical insights, and new materials and devices.[1] The π-electron excitations occurring on the individual molecular, or polymer chain, units are the basic origin of the observed nonresonant nonlinear optical coefficients which are oftentimes unusually large. The coefficients are broad band and ultrafast, and, as shown by theory and experiment, their sign, magnitude, and frequency dependence are determined by many-body electron correlation effects. This level of understanding, in turn, now makes viable computer-aided molecular design of new nonlinear optical chromophores.

In recent years as the field has naturally progressed toward technological applications, the main issues have focused on high-performance materials that comply with device manufacturing and end-use conditions. New challenges in materials synthesis are being presented, resulting in new methods of ultrastructure synthesis and the discovery of entirely new materials and high-performance compositions exhibiting high thermal, mechanical, and chemical stability. At the same time, there is increasing appreciation of the low-cost, practical materials processing associated with these materials, especially polymer thin films and fibers. These include ease and flexibility in synthesis, modification, and formulation; film deposition by spin coating, spraying, or dipping; and ease in fiber fabrication.

It has now been demonstrated on pilot plant scales that high-performance electro-optic polymer thin films can be routinely used in optoelectronic integrated circuit (OEIC) fabrication in existing microelectronic device manufacturing facilities. The key steps are standard, including spin coating, photolithography, etching, metallization, and multilayer assembly. High-optical-quality polymer optical fibers (POF) have long been available commercially for their linear optical properties, and the widespread use of optical-grade polymers in ophthalmic lenses, compact discs, and laser discs is well established.

I. MAXWELL'S EQUATIONS IN NONLINEAR OPTICAL MEDIA

To understand the nature of optical nonlinearity, one needs to start from the following Maxwell's equations, which are the general laws governing light (electromagnetic fields) interacting with matter.

$$\nabla \times \mathbf{E} = -\frac{1}{c}\frac{\partial \mathbf{B}}{\partial t}$$

$$\nabla \times \mathbf{B} = \frac{1}{c}\frac{\partial \mathbf{E}}{\partial t} + \frac{4\pi}{c}\mathbf{j} \tag{1}$$

$$\nabla \cdot \mathbf{E} = 4\pi\rho$$

$$\nabla \cdot \mathbf{B} = 0$$

0-8493-8923-2/97/$0.00+$.50

where $\mathbf{E}(\mathbf{r}, t)$ and $\mathbf{B}(\mathbf{r}, t)$ are the electric and magnetic fields, respectively. The relation between charge density ρ and current density $\mathbf{j}$ can be easily derived from Equation 1 as

$$\nabla \cdot \mathbf{j} + \frac{\partial \rho}{\partial t} = 0 \tag{2}$$

Equation 2 is commonly referred to as the law of charge conservation. In general, ρ and $\mathbf{j}$ can be expanded in terms of electric and magnetic multipole moments. Since optical media are generally nonmagnetic, the magnetic moment contribution to ρ and $\mathbf{j}$ can be neglected. Under electric dipole moment approximation, one can write ρ and $\mathbf{j}$ in the form

$$\rho = \rho_0 - \nabla \cdot \mathbf{P}$$

$$\mathbf{j} = \mathbf{j}_0 + \frac{\partial \mathbf{P}}{\partial t} \tag{3}$$

where ρ_0 and $\mathbf{j}_0$ are free electric charge and current densities, respectively, and $\mathbf{P}$ is the electric dipole moment per unit volume. We are primarily interested in a free medium, where there is no free electric charge and current so that

$$\rho_0 = 0$$

$$\mathbf{j}_0 = 0$$

The third Maxwell equation then becomes

$$\nabla \cdot \mathbf{E} + 4\pi \nabla \cdot \mathbf{P} = 0 \tag{4}$$

and the first two Maxwell equations simply combine into

$$\nabla^2 \mathbf{E} - \nabla(\nabla \cdot \mathbf{E}) - \frac{\partial^2 \mathbf{E}}{c^2 \partial t^2} - \frac{4\pi}{c^2} \frac{\partial^2 \mathbf{P}}{\partial t^2} = 0 \tag{5}$$

It is obvious that we need the information on the relationship between $\mathbf{P}$ and $\mathbf{E}$ to proceed further. This is where the optical nonlinearities are introduced. In general, the nonlinear responses are orders of magnitude smaller than the linear response and the electric polarization of a medium can be expanded, in frequency domain, according to the power of the applied $\mathbf{E}$ field:[2]

$$P_i(\omega) = \chi_i^{(0)} + \chi_{ij}^{(1)}(-\omega; \omega)E_j(\omega) + \chi_{ijk}^{(2)}(-\omega; \omega_1, \omega_2)E_j(\omega_1)E_k(\omega_2)$$
$$+ \chi_{ijkl}^{(3)}(-\omega; \omega_1, \omega_2, \omega_3)E_j(\omega_1)E_k(\omega_2)E_l(\omega_3) + \cdots \tag{6}$$

with $\omega = \omega_1 + \omega_2 + \omega_3 + \ldots$, where $\chi_{ijk}^{(n)} \ldots$ is the nth order optical susceptibility which is a $n + 1$ rank tensor. $\chi^{(0)}$ is the permanent zero-order susceptibility. $\chi^{(1)}$ is the linear susceptibility, often called the polarizability. All the higher order $\chi^{(n)}$ ($n > 1$) are the nonlinear optical responses of the medium. Summation over repeated indices (Einstein convention) is implied. The nonlinear optical responses of a medium are fully described by its nonlinear susceptibilities $\chi^{(n)}$, with $n > 1$. We need to point out that most of the materials in which we are interested do not have intrinsic dipole moments when no applied electric field is present. This means that $\chi_i^{(0)}$, the first term in Equation 6, usually vanishes. The medium responds to the applied optical/electric field in the following way. The optical field generates a nonlinear response of the material polarization $\mathbf{P}$ through Equation 6, which then, in turn, affects the optical field itself through Equation 4. With known $\chi^{(n)}$, the electrical field $\mathbf{E}$ is, at least in principle, solvable through Maxwell's equations.

It is worth mentioning that even-order nonlinear susceptibilities only exist in noncentrosymmetric media. This is a direct result of Equation 6. In centrosymmetric media, all of the susceptibility tensors are invariant under spatial inversion, and the polarization vector and applied field are all polar vectors which do change sign under spatial inversion operation. Let us pick the even-order nonlinear susceptibility and rewrite Equation 6 in the following form:

$$P_i^{(2n)}(\omega) = \chi_{ijk...2n}^{(2n)}(-\omega; \omega_1, \omega_2, \ldots, \omega_{2n})E_j(\omega_1)E_k(\omega_2) \cdots E_{2n}(\omega_{2n}) \tag{7}$$

Under spatial inversion, it becomes

$$
\begin{aligned}
-P_i^{(2n)}(\omega) &= \chi_{ijk...2n}^{(2n)}(-\omega; \omega_1, \omega_2, \ldots, \omega_{2n})[-E_j(\omega_1)][-E_k(\omega_2)] \cdots [-E_{2n}(\omega_{2n})] \\
&= (-1)^{2n}\chi_{ijk...2n}^{(2n)}(-\omega; \omega_1, \omega_2, \ldots, \omega_{2n})E_j(\omega_1)E_k(\omega_2) \cdots E_{2n}(\omega_{2n}) \\
&= \chi_{ijk...2n}^{(2n)}(-\omega; \omega_1, \omega_2, \ldots, \omega_{2n})E_j(\omega_1)E_k(\omega_2) \cdots E_{2n}(\omega_{2n}) \\
&= P_i^{(2n)}(\omega)
\end{aligned}
\tag{8}
$$

which indicates that $\chi_{ijk...2n}^{(2n)}(-\omega; \omega_1, \omega_2, \ldots \omega_{2n})$ must be identically zero. In other words, the even-order nonlinearities are forbidden in a medium with centroinversion symmetry. It should be clear that $\chi^{(n)}$ represents the macroscopic nonlinear susceptibility in a medium which consists of a large number of molecules that possess microscopic nonlinearities. While media with centrosymmetric molecules will certainly not have even-order optical nonlinear properties, media with noncentrosymmetric molecules exhibiting macroscopic centrosymmetry, such as cis-hexatriene (HT) in liquid solution, will also have no even-order nonlinear optical responses. Some experimental techniques can be used to break the macroscopic symmetry of the medium to measure its microscopic even-order nonlinearities due to individual molecules. An example of such techniques is the DC- (or electric field) induced second harmonic generation, commonly referred as DCSHG (or EFISH).[3-5]

Nonlinear optics research is mainly focused on the second- and third-order properties, with higher order effects only slightly touched. This is because higher order susceptibilities are usually several orders of magnitude smaller than those of lower orders, and are therefore very difficult to measure. For this reason, we would like to limit our discussions to up to third-order nonlinear responses. Although the second-order response is usually much stronger than the third-order response, $\chi^{(3)}$ will represent the leading nonlinearity in a medium with centroinversion symmetry.

We would like to discuss briefly the symmetry properties of nonlinear optical susceptibilities. The polarization induced through the second-order process can be written as

$$P_i^{(2)} = \chi_{ijk}^{(2)}(-\omega_3; \omega_1, \omega_2)E_j(\omega_1)E_k(\omega_2) \tag{9}$$

Since the product $E_j(\omega_1)E_k(\omega_2)$ is identical to $E_k(\omega_2)E_j(\omega_1)$, exchanging $E_j(\omega_1)$ and $E_k(\omega_2)$ in Equation 9 should have no effect on the induced polarization. This argument leads to the identity

$$\chi_{ijk}^{(2)}(-\omega_3; \omega_1, \omega_2) = \chi_{ikj}^{(2)}(-\omega_3; \omega_2, \omega_1) \tag{10}$$

In a lossless medium, the above symmetry can be extended to what is called *full* permutation symmetry, which states that *all* of the frequency arguments in a nonlinear susceptibility tensor are completely interchangeable, as long as the corresponding coordinate indices are interchanged accordingly. This full permutation symmetry is also referred to as Kleinman symmetry.[6] The Kleinman symmetry leads to the following additional identities for the second-order NLO susceptibility.

$$
\begin{aligned}
\chi_{ijk}^{(2)}(-\omega_3; \omega_1, \omega_2) &= \chi_{jik}^{(2)}(\omega_1; -\omega_3, \omega_2) = \chi_{jki}^{(2)}(\omega_1; \omega_2, -\omega_3) \\
&= \chi_{kij}^{(2)}(\omega_2; -\omega_3, \omega_1) = \chi_{kji}^{(2)}(\omega_2; \omega_1, -\omega_3)
\end{aligned}
\tag{11}
$$

Full permutation for the third-order NLO susceptibilities is listed below.

$$\chi_{ijkl}^{(3)}(-\omega_4; \omega_1, \omega_2, \omega_3) = \chi_{ikjl}^{(3)}(-\omega_4; \omega_2, \omega_1, \omega_3) = \chi_{ijlk}^{(3)}(-\omega_4; \omega_1, \omega_3, \omega_2)$$
$$= \chi_{ilkj}^{(3)}(-\omega_4; \omega_3, \omega_2, \omega_1) = \chi_{jikl}^{(3)}(\omega_1; -\omega_4, \omega_2, \omega_3)$$
$$= \chi_{kjil}^{(3)}(\omega_2; \omega_1, -\omega_4, \omega_3) = \chi_{ljki}^{(3)}(\omega_3; \omega_1, \omega_2, -\omega_4) \tag{12}$$
$$= \chi_{lkji}^{(3)}(\omega_3; \omega_2, \omega_1, -\omega_4)$$

Just as other physical quantities expressed in terms of tensors, nonlinear optical susceptibility tensors are closely related to material structure symmetry. Although tensors in general have a large number of components, the number of independent components are usually drastically reduced in reality due to material symmetry. A typical example is the proven result shown in Equation 8: all the even-order nonlinear susceptibilities are identical to zero in centrosymmetric media. Another common symmetry is isotropy for most materials in gas or liquid phase. In an isotropic medium, $\chi_{ijkl}^{(3)}$, for instance, has only three independent components from a total number of 81.

II. LIGHT PROPAGATION AND NONLINEAR INTERACTION IN OPTICAL MEDIA

Light propagation in an optical medium is essentially described by Equation 5. It is usually helpful to separate the polarization described in Equation 6 into a linear and a nonlinear part as

$$\mathbf{P} = \mathbf{P}^l + \mathbf{P}^{nl} \tag{13}$$

with

$$P_i^l(\omega) = \chi_i^{(0)} + \chi_{ij}^{(1)}(-\omega; \omega)E_j(\omega)$$

and

$$P_i^{nl}(\omega) = \chi_{ijk}^{(2)}(-\omega; \omega_1, \omega_2)E_j(\omega_1)E_k(\omega_2) + \chi_{ijkl}^{(3)}(-\omega; \omega_1, \omega_2, \omega_3)E_j(\omega_1)E_k(\omega_2)E_l(\omega_3) + \cdots$$

The linear part follows all the conventional treatments of linear electromagnetism

$$D_i = E_i + 4\pi P_i^{(1)} = E_i + 4\pi\chi_{ij}^{(1)}E_j = \varepsilon_{ij}E_j \tag{14}$$

and Equation 5, for linear optics and isotropic medium, is then reduced to

$$\nabla^2\mathbf{E} - \frac{\varepsilon}{c^2}\frac{\partial^2\mathbf{E}}{\partial t^2} = 0 \tag{15}$$

with $\varepsilon = 1 + 4\pi\chi^{(1)}$. This is a familiar linear wave equation which yields plane wave solutions

$$\mathbf{E} = \mathbf{E}(\omega)e^{i(\mathbf{k}\cdot\mathbf{r}-\omega t)} \tag{16}$$

The magnitude of the wave vector $\mathbf{k}$ is usually expressed as

$$\mathbf{k} = \frac{\omega}{c}[n_0(\omega) + i\gamma(\omega)]\mathbf{s} \equiv (k_{re} + ik_{im})\mathbf{s} \tag{17}$$

where $n_0(\omega) + i\gamma(\omega)$ is the linear index of refraction with its imaginary part γ representing linear absorption of the medium and $\mathbf{s}$ is a unit vector ($\mathbf{s} \cdot \mathbf{s} = 1$) pointing to the wave propagation direction. There should be no restriction on the electric field polarization direction and the wave propagation direction **s**. Wavelength λ is related to the wave vector **k** by

$$\lambda = \frac{2\pi}{k_{re}} = \frac{2\pi c}{\omega n_0(\omega)} \tag{18}$$

Since experiments are usually performed at wavelengths far away from the medium absorption bands, it is reasonable to assume $\gamma(\omega) \ll n_0(\omega)$, which leads to

$$\varepsilon = 1 + 4\pi\chi^{(1)}(\omega) = [n_0(\omega) + i\gamma(\omega)]^2 \approx n_0(\omega)[n_0(\omega) + 2i\gamma(\omega)] \tag{19}$$

so that

$$1 + 4\pi\mathrm{Re}[\chi^{(1)}(\omega)] = n_0^2(\omega)$$
$$4\pi\mathrm{Im}[\chi^{(1)}(\omega)] = 2n_0(\omega)\gamma(\omega) \tag{20}$$

In a nonlinear medium the wave Equation 5 becomes

$$\nabla^2\mathbf{E} - \nabla(\nabla\cdot\mathbf{E}) - \frac{\varepsilon}{c^2}\frac{\partial^2\mathbf{E}}{\partial t^2} = \frac{4\pi}{c^2}\frac{\partial^2\mathbf{P}^{nl}}{\partial t^2} \tag{21}$$

where the assumption of isotropic medium has already been applied for the sake of simplicity. Most optical media are dispersive and ε is a function of frequency. Therefore, Equation 16 should be solved for each frequency component. A summation of all the individual frequency components constitutes a general wave propagating in a dispersive medium.

Equation 21 describes specific nonlinear optical interactions, depending on the specific orders and types of nonlinear effects included in $\mathbf{P}^{nl}$. Of particular interest is a third-order effect called self-phase modulation, which is responsible for supercontinuum generation, a laboratory technique for generating short-pulsed wide color band for spectroscopic studies. Lasers are excellent sources for intensive light pulses that are coherent and monochromic. However, many dispersion studies require wide-band, short-pulsed light sources. A traditional approach is to run a dye laser pumped by a short-pulsed solid-state laser and use different dyes for frequency selection. The disadvantages of this approach include its inability to provide a continuous frequency band and its inefficiency due to complex procedures for dye change. A much more elegant method taking advantage of the third-order optical nonlinearity is the supercontinuum generation, in which short light pulses are focused on a third-order material and continuum light pulses, in addition to the fundamental input pulses, are recollimated at the other side of the material. Continuum pulses usually have their pulse width broadened compared with that of the fundamental seed pulses. This is essentially due to the group velocity dispersion occurring at the nonlinear medium used as continuum generator. The following example may give readers some idea as to how powerful the supercontinuum generation is as an ultrashort-pulsed color light source. Ultrashort light pulses, characterized by pulse widths of 100 femtoseconds (10^{-15} s), bandwidths of 10 nm centered at 630 nm and energy of 1 μJ per pulse, are capable of generating a wide frequency band of 350 nm to 900 nm at a 1-mm-thick ethylene-glycol cell. The fundamental 630-nm pulses can be obtained from an amplified CPM (colliding pulse mode locked) laser.

Let us first see how the third-order susceptibility is related to the intensity-dependent index of refraction. An applied field fast oscillating at an optical frequency ω in the following form

$$E(t) = E_0 \cos\omega t = \frac{E_0}{2}(e^{i\omega t} + e^{-i\omega t})$$

induces linear and nonlinear polarization

$$P = P^{(1)} + P^{(3)} = \chi^{(1)}E + \chi^{(3)}E^3 = \chi^{(1)}E_0 \cos(\omega t) + \chi^{(3)}E_0^3 \cos^3(\omega t)$$

$$= \chi^{(1)}E_0 \cos(\omega t) + \chi^{(3)}E_0^3 \left[\frac{3}{4}\cos(\omega t) + \frac{1}{4}\cos(3\omega t) \right] = P^{(1)}(\omega) + P^{(3)}(\omega) + P^{(3)}(3\omega)$$

(22)

where $\chi^{(1)}$ and $\chi^{(3)}$ represent the linear and third-order susceptibilities, respectively, in time domain. While $P^{(3)}(3\omega)$ is the induced polarization responsible for third harmonic generation, we are mainly interested in $P^{(3)}(\omega)$ since we are looking for the index of refraction at frequency ω. The total polarization at frequency ω can be extracted from Equation 22 and used for the index of refraction derivation.

$$P(\omega) = P^{(1)}(\omega) + P^{(3)}(\omega) = \left[\chi^{(1)} + \frac{3}{4}\chi^{(3)}E_0^2 \right] E_0 \cos(\omega t) = \chi_{eff}E_0 \cos(\omega t)$$

In analogy with Equation 19, one can obtain

$$n^2 = \varepsilon_{eff} = 1 + 4\pi\chi_{eff} = 1 + 4\pi \left[\chi^{(1)} + \frac{3}{4}\chi^{(3)}E_0^2 \right] = n_0^2 + 3\pi\chi^{(3)}E_0^2$$

which can be rewritten as

$$n^2 - n_0^2 = (n + n_0)(n - n_0) \approx 2n_0(n - n_0) = 3\pi\chi^{(3)}E_0^2$$

The intensity dependent index of refraction n can be derived as

$$n = n_0 + \frac{3\pi\chi^{(3)}E_0^2}{2n_0} = n_0 + n_2 E_0^2$$

(23)

where n_2 is defined as

$$n_2 = \frac{3\pi\chi^{(3)}}{2n_0}$$

(24)

Some authors prefer to express n in terms of light intensity. Since the field and the intensity in the cgs system are related by the expression

$$E_0^2 = \frac{8\pi}{cn_0} I$$

one can obtain

$$n = n_0 + n_2' I$$

with

$$n_2' = \frac{12\pi^2 \chi^{(3)}}{c n_0^2} \tag{25}$$

Equations 23 and 25 show that the nonlinear refractive index is directly related to the optical intensity. Since the beam intensity usually exhibits spatial variations, the index of refraction will also have geometric distributions. Let us assume that the intensity has a general Gaussian radius dependence. The index of refraction will then follow this radial Gaussian distribution, which, for the positive n_2 case, means higher index of refraction at the center of the beam and lower index of refraction at the edge of the beam. This lens-like refractive index distribution will give rise to the so-called self-focusing phenomenon. As the focused beam has higher intensity which simultaneously affects the nonlinear refractive index, this regenerative process may eventually confine the beam dimension to an order of its wavelength and result in material breakdown. Since self-focusing is an intensity-dependent third-order effect, the catastrophic optical damage is usually observed in optical fibers where light is confined to a very small region and propagates long distances in the medium so that the intensity is high to start with and even higher as the light travels along the fiber. Materials with negative $\chi^{(3)}$ have generated intensive research interest as the self-defocusing effect associated with the negative n_2 can successfully solve the optical medium damage problem due to self-focusing. Most natural materials and synthesized organic chemicals exhibit positive $\chi^{(3)}$. Among the very few materials which possess negative $\chi^{(3)}$ below all their optical resonance frequencies are a group of organic compounds, known as squaraines.[7–9] It has been discovered that only centrosymmetric squarylium molecules exhibit negative third-order nonlinear susceptibility. Much research is currently focused on the understanding of the origin of this phenomenon at the molecular level.[7–11] While self-focusing causes material damage in extreme cases, it also helps light propagation in the medium if the intensity is not too high. One example is that the self-focusing may compensate the spreading of the beam due to diffraction. Let us now come back to discuss our main topic, the mechanism for self-phase modulation and supercontinuum generation.

There are two approaches for us to see how short pulses can generate supercontinuum. One is to include the third-order susceptibility $\chi^{(3)}(\omega)$ in $\mathbf{P}^{nl}$ so that

$$\mathbf{P} = \chi^{(1)}(\omega)\mathbf{E} + \frac{1}{6}\chi^{(3)}(\omega)|\mathbf{E}|^2\mathbf{E} \tag{26}$$

and use Equation 21 with the specified $\mathbf{P}^{nl}$. By assuming a trial solution for $\mathbf{E}(\mathbf{r}, t)$, one can derive a dispersion relation and the propagation of electric field $\mathbf{E}$ in time. The frequency dependence of duly generated E through optical nonlinear processes can be analyzed by means of Fourier transform. Another approach is to determine the wave equation from the dispersion relation to the lowest order in one variable, either t or $\mathbf{r}$. We will try the second approach. Our treatment is similar to that of Whitham.[12]

We start with a new definition of the wave vector $\mathbf{k}$, which is an extension to Equation 17.

$$k = \frac{\omega}{c}\left[n_0(\omega) + i\gamma(\omega) + n_2 E^2\right] \tag{27}$$

which can be rewritten as

$$\omega = \frac{ck}{[n_0(\omega) + i\gamma(\omega) + n_2 E^2]} \approx \frac{ck}{n_0(\omega)}\left[1 - \frac{i\gamma(\omega)}{n_0(\omega)} - \frac{n_2}{n_0(\omega)}E^2\right] \tag{28}$$

where assumptions $\gamma \ll n_0(\omega)$ and $n_2 \ll n_0$ have been applied. We would like to expand the first factor in Equation 28 in terms of the incoming wave vector $\mathbf{k}_0$ so that

$$\frac{ck}{n_0(\omega)} = \omega_0 + \left.\frac{\partial\omega}{\partial k}\right|_{k_0}(k - k_0) + \frac{1}{2}\left.\frac{\partial^2\omega}{\partial k^2}\right|_{k_0}(k - k_0)^2 \tag{29}$$

with

$$\omega_0 = \frac{ck_0}{n_0(\omega)} \tag{30}$$

Substituting Equation 29 into Equation 28 yields

$$\omega = \omega_0 + \left.\frac{\partial\omega}{\partial k}\right|_{k_0}(k - k_0) + \frac{1}{2}\left.\frac{\partial^2\omega}{\partial k^2}\right|_{k_0}(k - k_0)^2 - \frac{i\gamma(\omega_0)\omega_0}{n_0(\omega_0)} - \frac{n_2\omega_0}{n_0(\omega_0)}E^2 \tag{31}$$

Assume the electric field is transverse and is of the following separable format in cylindrical coordinates

$$E(r, z, t) = R(r)Z(z, t)e^{i(k_0 z - \omega_0 t)} \tag{32}$$

One can write the electric field in terms of its Fourier components

$$E(r, z, t) = R(r)\int_{-\infty}^{\infty} F(k)e^{i(kz - \omega(k)t)}dk \tag{33}$$

where $F(k)$ is the Fourier transform of $Z(z, t)\exp[i(k_0 z - \omega_0 t)]$ so that

$$\begin{aligned}
Z(z, t) &= \int_{-\infty}^{\infty} F(k)e^{i[(k-k_0)z - (\omega(k)-\omega_0)t]}dk \\
&= \int_{-\infty}^{\infty} F(K + k_0)e^{i[Kz - (\omega(k)-\omega_0)t]}dK
\end{aligned} \tag{34}$$

where K is defined as $K = k - k_0$. The pulse shape function Z can be further expanded using Equation 31.

$$Z(z, t) = \int_{-\infty}^{\infty} F(K + k_0)e^{i[Kz - (\omega_0' K + (1/2)\omega_0'' K^2 - ((i\omega_0\gamma_0)/n_0) - ((\omega_0 n_2)/n_0)E^2)t]}dK \tag{35}$$

where parameters are defined as

$$\omega_0' = \left.\frac{\partial\omega}{\partial k}\right|_{k_0}, \qquad \omega_0'' = \left.\frac{\partial^2\omega}{\partial k^2}\right|_{k_0}, \qquad \gamma_0 = \gamma(\omega_0), \qquad n_0 = n_0(\omega_0)$$

The time rate of change for Z is given by

$$\frac{\partial Z}{\partial t} = -i\int_{-\infty}^{\infty}\left(\omega_0' K + \frac{1}{2}\omega_0'' K^2 - \frac{i\omega_0\gamma_0}{n_0} - \frac{\omega_0 n_2}{n_0}E^2\right) \times F(K + k_0)e^{i[Kz - (\omega(k)-\omega_0)t]}dK \tag{36}$$

This is equivalent to

$$\frac{\partial Z}{\partial t} = -\omega_0' \frac{\partial Z}{\partial z} + \frac{i}{2} \omega_0'' \frac{\partial^2 Z}{\partial z^2} - \frac{\omega_0 \gamma_0}{n_0} Z + \frac{i\omega_0 n_2}{n_0} E^2 Z \qquad (37)$$

which leads to the nonlinear Schrodinger equation

$$i\left(\frac{\partial Z}{\partial t} + \omega_0' \frac{\partial Z}{\partial z} + \frac{\omega_0 \gamma_0}{n_0} Z\right) + \frac{1}{2} \omega_0'' \frac{\partial^2 Z}{\partial z^2} + \frac{\omega_0 n_2}{n_0} |R|^2 |Z|^2 Z = 0 \qquad (38)$$

The second term in Equation 38 is related to the group velocity dispersion (GVD) of the medium as it is clear that the group velocity of propagation of the wave package is ω_0' and

$$\omega_0'' = \left.\frac{\partial^2 \omega}{\partial k^2}\right|_{\omega_0} = \left.\frac{\partial v_g}{\partial k}\right|_{\omega_0}$$

is a measure of GVD. The last term, involving n_2 in Equation 38 is related to the self-phase modulation (SPM), which we will discuss in detail. If the last term is sufficiently large, the GVD and SPM terms can cancel each other so that Equation 38 will be reduced to

$$\frac{\partial Z}{\partial t} + \omega_0' \frac{\partial Z}{\partial z} + \frac{\omega_0 \gamma_0}{n_0} Z = 0 \qquad (39)$$

The necessary conditions for the simplification are

$$\omega_0'' > 0 \quad \text{and} \quad \frac{\partial^2 Z}{\partial z^2} \frac{1}{Z} < 0$$

and

$$\omega_0'' < 0 \quad \text{and} \quad \frac{\partial^2 Z}{\partial z^2} \frac{1}{Z} > 0$$

with $\omega_0'' < 0$ and $\omega_0'' > 0$ referred to as natural and anomalous dispersion, respectively.

For the sake of simplicity, we would like to assume that the medium under consideration is lossless and that the group velocity dispersion is negligible. Since we are essentially interested in the pulse envelope, we further simplify the problem by neglecting the transverse field distribution $R(r)$. Equation 38 then becomes

$$\frac{\partial Z}{\partial z} + \frac{1}{v_g} \frac{\partial Z}{\partial t} = \frac{i\omega_0 n_2}{c} |Z|^2 Z \qquad (40)$$

with

$$v_g = \left.\frac{\partial \omega}{\partial k}\right|_{\omega_0} = \left.\frac{\partial}{\partial k}\left(k\frac{c}{n}\right)\right|_{\omega_0} \approx \frac{c}{n_0} = v_{ph} \qquad (41)$$

for the nondispersive medium approximation. One may write Z in terms of its amplitude A and phase ϕ so that $Z = Ae^{i\phi}$. Equation 40 then can be rewritten as two equations (42 and 43) to satisfy A and φ, respectively.

$$\frac{\partial A}{\partial z} + \frac{1}{v_g}\frac{\partial A}{\partial t} = 0 \tag{42}$$

and

$$\frac{\partial \phi}{\partial z} + \frac{1}{v_g}\frac{\partial \phi}{\partial t} = \frac{\omega_0}{c}n_2 A^2 \tag{43}$$

With an introduction of local time

$$\tau = t - \frac{z}{v_g} \tag{44}$$

it is straightforward to obtain

$$\frac{\partial A(z,\tau)}{\partial z} = \frac{\partial A(z,t)}{\partial z} + \frac{\partial A(z,t)}{\partial t}\frac{\partial t}{\partial z} = \frac{\partial A(z,t)}{\partial z} + \frac{1}{v_g}\frac{\partial A(z,t)}{\partial t} = 0 \tag{45}$$

which is equivalent to

$$A(z,\tau) = A(\tau)$$

Equation 43, therefore, can be simplified to

$$\frac{\partial \phi(z,\tau)}{\partial z} = \frac{\omega_0}{c}n_2 A^2 \tag{46}$$

which yields

$$\phi(z,\tau) = \frac{\omega_0}{c}n_2 \int_0^z A^2 dz' = \frac{\omega_0}{c}n_2 A^2(\tau)z \tag{47}$$

assuming that the medium starts at $z = 0$. After propagating a distance z, the electric field envelop should be

$$Z(z,\tau) = Ae^{i\phi} = A(\tau)e^{i(\omega_0 n_2/c)A^2(\tau)z} \tag{48}$$

This is a pulse whose intensity is unaltered during propagation and which experiences a phase shift φ, specified in Equation 47, due to self-phase modulation. Since the intensity $A^2(\tau)$ varies through the pulse [$A^2(\tau)$ varies with t at a fixed position z], various parts of the pulse undergo different phase shifts, leading to a frequency shift which is directly proportional to the distance traveled. As the pulse duration is much longer than the optical field period ($2\pi/\omega_0$), the electric field at each position τ within the pulse shape has a specific local and instantaneous frequency which is given by

$$\omega(\tau) = \omega_0 + \delta\omega(\tau) \tag{49}$$

where

$$\delta\omega(\tau) = -\frac{\partial\phi}{\partial\tau} = -\frac{\omega_0}{c}\,n_2 z\,\frac{\partial A^2}{\partial\tau} \tag{50}$$

The frequency shift $\delta\omega(\tau)$ is generated at the time location τ of the pulse shape. Suppose the initial pulse is given by the following chirp-free (constant-phase) envelop

$$A(t) = E_0\,\text{sech}(t/t_0) \tag{51}$$

where t_0 is a measure of the pulse duration. The frequency shift can be illustrated as in Figure 1. Assuming positive n_2, the instantaneous frequency shift in the leading edge is negative, whereas those in the trailing edge is positive. The total frequency excursion is

$$2|\delta\omega_{max}| = 1.54\,\frac{\omega_0 n_2 E_0^2 z}{c t_0} = 1.54\,\frac{\Delta\phi_{max}}{t_0} \tag{52}$$

where $\Delta\phi_{max}$ is the phase change between the peak and edges of the pulse. This spectral broadening due to SPM becomes significant when the frequency shift approaches and exceeds the initial bandwidth of the pulse. The experimental example of supercontinuum generation mentioned earlier certainly belongs to this category. Equation 52 clearly shows that pulses with short duration t_0 and higher peak intensity E_0 will generate broader continuum band, and obviously one would like to choose a material with large third-order susceptibility (therefore large n_2) as a medium for the continuum generator.

III. HARMONIC OSCILLATOR AND LINEAR OPTICAL SUSCEPTIBILITY

A model describing an electron bound in a harmonic potential and driven by an applied electric field at frequency ω offers a phenomenological explanation of linear optical susceptibility. Assume the medium consists of N atoms with one electron in each of these atoms. Each electron can oscillate at the intrinsic frequency ω_0 under the harmonic potential created by its own nuclei. The applied field is fast oscillating at an optical frequency ω in the following form

$$E(t) = E_0\cos\omega t = \frac{E_0}{2}\,(e^{i\omega t} + e^{-i\omega t}) \tag{53}$$

The equation of motion of such an electron under the influence of the driving field and the harmonic

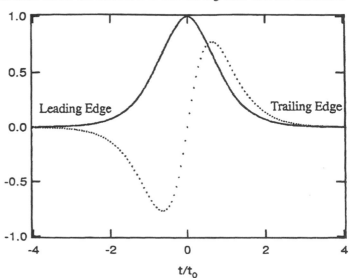

Figure 1 Initial pulse shape represented by sech(t/t_0) (solid curve) and the corresponding frequency shift represented by $-(1/t_0)d(\text{sech}^2(t/t_0))/dt$ (dotted curve).

potential can be obtained from Newton's second law as

$$\frac{d^2x}{dt^2} + \gamma \frac{dx}{dt} + \omega_0^2 x = -\frac{eE_0}{2m}(e^{i\omega t} + e^{-i\omega t}) \tag{54}$$

where x is the displacement of the electron from its equilibrium position, ω_0 is the intrinsic harmonic oscillator frequency, γ is the damping coefficient related to linear optical loss, and e and m are electron charge and mass, respectively. Inspection of Equation 54 suggests that its solution should be a superposition of $e^{i\omega t}$ and $e^{-i\omega t}$ and a trial solution is, therefore, assumed in the following form to guarantee a physical (real) x

$$x = Ae^{i\omega t} + \text{complex conjugate} \tag{55}$$

Substituting x in Equation 54 with Equation 55 and equating all the coefficients of $e^{i\omega t}$ and $e^{-i\omega t}$ separately yields

$$A = -\frac{eE_0}{2m}\frac{1}{i\gamma\omega + \omega_0^2 - \omega^2} \tag{56}$$

Since the polarization and displacement have a simple relation $P = -ex$, one can obtain

$$P = \frac{e^2E_0}{2m}\frac{e^{i\omega t}}{i\gamma\omega + \omega_0^2 - \omega^2} + cc \tag{57}$$

Where cc refers to complex conjugate. The definition of linear optical susceptibility

$$p^{(1)}(\omega) = \frac{1}{2}(\chi^{(1)}(\omega)e^{i\omega t} + cc)E_0 \tag{58}$$

leads to

$$\chi^{(1)}(\omega) = \frac{Ne^2}{m}\frac{1}{i\gamma\omega + \omega_0^2 - \omega^2} = -\frac{Ne^2}{m}\frac{(\omega^2 - \omega_0^2) + i\gamma\omega}{(\gamma\omega)^2 + (\omega^2 - \omega_0^2)^2} \tag{59}$$

where N is the number density of harmonic oscillators/atoms in the medium. This is a typical Lorentz shape. The dispersion clearly involves one resonance as the external optical frequency ω is tuned towards the oscillator intrinsic frequency ω_0. The real part of the $\chi^{(1)}$ changes its sign when the external driving field scans its frequency across the resonance while the imaginary part remains unchanged in sign. We notice that the induced dipole moment oscillates at the same frequency as that of the applied field but has some phase lag with respect to the driven field. Since oscillating dipoles radiate electric field, the net field is to be a superposition of the external field and the dipole radiated field with a mismatched phase.

IV. ANHARMONIC OSCILLATOR AND NONLINEAR OPTICAL SUSCEPTIBILITIES

The harmonic model offers a successful classic explanation of the origin of the linear susceptibility. However, it fails to predict any nonlinear response with respect to the applied optical field. A harmonic oscillator corresponds to the situation where the restoring force applied to an electron is proportional to its displacement from the equilibrium position. Equation 57 shows that the electron displacement is proportional to the amplitude of the applied optical field. If the optical field is strong enough then the displacement can be so large that the restoring force will be proportional to not only the displacement, but also to its second or third power. In this case, the harmonic oscillator model has to be extended to the anharmonic model, from which nonlinear optical susceptibilities can be deduced. Similar to the polarization expansion in Equation 6, it is assumed that the higher orders in the restoring force are always orders of magnitude smaller than the linear order.

Let us first add a quadric term to Equation 54 so that it becomes

$$\frac{d^2x}{dt^2} + \gamma\frac{dx}{dt} + \omega_0^2 x + Bx^2 = -\frac{eE_0}{2m}(e^{i\omega t} + e^{-i\omega t}) \tag{60}$$

where Bx^2 is the anharmonic restoring force. The trial solution for Equation 60 has to include some second harmonic terms in order to satisfy the newly added quadratic term.

$$x = A^{(1)}e^{i\omega t} + A^{(1)*}e^{-i\omega t} + A^{(2)}e^{i2\omega t} + A^{(2)*}e^{-i2\omega t} = x^{(1)} + x^{(2)} \tag{61}$$

The technique used to solve Equation 60 with trial solution (Equation 61) is analogous to the perturbation theory in quantum mechanics. We allow the first term and its complex conjugate in Equation 61 to satisfy Equation 54 so that $A^{(1)}$ has the same expression as A in Equation 56. Then the third term and its complex conjugate represented by $x^{(2)}$ in Equation 61 is to satisfy the "homogeneous form" of Equation 60, with its quadratic term replaced by $x^{(1)}$. This operation means rewriting Equation 60 as

$$\frac{d^2x^{(1)}}{dt^2} + \gamma\frac{dx^{(1)}}{dt} + \omega_0^2 x^{(1)} = -\frac{eE_0}{2m}(e^{i\omega t} + e^{-i\omega t})$$

and

$$\frac{d^2x^{(2)}}{dt^2} + \gamma\frac{dx^{(2)}}{dt} + \omega_0^2 x^{(2)} + B(x^{(1)})^2 = 0 \tag{62}$$

which directly lead to

$$A^{(2)} = \frac{e^2E_0^2}{4m^2}\frac{1}{[i\gamma\omega + \omega_0^2 - \omega^2]^2}\frac{-B}{2i\gamma\omega + \omega_0^2 - 4\omega^2} \tag{63}$$

From the relation between $P^{(2)}(2\omega)$ and $\chi^{(2)}(-2\omega; \omega, \omega)$

$$P^{(2)}(2\omega) = \frac{1}{2}(\chi^{(2)}(-2\omega; \omega, \omega)e^{i2\omega t} + cc)E_0^2 \tag{64}$$

One can obtain

$$\chi^{(2)}(-2\omega; \omega, \omega) = \frac{Ne^3}{2m^2}\frac{1}{[i\gamma\omega + \omega_0^2 - \omega^2]^2}\frac{B}{2i\gamma\omega + \omega_0^2 - 4\omega^2} \tag{65}$$

which is the susceptibility corresponding to the second harmonic ($2\omega = \omega + \omega$) generation, one of the second-order responses. It is straightforward to figure out the zero frequency ($0 = \omega - \omega$) response, $\chi^{(2)}(0; \omega, -\omega)$, if one includes a constant term in the trial solution Equation 61. A more general approach is to include two arbitrary (Fourier) frequency components, ω_1 and ω_2, in the applied field in Equation 60. The anharmonic model then will be able to show the sum frequency ($\omega_1 + \omega_2$) generation and the difference frequency ($\omega_1 - \omega_2$) generation. Equation 65 indicates resonances not only at the fundamental atomic resonance frequency ω_0 but also at $2\omega = \omega_0$. The latter is often referred to as two-photon resonance.

An empirical rule discovered by Miller in 1964[13] states that the quantity

$$\delta_{ijk}^{(2\omega)} = \frac{\chi_{ijk}^{(2)}(2\omega)}{\chi_{ii}^{(1)}(2\omega)\chi_{jj}^{(1)}(\omega)\chi_{kk}^{(1)}(\omega)} \tag{66}$$

almost remains unchanged for all noncentrosymmetric crystals, although $\chi_{ijk}^{(2)}(2\omega)$ can vary over several orders of magnitude. The quantity is also called Miller's delta. This Miller's rule can be understood in terms of the anharmonic oscillator model just presented. Let us reduce Equation 66 to its one dimensional analog

$$\delta^{(2\omega)} = \frac{\chi^{(2)}(2\omega)}{\chi^{(1)}(2\omega)[\chi^{(1)}(\omega)]^2} \tag{67}$$

By inspecting Equations 59 and 65, we can see that Miller's delta is actually an intrinsic parameter directly related to the anharmonic term coefficient B. The relation

$$\delta^{(2\omega)} = \frac{mB}{2e^3N^2} \tag{68}$$

which was first derived by Garrett and Robinson,[14] indicates that Miller's rule means a nearly constant B for all crystals since atomic number density N remains in the order of $10^{22}/cm^3$ for all condensed matters. The anharmonic model has been applied to inorganic crystals[15] to calculate $\delta^{(2\omega)}$. Within the model approximation, B is estimated by calculating the bond charge from the electron overlap and screened nuclear charges along the various bonds in the unit cell. The bond susceptibilities are then summed over the bonds in the unit cell to obtain the crystal susceptibility. The detailed calculation of the bond charge leads to surprisingly successful predictions for $\delta^{(2\omega)}$. The following estimate may serve as an example for calculating Miller's delta using the anharmonic oscillator model. Consider two charges, e and $2e$, separated by a distance r_0. We notice that the configuration is noncentrosymmetric, which guarantees a non-zero second-order susceptibility. Expanding the electric potential for this configuration in terms of its minimum position leads us to

$$V(x) = \frac{e^2}{4\pi r_0}\left(5.83 + 24.1\frac{x^2}{r_0^2} - 13.3\frac{x^3}{r_0^3} + \cdots\right) \tag{69}$$

which yields coefficient B from the first anharmonic term as

$$B = \frac{-39.9e^2}{4\pi mr_0^4} \tag{70}$$

Choosing $r_0 = 0.3$ nm and $N = 6 \times 10^{22}/cm^3$, one obtains, from Equation 68,

$$\delta^{(2\omega)} = 11.3 \times 10^{-7} \text{ esu}$$

which is within the range of observed Miller's deltas listed in Table 1 for inorganic crystals.

We notice in Table 1 that organic materials, in general, do not follow Miller's rule. Miller's delta for 2-methyl-4-nitroaniline (xxx component) is 620×10^{-7} esu, orders of magnitude larger than that of m-nitroaniline (zyy component) (12×10^{-7} esu). This is because organic systems are composed of highly conjugated molecules and π-electrons contributing to the optical nonlinearity are not located on localized bonds. The electron delocalization is the key characteristic which is responsible for exceptionally large nonlinear susceptibilities in conjugated organic polymers. The detailed microscopic origin is discussed in the following sections.

The quadratic anharmonic term in Equation 60 corresponds to a quadratic restoring force which is from a potential

$$U(x) = -\int F_{\text{restoring}}(x)dx = -\int Bx^2dx = -\frac{1}{3}Bx^3 \tag{71}$$

Since this potential does not satisfy centrosymmetric inversion $U(x) = U(-x)$, B has to vanish in centrosymmetric materials. This is consistent with our earlier argument that centrosymmetric materials

Table 1 Second Harmonic Susceptibilities and Miller's Deltas of Selected Crystals

Crystal	$\chi_{ijk}^{(2\omega)}$ (10^{-9} esu)	$\delta_{ijk}^{(2\omega)}$ (10^{-7} esu)
Quartz	$xxx = 1.2$	9.2
LiNbO$_3$	$zxx = 14.2$	4.4
	$yyy = 7.5$	2.7
LiIO$_3$	$zxx = 17$	2.7
NH$_4$H$_2$PO$_4$ (ADP)	$zxy = 1.8$	19
	$xyz = 1.6$	13
KH$_2$PO$_4$ (KDP)	$zxy = 1.5$	17
	$xyz = 1.1$	14
KD$_2$PO$_4$ (KD*P)	$zxy = 1.4$	12
	$xyz = 1.4$	12
CdS	$zzz = 93$	15
	$zxx = 97$	7.6
	$zxy = 107$	9.2
m-Nitroaniline	$zxx = 48$	130
	$zyy = 3.8$	12
	$zzz = 50$	190
2-Methyl-4-nitroaniline (MNA)	$xxx = 600$	620
	$xyy = 90$	230
Methyl-(2, 4 dinitrophenyl)-amino-2-proponate (MAP)	$yxx = 40$	120
	$yyy = 44$	190
	$yzz = 8.8$	36
	$yxz = -1.3$	-4.5

From Singer, K. D., Ph.D. Dissertation, University of Pennsylvania, Philadelphia, PA. With permission.

do not exhibit even-order nonlinear responses. Therefore, for centrosymmetric structures one has to include a higher order term in the restoring force and modify Equation 60 as follows:

$$\frac{d^2x}{dt^2} + \gamma\frac{dx}{dt} + \omega_0^2 x + Cx^3 = -\frac{eE_0}{2m}(e^{i\omega t} + e^{-i\omega t}) \qquad (72)$$

Similar procedures for solving Equation 60 can be adopted to obtain the third harmonic polarization. The trial solution should be in the form:

$$x = (A_\omega^{(1)}e^{i\omega t} + cc) + (A_\omega^{(3)}e^{i\omega t} + cc) + (A_{3\omega}^{(3)}e^{i3\omega t} + cc)$$

$$= x_\omega^{(1)} + x_\omega^{(3)} + x_{3\omega}^{(3)} = x^{(1)} + x^{(3)} \qquad (73)$$

with $x^{(1)}$ satisfying Equation 54 and $x^{(3)}$ satisfying

$$\frac{d^2x^{(3)}}{dt^2} + \gamma\frac{dx^{(3)}}{dt} + \omega_0^2 x^{(3)} + C(x^{(1)})^3 = 0 \qquad (74)$$

Equation 74 immediately yields

$$x_{3\omega}^{(3)} = -\frac{C(x^{(1)})^3|_{3\omega}}{3i\gamma\omega + \omega_0^2 - 9\omega^2}$$

$$= -\frac{C}{3i\gamma\omega + \omega_0^2 - 9\omega^2}\left(-\frac{eE_0}{2m}\right)^3\left(\frac{e^{i\omega t}}{i\gamma\omega + \omega_0^2 - \omega^2} + cc\right)^3\Bigg|_{3\omega} \qquad (75)$$

$$= \frac{1}{8}\left(\frac{eE_0}{m}\right)^3\frac{C}{3i\gamma\omega + \omega_0^2 - 9\omega^2}\left[\frac{e^{i3\omega t}}{(i\gamma\omega + \omega_0^2 - \omega^2)^3} + cc\right]$$

and

$$x_\omega^{(3)} = -\frac{(x^{(1)})^3 \mid_\omega}{i\gamma\omega + \omega_0^2 - \omega^2}$$

$$= -\frac{C}{i\gamma\omega + \omega_0^2 - \omega^2}\left(-\frac{eE_0}{2m}\right)^3\left(\frac{e^{i\omega t}}{i\gamma\omega + \omega_0^2 - \omega^2} + cc\right)^3\Bigg|_\omega \tag{76}$$

$$= \frac{1}{8}\left(\frac{eE_0}{m}\right)^3\frac{C}{i\gamma\omega + \omega_0^2 - \omega^2}\left[\frac{3e^{i\omega t}}{(i\gamma\omega + \omega_0^2 - \omega^2)^2(-i\gamma\omega + \omega_0^2 - \omega^2)} + cc\right]$$

The relation between the third-order susceptibilities and polarization given by

$$P^{(3)} = \frac{1}{2}\{[\chi^{(3)}(-3\omega; \omega, \omega, \omega)E_0^3 e^{i3\omega t} + cc] + [\chi^{(3)}(-\omega; \omega, -\omega, \omega)E_0^3 e^{i\omega t} + cc]\} \tag{77}$$

leads to the third harmonic susceptibility

$$\chi^{(3)}(-3\omega; \omega, \omega, \omega) = \frac{N}{4}\left(\frac{e^4}{m^3}\right)\frac{C}{[3i\gamma\omega + \omega_0^2 - (3\omega)^2](i\gamma\omega + \omega_0^2 - \omega^2)^3} \tag{78}$$

and the other form of the third-order response

$$\chi^{(3)}(-\omega; \omega, -\omega, \omega) = \frac{3N}{4}\left(\frac{e^4}{m^3}\right)\frac{C}{[(\gamma\omega)^2 + (\omega_0^2 - \omega^2)^2](i\gamma\omega + \omega_0^2 - \omega^2)^2} \tag{79}$$

Equation 78 points out that $\chi^{(3)}(-3\omega; \omega, \omega, \omega)$ has resonance behavior either at fundamental frequency $\omega = \omega_0$ or third harmonic frequency $3\omega = \omega_0$. The expression for $\chi^{(3)}(-3\omega; \omega, \omega, \omega)$ can be rewritten in terms of the linear response $\chi^{(1)}(\omega)$ by using Equation 59 to eliminate the denominator factor $i\gamma\omega + \omega_0^2 - \omega^2$. We therefore have

$$\chi^{(3)}(-3\omega; \omega, \omega, \omega) = \frac{m}{4N^3 e^4} C\chi^{(1)}(3\omega)[\chi^{(1)}(\omega)]^3 \tag{80}$$

Nonresonance third harmonic susceptibility $\chi^{(3)}(-3\omega; \omega, \omega, \omega)$ can be estimated using Equation 78. When the frequency is far below resonance, the two denominators in Equation 78 can be reasonably approximated by ω_0^8. To get an estimate for the anharmonic coefficient C, we would like to assume that when the displacement x and the atomic separation s are of the same order of magnitude, harmonic and anharmonic restoring forces have the same value so that $\omega_0^2 s = Cs^3$. Equation 78 then becomes

$$\chi^{(3)}(-3\omega, \omega, \omega, \omega) = \frac{N}{4}\left(\frac{e^4}{m^3}\right)\frac{C}{\omega_0^8} = \frac{N}{4s^2\omega_0^6}\left(\frac{e^4}{m^3}\right) \tag{81}$$

Using $s = 0.3$ nm, $\omega_0 = 10^{16}$ rad/s and $N = 6 \times 10^{22}/\text{cm}^3$, one can obtain

$$\chi^{(3)}(-3\omega; \omega, \omega, \omega)\mid_{\omega \to 0} = 1 \times 10^{-15} \text{ esu}$$

which is within reasonable range of measured third harmonic susceptibilities of various materials.

The third-order response expressed in Equation 79 corresponds to degenerate four-wave mixing process where two counter-propagating photons create a standing wave pattern grating in the medium and the third photon is scattered off by the grating. Its real and imaginary parts are responsible for self-actions such as self-focusing and two-photon absorption, respectively. Due to the limitation of the classic anharmonic oscillator model, it shows only one resonance at the model's characteristic frequency ω_0.

Although the classic harmonic and anharmonic oscillator models are successful in predicting some of the linear and nonlinear optical behaviors of a medium, it is far from sufficient to offer a complete explanation about the phenomena experimentally observed. One obvious problem with the classic model is that the model has only one characteristic frequency while a real system consists of molecules having a large number of excited states. One needs to treat the problem quantum mechanically and solve the Schrodinger equation with specific Hamiltonian reflecting the structure of a particular molecule. This is discussed in the following sections.

V. MICROSCOPIC THEORY

We have discussed the nonlinear optical responses for model systems such as anharmonic oscillators. Although these discussions give us some insights on how nonlinear optical responses can result from interaction with an electromagnetic field, these models are highly phenomenological and have omitted many important features pertaining to a real physical system. These simple models are therefore of no practical use in representing the responses of real material systems. Real material systems are composed of atoms, which interact with each other to form molecular or crystalline structures. To describe correctly the nonlinear optical response of these systems, rather than of a simple phenomenological model, it is necessary to consider every structural aspect of the material such as the positions of the atoms, electronic energy band structures, type of bonding, and so forth, and the computation of the nonlinear optical susceptibility must be done quantum mechanically.

Inorganic materials such as semiconductors are generally crystalline, where all the atoms within the crystalline structure interact to form a macroscopic network. To calculate the nonlinear optical responses of this practically infinite network of atoms, simplified models that utilize the structural symmetry of the crystal such as the energy bands and elementary excitations are needed in order to reduce the practically infinite degree of freedom of the system into a model system which is computationally manageable. Organic materials, on the other hand, are usually composed of microscopic entities, namely the molecules, which interact with each other to form molecular crystals. The intramolecular interaction energy is generally much larger than the intermolecular interaction, as the intermolecular interactions are the weaker van der Waal's interaction or hydrogen bond compared to intramolecular covalent bond. Therefore, unlike the case for most inorganic materials, organic materials can be view as collections of weakly interacting molecular units, each independently interacting with the external electromagnetic field. The intermolecular interactions can be adequately taken care of by considering the correction to the local field.

For organic materials, the macroscopic susceptibility $\chi^{(2)}_{ijk}(-\omega_3; \omega_1, \omega_2)$ and $\chi^{(3)}_{ijkl}(-\omega_4; \omega_1, \omega_2, \omega_3)$ can therefore be related to the microscopic molecular susceptibility $\beta_{ijk}(-\omega_3; \omega_1, \omega_2)$ and $\gamma_{ijkl}(-\omega_4; \omega_1, \omega_2, \omega_3)$ through simple relations:

$$\chi^{(2)}_{ijk}(-\omega_3; \omega_1, \omega_2) = N\langle R_{im'} R_{jn'} R_{ko'} f^{\omega_1}_{j'n'} f^{\omega_2}_{k'o'} f^{\omega_3}_{m'i'} \beta_{i'j'k'}(-\omega_3; \omega_1, \omega_2) \rangle \tag{82}$$

and

$$\chi^{(3)}_{ijkl}(-\omega_4; \omega_1, \omega_2, \omega_3) = N\langle R_{im'} R_{jn'} R_{ko'} R_{lp'} f^{\omega_1}_{j'n'} f^{\omega_2}_{k'o'} f^{\omega_3}_{l'p'} f^{\omega_4}_{m'i'} \gamma_{i'j'k'l'}(-\omega_4; \omega_1, \omega_2, \omega_3) \rangle \tag{83}$$

where N is the number of molecules per unit volume and $f^{\omega}_{i'j'}$'s are local field correction tensors pertaining to the local molecular reference frame. The unprimed(primed) coordinates designate the laboratory(molecular) fixed axes. R's are the rotation matrices that transform the molecular frame to the laboratory frame. The brackets $\langle \cdots \rangle$ represent the average over the possible orientations of the molecules within the macroscopic phase. The molecule's orientations can be fixed in space, as in the case of molecular crystals, or form continuous distributions, as in the cases of liquids, liquid crystals, Langmuir-Blodgett films, or in guest-host polymers. Thus the understanding of the macroscopic susceptibilities is reduced to the understanding of the corresponding microscopic molecular susceptibilities and the orientation distributions of the molecules within the macroscopic phase.

Since the number of atoms and electrons within a molecule is usually small, calculation of the molecular nonlinear optical susceptibility can be performed from first principle without relying on approximate models such as the energy bands. Sophisticated computational theories have been developed for the nonlinear optical responses of molecules and their accuracy in determining the nonlinear optical susceptibility has been confirmed experimentally.

To provide an accurate theoretical description at the molecular level of the nonlinear optical properties of organic molecules, the theory must explicitly account for the effect of electron correlation. At the microscopic level, the molecular second- and third-order susceptibility tensors $\beta_{ijk}(-\omega_3; \omega_1, \omega_2)$ and $\gamma_{ijkl}(-\omega_4; \omega_1, \omega_2, \omega_3)$ are defined through the equations

$$p_i^{\omega_3} = \beta_{ijk}(-\omega_3; \omega_1, \omega_2)E_j^{\omega_1}E_k^{\omega_2} \tag{84}$$

$$p_i^{\omega_4} = \gamma_{ijkl}(-\omega_4; \omega_1, \omega_2, \omega_3)E_j^{\omega_1}E_k^{\omega_2}E_l^{\omega_3} \tag{85}$$

where $p_i^{\omega_3}$ or $p_i^{\omega_4}$ are the ith component of the molecular polarization induced at frequency ω_3 or ω_4 in response to the specified components of applied electric fields at frequencies ω_1 and ω_2, or ω_1, ω_2, and ω_3. By considering only the electric field amplitudes of the incident light, we have made the generally valid approximation that the electric dipole interaction with the molecule is much stronger than the magnetic and higher order electric interactions.

Different combinations of the frequencies of the applied field result in various nonlinear optical phenomena. For example, for second-order processes, if the frequencies ω_1 and ω_2 are degenerate at frequency ω, then $\beta_{ijk}(-2\omega; \omega, \omega)$ is responsible for the creation of light at 2ω through second harmonic generation (SHG). If one of the applied frequencies is zero, then $\beta_{ijk}(-\omega; 0, \omega)$ represents the linear electro-optic effect. Likewise, for third-order processes, $\gamma_{ijkl}(-3\omega; \omega, \omega, \omega)$ is responsible for third harmonic generation (THG). Another important third-order optical phenomenon is the intensity-dependent refractive index that results from $\gamma_{ijkl}(-\omega; \omega, -\omega, \omega)$, where we have taken the complex conjugate of one of the incident fields. Further, if in addition to the optical field at frequency ω, one applies a DC electric field, the third-order process of DC-induced second harmonic generation (DCSHG) occurs via the susceptibility $\gamma_{ijkl}(-2\omega; \omega, \omega, 0)$. Thus it is clear that both $\beta_{ijk}(-\omega_3; \omega_1, \omega_2)$ and $\gamma_{ijkl}(-\omega_4; \omega_1, \omega_2, \omega_3)$ govern a multitude of fundamental nonlinear optical processes each determined by the conditions $\omega_3 = \omega_1 + \omega_2$ or $\omega_4 = \omega_1 + \omega_2 + \omega_3$, where each frequency ω_1, ω_2, and ω_3 may be taken as positive or negative.

Of the two principal methods for the calculation of molecular susceptibility, usually referred to as the summation-over-states[16–18] and finite field techniques,[19] the summation-over-states formalism offers several advantages. Primary among these is the ability to identify specific virtual excitation processes among the eigenstates of the system that make the most significant contributions to the molecular susceptibility. This results from the summation-over-states representation of the molecular susceptibility as a perturbation expansion over all possible virtual excitations. In contrast, the computationally simpler finite field technique, which involves taking derivatives of the perturbed ground state energy or dipole moment of the molecule as a function of applied field strength, yields only a final value for the molecular susceptibility with no information regarding its origin. Additionally, the summation-over-states method allows one to calculate the frequency dependence or dispersion of the molecular susceptibility since it is founded in time-dependent perturbation theory, while the finite field technique only calculates the zero-frequency limit. A third strength of the summation-over-states method is the capability to include the many-electron nature of the molecular wave functions through multiple-excited configuration interaction theory.

The summation-over-states method derives its name from the expression of the molecular susceptibility through summations over all energy eigenstates of the system of terms that involve the transition dipole moments and excitation energies of the eigenstates. The calculation of the molecular susceptibility therefore requires knowledge of the transition moments and excitation energies of the excited states of the molecule. The summation-over-states expression of the nonlinear optical susceptibility can be derived using time-dependent perturbation theory in quantum mechanics. Standard time-dependent perturbation, however, results in an expression of the nonlinear optical susceptibility that contains divergent terms. These divergences occur when any subset of the applied frequencies sums to zero. These divergences, known as secular divergence, result from the improper execution of the perturbation procedure. The secular divergences can be eliminated by employing a technique called the Method of Averages developed by Bogoliubov and Mitropolsky.[20] This method is a perturbation technique that involves separation of the wave function into slowly varying and rapidly varying components. The slowly varying component is responsible for the secular divergences that occur in standard perturbation theory and leads to shifts of energy levels while the rapidly varying component produces the polarization induced in response to the applied fields. Without going into the details of the derivation of the expressions for the nonlinear optical susceptibility using the Method of Averages, we simply present the results below.[21]

The susceptibility $\beta_{ijk}(-\omega_3; \omega_1, \omega_2)$ is given by

$$\beta_{ijk}(-\omega_3; \omega_1, \omega_2) = K(-\omega_3; \omega_1, \omega_2)\left(\frac{e^3}{\hbar^2}\right)I_{1,2}\sum_{m_1m_2}{}' \left[\frac{r^i_{gm_2}\bar{r}^k_{m_2m_1}r^j_{m_1g}}{(\omega_{m_2g} - \omega_3)(\omega_{m_1g} - \omega_1)} \right.$$
$$\left. + \frac{r^k_{gm_2}\bar{r}^i_{m_2m_1}r^j_{m_1g}}{(\omega_{m_2g} + \omega_2)(\omega_{m_1g} - \omega_1)} + \frac{r^k_{gm_2}\bar{r}^j_{m_2m_1}r^i_{m_1g}}{(\omega_{m_2g} + \omega_2)(\omega_{m_1g} + \omega_3)} \right]$$

(86)

and the susceptibility $\gamma_{ijkl}(\omega_4; \omega_1, \omega_2, \omega_3)$ is given by

$$\gamma_{ijkl}(-\omega_4; \omega_1, \omega_2, \omega_3)$$

$$= K(-\omega_4; \omega_1, \omega_2, \omega_3)\left(\frac{e^4}{\hbar^3}\right)I_{1,2,3}\left\{ \sum_{m_1m_2m_3}{}' \left[\frac{r^i_{gm_3}\bar{r}^l_{m_3m_2}\bar{r}^k_{m_2m_1}r^j_{m_1g}}{(\omega_{m_3g} - \omega_4)(\omega_{m_2g} - \omega_1 - \omega_2)(\omega_{m_1g} - \omega_1)} \right. \right.$$

$$+ \frac{r^l_{gm_3}\bar{r}^i_{m_3m_2}\bar{r}^k_{m_2m_1}r^j_{m_1g}}{(\omega_{m_3g} + \omega_3)(\omega_{m_2g} - \omega_1 - \omega_2)(\omega_{m_1g} - \omega_1)} + \frac{r^j_{gm_3}\bar{r}^k_{m_3m_2}\bar{r}^i_{m_2m_1}r^l_{m_1g}}{(\omega_{m_3g} + \omega_1)(\omega_{m_2g} + \omega_1 + \omega_2)(\omega_{m_1g} - \omega_3)}$$

$$\left. + \frac{r^j_{gm_3}\bar{r}^k_{m_3m_2}\bar{r}^l_{m_2m_1}r^i_{m_1g}}{(\omega_{m_3g} + \omega_1)(\omega_{m_2g} + \omega_1 + \omega_2)(\omega_{m_1g} + \omega_4)} \right] - \sum_{m_1m_2}{}' \left[\frac{r^i_{gm_2}r^l_{m_2g}r^k_{gm_1}r^j_{m_1g}}{(\omega_{m_2g} - \omega_4)(\omega_{m_2g} - \omega_3)(\omega_{m_1g} - \omega_1)} \right.$$

$$+ \frac{r^i_{gm_2}r^l_{m_2g}r^k_{gm_1}r^j_{m_1g}}{(\omega_{m_2g} - \omega_3)(\omega_{m_1g} + \omega_2)(\omega_{m_1g} - \omega_1)} + \frac{r^l_{gm_2}r^i_{m_2g}r^j_{gm_1}r^k_{m_1g}}{(\omega_{m_2g} + \omega_4)(\omega_{m_2g} + \omega_3)(\omega_{m_1g} + \omega_1)}$$

$$\left. \left. + \frac{r^l_{gm_2}r^i_{m_2g}r^j_{gm_1}r^k_{m_1g}}{(\omega_{m_2g} + \omega_3)(\omega_{m_1g} - \omega_2)(\omega_{m_1g} + \omega_1)} \right] \right\}$$

(87)

where $r^i_{m_1m_2}$ is the transition matrix element $< m_1|r^i|m_2 >$ between the unperturbed energy eigenstates $|m_1 >$ and $|m_2 >$ with $|g >$ being the ground state, $\bar{r}^i_{m_1m_2} = r^i_{m_1m_2} - r_{gg}\delta_{m_1m_2}$, and $\hbar\omega_{m_1g}$ is the excitation energy of state m_1. The prime on the summations indicates a sum over all states but $|g >$. The operators $I_{1,2}$ and $I_{1,2,3}$ denote the average of all terms generated by permutations of all the distinct input frequencies, that is, ω_1 and ω_2 for $I_{1,2}$, and ω_1, ω_2, and ω_3 for $I_{1,2,3}$. The factors $K(-\omega_3; \omega_1, \omega_2)$ and $K(-\omega_4; \omega_1, \omega_2, \omega_3)$ arise from the distinguishable arrangements of the input frequencies that can contribute to the susceptibility expression. The numerical value is given by $K = 2^mD$ where m is the number of nonzero input frequencies minus the number of nonzero output frequencies and D is the number of distinguishable ordering of the set of input frequencies. Therefore, given the transition matrix elements and the excitation energies of the electronic states for a particular molecular structure, the second- and third-order molecular nonlinear optical susceptibility tensors $\beta_{ijk}(-\omega_3; \omega_1, \omega_2)$ and $\gamma_{ijkl}(-\omega_4; \omega_1, \omega_2, \omega_3)$ for any process with any given set of frequencies can be calculated.

Organic molecules containing conjugated electronic structures are of particular interest because they exhibit much larger nonlinear optical responses compared to nonconjugated molecules such as saturated hydrocarbons. For the nonconjugated molecules, a simple empirical rule called the bond-additive rule can be used to describe their nonlinear optical responses. This rule states that the nonlinear optical response of a molecule can be regarded as the algebraic sum of the nonlinear optical responses of the individual chemical bonds it contains. In other words, for this rule to be valid, each chemical bond can be regarded as an independent entity as far as its nonlinear optical response is concerned. This rule is a reasonable description of the nonconjugated molecule since, for saturated bonding, the electrons within the molecule are localized at the site of the bond. For conjugated molecules, however, the bond-additive rule is no longer valid. The electrons in a conjugated structure are not localized at the bonding sites, but delocalized throughout the conjugated structure. Furthermore, the electron–electron interactions are significant in a conjugated molecule so that single-electron models would not be adequate for a satisfactory theoretical description. To describe the nonlinear response of a conjugated molecule, the theoretical model must be a many-electron theory in order to account for the electron correlation effects.

We discuss below a theoretical framework for the calculation of the nonlinear optical susceptibility that is based on multiple-excited configuration interaction theory applied to a self-consistent-field molecular orbital method that explicitly accounts for the electron–electron interaction and electron correlation. The all-valence electron, self-consistent-field, molecular orbital method is employed to calculate the ground-state wavefunction and the multiple-excited configuration interaction theory that incorporate electron correlation determine the excited-state properties.[22-24]

Under the Born-Oppenheimer approximation, which assumes that the electron within a molecule adjusts to the instantaneous nuclear configuration as if the nuclei were motionless, the electronic Hamiltonian is given by

$$H(1, 2, \ldots, n) = -\frac{\hbar^2}{2m} \sum_i \nabla_i^2 - \sum_A \sum_i \frac{e^2 Z_A}{r_{Ai}} + \sum_{i<j} \frac{e^2}{r_{ij}} \tag{88}$$

where n is the number of electrons, $Z_A e$ is the charge of nucleus A, m and e are the electron mass and charge, respectively, and r_{ij} is the distance between electrons i and j. The terms in Equation 88 correspond in order from left to right to the kinetic energy of the electrons, the Coulomb attraction between nuclei and electrons, and the Coulomb repulsion between electrons. It is understood that $H(1, 2, \ldots n)$ is also a function of the given nuclear configuration, which is assumed fixed. The many-electron wave function $\Psi_m(1, 2, \ldots n)$ is given by the Schrodinger equation

$$H(1, 2, \ldots, n)\Psi_m(1, 2, \ldots, n) = E_m \Psi_m(1, 2, \ldots, n) \tag{89}$$

Equation 89 is too complicated for exact solution because the third term in the electronic Hamiltonian couples the wavefunction of each electron to that of all the other electrons in the system. In the orbital approximation, the many-electron wave function is constructed from individual wave functions determined for each electron. To maintain antisymmetry, the wave function is taken as the Slater determinant of one electron orbitals Φ_m:

$$\Psi(1, 2, \ldots, n) = |\Phi_1 \overline{\Phi}_1 \Phi_2 \overline{\Phi}_2 \cdots \Phi_{n/2} \overline{\Phi}_{n/2}| \tag{90}$$

where Φ_m and $\overline{\Phi}_m$ are one-electron orbitals of opposite spin.

The orbitals are further approximated as linear combinations of atomic orbitals (LCAO):

$$\Phi_i = \sum_\mu c_{\mu i} \phi_\mu \tag{91}$$

where ϕ_μ is an atomic orbital. The atomic overlap integral $S_{\mu\nu}$ and density matrix $P_{\mu\nu}$ are defined as

$$S_{\mu\nu} = \int \phi_\mu^*(1)\phi_\nu(1)d\tau_1 \tag{92}$$

$$P_{\mu\nu} = 2 \sum_i c_{\mu i}^* c_{\nu i} \tag{93}$$

where the sum over i in $P_{\mu\nu}$ is over each of the doubly occupied molecular orbitals. The total electronic energy E is given by $E = \langle \Psi | H | \Psi \rangle$. This can be expressed in terms of the atomic orbitals as

$$E = \sum_{\mu\nu} P_{\mu\nu} H_{\mu\nu} + \frac{1}{2} \sum_{\mu\nu\lambda\sigma} P_{\mu\nu} P_{\lambda\sigma} \left[(\mu\nu|\lambda\sigma) - \frac{1}{2}(\mu\lambda|\nu\sigma) \right] \tag{94}$$

where

$$H_{\mu\nu} = \int \phi_\mu^*(1) \left[-\frac{\hbar^2}{2m} \nabla_1^2 - \sum_A \frac{e^2 Z_A}{r_{A1}} \right] \phi_\nu(1) d\tau_1 \tag{95}$$

and

$$(\mu\nu|\lambda\sigma) = \iint \phi_\mu^*(1)\phi_\nu(1) \frac{1}{r_{12}} \phi_\lambda^*(2)\phi_\sigma(2) d\tau_1 d\tau_2 \tag{96}$$

The coefficients $c_{\mu i}$ in the LCAO expansion Equation 91 that correspond to the ground-state wave function can be obtained by a variational solution of the Schrodinger equation using the method of Lagrange multiplier. The following set of algebraic equations, known as the Roothaans's equation,[25] is the results of the variational calculation:

$$\sum_\nu (F_{\mu\nu} - \epsilon_i S_{\mu\nu}) c_{\nu i} = 0 \tag{97}$$

where the Fock matrix $F_{\mu\nu}$ is defined by

$$F_{\mu\nu} = H_{\mu\nu} + \sum_{\lambda\sigma} P_{\lambda\sigma} \left[(\mu\nu|\lambda\sigma) - \frac{1}{2}(\mu\lambda|\nu\sigma) \right] \tag{98}$$

Solution of the algebraic set of equations (Equation 97) yields the molecular orbitals Φ_i in terms of the atomic orbital coefficients $c_{\nu i}$. Since Equation 97 involves the remaining set of coefficients $c_{\nu i}$ through $P_{\mu\nu}$, it requires an iterative solution.

Further approximations and parametrizations are required. Here, we discuss a specific parametrization procedure known as the CNDO/S.[26,27] In this parametrization procedure, the atomic orbitals ϕ_μ are approximated as Slater-type orbitals where the radial part is given by

$$R_{nl}(r) = (2\xi)^{n+(1/2)}[(2n)!]^{-1/2} r^{n-1} e^{-\zeta\tau} \tag{99}$$

where n is the radial quantum number and ζ is the Slater exponent. The angular dependence of the wave function is given by the spherical harmonics $Y_{lm}(\theta,\phi)$. In Roothaan's equation (Equation 97) the zero differential overlap (ZDO) approximation is made for the overlap integrals $S_{\mu\nu}$,

$$S_{\mu\nu} = \delta_{\mu\nu} \tag{100}$$

The ZDO approximation is also made for the two-electron integrals such that

$$(\mu\nu|\lambda\sigma) = \delta_{\mu\nu}\delta_{\lambda\sigma}(\mu\mu|\lambda\lambda) \tag{101}$$

We use μ_A to denote orbital μ on atomic site A. The one-center integral $(\mu_A\mu_A|\nu_A\nu_A)$ is taken as an input parameter that is independent of orbital type,

$$(\mu_A\mu_A|\nu_A\nu_A) = \gamma_{AA} \tag{102}$$

and the two-center integral is given by the Ohno potential

$$(\mu_A \mu_A \mid \nu_B \nu_B) = \gamma_{AB} = \cfrac{14.397 e V\mathring{A}}{\left\{ \left[\cfrac{28.794 e V\mathring{A}}{\gamma_{AA} + \gamma_{BB}} \right]^2 + [R_{AB}(\mathring{A})]^2 \right\}^{1/2}} \tag{103}$$

The resulting equations stating the CNDO approximations are

$$\sum_\nu (F_{\mu\nu} - \epsilon_i \delta_{\mu\nu}) c_{\nu i} = 0 \tag{104}$$

$$F_{\mu\mu} = I_\mu - (Z_A - 1)\gamma_{AA} + \left(P_{AA} - \frac{1}{2} P_{\mu\mu} \right)\gamma_{AA} + \sum_{B \neq A} (P_{BB} - Z_B)\gamma_{AB} \tag{105}$$

$$F_{\mu\nu} = -\frac{(\beta_\mu + \beta_\nu)}{2} S_{\mu\nu} - \frac{1}{2} P_{\mu\nu}\gamma_{AB} \qquad (\mu \neq \nu) \tag{106}$$

In the above equations, I_μ is the valence state ionization energy for atomic orbitals μ, Z_A is the core charge on atom A, P_{AA} is the net electron population on atom A, and β_μ (β_ν) is a hopping integral which refers to the atom on which $\mu(\nu)$ is centered. The ZDO approximation is not used for $S_{\mu\nu}$ in evaluating $F_{\mu\nu}$.

Solution of Equation 104 yields the ground-state electronic wave function for the molecular system of interest. To obtain the excited-state wave functions and energies the orbital coefficients obtained for the ground-state orbital configuration is employed. It is found that configuration interaction (CI) theory consisting of singly and doubly excited CI is often adequate for the computation of nonlinear optical susceptibility. In the singly and doubly excited CI theory, each state of the system Ψ_n is expanded as

$$\Psi_n = A_{n,0}\Psi_0 + \sum_{k=1}^{n/2} \sum_{p=n/2+1}^{m} A_{n,kp}\Psi_{kp} + \sum_{k=1}^{n/2} \sum_{l=1}^{n/2} \sum_{p=n/2+1}^{m} \sum_{q=n/2+1}^{m} A_{n,kplq}\Psi_{kplq} \tag{107}$$

where Ψ_0 is the ground-state wave function, Ψ_{kp} is a spin-singlet, single excited configuration given by

$$\begin{aligned} \Psi_{kp} = 2^{-1/2}(&\mid \Phi_1\overline{\Phi}_1 \cdots \Phi_k\Phi_{k+1}\overline{\Phi}_{k+1} \cdots \Phi_{n/2}\overline{\Phi}_{n/2}\overline{\Phi}_p \mid \\ &- \mid \Phi_1\overline{\Phi}_1 \cdots \overline{\Phi}_k\Phi_{k+1}\overline{\Phi}_{k+1} \cdots \Phi_{n/2}\overline{\Phi}_{n/2}\Phi_p \mid) \end{aligned} \tag{108}$$

where p is a virtual orbital that is unoccupied in Ψ_0, and Ψ_{kplq} is an analogous spin-singlet doubly excited configuration. The coefficients $A_{n,R}$, where R represents the set of molecular orbitals involved in the configuration, are determined by

$$\sum_R A_{n,R}(H_{RS} - E_n S_{RS}) = 0 \tag{109}$$

where $H_{RS} = \langle \Psi_R \mid H \mid \Psi_S \rangle$, E_n is the eigenvalue of H_{RS}, and S_{RS} is the overlap between Ψ_R and Ψ_S.

The molecular dipole moment operator μ is defined as

$$\mu = -e \sum_i r_i + e \sum_A Z_A r_A \tag{110}$$

where i sums over the valence electrons and A over the atomic cores. Within the Born-Oppenheimer approximation, the second term of Equation 110 is constant and only contributes to the diagonal elements of $\mu_{nn'}$ where

$$\mu_{nn'} = \langle \Psi_n | \mu | \Psi_{n'} \rangle \tag{111}$$

Rewriting the expansion of the state function Ψ_n as

$$\Psi_n = \sum_R A_{n,R} \Psi_R \tag{112}$$

where Ψ_R is either the ground, a singly excited, or a doubly excited configuration, we obtain

$$\mu_{nn'} = \sum_{R,S} A_{n,R} A_{n',S} \langle \Psi_R | \mu | \Psi_S \rangle \tag{113}$$

where $\langle \Psi_R | \mu | \Psi_S \rangle$ is the transition dipole moment between the configuration Ψ_R and Ψ_S. We can therefore evaluate $\beta_{ijk}(-\omega_3; \omega_1, \omega_2)$ and $\gamma_{ijkl}(-\omega_4; \omega_1, \omega_2, \omega_3)$ through Equations 86 and 87 with the electronic excitation energies and transition dipole moments for the molecular system obtained from the above procedure.

VI. SECOND-ORDER NONLINEAR OPTICAL SUSCEPTIBILITY

To illustrate the basic mechanisms of the second-order virtual excitation processes and the microscopic origin of the nonresonant susceptibility for noncentrosymmetric conjugated compounds, we will use the well-known compound 2-methyl-4-nitroaniline (MNA) as an example. MNA molecule is derived from a benzene molecule by substituting an amine (NH_2) electron donor group on one end of the benzene ring structure, a nitro (NO_2) acceptor group on the opposite end of the ring, and a methyl group next to the amine group. The resultant molecular ground-state dipole moment (6.94 D) points nearly axially across the ring from the amine group to the nitro group. MNA crystallizes in the solid state in a highly acentric monoclinic (C_c) structure. X-ray studies reveal a unique projection along the crystal polar axis where individual microscopic dipole moments of the molecular units are all aligned along the polar axis. The dominance of the π-electron excitations in MNA is demonstrated by the experimental finding that along the polar axis the macroscopic second harmonic susceptibility $\chi_{ijk}^{(2)}(-2\omega; \omega, \omega)$ is the same as the linear electrooptic susceptibility $\chi_{ijk}^{(2)}(-\omega; \omega, 0)$, thereby ruling out other contributions of nonelectronic origin. The finding that the electronic contribution dominates the frequency range from DC to optical frequency has been observed in other organic molecular solids and appears to be a property common to optically nonlinear organic solids. The value of the microscopic second harmonic susceptibility $\beta_{ijk}(-2\omega; \omega, \omega)$ and its dispersion can be experimentally determined by DC electric field-induced second harmonic generation (DCSHG) measurement of liquid solutions. The experimental arrangement in this measurement requires an applied DC electric field to remove the natural center of inversion symmetry of the solution phase. Theoretically, the value of $\beta_{ijk}(-2\omega; \omega, \omega)$ can be calculated from Equation 86. The compound p-nitroaniline (PNA), the high-symmetry parent molecular unit of MNA, presents an excellent example for both experimental and theoretical analysis. In the C_{2v} symmetry for PNA and second harmonic generation, the measured principal component β_x is given by

$$\beta_x = \beta_{xxx} + [\beta_{xyy} + \beta_{xzz} + 2\beta_{yyx} + 2\beta_{zzx}]/3 \tag{114}$$

where the x-direction is along the molecular dipole axis.

In the singly and doubly excited configuration interaction calculation for PNA, the β value is dominated by virtual excitations to a highly charge correlated π-electron excited state at $\hbar\omega_{ng}$ of 4.37 eV corresponding to the first major transition in the singlet–singlet excitation spectrum of PNA. This many-body excited state is composed mainly of an A_1 configuration in which an electron is promoted from the highest energy, single-electron orbital occupied in the ground state to the lowest energy, single-electron orbital that is unoccupied in the ground state. For second-order processes, description of the many-electron wave function by singly excited configuration interaction oftentimes yields acceptable results, although there exist cases in which doubly excited configuration interactions are required. On the contrary, for third-order processes, the inclusion of doubly excited configuration interaction is absolutely essential for the calculation, as we shall discuss later.

The corresponding electron density contour diagrams are shown in Figure 2. The many-body excited state clearly demonstrates the transfer of electron density from the region of the amine donor group across the ring to the nitro acceptor group and its neighboring sites, giving the excited state a highly charge-correlated nature along the polar axis of the molecule. The change in dipole moment of the excited state with respect to the ground state, $-e\Delta r_n^x$, was calculated to be 5.8 D, and the transition dipole moment, $-e\Delta r_{gn}^x$, 5.5 D. Virtual excitations to this state account for over 90% of the magnitude of β_x of PNA as well as determine its sign. Dispersion measurement of DCSHG results in frequency-dependent values of β_x. Comparison of the theoretical calculation with the experimental values shows excellent agreement, demonstrating that the origin of the second-order nonlinear optical response of conjugated organic molecules is essentially in terms of the highly charge-correlated π-electron states.

VII. THIRD-ORDER NONLINEAR OPTICAL SUSCEPTIBILITY

Experimental and theoretical studies of β_{ijk} in noncentrosymmetric conjugated structures and of one-photon and two-photon resonant processes in centrosymmetric conjugated linear chains such as finite polyenes have demonstrated that the π-electron states are dominated by electron correlation and that, correspondingly, single-particle descriptions are inadequate. One principal result observed in polyenes, for example, is that below the first optically allowed, dominant singlet 1^1B_u state is located a two-photon singlet 2^1A_g state.[28] Calculations based on the many-electron theory have obtained the correct state ordering and have shown that the 2^1A_g state is a highly correlated π-electron state that appears in the theoretical results only upon inclusion of at least doubly excited configurations. Here, using

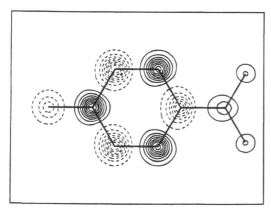

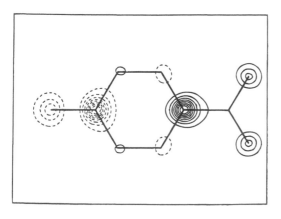

Figure 2 Electron density distribution difference contour diagram (top) and transition density contour diagram (bottom) for PNA. The contour is cut 0.4 Å above the molecular plane.

polyenes as an example, we will discuss the origin and mechanism responsible for the third-order nonlinearity of centrosymmetric conjugated structures.

Polyenes are important examples of organic conjugated linear chains, and represent the finite-chain limit to the infinite-chain polymer polyacetylene. Each carbon atom site in a polyene chain is sp^2 hybrid bonded to its three nearest neighbor carbon and hydrogen atom sites. The carbon sp^2 hybrid orbitals, together with the hydrogen 1s orbitals, combine to form bonding molecular orbitals of σ symmetry. The remaining electrons form a framework of delocalized, π-electron molecular orbitals. These weakly bonded π-electrons are responsible for low-lying electronic excitations of polyenes, and importantly, through their large transition dipole moments they dominate the molecular nonlinear susceptibility $\gamma_{ijkl}(-\omega_4; \omega_1, \omega_2, \omega_3)$ of these conjugated structures. Polyenes have a dimerized bonding structure where the short, double bonds are due to larger bonding electron density. Various structural conformations of polyene chains can be considered based on rotations about the carbon–carbon bonds. The two most common conformations, all-trans (trans) and cis-transoid (cis), are considered here. The multiple configuration interaction theory is employed to calculate the third harmonic generation (THG) susceptibility. The result of the calculation reveals that the γ_{xxxx} component of $\gamma_{ijkl}(-3\omega; \omega, \omega, \omega)$, with all fields along the direction of conjugation, is by far larger than the others. For example, the calculated values for the independent tensor components of $\gamma_{ijkl}(-3\omega; \omega, \omega, \omega)$ of trans-hexatriene (trans-HT), which is an all-trans polyene with six carbon atoms, at a nonresonant fundamental photon energy of 0.65 eV(λ = 1.907 μm) are γ_{xxxx} = 4.7, γ_{xyyx} = 0.4, γ_{yxxy} = 0.4, and γ_{yyyy} = 0.1 $\times 10^{-36}$ esu.

For centrosymmetric conjugated chain, the π-electron states have definite parity of A_g or B_u, and the one-photon transition moment vanishes between states of like parity. Since the ground state is always ^1A_g, it is evident from Equation 87 that the π states in a third-order process must be connected in the series $g \to {}^1B_u \to {}^1A_g \to {}^1B_u \to g$. Virtual transitions to both one-photon ^1B_u and two-photon ^1A_g states are necessarily involved. In the summation over intermediate states for trans-HT there are two major terms which constitute 87% of γ_{xxxx}. In both of these terms, the only ^1B_u state involved is the dominant low-lying one-photon 1^1B_u π-electron excited state. In addition to its low-energy, the importance of this state lies in the value of its transition-dipole moment with the ground state of 6.7 D, the largest among all low-lying states. For one of the two major terms, the intermediate ^1A_g state is the ground state itself; but for the other, it is the 5^1A_g state of HT. Since this state has a transition moment with 1^1B_u of 11.4 D, it is much more significant than the 2^1A_g that has a corresponding transition moment of only 2.42 D.

Important major features of the microscopic description of third-order virtual processes are contained in the transition density matrix $\rho_{nn'}$. The 2^1A_g and 5^1A_g states are both nearly 30% composed of doubly excited configurations, and the contributions to γ_{xxxx} of these two highly electron-correlated states are distinguished by $\rho_{nn'}$ defined by

$$\rho_{nn'}(\mathbf{r}_1) = \int \psi_n^*(\mathbf{r}_1, \mathbf{r}_2, \ldots, \mathbf{r}_M)\psi_{n'}(\mathbf{r}_1, \mathbf{r}_2, \ldots, \mathbf{r}_M)d\mathbf{r}_2 \cdots d\mathbf{r}_M \qquad (115)$$

so that

$$\langle \mu_{nn'} \rangle = -e \int \mathbf{r}\rho_{nn'}(\mathbf{r})d\mathbf{r} \qquad (116)$$

where M is the number of valence electrons included in the state wave function. Contour diagrams for $\rho_{nn'}$ of the ground, 2^1A_g, and 5^1A_g states with the 1^1B_u state are shown in Figure 3 where solid and dashed lines correspond to increased and decreased charge density. Whereas the virtual transition $2^1A_g \to 1^1B_u$ results in a modulated charge redistribution, the $5^1A_g \to 1^1B_u$ transition, in sharp contrast, produces a large charge separation that spans the entire chain length and results in a large transition moment. The same characteristic features are found in all other chain lengths of polyenes of both trans and cis conformations.

The importance of electron correlations to $\gamma_{ijkl}(-3\omega; \omega, \omega, \omega)$ of the conjugated linear chains is further illustrated by results obtained from calculations at the singly excited configuration interaction (SCI) level that purposely omit doubly excited configuration (DCI) but are otherwise identical. The values calculated for nonresonant $\gamma_{ijkl}(-3\omega; \omega, \omega, \omega)$ are negative in sign for all the polyene chains

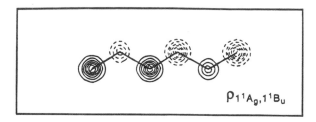

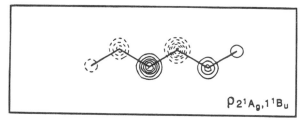

$\rho_{1^1A_g,1^1B_u}$

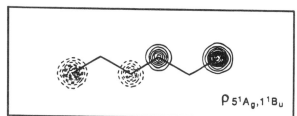

$\rho_{2^1A_g,1^1B_u}$

$\rho_{5^1A_g,1^1B_u}$

Figure 3 Transition density matrix diagrams of HT for the ground, 2^1A_g, and 5^1A_g states with the 1^1B_u state. The contour cut is taken at 0.4 Å above the HT molecular plane. The $2^1A_g \to 1^1B_u$ transition results in a charge redistribution concentrated at the center of the molecular structure which yields a small transition moment of 2.42 D. In sharp contrast, $\rho5^1A_g,1^1B_u$ for the virtual transition between the 5^1A_g and 1^1B_u states produces a large charge separation along the chain axis x-direction and an associated large transition moment of 11.40 D which dominates the contributing terms of $\gamma(-\omega_4; \omega_1, \omega_2, \omega_3)$ and is the key to understanding the nonlinear optical properties of conjugated linear chains. (From Helfin, J. R., Wong, K. Y., Zamani-Khamiri, O., and Garito, A. F., *Mol. Cryst. Liq. Cryst.*, 160, 37, 1988. With permission.)

which is contrary to experimental results.[29] This disagreement occurs because the SCI calculation improperly describes electron correlation which we have found to be of primary importance, such as in the illustrative case of the 5^1A_g state of HT. Instead, at the SCI level, $\gamma_{ijkl}(-3\omega; \omega, \omega, \omega)$ is predicted to be dominated solely by the virtual excitation process which involves only the ground and 1^1B_u states.

ACKNOWLEDGMENTS

The authors gratefully appreciate Dr. R. F. Shi for his important contribution and Dr. O. Zamani-Khamiri for her critical reading of the entire chapter.

REFERENCES

1. **Garito, A. F., Shi, R. F., and Wu, M. H.,** Nonlinear optics of organic and polymeric materials, *Physics Today,* 47, 51, 1994.
2. **Boyd, R. W.,** *Nonlinear Optics,* Academic Press, New York, 1991.
3. **Oudar, J. L.,** Optical nonlinearities of conjugated molecules, stilbene derivatives and highly polar aromatic compounds, *J. Chem. Phys.,* 64, 446, 1977.
4. **Singer, K. D., and Garito, A. F.,** Measurements of molecular second order optical susceptibilities using dc-induced second harmonic generation, *J. Chem. Phys.,* 75, 3572, 1981.
5. **Teng, C. C., and Garito, A. F.,** Dispersion of the second-order optical susceptibilities of organic systems, *Phys. Rev.,* B28, 6766, 1983.
6. **Kleinman, D. A.,** Nonlinear dielectric polarization in optical media, *Phys. Rev.,* 126, 1977, 1966.
7. **Zhou, Q. L., Shi, R. F., Zamani-Khamiri, O., and Garito, A. F.,** Negative third-order optical response in squaraines, *Nonlinear Opt.,* 6, 145, 1993.
8. **Chen, C. T., Marder, S. R., and Cheng, L. T.,** Molecular first hyperpolarizabilities of a new class of asymmetric squaraine dyes, *J. Chem. Soc. Chem. Commun.,* 259, 1994.
9. **Yu, Y. Z., Shi, R. F., Garito, A. F., and Grossman, C. H.,** Origin of negative $\chi^{(3)}$ in squaraines: Experimental observation of two-photon states, *Opt. Lett.,* 19, 786, 1994.

10. **Dirk, C. W., Cheng, L.-T., and Kuzyk, M. G.,** A three-level model useful for exploring structure/property relationships for molecular third-order optical polarizabilities, *Mater. Res. Soc. Symp. Proc.,* 247, 73, 1992.

11. **Dirk, C. W., Cheng, L.-T., and Kuzyk, M. G.,** A simplified three-level model describing the molecular third-order optical susceptibility, *Int. J. Quant. Chem.,* 43, 27, 1992.

12. **Whitham, G. B.,** *Lecture on Wave Propogation,* Springer-Verlag, New York, 1979.

13. **Miller, R. C.,** Optical second harmonic generation in piezoelectric crystals, *Appl. Phys. Lett.,* 5, 17, 1964.

14. **Garrett, C. G. B., and Robinson, F. N. H.,** Miller's phenomenological rule for computing nonlinear susceptibilities, *IEEE J. Quant. Electron.,* QE-2, 328, 1966.

15. **Levine, B. F.,** Bond charge calculation of nonlinear optical susceptibilities for various crystal structures, *Phys. Rev.,* B7, 2600, 1973.

16. **Lalama, S. J., and Garito, A. F.,** Origin of the nonlinear second-order optical susceptibilities of organic systems, *Phys. Rev.,* A20, 1179, 1979.

17. **Butcher, P. N., and Cotter, D.,** *The Elements of Nonlinear Optics,* Cambridge University Press, Cambridge, 1990.

18. **Helfin, J. R., and Garito, A. F.,** Third-order optical processes in linear chains: Electron correlation theory and experimental dispersion measurements, in *Polymers for Lightwave and Integrated Optics,* Hornak, L. A., Eds., Marcel Dekker, New York, 1992.

19. **Cohen, H. D., and Roothaan, C. C. J.,** Electric dipole polarizability of atoms by the Hartee-Fock method, *J. Chem. Phys.,* 43, S34, 1965.

20. **Bogoliubov, N. N., and Mitropolsky, Y. A.,** *Asymptotic Methods in the Theory of Nonlinear Oscillations,* Gordon & Breach, New York, 1961.

21. **Orr, B. J., and Ward, J. F.,** Perturbation theory of the nonlinear optical polarization of an isolated system, *Mol. Phys.,* 20, 513, 1971.

22. **Salem, L.,** *The Molecular Orbital Theory of Conjugated Systems,* Benjamin, New York, 1966.

23. **Pople, J. A., and Beveridge, D. L.,** *Approximate Molecular Orbital Theory,* McGraw-Hill, New York, 1970.

24. **Daudel, R., Elroy, G., Peters, D., and Sana, M.,** *Quantum Chemistry,* Wiley, New York, 1983.

25. **Roothaan, C. C. J.,** New developments in molecular orbital theory, *Rev. Mod. Phys.,* 23, 69, 1951.

26. **Delbene, J., and Jaffe, H. H.,** Use of the CNDO method in spectroscopy. I. Benzene, pyridine and the diazines, *J. Chem. Phys.,* 48, 1807, 1968.

27. **Lipari, N. O., and Duke, C. B.,** The electronic structure of dialkylbenzenes, *J. Chem. Phys.,* 63, 1748, 1975.

28. **Hudson, B. S., Kohler, B. E., and Schulten, K.,** *Excited States,* Vol. 6, Lim, E. C., Eds., Academic Press, New York, 1982.

29. **Helfin, J. R., Wong, K. Y., Zamani-Khamiri, O., and Garito, A. F.,** Symmetry-controlled electron correlation mechanism for third-order nonlinear optical properties of conjugated linear chains, *Mol. Cryst. Liq. Cryst.,* 160, 37, 1988.

Chapter 2

Computational Evaluation of Second-Order Optical Nonlinearities

John O. Morley and David Pugh

CONTENTS

I. INTRODUCTION

It is well established that conjugated organic molecules containing electron donor and acceptor groups exhibit extremely large optical nonlinear responses with applications in both second harmonic generation (SHG) and linear electro-optic modulation (LEO).[1–3] In an external electric field, the response of an organic molecule can be expressed in terms of the induced polarization (P_i) as:

$$P_i = \alpha_{ij}E_{ij} + \beta_{ijk}E_jE_k + \gamma_{ijkl}E_jE_kE_l + \cdots \tag{1}$$

where α, β, and γ are the polarizability, the first hyperpolarizability, and second hyperpolarizability, respectively. Molecules with large first hyperpolarizabilities, β_{ijk}, give rise to large second-order susceptibilities in noncentrosymmetric materials which can be utilised for applications in nonlinear devices. Experimentally, the first hyperpolarizability is evaluated by electric field-induced second harmonic generation (EFISH) in solution using a strong DC field for alignment.[1,3]

Theoretically, the first hyperpolarizability, β, of any molecular species can be evaluated by a number of quantum mechanical treatments though the present selective review is concerned mainly with the sum-over-states approach.[1–3] Historically, a number of groups[4–8] have used this approach to calculate the frequency-dependent hyperpolarizability using an all-valence electron or simpler π-electron treatment, to give values which have generally shown a good correlation with experimental data. Recently, Kanis et al.[9] have comprehensively reviewed the theoretical methods and results obtained on numerous organic compounds up to early 1993. For this reason, the present review has been directed at more recent studies which involve an examination of the role of single and double excitations in the evaluation of the hyperpolarizability tensor, a discussion of selected results on speculative organic structures for LEO and SHG applications, and an analysis of the influence of structure on the calculated hyperpolarizability.

II. THEORETICAL DISCUSSION

A. SUM-OVER-STATES METHOD

There are a number of derivations of the relevant formulae in the literature, where hyperpolarizabilities are often defined in slightly different ways in different publications. In the case of second-order formulae, the only possibilities for ambiguity relate to the presence or absence of factors of 2 or (1/2). The origins of this problem lie either in the alternative definitions of β through a power series,

$$\delta\mu = \alpha E + \beta E^2 + \cdots \tag{2}$$

or a Taylor series,

$$\delta\mu = \alpha E + (1/2)\beta E^2 + \cdots \tag{3}$$

or in confusion as to the definition of the input electromagnetic field, which may be represented as

$$E = E_\omega \sin(\omega t) \quad \text{or} \quad E = E_\omega e^{i\omega t} \tag{4}$$

It is therefore important to define carefully the quantities of interest. First consider the molecule subjected to an unspecified perturbation H'. The perturbed wave function after transient terms have decayed can be expressed, following Ward,[10] as

$$|\Psi^+\rangle = (1 + GH' + GH'GH' + \cdots)|g\rangle \tag{5}$$

with complex conjugate,

$$\langle\Psi^+| = \langle g|(1 + H'^*G^+ + H'^*G^+H'^*G^+ + \cdots) \tag{6}$$

where only those terms that may contribute in second order are retained. G is a propagator which enables the frequency denominators in the perturbation expansion to be written down automatically and $|g\rangle$ is the ground-state eigenfunction. The rules for the application of G are given in references 10 and 11. In second-order effects the perturbation, H', must act twice, so that the second-order-induced dipole is given by

$$\delta\mu^{(2)} = -\langle\Psi^+|er|\Psi^+\rangle$$
$$= -e[\langle g|rGH'GH'|g\rangle + \langle g|H'^*G^+H'^*G^+r|g\rangle + \langle g|H'^*G^+rGH'|g\rangle] \tag{7}$$

where $\mathbf{r}$ is a sum over all electronic coordinates.

In all the work described in this article, a molecule in its ground state is perturbed by electromagnetic fields. The unperturbed reference state is the ground state of the molecule and it greatly simplifies the analysis of the perturbation series if the large diagonal matrix elements $\langle g|r|g\rangle$ can be removed. Since it is the electronic part of the Hamiltonian which is of interest, this can be done formally by making a transformation of the coordinate system to the electronic charge centroid of the ground state. An equivalent transformation which can also be applied in more general cases is given in reference 12. Only diagonal matrix elements are affected by a translation of the origin of the coordinate system. Matrix elements in the charge centroid system are denoted by $\mathbf{r}_{n'n}$ and, in terms of those calculated in an arbitrarily translated system,

$$\mathbf{r}_{n'n} = \langle n' | \bar{\mathbf{r}} | n \rangle = \langle n' | \mathbf{r} | n \rangle - \langle g | \mathbf{r} | g \rangle \delta_{n'n} \tag{8}$$

Multiplying by the electronic charge and adding and subtracting the nuclear dipole also gives

$$\mathbf{r}_{nn} = -(1/e)(\mu_n - \mu_g) \tag{9}$$

leading to the important result that a large change in dipole moment between the ground and excited state is an essential requirement if the excited state is to contribute directly to the first hyperpolarizability.

Returning to Equation 7, terms can be picked out that lead to the term in $\delta\mu^{(2)}$ with a particular frequency dependence. The two cases of most general interest, which are the subjects of the investigations described later in the review, are second harmonic generation (SHG) and the linear electro-optic effect (LEO), also known as the Pockels effect.

B. SECOND HARMONIC GENERATION

In this case the input field will be at a single laser frequency and can be taken as

$$\mathbf{E} = \mathbf{E}_\omega \sin \omega t = (1/2i)\mathbf{E}_\omega(e^{i\omega t} - e^{i\omega t}) \tag{10}$$

and the perturbing Hamiltonian is

$$H' = -\frac{e\mathbf{r}}{2i} \mathbf{E}_\omega(e^{i\omega t} - e^{-i\omega t}) \tag{11}$$

This is substituted into Equation 7 and the terms in $\delta\mu^{(2)}$ leading to a 2ω response picked out. If

$$\delta\mu_i^{(2)}(2\omega) = \delta\mu_i^{(2)}(2\omega_+)e^{i\omega t} + \delta\mu_i^{(2)}(2\omega_-)e^{-i\omega t} \tag{12}$$

it is found that

$$\delta\mu_i^{(2)}(2\omega_+) = \beta_{ijk}(-2\omega; \omega, \omega)E_j E_k \tag{13}$$

and

$$\beta_{ijk}(2\omega; -\omega, -\omega) = \beta_{ijk}(-2\omega; \omega, \omega), \tag{14}$$

so that the complete 2ω response is

$$\delta\mu_i^{(2)}(2\omega) = 2\beta_{ijk}(-2\omega; \omega, \omega)E_j E_k \cos(2\beta\omega t) \tag{15}$$

$$= \beta_{ijk}^{SHG} E_j E_k \cos(2\omega t) \tag{16}$$

The quantity β_{ijk}^{SHG} gives the second harmonic response and is obtained from the formula (deduced from Equation 7):

$$\beta_{ijk}^{SHG} = \frac{e^3}{4\eta^2} \sum_{n,n'} \left[\begin{array}{l} r_{gn'}^j r_{n'n}^i r_{ng}^k \left\{ \dfrac{1}{(\omega_{n'g} - \omega)(\omega_{ng} + \omega)} + \dfrac{1}{(\omega_{n'g} + \omega)(\omega_{ng} - \omega)} \right\} \\[2em] + r_{gn'}^i r_{n'n}^j r_{ng}^k \left\{ \dfrac{1}{(\omega_{n'g} + 2\omega)(\omega_{ng} + \omega)} + \dfrac{1}{(\omega_{n'g} - 2\omega)(\omega_{ng} - \omega)} \right\} \\[2em] + r_{gn'}^j r_{n'n}^k r_{ng}^i \left\{ \dfrac{1}{(\omega_{n'g} - \omega)(\omega_{ng} - 2\omega)} + \dfrac{1}{(\omega_{n'g} + \omega)(\omega_{ng} + 2\omega)} \right\} \end{array} \right] \tag{17}$$

C. LINEAR ELECTRO-OPTIC EFFECT

In this case a large low-frequency field is applied across a material through which light of frequency ω is passing. The refractive index of the medium for radiation of frequency ω is changed by an amount proportional to the low-frequency field. The perturbation in this case is

$$H' = -e\mathbf{r} \cdot (\mathbf{E}_0 + \mathbf{E}_\omega \sin \omega t) \tag{18}$$

and the cross terms between $\mathbf{E}_0$ and $\mathbf{E}_0$ in the quadratic contributions to H' lead to an induced dipole contribution of frequency ω proportional to the products $E_0^j E_\omega^k$. It is this dipole which gives rise to the linear electro-optic effect and its magnitude is connected to the product of the field strengths by a hyperpolarizability defined by the equation

$$\delta\mu_i^{(2)}(2\omega) = \beta_{ijk}^{LEO}(-\omega; \omega, 0)E_\omega^j E_0^k \tag{19}$$

The hyperpolarizability is obtained from the perturbation theory as:

$$\beta_{ijk}^{LEO}(-\omega; \omega, 0) = -\frac{e^3}{\eta^2} \sum_{n',n} \left[\begin{array}{l} r_{gn'}^i r_{n'n}^j r_{ng}^k \left\{ \dfrac{1}{(\omega_{n'g} + \omega)(\omega_{ng} + \omega)} + \dfrac{1}{(\omega_{n'g} - \omega)(\omega_{ng} - \omega)} \right\} \\[2em] + r_{gn'}^j r_{n'n}^i r_{ng}^k \left\{ \dfrac{1}{\omega_{n'g}(\omega_{ng} + \omega)} + \dfrac{1}{\omega_{n'g}(\omega_{ng} - \omega)} \right\} \\[2em] + r_{gn'}^j r_{n'n}^k r_{ng}^i \left\{ \dfrac{1}{\omega_{n'g}(\omega_{ng} + \omega)} + \dfrac{1}{\omega_{n'g}(\omega_{ng} - \omega)} \right\} \end{array} \right] \tag{20}$$

D. IMPLEMENTATION OF SUM-OVER-STATES FORMULA USING SEMI-EMPIRICAL METHODS: CONFIGURATION INTERACTION

The sum-over-states (SOS) formulae are independent of any particular system for calculating excited states. They refer to the exact excited states of the molecule. In principle, the transition energies and most of the matrix elements are quantities that could be measured by various spectroscopic techniques. Except for the most simplified approximate treatments, such as the two-state model, such an approach is not practicable. For the simpler donor/acceptor structures such as 4-nitroaniline (p-NA) (1) and 2-methyl-4-nitroaniline (MNA) (2), an analysis of the first hyperpolarizability has been carried out using a direct calibration based on the experimental excitation energy of the charge-transfer state.[5,13]

1   2   (Me = CH_3)

There is good reason to suppose that the exceptionally large β values of interest in organic molecules arise from the contributions of only a few of the lowest excited states. The two-state (ground state and charge-transfer state) model is the extreme case. If the large β values found in unsaturated, conjugated systems can indeed be attributed to the response of the mobile π-electrons, then it is reasonable to suppose that a theory that copes with the lowest part of the electronic excitation spectrum should be adequate to identify the principal contributing terms in an SOS expansion.

The investigation of the electronic energy spectrum of conjugated molecules using semi-empirical quantum theory has a long history, particularly in connection with dyestuffs.[14] The PPP method has been successful where the π-electron system can be sharply isolated, but more general, all-valence electron methods are necessary in most cases, the most well known of these being the CNDO/S scheme for spectroscopic calculations.[14] More recently INDO-based methods have also come into prominence (see, for example, references 8 and 9).

The approach to β calculations based on the SOS method is, therefore, to select a semi-empirical scheme that gives good excitation energies for the lower excited states, as determined by a comparison with experiment, and to use it to calculate the quantities required in the SOS expressions. The parametrization of the semi-empirical method serves, essentially, as a calibration of the SOS calculation, since the SOS theory shows that the excitation energy of the lowest dipole-allowed excited states is of critical importance. The SOS series must be truncated at some point and this can be done in a number of ways, some of which are described below. When approaching the SOS method in this way, the aim is clearly to find a good approximation to the energies and matrix elements of the low-lying electronic states and therefore get close to reproducing the true values of the leading terms in the SOS expansion. If the SOS method is the starting point, then a reasonable approximation to the true excited states is the first requirement. It is here that the SOS approach differs markedly from the finite field method and its variants, where a modified Hartree-Fock ground state is calculated. If the excited states are adequately reproduced by the Hamiltonian and basis set employed, then the frequency dependence of β should be reasonably accounted for.

There are currently a number of semi-empirical schemes in use for calculating β. Details of most of these can be found in reference 8. Here the CNDOVSB system, developed from the CNDO/S semi-empirical scheme by the authors,[6] is used in the following paragraphs as a basis for a discussion of the effect of the different levels of approximation that can be used within a given semi-empirical system on the results of first hyperpolarizability calculations. The hyperpolarizability calculations are made using the SOS expansion and also through the related Correction Vector Method.[15] The CNDOVSB Hamiltonian has been described in a number of publications.[2,6,11] It is essentially a CNDO/S Hamiltonian adjusted to give good correlation with experimental transition energies and ground-state dipole moments for a range of conjugated (mainly aromatic) molecules, the great majority of which were of donor–acceptor substituted type.

First, the method used for singly excited configuration interaction (SCI) calculations, which form the basis for much of the work described later in this review, is described. The Hartree-Fock equations (HF) (in the Roothaan-Hall scheme)[14] are first solved to give the N HF orbitals (ψ_i, $i = 1, N$) with orbital energies, ε_i, from which the ground-state HF determinant, Φ_{HF} is constructed. The excited states, ψ_n, are expanded over a set of singly excited singlet configurations (Slater determinants) obtained from Φ_{HF} by exciting one electron to a virtual HF orbital:

$$\Psi_n = \sum_{i \to j} C_{n,i \to j} \Phi_{i \to j} \qquad (21)$$

There is no interaction between the singly excited configurations Φ_{i-j} and the HF ground state, so that the ground state is unmodified in this form of configuration interaction (CI). The CI matrix formed by the singly excited states interacting with each other is diagonalized to obtain the excited states, Ψ_l and their energies ε_l. The matrix elements for use in the SOS equation are given by

$$\mathbf{r}_{n'n} = \sum_{i\to j}\sum_{k\to l} C_{n',i\to j}C_{n,k\to l}[\langle j|\mathbf{r}|l\rangle\delta_{ik} - \langle i|\mathbf{r}|k\rangle\delta_{jl}]$$

$$\mathbf{r}_{gn} = \sqrt{2}\sum_{i\to j} C_{n,i\to j}\langle i|\mathbf{r}|j\rangle$$

(22)

Usually a basis set including the 50 lowest energy singly excited (singlet) configurations has been used. The values of the calculated excitation energies, adjusted in the parametrization of the method, are not sensitive to the size of the many-electron basis set. In the SOS calculation it is found that the major contributions come from a small number of states, often, but not always, dominated by a large contribution from a clearly identifiable charge-transfer state. Most of the significant contributions are clearly related to low-lying π-π* states, although the effects of nondiagonal terms involving factors of the type $r_{gn'}r_{n'n}r_{ng}$, where n' and n are different, are often significant. The use of 50 excited states therefore usually provides safe coverage within which significant contributions will be picked up. The method described in the previous paragraph has been extensively used and further applications are discussed later in the paper. The SCI level of approximation is adequate to reproduce the essential features of the frequency-dependent first hyperpolarizabilities. It represents a minimal level of approximation, in that multiply excited configurations have been rather arbitrarily excluded. The Hamiltonian in a semi-empirical calculation is parametrized through the adjustment of various key integrals to obtain agreement with experimental data. The adjustment of the integrals, which differ from values which could in some cases be computed, is certainly intended to take some account of the higher order correlations between different electrons which would be explicitly represented through the inclusion of multiply excited configurations. Accordingly, it is not apparent that it is valid to consider that the addition of higher order configurations necessarily improves the calculation. Over-elaboration of semi-empirical calculations is probably not desirable. Nevertheless there are reasons for exploring the effect of the inclusion of multiply excited configurations on calculations of β, which will be discussed as the extension of the CNDOVSB method to multiple CI is described in the next section.

E. MULTIPLY EXCITED CI: EFFECTS ON β

The linked matrix elements appearing in the SOS formula in first, second, and third order are, respectively,

$$\langle g|\bar{\mathbf{r}}|n'\rangle\langle n'|\bar{\mathbf{r}}|g\rangle, \quad \langle g|\bar{\mathbf{r}}|n'\rangle\langle n'|\bar{\mathbf{r}}|n\rangle\langle n|\bar{\mathbf{r}}|g\rangle, \quad \text{and} \quad \langle g|\mathbf{r}|n''\rangle\langle n''|\bar{\mathbf{r}}|n'\rangle\langle n'|\bar{\mathbf{r}}|n\rangle\langle n|\bar{\mathbf{r}}|g\rangle \quad (23)$$

If there were no interaction under the ground-state Hamiltonian between the individual configurations, then each excited state would be represented by a single configuration. States connected by the one-electron operator, $\mathbf{r}$, can differ only in one orbital. Hence, in the first- and second-order formulae, $|n\rangle$ and $|n'\rangle$ must be singly excited configurations. In the third-order formula, even in this simplified, noninteracting scheme, the intermediate state $|n'\rangle$ can be a doubly excited configuration. Double excitations therefore have a much more direct role in third-order calculations than in first- and second-order calculations.[11,16] Their effect has been investigated in a number of publications and it has been established that their presence is essential if reasonable estimates of γ are to be obtained. If only single excitations are included in such a calculation, the sign of the leading component of the γ tensor is usually incorrect. The general structure of the theory suggests, however, that the importance of multiply excited configurations in second-order calculations is much less since they do not appear directly in the perturbation theory terms. It is at least possible that their effect may be subsumed into the parametrization.

Nevertheless, there are reasons for pressing the question further. Doubly excited configurations mix under the unperturbed Hamiltonian, with Φ_{HF} and with the singly excited configurations, Φ_{i-j}. The singly excited configurations are thereby indirectly mixed with the ground state and every molecular electronic state may include contributions from terms of all types in the *singly and doubly excited configuration interaction* (SDCI) expansion:

$$\Psi_n = C_{n,\text{HR}}\Phi_{\text{HF}} + \sum_{i\to j} C_{n,i\to j}\Phi_{i\to j} + \sum_{i\to j,k\to l} C_{n;\to j,k\to l}\Phi_{i\to j,k\to l} \tag{24}$$

The excited states themselves may be modified significantly in ways that are not accounted for by changes of parametrization. One well-known example is the behavior of the lowest states in the fully conjugated polyenes where the ordering of the states is changed when multiply excited configurations are introduced into the calculation and the $2A_{1g}$ state appears at an energy lower than the $1B_{1u}$ state.[17] There is experimental evidence that the calculated ordering is correct. In fact this low-energy A_{1g} state does not have a significant effect even on the third-order calculations, but the result provides evidence of possible large changes in the structure of the excited states when multiply excited configurations are included. A systematic study of the effect of including double and higher order CI in β calculations is therefore necessary. It has been essential to develop procedures for carrying out such calculations in order to be able to study γ by the SOS method and the same techniques can be employed for the easier β calculations.

The central problem of multiply excited configuration interaction (MCI) work is that the number of configurations that must be considered increases very rapidly with the order of the CI and the size of the molecule. In the third-order work the triple summation in the expansion for γ in the SOS method and the calculation of the associated matrix elements can also absorb appreciable amounts of computer time. In the second order, the SOS summation itself does not usually pose any computational problems. The computational difficulty in the semi-empirical theory is therefore principally the diagonalization of the CI matrix which may have a dimension of many thousands. For the SOS method it is necessary to perform a complete diagonalization and compute all the eigenvalues and eigenvectors for the multiconfigurational basis chosen. An alternative approach, the correction vector method,[15,18-20] which avoids this difficulty, is described in the next subsection. The use of the correction vector method is extremely advantageous in the case of third-order calculations where the use of very large numbers of configurations is essential. In the case of second-order calculations it enables the effect of incorporating larger basis sets to be investigated and, although results obtained for β by this procedure have not, so far, been qualitatively different from those obtained by the SOS method with a more restricted set of configurations, it is probable that the method will become the standard procedure for larger molecules and also in future calculations using *ab initio* methods with large sets of polarizing functions. The correction vector method and its relation to the SOS method are described in the next section.

F. CORRECTION VECTOR METHOD[15, 18-20]

Suppose that a set of configurations (Φ_I, $I = 1$, M) formed by exciting electrons from the occupied HF orbitals to virtual orbitals has been defined. The set may include configurations of different degrees of multiple excitation.

In the SOS method the Φ_I's are used as a basis for the expansion of the molecular eigenstates,

$$\Psi_n = \sum_I C_{nI}\Phi_I \tag{25}$$

and the perturbed wave function expanded over the resulting Ψ_n. The fact that the Ψ_n's are eigenfunctions of the unperturbed molecular Hamiltonian, as represented in the semi-empirical model being used, leads to the particular features of the SOS series such as the introduction of the excitation energies in the denominators. A very useful feature of the SOS series is that the Ψ_n can be arranged in order of increasing energy, so that the largest terms, those with the smallest energy denominators, appear first. The series can then be truncated at some point in the expectation that the largest terms will have been retained. This is the basis of all the simpler treatments such as the two-state model. If, however, we consider the expansion of the perturbed ground-state wave function in terms of the whole of the set of M configurations, Φ_I, $I = 1$, M

$$\Psi'_g = \sum_n C_{gI}\Phi_I \tag{26}$$

then the result will be identical with that obtained by expanding in terms of the set of (Ψ_n, $n = 1$, M) into which they have been transformed by the CI calculation. The advantage of working with the noninteracted set of configurations, Ψ_I, is that the complete diagonalization of the CI matrix does not have to be carried out, as is explained below.

All second-order effects can be broken down into a superposition of the results of the action of two perturbations, $\mathbf{E}_1 \exp(i\omega_1 t)$ and $\mathbf{E}_2 \exp(i\omega_2 t)$, in succession. The ground state is first perturbed by the ω_1 field when it is modified by the addition of a term $\mathbf{E}_1 \cdot \varphi_1(\omega_1) \exp\{-i(\omega_g - \omega_1)\}$, where $\omega_g = E_g/\eta$ is the intrinsic frequency factor for the ground state. The field at ω_2 then acts on the first-order perturbation of the wave function to give the second-order correction $\phi_2 (\omega_2 + \omega_1; \omega_1):\mathbf{E}_2\mathbf{E}_1 \exp\{-i(\omega_g - \omega_2 - \omega_1)\}$. In the case of SHG the response to the real perturbation of Equation 11 is made up from contributions, $\varphi_2 (2\omega; \omega)$ and $\varphi_2(-2\omega; -\omega)$, each of which is produced by the action, in second order, of one of the complex components of the field on the first-order corrections, $\varphi_1(\omega)$ and $\varphi_1(-\omega)$. The first- and second-order correction vectors are found from the following equations:[17,19,20]

$$(H_0 + \eta\omega_g + \eta\omega_1)\varphi_1^i(\omega_1) = er_i\psi_g \tag{27}$$

$$\{H_0 - \eta\omega_g + \eta(\omega_1 + \omega_2)\}\varphi_2^{ij}(\omega_1 + \omega_2; \omega_1) = er_i\varphi_1^j(\omega_1) \tag{28}$$

The induced dipole at 2ω is calculated from the perturbed wave function and the hyperpolarizability extracted to give (for SHG):[18]

$$\beta_{ijk}(-2\omega; \omega, \omega) = \frac{e}{2} \hat{P}[\langle\psi_g|r_i|\varphi_2^{jk}(2\omega; \omega)\rangle + \langle\varphi_2^{jk}(-2\omega; -\omega)|r_i|\psi_g\rangle + \langle\varphi_1^j(-\omega)|r_i|\varphi_1^k(\omega)\rangle] \tag{29}$$

where the permutation operator, $\hat{P}$, implies that all terms resulting from the permutation of the jk coordinates must be added. The problem of calculating β, therefore, reduces to that of finding the correction vectors from Equations 27 and 28. The reorganization of the perturbed Schrodinger equations into the correction vector formulism is completely general and is not linked at this stage to any particular choice of multielectron basis set. The method of solution to be employed for Equations 27 and 28 is still to be selected. If the φ_1 and φ_2 were to be expanded in terms of the eigenfunctions of the unperturbed Hamiltonian, H_0, then the correction vector equations would reduce to the sum-over-states expansion and nothing would have been gained. The advantages of the correction vector method are realized when φ_1 and φ_2 are expanded directly in terms of the basis set of multiply excited configurations

$$\varphi_1^i(\omega) = \sum_I C_I^i(\omega)\Phi_I \tag{30}$$

$$\varphi_2^{ij}(\omega_2 + \omega_1; \omega_1) = \sum_I C_I^{ij}(\omega_2, \omega_1)\Phi_I \tag{31}$$

The only eigenfunction that must be known to apply Equations 27 and 28 is the ground-state function,

$$\psi_g = \sum_I g_I\Phi_I \tag{32}$$

which must be expanded over the same basis set. Substitution of these expansions into Equations 27 and 28, multiplication by one of the basis functions, and integration over all electron coordinates lead to a set of simultaneous linear equations for the expansion coefficients

$$\sum_I \{\langle J|H_0|I\rangle + \eta(\omega - \omega_g)\delta_{JI}\}C_I^i(\omega) = e \sum_I g_I\langle J|r_i|g\rangle \tag{33}$$

$$\sum_I \{\langle J|H_0|I\rangle + \eta(\omega_1 + \omega_2 - \omega_g)\delta_{JI}\}C_I^{ij}(\omega_2, \omega_1) = e \sum_I C_I^j(\omega_1)\langle J|r_i|I\rangle \tag{34}$$

These simultaneous equations are solved directly to find the expansion vectors, $\mathbf{C}_I (\omega)$ and $\mathbf{C}_I(\omega_2,\omega_1)$ for all components. The solutions of the linear inhomogeneous equations are easy to obtain if the associated matrix is positive definite since the Gauss-Seidel iteration procedure is efficient and sure to converge. If the associated matrix is nondefinite, the Gauss-Seidel iteration scheme does not converge and one has to resort to conjugate gradient methods which are slow but surer. In the latter case the recent algorithm of Ramasesha has been found to be more efficient. We have therefore used the Gauss-Seidel scheme for positive omegas and Ramasesha's method for negative values.[21,22] A knowledge of the expansion coefficients, g_I, of the ground state in terms of the same basis set is required to carry out the procedure. It is possible to carry out an exact diagonalization of the CI matrix for a few low-lying states, using the Davidson algorithm, even for very large basis sets and the expansion of the ground state is obtained in this way.

Provided the same basis set of configurations is employed for all expansions of eigenfunction, including the ground state, and correction vectors, the SOS method and the correction vector method should lead to identical results for β. The SOS method converts the basis set of configurations (Φ_I, $I = 1$, M) into the corresponding set of eigenfunctions of H_0 in the same Hilbert space (Ψ_n, $n = 1$, M). The advantage of the SOS method is that, since the Ψ_I are eigenfunctions of H_0, the terms of the SOS expansion can be arranged in order of increasing energy, giving a criterion for truncation of the series, and the energy eigenfunctions appear automatically in the expansion and the terms can sometimes be related directly to spectroscopic observations. The disadvantages are that the CI matrix has to be diagonalized for all the states used in the expansion. The correction vector avoids the necessity for extensive diagonalization of the CI method, but loses some of the physical directness of the SOS method. It is essential for large basis sets.

The correction vector method, since it is specially suited to the treatment of very large numbers of configurations, has been extensively used to carry out full CI calculations (FCI) within a given one-electron basis set. The number of configurations that arise in FCI[18,20] calculations is such that, even with the correction vector approach, such computations are restricted in practice to PPP models where the basis set consists of only one orbital from each heavy atom. The advantage of FCI is that it is *size consistent*, whereas in any less general form of CI it is necessary to reassess the degree of multiple excitation that is necessary as the size of the molecule is increased. Accounts of the meaning and implications of size consistency can be found in standard quantum chemistry references. When FCI is used, the initial step in all the procedures described above—the solution of the Roothaan-Hall equations to obtain the Hartree-Fock type orbitals from which the many-electron determinants are constructed—becomes unnecessary. Since *all* configurations constructed from the one-electron basis are to be used, there is no advantage in attempting to find a set of one-electron molecular orbitals that are arranged according to an energy criterion. The FCI calculations have therefore employed a *generalized valence bond* method which makes use of localized spin-adapted functions, with the advantage that very sparse matrices result, reducing computer storage requirements and generally leading to a mathematical structure that is well adapted to the correction vector approach.[17] In the work described here[17,19,20,23] we have attempted to use the correction vector method with a Hartree-Fock type one-electron basis in the context of an all-valence electron minimum basis set model. For most of the work the many-electron basis has consisted of single and double excitations from occupied to virtual HF orbitals. The advantages of sparseness and complete size consistency in the GVB, FCI treatments have been lost, but the correction vector method enables much larger numbers of configurations to be used in comparison with SOS calculations. In the following sections the results, for β calculations, of the SDCI correction vector approach are compared with that of the SCI models used earlier with the SOS method for some typical series of molecules.

G. POLYENES AND POLYPHENYLS[17]

The molecular geometries of the donor–acceptor polyenes (3) and polyphenyls (4) have been found either by modification of templates obtained for protoptype compounds from crystallographic data, or by molecular modeling using structural packages. The structures used in reference 17 are identical with those used in previous work employing the SCI method.

The Hartree-Fock molecular orbital calculations, which provide the basis for the CI expansions, have been calculated with the CNDOVSB method. Excited configurations have been constructed by promoting one or two electrons from the π-molecular orbitals only, since the major nonlinearities have been linked, both theoretically and experimentally to the response of the π-electron system. For smaller molecules it is possible to use a many-electron basis set that includes all singly and doubly excited configurations between π-molecular orbitals, but this becomes impracticable for the larger homologs.

Here an alternative strategy is employed which is analogous to the complete active subspace self-consistent field (CASSCF) approach often used in *ab initio* calculations of excited states. A set of molecular orbitals consisting of those of appropriate symmetry that lie closest to the energy gap between the occupied and unoccupied states is selected. For the larger molecules investigated here this set has been restricted to include only 10 molecular π-orbitals, which, because of the extra π-orbitals donated by the N atoms in the NMe_2 and NO_2 groups, are occupied by 12 electrons. Six of the chosen orbitals are then below the energy gap (HF occupied orbitals) and four are above (HF virtual orbitals). All single and double excitations between these orbitals have been included to give a total of 325 configurations.

3   4

The polyene (**3**, $n = 3$) and the polyphenyl (**4**, $n = 2$) have the same number of π-electrons and are the largest of the molecules treated where it has been possible to carry out a calculation in which all single and double excitations between π-orbitals are included (2080 configurations) The results of the complete π-SDCI calculation are included in this case for the sake of comparison. The analysis of the restricted versus the full SDCI treatment for both the polyene (**3**, $n = 3$) and the polyphenyl (**4**, $n = 2$), shows that most of the qualitative features are reproduced in the active space formed from the 10 π-orbitals and that the introduction of a larger active space only has the effect of producing a small overall reduction in the hyperpolarizability. It has usually been found that the introduction of more electron correlation into molecular calculations tends to reduce the hyperpolarizabilities.

The results of the SDCI calculations are also compared with those obtained using the SCI method which are similar, but not identical, to the earlier SCI work. The differences arise because only those singly excited configurations derived from promoting electrons within the set of 10 π-orbitals defined above have been used in the present studies, whereas in the earlier work a larger set was selected, out of all possible single excitations in the CNDO basis set, strictly in order of increasing energy of the configuration before interaction (see, for example, reference 6). At least 10 occupied and 10 virtual orbitals were involved in this set of configurations.

The ground-state dipole moments and the λ_{max} values calculated with the SCI and SDCI treatments for the polyenes and polyphenyls show a number of differences. For the polyphenyls, the change in λ_{max} between the SCI and SDCI values is marginal, only about 2 nm except for the first member of the series where it is 10 nm. The λ_{max} values from the SDCI calculation are always greater than those from the SCI calculation. In the case (2* in Table 1) in which all the π-orbitals configurations are included, the SDCI value of λ_{max} is reduced by about 10 nm. In the substituted polyenes the differences between the SCI and SDCI λ_{max} values are much greater, with the SDCI values being lower except in the first two homologs. The general trend with n is similar to that found in the unsubstituted polyenes, although in the latter the SDCI λ_{max} is greater than the SCI value. The larger discrepancies between the SCI and SDCI results in the unsubstituted polyenes are related to the strong correlation effects that lead to the appearance of a low-lying ^1A_g excited state below the lowest dipole allowed ^1B_u state. In the polar polyenes the lowest excited state, as given in Table 1, can be regarded as derived from a mixture of these two states. As such it would be exxpected to show some of the sensitivity to the inclusion of electron correlation found in the nonpolar molecules. It is also clear that the λ_{max} values for the polyphenyls approach a steady value, despite some irregularity, much more rapidly with increasing n than do those of the polyenes.

Table 1 gives the calculated results for the ground-state dipole moments and principal electronic transition wavelengths for the polyenes and polyphenyls as obtained from SCI and SDCI. Table 2 gives data for the first hyperpolarizabilities for a fundamental wavelength of 1.907 μm (0.65 eV). In the case of the polyenes, the change from SCI to SDCI has the effect of reducing the magnitude of the quadratic response. The divergence between the two sets of results increases with increasing chain length. For the longest chain ($n = 8$) the β values calculated with SDCI are reduced by about one third compared with the SCI values.

Table 1 Effect of Singly and Doubly Excited Configurations on the Transition Energies of the Polyenes **(3)** and Polyphenyls **(4)**

	Polyenes				Polyphenyls		
n	μ_g	λ(SCI)	λ(SDCI)	n	μ_g	λ(SCI)	λ(SDCI)
1	11.93	381.4	392.5	1	9.03	323.5	333.4
2	15.47	503.6	508.5	2	9.72	368.1	369.5
				2*	9.72	369.5	359.8
3	17.95	600.5	588.3	3	9.96	380.9	328.8
3*	17.95	601.3	547.2				
4	19.78	677.2	658.6	4	10.13	382.8	384.0
5	21.14	737.7	712.5	5	10.19	387.2	390.2
6	22.55	789.9	759.3	6	10.22	388.3	390.0
7	23.88	827.8	790.9	7	10.24	389.9	390.1
8	24.03	857.3	816.6	8	10.33	375.9	377.0
9	25.58	909.5		9	10.34	391.4	391.7
10	24.97	920.1					
11	25.27	929.8					
12	25.52	936.7					

Note: See Section IV for a definition of the terms used. The values of n are to be interpreted in terms of Structures **3** and **4** (SCI, SDCI: see text). The asterisk indicates that full π-π* CI has been used.

A measurement of β for the all-*trans* molecule, H_2N–$(CH=CH)_3$–NO_2 has been reported as (86×10^{-30}) esu,[24] which is consistent with the tabulated results.

The effect of the introduction of doubly excited configurations in the polyphenyls is less marked and, for the higher homologs, the divergence is in the opposite direction, amounting to an increase of a about 10%, although, as in previous work on this series, some irregularities have been found. The general trends are similar in the SCI and SDCI results and the much less rapid increase of β with n in the polyphenyl series, as contrasted with the polyenes, is confirmed.

Table 2 Effect of Singly and Doubly Excited Configurations on the Hyperpolarizability of the Polyenes **(3)** and Polyphenyls **(4)**

	Polyenes		Polyphenyls	
n	β^{SCI}	β^{SDCI}	β^{SCI}	β^{SDCI}
1	18.1	18.1	14.48	11.40
2	127.0	105.2	42.30	34.85 (28.32)
3	473.5	326.9 (274.3)	70.89	64.99
4	1,377	874.3	96.55	95.73
5	3,089	1,970	98.16	106.72
6	6,239	4,072	107.69	117.09
7	11,140	7,290	127.28	139.97
8	18,030	12,040	182.70	191.37
9	50,900		131.48	152.29
10	76,450			
11	114,500			
12	168,500			

Note: See Section IV for a definition of the terms used. The two results in parentheses have been calculated using full π-π* CI.

III. CALCULATED HYPERPOLARIZABILITIES

A considerable number of calculations have been reported for LEO and SHG applications on a variety of substituted benzenes and related structures.[1-3,9] Materials for the former application, where a signal is imposed on a carrier wave generally at a wavelength of around 1300 to 1500 nm, are required to be transparent in this region only and their electronic absorption bands can occur anywhere in the visible region of the spectrum. Recent calculations and experimental studies on materials with possible applications in this area have explored the substitution of one or more phenyl rings in a given substituted benzene, either by heterocyclic rings such as thiophene[25–27] and thiadiazole,[28] or by more complex rings such as those found in coumarins,[29] indoanilines,[30] and phenylazonaphthalenes.[31,32]

In SHG applications, such as optical data storage where the emission from a semiconductor laser is converted from the infrared region to the blue region of the spectrum (from around 830 to 415 nm), enabling up to four times as much data to be stored, the active organic molecule must be transparent at the SHG wavelength to avoid reabsorption of the converted light. Furthermore, because the first prototype organic nonlinear devices are likely to be fabricated from poled polymer films, the active molecule must possess a sufficiently large dipole moment to allow it to be orientated in the polymer by electric field poling. Unfortunately, the majority of molecules with large hyperpolarizabilities generally show some absorption in the critical 415 nm region of the spectrum. For example, N,N-dimethyl-4-nitroaniline (5) has an acceptable hyperpolarizability for SHG application,[33,34] a reasonable dipole moment[35] for orientation under electric field poling, and is transparent in dilute solution with an absorption maximum ranging from 351 to 390 nm depending on the solvent used for measurement.[36] However, in a concentrated medium such as a polymer film, the tail of the absorption band extends into the blue region of the spectrum and the molecule therefore reabsorbs the SHG emission.

NMe$_2$ — (benzene ring) — NO$_2$

5

OMe — (benzene ring) — CHO

6

The introduction of weaker substituents in place of the strong donor and acceptor groups usually results in a beneficial hypsochromic shift toward the ultraviolet region. Thus the replacement of the dimethylamino group by the weaker methoxyl group results in a large hypsochromic shift (84 nm) in the absorption maximum.[36] A similar shift occurs if the electron-attracting nitro group in the same molecule is replaced by the weaker aldehyde group, and if both are replaced, the resulting 4-methoxybenzaldehyde (6) is completely transparent with an absorption at 277 nm.[36] However, in moving from the aniline (5) to the aldehyde (6), the hyperpolarizability falls from an experimental value measured in dimethylsulfoxide ranging from 21.0 at 1356 nm[33] to 52.8 at 1890 nm[34] for the former (5), to around 0.33 for the latter (6) in the same solvent,[34] with a concomitant reduction in the dipole moment from 6.84 to 3.88 D.[34] As a consequence of these deficiencies, other molecular systems such as the squarates and related systems[37] have been investigated which are essentially transparent. Calculations on a related theme have examined the effect of weak donors and acceptors on the hyperpolarizability of modified polyenes.[38–40]

A number of other systems with potentially large dipole moments which would assist their orientation in poled polymer films for either SHG or for LEO applications have been explored: these systems include pyridinium cyclopentadienylides[41,42] and sydnones.[43] In other studies, the important resonance enhancement effects on the hyperpolarizability have been calculated for azothiophene dyes.[44] Intermolecular charge-transfer complexes between electron acceptors such as tetracyanoethylene and donors such as hexamethylbenzene have been calculated and compared with intramolecular complexes.[45] More fundamental calculations have explored the relationship between bond length alternation in donor–acceptor alkenes and merocyanines on the hyperpolarizability.[46–52] Finally, there have been recent studies on quinoline derivatives,[53] diaryls and polyaryls,[54] fluorinated stilbenes,[55,56] acylhydrazines,[57] conjugated systems containing azomethine bonds,[58] and the influence of substituents on chromophore architecture,[59] Selected results are presented below.

A. DONOR–ACCEPTOR THIOPHENES

The substitution of a benzene ring by thiophene in donor–acceptor stilbenes,[25,26] styrenes,[27] or biphenyls[27] has a considerable effect on the nonlinear properties. For example, the insertion of one or more vinylthienyl groups between the aromatic ring and double bond of 1-(4'-dimethylaminophenyl)-2,2-dicyanoethylene (**7**, $n = 0$) results in a substantial increase in the product of the hyperpolarizability and dipole moment, $\mu\beta$, from 300 for $n = 0$ to 3800 for $n = 3$ (Table 3).[25]

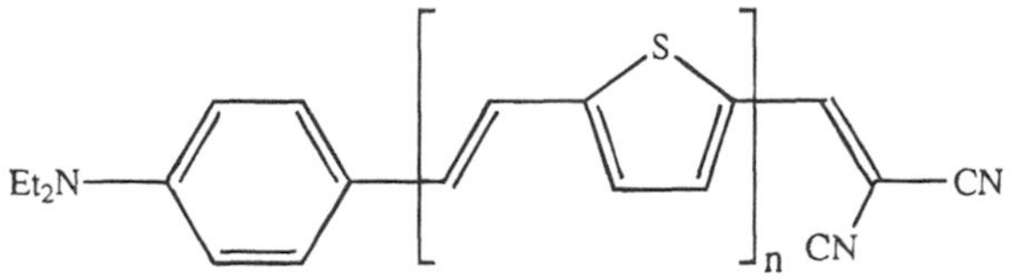

7

An even larger increase is found in the closely related 1-(4'-dimethylaminophenylthienyl)-1,2,2-tricyanoethylenes (**8**) where the product, $\mu\beta$, rises to 9100 for $n = 2$ (Table 3).[26] Furthermore, the substitution of the dimethylamino and nitro groups in 2-dimethylamino-5-nitrothiophene (**9a**) by the dimethylhydrazono and dicyanovinyl groups respectively to give (**9b**) and (**9c**) results in a sharp increase in both the experimental and calculated hyperpolarizability.[27] A further increase results when the conjugation path is extended by an additional thiophene ring (Table 3).[27]

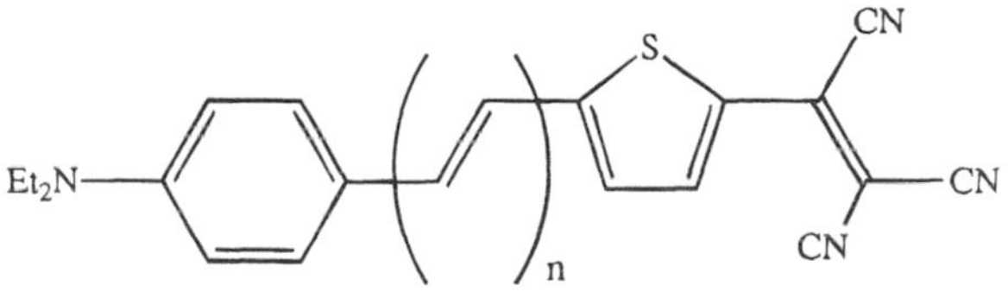

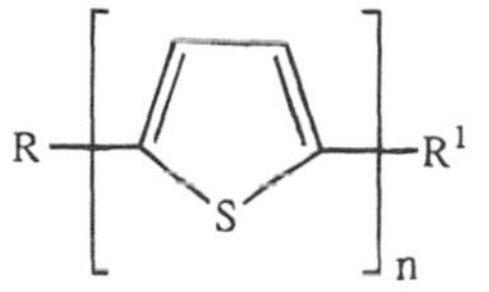

8

9

$$^{a}R = Me_2N; \; R^1 = NO_2; \; n = 1$$

$$^{b}R = Me_2NN{=}CH; \; R^1 = NO_2; \; n = 1$$

$$^{c}R = Me_2NN{=}CH; \; R^1 - CH{=}C(CN)_2; \; n - 1$$

$$^{d}R = Me_2NN{=}CH; \; R^1 = CH{=}C(CN)_2; \; n = 2$$

Table 3 Hyperpolarizabilities and Absorption Maxima of Donor–Acceptor Thiophenes (**7**, **8**, and **9**)

Molecule	n	λ	λ (nm)	$\mu\beta$	β_0	β_0 (calc)	Ref.
			Experimental Values				
7	0	417	1907	300			25
	1	513		1300			25
	2	547		2300			25
	3	556		3800			25
8	1	640	1907	6200			26
	2	662		9100			26
9a	1	388	1340	190	13	11	27
9b	1	445		330	25	27	27
9c	1	475		680	62	65	27
9d	2	494		1650	100	115	27

Note: See Section IV for a definition of the terms used.

B. THIADIAZOLES AND RELATED SYSTEMS

The potential of stable heterocycles such as the donor–acceptor thiadiazoles (10a), oxadiazoles (10b), and triazoles which can exist in three distinct tautomeric forms (10c–e) for nonlinear applications has been explored at the CNDOVSB level.[28]

All of these structures show a large change in the magnitude of the dipole moment on excitation to the first excited state (Table 4). The change is largest for the triazole (10d) which increases from 3.9 to 17 D, followed by the oxadiazole (10b) and triazole (10e), with each showing an increase of around 11 D. The calculated hyperpolarizabilities of these systems, however, are smaller than those expected on the basis of the change in dipole moments on excitation. This arises because the hyperpolarizability is not only dependent on the change in dipole moment on excitation, but also on the magnitude of the transition moment or oscillator strength, which in these cases is relatively small. The thiadiazole (10a) is predicted to have the largest value of this series. Of the triazoles, the 1,3,4-isomer (10c) has the largest value and the 1,2,4-isomer (10e) the smallest.[28]

10

$$^aR = NMe_2;\ R^1 = NO_2;\ Y1 = S;\ X3 = X4 = N$$

$$^bR = NMe_2;\ R^1 = NO_2;\ Y1 = O;\ X3 = X4 = N$$

$$^cR = NMe_2;\ R^1 = NO_2;\ Y1 = NH;\ X3 = X4 = N$$

$$^dR = NMe_2;\ R^1 = NO_2;\ Y1 = X3 = N;\ X4 = NH$$

$$^eR = NMe_2;\ R^1 = NO_2;\ Y_1 = X4 = N;\ X3 = NH$$

Table 4 Calculated Electronic Properties and Hyperpolarizabilities[28] of the Donor–Acceptor Heterocycles (10)

Molecule	μ_g	μ_e	λ	f	β_0	$\beta_{1.17}$
10a	8.40	15.7	366	0.35	9.29	19.9
10b	8.20	19.2	328	0.29	6.53	12.7
10c	9.09	19.2	339	0.31	7.70	15.7
10d	3.94	17.1	365	0.11	4.94	9.69
10e	10.8	22.1	303	0.12	2.50	3.86

Note: See Section IV for a definition of the terms used.

C. SUBSTITUTED COUMARINS

Experimental and theoretical studies have been reported on the nonlinear properties of substituted coumarins containing electron donors at the 7-position of the aromatic ring and a variety of electron attractors at the 3- and 4-positions of the other ring.[29]

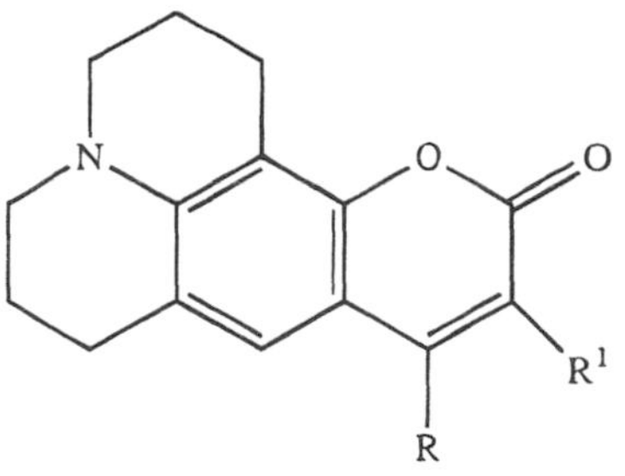

11

$^a R = R^1 = H$

$^b R = CF_3; R^1 = H$

$^c R = H; R^1 = CHO$

$^d R = H; R^1 = CN$

$^e R = H; R^1 = C_5H_4N$

Table 5 A Comparison Between Experimental and Calculated Dipole Moments and Hyperpolarizabilities for Substituted Coumarins (**11**)[29]

Structure	Experimental Results				Calculated	
	λ	μ	$\beta_{1.17}$	β_0	μ	β_0
11a	392	6.75	27.9	11.0	5.40	8.20
11b	418	6.55	49.1	15.8	5.21	18.8
11c	452	7.56	111.0	25.4	6.81	21.7
11d	446	10.2	132.0	32.4	9.20	28.6
11e	430	7.44	114.0	32.9	6.01	25.2

Note: See Section IV for a definition of the terms used.

Although the calculated dipole moments are underestimated, there is a reasonable correlation between the experimental and calculated hyperpolarizabilities (Table 5). The results clearly show that the introduction of an attractor at the 3- or 4-positions of the coumarin ring results in a substantial increase in the hyperpolarizability. Thus in moving from the parent of this series (**11a**) to the 4-trifluoromethyl- (**11b**) or the 3-aldehyde derivative (**11c**), the calculated hyperpolarizability more than doubles and follows the experimental trends.[29] The largest values are obtained with strong attractors such as the cyano group at the 3-position of the ring (Table 5).

D. SUBSTITUTED INDOANILINES

The conformations of a series of indoanilines have been calculated and the hyperpolarizabilities assessed at the PM3 level.[30] Two stable conformers are possible where the phenyl ring is twisted from the plane of the quininoid ring. The calculated results show that the insertion of an electron donor such as a methoxyl or dimethylamino group into the 4-position of the phenyl ring of the parent structure (**12a**) results in a large enhancement of the hyperpolarizability from 1.1 to 6.5 and 24.6, respectively, while the presence of the electron-attracting bromine atom reduces the value to −0.3 (Table 6). Overall the calculated hyperpolarizabilities are far too small by comparison with experimental results even allowing for the effect of solvent on the latter.[30]

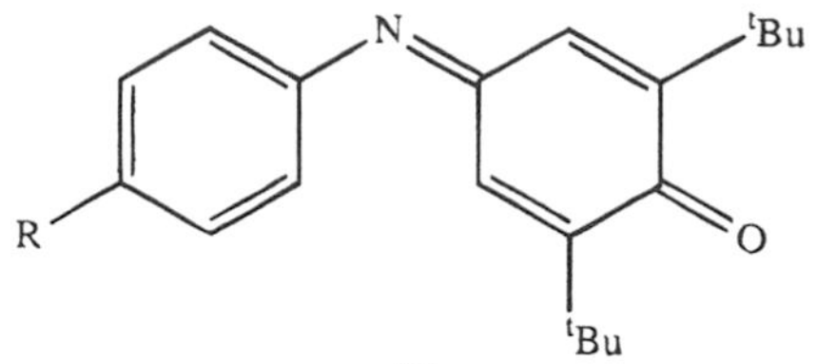

12

$$^aR = H$$

$$^bR = Br$$

$$^cR = OMe$$

$$^dR = NMe_2$$

Table 6 A Comparison Between Calculated and Experimental Hyperpolarizabilites for Substituted Indoanilines **(12)**[30]

Structure	μ^{calc}	β^{calc}	β^{expt}
12a	2.54	1.1	5.9
12b	2.04	−0.3	13.0
12c	2.98	6.50	17.0
12d	2.69	11.0	78.0

Note: See Section IV for a definition of the terms used; results are given for the conformer with the largest hyperpolarizability.

E. PHENYLAZOARENES AND AZOHETEROCYCLES

A detailed study of the change of hyperpolarizability resulting from the substitution of either one or both phenyl rings in 4-dimethylamino-4′-nitroazobenzene (**13a**) by a naphthyl, anthryl, furyl, or thienyl rings has been carried out using both the sum-over-states[31] and correction vector approach.[32]

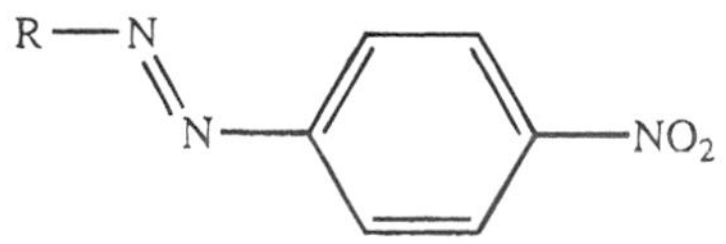

13

$$^aR = 4-C_6H_4NMe_2$$

$$^bR = 4-C_{10}H_7NHMe$$

$$^cR = 4-C_{14}H_9NHMe$$

In moving from the azobenzene (**13a**) to the phenylazonaphthalene (**13b**) and the phenylazoanthracene (**13c**), there is little change to the hyperpolarizability, calculated at 0.65 eV using both singly and

Table 7 Calculated Electronic Properties and Hyperpolarizabilities[32] of the Donor–Acceptor Azoarenes **(13)** and Azoheterocycles **(14,15)**

Molecule	μ_g	$\delta\mu_e$	λ	f	$\beta_{0.65}$	$\beta_{1.17}$
13a	9.59	12.2	441.7	1.01	75.5	247.2
13b	10.1	7.74	477.1	0.94	65.9	346.7
13c	10.1	5.11	538.4	0.76	81.0	1913.9
14a	11.0	6.53	498.3	0.91	70.0	547.3
14b	11.2	7.64	523.0	0.95	89.2	3243.8
15a	11.0	9.85	540.0	0.86	124.4	−3012.1
15b	11.1	4.86	562.0	0.98	76.1	−627.3

Note: See Section IV for a definition of the terms used.

doubly excited configurations, though there is a large increase at 1.17 eV which arises from resonance enhancement effects (Table 7).[32]

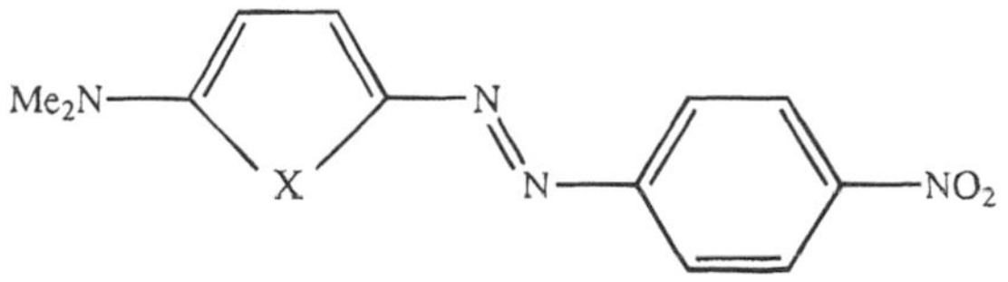

14

$^aX = S$

$^bX = O$

In contrast, the stepwise replacement of the two phenyl rings of the azobenzene by thienyl or furyl rings to give first the phenylazoheterocycle **(14a)** and **(14b)** and then the bisazoheterocycles **(15a)** and **(15b)**, respectively, results in a substantial increase in the hyperpolarizability, particularly for the value at 0.65 eV which rises from 75.5 in the azobenzene **(13a)** to 124.4 in the azothiophene **(15a)**.

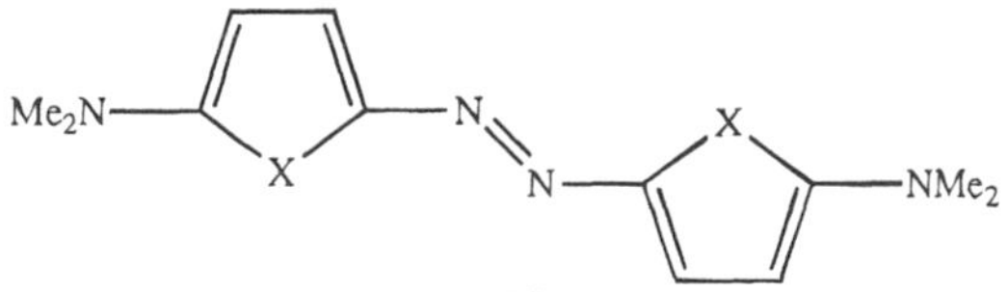

15

$^aX = S$

$^bX = O$

F. SQUARATES AND RELATED MOLECULES

Theoretical studies have been reported on the potential of small ring systems such as 2,3-dimethoxy-2-cyclopropenone or dimethyl deltate **(16)**, 3,4-disubstituted-3-cyclobutene-1,2-diones or squarates **(17a–c)**, 4,5-dimethoxy-4-cyclopentene-1,2,3-triones or dimethyl croconate **(18)**, and 5,6-dimethoxy-5-cyclohexene-1,2,3,4-tetraones or dimethyl rhodizonate **(19)** for SHG applications using both the CNDOVSB approach and a coupled Hartree-Fock method at the 3–21 G level.[37]

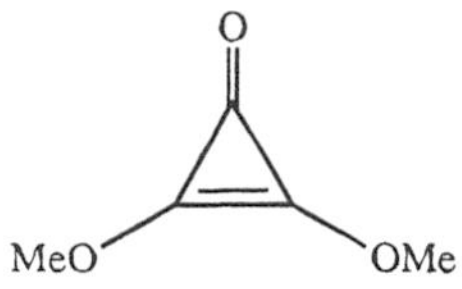

MeO OMe

16

O O

R R

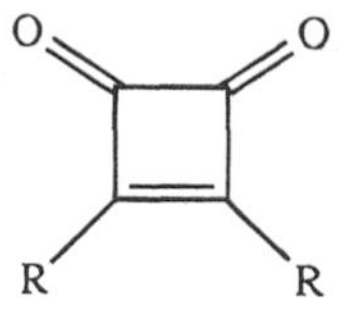

17

$^{a}R = OMe$
$^{b}R = NHMe$
$^{c}R = NMe_2$

The calculated hyperpolarizabilities of these systems increase with increasing ring size as expected though the values for dimethyl deltate **(16)** and dimethyl squarate **(17a)** are too small for practical application.

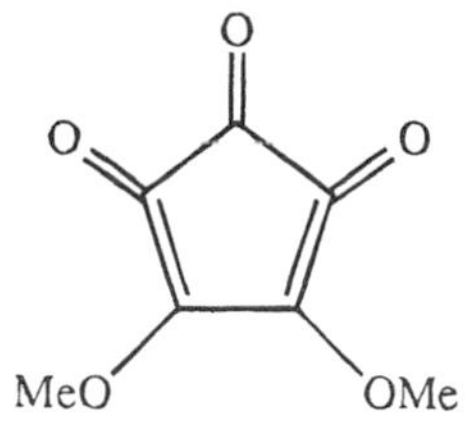

MeO OMe

18

MeO MeO

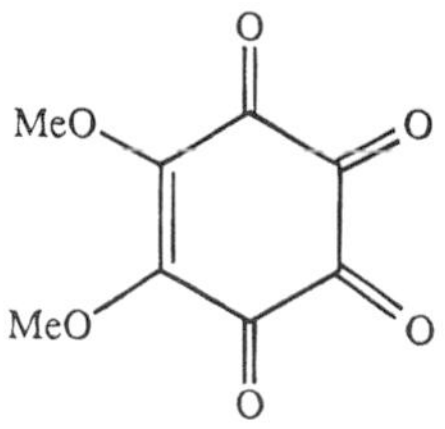

19

Table 8 Calculated Dipole Moments, Transition Energies, Oscillator Strengths and Hyperpolarizabilities of the Deltate **(16)**, Squarates **(17)**, Croconate **(18)**, and Rhodizonate **(19)** Compared with the Values for N,N-Dimethyl-4-nitroaniline **(5)** and 4-Methoxybenzaldehyde **(6)**[37]

Structure	CNDOVSB Results				3–21 G Results		Experimental	
	λ	f	β_0	$\beta_{1.17}$	μ	β_0	λ	μ
5	333	0.68	13.33	26.68	8.39		351	6.84
6	277	0.63	4.29	6.40	5.08		277	3.88
16	250	0.38	0.83	1.07	1.53	−0.30	244	
17a	302	0.55	6.68	10.70	2.92	−2.65	246	
17b	342	0.55	9.14	16.51	7.68		363	
17a	356	0.57	10.56	20.77	7.57			
18	346	0.33	12.97	25.56	6.32	6.68	292	5.85
19	386	0.51	16.31	39.18	6.19			

Note: See Section IV for a definition of the terms used.

However, the introduction of the more powerful methylamino **(17b)** and dimethylamino groups **(17c)** in place of the two methoxyl groups results in a significant enhancement of the hyperpolarizability; there is also a substantial boost to the dipole moment which is predicted to be comparable to that of the nitroaniline **(5)**; and most importantly, the absorption of these structures is predicted to occur in the ultraviolet region of the spectrum (Table 8). A greater enhancement occurs when the ring is expanded from the dimethyl squarate **(17a)** to dimethyl croconate **(18)** and then to the dimethyl rhodizonate **(19)**. While the latter **(19)** has the largest hyperpolarizability, it suffers from the disadvantage that its absorption is close to the visible region. The value for dimethyl croconate (18), however, is comparable to *N,N*-dimethyl-4-nitroaniline (5) and it also shows the necessary transparency at the SHG wavelength.

G. ARYLALKENES CONTAINING WEAK DONORS AND ACCEPTORS

Previous studies have shown that the hyperpolarizabilities of polyenes containing a strong donor and acceptor at either end of the chain[7,39] are substantially greater than those for simple aromatics though most systems of interest show strong absorption in the visible region of the spectrum and therefore cannot be utilized for SHG applications. However, the insertion of weaker groups at the end of a conjugation chain, such as the methoxyl and aldehyde group, has a much reduced effect on the absorption wavelength.

20

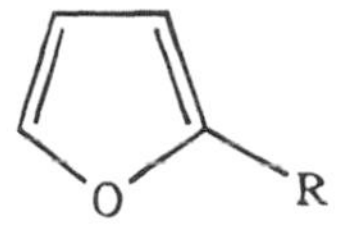

21

[a]R = CHO

[b]R = (CH=CH)CHO

[c]R = (CH=CH)$_2$CHO

[d]R = (CH=CH)$_3$CHO

[e]R = (CH=CH)$_4$CHO

The insertion of an alkenyl group between the aromatic ring and the aldehyde acceptor in 4-methoxybenzaldehyde (20a) to give 4-methoxycinnamaldehyde (20b) results in a substantial bathochromic shift in the predicted absorption band, an increase in the ground-state dipole moment, and a sharp increase in the hyperpolarizability (Table 9).[38]

Table 9 Calculated Transition Energies, Dipole Moments, and Hyperpolarizabilities of the Arylpolyenals (20 and 21)[38] and Styrylcyanoacetates (22 and 23)[40]

Structure	n	μ	λ	f	β_0	$\beta_{1.17}$
20a	0	4.14	268	0.72	4.28	6.29
b	1	4.88	321	1.16	11.3	21.4
c	2	5.24	366	1.60	21.7	55.9
d	3	5.46	404	2.05	38.7	136.2
e	4	5.61	437	2.46	57.8	296.1
21a	0	2.88	274	0.56	2.21	3.52
b	1	3.42	337	0.96	9.69	20.5
c	2	3.67	383	1.38	21.8	64.4
d	3	3.81	420	1.82	39.1	161.7
e	4	3.91	451	2.25	60.1	374.6
22a	cis	9.59	367	0.79	4.56	11.4
22b	cis	8.09	344	0.56	2.99	7.54
22c	cis	7.11	363	0.97	8.61	20.6
	trans	3.50	364	0.95	11.4	27.4
23	cis	8.22	399	0.81	14.6	49.0
	trans	4.47	400	0.79	19.3	61.2

Note: See Section IV for a definition of the terms used.

The insertion of further alkenyl groups results in a further bathochromic shift in the absorption band coupled with an approximate doubling of the hyperpolarizability with each alkenyl group (Table 9).

A similar pattern emerges in 2-furaldehyde (**21a**) from the insertion of alkenyl groups between the furan ring and the aldehyde group. Although the insertion of three or four alkenyl groups results in the largest hyperpolarizabilities, the molecules are no longer transparent at the critical SHG wavelengths, and the best compromise between transparency and hyperpolarizability appears to be the use of one or two alkenyl group as in the 4-methoxyphenylalkenals (**20b** and **20c**) and the furylalkenals (**21b** and **21c**).[38]

In another approach, the influence of the position and number of donor groups present in the phenyl ring has been explored for the related methoxystyrylcyanoacetates (**22**) and (**23**).[40]

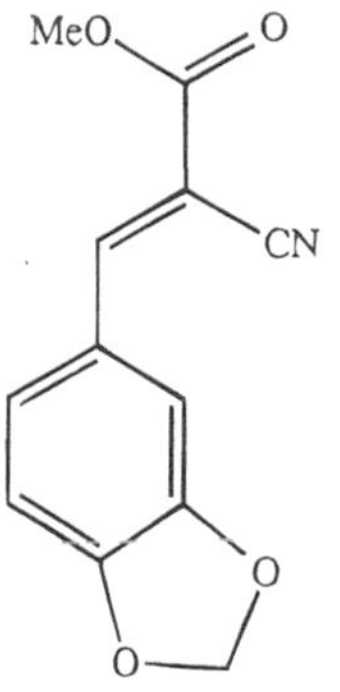

22

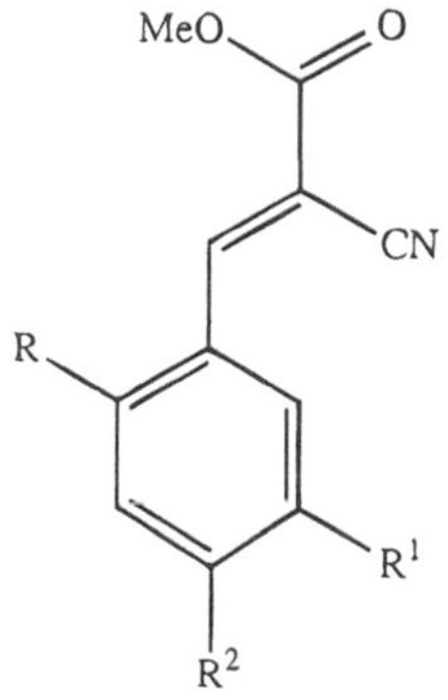

23

$$^{a}R = OMe; R^1 = R^2 = H$$
$$^{b}R^1 = OMe; R = R^2 = H$$
$$^{c}R^2 = OMe; R = R^2 = H$$

The calculated results on *cis*-conformers (where the carbonyl and cyano groups are on the same side of the double bond) show that the hyperpolarizability of the 3-methoxy derivative (**23b**) is relatively small as expected on the basis of the known weak electron-donating effects of meta-substituents of this type. The 5-methoxy isomer (**23a**) also shows small values, but the 4-methoxystyrylcyanoacetate (**23c**) has the largest hyperpolarizability of the three isomers considered (Table 9).

In the *trans*-conformer of the 4-methoxy derivative (**23c**) the dipole moment falls because the two electron-attracting groups are opposed to one another, but the vector component of the hyperpolarizability rises (Table 9). The calculated hyperpolarizability of the related piperonylidene ester (**22**) is surprisingly larger than any of the simple methoxy derivatives (**23a–c**), and larger than that expected from the simple additive effects of the 3-methoxy and 4-methoxy groups (Table 9). The overall values obtained for the *cis*- and *trans*-isomers (**22**) are approximately three times larger than those obtained for the simple arylalkenes (**20b**) and (**21b**).

H. CYCLOPENTADIENYLPYRIDINES

Theoretical investigations of the highly polar pyridinium cyclopentadienylide and some of its derivatives (**24**) using the finite field/PM3 method,[41] and the related cyclopentadienylidene 1,4-dihydropyridine (**25**) and the azolylidene 1,4-dihydropyridine (**26**) using the CNDOVSB method[42] have shown that the molecules possess fairly large hyperpolarizabilities and dipole moments which would facilitate their orientation in poled polymer films.

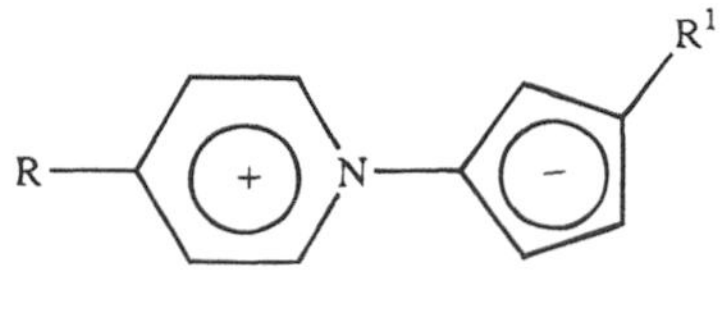

24

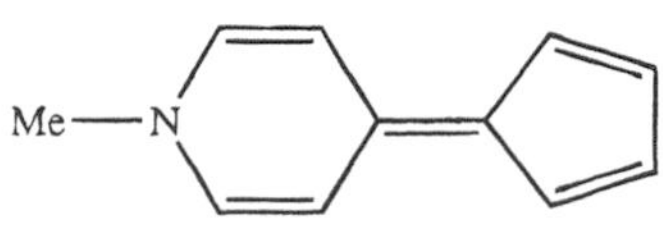

25

[a] $R = R^1 = H$
[b] $R^1 = NH_2; R = H$
[c] $R^1 = H; R = NO_2$
[d] $R^1 = NH_2; R = NO_2$

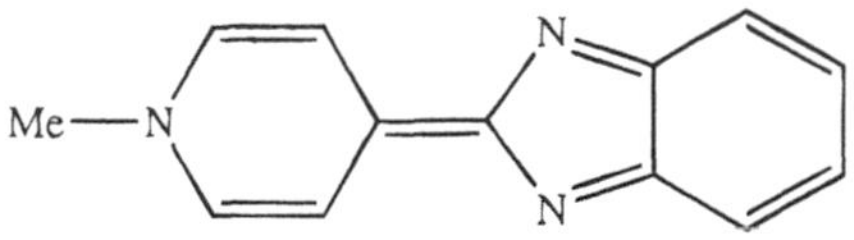

26

The results reported using the finite field/PM3 method,[41] however, differ from those obtained at the CNDOVSB level.[42] In the former,[41] the dipole moment of pyridinium cyclopentadienylide (**24a**) at 4.19 D is substantially underestimated relative to the experimental value of 13.3 D, but the latter[42] gives better result at 8.19 D (Table 10). The resulting hyperpolarizabilities at zero field strength are therefore different with values of -12.2 for the former and -55.8 for the latter (Table 10). The negative sign is a reflection of the change of direction of charge transfer which takes place on excitation from the ground state, which is polarized with electron donation from the left-hand pyridinium ring to the right-hand cyclopentadienylide ring, to the excited state which is polarized in such a way that the two rings are now broadly neutral.[42]

Table 10 Calculated Dipole Moments, Hyperpolarizabilities, and Excited-State Properties of the Cyclopentadienylpyridines (**24–26**)

Molecule	μ_{exp}	μ_g	μ_e	λ	f	β_0	$\beta_{0.95}$	Ref.
24a	13.3	4.19				-12.2		41
24b		3.76				-24.5		41
24c		4.01				11.1		41
24d		5.49				30.1		41
24a	13.3	8.19	0.18	515	0.73	-55.8	-188.5	42
25	10.3[a]	11.54	8.03	431	1.19	-23.3	-48.2	42
26	9.03	14.14	6.32	521	1.35	-93.9	-344.5	42

Note: See Section IV for a definition of the terms used.

[a] Data for the *N*-benzyl derivative.

The introduction of an amino group into the left-hand ring at the 9-position results in a doubling of the hyperpolarizability while the presence of a nitro group in the right-hand ring at the 3-position gives a similar value to the unsubstituted molecule but of the opposite sign.[41] A combination of the two groups results in a substantial enhancement of the value to 30.1 (Table 10). The values obtained with the CNDOVSB method both at zero field and at 0.95 eV for the cyclopentadienylides (**24**) and (**26**) are much larger than that for (**25**) and mirror the large changes of around -8 D in their dipole moments on excitation.[42] The approximate threefold resonance enhancement for the pyridinium cyclopentadienylide **(24)** is matched by the 4-(benzimidazolylidene)-1-methyl-1,4-dihydropyridine **(26)** as the transition energies of both molecules are very similar at 515 and 521 nm, respectively (Table 10).

I. PHENYLSYDNONES

Mesoionic compounds such as 3- and 4-phenylsydnones **(27)** and **(28)** have recently been evaluated as nonlinear materials both in terms of their hyperpolarizability and dipole moments.[43] Surprisingly, the hyperpolarizability of 3-phenylsydnone **(27a)** calculated either as the vector component along the dipole moment, $\beta 1$, or as the vector component in the transverse direction, $\beta 2$, is very small despite the very large reduction in the magnitude of the dipole moment from 8.26 to 1.78 D on excitation from the ground state to the first major excited state (Table 11).

An analysis of the contribution of individual excited states to the hyperpolarizability, however, shows that there is a wide oscillation in its value with an increasing number of excited states.

27   28

$^aR = H$

$^bR = Me_2N$

$^cR = NO_2$

Unlike the vast majority of donor–acceptor organic systems such as N,N-dimethyl-4-nitroaniline **(5)** where the hyperpolarizability is dominated by the transition of an electron from the ground to the first excited state, this does not apply to 3-phenylsydnone **(27a)**, because there are a number of excited states which contribute to the final value of the hyperpolarizability. These predicted excited states which have significant oscillator strengths are also found experimentally.

Table 11   Calculated Hyperpolarizabilities and Excited-State Properties of the Phenylsydnones **(27)** and **(28)**[43]

Molecule	μ_g	μ_e	λ	f	$\omega = 0$		$\omega = 1.17$	
					$\beta 1$	$\beta 2$	$\beta 1$	$\beta 2$
27a	8.26	1.78	388	0.28	−0.48	0.36	−3.33	2.31
27b	11.6	4.70	388	0.26	12.7	0.39	25.8	−0.57
27c	2.85	9.40	413	0.19	−3.04	1.20	−8.10	−4.78
28a	6.95	4.04	411	0.75	−6.08	5.41	−18.6	15.1
28b	9.34	9.19	425	0.92	4.31	−5.47	11.6	−20.3
28c	6.87	8.16	431	0.90	5.67	20.2	21.7	75.7

Note: See Section IV for a definition of the terms used. $\beta 1$ is the vector component of the molecular hyperpolarizability in the direction of the molecular dipole moment; $\beta 2$ is the same quantity in the transverse direction.

The introduction of the electron-donating dimethylamino group into the phenyl ring, to give structure **(27b)**, has a large effect on the hyperpolarizability with the vector component along the dipole moment, $\beta 1$, dominant (Table 11). The corresponding 3-(4′-nitrophenyl)sydnone **(27c)**, however, has a much smaller value as a consequence of the two competing electron attractors located at both ends of the molecule, although the value is larger than the parent **(27a)**. In both these cases, more than one excited state contribute to the final values.

4-Phenylsydnone (**28a**) is calculated to have a much larger hyperpolarizability than the 3-isomer (**27a**) both for the vector component along the dipole moment, β1, and for the component in the transverse direction, β2 (Table 11). Again, a number of excited states contribute to the final value though in this case the first excited state has the strongest oscillator strength. The introduction of the dimethylamino group into the phenyl ring to give (**28b**) has little effect on the magnitude of the hyperpolarizability, but the nitro group in (**28c**) shows a larger effect with a calculated value which is greater than that calculated for 3-(4′-dimethylamino) sydnone (**27b**).

J. FREQUENCY-DEPENDENT EFFECTS ON HYPERPOLARIZABILITY OF A BLUE AZOTHIOPHENE DYE

The structure of a commercial dye, 2-(4′-(*N,N*-diethylamino)-2′-acetamidophenylazo)-3,5-dinitrothiophene (**29**), has been calculated and shown to be essentially planar between donor and acceptor groups with a strong intramolecular hydrogen bond between the acetamido group and one nitrogen of the azo linkage.[44] The hyperpolarizability of this structure, calculated for the linear electro-optic effect (LEO), is predicted to be large and arises mainly from the transition of an electron from the ground state to the first excited state.[44]

29

The effect of a variation of the applied field on the experimental and calculated hyperpolarizability of a variety of donor–acceptor aromatics has been systematically explored for second harmonic generation where there are large resonance enhancement effects[9] arising from terms such as $\omega_{ng} - 2\omega$ in the denominator of Equation 17. Similar enhancements are found for the LEO effect in the azothiophene (**29**) where the magnitude of the hyperpolarizability is now partly dependent on the reciprocal of terms such as $\omega_{ng} - \omega$ (see Equation 20). The results show that the frequency-dependent values of the vector component of the hyperpolarizability increase with increasing field strength and become extremely large as the applied field, ω, approaches the transition energy, ω_{ng}. The predicted value at $\omega = 2.08$ eV, which is very close to the transition energy, ω_{ng} at 2.082 eV, is approximately 10^{11} (Table 12).[44]

Table 12 Calculated Vector Components of the Frequency-Dependent Hyperpolarizability of 2-(4′-(*N,N*-diethylamino)-2′-acetamidophenylazo)-3,5-dinitrothiophene (**29**) using the CNDOVSB method[44]

Applied Field (W)		LEO Effect		SHG Effect	
nm	eV	$(\omega_{ng} - \omega)$	β^{ω}	$(\omega_{ng} - 2\omega)$	$\beta^{2\omega}$
0	0	2.08	4.94×10^2	2.08	
1890	0.66	1.43	5.95×10^2	0.78	2.34×10^2
1300	0.95	1.14	7.50×10^2	0.19	9.71×10^2
1186	1.04	1.04	8.31×10^2	0.002	-1.81×10^6
1060	1.17	0.92	9.77×10^2	−0.52	-7.90×10^2
800	1.55	0.54	2.15×10^3	−1.01	1.24×10^4
700	1.77	0.32	5.21×10^3	−1.45	9.85×10^3
650	1.91	0.18	1.49×10^4	−1.73	6.50×10^1
600	2.07	0.02	1.18×10^7	−2.05	1.71×10^3
593	2.08	0.002	3.47×10^{11}	−2.08	5.12×10^6
550	2.25	−0.16	1.41×10^4	−2.41	3.02×10^3
500	2.48	−0.39	2.23×10^3	−2.87	2.05×10^2
450	2.76	−0.67	7.79×10^3	−3.43	-1.33×10^3
354	3.50	−1.41	4.42×10^3	−4.91	-4.94×10^1

Note: β^{ω} and $\beta^{2\omega}$ are the calculated hyperpolarizabilities for the LEO and SHG effects, respectively. The $(\omega_{ng} - \omega)$ and $(\omega_{ng} - 2\omega)$ values have been rounded to two decimal places except where stated otherwise.

In addition, large values of approximately 10^6 are calculated for the SHG effect both at the second harmonic frequency where ω_{ng} approaches 2ω, and at the fundamental frequency where ω_{ng} approaches ω, which demonstrate the almost equal importance of the $\omega_{ng} - 2\omega$ and $\omega_{ng} - \omega$ reciprocal terms in Equation 17.

K. INTERMOLECULAR CHARGE-TRANSFER COMPLEXES

The nonlinear optical response of model molecular 1:1 and 2:1 organic π-electron donor–acceptor complexes, such as those between arylamines and tetracyanoethylene, have been investigated using an INDO/S sum-over-states approach.[45] The hyperpolarizabilities of these systems arise from the large change in dipole moment that accompanies the intermolecular charge-transfer transition coupled with their low-lying excitation energies. For 1:1 complexes, the largest values are obtained from complexes between tetramethyl-*p*-phenylene-diamine (**30a**) and 7,7,8,8-tetracyanoquinodimethane (**31**) or 1,2,4,5-tetraaminobenzene (**30d**) and tetracyanoethylene (**32**). The hyperpolarizability drops as expected when the strength of the donor is reduced (Table 13). A similar pattern emerges for the 2:1 complexes between two molecules of the donor coupled with one of the acceptor, except that the hyperpolarizabilities here are more than twice as large as the 1:1 complexes.

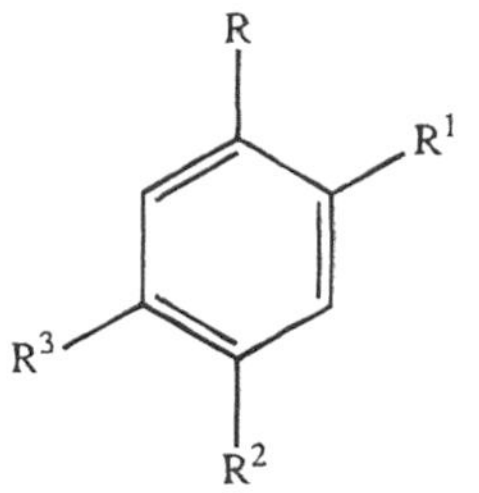

30

$$^{a}R = R^2 = NMe_2; R^1 = R^2 = H$$

$$^{b}R = R^2 = NH_2; R^1 = R^2 = H$$

$$^{c}R = R^2 = OMe; R^1 = R^2 = H$$

$$^{d}R = R^2 = R^1 = R^2 = NH_2$$

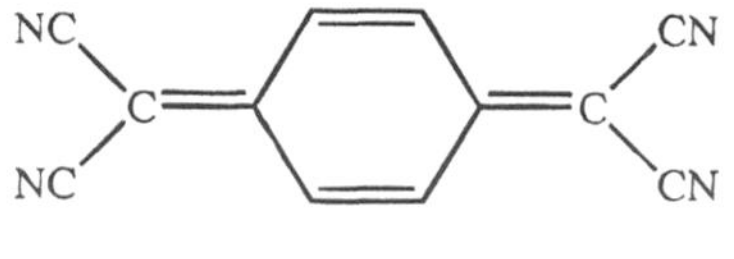

31

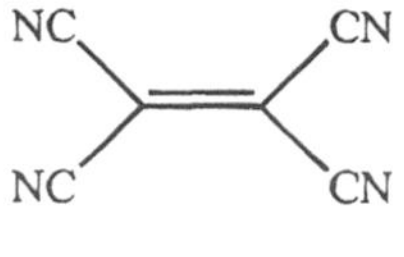

32

Table 13 Calculated Transition Energies, Dipole Moments, and Hyperpolarizabilities of Intermolecular Charge-Transfer Complexes[45]

| | Intermolecular Complex | | | | | | |
Type	Donor	Acceptor	μ	λ	f	$\delta\mu$	β_{zzz}
1:1	30a	31	1.07	2.04	0.14	14.43	68.36
	30b	31	1.04	2.33	0.14	14.46	39.01
	30c	31	0.40	3.02	0.08	14.29	8.39
2:1	30a	31	1.42	1.77	0.16	18.29	188.44
	30b	31	1.25	2.02	0.15	18.63	89.85
	30c	31	0.50	2.64	0.08	18.62	16.78
1:1	30d	32	1.33	2.04	0.14	14.54	68.51
	30b	32	0.83	2.45	0.10	14.94	23.22
	30c	32	0.54	3.25	0.09	14.98	8.23
2:1	30d	32	1.63	2.85	0.16	17.29	148.25
	30b	32	0.10	2.19	0.11	17.88	49.03
	30c	32	0.64	2.99	0.11	16.81	13.75

Note: See Section IV for a definition of the terms used. β_{zzz} is the calculated molecular hyperpolarizability at 0.65 eV.

L. INFLUENCE OF BOND LENGTH ALTERNATION ON MOLECULAR HYPERPOLARIZABILITY

Recent work has suggested that the degree of bond length alternation, expressed as the difference between the average length of carbon–carbon double and single bonds in a polymethine chain, found in donor–acceptor polyenes and merocyanines has a critical bearing on their calculated hyperpolarizabilities.[46,47] Electric field-dependent calculations with point charges on the dimethylaminopolyenal (**33**) show that with increasing applied electric field the geometry of the molecule changes from the neutral polyene structure (**N**) to a highly polar zwitterionic form (**Z**) via a cyanine-type structure which shows no appreciable bond length alternation. Here the calculated differences between the average double and single bond lengths between the neutral form (**N**) and the zwitterion (**Z**) change from -0.12 Å to 0.05 Å with the application of the external field. This change is accompanied by a dramatic change in the calculated value of the hyperpolarizability which changes sign, and shows only a small value at zero bond length alternation, with a maximum value of approximately 90 at a bond length alternation value of -0.055.[46,47]

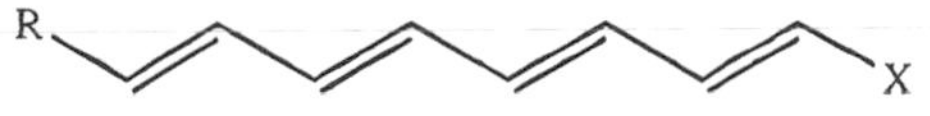

33

$^aR = NMe_2$; $X = CHO$

$^bR = NH_2$; $X = CH=C(CN)_2$

$^cR = OMe$; $X = NO_2$

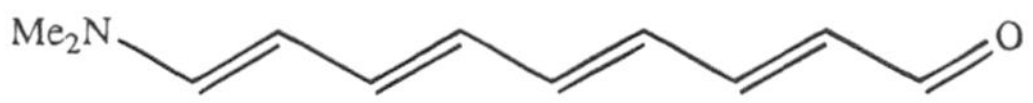

N

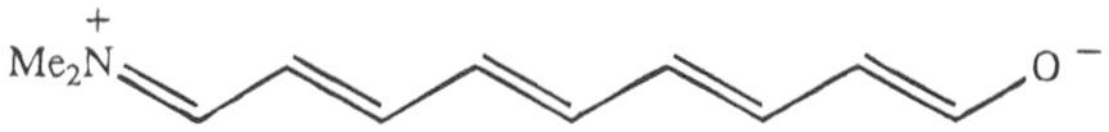

Z

Marder et al.[46–52] have proposed that the ground-state structures of polyenes and merocyanines are dramatically changed from a neutral polyenic or quininoid structure toward a zwitterionic form in moving from solvents with low dielectric constants to those with higher values. This change is reflected in the experimental hyperpolarizabilities evaluated in different solvents.[49] Recent calculations using a modified INDO sum-over-states method with structural optimization carried out in the presence of external homogeneous external field[52] again show that the dimethylaminopolyenal changes from the neutral polyene structure (**N**, with a dipole moment of around 10 D), to a highly polar zwitterionic form (**Z**, with a dipole moment of approximately 54 D) as the field strength is increased from zero to 10^8 V cm^{-1}. The hyperpolarizability of this polyene (**33a**) and other related polyenes (**33b**) and (**33c**) are found to be highly dependent on the strength of the applied field (Table 14).[52]

Table 14 Effect of Applied Field Strength on the Calculated Hyperpolarizabilities of Donor–Acceptor Polyenes (**33**)[52]

Structure		Field strength (10^7 V cm^{-1})		
		3	4	5
33a	β_{xxx}	497	587	191
33b	β_{xxx}	858	603	−458
33c	β_{xxx}	323	443	452

IV. DEFINING TERMS

μ is the dipole moment (in D).

μ_{exp}, μ_g, and μ_e are the experimental and calculated ground and excited state dipole moments, respectively (in D).

β_0, $\beta_{0.65}$, $\beta_{0.95}$, and $\beta_{1.17}$ are components of the hyperpolarizability tensor at zero field, 0.65, 0.95, and 1.17 eV, respectively (in units of 10^{-30} cm^5 esu^{-1}).

λ is the transition energy or absorption maximum (in nm).

f is the oscillator strength.

ω is the frequency of the applied field (eV).

$\mu\beta$ is the product of the dipole moment and hyperpolarizability at field strength ω (in units of 10^{-48} cm^5 esu^{-1}).

$\delta\mu$ is the difference between the ground- and first excited-state dipole moment (in D).

REFERENCES

1. **Prasad, P. N., and Williams, D. J.**, Eds., *Nonlinear Optical Effects in Molecules and Polymers,* Wiley, New York, 1991.
2. **Chemla, D. S., and Zyss, J.**, Eds., *Nonlinear Optical Properties of Organic Molecules and Crystals,* Academic Press, New York, 1987.
3. **Kobayashi, K.**, Ed., *Nonlinear Optics of Organics and Semiconductors,* Springer-Verlag, Tokyo, 1989.
4. **Morrell, J. A., and Albrecht, A. C.**, Second-order hyperpolarisabilities of p-nitroaniline calculated from perturbation theory based expression using CNDO/S generated electronic states, *Chem. Phys. Lett.,* 64(1), 46, 1979.
5. **Lalama, S. J., and Garito, A. F.**, Origin of the nonlinear second-order optical susceptibilities of organic systems, *Phys. Rev.,* A208, 1179, 1979.
6. **Docherty, V. J., Pugh, D., and Morley, J. O.**, Calculation of the second-order electronic polarisabilities of some organic molecules, *J. Chem. Soc., Faraday Trans. 2,* 81, 1179, 1985.
7. **Dirk, C. W., Tweig, R. J., and Wagniere, G.**, The contribution of π electrons to second harmonic generation in organic molecules, *J. Am. Chem. Soc.,* 108, 5387, 1986.
8. **Li, D., Ratner, M. A., and Marks, T. J.**, Molecular and macromolecular nonlinear optical materials. Probing architecture/electronic structure/frequency doubling relationships via an SCF-LCAO MECI π electron formulism, *J. Am. Chem. Soc.,* 110, 1707, 1988.

9. **Kanis, D. R., Ratner, M. A., and Marks, T. J.,** Design and construction of molecular assemblies with large second-order optical nonlinearities. Quantum chemical aspects, *Chem. Rev.,* 94, 195, 1994.

10. **Ward, J. F.,** Calculation of nonlinear optical susceptibilities using diagrammatic perturbation theory, *Rev. Mod. Phys.,* 37, 1, 1965.

11. **Morley, J. O., Pavlides, P., and Pugh, D.,** On the calculation of the hyperpolarizabilities of organic molecules by the sum over virtual excited states approximation, *Int. J. Quantum Chem.,* 43, 7, 1992.

12. **Orr, B. J., and Ward, J. F.,** Perturbation theory of the non-linear optical polarization of an isolated system, *Mol. Phys.,* 20, 513, 1971.

13. **Teng, C. C., and Garito, A. F.,** Dispersion of the nonlinear second-order susceptibility of organic systems, *Phys. Rev.,* B28, 6766, 1983.

14. **G. A. Segal,** *Semiempirical Methods of Electronic Structure Calculation,* Plenum Press, New York, 1976.

15. **Hameka, H. F.,** *J. Chem. Phys.,* 67, 2935, 1977.

16. **Garito, A. F., Heflin, J. R., Wong, K. Y., and Zamani-Khamiri,** Enhancement of nonlinear optical properties of conjugated linear chains through lowered symmetry, in *Organic Materials for Non-linear Optics,* Hann, R. A., and Bloor, D., Eds., Royal Society Chemistry Special Publication, 69, 1989.

17. **Albert, I. D. L., Pugh, D., and Morley, J. O.,** Further studies on the polarizabilities and hyperpolarizabilities of the substituted polyenes and polyphenyls, *J. Chem. Soc. Faraday Trans.,* 90, 2617, 1994.

18. **Ramasesha, S., and Soos, Z. G.,** Correlated states in linear polyenes, radicals and ions: exact PPP transition moments and spin densities, *J. Chem. Phys.,* 80, 3278, 1984.

19. **Albert, I. D. L., Pugh, D., and Morley, J. O.,** The correction vector approach to the linear and nonlinear properties of organic molecules, *J. Chem. Phys.,* 99, 5197, 1993.

20. **Albert, I. D. L., Morley, J. O., and Pugh, D.,** A π-CI approach to the study of correlation effects on the nonlinear optical properties of organic π-conjugated systems, *J. Chem. Phys.,* 102, 237, 1995.

21. **Rettrup, S.,** *J. Comp. Phys.,* 45, 100, 1982.

22. **Ramasesha, S., Albert, I. D. L., and Sinha, B. B.,** *Mol. Phys.,* 72, 537, 1991.

23. **Albert, I. D. L., Pugh, D., Morley, J. O., and Ramasesha, S.,** Linear and nonlinear optical properties of cumulenes and polyynes: a model exact study, *J. Phys. Chem.,* 96, 10160, 1992.

24. **Donald, D. S.,** in *New Aspects of Organic Chemistry II,* Yoshida, Z., and Ohshiro, Y., Eds., Kudamska, Tokyo, 1990, 40.

25. **Jen, A. K. Y., Rao, V. P., Wong, K. Y., and Drost, K. J.,** Functionalised thiophenes: second order nonlinear optical materials, *J. Chem. Soc., Chem. Commun.,* 1, 90, 1993.

26. **Rao, V. P., Jen, A. K. Y., Wong, K. Y., and Drost, K. J.,** Dramatically enhanced second-order nonlinear optical susceptibilities in (tricyanovinyl)-thiophene derivatives, *J. Chem. Soc., Chem. Commun.,* 14, 1118, 1993.

27. **Hutchings, M. G., Ferguson, I., McGeein, D. J., Morley, J. O., Zyss, J., and Ledoux, I.,** Quadratic nonlinear optical properties of some donor-acceptor substituted thiophenes, *J. Chem. Soc. Perkin Trans. 2,* 171, 1995.

28. **Morley, J. O.,** Non-linear optical properties of organic molecules. Part 20. Calculation of the structure, electronic properties and hyperpolarisabilities of donor-acceptor heterocycles containing sulphur, oxygen, and nitrogen, *J. Chem. Soc. Perkin Trans. 2,* 177, 1995.

29. **Moylan, C. R.,** Molecular hyperpolarisabilities of coumarin dyes, *J. Phys. Chem.,* 98, 13513, 1994.

30. **Matsuzawa, N., and Dixon, D. A.,** Theoretical prediction of the hyperpolarisabilities for 4-aminoindoaniline, *J. Phys. Chem.,* 98, 11677, 1994.

31. **Morley, J. O.,** Theoretical evaluation of novel organic materials for electro-optic modulation, *Proc. SPIE,* 1775, 2, 1993.

32. **Albert, I. D. L., Pugh, D., and Morley, J. O.,** Optical nonlinearities in azoarenes, *J. Phys. Chem.,* 99, 8024, 1995.

33. **Singer, K. D., Sohn, J. E., King, L. A., Gordon, H. M., Katz, H. E., and Dirk, C. W.,** Second-order nonlinear optical properties of donor- and acceptor-substituted aromatic compounds, *J. Opt. Soc. Am. B,* 6, 1339, 1989.

34. **Dulcic, A., and Sauteret, C.,** The regularities observed in the second order hyperpolarisabilities of variously disubstituted benzenes, *J. Chem. Phys.,* 69, 3453, 1978.

35. **McClelland, A. L.,** *Tables of Experimental Dipole Moments,* W. H. Freeman, San Francisco, 1963, Vol. I; Rahara Enterprises, San Francisco, 1974, Vol. II.

36. **Phillips, J. P., Freedman, L. D., and Craig, J. C.,** *Organic Electronic Spectral Data,* Interscience, New York, 1963, Vols. I–X.

37. **Dory, M., Andre, J-M., Delhalle, J., and Morley, J. O.,** Theoretical studies of the electronic structure, conformations, spectra and hyperpolarisabilities of squarates and related molecules, *J. Chem. Soc. Faraday Trans.,* 90, 2319, 1994.

38. **Morley, J. O.,** Non-linear optical properties of organic molecules. Part 15. Calculation of the structure and hyperpolarisabilities of arylalkenes containing weak donor and acceptors, *J. Chem. Soc. Perkin Trans. 2,* 1211, 1994.

39. **Morley, J. O., Docherty, V. J., and Pugh, D.,** Non-linear optical properties of organic molecules. Part 2. Effect of conjugation length and molecular volume on the calculated hyperpolarisabilities of polyphenyls and polyenes, *J. Chem. Soc. Perkin Trans. 2,* 1351, 1987.

40. **Morley, J. O.,** Unusual enhancement of the hyperpolarisability of a donor-acceptor styrene by an additional meta-substituent, *J. Phys. Chem.,* 98, 11818, 1994.

41. **Pranata, J., and Murray, C. J.,** Calculation of nonlinear optical properties of pyridinium cyclopentadienylides, *J. Phys. Org. Chem.,* 6, 531, 1993.
42. **Morley, J. O.,** Non-linear optical properties of organic molecules. Part 14. Calculations of the structure, electronic properties, and hyperpolarisabilities of cyclopentadienylpyridines, *J. Chem. Soc. Faraday Trans.,* 90, 1853, 1994.
43. **Morley, J. O.,** Non-linear optical properties of organic molecules. Part 19. Calculation of the structure, electronic properties and hyperpolarisabilities of phenylsydnones, *J. Phys. Chem.,* 99, 1923, 1995.
44. **Morley, J. O.,** Non-linear optical properties of organic molecules. Part 13. Calculation of the structure and frequency dependent hyperpolarisabilities of a blue azothiophene dye, *J. Chem. Soc. Faraday Trans.,* 90, 1849, 1994.
45. **Di Bella, S., Fragala, I. L., Ratner, M. A., and Marks, T. J.,** Electron donor–acceptor complexes as potential high-efficiency second order nonlinear optical materials. A computational investigation, *J. Am. Chem. Soc.,* 115, 682, 1993.
46. **Marder, S. R., Gorman, C. B., Tiemann, B. G., and Cheng, L-T.,** Optimising the second-order optical nonlinearities of organic molecules: asymmetric cyanines and highly polarised polyenes, *Proc. SPIE,* 1775, 19, 1993.
47. **Marder, S. R., Gorman, C. B., Tiemann, B. G., and Cheng, L-T.,** Stronger acceptors can diminish nonlinear optical response in simple donor-acceptor polyenes, *J. Am. Chem. Soc.,* 115, 3006, 1993.
48. **Marder, S. R., Perry, J. W., Tiemann, B. G., Gorman, C. B., Gilmour, S., Biddle, S. L., and Bourhill, G.,** Direct observation of reduced bond length alternation in donor/acceptor polyenes, *J. Am. Chem. Soc.,* 115, 2524, 1993.
49. **Bourhill, G., Bredas, J. L., Cheng, L-T., Marder, S. R., Meyers, F., Perry, J. W., and Tiemann,** Experimental demonstration of the dependence of the first hyperpolarisability of donor-acceptor polyenes on the ground state polarisation and bond length alternation, *J. Am. Chem. Soc.,* 116, 2619, 1994.
50. **Ortiz, R., Marder, S. R., Cheng, L-T., Tiemann, B. G., Cavagnero, S., and Ziller, J. W.,** The dependence of the molecular first hyperpolarisabilities of merocyanines on ground state polarisation and length, *J. Chem. Soc., Chem. Commun.,* 2263, 1994.
51. **Tiemann, B. G., Cheng, L-T., and Marder, S. R.,** The effect of varying ground state aromaticity on the first molecular electronic hyperpolarisabilities of organic donor-acceptor molecules, *J. Chem. Soc., Chem. Commun.,* 735, 1993.
52. **Meyers, F., Marder, S. R., Pierce, B. M., and Bredas, J. L.,** Electric field modulated nonlinear optical properties of donor-acceptor polyenes. Sum-over-states investigation of the relationship between molecular polarisabilities (α, β, and γ) and bond length alternation, *J. Am. Chem. Soc.,* 116, 10703, 1994.
53. **Ryu, U. C., Donghoon, K., Nakjoong, L., Yoon, S.,** Semiempirical calculations of hyperpolarisabilities of quinoline derivatives, *J. Korean Chem. Soc.,* 37, 1, 62, 1993.
54. **Barzoukas, M., Fort, A., Boy, P., Combellas, C., and Thiebault, A.,** Experimental and computational investigation of the quadratic hyperpolarisability of a series of diaryls and polyaryls, *Chem. Phys.,* 185, 65, 1994.
55. **Moylan, C. R., Betterton, K. M., Tweig, R. J., and Walsh, C. A.,** The hyperpolarisabilities of fluorinated 4-methoxystilbenes, *Mol. Cryst. Liq. Cryst. Sci. Technol.,* B-8, 69, 1994.
56. **Sugiyama, Y., Suzuki, Y., Mitamura, S., and Nishiyama, T.,** Second-order nonlinearities of substituted stilbenes and related compounds containing trifluoromethyl as the electron-withdrawing group, *Bull. Chem. Soc. Jpn.,* 66, 687, 1993.
57. **Tachikawa, Y., Kawaguchi, Y., Yonehara, H., Kawara, T., Maki, H., and Pac, C.,** Molecular orbital calculations of nonlinear optical properties of acylhydrazines, *Chem. Funct. Dyes., Proc. 2nd Int. Symp.,* 485, 58, 1993.
58. **Matsuura, A., and Hayano, T.,** Theoretical prediction of the donor/acceptor site-dependence on the first hyperpolarisabilities in conjugated systems containing azomethine bonds, *Mater. Res. Soc. Symp. Proc.,* 291, 1993.
59. **Kanis, D. R., Marks, T. J., and Ratner, M. A.,** Calculation and electronic description of molecular quadratic hyperpolarisabilities employing the ZINDO-SOS quantum chemical formulism. Chromophore architecture and substituent effects, *Mol. Cryst. Liq. Cryst. Sci. Technol.,* B6, 317, 1994.

Chapter 3

Measurement Techniques for Refractive Index and Second-Order Optical Nonlinearities

Toshiyuki Watanabe, Hari Singh Nalwa, and Seizo Miyata

CONTENTS

I. INTRODUCTION

Second-order nonlinear optical properties of organic materials are evaluated in terms of second-order nonlinear susceptibility $\chi^{(2)}$ and first hyperpolarizability β. Therefore the magnitude of $\chi^{(2)}$ and β provides an assessment on the second-order nonlinear optic (NLO) efficiency of organic molecules and polymers. Second harmonic generation (SHG) efficiency either in terms of microscopic or macroscopic optical nonlinearity can be determined in solid form using powders and thin films, or in liquid form by dissolving dyes in an appropriate organic solvent. Another important parameter related to SHG properties is the refractive index of a material. In this chapter, we have summarized various measurement techniques that have been employed for determining refractive index, $\chi^{(2)}$ and β. Therefore, the scope of this chapter is to discuss in detail various techniques for evaluating second-order optical nonlinearity. In the first part, refractive index measurements are summarized, while the second part of this chapter reports measurements of $\chi^{(2)}$ and β in thin films and liquids. The information provided in this chapter has been taken partially from our previously published monograph entitled "Optical second harmonic generation in organic molecular and polymeric materials: measurement techniques and materials" by H. S. Nalwa, T. Watanabe, and S. Miyata, in *Progress in Photochemistry and Photophysics,* Vol. 5, edited by J. F. Rabek, CRC Press, Boca Raton Florida, 1990, p. 103, with permission from CRC Press. The aim of this chapter is to furnish an exposition of updated measurement techniques for both the refractive index and SHG properties.

This chapter is an up-to-date survey of the majority of available and commercial techniques in the study of refractive index and second-order optical nonlinearity. The methods discussed here have been obtained from numerous publications and should be useful to newcomers in this field. Each technique is introduced separately. This chapter provides the most important information on theoretical fundamentals and experimental instrumentation of different methods which can be used in the study of second-order NLO materials.

II. REFRACTIVE INDEX MEASUREMENTS

Refractive index is one of the most important parameters in determining the nonlinear optical susceptibilities as well as in designing optical waveguides, because the propagation of waves is dominated by the refractive index of a substrate, waveguide layer and over layers. Furthermore, the conversion efficiency of frequency doubler and coupling efficiency into waveguide is also affected by these factors. A brief definition of refractive index is provided here. When light waves traveling through one medium at a velocity V_1 enter a second medium at a velocity V_2, their direction is changed. The index of refraction (n) of a medium is expressed as the ratio of the velocity of wave in vacuum c to their velocity in the medium and can be expressed as $c = n_1 V_1 = n_2 V_2$. The sine of angle formed by the incident wave and the normal to the surface of separation has the same ratio to the sine of the angle formed by the refracted wave and its normal. This ratio defines n which depends on frequency and can be written as $n_1 \sin \theta_1 = n_2 \sin \theta_2$ where θ_1 is the angle of incidence and θ_2 is the angle of refraction made with the normal. The Snell's law of refraction can be expressed by $n = \sin \theta_1 / \sin \theta_2$ when the standard medium is air; this is further discussed in the text. Each substance has a different value of refractive index under standard conditions of wavelength, temperature, and pressure. The instrument for measuring the refractive index is called a refractometer. In this section, several techniques used to measure the refractive index are discussed.

A. MINIMUM DEVIATION METHOD
The measurement configuration for the minimum deviation method[1] is shown in Figure 1, where α represents the aspect angle between two faces of prism. The B_1 and B_2 are the points of intersection of the incident and the emergent ray with two faces. ϕ_1 and ψ_1 are the angles of incidence and refraction at B_1, and ϕ_2 and ψ_2 are the inner and outer angles at B_2. C is the point of intersection of the normals to the prism at B_1 and B_2, and D is the point of intersection of the incident and emergent ray. If ε is the angle of deviation (i.e., the angle the emergent ray makes with the incident ray), then the following deviation can be given:

$$\phi_1 + \phi_2 = \varepsilon + \alpha \tag{1}$$

$$\psi_1 + \psi_2 = \alpha \tag{2}$$

From the law of refraction

$$\sin \phi_1 = n \sin \psi_1 \tag{3}$$

$$\sin \phi_2 = n \sin \psi_2 \tag{4}$$

where n is the refractive index of the material with respect to the surrounding medium, air. The deviation angle ε will have an extremum when

$$\frac{d\varepsilon}{d\phi_1} = 0 \tag{5}$$

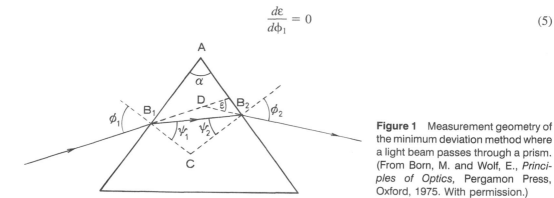

Figure 1 Measurement geometry of the minimum deviation method where a light beam passes through a prism. (From Born, M. and Wolf, E., *Principles of Optics*, Pergamon Press, Oxford, 1975. With permission.)

when the following relation is satisfied

$$\phi_1 = \phi_2 \tag{6}$$

$$\psi_1 = \psi_2 \tag{7}$$

then the deviation angle is minimum. Therefore, when the passage of the beam through the prism is symmetrical, the minimal value of the deviation is expressed by

$$\varepsilon_{min} = 2\phi_1 - \alpha \tag{8}$$

In terms of ε_{min} and α, the angles of incidence and reflection at the first face of the prism can be written as

$$\phi_1 = 0.5 \, (\varepsilon_{min} + \alpha) \tag{9}$$

$$\psi_1 = 0.5\alpha \tag{10}$$

From Equations 3, 4, 9, and 10, the refractive index can be given as

$$n = \frac{\sin \phi_1}{\sin \psi_1} = \frac{\sin[0.5(\varepsilon_{min} + \alpha)]}{\sin(0.5\alpha)} \tag{11}$$

This equation is generally used in determination of the refractive index of materials. The ε_{min} and α are measured with the help of a spectrometer and n is evaluated from Equation 11. The precision of this technique is better than 10^{-5}. In the case of a biaxial crystal, it is more complicated to measure the refractive indices. The prisms must be cut and polished. Figure 2 shows the prism configuration of a biaxial crystal.[2] At the minimum deviation, the value of the indices along the bisector plane are measured. The orientational discrepancy $\Delta\theta$ due to the cutting can be negligible, which is the angle between these planes and the corresponding crystallographic planes, since it can induce the measurement error which is proportional to $\Delta\theta^2$. For prism 1 (prism 2), the propagation of beam which is perpendicular to the bisector plane and polarized along the v and h axes will yield indices n_c and n_b (n_b and n_a).

By using the different kinds of lasers ranging in wavelengths from ultrashort wavelength to infrared (IR) wavelength, the dispersion of refractive index can be obtained. In general, these data are fitted by a least-mean-square method; the four coefficients of a Sellmeir expression, such as A, B, C, and D in following equation, are obtained. This Sellmeir equation can be derived from the Lorentz-Lorent equation.[1]

$$n^2 = A + \frac{B}{1 - c/\lambda^2} - D\lambda^2 \tag{12}$$

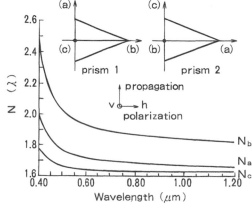

Figure 2 Prism configuration of a biaxial crystal to determine the refractive index. The bisector plane and prism 1 (prism 2) are the (1, 0, 0) plane [(0, 0, 1) plane] and the c axis (b axis). For prism 1 (prism 2), propagation perpendicular to the bisector plane and polarized along v and h will produce indices N_c and N_b (N_b and N_a). Refractive indices of 3-methyl-4-nitropyridine-1-oxide are plotted as a function of wavelength by using the Sellmeir-type equation. (From Zyss, J., et al., *J. Chem. Phys.* 74, 4800, 1981. With permission.)

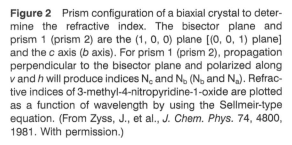

where n is the refractive index. A, B, C, and D are constants, and λ represents the wavelength. The minimum deviation method yields the most precise values compared with other techniques. However, in the case of polymers, it is difficult to make a good prism. So other techniques are more useful to measure the refractive indices of poled polymers.

B. IMMERSION METHOD (BECKE LINE METHOD)

The minimum deviation method requires bulk samples with large dimensions. It is sometimes difficult to obtain a good optical prism with large dimensions having an excellent optical quality and flatness. In this situation, the minimum deviation method cannot be applied. However, the refractive index of a thin single crystal which can not be grown into a large size and a polymer film can be determined by the immersion method.

The immersion method, also known as the Becke line method, depends on the fact that a transparent body is not visible when immersed in a liquid having exactly the same refractive index.[1] On the other hand, when there is a difference in the refractive index between the sample and liquid, the Becke line, which is a slightly brighter line than the other part, can be observed at a boundary between the sample and liquid. This line can be observed through a microscope. When the sample stage of a microscope is moved down, the Becke line shifts from low refractive index material to high refractive index material. On the contrary, Becke line shifts from high refractive index material to low refractive index material when the sample stage of a microscope is moved up.[3]

Using this procedure, the difference of the refractive index between liquid and sample can be estimated. Therefore, if the refractive index of the immersion liquid is known, one can easily obtain the refractive index of sample. When the refractive index of the immersion is unknown, then the value can be measured by refractometers such as an Abbe refractometer.[1] The refractive index of the sample is given by the following equations:

$$N = \sin \alpha / \sin \beta \tag{13}$$

$$\beta + \gamma = \delta \tag{14}$$

$$n = N \sin \gamma \tag{15}$$

$$n = \sin \delta \sqrt{N^2 - \sin^2 \alpha} - \cos \delta \sin \alpha \tag{16}$$

where n is the refractive index of the sample, N is the refractive index of the prism, and γ is the incident angle to the main prism. α and β are incident and refract angles from the prism to air interface, respectively, as shown in Figure 3. It is also possible to measure the refractive indices of polymers directly by pressing thin films onto the prism of the Abbe refractometer. For good optical contact, the matching liquid is filled between the film and prism. The refractive index of the matching liquid should be larger than that of the polymer. In the case of biaxial crystal, this procedure will be more complicated than with isotropic material, since a crystal has two refractive indices for any given direction of propagation. It is possible to measure the refractive index of a biaxial crystal using polarized light when the film is fixed so that light travels parallel to each dielectric axis. The immersion method resembles the technique that uses the Abbe instrument, discussed in the text, without an auxilliary prism and it provides greater range than the Abbe type.

C. PRISM COUPLING METHOD

This measurement is done by using the prism coupling technique with an optical waveguide. This method is also called a m-line technique. The prism coupling method is the most efficient technique for measuring the refractive index of poled polymers. It provides accurate results for spin-coated thin films. It is a simple and fast technique[4] and is used as discussed below. The coupling of a laser beam by a prism into a planner dielectric light guide is governed by the angle θ of incidence of the light onto the prism base (Figure 4). This angle θ determines the phase velocity in the x direction, $V(i) = (c/n_p)\sin\theta$, of the incident waves in the prism (index n_p) and in the gap. Strong coupling of light into the film occurs only when we choose θ so that $V(i)$ equals the phase velocity V_m of one of the characteristic modes of propagation in the guide ($m = 0, 1, 2$) Thus, by determining these synchronous

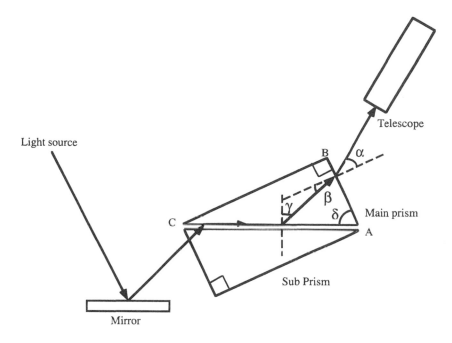

Figure 3 Measurement geometry of the Abbe refractometer. The incident light is illuminated from the AC plane.

angles θ_m of strongest coupling, we can find experimentally the characteristic propagation constants of a given film, relative to the propagation constants, $k_0 = \omega/c$, of free space.

$$N_m = c/V_m = n_p \sin \theta \tag{17}$$

On the other hand, we can calculate theoretical values N_m (effective refractive index) for the relative propagation constants from the known dispersion equation of planner dielectric light guide.

$$N_m = N(m, n, W, k, n_0, n_2, p) \tag{18}$$

In Equation 18, n_0 and n_2 are the refractive indices of the two media next to the film (Figure 4), and, p indicates the polarization of the laser beam ($p = 0$ for TE polarization and $p = 1$ for TM), where W is the thickness of the films. The propagation of light in the waveguide can be explained by a dispersion equation. In the case of TE, the dispersion equation is given by

$$k_x W = (m + 1)\pi - \tan^{-1}(k_x/\gamma_s) - \tan^{-1}(k_x/\gamma_c) \tag{19}$$

In the case of TM, the dispersion equation is given as

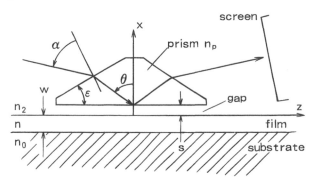

Figure 4 Schematic diagram of a prism film coupler. (From Ulrich, R. and Torge, R., *Appl. Opt.*, 12, 2901, 1973. With permission.)

62

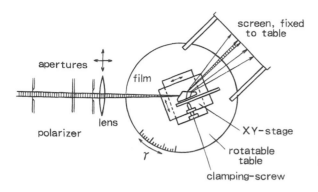

Figure 5 Schematic diagram of prism input optics and observation screen for determining the refractive index. (From Ulrich, R. and Torge, R., *Appl. Opt.*, 12, 2901, 1973. With permission.)

$$k_x W = (m + 1)\pi - \tan^{-1}[(n_0/n)^2((k_x/\gamma_s))] - \tan^{-1}[(n_2/n)^2((k_x/\gamma_c))] \tag{20}$$

where the propagation constants, γ_c, γ_s, and k_x are given by Equations 21, 22, and 23, respectively.

$$\gamma_c = k_0(N_m^2 - n_2^2)^{1/2} \tag{21}$$

$$k_x = k_0(n^2 - N_m^2)^{1/2} \tag{22}$$

$$\gamma_s = k_0(N_m^2 - n_0^2)^{1/2} \tag{23}$$

where n_2 and n_0 are the refractive index of overlayer and substrate of waveguide, respectively and n is the refractive index of material.

The experimental arrangement for the observation of the coupling and for the measurement of the coupling angle is shown schematically in Figure 5. By means of a spring-loaded clamp, the film on the substrate is pressed against the base of the coupling prism. The prism can be either symmetric (Figure 6a) or a half-prism (Figure 6b). It sits on a *xy* translation stage that is mounted on a precision (1 < min of arc) rotary table or goniometer. The laser beam is linearly polarized (TE or TM) and must be a TEM$_{00}$ cross section. A lens focuses the beam into the prism so that the beam waist coincides with the prism base. The point where the beam strikes the prism base is the coupling spot. At this point, the parameters n and W are measured. In the case of a symmetric prism the coupling spot is preferably near the center of the prism base; in a half-prism it must be close to the corner.

Besides the prism shape (Figure 6), the most important parameters of a coupling prism are its refractive index n_p and the prism angle ϵ. These parameters determine the range of propagation and the propagation constants $N = n_p \sin\theta$ of light along the prism. The parameter N can be described by the following equation:

$$N = \sin\alpha \cos\epsilon + (n_p^2 - \sin^2\alpha)^{1/2} \sin\epsilon \tag{24}$$

where α is the prism angle of incidence on the entrance face of the prism (Figure 4). When α varies from $-\pi/2$ to $\pi/2$, the range of possible N values with a given prism is obtained. The limits N_{min} and

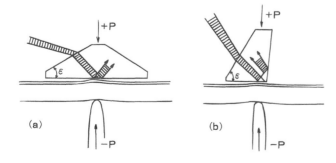

Figure 6 Prism shapes for prism coupling measurements. (a) Symmetry prism, useful for flexible films such as polymers. (b) Half prism, useful for thick samples such as single crystals. (From Ulrich, R. and Torge, R., *Appl. Opt.*, 12, 2901, 1973. With permission.)

N_{max} of this range in a film of index n, deposited on a substrate of index n_0, all guided modes ($m = 0,1,2, \ldots$) are in the interval $n_0 < N_m < n$. Therefore, in order to measure all modes of film, generally a prism is required whose N range covers the interval from n_0 to n. The result of the measurement is a list containing the observed modes of the film, identified by their mode numbers $m = 0,1,2, \ldots$ and for each mode the measured coupling angle θ_m. By solving Equation 19 or 20 from the measurement results of the coupling angle at each mode, refractive indices of films can be obtained.

D. ELLIPSOMETRY METHOD

Since the ellipsometry method[5,6] can measure the refractive index of the ultrathin film, it has attracted much attention. Ellipsometry is the measurement of the effect of the reflection on the state of polarization of light. The state of polarization is characterized by the phase and amplitude relationship between the two-component plane waves of the electric field vector in which the polarized oscillation may be resolved. An incident plane wave propagating in the z direction of local orthogonal system can be expressed by[5]

$$E_i(r, t) = Re[(\hat{x}E_x + \hat{y}E_y)\exp(ikz - i\omega t)] \tag{25}$$

where E_x and E_y are the complex field coefficients describing the amplitude and phase dependencies of the projections of $E_i(r,t)$ along the x and y axes. If the electromagnetic wave is reflected by a smooth surface, the outgoing wave can be presented in the absence of anisotropic effects in another local coordinate system as

$$E_i(r, t) - Re[\hat{x}r_pE_x + \hat{y}r_sE_y)\exp(ikz' - i\omega t)] \tag{26}$$

where one wave, designated p, is polarized in the plane of incidence, parallel to the x axis, and another wave, designated s, is polarized perpendicular to the plane of incidence, parallel to the y axis, and the z axes define the plane of incidence. In this approximation, the effect of the surface is described by two coefficients, r_p and r_s. In general, reflection causes a change in the relative phase of the p and s waves and change in the ratio of their amplitude. The four parameters (phase and amplitude of p and s waves) describing the plane wave of Equations 25 and 26 are represented by the concept of polarization ellipse. In order to know the properties of both incident and reflected waves, it is useful to introduce the polarization ellipse describing these four parameters. The polarization ellipse can be described by size and shape. The size $E = (|E_x|^2 + |E_y|^2)^{1/2}$ is a scalar whose square is proportional to the intensity I. The shape is an intensity-dependent quantity that clearly requires two parameters to specify, such as the minor/major axis ratio and the azimuth angle of the major axis of the polarization ellipse. One convenient representation of the shape is the polarization state $\chi = E_x/E_y$. Since E_x, E_y, and χ are complex, they therefore contain the required two parameters. For example, if E_x and E_y are in phase, then χ is real, both field components are strictly proportional at all times, and the light is linearly polarized with an azimuth angle $\psi = \tan^{-1}\chi$. If E_x and E_y are 90° out of phase, then χ is purely imaginary and one field component is at a maximum while the other is at zero; if, in addition, $|E_x| = |E_y|$ then $\chi = \pm i$, and the light is circularly polarized.

Given either intensity or polarization state for the incident and reflected light, the sample properties can be calculated by taking appropriate ratios. In reflectometry, we distinguish between p- and s-polarized light. The intensity rations are given by

$$(I_r/I_i)_p = |r_pE_x|^2/|E_x|^2 = |r_p|^2 = R_p \tag{27}$$

$$(I_r/I_i)_s = |r_sE_y|^2/|E_y|^2 = |r_s|^2 = R_s \tag{28}$$

The reflectances R_p and R_s are the absolute squares of the respective complex reflectance coefficients. At normal incidence both polarizations are equivalent and $R_p = R_s$. In ellipsometry, the ratio of polarization state yields a different perspective of the sample. From Equations 27 and 28, the following equation is derived

$$\chi_r/\chi_i = (r_p E_x/r_s E_y)/(E_x/E_y) = r_p/r_s = \rho = (\tan \psi \exp(i\Delta)) \tag{29}$$

In the case of a film-covered surface, the amplitude and phases of the component waves for reflection depend on interference between the beam reflected from the ambient–film interface and the beams refracted into the ambient interfaces. From this point of view, a relationship between Δ and ψ and the properties of the reflecting system are easily derived in terms of the Fresnel reflection coefficient for the two-component wave at the two interfaces and the optical thickness of the film. The resulting fundamental equation of ellipsometry is as follows[6]:

$$\tan \psi \exp(i\Delta) = \frac{r_{1p} + r_{2p} \exp(-2i\delta)}{1 + r_{1p} r_{2p} \exp(-2i\delta)} \times \frac{1 - r_{1s} r_{2s} \exp(-2i\delta)}{r_{1s} + r_{2s} \exp(-2i\delta)} \tag{30}$$

where δ is the change in phase of the vacuum wavelength λ caused by transversing the film of thickness d and index of refraction n_1, r_{1p} and r_{1s} are the Fresnel coefficients for the two-component waves for reflection into the ambient from the surface of the film at the ambient–film interface, and r_{2p} and r_{2s} are the coefficients for reflection into the film from the substrate at the film–substrate interface.

The relationship between δ, refractive index n, and thickness d is represented as

$$\delta = (360/\lambda)d(n^2 - \sin^2\phi)^{1/2} \text{ degree} \tag{31}$$

where ϕ shows the incident angle, and λ is the wavelength of the light source.

The general formulas for the Fresnel reflection coefficients for reflection from surface of medium b into medium a at the a–b interface are:

$$r_p = \frac{n_a \cos \phi_b - n_b \cos \phi_a}{n_a \cos \phi_b + n_b \cos \phi_a} \tag{32}$$

$$r_s = \frac{n_a \cos \phi_a - n_b \cos \phi_b}{n_a \cos \phi_a + n_b \cos \phi_b} \tag{33}$$

where ϕ_a is the angle of incidence and ϕ_b is the angle of refraction. Substitution of Equations 31, 32, and 33 into Equation 30 and separation of the resulting equation into its real and imaginary parts yields one equation for Δ and one for ψ. Both equations are the functions of the angle of incidence in the ambient medium, the wavelength of the light, the index of refraction of the substrate, and the thickness and index of refraction of the film. All of these quantities are independently determined or fixed constants except for the properties of the film which can be obtained from ellipsometer measurements by solving the equations.

A graphical representation of the dependence of Δ and ψ on the properties of transparent (i.e., nonabsorbing) isotropic films on silicon is shown in Figure 7. The plot was constructed by programming the expression for Δ and ψ. The computer yields tables of Δ and ψ as a function of film thickness, or δ, for any value for the index of refraction of film inserted in the program. A significant property of the dependence of Δ and ψ on the index of refraction of the film is that, for all practical cases, no two curves overlap or intersect; as a result, each point in the plane corresponds to a unique value of the index of refraction of the film. The parameter such as Δ, ψ can be measured according to the following procedure.

The ellipsometer is a specially made polarizing spectrometer with the optical arrangement shown in Figure 8. It employs Glan-Thompson polarizing prisms mounted in divided circles which can read up to $\pm0.01°$. The mica wave plate is similarly mounted and used as an approximately quarter-wave for the monochromatic light source. If the fast axis of the quarter-wave plate is set at $45°$, where angles are measured relative to a conventional coordinate system in which the positive z axis is the direction of propagation of the light beam and the xz plane is the plane of the incidence, the reflected beam can be extinguished by suitably adjusting the orientations of the polarizer and analyzer, respectively; the settings at extinction, P_0 and A_0, are related to Δ and ψ by

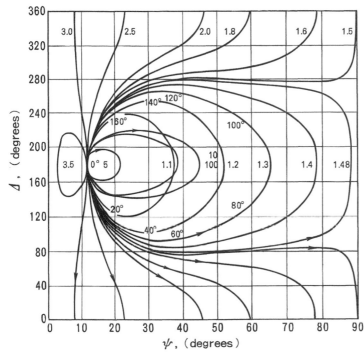

Figure 7 Nomograph for translating the measured values into thickness and index of refraction. The curves corresponds to a film of fixed refractive index and shows the increasing direction of the thickness. (From Archer, R. J., *J. Opt. Soc. Am.,* 52, 970, 1962. With permission.)

$$\tan \Delta = \sin \varepsilon \tan(90° - 2P_0) \tag{34}$$

$$\tan \psi = \cot L \tan(-A_0) \tag{35}$$

where

$$\cos 2L = -\cos \varepsilon \cos(2P_0) \tag{36}$$

and ε is the relative retardation of the wave plate. Using as a detector, a photomultiplier microphotometer, approximate extinction settings are made by adjusting P and A for approximately minimal transmission. The exact extinction settings for the polarizer are determined by measuring P at equal intensity on each side of the minimum. The exact minimum P_0 is the average of the two values. The polarizer is then set at P_0, and A_0 is obtained by the same method of measuring analyzer setting giving equal intensity on each side of the minimum. The experimental results obtained from this procedure can be plotted to the Δ-ψ curves in Figure 7, and the curves give the index of refraction of films and the thickness.

E. OPTICAL RETARDATION MEASUREMENTS

The optical retardation R can be measured using ellipsometry.[7] A photoelastic modulator improves the measurement accuracy coupled with a lock-in amplifier. The schematic apparatus of a photoelastic

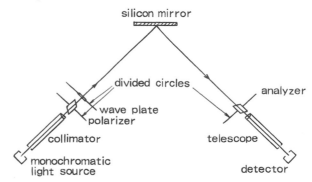

Figure 8 Ellipsometry technique of determining refractive indices. (From Archer, R. J., *J. Opt. Soc. Am.,* 52, 970, 1962. With permission.)

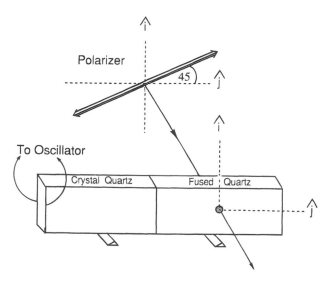

Figure 9 Schematic diagram of polarization modulator. (From Jasperson, S. N. and Schnatterly, S. E., Rev. Sci. Instrum., 40, 761, 1969. With permission.)

modulator is shown in Figure 9. The schematic diagram of the measurement apparatus is shown in Figure 10.

The electric field detected in the front of the photodiode can be represented as follows[8]:

$$E_f = A \times T(-\theta)ST(\theta) \times M \times T(-4/\pi)P \times L = \frac{1}{\sqrt{2}} [\cos^2\theta + \rho \sin^2\theta \cdot \exp(i\Delta)$$

$$- \sin\theta \cos\theta \cdot \exp(i\delta) \cdot (1 - \rho \exp(i\Delta)] \quad (37)$$

where $L, P, A, M, S, T(\theta)$ are the Jones matrix and represent light source, polarizer, analyzer, modulator, sample, and rotation stage, respectively. Each matrix is given by

$$E_f = \begin{bmatrix} E_x \\ E_y \end{bmatrix} \quad (38)$$

$$L = \begin{bmatrix} 1 \\ 1 \end{bmatrix} \quad (39)$$

$$P, A = \begin{bmatrix} 1 & 0 \\ 0 & 0 \end{bmatrix} \quad (40)$$

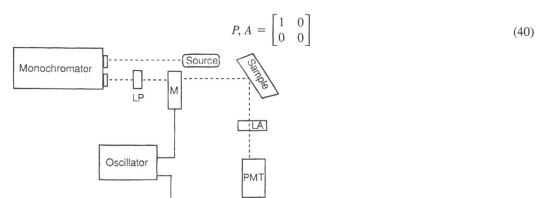

Figure 10 Schematic diagram of the ellipsometry system. LP and LA are linear polarizer, M is the polarization modulator, and PMT is the photomultipler tube. (From Jasperson, S. N. and Schnatterly, S. E., *Rev. Sci. Instrum.*, 40, 761, 1969. With permission.)

$$M = \begin{bmatrix} 1 & 0 \\ 0 & \exp(i\delta) \end{bmatrix} \tag{41}$$

$$S = \begin{bmatrix} 1 & 0 \\ 0 & \exp(i\Delta) \end{bmatrix} \begin{bmatrix} 1 & 0 \\ 0 & \rho \end{bmatrix} \tag{42}$$

$$T(\theta) = \begin{bmatrix} \cos\theta & \sin\theta \\ -\sin\theta & \cos\theta \end{bmatrix} \tag{43}$$

where δ is the phase of modulator, Δ is the phase of sample, ρ is T_{max}/T_{min}, and θ is the angle between fast axis of sample and the optical axis of measurement system, T_{max} and T_{min} represent maximum transmittance and minimum transmittance. Optical intensity detected by the photodiode is

$$I = E_f \times E_f^* = 1/2(\alpha + \beta + \gamma) \tag{44}$$

where α, β, γ are defined as

$$\begin{aligned} \alpha &= (\rho^2 - 1)\sin^2\theta + 1 \\ \beta &= (\sin 2\theta \cdot ((\rho^2 + 1)\sin^2\theta + \rho\cos 2\theta \cdot \cos\Delta - 1) \cdot \cos\delta \\ \gamma &= -\rho \sin 2\theta \cdot \sin\Delta \cdot \sin\delta \end{aligned} \tag{45}$$

To simplify these equations, the amplitude is represented as A.

$$\delta - A\sin(\omega t) \tag{46}$$

where ω is the modulated frequency,

Using Bessel function, β and γ are given as follows:

$$\begin{aligned} \beta &= \sin 2\theta \cdot ((\rho^2 + 1)\sin^2\theta + \rho\cos 2\theta \cdot \cos\Delta - 1) \cdot (J_0(A) + 2J_2(A)\cos(2\omega t)) \\ \gamma &= -\rho\sin 2\theta \cdot \sin\Delta \cdot 2J_1(A)\sin(\omega t) \end{aligned} \tag{47}$$

$J_0(A)$ becomes zero by controlling modulation depth. Then α, β, γ become the DC component, fundamental component, and second harmonic component, respectively.

Each component presented as

$$\begin{aligned} I_{DC} &= 0.5((\rho^2 - 1)\sin^2\theta + 1) \\ I_\omega &= \rho\sin 2\theta \cdot \sin\Delta \cdot J_1(A) \\ I_{2\omega} &= \sin 2\theta \cdot ((\rho^2 + 1)\sin^2\theta + \rho\cos 2\theta \cdot \cos\Delta - 1) \cdot J_2(A) \end{aligned} \tag{48}$$

When $\theta = 0$,

$$\begin{aligned} I_{DC}(0) &= 0.5 \\ I_\omega(0) &= 0 \end{aligned} \tag{49}$$

When $\theta = 45°$,

$$\begin{aligned} I_{DC}(45°) &= 0.25(\rho^2 + 1) \\ I_\omega(45°) &= \rho\sin\Delta \cdot J_1(A) \end{aligned} \tag{50}$$

From these equations, the following equation can be derived

$$\Delta = \sin^{-1}[I_\omega(45°)/2\rho \cdot J_1(A) \cdot I_{DC}(0)](\text{rad}) \qquad (51)$$

Therefore,

$$R = \Delta \times \lambda/2\pi \qquad (52)$$

By changing the incident angle, the refractive index of the sample which is out of the film plane also can be obtained from the retardation. The refractive indices of three axes n_x, n_y, n_z, can be derived by solving the following equations:

$$R_0 = (n_x - n_y)d \qquad (53)$$

$$R_{t1} = (n_{e1} - n_y)d/\cos \alpha_1 \qquad (54)$$

$$n_{e1}^2 = n_x^2 n_z^2/(n_x^2 \sin^2 \alpha_1 + n_z^2 \cos^2\alpha_1) \qquad (55)$$

$$n_{e1} \sin \alpha_1 = \sin \theta \qquad (56)$$

$$R_{t2} = (n_{e2} - n_y)d/\cos \alpha_2 \qquad (57)$$

$$n_{e2}^2 = n_x^2 n_z^2/(n_x^2 \sin^2 \alpha_2 + n_z^2 \cos^2 \alpha_2) \qquad (58)$$

$$n_{e2} \sin \alpha_2 = \sin \theta_2 \qquad (59)$$

where d is the film thickness, α is the refractive angle given by Snell's law. R_0 is the retardation of film for normal to the film. R_{t1} and R_{t2} are the retardation when film is tilted around the first axis. n_{e1} and n_{e2} are the refractive index of material for p-polarization at different tilt angle.

F. TRANSMITTANCE MEASUREMENT (INTERFERENCE METHOD)

Transmittance measurement is advantageous because it can measure the dispersion of the refractive index. By measuring the optical spectra of film of a specific thickness, the refractive index can be obtained. The transmittance spectra of 20 to 100 μm-thick film appear as an interference fringe pattern. This fringe is due to the interference of incident light between the front and back sides of the film. The reflectivity R and transmittance T are oscillated as a function of nd/λ, where d is the film thickness and n is the refractive index of film. Let us suppose that the reflectivity has a maximum at λ_{2m} and a minimum at $\lambda_{2m} + 1$, where m represents the order of interference. If we assume that the refractive index is constant in this range, the following relations are derived[9]:

$$2m = \frac{4n_1 d}{\lambda_{2m}}, \qquad 2m + 1 = \frac{4n_1 d}{\lambda_{2m+1}} \qquad (60)$$

From these equations, the retardation can be derived by erasing the order of interference m.

$$n_1 d = \frac{\lambda_{2m} \cdot \lambda_{2m+1}}{4(\lambda_{2m} - \lambda_{2m+1})} \qquad (61)$$

In general, the above relation cannot be satisfied completely because of the dispersion of the refractive index. Therefore, plotting the order of interference as a function of $1/\lambda$ and interpolating to zero, the order of interference can be precisely determined. By using Equation 61, the refractive index can be obtained. The interference method can be used for determining the refractive index of gases as well as liquids and some solids. With an interference-type instrument, the difference in the refractive index of one part of a million can be estimated with great sensitivity.

G. ATTENUATED TOTAL REFLECTION (ATR) METHOD

The attenuated total reflection method (ATR)[10] permits the determination of film thickness and two components of the anisotropic refractive index. The ATR technique does not require the thick films which permit the propagation of light in the guided mode like a prism coupling method. The ATR technique can measure the refractive index of monolayer LB films incorporating with surface plasmon. This technique is a generalization of the Kreschtmann method for observing the plasma surface waves in the metallic layer.

A glass slide, coated with a thin semi-transparent silver layer (=50 nm) on which dielectric film is deposited, is put in optical contact with the base of a prism. A collimated monochromatic light beam is reflected on the multilayered structure through the prism (Figure 11).[10] The reflectivity $R(\varphi)$ is recorded as a function of the incident angle. In the real experiment, the goniometer angle Φ is measured and transferred easily into the angle φ defined inside the prism. SPW waves propagate along the interface between a metal and dielectric material, whereas the guided waves propagate in the dielectric layers. The specific features of surface waves are extreme sensitivity to the optical properties of the dielectric medium and a strong electric field near the boundary or in the guided layers. The enhancement of the electric field is also necessary to apply the second harmonic generation and Raman effects. This high sensitivity to the refractive index and the absorption coefficient was used for the measurement of a complex refractive index.

Let us consider the principle of ATR techniques. A plane boundary between two mediums which correspond to ε_1 and ε_2 is represented in Figure 12. A surface wave is propagating along the interface between two layers, and its electric field decreases exponentially on both layers. x is the propagation direction. The extinction coefficients are presented as α. The electric field in each layer is described as follows:

$$E_1 = E_1 \exp(-i\omega t + ik_x x + \alpha_1 z) \tag{62}$$

$$E_2 = E_2 \exp(-i\omega t + ik_x x - \alpha_2 z) \tag{63}$$

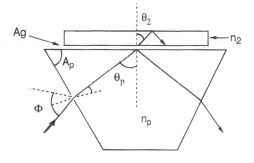

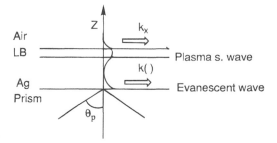

Figure 11 Kretschmann configuration for producing guided wave and surface plasmon waves (SPW). (From Dumont, M. and Levy Y. *Nonlinear Optics of Organics and Semiconductors, Springer-Verlag,* Berlin, 1989. With permission.)

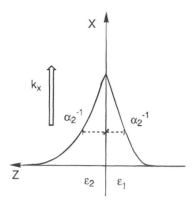

Figure 12 Structure of surface plasmon waves. (From Dumont, M. and Levy Y. *Nonlinear Optics of Organics and Semiconductors, Springer-Verlag,* Berlin, 1989. With permission.)

with α_1 and $\alpha_2 > 0$. The Maxwell equations and the continuity conditions at the boundary show that this type of wave exists if light is TM polarized ($H_x = H_z = E_y = 0$; $k_x E_{ix} = -k_{iz} E_{iz}$, where $k_{iz} = i\alpha_i$ is the imaginary z component of the wave vector in medium i: $k_i^2 = \varepsilon_i \omega^2/c^2 = k_x^2 - \alpha_i^2$) and if

$$\varepsilon_1 < -\varepsilon_2 < 0 \tag{64}$$

In addition, one finds the well-known dispersion relation

$$k_x^2 = \varepsilon_1 \varepsilon_2 (\varepsilon_1 + \varepsilon_2)^{-1} k_0^2 \tag{65}$$

with $k_0 = \omega/c$. A large negative ε (i.e., imaginary index) is possible in a perfect metal described by the free-electron model. This model gives

$$\varepsilon_1 = 1 - \omega_p^2/\omega^2 \tag{66}$$

where ω_p is the plasma frequency of free electrons. By using this value in Equation 66, one gets the dispersion curves shown in Figure 13. ω must be smaller than $\omega_p(1 + \varepsilon_2)^{-1/2}$. Real metals are not perfect: their dielectric constants have an imaginary part which produces absorption and, hence, fast dumping of surface waves. The penetration length of the electric field in each medium is given by

$$l_i = \alpha_i^{-1} = \sqrt{-(\varepsilon_1 + \varepsilon_2)/\varepsilon_1^2} \lambda_0/2\pi \tag{67}$$

In the case of air and silver (for $\lambda_0 = 600$ nm), one has $l_{air} = 400$ nm and $l_{Ag} = 25$ nm. Let us consider an absorption-free dielectric film of thickness t_2 and index of refraction n_2, where $n_3 < n_2 < n_1$. It is well known that guided waves may propagate inside this film with the general structure as shown in

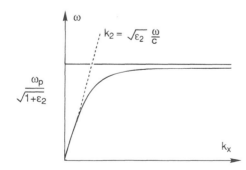

Figure 13 Dispersion function of surface plasmon waves (SPW). (From Dumont, M. and Levy Y. *Nonlinear Optics of Organics and Semiconductors, Springer-Verlag,* Berlin, 1989. With permission.)

Figure 14. Finally, the Fabry-Perot interference condition imposes well-defined values of θ_2 which correspond to the modes of the wave guide. By solving the mode according to Equation 19 or 20, one can obtain the refractive index of the medium.

III. SECOND-ORDER OPTICAL NONLINEARITY MEASUREMENTS

The second part of this chapter provides a general description of commonly used techniques for measuring second-order nonlinear optical properties of powders, thin films, and liquids. Recently, new techniques have been developed to measure SHG efficiency of powders with accuracies as well as β values. The NLO coefficients are generally measured by the Maker fringe technique using a reference material such as quartz. In powder measurements, urea or ADP are used as the reference standards.

A. NONLINEAR OPTICAL COEFFICIENTS

The general description of microscopic and macrocopic second-order optical nonlinearity has been provided in Chapters 1 and 2. Second-order polarization is given by the following equation[11]:

$$P(2\omega) = \varepsilon_0 \sum (1/2)\chi_{ijk}(-2\omega; \omega, \omega)E_j(\omega)E_k(\omega) \text{ (in MKS units)} \tag{68}$$

where $P(2\omega)$ is the second harmonic power, and ω is the angular frequency at the fundamental wavelength. Similarly, the induced polarization is presented in terms of d, which is a $3 \times 3 \times 3$ third-rank tensor. The SHG coefficients satisfy the permutation symmetry, since the electric field components in Equation 68 are not significant. For SHG where $\omega_1 = \omega_2 \equiv \omega$, d_{ijk} is related to second-order NLO susceptibility $\chi^{(2)}$ as follows:

$$\chi^{(2)}_{ijk}(-2\omega; \omega, \omega) = 2d_{ijk}(-2\omega; \omega, \omega) \tag{69}$$

Furthermore, in the contracted form, the suffix j and k can be replaced by single suffix m, with values 1 to 6. The relation between the contracted and uncontracted suffixes can be expressed by

$$
\begin{array}{rl}
jk & m \\
11 & \to 1 \\
22 & \to 2 \\
33 & \to 3 \\
23 \text{ or } 32 & \to 4 \\
31 \text{ or } 13 & \to 5 \\
12 \text{ or } 21 & \to 6
\end{array}
\tag{70}
$$

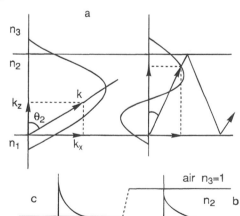

Figure 14 Structure of modes in wave guide. (a) Guided mode in a dielectric film (TE_0 and TE_1). (b) TM_0 mode when medium 1 is a metal: metal/dielectric SPW. (c) TM_0 mode when medium 1 is a metal: metal/air SPW. (From Dumont, M. and Levy Y. *Nonlinear Optics of Organics and Semiconductors*, Springer-Verlag, Berlin, 1989. With permission.)

In the contracted form, the 18 elements of dim and the components of the second-order polarization at frequency 2ω can be written in the matrix form:

$$\begin{pmatrix} P_x \\ P_y \\ P_z \end{pmatrix} = \varepsilon \begin{pmatrix} d_{11} & d_{12} & d_{13} & d_{14} & d_{15} & d_{16} \\ d_{21} & d_{22} & d_{23} & d_{24} & d_{25} & d_{26} \\ d_{31} & d_{32} & d_{33} & d_{34} & d_{35} & d_{36} \end{pmatrix} \begin{pmatrix} E_x^2 \\ E_y^2 \\ E_z^2 \\ 2E_y E_z \\ 2E_x E_z \\ 2E_x E_y \end{pmatrix} \tag{71}$$

When there is no absorption at harmonic wavelength and fundamental wavelength, the following equalities are obtained according to the Kleinman's symmetry:

$$d_{12} = d_{26}; \quad d_{13} = d_{35}; \quad d_{14} = d_{25} = d_{36}; \quad d_{15} = d_{31}; \quad d_{16} = d_{21}; \quad d_{24} = d_{32} \tag{72}$$

Since the d coefficient is a polar tensor of odd rank, its elements are identically zero for any medium possessing a center of inversion. The number of nonvanishing elements of a third-rank tensor are obtained by the point group symmetry of a crystal. The d_{ijk} coefficient and the electro-optic coefficient r_{ijk} are related to each other as follows through the two-level model:[12]

$$r_{ijk}(-\omega; \omega, 0) = -\frac{4d_{kij}}{n_i^2(\omega)n_j^2(\omega)} \frac{f_{ii}^\omega f_{jj}^\omega f_{kk}^0}{f_{kk}^{2\omega} f_{ii}^\omega f_{jj}^\omega} \times \frac{(3\omega_0^2 - \omega^2)(\omega_0^2 - \omega'^2)(\omega_0^2 - 4\omega'^2)}{3\omega_0^2(\omega_0^2 - \omega^2)^2} \tag{73}$$

where ω_0 is the frequency of the first strongly absorbing electronic transition in the molecule, and ω' and ω are the fundamental wavelengths in the second harmonic generation and for electro-optic coefficient measurements, respectively. By evaluating the SHG coefficients, one can also roughly estimate the electro-optic coefficients. The SHG coefficients form a third-rank tensor, therefore d_{ij} tensors follow the same symmetry rules as the piezoelectric tensors. Second-order nonlinear optical tensors for 21 non-centrosymmetric classes are listed below:[13,14]

1. Tricilinic system
   Class l-C_1

$$\begin{pmatrix} d_{11} & d_{12} & d_{13} & d_{14} & d_{15} & d_{16} \\ d_{21} & d_{22} & d_{23} & d_{24} & d_{25} & d_{26} \\ d_{31} & d_{32} & d_{33} & d_{34} & d_{35} & d_{36} \end{pmatrix}$$

2. Monoclinic system
   Class m-C_{1h}

$$\begin{pmatrix} d_{11} & d_{12} & d_{13} & 0 & 0 & d_{16} \\ d_{21} & d_{22} & d_{23} & 0 & 0 & d_{26} \\ 0 & 0 & 0 & d_{34} & d_{35} & 0 \end{pmatrix} m \perp Z$$

   Class m-C_{1h}

$$\begin{pmatrix} d_{11} & d_{12} & d_{13} & 0 & d_{15} & 0 \\ 0 & 0 & 0 & d_{24} & 0 & d_{26} \\ d_{31} & d_{32} & d_{33} & 0 & d_{35} & 0 \end{pmatrix} m \perp Y \text{(IRE convention)}$$

Class 2-C$_2$

$$\begin{pmatrix} 0 & 0 & 0 & d_{14} & d_{15} & 0 \\ 0 & 0 & 0 & d_{24} & d_{25} & 0 \\ d_{31} & d_{32} & d_{33} & 0 & 0 & d_{36} \end{pmatrix} 2//Z$$

Class 2-C$_2$

$$\begin{pmatrix} 0 & 0 & 0 & d_{14} & 0 & d_{16} \\ d_{21} & d_{22} & d_{23} & 0 & d_{25} & 0 \\ 0 & 0 & 0 & d_{34} & 0 & d_{36} \end{pmatrix} 2//Y\text{(IRE convention)}$$

3. Orthorhombic system
   Class $mm2$-C$_{2v}$

$$\begin{pmatrix} 0 & 0 & 0 & 0 & d_{15} & 0 \\ 0 & 0 & 0 & d_{24} & 0 & 0 \\ d_{31} & d_{32} & d_{33} & 0 & 0 & 0 \end{pmatrix} \text{(Stretched films of poled polymers)}$$

   Class 222-D$_2$

$$\begin{pmatrix} 0 & 0 & 0 & d_{14} & 0 & 0 \\ 0 & 0 & 0 & 0 & d_{25} & 0 \\ 0 & 0 & 0 & 0 & 0 & d_{36} \end{pmatrix}$$

4. Tetragonal system
   Class 4-C$_4$

$$\begin{pmatrix} 0 & 0 & 0 & d_{14} & d_{15} & 0 \\ 0 & 0 & 0 & d_{15} & -d_{14} & 0 \\ d_{31} & d_{31} & d_{33} & 0 & 0 & 0 \end{pmatrix}$$

   Class $\bar{4}$-S$_4$

$$\begin{pmatrix} 0 & 0 & 0 & d_{14} & d_{15} & 0 \\ 0 & 0 & 0 & -d_{15} & d_{14} & 0 \\ d_{31} & -d_{31} & 0 & 0 & 0 & d_{36} \end{pmatrix}$$

   Class $4mm$-C$_{4v}$

$$\begin{pmatrix} 0 & 0 & 0 & 0 & d_{15} & 0 \\ 0 & 0 & 0 & d_{15} & 0 & 0 \\ d_{31} & d_{31} & d_{33} & 0 & 0 & 0 \end{pmatrix}$$

Class $\bar{4}2m$-D$_{2d}$

$$\begin{pmatrix} 0 & 0 & 0 & d_{14} & 0 & 0 \\ 0 & 0 & 0 & 0 & d_{14} & 0 \\ 0 & 0 & 0 & 0 & 0 & d_{36} \end{pmatrix}$$

Class 422-D$_4$

$$\begin{pmatrix} 0 & 0 & 0 & d_{14} & 0 & 0 \\ 0 & 0 & 0 & 0 & -d_{14} & 0 \\ 0 & 0 & 0 & 0 & 0 & 0 \end{pmatrix}$$

5. Cubic system
   Class $\bar{4}3m$, 23-T and T$_4$

$$\begin{pmatrix} 0 & 0 & 0 & d_{14} & 0 & 0 \\ 0 & 0 & 0 & 0 & d_{14} & 0 \\ 0 & 0 & 0 & 0 & 0 & d_{14} \end{pmatrix}$$

   Class 432-O All components are zero

6. Trigonal system
   Class 3-C$_3$

$$\begin{pmatrix} d_{11} & -d_{11} & 0 & d_{14} & d_{15} & -d_{22} \\ -d_{22} & d_{22} & 0 & d_{15} & -d_{14} & -d_{11} \\ d_{31} & d_{31} & d_{33} & 0 & 0 & 0 \end{pmatrix}$$

   Class 3m-C$_{3v}$

$$\begin{pmatrix} 0 & 0 & 0 & 0 & d_{15} & -d_{22} \\ -d_{22} & d_{22} & 0 & d_{15} & 0 & 0 \\ d_{31} & d_{31} & d_{33} & 0 & 0 & 0 \end{pmatrix} m \perp X(\text{IRE convention})$$

   Class 3m-C$_{3v}$

$$\begin{pmatrix} d_{11} & -d_{11} & 0 & 0 & d_{15} & 0 \\ 0 & 0 & 0 & d_{15} & 0 & -d_{11} \\ d_{31} & d_{31} & d_{33} & 0 & 0 & 0 \end{pmatrix} m \perp Y$$

   Class 32-D$_3$

$$\begin{pmatrix} d_{11} & -d_{11} & 0 & d_{14} & 0 & 0 \\ 0 & 0 & 0 & 0 & -d_{14} & -d_{11} \\ 0 & 0 & 0 & 0 & 0 & 0 \end{pmatrix}$$

7. Hexagonal system
   Class 6-C$_6$

$$\begin{pmatrix} 0 & 0 & 0 & d_{14} & d_{15} & 0 \\ 0 & 0 & 0 & d_{15} & -d_{14} & 0 \\ d_{31} & d_{31} & d_{33} & 0 & 0 & 0 \end{pmatrix}$$

   Same as Class 4

Class $\bar{6}$-C_{3h}

$$\begin{pmatrix} d_{11} & -d_{11} & 0 & 0 & 0 & -d_{22} \\ -d_{22} & d_{22} & 0 & 0 & 0 & -d_{11} \\ 0 & 0 & 0 & 0 & 0 & 0 \end{pmatrix}$$

Class $\bar{6}m2$-D_{3h}

$$\begin{pmatrix} 0 & 0 & 0 & 0 & 0 & -d_{22} \\ -d_{22} & d_{22} & 0 & 0 & 0 & 0 \\ 0 & 0 & 0 & 0 & 0 & 0 \end{pmatrix} m \perp X \text{(IRE convention)}$$

Class $\bar{6}m2$-D_{3h}

$$\begin{pmatrix} d_{11} & -d_{11} & 0 & 0 & 0 & 0 \\ 0 & 0 & 0 & 0 & 0 & -d_{11} \\ 0 & 0 & 0 & 0 & 0 & 0 \end{pmatrix} m \perp Y$$

Class $6mm$-C_{6v}

$$\begin{pmatrix} 0 & 0 & 0 & 0 & d_{15} & 0 \\ 0 & 0 & 0 & d_{15} & 0 & 0 \\ d_{31} & d_{31} & d_{33} & 0 & 0 & 0 \end{pmatrix}$$
Same as Class $4mm$

Class 622-D_6

$$\begin{pmatrix} 0 & 0 & 0 & d_{14} & 0 & 0 \\ 0 & 0 & 0 & 0 & -d_{14} & 0 \\ 0 & 0 & 0 & 0 & 0 & 0 \end{pmatrix}$$
Same as Class 422

The symmetry classes of a very wide variety of organic single crystals are discussed in Chapter 4. For the poled polymer system, second-order NLO tensor components are:

Class ∞mm

$$\begin{pmatrix} 0 & 0 & 0 & 0 & d_{15} & 0 \\ 0 & 0 & 0 & d_{24} & 0 & 0 \\ d_{31} & d_{31} & d_{31} & 0 & 0 & 0 \end{pmatrix} \quad \text{(Poled polymers)}$$

B. POWDER METHOD

The powder technique is the easiest measurement for the quick evaluation of second-order nonlinear optical properties of materials.[15] The experimental setup used for this powder measurement is shown

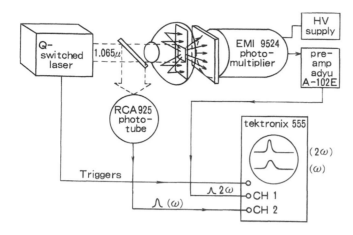

Figure 15 Schematic diagram for evaluating powder samples for SHG. (From Kurtz, S. K. and Perry, T. T., *J. Appl. Phys.*, 39, 3798, 1968. With permission.)

in Figure 15.[15] It consists of a Q-switched laser, and is illuminated to the sample through the UV-VIS cut filter. The second harmonic generation is detected by a photomultiplier tube and displayed on an oscilloscope. The system permitted the insertion of narrow-band pass filters at the second harmonic wavelength between the short filter and photomultiplier to eliminate spurious signals. In order to obtain the efficiency of a second harmonic collection at the detector, a parabolic reflector is placed directly in the front of the sample with a small access hole for the laser beam. Samples were prepared and mounted using several techniques. For qualitative results, a thin layer (−0.2 mm) of ungraded powder is placed on a microscope slide and held with transparent tape. For quantitative work, powder is graded by use of standard sieves for desired range of particle size (usually 75–100 μm) and loaded into a fused silica cell of known thickness.

This powder technique enables us to classify materials into three categories;

1. Materials in which the phase-matchable component is larger than the other component: noncentrosymmetric crystal structure
2. Materials in which the non-phase-matchable component is larger than the other component: noncentrosymmetric crystal structure
3. SHG inactive: centrosymmetric crystal structure

It is easy to classify a material in class 3, since these materials do not exhibit any SHG in powder form. Furthermore, it is possible to determine whether material belongs to either class 1 or class 2 by observing the powder SHG intensity as a function of particle size. Figure 16 shows the powder efficiency of phase-matchable material and non-phase-matchable material.

The SHG intensity of a non-phase-matchable material is described by the following approximation:

$$I(2\omega) \propto d^2 l_c^2 I(\omega)/2\langle r \rangle \qquad (\langle r \rangle >> l_c, \text{ not phase matchable})$$

$$I(2\omega) \propto d^2 l I(\omega) \qquad (\langle r \rangle >> k, \text{ phase matchable})$$

$$(74)$$

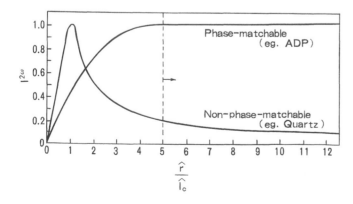

Figure 16 Schematic diagram showing particle size dependence of phase-matching and nonmatching materials. (From Kurtz, S. K. and Perry, T. T., *J. Appl. Phys.*, 39, 3798, 1968. With permission.)

The SHG intensity of non-phase-matchable material decreases with the increase of averaging particle size $\langle r \rangle$ when $\langle r \rangle$ is larger than coherence length l_c. In the case of phase-matchable material, the SHG intensity does not decrease when $\langle r \rangle$ is larger than l_c. Therefore, this difference can evaluate whether the material is phase matchable or not in bulk crystal.

C. SHEW METHOD

In the case of powder, SHG measurement depends on the particle size and is not proportional to the tensor component. The second harmonic evanescent wave (SHEW) method,[16,17] a newly developed powder technique by the Hitachi research group, gives a more accurate value for non-phase-matchable materials than the conventional powder measurement. In this method, the phase-matched directional SH in total reflection from the sample surface is detected through a prism (Figure 17). The sample holder and experiemtal setup used in the SHEW method are shown in Figures 18 and 19. A rutile (TiO$_2$) crystal is used as the prism material because it belongs to the centrosymmetric group and it is

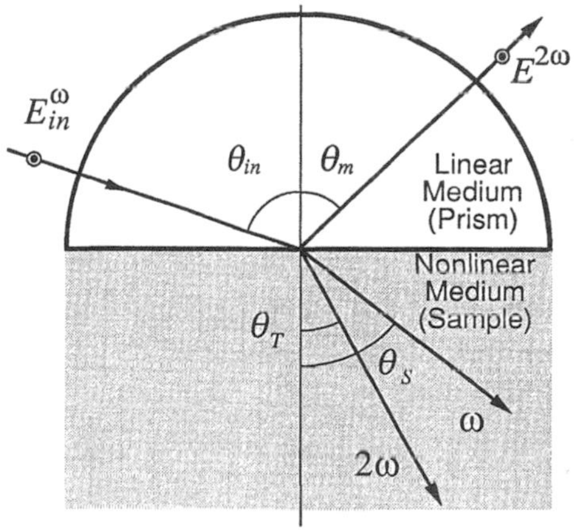

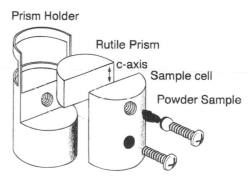

Figure 17 Schematic of the phase-matched SHG with the total reflection. The phase-matched SH propagates at the angle θ_m depending on incident angle θ_{in} and the prism refractive index n_p. The incident wave is polarized parallel to the c axis of the prism. (From Kiguchi, M., et al., *J. Appl. Phys.*, 75, 4332, 1994. With permission.)

Figure 18 Sample holder for SHEW method. The powder sample is pushed into contact with the hemicylinderical rutile prism by a screw. (From Kiguchi, M., et al., *J. Appl. Phys.*, 75, 4332, 1994. With permission.)

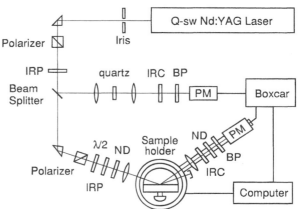

Figure 19 Experimental setup for SHEW technique. IRP, IRC, BR, and ND refer to IR pass, IR cut, bandpass, and neutral density filters, respectively. (From Kiguchi, M., et al., *J. Appl. Phys.*, 75, 4332, 1994. With permission.)

a transparent material with large refractive indexes. The strong nondirectional SH from bulk crystals is not detected because of the total reflection geometry. The phase velocity of the evanescent wave propagating at the interface is given by

$$v_e = \omega/(n_P^\omega k_0 \sin \theta_{in}) \qquad (75)$$

where $n_P\omega$ is the refractive index of the prism at the frequency ω, k_0 is the wave number of the incident wave, and θ_{in} is its incident angle. The nonlinear dipoles, which are induced by the evanescent wave, propagate with the same velocity along the x axis, and generate the SH; every phase of the SH is matched when $v_e \sin\theta$ is equal to the phase velocity of SH. The phase-matching condition is therefore given by

$$n_P^\omega \sin \theta_{in} = n_P^{2\omega} \sin \theta_m \qquad (76)$$

The SH wave propagates in the direction of the phase-matching angle θ_m. This is the Cerenkov-radiation-type phase matching. The SH for every element of nonlinear coefficient tensor can be phase matched. The SH intensity is dependent on all elements. The largest element will be dominant, whether it is phase matchable in a bulk crystal or not. The SH field at z is written as

$$E^{2\omega} \propto dE^\omega E^\omega \exp(-2z/\zeta) \qquad (77)$$

with

$$\frac{1}{\zeta} = \frac{2\pi n_p}{\lambda} \sqrt{\sin^2\theta_{in} - \left(\frac{n_s}{n_p}\right)^2} \qquad (78)$$

where z is the distance from the sample surface, d is the SH coefficient, E^ω is the fundamental field, and ξ is the penetration depth. As ξ is smaller than the wavelength, the simple field summation along the z axis is an adequate approximation. The observed SH power $P^{2\omega}$ is therefore given by

$$p^{2\omega} \propto M_{eff}\zeta^2 \cos \theta_{in}(p^\omega)^2 \qquad (79)$$

Here, p^ω is the fundamental beam power, and $\cos \theta_{in}$ is the compensative factor for an irradiated area changing with the incident angle θ_{in}. Figure 20 shows the effective figure of merit obtained using the

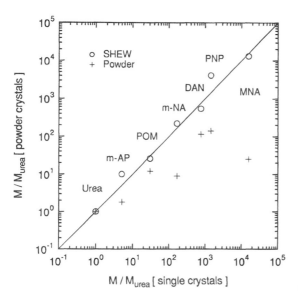

Figure 20 Effective figure of merit obtained using SHEW method and conventional powder method, compared with the literature values precisely measured using single crystals. Each value is normalized by the appropriate value of urea. MNA = 2-methyl-4-nitroaniline; PNP = 2-(N-prolinol)-5-nitro-pyridine, DAN = 4-(N,N-dimethylamino-3-acet-amido-nitrobenzene, m-NA = meta-nitroaniline, POM = 3-methyl-4-nitropyridine-1-oxide, m-AP = meta-aminophenol. (From Kiguchi, M., et al., J. Appl. Phys., 75, 4332, 1994. With permission.)

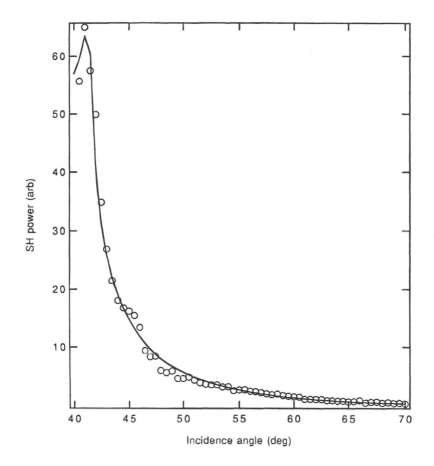

Figure 21 SHEW signal for *m*-nitroaniline powder crystals. Curve is fitted using the least squares method. (Courtesy of Dr. M. Kiguchi, Advance Research Laboratory, Hitachi Ltd.)

SHEW method to show the reliability of this technique. The vertical axis represents the figure of merit for powder crystals using the SHEW method and the powder method, while the horizontal axis shows the literature value of the largest tensor element in the figure-of-merit tensor for each singe crystal determined by the Maker fringe or other measurements. It can be seen that data obtained by the SHEW method (shown by circles) are closer to the line than those of the powder method, (represented by crosses). The accuracy of the powder method is beyond an order of magnitude, while the SHEW method yields adequate results. The discrepancies between the figure of merit by SHEW and literature value is within a factor of 2 to 3 considering all possible errors. Therefore the results from the SHEW method are in better agreement with those of the conventional powder method. Figure 21 shows the SHEW signal from *m*-nitroaniline powder crystals. There is excellent agreement where the curve is fitted to the experimental data using the least squares method. This shows that the SHEW method is far more reliable than the powder method. It is a simple technique applicable for a wide range of materials.

D. PHASE-MATCHING METHOD

The nonlinear optical coefficients of materials can be obtained using the phase-matching method.[18] This technique requires the phase-matching condition to measure the conversion efficiency directly using a power meter. Therefore, materials must have the proper birefringence and also off-diagonal tensor components such as d_{31}. Diagonal tensor components cannot be measured using this technique. The second harmonic power, $P(2\omega)$, generated by a single-mode Gaussian beam (TEM$_{oo}$) of power $P(\omega)$ incident on a plane parallel slab of thickness L of a nonlinear crystal, is given by[18]:

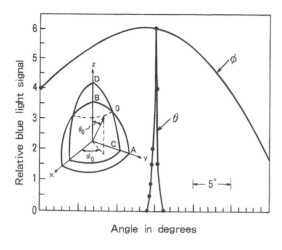

Relative blue light signal

Angle in degrees

Figure 22 Phase-matching configuration of KDP crystal. Maximum output occurs at $\theta_0 = 52 \pm 2°$. The laser beam was collimated to within $\pm 1/4°$. AOB is an arc on the refractive index surface for red ordinary rays, and COD is for blue extraordinary rays. (From Maker, P. D., et al., *Phys. Rev. Lett.*, 8, 21, 1962. With permission.)

$$P(2\omega) = \frac{128\pi^3\omega^2 L^2 P(\omega)^2 d^2}{c^3 w_o^2 n(\omega)^2 n(2\omega)} \times \frac{\sin^2(\Delta kL/2)}{(\Delta kL/2)} \text{ (in CGS units)} \qquad (80)$$

where ω is the angular frequency of the fundamental wave, W_0 is the spot radius of fundamental beam, d is the SHG coefficient, $n(\omega)$ is the refractive index of material at the fundamental wavelength, $n(2\omega)$ is the refractive index of material at the second harmonic wavelength, c is the velocity of light in a vacuum, $\Delta k = 2k(\omega) - k(2\omega)$ is the wave vector mismatch, and L is the optical pass length of material.

In general, when the light at ω and 2ω is polarized in the same direction, the wave vector Δk is not zero because of the dispersion of the refractive index. However, if material has the appropriate birefringence compared to the its dispersion, phase matching is possible. In order to obtain the phase matching, a condition to satisfy $\Delta k = 0$ should be considered. Figure 22 shows the SHG intensity of the KDP crystal as a function of crystal orientation.[18] In the KDP crystal, only at the appropriate angle θ(angle between the wave vector and optic axis), is phase matching attained. The SHG intensity also slowly changes when the crystal rotates, since the projection of the effective nonlinear coefficient is reduced. In the case of KDP, ordinary rays are used as fundamentals and extraordinary rays are used as second harmonics. However, in the case of positive crystal this procedure would have been reversed— extraordinary fundamentals and ordinary second harmonics are used.

And when the polarization of the two input beams is mixed (i.e., an ordinary ray and an extraordinary ray are used as fundamentals), an ordinary second harmonic in a positive crystal and an extraordinary second harmonic in a negative crystal can be produced. If both fundamentals have the same polarization, we speak of type I phase matching, and if their polarizations are orthogonal, the phase matching is called type II.

The details of a phase-matching condition for biaxial crystals were described by Hobden[19] and Ito et al.[20] The type of phase matching featuring the birefringence of the crystal is the one most universally employed. However, a variety of other methods have been used for the phase matching. In those cases where $k_1 + k_2 > k_3$, phase matching can be achieved by making the interaction take place between noncolinear beams. Noncolinear phase matching can be used to determine the d constant of materials.

In the phase-matching condition, conversion efficiency is proportional to the square of the optical pathlength. By using this condition, one can determine the effective nonlinear coefficient (d_{eff}) of material relative to reference. Sometimes this procedure can be employed to determined the absolute value of d_{eff} of materials. When a reference sample is used for the determination of d constant, one must take care of the linearity of the photomultiplier tube. Usually neutral density filters are used to protect the saturation of the photomultiplier tube.

E. MAKER FRINGE METHOD

One of the most popular and useful techniques is the Maker fringe method.[21,22] This technique does not require the measurement of absolute value of SHG intensity, since the SHG intensity always compares with that of the reference. The d_{11} of fused silica or d_{36} of KDP is used as a reference. The nonlinear optical coefficients of the reference were measured using the phase-matching technique

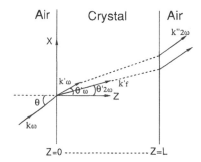

Figure 23 Propagation of the fundamental and second harmonic waves. The y axis is perpendicular to the plane of the figure. (From Jerphagnon, J. and Kurtz, S. K., *J. Appl. Phys.*, 39, 3798, 1970; 41, 1667, 1970. With permission.)

mentioned earlier and the absolute power of SHG was measured. In this technique the parallel slab plate is rotated about an axis perpendicular to the laser beam, giving the Maker fringe pattern. This pattern ascribes to the interference between the nonlinear polarization and free harmonic wave which is termed as phase-mismatch. The phase-mismatch Δk is given as

$$\Delta k = 2k_\omega - k_\omega = \frac{4\pi}{\lambda(n_\omega \cos \theta_\omega - n_{2\omega} \cos \theta_{2\omega})} \tag{81}$$

where θ_ω and $\theta_{2\omega}$ are the angles of refraction at fundamental and harmonic wavelengths. θ is the incident angle of the fundamental beam. The relationship between these angles is shown in Figure 23.

When Δk is not zero, the periodic intensity is obtained as a function of the incident angle. This period is known as a coherence length (l_c). For normal incidence of the fundamental beam on a sample of thickness L, the second harmonic power varies as

$$P(2\omega) \propto \sin^2\left(\frac{\Delta\psi}{2}\right) \tag{82}$$

$$\psi = \frac{4\pi L(0)}{\lambda_\omega}(n_\omega - n_{2\omega}) \tag{83}$$

$$l_c = \frac{\lambda}{4(n_\omega \cos \theta_\omega - n_{2\omega} \cos \theta_{2\omega})} \tag{84}$$

This oscillation depends only on the dispersion of the refractive index and the thickness of the samples. This dephasing limits the effective crystal thickness.

From Equation 82, the condition for the mth minimum of fringe is

$$\psi = \frac{\pi L}{2l_c(\psi_m)} = 2m \tag{85}$$

By fitting the minima of the experimental fringe to Equation 85, the coherence length can be determined. In the case of uniaxial crystal, the coherence length at normal incidence can be presented by Equation 86

$$l_c = L(\sin^2 \theta_{m+1} - \sin^2\theta_m)/4n_{2\omega}n_\omega \tag{86}$$

where θ_{m+1}, θ_m are the angular separation of $(m + 1)$ and mth minima.

The precise analysis of the Maker fringe method is demonstrated by Jerphagnon and Kurtz.[21,22] The SH intensity is a function of the incident angle as[23]

$$I_{2\omega} = \frac{128\pi^3}{c^2} T_Q^2 (t_\omega^{0Q})^4 \left(\frac{\chi_{ijk}^{(2)}}{\Delta\varepsilon_Q} \right)^2 I_\omega^2 \sin^2\left(\frac{\Delta\psi_Q}{2} \right) \tag{87}$$

$$\Delta\varepsilon_Q = n_{2\omega}^2 - n_\omega^2 \tag{88}$$

The transmission factor depends on the polarization. When s-polarized fundamental and s-polarized SH beams are detected, the transmission factor T_Q and t^Q are described as follows:

$$T_Q = (n_\omega + n_{2\omega})/(1 + n_{2\omega}) \tag{89}$$

$$t_\omega^{0Q} = \frac{2}{1 + n_\omega} \tag{90}$$

The polar axis of the sample must be parallel to the electric field. Only diagonal tensor components can be obtained without considering the projection factor when s-polarized fundamental beam is used and s-polarized SH is detected. For example, for the measurement of d_{33}, it is better to use z axis as the rotation axis. In other cases the projection of the electric field to each axis of tensor components must be considered.

Figure 24 shows the experimental setup for the Maker fringe method. In order to increase the second harmonic power, the laser beam is focused by the lenses F1 and F3. Two Glan-Thomson polarizers or half-wave plates are used to polarize the beam in such a way that the electric field is horizontal or vertical. The rotation speed must be considered, since the thick sample or short coherence material drastically changes the second harmonic power as a function of the rotation angle. Both reference and signal are detected using a boxcar integrator and recorded.

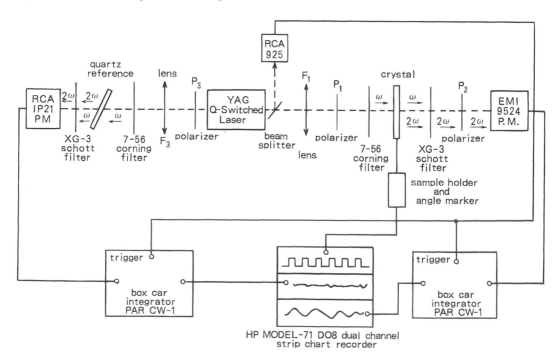

Figure 24 Experimental configuration of Maker fringe measurements. The electrical connections are represented by continuous lines; light propagation is represented by dashed lines. (From Jerphagnon, J. and Kurtz, S. K., *J. Appl. Phys.*, 41, 1667, 1970. With permission.)

F. EFISH MEASUREMENTS

Electric field-induced second harmonic generation (EFISH) can be applied in liquids to obtain the β and γ values.[24-28] When one considers an electroded glass cell such as shown in Figure 25, the liquid is enclosed between the glass walls, which forms a wedge angle α. The optical wave incident on the nonlinear glass–liquid boundaries will introduce a nonlinear polarization

$$P(2\omega) = \chi^{(3)}E_0 E(\omega)^2 \equiv d_{eff}E(\omega)^2 \tag{91}$$

where Γ is the macroscopic third-order optical nonlinearity of the liquid, and d_{eff} is the effective nonlinear coefficient. The microscopic third-order nonlinearity γ for a single molecule can be determined from

$$\chi^{(3)} = N <\gamma_{ijkl}> f_0 f_\omega^2 f_{2\omega} \tag{92}$$

where N is the molecular density of the liquid, and the f's are the local field factors at their indicated frequencies.

$$f(\omega) = [n(\omega)^2 + 2]/3 \tag{93}$$

$$f_0 = \frac{\varepsilon_0(\varepsilon_\infty + 2)}{\varepsilon_\infty + 2\varepsilon_0} \tag{94}$$

$\langle\gamma\rangle$ represents the average of overall molecular orientations. The γ has three contributions, such as γ_e, γ_v, γ_r. Where γ_e is the average value electronic contribution, γ_v is the average vibrational nonlinearity, and γ_r is the average dipolar rotational contribution induced by the electric field.[24-28]

$$\gamma \approx \gamma_r = \mu\beta/5kT \tag{95}$$

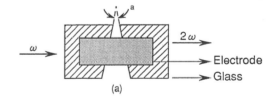
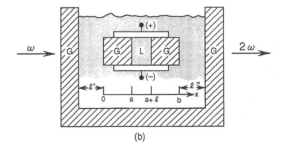
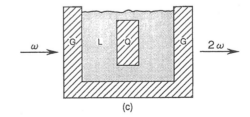

Figure 25 (a) Top view of the small open electroded liquid cell. The wedge angle α is typically 10–20 mrad. (b) Side view (not to scale) of entire assembly of small open electroded cell. (c) Reference cell containing the quartz crystal Q and an index matching liquid. (From Levine, B. F. and Bethea, C. G., *J. Chem. Phys.* 63, 2666, 1975. With permission.)

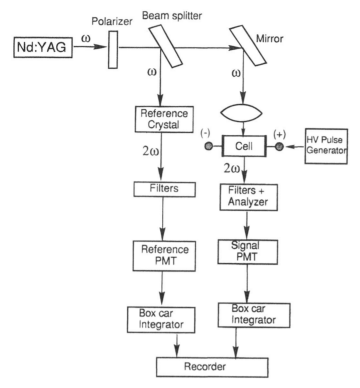

Figure 26 Schematic of the experimental apparatus of EFISH. (From Levine, B. F. and Bethea, C. G., *J. Chem. Phys.* 63, 2666, 1975. With permission.)

In the case of EFISH measurement, the averaging β can be obtained, where β is represented by three independent components such as β_{xxx}, β_{xyy}, and β_{xzz}. The x axis is defined as a direction parallel to the CT axis of the molecule.

$$\beta_x = \beta_{xxx} + \beta_{xyy} + \beta_{xzz} \tag{96}$$

Both the solute and liquid generate SHG, then measured $\chi^{(3)}$ is represented as

$$\chi^{(3)}(-2\omega; \omega, \omega, 0) = N_p f_p(2\omega, \omega, \omega, 0)\gamma_p(-2\omega; \omega, \omega, 0)$$
$$+ N_S f_S(2\omega, \omega, \omega, 0)\gamma_p(-2\omega; \omega, \omega, 0) \tag{97}$$

where p and s denote the solute and solvent.

In this EFISH experiment, jitter of laser pulse presents a problem, so a special electronic part (pulse delay generator) was designed to trigger all the electronic components of this system at the same frequency as shown in Figure 26. The *S/N* ratio limitation is mainly due to the fluorescence from the glass walls and burning effect. When quartz is used as the reference, the effective second-order coefficient is given by[23]:

$$I(2\omega) = \frac{288\pi^3}{(cg)^2}\left(\frac{\chi_G^{(3)}}{\Delta\varepsilon_G}\right)[E_0 I(\omega)(t_\omega^{G0})^2 t_\omega^{G0} T_L]^2 \left\{ \left| Y\left(\frac{\gamma_s^{E'}}{M_s} + \frac{C\gamma_p^{E'}}{M_p}\right)[1 - W\exp(i\Delta\psi_E)] \right. \right.$$
$$\left. \left. - [1 - V\exp(i\Delta\psi_E)] \right|^2 + Y^2 C^2\left(\frac{\gamma_p^{E''}}{M_p}\right)^2 |1 - E\exp(i\Delta\psi_E)|^2 \right\} \tag{98}$$

where g takes into account absorption that may occur at the second harmonic frequency. E_0 is the externally applied DC electric field. The following quantities are defined:

$$g = \exp[2\omega K_{2\omega} l(x)/c] \tag{99}$$

$$V = T_G (t_\omega^{GL} t_\omega^{LG})^2 T_1 t_{2\omega}^{LG} g \tag{100}$$

$$W = T_L g/(T_2 t_{2\omega}^{LG}) \tag{101}$$

$$Y = \frac{T_2}{T_1} \frac{\Delta\varepsilon_G}{\Delta\varepsilon_L} \frac{N_A f(C) d(C)}{(1 + C)\chi_G^{(3)}} \tag{102}$$

$$T_1 = (n_\omega^G + n_\omega^G)/(n_{2\omega}^G + n_{2\omega}^L) \tag{103}$$

$$T_2 = (n_{2\omega}^G + n_\omega^L)/(n_{2\omega}^G + n_{2\omega}^L) \tag{104}$$

$$T_L = (n_\omega^L + n_{2\omega}^L)/(n_{2\omega}^L + n_{2\omega}^G) \tag{105}$$

$$T_G = (n_{2\omega}^L + n_\omega^G)/(n_{2\omega}^L + n_{2\omega}^G) \tag{106}$$

The transmission factors t_ω, 2_ω^{ij} are defined as follows:

$$t_{\omega(2\omega)}^{ij} = \frac{2n_{\omega(2\omega)}^i}{n_{\omega(2\omega)}^i + n_{\omega(2\omega)}^j} \tag{107}$$

Also, G refers to glass, L to liquid, and O to air.

The separation of the corresponding contributions to γ in the EFISH technique is not evident. In general, γ_r is estimated from the temperature dependence of γ, since γ_r dominates in organic liquid. Hence using the Equation 95, the β of materials can be obtained.

G. HYPER-RAYLEIGH SCATTERING

Hyper-Rayleigh scattering (HRS),[29,30] which is the incoherent scattered second harmonic generation, can be observed in isotropic samples even from centrosymmetric molecules in the liquid phase.[29] The HRS technique has definite advantages over the EFISH method. In the case of the EFISH technique, the value of β cannot be obtained without the measurements of the molecular dipole moment and second hyperpolarizability γ. HRS can also measure a value of the first hyperpolarizability of molecules with their dipole moment perpendicular to β_{vec}. Since HRS does not require an orientation of the NLO molecules, polar molecules in which β may originate from octopolar charge distributions can be studied by the technique. The absence of molecular orientation induced by an electric field is a major advantage of HRS allowing for the determination of the β of ionic species. The polarization dependence of the scattered light for HRS is a very important factor in determining the ratio β_{xxx}/β_{xyy} of the molecules, where EFISH cannot distinguish between these two components.

The coherent SHG intensity is proportional to the square of $\langle\cos^3\theta\rangle$, where $\langle\cos^3\theta\rangle$ represents the molecular orientation. If the molecular distribution is completely random, then $\langle\cos^3\theta\rangle$ is 0. On the contrary, $\langle\cos^3\theta\rangle$ is 1 in the case of perfect alignment. In the scattered SHG, the SH intensity is proportional to the square of the $\langle\cos^2\theta\rangle$. $\langle\cos^2\theta\rangle$ is related to the order parameter and is not equal to 0 even in the random orientation. In linear light scattering, the scattering is mainly due to density fluctuations in the solution governed by transitional diffusion processes. Because of the difference in symmetry requirements, fluctuations in molecular orientation are the main cause for second-order light scattering. If we assume that correlation between β_{ijk} for two different volumes 1 and 2 only exists over distances shorter than the wavelength, the intensity of the second harmonic light is proportional to $\langle\beta_{ijk},1\,\beta_{ijk},2\rangle$ ($\langle\rangle$indicates orientational average). In the case of randomly oriented and independent solutions, the scattered SH intensity is proportional to the number of molecules N (in the case of a solvent–solute system N_S and N_s) and to $\langle\beta_{ijk},1\,\beta_{ijk},2\rangle$. In general, SHG can be observed in a noncentrosymmetric medium. The incident light propagates in the Y direction with polarization in the X direction; the scattered SH intensity in the Z direction is written as follows[30]:

$$I_Z(2\omega) = g[N_S\langle\beta_{ZZZ}^2\rangle_S + N_s\langle\beta_{ZZZ}^2\rangle_s]I^2(\omega) \tag{108}$$

$$I_X(2\omega) = g[N_S\langle\beta_{ZXX}^2\rangle_S + N_s\langle\beta_{ZXX}^2\rangle_s]I^2(\omega) \tag{109}$$

where

$$\langle\beta_{ZZZ}^2\rangle = \frac{1}{7}\sum_x \beta_{xxx}^2 + \frac{6}{35}\sum_{x\neq y}\beta_{xxx}\beta_{xyy} + \frac{9}{35}\sum_{x\neq y}\beta_{xyy}^2 + \frac{6}{35}\sum_{xyz,cyclic}\beta_{xxy}\beta_{yzz} + \frac{12}{35}\beta_{xyz}^2 \tag{110}$$

$$\langle\beta_{ZXX}^2\rangle = \frac{1}{35}\sum_x \beta_{xxx}^2 - \frac{2}{105}\sum_{x\neq y}\beta_{xxx}\beta_{xyy} + \frac{11}{105}\sum_{x\neq y}\beta_{xyy}^2 - \frac{2}{105}\sum_{xyz,cyclic}\beta_{xxy}\beta_{yzz} + \frac{8}{35}\beta_{xyz}^2 \tag{111}$$

where $I_z(x)(2\omega)$ is the experimentally observed scattered light intensity polarized in the $z(x)$ direction. The proportionality factor g depends upon the scattering geometry, instrumental factors, and the local field corrections. The g is determined as a function of fundametal power beam and calibrated using well-known samples. The indices XYZ and xyz indicate the laboratory and molecular coordinates, respectively. At low solute concentration N_S can be taken as constant and $I(2\omega)$ will linearly depend upon N_S. From the intercept and the slope of the plot $I(2\omega)/I(\omega)^2$ vs. N_s we calculate β_s, if β_S is known, or vice versa. The internal reference method (IRM) effectively eliminates the requirement of local field correction factors since these factors not significantly change the refractive index of the solution.

For systems in which the solute molecules show absorption at 2ω, Equation 109 has to be modified to take this absorption into account:

$$I_Z(2\omega) = g[N_S\langle\beta_{ZZZ}^2\rangle_S + N_s\langle\beta_{ZZZ}^2\rangle_s]e^{-\varepsilon(2\omega)l\cdot N_S}I^2(\omega) \tag{112}$$

$$I_X(2\omega) = g[N_S\langle\beta_{ZXX}^2\rangle_S + N_s\langle\beta_{ZXX}^2\rangle_s]e^{-\varepsilon(2\omega)l\cdot N_S}I^2(\omega) \tag{113}$$

where $\varepsilon(2\omega)$ is the extinction coefficient of solute molecules and 1 is an effective optical path length.

Figure 27 shows the experimental setup for measuring the HRS. The laser is focused into a small cuvette containing the sample. The energy of incident beam was varied by using the polarizer and a half-wave plate. The scattered light is collected by an efficient condensor lens under 90° and detected by photomultiplier tube. A low-pass optical filter and an interference filter (or monochromator) are

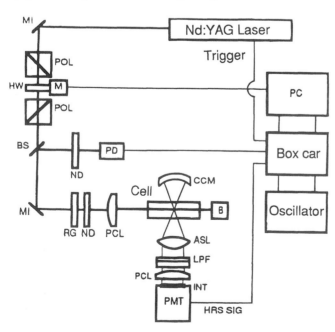

Figure 27 Experimental setup for Hyper-Rayleigh scattering in solution. (From Verbiest, T., et al. *Nonlinear Optical Properties of Organic Materials*, SPIE, Washington, DC, 1991. With permission.)

used to isolate the second harmonic scattered light. Gated integrators are used to retrieve actual values for the intensities of the incident and second-order scattered light pulses. An important and exciting possibility of the HRS technique, combining several advantages over EFISH, is the measurement of β for molecular ions with octopolar charge distribution. The dye molecule crystal violet contains the tris(p-dimethylaminophenyl)methyl cation, has D_{3h} symmetry, and presents therefore an octopolar ionic species. For crystal violet, the large value of $\beta_{xxx} = 3060 \times 10^{-30}$ esu in methanol solution was reported. Unique to the HRS is the possibility of analyzing depolarization effects in the second harmonic scattered light. As a first approach, the polarization states of the HRS signal were obtained from solutions of crystal violet in chloroform. From Equations 112 and 113, one can expect a depolarization ratio of (I_{vv}/I_{vh}) 1.5 for crystal violet since the symmetry of this cation belongs to the D_{3h} space group and that it has therefore only two independent β coefficients (i.e., $\beta_{xxx} = \beta_{xyy}$).

In summary, theoretical principles and experimental instruments of different techniques which should be useful for newcomers in this field have been discussed. Various techniques for measuring the refractive index and SHG efficiency of organic materials have been summarized. Two techniques, the SHEW method and HRS, are new additions to this literature. This chapter has provided an up-to-date overview for researchers who are interested in this exciting new field.

ACKNOWLEDGMENTS

The authors are grateful to Ms. Barbara Caras of CRC Press for permission to use some of the contents from our published monograph. The authors would also like to thank Dr. M. Kiguchi and Miss Midori Kato of Hitachi Advanced Research Laboratory for providing original figures on the SHEW method and useful discussions.

REFERENCES

1. **Born, M. and Wolf, E.,** *Principles of Optics,* Pergamon Press, Oxford, 1975.
2. **Zyss, J., Chemla, D. S., and Nicoud, J. F.,** Demonstration of efficient nonlinear optical crystals with vanishing molecular dipole moment: second-harmonic generation in 3-methyl-4-nitropyridine-1-oxide, *J. Chem. Phys.,* 74, 4800, 1981.
3. **Tsuboi, S.,** *Henko Kenbikyo (Polarizing Microscope),* Iwanami, Tokyo, 1965. (In Japanese).
4. **Ulrich, R. and Torge, R.,** Measurement of thin film parameters with a prism coupler, *Appl. Opt.,* 12, 2901, 1973.
5. **Aspnes, D. E.,** The accurate determination of optical properties by ellipsometry, in *Handbook of Optical Constants in Solid,* Palik, E. D., Ed., Academic Press, Orlando, FL, 1985, chap. 5.
6. **Archer, R. J.,** Determination of the properties of films on silicon by the method of ellipsometry, *J. Opt. Soc. Am.,* 52, 970, 1962.
7. **Jasperson, S. N. and Schnatterly, S. E.,** An improved method for high reflectivity ellipsometry based on a new polarization modulation technique, *Rev. Sci. Instrum.,* 40, 761, 1969.
8. **Jellison, G. E. and Modine, F. A.,** Two-channel polarization modulation ellipsometer, *Appl. Opt.,* 29, 959, 1990.
9. **Guenther, R. D.,** *Modern Optics,* John Wiley & Sons, New York, 1990.
10. **Dumont, M. and Levy, Y.,** Measurement of electronic properties of organic thin films by attenuated total reflection, in *Nonlinear Optics of Organics and Semiconductors,* Springer Proceedings in Physics, Vol. 36, Kobayashi, T., Ed., Springer-Verlag, Berlin, 1989, 256.
11. **Weber, M. J.,** Ed., *Handbook of Laser Science and Technology: Vol. 3. Optical Materials,* CRC Press, Boca Raton, FL, 1986.
12. **Burland, D. M., Miller, R. D., and Walsh, C. A.,** Second-order nonlinearity in poled-polymer systems, *Chem. Rev.,* 94, 31, 1994.
13. **Nye, J. F.,** *Physical Properties of Crystals,* Oxford University Press, London, 1957.
14. **Shen, Y. R.,** *The Principles of Nonlinear Optics,* John Wiley & Sons, New York, 1984.
15. **Kurtz, S. K. and Perry, T. T.,** A powder technique for the evaluation of nonlinear optical materials, *J. Appl. Phys.,* 39, 3798, 1968.
16. **Kiguchi, M., Kato, M., and Taniguchi, Y.,** Measurement of second-harmonic generation from colored powders, *Appl. Phys. Lett.,* 63, 2165, 1993.
17. **Kiguchi, M., Kato, M., Kumegawa, N., and Taniguchi, Y.,** Technique for evaluating second-order nonlinear optical materials in powder form, *J. Appl. Phys.,* 75, 4332, 1994.
18. **Maker, P. D., Terhune, R. W., Nisenoff, M., and Savage, C. M.,** Effects of dispersion and focusing on the production of optical harmonics, *Phys. Rev. Lett.,* 8, 21, 1962.
19. **Hobden, M. V.,** Phase-matched second-harmonic generation in biaxial crystals, *J. Appl. Phys.,* 38, 4365, 1967.

20. **Ito, H., Naito, H., and Inaba, H.,** Generalized study on angular dependence of induced second-order nonlinear optical polarizations and phase-matching in biaxial crystals, *J. Appl. Phys.,* 46, 3992, 1975.
21. **Jerphagnon, J. and Kurtz, S. K.,** A detailed comparison of theory and experiment for isotropic and uniaxial crystals, *J. Appl. Phys.,* 39, 3798, 1970.
22. **Jerphagnon, J. and Kurtz, S. K.,** Maker fringes: a detailed comparison of theory and experiment for isotropic and uniaxial crystals, *J. Appl. Phys.,* 41, 1667, 1970.
23. **Burland, D. M., Walsh, C. A., Kajzar, F., and Sentein, C.,** Comparison of hyperpolarizabilities obtained with different experimental methods and theoretical techniques, *J. Opt. Soc. Am.* B8, 2269, 1991.
24. **Oudar, J. L. and Person, H. L.,** Second-order polarizabilities of some aromatic molecules, *Opt. Commun.,* 15, 258, 1975.
25. **Bethea, C. G.,** Experimental technique of dc induced SHG in liquids: measurement of the nonlinearity of CH_2I_2, *Appl. Opt.,* 14, 1447, 1975.
26. **Levine, B. F. and Bethea, C. G.,** Absolute signs of hyperpolarizabilities in the liquid state, *J. Chem. Phys.,* 60, 3856, 1974.
27. **Levine, B. F. and Bethea, C. G.,** Second and third order hyperpolarizabilities of organic molecules, *J. Chem. Phys.,* 63, 2666, 1975.
28. **Levine, B. F.,** Donor-acceptor charge transfer contributions to the second order hyperpolarizability, *Chem. Phys. Lett.,* 37, 516, 1976.
29. **Clays, K. and Persons, A.,** Hyper-Rayleigh scattering in solutions, *Phys. Rev. Lett.,* 66, 2980, 1991.
30. **Verbiest, T., Hendrickx, E., Persons, A., and Clays, K.,** Measurement of molecular hyperpolarizabilities using Hyper-Rayleigh scattering, in *Nonlinear Optical Properties of Organic Materials V,* Vol. 1775, Williams, D. J., Ed., SPIE, Washington, DC, 1991, 206.

Chapter 4

Organic Materials for Second-Order Nonlinear Optics

Hari Singh Nalwa, Toshiyuki Watanabe, and Seizo Miyata

CONTENTS

I. INTRODUCTION

Nonlinear optical organic materials will be the key elements for future photonic technologies. The field of nonlinear optics is interdisciplinary, as it needs expertise in fields such as mathematics, chemistry, materials science, physics, electrical engineering, and optics. Photonic technologies are analogous to the field of electronics; they can even be hybridized for making devices and interconnects. It is appropriate to mention that current research activities are aimed for the "photonics versus electronics" theme. Photonic technologies are based on the fact that photons are capable of processing information with the speed of light. In electronic devices, electrons are the key elements performing multifunctions, whereas in photonic devices photons will perform the same functions at a much faster speed and in a cleaner, easier way. Such dreams may come true because lasers are expected to be produced as inexpensively as a dollar a piece.

The word laser is an acronym that stands for light amplification by stimulated emission of radiation. Lasers are sources of electromagnetic radiation which are characteristically different from an ordinary light source. Laser light possesses coherence, collimation, and a narrow spectral width

0-8493-8923-2/97/$0.00+$.50
© 1997 by CRC Press, Inc.

89

and can focus at a possible minimum spot size. The combination of these unique features of laser radiation makes lasers highly desirable for a wide range of applications in optics. The first laser, developed by Maiman[1] in 1960, opened a new era for laser research. After this discovery, Javan et al.[2] designed the first continuous helium-neon laser. Laser engineering has matured and recently more durable, reliable, and sophisticated lasers have been developed. A variety of lasers, such as gas lasers, solid-state lasers, organic-dye lasers, semiconductor lasers, chemical lasers, excimer lasers, and tuneable lasers, are currently available.[3,4] The importance of lasers was realized long ago with the invention of optical fiber systems widely used in telecommunications. It is now almost three decades since the first nonlinear optical phenomenon in a quartz crystal was reported by Franken et al.[5] In the beginning, studies were concentrated mainly on inorganic materials such as quartz, potassium dihydrogen phosphate (KDP), lithium niobate (LiNbO$_3$), and semiconductors such as cadmium sulfide, selenium, tellurium, cadmium germanium arsenide; the reports on organic materials were very rare. Several excellent reference texts are available on the physics aspects of nonlinear optics, and are related to inorganic materials; they are Bloembergen,[6] Shen,[7] Yariv,[8] Reintjes,[9] Hanna et al.,[10] and others.[11–14]

An important development in nonlinear optical material occurred in 1970, when Davydov et al.[15] reported a strong second harmonic generation (SHG) in organic molecules having electron donor and acceptor groups connected with a benzene ring. This discovery led to an entirely new concept of molecular engineering to synthesize new organic materials for the SHG studies. In the 1980s, tremendous growth occurred in design and development of organic materials for second-order nonlinear optics. The topic of organic nonlinear optics has been reviewed by Zyss,[16] Williams,[17] and Burland et al.[18] and has also been compiled in several conference proceedings.[19–22] The first text to investigate the various aspects of organic nonlinear optics was edited by Chemla and Zyss.[23] The importance of this field is demonstrated by the launching of several periodicals, namely *Nonlinear Optics*, *Optical Materials*, and *International Journal of Nonlinear Optical Physics*, exclusively dealing with the field of nonlinear optics. In addition, papers on aspects of nonlinear optics have been published in other lead journals. A few international conferences deal with nonlinear optical aspects of inorganic and organic materials. Of particular importance are conferences held annually by the Society of Photo-Optical Instrumentation Engineers (SPIE) and Optical Society of America (OSA). The relatively rapid developments in the subject reflect continued growth of interest. As a consequence, enormous data are available to make useful comparisons. The chapter presented here benefits from the knowledge of tailor-made organic materials for second-order nonlinear optical processes, especially for SHG. A meaningful insight can be achieved through a comprehensive review of emerging new organic materials. Therefore, this review chapter has been prepared to assess the present status of organic materials in developing new technologies for laser diodes, modulators, optical interconnects, laser printers, copiers, waveguides, and other applications. The organic materials described here range from nanoscale molecules to single crystals to giant organic polymers. Organic materials that have been investigated for second-order nonlinear optics can be summarized into several categories such as single crystals, Langmuir-Blodgett (LB) films, polar polymers, guest (NLO-dye)-host (polymer matrix) systems, NLO-chromophore functionalized polymers, self-assembled systems, and liquid crystals. During the past few years, growth has been rapid in the development of organometallic molecules and polymers for nonlinear optics in which the role of metal to ligand bonding is used to optimize optical nonlinearities. This review chapter aims to present the detailed description of organic molecules and polymers related to second-order nonlinear optics.

II. SECOND-ORDER NONLINEAR OPTICAL PHENOMENON

Although a quantitative description of molecular nonlinear optical responses and their relationship to chemical structures have been described in another chapter, it is necessary to introduce very briefly the fundamental aspects here again. A laser is a source of light and the light emitted by a laser has a very high degree of collimation and coherence. The light of a laser beam is highly monochromatic and can be focused to a minimum spot size on a material. Therefore, a focused laser beam can provide a high power per unit area. The laser light can stimulate the atomic or molecular systems to produce transition and release the stored energy also as light. This emission of light requires two energy levels that are separated by the photon energy of the light to be emitted. Therefore what we have is a process of controlling the light by light in the atomic or molecular system. When the electromagnetic field of a laser beam is illuminated on an atom or a molecule, it induces electrical polarization that gives rise to many of the unusual and interesting properties that are optically nonlinear. A relationship between the polarization p induced in a molecule and the applied electric field E of incident electromagnetic wave at frequency ω can be written by[23–28]:

$$p_i(\omega_1) = \sum_j \alpha_{ij}(-\omega_1; \omega_2)E_j(\omega_2) + \sum_{jk} \beta_{ijk}(-\omega_1; \omega_2, \omega_3)E_j(\omega_2)E_k(\omega_3)$$

$$+ \sum_{jkl} \gamma_{ijkl}(-\omega_1; \omega_2, \omega_3, \omega_4)E_j(\omega_2)E_k(\omega_3)E_l(\omega_4) + \cdots \tag{1}$$

where $p_i(\omega_1)$ is the induced polarization in a microscopic medium at laser frequency ω_1 along the ith molecular axis, α is the linear polarizability, β is the first hyperpolarizability, and γ is the second hyperpolarizability, and E_j is the applied electric field component along the jth direction. Analogously, the macroscopic polarization induced in bulk media under high electromagnetic fields of a laser beam can be expressed in a power series as:

$$P_i(\omega_1) = \sum_J \chi_{IJ}^{(1)}(-\omega_1; \omega_2)E_J(\omega_2) + \sum_{JK} \chi_{IJK}^{(2)}(-\omega_1; \omega_2, \omega_3)E_J(\omega_2)E_K(\omega_3)$$

$$+ \sum_{JKL} \chi_{IJKL}^{(3)}(-\omega_1; \omega_2, \omega_3, \omega_4)E_J(\omega_2)E_K(\omega_3)E_L(\omega_4) + \cdots \tag{2}$$

Here $\chi^{(1)}$, $\chi^{(2)}$, and $\chi^{(3)}$ are the first-, second-, and third-order nonlinear optical susceptibilities, respectively. The even-order tensors vanish in a centrosymmetric medium, whereas odd-order tensors do not have any symmetry restrictions. The macroscopic second-order optical nonlinearities are related to their corresponding microscopic terms β as follows:

$$\chi_{IJK}^{(2)} = N\beta_{ijk}f_I f_J f_K \tag{3}$$

where N is the molecular number density and f is the local field factor due to intermolecular interactions. By knowing the magnitude of β either theoretically or experimentally, a general trend of the corresponding macroscopic second-order NLO coefficients can be estimated for a molecular system. The β values of organic molecules can be calculated theoretically by quantum chemical approaches such as *ab initio*, Pariser-Parr-Pople (PPP) method, complete neglect of differential overlap (CNDO), intermediate neglect of differential overlap (INDO), and modified neglect of differential overlap (MNDO). The individual tensor components for SHG process discussed here follow the work of Bloembergen[25,26] and Ward.[27,28] The quantum mechanical formula derived by time-dependent perturbation theory can be expressed as follows:

$$\beta_{ijk}(-2\omega; \omega, \omega) = -\frac{e^3}{8\hbar^2} \sum_{\substack{n' \neq g}} \sum_{\substack{n \neq g \\ n = n}}$$

$$\left[\left(r_{gn'}^j r_{n'n}^i r_{gn}^k + r_{gn'}^k r_{n'n}^i r_{gn}^j \right) \times \left(\frac{1}{(\omega_{n'g}-\omega)(\omega_{ng}+\omega)} + \frac{1}{(\omega_{n'g}+\omega)(\omega_{ng}-\omega)} \right) \right.$$

$$+ \left(r_{gn'}^j r_{n'n}^j r_{gn}^k + r_{gn'}^i r_{n'n}^k r_{gn}^j \right) \times \left(\frac{1}{(\omega_{n'g}+2\omega)(\omega_{ng}+\omega)} + \frac{1}{(\omega_{n'g} - 2\omega)(\omega_{ng}-\omega)} \right)$$

$$\left. + \left(r_{gn'}^j r_{n'n}^k r_{gn}^i + r_{gn'}^k r_{n'n}^j r_{gn}^i \right) \times \left(\frac{1}{(\omega_{n'g}-\omega)(\omega_{ng}-2\omega)} + \frac{1}{(\omega_{n'g}+\omega)(\omega_{ng}+2\omega)} \right) \right] \tag{4}$$

where r_{gn}^i and $r_{n'n}^j$ are matrix elements of the ith components of the dipole operator for the molecule between the unperturbed ground and excited states and between two excited states, respectively, ω is the incident laser frequency, $\hbar\omega_{ng}$ corresponds to the excitation energy from the ground state ($|g\rangle$) to the excited state ($|n\rangle$). The orientationally averaged first hyperpolarizability can be written as:

$$\beta = \beta_{xxx} + \beta_{xyy} + \beta_{xzz} \tag{5}$$

For SHG where $\omega_1 = \omega_2 \equiv \omega$, the nonlinear optical coefficient d_{ijk} is related to second-order NLO susceptibility $\chi^{(2)}$ as follows:

$$d_{ijk}(-2\omega; \omega, \omega) = \frac{1}{2} \chi^{(2)}_{ijk}(-2\omega; \omega, \omega) \tag{6}$$

Now the polarization is expressed by

$$P_i^{2\omega} = d_{ijk}(-2\omega; \omega, \omega)E_j(\omega)E_k(\omega) \tag{7}$$

By interchanging j and k in the equation above, one can replace the subscripts kj and jk by the contracted indices[29]

$$xx = 1, \quad yy = 2, \quad zz = 3,$$

$$yz = zy = 4, \quad xz = zx = 5, \quad xy = yx = 6$$

The resulting d_{ij} components form a 3×6 matrix that operates on the electric field E tensor to generate polarization P as follows. The d is a third-rank tensor that contains 18 elements and can be written as:

$$\begin{pmatrix} P_x \\ P_y \\ P_z \end{pmatrix} = \begin{pmatrix} d_{11} & d_{12} & d_{13} & d_{14} & d_{15} & d_{16} \\ d_{21} & d_{22} & d_{23} & d_{24} & d_{25} & d_{26} \\ d_{31} & d_{32} & d_{33} & d_{34} & d_{35} & d_{36} \end{pmatrix} \begin{pmatrix} E_x^2 \\ E_y^2 \\ E_z^2 \\ 2E_zE_y \\ 2E_zE_x \\ 2E_xE_y \end{pmatrix} \tag{8}$$

The contracted d_{ij} tensor obeys the symmetry restrictions as the piezoelectric tensor, hence they are reduced to only a few independent elements but have the same form in a given point-group symmetry. Only crystals that lack a center of symmetry can possess a nonvanishing d_{ijk} tensor that can be determined from SHG measurements. For SHG $(-2\omega; \omega, \omega)$ experiments, the tensor is given as:

$$P_i^{2\omega} = d_{ijk}(-2\omega; \omega, \omega)E_j^\omega E_k^\omega \tag{9}$$

The subscripts jk and kj of the d coefficient can be interchanged and written as:

$$d_{ijk}(-2\omega; \omega, \omega) = d_{ikj}(-2\omega; \omega, \omega) \tag{10}$$

The subscripts jk and kj can be replaced by the contracted indices as follows;

$$\begin{array}{lll} d_{xxx} = d_{11} & d_{xyy} = d_{12} & d_{xzz} = d_{13} \\ d_{xyz} = d_{xzy} = d_{14} & d_{xxz} = d_{xzx} = d_{15} & d_{xxy} = d_{xyx} = d_{16} \end{array} \tag{11}$$

For example, the d_{ij} tensor for KDP that belongs to 42m point-group can be written as:

$$d_{ij} = \begin{bmatrix} 0 & 0 & 0 & d_{14} & 0 & 0 \\ 0 & 0 & 0 & 0 & d_{14} & 0 \\ 0 & 0 & 0 & 0 & 0 & d_{36} \end{bmatrix} \tag{12}$$

In that case the component of nonlinear polarization will be:

$$P_x = 2d_{14}E_zE_y$$
$$P_y = 2d_{14}E_zE_x$$
$$P_z = 2d_{36}E_xE_y$$

A relationship between the β and second-order nonlinear coefficient d_{IJK} for organic molecules has been developed by Zyss and Oudar[30]

$$d_{IJK}(-2\omega; \omega, \omega) = Nf_I(2\omega)f_J(\omega)f_K(\omega)b_{IJK}(-2\omega; \omega, \omega) \tag{13}$$

where

$$b_{IJK}(-2\omega; \omega, \omega) = \frac{1}{N_g}\sum_{ijk}\left(\sum_{s=1}^{N}\cos\theta_{Ii}^{(s)}\cos\theta_{Jj}^{(s)}\cos\theta_{Kk}^{(s)}\right)\beta_{ijk}(-2\omega; \omega, \omega) \tag{14}$$

where N is the number density of molecules in crystal unit volume and N_g is the number of equivalent positions in the unit cell, f terms are the local field factors, and b_{IJK} refers to the unit cell nonlinearity per molecule. The second-order NLO susceptibility depends strongly on both β value and its orientation in a crystal. The total β of a molecule can be determined from electric field induced second harmonic (EFISH) measurements. The total β is a sum of two parts; β_{add} resulting from the interaction between the substituent and the conjugated system and β_{CT} the predominating part arising from the donor-acceptor charge transfer contribution.

$$\beta_x = \beta_{add} + \beta_{CT}, \tag{15}$$

In a quantum two-level model, the β_{CT} can be expressed as[31]:

$$\beta_{CT} = \frac{3e^2h^2}{2m}\frac{W}{[W^2 - (2h\omega)^2][W^2 - (h\omega)^2]}f\Delta\mu \tag{16}$$

where f is the oscillator strength of the optical transition involved in the two-state process, and $\Delta\mu$ is the difference between excited-state and ground-state dipole moments, W and $h\omega$ are the energy gap and the energy of the laser radiation, respectively. According to this relation, β_{CT} increases for harmonic frequencies approaching absorption edge. The β_{add} has been found to be only 10% of the total β_x for stilbene compounds. To optimize β_x, large values of oscillator strength f and induced dipole moment μ are important factors. Oudar and Chemla[31,32] demonstrated theoretically as well as experimentally the validity of two-level model for a series of stilbene derivatives. The β_{CT} value of stilbene derivatives increased remarkably with the increase of the π-conjugation length

Table 1 Theoretically Calculated Charge-Transfer Contribution β_{CT} from Two-Level Model and Experimental β_{exp} for a Series of Stilbene Derivatives and p-nitroaniline

Molecules	W (eV)	f	$\Delta\mu$ (D)	β_{CT} (10^{-30} esu)	β_{exp} (10^{-30} esu)
H_2N—〇—NO_2	3.89	0.51	12.35	19	34.5
H_3C,H_3C>N—〇—〇—NO_2	2.85	0.56	24	217	220
H_2N—〇—〇—NO_2	3.06	0.54	42	227	260
H_3C,H_3C>N—〇—〇—NO_2	2.95	0.72	42	383	450
H_3C,H_3C>N—〇—NO_2	2.65	0.73	35	715	650

From Oudar, J. L., *J. Chem. Phys.*, 67, 446, 1997; Oudar, J. L. and Chemla, D. S., *J. Chem. Phys.*, 66, 2664, 1977. With permission.

and the strength of donor and acceptor groups (Table 1). Figure 1 indicates that the two-level model is appropriate for describing optical nonlinearities of conjugated molecules which possess strong intramolecular charge-transfer interactions.

Both microscopic and macroscopic nonlinear properties are correlated as discussed above. Another important NLO organic system is poled polymers. In the presence of electrical poling, the $\langle\beta_{IJK}\rangle$ can be deduced from the orientational average of the local-field-corrected molecular second-order optical nonlinearity $(\beta_{ijk})^{33}$;

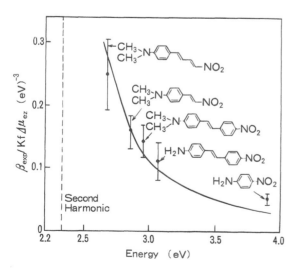

Figure 1 Second-order molecular hyperpolarizability as a function of conjugation length for a number of benzene and stilbene derivatives. (From Oudar, J. L., *J. Chem. Phys.*, 67, 446, 1977. With permission.)

$$\langle \beta_{IJK} \rangle = \int d\phi \int \sin\theta \, d\theta \int d\varphi b_{IiaJjaKk} \, \beta_{ijk} G(\theta) \tag{17}$$

where the rotation matrices aIL between the molecular frame and the laboratory frame is given by the Euler angles (ϕ, θ, and φ), the bIL is the inverse matrices of aIL. The $G(\theta)$ is the distribution function. The macroscopic second-order susceptibility can be derived from the following expression:

$$\beta_{333} = L_3(p)\beta_{xxx} + 1.5[L_1(p) - L_3(p)]\,\beta_{xyy} \tag{18}$$

where $L_1(p)$ and $L_3(p)$ are the first- and third-order Langevin functions, respectively, and p is expressed by the local field correction factor of Onsager type;

$$p = \frac{\epsilon(n^2 + 2)\mu}{n^2 + 2\epsilon} \frac{E_p}{kT} \tag{19}$$

where ϵ is the dielectric constant, n is the refractive index, and μ is the molecular dipole moment. Third-order Langevin function can be estimated and transferred to the order parameter ϕ according to the following expression:

$$\phi = [3\langle \cos^2 \theta \rangle - 1]/2 \tag{20}$$

where $\langle \cos^2 \theta \rangle$ is a second-order Langevin function.

The second-order NLO coefficients d_{ijk} are related to a linear electro-optic coefficients r_{ijk} by the following equation:[36]

$$d_{ijk} = -(\epsilon_{ii}\epsilon_{jj}/4\epsilon_0)r_{ijk} \tag{21}$$

Another relationship between electro-optic coefficient (r) and NLO coefficient (d) can be written through the two-level model:[33]

$$r_{IJK}(-\omega; \omega, 0) = -\frac{4d_{KIJ}}{n_I^2(\omega)n_J^2(\omega)} \frac{f_{II}^\omega f_{JJ}^\omega f_{KK}^0}{f_{KK}^{2\omega'} f_{II}^{\omega'} f_{JJ}^{\omega'}} \times \frac{(3\omega_0^2 - \omega^2)(\omega_0^2 - \omega'^2)(\omega_0^2 - 4\omega'^2)}{3\omega_0^2(\omega_0^2 - \omega^2)^2} \tag{22}$$

where ω_0 is the frequency of the first strongly absorbing electronic transition in the molecule, and ω' and ω are the fundamental wavelengths in SHG and for electro-optic coefficient measurements, respectively. By evaluating the SHG coefficients, one can also roughly estimate the electro-optic coefficients. Moreover, through the oriented gas model, the NLO and electro-optic coefficients can be compared fairly with β values. The d_{ij} tensors follow the same symmetry rules as the piezoelectric tensors and the electro-optic tensors. For example, electro-optic tensor elements of point-group symmetry 2 are as follows. The organic single crystal 2-(N-prolinol)-5-nitropyridine (PNP) belongs to this point group and its details can be found in section VIIA and Chapter 6.

$$\begin{bmatrix} 0 & r_{12} & 0 \\ 0 & r_{22} & 0 \\ 0 & r_{32} & 0 \\ r_{41} & 0 & r_{43} \\ 0 & r_{52} & 0 \\ r_{61} & 0 & r_{62} \end{bmatrix} \tag{23}$$

Second-order nonlinear susceptibility tensors for 21 classes lacking inversion symmetry are summarized in Chapter 3. For more details, see reference 34. Bosshard et al.[35] have provided an excellent description of electro-optic coefficients and nonlinear optical coefficients in organic single crystals.

III. ORIGIN OF THE SECOND-ORDER NLO EFFECTS IN ORGANIC MOLECULES

Here we analyze the relationship between the chemical structures and second-order nonlinear optical properties by providing several examples dealing with molecular engineering. Virtually second-order NLO effects in organic molecules originate from a strong donor–acceptor intramolecular interaction. This concept was demonstrated by Davydov and co-workers[15] in 1970 while screening SHG activity in a wide variety of substituted benzenes. They concluded that dipolar aromatic molecules possessing an electron donor group and an electron acceptor group contribute to large second-order optical nonlinearity arising from the intramolecular charge transfer between the two groups of opposite nature. Therefore, a typical SHG active molecule can be presented as shown below if it lacks a center of symmetry. On the other hand, π-conjugated molecules with a donor and an acceptor will not display SHG activity if they possess a center of symmetry. This symmetry requirement eliminates many materials from being SHG active and at the early stage of designing and synthesizing novel materials, one has to consider ways of introducing noncentrosymmetry in the molecular structures.

Donor ▬ $\boxed{\text{π-conjugated system}}$ ▬ **Acceptor**

Large second-order optical nonlinearity originates from organic conjugated molecules having an electron acceptor group at one end and a donor group at the opposite end. The π-conjugated systems could be benzene, azobenzene, stilbene, tolans, biphenyl, benzylidene, heterocycle, polyenes, etc. The electron acceptor and donor groups that can be attached to a π-conjugated system are as follows: (1) Acceptor groups: NO_2, NO, CN, COOH, COO^-, $CONH_2$, CONHR, $CONR_2$, CHO, SSI, SO_2R, $SO_2C_3F_7$, SO_2CH_3, COR, $COCF_3$, CF_3, $COCH_3$, $CH=C(CN)_2$, $C_2(CN)_3$, SO_2NH_2, N_2^+, NH_3^+, $N(CH_3)_3^+$ and aromatic*. (R is an alkyl group). (2) Donor groups: NH_2, $NHCH_3$, $N(CH_3)_2$, NHR, N_2H_3, F, Cl, Br, I, SH, SR, OR, CH_3, OH, $NHCOCH_3$, OCH_3, SCH_3, OC_6H_5, $C(CH_3)_3$, $COOCH_3$, O^-, S^-, and aromatic*. (R is an alkyl group). (*Aromatic group is capable of both kinds of effects.)

The absorption band of an organic dye can be tailored by either increasing the π-conjugation length or by substituting donor–acceptor groups to a conjugated system. As a result the absorption band of the UV-visible spectrum can be shifted and will have either bathochromic (red shift toward a longer wavelength) or hypsochromic (blue shift toward a shorter wavelength) features. For example, long-chain conjugated polyene molecule will exhibit color evidenced by an absorption band in the visible region. The position of an absorption band could be changed among stilbene, azobenzene, benzylidene, heterocycle, and tolan π-backbones. Electron donor–acceptor groups are of particular interest because they tend to cause bathochromic shifts which also increase the intensity of absorption bands. A donor group can provide additional electrons into the π-conjugated system leading to a strong interaction from a donor–acceptor combination. The donor–acceptor interaction is not only affected by the length of conjugation but also by the relative positions at the π-conjugated system; there are many such examples. Because organic structures offer tremendous possibilities of chemical modification, their optical and nonlinear optical properties can be tailored. The 1970s discovery that large second-order optical nonlinearity originates from π-conjugated molecules having an electron acceptor group at one end and a donor group at the opposite end of a benzene ring provided the basic understanding of a structure–property relationship. Now, three decades after the discovery, this simple notion has given birth to a new field of science, nonlinear optics, for which functional organic dyes are the key elements. The optically induced polarization in these dyes generates nonlinear optical properties. Certain dye chromophoric systems restrict their application to NLO devices, although they have parallel potentials in applications associated with light absorption, light emission, photoelectrical activity, chemical and photochemical activities. Functional dye molecules with highly polarizable π-electron systems are of great interest in nonlinear optics. The past decade

Table 2 Molecular Hyperpolarizability of Nitrobenzene, Aniline, and *p*-Nitroaniline Showing the Influence of Donor–Acceptor Intramolecular Charge-Transfer Interaction

Molecule	Dipole Moment (D)	β $(10^{-30}$ esu)	Wavelength (μm)	Solvent	Ref.
Nitrobenzene	3.93	1.97	1.318	Neat	42
		2.20	1.06	Neat	43
Aniline	1.56	7.9	1.318	Neat	42
		1.10	1.06	Neat	43
p-Nitroaniline	6.29	21.1	1.318	Neat	42
		34.5	1.06	Methanol	43

has seen a dramatic increase in NLO dyes for specialized opto-electronic applications. The field of nonlinear optics is a uniquely concerned with the molecular design of functional dyes in which NLO response is governed largely by the chromophores involving interactions with light. Either of the groups in a combination are suitable. More complex NLO chromophores can be envisaged using multiple donor and acceptor groups linked to different π-conjugations either by length or elements. Although there are a few exceptions, a group may be an electron acceptor in one condition and an electron donor in another, depending on the situation. These acceptor–donor groups are generally attached to conjugated systems such as benzene, azobenzene, stilbene, tolans, heterocycles, and benzylidene to generate NLO materials with large molecular hyperpolarizability and to tailor their transparency. Therefore a wide variety of organic materials having electron acceptor and donor groups have been synthesized particularly for second-order nonlinear optics; those can be used for electro-optics as well as for third-order nonlinear optics.

Zyss[37-39] used quantum chemistry calculations to establish a relationship between various molecular structures and optical nonlinearities. Molecular hyperpolarizabilities of monosubstituted benzenes and donor–acceptor-disubstituted phenyl-polyenes were evaluated from the finite field perturbed INDO approach, and the roles of charge-transfer and chain length in second-order nonlinear optical processes of conjugated molecules were confirmed. The origin of the second-order optical nonlinearity in a donor–acceptor-substituted benzene (e.g. *p*-nitroaniline) was described excellently by Chemla et al.[40] Lalama and Garito[41] described a simplified two-level model of the molecular charge-transfer contribution to the hyperpolarizability using *p*-nitroaniline. Levine[42] and Oudar and Chemla[43] showed the importance of charge-transfer interactions while calculating the magnitude and sign of the molecular hyperpolarizability for nitrobenzene *(1)*, aniline *(2)*, and *p*-nitroaniline *(3)*. The chemical structures of these three molecules are depicted below. Table 2 lists the β values of nitrobenzene, aniline, and *p*-nitroaniline to show the influence of donor–acceptor intramolecular charge-transfer interaction.

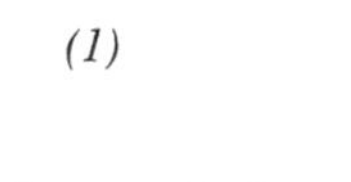

(1)    *(2)*    *(3)*

It can be seen that much interest in second-order NLO properties of substituted benzene derivative systems stems from the basic studies of the chemistry and physics of *p*-nitroaniline (*p*-NA) molecule. The *p*-NA molecule consists of a benzene ring in which an electron donor amino (NH_2) group is substituted in a *para*-position to an electron acceptor nitro (NO_2) group. These opposite ends of the conjugated system lead to maximum acentricity in the molecule. The π-electron donor and acceptor configuration composed of NH_2 and NO_2 groups, respectively, in *p*-NA results in intramolecular charge-transfer interactions. Therefore, *p*-NA exhibits extremely large first hyperpolarizability (β) due to the highly asymmetric charge distribution arising from the π-electronic structure of the molecule. The origin of first hyperpolarizability in *p*-NA can be explained as follows. In a benzene molecule, β disappears due to the symmetry with a center of inversion. The electron donor or electron

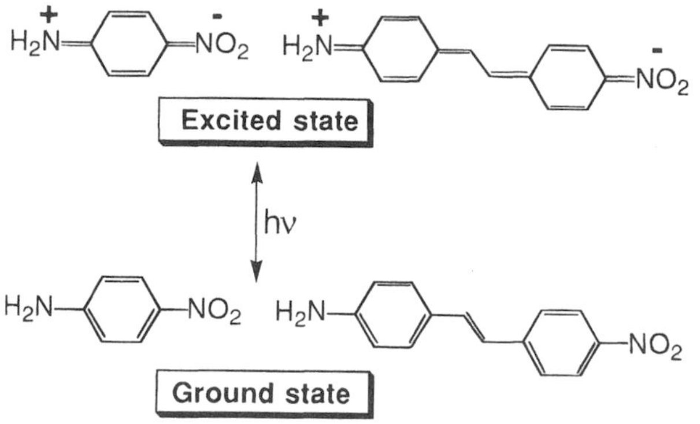

Figure 2. Molecular structures of the ground-state and lowest energy excited states for p-NA and 4-amino-4'-nitrostilbene.

acceptor substitution in the benzene ring removes such a restriction and thus generates β. This can be better understood from the comparison of the β values of the nitrobenzene, aniline, and p-NA. The β values reported by Levine[42] show that the β factor of p-NA is about 11 times larger than nitrobenzene and about 26 times larger than aniline (Table 2). The strong donor–acceptor charge-transfer interactions contribute to large β value of p-NA molecule. Let us compare p-NA with o-nitroaniline and m-nitroaniline. The β values of corresponding o-nitroaniline and m-nitroaniline are 6.4×10^{-30} and 4.2×10^{-30} esu, respectively, much lower than that of p-NA molecule. The donor–acceptor charge-transfer interactions in o- and m-nitroanilines are negligible because β predominates in a mesomeric manner. The delocalization of π-electron in p- and o-nitroanilines contributes to a large hyperpolarizability. A resonance structure having alternating single and double bonds can only exist in p- and o-isomers but not for the m-nitroaniline. As a result, the delocalization of π-electron in p- and o-nitroanilines contributes to a large nonlinearity. The β changes in an order p-NA > o-NA > m-NA because of the resonance structures. β is larger for p-NA molecule because the donor–acceptor charge-transfer intramolecular interactions are prominent by attaining the most desirable resonance structure. p-Nitroaniline exhibits very large first hyperpolarizability due to its highly asymmetric charge-correlated excited states of the π-electronic structure of the molecule.[41] Figure 2 shows the molecular structures of the ground-state and resonance excited state for p-nitroaniline and 4-amino-4'-nitrostilbene, both of them show large β values.

Stahelin et al.[44] investigated the values of β for p-NA in solvents of varying polarity using the EFISH technique. Table 3 lists the optical absorption peaks and hyperpolarizability for p-NA in solvents of different polarities. The β_0 extrapolated from a log-log plot of the hyperpolarizability to zero-frequency was found to be correlated with the solvent-induced shift of the p-NA charge transfer absorption maximum (λ_{max}). The measured quantity $\mu\beta_z$ changes by a factor of ~2.5 on going from the least polar solvent (chloroform) to one of the most polar solvents (N-methylpyrollidone). The β_0 for p-NA depends on the solvent polarity and ranges from 8.1×10^{-30} to 15.8×10^{-30} esu on going from nonpolar to polar solvents. The correlation between λ_{max} and β_0 is the same as that of donor–acceptor p-substituted benzenes with different substituents. The observed relationship between λ_{max} and β_0 was analyzed by a simple four-level Huckel model and the two-level hyperpolarizability model and it was found that both λ_{max} and β_0 depend on the relative position of the electron acceptor and donor energy levels, which can be shifted either by polarity of the solvent or by changing the chemical nature of the substituents. Therefore, a relationship between λ_{max} and β_0 has been considered as a fundamental property of π-conjugated donor–acceptor systems. p-Nitroaniline represents a model for one-dimensional donor–acceptor charge-transfer interactions. The p-NA molecule crystallizes in a centrosymmetric space group, which restricts the observation of any macroscopic second-order optical effects. Therefore, closely related p-NA derivatives, which do not have a center of symmetry in the crystalline state, have been found to display interesting NLO properties and are gaining considerable attention.

IV. MOLECULAR ENGINEERING OF SECOND-ORDER NONLINEAR OPTICAL MATERIALS

The second-order NLO properties of organic materials, which are currently of intense interest, originate from the basic concept of the p-NA molecule. Now let us look at a 2-methyl-4-nitroaniline (4) molecule

(4)

in which a donor amine (NH_2) group is located at the first carbon atom of the benzene ring, a methyl (CH_3) group at the second carbon atom, and an acceptor nitro (NO_2) group at the fourth carbon atom; this chemical substitution provides the maximum acentricity to the molecule and gives rise to a noncentrosymmetric crystal structure. In 2-methyl-4-nitroaniline (MNA), the simple substitution of a methyl group at second carbon atom of a p-NA molecule changes the whole physical aspects of optical nonlinearity. We have acquired the concept of generating large molecular hyperpolarizability from the p-NA molecule; however, being centrosymmetric, it can not display SHG activity as stated earlier. Using the same concept of donor–acceptor intramolecular interactions and substituting a methyl group to p-NA provides us with a novel chemical species, the MNA molecule that exhibits exceptionally large SHG activity being noncentrosymmetric. Therefore, the molecular engineering approach offers an avenue to design and tailor second-order NLO materials of interest. Currently, chemical species having charge-transfer characteristics analogous to a p-NA molecule are gaining considerable attention.

Quantum chemistry approaches assist in establishing a structure–property relationship between chemical species and NLO interactions by calculating molecular hyperpolarizabilities of organic structures. Here a very selective example is provided to show how nonlinear optical effects are changed as the chemical structure is modified. First let us consider these three molecules; p-NA, 1,5-diamino-2,4-dinitrobenzene (DADB), and 1,3,5-triamino-2,4,6-trinitrobenzene (TATB) which are unique due to their dimensionality and charge-transfer interactions. TATB was first reported by

Table 3 Optical Absorption Peaks and Hyperpolarizability for p-NA in Solvents of Different Polarities

Solvent	λ_{max} (nm)	$\mu\beta_z$ (10^{-30} esu D)	μ (D)	β_z (10^{-30} esu)	β_0 (10^{-30} esu)
Acetone	368	189	7.3	25.9	11.9
Acetonitrile	366	181	6.2	29.2	13.6
Chloroform	348	107	6.4	16.8	8.6
1,4-Dioxane	354	114	7.0	16.3	8.1
Dichloromethane	350	105	6.2	16.6	8.5
Dimethoxyethane	366	150	7.7	19.5	9.1
DMF	382	219	7.3	30.0	12.6
Ethylacetate	358	152	6.7	22.6	11.0
Methanol	370	195	6.1	32.0	14.5
MME	368	186	7.8	23.8	11.0
NMP	386	260	6.8	38.4	15.8
Tetrahydrofuran	362	152	7.1	21.4	10.2
Tetrahydropyran	360	160	6.7	22.6	11.0

DMF = dimethylformamide; MME = 2-methoxymethyl ether, NMP = N-methyl pyrollidone.
From Stahelin, M., et al., *Chem. Phys. Lett.,* 191, 245, 1992. With permission.

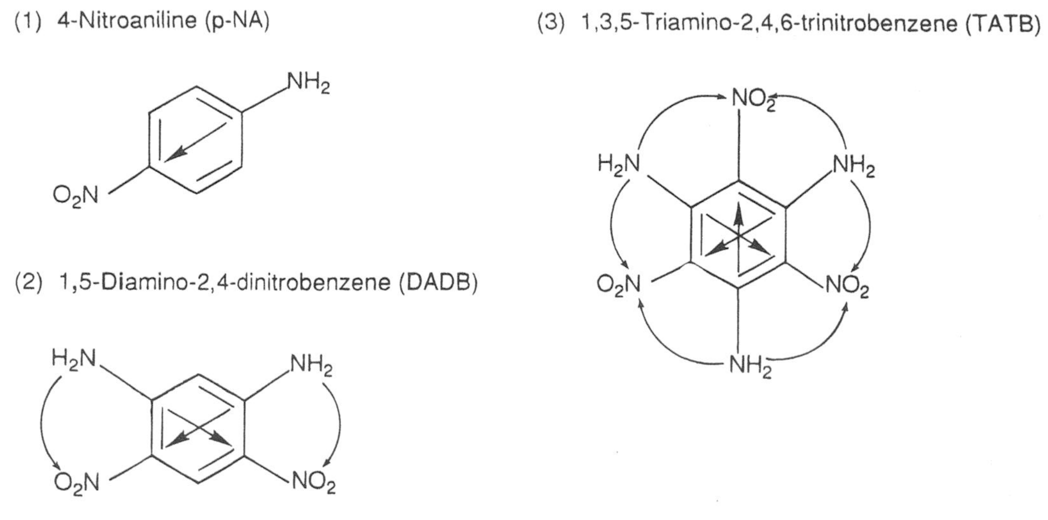

(1) 4-Nitroaniline (p-NA)

(3) 1,3,5-Triamino-2,4,6-trinitrobenzene (TATB)

(2) 1,5-Diamino-2,4-dinitrobenzene (DADB)

Figure 3. Charge-transfer interactions of p-NA, DADB, and TATB. (From Nalwa, H. S., et al., *Opt. Mater.,* 2, 73, 1993. With permission.)

Cady and Larson[45] in 1965 and its NLO properties were first described by Zyss.[46] On the other hand, DADB and its derivatives were first synthesized and their second-order NLO properties were first reported by Nalwa et al.[47] A comparative theoretical study of the p-NA, DADB, and TATB molecules was also carried out by Nalwa et al.[48] Figure 3 shows the chemical structures and charge-transfer interactions of p-NA, DADB, and TATB. Here the p-NA is a one-dimensional charge-transfer molecule, whereas the DADB and TATB are two-dimensional charge-transfer systems. Although these molecules are closely related, the number and location of electron acceptor nitro groups and donor amino groups on the benzene ring significantly change their charge-transfer characteristics; as a result, their molecular hyperpolarizabilities are changed. The geometry optimization and finite-field molecular hyperpolarizability calculations of p-NA, DADB, and TATB were performed with the semiempirical Molecular Orbital Package (MOPAC) AM1 method.[48]

In p-NA, an NH_2 group is located at the *para* to an NO_2 group and this system represents a model for one-dimensional charge-transfer molecules. In the case of DADB, amino groups are located at *para* as well as *ortho* positions to nitro groups, therefore its dominantly *para* and quasi-*ortho* interactions lead to a two-dimensional charge-transfer character. TATB is an octupolar molecule having an alternating sequence of acceptor nitro and donor amino groups at benzene ring which interact in both *para* and *ortho* positions. These molecules present a beautiful example where DADB forms a perfect bridge between p-NA and TATB. The orientationally averaged hyperpolarizability $\beta_x = \beta_{xxx} + \beta_{xyy} + \beta_{xzz}$ of p-NA, DADB, and TATB calculated by using MOPAC-AM1 method were estimated to be 12.017×10^{-30}, 6.793×10^{-30}, and -0.196×10^{-30} esu, respectively.[48] One can see that moving from p-NA to DADB to TATB yields large changes in the β components not only by their sign but also in their magnitude because of the different process of charge-transfer interaction caused by the number and location of donor–acceptor substituents on the benzene ring. It becomes clear here that the β changes in the sequence p-NA > DADB > TATB because of its sensitivity to the substitution. On the other hand, the α and γ change in the sequence TATB > DADB > p-NA. The γ values of TATB and DADB are 2.4 times larger than p-NA. The β of p-NA is about 1.76 times larger that of DADB molecule, which is related to the change in dipole moment that decreases with the substitution. The dipole moment of 7.99, 7.36, and 0.3099 debye were obtained for p-NA, DADB, and TATB, respectively. In p-NA, the diagonal component β_{xxx} is the largest, whereas for DADB, the off-diagonal component β_{xyy} is the largest. This indicates that a remarkable change in β components occurs as one goes from p-NA to DADB to TATB. The TATB has an octupolar symmetry, and a nonvanishing β value was demonstrated by Zyss[46] on the basis of group theory. Bredas et al.[49,50] reported the β_{xxx}, β_{xyy}, and β_{xzz} components as -8.01×10^{-30}, 8.01×10^{-30}, and zero esu, respectively, for TATB by *ab initio* 3-21G CPHF method. The INDO/SDCI calculation gave β_{xxx}, β_{xyy}, and β_{xzz} of -7.68×10^{-30}, 7.68×10^{-30}, and zero

Table 4 Physical Properties of p-NA ($M_w = 138.12$ g/mol), DADB ($M_w = 198.13$ g/mol), and TATB ($M_w = 258.15$ g/mol)

Molecule	m.p. (°C)	λ_{max} (nm)	α (10^{-23} esu)	β_x (10^{-30} esu)	γ (10^{-36} esu)	Dipole Moment (D)	SHG (× Urea)
p-NA	149	352	2.322	12.017	12.71	7.99	0
DADB	298	386	3.071	6.793	29.71	7.36	0
TATB	300	351	3.766	−0.196	30.65	0.309	3

Optical absorption spectra were recorded for a concentration of 10^{-5} M for both p-NA and DADB in methylene chloride. For TATB, the melting point is taken from Cady, and Larson, Acta *Crystallogr.*, 18, 485, 1965, whereas the data of absorption maximum and SHG are those reported by Ledoux et al., *Chem. Phys. Lett.*, 172, 440, 1990. (From Nalwa, H. S., et al., *Opt. Mater.*, 2, 73, 1993. With permission.)

esu, respectively. Table 4 lists the physical properties of p-NA, DADB, and TATB molecules. Both DADB and TATB show high melting temperature at around 300 °C due to the inter- and intramolecular hydrogen bonding between the hydrogen of amino groups and adjacent oxygen of nitro groups. Both p-NA and DADB show no powder SHG, whereas the TATB exhibits a SHG intensity 3 times that of standard urea as reported by Ledoux et. al.[51] Therefore, these three molecules represent a very interesting structural-property relationship.

Zyss[46,53] analyzed the role of dipolar and octupolar molecules for second-order nonlinear optics and the following discussion is based exclusively on those descriptions. The relative deviation α is considered as the relative amount of out-of-plane contribution β, namely,

$$\alpha = \frac{\|T_{J=3}\|^2}{\|\beta\|^2} = \frac{\|\beta_{J=3}^{3D}\|^2 - \beta_{J=3}^{2D}\|^2}{\|\beta\|^2} \tag{23}$$

$$\alpha = \frac{1}{5}\frac{1}{[1 + (\rho^{2D})^2]} \tag{24}$$

where ρ is ratio of the relative magnitudes of the octupolar over dipolar components of a molecular quadratic hyperpolarizability in the case of a C_{2v} (mm2) molecule[54] and ρ^{2D} and ρ^{3D} are defined as follows:

$$\rho^{2D} = \frac{\|\beta_{J=3}^{2D}\|}{\|\beta_{J=1}^{2D}\|} \tag{25}$$

$$\rho^{2D} = \frac{1}{3}\frac{(\beta_{xxx} - 3\beta_{xyy})^2 + (\beta_{yyy} - 3\beta_{yxx})^2}{(\beta_{xxx} + \beta_{xyy})^2 + (\beta_{yyy} + \beta_{yxx})^2} \tag{26}$$

$$\rho^{3D} = \sqrt{\frac{2}{3}}\left|\frac{1 - 3u + 6u^2}{1 + u}\right| \tag{27}$$

$$\rho^{3D} = \frac{\|\beta_{J=3}^{3D}\|}{\|\beta_{J=1}^{3D}\|} \tag{28}$$

$$\rho^{3D} = \frac{2}{3}\frac{\beta_{xxx}^2 + \beta_{yyy}^2 + 6\beta_{xyy}^2 + 6\beta_{yxx}^2 - 3\beta_{xyy}\beta_{xxx} - 3\beta_{yxx}\beta_{yyy}}{(\beta_{xxx} + \beta_{xyy})^2 + (\beta_{yyy} + \beta_{yxx})^2} \tag{29}$$

$$\rho = \frac{\|\beta_{J=3}\|}{\|\beta_{J=1}\|} \tag{30}$$

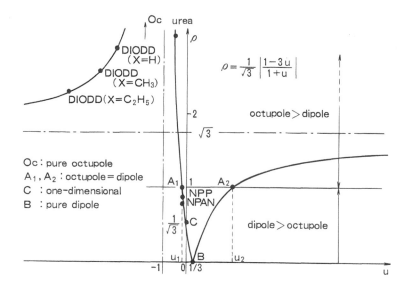

Figure 4 Dependence of ρ (ratio of octupolar over dipolar contributions) on parameter $u = \beta_{XYY}/\beta_{XXX}$ measuring the in-plane anisotropy of the β tensor in the case of a C_{2v} molecules. DIODD is 3-amino-4,6-dinitroaniline[47] NPP,[57,58] NPAN,[58] and urea. (From Zyss, J., *J. Chem. Phys.*, 98, 6583, 1993. With permission.)

$$\rho = 1/\sqrt{3} \left| \frac{(\beta_{xxx} - 3\beta_{xyy})}{(\beta_{xxx} + \beta_{xyy})} \right| \tag{31}$$

$$\rho = 1/\sqrt{3} \left| \frac{1 - 3u}{1 + u} \right| \tag{32}$$

where the variable u reflecting the anisotropy of the in-plane nonlinearity is expressed as:

$$u = \frac{\beta_{xyy}}{\beta_{xxx}} \tag{33}$$

In accordance with Equations 30 through 33, molecular systems can be classified in terms of their octupolar vs. dipolar character. Figure 4 shows the variation of ρ with respect to u. The octupolar and dipolar contribution peaks for $u = -1$ and $u = 1/3$ correspond to no dipole and octupole, respectively, and therefore a pure octupolar and dipolar system, respectively.[53] Urea and DIODD (derived from DADB) analogs are the most octupolar-like species and both have mm2 symmetry. The angular dependence of macroscopic second-order susceptibility β of the 2-D charge-transfer molecules has been deduced, whereas off-diagonal hyperpolarizability component β_{xyy} is considered as the largest β component contrary to 1-D charge-transfer molecules where the X axis of the molecule is directed along a dipole moment and the Y axis is perpendicular to the dipole moment within the molecular plane. The average molecular hyperpolarizability β was obtained by averaging overall molecular orientation using the standard transformation equation. The macroscopic second-order nonlinear susceptibilities β_{111} and β_{133} were derived from the following equations[47]:

$$\beta_{111} = \cos^3\theta \, \beta_{xxx} + 3 \cos\theta \sin^2\theta \, \beta_{xyy} + 3 \cos^2\theta \sin\theta \, \beta_{yxx} + \sin^3\theta \, \beta_{yyy} \tag{34}$$

$$\beta_{133} = 0.5 \cos\theta \sin^2\theta \, \beta_{xxx} + (0.5 \cos^3\theta + \sin^2\theta \cos\theta)\beta_{xyy} + 0.5 \cos\theta \, \beta_{xzz}$$

$$+ (\cos^2\theta \sin\theta + 0.5 \sin^3\theta) \, \beta_{yxx} + 0.5 \sin\theta \cos^2\theta \, \beta_{yyy} + 0.5 \sin\theta \, \beta_{yzz} \tag{35}$$

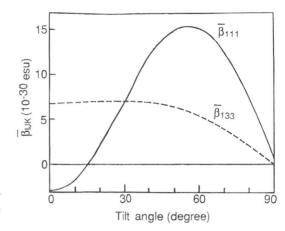

Figure 5 Angular dependence of β_{xyy} for 2-D molecules. (From Nalwa, H. S., Watanabe, T., and Miyata, S., *Adv. Mater.*, 7, 1001, 1995. With permission.)

Using these equations, average macroscopic molecular hyperpolarizabilities β_{111} and β_{133} were calculated for the molecular tilt angle θ ranging from 0 to 90°. Figure 5 shows the angular dependence β_{IJK} of the 2-D charge-transfer molecules where the main contributing component β_{xyy} is always larger by several factors than the β_{xxx} component. The macroscopic molecular hyperpolarizability strongly depends on the tilt angle and shows a maximum at an angle of 56°. Table 5 lists the β_{xyy} and β_{xxx} components of 1-D and 2-D molecules. Urea and DIODD analogs are the most octupolar-like species and both have mm2 symmetry. The DIODD analogs are quite interesting from the viewpoint of their higher off-diagonal β_{xyy} component which is three to four times larger than that of the β_{xxx} component. In some cases, the β_{xyy} has been found to be as much as 30 times larger than the β_{xxx} component for 2-D molecules.[55,56] Similarly, for DNCB and MANN molecules, the β_{xyy} is 11 and 32 times larger, respectively, than β_{xxx} component. The ratio between

Table 5 Values of Various Parameters u, α, ρ^{2D}, ρ^{3D} for Various 1-D and 2-D Molecules

Molecule	β_{xyy}	β_{xxx}	$u = \beta_{xyy}/\beta_{xxx}$	ρ^{2D}	α	ρ^{3D}
Urea	0.60	−1.14	−0.52	3.09	0.019	7.11
DIODD-type						
1a. ($X = F$)	−3.50	1.93				
1b. ($X = Cl$)	−4.55	1.03				
1c. ($X = NH_2$)	9.87	−3.11	−3.17	2.80	0.0023	26.63
1d. ($X = NHCH_3$)	12.19	−3.09	−3.94	2.52	–	–
1e. ($X = NHC_2H_5$)	13.64	−2.79	4.89	2.32	–	–
DDCV	20.30	−2.85	7.12			
p-NMDA	11.35	−1.63	6.96			
DABNP	47.30	9.40	5.03			
MANN	11.86	0.37	32.05			
OANN	11.87	0.03	395.66			
DNCB	7.40	−0.66	11.21			
NPP	2.3	−17.2	−0.134	0.93	0.11	1.42
NPAN	2.3	−11.9	−0.193	0.912	–	–
MAP	−0.48	−27.0	–	0.744	–	–
p-NA (1-D)	2.23	14.26	0.156			
ANST (1-D)	1.91	46.27	0.041			
ODDFA-type						
2.a ($X = NHCH_3$)	−1.30	7.88				
2b. ($X = NHCH_2CH_3$)	−3.61	13.31				
2c. ($X = NH(CH_2)_2CH_3$)	−3.59	14.86				
2d. ($X = NH(CH_2)_8CH_3$)	0.36	13.16				
TATB			−1	∞	0	∞

Partly adopted from Zyss, J. *J. Chem. Phys.*, 98, 6583, 1993. With permission.

β_{xyy} and β_{xxx} components grows much larger for OANN molecule. For 1-D p-NA and ANST molecules, the β_{xxx} is 6 and 23 times larger than the β_{xyy} component, respectively. It is amazing to have such large off-diagonal components for 2-D NLO molecules. Both N-4-nitrophenyl-(L)-prolinol (NPP)[57] and N-4-nitrophenyl-N-methylacetonitrile (NPAN)[58] have relatively balanced octupolar and dipolar contributions and MAP[59] does not satisfy mm2 symmetry. An excellent description of multipolar nonlinear optical media can be found in reference 60. In the case of TATB which is a planar system with D_{3h} symmetry, $\beta_{xyy} = -\beta_{xxx}$ which cancels the vectorial component.

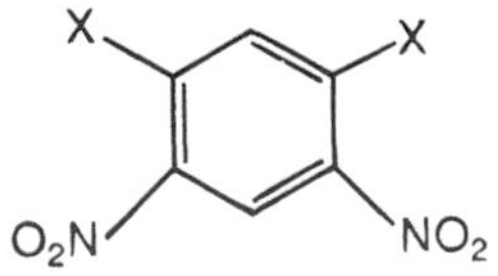

1. DIODD-type
1a. X = F
1b. X = Cl
1c. X = NH$_2$
1d. X = NHCH$_3$
1e. X = NHCH$_2$CH$_3$

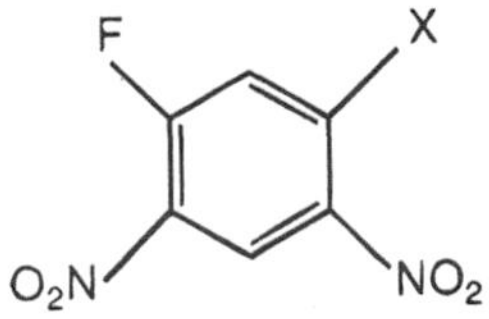

2. ODDFA-type
2a. X = NHCH$_3$
2b. X = NHCH$_2$CH$_3$
2c. X = NH(CH$_2$)$_2$CH$_3$
2d. X = NH(CH$_2$)$_8$CH$_3$

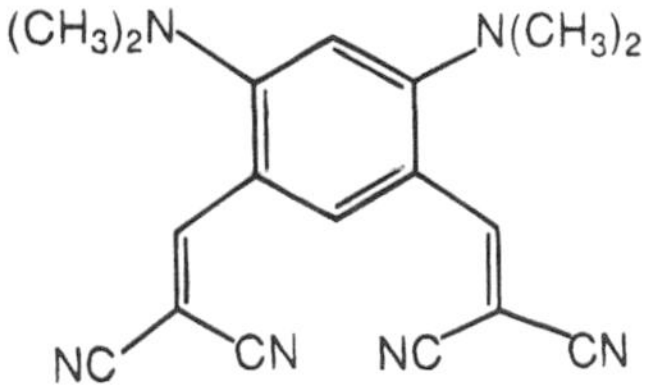

(DDCV)

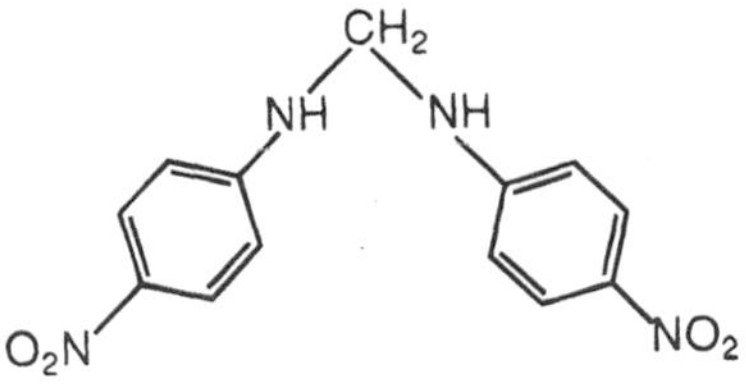

(p-NMDA)

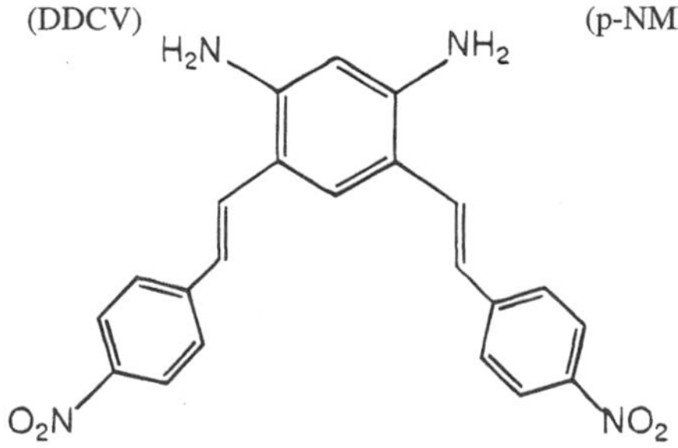

(DABNP)

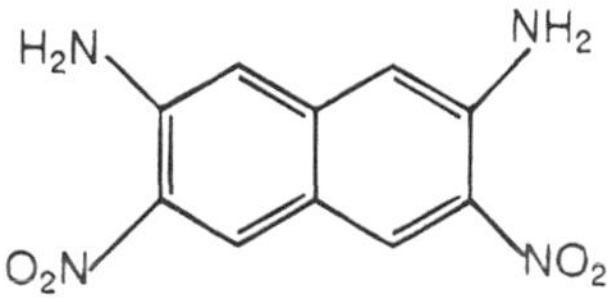

(MANN)

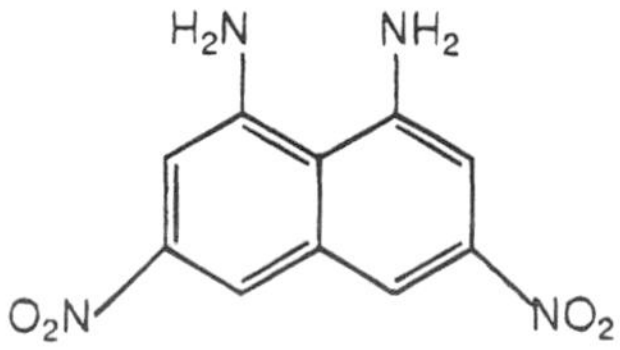

(OANN)

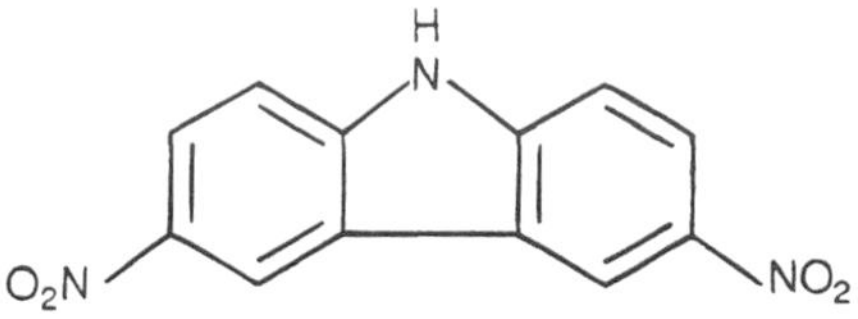

(DNCB)

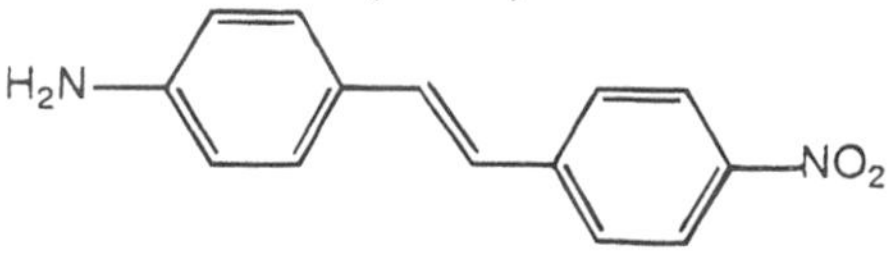

(ANST)

The second-order NLO properties of the alkyl chain derivatives of DADB consist of alkyl chain C_nH_{2n+1}, where $n = 1, 2, 3, 6, 8, 10, 11$ and 18, were also evaluated by using computational and experimental techniques to determine the possible role of alkyl chain length in the physical and NLO properties.[47,52,61] Suslick et al.[62] measured the hyperpolarizabilities of porphyrins having both electron donor and acceptor substituents which are also two-dimensional materials. The EFISH measurements were conducted at 1.19 μm by dissolving porphyrins in chloroform. The dipole moment and β values are affected by the position of donor and acceptor groups. The β value of porphyrin with $R_1 = R_2 = R_3 = NO_2$, $R_4 = NH_2$ was $\geq 10 \times 10^{-30}$ esu. In this case, only amino group pushes charge into

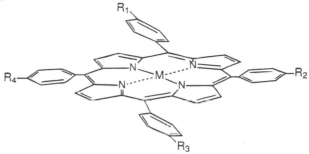

the macrocycle and is drained by three nitro groups, which leads to a lower β value. The porphyrin with $R_1 = R_2 = NO_2$, $R_3 = R_4 = NH_2$ exhibited a β value of 30×10^{-30} esu and dipole moment of 7×10^{-18} esu. The porphyrin with $R_1 = NO_2$, $R_2 = R_3 = R_4 = NH_2$ exhibited a β value of 20×10^{-30} esu and dipole moment of 5×10^{-18} esu. In this case, the charge is pushed into the π-conjugated ring by three amino groups. The charge transfer being well aligned yielded a large β value.

Moylan et al.[63] synthesized a variety of triaryl azole chromophores including triaryl-substituted imidazole, oxazole, thiazole, and phenanthro[9,10-*d*]imidazoled. The homophores, 2,4,5-triarylimidazoles 1a–i, 2,4,5-triaryloxazoles 2a c,j , and 2,4,5-triarylthiazoles 3a, e, are substituted with a donor group (D) at the *para*-position of both 4-aryl and 5-aryl rings and an acceptor group (A) at the *para*-position of the 2-aryl ring (Table 6). These triarylazole derivatives exhibit excellent thermal stability because weight loss at 300 °C was quite low. The amino donor groups cause a red-shift in the absorption spectra as well as lead to larger molecular hyperpolarizability compared with alkoxy and arylthio groups. A phenylacetylene linkage probably also leads to a red-shift in the absorption and to larger molecular hyperpolarizability. On the other hand, introduction of a phenylsulfonyl group causes a blue-shift in the absorption spectrum and decreases the hyperpolarizability relative to the nitro group. The oxazoles also showed the similar effect of the acceptor group as to imidazoles. Chromophore 2n, which contains both phenylsulfone and perfluoroalkyl acceptors, shows large hyperpolarizability and very high thermal stability up to 432 °C. The ground-state dipole moment and hyperpolarizability of thiazole derivatives were larger than substituted imidazoles and oxazoles. The molecular hyperpolarizabilities of the triaryl-substituted azole derivatives increased in an order imidazoles < oxazoles < thiazoles; however, the difference was not significantly large. This series of chromophores provides a good example of structure–property relationships.

Allen et al.[64] reported the hyperpolarizabilities of a family of molecules based on substituted dihydropyrazoles. The β values change significantly with substituent groups and increased conjugation (Table 7). The details of a single crystal 3-(1,1-dicyanoethenyl)-1-phenyl-4,5-dihydro-1*H*-pyrazole (DCNP) are mentioned in VII. This compound show a very large β and electro-optical coefficient with electro-optical figure of merit as 8.6×10^{-10} m/V and has a very low, reduced half-wave voltage of 370 V, about eight times smaller than lithium niobate. The wavelength dependences show that the electro-optical coefficient at 633 nm has advantage over lithium niobate although it reduces at longer wavelengths.

An example of structure–property relationship here is that provided by Nicoud and Twieg[65] for benzene derivatives showing the effect of substituents and chiral auxiliary. It is apparant that second-order NLO properties can be tailored by applying different donor groups and substituents (Table 8). In particular, additional groups such as COOH, OH, CN, and amide derivatives assist crystal packing by forming hydrogen bonding. Nitrobenzene and nitropyridine derivatives crystallize in noncentrosymmetric groups and hence are SHG active. Authors reported that nitropyridine derivatives from alaninol and phenylalaninol also exhibit very large powder SHG efficiencies in the range of NPP and PNP molecules. Molecular engineering plays an important role in designing new materials of interest.

Rao and his co-workers[66–68] reported the synthesis, electronic absorption, and first hyperpolarizabilities

Table 6 Chemical Structures, λ_{max} (nm), Dipole Moment and β Values (10^{-30} esu) of Donor–Acceptor-Substituted Triaryl Azole Derivatives

No.	X	Donor (D)	Acceptor (A)	Solvent	μ (D)	λ_{max}	β	λ (μm)
1a.	NH	OCH_3	NO_2	Dioxane	7.0	410	52.3	1.064
1b.	NH	OCH_3	$C\equiv C\text{-}PhNO_2$	Dioxane	8.1	400	69.1	1.064
1c.	NH	OCH_3	SO_2Ph	Dioxane	8.0	362	10.1	1.064
1d.	NH	$C\text{-}(CH_2)_5N$	NO_2	$CHCl_3$	7.2	438	53.2	1.064
1e.	NII	$OCH_2(C_2H_5)CHC_4H_9$	NO_2	$CHCl_3$	6.3	416	24.5	1.907
1f.	NH	$OCH_2(C_2H_5)CHC_4H_9$	$SO_2C_4F_9$	$CHCl_3$	6.5	384	13.9	1.064
1g.	NH	$C\equiv C\text{-}C_6H_4OCH_3$	NO_2	Dioxane	8.0	344	53.2	1.064
1h.	NH	$SPhOCH_3$	NO_2	Dioxane	6.6	398	55.9	1.064
1i.	NH	$C\text{-}(CH_2)_6N$	NO_2	$CHCl_3$	8.3	476	78.7	1.907
2a.	O	OCH_3	NO_2	Dioxane	6.3	390	47.3	1.064
2b.	O	OCH_3	$C\equiv C\text{-}PhNO_2$	Dioxane	7.1	378	63.9	1.064
2c.	O	OCH_3	SO_2Ph	Dioxane	7.0	358	17.1	1.064
2j.	O	OCH_3	C_6F_{13}	Dioxane	5.7	344	13.7	1.064
2k.	O	OCH_3	$SO_2C_4F_9$	Dioxane	7.9	378	31.1	1.064
2l.	O	OCH_3	CF_3	Dioxane	5.3	338	9.2	1.064
2m.	O	OCH_3	SO_2CF_3	Dioxane	7.0	352	15.6	1.064
2n.	O	OCH_3	SO_2PhCF_3	$CHCl_3$	8.0	370	31.5	1.064
3a.	S	OCH_3	NO_2	Dioxane	8.1	400	$\beta_0 = 21.1$	1.064

From Maylan, C. R., et al., *Chem. Mater.*, 5, 1499, 1993. With permission.

Table 7 Ground State Dipole Moment and β Values of Substituted Dihydropyrazoles

X	Y	μ_g (debye)	λ_{max} (nm)	β at 1.06 μm (10^{-30} esu)
C_6H_5	C_6H_5	3.6	438	100.5
NH_2	NO_2	8.7	415	49.5
$CH=CH\text{-}NH_2$	$CH=CH\text{-}NO_2$	9.7	506	1554
$C_6H_5\text{-}OCH_3$	$C_6H_5\text{-}NO_2$	7.2	485	624
$C_6H_5\text{-}NO_2$	$C_6H_5\text{-}OCH_3$	9.1	441	37.1
C_6H_5	$CH=C(CN)_2$	7.0	499	777.9

From Allen, S., et al., *J. Appl. Phys.*, 64, 2583, 1988. With permission.

Table 8 Power SHG Activities of Nitrobenzenes and Analogous Nitropyridine Derivatives

Donor	Z	R_1	R_2	SHG ($\times$ Urea)	Abbreviated Name[a]
NH_2	CH	H	H	0	*p*-NA
	CH	H	CH_3	22	MNA
	CH	H	Cl	40	CNA
	CH	H	Br	4	BNA
	CH	H	$NHCOCH_3$	20	
$NHCH_3$	CH	H	CH_3	80	MNMA
	CH	H	Cl	20	CNMA
$NH\text{-}(CH_2)_2\text{-}OH$	CH	H	COOH	15	
	CH	H	H	80	APNP
	N	H	H	3	
$NH\text{-}(CH_2)_3\text{-}OH$	CH	H	CH_3	8	
	CH	H	CF_3	30	
	CH	CF_3	H	80	
$N(CH_3)_2$	CH	H	$NHCOCH_3$	115	DAN
	CH	H	$NHCOCF_3$	70	
Cl	N	H	H	1	
	N	H	NO_2	5	
OCH_3	N	H	H	2	
$NH\text{-}(CH_2)_3\text{-}COOH$	CH	H	H	115	BANP
$NCH_3\text{-}CH_2\text{-}CN$	CH	H	H	140	NPAN
$NCH_3\text{-}(CH_2)_2\text{-}CN$	CH	H	H	85	NPPN

[a] MNA, 2-methyl-4-nitroaniline; CNA, 2-chloro-4-nitroaniline; BNA, 2-bromo-4-nitroaniline; MNMA, 2-methyl-4-nitro-*N*-methylaniline; CNMA, 2-chloro-4-nitro-*N*-methylaniline; APNP, *N*-(4-nitrophenyl)-3-amino-1-propanol; DAN, 3-acetamido-4-dimethylaminonitrobenzene; BANP, *N*-(4-nitrophenyl)-3-aminobutanoic acid; NPAN, *N*-(4-nitrophenyl)-3*N*-methylaminoacetonitrile.

From Nicoud, J. F. and Twieg, R. I., *Nonlinear Optical Properties of Organic Molecules and Crystals,* Chemla, D. S. and Zyss, J., eds. Academic, 1987. With permission.

of two series of donor–acceptor-substituted conjugated thiophene compounds. These donor–acceptor-substituted thiophene compounds exhibit two absorption bands. An intense lowest absorption band was observed in the visible region which is associated with the intramolecular charge-transfer transition caused by excitation from the donor to the acceptor group. This band denotes a large dipole moment change between ground and excited states and is solvatochromic in nature. The second weak absorption band was observed in the UV-visible region between 250 and 700 nm. The lowest absorption charge-transfer bands of various compounds and their chemical structures are listed in Table 9.

The β values of these compounds were measured by the EFISH technique in 1,4-dioxane at 1.907 μm. The compounds containing dicyanovinyl groups show larger $\beta\mu$ than those possessing nitro groups, because dicyanovinyl group is a strong electron acceptor. A comparison of dialkylamino-stilbenes indicates that the replacement of one benzene does not affect the $\beta\mu$ values, whereas the replacement of both benzene rings with thiophene rings causes a 2-fold increase in $\beta\mu$. The dialkylamino-dicyanovinyl stilbene compounds also show a similar trend. The replacement of both benzene rings in dicyanovinyl- and nitro-substituted stilbene compounds with thiophene rings drastically increases the magnitude of second-order optical nonlinearity. A comparison of thiophene-based compounds with stilbenes indicates that the $\beta\mu$ value is quite sensitive to the conjugating moieties that connect the donor and acceptor groups. The $\beta\mu$ values of thiophene compounds are 1.8 to 2.5 times larger than the corresponding

Table 9 The $\beta\mu$ Measured by EFISH Technique at 1.907 μm and Absorption Maxima Recorded in Dioxane of Some Thiophene-Based Donor–Acceptor Compounds and their Benzene Analogs

Compound	λ_{max} (nm)	$\beta\mu$ (10^{-48} esu)
Me_2N—⟨benzene⟩—NO_2	370	110
Me_2N—⟨benzene⟩—CH=CH—⟨benzene⟩—NO_2	424	580
Et_2N—⟨benzene⟩—CH=CH—⟨thiophene-S⟩—NO_2	478	600
Et_2N—⟨benzene⟩—(CH=CH—⟨thiophene-S⟩)$_2$—NO_2	506	1400
piperidino-⟨thiophene-S⟩—CH=CH—⟨benzene⟩—NO_2	460	660
piperidino-⟨thiophene-S⟩—CH=CH—⟨benzene⟩—NO_2	500	563
diethylamino-⟨thiophene-S⟩—CH=CH—⟨thiophene-S⟩—NO_2	516	1040
Et_2N—⟨benzene⟩—CH=C(CN)CN	419	300
Et_2N—⟨benzene⟩—CH=CH—⟨benzene⟩—CH=C(CN)CN	468	1100
Et_2N—⟨benzene⟩—CH=CH—⟨benzene⟩—C(CN)=C(CN)CN	594	2700
Et_2N—⟨benzene⟩—CH=CH—⟨thiophene-S⟩—CH=C(CN)CN	513	1300

Table 9—_Continued_

Compound	λ_{max} (nm)	$\beta\mu$ $(10^{-48}$ esu)
	640	6200
	584	2600
	558	1250
	650	2850
	718	1900
	662	9100
	547	2300
	653	7400

Table 9—*Continued*

Compound	λ_{max} (nm)	$\beta\mu$ $(10^{-48}$ esu)
	556	3800
	556	940
	594	1500
	625	1600
	660	3800
	604	1350
	635	330
	638	2400

From Rao, J. P., et al., *SPIE Proc.*, 1775, 32, 1992; Wong, K. Y., et al., *SPIE Proc.*, 1775, 74, 1992; Jen, A. K., et al., *J. Chem. Soc. Chem. Commun.*, 90, 1993; Jen, A. K., et al., *SPIE Proc.*, 2143, 30, 1994. With permission.

benzenoid stilbene compounds. Furthermore, a thiophene ring also causes a red-shift of about 100 nm in the absorption spectra. The thiophene-substituted donor–acceptor compounds with different conjugation lengths show a power law dependence of $\beta\mu$ on the conjugation length. The plot of $\beta\mu$ against the number of conjugated bonds for thiophene substituents show no saturation and the linear fit of the data yields a power law exponent of 1.8 for dicyanovinyl group series and 2.0 for the nitro group series. Jen et al.[69] reported a series of dithiolyldinemethyl derivatives for improving electron donor activity and thermal stability. The presence of sulfur substituents leads to greater donor ability of the dithiolyldinemethyl group which results into larger $\mu\beta$ values. The incorporation of a benzenoid ring in the dithiolyldinemethyl group causes a decrease in $\mu\beta$ values. A comparison of the $\mu\beta$ values suggest that the thiophene ring introduces more electron delocalization in donor–acceptor compounds than that of a benzenoid ring. Bithiophene compounds show better thermal stability at least higher by 50 °C than the stilbene compounds.

We have provided a few examples of structure–property relationships in Tables 4 through 9. Both physical and nonlinear optical properties of organic materials can be modified successfully by molecular engineering; many such examples are the basis of this chapter and can be found in preceding sections.

V. FACTORS AFFECTING SECOND-ORDER OPTICAL NONLINEARITY

The trade-off between optical transparency and SHG efficiency is an important issue while designing devices using NLO materials; this situation has been dealt by researchers in past years. As discussed earlier, hyperpolarizability of an organic molecule is related to the charge-transfer characteristics eventually governed by the increasing conjugation length and strength of donor and acceptor groups. Therefore a trade-off should exist between hyperpolarizability and optical transparency such as large β values can be obtained from highly polar molecules but that leads to a loss in optical transparency. Both second-order optical nonlinearity and optical transparency are affected by the nature of the conjugated bonds, length of π-conjugation, strength of electron donor and acceptor substituents and conformation, such factors are discussed below.

A. CONJUGATION LENGTH

Huijts and Hesselink[70] performed a systematic study on a series of disubstituted conjugated organic molecules to establish a relationship between the conjugation length and hyperpolarizability. The substituent groups for all molecules were the same, the methoxy (CH_3O) group as a donor and the nitro (NO_2) as an acceptor. EFISH studies were performed on *p*-methoxynitrobenzene, α-*p*-methoxyphenyl-ω-*p*-nitrophenyl polyene (MPNP with $n = 1, 2, 3, 4,$ and 5 for ethene to decapentaene). As one can see from the chemical structures, the length of conjugation considering the number of π-bonds between the two substituents CH_3O and NO_2 varies from two to nine where the phenyl ring was counted as two. The conjugation length dependence of $\mu\beta$ measured from the EFISH showed that $\mu\beta$ varies with a 3.4 power of the number of π-bonds, whereas β_0 corrected for resonance effects varies as the third power of the length of the conjugated π-electron system over the entire range of measurements. The β_0 of these compounds was estimated according to the two-level model. Therefore the length of π-conjugation has a very significant effect on the magnitude of hyperpolarizability (Table 10). From the

Table 10 Dipole Moment (μ), Electronic Transition Energy (ω_0), and Hyperpolarizabilities of CH_3O- and NO_2-substituted Compounds Showing the Effect of π-Conjugation Length on Optical Nonlinearity

Compound	Number of π-bonds	ω_0 (eV)	μ (D)	$\mu\beta$ (10^{-46} esu)	β (10^{-30} esu)	β_0 (10^{-30} esu)
1	2	4.08	4.9	0.30	6	8
2	3	3.58	5.5	0.64	12	12
3	5	3.45	5.7	4.6	81	78
4	6	3.16	6.0	8.1	135	106
5	7	3.02	6.6	18.1	274	188
6	8	2.88	6.7	24.6	367	211
7	9	2.85	7.0	43.6	623	340

From Huijts, R. H. and Hesselink, G. L. S., *Chem. Phys. Lett.,* 156, 209, 1989. With permission

viewpoint of an efficiency-transparency trade-off, although β increases remarkably as the length of π-conjugation increases, increased conjugation length causes a loss in optical transparency.

$$CH_3O-\langle\bigcirc\rangle-NO_2 \tag{1}$$

$$CH_3O-\langle\bigcirc\rangle-CH=CH-NO_2 \tag{2}$$

$$CH_3O-\langle\bigcirc\rangle-CH=CH-\langle\bigcirc\rangle-NO_2 \tag{3}$$

$$CH_3O-\langle\bigcirc\rangle-(CH=CH)_2-\langle\bigcirc\rangle-NO_2 \tag{4}$$

$$CH_3O-\langle\bigcirc\rangle-(CH=CH)_3-\langle\bigcirc\rangle-NO_2 \tag{5}$$

$$CH_3O-\langle\bigcirc\rangle-(CH=CH)_4-\langle\bigcirc\rangle-NO_2 \tag{6}$$

$$CH_3O-\langle\bigcirc\rangle-(CH=CH)_5-\langle\bigcirc\rangle-NO_2 \tag{7}$$

The π-conjugation length dependence of β has been investigated in push-pull polyenes and carotenoids of increasing π-conjugation length and substituted with donor and acceptor end groups.[71–73] The donor groups are dimethylaminophenyl (I-1 through I-5), benzodithia (II-1 through II-5), julolidinyl (III-1 through III-5), and ferrocenyl (IV-1 through IV-5); four different acceptor groups are formyl (CHO), dicyanovinyl [$CH=CH(CN)_2$], p-cyanophenylvinyl ($CH=CHC_6H_4CN$), and p-nitrophenylvinyl ($CH=CHC_6H_4NO_2$). The μβ were measured by the EFISH technique in chloroform whereas μβ(0) were calculated using the two-level model. The β value increases remarkably as the conjugation length of polyenic chain increases; the longest molecules showed exceptionally large μβ(0) (Table 11). The enhancement of μβ(0) as a function of the number of double bonds yields an exponent of 2.3. These molecules have 1-D delocalization of π-electrons along the chain, therefore the β tensor is 1-D mainly along the charge-transfer axis. Viewing in terms of nonlinearity–transparency trade-off, the polyenic conjugated chain is advantageous in the case of weak donor and acceptor end groups because β is affected by the strength of end groups. Two series of molecules show a bathochromic effect and hyperchromic effect as the length of polyene chain increases along with a substantial increase of μβ(0). The off-resonance μβ(0) of I and II molecules are 50 times larger than that of p-NA. The μβ values are significantly larger for longer chain conjugated molecules. The μβ(0) of III-5 having dicyanovinyl group is 17 times larger than that of DANS and one of the highest off-resonance values known so far. The incorporation of a triple bond in the middle of polyenic bridge leads to lower μβ values as shown by the p-cynaophenyl and p-nitrophenyl derivatives. The effect of triple bond depends on the end group

Table 11 Absorption Maximum Wavelength (λ) Recorded in Chloroform, Hyperpolarizabilities Measured by EFISH of Polyene Compounds Showing the Effect of π-Conjugation Length and Donor–acceptor Groups on β

Compound	Acceptor	λ_{max} (nm)	$\mu\beta(2\omega)$ (10^{-48} esu)	$\mu\beta(0)$ (10^{-48} esu)	Measurement Wavelength (μm)
I-1	CHO	372	30	20	1.34
	CH=CH(CN)$_2$	446	460	230	1.34
		446	230	170	1.91
	CH=CHC$_6$H$_4$CN	410	250	140	1.34
	CH=CHC$_6$H$_4$NO$_2$	452	1000	480	1.34
I-2	CHO	456	1200	570	1.34
	CH=CH(CN)$_2$	562	2850	1700	1.91
	CH=CHC$_6$H$_4$CN	465	1950	900	1.34
	CH=CHC$_6$H$_4$NO$_2$	488	2200	900	1.34
I-3	CHO	466	2200	1000	1.34
	CH=CH(CN)$_2$	540	3700	2300	1.91
	CH=CHC$_6$H$_4$CN	457	1600	750	1.34
	CH=CHC$_6$H$_4$NO$_2$	467	1660	750	1.34
I-4	CHO	485	2700	1100	1.34
I-5	CHO	500	7250	2800	1.34
II-1	CHO	384	320	200	1.34
	CH=CH(CN)$_2$	489	1030	710	1.91
II-2	CHO	450	2000	1000	1.34
	CH=CH(CN)$_2$	560	5500	3300	1.91
II-3	CHO	461	4200	1950	1.34
	CH=CH(CN)$_2$	538	4300	2700	1.91
II-4	CHO	474	4300	1900	1.34
	CH=CH(CN)$_2$	574	9400	5450	1.91
II-5	CHO	498	8900	3400	1.34
	CH=CH(CN)$_2$	588	13400	7500	1.91
III-1	CH=CH(CN)$_2$	520	1500	980	1.91
III-2	CH=CH(CN)$_2$	570	5500	3200	1.91
III-3	CH=CH(CN)$_2$	536	5800	3650	1.91
III-4	CH=CH(CN)$_2$	572	11000	6400	1.91
III-5	CH=CH(CN)$_2$	572	13700	8000	1.91
IV-1	CH=CH(CN)$_2$	320,526	90	60~80	1.91
IV-2	CH=CH(CN)$_2$	370,556	420	250~340	1.91
IV-3	CH=CH(CN)$_2$	456,568	1120	660~875	1.91
IV-4	CH=CH(CN)$_2$	458	4600	3300	1.91

From Barzoukas M., et al., *Chem. Phys.*, 133, 323, 1989; Blanchard-Desce, M., *Condensed Matter News*, 2, 12, 1993; Blanchard-Desce, M., et al., *SPIE Proc.*, 2143, 20, 1994. With permission.

and it may be useful from transparency point of view while retaining almost the same optical nonlinearity. This is somewhat visible in push-pull carotenoid containing formyl acceptor end group. The enhancement of $\mu\beta(0)$ is mainly contributed to an increase of $\beta(0)$ because the ground-state dipole moment may not be affected significantly with number of double bonds. The benzodithia group seems to be less effective for enhancing β than the dimethylamino group as an electron donor. For these donor–acceptor-substituted polyenes, the β does not level off for the length of up to 30 Å.

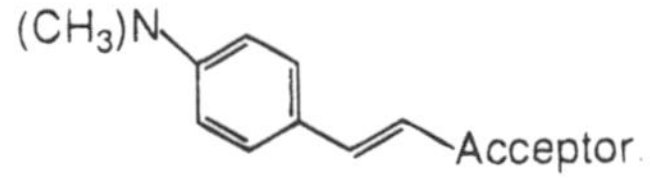

I-1

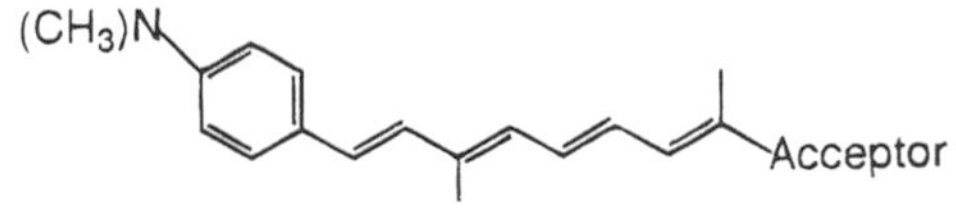

I-2

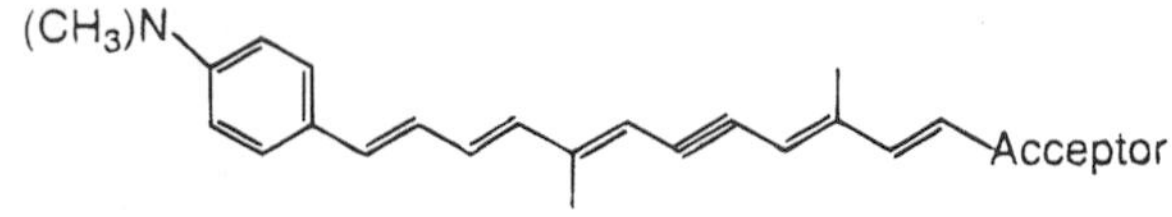

114

I-3

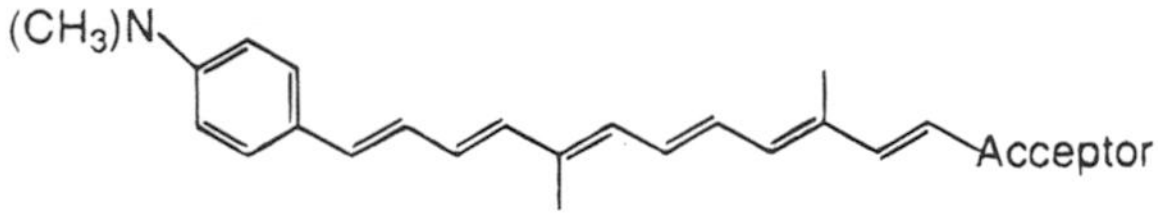

I-4

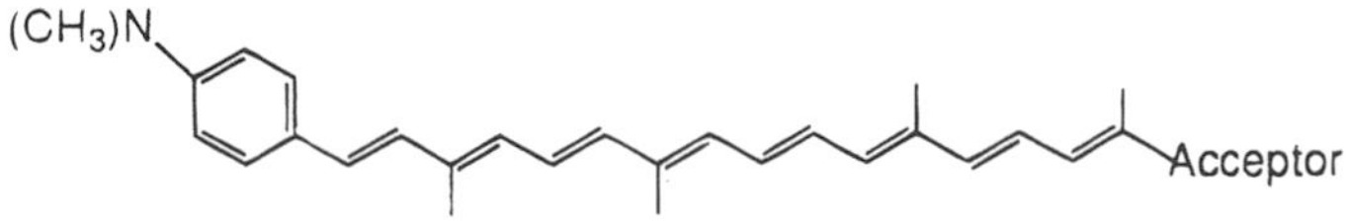

I-5

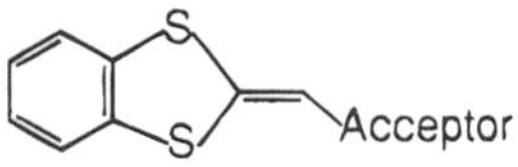

II-1

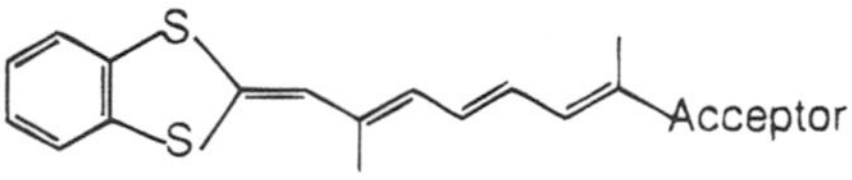

II-2

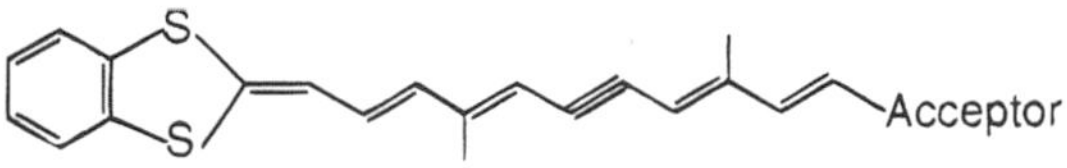

II-3

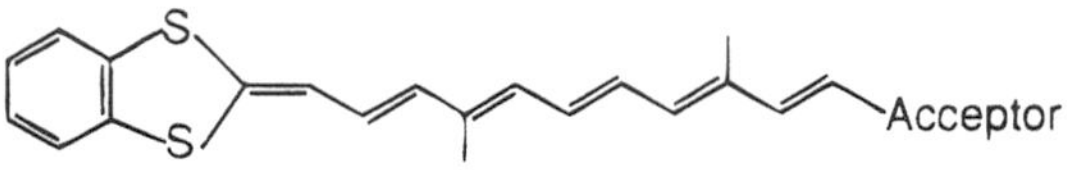

II-4

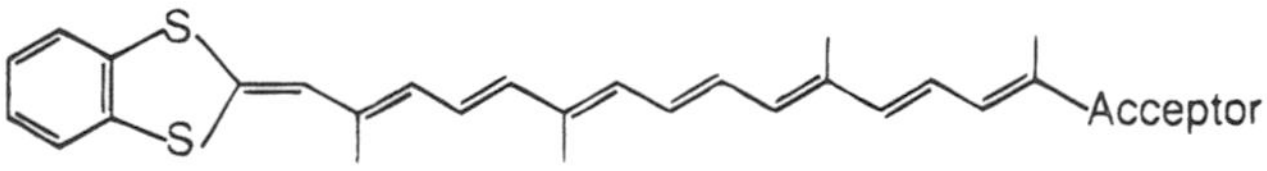

II-5

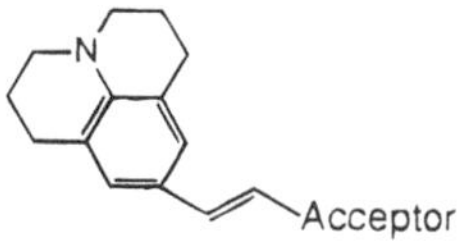

III-1

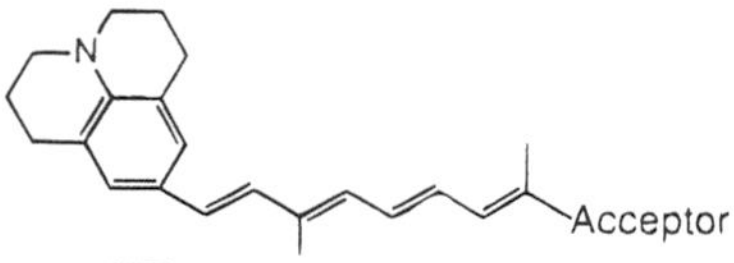

III-2

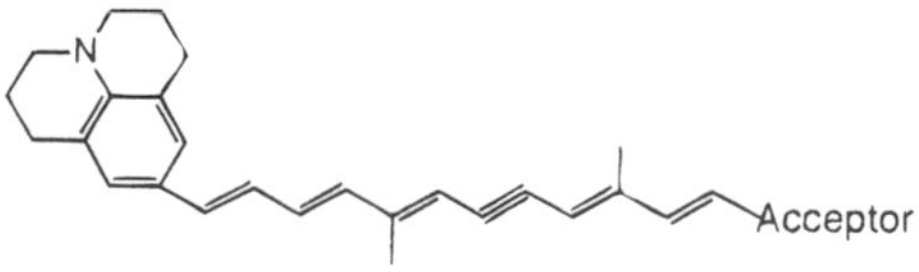

III-3

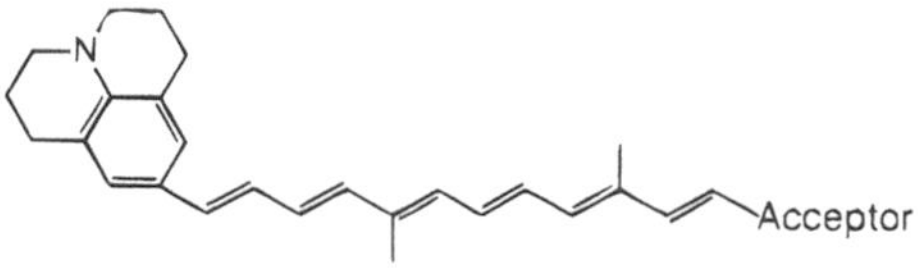

III-4

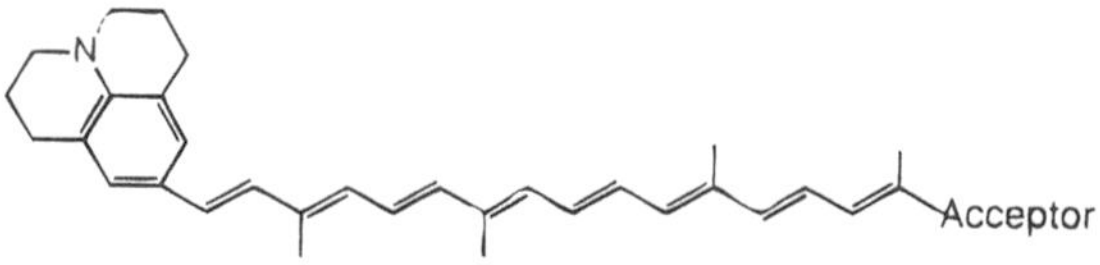

III-5

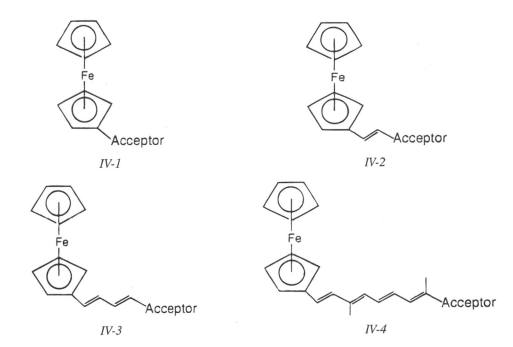

IV-1

IV-2

IV-3

IV-4

Chang et al.[74] reported the conjugation dependences of molecular hyperpolarizabilities for a wide range of donor–acceptor-substituted π-conjugated systems. Table 12 lists the dipole moment, α, β, and γ values of *para*-disubstituted benzenes. For polyphenyls β optimizes with $n = 2$–3 and biphenyl induces coplanarity, and a good combination of donor–acceptor groups gives best coplanarity.

Morley et al.[75] calculated hyperpolarizabilities of polyphenyls and polyenes containing a nitro group and a dimethylamino group located at opposite ends of the conjugated systems (Table 13). For polyphenyls, the β increases slowly with an increasing number of phenyl units at zero frequency (ω, o) and at an applied frequency of 1.17 eV. The β_x increases up to six phenyl rings although the effect per unit volume is optimized for 4-dimethylamino-4′-nitroterphenyl. Chang et al.[76] reported that measured β value of polyphenyls maximizes at two phenyl units and decreases as the number of phenyl rings increases. Therefore, both theoretical and experimental data support that large β values are difficult to obtain in polyphenyls compared with other π-conjugated systems. On the other hand, β increases rapidly with increasing ethenyl units for polyenes having the same donor and acceptor groups and the effect per unit volume is optimized for 20 units. Polyenes show β values 20 times larger than the β value of the polyphenyls.

The exponent (x) was determined for the conjugation length of donor–acceptor-substituted systems, which describes the power law dependence of β proportial L^x, where L is the delocalization length. The exponents of all these systems range between 1.9 and 3.5 (Table 14). The difference may occur because of computational and experimental techniques. The exponents of the α-phenylpolyene and α,ω-diphenylpolyene were found to be different. The exponents also depend on the combination of donor and acceptor groups and their value is related to the magnitude of the β values.

B. STRENGTH OF DONOR AND ACCEPTOR GROUPS

The research team at Du Pont has performed excellent studies to develop the relationship between organic molecular structures and intrinsic molecular optical nonlinearities and transparency trade-off. The effects of various factors such as strength of donor and acceptor groups, nature of conjugated systems, and π-conjugation length on polarizabilities and hyperpolarizabilities have been investigated using solution-phase dc EFISH and THG techniques.[74,76] The effect of various donor and acceptor groups on the magnitude of β and γ values was studied for different conjugated systems such as benzene, stilbene, styrene, biphenyl, tolane, and fluorene. The μ, α, β, and γ results on a variety of donor–acceptor-substituted π-conjugated systems are listed in Table 15. The following conclusions can be drawn from these experimental results: (1) The *para*-disubstituted benzenes show significant increase in β over the sum of monosubstituted benzenes due to charge-transfer; (2) the efficacies

Table 12 Absorption Maximum, Dipole Moment, α, β, and γ Values (in esu) of Para-disubstituted Benzenes

Acceptor	Donor	Solvent	n	λ_{max} (nm)	μ (10^{-18})	α (10^{-23})	β (10^{-30})	γ (10^{-36})
		Donor—[benzene]—(CH=CH)$_n$—[benzene]—Acceptor						
CN	OCH$_3$	Chloroform	1	340	3.8	3.4	19	54
			2	360	4.5	3.8	27	122
			3	380	4.4	4.4	40	234
NO$_2$	OCH$_3$	Chloroform	1	376	4.5	3.4	34	93
			2	397	4.8	4.0	47	130
			3	414	5.1	4.2	76	230
			4	430	5.8	4.8	101	
NO$_2$	N(CH$_3$)$_2$	Chloroform	1	430	6.6	3.4	73	225
			2	442	7.6	4.0	107	
			3	458	8.2	4.2	131	
		Donor—[benzene]—(C≡C)$_n$—[benzene]—Acceptor						
CN	SCH$_3$	Chloroform	1	333	4.0	3.5	15	35
			2	330	4.7	3.8	17	42
NO$_2$	SCH$_3$	Chloroform	1	362	4.0	3.8	20	95
			2	338	3.9	4.0	17	61
		Donor—[benzene]$_n$—Acceptor						
NO$_2$	NH$_2$	NMP	1	370	7.8	1.6	10	21
			2	372	7.8	2.6	24	96
			3	360	7.6	3.5	16	124
			4	344	10	3.3	11	133
NO$_2$	OCH$_3$	p-Dioxane	1	302	4.6	1.5	5.1	10
			2	332	4.5	2.8	9.2	39
			3	340	5.0	3.8	11	
		Donor—[benzene]—(CH=CH)$_n$—Acceptor						
CHO	OCH$_3$	Chloroform	1	318	4.0	2.5	12	28
			2	350	4.3	3.0	28	43
			3	376	4.6	3.5	42	120
CHO	N(CH$_3$)$_2$	Chloroform	1	384	5.6	2.6	30	63
			2	412	6.0	3.3	52	140
			3	434	6.3	4.0	88	257
CHC(CN)$_2$	N(CH$_3$)$_2$	Chloroform	1	486	8.4	3.2	82	
			2	520	9.0	3.6	163	

From Chang, L. T., et al., *J. Phys. Chem.*, 95, 10643, 1991. With permission.

Table 13 CNDO Calculated Dipole Moments and Hyperpolarizabilities (10^{-30} cm^5 esu^{-1}) of Polyphenyls and Polyenes

Chain Length (n)	O_2N–⬡$_n$–$N(CH_3)_2$		O_2N–polyene$_n$–$N(CH_3)_2$	
	$\beta_x(\omega,0)$	$\beta_x(\omega, 1.17\text{ eV})$	$\beta_x(\omega,0)$	$\beta_x(\omega, 0.656\text{ eV})$
1	12.29	24.56		
2	31.35	83.91	14.80	19.49
3	42.17	130.47		
4	47.74	155.45	80.58	133.89
5	49.36	163.81		
6	52.95	176.30	237.8	520.0
7	53.0	180.21		
8	54.07	182.86	491.8	1445.2
9	53.51	179.17		
10			827.0	3307.7
12			1211.4	6809.8
16			1888.4	21475.3
20			2468.2	66960.1
24			2823.5	175409.0

From Morley, et al., *J. Chem. Soc. Perkin Trans. II,* 1351, 1987. With permission.

Table 14 Exponents of Different Donor–Acceptor Systems

Donor—⬡–(polyene)$_n$–⬡—Acceptor

Conjugated System	Donor	Acceptor
1	$N(CH_3)_2$	NO_2
2 and 7	OCH_3	NO_2
3	OCH_3	CN
4	$N(CH_3)_2$	$CH(CN)_2$
5	$N(CH_3)_2$	CHO
6	OCH_3	CHO
8 and 9	$N(CH_3)_2$	CHO
10	$C_6H_4S_2$	CHO

Conjugated System	Length (n)	Wavelength (μm)	X				Ref.
			$\beta(\omega)$	β_0	$\mu\beta(\omega)$	$\mu\beta_0$	
1	2, 3, 5, 6, 7, 8	1.907	1.9	1.9	2.2	2.1	74
2	2, 3, 5, 6, 7, 8	1.907	2.2	2.0	2.3	2.2	74
3	2, 3, 5, 6, 7, 8	1.907	2.5	2.3	2.4	2.3	74
4	2, 3, 5, 6, 7, 8	1.907	2.4	2.1	2.6	2.3	74
5	2, 3, 5, 6, 7, 8	1.907	2.8	2.7	3.1	2.9	74
6	2, 3, 5, 6, 7, 8	1.907	3.3	3.2	3.6	3.5	74
7	2, 3, 5, 6, 7, 8	1.06	3.1	2.6	3.3	2.8	70
8	2, 3, 4, 5, 6	–	2.7				75
9	3, 6, 8, 10	1.34			2.7	2.4	71
10	1, 4, 6, 8	1.34			2.6	2.3	71

Table 15 Absorption Maximum, Dipole Moment, α, β, and γ Values (in esu) of *para*-Disubstituted Benzenes

Acceptor	Donor	Solvent	λ_{max} (nm)	μ (10^{-18})	α (10^{-23})	β (10^{-30})	γ (10^{-36})
		Donor—⬡—Acceptor					
SO_2CH_3	OH	*p*-Dioxane	290	3.4	1.7	1.3	3
CN	CH_3	Neat		4.4	1.5	0.7	5
CN	Cl	*p*-Dioxane		2.3	1.6	0.8	5
CN	Br	*p*-Dioxane		2.4	1.8	1.1	7
CN	OC_6H_5	*p*-Dioxane		4.1	2.6	1.2	9
CN	OCH_3	*p*-Dioxane	248	4.8	1.7	1.9	4
CN	SCH_3	*p*-Dioxane		4.4	2.0	2.8	9
CN	NH_2	*p*-Dioxane	270	5.0	1.6	3.1	6
CN	$N(CH_3)_2$	*p*-Dioxane	290	5.6	2.1	5.0	10
CHO	CH_3	Neat		3.0	1.6	1.7	7
CHO	OC_6H_5	Neat	269	2.8	2.5	1.9	12
CHO	OCH_3	Neat	269	3.5	1.7	2.2	8
CHO	SCH_3	Neat	310	3.1	1.9	2.6	13
CHO	$N(CH_3)_2$	*p*-Dioxane	326	5.1	2.0	6.3	18
i-$SO_2C_3H_7$	OCH_3	Chloroform	290	5.4	2.7	3.3	5
$COCF_3$	OCH_3	*p*-Dioxane	292	3.5	2.9	3.6	12
$COCF_3$	OC_6H_5	*p*-Dioxane	292	4.0	2.0	3.6	7
$COCF_3$	$N(CH_3)_2$	*p*-Dioxane	356	5.9	2.4	10	16
NO	$N(CH_3)_2$	*p*-Dioxane	407	6.2	2.1	12	
NO_2	CH_3	*p*-Dioxane	272	4.2	1.6	2.1	8
NO_2	Br	*p*-Diaxane	274	3.0	1.8	3.3	
NO_2	OH	*p*-Dioxane	304	5.0	1.5	3.0	8
NO_2	OC_6H_5	*p*-Dioxane	294	4.2	2.6	4.0	9
NO_2	OCH_3	*p*-Dioxane	302	4.6	1.5	5.1	10
NO_2	SCH_3	*p*-Dioxane	322	4.4	1.9	6.1	17
NO_2	N_2H_3	*p*-Dioxane	366	6.3	1.8	7.6	9
NO_2	NH_2	Acetone	365	6.2	1.7	9.2	1.5
NO_2	$N(CH_3)_2$	Acetone	376	6.4	2.2	12	28
NO_2	CN	*p*-Dioxane		0.9	1.7	0.6	7
NO_2	CHO	*p*-Dioxane	376	2.5	1.7	0.2	7
$CHC(CN)_2$	OCH_3	*p*-Dioxane	345	5.5	2.4	9.8	30
$CHC(CN)_2$	$N(CH_3)_2$	Chloroform	420	7.8	2.8	32	
$CHC(CN)_2$	Julolidine	CH_2Cl_2	458	8.0	3.0	44	
$C_2(CN)_3$	NH_2	CH_2Cl_2	498	7.8	3.4	39	
$C_2(CN)_3$	$N(CH_3)_2$	CH_2Cl_2	516	8.2	3.7	50	
$C_2(CN)_3$	Julolidine	CH_2Cl_2	556	8.5	3.9	60	
		Donor—⬡—〓—Acceptor					
CN	OCH_3	Chloroform	304	4.2	2.3	7.0	11
CN	$N(CH_3)_2$	Chloroform	364	6.0	2.8	23	29
CHO	Br	Chloroform	298	2.0	2.3	6.5	26
CHO	OCH_3	Chloroform	318	4.2	2.5	11	28
CHO	$N(CH_3)_2$	Chloroform	384	5.6	2.6	30	63
NO_2	OH	Chloroform	312	5.1	2.4	18	52
NO_2	OCH_3	Chloroform	352	4.6	2.6	17	35
NO_2	$N(CH_3)_2$	Chloroform	438	6.5	3.2	50	

Table 15—*Continued*

Acceptor	Donor	Solvent	λ_{max} (nm)	μ (10^{-18})	α (10^{-23})	β (10^{-30})	γ (10^{-36})

Donor⟶[stilbene: two phenyl rings linked by a vinylene bridge]⟶Acceptor

Acceptor	Donor	Solvent	λ_{max} (nm)	μ (10^{-18})	α (10^{-23})	β (10^{-30})	γ (10^{-36})
$SO_2C_6F_{13}$	OCH_3	p-Dioxane	347	7.8	4.8	14	93
$COCF_3$	OCH_3	p-Dioxane	368	4.2	3.9	16.4	83
CN	OH	p-Dioxane	344	4.5	3.2	13	52
CN	OCH_3	Chloroform	340	3.8	3.4	19	54
CN	$N(CH_3)_2$	Chloroform	382	5.7	3.9	36	125
NO_2	OH	p-Dioxane	370	5.5	3.3	17	104
NO_2	OCH_3	p-Dioxane	364	4.5	3.4	28	79
		Chloroform	370	4.5	3.4	34	93
NO_2	SCH_3	Chloroform	380	4.3	3.8	34	100
NO_2	$N(CH_3)_2$	Chloroform	427	6.6	3.4	73	225
NO_2	Julolidinamine	Chloroform	438	7	4.5	96	

Donor⟶[tolane: two phenyl rings linked by an acetylene (triple bond) bridge]⟶Acceptor

Acceptor	Donor	Solvent	λ_{max} (nm)	μ (10^{-18})	α (10^{-23})	β (10^{-30})	γ (10^{-36})
SO_2CH_3	NH_2	Chloroform	338	6.5	3.4	13	59
CO_2CH_3	NH_2	Chloroform	332	3.8	3.7	15	62
$COCH_3$	SO_2CH_3	Chloroform	334	3.3	3.2	12	29
CN	NH_2	Chloroform	342	5.2	3.2	20	55
CN	$NHCH_3$	Chloroform	358	5.7	3.4	27	90
CN	$N(CH_3)_2$	Chloroform	372	6.1	3.7	29	99
NO_2	OCH_3	p-Dioxane	356	4.4	3.9	14	52
NO_2	SCH_3	Chloroform	362	4.0	3.8	20	95
NO_2	NH_2	Chloroform	380	5.5	3.2	24	120
		NMP	410	5.5	3.6	40	140
NO_2	$NHCH_3$	Chloroform	400	5.7	4.0	46	130
NO_2	$N(CH_3)_2$	Chloroform	415	6.1	4.1	46	151

Donor⟶[biphenyl: two phenyl rings directly linked]⟶Acceptor

Acceptor	Donor	Solvent	λ_{max} (nm)	μ (10^{-18})	α (10^{-23})	β (10^{-30})	γ (10^{-36})
$SO_2C_6H_{12}OH$	$N(CH_3)_2$	Chloroform	340	6.0	4.6	13	38
CN	OH	p-Dioxane	292	4.8	2.5	6.3	10
NO_2	OCH_3	p-Dioxane	332	4.5	2.8	9.2	39
NO_2	NH_2	Chloroform	372	5.0	2.8	24	70
NO_2	$N(CH_3)_2$	Chloroform	390	5.5	3.4	50	130

Donor⟶[fluorene ring system]⟶Acceptor

Acceptor	Donor	Solvent	λ_{max} (nm)	μ (10^{-18})	α (10^{-23})	β (10^{-30})	γ (10^{-36})
NO_2	Br	p-Dioxane	330	2.8	2.6	6.0	30
NO_2	$N(CH_3)_2$	p-Dioxane	410	5.6	3.6	40	95
		Chloroform	417	6.0	3.6	55	
NO_2	OCH_3	p-Dioxane	356	4.7	2.7	11	28

From Chang L. T., et al., *J. Phys. Chem.*, 95, 10643, 1991; *J. Phys. Chem*, 95, 10631, 1991. With permission.

Table 16 Optical Maxima (λ_{max}) in Nanometer, Dipole Moments (μ) in Debye, and EFISH-Measured β Values in 10^{-30} esu of Donor–Acceptor Substituted Stilbene and Tolane Chromophores

Donor (D)	Acceptor (A)	λ_{max}	μ	β	β_0	λ_{max}	μ	β	β_0
CH_3O	NO_2	364	4.5	59.6	27.0	–	–	16.0	–
CH_3O	SO_2CH_3	335	6.1	9.1	5.0	310	5.9	10.8	7.1
CH_3O	SO_2CF_3	350	6.6	34.1	17.3	327	6.2	20.8	11.7
$(CH_3)_2N$	NO_2	424	6.2	83.0	52	402	7.1	102	37.5
$(CH_3)_2N$	SO_2CH_3	376	6.9	65.6	28.7	358	7.5	39.8	19.3
$(CH_3)_2N$	SO_2CF_3	–	–	–	–	388	8.4	5.6	34.8
CH_3S	NO_2	378	5.1	68.0	29.4	–	–	–	–
CH_3S	SO_2CH_3	–	–	–	–	320	5.2	15.9	9.2
C_6H_5S	SO_2CH_3	344	4.4	19.4	10.1	–	–	–	–
CH_3S	NO_2	364	4.4	57.6	27.0	–	–	–	–
H	NO_2	–	–	–	–	326	4.6	16.0	9.1
H	SO_2CH_3	–	–	–	–	310	5.3	3.75	2.27

From Stiegman, A. E., et al., *J. Am. Chem. Soc.*, 113, 7658, 1991; Burland, D. M., et al., *J. Appl. Phys.*, 71, 410, 1992. With permission.

of acceptor groups increase in the order of SO_2CH_3, CN, CHO, $COCF_3$, NO, NO_2, $CHC(CN)_2$, and $C_2(CN)_2$; (3) with nitro acceptor group, the relative effectiveness of various donor groups in an increasing order are: OCH_3 < OH < Br < OC_6H_5 < OCH_3 < SCH_3 < N_2H_3 < NH_2 < $N(CH_3)_2$, and the julilodine amine; (4) the magnitude of optical nonlinearities depends on the strength of donor–acceptor groups and the best combination of donor–acceptor groups provides about 10 times enhancement, the aldehyde group is better than cyano whereas NO and nitro have almost the same effect; (5) the β values are quite large for disubstituted benzenes with $CHC(CN)_2$, $C_2(CN)_2$, dimethylamino, and julolidine groups; (6) the β value of a donor–acceptor-substituted tolan is about half that of the respective stilbenes because acetylene introduces strong bond alternation and because of the lack of conjugation between donor and acceptor groups; (7) the charge-transfer through the biphenyl promotes coplanarity; (8) the β values of donor–acceptor-substituted fluorenes are enhanced relative to biphenyls; (9) the β values of donor–acceptor-substituted styrenes fall between those of benzene and stilbenes.

Stiegman et al.[77] synthesized a series of donor–acceptor-substituted acetylene compounds by systematically varying both the conjugation length and the strength of donor–acceptor groups. Although the strength of donor–acceptor groups increases the magnitude of β, it does not increase while increasing the conjugation length from one to two acetylene linkers; however, it increases sharply from two to three acetylene linkers. Burland et al.[78] evaluated the role of double bond in stilbene (CH=CH) system vs. triple bond in tolane (C≡C) system. The trade-off between second-order optical nonlinearity and transparency was examined by comparing the effect of conjugation and donor–acceptor groups on hyperpolarizability and optical transparency. The β values were measured by the EFISH technique at 1.064 μm and dipole moments in p-dioxane solution. Table 16 lists the optical maxima, dipole moments, and β values of donor–acceptor-substituted stilbene and tolane chromophores. Tolane with methoxy and sulfone trifluromethyl seems the most promising NLO chromophore, and sulfone-substituted tolanes nearly satisfy the material requirements for SHG.

Kondo et al.[79] evaluated the SHG activities of the phenylpyridylacetylenes and indicated the role of counter anion and donor–acceptor groups (Table 17). The compounds having week electron donor and acceptor groups at the 4-position of benzene ring resulted into high SHG efficiencies. For example, the powder SHG efficiency of (4-iodophenyl)(4-pyridyl) acetylene and (4-methoxyphenyl)(4-pyridyl) acetylene were 15 and 5 times larger that of urea at 1.064 μm. The SHG efficiency of (4-chlorophenyl)(4-methylpyridiniumyl) acetylene methyl sulfate was 7 times larger than that of urea. Authors pointed out the role of counter anion in controlling the noncentrosymmetric crystal packing. The λ_{cutoff} wavelengths

Table 17 Powder SHG activities and λ_{cutoff} (nm) of Phenylpyridylacetylene Compounds

Compound	X	Y	SHG (× urea)	λ_{cutoff} (nm)
	OCH$_3$	–	5.1	345
	I	–	15	352
	Br	–	2.0	320
	Cl	–	1.0	320
	OCH$_3$	–	<0.1	377
	Cl	–	<0.1	361
	NO$_2$	–	0	410
	Br	C$_7$H$_7$SO$_3^-$	<0.1	380
	Cl	C$_7$H$_7$SO$_7^-$	<0.1	378
	Br	CH$_3$SO$_4^-$	1.0	383
	Cl	CH$_3$SO$_4^-$	7.0	380
	Cl	C$_6$H$_2$N$_3$O$_7^-$	<0.1	478
	OCH$_3$	C$_6$H$_2$N$_3$O$_7^-$	<0.1	478
	Cl	Cl	0	370

From Kondo., et al., *J. Org. Chem.*, 57, 1623, 1992. With permission.

of phenylpyridylacetylenes were as low as 320 nm for the weak donor–acceptor and as high as 410 nm for strong donor–acceptor groups. The picrate anion raised λ_{cutoff} wavelengths to 487 nm indicating the role of counter anion. These compounds have rather low λ_{cutoff} wavelengths and could be useful for blue-green laser diodes.

The effect of donors and acceptors can also be seen from a comparative study of benzenoid stilbene compounds with corresponding thiophene analogs.[66-69] Of great importance were found to be the dithiolyldinemethyl derivatives for enlarging $\mu\beta$ values as well as thermal stabilities.

C. NATURE OF π-BONDING SEQUENCE

The nature of π-conjugated bond also significantly affects the magnitude of optical nonlinearities. The effect of orbital hybridization and bond alternation on μ, α, β and γ values has been examined for series of α,ω-diphenylpolyene charge-transfer deratives.[74,76] Table 18 compares the optical nonlinearities of the acetylenic and ethylenic linkages having identical donor–acceptor groups. For all combinations of donor and acceptor groups, the *trans*-stilbene derivatives show significantly large optical nonlinearities, with β and γ values about 40% to 50% larger than those of the diphenylacetylenic analogs. The electron-rich π-orbitals of the sp-hybridized acetylenic carbons may be responsible for the less effective π-delocalization due to the orbital energy mismatch with the π-orbitals of the sp^2-hybridized phenyl carbons. In addition, the π-orbital overlap is optimized due to the short triple bond length in the acetylenic resonance structure. Contrary to this, all carbons are sp^2 hybridized in the *trans*-stilbene which permit effective delocalization of all the π-electrons.

The EFISH measurements of two series of push-pull polyenes; 4-dimethylaminophenyl-polyene aldehydes and benzodithia-polyene aldehydes showed sharp increase in β values.[71] When a triple bond is introduced in a polyene chain having the same donor and acceptor groups, the β values decrease for the corresponding π-conjugated molecules. The β values of the following molecules can be compared with the indentical molecules. This indicates that the triple bond introduces a structural and electronic inhomogeneity in the π-conjugated system which causes a decrease in the β values.

Table 18 Optical Nonlinearities of the Acetylenic and Ethylenic Linkages Having Identical Donor–Acceptor Groups; λ_{max} (nm), μ(Debye), $\beta(10^{-30}$ esu), $\gamma(10^{-36}$ esu)

		D—⬡—=—⬡—A				D—⬡—≡—⬡—A			
Acceptor (A)	Donor (D)	λ_{max}	μ	β	γ	λ_{max}	μ	β	γ
CN	$N(CH_3)_2$	382	5.7	36	125	372	6.1	29	99
NO_2	Br	344	3.2	14	98	335	3.0	10	50
NO_2	OCH_3	364	4.5	28	79	356	4.4	14	52
NO_2	SCH_3	374	4.3	26	113	362	4.0	20	95
NO_2	$N(CH_3)_2$	427	6.6	73	225	415	6.1	46	151

From Chang L. T., et al., *J. Phys. Chem.*, 95, 10643, 1991; Chang, L. T. et al., *J. Phys. Chem.*, 95, 10631, 1991. With permission.

$$\lambda_{max} = 461 \text{ nm, } \mu\beta(2\omega) = 4200 \times 10^{-48} \text{ esu, } \mu\beta(0) = 2000 \times 10^{-48} \text{ esu}$$

$$\lambda_{max} = 466 \text{ nm, } \mu\beta(2\omega) = 2200 \times 10^{-48} \text{ esu, } \mu\beta(0) = 1000 \times 10^{-48} \text{ esu}$$

A comparative study of benzenoid stilbene compounds with corresponding thiophene analogs showed that a thiophene moiety results to more electron delocalization in donor–acceptor compounds than that of a benzenoid moiety.[66–69] The $\mu\beta$ values of thiophene compounds were up to 2.5 times larger than that of the corresponding benzenoid stilbene compounds.

D. SUBSTITUTION

It is also very important to know on how molecular substitutions affect optical nonlinearity. The following examples show the effects of heteroatoms, electron donor and acceptor group substitution in benzene and stilbene systems.[80] Table 19 lists the dipole moment, α, β, and γ values of substituted benzenes and stilbenes. The drastic perturbations of the π-conjugation such as aza and perfluoro substitution on benzenes and azo and azomethine substitution for stilbenes significantly lowered β values. The differences in bond length and hybridization between the acetylenic carbons and the phenyl carbons cause a decrease in optical nonlinearities when compared with analogous stilbene derivatives. Therefore heteroatoms and side-group-substituted benzene and stilbene derivatives show reduced optical nonlinearities due to the substitution.

Nicoud and Twieg[65] reported the hyperpolarizabilities of heterocyclic analogs of donor–acceptor benzene and stilbene (Table 20). Hyperpolarizabilities of these derivatives were found to be influenced by the location of nitrogen atom. A hypsochromic shift of about 30 nm/N occurred for analogs of pyridine or pyrimidine derivatives derived from p-NA. The β values of stilbene analogs change significantly with the incorporation of nitrogen atom either in the phenyl ring or in π-conjugated bridge and absorption spectral features and dipole moments are affected similarly. Therefore second-order nonlinear optical properties can be tailored by exercising chemical modification strategies.

VI. STRATEGIES FOR GENERATING NONCENTROSYMMETRIC STRUCTURES

As indicated earlier, the majority of organic materials crystallize in centrosymmetric space groups and, consequently they fail to exhibit second harmonic generation. In particular, most achiral organic molecules possessing large ground-state dipole moments tend to crystallize into a centrosymmetric space group restricting their use for second-order nonlinear optics. To generate SHG active organic structures, different techniques have been used to remove the center of symmetry. Various chemical as well as physical strategies have been implemented to introduce noncentrosymmetric crystal structures capable of displaying SHG activity through intermolecular forces. Therefore while designing organic materials for SHG, one has to consider chemical and physical means to engineer molecules into noncentrosymmetric structures. The various strategies that have been used to form acentric crystal structures are as follows: (1) chirality[57,59]; (2) hydrogen bonding[57,82,83]; (3) Steric hinderance[84]; (4) Langmuir-Blodgett (LB) technique[85–87]; (5) guest–host systems[88–90]; (6) electrical poling[91,92]; (7) co-crystallization[93]; (8) crystal growth technique[94]; (9) solvents for recrystallization[95,96]; (10) lamda (Λ)-type conformation[97,98]; (11) reduced dipole-dipole interaction[100]; and, (12) organometallic structures.[101]

These methods have been useful for ensuring a dipolar alignment favorable to generating SHG activity. Several techniques will be discussed here that deal with poled polymers and LB films. Although molecular interactions are large in single crystals, unfortunately, the majority of organic materials tend to crystallize with a center of symmetry which precludes second-order nonlinear optical effects. A survey of organic materials in connection with noncentrosymmetric space group population has been tabulated by Nicoud and Twieg.[65] The most populated noncentrosymmetric

Table 19 Absorption Maximum, Dipole Moment, α, β, and γ Values of Substituted Benzenes and Stilbenes

X	Substitution	λ_{max} (nm)	μ (10^{-18} esu)	α (10^{-23} esu)	β (10^{-30} esu)	γ (10^{-36} esu)
	3,Aza; 4,NH$_2$		6.5	1.7	3.7	11
	2,Aza; 4,OCH$_3$		3.5	1.5	2.2	10
	3,CH3; 4,NH$_2$		6.2	1.9	8.7	17
	2,Cl; 4,NH$_2$	350	5.9	1.8	6.8	12
	3,OCH3; 4,NH$_2$		6.0	1.8	8.7	19
	2,F; 4,OCH$_3$	304	4.4	1.8	2.5	10
	2,5,F; 4,OCH$_3$	304	4.9	1.8	2.6	12
	2,3,5,6F; 4,OCH$_3$	270	4.2	1.8	1.7	7
H	1,Aza	346	4.4	2.7	4.9	26
CH$_3$	1,Aza	351	4.7	3.5	15	77
OCH$_3$	1,Aza	376	4.4	3.3	14	76
OCH$_3$	1',Aza	349	5.4	2.5	6.6	30
NH$_2$	1,1',Azo	420	5.8	3.5	29	170
NEtEtOH	1,1'Azo	455	6.0	3.8	32	190
OCH$_3$	3,CH$_3$	366	5.2	3.7	26	96
OCH$_3$	3,OCH$_3$	380	4.7	3.7	23	117
OCH$_3$	3,F	363	4.1	3.4	18	92
OCH$_3$	2,OCH$_3$	395	4.8	3.9	32	11
OCH$_3$	1'CN	361	5.3	3.8	21	86
Br	1',CN	340	4.6	2.9	8.0	30
Br	1,Aza;1',CN	382	4.1	3.6	2.1	59
OCH$_3$	2'NO$_2$	390	4.7	3.7	22	110

The Solvent for all these Compounds was *p*-dioxane except 3, Aza; 4, NH$_2$ where the solvent was Acetone (From Cheng, L. T., et al., *SPIE Proc.*, 1147, 61, 1989. With permission.)

Table 20 Calculated Absorption Maxima, Dipole Moment, and Hyperpolarizability of Heterocyclic Analogs of Donor–Acceptor Substituted Benzene and Stilbene

Compound	λ_{max} (nm)	μ_o (D)	β (10^{-30} esu)
	325	8.2	35.3
	310	8.2	33.9
	355	8.7	46.7
	283	7.2	25.6
	354	8.2	47.6
	349	9.5	34.8
	384	10.0	248.0
	410	6.0	371.0
	337	13.3	77.0
	335	9.4	106.0

From Nicoud J. F. and Twieg, R. J., *Nonlinear Optical Properties of Organic Molecules and Crystals,* Chemla, D. S. and Zyss, J., Eds., Vol. 1, Academic, New York, 1987, Chap. II-3, p. 277. With permission.

space groups include P2$_1$, Pna2$_1$, C$_2$, Cc, and P2$_1$2$_1$2$_1$. A comparison of noncentrosymmetric space group population surveys reveals that approximately 70% of the total number of compounds investigated, crystallize into centrosymmetric space groups. As a result, a wide variety of organic compounds are unable to display second-order nonlinear optical properties. To overcome this major problem, a number of molecular engineering approaches have been suggested because this is still an important issue that should be addressed by researchers involved in the field of nonlinear optics. A few molecular concepts have been implemented during the evolution of organic materials for nonlinear optics. Oudar and Hierle[59] pointed out that the addition of a chiral radical ensures a noncentrosymmetric crystal structure and methyl-(2,4-dinitrophenyl)-aminopropanoate (MAP) is a conceptual structure. This hypothesis was supported further by Zyss et al.[57] with the discovery of N-(4-nitrophenyl)-l-prolinol (NPP). The simultaneous chiral and hydrogen-bonding character of the prolinol moiety which acts as electron donor results in a quasioptimal angle. Barzoukas et al.[58] reported that a highly polar group grafted to an achiral molecule at a position away from the donor–acceptor charge-transfer polar axis helps in introducing noncentrosymmetricity. Both chirality and hydrogen bonding are effective in designing noncentrosymmetric crystal structures. Steric hindrance is an another important molecular engineering strategy as demonstrated by Levine et al.[84] Two molecular designing approaches of co-crystallization and lambda (Λ)-shape molecules as discussed below have also been invented. Langmuir-Blodgett technique and electrical poling of a NLO dye in glassy polymer matrix, which ensure the noncentrosymmetric structure desired for quadratic nonlinear properties, have been discussed in detail in preceding sections.

Yamamoto et al.[97] reported that organic molecules which attain a lambda (Λ)-shape conformation tend to crystallize into noncentrosymmetric space groups. This new concept has been substantiated by synthesizing a series of methanediamine derivatives. Table 21 lists the chemical structures, physical properties, and powder SHG efficiencies of a variety of Λ-shape molecules. In these molecules, two aniline derivatives are bonded together via methylene bridge (−CH$_2$−). For example, in N, N'-bis-(4-nitrophenyl)methane-diamine (NMDA), two p-NA units are linked by a methylene linkage. Interestingly, the parent compounds of 4-nitroaniline (1), N-methyl-4-nitroaniline (3), 2-methoxy-4-nitroaniline (4), 2-amino-4-nitrotoluene (5), and 4-aminobenzoic acid methyl ester (6) show no SHG activity, whereas, 2-methyl-4-nitroaniline (2) and 4-aminobenzoic acid ethylester (7) show SHG efficiencies of 60 and 8 times that of urea. Condensation of these parent compounds with formaldehyde in methanol yields corresponding methanediamine derivatives. All of these methanediamine derivatives are SHG active. The p-NA derivative abbreviated as NMDA shows powder efficiency as high as 80 times larger than of urea. Theoretical calculations by the MOPAC AM1 method demonstrate that the optimum conformation of NMDA folds down and two aromatic rings are twisted, therefore NMDA molecule attains a Λ-shape conformation as shown in Figure 6a. Λ-shape molecule crystallizes into a noncentrosymmetric crystal structure because it easily stacks along one direction, which means that all the molecular dipoles are parallel to the crystal axis (Figure 6b). Theoretical calculation of N, N'-bis-(4-ethoxycarbonylphenyl)methanediamine (ECPMDA) yields d$_{36}$ coefficients of 48.5 × 10^{-9} esu (20.3 pm/V) and molecules attain a Λ-shape configuration. The present study extended to other Λ-shape molecules confirms this novel idea of introducing noncentrosymmetric crystal structures. Out of all the investigated compounds having a Λ-shape conformation, more than 75% crystallize with the lack of center of symmetry, hence they are SHG active. A noncentrosymmetric structure can also be generated by poling of guest–host systems. Langmuir-Blodgett technique is another way of assembling organic structures into a desirable configuration capable of displaying SHG. These two techniques have been discussed in section VII.

Okamoto et al.[93] investigated that p-NA on mixing with its N-alkyl derivatives shows SHG from the powders. The parent p-NA derivatives in which the hydrogen atom of the amino group is substituted by methyl, ethyl, propyl, butyl, isopropyl, hexyl, dimethyl, diethyl, and phenyl functionalities show either very weak SHG or no SHG itself. When p-NA is mixed with its N-alkyl derivatives discussed above, the co-crystals exhibit large SHG efficiency. SHG efficiency of the mixtures depends strongly on the weight ratio of the compounds. For example, the mixtures having weight ratios of 0:1, 1:1, 1:0,1, 1:0.2, and 1:0.5 of p-NA to N-isopropyl-4-nitroaniline show powder efficiencies of 1.5, 250, 163, 280, and 416 times larger than that of urea, respectively.[93] Tasaka et al.[98] reported SHG in a mixture of p-NA and its N-alkyl derivatives. In general, mixtures having a typical eutectic phase diagram showed large SHG activities. The powder SHG of the p-NA/isopropyl-NA mixture was 1670 times larger than that of urea. Table 22 lists the maximum powder SHG efficiencies of other derivatives and a similar

Table 21 Chemical Structures, Physical Properties, and Powder SHG Efficiencies of a Variety of Λ-Shape Molecules

Aniline Derivative	T_m (°C)	λ_{Cutoff} (nm)	SHG (for urea)	Λ-Shaped Methanediamine Derivative	T_m (°C)	λ_{Cutoff} (nm)	SHG (× urea)
O_2N–C$_6$H$_4$–NH_2	147	470	0	$(O_2N$–C$_6$H$_4$–$NH)_2$–CH_2	240	460	80
O_2N–C$_6$H$_3$(CH_3)–NH_2	131	480	60–150	$(O_2N$–C$_6$H$_3$(CH_3)–$NH)_2$–CH_2	260	450	6.3
O_2N–C$_6$H$_4$–NH_2–CH_3	106	490	0	$(O_2N$–C$_6$H$_4$–$N(CH_3))_2$–CH_2	242	470	6
O_2N–C$_6$H$_4$–NH_2, OCH_3	140	490	0	$(O_2N$–C$_6$H$_3$(OCH_3)–$NH)_2$–CH_2	254	480	2.5
O_2N–C$_6$H$_3$(NH_2)(CH_3)	118	510	0	(O_2N, CH_3-phenyl–$NH)_2$–CH_2	241	480	3
CH_3O–C(=O)–C$_6$H$_4$–NH_2	114	330	0	$(CH_3O$–C(=O)–C$_6$H$_4$–$NH)_2$–CH_2	215	330	30 / 0
C_2H_5O–C(=O)–C$_6$H$_4$–NH_2	90	330	8	$(C_2H_5O$–C(=O)–C$_6$H$_4$–$NH)_2$–CH_2	195	330	25
C$_6$H$_4$(NH_2)(NO_2)	71	500	0	(C$_6$H$_4$(NO_2)–$NH)_2$–CH_2	195	490	0

From Yamamoto H., et al., *J. Chem. Soc. Jpn. Chem. Indust. Chem.*, 7, 789, 1990. With permission.

behavior has been observed. The X-ray diffraction studies demonstrate the formation of a new noncentrosymmetric crystal structure entirely different from the parent compounds. These findings suggest a new avenue of introducing acentric structures by co-crystallization.

Nalwa et al.[99] applied the co-crystallization concept to *N*-9-octadecenyl-4-nitroaminobenzene (ODNA) and *p*-NA mixtures using chloroform as a solvent. A 2:1 mixture of *N*-9-octadecenyl-4-nitroaminobenzene to *p*-NA show powder efficiency as high as 135 times larger that of urea. The SHG efficiency of ODAN:*p*-NA mixtures of 1:1, 1:3, and 1:4 were 30, 38, and 77 times that of urea, respectively. The 1:2 and 1:10 mixtures showed weak SHG efficiency. On the other hand, 3:1 mixture

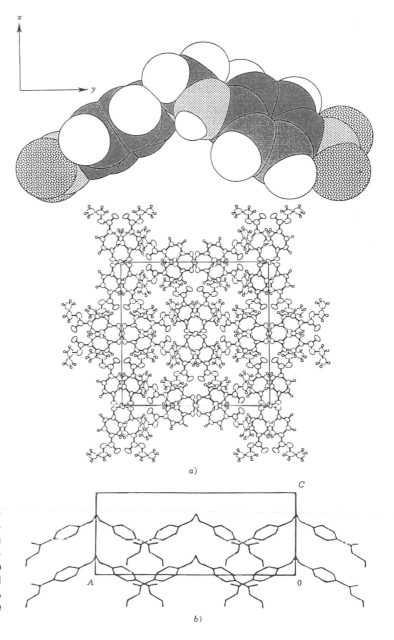

Figure 6 (A) Optimized conformation of N, N'-bis (p-nitrophenyl)-methanedlamlne (p-NMDA) that requires a Λ-shape with permanent dipole moment parallel to the molecular X axis. (B) Crystal structure of a Λ-shaped ECPMDA molecules (A) as viewed along the C axis, (B) along the B axis.

of ODAN:p-NA showed SHG efficiency 121 times that of urea. The SHG efficiency of the mixture decays with time as shown in Figure 7. The mixtures having higher ODNA contents than that of p-NA in the mixtures were found to be better candidates. The SHG activities of 2:1 and 3:1 mixtures of ODAN:p-NA stabilize after a day or two. The SHG activity begins to increase for 1:4 and 1:3 mixtures of ODAN:p-NA after storing for a few days after an initial decrease. Similar results have been confirmed for other nitroaniline derivatives such as N, N'-dialkyl-4,6-dinitro-1,3-diaminobenzenes although their SHG efficiencies were much lower. Therefore, the co-crystallization technique seems to be an effective way to generate noncentrosymmetric crystal structures exhibiting SHG activities.

$$O_2N-\!\!\!\!\bigcirc\!\!\!\!-N(CH_2)_8CH=CH(CH_2)_7CH_3 \quad + \quad \text{p-NA}$$
$$\underset{H}{|}$$

SHG active

Table 22 Powder SHG Efficiencies of Co-crystals of *p*-NA and Its *N*-Alkyl Derivatives

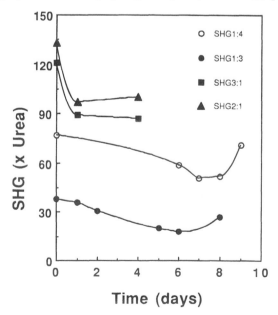

No.	N-Alkyl Derivatives		Powder SHG N-Alkyl p-NA (× urea)	SHG of 1:1 Mixture of p-NA/N-Alkyl p-NA (× urea)
	X	Y		
Methyl	H	CH_3	0.01	0.0
Ethyl	H	C_2H_5	0.03	0.1
n-Propyl	H	C_3H_7	0.05	25
isoPropyl	H	$CH(CH_3)_2$	2.5	1000
n-Butyl	H	C_4H_9	0.4	28
n-Hexyl	H	C_6H_{12}	0.15	39
Dimethyl	CH_3	CH_3	0.3	58
Diethyl	C_2H_5	C_2H_5	0.0	8
Phenyl	H	C_6H_5	0.0	0.0

From Tasaka, S., et al., *Jpn. J. Appl. Phys.*, 30, 296, 1991. With permission.

VII. SECOND-ORDER NONLINEAR OPTICAL MATERIALS

In recent years, many significant achievements have occurred in the field of nonlinear optics because of the development of laser technology and new nonlinear optical materials of both inorganic and organic types. Most of the scientific developments applied to the organic molecules were borrowed from their counterpart inorganic materials and, by the virtue of its nature, the field of nonlinear optics emerged along the same route. Although it is an additional option, it seems appropriate to provide first a brief introduction to the inorganic second-order NLO materials and their NLO devices which would be the basis of developing photonic technologies from organic molecules and polymers. The phenomenon of SHG in inorganic materials, first reported in 1961, led to the development of recent NLO materials such as lithium niobate (LNB; $LiNbO_3$), potassium niobate ($KNbO_3$), barium titanate ($BaTiO_3$), potassium titanyl phosphate (KTP; $KTiOPO_4$), potassium dihydrogen phosphate

Figure 7 Decay of the SHG efficiency of various mixtures of ODNA and *p*-NA. Mixtures 1:3 and 1:4 are for *p*-NA:ODNA whereas 3:1 and 2:1 are for ODNA: *p*-NA, respectively.

(KDP; KH_2PO_4), potassium didueterium phosphate (KD*P; KD_2PO_4), lithium iodate ($LiIO_3$), β-barium borate (BBO; β-BaB_2O_4), $Ba_2NaNb_5O_{15}$, etc. and a variety of semiconductors such as gallium arsenide (GaAs), cadmium sulfide (CdS), cadmium selenide (CdSe), cadmium telluide (CdTe), cadmium germanium arsenide ($CdGeAs_2$), gallium phosphide (GaP), zinc oxide (ZnO), zinc germinium phosphide ($ZnGeP_2$), and tellurium (Te). The $CdGeAs_2$ and $AgGaSe_2$ are appropriate infrared NLO materials, although they suffer from low laser damage threshold ($\sim$10 MW cm^{-2}) and poor optical transparency.[102] From application view points, $LiNbO_3$ has emerged as one of the most interesting SHG materials because optical waveguide SHG devices compatible with laser diodes and electro-optic waveguide devices for light switching and fast modulation have been fabricated using it. An excellent review on the crystal structure, crystal growth, linear and nonlinear optic, electro-optic, and dielectric proprties as well as on emerging new applications of KTP has been written by Bierlein and Vanherzeele.[103] KTP is also interesting for various sum and difference frequency and optical parametric applications from 0.35 to 4.5 μm and as an optical waveguide modulator. Despite the moderate level of optical damage threshold ($\sim$100 MW cm^{-2}) for KTP and $LiNbO_3$, they have emerged as the superior NLO materials for nonlinear optic and electro-optic devices including modulators, parametric oscillators, and optical switches. The single crystals of BBO and LBO have been used for efficient generation of sum and difference frequencies, higher harmonic generation, and optical parametric oscillators. Mei et al.[104] reported SHG properties of $MBe_2BO_3F_2$ (MBBF) where M is sodium and potassium. The KBBF belongs to R32 space group and has $d_{11} = -d_{22} = 2.5\ d_{36}$ (KDP) and cutoff wavelength as short as 155 nm. NBBF has a powder SHG of 1.7 times that of ADP and shows a cutoff wavelength of 160 nm. Both KBBF and NBBF are promising NLO crystals for vacuum UV applications.

The ideal material that could have potential applications in nonlinear optical devices should possess a combination of physicochemical properties as summarized below. In recent years, interest has been rapidly growing in organic NLO materials. The superiority of organic NLO materials results from their versatility and the possibility of tailoring them for a particular end use. An ideal nonlinear optical material should possess the following characteristics;

1. Large nonlinear figure of merit for frequency conversion
2. High laser damage threshold
3. Fast optical response time
4. Wide phase matchable angle
5. Architectural flexibility for molecular design and morphology
6. Ability to process into crystals, thin films
7. Optical transparency (no absorption at fundamental and SH wavelengths)
8. Ease of fabrication
9. Nontoxicity and good environmental stability.
10. High mechanical strength and thermal stability.

Compared with inorganic NLO materials, organic materials may fulfill many of these requirements, but there are also some drawbacks to organic NLO materials, such as environmental stability, mechanical strength, performance at low and high temperatures, and economic aspects. For example, the classical KTP has a melting point of 1180 °C, whereas organic materials have much lower melting temperature. Because organic materials offer a very wide range of chemical modification, some of these problems can easily be overcome. Recent growing efforts in molecular engineering suggest that organic NLO materials possess comparably better NLO properties than inorganic materials, hence, their tremendous practical potentials have been anticipated.

Optical SHG in an organic material was reported for the first time in 1964 by two research groups, in 3,4-benzopyrene by Rentzepis and Pao,[105] and in hexamethylenetetraamine single crystals by Heilmer et al.[106] Two years later, Orlov[107] reported SHG from the hippuric acid. Further research in organic NLO was stimulated in 1970, when Davydov and co-workers[15] reported strong SHG from organic compounds having electron donor and acceptor groups attached to a benzene ring while screening a large variety of organic compounds using the Kurtz and Perry powder method. This discovery created a ground for the basic understanding of the organic structure that may display large SHG effects and it aroused considerable interest in searching for new organic materials that possess large second-order optical nonlinearity, which are the most prevalent nowadays. Nonlinear organic materials have been a subject of intense research in the past decade. Depending on the habits, second-order NLO materials can be classified into several different categories such as: (1) single crystals; (2) organometallic compounds; (3) Langmuir-Blodgett films; (4) self-assembled systems; (5) liquid crystals; (6) poled polymers.

(3,4-benzopyrene)

(hippuric acid)

Every effort has been made to include all materials reported in the literature, although some may have been left out to which the authors did not have access. A majority of bulk materials are summarized along with single crystals because they are structurally related and are helpful in developing structure–property relationships. Likewise, some of the organometallic compounds have been listed under single crystals category whereas others are under a separate heading. Poled polymers have been divided into NLO-dye functionalized polymers, cross-linked and ferroelectric polymers categories being structurally different. Liquid crystals and ultrathin materials for SHG are characteristically different due to their chemical structures and way of processing.

A. SINGLE CRYSTALS AND RELATED BULK MATERIALS

Many NLO single crystals have been identified as potential candidates in optical and electro-optical devices. Currently NLO inorganic single crystals have been used in a variety of photonic applications and similar uses of NLO organic single crystals have been expected in industry. In this section, we will describe crystal growth of NLO organic materials, briefly giving the various processes of their purification, crystal growth, and characterization. Excellent monographs describing growth and characterization of molecular crystals have been written by Badan et al.,[108] Penn et al.,[109] and Karl.[110] Various techniques that have been used in growing bulk single crystals are described here. In growing bulk single crystals of organic materials, various steps involving ultrapurification of crude product, crystal growth, cutting, and polishing are involved.

1. Purification of Materials

High purity of the material is an essential prerequisite for crystal growth. Therefore the first step in the crystal growth is the purification of an organic material in appropriate solvents. Impurities as low as possible at the scale of $10 \sim 100$ parts per million are required. The high solubility of the material is important and purification needs repetition of the recrystallization process in an appropriate solvent. Although the chromatographic techniques of high-performance liquid chromatography (HPLC) or gas chromatograpgy (GC) can be used for purification, they yield a very small quantity of purified product per cycle. In particular, zone refining, sublimation, and distillation methods have been frequently used for ultrapurification to obtain large quantities of materials. In a recrystallization process, the material is dissolved in a hot solvent, then gradually cooled down to crystallize. Of great importance is the appropriate selection of organic solvents used in the recrystallization process. Recrystallization is the most common technique of purifying organic materials. Organic solvents generally used in the purification process of organic NLO materials such as 2-cyclooctylamino-5-nitropyridine (COANP) from 4:1 isoctane/toluene mixture,[111] 4-(N,N′-dimethylamino)-3-acetamidonitrobenzene (DAN) from tetrahydrofuran,[112] 2-(α-methylbenzylamino)-5-nitropyridine (MBA-NP) from toluene/isooctrane mixture,[113] methyl-3-nitro-4-pyridine-1-oxide (POM) from acetone and water,[114] m-nitroaniline (m-NA) from acetone,[115] NPP from 2:1 chloroform/cyclohexane or toleune[57] and chloronitrobenzene (CNB) from acetone.[116] Zone refining method has been found to be the most effective in producing ultrapure materials at a scale of parts per billion. Sublimation is used for thermally unstable solid materials that possess good volatility at temperatures below their decomposition points. Sublimation is generally performed at 10^{-5} to 10^{-6} torr to avoid decomposition.

2. Crystal Growth Techniques
a. Vapor Growth Methods

Sublimation is one of the simplest techniques to obtain bulk crystals of high purity for NLO devices. This method is suitable for obtaining impurity-free single crystals because the crystals are grown either in a vacuum or inert gas atmosphere. The details of growing single crystals by sublimation have been given by Karl.[110] High optical quality single crystals of various NLO materials were grown by this method.

Table 23 Solvent Effect on the Morphology of MBA-NP Crystals

Solvent	Dipole Moment $(10^{-30}$ cm)	Solubility	Morphology	Quality
Acetone	10.01	High	Prismatic	Fair
Ethanol	5.7	Medium	Prismatic	Excellent
Methanol	5.5	Medium	Prismatic	Excellent
Toluene	1.3	Low	Triangular	Good
o-Xylene	2.0	Low	Plate	Good

From Sherwood, J. N., *Proceedings of the First International Workshop on Crystal Growth in Organic Materials,* 1989, p. 180. With permission.

b. Melt Growth Methods

This is one of the most common methods used to obtain organic single crystals and a necessary condition is the thermal stability of the material to be used. Various techniques such as Bridgman-Stockbarger (BS), Kyropoulos, and Czochralsky methods are well known. Single crystals of *m*-nitroaniline, *m*-dinitrobenzene, NPP, MAP, and other materials have been grown from melt growth methods. The Bridgman-Stockbarger method has been used to grow single crystals of *m*-nitroaniline, *m*-dinitrobenzene, 2-methyl-4-nitro-*N*-methylaniline (MNMA), MAP, and NPP.

c. Solution Growth Methods

This technique is useful in providing constraint-free single crystals, and identification of a suitable solvent is the primary step in this procedure. A supersaturated solution containing a solute concentration exceeding the equilibrium value on the solubility curve and with metastable domains is used. In particular, slow cooling of saturated solutions, solvent evaporation, and temperature difference between regions of growth and dissolution processes have been used to grow single crystals from solutions. High-quality POM, MBA-NP, MAP, 4-aminobenzophenone, and *m*-NA single crystals were grown by this technique; more details are given by Penn et al.[109] Sherwood[117] reported the effect of solvent on the morphology of MBA-NP crystals (Table 23). Methanol was found to be the most suitable solvent for producing excellent quality crystals with some inperfection in the (010), polar, sectors. Crystals of MBA-NP prepared by the slow cooling of saturated methonal solutions yielded crystals in sizes up to $7 \times 7 \times 10$ cm^3 of high visual quality.

Crystals grown by the melt and vapor techniques suffer from mechanical damage introduced by the contacts of crystals with the walls of the container and this is a barrier to increased crystal perfection. In that case, solution growth technique is free of constraints. Organic crystals are mechanically soft compared with inorganic crystals because they are bound by weak Van der Waals forces, therefore, their cleavage, cutting and polishing is relatively difficult.

3. Spectroscopic Characterization

Single crystals obtained by different crystal growth techniques are spectroscopically characterized using polarization microscopy, X-ray diffraction, optical absorption spectroscopy, refractive index, and NLO susceptibility measurements. An excellent text on nonlinear optical single crystals covering both inorganic and organic crystals was published in 1991 by Dmitriev et al.[118] Substantial progress has been made in the development of organic nonlinear optical crystals because organic materials offer more possibilities of chemical modification and structural tailoring than inorganic materials. In this section, an attempt has been made to provide the details on organic crystals that have been reported up to 1995. Because the number of single crystals is so large, compiling them required a tremendous effort. In this section, the physical and second-order NLO properties of single crystals and related derivatives are described.

4. Single Crystal Materials
(1) Urea

Chemical structure[119]: $H_2N\text{-}CO\text{-}NH_2$
Point group: P42$_1$m Crystal system = tetragonal, positive uniaxial crystal: $n_e < n_o$
Melting point = 133 ~ 135 °C; transparency range: 0.2 ~ 1.8 μm, λ_{cutoff} = 200 nm

Table 24 Refractive Indexes,[122,123] Breakdown Threshold,[124] and Linear Absorption[125] of Urea Crystals

λ (μm)	n_o	n_e	α (cm^{-1})	tp (ns)	I (10^9 W cm^{-2})
0.213	1.7308	2.0155		10	0.5
0.266	1.5777	1.7575	0.04		
0.355	1.5207	1.6580		10	1.4
0.532	1.4939	1.6098		10	3.0
1.064	1.4811	1.5830	0.02	10	5

Powder SHG = 2.5 times that of ADP, d_{14} = 1.40 pm/V; NLO coefficients: d_{36} = 1.30 pm/V at 1.06 μm[120] Hyperpolarizability (β) = 0.36 × 10^{-30} esu; electro-optical coefficients: r_{63} = 0.83 pm/V, r_{41} = 1.9 pm/V[121]

Table 24 lists more details of urea crystals.

Twieg and Jain[81] synthesized new urea derivatives. The powder SHG efficiency of both DNP-Sc and DNPU was 8.8 times that of urea. The DNPU probably acquires a Λ-shape conformation similar to that reported by Yamamoto et al.[97] Urea can also be represented by a Λ-type conformation. Pyridylurea showed no SHG. Dimethylaminophenylurea showed SHG efficiency 10 times that of urea.[81]

(DNP-SC)   (DNPU)

Ledoux and Zyss[126] studied β values of urea, monomethylurea, dimethylurea, and tetramethylurea in different solvents. The β values were found to be solvent dependent. The β values of urea in water, DMF, and dimethyl sulfoxide (DMSO) were 0.45 × 10^{-30}, 0.3 × 10^{-30}, and 0.29 × 10^{-30} esu, respectively. The β values of monomethylurea in water, DMF, and DMSO were 0.45 × 10^{-30}, 0.65 × 10^{-30}, and 0.43 × 10^{-30} esu, respectively. The dimethylurea showed β values of 0.6 × 10^{-30}, 0.28 × 10^{-30}, and 0.18 × 10^{-30} esu in water, DMF, and DMSO, respectively. The β values of tetramethylurea was 0.41 × 10^{-30} esu.

(2) N-(2-Butenyl)-p-nitroaniline and Related Derivatives

Chemical structure[127]

Point group: orthorhombic: space group = Pna2$_1$
Unit cell parameters: a = 7.457 Å, b = 18.282 Å, c = 7.347 Å, V = 1001 Å^3
Melting point = 66.5 °C, powder SHG = 172 × urea

SHG activities of single crystals grown by slow crystallization from different solvents were 172 (ethanol), 160 (toluene), 195 (ethyl acetate), 166 (1-propanol), 177 (1,2-dimethoxyethane), 205 (pyridine), 157 (chloroform), 175 (methyl ethyl ketone), 171 (acetonitrile), and 255 (ethyl acetate/hexane) times that of urea. The SHG activities tend to increase on storing at room temperature for several months. The SHG activities were also found to change with annealing time. The mixed crystals of p-NA with N-(2-butenyl)-p-nitroaniline showed SHG activities of 177, 222, 209, 1190, 1650, 1770, and 2010 times that of urea for the 0.0, 0.2, 0.4, 0.6, 0.7, 0.8, and 0.9 mole fraction of p-NA, respectively, mixed in solution rather than in mixed crystals. The SHG activities measured after 8 days on standing at room temperature decreased. The SHG activities also changed with state of mixing p-NA in solution or in solid state, because the crystals were found to be different from those crystallized from solution.

Table 25 Refractive Indexes[130] and Breakdown Threshold[131] of *m*-NA Crystals

λ (μm)	n_x	n_y	n_z	tp (ns)	I (10^8 W cm^{-2})
0.5123	0.718	1.750	1.810		
0.530	1.705	1.738	1.798		
0.6274	1.670	1.709	1.758		
1.06	1.631	1.678	1.719	20	2.0
1.54	1.616	1.667	1.700		

The single crystals of *N*-(2-propenyl)-*p*-nitroaniline, *N*-(3-methyl-2-butenyl)-*p*-nitroaniline, and *N*-butyl-*p*-nitroaniline showed no SHG or very low SHG activities. The addition or subtraction of a methyl group and butyl group seem unfavorable for packing, whereas the 2-butenyl group plays an important role in packing. The SHG activities of the mixed crystals of *p*-NA with *N*-(2-propenyl)-*p*-nitroaniline, *N*-(3-methyl-2-butenyl)-*p*-nitroaniline, and *N*-butyl-*p*-nitroaniline were 377, 877, and 39 for 0.5-mol fraction, 1440, 1630, and 54 for 0.7-mol fraction, and 1,420, 2,150 and 540 times that of urea for 0.9-mol fraction of *p*-NA, respectively.

Okamoto et al.[128] reported large SHG efficiency from mixed powder and *p*-NA thin films deposited on isopropyl nitroaniline (IPNA). The mixed powder of *p*-NA and IPNA (10:1) showed d_{eff}(*p*-NA + IPNA) = 0.9 d_{eff}(MNA) = 230 pm/V. The refractive indexes of the composite were n_ω = 1.71 and $n_{2\omega}$ = 1.79. Thin film of *p*-NA deposited on IPNA layer by vacuum evaporation also showed large SHG efficiency, d_{eff}(*p*-NA on IPNA) = 1.1 d_{eff}(MNA) = 270 pm/V. The figure of merit was 1.5 × FOM d_{eff}(MNA) = 13,210 pm^2/V^2. The SHG intensity of *p*-NA on IPNA was stable for 1200 h at room temperature; however, for the mixed *p*-NA + IPNA system it decreased rapidly.

(3) meta-Nitroaniline (m-NA)

Chemical structure[129]:

Point group: mm2 (orthorhombic); space group = Pca2$_1$; negative biaxial crystal: 2V$_z$ = 104° (λ = 0.532 μm)

Unit cell parameters: a = 19.23 Å, b = 6.48 Å, c = 5.06 Å; Z = 4; density: 1.430 g/cm^3

Melting point = 114 °C, λ$_{max}$ = 372 nm; λ$_{cutoff}$ = 500 nm

NLO coefficients: d_{31} = 20 pm/V, d_{32} = 1.6 pm/V, d_{33} = 21 pm/V at 1.06 μm.[132] d_{31} = 19.5 pm/V, d_{32} = 1.6 pm/V, d_{33} = 20.5 pm/V at 1.06 μm.[133]

Electro-optical coefficients: r_{13} = 7.4 pm/V, r_{23} = 0.1 pm/V, r_{33} = 16.7 pm/V.[134]

Table 25 lists the refractive indexes and breakdown threshold of m-NA crystals.

(4) meta-Aminophenol

Chemical structure[134]:

Crystal class = orthorhombic, Space group = Pmc2$_1$.

Unit cell parameters: a = 6.12 Å, b = 11.23 Å, c = 8.31 Å; transparency range = 0.32 ~ 1.7 μm; Melting point = 122 °C

NLO coefficients: d_{31} = 0.75 pm/V, d_{32} = 2.87 pm/V, d_{33} = 3.35 pm/V at 1.06 μm

(5) meta-Nitrobenzene Derivatives

	X = NO₂ (m-DNB)[135]	X = Cl (m-CNB)[133]	X = Br (m-BNB)[133]
Formula	$C_6H_4N_2O_4$	$C_6H_4NO_2Cl$	$C_6H_4NO_2Br$
Melting point (°C)	90	46	56
Transparency (μm)	—	0.4 ~ 2.0	0.42 ~ 2.1
Crystal system	Orthorhombic	Orthorhombic	Orthorhombic
Space group	—	$Pbn2_1$	$Pbn2_1$
a (Å)	13.257	6.0	5.92
b (Å)	14.048	21.40	21.52
c (Å)	3.806	5.35	5.34
Z value	4	4	4
Density (g/cm³)	1.565	1.582	1.969
NLO coefficients (pm/V)	$d_{31} = 1.78$	$d_{31} = 4.6$	$d_{31} = 4.0$
	$d_{32} = 2.7$	$d_{32} = 4.0$	$d_{32} = 4.6$
	$d_{33} = 0.74$	$d_{33} = 7.75$	$d_{33} = 8.0$

Refractive indexes of m-DNB crystals[135]

Wavelength (μm)	n_x	n_y	n_z
0.436	1.803	1.736	1.508
0.532	1.759	1.698	1.491
0.633	1.738	1.680	1.483
1.064	1.709	1.654	1.471

(6) 2-Methyl-4-nitroaniline (MNA)

MNA is a derivative of p-NA in which a methyl (CH_3) group is substituted on an *ortho*-position to the donor–amino group. The substituted methyl group is a weak electron donor; therefore, it contributes in the enhancement of the donating activity of the amino group. MNA molecule has a noncentrosymmetric crystal structure; hence, it exhibited SHG activity. This observation has been confirmed by Levine et al.[136] in crystalline MNA molecule. The optical quality single crystal of MNA can be grown by various techniques: (1) solution, (2) melt, (3) vapor, and (4) sublimation. Levine et al.[136] reported that most suitable MNA crystals are obtained by the vapor growth method. Lipscomb et al.[137] obtained single crystals of MNA by using a mixture of 50% methanol and 50% *ortho*-xylene. The MNA crystals of different sizes ranging from microscopic to about $0.5 \times 0.5 \times 0.03$ cm³ grow from the solvent mixture. Grossman[138] reported that the good optical quality large MNA crystals are grown by the sublimation and the liquid solution methods. In solution growth, acetone and toluene vapor exchange forms large MNA crystals. The dimensions of these crystals are several millimeters thick and 1 cm in length and width. The MNA crystal has a planar structure. The dihedral angle between the plane of the amino group and the plane of the benzene ring is 7.2°, whereas the dihedral angle between the plane of the nitro group and the plane of the ring is 1.1°. The single crystal X-ray diffraction studies evidence monoclinic Cc space groups with point group m. The MNA crystal is almost transparent over an optical range of about 500 to 1900 nm.

The d_{11} coefficient of MNA in solution was determined by an electric-field-induced second harmonic generation technique to be 500 ± 25% over that of d_{11} coefficient of quartz, which is 2000 times larger than d_{11} coefficient of LiNbO₃. Lipscomb et al. [137] reported the calculated nonlinear second-order susceptibility of 600×10^{-9} esu in MNA using the 1.20×10^{-9} esu d_{11} value of quartz.

Chemical structure:

Table 26 Refractive Indexes, Breakdown Threshold, and Linear Absorption of MNA Crystals

Wavelength (μm)	n_x	n_y	α (cm^{-1})	tp (ns)	I (10^8 W cm^{-2})
0.532	2.2		1.0	10	
0.6328	2.0	1.6			
1.064	1.8			20	2.0

From Levine B. F., et al., *J. Appl. Phys.*, 50, 2523, 1979. With permission.

Point group: monoclinic; space group: Cc; transparency range: 0.5 ~ 2.5 μm, λ_{max} = 360 nm, λ_{cutoff} = 500 nm
Positive biaxial crystal: $2V_z$ = 138° unit cell dimensions: a = 11.57 Å, b = 11.62 Å, and c = 8.22 Å., α = 90°, β = 137°, and γ = 90°, density = 1.39 g/cm^3
Hyperpolarizability β = (10^{-40} m^4/V) (reference 139):

Wavelength (μm)	β_z
1.064	77 ± 16
1.318	38 ± 7
1.907	32 ± 5 (measured 32 ± 5)

β_o extrapolated to infinite wavelengths = 34 ± 6 × 10^{-40} m^4/V; ground-state dipole moments (μ_g) = 2.47 × 10^{-29} cm; transition dipole moments (μ_{eg}) = 1.61 × 10^{-29} cm; excited state dipole moments (μ_e) = 6.3 × 10^{-29} cm; powder SHG = 22 times of urea

NLO coefficients (pm/V)[139]:

Measured	Calculated
d_{12} = 26.7	38 + 10 (calc. 30 ± 8)
d_{11} = 184	
d_{11} = 250	250 ± 50 (calc. 286 ± 61) (λ = 1.06 μm)

Electro-optical coefficients: r_{11} = 70 pm/V

Twieg and Jain[81] studied SHG activities of MNA analogs from both racemic and optically active amines. Replacement of methylamino instead of amino yielded an achiral derivative MNNA which showed powder efficiency 80 times that of urea or 4 times that of MNA. 2-Trifluoromethyl and methyl analogs of MNA showed little or no powder SHG.

(7) Chloronitroanilines

The replacement of the methyl group with chloro groups leads to 2-chloro-4-nitroaniline (CNA). Single crystals of *m*-nitroaniline and 2-chloro-4-nitroaniline (CNA) alloy grown from the melt have been reported by Singh et al.[140] The alloy crystal shows damage threshold of 100 MW/cm^2 without any sign of deterioration.

Chemical structure:

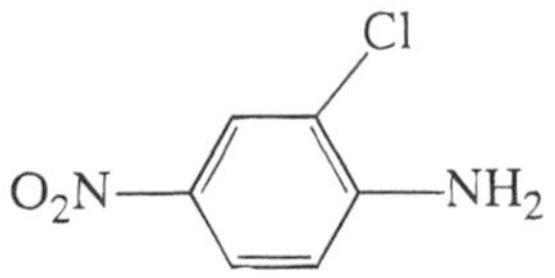

Point group: mm 2; space group = Pca2$_1$; negative biaxial crystal
Unit cell parameters: a = 4.109 Å, b = 13.515 Å, c = 14.183 Å, Z = 4
The calculated d_{eff} coefficient for the alloy was found to be 14 pm/V. The SHG efficiency is larger than that of *m*-NA alone.

Gerbi et al.[141] reported SHG properties of a series of chloronitroanilines as bulk crystals and in thin films. The SHG efficiency of 2-chloro-4-nitroaniline was two times that of urea.

(i) 5-Chloro-2-nitroaniline
   Space group; Pna2$_1$; unit cell parameters: a = 30.844 Å, b = 3.852 Å, c = 6.004 Å, Z = 4, α = β = γ = 90°; The powder SHG efficiency was 20 times that of urea.
(ii) 3-Chloro-2-nitroaniline
   Space group: P2$_1$/c; unit cell parameters: a = 3.813 Å, b = 12.531 Å, c = 14.467 Å, Z = 4, β = 91.20°; The powder SHG efficiency >0.001 times that of urea, thin film of 3-chloro-4-nitroaniline in PMMA showed $\chi^{(2)}$ of 1 × 10^{-10} esu.

136

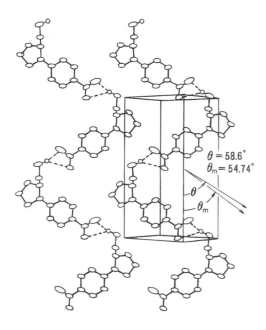

Figure 8 ORTEP of *N*-(4-nitrophenyl)-(L)-prolinol (NPP) where molecules are held head to tail by intermolecular hydrogen bonding. The molecular dipole is dispersed within 4° of the optimum alignment for bulk phase matching. (From Ledoux, I., Josse, D., Vidakovic, P., and Zyss, J., *Opt. Eng.*, 25, 202, 1986. With permission.)

(8) *N-(4-Nitrophenyl)-*(L)-prolinol (NPP)

NPP is synthesized from 1-fluoro-4-nitrobenzene with L-prolinol in DMSO using potassium carbonate. NPP was studied by Zyss and co-workers.[57,142] A yellow solid appears after heating the mixture for 24 hr at 50 °C; NPP melts at 100 °C and is soluble in common organic solvents. NPP shows a strong charge transfer absorption band at 390 nm in ethanol and it has a transparency range from 0.5 to 2.0 μm. In the NPP molecule, the benzene ring is planar, and both nitrogen atoms lie slightly out of the aromatic plane. The nitro group is coplanar and it twists about the C−N bond by 5.8°. The angle between the mean plane of the molecule and crystallographic plane (101) is 11°. The nonplanarity of the amino group demonstrates pyramidal hybridization of the amino nitrogen. The bond lengths data indicate a strong relationship between NPP and *p*-NA molecules. Molecular packing of NPP has one of the most interesting features. Figure 8 shows the molecular packing of NPP in the crystal unit cell. The molecules are held together by head-to-tail intermolecular hydrogen bonding between the alcohol of one molecule and the nitro group of the next molecule. In NPP, the crystallographic plane (101) is a possible cleavage plane. The angle between the charge-transfer axis (*a*) and the binary symmetry axis (*Y*) corresponds to 58.6°. Figure 9 shows the best polarization configuration of the NPP molecule. The mean molecular plane *XY* (110) is coincident with the dielectric axis *XY* plane of the crystal.[143] The nonlinear optical coefficients of NPP are d_{21} and d_{22}, and from an oriented gas description, these coefficients can be written

$$d_{21} = N\,\beta_{\alpha\alpha\alpha}\,\cos\theta\,\sin^2\theta \tag{36}$$

$$d_{22} = N\,\beta_{\alpha\alpha\alpha}\,\cos^3\theta \tag{37}$$

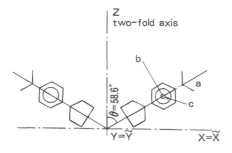

Figure 9 Molecular alignment of NPP showing molecular stacking parallel to the cleavage plane *YX*, *Z* being normal to the cleavage plane. Maximum d_{eff} is obtained when light propagates perpendicular to the molecular planes along *Z*. For phase matching the optimum angle is 54.7°, whereas in NPP molecule the optimum angle is 58.6°, hence, it exhibits the maximum d_{21} coefficient. (From Ledoux, I., Josse, D., Vidakovic, P., and Zyss, J., *Opt. Eng.*, 25, 202, 1986. With permission.)

The third-rank tensor β has only one coefficient $\beta_{\alpha\alpha\alpha}$ along α. The angular projection factor $\beta_{\alpha\alpha\alpha}$ is maximized for $\theta = 54.74°$, which optimizes the phase-matching d_{21} coefficient. For NPP, the actual θ value is 58.6°, close to the optimal angle; hence d_{21} may be considered an optimized NLO coefficient. The ratio of d_{22}/d_{21} calculated for $\theta = 58.6°$ (i.e., $\psi = \cos^2/\sin^2$) corresponds to 0.3726. The d_{21} and d_{22} coefficients of NPP are 83.78 pm/V and 30.58 pm/V, respectively. The optimized crystalline structure for phase-matched bulk quadratic nonlinear interactions occurs in conjunction of chirality and hydrogen bonding, and the electron-donating amino alcohol group induces a favorable crystalline packing. Molecules are linked together by hydrogen bonds in quasiplanar chains stacked in parallel crystalline planes parallel to the 2-fold symmetry. Ledoux et al.[144] reported the generation of high-intensity subpicosecond pulses tuneable in the 1- to 1.6-μm range by parametric amplification of a femtosecond continuum white light in the NPP crystal. Parametric emission in NPP at degeneracy at a pump intensity greater than 3 GW/cm² has been observed.

Chemical structure:

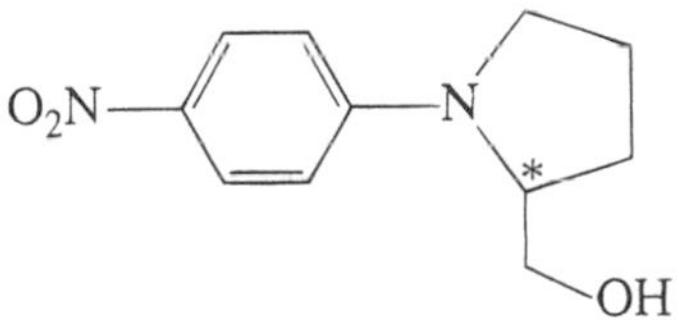

Chemical formula: $C_{11}H_{14}N_2O_3$; molecular weight: 222; crystal class: monoclinic; space group: P2$_1$
Unit cell parameters: $a = 5.261$ Å, $b - 14.908$ Å, $c = 7.185$ Å, density is 1.36 g/cm³.
Melting point: 116 °C; transparency range: 0.48 ~ 2.0 μm; $\lambda_{max} = 397$ nm, $\lambda_{cutoff} = 500$ nm, ($\epsilon_{max} = 18{,}700$); Powder SHG = 154 times that of urea
Hyperpolarizability $\beta = 17 \times 10^{-30}$ esu[57]
NLO coefficients: $d_{21} = 81$ pm/V, $d_{22} = 31$ pm/V, $d_{eff} = 102$ pm/V at 1.064 μm
Figure of merit $d_{eff}/n^3 = 834$ (pm/V)² at 1.064 μm, damage threshold: > 3GW/cm²

Refractive indexes of NPP:

	λ (nm)									
	509	532	546	577	589	633	644	690	1064	1340
n_x	2.355	2.277	2.231	2.153	2.128	2.066	2.055	2.051	1.926	1.917
n_y	2.116	2.024	1.982	1.927	1.911	1.876	1.857	1.857	1.774	1.757
n_z	1.497	–	1.491	1.495	1.484	1.478	1.474	1.474	1.457	1.440

Davis et al.[145] reported powder SHG results on a series of NPP-related materials as shown in Table 27. The most promising material was compound 12, a close analog of NPP and DAN that has high SHG activity. Compound 12 is orthorhombic and belongs to P2$_1$2$_1$2$_1$ space groups. Its unit cell parameters are; $a = 7.717$ Å, $b = 20.462$ Å, $c = 8.517$ Å, $\alpha = \beta = \gamma = 90°$, V = 1345 Å³, Z = 4. The replacement of nitroaniline system by 2-nitropyridine in compounds 6 and 7 yielded novel materials having powder SHG of 36 and 8 times that of urea, respectively.

Lequan et al.[146] reported β values of N-(4-nitrostilbenyl)(s) prolinol (NSP). NSP belongs to noncentrosymmetric space group P2$_1$2$_1$2$_1$ and has an absorption maximum at 441 nm. The dipole moment determined experimentally was 7.1 D, whereas calculated from AM1 and MNDO were 8.34 and 7.93 D, respectively. The EFISH measured $\beta_\mu(2\omega)$ and $\beta_\mu(0)$ values of NSP were 113×10^{-30} and 84.5×10^{-30} esu, respectively. The $\beta_\mu(0)$ values determined from AM1 and MNDO were 72×10^{-30} and 66.4×10^{-30} esu, respectively. Nonoptimal crystal packing of NSP caused weak powder SHG comparable to that of urea.

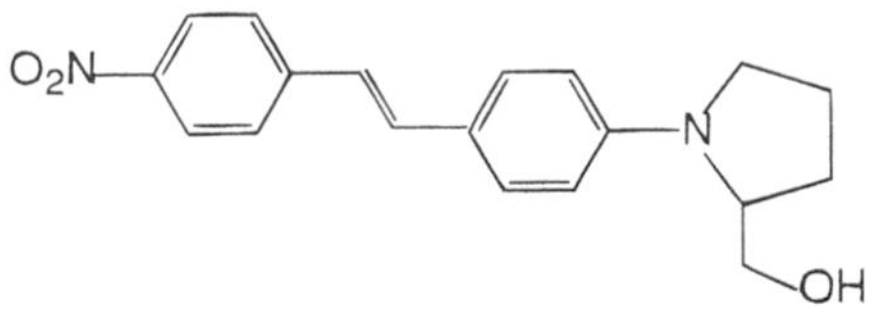

Table 27 Powder SHG of NPP-related Materials Showing the Effect of Substituents

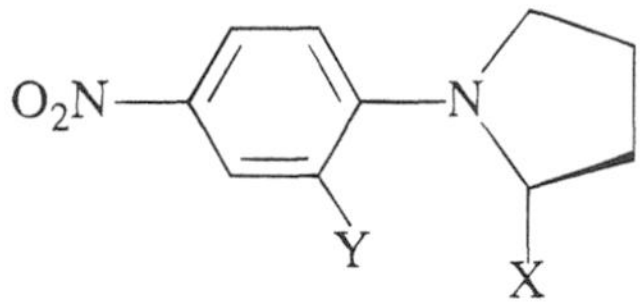

No.	X	Y	m.p. (°C)	Powder SHG (× urea)
1	CH$_2$NH$_2$	H	89.5 ~ 90.5	0.68
2	CH$_2$CN	H	164 ~ 165	0.02
3	CH$_2$NH$_3$Cl	H	~ 203	1.7
4	CH$_2$OTs	H	97 ~ 99	0.01
5	CH$_2$NPhth	H	245 ~ 246	15.5
6	COOH	H	138 ~ 139	7.0
7	COOCH$_3$	H	79 ~ 79.5	26
8	CONH$_2$	H	202 ~ 205	17
9	CONHNH$_2$	H	209 ~ 211	1.66
10	CONHC$_2$H$_5$	H	139 ~ 141	0.13
11	CH$_2$OH	H	115 ~ 116	35.5
12	CH$_2$OH	NHCOCH$_3$	110 ~ 112	204

From Davis, P.J., et al. *Organic Materials for Nonlinear Optics*, Special Publication No. 69, 1969, p. 163. With permission.

(9) 2-(N-Prolinol)-5-nitropyridine (PNP)

Chemical structure[147]:

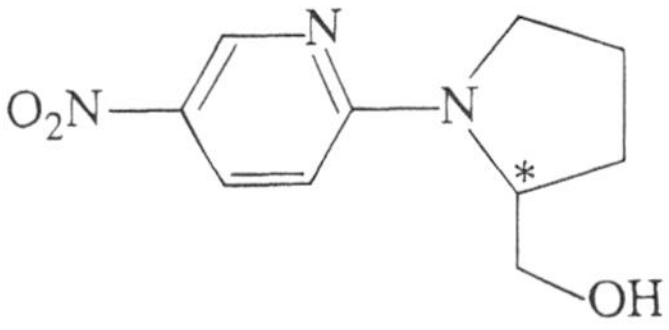

Point group: 2; space group: P2$_1$
Cell parameters: a = 5.182 Å, b = 14.964 Å, c = 7.045 Å, β = 106.76°, V = 523.1 Å^3, Z = 2, density = 1.42 g/cm^3, α = 9.0 cm^{-1}, melting point = 116 °C, transparency = 0.5 ~ 2.0 μm
Powder SHG = 140 times that of urea

Hyperpolarizability β (10^{-40} m^4/V)[139,148,149]

Wavelength (μm)	β_z
1.064	76 ± 15
1.318	47 ± 9
1.907	74 ± 7 (measured 47 ± 7)

β_0 extrapolated to infinite wavelengths = 38 ± 7 × 10^{-40} m^4/V; ground-state diople moments (μ_g) = 2.4 × 10^{-29} cm; transition dipole moments (μ_{eg}) = 1.83 × 10^{-29} cm; excited state dipole moments (μ_e) = 5.6 × 10^{-29} cm

Nonlinear optical coefficients:

At 1.064 μm
d_{yxx} = 48 pm/V
d_{yyy} = 17 pm/V

At 1.907 μm (pm/V)
d_{16} = 24 (calc. 38)
d_{21} = 29 (calc. 37)
d_{22} = 9 (calc. 11)

Electro-optical coefficients r (pm/V)[149]:

Wavelength (nm)	514.5	632.8	1064
r_{12}	20.2	13.1 (calc. 22.0)	8.7
r_{22}	28.3	12.8 (calc. 9.0)	7.6

Refractive indexes of PNP:

$\lambda(\mu m)$	n_x	n_y	n_z
0.488	2.239	1.929	1.477
0.5145	2.164	1.873	1.474
0.580	2.040	1.813	1.468
0.6328	1.990	1.788	1.467
1.064	1.880	1.732	1.456

10. (−)-2-(α-Methylbenzylamino)-5-nitropyridine (MBA-NP)

Chemical structure[150,151]:

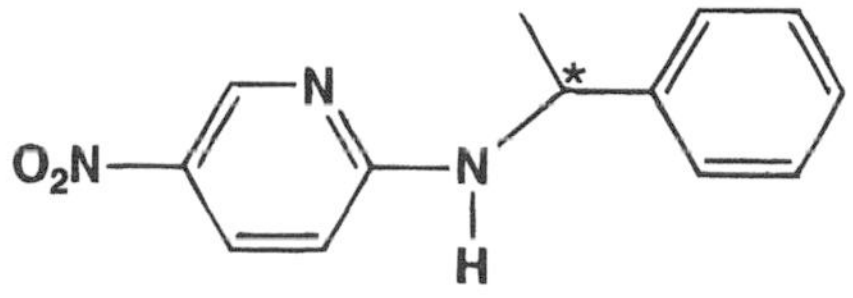

Space group: $P2_1$ point group = 2 (monoclinic)
Cell parameters: a = 5.392 Å, b = 6.354 Å, c = 17.924 Å, β = 94.6°
Melting point = 83.2 °C transparency range: 0.45 ∼ 1.6 μm, λ_{max} = 359 nm, λ_{cutoff} = 500 nm
Hyperpolarizability (β) = 15 × 10^{-30} esu, β_{calc} = 15.4 × 10^{-30} esu, Dipole moment = 6.07 debye;
Powder SHG = 25 times that of urea
Nonlinear optical coefficients: d_{22} = 60 pm/V; damage threshold = 1 GW cm^{-2}

Refractive index[150]:

$\lambda(\mu m)$	n_x	n_y	n_z
0.532	1.6895	1.8584	1.7632
1.064	1.65	1.7144	1.6882

(11) N-(4-Nitrophenyl)-4-methylamino acetonitrile (NPAN)

A new p-NA-like molecular compound, NPAN has been reported by Barzoukas et al.[58] The NPAN is obtained through a Mannich reaction of N-methyl 4-nitroaniline. The achiral NPAN molecule possesses a highly polar substituent. The polar cyano (CN) group is attached far away from the donor–acceptor π-electron conjugated system. It is electrically as well as sterically independent of the charge-transfer polar part. The cyano group has high dipole moment (μ = 3.47 debye) which favors the crystal packing owing to dipole-dipole interactions. This newly tailored noncentrosymmetric molecular compound also exhibits high quadratic nonlinear efficiency. On the other hand, its 5-nitropyridine analogs do not show SHG efficiency because of their centrosymmetric crystal structure.

The crystal structure of NPAN molecule has been investigated by Vidakovic et al.[152] The aromatic ring is planar. Both nitrogen atoms, i.e., the nitro group and the amino group, are slightly twisted out of the aromatic plane in the same direction. The nitro group is rigid planar whereas the amino group is quasiplanar. The quasiplanarity of the amino group gives a rigid conformation to the molecule because of the steric constraint on the atoms in close proximity of the nitro atom. The orientation of the methyl group is nearly symmetrical, which is sterically hindered, whereas the cyanomethyl group is oriented in a noncentrosymmetric fashion. The NPAN molecule has a blocked chiral conformation favoring the crystal packing, and the molecular packing plays a key role in the SHG activity.

The electronic absorption spectra of NPAN recorded in ethanol exhibit an intense charge-transfer absorption band at about 358 nm. Under similar conditions, the NPP molecule shows an absorption maximum at 390 nm, therefore a blue-shift is observed in NPAN (i.e., 30 nm). The increased energy gap in NPAN is associated with the cyano group and is of much interest because of the broader transparency range than those of common p-NA derivatives. The powder SHG efficiency of NPAN molecule is more than two orders of magnitude larger than that of urea. The optimized NLO coefficient d_{zyy} of crystalline NPAN is in the range of 140 × 10^{-9} esu. Vidakovic et al. [152] reported the crystal

Table 28 Structural Formula, Electronic Absorption Spectral Data, and SHG Powder Test (λ = 1.06 μm) for NPAN and Its Related Compounds

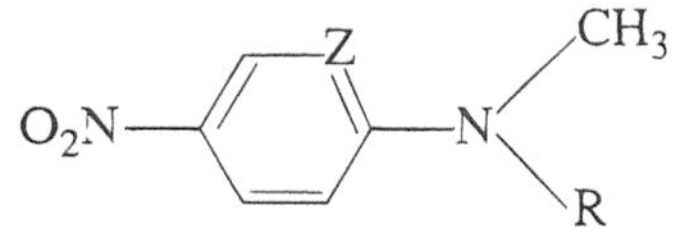

Compound	R	Z	λ_{max} (nm)	ε_{max}	SHG ($\times$ urea)
NPAN	$-CH_2-CN$	CH	358	18,440	140
NPPN	$-CH_2-CH_2-CN$	CH	374	19,920	85
NPAP	$-CH_2-C\equiv CH$	CH			0
NPAA	$-CH_2-CONH_2$	CH			0
DMNA	$-H$	N	387	20,000	1

From Barzoukas, M., et al., *J. Opt. Soc. Am. B.*, 4, 977, 1987. With permission.

growth of NPAN bulk single crystal by Bridgman-Stockbarger (BS) and NPAN fibers by inverted BS techniques. Morita and Vidakovic[153] reported angle and temperature tuning of phase-matched SHG in NPAN. Angularly phase-matched SHG efficiencies as 1.8×10^{-3} at 1.064 μm, 2.0×10^{-4} at 1.21 μm, and 1.2×10^{-4} at 1.30 μm have been measured. Temperature-tuning sensitivities of phase-matched SHG were -1.2, -1.2, and -1.4 mrad/°C at 1.064, 1.21, and 1.30 μm, respectively.

In the case of NPAN fibers, the SHG signal was very weak. The powder test and electric field-induced second-harmonic measurements were evidence that NPAN is a good organic NLO material. Although the nonlinearity of NPAN molecule is about half that of NPP, its transparency range is of interest.

Crystal class: orthorhombic; space group: Fdd2; positive biaxial crystal: $2V_z = 45.4 + 2.0°$ ($\lambda = 0.547$ μm)
$\lambda_{max} = 360$ nm, $\lambda_{cutoff} = 490$ nm; melting point = 114 °C
Unit cell parameters: a = 4.33 Å, b = 26.2 Å, c = 34.2 Å, Z = 2, Density = 1.31 g/cm³
NLO coefficients: $d_{eff} = 58.65$ pm/V[153b]; $d_{32} = 59$ pm/V; $d_{33} = 25$ pm/V at 1.064 μm; $d_{32} = 48$ pm/V; $d_{33} = 24$ pm/V at 1.34 μm; hyperpolarizability (β) = 12×10^{-30} esu

Nicoud[154] used chirality as an efficient strategy for the tentative control of noncentrosymmetry in organic crystals. These compounds have large SHG efficiencies but their crystal structures are not yet known.

(12) 2-Cyclooctylamino-5-nitropyridine (COANP)

COANP is an achiral molecule. The material is synthesized by reacting 2-chloro-5-nitropyridine with cyclooctylamine using *N*-methyl pyrrolidinone and triethylamine. Optical and nonlinear optical properties of COANP were reported by Gunter and co-workers.[111,155,156] The crystals of COANP are grown from ethylacetate and acetonitrile. The structure of COANP molecule has been analyzed using single crystal X-ray diffraction, computed by X-ray system, and also solved by direct methods. The COANP is a biaxial crystal that belongs to point group symmetry mm2. The hydrogen bonding with a distance of 2.9 Å takes place between the oxygen atoms of the nitro groups and the hydrogen atoms of the amino groups. The distance between oxygen atom and nitrogen atom of pyridine ring is 3.9 Å and the interaction between these two atoms (N and O) is very weak but it involves no hydrogen bonds. The cyclooctylamine part of the COANP molecule affects the noncentrosymmetric crystal packing and the amphiphilicity leads to packing of the molecule. The related derivatives obtained from smaller cycloaliphatic amines starting from cyclopropylamine through cycloheptylamine do not exhibit any SHG activity.

Chemical structure:

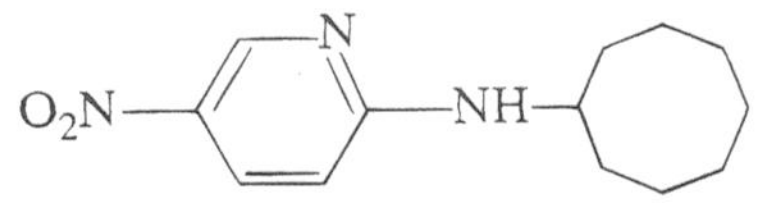

Point group: mm2; space group: Pca2$_1$; positive biaxial crystal: $2V_z = 45.4 + 2.0°$ ($\lambda = 0.547$ μm); transparency range: 0.5 ~ 1.4 μm, $\lambda_{cutoff} = 500$ nm
Cell parameters: $a = 26.281$ Å, $b = 6.655$ Å, $c = 17.6307$ Å, unit volume $V = 1334.5$ Å³, $Z = 4$,

Table 29 Chiral Amino-Alcohol 4-nitroaniline and 2-Amino-5-nitropyridine Derivatives and Achiral *p*-NA Derivatives

D	Z	R$_1$	R$_2$	SHG (× urea)
HOH$_2$C-CH-NH- (with CH$_3$)	CH	H	H	22
	N	H	H	92 (NPA)
HOH$_2$C-CH-NH- (with CH$_2$–phenyl)	N	H	H	130 (NPPA)
pyrrole-N- with CH$_2$OH	CH	H	H	150 (NPP)
	CH	CN	H	22 (CNPP)
	CH	H	OH	140 (HNPP)
HOH$_2$C-CH$_2$-CH$_2$-CH$_2$-NH-	CH	H	H	80 (APNP)
	CH	CF$_3$	H	30
	CH	H	CF$_3$	80
HOOC-CH$_2$-CH$_2$-CH$_2$-NH-	CH	H	H	115 (BANP)
NC-CH$_2$-N- (with CH$_3$)	CH	H	H	140 (NPAN)
NC-CH$_2$-CH$_2$-NH- (with CH$_3$)	CH	H	H	85 (PNPP)

From Nicoud, J.F., *Organic Materials for Nonlinear Optics*, Special Publication No. 69, 1989. With permission.

Density = 11.24 g/cm^3; melting point = 87.9 °C;

Hyperpolarizability β (10^{-40} m^4/V)[139]:

Wavelength (μm)	β$_z$
1.064	92 ± 18
1.318	50 ± 9
1.907	46 ± 7

β$_0$ extrapolated to infinite wavelengths = 34 ± 6 × 10^{-40} m^4/V

Ground-state diople moments (μ$_g$) = 2.28 × 10^{-29} cm; transition dipole moments (μ$_{eg}$) = 1.75 × 10^{-29} cm; excited state dipole moments (μ$_e$) = 6.6 × 10^{-29} cm; powder SHG efficiency: 30 × urea

Nonlinear optical coefficients (pm/V):

$$d_{31} = 15.0$$

$$d_{32} = 26.0 \qquad 16 \text{ (cals 19) at 1.907 } \mu m$$

$$d_{33} = 10.0 \qquad 8.5 \text{ (cals 6.2) at 1.907 } \mu m$$

$$d_{eff} = 24$$

SHG conversion efficiency: 6.4×10^{-3} W^{-1}; linear absorption coefficient $\alpha < 1$ cm^{-1} at 1.35 μm; Electro-optical coefficients r (pm/V)[149]

Wavelength (μm)	514.5	632.8	1.064
r_{13}	6.8	3.4	0.9
r_{23}	26	13 (calc 14)	6.3
r_{33}	28	15 (calc 6.0)	7.7

Refractive indexes[149]

λ (μm)	n_x	n_y	n_z
0.480	1.766	2.505	1.776
0.547	1.700	1.839	1.687
0.577	1.690	1.824	1.663
0.650	1.668	1.772	1.643
1.064	1.636	1.715	1.604

(13) 4-(N,N-Dimethylamino)-3-acetamidonitrobenzene (DAN)

A reaction of dimethylamine and 3-acetamido-4-fluoronitrobenzene in DMSO yields DAN as reported by Baumert et al.[112] DAN is recrystallized either from an ethanol-water or tetrahydrofuran-water solution. High optical quality crystals of dimensions of 10 mm $\times$ 5 mm $\times$ 2 mm can be grown. The DAN crystal is optically biaxial. In DAN, an intermolecular hydrogen bonding occurs between the acetamido hydrogen and adjacent nitro group. The acetamido group promotes noncentrosymmetric crystal structure by forming hydrogen bonding. Both the acetamido group and nitrogen electron pair are twisted with respect to the plane of the benzene ring.

The charge-transfer interaction between dimethylamino donor and nitro acceptor groups yields β of 30×10^{-30} esu determined using the electric field-induced SHG (EFISH) technique. Baumert et al.[112] reported the effective NLO coefficient

$$d_{eff} = d_{21} \cos^2 \Phi + d_{23} \sin^2 \Phi + d_{25} \cos \Phi \sin \Phi = 64.26 \times 10^{-9} \text{ esu} \qquad (38)$$

for the phase-matched SHG (where Φ is the internal phase matching angle).

Kerkoc et al.[157] measured the d_{eff} in several phase-matched configurations to determine all different NLO coefficients allowed by Kleinman's symmetry. The nonlinear figure of merit was estimated to be 140 (pm/V)^2, more than 10 times larger than KTP crystal. DAN exhibits a laser damage threshold of 5 GW/cm^2 for pulses of 100-psec width. The higher nonlinear efficiency in DAN originates from the molecular packing. Crystal data indicate that the tilt angle for DAN is 70.6°, away from the optimal angle of 54.74°, still the macroscopic nonlinearity is 78% of the maximum value.[158] DAN has been considered a potential material for parametric oscillators and frequency converters.

Chemical structure:

Space group: $P2_1$; Point group monoclinic; transparency range: $0.51 \sim 1.7$ μm, $\lambda_{cutoff} = 490$ nm
Cell parameters: $a = 4.786$ Å, $b = 13.053$ Å, $c = 8.736$ Å, $\beta = 94.43°$[157]; melting point $= 163$ °C; powder SHG $= 115$ times that of urea.
NLO coefficients at 1.064 μm (pm/V): $d_{23} = 50$, $d_{22} = 5.2$, $d_{21} = 1.5$, $d_{eff} = 62.84$; SHG conversion efficiency: 1.1×10^{-3} W^{-1}, damage threshold $= 5$ GW/cm^2

Refractive index:

λ (μm)	n_x	n_y	n_z
0.4965	1.574	1.779	2.243
0.5145	1.557	1.748	2.165
0.532	1.554	1.732	2.107
0.546	1.552	1.723	2.072
0.585	1.545	1.701	2.005
0.6328	1.539	1.682	1.949
1.064	1.517	1.636	1.843

(14) 4-Nitro-4'-methylbenzylideneaniline Derivatives

NMBA is synthesized by the condensation reaction of 4-nitrobenzaldehyde with 4-aminotoluene. The crude product is recrystallized from *n*-hexane and is zone refined to reduce the impurities.[159] NMBA single crystals as large as 50 mm × 30 mm × 5 mm grow in a 4-week period. NMBA shows polymorphism. The first phase (triclinic P1) is centrosymmetric, so it exhibits no SHG, whereas the second phase (monoclinic, space group Pc, point group m) is noncentrosymmetric. Unit cell parameters are: $a = 7.419$ Å, $b = 11.679$ Å, $c = 7.447$ Å and $\beta = 110.35°$. The molecules are planar and lie with the molecule dipole 24° apart.[160] The d_{11} and d_{33} coefficients of NMBA were 4 and 143 times that of quartz d_{11}. Since the half angle between the molecular dipole is 12°, quite far from the optimal angle of 54.774°, therefore low phase-matching efficiency has been anticipated. Gotoh et al.[161] modified NMBA and prepared new derivatives:

a. 4-Nitrobenzylidene-3-acetamino-4-methoxyaniline

Crystal class: monoclinic; space group $=$ Cc; unit cell parameters: $a = 8.3814$ Å, $b = 26.838$ Å, $c = 7.4871$ Å, $\beta = 115.336°$; melting point $= 193$ °C; cutoff wavelength $= 505$ nm; NLO coefficients: $d_{11} = 1080 \times 10^{-9}$ esu (454 pm/V); damage threshold $= 1$ GW/cm^2; refractive index; $n = 1.901$ at 1.064 μm and $n = 2.164$ at 532 nm.

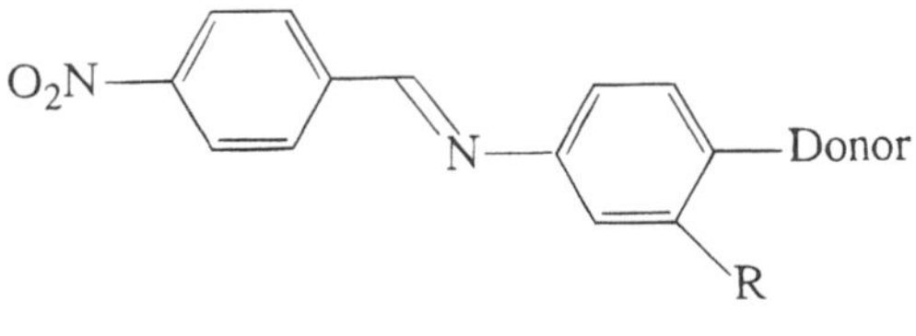

Crystal	Donor	Regulating group (R)	SHG (× urea)
MNBA	OCH_3	$NHCOCH_3$	230
MNBA-Et	OCH_3	$NHCOC_2H_5$	67
MNBA-Cl	OCH_3	$NHCOCH_2Cl$	33
MNBA-Br	OCH_3	$NHCOCH_2Br$	67
HNBC	OH	Cl	35
HNBM	OH	CH_3	29
MNBH	OCH_3	OH	52
EONB	OCH_2CH_5O	OCH_2CH_5O	130
MNA			15

From Gotoh, T., et al., *Proceedings of the First International Workshop on Crystal Growth of Organic Materials*, 1989, p. 234.

b. 4-Nitro-4′-methylbenzylideneaniline (NMBA)

Chemical structure[162,163]:

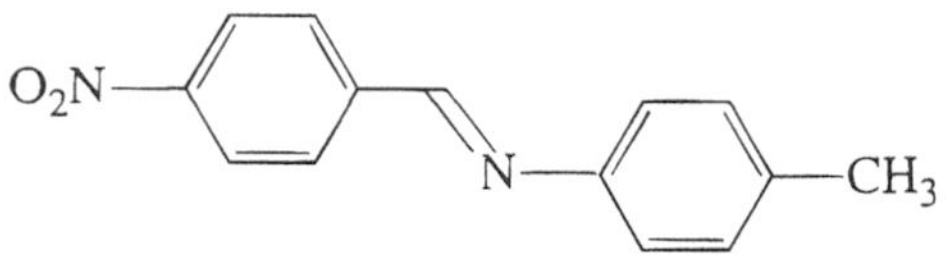

Crystal class: monoclinic; space group = Pb; unit cell parameters: a = 7.419 Å, b = 11.679 Å, c = 7.447 Å, β = 110.583°, Z = 2; cutoff wavelength = 450 nm; transparency = 500 ~ 1600 nm; melting point = 119 °C; powder SHG = 16 times that of urea/KCl disk

Table 30 lists refractive indexes of NMBA crystals.

NLO coefficients[162]

Wavelngth (μm)	d_{11} (pm/V)	d_{33} (pm/V)	d_{31} (pm/V)
1.00	174.15	0.782	44.91
1.064	138.77	0.618	40.87
1.30	70.81	0.439	34.75

Linear electro-optic coefficient:

Wavelength (nm)	r_{11} (pm/V)
632.8	25.2
514.5	36.4
488.0	37.2

Phase matching (001):

Wavelength (μm)	Incidence angle (degree)	
	Type I	Type II
1.00	35.4	−12.5
1.064	31.0	−6.1
1.30	28.20	1.20

(15) 3-Methyl-4-methoxy-4′-nitrostilbene (MMONS)

Tam et al.[164] prepared MMONS by a Wittig reaction of diethyl-p-nitrobenzyl phosponate and 3-methyl-p-anisaldehyde. MMONS crystal can be obtained from a chloroform-ethanol solution. MMONS shows powder SHG efficiency of 1250 times larger that of urea. The nonlinear figure of merit is 850 $(pm/V)^2$ as reported by Bierlein et al.[165]

Table 30 Refractive Indexes Derived from Sellmeier Equation for NMBA

$$n_i^2 = A + (B\lambda^2/\lambda^2 - C) - D\lambda^2$$

where i = x, y, or z and λ is the wavelength and dn is the average residual refractive index.

n	A	B	C	D	dn
n_x	2.7368	1.0472	0.1378	0.0406	0.00083
n_y	2.4700	0.2824	0.1261	0.0210	0.00057
n_z	2.0419	0.2400	0.0662	0.0126	0.00054

Adapted from Halfpenny, P. J., et al. *SPIE Proc.*, 2025, 171, 1993. With permission.

Chemical structure:

$$O_2N-\text{(aryl)}-CH=CH-\text{(aryl)}\begin{cases}-CH_3 \\ -OCH_3\end{cases}$$

Point group: mm 2; space group: Aba2; transparency range: 0.51 ~ 1.7 μm
Unit cell parameters: $a = 15.584$ Å, $b = 13.463$ Å and $c = 13.299$ Å, density = 1.282 g/cm^3, Z = 8
NLO coefficients at 1.064 μm: $d_{33} = 184$ pm/V, $d_{32} = 41$ pm/V, $d_{24} = 55$ pm/V, $d_{24} = 71$ pm/V (phase-matched SHG)
Electro-optic coefficients: $r_{33} = 39.9$ pm/V, $r_{23} = 19.3$ pm/V, $r_{c2} = 30.0$ pm/V
Molecular hyperpolarizability (β): 18×10^{-30} esu measured by EFISH
Temperature bandwidth: $\Delta T = 0.17$ °C (1.064 μm)
Walkoff: $\rho = 9.6°$ (1.064 μm), $\rho = 7.50$ (1.047 μm); acceptance angles: $\Delta\theta = 0.035°$ (1.064 μm), $\Delta\theta = 0.047°$ (1.047 μm)

Type II phase-matching angles:
$\phi = 0°$, $\theta = 73.2°$ (56.3° external) 1.064 μm
$\phi = 0°$, $\theta = 77.6°$ (65.4° external) 1.047 μm

Refractive index of MMONS:

λ (μm)	n_x	n_y	n_z
0.543	1.597	1.756	2.312
0.6328	1.569	1.693	2.129
1.064	1.530	1.630	1.961
1.319	1.525	1.622	1.940

(16) N-Methoxymethyl-4-nitroaniline (MMNA)

Hosomi et al.[166] synthesized MMNA by a reaction of p-NA with formaldehyde and methanol in sodium hydroxide at room temperature. MMNA crystals grown by recrystallizing from alcohol melt at 115 °C. MMNA is soluble in common organic solvents. Optical absorption spectrum recorded in ethanol indicates a cutoff wavelength of 420 nm. Good optical quality, pale yellow MMNA crystals with dimensions as large as 40 mm × 5 mm × 3 mm grew from a tetrahydrofuran solution. MMNA is a very attractive material because phase matching is possible over a wide range of temperatures and wavelengths without cutting.

Chemical structure:

$$O_2N-\text{(aryl)}-NH\text{-}CH_2\text{-}COCH_3$$

Point group: 222; space group: P2$_1$2$_1$2$_1$
Transparency range: 0.42 ~ 2.0 μm, $d_{36} = 25 \times d_{11}$ (SiO$_2$) at 1.064 μm, $d_{eff} = 29 \times 10^{-9}$ esu
Hyperpolarizability = 14.3×10^{-30} esu (MOPAC method)

Refractive indexes of MMNA:

λ (μm)	n_x	n_y	n_z
0.532	1.825	1.626	1.593
1.064	–	1.635	1.580

Table 31 Refractive Indexes and Breakdown Threshold of MAP Crystals[59]

Wavelength (μm)	n_x	n_y	n_z	tp (ns)	I (10^9 W cm^{-2})
0.532	1.5568	1.7100	2.0353	10	0.15
1.064	1.5078	1.5991	1.8439	10	3.0

(17) 2-Methyl-4-nitro-N-methylaniline (MNMA)

MNMA is synthesized by a reaction of 2-fluoro-5-nitrotoluene with methylamine. MNMA crystals with dimensions 0.05 mm × 0.13 mm × 0.15 mm were obtained by the Bridgman method.[167] Vizgert et, al.[168] reported powder efficiency of 80 and 10 times that of urea and KDP, respectively. MNMA shows a figure of merit of 26 (pm/V)^2 for d_{31} coefficient comparable to KNbO$_3$. Hydrogen bonding between the nitro group and the amino hydrogen of an adjacent molecule yields to a SHG-active structure and crystal packing. Twieg and Jain[81] also reported a powder SHG 80 times larger than urea or 4 times larger than that of MNA.

Chemical structure:

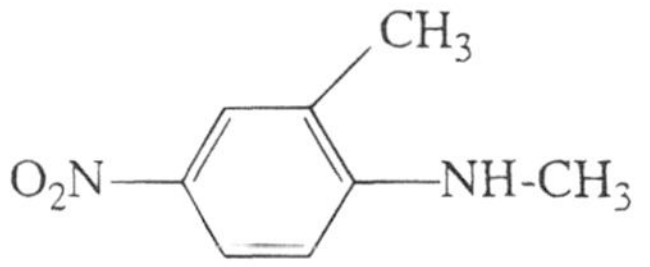

Point group: mm2; space group: Pna2$_1$ (Z = 4); transparency range: 0.51 ~ 2.0 μm
NLO coefficients: d_{31} = 13 pm/V, d_{33} = 2.6 pm/V, d_{15} = 12 pm/V
Electro-optical coefficients: r_{31} = 8 pm/V, r_{33} = 7.5 pm/V

Refractive indexes of MNMA:

Wavelength (μm)	n_x	n_z	$n_x - n_z$
0.5145	2.454	1.548	0.906
0.532	2.385	1.541	0.844
0.633	2.148	1.520	0.628
1.064	1.936	1.506	0.430

(18) Methyl-(2,4-dinitrophenyl)-aminopropanoate (MAP)

Oudar and Hierle[59] obtained a good optical quality, single crystal of MAP from 34% ethylacetate in n-hexane. MAP shows a transparency range of 0.5 ~ 2.0 μm. The figure of merit for SHG was 15 times larger that than of LiNbO$_3$. In MAP, two phase-matching configurations are possible. In type I phase-matching configuration, the efficiency increases to 15 times that of LiNbO$_3$. MAP shows a damage threshold of 1 GW/cm^2 for 10^{-8} sec pulses at 1.064 μm. Strong intermolecular charge-transfer interactions between chiral radical (amino donor group) and the acceptor nitro group in para position leads to large NLO coefficients. Refractive indexes and breakdown threshold of MAP are listed in Table 31.

Chemical structure:

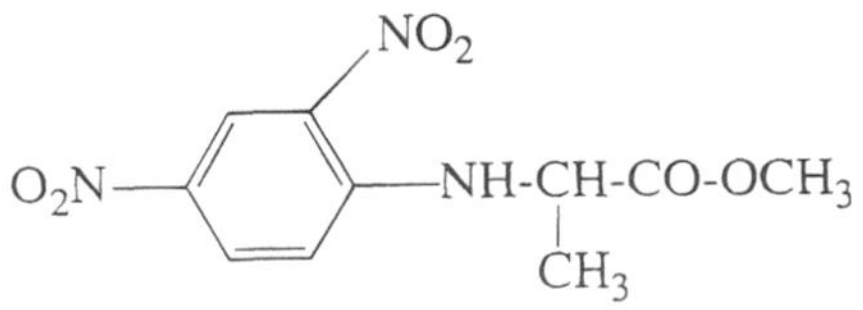

Point group: 2 (monoclinic); space group = P2$_1$; transparency range: 0.5 ~ 2.5 μm, λ_{cutoff} = 490 nm in solution and 550 nm in crystal; melting point = 80.9 °C
Unit cell parameters: a = 6.829 Å, b = 11.121 Å, c = 8.116 Å, β = 95.59°, Z = 2, Density: 1.46 g/cm^3, positive biaxial crystal: 2V$_z$ = 74°

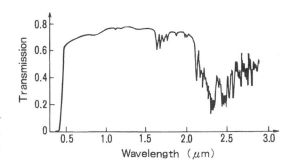

Figure 10 Optical transmission spectrum of 3-methyl-4-nitropyridine-1-oxide, POM. (From Zyss, J., and Chemla, D. S., *J. Chem. Phys.*, 74, 4800, 1981. With permission.)

Hyperpolarizability $(\beta) = 22 \times 10^{-30}$ esu, powder SHG = 10 times that of urea
NLO coefficients: $d_{21} = d_{16} = 16.7$ pm/V, $d_{23} = d_{34} = 3.68$ pm/V, $d_{25} = d_{36} = d_{14} = -0.544$ pm/V, $d_{22} = 18.4$ pm/V.

(19) Mixed MAP:MNA Crystals

Rao et al.[169] reported SHG from the mixed MAP:MNA crystals grown from organic solution containing equimolar portions of MAP and MNA using a 40:60 mixture of ethanol and ethyl acetate. The X-ray diffraction studies showed that molecules stack along the *b* axis with MAP atop MNA atop MAP. The optical transparency of the mixed single crystal is quite different than MAP and MNA because MAP and MNA transmit below 500 nm. The mixed crystal details are as follows: Point group: monoclinic; space group = P2$_1$; transparency range: 0.55 ~ 2.0 µm; melting point = 104.30 °C; unit cell parameters: $a = 6.9196$ Å, $b = 7.673$ Å, $c = 18.554$ Å, $\beta = 92.547°$, $Z = 2$; powder SHG efficiency of MAP:MNA was the same as that of MAP, but lower than MNA.

(20) 3-Methyl-4-nitropyridine-1-oxide (POM)

POM single crystals are commercially available from Quartz et Silice Company of France. The parent compound pyridine-1-oxide shows an electronic "push-pull" property caused by the N-oxide bond which facilitates substitution of donors as well as acceptors in the 4-position. Zyss et al.[100,170] investigated the charge-transfer activity of N-oxide bond with respect to the substituents at the 4-position. In POM, a methyl group is substituted at 3-position of a 4-nitropyridine-1-oxide. Hierle et al.[114] studied the growth kinetics of POM single crystals in three different solvents: acetonitrile, methylacetate, and 1,2-dichloroethane. Single crystals as large as 8 mm × 15 mm × 25 mm and weighing 2 to 5 g can be obtained from the seeds. Figure 10 shows the transparency of a 400-µm-thick plate of POM as a function of wavelength. POM exhibits a transparency range of 0.5 to 1.6 µm. POM shows a figure of merit of 350 times larger than that of KDP. It shows an optical damage threshold of 250 MW/cm^2 at 532 nm for 10-psec pulse width (10 Hz). Zyss et al.[170] reported the parametric gain phase-matched SHG orientation of POM single crystals. POM single crystals can be used for parametric gain phase-matching when pumping at 53 nm. It is often possible to convert an infrared beam at 850 nm into 1.4 µm. POM has three conversion processes corresponding to phase matching: 0.85 µm → 1.3 µm, 0.85 µm → 1.55 µm, and 1.3 µm → 1.55 µm at pump wavelengths of 0.549, 0.54g, and 0.707 µm with phase-matching angles of 55.7, 51.1, and 43.9°, respectively. Parametric emission can be achieved in POM crystals by using a high-intensity (200–260 MW/cm^2) pump beam at 0.532 µm. The emission intensity is in the range of a few kilowatts per square centimeter close to degeneracy (1.06 µm). Table 32 lists the refractive indexes and breakdown threshold of POM crystals.

Table 32 Refractive Indexes and Breakdown Threshold of POM Crystals

Wavelength (µm)	n_x	n_y	n_z	tp (ns)	I (10^9 W cm^{-2})
0.468	1.6875	1.8085	2.1134		
0.509	1.6665	1.7656	2.0279		
0.532	1.6591	1.7500	1.9969	0.025	0.27
0.579	1.6487	1.7281	1.9530		
0.644	1.6402	1.7091	1.9153		
1.064	1.6242	1.6633	1.8287	0.02	2.0

Table 33 Refractive Indexes and Electro-optical Coefficients r (pm/V) of DBNMNA Crystals[172,173]

λ (nm)	n_x	n_y	n_z	r_{42}	r_{51}	$n_a^3 r_{13} - n_c^3 r_{33}$
514.5	2.18	1.72	1.54	86	83	148
532	2.12		1.53			
632.8	1.90	1.62	1.48	20.4	41.4	32
810	1.81	1.59	1.46	11.5	31	8.3
1064	1.82		1.47			

Chemical structure:

Point group: 222 (orthorhombic); space group = $P2_12_12_1$

Transparency range: $0.4 \sim 3.0$ μm, $\lambda_{max} = 325$ nm, $\lambda_{cutoff} = 420$ nm in solution and 485 nm in crystal; melting point = 136 °C; positive biaxial crystal: $2V_z = 69.7°$ ($\lambda = 0.532$ μm)

Unit cell parameters: $a = 21.359$ Å, $b = 6.111$ Å, $c = 5.132$ Å, $\beta = 90°$ density = 1.53 g/cm³; hyperpolarizability (β) = 8×10^{-30} esu; powder SHG = 13 times that of urea

Nonlinear optical coefficients: $d_{14} = d_{25} = d_{36} = 10.0$ pm/V at 1.064 μm

Electro-optical coefficients: $r_{52} = 5.1$ pm/V, $r_{41} = 3.6$ pm/V, $r_{63} = 2.6$ pm/V[171]

Damage threshold = 0.25 GW/cm² at 532 nm.

(21) 2,6-Dibromo-N-methyl-4-nitroaniline (DBNMNA)

Nahata et al.[172,173] prepared a new organic electro-optical crystal material DBNMNA by the reaction of bromine with N,N'-dimethyl-4-nitroaniline in methanol. Orthorhombic b-platelets with dimensions of $2 \times 2 \times 0.2$ mm³ were grown by a temperature-difference solution growth technique within 2 weeks. The donor–acceptor axes lie 19.8° off the axis in the a-c plane. Table 33 lists refractive indexes and EO coefficients of DBNMNA crystals.

Chemical structure:

Point group: mm 2; space group: Fdd2 ($Z = 16$); positive biaxial crystal

Unit cell parameters: $a = 11.745$ Å, $b = 29.640$ Å, $c = 10.807$ Å, $V = 3762.1$ Å³ density = 2.189 g/cm³; transparency range: $0.42 \sim 1.65$ μm, $\lambda_{cutoff} = 420$ nm

(22) Thiourea Crystals

a. Thiosemicarbazide Cadmium Thiocyante (TSCCTC)

Point group[174]: 222; space group: $P2_12_12_1$; transparency range: $0.30 \sim 2.0$ μm; unit cell parameters: $a = 6.976$ Å, $b = 10.7655$ Å, $c = 12.9780$ Å, $Z = 4$, $V = 974.7$ Å³; density = 2.18 g/cm³

Refractive indexes:

Wavelength (μm)	n_x	n_y	n_z
0.4358	1.7609	1.7892	1.918
0.532	1.7317	1.7588	1.88455
0.5461	1.7290	1.7547	1.8810
0.5893	1.7219	1.7469	1.8726
0.6358	1.7159	1.7410	1.8630
0.7065	1.7094	1.7293	1.8568
1.064	1.6948	1.7104	1.8384

NLO coefficients: $d_{14} = 0.64$ pm/V

b. Tri-allylthiourea Cadmium Chloride (ATCC)

Point group[175]: C3v; space group: R3c; transparency range: $0.30 \sim 1.5$ μm; unit cell parameters:
$a =$
$b = 11.527$ Å, $c = 27.992$ Å, $Z = 6$, $V = 3332$ Å^3, density $= 1.65$ g/cm^3

Refractive indexes:

Wavelength (μm)	n_0	n_e
0.415	1.77889	1.69868
0.532	1.73050	1.66977
0.66	1.72032	1.65573
0.83	1.70817	1.64646
1.064	1.69963	1.63986
1.32	1.69404	1.63548

NLO coefficients: $d_{31} = 0.63$ pm/V, $d_{33} = 0.8$ pm/V, $d_{22} = 0.29$ pm/V

c. Tri-allylthiourea Cadmium Bromide (ATCB)

Point group[176]: C3v; space group: R3c; negative uniaxial crystal; transparency range: $0.3 \sim 1.5$ μm; unit cell parameters: $a = b = 11.6207$ Å, $c = 28.5694$ Å, $Z = 6$, $V = 3352$ Å^3; density $= 1.85$ g/cm^3; powder SHG efficiency: $1.5 \times$ urea, $12.5 \times$ KDP crystal.

Refractive indexes:

Wavelength (μm)	n_0	n_e
0.532	1.7485	1.6882
0.5461	1.7449	1.6859
0.5875	1.7376	1.6800
0.5893	1.7373	1.6798
0.6562	1.7292	1.6718
0.7065	1.7237	1.6689
1.064	1.7198	1.6739

d. Zinc Tris(thiourea) Sulfate (ZTS)

Point group[177]: mm2; space group: Pca2$_1$; melting point $= 200$ °C; transparency range: $0.30 \sim 1.1$ μm, $\lambda_{cutoff} = 280$ nm; powder SHG efficiency: $1.2 \times$ KDP crystal; NLO coefficients: $d_{31} = 0.31$ pm/V, $d_{32} = 0.35$ pm/V, $d_{33} = 0.23$ pm/V, $d_{eff} = 0.13$ pm/V

Refractive indexes:

λ (μm)	n_a	n_b	n_c
0.420	1.6907	1.7613	1.7684
0.440	1.6834	1.7535	1.7606
0.450	1.6805	1.7499	1.7565
0.475	1.6736	1.7417	1.7489
0.500	1.6682	1.7359	1.7422
0.520	1.6647	1.7313	1.7381
0.532	1.6625	1.7292	1.7357
0.570	1.6569	1.7228	1.7294
0.600	1.6538	1.7192	1.7258
0.650	1.6591	1.7136	1.7190
0.700	1.6455	1.7089	1.7154
0.800	1.6391	1.7025	1.7041
0.900	1.6359	1.6983	1.7041
1.000	1.6328	1.6944	1.7007
1.064	1.6306	1.6936	1.6984
1.100	1.6301	1.6926	1.6972
1.200	1.6277	1.6902	1.6942

Table 34 NLO and Miller (δ) Coefficients of NPPA

λ (μm)	NLO Coefficients (pm/V)					δ Coefficients (10^{-2} m²/C)				
	d_{22}	d_{21}	d_{16}	d_{23}	d_{34}	d_{22}	d_{21}	d_{16}	d_{23}	d_{34}
1.064	2.6	0.37	0.53	31	24.9	4.6	1.3	2.0	34	22
1.318	1.7	0.25	0.45	28	22.8	3.4	0.9	1.7	34	25

From Sutter, K., et al. *J. Opt. Soc. Am. B*, 8, 1483, 1991. With permission.

(23) 2-Adamantylamino-5-nitropyridine (AANP)

Chemical structure[178]:

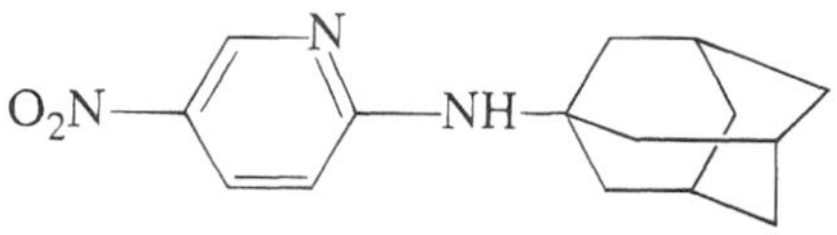

Point group: mm2; space group: Pna2₁
Unit cell parameters: a = 7.990 Å, b = 26.316 Å, c = 6.594 Å; transparency range: 0.46 ~ 1.7 μm
NLO coefficients at 1.064 μm: d_{31} = 80 pm/V, d_{33} = 60 pm/V; SHG conversion efficiency: 2.0 × 10^{-3} W^{-1}

Refractive indexes of AANP:

λ (μm)	n_x	n_y	n_z
0.53	1.77	1.61	1.86
1.064	1.67	1.59	1.71

(24) 4-(2,2-Dicyanovinyl)-anisole (DIVA)

Sasabe and coworkers[179] reported crystal growth and phase-matched SHG in DIVA. DIVA possesses an electron donor methoxy group and an electron acceptor dicyanovinyl group. Crystals, 10 × 10 × 5 mm³, were obtained by slow evaporation of acetone at room temperature over 2 to 3 days. Type I phase matching was observed at 30° from the normal of the (100) face and Type II was observed near-normal incidence in (001) face with the fundamental polarized at 45° from the polar axis. Blue laser light was obtained from Type I phase-matching SHG at 13° from the normal of the (100) face at a wavelength of 812 nm.

Chemical structure:

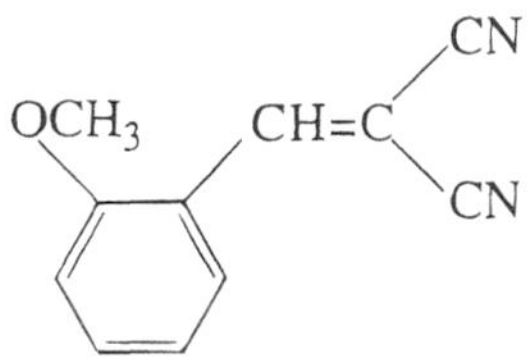

Space group: P2₁; transparency range ≤ 406 nm; NLO coefficients: d_{22} = 10.9 pm/V, d_{eff} = 20.5 pm/V.

(25) N-(4-Nitro-2-pyridinyl)-phenylalaninol (NPPA)

Chemical structure[180]:

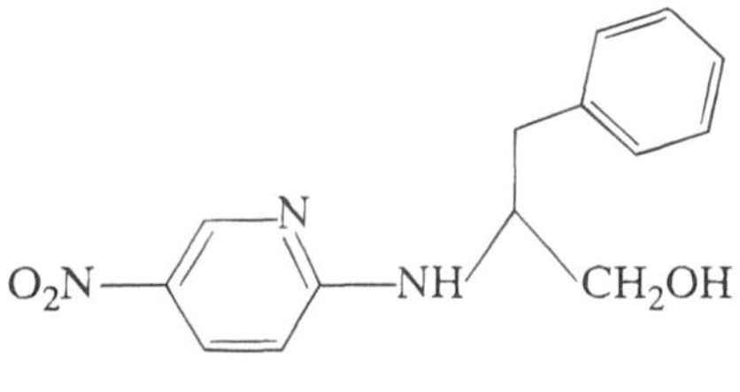

Point group: 2; space group; $P2_1$; transparency range: $0.5 \sim 2.0$ μm with absorptions at 1.150, 1.500, and 1.7090 μm

Type I phase-matching angles: $\phi = 55°$ at 1.064 μm, $\phi = 75°$ at 1.318 μm

Type II phase-matching angles: $\phi = 20°$, $\theta = 0°$ at 1.064 μm

Electric-optic coefficient: $r_{22} = 10$ pm/V at 0.633 μm

Table 34 and 35 lists the NLO and refractive indexs data of NPPA crystals, respectively.

(26) Chalcones

Goto et al.[181,182] reported SHG activity of numerous chalcones having different donor and acceptor groups. The chalcones with alkoxy, alkylthio, and a halogen showed relatively large SHG efficiencies and optical transparencies in the blue region. Table 36 lists the optical and NLO properties of a wide variety of chalcone compounds.

a. Chalcone Single Crystals[183]

i. X = 4-C$_2$H$_5$O Y = 4'-OCH$_3$

Chemical formula (see reference 183): $C_{18}H_{18}O_3$; space group: Pna2$_1$
point group = mm2
Cell parameters: $a = 6.321$ Å, $b = 14.704$ Å, $c = 16.259$ Å, $Z = 4$; volume $= 1511$ Å^3;
density $= 1.241$ g/cm^3

Refractive indexes:

λ (μm)	n_x	n_y	n_z
0.532	1.493	1.710	1.983
1.064	1.473	1.663	1.859

d_{eff} at 1.064 μm $= 3.5$ pm/V for type I and 5.7 pm/V for type II; damage threshold $= 30$ GW/cm^2 at 1.053 μm laser with 1-ns pulse width

ii. X = 4-SCH$_3$ Y = 4'-Cl

Chemical formula: $C_{16}H_{13}ClOS$; space group: Pc; point group = monoclinic
Cell parameters: $a = 33.389$ Å, $b = 7.112$ Å, $c = 5.869$ Å, $\alpha = 90°$; $\gamma = 90°$; $\beta = 91°$ $Z = 4$
Volume $= 1394$ Å^3; density $= 1.38$ g/cm^3

Table 35 Refractive Indexes and Dielectric Axes of NPPA Crystals

λ (μm)	n_x	n_y	n_z	k (degree)	d (degree)
0.465	1.560	1.809	2.252	4.3	14.8
0.488	1.542	1.767	2.151	3.9	14.4
0.5017	1.537	1.750	2.099	3.4	13.9
0.5145	1.537	1.740	2.062	3.1	13.6
0.532	1.534	1.730	2.026	2.0	12.5
0.566	1.531	1.715	1.975	1.6	12.1
0.5882	1.528	1.707	1.949	1.3	11.8
0.6094	1.526	1.701	1.930	1.0	11.5
0.632.8	1.524	1.694	1.887	0.5	11.0
0.6764	1.520	1.686	1.887	−0.2	10.3
1.064	1.511	1.663	1.813	−1.4	9.1
1.318	1.513	1.6655	1.799	–	–

From Sutter, K., et al., *J. Opt. Soc. Am. B*, 8, 1483, 1991. With permission.

Table 36 Powder SHG, Optical Absorption, Dipole Moment, Hyperpolarizability, and Melting Points of Chalcone Compounds

X	Y	SHG (× urea)	λ_{max} (nm)	λ_{cutoff} (nm)	$\Delta\mu_{ge}$ (D)	β (10^{-30} esu)	Tm (°C)
4-H	H	4.5	308	370	18.7	24.9	60
4-$(CH_3)_2$N	H	0	418	500	11.4	105.2	115
4-OCH_3	H	5.5	341	410	15.5	40.0	76.8
4-C_4H_9O	H	1.9	344	410	16.2	55.7	66
4-$C_6H_{13}O$	H	0	342	410	16.7	62.3	73.5
4-$C_8H_{17}O$	H	0	348	410	16.7	47.7	84
4-CH_3S	H	5.5	350	415	13.7	30.4	82
4-Cl	H	0	313	370	16.3	39.1	116
4-Br	H	7.1	314	380	17.8	46.8	121
4-NO_2	H	4.9	310	370	18.0	36.2	167
4-NC	H	0	304	340	17.6	42.5	158.5
H	4′-$N(CH_3)_2$	0	387	470	13.2	62.3	168
H	4′-NH_2	0	350	430	15.2	57.5	152.8
H	4′-OCH_3	0	317	380	13.1	111.9	108
H	4′-$OC_{18}H_{37}$	0	318	380	16.6	48.6	90
H	4′-SCH_3	4	332	400	16.3	34.8	123
H	4′-Br	0	313	370	14.3	67.6	105
H	4′-Cl	0	312	370	13.9	79.3	99
H	4′-NO_2	0	315	400	17.6	27.7	152
4-$N(CH_3)_2$	4′-H	0	418	500	11.4	105.2	115
4-$N(CH_3)_2$	4′-$N(CH_3)_2$	0	421	500	10.2	114.2	147.5
4-$N(CH_3)_2$	4′-NH_2	0	415	510	10.7	120.8	172.8
4-$N(CH_3)_2$	4′-OCH_3	10.1	415	500	11.0	100.6	131
4-$N(CH_3)_2$	4′-SCH_3	0.2	424	510	1.2	59.1	109
4-$N(CH_3)_2$	4′-C_9H_{19}	0	390	445	–	–	–
4-$N(CH_3)_2$	4′-Cl	0	425	500	11.4	103.4	–
4-$N(CH_3)_2$	4′-Br	0	426	510	11.5	112.6	143
4-$N(CH_3)_2$	4′-NO_2	0.3	450	560	–	–	226.2
4-OCH_3	4′-NH_2	8	346	430	15	78	150.5
4-OCH_3	4′-OCH_3	1.0	342	400	16	53.6	103.4
4-OCH_3	4′-$OC_{18}H_{37}$	2.3	346	410	14.6	55.5	102
4-OCH_3	4′-SCH_3	8	344	420	16.6	42.6	137
4-OCH_3	4-F	1.6	340	400	16.9	34.6	112.5
4-OCH_3	4′-Cl	1.5	346	410	–	–	123.8
4-OCH_3	4′-Br	7.1	346	410	16.7	40.0	145
4-SCH_3	4′-$N(CH_3)_2$	0	390	480	13.7	115.7	150
4-SCH_3	4′-OCH_3	0	360	410	13.9	68.7	111.5
4-SCH_3	4′-SCH_3	0	356	440	13.9	68.7	134
4-SCH_3	4′-F	2.4	356	425	13.6	37.6	112
4-SCH_3	4′-Cl	16	360	430	13.5	32.9	151.5
4-SCH_3	4′-Br	15.4	360	430	13.6	13.6	165
4-SCH_3	4′-NO_2	22.6	360	430	13.6	28.9	181.5
4-SCH_3	4′-NH_2	0.8	370	460	16.1	58.9	106
4-SCH_3	3′Cl	35	360	420	13.6	51.6	100.5
4-C_2H_5O	4′-OCH_3	8.9	344	410	16.2	60	111
4-C_4H_9	4′-OCH_3	0	345	400	15.4	56.9	92
4-C_4H_9O	4′-H	1.9	344	410	16.2	55.7	66
4-$C_6H_{13}O$	4′-H	0	342	410	16.7	62.3	73.5
4-$C_8H_{17}O$	4′-H	0	348	410	16.7	47.7	84
4-$C_{18}H_{37}O$	4′-H	0	344	410	15.9	36.9	77.5
4-$C_{18}H_{37}O$	4′-OCH_3	1	344	410	16.7	66.2	99.5
4-$C_{18}H_{37}O$	4′-$N(CH_3)_2$	5.5	392	470	12.7	127.2	110.5
4-$C_{18}H_{37}O$	4′-NO_2	0	356	450	14.9	44.9	106
4-$C_{18}H_{37}O$	4′-NH_2	0	324	420	–	–	100.5
4-$C_{18}H_{37}O$	4′-SCH_3	0	356	440	15.7	40.9	111
H	H	4.5	308	370	18.7	24.9	60

Table 36   *Continued*

X	Y	SHG (× urea)	λ_{max} (nm)	λ_{cutoff} (nm)	$\Delta\mu_{ge}$ (D)	β (10^{-30} esu)	Tm (°C)
4-SCH$_3$	H	5.5	350	415	13.7	30.4	58
4-SCH$_3$	4'-N(CH$_3$)$_2$	0	390	480	13.7	115.7	150
4-SCH$_3$	3'-N(CH$_3$)$_2$	0	351	470	14.8	41	142
4-SCH$_3$	4'-NH$_2$	11	370	440	13.2	19.2	168
4-SCH$_3$	3'-NH$_2$	1	350	420	14.1	48.2	124
4-SCH$_3$	4'-OCH$_3$	0	360	410	12.7	46.3	111.5
4-SCH$_3$	3'-OCH$_3$	0	355	430	13.7	42.6	86.5
4-SCH$_3$	4'-SCH$_3$	0	353	430	13.9	68.7	134
4-SCH$_3$	4'-F	2.4	356	425	13.6	35.3	112
4-SCH$_3$	4'-Cl	16	360	430	13.5	32.9	151.5
4-SCH$_3$	3'-Cl	35	360	430	13.6	51.8	100.5
4-SCH$_3$	4'-Br	15.4	360	430	13.6	13.6	165
4-SCH$_3$	3'Br	8	356	430	14.0	44.8	97
4-SCH$_3$	4'-I	22.6	360	430	13.6	28.9	181.5
4-SCH$_3$	4'-CH$_3$	0	356	420	12.9	48.4	104.5
4-SCH$_3$	3'-CH$_3$	0	355	420	13.7	39.8	89.5
4-SCH$_3$	4'-C$_2$H$_5$	0	356	420	13.1	43.4	103
4-SCH$_3$	3'-C$_5$H$_5$	0.1	360	420	13.5	22.1	153
4-SCH$_3$	4'-OH	0.8	350	430	13.7	18.3	170
4-SCH$_3$	4'-NO$_2$	0.8	370	460	15.6	28.4	160.5
4-SCH$_3$	2'-NO$_2$	0	350	435	15.2	110.6	134.5
4-SCH$_3$	3'-NO$_2$	0	350	430	15.4	58.8	148
4-SCH$_3$	4'-COCH$_3$	12.8	360	470	14.8	37.6	160
4-SCH$_3$	3',4'-(OCH$_3$)$_2$	0	356	420	13.0	37.1	102
4-SCH$_3$	3',4',5-(OCH$_3$)$_3$	0	360	420	13.1	54.6	97
4-SCH$_3$	4'-OC$_6$H$_{13}$	0	354	420	13.1	53.6	100
H	4'-SCH$_3$	4	332	400	16.3	34.8	123
4-F	4'-SCH$_3$	0	344	400	14.7	54.1	126
4-Cl	4'-SCH$_3$	8	352	420	13.3	47.9	161.5
4-Br	4'-SCH$_3$	14	330	400	17.5	54.7	180
3-Br	4'-SCH$_3$	0	336	400	14.9	39.3	142
4-C$_2$H$_5$S	4'-F	20	358	420	13.6	32.3	100
4-C$_2$H$_5$S	4'-Cl	5.1	359	430	–	–	122.3
4-C$_2$H$_5$S	3'Cl	0.1	360	430	13.9	23.5	108.7
4-C$_2$H$_5$S	4'-Br	5.9	360	430	13.8	28.4	130
4-C$_2$H$_5$S	3'-Br	12.7	360	430	13.9	30.8	104.9
4-C$_2$H$_5$S	4'-I	15.1	360	430	13.8	27.0	127.7
4-C$_4$H$_9$S	4'-Cl	0	363	420	13.6	41.9	115.9
4-C$_4$H$_9$S	3'Cl	0	360	430	13.9	33.7	85
4-C$_4$H$_9$S	4'-Br	12.6	360	420	13.8	37.9	126
4-C$_4$H$_9$S	4'-I	12.3	360	430	13.8	32.9	142.7
4-C$_8$H$_{17}$S	4'-F	14.6	360	430	13.5	54.9	66
4-C$_8$H$_{17}$S	4'-Cl	0	362	430	13.6	37.7	88
4-C$_8$H$_{17}$S	3'Cl	16	363	430	13.7	43.7	69.5
4-C$_8$H$_{17}$S	4'-Br	0	362	435	13.6	40.0	69.2
4-C$_8$H$_{17}$S	3'Br	22.3	363	435	13.7	41.4	66
4-C$_8$H$_{17}$S	4'-I	0	362	435	13.7	35.1	102.5

From Goto, Y., et al. *Kobunshi Ronbunshu*, 47, 791, 1990; Goto, Y., et al. *Nippon Kagaku Kaishi*, 968, 1990. With permission.

iii. X = 4-C₂H₅O   Y = 4'-OCH₃

Chemical formula: $C_{18}H_{18}O_3$; space group: Pna2$_1$; point group = mm2
Cell parameters: a = 6.321 Å, b = 14.704 Å, c = 16.259 Å, Z = 4
Volume = 1511 Å^3; density = 1.241 g/cm^3

Refractive indexes:

λ (μm)	n_x	n_y	n_z
0.532	1.493	1.710	1.983
1.064	1.473	1.663	1.859

d_{eff} at 1.064 μm = 3.5 pm/V for type I and 5.7 pm/V for type II
Damage threshold = 30 GW/cm^2 at 1.053 μm laser with 1-ns pulse width

iv. X = 4-Br   Y = 4'-OCH₃

4-Bromo-4'-methoxychalcone (BMC)[184]
Melting point = 160 °C, λ_{cutoff} = 380 nm in solution and 430 nm in crystal; Point group = m
Cell parameters: a = 15.898 Å, b = 7.158 Å, c = 5.983 Å, β = 97.19°

Refractive indexes:

λ (μm)	n_x	n_y	n_z
0.532	1.58	1.50	1.92
0.632	1.55	1.47	1.90
1.064	1.52	1.42	1.83

Powder SHG = 11.5 times that of urea; d_{13} = 27 pm/V, d_{33} = 6.5 pm/V; damage threshold = 30 GW/cm^2; Type I phase matching

v. Thienylchalcones

Chemical structures[185,186]:

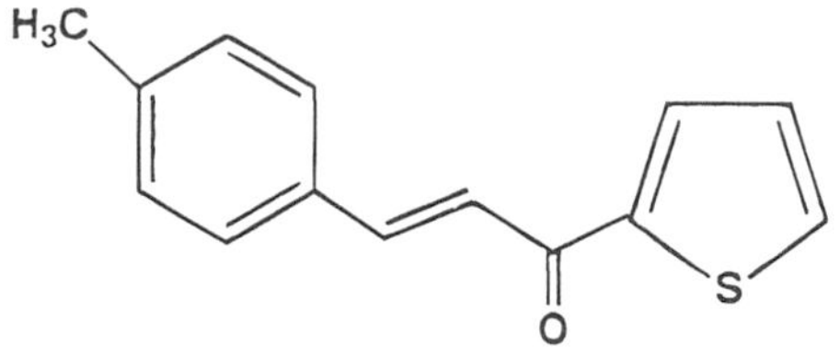

Point group: mm 2; space group: P2$_1$
Cell parameters: a = 12.3209 Å, b = 5.8782 Å, c = 17.5308 Å, β = 109.867°, Z = 4
Volume = 1194.1 Å^3; density = 1.270 g/cm^3; λ_{max} = 331 nm, λ_{cutoff} = 390 nm; melting point = 116.3 °C
Powder SHG = 15 times that of urea; hyperpolarizability = 44.2 × 10^{-30} esu; NLO coefficients at 1.064 μm: d_{eff} = 7.1 pm/V
External temperature acceptance: ΔT = 2.2 °C; walk-off angle: ρ = 3.6° (1.064 μm); Vickers hardness: 17
Acceptance angles: Δθ = 0.053° cm, Δθ = 1.2° cm$^{1/2}$ (experimental)
Δθ = 0.04° cm, Δθ = 4.7° cm$^{1/2}$ (calculated)
Type I phase-matching angles: ф = 0°, θ = 61.6°

Goto et al.[186] reported the SHG of 29 types of thienyl chalcone derivatives having a thiophene ring. Nonsubstituted chalcone showed a SHG 4.5 times greater than that of urea. It had a λ_{max} of 308 nm and λ_{cutoff} of 370 nm. Its β value was 24.9 × 10^{-30} esu. 1,3-Dithienylpropenone showed powder SHG 52 times that of urea and its β value was 34.8 × 10^{-30} esu. It had a λ_{max} of 349 nm and λ_{cutoff} of 300 nm. The SHG activity of thienyl chalcones depends on the position of thiophene ring to carbonyl group and the nature of substituents on benzene ring. Chalcones having a bromine-substituted thienyl and another substituted phenyl ring were 50% SHG active. The absorption features of chlorine, bromine, and iodine substituted compounds were similar. The effect of a substituted phenyl ring can seen below.

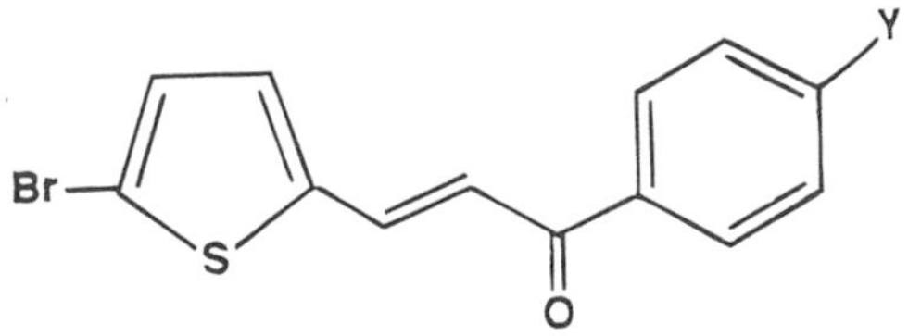

Y	Tm (°C)	SHG (× urea)	λ_{max} (nm)	λ_{cutoff} (nm)	β $(10^{-30}$ esu)
NH_2	171.2	8	367	440	48.9
OCH_3	140.8	26	351	415	44.0
Cl	131.0	22	351	430	43.1
Br	147.5	7	354	430	42.7
I	146.3	3	353	435	56.2
NO_2	166.5	15	363	435	40.6

(27) 3-Methoxy-4-hydroxybenzaldehyde (MHBA)

Chemical structure[187]:

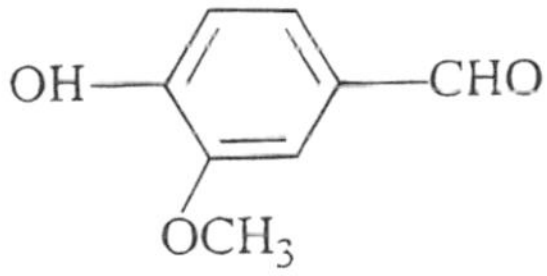

The carbonyl group acts as a strong acceptor and the hydroxy group as an electron donor. The single crystal can be grown with dimensions of $50 \times 25 \times 10$ mm^3 by a solution growth method. The type I and type II phase matching for SHG in MHBA was in wide-wavelength ranges, including 1064 and 830 nm. The MHBA yields blue-violet out.

Chemical formula: $C_8H_8O_3$; melting point = 82 °C; space group: P2$_1$; point group = 2

Cell parameters: a = 14.057 Å, b = 7.875Å, c = 15.037Å, β = 115.45°; Volume = 1503.30 Å^3, Z = 8; density = 1.34 g/cm^3

Transparency range: 0.37 ~ 1.6 μm, λ_{cutoff} = 370 nm; powder SHG efficiency: 30 × urea

NLO coefficients: d_{11} = 15.75 pm/V, d_{13} = 21 pm/V, d_{14} = 5.2 pm/V, d_{eff} = 8.47 pm/V at 1.064 μm and 10 pm/V at 0.830 μm

Hardness = 1.6 Mohs scale; type I phase-matching angles: ϕ = 0°, θ = 68.13° at 1.064 μm; walk-off angle; ρ = 6.1°; angular acceptance = 9.05 mrad mm

Refractive indexes of MHBA:

Wavelength (nm)	n_x	n_y	n_z
415.0	1.62183	1.76697	1.96996
532.0	1.56246	1.70331	1.81376
546.1	1.55840	1.70018	1.89349
587.5	1.55143	1.69045	1.80896
589.3	1.55127	1.69039	1.79235
667.8	1.53996	1.67963	1.77105
705.7	1.53673	1.67668	1.76812
830.0	1.52717	1.66944	1.75470
1064	1.51438	1.66261	1.74390

Table 37 SHG Efficiency of Stilbene Derivatives

Compound	Donor	Acceptor	X	Y	SHG (× urea)	Solvent
CMONS	CH$_3$O	NO$_2$	CN	H	300	Ethyl acetate
					250	Toluene
					90	Melt
					4	Tetrahydrofuran
	CH$_3$O	Br	H	H	7	Ethyl acetate
MONS	CH$_3$O	NO$_2$	H	H	90	Ethyl acetate
	CH$_3$O	NO$_2$	CH$_3$	H	0.4	Ethyl acetate
BMONS	CH$_3$O	NO$_2$	Br	H	168	Methanol
					37	From reaction
BONS	Br	NO$_2$	H	H	177	Hot toluene
					73	Methylene chloride
					33	Hot acetone
	Br	NO$_2$	H	CN	0.1	Chloroform/hexane
	Br	NO$_2$	CN	H	27	Chloroform
					23	Acetone
					22	Acetonitrile
					19	Toluene
					18	Ethyl acetate
BBONS	Br	NO$_2$	Br	H	93	Chloroform/pentane
					77	Methanol
					11	From reaction
	Cl	NO$_2$	H	H	11	Ethyl acetate
					4.4	Toluene
	Cl	NO$_2$	H	CN	0.5	Methanol
	F	NO$_2$	H	H	7.7	Toluene
IONS	I	NO$_2$	H	H	117	Chloroform
					86	Toluene
					67	DMSO
					72	Ethanol
					50	Ethyl acetate
	CHO	NO$_2$	H	H	28	Methylene chloride/pentane

CMONS = 2-Cyano-*p*-methoxy-*p'*-nitrostilbene; MONS = *p*-methoxy-*p'*-nitrostilbene; BMONS = 2-bromo-*p*-methoxy-*p'*-nitrostilbene; BONS = *p*-bromo-*p'*-nitrostilbene; BBONS = 2-bromo-*p*-bromo-*p'*-nitrostilbene; IONS = 4-iodo-4'-nitrostilbene. From Wang, Y., et al., *Chem. Phys. Lett.*, 148, 136, 1988. With permission.

(28) α-[4'-Methoxyphenyl)methylene-4-nitrobenzenacetonitrile (CMONS)

Wang et al.[188] reported on a series of new organic nonlinear optical materials of stilbene derivatives (Table 37). These molecules exhibited polymorphism associated with the crystal growth media. The CMONS crystals grown from different solvents varied remarkably in their SHG efficiency. A similar effect was also noticed for other compounds. The ethyllactate grown crystal showed the highest powder SHG efficiency of 300 times that of urea. The highest powder SHG efficiencies of MONS, BMONS, BONS, BBONS, IONS were 90, 168, 177, 93, and 117, respectively, relative to urea. Several weak donors such as methoxy, bromo, and iodo, which provide better optical transparency, were found to be effective in enhancing the optical nonlinearity.

Table 38 Powder SHG Efficiencies of Solution-Grown CMONS Crystals

			Powder SHG (× MNA)	
Polymorph	Solvent	Color	1.06 μm	1.32 μm
CMONS-a	Toluene	Yellow	5 ~ 7	3 ~ 4
CMONS-b	Acetic acid	Orange	0.5 ~ 0.7	2.5 ~ 8
CMONS-c	Dichloromethane/ methylated spirit		0	0 ~ 0.01

From Oliver, S. N., et al., *Appl. Phys. Lett.*, 56, 307, 1990. With permission.

Chemical structure:

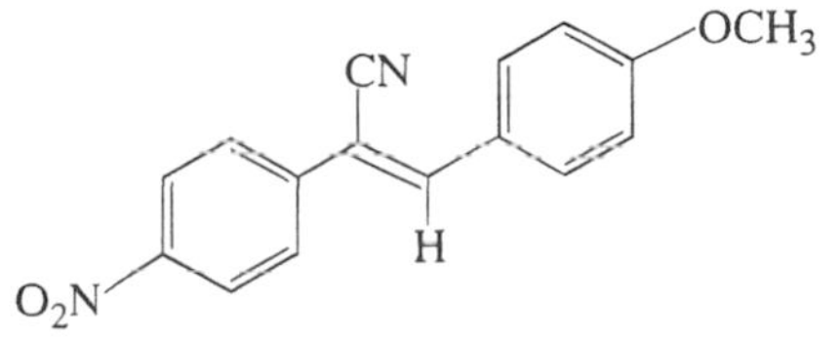

Oliver et al.[189] carried out SHG studies in CMONS powders under 1.06 and 1.32 μm irradiation to examine the influence of crystal growth conditions and thermal history of the crystals. Three distinct solution grown polymorphs (CMONS-a, CMONS-b, and CMONS-c) were indentified showing tremendous variations in their SHG activity. One of the polymorphs (CMONS-b) exhibited a 25-fold increase in SHG activity when the irradiation was changed to 1.32 μm whereas other polymorphs show a small or no increase. The melt grown CMONS-b showed SHG activity 8 times that of MNA at 1.32 μm, whereas this efficiency was only 0.3 times at 1.06 μm (Table 38).

Sugiyama et al.[190] introduced a trifluoromethyl group as the electron-withdrawing group instead of nitro to optimize NLO efficiency-transparency trade-off. Table 39 lists the details of optical and NLO properties of trifluoromethyl-containing stilbene derivatives. The β values were measured from EFISH technique and SHG on powder samples from 75 to 125 μm were determined as urea reference. These compounds have β values comparable with that of *p*-NA while their cut-off wavelength is shorter by

Table 39 Optical and NLO Properties of Trifluoromethyl Containing Stilbene Derivatives

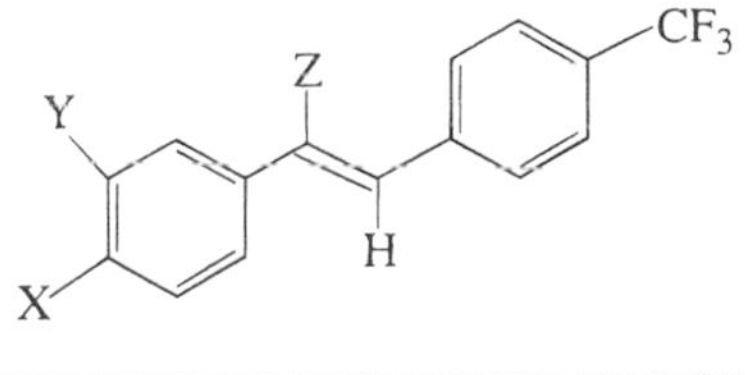

X	Y	Z	λ_{max} (nm)	λ_{cutoff} (nm)	β (10^{-30} esu)	Powder SHG (× urea)
OH	H	H	327	406	12.7	0.0
OH	CH₃	H	330	395	–	0.08
OH	OCH₃	H	333	423	–	0.31
OH	OC₂H₅	H	333	413	–	0.17
OCH₃	H	H	323	375	12.2	0.18
OCH₃	CH₃	H	326	379	12.8	0.0
H	H	CN	314	372	4.5	0.0
Cl	H	CN	318	382	5.1	0.0
Br	H	CN	320	380	8.1	3.0
CH₃	H	CN	323	385	7.4	0.0
OCH₃	H	CN	340	412	14.6	0.0
OH	H	CN	348	484	15.8	0.0
SCH₃	H	CN	362	437	15.5	1.5

From Sugiyama, Y., et al., *Sen-i-Gakkai Symp.*, B88, 1992. With permission.

90 nm. Introduction of bromo and methylthio groups with a cyano group to the double-bond position leads to greater SHG efficiency.

a. 2,4-dinitro-3′,4′,5′-trimethoxystilbene

Chemical structure[191]:

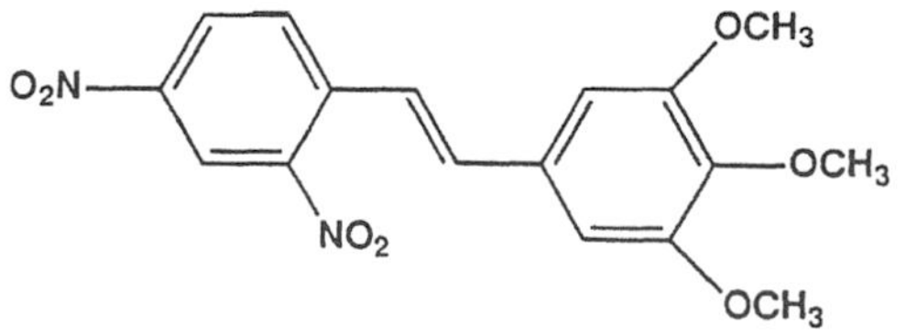

Crystal system: orthorhombic; space group: $P2_12_12_1$; cell parameters: $a = 7.345$ Å, $b = 13.425$ Å, $c = 16.956$ Å, volume = 1672 Å^3; $Z = 4$; density: 1.42 g/cm^3, $R = 0.039$; melting point = 189 °C, $\lambda_{max} = 383.1$ nm in methanol; powder SHG = 0.2 times that of urea.

(29) Hexamethylenetetramine (HMT)

HMT was the first organic crystal whose NLO and electro-optical coefficients were determined. These have been studied by many workers.

Chemical structure[106,192]:

Point group I43m; NLO coefficients: $\chi^{(2)} = 30 \times 10^{-9}$ esu, electro-optical coefficient = 12.6×10^{-8} esu[106]; electro-optical coefficients: $r_{41} = 0.8$ pm/V[192]

(30) Sucrose

Point group[193]:2; Crystal system = monoclinic; negative biaxial crystal
Melting point = 133 ∼ 135 °C, hygroscopic; transparency range: 0.192 ∼ 1.835 μm

Refractive indexes of Sucrose:

λ (μm)	n_x	n_y	n_z
0.5123	1.5404	1.5681	1.5737
1.064	1.75278	1.5552	1.5592

NLO coefficients: $d_{eff} = 0.2\ d_{eff}$ ADP at 1.064 μm; damage threshold: 500 MW cm^{-2} at 1.064 μm for 10 ns.

(31) (−)4-(4-Dimethylaminophenyl)-3-(2′-hydroxypropylamino) cyclobutane-1,2-dione (DAD)

Pu et al.[194,195] reported a new electron acceptor, cyclobutanedione, for NLO materials to prevent centrosymmetric crystal structures by the introduction of chirality and hydrogen bonding. Further chemical modifications gave a variety of novel materials. The intermolecular charge-transfer band of cyclobutanedione is at the same position to nitro analog. Tables 40 and 41 list physical and NLO properties of cyclobutanedione compounds.

Table 40 Chemical Structures, Physical and NLO Properties of Cyclobutanediones

	Material	
Data	DAD (R=CH₃)	MACD (R=CH₃OC₂H₅)
Formula	$C_{15}H_{18}N_2O_3$	$C_{17}H_{22}N_2O_4$
Crystal system	triclinic	triclinic
Space group	P1	P1
a (Å)	5.69	5.914
b (Å)	12.68	14.15
c (Å)	5.25	5.254
a (°)	93.8	96.49
b (°)	103.8	104.86
g (°)	102.0	99.0
V (Å³)	357.3	414
Z value	1	1
Density (g/cm³)	1.27	1.276
λ_{cutoff} (nm)	460	461
λ_{max} (nm)	397	397
β (10^{-30} esu)	116	126
$D\mu_e$	23	
Powder SHG (× urea)	64	58
d_{11} (pm/V)	200 (exp)	
	450 (calc)	

From Pu, L. S. and Tomono, T., Sen-i-Gakkai Symp. Preprints, B76, 1992; Pu, L. S., *Nonlinear Opt.* 3, 233, 1992.

Table 41 Chemical Structures, Physical and NLO Properties of Modified Cyclobutanediones

X	Y	Z	λ_{cutoff} (nm)	SHG (× urea)	β (10^{-30} esu)
CH₃	H	OH	426.7	2.5	
N(CH₃)₂	H	OH	445.6	0.0	
N(CH₃)₂	H	NHCH₂CH(OH)CH₃	459.8	64	171
N(CH₃)₂	H	NHCHC₂H₅CH₂OH	459	6	–
N-(CH₃)CH₂)₂OCH₃	H	NHCH₂CH(OH)CH₃	461	58	164
N(CH₃)₂	C₂H₅	NHCHC₂H₅CH₂OH	465.6	26	–
N(CH₃)₂	H	P5	476.8	8	

From Pu, L. S. and Tomono, T., Sen-i-Gakkai Symp. Preprints, B76, 1992; Pu, L. S., *Nonlinear Opt.* 3, 233, 1992.

(32) 2-Methoxy-5-nitrophenol (MNP)

Chemical structure[196]:

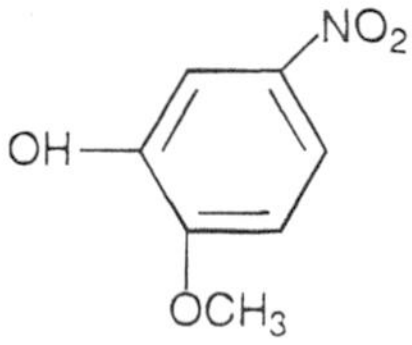

Crystal class = monoclinic; space group = $P2_1$
Unit cell parameters: $a = 11.959$ Å, $b = 16.574$ Å, $c = 3.857$ Å, $\beta = 90.06°$ Volume = 764 Å^3, $Z = 4$;
   density = 1.47 g/cm^3
$\lambda_{cutoff} = 460$ nm; melting point = 106 °C; hyperpolarizability (β) = 16×10^{-30} esu (MNDO-PM3 method)

NLO coefficients[197]:

Wavelength (μm)	d_{22} coefficient (pm/V)
1.06	43.3
1.5	95
0.948	47.8

(33) 4-Hydroxy-3-methoxybenzonitrile (HMB)

Chemical structure[198]:

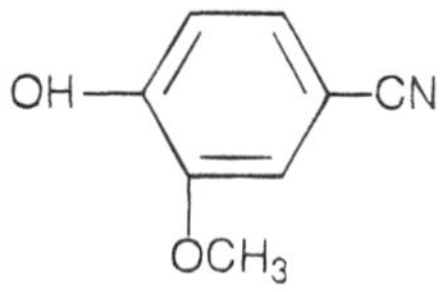

Crystal class = orthorhombic; space group = Fdd2
Unit cell parameters: $a = 29.78$ Å, $b = 49.20$ Å, $c = 4.07$ Å, Volume = 5961 Å^3, $Z = 32$, density =
   1.33 g/cm^3
$\lambda_{cutoff} = 330$ nm, nm in methanol solution; melting point = 88 °C; powder SHG = 2 times that of urea
NLO coefficient: $d_{11} = 4.4$ pm/V (theoretical), $d_{11} = 1.4$ pm/V (measured)
Refractive indexes at 550 nm: $a = 1.498$, $b = 1.670$, $c = 1.735$.

(34) 1-(4-nitrophenyl)pyrrole (NPRO)

Chemical structure[199,200]:

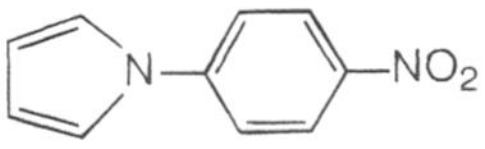

Crystal class = orthorhombic   Point group = *mm*2; space group = Fdd2
Packing density = 4.57×10^{21}/cm^{-3}; $\lambda_{max} = 323$ nm, $\lambda_{cutoff} = 410$ nm; melting point = 189 °C
Hyperpolarizability (β) = 23.75×10^{-30} esu; NLO coefficient: $d_{33} = 275.2$ pm/V (calculated)
Table 42 shows NLO properties of NPRO derivatives.

Table 42 Physical and NLO Properties of NPRO and Related Materials

Material	ENIM	DMNI	DMNT	DCNT
Crystal class	Orthorhombic	Orthorhombic	Orthorhombic	Monoclinic
Space group	Pca2$_1$	P2$_1$ab	Pca2$_1$	P2$_1$
Melting point(°C)	162	185	156	160
Packing density (10^{21}/cm^{-3})	3.85	3.82	3.97	3.97
λ_{max}(nm)	276	269	285	275
β (10^{-30} esu)	32.36	32.36	19.31	19.31
d_{ij}(pm/V)	(d_{33}) = 57	(d_{32}) = 67	(d_{32}) = 46	(d_{21}) = 71

NLO coefficients were calculated from oriented gas model and the number in parentheses indicates *ij*. Absorption spectra were recorded in ethanol solution.

From Katoh, T., et al. *Extended abstracts*, No. 30a-Q-2, Autumn Meeting 1992, Japanese Society of Applied Physics; Okazaki, M., et al. *Nippon Kagaku Kaishi*, 10, 1237, 1992. With permission.

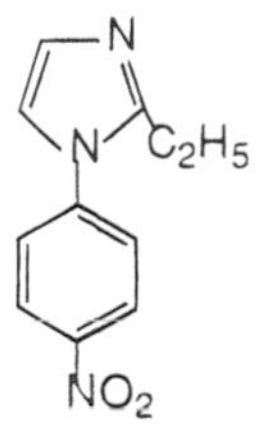

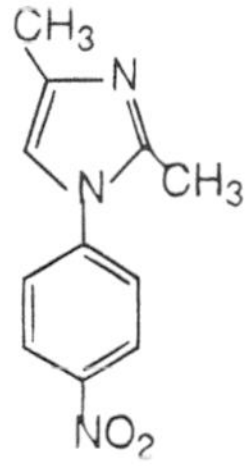

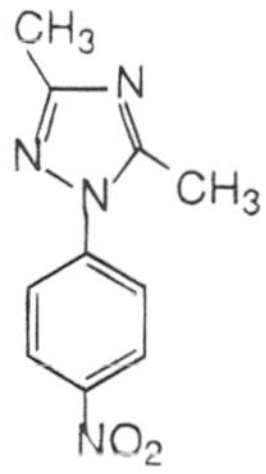

ENMI = 2-ethyl-1-(4-nitrophenyl)imidazole; DMNI = 2,4-dimethyl-1-(4-nitrophenyl)imidazole; DMNT = 3,5-dimethyl-1-(4-nitrophenyl)-1,2,4-triazole; DCNT = 3,5-dichloro-1-(4-nitrophenyl)-1,2,4-triazole; DMNP = 3,5-dimethyl-1-(4-nitrophenyl)pyrazole.

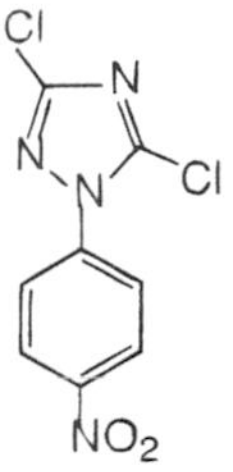

(35) 3,5-Dimethyl-1-(4-nitrophenyl)pyrazole (DMNP)

Chemical structure[201]:

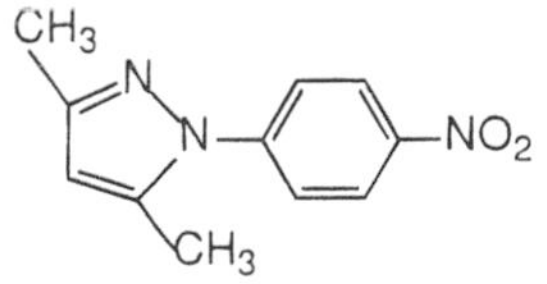

Crystal class = *mm*2; space group = P2$_1$ab
Unit cell parameters: *a* = 12.093 Å, *b* = 21.374 Å, *c* = 3.956 Å, Z = 4; packing density = 3.75 × 10^{21}/cm^{-3}, λ_{max} = 312 nm, λ_{cutoff} = 402 nm in ethanol
Melting point = 102 °C hyperpolarizability (β) = 29.16 × 10^{-30} esu

NLO coefficients of DMNP:

Theoretical	Experimental
d_{15} = 10.89 pm/V	
d_{24} = 69.96 pm/V	100.54 ± 58.65 pm/V
d_{31} = 10.89 pm/V	
d_{32} = 69.96 pm/V	
d_{33} = 28.07 pm/V	29.32 ± 4.19 pm/V;

Refractive indexes of DMNP:

λ (μm)	n_x	n_y	n_z
0.532	1.52	1.77	1.69
1.06	1.53	1.86	1.75

Okazaki et al.[202] reported pyrrolyl, diazolyl, and triazolyl-substituted 5-nitropyridines and only 2-(3,5-dimethyltriazolyl)-5-nitropyridine was found to be SHG active out of five different compounds. These heterocyclic compounds seem promising because of their large SHG activity and lower cutoff wavelength.

Chemical structure:

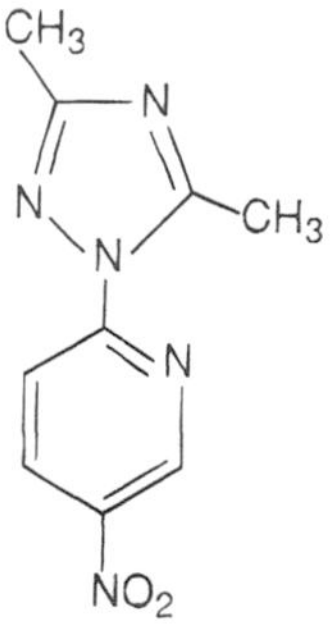

Crystal class = $mm2$; space group = Pna2$_1$
Unit cell parameters: a = 8.054 Å, b = 18.790 Å, c = 6.557 Å, V = 992.3 Å^3 Z = 4, density = 1.47 g/cm^3
λ_{max} = 298 nm, λ_{cutoff} = 389 nm; powder SHG = 14 × that of urea
Calculated NLO coefficients: d_{13} = 8.4 pm/V, d_{32} = 17.3 pm/V, d_{33} = 46.1 pm/V

(36) 2-(Aminomethylene)-benzo[b]thiophe-3(2H)-ones

Nakazumi et al.[203] prepared 2-(aminomethylene)-benzo[b]thiophe-3(2H)-ones and their sulfoxides by ring contraction of 3-bromo-4H-1-benzothiopyran-4-one and 3-bromothiochroman-4-one, respectively, with amines. The relationship between the structure and SHG activity indicates that transconfiguration between the amino groups and the carbonyl groups and the presence of an amino proton to form intermolecular hydrogen bonding between the CO and the NH are required for SHG activities. The sulfoxides or sulfones of 2-(aminomethylene)-benzo[b]thiophe-3(2H)-ones have lower SHG activity, but are advantageous in bringing about the blue-shift of absorption maxima for blue-region transparency (below 400 nm).

a. Methoxyphenyl-4H-1-benzothiopyran-4-one

Chemical structure:

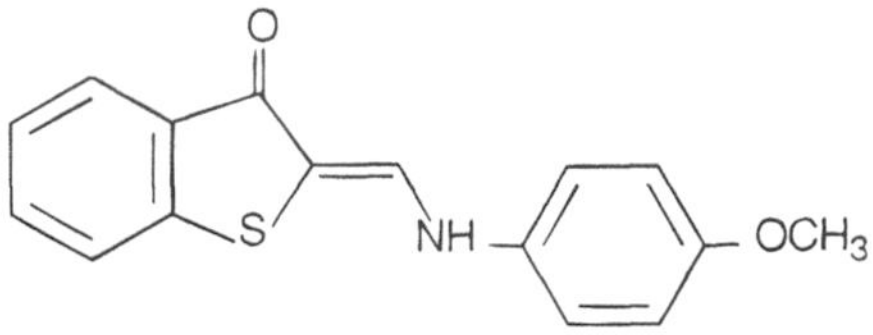

Formula = $C_{16}H_{13}NO_2S$; molecular weight = 283.34; crystal class = monoclinc; space group = Pa
Unit cell parameters: a = 11.670 Å, b = 4.964 Å, c = 11.473 Å, b = 92.79°, Volume = 663.8(7) Å^3;
    Z = 2; density: 1.42 g/cm^3
λ_{max} = 457 nm in chloroform
Powder SHG = 7 times that of urea (The replacement of 4-methoxyphenyl group by phenyl and methyl yields powder SHG efficiency of 0.006 and 0.013 times that of urea, respectively.)

Hyperpolarizability $(\beta) = 55.8 \times 10^{-30}$ esu (PPP-MO method)
The methyl substitution in place of 4-methoxyphenyl gives lower $(\beta) = 38.0 \times 10^{-30}$ esu.

b. Nitrophenylthiochroman-4-one

Chemical structure:

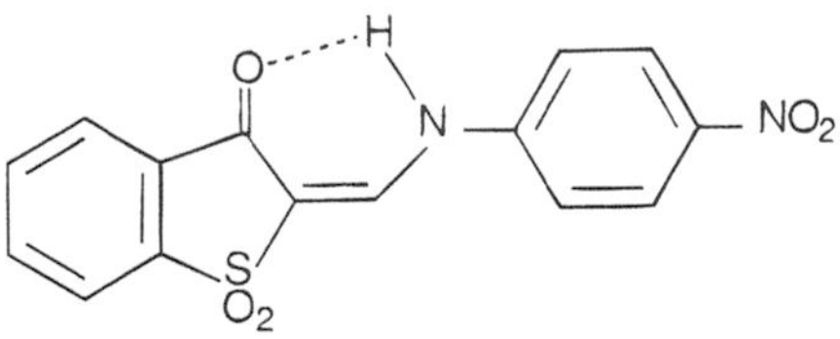

Formula $= C_{15}H_{10}N_2O_5S$; molecular weight $= 330.31$; crystal class $=$ monoclinc; space group $= P2_1$
Unit cell parameters: $a = 14.403$ Å, $b = 6.850$ Å, $c = 7.202$ Å, $\beta = 94.35°$, volume $= 708.5(5)$ Å³;
$Z = 2$; density: 1.55 g/cm³
$\lambda_{max} = 382$ nm in chloroform
Powder SHG ≤ 0.002 times that of urea (The replacement of nitro group by t-butyl and n-butoxy yields powder SHG efficiency of 3.0 and 0.016 times that of urea, respectively.)

(37) Carbamyl Compounds

Francis and Tiers[204] reported the powder SHG efficiencies of straight-chain alkyl and analogous poly-methylene esters and amides of nitrophenyl carbamic acids. These compounds exhibit SHG between 0 and 154 times that of urea at 1.064 μm, which strongly depends on the crystallization solvent and crystallizing conditions. The SHG activity for alkyl chain lengths between C_4 and C_{11} is high but starts leveling off after C_{11}, though not to a zero value. Crystallization solvent has a remarkable effect on SHG efficiency, which may be associated with the orientation of the NLO moiety. Attaching straight-chain polymethylene chains via urethane or urea bridges seems effective in yielding noncentrosymmetric structures. Table 43 lists SHG efficiency of urethane and urea compounds.

(38) Phenylureas

Miyata and co-workers[205] investigated SHG properties of phenyurea compounds (Table 44).

a. N-(2-Alkyl)-N'-(4-nitrophenylurea)

Chemical structure[205a,b]:

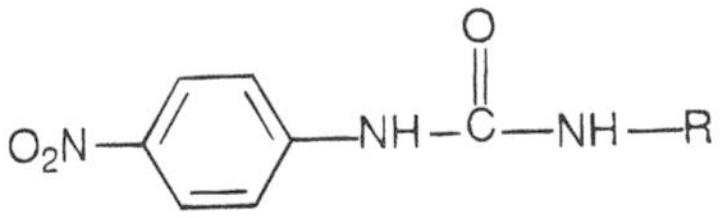

$$R = CH(CH_3)_2 \quad \text{(IPNPU)}$$

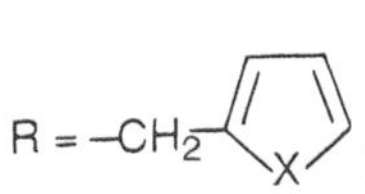

$X = O$; (FNPU), $X = S$ (TNPU)

b. Isopropyl-4-acetylphenylurea (IAPU)

Space group[205c]: $P2_1$; crystal system $=$ monoclinic; melting point $= 150$ °C
Unit cell parameters: $a = 18.645$ Å, $b = 6.539$ Å, $c = 4.90$ Å, $\beta = 96.54°$, $Z = 2$, density $= 1.23$ g cm⁻³
Transparency $= 0.380 \sim 1.4$ μm; Powder SHG $= 11$ times that of urea at 830 nm
NLO coefficients: $d_{22} = 35$ pm/V at 1.064 μm, Vicker's hardness $= 20$.

Table 43 Powder SHG Efficiency of Straight-Chain Urethanes and Ureas

R	Crystallization Solvent	mp (°C)	SHG (× urea)
CH_3	Methanol	178–179	0.001
C_2H_5	Ethanol (hot)	127–128	23
C_2H_5	Ethanol (cool)	130	5.5
$n\text{-}C_3H_7$	n-Propanol	116-117	0.0
$n\text{-}C_4H_9$	n-Butanol	96	85
$n\text{-}C_5H_{11}$	Ethanol (hot, aq)	93	46
$n\text{-}C_5H_{11}$	Ethanol (cool, aq)	93	0.03
$n\text{-}C_6H_{13}$	Ethanol	104	154
$n\text{-}C_7H_{15}$	Ethanol	103–104	110
$n\text{-}C_8H_{17}$	Ethanol	110	62
$n\text{-}C_9H_{19}$	Ethanol	106.5	13
$n\text{-}C_9H_{19}$	Ethanol (hot)	108	80
$n\text{-}C_{10}H_{21}$	Ethanol	112–114	99
$n\text{-}C_{11}H_{23}$	Ethanol	113	20
$n\text{-}C_{12}H_{25}$	Ethanol	116–118	8
$n\text{-}C_{13}H_{27}$	Heptane	115	7.6
$n\text{-}C_{13}H_{27}$	Ethanol	115.5	7.1
$CH_2{=}CH(CH_2)_9$	Ethanol	102.5	0.8
$HC{\equiv}C(CH_2)_9$	Ethanol	113	32
$n\text{-}C_{16}H_{33}$	Ethanol	120–121	60
$C_6H_5(CH_2)_5$	Ethanol	74	9
$n\text{-}C_{18}H_{37}$	Dichloromethane	121.5	14
$n\text{-}C_{16}H_{33}$	Ethanol	119–120	0.5
$cis\text{-}n\text{-}C_8H_{17}CH{=}CH(CH_2)_8$	Ethanol	88–93	0.5
$cis\text{-}n\text{-}C_8H_{17}CH{=}CH(CH_2)_{12}$	Isooctane	80	3.6
$n\text{-}C_{22}H_{45}$	Ethanol	123–124	0.6
$Cl(CH_2)_6$	n-Propanol	88	24
$Cl(CH_2)_6$	Ethanol	89	18
$Br(CH_2)_{11}$	Ethanol	122	67
$Br(CH_2)_{11}$	Acetone	123	37
$Br(CH_2)_{12}$	Ethanol	101	44
$Br(CH_2)_{16}$	Ethanol	110	23
$I(CH_2)_{11}$	Ethanol	127–129	17
$HO(CH_2)_8$	Ethanol (aq)	115	10
$CH_3S(CH_2)_2$	Ethanol (aq)	75	18

R	Crystallization Solvent	mp (°C)	SHG (× urea)
$n\text{-}C_5H_{11}$	Ethanol	131	0.001
$n\text{-}C_5H_{11}$	Acetone/isooctane	133	0.003
$n\text{-}C_6H_{13}$	n-$C_6H_{13}NH_2$	112	0.9
$n\text{-}C_{10}H_{21}$	Ethanol	118	38
$n\text{-}C_{12}H_{25}$	Ethanol	118	8
$n\text{-}C_{14}H_{29}$	Ethanol	121	11
$n\text{-}C_{16}H_{33}$	Ethanol	120	6
$n\text{-}C_{18}H_{37}$	Acetone	122	5
$n\text{-}C_{18}H_{37}$	Tetrahydrofuran	123	0.9
$cis\text{-}n\text{-}C_8H_{17}CH{=}CH(CH_2)_8$	Ethanol	105	1.6

From Francis, C. V. and Tiers, G. V. D., *Chem Mater.*, 4, 353, 1992. With permission.

Table 44 Physical and NLO Properties of Phenyurea Compounds

Material	R = i-propyl (IPNPU)	R = n-propyl (NPNPU)	R = n-butyl (NBNPU)	R = furyl (FNPU)	R = thienyl (TNPU)
Crystal class	Monoclinic	Orthorhombic	Monoclinic	Monoclinic	Monoclinic
Space group	$P2_1$	$P2_12_12_1$	$P2_1$	Pa	Pc
a (Å)	4.897	6.491	18.530	19.848	4.713
b (Å)	6.371	34.019	6.512	6.477	6.481
c (Å)	17.819	4.842	4.872	4.672	20.110
β (°)	97.121	–	83.989	90.73	90.267
Z value	2	4	2	2	2
R (%)	8.74	9.37	11.30	4.03	6.93
Melting point (°C)	192	152	148	199.8	230.4
Density (g/cm³)	1.343	1.385	1.347	1.445	1.50
λ_{cutoff} (nm)	375	375	380	380	380
β_{xxx} (10^{-30} esu)		15.1	16.0	17.07	20.01
Dipole moment (D)		9.24	9.31		
Powder SHG (× urea)	130	65	130	90	95
d_{ij} (pm/V) calculated	$d_{21} = 30$	$d_{14} = 38$	$d_{21} = 22$	$d_{11} = -19.97$	$d_{11} = -16.30$
	$d_{22} = 77$		$d_{22} = 58$	$d_{33} = 57.83$	$d_{33} = 70.76$
				$d_{13} = 40.14$	$d_{13} = 44.58$
				$d_{31} = 8.72$	$d_{31} = 14.78$

The structure shown in Section 38.a.

From Takashima, M., et al., *Extended Abstracts*, No. 16a-Z-3, Japanese Society of Applied Physics 1991; Katogi, Masters thesis. With permission.

(39) Tolane Derivatives

Chemical structure[206]:

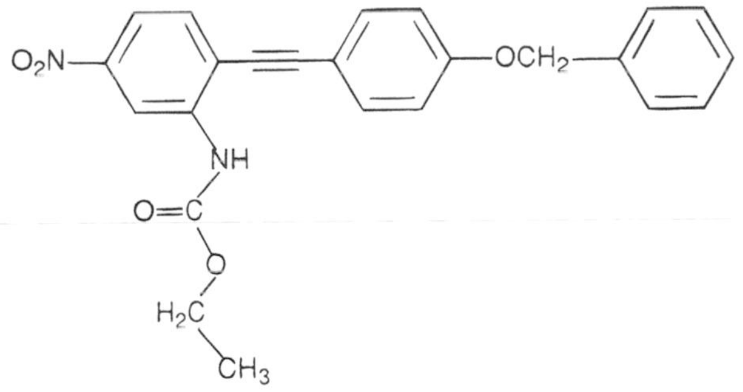

i. α-form

Crystal system: monoclinic; space group: Cc, $a = 25.828$ Å, $b = 4.947$ Å, $c = 21.090$ Å, $\beta = 127.78°$, $Z = 4$; density = 1.299 (g/cm³), $d_{eff} = 48$ pm/V.

ii. β-form

Crystal system: monoclinic; space group: $P2_1/C$; $a = 10.514$ Å, $b = 42.385$ Å, $c = 14.185$ Å, $\beta = 90.47°$, $Z = 2$; density = 1.313 g/cm³.

(40) 2-Amino-5-nitropyridine-(+)-tartarate (ANPT)

Chemical structure[207]:

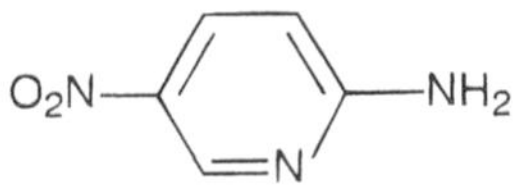

Crystal class = monoclinic; space group = $P2_1$
Unit cell parameters: $a = 7.6105$ Å, $b = 9.2019$ Å, $c = 8.2413$ Å, $\beta = 96.48°$
NLO coefficients (theoretical): $d_{33} = 41$ pm/V

Table 45 Powder SHG of *N*-methylstilbazolium Salts Measured at 1.097 and 1.064 μm Wavelengths Relative to That of a Urea Powder Standard

R	Wavelength (μm)	X=CF₃SO₃	X=BF₄	X=CH₃C₆H₄SO₃	X=Cl
4-CH₃OC₆H₄	1.907	50	0	120	60
	1.064	54	0	100	270
4-CH₃OC₆H₄-CH=CH	1.907	0	1.0	28	48
	1.064	0	2.2	50	4.3
4-CH₃SC₆H₄	1.907	0	0	–	0
	1.064	0	0	1	0
2,4-(CH₃O)₂C₆H₃	1.907	40	5.5	0	0.4
	1.064	67	2.9	0.08	0.7
C₁₀H₈-(pyrenyl)	1.907	0.8	–	37	–
	1.064	1.1	–	14	–
4-BrC₆H₄	1.907	0	0	1.7	22
	1.064	0	0.02	5.0	100
4-(CH₃)₂NC₆H₄	1.907	0	75	2000	0
	1.064	0	–	15	0
4-(CH₃)₂NC₆H₄CH=CH	1.907	500	350	115	0
	1.064	5	4.2	5	0
4-(CH₂CH₂CH₂CH₂N)C₆H₄	1.097	0.5	5.2	0.2	1.1
	1.064	0.06	0.05	0.03	0

From Marder, S.R., et al. *SPIE Proc.*, 1147, 108, 1989. With permission.

(41) Stilbazolium Derivatives

Marder et al.[208] reported the powder SHG of a series of *N*-methylstilbazolium salts with various donor groups and counterions. These compounds show very large powder SHG efficiencies (Table 45). The compounds 4-CH₃OC₆H₄−CH=CH−C₅H₄N(CH₃)⁺Cl⁻, (CH₃)₂NC₆H₄−CH=CH−CH=CH−C₅H₄N-(CH₃)⁺CF₃SO₃⁻, and (CH₃)₂NC₆H₄−CH=CH−C₅H₄N(CH₃)⁺CH₃C₆H₄SO₃⁻, show SHG efficiencies 270, 500, and 2000 times that of urea standard. This study indicates that dipolar ionic compounds have a higher tendency to crystallize in noncentrosymmetric structures than dipolar covalent compounds.

a. N,N-Dimethylamino-N'-methylstilbazolium p-toluenesulfonate (DAST)

The crystal structure of *N,N*-dimethylamino-*N'*-methylstilbazolium *p*-toluenesulfonate (DAST) (CH₃)₂NC₆H₄-CH=CH-C₅H₄N(CH₃)⁺*p*-CH₃C₆H₄SO₃, which shows exceptionally large SHG efficiency, was analyzed.[209] The crystal details are crystal class = monoclinic, space group = Cc, a = 10.365 Å, b = 11.322 Å, c = 317.897 Å, β = 92.24°, volume = 2098.2 Å³, Z = 4, density = 1.3 g cm⁻³. The λ_{max} = 474 nm in solution and 540 nm for single crystal. NLO coefficients reported by Lawrence[210] are: d_{11} = 475 ± 150 pm/V at 1.907 μm, 600 ± 200 pm/V at 1.6 μm, 1050 ± 250 pm/V at 1.3 μm, and 1900 ± 500 pm/V at 1.2 μm, d_{12} = 200 ± 50 pm/V 1.907 μm, d_{26} = 30 ± 10pm/V 1.907 μm. Pan et al.[220b] reported R_{11} = 47, R_{22} = 21, R_{13} = 5 pm/V at 1535 nm, R_{11} = 53, R_{21} = 25, R_{13} = 6.2 pm/V at 1313 nm, R_{11} = 77, R_{21} = 42, R_{13} = 15 pm/V at 800 nm and R_{11} = 92, R_{21} = 60, R_{13} = 16 pm/V at 720 nm and low dielectric constant ε_1 = 5.2 for DAST crystals.

Table 46 Crystallographic Data and NLO Properties of Stilbazolium Salts

| Data | X=OCH₃ | | X=OH[212] | X=CN[213] |
	Type 1	Type 2		
Formula	$C_{22}H_{23}NO_4S$	$C_{22}H_{23}NO_4S \cdot H_2O$	$C_{21}H_{21}NO_4S$	$C_{22}H_{20}N_2O_3S$
Formula weight	397.5	415.5	383.5	392.5
Crystal system	Triclinic	Triclinic	Triclinic	Orthorhombic
Space group	P1	P1	P1	$Pna2_1$
a (Å)	6.757	9.6720	6.615	35.115
b (Å)	7.911	10.1782	7.971	8.899
c (Å)	9.619	11.6910	9.359	6.373
α (°)	78.86	106.357	105.63	90
β (°)	80.62	109.299	97.39	90
γ (°)	83.49	90.273	95.40	90
Volume (Å³)	496(2)	1036(6)	467.0	1991.5
Z value	1	2	1	4
Density (g/cm³)	1.33	1.33	1.36	1.31
λ_{max} (nm)	392		383	335
λ_{cutoff} (nm)	480		470	390
SHG (× urea)	29.5	Inactive	8.9	3.7
NLO coefficient (pm/V)	$d_{eff} = 190$		$d_{11} = 314$ $d_{11} = 500$	$d_{33} = 20.5$

Table 47 Crystallographic Data and NLO Properties of MTSPS·H₂O

Data	Phase I	Phase II
Formula	$C_{17}H_{19}NO_3S_2 \cdot H_2O$	$C_{17}H_{19}NO_3S_2 \cdot H_2O$
Formula weight	367.48	367.47
Crystal system	Monoclinic	Orthorhombic
Space group	$P2_1$	$Pc2_1b$
a (Å)	10.380	6.989
b (Å)	11.124	12.430
c (Å)	6.968	20.59
β (°)	99.93	
Volume (Å³)	863.8	
Z value	2	4
Density (g/cm³)	1.413	1.365
λ_{max} (nm)	398	
λ_{cutoff} (nm)	490	
$\beta(0)$	37×10^{-30} esu	
β(EFISH at 1.34 μm)	61×10^{-30} esu	
Powder SHG	370 × urea (at 1.064 μm)	200 × urea (at 1.064 μm)
NLO Coefficient (d_{IJK})	d_{YXX} d_{YYY} d_{YYY}	d_{YXX} d_{YZZ}
$\lambda = 1.34$ μm (pm/V)	−100 8.9 8	70 19
$\lambda = 0$ (static) (pm/V)	60 5.2 4.8	42 11
$r_{IJK}(\lambda = 1.34$ μm) (pm/V)	−26 −2.3 −2	−18 −4.8
r_{IJK}(static) (pm/V)	−24 −2.2 −1.9	−17 −4.6

From Sebutoviez, C., et al., *Chem. Mater.*, 6. 1358, 1994. With permission.

b. Stilbazolium p-toluenesulfonates

The stilbazolium *p*-toluenesulfonates are: 1-methyl-4-(2-(4-methoxyphenyl)vinyl)pyridinium-4-toluene-sulfonate, 1-methyl-4-(2-(4-hydroxyphenyl)vinyl)pyridinium-4-toluenesulfonate (Mc-pTS), and 1-methyl-4-(2-(4-cyanophenyl)vinyl)pyridinium-4-toluenesulfonate.[211] Okada et al.[212,213] reported SHG properties of stilbazolium salts. Their crystallographic data and NLO properties are listed in Tables 46 and 47.

Chemical structure:

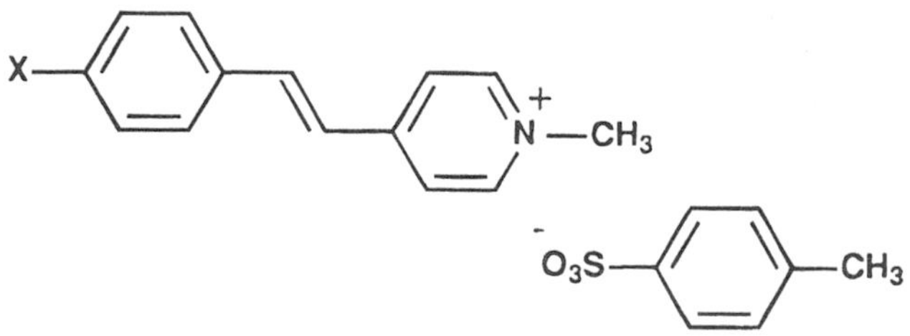

c. 4-(Methylthio)-4′-(3-sulfonatopropyl)stilbazolium (MTSPS)

Chemical structure[214]:

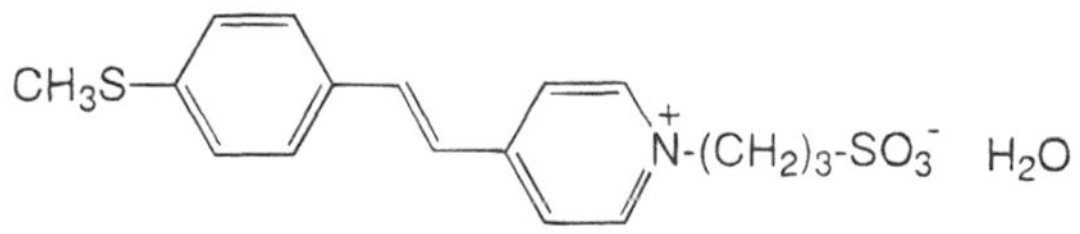

In this series, out of 33 derivatives, only six compounds showed SHG activity. The power SHG of 4-*n*-butyl′-(3-sulfonatopropyl)stilbazolium (BSPS) and 4-(3-sulfonatopropyl)stilbazolium (SPS-phase II) were 10 and 15 times that of urea at 1.064 μm.

d. Stilbazolium-PMS₃ Intercalation Compounds

The results above indicate that dimethylamino-*N*-methyl stilbazolium (DAMS$^+$) cation shows the largest SHG efficiency. Lacroix et al.[215] reported large SHG efficiency of intercalated layers materials based on organic cationic chromophores and inorganic host MPS$_3$ (M = Mn^{II}, Cd^{II}) phases. The intercalation process induces a spontaneous poling of cationic chromophore 4-[4-(dimethylamino)-α-styryl]-1 methyl-pyridinium-(DMAS$^+$) into the host MPS$_3$ matrix. The nanocomposite materials Cd$_{0.86}$ PS$_3$ (DAMS)$_{0.28}$ and Mn$_{0.86}$PS$_3$(DAMS)$_{0.28}$ exhibited a SHG efficiency 750 and 300 times that of urea at 1.34 μm, respectively. The SHG efficiency at 1.064 μm was 5 and 2 times that of urea for these Cd and Mn derivatives, respectively. In addition to large SHG efficiency, the Mn$_{0.86}$PS$_3$(DAMS)$_{0.28}$ also showed a permanent magnetization below 40 K. Therefore, the possibility to correlate NLO properties with magnetization of the host lattice in the case of MnPS$_3$ was postulated.

(42) Dimethylaminocyanobiphenyl (DMACB)

Ledoux et al.[216,217] synthesized and measured β values of a series of dimethylaminocyanopolyphenyl oligomers CN-(C$_6$H$_4$)$_n$-N(CH$_3$)$_2$ where *n* is 1,2,3, and 4 by EFISH technique. The compound with *n* = 1 showed no powder SHG whereas the compound with *n* = 2 had·SHG efficiency 20 times that of urea. For compounds *N* = 3 and *n* = 4, the SHG efficiency was rather small <KDP. The β values were 7.5×10^{-30}, 47×10^{-30}, 70×10^{-30}, and 78×10^{-30} esu at 1.34 μm and 8.3×10^{-30}, 62×10^{-10}, 93×10^{-30}, and 101×10^{-30} esu at 1.064 μm for *n* = 1, 2, 3, and 4, respectively. The static β value was found to increase from *n* = 1 to *n* = 3 but levels off for *n* = 4 and the highest β value of 55×10^{-30} esu at zero frequency was for *n* = 4.

Chemical structure:

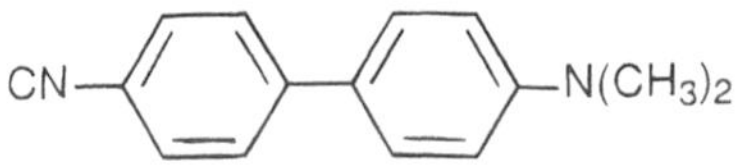

Formula = $C_{15}H_{14}N_2$; formula weight = 222; crystal class = monoclinic; space group = Cc
Unit cell parameters: a = 9.503 Å, b = 16.429 Å, c = 8.954 Å, β = 122.04°, volume = 1185.8(8)
Å³; Z = 4; density = 1.40 g/cm³
λ_{cutoff} = 380 nm; hyperpolarizability (β) = 47 × 10^{-30} esu at 1.34 μm, 2 × 10^{-30} esu at 1.064 μm,
32 × 10^{-30} esu at ω = 0 μm.
Powder SHG = 20 × that of urea; NLO coefficients (theoretical) d_{33} = 460 pm/V at 1.064 μm, d_{33} =
240 pm/V at ω = 0, r_{33} = 570 pm/V at ω = 0.

Takagi et al.[218] reported SHG in crystals of biphenyls containing bulky substituents. A nitro group
and a 3,5-di-*tert*-butylphenol moiety were found to be effective in controlling molecular arrangements
in crystals. The X-ray analysis of 4-hydroxy-3,5-di-(*t*-butyl)-2′,4′-dinitrobiphenyl indicated that the
nitro group was twisted about 40°. The 4-hydroxy-3,5-dimethyl-2′,4′-dinitrobiphenyl yielded two types
of crystals from different solvents. The OH group at the 4-position in this molecule formed a strong
hydrogen bond with alcoholic solvents which affects the molecular packing in crystals.

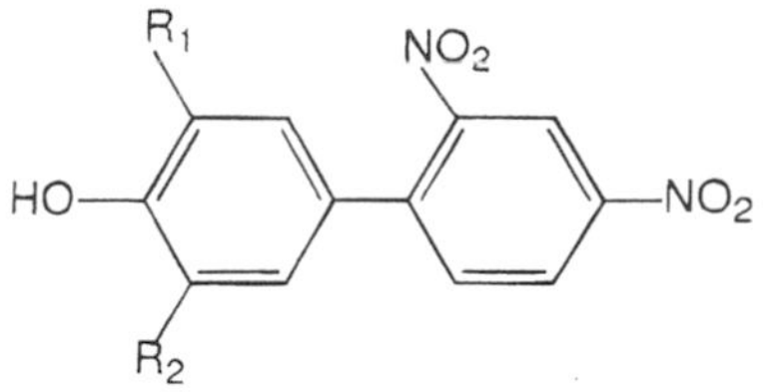

a. R₁ = R₂ = t-butyl
Crystal system = orthorhombic; space group = Pca2₁; a = 11.490 Å, b = 9.704 Å, c = 17.482 Å,
volume = 1949.2 Å³, Z = 4; density = 1.267 g cm⁻³; λ_{max} = 350 nm; powder SHG = 4.4 times that
of urea.

b. R₁ = R₂ = CH₃
Crystal system = triclinic; space group = P1; a = 7.8453 Å, b = 8.3575 Å, c = 11.3785 Å, α =
94.504°, β = 112.447°, γ = 71.306°, volume = 652.33 Å³, Z = 2; density = 1.417 g cm⁻³; λ_{max} =
345 nm; powder SHG = 2.1 times that of urea.

(43) 16-(p-Butoxybenzylidene)androsta-1,4-diene-3,17-dione

Chemical structure[219]:

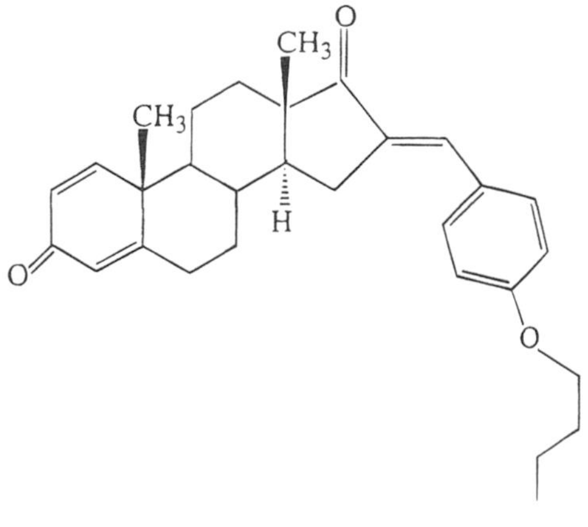

Formula = $C_{30}H_{36}O_3$; formula weight = 444.61; Crystal class = orthorhombic; space group = P2₁2₁2₁
Unit cell parameters: a = 10.928 Å, b = 37.398 Å, c = 6.139 Å, unit volume v = 2509(2) Å³; Z = 4;
density = 1.177 g/cm³

Table 48 Refractive Indexes of FMA

Refractive Indexes	Wavelength λ (μm)								
	0.4305	0.4535	0.488	0.5145	0.532	0.6328	0.838	1.064	1.152
n_o	1.751	1.721	1.691	1.685	1.672	1.641	1.619	1.612	1.617
n_e	2.137	2.064	2.007	1.983	1.958	1.887	1.841	1.821	1.811

From Kinoshita, T., et al. *J. Opt. Soc. Am.*, A11, 986, 1994. With permission.

λ_{max} = 320 nm; λ_{cutoff} = 408 nm in methylene dichloride
Dipole moment = 6.02 debye; hyperpolarizability (β_{xxx}) = 295 $\times$ 10^{-30} esu at 1.77 eV
Powder SHG = 6.7 times that of urea; NLO coefficients (theoretical): d_{36} = 2.6 pm/V

(44) 2-Furyl Methacrylic Anhydride (FMA)

Chemical structure[220,221]:

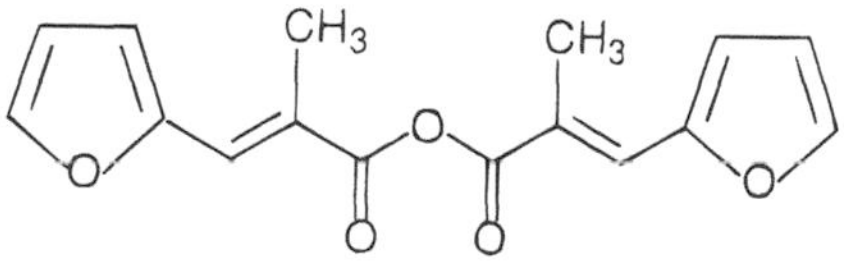

Crystal class = tetragonal 4 mm; space group = I4$_1$cd, positive uniaxial crystal
Unit cell parameters: $a = b = 16.103$ Å, $c = 10.679$ Å; $Z = 8$; volume = 2.7691 nm^3; density =
   2.7691 g/cm^3; transparency range: 0.45 ~ 1.09 μm; λ_{cutoff} = 380 nm
Melting point = 135 °C; NLO coefficients: d_{33} = 24 pm/V, d_{31} = 16 pm/V
Walkoff angle at 1.064 μm = 6.58°, Type II phase matching, phase matchable at 908 nm
Refractive indexes of FMA are listed in Table 48.

(45) Methyl-3-(p-nitrophenyl) carbazate (MNCZ)

Chemical structure[222,223]:

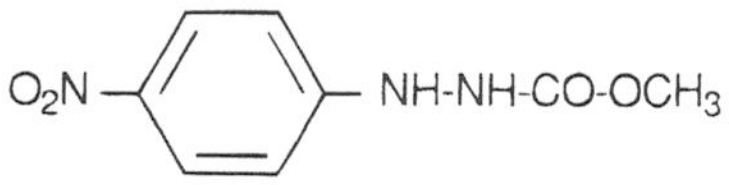

Crystal class = orthorhombic; space group = P2$_1$2$_1$2$_1$
Unit cell parameters: $a = 11.48$ Å, $b = 18.51$ Å, $c = 4.70$ Å, $Z = 4$, $\beta = 90°$; λ_{cutoff} = 470 nm
Refractive indexes: at 1.064 μm, n_a = 1.71; n_b = 1.557; n_c = 1.58;
   at 0.532 μm, n_a = 1.81; n_b = 1.63; n_c = 1.65;
Powder SHG = 1.4 times that of MNA; nonlinear optical coefficients: d_{eff} = 27 pm/V
Type I phase-matching angles: ϕ = 19° and θ = 0° at 1.064 μm; type II phase-matching angles: ϕ =
   26°, θ = 90° at 1.064 μm

(46) Dihydroxydiphenyl Sulfone (DHDPS)

Crystal class = *mm*2; space group = Aba2[224]
Unit cell parameters: $a = 1.5042$ nm, $b = 1.9280$ pm, $s = 816.0$ pm; transparency range: 0.43 ~ 1.15
   μm; λ_{cutoff} = 430 nm
NLO coefficients: d_{33} = 14 $\times$ d_{11} SiO$_2$, d_{11} = 5.2 $\times$ d_{11} SiO$_2$, d_{32} = 0.7 $\times$ d_{11} SiO$_2$

Sellmeir coefficients for DHDPS

n	A	B	C	D (10^3)
n_x	1.70764	0.99256	0.04793	-12.083
n_y	1.77690	0.75585	0.04172	-2.0940
n_z	1.64774	0.87583	0.04187	-5.7908

(47) 3,9-Dinitro-5a,6,11a,12-tetrahydro[1,4]-benzoxadino [3,2-b]benzoxadine (DNBB)

Chemical structures[225]:

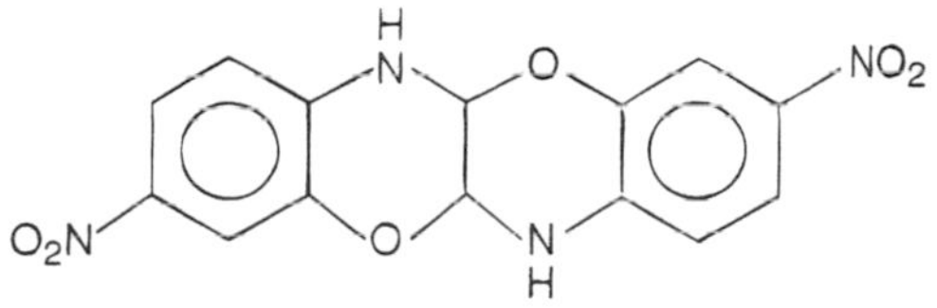

Point group = monoclinic; space group: Cc
Unit cell parameters: $a = 10.192$ Å, $b = 2.494$ Å, $c = 6.864$ Å, $\beta = 122.65°$ unit volume = 1384.0 Å^3; density: 1.58 g/cm^3
λ_{cutoff} = 500 nm; melting point = 280 °C; powder SHG = 200 times that of urea.

(48) Benzophenone Derivatives

Chemical structure[226]:

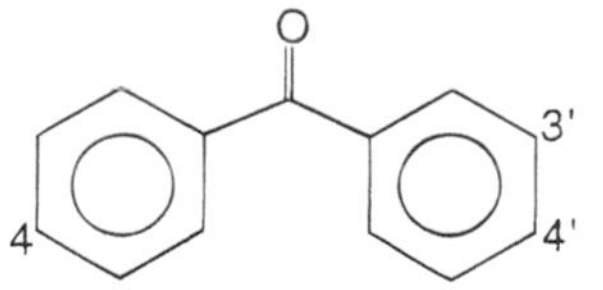

Substituents		Powder SHG (× urea)	λ_{cutoff} (nm)
4-OCH$_3$	4'-NO$_2$	5.3	388
4-OCH$_3$	4'-NH$_2$	3.7	400
4-OCH$_3$	4'-NHCH$_3$	5.4	410
4-OCH$_3$	4'-NH(CH$_3$)$_2$	4.1	410
4-OCH$_3$	4'-NH-C$_2$H$_5$	0	400
4-SCH$_3$	4'-NO$_2$	0	420
4-SCH$_3$	4'-NH$_2$	0	400
4-SCH$_3$	4'-NHCH$_3$	9.7	420
4-SCH$_3$	4'-NH(CH$_3$)$_2$	4.1	420
4-SCH$_3$	4'-NH-C$_2$H$_5$	5.4	410
4-NH$_2$	4'-NO$_2$	0	430
4-NHCH$_3$	4'-NO$_2$	2	460
4-OCH$_3$	3'-NO$_2$	7.4	350

a. 4-Methoxy-4'-nitrobenzophenone

Symmetry class[226]: monoclinic; space group: P2$_1$/C; unit cell parameters: $a = 13.414$ Å, $b = 10.867$ Å, $c = 8.436$ Å, $\beta = 90.74°$, volume = 1229.7 Å^3; λ_{cutoff} = 388 nm; powder SHG = 5.3 times that of urea.

b. 4-Aminobenzophenone (ABP)

Symmetry class[227]: monoclinic; space group: P2$_1$; unit cell parameters: $a = 8.304$ Å, $b = 5.460$ Å, $c = 12.074$ Å, $\beta = 98.87°$, $Z = 2$; transparency range: 420–2000 nm; λ_{cutoff} = 411 nm; powder SHG = 360 times that of ADP; damage threshold = 30 GW/cm^2.

(49) Xanthone Derivatives[228]

Substituents		Powder SHG ($\times$ urea)	λ_{cutoff} (nm)
H		2.2	360
1-NO$_2$		0	368
2-NO$_2$		4.5	392
3-NO$_2$		0.7	380
4-NO$_2$		0	370
1-NH$_2$		0	460
2-NH$_2$		0.4	460
3-NH$_2$		13.5	400
4-NH$_2$		0	410
3-OH		4.9	
3-NO$_2$		4.5	392
3-NH$_2$		13.0	400
3-NHCH$_3$		0	403
3-NH(CH$_3$)$_2$		0	402
3-NH-C$_2$H$_5$		0.5	418
3-N(C$_2$H$_5$)$_2$		0.7	420
3-NHCONH$_2$		7.5	352
3-NHCOCH$_3$		1.4	360
3-NHCOC$_6$H$_5$		0.4	367
3-NHCOC(CH$_3$)(CF$_2$)C$_6$H$_2$		3.4	358
2-OH		4.6	
2-NO$_2$		4.5	392
2-NH$_2$		0	460
2-NHCONH$_2$		4.3	420
2-NHCOCH$_2$O-R$_1$		4.8	390
2-NHCOC(CH$_3$)(CF$_3$)C$_6$H$_5$		0	
2-CH$_3$	7-NO$_2$	3.9	390
4-CH$_3$	7-NO$_2$	2.8	385
1-NH$_2$	6-NO$_2$	12.9	460
2-NH$_2$	7-NO$_2$	0	480
1-NO$_2$	6-NO$_2$	0	400
2-NO$_2$	7-NO$_2$	5.9	391

From Imanishi, Y., et al. *SPIE Proc.*, 1361, 570, 1990. With permission.

a. 3-Aminoxanthone

Symmetry class: monoclinic; space group: P2$_1$; unit cell parameters: a = 13.394 Å, b = 4.718 Å, c = 8.084 Å, β = 103.12°, Z = 2; unit volume = 497.5 Å^3; λ_{cutoff} = 430 nm; melting point = 232 °C; dipole moment = 6.3 debye at 1.34 μm; powder SHG = 13.5 times that of urea.

(50) 9-Hydroxyethylcarbazole Derivatives

Chemical structure:

Sasabe and co-workers[229] reported on crystal structures of carbazole derivatives (Table 49).

Table 49 Crystallographic Data of 9-Hydroxyethylcarbazole Derivatives[229]

Data	$R_1 = R_2 = H$	$R_1 = NO_2, R_2 = H$	$R_1 = R_2 = NO_2$
Crystal system	Tetragonal	Monoclinic	Triclinic
Space group	I4$_1$	P2$_1$	P$\bar{1}$
a (Å)	27.545	13.489	8.317
b (Å)	27.545	4.764	10.634
c (Å)	6.057	9.636	8.125
α (°)	90	90	107.20
β (°)	90	102.87	100.10
γ (°)	90	90	71.63
Volume (Å³)	4595.6	603.7	648.8
Z value	16	2	2
λ_{max} (nm)	330	374	370

(51) Carbamic Acid Alkylesters

a. 4'-Nitrophenylmethyl-4-methoxybenzoate
Chemical structure[230]:

Space group: Pna2$_1$; crystal system = orthorhombic; unit cell parameters: a = 13.678 Å, b = 24.931 Å, c = 3.913 Å, Z = 4, R = 5.11%; density = 1.430 g cm^{-3}; λ_{max} = 268 nm; λ_{cutoff} = 325 nm; powder SHG = 10 times that of urea; NLO coefficients: d_{33} = 1.8 pm/V, d_{32} = 10 pm/V, d_{31} = 8.6 pm/V.

b. Isopropyl-4-(4methoxyphenylmethylamino) benzoate
Chemical structure[230]:

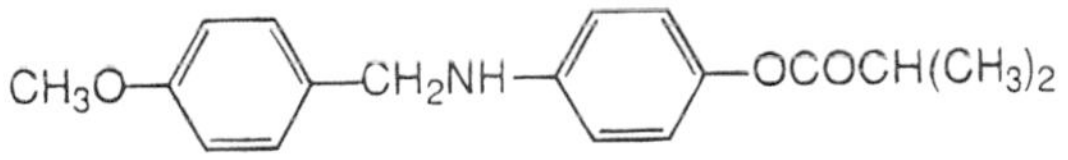

Space group: P2$_1$/C; crystal system = monoclinic; unit cell parameters: a = 8.118 Å, b = 27.657 Å, c = 7.823 Å, β = 112.40°; R = 7.0%, Z = 4; density = 1.223 g cm^{-3}; λ_{max} = 275 nm, λ_{cutoff} = 335 nm; powder SHG = 5 times that of urea; NLO coefficients: d_{33} = 57.6 pm/V, d_{31} = 12.3 pm/V.

c. 4-Nitrophenylcarbamic Acid Ethylester (ECNB)
Chemical structure[230]:

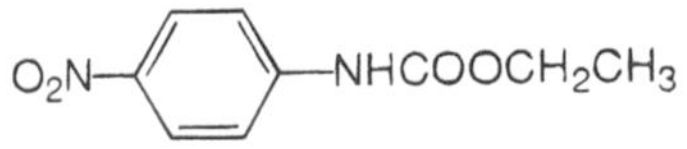

Space group: P2$_1$; unit cell parameters: a = 6.133 Å, b = 5.031 Å, c = 15.804 Å, β = 91.03°, R = 4.2%; Z = 2; density = 1.424 g cm^{-3}, λ_{max} = 390 nm, λ_{cutoff} = 480 nm; powder SHG = 50 times that of urea; NLO coefficients: d_{22} = 32 pm/V, d_{21} = 30 pm/V.

(52) 5-Nitrouracil
Chemical structure[231,232]:

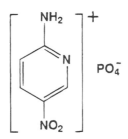

Point group: orthrohombic; space group: $P2_12_12_1$
Cell parameters: $a = 9.94$ Å, $b = 10.30$ Å, $c = 5.47$ Å; density: 1.83 g/cm³; melting point = 300 °C; transparency range: 0.5 ~ 1.5 μm.

Refractive indexes of 5-nitrouracil:

λ (μm)	n_x	n_y	n_z
0.53	1.57	2.00	1.71
1.064	1.54	1.97	1.67

NLO coefficients (pm/V): $d_{36} = (1.0 + 0.2)$ (d_{33} (LiIO₃), $d_{14} = 8.7$ pm/V[232]; damage threshold = 3 GW/Cm² at 1.06 μm and 1 GW/cm² at 532 nm for 10 ns.

(53) 2-Amino-5-nitropyridinium Dihydrogen Phosphate (2A5NPDP)
Chemical structure[233,234]:

Space group: Pna2₁; point group = *mm2*
Unit cell parameters: $a = 25.645$ Å, $b = 6.228$ Å, $c = 5.675$ Å; unit volume = 906.4 (2) Å³; density: 1.7376 g/cm³
Transparency range: 0.42 ~ 1.65 μm, $\lambda_{cutoff} = 420$ nm; dipole moment = 6.3 debye at 1.06 μm
Hyperpolarizability (β) = 20.3×10^{-30} esu (acetone) at 1.34 μm; 27.3×10^{-30} esu (ether) at 1.34 μm; 27.0×10^{-30} esu (acetone) at 1.06 μm; 33.9×10^{-30} esu (ether) at 1.06 μm.

Nonlinear optical coefficients of 2A5NPDP:

NLO Coefficient	At 1.06 μm (pm/V)		At 1.34 μm (pm/V)	
	Theoretical	Experimental	Theoretical	Experimental
d_{15}	3.53	7.2	27.8	6
d_{24}	15.9	1.3	12.5	—
d_{31}	35.6	—	28.3	—
d_{23}	16.3	—	12.8	—
d_{33}	122.5	—	87.1	12

Horiuchi et al.[235] reported the growth of 2A5NPDP single crystal by spontaneous nucleation in tetramethoxysilane gel by slow cooling. The refractive indexes of gel-grown crystals were: $n_x = 1.6156$, $n_y = 1.6411$, and $n_z = 1.7330$ at 0.76 μm and $n_x = 1.5788$, $n_y = 1.6126$, and $n_z = 1.6816$ at 1.34 μm, almost the same as reported earlier. The largest NLO coefficients at 1.34 mm were: $d_{15} = 15.42$ pm/V, $d_{eff} = 11.39$ pm/V for A platelets and $d_{15} = 2.33$ pm/V, $d_{eff} = 2.23$ pm/V for B platelets cut for both phase matchings when the sample was rotated around a vertical axis. The higher NLO coefficients were considered to be related with the better quality of single crystals. The better optical quality also led to a figure of merit of 88% for gel-grown crystals.

Masse et al.[236] reported quadratic NLO properties of a series of organic-inorganic polar crystals where inorganic bulky anions or polyanions assisted in forming polar packing.

Crystal	Space group	a (Å)	b (Å)	c (Å)	β (°)	Powder SHG
$C_5H_6N_3O_2^+NO_2^-$	$P2_1/n$	13.672	7.762	7.577	101.45	−
$(C_5H_6N_3O_2^+)_2CuCl_4^-$	$P2_1/c$	7.701	31.99	7.107	92.65	−
$(C_5H_6N_3O_2^-)_2Cr_2O_7^-$	$P2_1/nb$	22.454	15.129	5.189	−	+
$C_5H_6N_3O_2^-H_2PO_4^-$	$Pna2_1$	25.645	6.228	5.675	−	+
$C_5H_6N_3O_2^-HSO_4^-$	$Pcab$	29.617	13.185	9.013	−	−
$C_5H_6N_3O_2^-1\text{-}C_4H_5O_6^-$	$P2_1$	8.248	9.199	7.611	96.5	+
$2A5NP\cdot CCl_3\text{-}COO^-$	$P2_1/\gamma$	14.136	6.876	12.316	103.49	+
$2A5NP\cdot CHCl_2\cdot COO^-$	$P2_12_12_1$	6.010	10.841	17.416	−	(Equal to POM)
$2A5NP\cdot CH_2Cl\text{-}COOH$	Cc	4.857	21.755	9.428	104.08	(Equal to POM)

(54) 1,1'-(1,8-Naphthylene)-bis(1H-1,2,3-triazole) (BTN)
Chemical structure[237]:

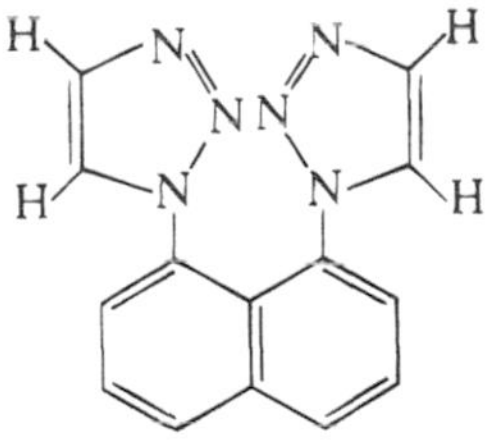

Formula = $C_{14}H_{10}N_6$; molecular weight = 262.3; space group: Fdd2; crystal system = orthorhombic
Unit cell parameters: $a = 5.489$ Å, $b = 25.241$ Å, $c = 17.675$ Å; unit volume = 2448.6 (2) Å³; Z = 8; density: 1.42 g/cm³ $\lambda_{cutoff} = 325$ nm; dipole moment = 4.445 debye; hyperpolarizability (β) = -1.682×10^{-30} esu (CNDO/S-CI calculation).

(55) L-Arginine phosphate (LAP) and Tris-hydroxymethyl-aminomethane Phosphate (THAMP)

a. L-Arginine Phosphate (LAP)

Chemical structure[238]: $(H_2N)_2CNH(CH_2)_3(NH_3)^+COO^- \cdot H_2PO_4^- \cdot H_2O$
Point group: 2, monoclinic; space group = $P2_1$; negative biaxial crystal: $2V_z = 141°$
Transparency range: 0.25 ~ 1.3 μm; Melting point = 130 °C

Refractive indexes of LAP:

λ (μm)	n_x	n_y	n_z
0.355	1.5332	1.6078	1.6211
0.532	1.5116	1.5792	1.5662
1.064	1.4973	1.5589	1.5674

Table 50 Crystallographic Data of LAP, DLAP, THAMP, and THAMS

Data	LAP	DLAP	THAMP	THAMS
Crystal system	Monoclinic (P2$_1$)	Monoclinic	Monoclinic	Monoclinic
λ_{cutoff} (nm)	260	260	260	210
Melting point (°C)	140	–	169	167
Hardness (Vickers)	56	–	58	54
Angular acceptance from SHG (mrad cm)				
Type I	–	0.5	2.1	
Type II	0.84	1.1	–	–
Temperature acceptance from SHG (°C cm)				
Type I	14	–	–	–
Type II	27	–	–	–
d_{eff}(pm/V)				
Type I	0.69		0.16	0.1
Type II	0.87	0.83	–	–

From Sasaki, T., et al. *Proceedings of First International Workshop on Crystal Growth of Organic Materials (CGOM)*. 1989:2216. With permission.

Linear absorption coefficient: α (cm^{-1}) = <0.001 at 532 nm and 0.1 ~ 0.18 at 1.064 μm
Breakdown threshold: tp = 0.6 ns at 526 nm, 25 ns at 1.053 μm, and 1 ns at 1.064 μm
I (10^9 W cm^{-2}) = 60 at 526 nm, 13 at 1.053 μm, and 9.8 ~ 13.4 at 1.064 μm[239]
NLO coefficients: d_{21} = 0.931 pm/V, d_{22} = 0.413 pm/V, d_{23} = 0.448 pm/V, d_{36} = 0.635 pm/V

b. Deuterated L-Arginine Phosphate (DLAP)[238]

Point group: 2; negative biaxial crystal: 2V$_z$ = 142°; Transparency range: 0.25 ~ 1.3 μm

Refractive indexes:

λ (μm)	n_x	n_y	n_z
0.355	1.5298	1.6043	1.6167
0.532	1.5090	1.5764	1.5847
1.064	1.4960	1.5584	1.5655

Linear absorption coefficient α (cm^{-1}) = 0.07 ~ 0.12 at 266 nm, <0.001 at 532 nm, and 0.01 at 1.064 μm
Breakdown threshold: tp = 0.6 ns at 526 nm, 25 ns at 1.053 μm and 1 ns at 1.064 μm
I (10^9 W cm^{-2}) = 67 at 526 nm, 87 at 1.053 μm, and 9.2 ~ 13 at 1.064 μm (ref. 239).

Sasaki et al.[240] reported crystal growth and NLO properties of LAP and deuterated LAP crystals. LAP is an organic salt consisting of L-arginine and phosphoric acid. LAP is monoclinic and biaxial having absorptions at 1.06 μm due to the higher harmonics of the molecular vibration of the functional groups such OH, NH, $-NH_3$, NH_2, and $-NH_2^+$. LAP's molecular structure is $(H_2N)_2CNH(CH_2)_3$-$(NH_3)^+COO^-\cdot H_2PO_{i4}^-\cdot H_2O$. Absorption was reduced by the deuteration of these functional groups; deuterated LAP had low absorption loss. THAMP having molecular structure $(NH_3)^+C(CH_2OH)_3\cdot H_2PO_4{}^-$ and THAM-sulfate (THAMS) are both monoclinic and biaxial. Their crystals were grown by the slow-cooling method. Table 50 lists the details of these four single crystals.

Haussuhl et al.[241] reported SHG properties of single crystals with dimensions up to $80 \times 60 \times 30$ mm of L-arginine HCl·H_2O and L-arginine HBr·H_2O.

(56) Potassium Malate

Chemical structure[242]: COOK-CHOH-CH_2COOK·1.5H_2O
Point group: m; negative biaxial crystal: $2V_2 = 131.1°$; transparency range: $0.24 \sim 1.3$ μm
NLO coefficients: $d_{31} = 0.61$ pm/V; $d_{32} = 2.3$ pm/V

Refractive indices:

λ (μm)	n_x	n_y	n_z
0.532	1.5606	1.5060	1.4954
1.064	1.5450	1.4931	1.4838

(57) Benzalbarbituric Acid Derivatives

Kondo et al.[243] reported powder SHG of a number of p-substituted benzalbarbituric acid derivatives.

Compound	R_1	R_2	R_3	SHG (× urea)
1	OH	H	H	11 (512 nm)
2	OH	H	CH_3	0
3	OH	CH_3	CH_3	0.6
4	OH	H	C_6H_5	0
5	OH	C_6H_5	C_6H_5	1
6	NHCOCH$_3$	H	H	12
7	NHCOCH$_3$	H	CH_3	0
8	NHCOCH$_3$	CH_3	CH_3	21
8	NHCOCH$_3$	C_6H_5	C_6H_5	1
9	NHCOCH$_3$	H	C_6H_5	0

Compound 8 shows SHG activity that is the same as of MNA. The details of its crystal structure are as follows: melting point = 280 °C; space group: Pc; crystal system = monoclinic; unit cell parameters: $a = 5.472$ Å, $b = 8.500$ Å, $c = 16.618$ Å, $\beta = 118.36°$; powder SHG = 21 (× urea) at 459 nm.

(58) Methanediamine Derivatives

Yamamoto et al.[97] reported large SHG from several methanediamine derivatives forming lambda (Λ) shape conformation by bonding two aniline derivatives via the methylene bridge (Table 51). Crystal structures of N, N′-bis-(2-methylcarbonylphenyl) methanediamine (MNPMDA), N,N′-bis-(2-methoxy-carbonylphenyl)methanediamine (MCPMDA), and N,N′-bis-(2-ethoxycarbonylphenyl)methanediamine (ECPMDA) have been analyzed. The crystal details are listed below. More details of Λ-shape molecules can be found in reference 97.

Table 51 Crystallographic Data of Λ-Shaped Methanediamine Derivatives

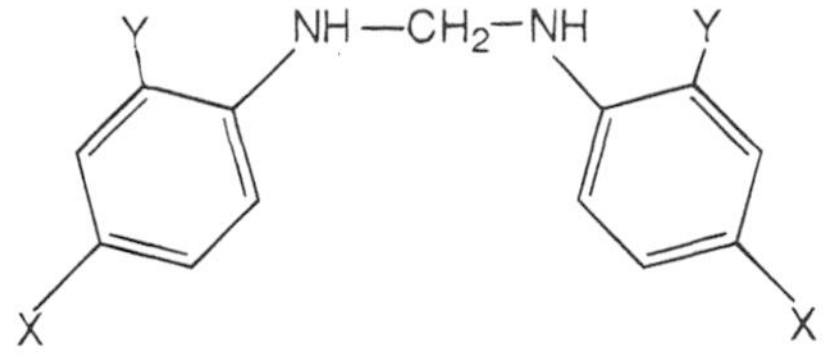

p-MNDA	MNPMDA	MCPMDA	ECPMDA
X = NO$_2$	X = NO$_2$	X = COOCH$_3$	X = COOC$_2$H$_5$
Y = CH$_3$	Y = H	Y = H	Y = H

Data	MNPMDA	MCPMDA	ECPMDA	p-NMDA	o-NMDA
Melting point (°C)	240	215	195	260	241
Crystal system	Monoclinic	Monoclinic	Tetragonal	Trigonal	Monoclinic
Space group	C2	Cc	I4$_1$cd	P3	Cc
a (Å)	16.795	16.275	20.472	18.791	23.237
b (Å)	5.233	7.779	20.475	18.788	4.111
c (Å)	9.802	14.029	8.488	4.147	14.635
α (°)	90	90	107.20		
β (°)	120.6	119.5	90	120	111.4
γ (°)	90	90	90		
Volume (Å^3)	595.6	603.7	648.8		
Z value	2	4	8	3	4
R (%)	3.9	5.1	5.8	5.8	6.0
Density (g cm^{-3})	1.42	1.35	1.28	1.36	1.47
λ$_{cutoff}$ (nm)	460	330	330	450	480
Λ-angle (°)	110	125	120	120	125
Dipole moment (D)	0.6	1.7	1.5	5.4	3.0
SHG (× urea)	80	30	25	6.3	3
β$_{xxx}$ (10^{-30} esu)	2.1	2.1	1.1	1.7	0.211
β$_{xyy}$ (10^{-30} esu)	2.7	10.2	10.2	12.2	2.7
Refractive index	$n_y = 1.630(\omega)$		$n_y = 1.761(\omega)$		
	$n_y = 1.687(2\omega)$		$n_y = 1.791(2\omega)$		
d$_{ij}$ (pm/V)	$d_{yyy} = 9.3$		$d_{zzz} = 4.7$		
	$d_{yxx} = 13.1$		$d_{zzz} = 20.4$		

From Miyata, S., Watanabe, T., and Nalwa, H. S., *Proceedings of the International School of Physics "Enrico Fermi", Nonlinear Optical Materials: Principles and Applications,* Degiorgio, V. and Flytzanis, C., eds., IOS Press, Amsterdam 1995, p. 225. With permission.

(59) 2-Cyano-2-propenoic Acid Derivatives

Sano et al.[245] reported of 2-cyano-2-propenoic acid derivatives. The chemical structures of two derivatives, 2-cyano-3-(3,4-methylenedioxyphenyl-2-propenoic acid ethyl ester (CMPE) and 2-cyano-3-(3,-(3,4-dimethoxyphenyl-2-prpenoic acid ethyl ester (CDPE), are shown below. The CMPE shows polymorphism and crsytals grown from toluene were SHG inactive. The CMPE shows SHG activity 16 times that of urea whereas CDPE was found to be SHG inactive. In CMPE CO and CN groups are oriented in the same direction on the conjugated plane consisting of the C=C double bond and the phenyl ring. In CDPE, the dipole moments of CO and CN groups cancel each other out and the adjacent molecular planes are not aligned in parallel. The *cis* conformation for CO and CN groups is important for SHG activity in CMPE.

Nakatani et al.[246] reported SHG activities of α-cyano cinnamic acid ester derivatives. Of 13 compounds, only 5 were SHG active. The powder SHG activities of (i) R=CH$_3$, R_1=OC$_2$H$_5$, R_2=R_3=H, (ii) R=C$_2$H$_5$, R_1=OC$_2$H$_5$, R_2=R_3=H, (iii) R=CH$_3$, R_1=H, R_2=R_3=OCH$_3$, (iv) R=CH$_3$, R_1=R_2=H, R_3=OCH$_3$, and (v) R=C$_2$H$_5$, R_1=R_2=H, R_3=OC$_2$H$_5$ were 0.02, 0.22, 5, 10, and 0.5 times that of urea, respectively. These compounds showed good optical transparency. Compounds having R_1=R_3=H, R_2=NH$_2$ and R=CH$_3$, showed powder SHG of 20 times that of urea while compounds R_1=R_3=H, R_2=NH−CO−CH(CH$_3$)−C$_6$H$_5$ and R=CH$_3$ (CPAC) showed SHG of 46 times that of urea.

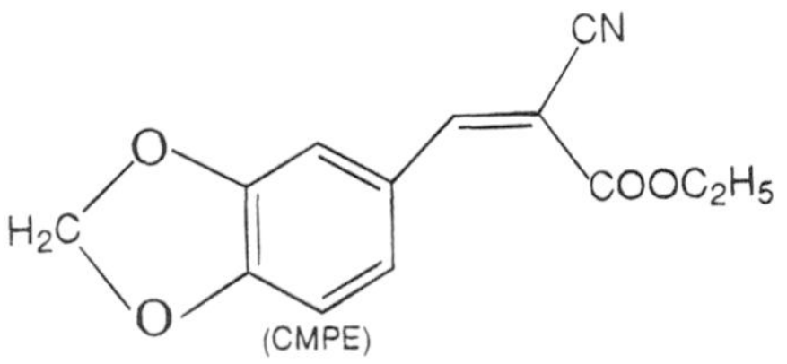

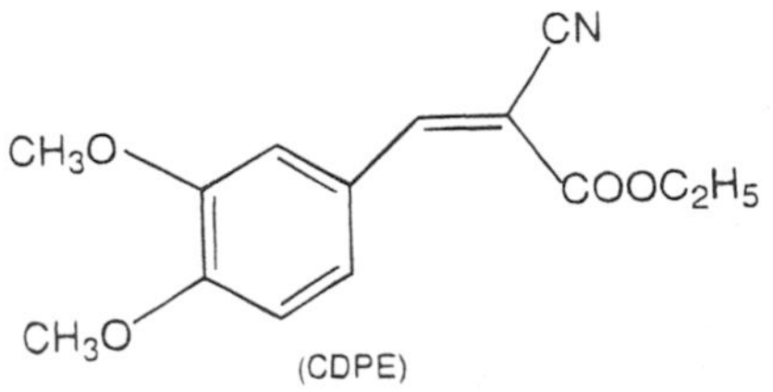

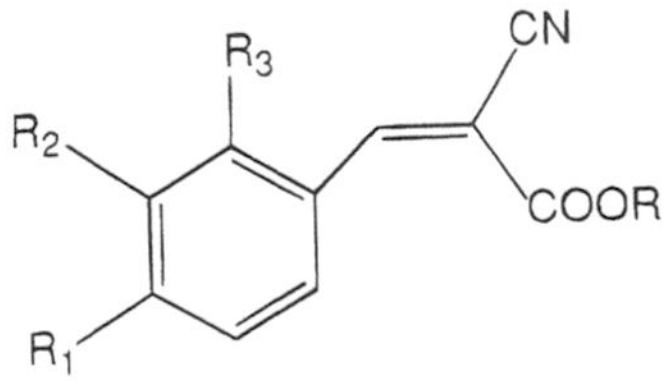

R=CH₃, R₁=R₂=H, R₃=OCH₃, (CMP)

Rendered: R=CH$_3$, R$_1$=R$_2$=H, R$_3$=OCH$_3$, (CMP)

Data	CMPE[a]	CDPE[b]	CMP[b]
Melting point (°C)	105	153	110
Color	Greenish yellow	Yellow	Yellow
Crystal system	Monoclinic	Monoclinic	Monoclinic
Space group	C2	P2$_1$/n	P2$_1$
a(Å)	14.357	13.849	4.020
b (Å)	7.063	8.0258	9.984
c (Å)	12.548	13.122	13.757
β (°)	111.01	114.22	92.25
Volume (Å^3)	1187.9	1330.2	–
Z value	4	4	2
Density (g cm^{-3})	1.37	1.31	1.307
SHG ($\times$ urea)	16	none	10

For CMP, λ_{cutoff} = 410 nm, d_{22} = 29 pm/V, β = 4.6 $\times$ 10^{-30} esu.

[a] From Sano, K., et al. *Nonlinear Optics,* 2, 89, 1992. With permission.
[b] From Nakatani, et al. *Jpn. J. Appl. Phys.,* 31, 1802, 1992. With permission.

(60) 4-Amino-4'-nitrodiphenyl Sulfide (ANDS)

a. Chemical structure[247]:

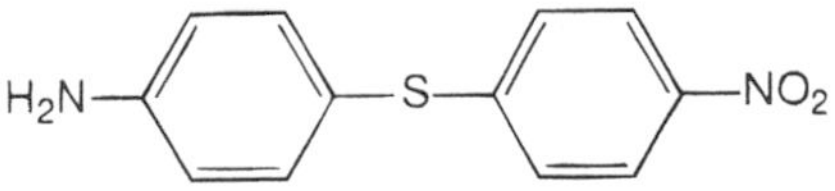

(ANDS)

Formula = C$_{12}$H$_{10}$N$_2$O$_2$S; space group: Cmc2$_1$; Crystal system = m2; unit cell parameters: a = 6.688 Å, b = 7.367 Å, c = 23.380 Å; unit volume = 1152 Å^3; Z = 4; density = 1.42 g/cm^3; λ_{max} = 341 nm; λ_{cutoff} = ~500 nm; powder SHG = 20 times that of urea; β = 15 $\times$ 10^{-30} esu.

b. Methyl-4-nitrophenylsulfide (MNS)[247b]

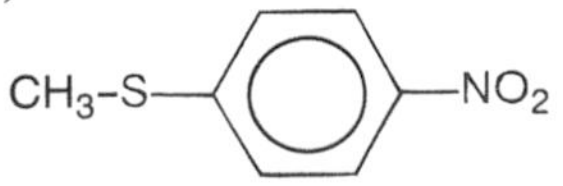

(MNS)

β_0 = 10 $\times$ 10^{-30} esu, λ_{max} = 335 nm

(61) 1,3,5-Triamino-2,4,6-trinitrobenzene (TATB)

The crystal structure of TATB was reported in 1965 by Cady and Larson[45] TATB is an unique molecule, an octupolar organic system, and its quadratic properties were reported first by Zyss.[46]

In TATB, the acceptor nitro and donor amino groups are located at alternate positions on the benzene ring. Because of strong intra- and intermolecular hydrogen bonding, the TATB molecule is planar. It is noncentrosymmetric and adopts D_{3h} symmetry. A complete cancellation of the β tensor takes place due to the crossed character of the intramolecular charge transfer between the amino and nitro groups. It is important that TATB exhibits powder SHG three times larger than urea. Discovery of SHG in the octupolar TATB molecule has opened a new concept in the field of nonlinear optics, as discussed exclusively by Zyss.[46]

Chemical structure:

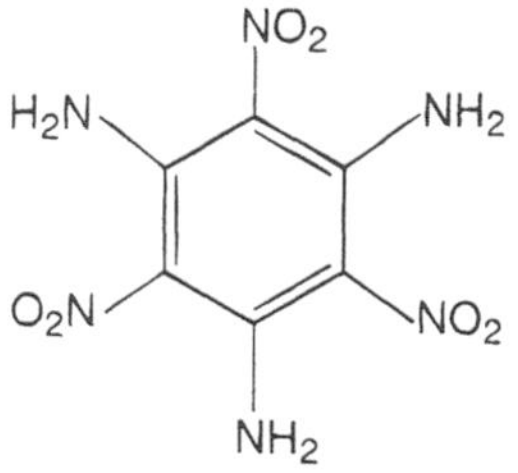

Formula = $C_6H_6N_6O_6$, melting point = 300 °C; space group: PI; crystal system = triclinic
Unit cell parameters: a = 9.010 Å, b = 9.028 Å, c = 6.812 Å, α = 108, 59°, β = 91.82°, γ = 119.97°, Z = 2; density: 1.937 g/cm³
Crystals are strongly pleochroic and biaxial negative; refractive indexes: n_x = 1.45, n_y = 2.3, n_z = 3.1 showing large birefringence
λ_{max} = 351 nm in methylene chloride. Powder SHG = 3 times that of urea.[51]
1,3,5-Tridimethylamino-2,4,6-trinitrobenzene and 1,3,5-ributylamino-2,4-6-trinitrobenzeol showed no SHG.

β values of TATB (calculated)[50]:

β values	AM1 Method	3–21 G Method	INDO/SDCI Method
β_{xxx} (10^{-30} esu)	−11.08	−8.01	−7.68
β_{xyy} (10^{-30} esu)	11.08	8.01	7.68
β_{vec} (10^{-30} esu)	0.0	0.0	0.0
β (10^{-30} esu)	22.16	16.02	15.36

In TATB, D_{3h} symmetry imposes $\beta_{xyy} = \beta_{xxx}$.

Verbiest et al.[248] reported the β values of the octapolar molecules trinitrobenzene and tricyanobenzene. The β values measured by Hyper Rayleigh Scattering (HRS) method were β_{333} = 0.9 × 10^{-30}, β_{311} = 0.9 × 10^{-30}, and $|\beta|$ = 1.8 × 10^{-30} esu for tricyanobenzene and β_{333} = 2.3 × 10^{-30} esu, β_{311} = 2.3 × 10^{-30} esu, and $|\beta|$ = 4.6 × 10^{-30} esu for trinitrobenzene. The calculated β_{333} = 0.42 × 10^{-30}, β_{311} = 0.42 × 10^{-30}, and $|\beta|$ = 0.84 × 10^{-30} esu for tricyanobenzene, whereas β_{333}, β_{311} and $|\beta|$ for trinitrobenzene were same as experimental values. The HRS measured β_{333} (tricyanobenzene) $\simeq$ β_{333}(benzonitrile) whereas $|\beta|$(tricyanobenzene) = $2|\beta|$(benzonitrile). The calculated β_{333} and $|\beta|$ of nitrobenzene were −1.91 × 10^{-30} esu and 2.35 × 10^{-30} esu, respectively; therefore trinitrobenzene showed a $|\beta|$ value twice that of nitrobenzene.

The β values of the octupolar vinamidinium salts made from symmetric cyanine dyes were measured between 18 ~ 56 × 10^{-30} esu by Stadler et al.[248b]

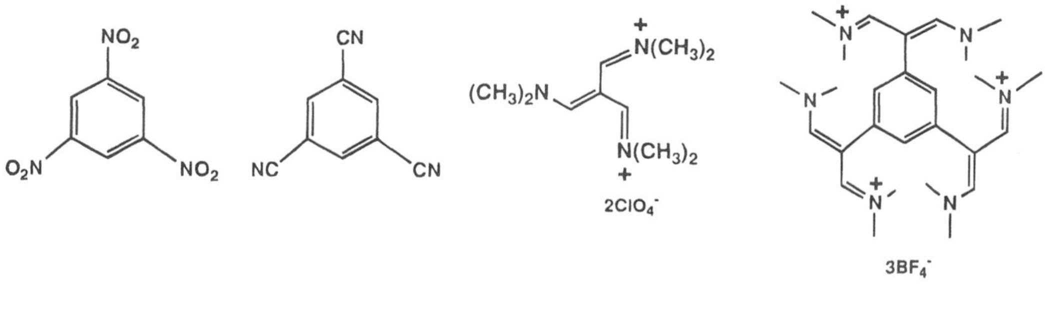

(62) Trinenzamide Derivatives

Chemical structure:

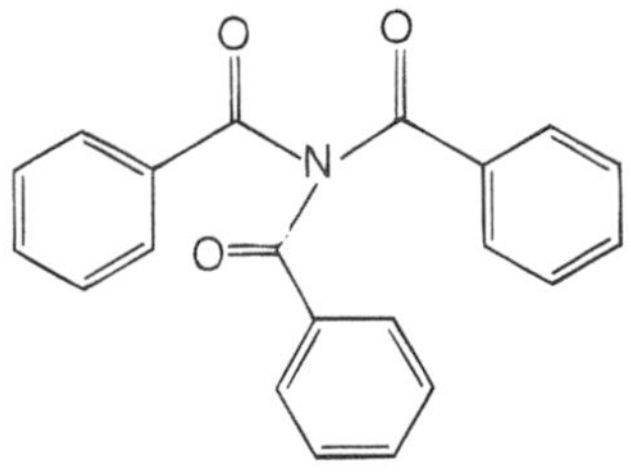

Formula = $C_{21}H_{15}NO_3$; space group: P2$_1$; crystal system = monoclinic
Unit cell parameters: a = 5.388 Å, b = 16.053 Å, c = 9.987 Å, b = 102.1°, Z = 2
λ_{cutoff} = 330 nm; powder SHG = 1.5 times that of urea
Hyperpolarizability (β) = 25.1 × 10^{-30} esu using *ab initio* STO-3G method

(63) Phenoxide-pyridinium Derivatives

Bacquet et al.[250] reported second-order NLO properties of phenoxide-pyridinium derivatives. Cubic single crystals of methylpyridium-3′,5′-(di-*t*-butyl)-4′-phenoxide (MPDBP) were characterized by X-ray diffraction and this belongs to space group P2$_1$2$_1$2$_1$. Large single crystals were grown by recrystallization in ethanol-heptane solution. The ground state was intermediate between the zwitterionic and the quinoid structures. In the unit cell, the molecules are packed head-to-tail. The MPDBP showed λ_{max} at 520 nm in methylene chloride and the SHG from cubic single crystals was observed at 1.32 μm. The dipole moment of 0.8 debye and β_{xxx} of 0.5 × 10^{-30} esu was determined by solvatochromism.

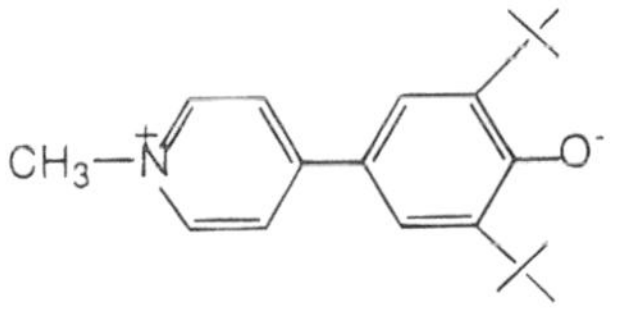

The following phenoxide-pyridinium derivatives have β values of 12 × 10^{-30} and 11 × 10^{-30} esu for 2a and 2b, respectively. Both compounds showed the λ_{max} at almost same wavelength around 292 nm in methylene chloride. In 2a, the excited state dipole moment (μ_e) of −5.8 debye is in the opposite direction to the ground-state dipole moment (μ_g) of 4.0 debye. On the other hand, 2b has both μ_g of 6.2 debye and μ_e of 16.2 debye in the same direction.

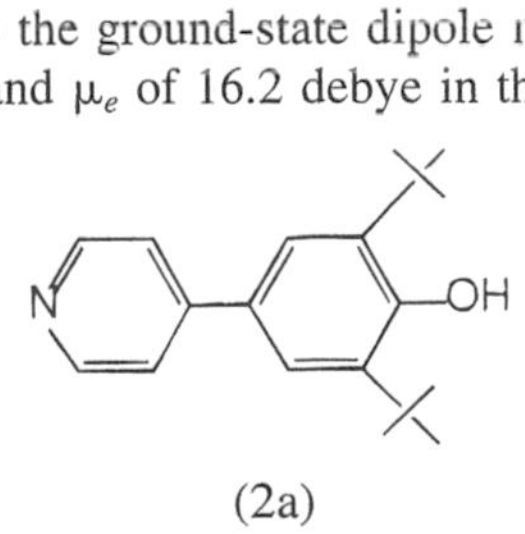

(2a) 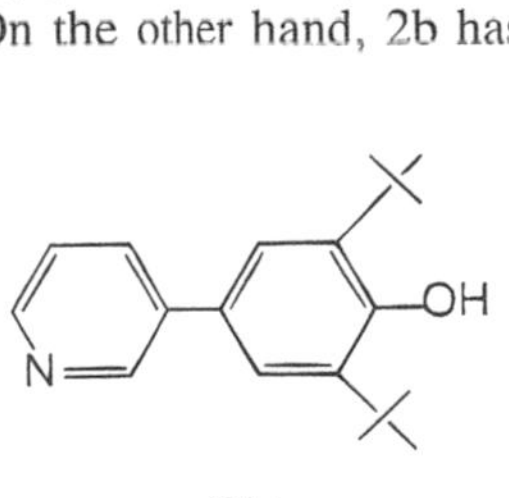   (2b)

(64) Dihydropyrazole Derivatives

Allen et al.[64] reported the hyperpolarizabilities of a family of molecules based on substituted dihydropyrazoles. The β values changed significantly with substituent groups and increased conjugation (Table 7). The details of a single-crystal DCNP are mentioned below. This compound shows very large β and electro-optical coefficients with an electro-optical figure of merit of 8.6 × 10^{-10} m/V and has a very low reduced half-wave voltage of 370 V, about 8 times smaller than lithium niobate. The wavelength dependence shows that electro-optical coefficient at 633 nm has an advantage over lithium niobate, although it reduces at longer wavelengths.

a. 3-(1,1-Dicyanoethenyl)-1-phenyl-4,5-dihydro-1H-pyrazole (DCNP)[64]

Formula = $C_{13}H_{10}N4$; molecular weight = 222.25; space group: Cc; crystal system = monoclinic; unit cell parameters: a = 11.842 Å, b = 12.346 Å, c = 7.865 Å, β = 90.34°, α = γ = 90°; volume =

1150 Å^3, $Z = 4$; density = 1.28 g cm^{-3}; refractive indexes: $n_x = 1.9$, $n_z = 2.7$; powder SHG = 16 times that of urea at 1.06 μm, lower due to reabsorption of the 532 nm harmonic radiation, and 100 times larger than urea at 1.90 μm; hyperpolarizability: β = 7.779×10^{-28} esu at 1.17 eV; electro-optical coefficient: $r_{33} = 87$ pm/V.

(65) Cyano-(alkoxycarbonyl)methylene-2-ylidene-4,5-dimethyl-1,3-dithiole

Uemiya and Nogami[251] reported on SHG properties of dithiole compounds. Their SHG efficiency as high as 24 times of urea was obtained.

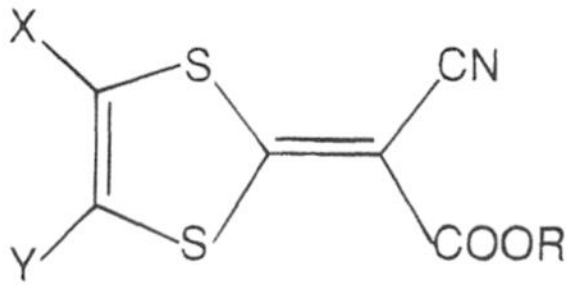

X	Y	R	Crystallization solvent	SHG Activity (× urea)
H	H	CH$_3$	Methanol	0.1
CH$_3$	H	C$_2$H$_5$	Ethanol	10
CH$_3$	H	n-C$_3$H$_7$	n-Propanol	0.1
CH$_3$	H	i-C$_3$H$_7$	Methanol	0.5
CH$_3$	CH$_3$	C$_2$H$_5$	Methanol	24

From Uemiya, T. and Nogami, T., *Advanced Nonlinear Optical Organic Materials I*, 1991 (in Japanese). With permission

a. Cyano-(ethoxycarbonyl)methylene-2-ylidene-4,5-dimethyl-1,3-dithiole[251]

Space group: P1-; crystal system = triclinic; unit cell parameters: $a = 8.711$ Å, $b = 9.748$ Å, $c = 7.800$ Å, α = 95.50°, β = 107.67°, γ = 109.14°; volume = 582.0 Å^3, $Z = 2$; density = 1.37 g cm^{-3}; powder SHG. Single crystals obtained from acetone were SHG inactive whereas those grown from methanol showed SHG activity 24 times that of urea, $d_{11} = 90$ pm/V (calculated).

(66) Indole-3-aldehyde

Youping et al.[252] reported crystal growth and characterization of single crystals of indole-3-aldehyde. The high-quality single crystals having dimensions of $18 \times 8 \times 35$ mm^3 were grown in acetone from seeds by the evaporation method. The indole-3-aldehyde molecule is planar, and the dihedral angle between the indole ring and the plane of the aldehyde group is 16.19°. The molecules are linked by hydrogen bonds between N−H and oxygen.

Chemical structure:

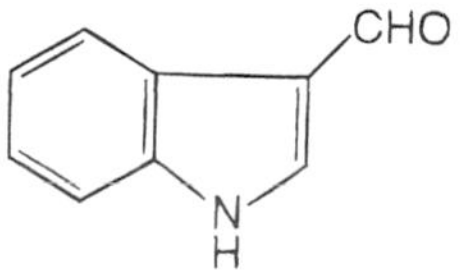

Melting point: 198 °C; space group: Pca2$_1$; crystal system = orthorhombic
Unit cell parameters: $a = 14.154$ Å, $b = 5.836$ Å, $c = 8.724$ Å; volume = 720.6 Å^3, $Z = 8$; density = 1.34 g cm^{-3}; $\lambda_{cutoff} = 340$ nm; powder SHG = 38 times that of KDP.

(67) 3-(4-tert-Butylphenyl)-3-hydroxy-1-(4-methoxyphenyl)-2-propen-1-one (BPMP)

Chemical structure[253]:

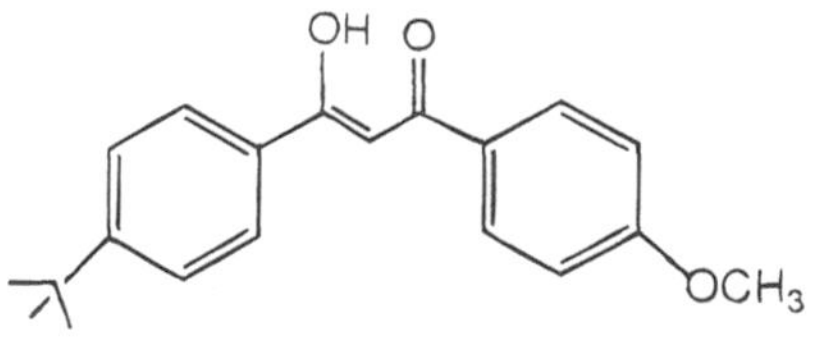

Formula = $C_{20}H_{22}O_3$ (molecular weight = 310.39); λ_{cutoff} = 410 nm in solution; space group: Pna2$_1$; crystal system = orthrohmbic

Unit cell parameters: a = 10.6819 Å, b = 9.964 Å, c = 16.2694 Å; volume = 1731.6 Å^3, Z = 4; density = 1.191 g cm^{-3}; powder SHG = 10 times that of urea at 1.064 μm.

BPMP single crystals were grown from acetonitrile, chloroform, ethylacetate, n-hexane, methanol, and toluene solutions.

(68) 8-(4′-Acetylphenyl)-1,4-dioxa-8-azaspiro[4,5]decane (APDA)

Chemical structure[254–256]:

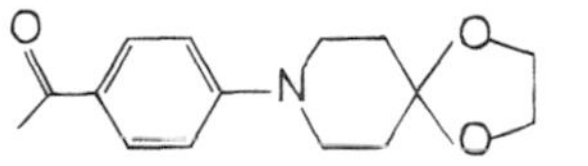

Formula = $C_{15}H_{19}NO_3$ (molecular weight = 261.32); space group: Pna2$_1$; crystal system = orthrohmbic; melting point = 123 °C

Unit cell parameters: a = 6.875 Å, b = 9.036 Å, c = 21.676 Å; volume = 1346.5 Å^3, Z = 4; density = 1.1289 g cm^{-3}; λ_{max} = 350 nm in CH$_2$Cl$_2$; λ_{cutoff} = 400 nm

Refractive indices:

n_a = 1.637 n_b = 1.549 n_c = 1.658 at 1.064 μm

n_a = 1.679 n_b = 1.565 n_c = 1.702 at 0.532 μm

Powder SHG = 12 times that of urea at 1.064 μm; hyperpolarizability β = 2.02 × 10^{-30} esu (calculated)

NLO coefficients of APDA:

Calculated	Measured at 1.064 μm
d_{33} = 61 pm/V	d_{33} = 50 pm/V
d_{32} = 8.7 pm/V	d_{32} = 7 pm/V

Type I phase matching, d_{eff} = 14.9 pm/V.

(69) N-(4-Aminobenzenesulfonyl)acetamide (ASBA)

Chemical structure[257,258]:

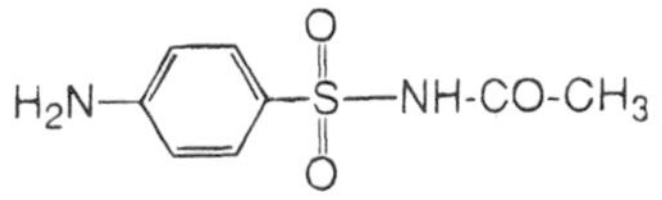

Space group: P4$_1$; crystal system = tetragonal; melting point = 182 °C (uniaxial crystal)

Unit cell parameters: a = b = 7.921 Å, c = 16.452 Å, Z = 4; density = 1.38 g cm^{-3}

Absorption maximum = 269 nm for single crystal of thickness 11 mm; λ_{cutoff} = 350 nm

Powder SHG = 6.4 times that of urea at 830 nm; Hyperpolarizability β = 5.36 × 10^{-30} esu by MOPAC-AM1 method.

NLO coefficients: of ASBA:[258] d_{33} = 6.2 pm/V, d_{31} = 3.3 pm/V, d_{15} = 2.8 pm/V, d_{eff} = 1.9 pm/V

Refractive indexes of ABSA crystal:

λ (nm)	441.5	488.0	632.8	1064
n_o	1.5947	1.5809	1.5622	1.5441
n_e	1.7737	1.7481	1.7088	1.6797

Type II phase matching, d_{eff} = 2.4 pm/V; optical damage ≥ GW/cm^2; Vicker's hardness = 32.

(70) L-Pyrrolidone-2-carboxylic acid (L-PCA)

Chemical structure[259]:

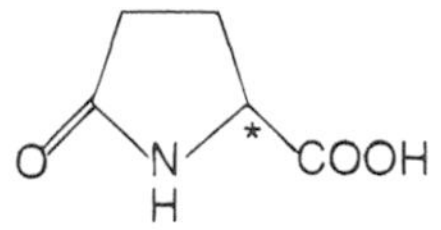

Space group: $P2_12_12_1$; crystal system = orthorhombic; melting point = 162 °C

Unit cell parameters: $a = 9.010$ Å, $b = 13.422$ Å, $c = 14.640$ Å, $Z = 12$; density = 1.44 g cm^{-3}

Absorption cutoff wavelength = 260 nm; refractive indices (uniaxial crystal): $n_o = 1.67$; $n_e = 1.54$ at 1.064 μm

Powder SHG = same as of urea; NLO coefficients: $d_{14} = 0.32$ pm/V at 532 nm

Type II phase matching, $d_{eff} = 0.28$ pm/V; Vicker's hardness = 33.

(71) 4-Nitrophenylhydrazines

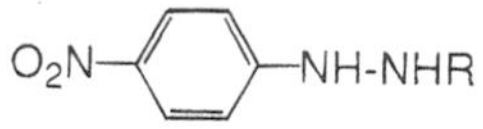

Parameter	R = COH	R = H
Space group	Pca2$_1$	Pca2$_1$
Crystal system	Orthorhombic	
a (Å)	10.352	12.5
b (Å)	9.587	3.5
c (Å)	8.065	14.7
Z	4	4
Volume (Å^3)	800.4	–
Density (g cm^{-3})	1.505	–
λ_{max} (nm)	348	381
SHG ($\times$ urea)	17	24
β (10^{-30} esu)	4.90 (AM1)	16.4
	14.7 (CNDO/S-CI)	–
	12.2 (DC-SHG)	44

See references 260 and 261.

Serbutoviez et al.[262] reported synthesis, characterization, and second-order NLO properties of hydrazone derivatives. The β values were measured by EFISH at 1.907 μm in 1,4-dioxane, β$_o$ was extrapolated, and absorption peaks are shown below. The β value of donor groups of benzaldehyde 4-nitrophenylhydrazone derivatives changed as N(CH$_3$)$_2$ > OCH$_3$ > Br > CH$_3$. The β value increased rapidly with replacement of one phenyl ring with a thiophene ring showing thiophene moietiy is important. The β value as high as 220×10^{-40} m^4/V was for thiophene-substituted derivatives at 1.907 μm. Powder measurements showed large SHG efficiency, which varied depending on the chemical structure between a few times and two orders of magnitude larger than that of urea.

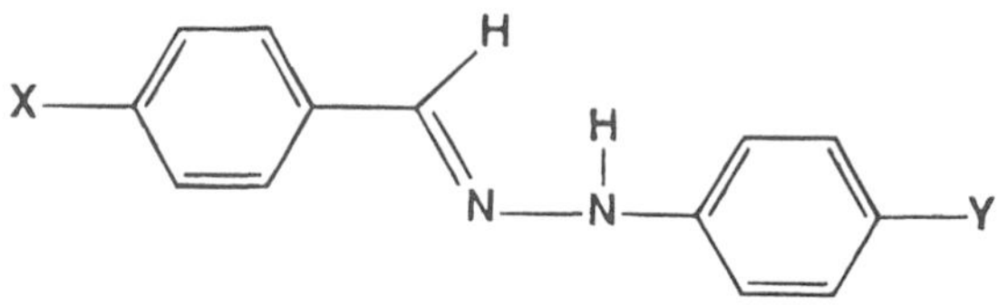

		λ_{max}(nm)	β (10^{-40} m^4/V)	β$_o$ (10^{-40} m^4/V)
1. X = N (CH$_3$)$_2$	Y = NO$_2$	420	200	153
2. X = NO$_2$	Y = N(CH$_3$)$_2$	464	–	–
3. X = OCH$_3$	Y = NO$_2$	399	130	103
4. X = NO$_2$	Y = OCH$_3$	433	160	121
5. X = CH$_3$	Y = NO$_2$	395	70	56
6. X = NO$_2$	Y = CH$_3$	423	85	65
7. X = Br	Y = NO$_2$	391	80	64

| 8. $X = NO_2$ | $Y = OCH_3$ | 482 | 220 | 153 |
| 9. $X = NO_2$ | $Y = CH_3$ | 470 | 140 | 100 |

| 10. | | 441 | 220 | 164 |

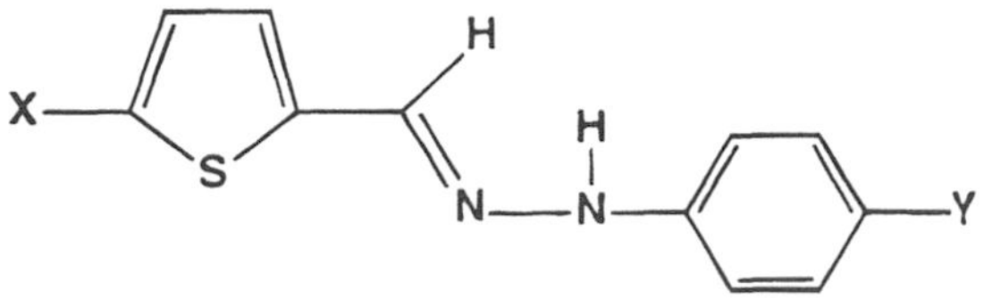

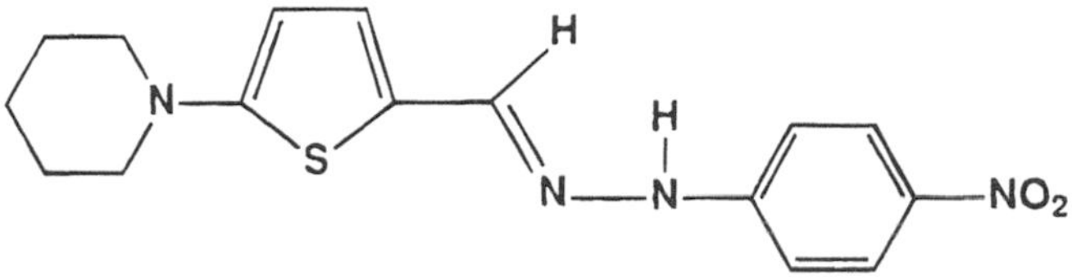

a. 4-(Dimethylamino)-benzaldehyde-4′-nitrophenylhydrazone (DANPH)

Space group: Cc; crystal system = monoclinic; unit cell parameters: $a = 6.281$ Å, $b = 28.133$ Å, $c = 8.380$ Å, $\beta = 97.87°$, $Z = 4$; density = 1.287 g cm^{-3}; color = red green; powder SHG = about 100 times that of urea at 1.3 μm; NLO coefficient d_{12} was calculated as 250 pm/V at 1.5 μm.

b. 4-(1-Azacycloheptyl)-benzaldehyde-4′-nitrophenylhydrazone (ACNPH)

Space group: Pca2$_1$; crystal system = orthorhombic; unit cell parameters: $a = 31.878$ Å, $b = 7.1440$ Å, $c = 7.4310$ Å, $Z = 4$; density = 1.328 g cm^{-3}; color = dark red.

(72) 2-(4-Nitroanilino)-1,3,5-triazine Derivatives

Yonehara et al.[263] synthesized a series of 2-(4-nitroanilino)-1,3,5-triazine derivatives. The SHG activity and absorption features of these compounds depended on chemical structures. The absorption maxima were at 350 nm, shorter by 30 to 50 nm than those of nitroanilines. 2-(4-Nitroanilino)-4,6-diphenyl-1,3,5-triazine recrystallized from toluene showed powder SHG 104 times that of urea. Its recrystallization with DMF yielded SHG-inactive crystals that were soluble in a 1:1 ratio by hydrogen bonding. The chemical structure, physical, and NLO properties of various derivatives are shown below:

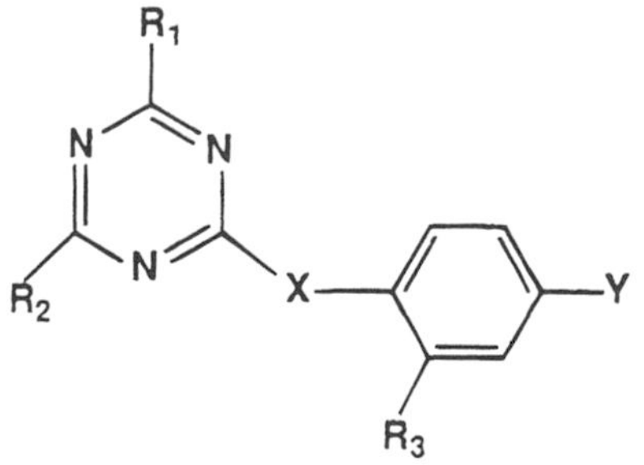

R_1	R_2	R_3	X	Y	Melting Point (°C)	λ_{max} (nm)	SHG (× urea)
Cl	Cl	H	NH	NO_2	310	323	20
Cl	Cl	CH_3	NH	NO_2	239	304	0.007
Cl	Cl	Cl	NH	NO_2	200	309	1
Cl	Cl	H	O	NO_2	185	267	4.1
Cl	Cl	H	S	NO_2	118	255	0.8
Cl	Cl	H	NH	3-NO_2	191	268	10

Cl	OCH$_3$	H	NH	NO$_2$	261	327	0.37
Cl	N(C$_2$H$_5$)$_2$	H	NH	NO$_2$	194	337	0.02
Cl	CH$_3$COCHCOOC$_2$H$_5$	H	NH	NO$_2$	246	326	9.4
Cl	OCH$_3$	CH$_3$	NH	NO$_2$	200	314	0.003
OCH$_3$	OCH$_3$	–	NH	NO$_2$	223	331	0.1
OH	OH	–	NH	NO$_2$	>320	328	0.04
OCH$_3$	OCH$_3$	–	O	CHO	149.5	–	0.06
OCH$_3$	OCH$_3$	–	O	CH=C(CN)$_2$	–	–	0.0004
C$_6$H$_5$	C6H$_5$	–	NH	NO$_2$	268	342	104

Cl	Cl	–	NH	NO$_2$	>300	370	11
OH	OH	–	NH	NO$_2$	>300	368	0.5
OCH$_3$	OCH$_3$	–	NH	NO$_2$	>300	376	0.6
CH$_3$	CH$_3$	–	NH	NO$_2$	>300	379	8.9
Cl	ANS[a]	–	NH	NO$_2$	>300	393	6.5

[a] ANS = 4-amino-4′-nitrostilbene.

(73) 2-Arylideneindan-1,3-diones

The derivatives of 2-benzylideneindan-1,3-dione with different substituents on the 2-benzylidene ring showed powder SHG efficiencies 50 ~ 100 times that of urea at 1.064 μm, as reported by Matsushima et al.[264] The SHG efficiency depends on the crystallization solvent. The derivatives with 4′-methyl and 4′-hydroxy have phase-matchable and large SHG activities. With the change of crystallization conditions, such as temperature and solvent, both SHG efficiency and crystal morphology changed. Yellow crystals of 4′-methyl, 20 mm × 2.5 mm × 0.4 mm, were grown from a hot methanol solution by solvent evaporation at room temperature for 2 months, which showed powder SHG activity 63.7 times that of urea. Also yellow crystals of 4′-hydroxy, 35 mm × 3 mm × 0.6 mm, were obtained from a 2:1 mixture of hexanetetrahydrofuran by slow evaporation of the solvent at room temperature for 1 month. The powder crystals of this compound obtained from ethanol, hot ethanol, and hot hexanetetrahydrofuran mixture exhibited SHG efficiencies of 0.62, 44.5, and 65.4 times that of urea, respectively (Table 52).

(74) Dinitrophenyl-chlorophenyl-propanamine (DNCPA)

The high optical quality crystals of N-(2,4-dinitrophenyl)-1-chloro-3-phenyl-2-propanamine of 19 mm × 16 mm × 5 mm were grown for 1 month.[265] The DNCPA forms orthorhombic crystals with space group P2$_1$2$_1$2$_1$. The unit cell parameters: a = 14.09 Å, b = 15.61 Å, c = 7.15 Å; volume = 1573 Å^3; β = 90°; Z = 4; density = 1.418 g cm^{-3}; λ_{cutoff} = 530 nm; and Vicker's hardness = 25.

Refractive indexes:

Indexes/λ	514 nm	532 nm	633 nm	1064 nm
n_x	1.523	1.517	1.511	1.499
n_y	1.909	1.887	1.825	1.764
n_z	1.990	1.945	1.853	1.786

Table 52 Effect of Crystallization Solvent on SHG of 2-arylideneindan-1,3-diones

		Absorption Characteristics		SHG Intensity (× urea)	
R	Solvent	λ_{max} (nm)	λ_{cutoff} (nm)	(crystallization solvent)	
H	1,4-Dioxane	339	442	0.88 (ethanol)	1.31 (hexane)
2-CH$_3$	Benzene	352	451	2.61 (ethanol)	4.28 (benzene)
3-CH$_3$	Benzene	348	447	9.03 (ethanol)	9.23 (benzene/hexane)
4-CH$_3$	1,4-Dioxane	350	440	78.2 (ethanol)	70.6 (benzene)
2-OH	1,4-Dioxane	348	450	0.01 (ethanol)	0.0 (benzene)
3-OH	Ethanol	338	460	0.27 (ethanol)	0.3 (acetone)
4-OH	1,4-Dioxane	382	460	58.6 (ethanol)	120 (hexane/THF)
2-OCH$_3$	Benzene	390	453	5.99 (ethanol)	0.0 (benzene)
3-OCH$_3$	Benzene	340	443	0.0 (ethanol)	0.0 (benzene)
4-OCH$_3$	1,4-Dioxane	380	450	8.3 (ethanol)	14.6 (benzene/hexane)
4-N(CH$_3$)$_2$	1,4-Dioxane	469	553	1.03 (ethanol)	32.7 (acetone)
4-N(C$_2$H$_5$)$_2$	1,4-Dioxane	475	553	88.9 (ethanol)	93.6 (benzene/hexane)
4-NHAc	1,4-Dioxane	387	470	0.10 (ethanol)	0.10 (acetone)
4-C$_2$H$_5$	1,4-Dioxane	350	450	0.02 (ethanol)	0.05 (hexane)
4-OPr	Ethanol	385	470	0.0 (ethanol)	0.0 (benzene/hexane)
4-F	1,4-Dioxane	341	430	0.12 (ethanol)	0.05 (benzene/hexane)
4-Cl	1,4-Dioxane	344	450	0.0 (ethanol)	0.0 (benzene/hexane)
4-Br	1,4-Dioxane	347	442	0.1 (ethanol)	0.37 (hexane)
4-SCH$_3$	Ethanol	406	520	0.0 (ethanol)	0.0 (benzene/hexane)
3,5-(OCH$_3$)$_2$	1,4-Dioxane	350	449	0.3 (ethanol)	0.01 (benzene)

From Matushima, R., et al., *J. Mater. Chem.* 2, 507, 1992. With permission.

$\beta_z = 5.86 \times 10^{-30}$ esu; dipole moment = 7.87 D by the CNDO/S method
Effective NLO coefficients given by:
Type I: $d_{eff} = -d_{36} \sin \theta \sin 2\phi = 0.13$ pm/V, $\Phi_{PM} = 27.8°$; walkoff angle = 0.20 rad
Type II: $d_{eff} = -0.5(d_{14} + d_{25}) \sin \theta \sin 2\phi = 0.11$ pm/V, $\Phi_{.} = 38.3°$; walkoff angle = 0.22 rad.

(75) Nitropyridyl-leucinol (NPLO)

The optical quality crystals of *N*-(5-nitro-2-pyridyl) leucinol with dimensions of 22.4 mm × 112.3 mm × 6.3 mm were grown by slow cooling ethanol/toluene-mixed solution for 2 months.[265,266] The NPLO crystals were monoclinic with space group P2$_1$. The unit cell parameters: $a = 10.04$ Å, $b = 5.883$ Å, $c = 11.60$ Å; volume = 641.6 Å^3; $\beta = 110.43°$; Z = 2; density = 1.240 g cm^{-3}; $\lambda_{cutoff} = 460$ nm and Vicker's hardness = 18.

Refractive indexes:

Indexes/λ	488 nm	514 mm	633 nm	1064 nm
n_x	1.470	1.463	1.457	1.451
n_y	1.712	1.681	1.631	1.598
n_z	2.218	2.116	1.933	1.812

$\beta_z = 6.39 \times 10^{-30}$ esu; dipole moment = 7.27 D by the CNDO/S method
Effective NLO coefficients were given by:
Type I: $d_{eff} = -d_{21} \cos^2 \theta - d_{23} \sin^2 \theta + d_{25} \sin 2\theta = 37$ pm/V; $\Phi_{PM} = 52°$, walkoff angle = 0.22 rad

Type II: $d_{eff} = d_{14} \sin\theta\cos\theta - d_{16}\cos^2\theta - d_{34}\sin^2\theta + d_{36}\sin\theta\cos = 3$ pm/V, $\Phi_{PM} = 33°$; walkoff angle $= 0.22$ rad

Figure of merit $= 296$ pm^2/V^2; damage threshold $= 6$ GW/cm^2 with a peak power of 29 MW and a pulse duration of 8 nsec.

(76) Ethylene Derivatives

Osaka et al.[267] reported optical and NLO properties of ethylene and methanimine derivatives possessing electron-donor and acceptor groups. Three ethylene derivatives were found to be SHG active with transparency in the blue wavelength region. Ethylenes with stronger donor and acceptor groups tend to acquire a centrosymmetric space group because of larger, strong intermolecular charge-transfer interactions. The central C=C bond twisting results in smaller β values than those of nontwisted molecular structures.

Parameter	$R_1 = R_2 = CH_3S$	$R_1 = CH_3S$ $R_2 = (CH_3)_2N$
Space group	$P2_1$	$P2_12_12_1$
Crystal system	Monoclinic	Orthorhombic
Melting point (°C)	57~59	92~93
λ_{max} (nm)	330	320
λ_{cutoff} (nm)	370	355
a (Å)	8.754	18.443
b (Å)	14.399	8.794
c (Å)	4.156	6.731
Z	2	4
Volume (Å^3)	523.7	1091.6
Density (g cm^{-3})	1.378	1.304
λ_{max} (nm)	330	320
λ_{cutoff} (nm)	370	355
SHG ($\times$ urea)	5.5	0.8
β_{EFISH} (10^{-40} m^4/V)	29	2.5

(77) Silicon-Based Donor–Acceptor Compounds

Magnani et al.[268,269] reported the NLO properties of donor–acceptor-substituted silicon compounds. The β values increased as the number of silicon atoms increased because of the delocalization of α-electrons along the Si-Si backbone. The β values of (4-(2,2-dicyanoethenyl)phenyl][4-(dimethylamino)phenyl]dimethylsilane ($n = 1$), 1-(4-(2,2-dicyanoethenyl)phenyl]-2-[4(dimethylamino)phenyl]-1,1,2,2-tetramethylsilane ($n=2$), and 1-(4-(2,2-dicyanoethenyl)phenyl]-6-[4;-(dimethylamino)phenyl]-1,1,2,2,3,3,4,4,5,5,6,6- dodecamethylsilane ($n = 6$) measured by EFISH were 16×10^{-30}, 22×10^{-30}, and 38×10^{-30} esu, and their λ_{max} values recorded in chloroform were 320, 334, and 276 nm, respectively.

Hissink et al.[270] reported optical and NLO properties of several donor–acceptor silicon compounds. The β values measured by the HRS method changed depending on the nature of the donor–acceptor groups. The β values were 80×10^{-30}, 20×10^{-30}, and 15×10^{-30} esu, having the donor dimethylamino group on one side, and the acceptor $CH = C(CN)_2$, $SO_2C_4H_9$, and $SO_2C_6H_5$ groups on another side, respectively. Their λ_{max} recorded in chloroform were 320, 334, and 276 nm, respectively. Perfluorobutylsulfonyl phenyl-thiomethylphenyl-tetramethyldisilane, perfluorobutylsulfonyl-phenyl-methoxyphenyltetramethyldisilane showed β values of 11×10^{-30} and 10×10^{-30} esu, respectively. The β values for dicyanoethylphenyl-dimethylaminophenyl-tetramethylsilane was the largest at 80×10^{-30} esu. Tetramethylsilane with acceptor dicyanoethenyl group and donor thiomethyl, methoxy, and fluoro groups showed a β of 55×10^{-30}, 53×10^{-30}, and 38×10^{-30} esu, respectively. Similarly, optical absorption bands changed with different donor–acceptor groups. The silicon compound having perfluorobutylsulfone and dimethylamino groups showed a β value little lower than p-NA, although its optical transparency was much better. The details of the single crystals of perfluorobutylsulfonyl-phenyl-dimethylaminophenyl-tetramethyldisilane are listed below.

Formula = $C_{22}H_{26}F_9NO_2SSi_2$; space group: P1; crystal system = triclinic
Unit cell parameters: $a = 14.032$ Å, $b = 14.052$ Å, $c = 15.898$ Å, $\alpha = 113.11°$, $\beta = 94.31°$, $\gamma = 100.57°$; volume = 2796.5 Å^3; Z = 4, density = 1.415 g cm^{-3}
λ_{max} = 330 nm; λ_{cutoff} = 365 nm; β = 20×10^{-30} esu at 1.064 μm by HRS method.

(78) Pyrrolo [1,2-a]quinolines

Chemical structure[271]:

R	μ (D)	EFISH β$_z$ = $(10^{-30}$ esu)		λ_{max} (nm)	SHG (x urea)
		at 1.064 μm	at 1.91 μm		
1. H	3.4	1.7	1.2	306	0.17
2. CHO	4.3	17	11	331	3.2
3. NO$_2$	4.7	47	26	373	1.7
4. (CN)$_2$C=CH	7.6	84	35	418	10
5. NO$_2$C$_6$H$_4$CH=CH	6.3	326	137	417	0.067

The powder SHG efficiencies of quinoline compounds are comparable to that of urea. The SHG activity of MNA being 6.5 times that of urea shows that the SHG of compound 4 is few times larger than that of MNA. Polymorphism also affected the SHG. The crystal data of compounds 2 and 3 are as follows: Compound 2 crystal class = orthorhombic; space group = P2$_1$2$_1$2$_1$; unit cell parameters: $a = 9.738$ Å, $b = 10.078$ Å, $c = 15.387$ Å; volume = 1510 Å^3; Z = 4; calculated density = 1.30 g cm^{-3}. Compound 3 crystal class = monoclinic; space group = P2$_1$; unit cell parameters: $a = 8.796$ Å, $b = 12.761$ Å, $c = 14.010$ Å; volume = 1555 Å^3, Z = 4; calculated density = 1.34 g cm^{-3}.

(79) 2-Dicyanomethylene-1,3-dioxolane

The optical quality, pale yellow crystals of 2-dicyanomethylene-1,3-dioxolane with dimensions of 1.0 mm × 0.5 mm × 0.5 mm were grown by slow evaporation of an alcohol-water solution.[272] The crystals were monoclinic with space group Cc. The unit cell parameters: $a = 5.288$ Å, $b = 15.044$ Å, $c = 16.356$ Å; volume = 1284.4 Å^3; $\beta = 99.23°$; Z = 8; density = 1.406 g cm^{-3}; powder SHG = 2 times that of urea.

(80) Diphenylmethyl(Z)-1-(1-methylthio-2-nitrovinyl) tetrahydropyrrole-2-carboxylate

Chemical structure[273]:

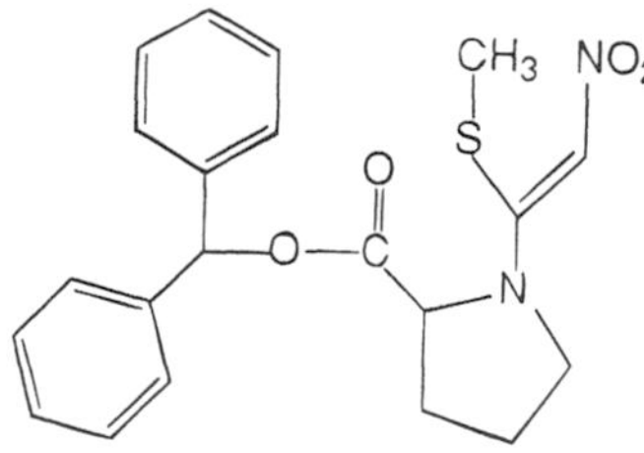

Crystal class = orthorhombic; space group = $P2_12_12_1$; unit cell parameters: a = 9.799 Å, b = 11.853 Å, c = 17.316 Å; volume = 2011.4 Å^3; Z = 4; density = 1.320 g cm^{-3}; powder SHG = 0.25 times that of urea.

(81) Adducts of triiodomethane

a. Chemical formula = $CH_3I \cdot 3S_8$; a = b = 24.36 Å, c = 4.44 Å, α = β = 90°; γ = 120°; d_{33} = 2.6 pm/V, r_{33} = 0.74 pm/V, r_{13} = 0.84 pm/V, r_{22} = 4.0 pm/V; β_{EFISH} = 25 × 10^{-40} m^4/V (see reference 274).
b. Chemical formula = $CH_3I \cdot 3C_9H_7N$; a = b = 22.40 Å, c = 4.59Å, α = β = 90°, γ = 120°, d_{33} = 2.0 pm/V, r_{33} = 0.66 pm/V, r_{13} = 0.97 pm/V, r_{22} = 1.86 pm/V, β_{EFISH} = 17.5 × 10^{-40} m^4/V.

(82) Ferroelectric Liquid Crystalline Thiadiazole Derivatives

Chemical structure[275]:

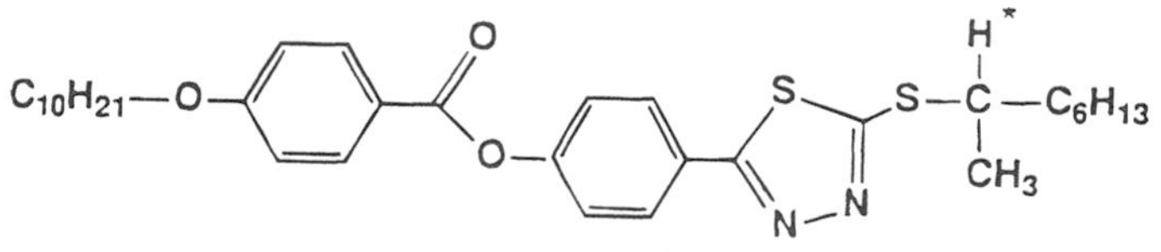

Space group = PnC2; unit cell parameters: a = 42.73 Å, b = 15.14 Å, c = 5.73 Å, α = β = γ = 90°; SHG active, β = − 0.373 × 10^{-30} esu.

(83) Tris(pyrocatecholato)stannate (IV) Compounds

Chemical structure[276]: $[(O-4-NO_2-C_6H_3O)_3Sn]$ $[NH(C_2H_5)3]_2$; space group = Pbca; unit cell parameters: a = 13.655 Å, b = 18.299 Å, c = 36.030 Å, α = β = γ = 90°; volume = 7003 Å^3; Z = 8; density = 1.48 g cm^{-3}; λ_{max} = 323, 396 nm (yellow-orange); powder SHG = 1.33 times that of urea at 1.064 μm and 0.44 times that of urea at 1.9 μm.

The powder SHG efficiency of compounds $[L_2L'Sn]^{2-}$ where L = (OC_6H_4O), L' = (OC_6H_4O), (O-4-NO$_2$-C$_6$H$_3$O), (O-3-CH$_3$O-C$_6$H$_3$O), and (OC_6Cl_4O) were 0.40, 0.28, 0.19, and 0.07 times that of urea at 1.064 μm, respectively. The SHG efficiency of compound, L = (O-4-NO$_2$-C$_6$H$_3$O) and L = (OC_6H_4O) was 0.71 times and L = (O-3-CH$_3$O-C$_6$H$_3$O) and L' = (O-4-NO$_2$-C$_6$H$_3$O) was 0.32 times that of urea at 1.9 μm.

(84) Pyridine and Bipyridine Complexes

Chemical structure[277]:

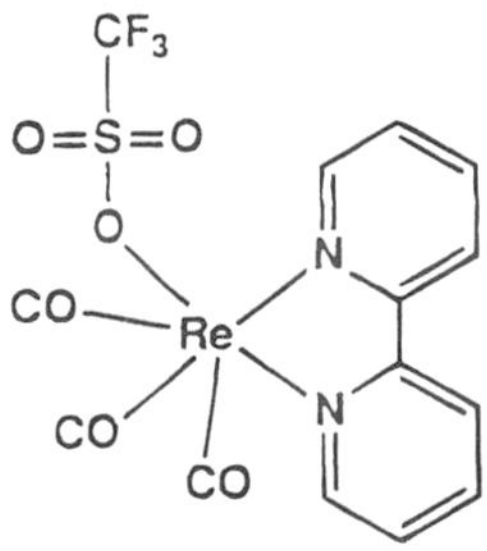

Chemical formula: $ReSF_3O_6N_2C_{14}H_8$; Space group = $P2_1$; crystal class = monoclinic; unit cell parameters: $a = 8.993$ Å, $b = 14.282$ Å, $c = 6.554$ Å, $\beta = 91.59°$, volume = 841.5 Å^3, Z = 2; density = 2.271 g cm^{-3}; yellow; powder SHG = 1.7 ~ 2.0 times that of urea at 1.06 μm. The powder SHG efficiency of analogous compounds, $ClRe(bipyridine)(CO)_3$, $BrRe(pyridine)_2(CO)_3$, $BrRe(4\text{-styrylpyridine})_2(CO)_3$, $Pt(bipyridine)Cl_4$, $Pd(bipyridine)Cl_4$, $Mo(bipyridine)(CO)_4$, $W(bipyridine)(CO)_4$, and $Cr(bipyridine)(CO)_4$ were 3.0, 0.2, 0.1, 1.2, 0.1, 0.5, 0.3, and 0.2 times of urea at 1.064 μm, respectively.

(85) Pt(PEt₃)₂(I)(p-nitrophenyl) and Related Compounds

Tam and Calabrese[278] reported SHG activity for $M(PEt_3)_2(X)$ where M = Pd, Pt; X = Br, I as donors in aromatic donor–acceptor compounds. The SHG efficiency of $trans\text{-}M(PEt_3)_2$ $(X)(2\text{-}Z\text{-}4\text{-}Y\text{-phenyl})$ where M = Pd, Pt and X = Br, I, Y = CHO, NO_2, CN, Z = H, CH_3, NO_2, were 0.4 to 14 times that of urea. All these compounds were white and air stable. The large crystal of $trans\text{-}Pt(PEt_3)_2(I)(p\text{-}nitrophenyl)$ was pale yellow. Crystals of these compounds were grown from dichloromethane/pentane and crystallize in noncentrosymmetric space group to exhibit SHG activity.

$trans\text{-}Pt(PEt_3)_2(I)(p\text{-}nitrophenyl)$; chemical formula: $PtIP_2O_2NC_{18}H_{34}$; space group = $Cmc2_1$; crystal class = orthorhombic; unit cell parameters: $a = 12.705$ Å, $b = 13.208$ Å, $c = 14.302$ Å, $V = 2400$ Å^3, Z = 4; density = 1.883 g cm^{-3}; colorless, powder SHG = 8.0 times of urea at 1.06 μm.

(86) cis-1-Ferrocenyl-2-(4-nitrophenyl)ethylene

Chemical structure[279]:

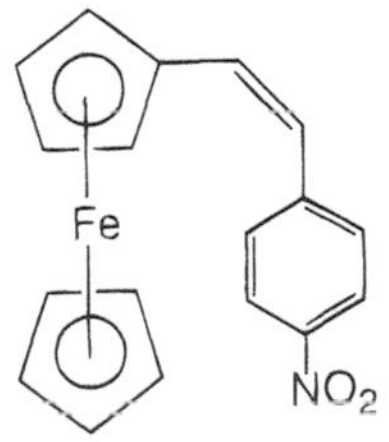

Formula = $C_{18}H_{15}FeNO_2$ (molecular weight = 333.2); space group: Cc; crystal system = monoclinic; unit cell parameters: $a = 6.015$ Å, $b = 44.547$ Å, $c = 1.552$ Å, $\beta = 9105.11°$; volume = 2988.4 Å^3, Z = 8; density = 1.48 g cm^{-3}.

Powder SHG = 62 times of urea at 1.907 μm (no SHG from *trans* isomer). Single crystals were grown from 2-propanol. This compound exhibits solvatochromic behavior, and the absorption peaks were observed at 320, 406, and 462 nm in heptane and 340 and 492 nm in dimethylformamide.

(87) Di-iron alkenylidyne Compounds

Chemical structure[280]:

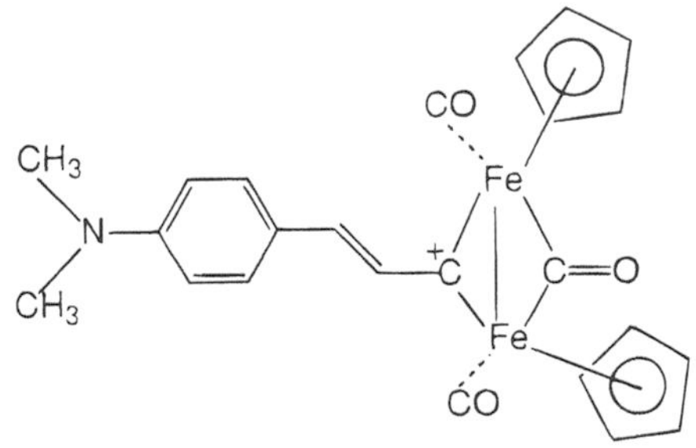

Formula = $C_{24}H_{22}Fe_2NO_3BF_4$; space group: $P2_1/n$; crystal system = monoclinic
Unit cell parameters: $a = 10.702$ Å, $b = 13.349$ Å, $c = 16.932$ Å, $\beta = 93.42°$; volume = 2414.6 Å^3, Z = 4; density = 1.57 g cm^{-3}; powder SHG = 0.77 times that of urea at 1.907 μm

(88) Ferrocenyl Derivatives

Coe et al.[281] found large SHG efficiency for ferrocenyl derivatives at 1.907 μm, although no SHG was observed for the derivatives having dimethylamino or ethoxy groups. The tungsten derivatives exhibited lower SHG values by a factor of 2 to 6 than their molybdenum counterparts.

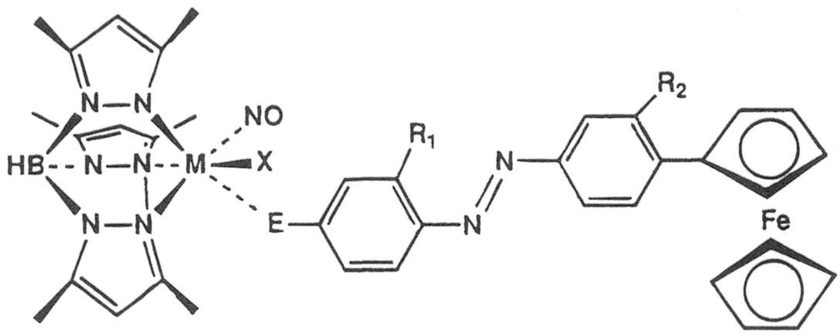

M	X	E	R_1	R_2	λ_{max} (nm) in CH_2Cl_2	SHG ($\times$ urea) at 1.907 μm
Mo	Cl	O	H	H	500	Very weak
Mo	I	O	H	H	482	Very weak
Mo	OH	O	H	H	468	–
Mo	Cl	NH	H	H	480	30
W	Cl	NH	H	H	422	12
Mo	Cl	NH	CH_3	H	488	50
Mo	I	NH	CH_3	H	511	1
Mo	F	NH	CH_3	H	509	–
Mo	Br	NH	CH_3	H	–	25
W	Cl	NH	CH_3	H	427	8
Mo	Cl	NH	CH_3	CH_3	–	1

The crystal structure of compound M = W, X = Cl, E = NH, R_1 = CH_3, and R_2 = H was analyzed: formula = $C_{38}H_{42}BClFeNOW \cdot CH_2Cl_2$; space group: $P2_1$; crystal system = monoclinic; unit cell parameters: a = 10.151 Å, b = 20.470 Å, c = 10.367 Å, β = 91.72°; volume = 2153 Å^3; Z = 2; density: 1.582 g cm^{-3}.

(89) Bimetallic Complexes

Coe et al.[282] reported SHG efficiency of several molybdenum or tungsten mononitrosyl complexes with a general formula [M(NO)LX where M = Mo, W, X = Cl, I groups were linked to chiral ligand. The general chemical structure of the complex is shown below. Three complexes [Mo(NO)LCl(NHC$_6$H$_4$$-$N = N$-$4'$-$Fc)] [Fc = [$(\eta)^5$ C$_5$H$_5$) Fe $(\eta)^5$C$_5$H$_4$)], L = HB (3,5$-$ (CH$_3$)$_2$C$_3$N$_2$H)$_3$}, [W(NO)LCl(NHC$_6$H$_3$-(3-CH$_3$)-4$-$N=N$-$C$_6$H$_4$'$-$Fc)], and [Mo(NOL)LCl(NHC$_6$H$_3$-(3-CH$_3$)-4$-$N=N$-$C$_6$H$_4$$-$4'$-$Fc)] showed SHG efficiency of 59, 53, and 123 times that of urea at 1.906 μm. Either very small or no SHG efficiency was observed without [M(NO)LCl]$^-$ and Fc$-$C$_6$H$_4$$-$N=N$-C_6H_4$ fragments.

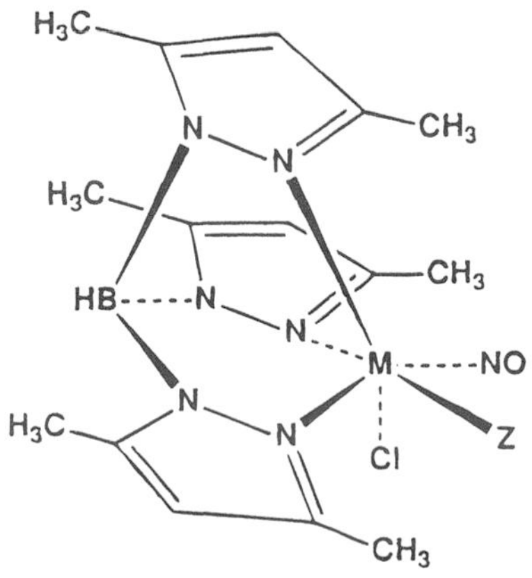

The crystal structures of the following complexes have been resolved by Coe et al.[283]

Chemical formula: [Mo(NO)LCl(NHC$_6$H$_3$-(3-CH$_3$)-4-N=N-C$_6$H$_4$-4'-Fc)]; Space group: P2$_1$; crystal system = monoclinic; unit cell parameters: a = 10.0954 Å, b = 20.6345 Å, c = 10.3965 Å, β = 91.9109°.

Chemical formula: [W(NO)LCl(NHC$_6$H$_3$-(3-CH$_3$)-4-N=N-C$_6$H$_4$-4'-Fc)]; space group: P2$_1$; crystal system = monoclinic; unit cell parameters: a = 10.147 Å, b = 20.455 Å, c = 10.366 Å, β = 91.79°.

Peripherally metallated porphyrin complex [Mo(NO)LCl(p-O-TPPH$_2$) crystallizes in the centrosymmetric space group PI with unit cell parameters: a = 13.075 Å, b = 15.181 Å, c = 17.010 Å, β = 90.85°. The SHG efficiency was 1.93 times that of urea at 1.906 μm, which may result either from noncentrosymmetric crystals present in powder sample or the interaction of circularly polarized light with the centrosymmetric crystallites. Another porphyrin complex [Mo(NO)LCl(o-O-TPPH$_2$) crystallizes in the noncentrosymmetric orthorhombic space group Pna2$_1$; unit cell parameters: a = 16.430 Å, b = 13.244 Å, c = 25.055 Å; α = β = γ = 90°. The compound was SHG active, but details are unknown.

(90) Platinum Acetylides

Chemical structure[284]:

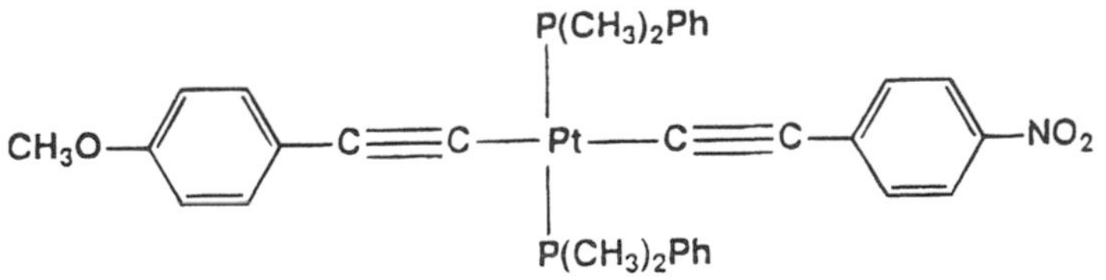

Space group: P1; crystal system = triclinic; unit cell parameters: a = 5.985 Å, b = 9.065 Å, c = 14.105 Å, α = 84.27°, β = 88.12°, γ = 878.35°, volume = 745.7 Å^3; Z = 1, λ_{max} = 386 nm in acetonitrile; SHG activity = same as that of quartz at 1.064 μm.

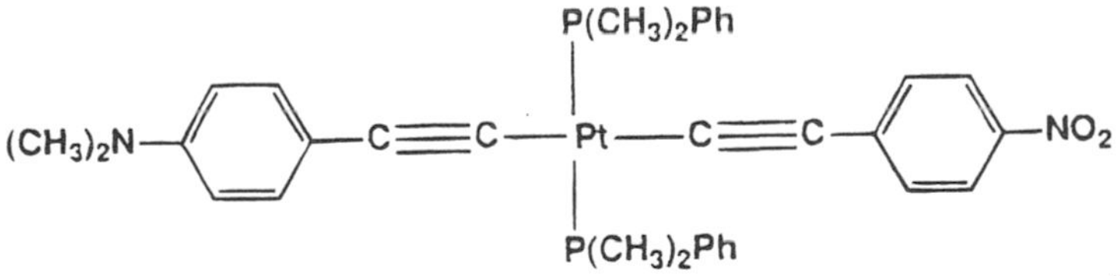

Space group: Cc or C/2c; crystal system = monoclinic; unit cell parameters: a = 20.961 Å, b = 5.693 Å, c = 28.551 Å, β = 110.29°, volume = 3196 Å^3, Z = 4; λ_{max} = 378 nm in acetonitrile; SHG activity = 0.4 times that of quartz at 1.064 μm.

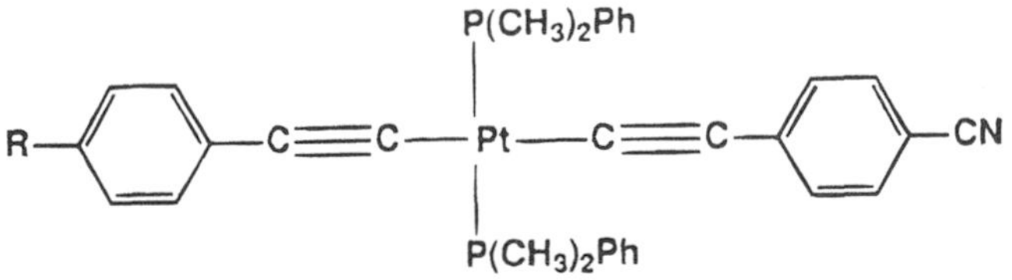

The compounds R = CH$_3$O; λ_{max} = 350 nm in acetonitrile; SHG activity = 0.5 times that of quartz; R = CH$_3$S; λ_{max} = 342 nm in acetonitrile; SHG activity = 0.33 times that of quartz, and R = H; λ_{max} = 338 nm in acetonitrile; SHG activity = 0.5 times that of quartz. The SHG activity of R = H is comparable with those compounds having donor group R = CH$_3$O or CH$_3$S in the para-position. All asymmetrically substituted platinum acetylides showed SHG activity.

(91) Bisbenzylidene Derivatives

a. 2,5-Bis(benzylidene)cyclopentanone (BBCP)
Space group: C222$_1$; d_{14} = 7 pm/V; d_{eff} = 12 pm/V at 1.064 μm (see reference 285).

b. 2,6-bis(4-methylbenzylidene)-4-tert-butylcyclohexanone (MBBCH)
Space group: Pnm2$_1$; d_{31} = 15 pm/V; d_{32} = 12 pm/V at 1.064 μm.

Table 53 Cutoff Wavelength in Cyclohexane and Powder SHG Efficiencies of Camphor Derivatives

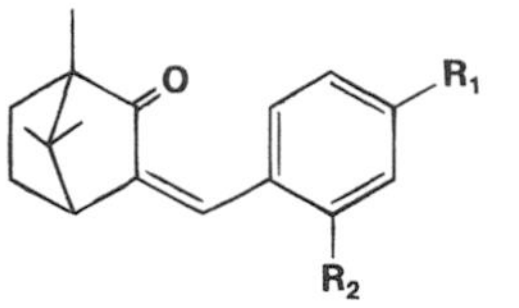

R_1	R_2	Melting Point (°C)	λ_{cutoff} (nm)	SHG (× urea)	β (10–30 esu)
H	H	96	320	2.0	10.0
CH$_3$	H	99	330	0.5	
CH$_3$O	H	126	340	0.13	
F	H	94	320	3.0	
Cl	H	106	320	2.1	
Br	H	131	320	1.9	
H	Cl	81	320	0.62	
CF$_3$	H	99	310	0.13	
N(CH$_3$)$_2$	H	144	380	<0.10	
CH$_3$S	H	88	350	0.10	
CDC compound		88	350	10.0	30

From Kawamata, J., et al., *Chem Lett.*, 921, 1993. With permission.

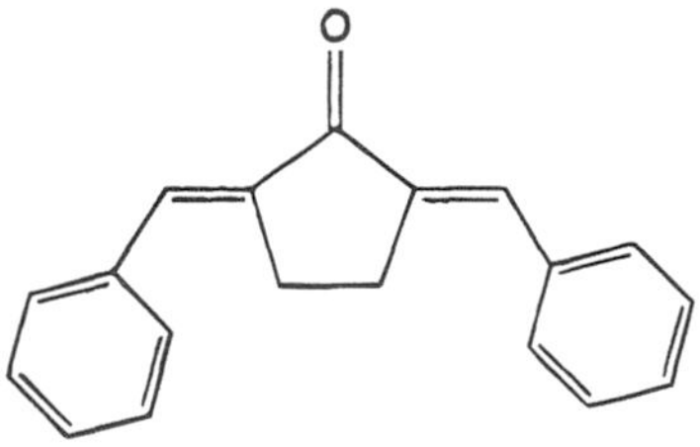

(BBCP)

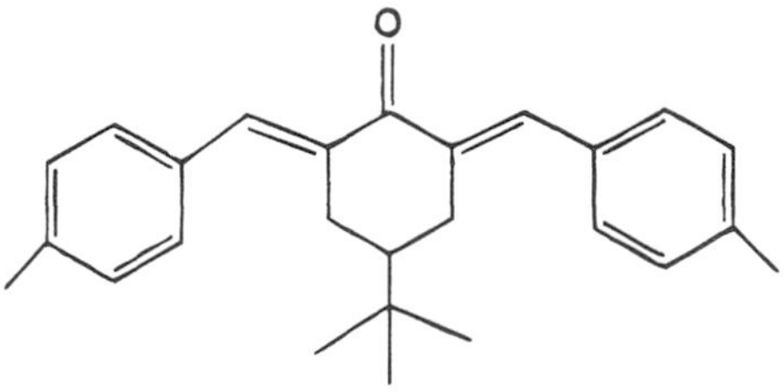

(MBBCH)

Kawamata and Inoue[286] reported powder SHG efficiency for a series of 3-benzylidene-D-camphor derivatives. 3-Cinnamylidene-D-camphor (CDC), which has a cutoff wavelength of 350 nm, showed SHG efficiency 10 times larger than that of urea standard at 1.06 μm. These compounds are interesting because of their optical transparency; it can be applied for SHG of a GaAs diode laser (750 ~ 830 nm). Hendricks et al.[287] measured β values of retinal and related derivatives by using HRS technique in a methanol solution at 1.064 μm. The β values of 140×10^{-30}, 310×10^{-30}, 470×10^{-30}, 730×10^{-30}, and 3600×10^{-30} esu were obtained for vitamin A acetate, retinoic acid, retinal Schiff base, retinal, and RPSB, respectively. These molecules show significantly large β values. The β values of 5×10^{-30}, 10×10^{-30}, 270×10^{-30}, and 1040×10^{-30} esu were obtained for β-cyclocitral, β-ionone, retinal, and carotenal, respectively, in a chloroform solution at 1.064 μm. These polyene derivatives have 2, 3, 6, and 10 double bonds, respectively. Therefore, the β values of these polyenes increases with the increase of π-conjugation length. A fraction of β value was found to be lost during *trans* to 13-*cis* or 9-*cis* isomerization of a retinal double bond. On the other hand, the β value increased by about an order of magnitude in going from the neutral form of the retinal Schiff base to the protonated form which exists in the bacteriorhodopsin protein.

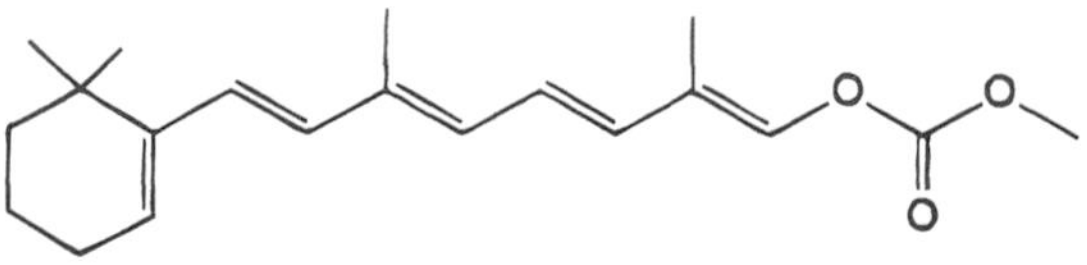

(Vitamin A acetate)

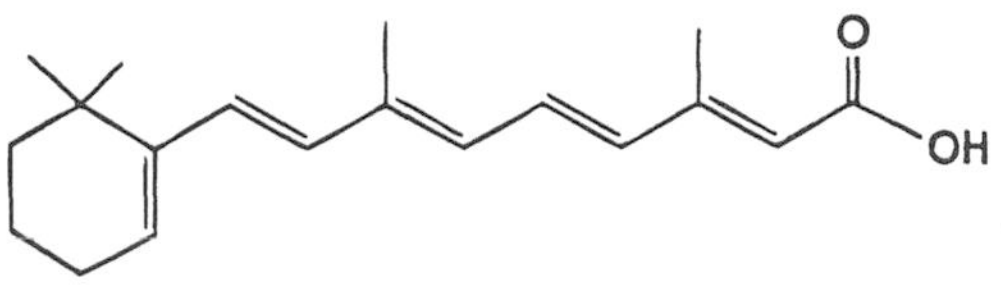

(Retinoic acid)

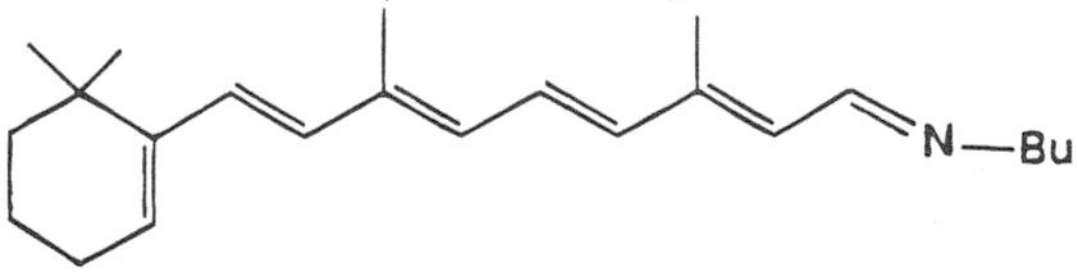

(Retinal Schiff base)

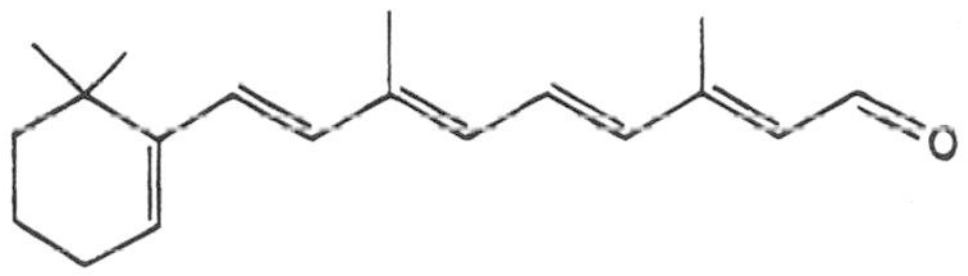

(Retinal)

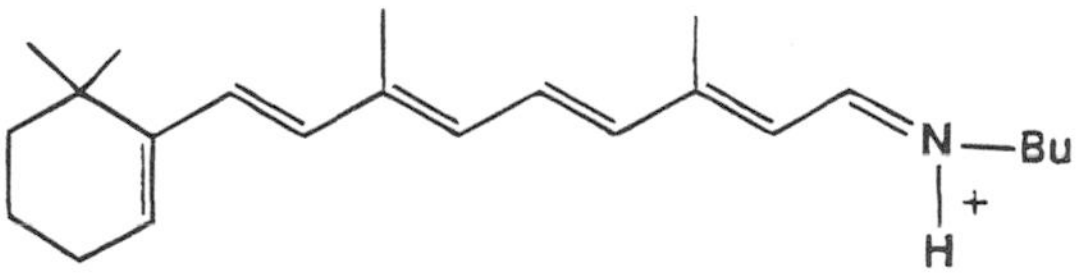

(RPSB)

(92) Salts of Dibenzoyltartaric Acid and 4-Aminopyridine

a. 1.5:1 (acid:base) monohydrate salt

Space group[288]: $P2_1$; crystal system = monoclinic; unit cell parameters: $a = 30.339$ Å, $b = 7.881$ Å, $c = 14.355$ Å; $\beta = 97.48°$; volume = 3403.1 Å^3; $Z = 4$, $R = 0.058$; density = 1.267 g cm^{-3}; colorless; melting point = 138 °C; powder SHG = 1.4 to 1.6 times that of urea.

b. 1:1 (acid:base) salt[288]

Space group: C2; crystal system = monoclinic; unit cell parameters: $a = 7.500$ Å, $b = 14.968$ Å, $c = 10.3705$ Å; $\beta = 102.67°$; volume = 1135.9 Å^3; $Z = 2$, $R = 0.048$, density = 1.322 g cm^{-3}; colorless; melting point = 141°C; powder SHG = 1.4 to 1.6 times that of urea

(93) Channel Inclusion Complexes

Tam et al.[289] reported a new avenue of introducing noncentrosymmetry for SHG activity by forming guest–host channel inclusion complexes between hosts thiourea or tris-o-thymotide (TOT) and organometallic guests. Neither hosts nor guests showed significant SHG themselves. Among thiourea organometallic inclusion complexes, 69% were SHG active whereas only 50% TOT complexes were SHG active. Authors suggested that if channel dimensions are maintained smaller than the guest dimensions then head-to-tail dipolar orientation by guest molecules is achieved along the channel directions. Contrary to this, if dipolar orientation of guest molecules becomes difficult for large channels because they accomodate a head-to-tail antiparallel pair, Thiourea-organmetallic complexes crystallize in a noncentrosymmetric space group and their SHG activity is governed by the nature of the organometallic species.

a. 3:1 Thiourea:(η^5-benzene)Cr(CO)$_3$ Complex

Space group: R3c; unit cell parameters: $a = b = 16.130$ Å, $c = 12.569$ Å, $\alpha = \beta = 90°$, $\gamma = 120°$, volume = 2832 Å^3; $Z = 6$; $R = 0.030$; density = 1.556 g cm^{-3}; powder SHG = 2.3 times that of urea.

b. 3:1 Thiourea:(η^4-1,3-cyclohexadiene)Fe(CO)$_3$ Complex

Space group: Pna2$_1$; unit cell parameters: $a = 12.562$ Å, $b = 16.128$ Å, $c = 9.536$Å, $\alpha = \beta = \gamma = 90°$, volume = 1932 Å^3; $Z = 4$; $R = 0.027$; density = 1.541 g cm^{-3}; powder SHG = 0.4 times that of urea.

c. 3:1 Thiourea:(η^4-trimethylenemethane)Fe(CO)$_3$ Complex
Space group: R3c; unit cell parameters: $a = b = 16.234$ Å, $c = 12.488$ Å, $\alpha = \beta = 90°$, $\gamma = 120°$; volume = 2850.2 Å^3; $Z = 6$; $R = 0.021$; density = 1.476 g cm^{-3}; powder SHG = 0.3 times that of urea.

d. 3:1 Thiourea:(η^5-cyclohexadienyl)Mn(CO)$_3$ Complex
Space group: R3c; unit cell parameters: $a = b = 16.307$ Å, $c = 12.4653$ Å, $\alpha = \beta = 90°$, $\gamma = 120°$; volume = 2913.9 Å^3; $Z = 6$; $R = 0.0245$; density = 1.530 g cm^{-3}; powder SHG = 0.4 times that of urea.

Inclusion complexes of 3:1 thiourea:(η^6-fluorobenzene)Cr(CO)$_3$, 3:1 thiourea:(η^5-Cp)Mn(CO)$_3$, thiourea:(η^5-Cp)Re(CO)$_3$, 3:1 thiourea:(η^5-CpCr)(CO)$_2$(NO), 3:1 thiourea:(η^4-1,3-butadiene)Fe(CO)$_3$, 3:1 thiourea:(η^4-pyrrolyl)Mn(CO)$_3$, and 3:1 thiourea:(η^4-thiophene)Cr(CO)$_3$ exhibited SHG activities 2.0, 0.3, 0.5, 0.1, 1.0, 0.2, and 0.1 times that of urea, respectively. Inclusion complexes with tris-o-thymotide, 1:1 TOT:(p-cynobenzoyl)Mn(CO)$_3$ showed SHG activity 0.2 times that of urea, whereas SHG activity of all 1:1 complexes of TOT:η^6-indan)Cr(CO)$_3$, TOT:η^6-anisole)Cr(CO)$_3$, and TOT:η^6-tetralin)Cr(CO)$_3$ were 0.1 times that of urea. The SHG activity of TOT:η^6-(dimethylamino)benzeneCr(CO)$_3$, and TOT:[η^6-benzene)Mn(CO)$_3$]BF$_4$ were 0.13 and 0.1 times that of urea, respectively.

(94) 4,5-Dimercapto-1,3-dithiole-2-thionato Compounds

a. X = O (4,5-Bis(2',4'-dinitrophenylthio)-1,3-dithiole-2-one) β-BNPT-DTO
Space group[290]: P2$_1$2$_1$2$_1$; crystal system = orthorhombic; unit cell parameters: $a = 10.078$ Å, $b = 19.352$ Å, $c = 9.926$ Å; volume = 1936 Å^3; $Z = 4$; $R = 0.027$; density = 1.77 g cm^{-3}; transparency range = 0.460–2.1 μm; melting point = 189 °C; powder SHG = 5 times that of urea; α-phase BNPT-DTT showed no SHG activity.

b. X = S (4,5-bis(2',4'-dinitrophenylthio)-1,3-dithiole-2-thione) β-BNPT-DTT
Space group: P2$_1$2$_1$2$_1$; crystal system = orthorhombic; unit cell parameters: $a = 10.143$ Å, $b = 20.488$ Å, $c = 9.656$ Å; volume = 2005 Å^3; $Z = 4$; $R = 0.038$; density = 1.76 g cm^{-3}; color = orange; melting point = 212 °C; powder SHG = 0.1 that of KDP; α-BNPT-DTT obtained by passing through a silica gel chromatographic column with chloroform showed no SHG activity.

(95) 1-(3-thienyl)-3-(4-chlorophenyl)-propen-1-one
Space group = P2$_1$; crystal class = monoclinic; unit cell parameters: $a = 5.942$ Å, $b = 4.868$ Å, $c = 20.104$ Å, $\beta = 95.94°$, volume = 578.4 Å^3; $Z = 2$; density = 1.43 g/cm^3; $\lambda_{cutoff} = 410$ nm; transparency range = 0.41–1.5 μm; melting point = 123.9 °C; powder SHG = 3 times that of KTP.[291]

(96) Halobenzonitrile and Halo-Dicyano-Pyrazine Compounds
Donald et al.[292] studied the SHG properties of halobenzonitrile and halo-dicyano-pyrazine compounds. The 4-chloro (X = Cl), 4-bromo (X = Br), and 4-iodo (X = 1) benzonitrile crystals were found to have similar structures. The arrays of molecules are antiparallel in 4-chlorobenzonitrile and 4-iodobenzonitrile, which leads to centrosymmetric structure and hence no significant SHG activity. On the other hand, the arrays in 4-bromobenzonitrile are parallel, which results in SHG activity. The crystal details of 4-bromobenzonitrile are as follows: unit cell parameters: $a = 4.097$ Å, $b = 8.538$ Å, $c = 9.049$ Å, $\beta = 90.83°$; $Z = 2$; β measured by EFISH at 1.19 μm = 1.1 × 10^{-30} esu. The β measured of 4-chlorobenzonitrile by EFISH at 1.19 μm was 0.8 × 10^{-30} esu.

Donald et al.[292] also reported another interesting class of halogen- and cyano-containing pyrazines. Because the halo-cyano interaction gives rise to favorable crystal packing, most of the material in these compounds shows very large powder SHG activity. The powder SHG activity depends on substituents; as high as 10,000 times that of quartz has been observed.

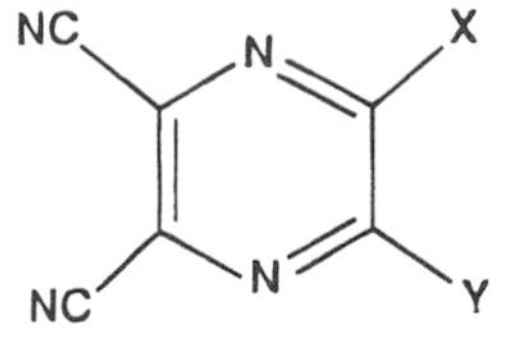

X	Y	SHG (× quartz)
Cl	Cl	500
4-Cl-C$_6$H$_4$-NH	Cl	6,500
(C$_6$H$_5$)$_2$CHNH	Cl	4,200
4-F-C$_6$H$_4$-NH	Cl	2,900
4-Br-C$_6$H$_4$-NH	Cl	10,000
4-CN-C$_6$H$_4$-NH	Cl	60
CH$_3$(CH$_2$)$_4$NH	Cl	170
Br	Br	100
4-Cl-C$_6$H$_4$-NH	Br	1,700
(C$_6$H$_5$)$_2$CHNH	Br	500
4-Cl-C$_6$H$_4$-NH	CN	0
4-CN-C$_6$H$_4$-NH	Br	100
4-Br-C$_6$H$_4$-NH	Br	4
4-Br-C$_6$H$_4$-NH	CN	0
4-CN-C$_6$H$_4$-NH	CN	0

a. X = Y = Cl (2,3-Dichloro-5,6-dicyanopyrazine)

Space group = P2$_1$; monoclinic; unit cell parameters: $a = 7.383$ Å, $b = 6.969$ Å, $c = 7.445$ Å, $\beta = 105.93°$; Z = 2; β measured by EFISH at 1.19 μm = 1.7×10^{-30} esu; μ = 4.4×10^{-18} esu

(97) L-Histidine Tetrafluoroborate

Marcy et al.[293] reported crystal structure and NLO properties of solution grown L-histidine tetrafluoroborate (HFB), which is a semiorganic material. The HFB was prepared by dissolving both L-histidine and tetrafluoroboric acid in one equivalent ratio in water. Rod-like single crystals were obtained from slow water evaporation at ambient temperature. The imidazole and amino (NH3)$^+$ groups are protonated and thus counterbalanced the negative charges of the carboxylate (COO)$^-$ and tetrafluoroborate (BF$_4$)$^-$ ions. The SHG originate from delocalized electrons related to the histidine imidazolium and carboxylate groups.

Chemical structure:

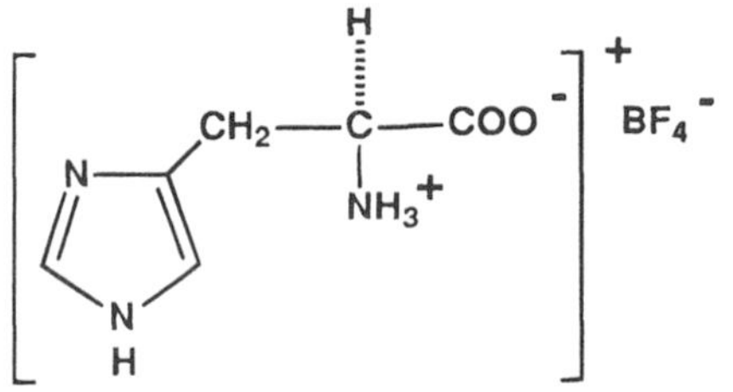

Formula: [(C$_3$N$_2$H$_4$)CH$_2$CH(NH$_3$)(CO$_2$)]$^+$BF4$^-$; space group = P2$_1$, crystal class-monoclinic
Unit cell parameters: $a = 5.022$ Å, $b = 9.090$ Å, $c = 10.216$ Å, $\beta = 93.484°$, transparency = 0.250–1.3 μm
Decomposition temperature = 205 °C, d_{eff} = 2 pm/V from type II phase matching.

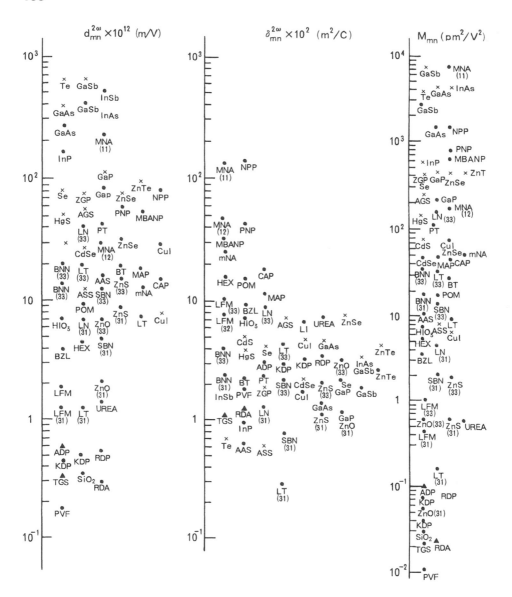

Figure 11 A comparative representation of the NLO coefficients, Miller's δ_{mn} and the figure-of-merit of well-known inorganic and organic second-order NLO materials. •, ▲, × refer to fundamental wavelengths of 1.06, 0.6943, and 10.6 μm, respectively. The number in parentheses ( ) represents the d_{mn} coefficients such as d_{11}, d_{14}, d_{31}, d_{33}, etc. mNA(31) = meta-nitroaniline. MAP(22) = 3-methyl-2(2-4-dinitrophenyl)-aminopropanoate, POM(36) = 3-methyl-4-nitropyridine-1-oxide, HEX(14) = hexamine, BZL(11) = benzil, MNA = 2-methyl-4-nitroaniline, CAP(31) = 2-cyclooctylamino-5-nitropyridine, PVF(31) = $(CH_2CF_2)_n$, NPP(21) = N-(4-nitro-phenyl)-(L)-prolinol, PNP(21) = 2-N-prolinol)-5-nitropyridine, MBAND(22) = 2-(α-methylbenzylamino)-s-nitropyridine, BZL(11) = benzil, UREA(36), TGS(23) = $(NH_2CH_2COOH)_3 \cdot H_2SO_4$, LN = $LiNbO_3$, LT = $LiTaO_3$, BT(15) = $BaTiO_3$, PT(31) = $PbTiO_3$, SBN = $Sr_{0.5}Ba_{0.5}Nb_2O_6$, BNN = $Ba_2NaNb_5O_{15}$, $HIO_3(14)$, LI(31) = $LiIO_3$, $SiO_2(11)$, LFM = $LiCHO_2 \cdot H_2O$, ADP(36) = $NH_4H_2PO_4$, KDP(36) = KH_2PO_4, KD*P(36) = KD_2PO_4, RDA(36) = RbH_2AsO_4, RDP(14) = RbH_2PO_4, HgS(11), Se(11), Te(11), AAS(31) = Ag_3AsS_3, ASS(31) = Ag_3SbS_3, AGS(14) = $AgGaS_2$, ZGP(14) = $ZnGeP_2$, CdSe(31), CdS(15), CuI(14), GaSb(14), GaAs(14), GaP(14), InSb(14), InAs(14), InP(14), ZnSe(36), ZnTe(14). (From Umegaki, S., *Organic Nonlinear Optical Materials*, Bunshin Press, Tokyo, 1990. With permission.)

Umegaki[294] compiled inorganic and organic single crystals for a significant comparison. Figure 11 shows NLO coefficients and NLO figures of merit (d^2/n^3) for a very wide variety of inorganic and organic materials. Organic single crystals have a much larger figure of merit (FOM) than the inorganic counterparts as depicted in the right-hand column. For example, MNA, NPP, PNP, and

MBANP exhibit remarkably large FOMs. Umegaki[294] calculated the FOM ($d_{ij}{}^2/n(2\omega)n(\omega)^2$) of 26,000, 16,000, 8,700, 3,700, 3,100, 1,400, 600, 50, and 20 (pm/V)^2 for 4'-nitrobenzylidene-3-acetamino-4-methoxyaniline (MNBA), merocyanine (MC), MNA, MMONS, 4-nitro-4'-methylbenzylideneaniline (NMBA), DMNP, MBANP, DAN, and POM single crystals, respectively, whereas the FOM of classical inorganic material LiNbO$_3$ was calculated as 150 (pm/V)^2. The FOM ($d_{eff}{}^2/n^3$) of MMONS is 850 (pm/V)^2. The FOM of MNMA is moderate, 26 (pm/V)^2 for the d_{31} coefficient. DAST single crystals exhibit NLO coefficients as high as 1900 pm/V at 1.2 μm.[210] The cutoff wavelength below 350 nm has likely been achieved for many single crystal materials. Organic crystals have been used for fabrication of different waveguides such as channel waveguide, planar and slab-type waveguides and cored fibers. (Such aspects are discussed in chapter 5.) Besides material performance characteristics such as thermal, chemical, and mechanical stability, high purity, and optical quality, NLO conditions that should be satisfied by a single crystal material for fabricating waveguides are relatively low refractive index, low absorption and scattering losses, high optical damage threshold (GW/cm^2), and high NLO and electro-optical coefficients. No single crystal possess all these features. Waveguide has been studied extensively in a variety of organic single crystals such as channel waveguides in *m*-nitroaniline and NPP, cored fiber waveguides in *m*-nitroaniline, *m*-dinitrobenzene, 2-bromo-4-nitroaniline, and NPAN, planar waveguide in *p*-chlorophenylurea, and tapered slab-type waveguide in MNA. Umegaki[295] pointed out that MNBA, merocyanine, MNA, MMNOS, NMBA, and POM are well suited for the Cerenkov-radiation type SHG because their NLO coefficients are diagonal, having parallel or nearly parallel alignment of molecules. However they can be applied for channel waveguides and fibers under the conditions of Cerenkov-radiation type phase matching. On the other hand, DMNP and DAN were classified as promising crystals. As a whole, single crystals are very promising materials for frequency doubling and other applications.

B. ORGANOMETALLIC COMPOUNDS

Organometallic compounds are different than organic materials because they possess a metal atom in the structural unit. Organometallics are important because they are expected to exhibit large molecular hyperpolarizabilities due to the transfer of electron density between the metal atom and the ligands. There are two types of charge-transfer transitions, metal-to-ligand and ligand-to-metal. The charge-transfer interactions can be increased further by using central metal atoms with diversed oxidation states and ligands. A wide variety of central metal atoms varying in size, nature, and oxidation states, as well as ligands with different nature and size, facilitate the development of relationships between molecular structures and NLO properties. The chirality can be introduced by metal centers which help in generating noncentrosymmetric molecules. The first review of NLO properties of organometallic materials was published in 1991 by Nalwa.[101] Since then some developments have occurred. The SHG properties of the single crystals of organometallic materials, are documented in section VII.A and the bulk organometallic compounds are discussed here.

1. Metallocenes

Cheng et al.[296] and Calabrese et al.[297] measured hyperpolarizabilities of metallocene derivatives with EFISH technique at 1.91 μm. The ferrocene (M = Fe) derivative where X = H and Y = COCH$_3$ showed a μ value of 3.0 × 10^{-18} esu and a β value of 0.3 × 10^{-30} esu. The ruthenocene (M = Re) derivative where X = CH$_3$ and Y = NO$_2$ showed a μ value of 3.5 × 10^{-18} esu and a β value of 0.6 × 10^{-30} esu. The π-conjugation of metallocenes was increased further; the resulting compounds structurally resemble some nitrostilbene compounds. A large bathochromic shift in the absorption bands was observed for pentamethyl, ethylenic cyano, and 2,4-dinitro substituents, while the 4-nitrophenyl butadiene ferrocene was red-shifted. The ruthenocenes showed a larger dipole moment, and the replacement of Fe with Ru reduced β values because Ru has a higher ionization potential. The second-order optical nonlinearities of the metallocenes were influenced by metal atom, *cis* and *trans* conformations, π-electron conjugation, and terminal substituents. The *trans*-ferrocene derivative showed larger optical nonlinearities than the *cis* compound. Pentamethyl substitution gave rise to a large dipole moment and optical nonlinearity. The replacement of a cyclopentadienyl ring with a pentamethylcyclopentadienyl ring also led to large optical nonlinearity. The β value of (*E*)-ferrocene-4-(4-nitrophenyl)ethylene was 31 × 10^{-30} esu, slightly larger that than 4-methoxy-4'-nitrostilbene (β of 29 × 10^{-30} esu). Table 54 lists NLO properties of metallocenes.

Table 54 NLO Parameters of Metallocene Derivatives Recorded in *p*-Dioxane Solvent

M	X	Y	μ (10^{-18} esu)	α (10^{-23} esu)	β (10^{-30} esu)
Fe	H	H (*trans*)	4.5	3.9	31
Fe	H	H (*cis*)	4.0	3.8	13
Fe	CH$_3$	H	4.4	5.3	40
Fe	H	CN	5.3	4.2	21
Fe	CH$_3$	CN	6.0	5.6	35
Fe[a]	H	H	4.9	4.5	23
Fe[b]	H	H	4.5	4.6	66
Ru	H	H	5.3	4.2	12
Ru	CH$_3$	H	5.1	5.0	24
Ru	CH$_3$	CN	5.9	5.5	24

[a] Fe = 2,4-dinitro derivative.

[b] Fe − 4-nitrophenyl butadienyl.

From Cheng, C. T., et al. *Mol. Cryst. Liq. Cryst.*, 198, 137, 1990; Calabrese, I. C., et al., *J. Am. Chem. Soc.*, 113, 7227, 1991. With permission.

Marder et al.[298] reported powder SHG of ferrocenyl derivatives at 1.907 μm. The powder SHG efficiency of a ferrocenyl compound where the counterion $X = I^-$ was as high as 220 times larger than that of urea, whereas for Br$^-$ the counterion was 170 times larger. On the other hand, no SHG signal was observed for Cl$^-$ and CF$_3$SO$_3^-$ counterions. The SHG efficiencies of ferrocenyl derivatives having counterions BF$_4^-$, PF$_6^-$, and NO$_3^-$ were 50, 0.05, and 75 times that of urea, respectively. For both B(C$_6$H$_5$)$_4^-$ and (*p*)-CH$_3$-C$_6$H$_4$SO$_3^-$ counterions, the SHG activity was 13 times that of urea.

Doisneau et al.[299,300] reported β values of bimetallic ferrocenyl derivatives measured by EFISH at 1.34 and 1.9 μm. These complexes are linked via an ethylenic bridge, where the ferrocenyl group acts as a donor and the oxazoline acts as a acceptor group. The sulfur to the oxazoline ring provides a bidentate ligand for further complex formation. Chirality is introduced in the ligand either at the sulfur or in the oxazoline ring that leads to noncentrosymmetric crystals for SHG. The measured and calculated β values are listed in Table 55. According to these results, the β values are enhanced significantly upon complexation. The highest β value of 1.23×10^{-28} esu was obtained for a Pd complex. Both Pd and Pt complexes showed larger β values than Ni complexes. A copolymer of PMMA having ferrocenyl-based chromophore (2) showed $d_{33} = 4.39$ pm/V and $d_{31} = 1.38$ pm/V.

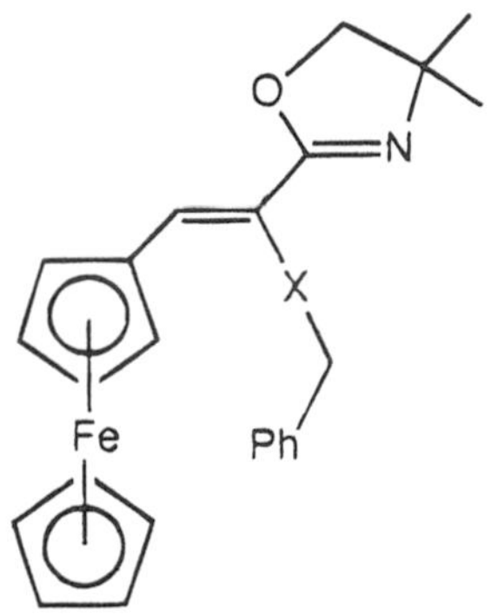

(1)

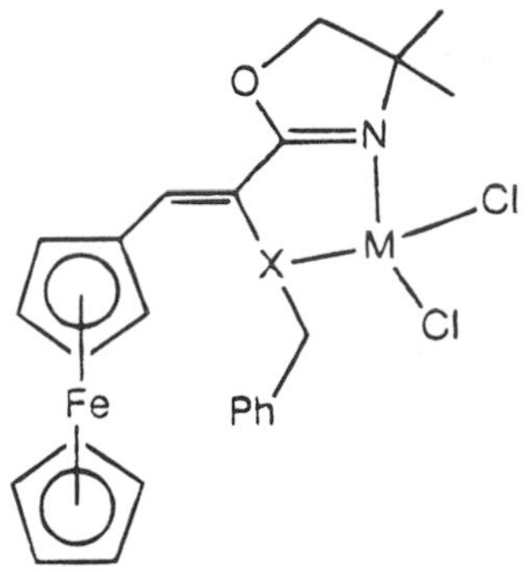

(2)

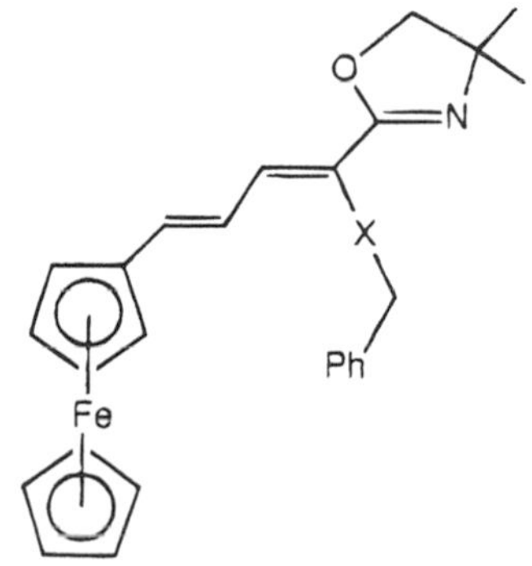

(3)

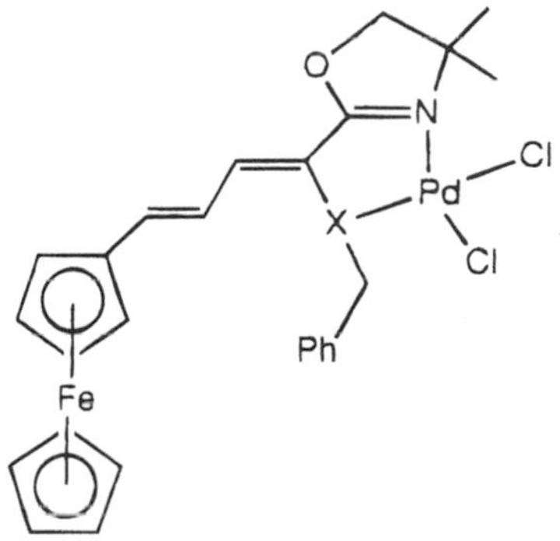

(4)

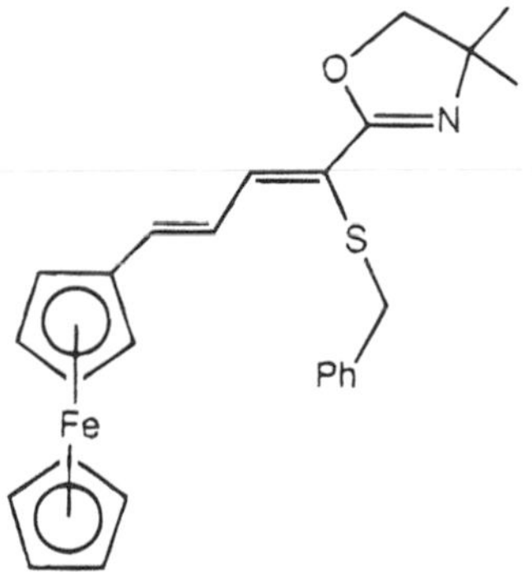

(5)

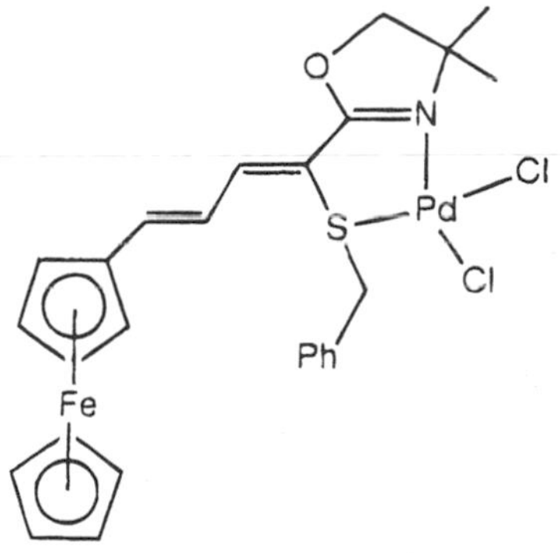

(6)

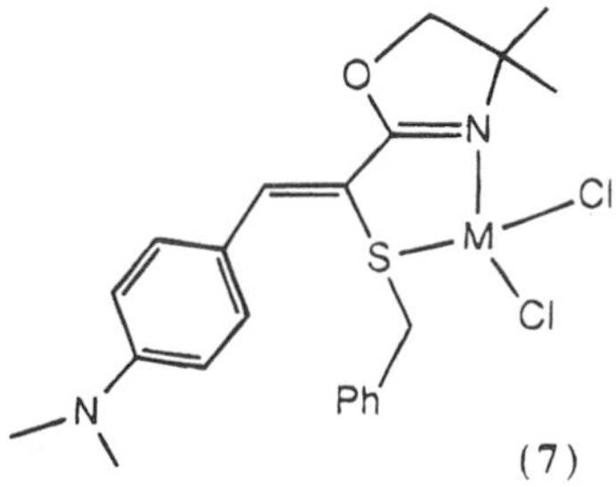

(7)

Table 55 Absorption Peaks and Dipole Moment Measured in Chloroform Solution, β_{EFISH} Measured in Chloroform Solution at 1.34 μm, and β_0 Calculated from a Two-Level Model

Compound	X	M	λ_{MLCT} (nm)	μ_g (D)	β_{EFISH} (10^{-30} esu)	β_0 (10^{-30} esu)
1	S	–	472	3.85	4.5	2.0
	SO	–	470	5.7	8.5	3.7
	SO$_2$		471	6.45	6.3	1.6
2	S	Ni	500	14.0	10.3	3.92
	S	Pd	530	13.2	12.2	4.13
	S	Pt	523	11.6	16.0	5.2
	SO	Ni	498	14.0	15.8	6.0
	SO	Pd	540	12.2	30.2	7.6
	SO	Pt	548	11.3	28.0	7.7
3	S	–	482	3.22	13.3	5.6
	SO	–	494	6.65	11.0	4.6
4	S	Pd	567	13.7	58.0	13.5
	SO	Pd	602	13.5	123.5	19.0
5			460	5.55	2.6	1.2
6			558	13.6	33.6	8.5
7		Pd	442	13.5	37.0	20.0

From Loucif, R. et al., *Mol. Eng.*, 2, 221, 1992. With permission.

Wright et al.[301] prepared organometallic NLO polymers consisting of pendant ferrocene chromophores in a poly(methylmethacrylate) copolymer. This polymer shows a T_g of 120 °C and a T_m of 225 °C. The SHG efficiency of this copolymer was 4 times that of the quartz reference and it showed a d_{33} coefficient of 1.72 pm/V at 1.064 μm.

Qin et al.[302] reported the powder SHG of some of the organo-zirconium and titanium compounds having the general formula $(h\text{-}C_5H_5)_2M(Cl)_n(EAr)_{2-n}$ where M = Ti or Zr; E = O or S; n = 1 or 0. Table 56 lists the powder SHG efficiency and absorption characteristics of zirconium compounds. Of the eleven monosubstituted zirconium derivatives synthesized, 91% were SHG active. Compound 10 showed the largest SHG, 4.5 times that of KDP. Most of these compounds were white. Both mono- (n = 1) and disubstituted (n = 0) titanocene derivatives were also examined. Disubstituted titanium compounds did not show any SHG although several of monosubstituted compounds were found to be SHG active.

2. Metal Nitrido Compounds

Pollagi et al.[303] reported SHG properties of one-dimensional nitrido conjugated polymers having the following chemical structure. The $[N \equiv M(OR)_3]_n$ structure had triple and single bonds where M = Mo and W and R = CMe$_3$ and CMe$_2$CF$_3$. The $[MoN(OCMe_3)_3]_n$ was noncentrosymmetric and crystallized into space group P6$_3$cm and showed SHG efficiency 0.25 times that of urea. The SHG efficiencies of $[WN(OCMe_3)_3]_n$ and $[MoN(OCMe_2CF_3)_3]_n$ were 0.20 and 0.35 times that of urea at 1.064 μm, respectively. On the other hand, no SHG was observed for $[MoN(OCMe_2Et)_3]_n$. The SHG efficiency and emission energy were affected by the nature of alkoxide ligands and the effective conjugation length of the $[MN]_n$ backbone.

3. Metal Carbonyl Complexes

Cheng et al.[296] measured hyperpolarizabilities of arene metal carbonyl complexes with the EFISH technique at 1.91 μm. Table 57 lists the dipole moment, polarizability, and hyperpolarizabilities of chromium tricarbonyl benzene π-complexes (π-CTCB) and tungsten pentacarbonyl pyridine σ-complexes (σ-TPCB). The optical nonlinearities of σ-TPCB complexes are larger than π-CTCB complexes, which are affected significantly by the 4-pyridine substituents. The higher dipole moment of this series results from the pyridine fragment and the metal cluster contributions. The sensitivity at 4-position substituents is related to several factors such as additive effect, the strength of the pentacarbonyl tungsten acceptor, and d-π overlapping. The β values of metal carbonyl derivatives were negative showing a reversal of sign for charge transfer in the ground and excited states.

Frazier et al.[304] evaluated SHG efficiency of about 60 transition organometallic compounds of Group VI metal carbonyl arene, pyridyl, and chiral phosphine complexes. The general formula of the metal carbonyls is $M(CO)_nL_{3-n}$, where M = Cr, Mo, W, L = phosphine ligand, n = 0, 1, 2, or 3; and the arene is a benzene ring. The arenecarbonylchromium(0) complexes; Cr(m-anisidine)$(CO)_3$, Cr[(−)-a-ethylphenethyl alcohol]$(CO)_3$, Cr(S)-(+)-(2-methylbutyl)benzene$(CO)_3$, Cr(nopolbenzyl ether)$(CO)_3$, Cr(L-phenylaniline ethyl ester hydrochloride)$(CO)_3$, Cr(styrene)$(Co)_3$, and Cr(p-xylene)$(CO)_2$nmdpp showed SHG efficiency of 0.3, 0.01, 1.7, 1.3, 0.6, 1.8, and 0.01 times that of ammonium dihydrogen phosphate (ADP) respectively. The nmdpp represents neomenthyldiphenylphosphine. The evidence showed that an electron acceptor group close to the benzene ring led to SHG-inactive materials. The metal atom interaction resulted in a positive effect; for example, the SHG efficiency of Cr(L-phenylaniline ethyl ester hydrochloride)$(CO)_3$ complex is about 30 times larger than that of the parent compound L-phenylaniline ethyl ester hydrochloride. Chiral phosphine carbonyl metal complexes; Fe(CO)nmdpp,

Table 56 Powder SHG Efficiency and Optical Absorption Maxima of Zirconium Compounds

Compounds	E	A	B	F	C	D	λ_{max} (nm)	SHG (× KDP)
1	O	H	H	NO$_2$	H	H	328	2.0
2	O	NO$_2$	H	H	H	H		1.4
3	O	NO$_2$	H	NO$_2$	H	H		0.0
4	O	H	H	F	H	H	315	0.15
5	O	F	H	H	H	H	338	2.0
6	O	H	H	Br	H	H	334	0.3
7	O	Cl	H	H	H	H	344	2.0
8	O	H	H	CO$_2$Pr	H	H	330	1.2
9	S	H	H	F	H	H	316	1.2
10	S	F	F	H	F	F	336	4.5
11	S	F	F	F	F	F	363	0.15

From Qin J., et al., Sen-i-Gakkai Symposium Preprints, p. A46, 1991. With permission.

Table 57 Dipole Moment, Polarizability, and Hyperpolarizabilities of Arene Metal Carbonyl Complexes, π-CTCB, and σ-TPCB

Complex	X	Solvent	μ (10^{-18} esu)	α (10^{-23} esu)	β (10^{-30} esu)
π-CTCB	H	Toluene	4.4	2.3	−0.8
π-CTCB	OCH$_3$	Toluene	4.7	2.7	−0.9
π-CTCB	NH$_2$	p-Dioxane	5.5	2.7	−0.6
π-CTCB	N(CH$_3$)$_2$	Toluene	5.5	2.9	−0.4
π-CTCB	COOOCH$_3$	p-Dioxane	4.0	2.9	−0.7
σ-TPCB	H	Toluene	6.0	3.3	−4.4
σ-TPCB	C$_6$H$_5$	Chloroform	6.0	4.6	−4.5
σ-TPCB	C$_4$H$_9$	p-Dioxane	7.3	4.2	−3.4
σ-TPCB	NH$_2$	DMSO	8	3.5	−2.1
σ-TPCB	COCH$_3$	Chloroform	4.5	4.0	−9.3
σ-TPCB	COH	Chloroform	4.6	3.7	−12

From Cheng, C. T., et al., *Mol. Cryst. Liq. Cryst.*, 198, 137, 1990. With permission.

Mo(CO)nmdpp, and W(CO)nmdpp showed SHG efficiency of 0.01, <0.01, and 0.06 times that of ADP, respectively, whereas the neomenthyldiphenylphosphine (nmdpp) showed larger SHG than metal complexes. The SHG activity was affected by the nature of the transition metal atom.

The pyridyl Group VI metal carbonyl complexes have a general formula M(CO)$_{5-n}$L$_n$, where M = Cr, Mo, W; L = phosphine ligand; and n = 0 or 1. The pyridine ring is attached to the metal through the nitrogen atom. The SHG efficiency of these complexes was rather small, between 0.4 and 1.0 times that of ADP. However, one of the parent compounds 4-benzoylpyridine showed SHG efficiency 14 times that of ADP, although its Cr and Mo complexes were SHG inactive, whereas the W complex W(CO)$_5$(4-benzoylpyridine) exhibited SHG efficiency of 0.2 relative to ADP. The SHG efficiency of Fe(CS$_2$)(CO)$_2$[P(OCH$_3$)$_3$]$_2$ was 0.03 whereas that of Mo(CO)$_4$(PPh$_3$)$_2$ was <0.01.

4. Square-planar Metal Complexes

Cheng et al.[296] measured hyperpolarizabilities of square-planar metal benzenes, platinum and palladium benzene derivatives (Table 58). These complexes are analogous to the donor–acceptor *para*-disubstituted benzenes. The optical nonlinearities of platinum compounds were higher than the palladium compounds because of the difference in inductive donating strength between the two metal atoms. The dipole moment and hyperpolarizabilities increased with acceptor group strength and, as a result, the nitro derivatives had larger optical nonlinearities.

Kanis et al.[305,306] used INDO/SOS (intermediate neglect of differential overlap) electronic structure model (ZINDO) to calculate the second-order NLO response of transition metal-containing chromophores. The synthesis, characterization, and β values of (4-dimethylamino-4′-stilbazole)W(CO)$_5$ were reported. The ZINDO-derived calculation for formyl (X = CHO)-substituted, pyridine-based chromium coordination complexes yielded β_{vec} of −44.6, −29.4, and −22.1 × 10^{-30} cm^5 esu^{-1} for n = 1, 2, 3, respectively.[296] Although the β_{vec} of analogous aminonitrostilbene increased significantly with the increase of π-conjugation, the chromium pentacarbonyl derivative showed no such behavior. Instead,

Table 58 Dipole Moment, Polarizability, and Hyperpolarizabilities of Square-Planar Metal Benzene Complexes

A	M	L	X	μ $(10^{-18}$ esu)	α $(10^{-23}$ esu)	β $(10^{-30}$ esu)
CHO	Pt	PEt$_3$	Br	2.5	5.8	2.1
NO$_2$	Pd	PEt$_3$	I	3.6	6.2	0.5
NO$_2$	Pd	PPh$_3$	I	5.5	1.05	1.5
NO$_2$	Pt	PEt$_3$	I	3.0	6.2	1.7
NO$_2$	Pt	PEt$_3$	Br	3.4	6.1	3.8

From Cheng, C. T., et al., *Mol. Cryst. Liq. Cryst.*, 198, 137, 1990. With permission.

it decreased with increasing conjugation length. The β values of other related organometallic structures were also estimated.

5. Phthalocyanines

Because metallophthalocyanines are centrosymmetric materials, they are not expected to exhibit SHG, although the observation of SHG in phthalocyanines has become possible through fabrication techniques from the break in symmetry. SHG activity in copper phthalocyanine (CuPc) where M = Cu and R = H, has been reported by several research groups. Chollet et al.[307] were the first to report SHG from vacuum deposited thin films of CuPc.

Using the LB approach, the d_{33} coefficients of 11.5×10^{-9} esu for p-p polarization and 65×10^{-9} esu for s-p polarization were estimated for a 162 nm thick CuPc film. This approach was not found to be suitable. Using the poled polymer film approach, the d_{33} coefficients of 31×10^{-9} esu for 162 nm thick film and 11.3×10^{-9} esu for 87-nm-thick film for *p-p* polarization were obtained. The authors did not observe any SHG from metal-free and zinc phthalocyanines.

Kumagai et al.[307] reported strong SHG efficiency from evaporated CuPc thin films with thicknesses between 40 and 200 nm. The second-order nonlinear optical susceptibility $\chi^{(2)}_{ZYY}$ of CuPc film was estimated as 4.51×10^{-8} esu, where Z and Y are the perpendicular and parallel directions to the film surface, respectively. Yamada et al.[308] analyzed the origin of SHG in vacuum-deposited CuPc thin films by performing *in situ* SHG experiments during the evaporation process as a function of film thickness. The SH intensity exhibited quadratic dependence between the 70 and 200 nm region, but it deviated for thinner and thicker regions. Kumagai et al.[307] also observed a deviation from the quadratic dependence in thicker regions in a reflection geometry. The origin of SHG in CuPc vacuum-deposited films was considered from the electric quadrupolar mechanism or preferably from a magnetic dipole mechanism. Hoshi et al.[309] demonstrated that the SH intensity of vanadyl phthalocyanines (VOPc), where M = VO and R = H, thin films increased quadratically with the increase of film thickness up to 64 nm and showed a saturation behavior at about 110 nm. The origin of SHG was considered a polar VOPc orientation in the epitaxially grown 4mm structure.

Liu et al.[310,311] reported SHG from the LB films of an asymmetrically substituted metal-free phthalocyanine; nitro-tri-*tert*-butylphthalocyanine as shown in Langmuir-Blodgett section. The SH intensities of the monolayer were 4.0×10^{-7} and 7.5×10^{-8} relative to a Z-cut quartz wedge for *p-p* and *s-p* configurations, respectively. The β value of $2–3 \times 10^{-30}$ esu and $\chi^{(2)}$ of $20–30 \times 10^{-9}$ esu was estimated for the monolayer. Neuman et al.[312] reported SHG from tetracumylphenoxy nickelphthalocyanine abbreviated as NiPc(CP)$_4$ where M = Ni and R = OC$_6$H$_4$-C(CH$_3$)$_2$-C$_6$H$_5$ (cumylphenoxy) films developed by the LB technique. The alignment of the LB film covered quartz substrate and the angle of incidence were found to be critical for SHG activity. The 35-layer LB films of NiPc(CP)$_4$ and octadecanol showed a nonlinear grating behavior with the same periodicity and groove depth over the entire surface.

Hoshi et al.[313] reported SHG in ultrathin films of fluoro-bridged aluminum phthalocyanine polymer (AlPcF)$_n$. The SHG intensity from (AlPcF)$_n$/KBr was found to be 2.38 times larger than that of (AlPcF)$_n$/silica for film of the same thickness (20 nm).

The centrosymmetric phthalocyanines can be insinuated by breaking inversion symmetry through physical means, such as film growth processes of MBE and LB techniques to observe SHG activity. The SHG in CuPc is a deformation-induced property, although the probability of quadrupolar and magnetic dipole mechanisms also exists. A polar orientation of VOPc molecules in epitaxial films seems to be responsible for the SHG appearance. SHG in phthalocyanines seems to be a highly surface-selective optical phenomenon that can be induced from the break in symmetry during fabrication processes.

6. Miscellaneous Organometallics

Lequan et al.[314] measured hyperpolarizabilities of tetraorganotin derivatives in dichloromethane using HRS. The β_{xyz} values of compounds 1, 2, and 3 were 13×10^{-30}, 24×10^{-30}, and 159×10^{-30} esu, respectively, which indicates that the β value increases with the increase of conjugation length. The β_{xyz} value of azo derivatives are larger than the biphenyl analogs. The value of β_{zzz} component was 181×10^{-30} esu for compound 4 which has C$_{3v}$ symmetry; its β, value is comparable to the tetrahedral complex 3. The organotin compounds with quasi-tetrahedral (T_d) geometry are nonpolar and they have a three-dimensional transition moment.

(1)

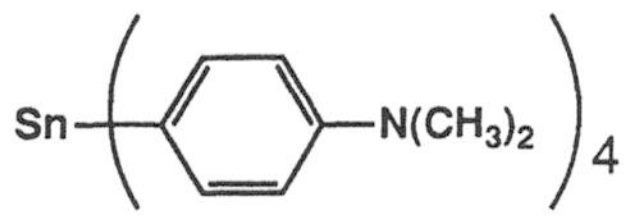

(2)

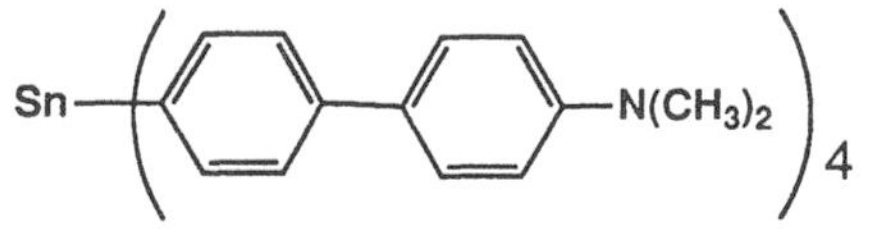

(3)

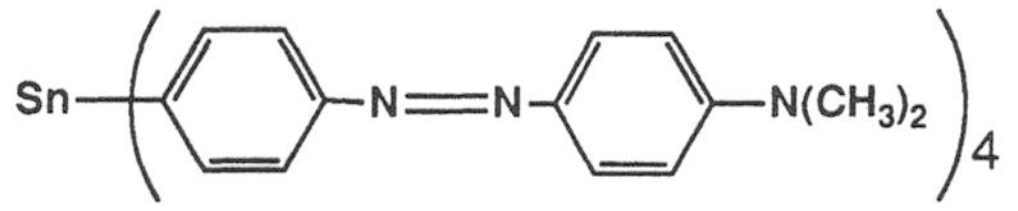

(4)

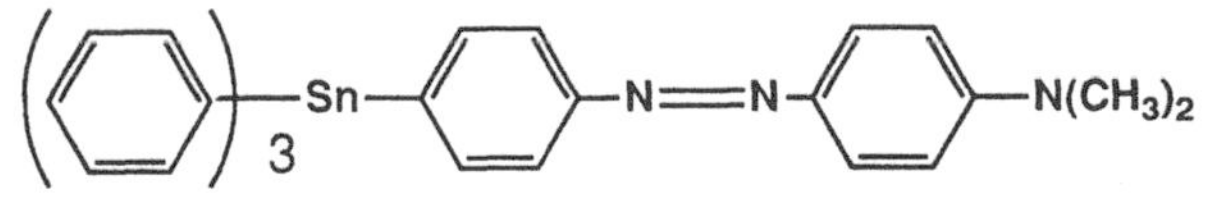

Chiang et al.[315] reported SHG from a class of dipolar coordination polymers having SALEN (a planar dianionic tetradendate ligand) in the backbone. Polar polymers were prepared from (SALEN)MX compounds where M = Cr, Mn, or Fe and X = CO_2 or SO_3. The acceptor trivalent metal atom is linked to the amine donor via the coordinated pyridine and insulated from the donor in the opposite direction by a methylene group between X and the amine. Compounds of Cr, Mn, and Fe crystallized from methanol were not SHG active. The compounds of Cr and Mn, recrystallized from a mixture of triethylamine and methanol, yielded a weak SHG at 1.064 μm, although their SHG efficiencies at 1.907 μm were 0.18 and 0.012 times that of urea, respectively. Both Cr and Mn compounds attained noncentrosymmetric space groups.

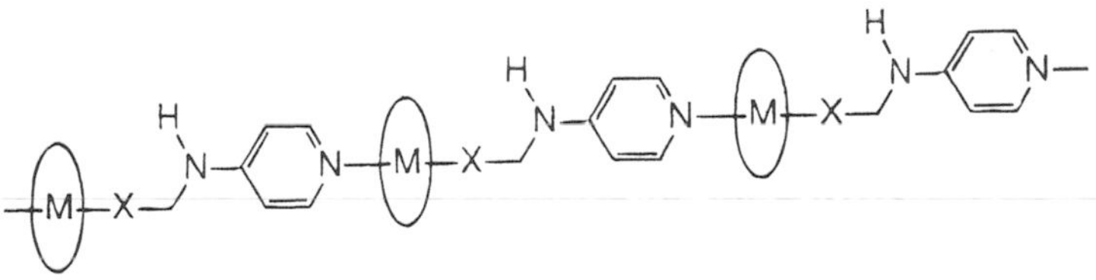

Bella et al.[316] estimated β values of bis(salicylaldiminato)nickel(II) complexes with ZINDO-SOS-derived calculations and EFISH technique. Theoretically estimated β_{vec} were −10.1, −12.9, −17.3, −38.7, and −55.2 × 10^{-30} cm^5 esu^{-1} for Compounds 1, 2, 3, 4, and 5, respectively. On the other hand, β_{vec} values measured by EFISH technique at 0.92 eV were −9.3, −9.6, −20.5, −43.0 and −55.0 × 10^{-30} cm^5 esu^{-1} for Compounds 1, 2, 3, 4, and 5, respectively. The β_{vec} values of 235, 200, and 400 × 10^{-30} cm^5 esu^{-1} for the nickel, copper, and zinc Schiff-base complexes at 1.34 μm have been reported.[316b]

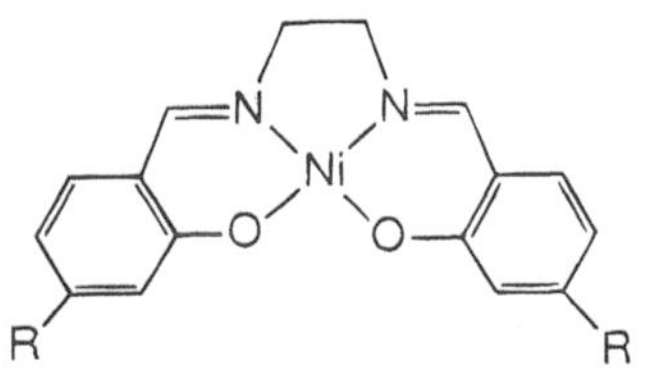

(1) R = H
(2) R = OCH_3

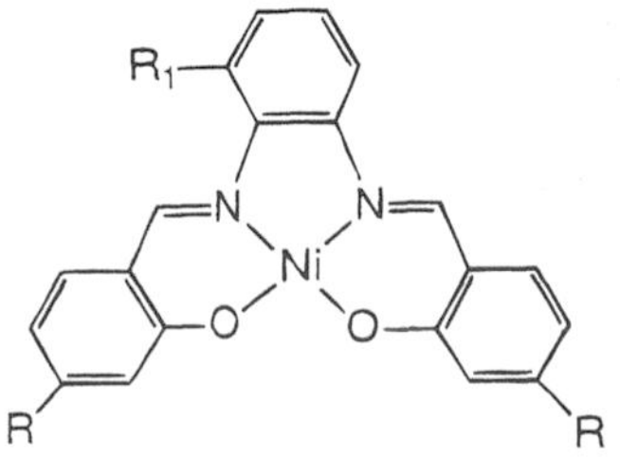

(3) $R = R_1$ = H
(4) R = H, R_1 = NO_2
(5) R = OCH_3, R_1 = NO_2

Laidlaw et al.[317] reported molecular hyperpolarizabilities of mixed-valence metal chromophores with mixed valences measured by HRS technique. The β values at 1.064 μm were 209 $\times$ 10^{-30} esu for [(CN)$_5$Ru^{II}(μ − CN)Ru^{III}(NH$_3$)$_5$]$^-$ (**I**), 157 $\times$ 10^{-30} esu for [η^5(C$_5$H$_5$)Ru^{II}(PPh$_3$)$_2$(μ − CN)Ru^{III}(NH$_3$)$_5$]$^{3+}$ (**II**), 65 $\times$ 10^{-30} esu for [η^5(C$_5$H$_5$)Ru^{II}(PPh$_3$)$_2$(μ − CN)Os^{III}(NH$_3$)$_5$]$^{3+}$, 255 $\times$ 10^{-30} esu for [(CO)$_5$Mo^0(μ − CN)Ru^{III}(NH$_3$)$_5$]$^{2+}$, and 130 $\times$ 10^{-30} esu for [(CO)$_5$W^0(μ − CN)Ru^{III}(NH$_3$)$_5$]$^{2+}$, respectively. The β values of these compounds were similar to those of *p*-NA and DANS.

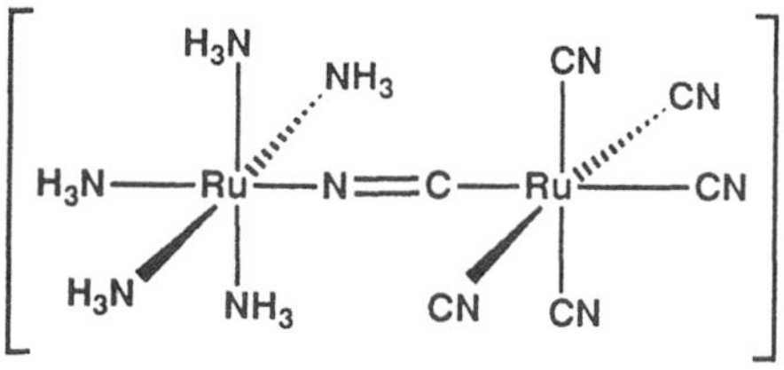

(I)

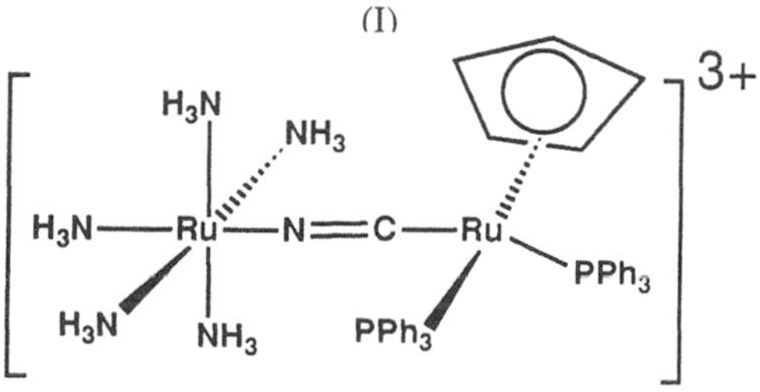

(II)

C. LANGMUIR-BLODGETT FILMS

Material processing for fabricating devices is aided by techniques such as sputtering, vacuum deposition, spin-coating, and casting. These methods do not yield films of precise thickness and the molecules remain randomly oriented onto the substrate. Although it is possible to orient free-standing film samples by mechanical stretching followed by an electrical poling-freezing process,[318,319] a similar poling technique can also be used for introducing a preferential orientation in spin-coated films. The details of such techniques are discussed in the sections dealing with guest–host systems and poled polymers. The degree of molecular orientation attained by the poling process may not be high enough. The increasing importance of ultrathin films in electronics, nonlinear optics, and bioengineering technologies has created demands for organic materials having precise films thickness and highly ordered microstructures. The LB technique is one of the most powerful tools to offer closely packed, uniformly oriented and controlled thickness monolayer and multilayers of desired amphiphiles. The history and experimental knowledge of LB technique can be found elsewhere.[320]

The fatty acids such as hexadecanoic acid [CH$_3$-(CH$_2$)$_{14}$-COOH], stearic acid [CH$_3$-(CH$_2$)$_{16}$-COOH], arachidic acid [CH$_3$-(CH$_2$)$_{18}$-COOH], ω-tricosenoic acid [CH$_2$=CH(CH$_2$)$_{20}$-COOH], docosanoic (behenic) acid [CH$_3$-(CH$_2$)$_{20}$-COOH] which contain long alkyl chains and a highly polar COOH end group, are the most suitable amphiphiles for developing LB films. The long alkyl chain is a hydrophobic part whereas the COOH group is a hydrophilic part and such a structural amphiphilicity is a prerequisite for a LB molecule. The amphiphilic molecules cannot dissolve in water because of the hydrophobic aliphatic chain, and they can not leave the water surface because of the hydrophilic COOH part. The deposition process by the LB technique is not straightforward and requires some preparation before experiments were conducted. The fabrication of the monolayer is affected by concentration of solvent, temperature, surface pressure, barrier speed, subphase, and its pH. Generally, the subphase is water, but mercury, glycerol, and other liquid subphases can be used. The monolayer deposited on a subphase is transferred to a clean, solid substrate such as glass slides, quartz plate, metal-coated slides, or mica. Organic solvents such as chloroform, acetone, hexane, benzene, toluene, and xylene are used for spreading the monolayer in accordance with the solubility of the amphiphiles. A typical LB film deposition procedure involves three steps:

(i) Spreading the monolayer of an amphiphilic molecule over an aqueous subphase of a trough where the evaporation of the solvent leaves the monolayer of an amphiphile on the subphase; (ii) starting compression after incubation for a few minutes. (By moving a barrier, the surface pressure increases and molecules are closely packed and uniformly aligned); and (iii) transferring the monolayer from the aqueous subphase onto a solid substrate at a particular deposition rate and at a defined surface pressure, depending on the nature of amphiphilic molecule. Figure 12 shows the sequential steps of the LB film formation process. Sometimes fatty acids are mixed with the amphiphiles to improve monolayer stability and they also act as a transfer promoter. LB monolayers on a substrate stack in three different ways depending on the nature of the LB molecule: (1) X-type stacking having the tail-to-head and tail-to-head configuration; (2) Z-type stacking having head-to-tail and head-to-tail modes; and (3) Y-type stacking having head-to-head and tail-to-tail modes. Figure 13 shows the stacking features of LB films. The head is the hydrophilic end and the tail is the hydrophobic end. The alternate layers deposited in a Y-mode may give rise to noncentrosymmetric structures needed for SHG. The LB films offer the following advantages:

1. High degree of molecular orientation
2. Precise control of film thickness at the molecular level
3. Control of layer architecture
4. High optical stability for optical processes
5. Building up noncentrosymmetric structures for second-order nonlinear optics.

Because LB technique facilitates the way to align molecules into a noncentrosymmetric structure, organic materials are designed and synthesized considering both the requirements of an amphiphile and a chromophore. A representative example of a highly polarizable LB chromophore suitable for SHG and electro-optics is shown below.

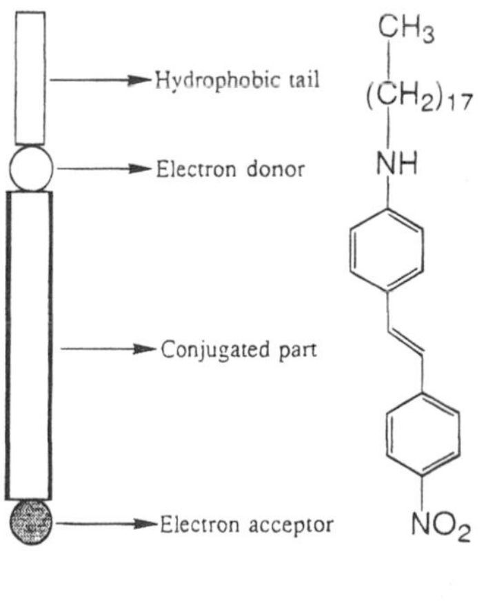

Investigation of SHG in LB films dates back to 1983 when Aktsipetrov et al.[321] first reported SHG properties from the LB monolayer of 4-nitro-4-N-octadecylazobenzene. The $\chi^{(2)}_{yyy}$ value of 4.2 $\times$ 10^{-8} esu for a Langmuir layer of 20 to 30 Å thickness was measured. Later Girling et al.[322,323] observed SHG from a LB monolayer of a merocyanine dye and from a noncentrosymmetric alternating LB film of ω-tricosenic acid and merocyanine dye. Because merocyanine dye undergoes chemical changes, the use of poly(methyl methacrylate) was proposed by Stroeve et al.[324] to improve the long-term stability of the merocyanine multilayers. The stacking of LB monolayers was found to be related to the contents of PMMA and dyes in the mixture. The mixture with a high content of PMMA formed Z-type multilayer, whereas that with high dye content acquired a Y-type structure at a surface pressure of 12 mN/m. The Z-type structures were SHG active. Cross et al.[325] reported the SHG of two hemicyanine dyes: E-1-methyl-4-[2-(4-octadecylaminophenyl) ethenyl]pyridinium iodide (N-stilbazene) and E-1-methyl-4-[2-(4-octadecyloxyphenyl)ethenyl] pyridinium iodide (O-

(1) Spreading

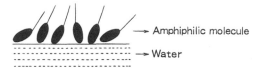

→ Amphiphilic molecule

→ Water

(2) Compression

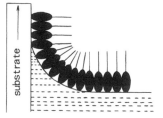

(3) Deposition (First layer)

substrate

(3) Deposition (Second layer)

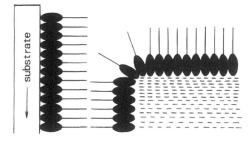

substrate

(3) Deposition (Third layer)

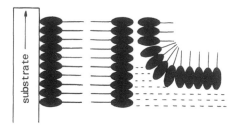

substrate

Figure 12 Langmuir-Blodgett film forming process showing spreading, compression, and deposition of monolayer and multilayers.

stilbazene). The SHG of *N*-stilbazene was about 10 times larger than that of *O*-stilbazene. Although SHG was observed from a monolayer but with deposition of the second monolayer of hemicyanine, the SHG signal disappeared because of the introduction of symmetry.

Hayden et al.[326] developed the multilayers of the hemicyanine dye and hemicyanine-PMMA mixture. In these LB films, the SHG signal increased as the number of the monolayer increased. Hayden et al.[327]

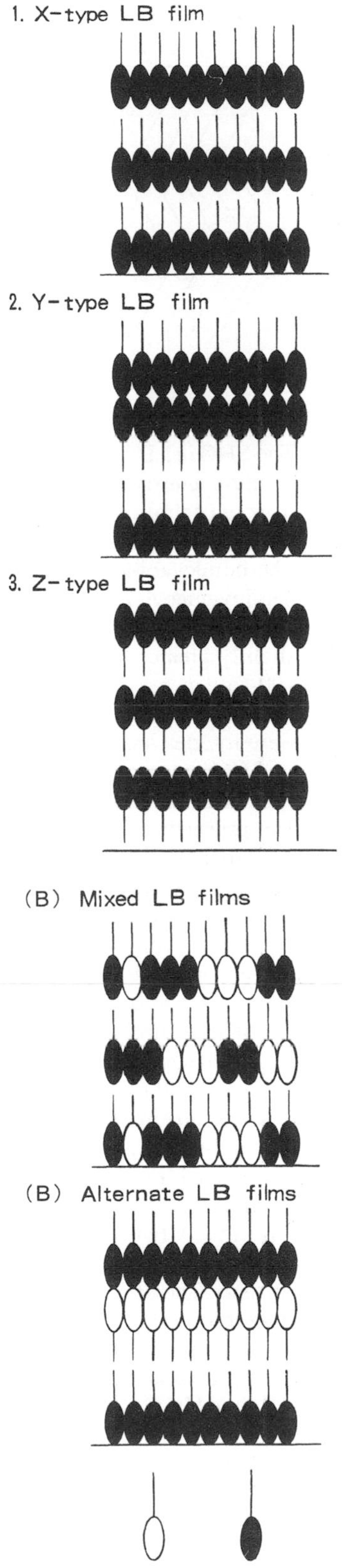

1. X-type LB film

2. Y-type LB film

3. Z-type LB film

(B) Mixed LB films

(B) Alternate LB films

Figure 13 (A) Stacking of LB multilayers into *X-*, *Y-*, and *Z*-type structures. (B) Mixed and alternate LB films. The filled refer to NLO chromophore molecules, whereas empty is fatty acid molecules.

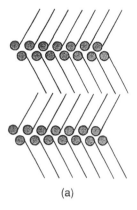 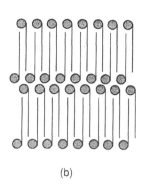

Figure 14 Structural models for noncentro-symmetric LB films: (a) a tilted Y-type bilayer and (b) an intercalated structure. (From Decher, G., Tieke, B., Bosshard, G., and Gunter, P., *Ferroelectrics*, 91, 193, 1989. With permission.)

also reported SHG from multilayered LB films of a mixture containing 52% hemicyanine and 42% PMMA. The SHG in LB films of the mixtures of a hemicyanine dye and poly(octadecyl methacrylate) PODMA, and hemicyanine dye and behenic acid, were also observed. The normalized SHG was found to be independent of dye concentration for the hemicyanine-PODMA mixture, whereas the signal increased as behenic acid was added to the dye. The results with hemicyanine-behenic acid were similar to those reported by Neal et al.[328] The hemicyanine-PODMA system was immiscible and complete Y-type LB multilayers of 54 molar% hemicyanine in PODMA showed an increase in SHG with the number of layers deposited on the upstroke. Multilayer LB films of hemicyanine interleaved with behenic acid deposited on the downstroke exhibited slightly greater than a quadratic increase in SHG as a function of the number of hemicyanine layers. A 100-fold increase in SHG activity was observed for hemicyanine dye-grafted poly(epichlorohydrin) backbone.[329] The quadratic enhancement of SHG was observed when the optically active bilayers of these dyes were interleaved with an optically inert material.

SHG properties of stilbazium salt and phenylhydrazine dye LB materials were reported by Lupo et al.[330] The LB molecules exhibited $\chi^{(2)}$ values on the order of 10^{-6} esu and the β values of 10^{-27} esu. The presence of two long aliphatic chains contributes to the stability and ordering of LB monolayers. A proper combination of an electron-donor and an acceptor group resulted into great SHG activity in LB monolayers. Ledoux et al.[331] reported the SHG activity in LB films of 4-[4-(N-n-dodecyl-N-,methylamino)phenyl-azo]-3-nitrobenzoic acid (DPNA). Although LB monolayers of DPNA can be developed with or without adding cations in the subphase, the divalent cations such as Ca^{+2} or Cd^{+2} proved advantageous for improving the quality of deposition and for stabilizing monolayers. Both monolayers and multilayers were SHG active and LB multilayers attained a Z-type arrangement. The $\chi^{(2)}$ of LB films were 2.1×10^{-20} and 2.5×10^{-20} $C^3 J^{-2}$ without and with Cd^{+2} ions, respectively.

The SHG properties of a series of amphiphilic dyes containing substituted nitropyridine and nitroaniline were reported by Decher et al.[332,333] Optical absorption spectroscopy indicated the cutoff wavelengths of 450 and 490 nm for pyridine and benzene derivatives, respectively. The LB films do not form an asymmetrical head-to-tail arrangement (X- or Y-type); instead, a head-to-head and tail-to-tail arrangement (Y-type) was formed (Fig. 14). The SH signal plotted as a function of the alkyl chain lengths for pyridine derivatives revealed an increase of SH intensity with the increase of chain length which optimized for a docosyl chain (Fig. 15). A better alignment of the chromophores in longer alkyl chain pyridine derivatives due to the hydrogen bridging between the amino and nitro substituents of adjacent molecules resulted in great SH intensity. The SH power increased quadratically with the number of layers. Figure 16 represents SH intensity plotted for 5, 10, 15, 30, 45, 60, 90, 135, and 270 bilayers for 2-docosylamino-5-nitropyridine (DCANP). The d_{31} and d_{33} coefficients of a DCANP film were approximately 0.9 and 6.8 pm/V, respectively. The head-to-head and tail-to-tail modes of DCANP molecules suggested a tilt angle of approximately 370°. The SHG and UV-visible spectroscopy measurements evidenced anisotropy with respect to the dipping direction.[334] In the case of benzene derivatives, no SH signal was detected due to the centrosymmetric arrangement of the molecules, therefore alternating layers were required to obtain an asymmetric arrangement.

Studies on second-order NLO properties have been focused mainly on the LB films of highly polarizable, one-dimensional charge-transfer molecules such as merocyanine, azobenzene derivatives, amidonitrostilbenes, hemicyanines, phenylhydrazone, stilbazium salt dyes, diazostilbene, and polyenic molecules, as discussed above. The first example of a two-dimensional charge-transfer LB molecule, namely *N,N′*-dioctadecyl-1,5-diamino-2,4-dinitrobenzene (DIODD) for SHG was reported by Nalwa et al.[47] Pure DIODD

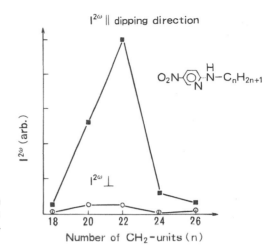

Figure 15 Second harmonic intensity dependence on the carbon chain length of nitropyridine derivatives. (From Decher, G., Tieke, B., Bosshard, C., and Gunter, P., *Ferroelectrics*, 91, 193, 1989. With permission.)

powder was SHG inactive, showing the presence of a center of symmetry in the molecule. SHG signals were detected for a LB monolayer. SHG experiments performed at a wavelength of 1.064 μm showed a SHG signal from the LB monolayer of a 1:1 mixture of DIODD and arachidic acid, whereas no SHG was noticed for 1:3 or 3:1 mixtures. The d_{11} of the monolayer was 11×10^{-9} esu and the d_{13} component was 3.85×10^{-9} esu. An octadecylamino chain in DIODD molecule was replaced by a fluorine atom resulting in a new amphiphile N-octadecyl-2,4-dinitro-5-fluoroaniline (ODDFA) with a one-dimensional charge-transfer character. The ODDFA itself formed stable monolayers on the water surface, which can be transferred easily onto a solid substrate. The LB monolayer of pure ODDFA exhibited d_{11} and d_{13} coefficients of 19×10^{-9} and 6.7×10^{-9} esu, respectively, at a tilt angle of 40°.

Richardson et al.[335] reported the β values the LB films of a series of organoruthenium complexes which had the ruthenium (cyclopentadienyl) bis(triphenylphosphine) head group; measured by the EFISH technique. The β increased with the increase of conjugation length and, for a two-benzene ring system, the β was about 6 times greater than that of a single benzene ring. The incorporation of an acetylenic moiety between two benzene rings also increased the β value. The *trans*-stilbene bridge system showed the largest β value of 37×10^{-30} esu, which is more than an order of magnitude higher from the ruthenium complex with the benzene ring. The LB films of ruthenium, Fe, or Co compounds show large β values and good chemical stability.[336]

Sakaguchi et al.[337] reported the second-harmonic light emission from the LB films of the ruthenium(II)-pyridine complex. The SH light intensity from RuC18B-2C18NB LB films decreased to ~50% for samples that irradiated either 355 nm or 460 nm just before Nd:YAG laser irradiation at 1.064 μm. The LB films of 1-methyl-4-(4-(N-octadecyl-N-methylamino)styryl)-pyridinium iodide (C18AStZ) and C18NB in a 2:3 molar ratio showed strong SHG; in this case, the SH light intensity was not affected

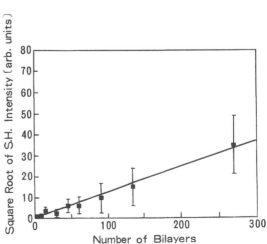

Figure 16 Square root of the second harmonic intensity as a function of the number of bilayers for a nitropyridine derivative. (From Decher, G., Tieke, B., Bosshard, G., and Gunter, P., *Ferroelectrics*, 91, 193, 1989. With permission.)

Table 59 Absorption Peak, Types of Deposition, and Second-Order Optical Nonlinearities of Langmuir-Blodgett Films

No.	Material	λ_{max} (nm)	Type of Deposition	No. of Monolayers	NLO Coefficients (1.064 μm)	Ref.
(1)	$C_{60}(C_4H_8N_2)$	220, 270, 328	Z-type	1,3,5	$\beta = 3.6 \times 10^{-29}$ esu $\chi^{(2)} = 1.8 \times 10^{-7}$ esu	341
	$C_{60}(C_6H_9NO_4)$	227, 278, 337	Z-type		$\beta = 2.3 \times 10^{-29}$ esu $\chi^{(2)} = 1.5 \times 10^{-7}$ esu	342
	$C_{60}(C_8H_{13}NO_4)$	225, 275, 336	Z-type		$\beta = 1.9 \times 10^{-29}$ esu $\chi^{(2)} = 1.1 \times 10^{-7}$ esu	342

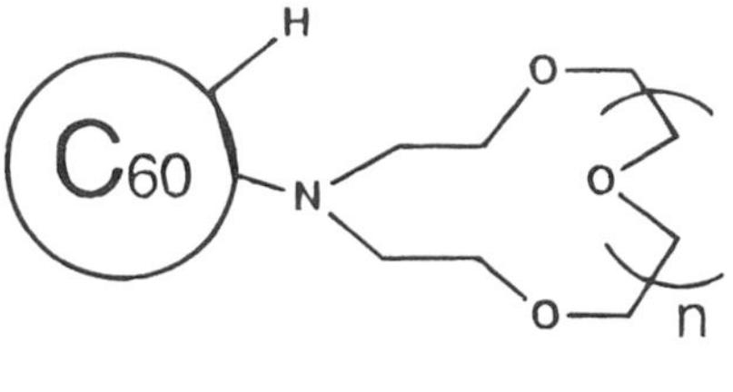

343

C_{60}-1-aza-12-crown-4
($n = 1$)
$\quad\quad\quad\quad\quad\quad\quad\quad\quad\quad$ $\chi^{(2)} = 2.3$ pm/V

C_{60}-1-aza-15-crown-5
($n = 2$)
$\quad\quad\quad\quad\quad\quad\quad\quad\quad\quad$ $\chi^{(2)} = 3.6$ pm/V

C_{60}-1-aza-18-crown-6
($n = 3$)
$\quad\quad\quad\quad\quad\quad\quad\quad\quad\quad$ $\chi^{(2)} = 3.2$ pm/V

$\quad\quad\quad\quad\quad\quad\quad\quad\quad\quad$ $\chi^{(2)} = 1$ pm/V
$\quad\quad\quad\quad\quad\quad\quad\quad\quad\quad$ (with K^+ ions)

(2)

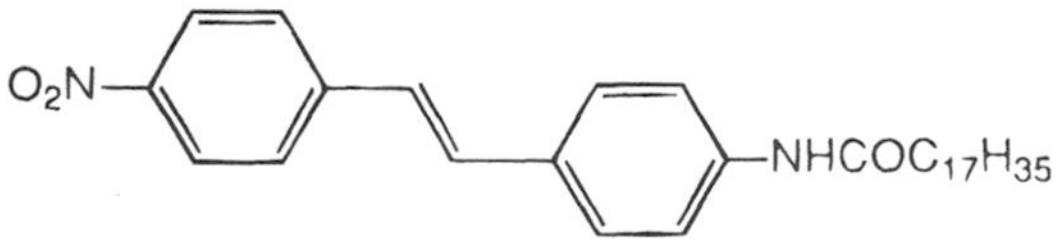

4-*n*-Heptadecylamido-4'-nitrostilbene (4-HANS)

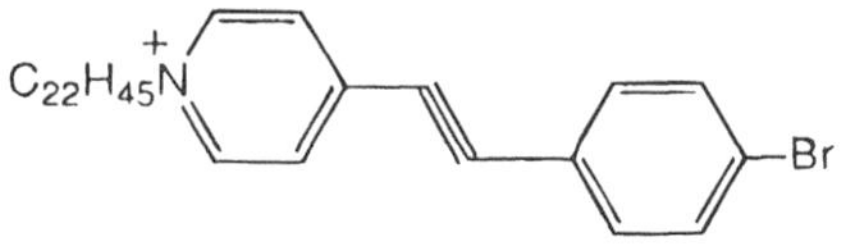

Functionalized diarylalkyne (JT11)

JT11/4-HANS		Y-type	1 (bilayer)	$\chi^{(2)} = 19$ pm/V	344
			7 (bilayers)	$\chi^{(2)} = 9.6$ pm/V	
			49 (bilayers)	$\chi^{(2)} = 1.7$ pm/V	

(3)

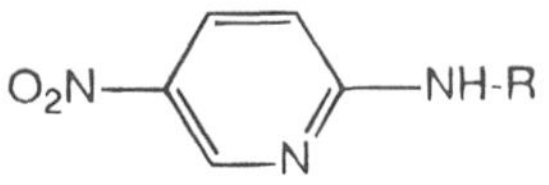

(i) $R = C_{18}H_{37}$ = 2-octadecylamino-5-nitropyridine (ODANP)
(ii) $R = C_{20}H_{41}$ = 2-eicosylamino-5-nitropyridine (ECANP)
(iii) $R = C_{22}H_{45}$ = 2-docosylamino-5-nitropyridine (DCANP)
(iv) $R = C_{24}H_{49}$ = 2-tetracosylamino-5-nitropyridine (TCANP)
(v) $R = C_{26}H_{53}$ = 2-hexacosylamino-5-nitropyridine (HCANP)

(i) ODANP			40 layers	$d_{33} = 5.8$ pm/V $d_{31} = 2.7$ pm/V $d_{32} = 2.5$ pm/V	345

Table 59 *Continued*

No.	Material	λ_{max} (nm)	Type of Deposition	No. of Monolayers	NLO Coefficients (1.064 μm)	Ref.
	(ii) ECANP			40 layers	$d_{33} = 4.8$ pm/V $d_{31} = 1.5$ pm/V $d_{32} = 1.9$ pm/V	345
	(iii) DCANP	375	Y-type	1–540	$d_{33} = 7.8$ pm/V at 1.064 μm $d_{33} = 5.6$ pm/V at 1.318 μm $d_{33} = 3.7$ pm/V at 1.907 μm	346
				40	$d_{31} = 2.0$ pm/V at 1.064 μm $d_{32} = 2.6$ pm/V at 1.064 μm	
	(vi) TCANP			40 layers	$d_{33} = 4.2$ pm/V $d_{31} = 0.7$ pm/V $d_{32} = 1.9$ pm/V	345
	(v) HCANP			40 layers	$d_{33} = 3.4$ pm/V $d_{31} = 0.4$ pm/V $d_{32} = 1.3$ pm/V	345

(4)

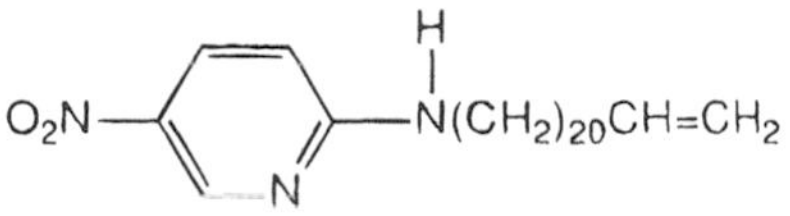

2-(21-Docosenyl)amino-5-nitropyridine (VECANP)

No.	Material	λ_{max} (nm)	Type of Deposition	No. of Monolayers	NLO Coefficients (1.064 μm)	Ref.
	VECANP	378	Y-type	1–40	$d_{33} = 8.5$ pm/V	347
	Polymerized films			20 bilayers	$d_{33} = 9.1$ pm/V $d_{31} = 1.8$ pm/V	348
	Polymerized films			40 bilayers	$d_{33} = 5.3$ pm/V $d_{31} = 1.2$ pm/V	348

(5)

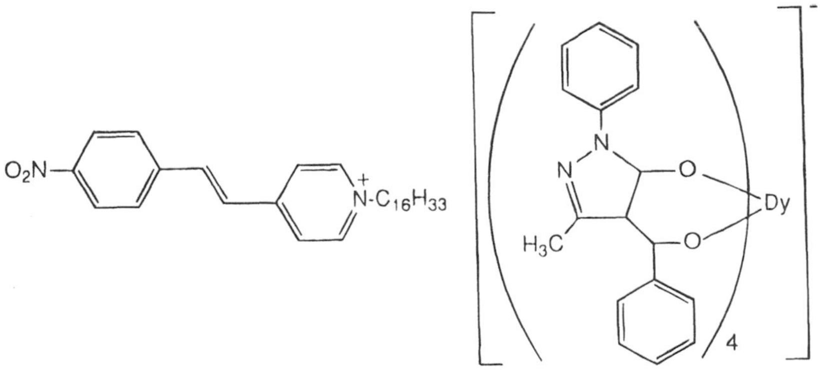

(*E*)-*N*-Hexadecyl-4-(2-(4-(dimethylamino)phenyl)ethenyl)pyridinium tetrakis(1-phenyl-3-methyl-4-benzoyl-5-pyrazolonato)-dysprosium (III) = ADy(PMBP)$_4$

No.	Material	λ_{max} (nm)	Type of Deposition	No. of Monolayers	NLO Coefficients (1.064 μm)	Ref.
	(i) ADy(PMBP)$_4$	496		1	$\beta = 6.6 \sim 9.3 \times 10^{-28}$ esu $\chi^{(2)} = 5.9 \sim 8.7 \times 10^{-7}$ esu	349

Table 59   *Continued*

No.	Material	λ_{max} (nm)	Type of Deposition	No. of Monolayers	NLO Coefficients (1.064 μm)	Ref.
	(ii) Hemicyanine-iodide			1	$\beta = 2.4 \sim 3.6 \times 10^{-28}$ esu $\chi^{(2)} = 3.5 \sim 5.0 \times 10^{-7}$ esu	

(6)

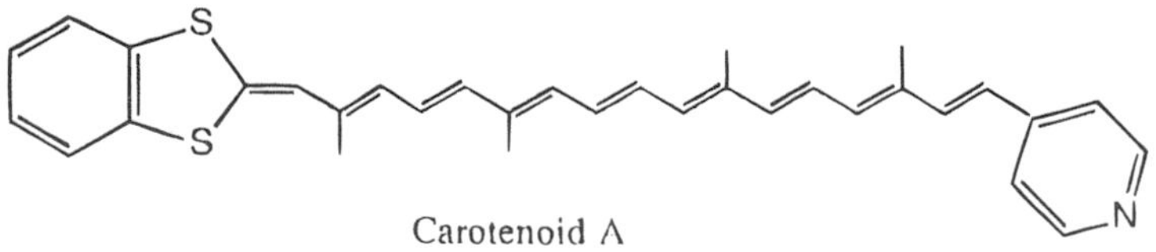

Carotenoid A

No.	Material	λ_{max} (nm)	Type of Deposition	No. of Monolayers	NLO Coefficients (1.064 μm)	Ref.
	(i) Carotenoid A + ω-tri-cosenic acid			1	$\beta = 92 \times 10^{-30}$ esu $\chi^{(2)} = 100 \times 10^{-9}$ esu	350
				10	$\beta = 50 \times 10^{-30}$ esu $\chi^{(2)} = 57 \times 10^{-9}$ esu	
	(ii) Carotenoid A + cyclodextrin			1	$\beta = 28 \times 10^{-30}$ esu $\chi^{(2)} = 22 \times 10^{-9}$ esu	

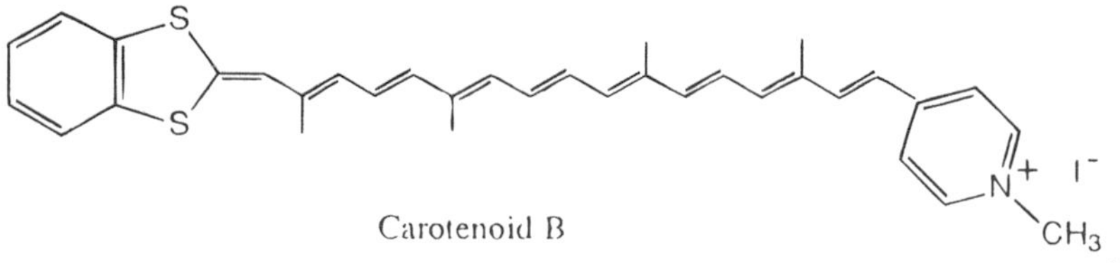

Carotenoid B

No.	Material	λ_{max} (nm)	Type of Deposition	No. of Monolayers	NLO Coefficients (1.064 μm)	Ref.
	(iii) Carotenoid B + ω-tricosenic acid			1	$\beta = 730 \times 10^{-30}$ esu $\chi^{(2)} = 760 \times 10^{-9}$ esu	
	(iv) Carotenoid B + cyclodextrin			1	$\beta = 400 \times 10^{-30}$ esu $\chi^{(2)} = 340 \times 10^{-9}$ esu	
				10	$\beta = 245 \times 10^{-30}$ esu $\chi^{(2)} = 175 \times 10^{-9}$ esu	

(7)

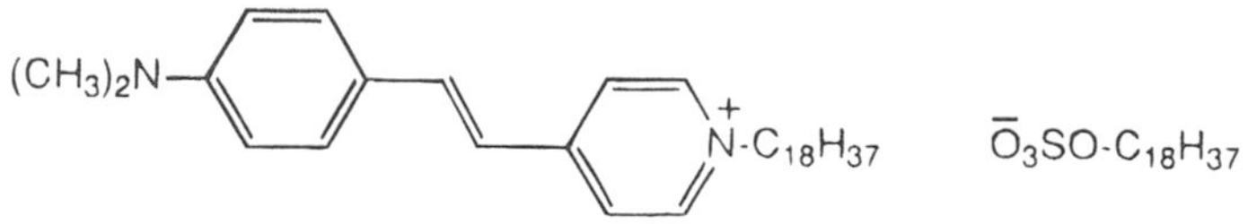

E-N-Octadecyl-4-[2-(4-dimethylaminophenyl)ethenyl]-pyridinium octadecylsulfate (Dye I-1), iodide (Dye I-2), and bromide (Dye I-3)

No.	Material	λ_{max} (nm)	Type of Deposition	No. of Monolayers	NLO Coefficients (1.064 μm)	Ref.
	(i) Dye I-1 (sulfate)	425	Y-type	41 bilayers	$\chi^{(2)}_{zzz} = 50$ pm/V $\chi^{(2)}_{zxx} = 8.3$ pm/V	351
					$\chi^{(2)}_{zzz} = 500 \sim 1500$ pm/V	352, 353
	Dye I-2 (iodide)	460			$\chi^{(2)}_{zzz} = 90 \sim 150$ pm/V	353
	Dye I-3 (bromide)	460			$\chi^{(2)}_{zzz} = 70 \sim 90$ pm/V	

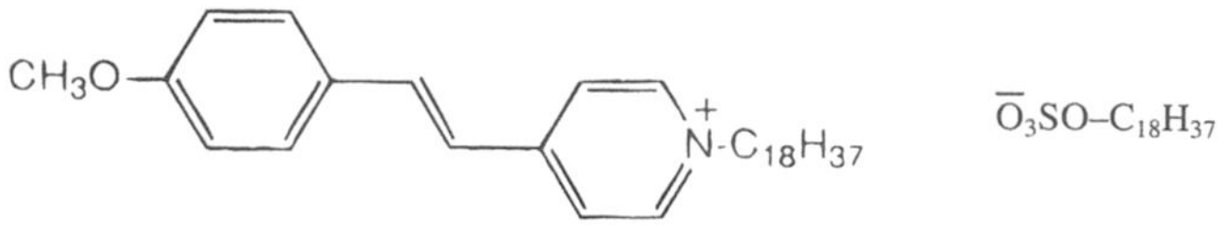

No.	Material	λ_{max} (nm)	Type of Deposition	No. of Monolayers	NLO Coefficients (1.064 μm)	Ref.
		398	Y-type	78 bilayers	$\chi^{(2)}_{zzz} = 12$ pm/V $\chi^{(2)}_{zxx} = 2.9$ pm/V	

Table 59 *Continued*

No.	Material	λ_{max} (nm)	Type of Deposition	No. of Monolayers	NLO Coefficients (1.064 μm)	Ref.
(8)						

E-N-octadecyl-4-[2-(4-dibutylaminophenyl)ethenyl]-quinolinium $^+$I (Dye III)

| | (i) Dye III | 530 | Z-type | | $\chi^{(2)}_{zzz} = 120$ pm/V | 354 |

E-N-octadecyl-4-[2-(4-dibutylaminophenyl)ethenyl]-pyridinium octadecylsulfate (Dye IV)

| | (i) Dye IV | 460 | Z-type | | $\chi^{(2)}_{zzz} = 50$ pm/V | 352, 354 |

E-N-dodecyl-4-[2-(4-docosyloxyphenyl)ethenyl]-pyridinium dodecyliodide (Dye V)

| | (i) Dye V | 360 | Z-type | | $\chi^{(2)}_{zzz} = 15$ pm/V | 352, 354 |
| (9) | | | | | | |

Z-β-(1-Octadecyl-4-pyridinium)-α-cyano-4-styryl-dicyanomethanide (Dye VI)

| | Dye VI | | | | $\chi^{(2)}_{zzz} = 500 \sim 1500$ pm/V $\chi^{(2)}_{zxx} = 30 \sim 80$ pm/V $\beta = 2 \sim 5 \times 10^{-37}$ m^{4}V^{-1} | 351 |
| (10) | | | | | | |

E-N-Octadecyl-4-[2-(4-octadecyloxyphenyl)ethenyl]-pyridinium octadecyliodide (Dye VII)

| | Dye VII | 360 | Z-type | | $\chi^{(2)}_{zzz} = 5$ pm/V | |

Table 59 *Continued*

No.	Material	λ_{max} (nm)	Type of Deposition	No. of Monolayers	NLO Coefficients (1.064 μm)	Ref.
(11)						

$$(H_{37}C_{18})_2N\text{—}\!\!\bigcirc\!\!\text{—}\!\!/\!\!=\!\!\backslash\!\!\text{—}\!\!\bigcirc\!\!\overset{+}{N}\text{-}CH_3 \qquad H_3COSO_3^-$$

4-[2-(*N,N'*-octadecy-4-aminophenyl)ethenyl]-*N*-methylpyridinium methoxysulfonate (Dye VIII)

No.	Material	λ_{max} (nm)	Type of Deposition	No. of Monolayers	NLO Coefficients (1.064 μm)	Ref.
	Dye VIII	325, 495			$\beta = 2.5 \times 10^{-27}$ esu at 1.06 μm $\beta = 6.3 \times 10^{-27}$ esu at 0.915 μm	355
(12)						

$$H_3C\text{-}(CH_2)_{21}\text{-}HN\text{—}\!\!\bigcirc\!\!\text{—}NO_2$$
$$\underset{CH_3}{}$$

n-Docosyl-2-methyl-4-nitroaniline (DCMNA)

No.	Material	λ_{max} (nm)	Type of Deposition	No. of Monolayers	NLO Coefficients (1.064 μm)	Ref.
	DCMNA	390	Y-type	50 layers	$\chi^{(2)}_{zzz} = 1.6$ pm/V $\chi^{(2)}_{zxx} = 0.08$ pm/V	356
(13)						

$$H_{37}C_{18}O\text{—}\!\!\bigcirc\!\!\text{—}N\!\!=\!\!N\text{—}\!\!\bigcirc\!\!\text{—}NO_2$$

N-Octadecyloxy-4-nitroazobenzene

$$+$$

$$H_{37}C_{18}O\text{—}\!\!\bigcirc\!\!\text{—}N\!\!=\!\!N\text{—}\!\!\bigcirc\!\!\text{—}COOH$$

N-octadecyloxy-4-carboxyazobenzene

No.	Material	λ_{max} (nm)	Type of Deposition	No. of Monolayers	NLO Coefficients (1.064 μm)	Ref.
	Homologous mixture	305			$\chi^{(2)} = 9 \times 10^{-8}$ esu	357
(14)						

$$H_{25}C_{12}O\text{—}\!\!\bigcirc\!\!\text{—}/\!\!=\!\!\backslash\!\!\text{—}\!\!\bigcirc\!\!\text{—}COOH \qquad (C12OPPy)$$

No.	Material	λ_{max} (nm)	Type of Deposition	No. of Monolayers	NLO Coefficients (1.064 μm)	Ref.
	C12OPPy	359		50 layers	$\chi^{(2)}_{eff} = 96$ pm/V (Cd salt) $\chi^{(2)}_{eff} = 34$ pm/V (Ba salt) $\beta = 8.27 \times 10^{-51}$ C^3 m^3 J^{-2}	358

$$H_{25}C_{12}S\text{—}\!\!\bigcirc\!\!\text{—}/\!\!=\!\!\backslash\!\!\text{—}\!\!\bigcirc\!\!\text{—}COOH \qquad (C12SPPy)$$

Table 59 *Continued*

No.	Material	λ_{max} (nm)	Type of Deposition	No. of Monolayers	NLO Coefficients (1.064 μm)	Ref.
	C12SPPy			50 layers	$\chi^{(2)}_{eff} = 42$ pm/V (Cd salt) $\chi^{(2)}_{eff} = 29$ pm/V (Ba salt) $\beta = 17.9 \times 10^{-51}$ C^3 m^3 J^{-2}	358

No.	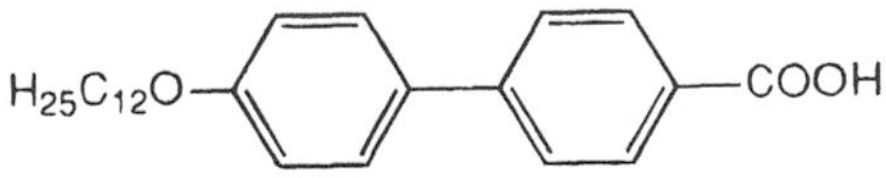 Material	λ_{max} (nm)	Type of Deposition	No. of Monolayers	NLO Coefficients (1.064 μm)	Ref.
	C12PPy			50 layers	$\chi^{(2)}_{eff} = 42$ pm/V (Ba salt) $\beta = 6.95 \times 10^{-51}$ C^3 m^3 J^{-2}	358

(15) Crown ether derivatives

| | 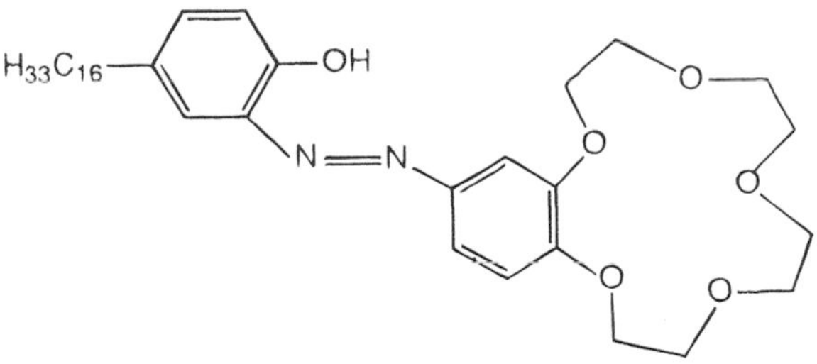 | | | | $\beta = 4 \times 10^{-29}$ esu | 359 |

| | 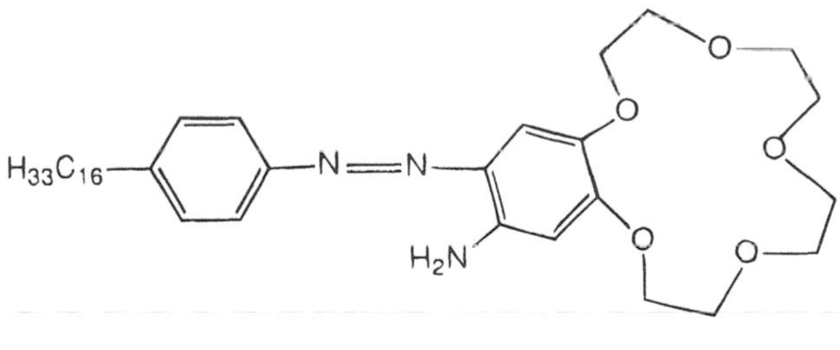 | | | | $\beta = 4 \times 10^{-28}$ esu | |

(16)

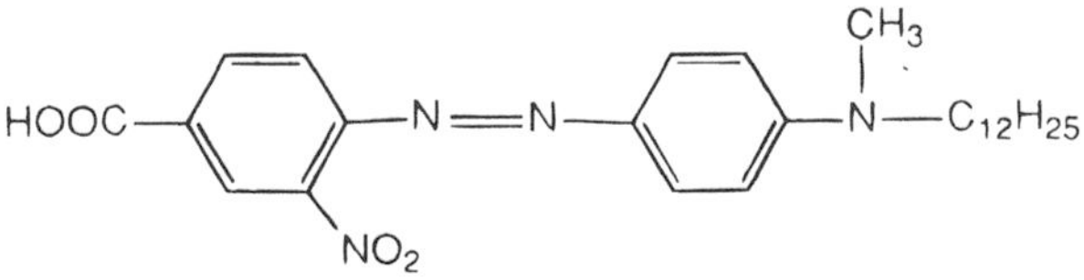

4-[4-(*N-n*-dodecyl-*N*-methylamino)phenylazo]-3-nitrobenzoic acid (DPNA)

No.	Material	λ_{max} (nm)	Type of Deposition	No. of Monolayers	NLO Coefficients (1.064 μm)	Ref.
	DPNA			Monolayer	$\beta = 2.8 \times 10^{-28}$ esu $\chi^{(2)} = 5.5 \times 10^{-6}$ esu	360
			Z-type		$\chi^{(2)} = 2.5 \times 10^{-20}$ C^3 J^{-2} (with Cd^{+2}) $\beta = 4.5 \times 10^{-48}$ C^3 m^3 J^{-2}	331
			Z-type		$\chi^{(2)} = 2.1 \times 10^{-20}$ C^3 J^{-2} (without Cd^{+2}) $\beta = 3.3 \times 10^{-48}$ C^3 m^3 J^{-2}	331

Table 59 *Continued*

No. Material	λ_{max} (nm)	Type of Deposition	No. of Monolayers	NLO Coefficients (1.064 μm)	Ref.

Polyene-1		Z-type	Bilayer	$\beta = 8.3 \times 10^{-28}$ esu	360
				$\chi^{(2)} = 6.4 \times 10^{-6}$ esu	
			Monolayer	$\chi^{(2)}_{zzz} = 6.0 \times 10^{-7}$ esu at	361
			Monolayer	$\phi = 37°$	
				$\chi^{(2)}_{zzz} = 11.3 \times 10^{-6}$ esu at $\phi = 38°$	
				(on hydrophilic substrate)	
		Y-type	Bilayer	$\chi^{(2)}_{zzz} = 2.0 \times 10^{-7}$ esu at	
				$\phi = 30°$	
(17)					

| (i) | 440 | | Monolayer | $\chi^{(2)} = 300$ pm/V | 362 |
| | | | | $\beta = 99 \times 10^{-50}$ C^3 m^3 J^{-2} | |

(ii)	450		Monolayer	$\chi^{(2)} = 670$ pm/V	
(i) + (ii)	420	Y-type	Alternating	$\chi^{(2)} = 340$ pm/V	
(18)					

Polyene-2			Monolayer	$\chi^{(2)}_{zzz} = 1.3 \times 10^{-6}$ esu at	361
				$\phi = 29°$	
Polyene-1 +			Bilayer	$\chi^{(2)}_{zzz} = 4.8 \times 10^{-6}$ esu at	
Polyene-2				$\phi = 40°$	
(19) Retinals					

| | | | | $\beta = 51 \times 10^{-50}$ C^3 m^3 J^{-2} | 363 |

Table 59 *Continued*

No.	Material	λ_{max} (nm)	Type of Deposition	No. of Monolayers	NLO Coefficients (1.064 μm)	Ref.

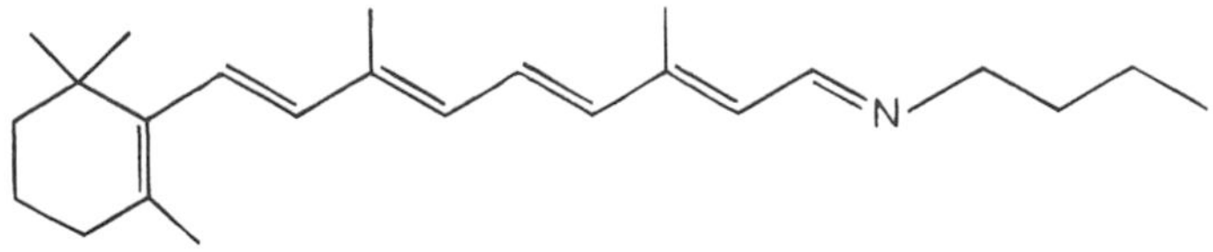

$\beta = 46 \times 10^{-50}$ C^3 m^3 J^{-2}

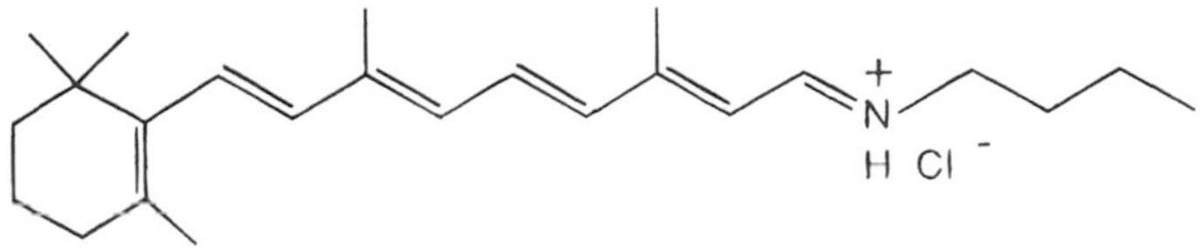

$\beta = 85 \times 10^{-50}$ C^3 m^3 J^{-2}

(20)

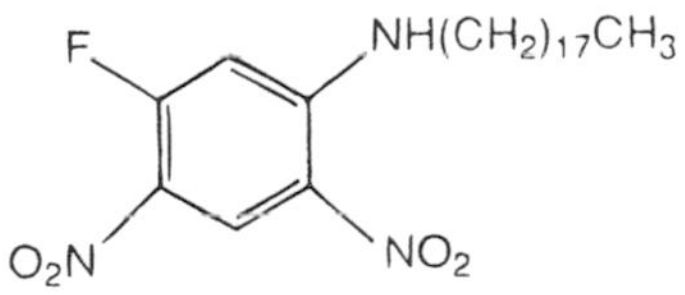

N-Octadecyl-2,4-dinitro-5-fluoroaniline (ODDFA)

(ODDFA)		365	Monolayer		$\chi_{zzz}^{(2)} = 3.8 \times 10^{-8}$ esu $\chi_{zxx}^{(2)} = 1.34 \times 10^{-8}$ esu	364

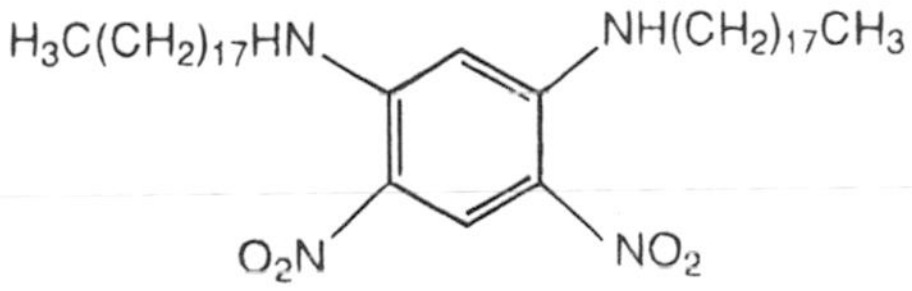

N,N'-Diocadecyl-1,5-diamino-2,4-dinitrobenzene (DIODD)

DIODD+ Arachidic acid		320,420	Monolayer		$\chi_{zzz}^{(2)} = 2.2 \times 10^{-8}$ esu $\chi_{zxx}^{(2)} = 7.7 \times 10^{-9}$ esu	47, 365

(21)

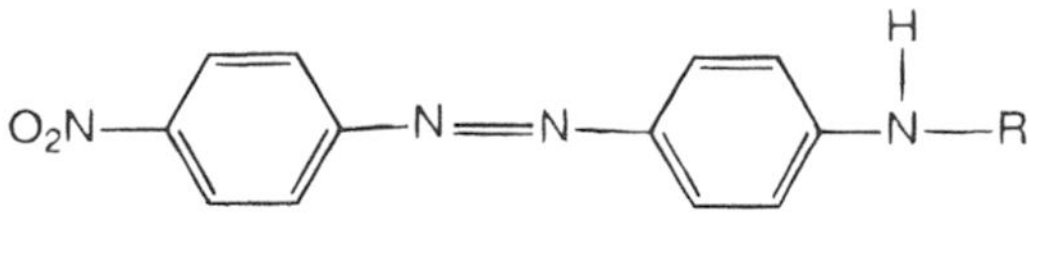

(i) $R = C_6H_{12}OH$ (ii) $R = C_{12}H_{24}OH$

(i)					$\beta = 13.4 \times 10^{-50}$ C^3 m^3 J^{-2} at $\phi = 15°$	366
(ii)					$\beta = 23.8 \times 10^{-50}$ C^3 m^3 J^{-2} at $\phi = 15°$	

Table 59 *Continued*

No.	Material	λ_{max} (nm)	Type of Deposition	No. of Monolayers	NLO Coefficients (1.064 μm)	Ref.
(22)						

| | | 475 | mixed with polymer | | $\chi^{(2)}_{zzz} = 2.2 \times 10^{-7}$ esu
$\beta = 2.4 \times 10^{-28}$ esu | 367 |

| | | | mixed with polymer | | $\chi^{(2)}_{zzz} = 1.5 \times 10^{-7}$ esu
$\beta = 1.2 \times 10^{-28}$ esu | |

| | | | mixed with polymer | | $\chi^{(2)}_{zzz} = 8.0 \times 10^{-8}$ esu
$\beta = 7.0 \times 10^{-29}$ esu | |

| (23) | | | | | | |

| | | 430 | | | $\chi^{(2)}_{zzz} = 15$ pm/V at $\phi = 43°$
$\chi^{(2)}_{zxx} = 6.6$ pm/V
$\beta = 44 \times 10^{-30}$ esu | 368 |

| | | 437 | | | $\chi^{(2)}_{zzz} = 4.0$ pm/V at $\phi = 55°$
$\chi^{(2)}_{zxx} = 4.0$ pm/V
$\beta = 21 \times 10^{-30}$ esu | 368 |

Table 59 *Continued*

No.	Material	λ_{max} (nm)	Type of Deposition	No. of Monolayers	NLO Coefficients (1.064 μm)	Ref.

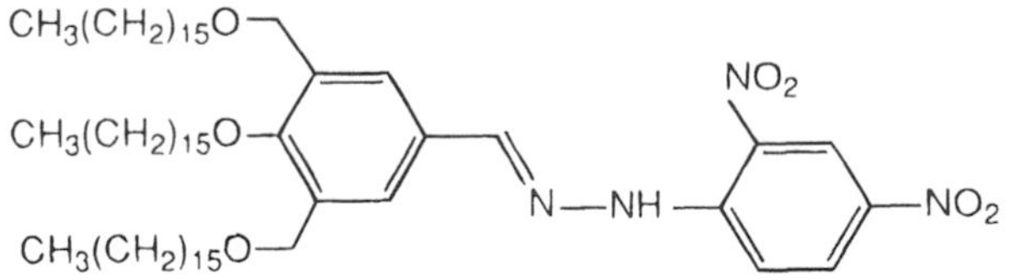

407

$\chi_{zzz}^{(2)} = 16$ pm/V at $\phi = 48°$
$\chi_{zxx}^{(2)} = 10$ pm/V
$\beta = 64 \times 10^{-30}$ esu

368

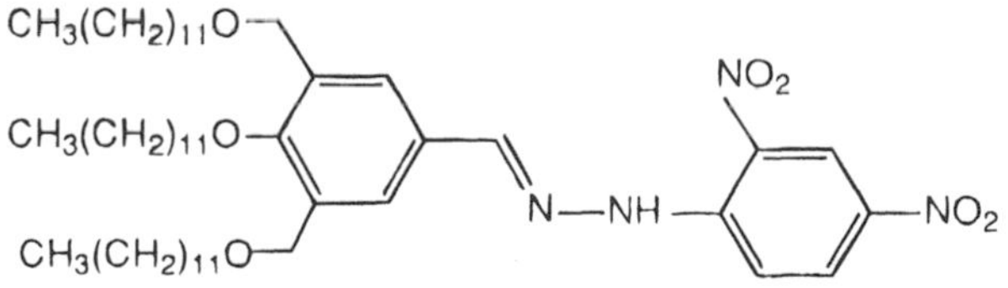

407

$\chi_{zzz}^{(2)} = 2.3$ pm/V at $\phi = 56°$
$\chi_{zxx}^{(2)} = 2.5$ pm/V
$\beta = 13 \times 10^{-30}$ esu

368

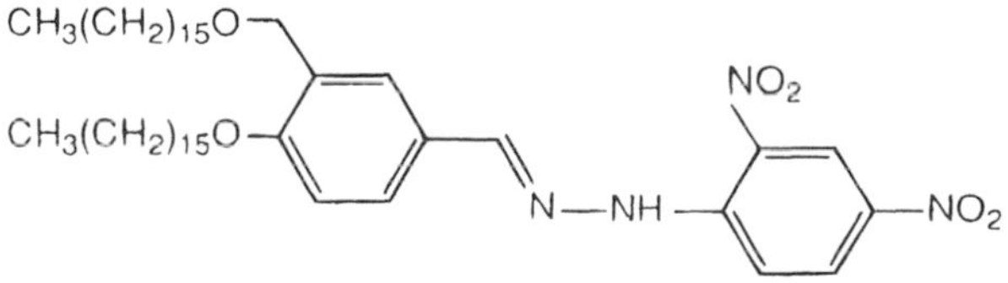

437

$\chi_{zzz}^{(2)} = 14$ pm/V at $\phi = 56°$
$\chi_{zxx}^{(2)} = 15$ pm/V
$\beta = 60 \times 10^{-30}$ esu

368

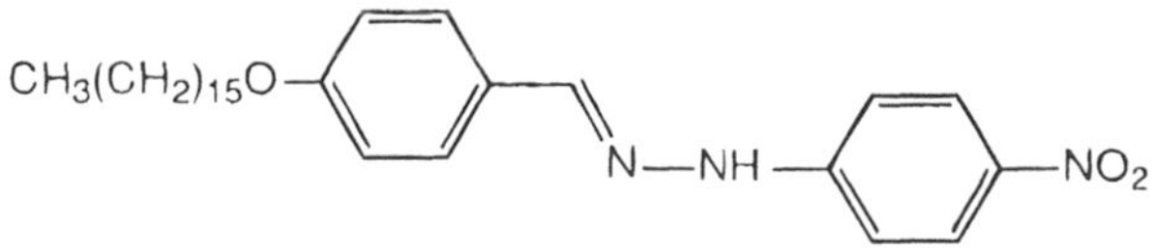

420

$\chi_{zzz}^{(2)} = 30$ pm/V at $\phi = 50°$
$\chi_{zxx}^{(2)} = 21$ pm/V
$\beta = 57 \times 10^{-30}$ esu

368

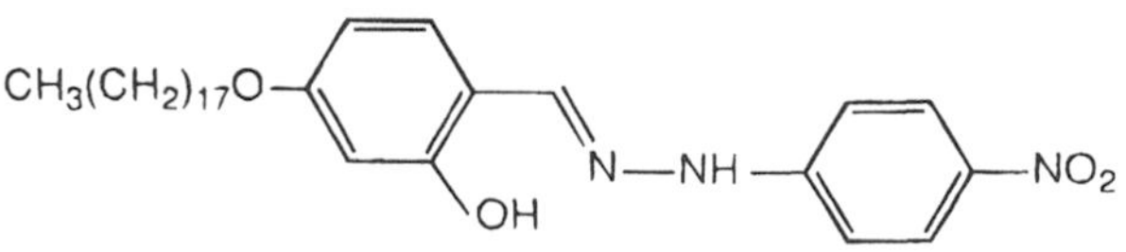

420

$\chi_{zzz}^{(2)} = 43$ pm/V at $\phi = 49°$
$\chi_{zxx}^{(2)} = 28$ pm/V
$\beta = 66 \times 10^{-30}$ esu

368

Table 59 *Continued*

No.	Material	λ_{max} (nm)	Type of Deposition	No. of Monolayers	NLO Coefficients (1.064 μm)	Ref.

$CH_3(CH_2)_{15}O-$ ... $CH_3(CH_2)_{15}O-$ ... $N-NH-$ ... NO_2 ... NO_2

417 — $\chi^{(2)}_{zzz} = 25$ pm/V at $\phi = 47°$; $\chi^{(2)}_{zxx} = 14$ pm/V; $\beta = 62 \times 10^{-30}$ esu — 368

$CH_3(CH_2)_{17}O-$ ... N^+-CH_3 ... $\bar{I}$

360 — $\chi^{(2)}_{zzz} = 36$ pm/V at $\phi = 35°$; $\chi^{(2)}_{zxx} = 8.4$ pm/V; $\beta = 43 \times 10^{-30}$ esu — 368

$CH_3(CH_2)_{17}S-$ ... N^+-CH_3 ... $\bar{I}$

380 — $\chi^{(2)}_{zzz} = 76$ pm/V at $\phi = 26°$; $\chi^{(2)}_{zxx} = 9.0$ pm/V; $\beta = 65 \times 10^{-30}$ esu — 368

$CH_3(CH_2)_{15}$ ... N ... $CH_3(CH_2)_{15}$ ... N^+-CH_3 ... $\bar{I}$

475 — $\chi^{(2)}_{zzz} = 146$ pm/V at $\phi = 43°$; $\chi^{(2)}_{zxx} = 63$ pm/V; $\beta = 304 \times 10^{-30}$ esu — 368

$CH_3(CH_2)_{15}$ ... N ... $CH_3(CH_2)_{15}$ ... $N^+-(CH_2)_3OH$ ... Br^-

475 — $\chi^{(2)}_{zzz} = 146$ pm/V at $\phi = 35°$; $\chi^{(2)}_{zxx} = 36$ pm/V; $\beta = 273 \times 10^{-30}$ esu — 368

Table 59 *Continued*

No.	Material	λ_{max} (nm)	Type of Deposition	No. of Monolayers	NLO Coefficients (1.064 μm)	Ref.

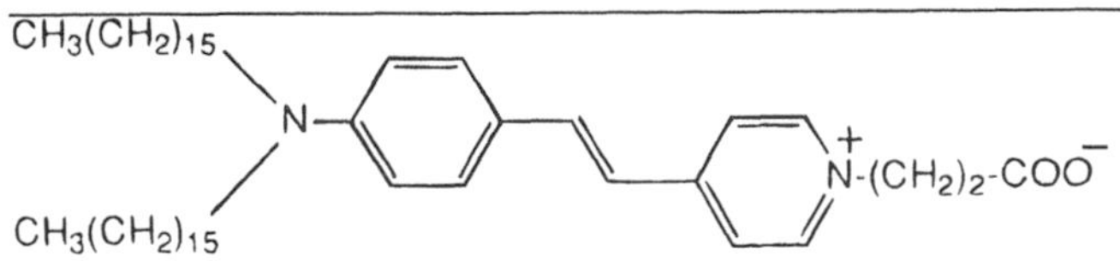

CH₃(CH₂)₁₅ ... N ... =N⁺-(CH₂)₂-COO⁻

475

$\chi^{(2)}_{zzz} = 129$ pm/V at $\phi = 38°$
$\chi^{(2)}_{zxx} = 39$ pm/V
$\beta = 197 \times 10^{-30}$ esu

368

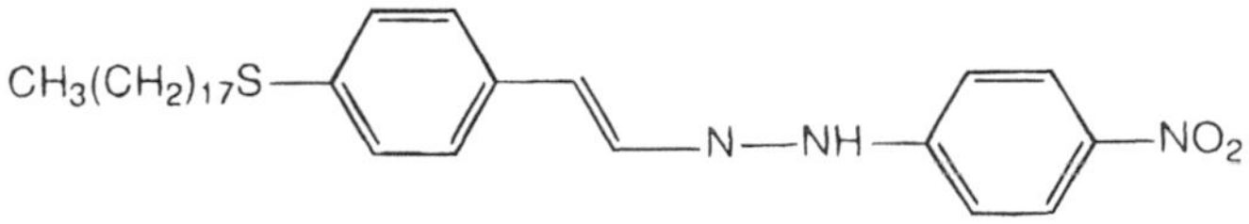

CH₃(CH₂)₁₇S—...—N—NH—...—NO₂

405

$\chi^{(2)}_{zzz} = 63$ pm/V at $\phi = 41°$
$\chi^{(2)}_{zxx} = 24$ pm/V
$\beta = 81 \times 10^{-30}$ esu

368

(24)

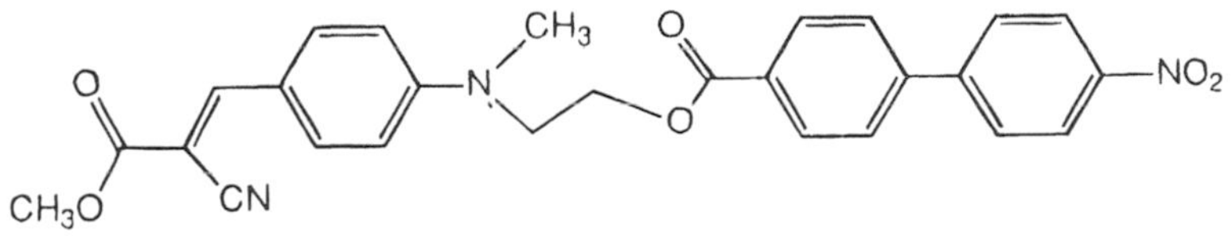

CH₃O, CN, N–CH₃ ... O ... —NO₂

417

$\chi^{(2)}_{zzz} = -1.02$ pm/V
$\chi^{(2)}_{zxx} = 1.89$ pm/V

369

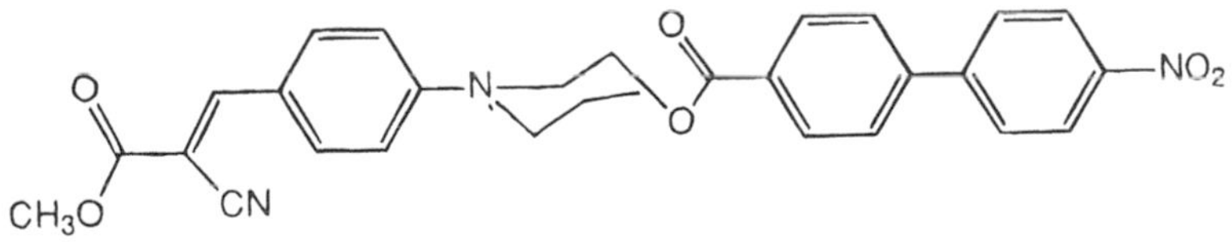

CH₃O, CN ... N ... O ... —NO₂

407

$\chi^{(2)}_{zzz} = 1.31$ pm/V
$\chi^{(2)}_{zxx} = 0.99$ pm/V

369

(25)

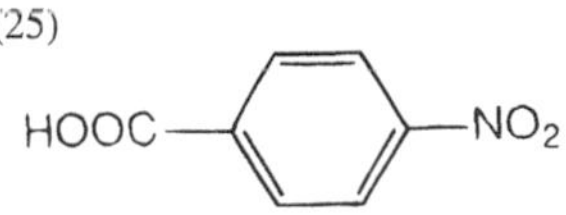

HOOC—...—NO₂

$\beta = 1.48 \times 10^{-50}$ C³ m³ J⁻²

370

(26)

H₂N—...—=NH₂⁺ Cl⁻ ... COOH

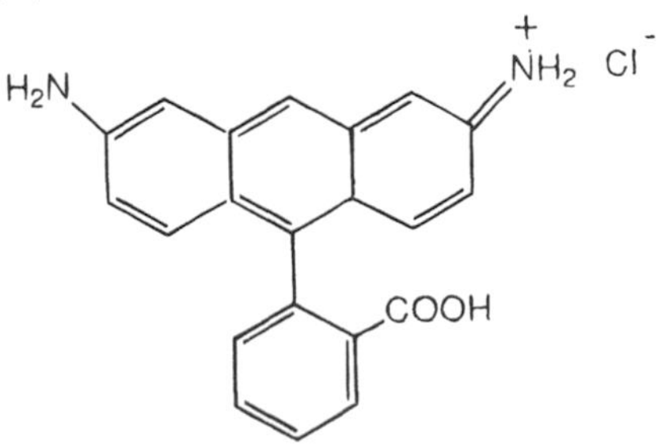

$\beta = 11 \times 10^{-50}$ C³ m³ J⁻² at $\phi = 34°$

371

226

Table 59 *Continued*

No.	Material	λ_{max} (nm)	Type of Deposition	No. of Monolayers	NLO Coefficients (1.064 μm)	Ref.
(27)						

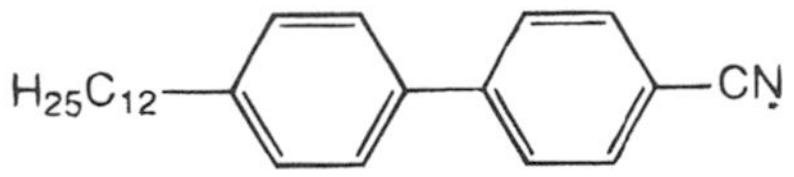

$\beta = 9.28 \times 10^{-50}$ C^3 m^3 J^{-2} at $\phi = 71°$ 372

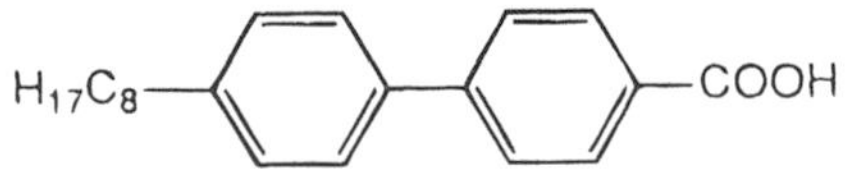

$\beta = 2.23 \times 10^{-50}$ C^3 m^3 J^{-2} at $\phi = 63°$

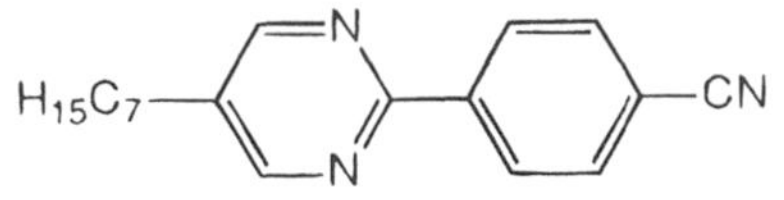

$\beta = 2.97 \times 10^{-50}$ C^3 m^3 J^{-2} at $\phi = 79°$

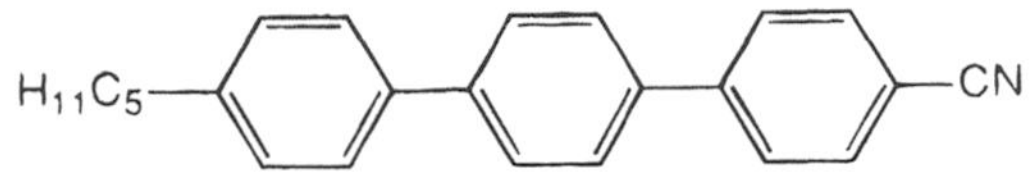

$\beta = 2.78 \times 10^{-50}$ C^3 m^3 J^{-2} at $\phi = 60°$

(28) *N*-acyl-p-nitroaniline

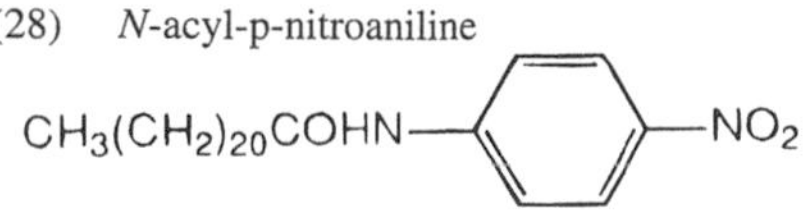

$CH_3(CH_2)_{20}COHN-$⟨benzene⟩$-NO_2$

| | | 330 | Y-type | | $\chi^{(2)}_{zzz} = 21.2$ pm/V | 373 |

(29) DR1-PMMA polymer

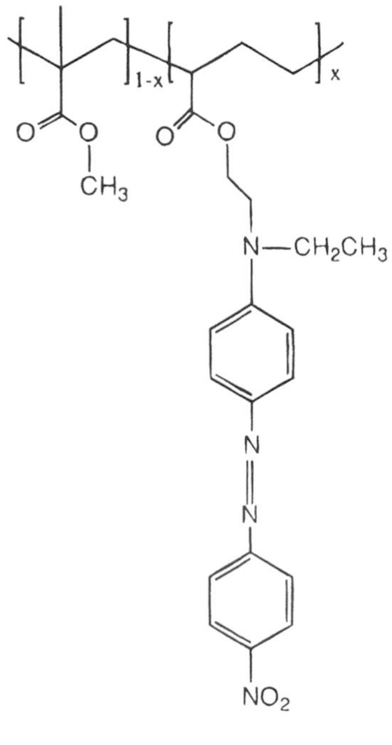

$\chi^{(2)}_{zzz} = 41.9$ pm/V 374
$\chi^{(2)}_{zxx} = 7.33$ pm/V

Table 59 *Continued*

No.	Material	λ_{max} (nm)	Type of Deposition	No. of Monolayers	NLO Coefficients (1.064 μm)	Ref.
(30)	Polysiloxane copolymer		X-type		$\chi^{(2)}_{zzz} = 10.6 \times 10^{-9}$ esu $\beta = 9.8 \times 10^{-30}$ esu	375

No.	Material	λ_{max} (nm)	Type of Deposition	No. of Monolayers	NLO Coefficients (1.064 μm)	Ref.
(31)	NLO polymer		Alternating		$\chi^{(2)}_{zzz} = 42$ pm/V at 1.097 μm	376

Table 59 *Continued*

No.	Material	λ_{max} (nm)	Type of Deposition	No. of Monolayers	NLO Coefficients (1.064 μm)	Ref.
(32)	Azobenzene DMA-copolymer 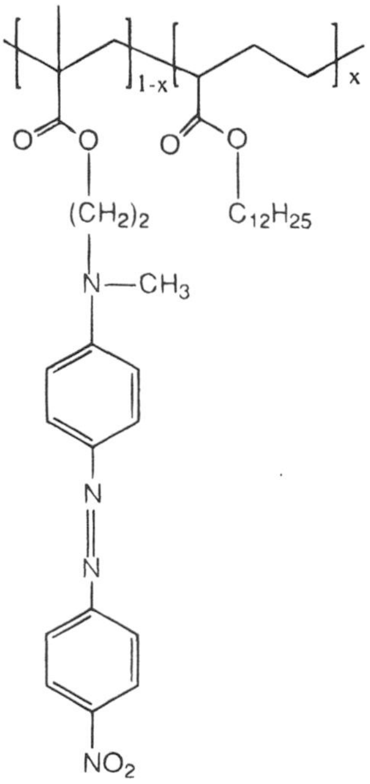	470	Z-type	monolayer	$\beta = 8.55 \times 10^{-30}$ esu at $\phi = 23°$	377

(33)

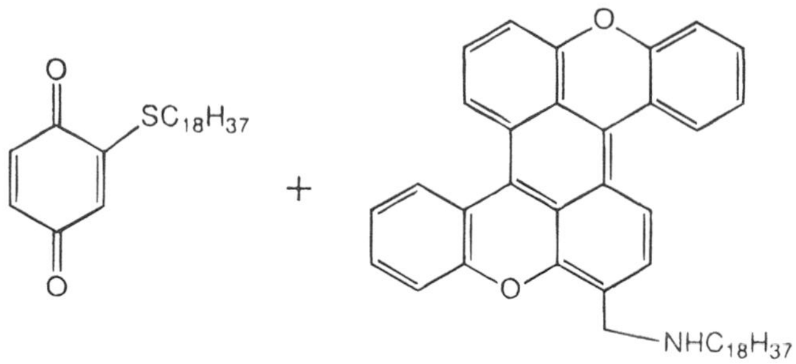

Octadecylthiobenzoquinone (acceptor)

7-(*n*-octadecylaminomethyl)-8-,16-dioxadibenzo[f,g]perylene (donor)

(34)	(i) + (ii)			Bilayer	$\chi^{(2)}_{zzz} = 0.5$ pm/V	378

| | 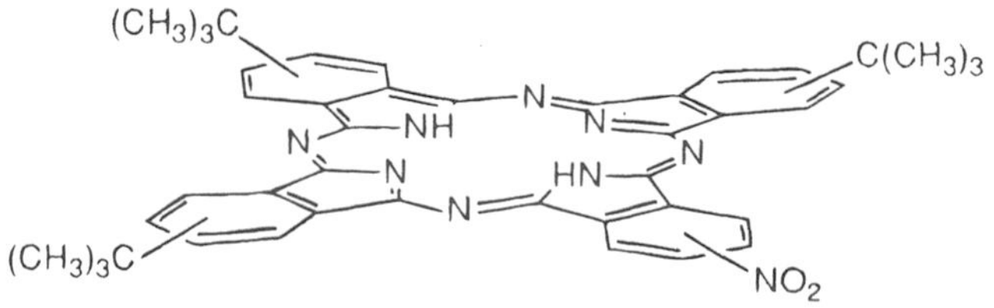 | 630 nm | Z-type | Monolayer | $\beta = 2\text{–}3 \times 10^{-30}$ esu $\chi^{(2)} = 20\text{–}30 \times 10^{-9}$ esu | 310,311 |

Table 59 *Continued*

No.	Material	λ_{max} (nm)	Type of Deposition	No. of Monolayers	NLO Coefficients (1.064 μm)	Ref.

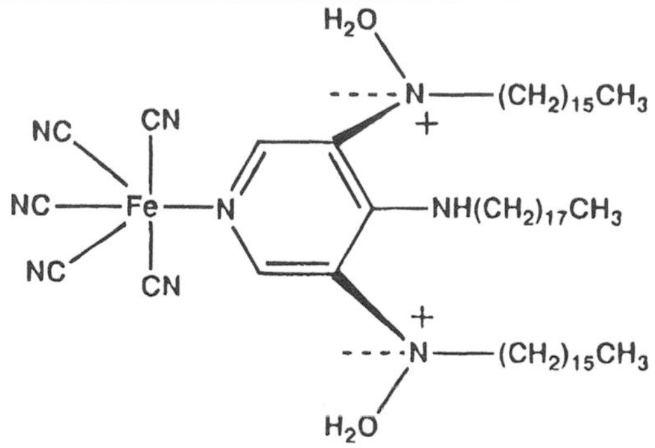

$\beta_{zzz} = 160 \times 10^{-30}$ esu   379

by the prior irradiation of dye laser pulses unlike RuC18C-2C18NB LB films. The change in the β is associated with the ground state to a metal-to-ligand charge-transfer excited states.

Carr et al.[338] reported SHG in a monomolecular LB film of a polysiloxane consisting of poly(dimethyl-silane) and poly(methylsilane) units of unspecified distribution. Polysiloxane backbone contained only ~50% pendant chromophoric groups. For SHG measurements, LB monolayer was deposited onto a hydrophilic glass substrate. The SH intensity from a monolayer was evaluated relative to hemicyanine and a β value of 3.5×10^{-49} C³ m³ J⁻² was obtained where β of merocyanine was taken as 9×10^{-49} C³ m³ J⁻², assuming that both materials have the same refractive indices.

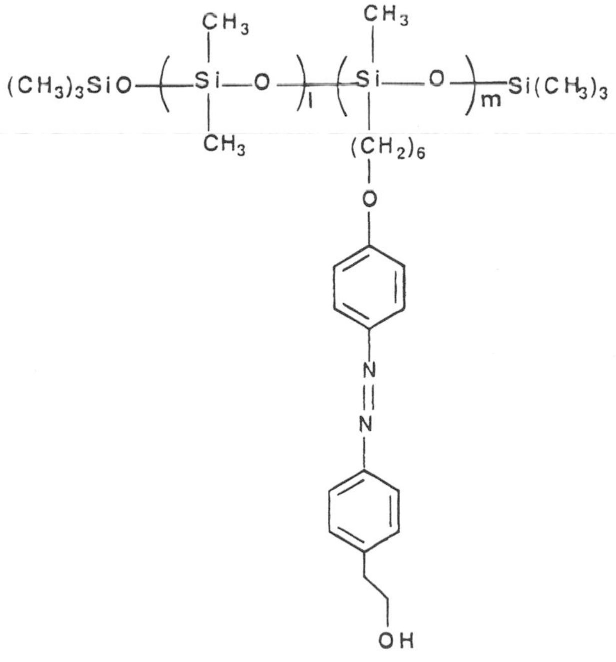

$1 = 9 \pm 2;\ m = 8 \pm 2$

Highly stable icosahedral-cage molecule C_{60} fullerene is a new form of carbon which has a soccer ball structure (see structure below). The C_{60} molecule has a highly aromatic character where all the atoms are connected by sp^2 bonds and the remaining 60 π electrons are distributed. The C_{60} seems to be an ideal candidate for third-order nonlinear optics because of its interesting π-electron conjugation. There are also some reports on the SHG activity of C_{60} from vapor-deposited thin films despite the symmetry. Wang et al.[339] measured a $\chi^{(2)}$ of 2.1×10^{-9} esu for a C_{60} thin films at room temperature. The SHG showed temporal decay on exposure to air. The SHG was temperature dependent. It showed a maximum at 140 °C where the $\chi^{(2)}$ value was 2.0×10^{-8} esu, about 10 times larger than room temperature. Kajzar et al.[340] reported the resonant $\chi^{(2)}$ value of 3.8×10^{-9} esu for C_{60} thin films at 1.064 μm. This $\chi^{(2)}$ value was about two times larger for C_{60} films than that reported by Wang et al.[339] at room temperature.

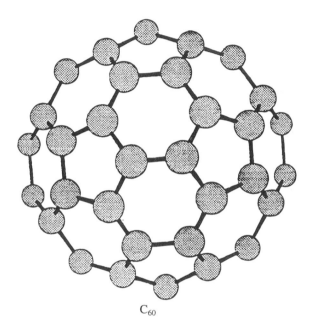

C_{60}

There are a few reports on SHG properties of LB films of C_{60} derivatives. SHG has been observed from the LB film of the $C_{60}(C_4H_8N_2)$ derivative[341] and C_{60}-glycine ester derivatives.[342]

Chemical structures, absorption maxima, type of film deposition, hyperpolarizabilities, and NLO coefficient of LB materials are listed in Table 59. As can be seen, SHG efficiency of LB films varies by three orders of magnitude depending on the chemical species. In particular, stilbazolium salts, pyridinium derivatives, azobenzene, and polyene compounds exhibit remarkably large NLO coefficients. In fact, LB films show one of the largest NLO coefficients ever reported for organic chemical species because of the fine molecular tuning and desired alignment.

D. SELF-ASSEMBLED MULTILAYERS

Self-assembled multilayers is another interesting approach to preparing covalently linked chromophores using the siloxane technology. Marks and co-workers[380–386] have suggested a new organic superlattice approach for developing covalently linked self-assembled, chromophore-containing multilayer systems for second-order nonlinear optics. In this new approach, self-assemblies of intrinsically acentric multilayers of high β-chromophores on inorganic oxide substrates are formed. The synthetic strategy for covalently linked NLO multilayers is shown in Figure 17. The general synthetic strategy using a NLO chromophore is shown in Figure 18. This strategy provides organic materials containing higher chromophore density and better alignment than poled polymers. A stilbazole chromophore precursor that provides the NLO activity causes the sequential construction of multilayers. These multilayers

(i) Defined substrate

(ii) Formation of coupling layer

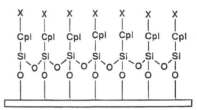

(iii) Incorporation of NLO chromophore

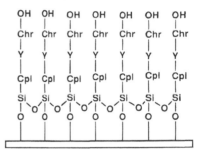

(iv) Formation of coupling layer

Figure 17 General strategies for constructing self-assembled multilayers having NLO chromophores. Cpl = spacer for the coupling reaction; Chr = NLO chromophore. (Reprinted with permission from Li, D., et al., *J. Am. Chem. Soc.*, 112, 7389, 1990. Copyright 1990, American Chemical Society.)

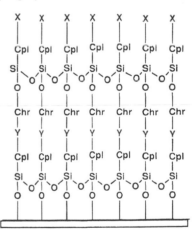

show excellent adhering properties and are insoluble in common organic solvents as well as in strong acids. The superlattice films showed $\chi^{(2)}_{zzz}$ of 2×10^{-7} esu, while the estimated value for a 22-Å thick single layer was 6×10^{-7} esu at 1.064 μm. The SHG efficiency of a five-layer structure decreased by 10% over a 1-month period. The β_{vec} values of chromophores SA-1 through SA-4 for self-assembled NLO materials are listed in Table 60. The β_{vec} values of chromophores were significantly influenced by the nature of the π-bonding sequence where SA-4 showed the largest hyperpolarizability on the order of 3.43×10^{-28} cm^5 esu^{-1}. Lundquist et al.[384] studied wavelength dependence of SHG for a stibazolium-based self-assembled monolayer films (Fig. 19). The SHG follows the absorption spectrum (λ_{max} at 480 nm) and resonant $\chi^{(2)}_{zzz}$ peaked around this region with a value of 9×10^{-7} esu. The $\chi^{(2)}_{max}$ coincides with a low-energy, chromophore-centered, charge-transfer one-photon excitation in the optical spectrum at 480 nm.

232

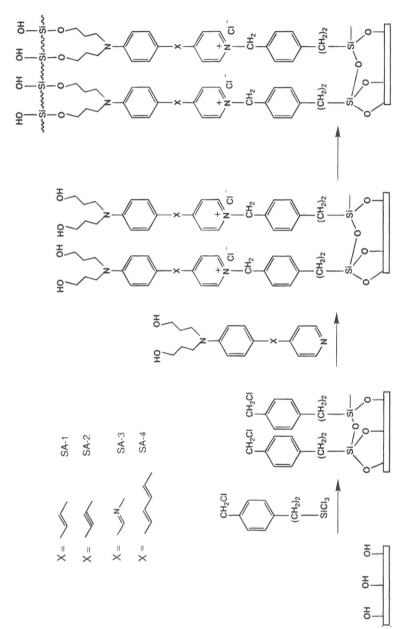

Figure 18 Schematic sequence of self-assembled multilayer formation. Multilayer formation with deposit of 2-trichlorosilyl-1-(4-chloromethyl)phenylethane on both sides of glass slides. The final structure consists of coupling layer, NLO chromophore layer, and capping layer. (Reprinted with permission from Kakkar, A. K., et al., *Langmuir*, 9, 388, 1993. Copyright 1993, American Chemical Society.)

Table 60 Hyperpolarizabilities and NLO Data of Self-assembled Thin Films

Materials	β Value	NLO Coefficient	Ref.
SA-1	$\beta_{vec} = 178 \times 10^{-30}$ cm^5 esu^{-1} at 0.65 eV	$\chi^{(2)} = 9 \times 10^{-7}$ esu	380
SA-2	$\beta_{vec} = 171 \times 10^{-30}$ cm^5 esu^{-1} at 0.65 eV		380
SA-3	$\beta_{vec} = 92.6 \times 10^{-30}$ cm^5 esu^{-1} at 0.65 eV		380
SA-4	$\beta_{vec} = 343 \times 10^{-30}$ cm^5 esu^{-1} at 0.65 eV		380
SA-5	$\beta = 10^{-30}$ esu at 1.06 μm	$d_{zzz} = 19$ pm/V	388,389
SA-6	$\beta = 1060 \times 10^{-30}$ esu at 1.06 μm	$d_{zzz} = 110$ pm/V	388,389
SA-7	$\beta = 150 \times 10^{-30}$ esu at 1.06 μm	$d_{zzz} = 6$ pm/V	388,389

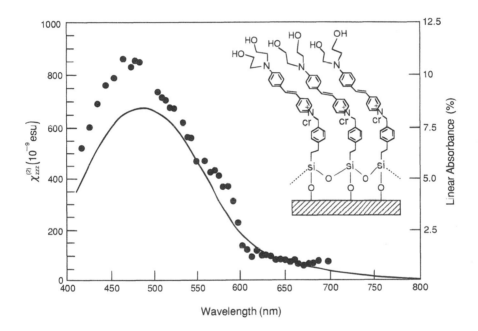

Figure 19 Wavelength dependence of $\chi^{(2)}$ for a self-assembled film (filled circle) and optical absorption spectrum (solid line). The $\chi^{(2)}$ dispersion is plotted as a function of the output wavelength. The 480 nm resonance in the optical spectrum corresponds to the 960 nm resonance in the $\chi^{(2)}$ dispersion. The chemical structure of a stilbazolium-based chromophore used for self-assembled monolayer is shown in the inset. (From Lundquist, P. M., et al., *Appl. Phys. Lett.*, 64, 2194, 1994, American Institute of Physics. With permission.)

Roscoe et al.[386] reported the effect of anion exchange on the SHG properties of self-assembled multilayers comprising stilbazolium chromophore. The $\chi^{(2)}_{zzz}$ values of self-assembled monolayers before ion exchange were 2.22×10^{-7}, 2.57×10^{-7}, 1.46×10^{-7}, and 1.41×10^{-7} esu for 1a(1), 1a(3), 1b(2), and 1b(3), respectively. The self-assembled monolayers were subjected to ion exchange anaerobically by exposing functionalized glass substrate to solutions of $(n\text{-Bu})_4N^-I^-$ (1), sodium salts of p-aminobenzensulfonate (2), and 4-[4-[(diethylamino)phenyl]azo]benzenesulfonate (3). The exchange of Cl^- monolayers with anions was monitored by XPS technique. The $\chi^{(2)}_{zzz}$ values after ion exchange increased to 2.98×10^{-7}, 2.99×10^{-7}, 1.83×10^{-7}, and 2.03×10^{-7} esu for 1a(1), 1a(3), and 1b(2), and 1b(3), respectively. A comparison of the $\chi^{(2)}_{zzz}$ value from these monolayers before and after ion exchange showed that ion exchange increased the $\chi^{(2)}_{zzz}$ value by 34%, 25%, and 44% for I^- (1), p-aminobenzensulfonate (2), and 4-[4-[(diethylamino)phenyl]azo]benzenesulfonate (3), respectively. The large enhancement in $\chi^{(2)}_{zzz}$ values occurs from changes in anion–cation packing configurations. After ion exchange of 1a with bromide ions, the $\chi^{(2)}_{zzz}$ value of 5×10^{-7} esu (~ 200 pm/V) was measured.[387]

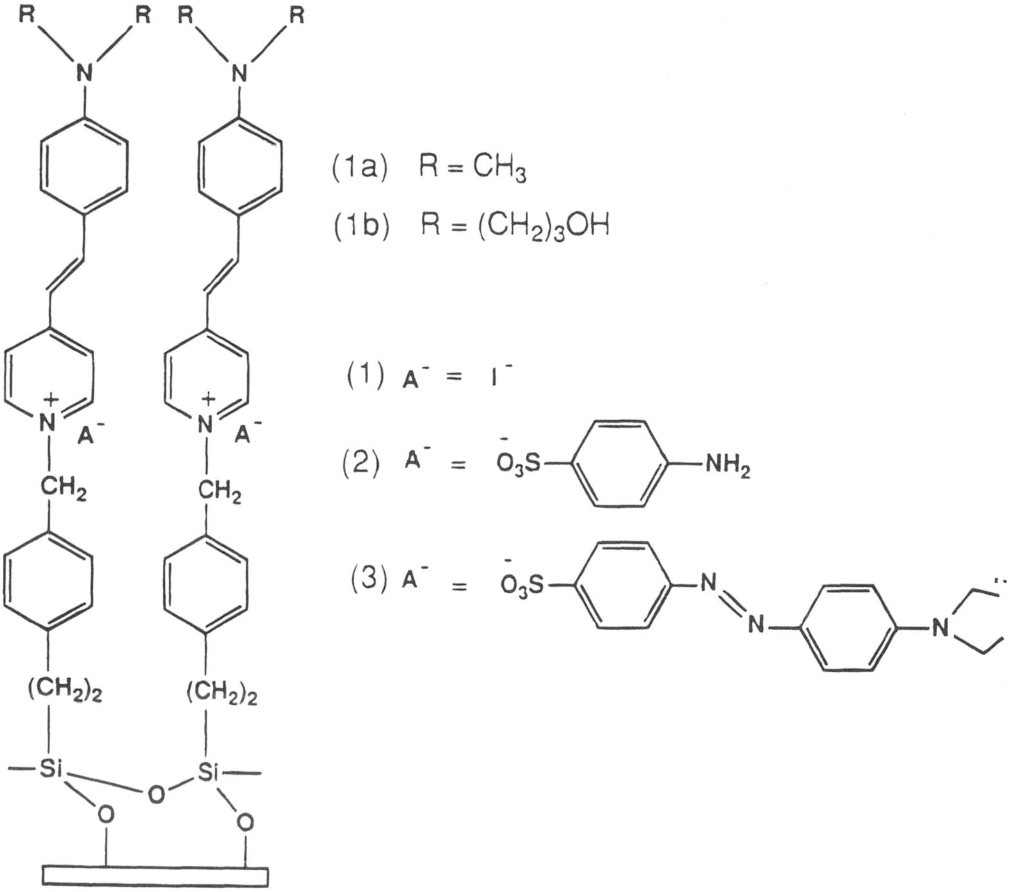

(1a) R = CH$_3$

(1b) R = (CH$_2$)$_3$OH

(1) A$^-$ = I$^-$

(2) A$^-$ = ^-O_3S—⬡—NH$_2$

(3) A$^-$ = ^-O_3S—⬡—N≡N—⬡—N

Katz et al.[388,389] reported SHG from acentric self-assembled multilayers of donor–acceptor phosphate-terminated polar azo dyes incorporatd into zirconium-based surface multilayers. This approach involved three steps: polar deposition sequence of surface phosphorylation, zirconation, and dye absorption. A schematic diagram of the resulting structure is shown below; the phosphate-terminated three NLO chromophores used in this study were SA-5, SA-6, and SA-7. The SHG values for SA-5, SA-6, and SA-7 are listed in Table 60. There was a significant change in NLO activity with the change of donor–acceptor chromophore structure. The resonantly enhanced d_{zzz} value for SA-6 was 110 pm/V. The order parameters $<\cos^3\theta)$ were >0.2 for SA-5 and ~0.6 for SA-6. These films were stable to solvents above 80°C. The SHG activity was not affected by the insertion of a nonpolar chromophore layer among the polar layers during the deposition sequence. Self-assembled multilayers showed SHG activities as large as that of LB films. As seen in the studies above, this is an interesting approach for generating SHG-active multilayers exhibiting very large second-order optical nonlinearity.

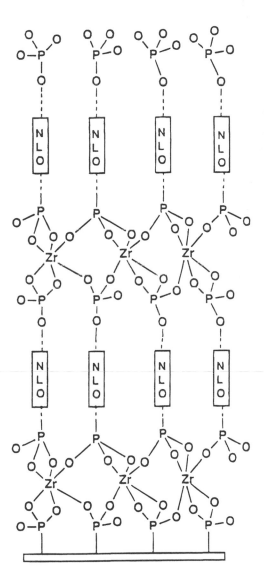

Phosphate-terminated donor–acceptor chromophores (NLO) (From Katz, H. E., et al., *Science*, 254, 1485, 1991; *J. Am. Chem. Soc.*, 116, 6636, 1994. With permission.)

SA-5   SA-6   SA-7

E. LIQUID CRYSTALS

Ferroelectric liquid crystals (FLCs) also have attracted attention for SHG. An advantage of these materials is that S_c^* phase and other chiral-tilted phase FLCs are noncentrosymmetric. Several research groups have explored the SHG efficiency of FLC compounds and their mixtures. Chemical strategies such as incorporating a NLO chromophore into the core of the FLC molecules or doping with NLO dyes have been found useful in enhancing the SHG efficiency.

Schmitt et al.[390] reported the switching behavior, spontaneous polarization, and SHG efficiency of a series of FLCs. The FLCs possessing large spontaneous polarization exhibited large SHG efficiency. A d_{22} as large as 5 pm/V was measured for a biphenyl compound containing NLO chromophore. These FLCs allowed the NLO moieties to freeze in the S_c^* phase by cooling to less than T_g under an electric field. SHG activity was stable even after removing the electric field below T_g. Angular phase matching was also demonstrated. Yoshino et al.[391] reported SHG efficiency of various FLCs and NLO dye-doped FLC mixtures. The SHG activity was enhanced by doping of FLCs with NLO dyes. The SHG efficiency of 3M2CPOOB, 1BC1EPOPB, and 1HpC1ECPOPB was enhanced as a function of the concentration of p-NA. The SHG intensity of o-NA/1BC1EPOPB, 4-methoxy-2-NA/1BC1EPOPB, and p-NA/1BC1EPOPB guest–host systems were 2.1, 2.3, and 6.5 times that of pure 1BC1EPOPB. The SHG intensity of m-NA/1BC1EPOPB was the same as that of pure 1BC1EPOPB, so m-NA does not lead to the enhancement of SHG. The NLO coefficients of FLCs are rather small and some improvement has been achieved by modifying chemical structures by attaching donor–acceptor groups. Table 61 lists the chemical structure, NLO coefficients, and spontaneous polarization (Ps) of FLCs. Over all, the SHG efficiencies of FLCs are lower compared with other organic NLO molecules and polymers. Much more research work is needed to enhance the SHG efficiency of FLCs.

Smith and Coles[399] reported $\chi^{(2)}$ values of liquid crystalline polymers having donor–acceptor-substituted side chains measured at fundamental wavelengths of 1.064 and 1.579 μm. The $\chi^{(2)}$ of liquid crystalline polymers measured at 1.064 μm are shown along with chemical structures (below). The LC polymers

Table 61 Chemical Structure, NLO Coefficients, and Spontaneous Polarization (Ps) of FLCs*

FLCs	d_{ijk} (pm/V)	Ref.

(1)

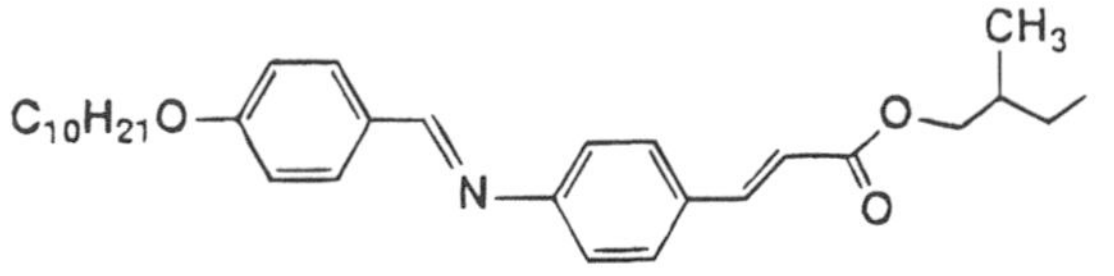

$C_{10}H_{21}O$—...—N=...—...—...—O—CH_3 (O)

(DOBAMBC)

	$d_{eff} = 0.0008$	392
	$d_{eff} = 0.0011$	393
	Ps $= -3$ nC cm^{-2}	394

(2)

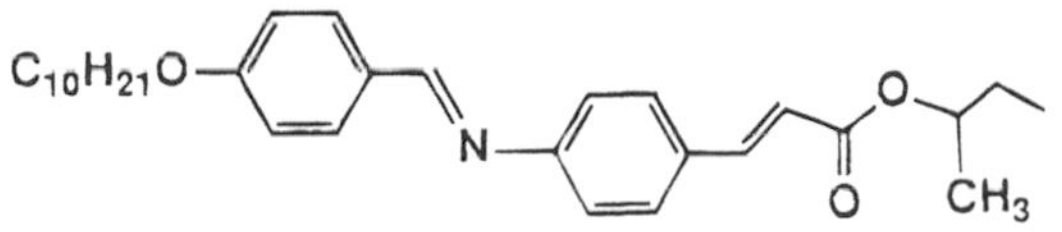

$C_{10}H_{21}O$—...—N=...—...—O—...—CH_3

(DOBA-1-MPC)

(3) ZL13654 (mixture having phenyl pyrimidines)

(4) SCE9 (contains phenyl benzoates)

	$d_{eff} = 0.0006$	
	$d_{eff} = 0.005$	393
	Ps $= -29$ nC cm^{-2}	391,394
	$d_{21} = 0.0026$	
	$d_{22} = 0.027$	
	$d_{23} = 0.073$	
	$d_{25} = 0.0009$	
	$d_{eff} = 0.0037$	395
	Ps $= 33.6$ nC cm^{-2}	394

(5)

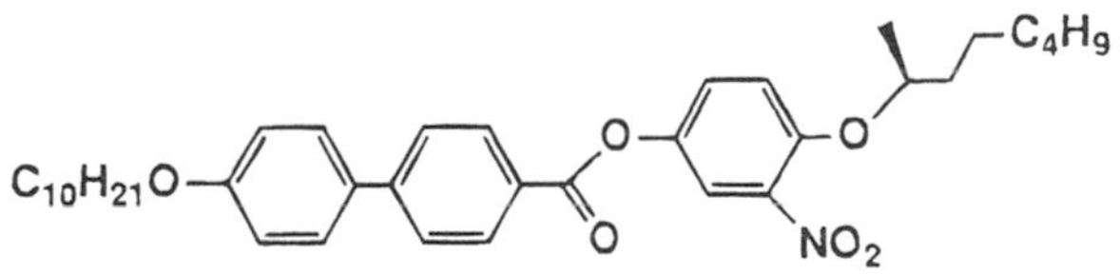

$C_{10}H_{21}O$—...—...—...—O—...—O—...—C_4H_9, NO_2

(W314)

	$d_{21} = 0.08$	
	$d_{22} = 0.60$	
	$d_{23} = 0.63$	
	$d_{25} = 0.16$	394
	$d_{eff} = 0.23$	
	Ps $= -420$ nC cm^{-2}	

(6)

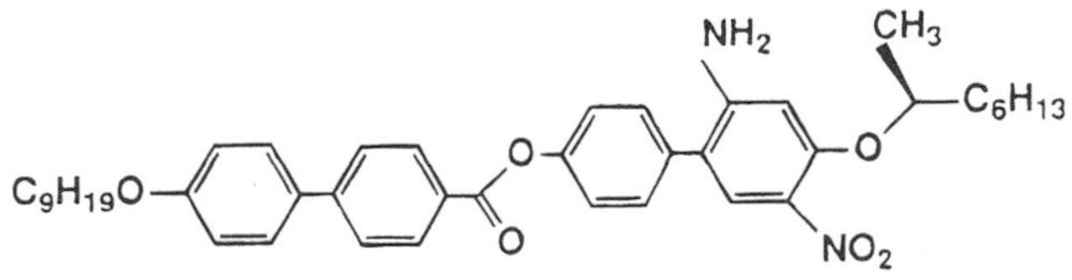

$C_9H_{19}O$—...—...—...—O—...—...—NH_2, CH_3, C_6H_{13}, O—, NO_2

	$d_{22} = 5.0$	
	$d_{eff} = 0.78$	
	$d_{14} = 0.49$	
	$d_{16} = 1.13$	
	$d_{23} = 1.46$	396
	Ps $= 306$ nC cm^{-2}	390

Table 61 *Continued*

FLCs	d_{ijk} (pm/V)	Ref.

(7) 1:1 mixture

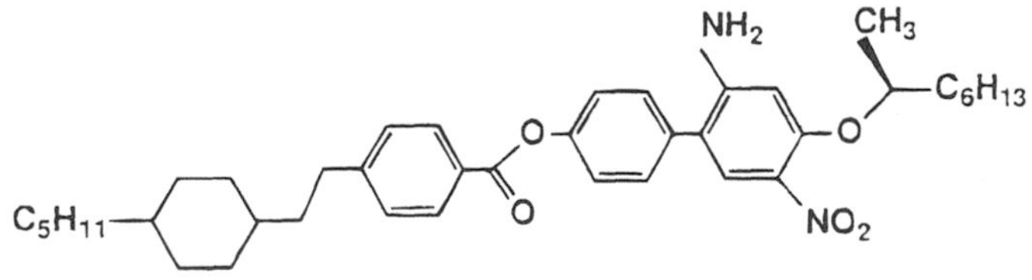

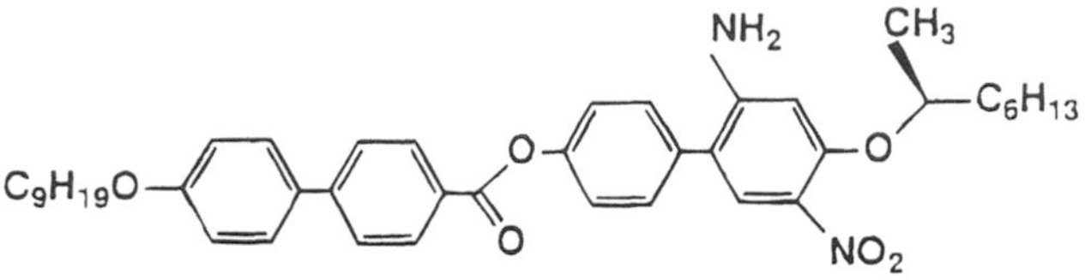

C_5H_{11} — cyclohexyl —CH₂CH₂— benzene —C(=O)—O— phenyl with NH₂, CH₃, C_6H_{13}, O, NO₂

$C_9H_{19}O$ — biphenyl —C(=O)—O— phenyl with NH₂, CH₃, C_6H_{13}, O, NO₂

$d_{eff} = 0.68$
$d_{22} = 2.1$
Ps = 179 nC cm^{-2} ... 390

(8)

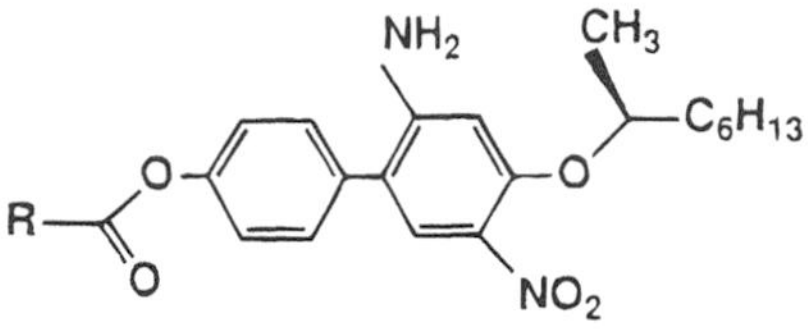

R—C(=O)—O— phenyl with NH₂, CH₃, C_6H_{13}, O, NO₂

SHG active
Ps = 139 nC cm^{-2} ... 390
SHG active ... 390

$C_8H_{17}O$ — biphenyl —C(=O)—O— phenyl —O— with CH₃, C(=O)O—C_4H_9

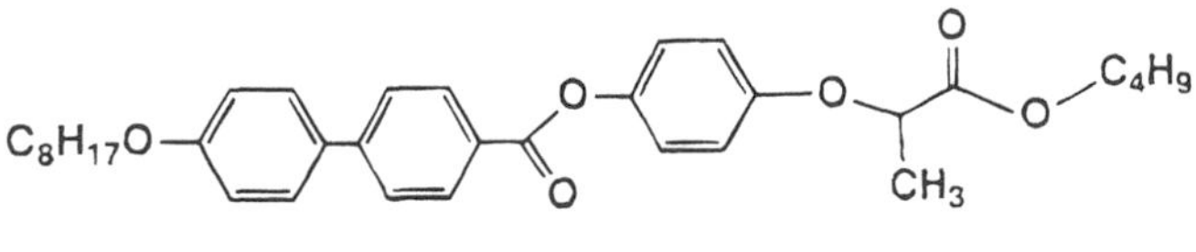

Ps = 208 nC cm^{-2}
SHG active
Ps = 207 nC cm^{-2} ... 390
SHG active
Ps = 141 nC cm^{-2} ... 390

(9)

$C_9H_{19}O$ — biphenyl —C(=O)—O— phenyl with R_1, CH₃, C_6H_{13}, O*, R_2

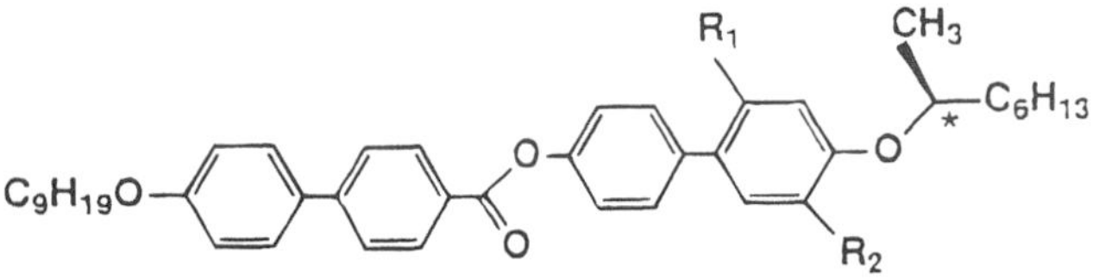

(i) $R_1 = NH_2$, $R_2 = NO_2$
    SHG active
    Ps = 58 nC cm^{-2} ... 390

(ii) $R_1 = NO_2$, $R_2 = NH_2$
    SHG active
    Ps = 131 nC cm^{-2} ... 390

(iii) $R_1 = NO_2$, $R_2 = N(CH_3)_2$
    SHG active
    Ps = 5 nC cm^{-2} ... 390

Table 61 *Continued*

FLCs	d_{ijk} (pm/V)	Ref.
(10)		
(1BC1EPOPB)	$d_{eff} = 0.0033$ Ps = 240 nC cm^{-2}	393
(11)		
(3M2CPOOB)	$d_{eff} = 0.01$ $d_{21} = 0.021$ $d_{22} = 0.009$ $d_{23} = 0.053$ $d_{25} = 0.023$ Ps = -200 nC cm^{-2}	393 391
(12)		
(ETFPPOPB)	$21.0 \times$ 3M2CPOOB* Ps = -150 nC cm^{-2}	391
(OTFPPOPB)	$6.0 \times$ 3M2CPOOB Ps = -90 nC cm^{-2}	391
(1HpC1EPOPB)	$3.0 \times$ 3M2CPOOB Ps = 130 nC cm^{-2}	391
(1HpC1ECPOPB)	$6.0 \times$ 3M2CPOOB Ps = 100 nC cm^{-2}	391

Table 61 *Continued*

FLCs 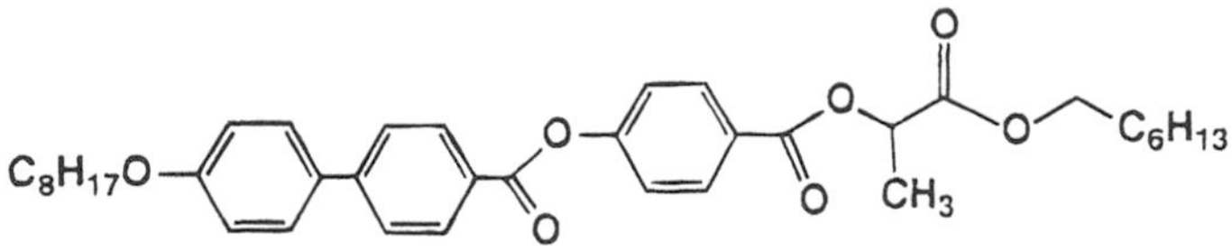	d_{ijk} (pm/V)	Ref.

C$_8$H$_{17}$O— ... —O—C$_6$H$_{13}$

(1HpC1ECPOPB)

3.0 × 3M2CPOOB
Ps = −55 nC cm^{-2}

391

(13)

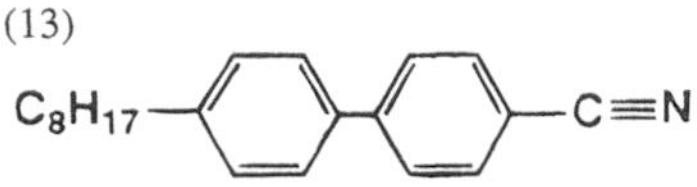

C$_8$H$_{17}$— ... —C≡N

(unrubbed substrate)

$\chi^{(2)} = 1.0 \times 10^{-15}$ esu
$\chi^{(2)}_{zzz} = 1.0 \times 10^{-15}$ esu
$\chi^{(2)}_{zyy} = 3.7 \times 10^{-15}$ esu

397

398

(rubbed substrate)

$\chi^{(2)}_{zzz} = 0.8 \times 10^{-15}$ esu
$\chi^{(2)}_{xxx} = 1.7 \times 10^{-15}$ esu
$\chi^{(2)}_{zyy} = 1.8 \times 10^{-15}$ esu
$\chi^{(2)}_{zxx} = 3.7 \times 10^{-15}$ esu
$\chi^{(2)}_{zxz} = 0.15 \times 10^{-15}$ esu
$\chi^{(2)}_{xyy} = 0.37 \times 10^{-15}$ esu

398

* The SHG intensity is expressed relative to the phase-matched SHG intensity of 3MQCPOOB on a scale of 100 for 3M2CPOOB.

containing a nitro acceptor group show larger $\chi^{(2)}$ values than those containing a cyano group. The $\chi^{(2)}$ value also increased with the increase of conjugation length and planarity that exists in the azo- and stilbene-based chromophores compared with those having a biphenyl side group.

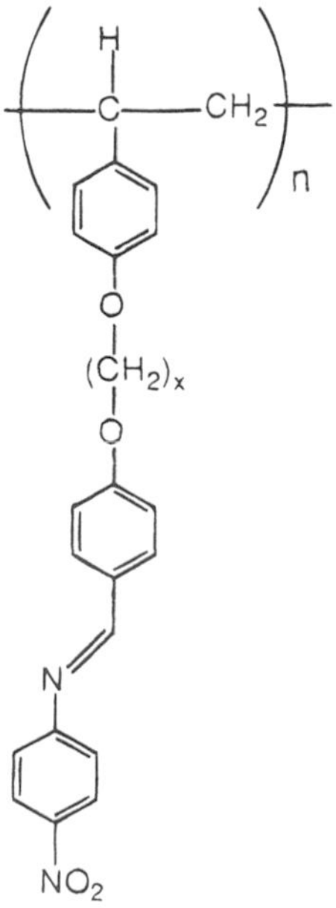

$\chi^{(2)}_{33} = 6.30 \times 10^{-9}$ esu
$\chi^{(2)}_{31} = 1.90 \times 10^{-9}$ esu

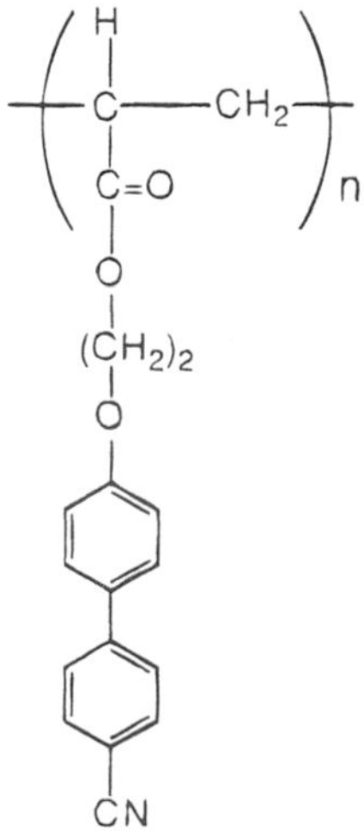

$\chi^{(2)}_{33} = 1.30 \times 10^{-9}$ esu
$\chi^{(2)}_{31} = 0.23 \times 10^{-9}$ esu

$\chi_{33}^{(2)} = 3.36 \times 10^{-9}$ esu
$\chi_{31}^{(2)} = 0.62 \times 10^{-9}$ esu

$\chi_{33}^{(2)} = 2.55 \times 10^{-9}$ esu
$\chi_{31}^{(2)} = 0.86 \times 10^{-9}$ esu

$\chi_{33}^{(2)} = 1.89 \times 10^{-9}$ esu
$\chi_{31}^{(2)} = 0.47 \times 10^{-9}$ esu

$\chi_{33}^{(2)} = 0.55 \times 10^{-9}$ esu
$\chi_{31}^{(2)} = 0.13 \times 10^{-9}$ esu

$$\chi_{33}^{(2)} = 0.25 \times 10^{-9} \text{ esu}$$
$$\chi_{31}^{(2)} = 0.06 \times 10^{-9} \text{ esu}$$

$$\chi_{33}^{(2)} = 3.00 \times 10^{-9} \text{ esu}$$
$$\chi_{31}^{(2)} = 0.87 \times 10^{-9} \text{ esu}$$

$$\chi_{33}^{(2)} = 6.00 \times 10^{-9} \text{ esu}$$
$$\chi_{31}^{(2)} = 1.24 \times 10^{-9} \text{ esu}$$

Yitzchaik et al.[400] reported SHG in polymers and blends of liquid crystal polymers (LCP), photochromic liquid crystal polymers (PCLC), quasi liquid crystals (QLC) with 4-(dimethylamino)-4′-nitrostilbene (DANS) and 4-[4-methoxybenzoyl)oxy]benzylideneamino(p-nitrostilbene) (MBANS). The chemical structures of PLCP, LCPs, and QLCs are shown below. The spiropyran comonomer content in the PLCP(1) and PLCP(2) is 15%. In PLCP(3), about 705 of the molecules are in merocyanine form in an ethanol solution. The $\chi^{(2)}_{zzz}$ values of PLCP(1), PLCP(2), QLC(1), and QLC(3) were 0.6, 0.4, 1.1, and 0.1×10^{-9} esu/cm^3, respectively, 4 days after the sample was prepared and poled at elevated temperatures. QLC(1)-PLCP(3) in 1:4 ratio and 10% MBANS in QLC(1) showed $\chi^{(2)}_{zzz}$ coefficients of 4.4×10^{-9} esu/cm^3 and 2.6×10^{-9} esu/cm^3, respectively, and these samples also showed long-term stability of SHG activity. The $\chi^{(2)}_{zzz}$ values of PLCPs and QLCs were significantly enhanced by doping with DANS or MBANS. For example, 2% DANS in QLC(3), LCP(1), LCP(2), PLCP(1), and PLCP(2) showed $\chi^{(2)}_{zzz}$ values of 5.2, 2.6, 2.2, 2.6, and 0.1×10^{-9} esu/cm^3, respectively. The $\chi^{(2)}_{zzz}$ of blends increased with an increasing poling field. For example, the $\chi^{(2)}_{zzz}$ of QLC(1)-PLCP(2) 1:4, 2% DANS in LCP(1), and 2% DANS in QLC(1)-PLCP(1) 1:4 were 44, 23, and 60×10^{-9} esu/cm^3 at 50 kV/cm, respectively. $\chi^{(2)}_{zzz}$ values increased about 10 fold with an applied field of 50 kV/cm.

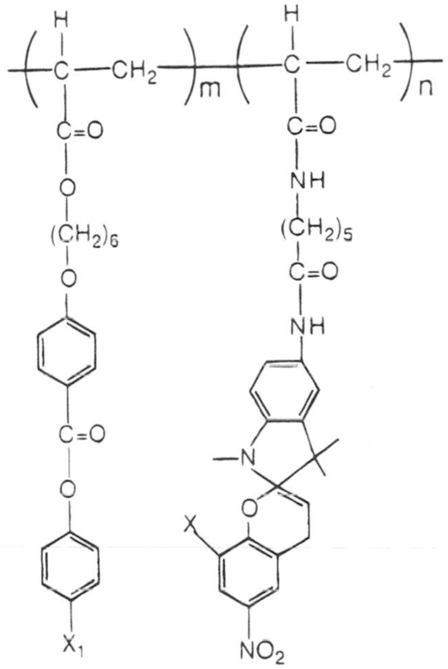

LCP(1)	X_1 = CN	$Z = n/(m + n) = 0$
LCP(2)	X_1 = OCH$_3$	$Z = n/(m + n) = 0$
PLCP(1)	X_1 = CN, X = H	$Z = n/(m + n) = 0.15$
PLCP(2)	X_1 = OCH$_3$, X = H	$Z = n/(m + n) = 0.15$
PLCP(3)	X_1 = CN, X = NO$_2$	$Z = n/(m + n) = 0.05$

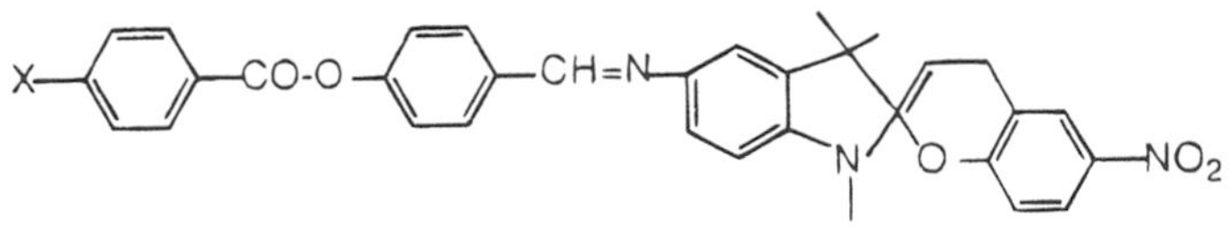

QLC(1) = X = OC$_6$H$_{13}$ QLC(2) = X = OCH$_3$

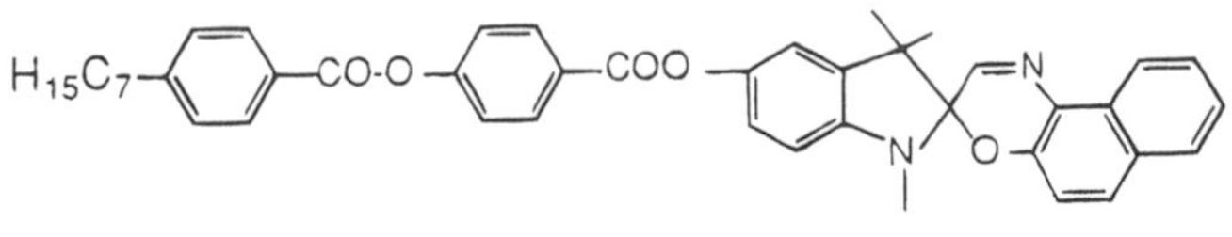

(QLC-3)

The SHG efficiencies of FLC molecules and LC polymers were very small compared with the traditional NLO materials such as barium titanate, NLO-chromophore-functionalized poled polymers, single crystals, or LB films. To increase SHG efficiency of these materials, more work must be done on chemical strategies developing novel FLCs and LCs material with large second-order optical nonlinearity.

244

F. POLED POLYMERS

The organic nonconjugated polymers are inherently insulating materials which possess the basic property of a dielectric to be polarized in the presence of an electric field and retain the induced electrostatic charges for a long time. The most important parameter that contributes to various electrophysical properties in a polymer system is the dipole moment. A highly polar polymer should have a permanent dipole moment and an asymmetric structure, whereas a nonpolar polymer lacks net permanent dipole moment and is structurally symmetrical. Therefore it is apparent that introducing a polar bond into the polymer molecule would assist in imparting a dipole moment, but one has to remember that, if these polar bonds are arranged symmetrically to neutralize each other's electrical field, the resulting dipole moment of the polymeric system would be zero. This situation can be understood best by comparing poly(tetrafluoroethylene) (PTtFE) and poly(trifluoro chloroethylene) (PTrFE) systems, both of which have carbon-fluorine polar bonds. PTtFE is nonpolar despite a large number of C−F bonds, because the bonds are arranged symmetrically, whereas PTrFE is polar because of its asymmetric structure. Therefore a rough idea about the polarity of a polymeric system can be obtained from the polarity of the imparted groups which are also responsible for molecular alignment. The concept of polarity confirms that poly(methyl methacryate), poly(vinyl chloride), poly(vinyl alcohol), polycarbonates, and poly(vinylide fluoride) are highly polar materials.

The dipoles in a polymer are randomly distributed when no electric field exists. The dipoles can be aligned in the direction of the field in a strong electric field although intermolecular forces in a polymeric system oppose such an alignment. If the polymeric system is brought to a glassy state by raising the temperature while still in the electric field, the opposing internal molecular forces decrease and a desired dipolar alignment can be secured (Figure 20(a)). The poled polymers obtained by this technique of

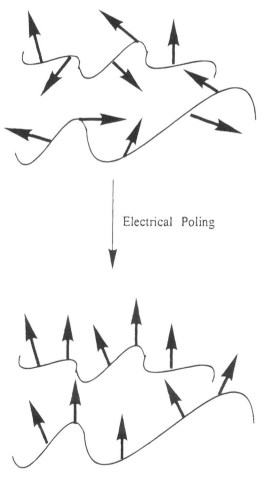

Electrical Poling

Figure 20 (a) Orientation of molecular dipoles under electrical poling.

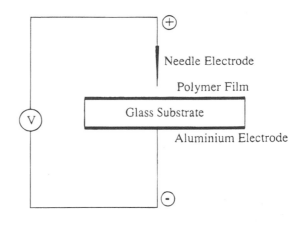

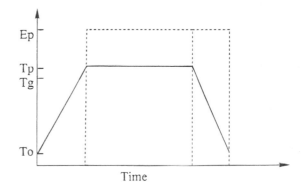

Figure 20 (b) Schematic representation of corona poling. The abbreviations used are Ep = poling field, Tp = poling temperature, Tg = glass transition temperature, and To = room temperature.

aligning molecular dipoles under the influence of an electric field and temperature is called thermoelectrets. This technique of thermoelectret formation has been known for decades.[318,319,401–404]

1. Poling Techniques

A poling temperature for a guest–host system is usually selected above T_g because it facilitates the easy rotation of dipoles. In a typical poling experiment, thin polymer films are subjected to a high electrical field for a period of a few minutes to several hours. After poling for a certain period, the thin film sample still under the influence of an electric field is cooled down to room temperature, which allows freezing of the dipoles in a polar alignment (i.e., noncentrosymmetric fashion) below T_g of the composite. This system is known as "poled polymer." This procedure is called thermoelectret formation.[402] A well-known fact about electret suggests that the dipolar alignment persists for a very long time (e.g., several years). The dipolar alignment induced by the electric field is characteristically different than that inherited by the noncentrosymmetric crystals. The magnitude of SHG activity is related to the dipolar alignment; therefore, poling conditions as well as the dipole moment of the NLO moieties are the important factors. Either parallel-plate or corona poling at elevated temperatures is used.[403,404] It depends on the strength of an individual polymeric material, but in general the dye–polymer system cannot sustain very high electric voltages; hence, the breakdown electric field must be watched carefully during the poling procedure. The dipolar orientation in organic crystals is of a much higher quality than the guest–host system.

NLO properties such as SHG and electro-optic Pockel's effect are associated closely with the chemical and crystalline structures of the polymeric materials. Of the 32 different crystal point group symmetries sketched below in which all crystalline materials have been classified, 11 possess a center of symmetry; hence, they are called "centrosymmetric" materials. On the other hand, the 21 that do not have a center of symmetry are called "noncentrosymmetric" crystal structures. Of the noncentrosymmetric classes, 20 exhibit piezoelectric behavior; hence, they are inherently active to the nonlinear optical SHG and electro-optic Pockel's effects. The NLO properties of ferroelectric polymers will be discussed in detail in another section. If molecular dipoles are aligned in an organic polymer in such a fashion that self-cancellation does not occur, then spontaneous polarization appears because of its unique chemical and morphological structures. As discussed earlier, polarization can be induced by applying a poling technique, although that polarization is electrically quite different from spontaneous polarization. Based on the dielectric properties of the poled polymer, SHG and electro-optic Pockel's effects are also affected by its elastic behavior, because these NLO parameters increase rapidly near the glass transition temperature (T_g) with an increase in the electrostriction constant and a decrease in the dynamic modulus. Therefore, it is important to study the effect of the poling field and temperature on the SHG and electro-optic properties of polymers and to optimize them because they depend strongly on poling conditions.

Several chemical strategies are available to prepare NLO polymers via poling technique. Take as an example where NLO chromophores are covalently attached to a polymer backbone. In a cast film, the NLO chromophores are randomly oriented. To generate, second-harmonic generation and electro-optical effects, the cast film is electrically poled at elevated temperatures under an external field that results in preferential orientation to the NLO chromophores. The electrical poling can be performed nearly at or above the glass-transition temperature by strong electric field via electrodes or corona poling techniques. Ordering parameter of $f = 0.30$ is generally obtained, where f ranges from 0 to 1 corresponding to the degree of orientation. The random alignment of dipoles and preferential alignment induced by electrical poling can be illustrated simply as follows:

Figure 20(b) shows the schematic representation of corona poling generally used for generating SHG activity in polymers. The poling process is as follows: The sample with electric field E_p is heated to temperature T_p, and the temperature is kept constant. The sample then is cooled to room temperature after poling for a specified time t_p and the electric field is switched off. This thermoelectric treatment generates an electret which is the electrostatic analog of a magnet. More details on poling techniques can be found in reference 318. The corona-poling technique is advantageous because of the simplicity of the setup and fast charging process. In corona poling, an electric field of a few kilovolts against the transparent thin film is applied through a sharp needle electrode. The charge distribution in corona poling is not as uniform as that of the parallel plate poling method. However, the corona poling can apply the strong electric field to the film compared with contact poling. The essential components of the poling experiment are a high-voltage power supply and a heating unit for variable temperature. The specimens used in a guest–host system must be transparent thin films. Good electrical contacts between the specimen and electrodes must be ensured during the poling experiment. Transparent thin films of the composite are deposited onto indium tin oxide (ITO) electrode-coated glass slides by spin coating or by the solvent evaporation technique. The corona discharge is a partial breakdown of air at atmospheric pressure. The corona discharge is induced by the inhomogeneous electric field around the needle or vary thin wire electrode. The corona poling can apply the relatively high voltage compared with contact poling. By selecting the thin wire, one can reduce on-set voltage for corona discharge. The corona ions have very low lateral conductivity, only charges can leak through the defect site. Localizing the effects helps the films avoid a breakdown, a common cause of failure during contact poling. It is better to keep the poling current within 2 μA, otherwise the sample will be degradated by electrical breakdown. Therefore, an electret possesses persistent polarization due to the orientation of the dipoles in the presence of an electric field. Thus the orientation of dipoles generates SHG activity in polymers after poling. In fact, poling induces molecular alignment in order to increase molecular interactions. The electret formation technique has been widely used for generating piezo-, pyro-, and ferro-electric properties in a broad range of polar polymers such as poly(vinylidene fluoride) and its copolymers, polyamides, etc. Based on the same fundamentals, this electret formation method has been employed for generating electro-optical and SHG effects.

The formation of polymer thermoelectrets by using electrical field and temperature is a conventional method. Besides poling at elevated temperatures, organic polymers can be poled by other physical methods which are not common. The details of various types of polymeric electrets can be found in reference 318. Polymer electrets prepared by various treatments are as follows:

1. Thermoelectrets (electric field and temperature): useful for NLO guest–host systems, NLO-dye attached side-chain and main-chain polymers.
2. Photoelectrets (electric field and radiation): useful for forming a cross-linked network of NLO polymers.
3. Thermophotoelectrets (electric field, radiation, and temperature): useful for forming a cross-linked network of NLO polymers by photocross-linking as well as by thermal curing.
4. Autoelectrets (radiation only): useful for forming a photocross-linked network of NLO polymers.

5. Radioelectrets (α-, β-, and γ-radiation with and without electric field and temperature): useful for forming a network of polymer from acetylenic type NLO dyes.

6. Electroelectrets (electric field only): useful for aligning NLO liquid crystal polymers.

7. Magnetoelectrets (magnetic field and temperature): specially useful for aligning NLO liquid crystal polymers.

The poling technique using temperature and electric field is currently the most commonly used to achieve a preferred polar order of NLO chromophores, either dispersed in a polymer matrix or covalently attached to the polymer backbone. If we look at the decades-old history of electrets, other poling techniques are of considerable interest depending on the chemical species of a NLO polymer. These various electret-forming techniques are not only useful for the dipolar alignment desired to induce an acentric space group into guest–host systems and NLO-dye attached polymers, but could be of great importance for long-term stability of SHG activity locked through photocross-linking or thermal curing processes which essentially need more study. The first step in this direction was taken by Sekkat and Dumont,[405] who applied the photo-induced poling technique or, more appropriately, the technique of photoelectret formation in which there is no need to heat the sample because poling is performed at room temperature simultaneously using irradiation at a particular wavelength and electric field. The poling process here uses the *trans-cis-trans* isomerization cycle in which irradiation with a suitable light source within the absorption band of the *trans*-isomer causes an increase in the population of *cis*-isomer. The *cis*-isomer rotates through thermal excitation and finally decay back to the *trans*-configuration. Therefore orientational order induced by this technique is quite different than the thermoelectret-forming technique. Photo-induced poling offers another advantage of localized poling because poling can be done precisely within a localized region of the polymer film by selectivity irradiating at the specified regions through a mask; therefore, it helps in producing poled as well as unpoled patterns.[406] This technique is promising for fabricating waveguides or other photonic devices using conventional lithographic techniques. A comparative study of the SHG activity induced by photo-induced poling and thermal poling of azo-dye doped poly(methyl methacrylate) films showed that, although the levels of polar order obtained by these two techniques were similar, stability of the polar order was reduced by up to 20 times using the photo-induced technique in terms of the decay time constant, which also depends on the poling temperature.[406]

Dao and Williams[407] reported the effect of gas on corona poling. To prepare the selective poled layers for frequency doubler, electron beam poling was demonstrated by Yang et al.[408] The electron beam can pole the lower part of films, which is useful to increase the overlap integral for 0-1 mode conversion without introducing the another waveguide layer. The DR1 doped PMMA was used in this experiment. Electron beam poling of polymer films was performed using a electron-scanning microscope under high vacuum with a beam energy of 10 keV and a current density of approximately 0.4 mA/m^2. The sample was heated to 90 °C and the electron beam was illuminated. During the irradiation, negative space charge is built up in the films. Although this technique can not give large nonlinearity to the films, the desired polarization profile was formed which was confirmed by thermally stimulated depolarization, laser-induced pressure-pulse experiment, and thermal wave-response measurements.

To obtain the large nonlinear coefficients for TE configuration (polarized in the plane), the effective in-plane poling technique was developed by Otomo et al.[409] In general, charge injection causes the electrical breakdown that prevents efficient in-plane poling. The arcing which induced the electrical breakdown occurred in the air-film interface. Therefore the cladding layer with a lower refractive index than the NLO active layer and a similar T_g was added to the NLO polymers. As a result of the introduction of the buffer layer, the poling voltage was increased up to 370 V/μm and stable as well as reproducible poling was performed at the fields in excess of 300 V/μm. These results suggest that the high resistivity and a thick cover layer are essentially required for the in-plane poling.

Corona poling sometimes induces the surface roughness which increases the optical loss. To improve the optical loss during the poling, protection layer was coated on the NLO polymers by Hill and Knoesen.[410] The requirements for the protective layer are good adhesion with high conductivity during the poling and the ability to be removed from the films without damaging the NLO polymers. The material which satisfies the above condition is poly(acrylic acid) (PAA). This polymer can eliminate the surface roughness damage which is induced by corona poling and pinholes.

In this section, we mention the relationship between $\chi^{(2)}$ and the first hyperpolarizability. For a guest molecule dissolved in polymeric materials, details of molecular shape and guest–host interactions determine the order parameters of NLO chromophore. Of course, the guest molecules may increase the free volume of host polymers and guest–guest interactions may prevent or induce the orientation of

NLO chromophore. Let us consider the rodlike guest molecule, such as 4-dimethylamino-4'-nitrostilbene (DANS); we expect it to associate with host mesogenic units or other guests. A mean-field description $U(\theta)$ determines the orientational distribution function, $P(\theta)$, of guest molecules in a nematic domain. $P(\theta)$ is the relative probability that a guest molecule will lie with its X axis oriented at angle θ to the polar director (or laboratory reference axis). We assumed that P is independent of ϕ in spherical polar coordinates. Also it is assumed that the "long axis", X, which is also herein assumed to be a good approximation to the direction of the permanent dipole moment μ, is parallel to the charge-transfer direction of molecules such as β_{xxx}. The normalized distribution function, presented by[411]

$$P'(\theta) = P(\theta)/\sin\theta \tag{37}$$

is constant, i.e., there is no preferred orientational direction. On the other hand, the nematic potential causes a distribution that peaks at $\theta = 0$ and $\theta = \pi$ and which is symmetric about $\theta = \pi/2$. The $P'(\theta)$ are determined by relative values of $\exp\{-U(\theta)/kT\}$. The nature of these distributions is easy to confirm using the dichroism. The order parameter is given by:

$$S = \langle 3\cos^2\theta - 1)/2 \rangle \tag{38}$$

where S is zero in the isotropic medium. A poling field in the isotropic case, even considering the very large electric field and large dipole moments μ in DANS (7.4 debye), can induce a small variation of $P'(\theta)$ because the μE electrostatic energy is small compared with kT at or above 300 °K. A similar effect occurs in the liquid crystal when

$$U(\theta)' = U(\theta) - \mu \cdot E \tag{39}$$

replaces $U(\theta)$ in determining $P'(\theta)$. The net polar alignment $\langle\cos\theta\rangle$ can be calculated by

$$\langle\cos\theta\rangle = \int_0^\pi \cos\theta \exp\{-U'(\theta)/kT\}\sin\theta \, d\theta \Big/ \int_0^\pi \exp\{-U'(\theta)/kT\}\sin\theta \, d\theta \tag{40}$$

The order of net polar alignment can be enhanced in the liquid crystal medium over the isotropic case. The limiting cases are isotropic distributions and the Ising model (in which only $\theta = 0$ and π and 2π are allowed orientations). The order of dipolar alignment is:

$$\langle\cos\theta\rangle_{\text{isotropic}} = \mu E/3kT \tag{41}$$

$$\langle\cos\theta\rangle_{\text{Ising}} = \mu E/kT \tag{42}$$

In contact poling we can not apply a high electric field. Therefore orientational order is restricted to $\mu E/kT \ll 1$. Thus

$$\langle\cos\theta\rangle = (\mu E/kT)\langle\cos^2\theta\rangle_{E=0} \tag{43}$$

We are interested in $\chi^{(2)}$ tensors in a poled polymer system. These may be calculated by considering the average molecular distribution as $\langle Z\rangle = \langle\cos\theta\rangle$. The poled polymers belong to $C_{\infty v}$ symmetry, the resultant $\chi^{(2)}$ tensor may be shown to have the following form in contracted index notation.

$$\begin{pmatrix} 0 & 0 & 0 & 0 & d_{15} & 0 \\ 0 & 0 & 0 & d_{24} & 0 & 0 \\ d_{31} & d_{32} & d_{33} & 0 & 0 & 0 \end{pmatrix} \tag{44}$$

If there is no absorption of materials, there are only two independent components for a poled polymer system in accordance with the Kleinman symmetry. By evaluating the NLO coefficients, one can also roughly estimate electro-optic coefficients. Through the oriented gas model, the NLO and electro-optic coefficients can be compared fairly well with β values. Poled polymeric systems have the following electro-optic tensor elements.

$$
\begin{bmatrix}
0 & 0 & r_{13} \\
0 & 0 & r_{23} \\
0 & 0 & r_{33} \\
0 & r_{42} & 0 \\
r_{51} & 0 & 0 \\
0 & 0 & 0
\end{bmatrix}
\tag{45}
$$

Due to symmetry, $r_{13} = r_{23}$ and $r_{42} = r_{51}$. After applying Kleinman symmetry, $r_{13} = r_{23} = r_{42} = r_{51}$ which leaves two independent components.

The purely electronic contribution of $\chi^{(2)}$ can be related to the purely electronic β tensors of the molecules by the oriented gas description in which their cancellation in the absence of electric field is negated by poling. Because $\mu\beta$ is very large in DANS, for description of high concentration samples, it is justifiable to neglect the contribution of the host liquid crystalline polymer to $\chi^{(2)}$. Also because β_{xxx} is the only significant element of β in DANS, we only need to know the distribution of molecular X axes to approximately describe $\chi^{(2)}$. For this purpose, because only the two sets $\chi^{(2)}_{zzz}$ and $\chi^{(2)}_{zxx}$ are nonvanishing as shown above, one requires knowledge only of

$$
\langle (K \cdot i)(K \cdot i)(K \cdot i) \rangle = \langle \cos^3 \theta \rangle
\tag{46}
$$

and

$$
\langle (K \cdot i)(I \cdot i)(I \cdot i) \rangle = \langle (\cos \theta)(\sin^2 \theta)(\cos^2 \phi) \rangle
\tag{47}
$$

where i, j, k (I, J, K) are the molecular (laboratory) cartesian unit vectors. Remembering the independence of a nematic system from the spherical polar coordinate ϕ,

$$
\langle (\cos \theta)(\sin^2 \theta)(\cos^2 \phi) \rangle = \int (\cos \theta)(\sin^2 \theta)P(\theta)d\theta \int (\cos^2 \phi)d\phi
$$

$$
= \int (\cos \theta - \cos^3 \theta)P(\theta)d\theta/2
\tag{48}
$$

$$
= (\langle \cos \theta \rangle - \langle \cos^3 \theta \rangle)/2
$$

For isotropic media perturbed by the $-\mu E$ electrostatic energy at high temperature, or nonsaturated alignment ($\mu E/kT \ll 1$) limit, values are calculated to be

$$
\langle \cos \theta \rangle_{\text{isotropic}} = \mu E/3kT
\tag{49}
$$

$$
\langle \cos^3 \theta \rangle_{\text{isotropic}} = \mu E/5kT
\tag{50}
$$

$$
\langle (\cos \theta)(\sin^2 \theta)(\cos^2 \phi) \rangle_{\text{isotropic}} = \mu E/15kT
\tag{51}
$$

The associated poled frozen polymer susceptibilities are given as:

$$
\chi^{(2)}_{zzz\text{isotropic}} = F\beta_{xxx}\mu E/5kT
\tag{52}
$$

$$
\chi^{(2)}_{zxx\text{isotropic}} = \chi^{(2)}_{zzz\text{isotropic}}/3
\tag{53}
$$

where F is an unspecified local field correction factor.

In the Ising model similar results are

$$\langle \cos \theta \rangle_{Ising} = \mu E / kT \tag{54}$$

$$\langle \cos^3 \theta \rangle_{Ising} = \mu E / kT \tag{55}$$

with the predictions

$$\chi^{(2)}_{zzz_{Ising}} = F\beta_{xxx}\mu E / kT \tag{56}$$

$$\chi^{(2)}_{zxx_{Ising}} = 0 \tag{57}$$

Certainly for real samples where the distribution function is intermediate between the isotropic and Ising limits, the susceptibilities also lie between the isotropic and Ising models. The macroscopic nonlinearity can be enhanced by up to a factor of five over the maximum achievable in isotropic media by use of liquid crystal host. High electric field poling such as $\mu E / kT > 1$ can be performed using corona poling technique. The dependence of $\langle \cos \theta \rangle$ and $\langle \cos^3 \theta \rangle$ on $\mu E / kT$ can be plotted.[412] Similarly the dependence of $\langle \cos^3 \theta \rangle$ and $(\langle \cos \theta \rangle - \langle \cos^3 \theta \rangle)/2$ on $\mu E / kT$ can also be shown. Only the Ising model can attain the complete alignment of the NLO chromophore at high electric field.

2. Guest–Host Systems

A guest–host system is generally prepared by mixing a NLO dye possessing large molecular hyperpolarizability β into an amorphous polymer matrix. The simplest example is the dispersion of the centrosymmetric NLO chromophores such as p-NA into poly(methyl methacrylate) PMMA matrix. A variety of guest–host systems can be prepared simply by mixing two components together. It is important to point out here that, although many organic polymers can be used as a host matrix, it is necessary to select an amorphous polymer of the high optical transparency. A NLO dye is dissolved into the polymer host using a common solvent, and thin films onto a substrate are prepared either by casting, spin-coating, or dipping. The guest–host thin films should be dried under a vacuum for evaporating the remaining solvent, otherwise it may cause severe problems such as blistering or air-voids. To generate second-order NLO activity, the guest–host system is poled under an external electric field at elevated temperatures which induces a degree of orientational order. The details of such poling technique have already been described. The advantages and disadvantages of a guest–host system can be summarized as follows:

A. Advantages
1. Advantages of using a wide variety of centrosymmetric NLO chromophores for SHG activity
2. Ease of processing into thin film by coating, dipping, and lithography
3. Ease of fabrication onto a very large area of substrate of any dimensions
4. High mechanical strength
5. Inexpensive mass production of NLO materials
6. Unlimited selection of desired NLO guest and polymer host components
7. Wide range of operating frequencies
8. Low dielectric constant

B. Disadvantages
1. Inherits the problems of the decay of the NLO activity due to orientational relaxation
2. Low NLO activity due to the limited solubility of NLO dye in the polymer matrix and dilution of NLO activity from inactive host polymer (The segregation between guest and host molecules occurs at higher concentrations.)
3. Inhomogeneity leads to scattering losses
4. Sublimation of NLO dye at elevated temperatures during the fabrication of a device
5. Thermal and environmental instability.

The earliest work on SHG-active guest–host systems was reported by Meredith et al.[413] in 1982 by doping a side-chain liquid crystal polymer with DANS. This work stimulated studies on guest–host systems and other reports were on SHG properties of Dispersed Red 1 (DR1) using as a guest NLO dye into PMMA host.[414] The second-order optical nonlinearity in NLO dye incorporated polymers decay as a result of the orientational relaxation, because such a guest–host system does not have a state of thermodynamic equilibrium. Molecular relaxation occurs gradually after the electrical field is removed leading to a decrease in SHG activity. Jeng et al.[415] proposed a new guest–host system by incorporating the DR1 into a phenyl siloxane polymer (PSP) and then forming a cross-linked network at higher temperatures. The SHG activity of the polymer cured at 200 °C was found to be stable at room temperature, although decayed to 60% of its original value after being maintained at 100 °C for about 30 h. Viewing various guest–host systems, it has become clear that the polymer host plays a very important role not only in inducing an acentric space group but also in stabilizing the SHG activity for an extended time period at room temperature. The time dependence of SHG intensity for p-NA guest molecules in two different polymer hosts indicated that the stability of second-order optical nonlinearity also depends on the host polymer matrix. Interestingly highly efficient SHG activity is observed in both p-NA/poly(ϵ-caprolactone) PCL and p-NA/poly(oxyethylene) POE systems, but the p-NA/PCL system shows no decay of SHG over a period of 1000 hours.[416] The p-NA/POE system exhibits SHG activity the same as that of MNA, whereas p-NA/PCL shows SHG activity 3.3 times that of MNA. One of the most interesting features of the p-NA PCL system is the spontaneous alignment of p-NA guest molecules in PCl matrix during crystallization without applying an electric field.

Hirota et al.[429] reported the SHG properties of polycarbonate (PC) and PMMA doped with novel disazo dyes which have a perfluoroalkylsulfonyl groups. The d_{33} as large as 67 pm/V was measured for polycarbonate doped with 4 mol% of 4-(4-(4-perfluorobutylsulfonylphenylazo)naphthylazo)-N-ethyl-N-hydroxyethylaniline and this dye retains more than 50% SHG activity after 1000 h at 50 °C and 35% after 500 h at 80 °C. The decay of d_{33} of polycarbonate-doped disazo dyes at 50 °C was found to be slow due to the molecular arrangement achieved at higher temperatures. Interestingly, the decay of polycarbonate-doped samples was less than that of PMMA-doped films and dyes with perfluoroalkylsulfonyl group showed larger d_{33} values than a nitro group.

R	H	NO₂	C₄F₉SO₂	NO₂	C₄F₉SO₂
d_{33} (pm/V)					
in PMMA)	31	31	33	29	34
in PC)	35	50	67	23	43

The initial NLO coefficients of (2-(4-(4-dimethylaminostyryl)phenyl methylene) propanedinitrile), (3-dicyanomethylene-5,5-dimethyl-1-(p-dimethylamino styryl)cyclohexene), and (3-dicyanomethylene-5,5-dimethyl-((5-dimethylamino-2-thienyl)vinyl cyclohexene), which were abbreviated as DPMP, DDDSC, and DDDTC, respectively, in PMMA matrices were reported by Gadret et al.[430] DPMP was synthesized according to the Singer's method. DDDSC was prepared by one-pot reaction of malononitrile, isophorone, and 4-dimethylaminobenzaldehyde and purified by column chromatography. DDDTC was synthesized in a way similar to the DDDSC molecule.

Thin films were prepared by spinning of doped polymer solutions on a glass substrate. The sample was poled according to the corona poling technique and the atmosphere of chamber was controlled to allow corona discharge. The order parameter of these guest–host systems was investigated and compared with poling field.

The linear optical and SHG properties of new guest–host poled polymeric thin films are reported based on combined ellipsometric and modulated reflection measurement by Levenson et al.[431] PMMA were doped with two different molecules which exhibit a large molecular hyperpolarizability β as measured by the EFISH technique.

Thin films of doped polymers have been deposited by spin-coating technique on different substrates. The blue molecules S and TP stand out in view of their high nonlinear coefficient as well as their low relaxation rate.

Second-order NLO properties of a series of alkyl derivatives of 2,4-dinitrobenzene derivatives in LB films and guest-host systems have been investigated.[432] These chromophores have interactions in *ortho* as well as in *para* directions. The SHG of a few guest–host systems prepared from PMMA matrix are listed in Table 62. The NLO coefficients of these systems are rather low, the largest being approxi-

Table 62 Refractive Indices and NLO Coefficients of Dyes Dispersed in the PMMA Matrix[432]

X	Y	Weight % in PMMA	Refractive Indexes		NLO Coefficients (pm/V)
			0.523 μm	1.064 μm	
Cl	NH-(CH$_2$)$_9$-CH$_3$	16.62	1.53	1.484	$d_{31} = 0.88$ $d_{33} = 2.8$
F	NH-(CH$_2$)$_{12}$-CH$_3$	16.32	1.507	1.497	$d_{31} = 0.70$ $d_{33} = 2.64$
F	NH-(CH$_2$)$_{15}$-CH$_3$	15.61	1.51	1.49	$d_{31} = 0.84$ $d_{33} = 1.8$
X = Y = NH-(CH$_2$)$_2$-CH$_3$		16.66	1.498	1.488	$d_{31} = 0.31$ $d_{33} = 2.4$
X = Y = NH-(CH$_2$)$_5$-CH$_3$		16.62	1.513	1.482	$d_{31} = 1.12$ $d_{33} = 3.78$
X = Y = NH-(CH$_2$)$_7$-CH$_3$		16.6	1.50	1.48	$d_{31} = 0.28$ $d_{33} = 1.45$

mately 3.78 pm/V at 1.064 μm. The d_{33}/d_{31} ratio is 3 for poled polymeric systems, but here it is quite different. The NLO coefficient of these systems depends on substituents.

Mortazavi et al.[403] reported the SHG in PMMA and azo dye, 4-[N-ethyl-N-(2-hydroxyethyl)] amino-4-nitroazobenzene (i.e., Disperse Red I or DR1). The polymer–dye films exhibit a change in color from light orange to red-violet during the corona poling close to the T_g. This effect is a reversible process because after completion of the poling process, the red-violet color changes to the light orange of the unpoled film. This unique characteristic results from the induced dipole moments and dipole orientation occurring during corona poling. The electronic absorption studies performed on corona-poled and unpoled samples indicate that no chemical changes occurred during the poling process. The d_{33} coefficient of PMMA-DR1 films poled by parallel-plate poling was 2.5 pm/V at 1.58 μm, whereas that of corona-poled film is 9.6 pm/V at the same fundamental wavelength. It suggests that the corona poling at elevated temperatures is a more effective in inducing the internal electric field than parallel-plate poling. The SHG properties of DR1/PMMA system stabilized after 10 days and remained constant up to about 8 months. The composite showed an optical damage threshold of 60 MV/cm.

Meredith et al.[413] reported the thermotropic nematic liquid crystalline copolymer doped with a 2% DANS.

DANS has an extremely large β value of 4.5 × 10^{-28} esu. A preferential dipolar alignment of the DANS in a host polymer was achieved by an electric field. A strong SHG signal was observed when the films were poled under an electric field of 1.3 V/μm at 293 °C for 6 h, which led to the maximum SHG. Poling of the molecularly doped thermotropic liquid crystalline polymer films below T_g yielded d_{11} of 0.63 pm/V, significantly larger (100-fold) than that of a 2% DANS doped PMMA sample. The polar alignment of the guest–host system persisted for a long time, but the SHG signal began to decrease if the laser power was increased due to the local heating. The time profile of the SHG intensity under various applied voltages was also studied. Upon zeroing of the applied voltage, SHG intensity decayed rapidly, but reapplication of the voltage again generated SHG and exceeded the initial value by 30% after only 6 min. Continued cycling lead to a further increase to a level 50% higher compared with the initial value. The increase in SHG intensity on repeated cycling probably restrains local cage barriers which hinders the rotation of DANS molecules.

Copolymers of vinylidene fluoride (VDF) and trifluoroethylene (TrFE) have also been used as a host. Hill et al.[427a] dissolved a NLO guest in the ferroelectric copolymer and the composite films containing up to 10% by weight of the guest molecule were obtained from an acetone solution. The copolymer (Forafon-7030) has a 7:3 molar ratio of VDF and TrFE, and acetone-casted thin films were transparent. The host was a copolymer P(VDF/TrFE) in the molar ratios of 7:3 and the guest was a NLO chromophore 4-(4'-cyanophenylazo)-N,N-bis-(methoxycarbonylmethyl)-aniline (2). This system showed a d_{33} coefficient of 2.6 pm/V at 1.064 μm. Pentalis et al.[427b] used 2% of chromophore (1) in the P(VDF-TrFE) matrix for frequency doubling. The SHG efficiency of the sample thermally annealed before poling was 1.4 times larger than non-annealed sample. The substitution of 5% of chromophore (2) by 5% of 4-amino-N,N-bis-(methylcarbonylmethyl)-4'-nitrostilbene (3) increased SHG efficiency by an average ratio of 1.3 as a result of the large dipole moment and β value of the stilbene derivative than that of aminoazo compound. The d_{33} of 2.55 pm/V was determined for a composite having 10% of the aminoazo compound. On applying an electric field, the carbon–fluorine groups of the host and the guest molecules tend to orient with the field. The NLO coefficient, as a function of chromophore concentration, became zero at 0.5 wt% due to the difference of sign of β between host polymer and

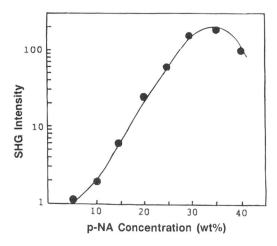

Figure 21 The variation of SHG intensity as a function of *p*-NA concentration in a POE system (Watanabe, T., et al., *SPIE Proc.*, 1147, 101, 1989. With permission.)

NLO dye. The NLO properties were mainly governed by the guest concentration. The poled composite had the bulk second-harmonic coefficient d_{33} as high as 2.6 pm/V at 1.064 μm. The waveguide structures constructed from the composite exhibited an optical loss of 5 ± 0.5 dB/cm at 633 nm. The bulk losses at the longer wavelengths (1.3 and 1.55 μm) were found to be significantly less (1.5 dB/cm) useful for telecommunications.

In the case of the poled polymer system, the molecular orientation induced by poling is poor compared with good single crystals such as MNA, because the electric field cannot overcome the thermal disorientation effect caused by the electrical breakdown of the host polymer film. To obtain the highly aligned chromophores in a host–guest system, the electric field-induced crystallization process was developed. In general, *p*-NA crystallizes into a centrosymmetric space group and hence does not exhibit SHG. It is possible to obtain the highly oriented *p*-NA crystals showing SHG activity by using the electric field-induced crystallization process. The first results of this approach for a *p*-NA and poly(oxyethylene) (POE) system was reported by Watanabe et al.[433] Both *p*-NA and POE are soluble in acetonitrile and benzene, and *p*-NA is miscible in POE. The *p*-NA and POE in molar compositions of 1:4, 1:8, and 1:16 were prepared. The 1:8 system exhibits SHG intensity 20 to 30 times that of urea; it decreases sharply for a 1:16 system. The d_{33} coefficient was $51 \times d_{11}$ (SiO_2) and the ratio of d_{33}/d_{31} was 7.5 which is larger than the ordinary poled polymer system ($d_{33}/d_{31} = 3$). This result suggests that the molecular orientation of *p*-NA induced under an electric field is excellent compared with the ordinary poled polymer system. Figure 21 shows the SHG intensity as a function of *p*-NA concentration. The maximum SHG was observed at 35 wt% of *p*-NA and decreased above this concentration because of the formation of SHG-inactive *p*-NA.

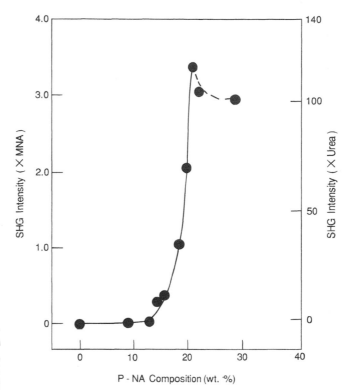

Figure 22 The variation of SHG intensity as a function of *p*-NA concentration in a PCL system. (From Miyazaki, T., Watanabe, T., and Miyata, S., *Jpn. J. Appl. Phys.*, 27, L1724, 1988. With permission.)

A more interesting *p*-NA/polymer system consists of poly (ε-carprolactone) (PCL). When *p*-NA is mixed with PCL, a large SHG was observed even without the poling process.[416] In a PCL matrix, *p*-NA molecules spontaneously crystallize into a noncentrosymmetric crystal structure. The variation of SHG intensity as a function of *p*-NA concentration in a PCL system is shown in Figure 22. A 21.6 wt% *p*-NA concentration showed SHG intensity as high as 115 times that of urea or 3.3 times that of MNA crystals. The PCL/*p*-NA system showed much higher stability of the SHG intensity compared with POE/*p*-NA system (Fig. 23). The PCL/*p*-NA system exhibited no decay over a period of 1000 h.[434] It means that the new acentric crystals that formed spontaneously

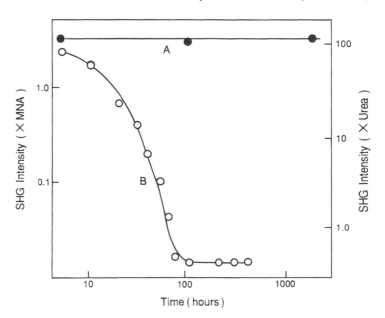

Figure 23 The temporal stability of SHG activity of POE/*p*-NA and PCL/*p*-NA systems. (From Miyazaki, T., Watanabe, T., and Miyata, S., *Jpn. J. Appl. Phys.*, 27, L1724, 1988. With permission.)

Table 63 SHG Properties of Polyester/*p*-NA Composite Systems

Polymer	*n*	*m*	CH$_2$/COO	*p*-NA wt%	SHG (× urea)
Poly(elhylene succinate)	2	2	2	17.2	65
Poly(ethylene adipate)	2	4	3	16.1	39
Poly(butylene adipate)	4	4	4	46.9	117
Poly(hexamethylene adipate)	6	4	5	22.3	77
Poly(ϵ caprolactone)	0	5	5	17.3	115
Poly(butylene sebacate)	4	8	6	27.4	129
Poly(hexamethylene subarate)	6	6	6	28.6	125
Poly(hexamethylene sebacate)	6	8	7	32.6	7
Poly(decamethylene sebacate)	10	8	9	28.2	Very weak
Poly(ethylene)	∞	∞	∞	–	0

From Nakanishi, H., et al., *SPIE Proc.*, 1147, 84, 1989.

in the PCL/*p*-NA system were more stable than those formed by electrical poling in the POE/*p*-NA system. The decomposition of the acentric crystal in POE/*p*-NA system occurred over time. The X-ray diffraction studies indicate that the SHG activity of POE/*p*-NA and PCL/*p*-NA systems are not related to the similar orientation of *p*-NA molecules occurring during electric poling, but instead appear as a result of the formation of an entirely new acentric crystal structure.[435] In the POE/*p*-NA system, acentric crystals are formed with the influence of the applied electric field, whereas in the PCL/*p*-NA system, acentric crystals are formed spontaneously due to the interaction between the PCL and *p*-NA molecules in the absence of an electric field.

The quadratic NLO properties of aliphatic polyesters and *p*-NA systems have also been investigated.[436] Both polyesters as well as *p*-NA belong to the centrosymmetric space groups; hence, they are not SHG active individually. When *p*-NA is mixed with polyesters, the composites attain a noncentrosymmetric crystal structure. The SHG intensity of these composites depends both on the *p*-NA concentration and on the ratio of methylene (CH$_2$) and ester groups, which is related to an interaction strength between a host polymer and *p*-NA. A poly(butylene sebacate) system showed the largest SHG activity about 130 times that of urea standard. The SHG intensities of various polyester/*p*-NA systems are listed in Table 63. These SHG-active systems acquire a new acentric phase as evidenced by the differential scanning calorimetry technique. The molecular interactions occur through the formation of the hydrogen bonding between the hydrogen of the amino group in *p*-NA and the oxygen of the carbonyl group in polyester. To generate a new acentric crystal, no electric field is required in the polyester/*p*-NA system, similar to the PCL/*p*-NA system. These studies demonstrate that the processible SHG-active composites can be prepared simply by mixing the two components in an appropriate ratio. The similar type SHG activity of *p*-NA and poly(γ-benzyl-L-glutamate) (PBLG) was reported by Kamino et al.[437] The SHG-active *p*-NA was formed by quenching the polymers to room temperature from its lyotropic liquid crystalline phase. The SHG activity of this system was 3.6 times larger than that of MNA powder.

Boyd et al.[438] reported the mobility of dyes in the polymer matrix. The rotational mobility of dye can be written as follows:

$$\rho = \frac{\langle \cos^3 \theta \rangle_T}{\langle \cos^3 \theta \rangle_{\text{free}}} \tag{58}$$

The $\langle \cos^3 \theta \rangle$ represents the third-order Langevin function. In a low electric field such as $\rho = \mu E/5kT$ < 1, ρ is independent of dipole moment of dye and electric field. The ρ can be determined by measuring the SHG intensity of the sample. Figure 24 shows the parameter ρ as a function of dye volume. The dyes were doped in both PMMA and polycarbonate matrix. The chemical structures of the polymer and dyes are shown below.

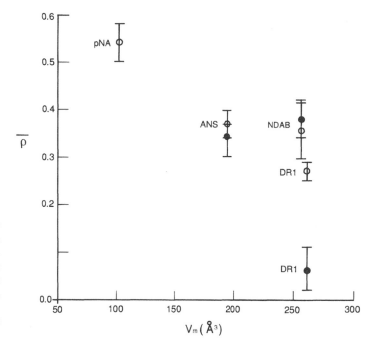

Figure 24 Measured mobility parameter ρ at room temperature for doped PMMA (open circles) and PC films (filled circles) as a function of dopant volume. (From Boyd, G. T., Francis, C. V., Trend, J. E., and Ender, D. A., *J. Opt. Soc. Am. B*, 8, 887, 1991. With permission.)

Rotational mobility decreased with increasing the dopant length in PMMA. By increasing the temperature, the rotational mobility increased and became 1 near the glass transition temperature. To constrain the decay of SHG activity, the large-volume dye was found to be a better dopant.

To prevent the decay of SHG, physical aging of guest–host systems was demonstrated by Lee et al.[439] The poly(4-vinyl pyridine-co-styrene) was used as a matrix and p-NMDA was used as a dopant. The film was first corona-poled about 5 °C above the T_g for 30 min, and cooled down to room temperature in the presence of electric field. In the subsequent aging process, the poled polymer was left at 5 °C below T_g. The aged film showed the slow decay compared with the unaged sample (Fig. 25). Physical aging can reduce the free volume of polymer or change the distribution of free volume and size. Therefore physical aging seems to be a good technique to improve the stability of poled polymers.

As suggested by Boyd et al.[438] the large dopant molecules seemed effective in preventing the decay of SHG activity. The large dopant molecules with high thermal stability were demonstrated by the IBM group.[440] The NLO dyes have the following structures and some other related dyes were also studied.

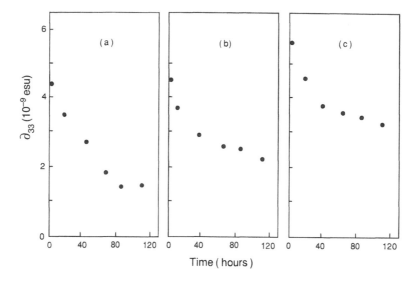

Figure 25 The decay of nonlinear optical coefficients for films containing 20 wt% *p*-NMDA. (a) 0 h, (b) 2 h, (c) 5 h aged sample. (From Lee, S. C., Kidoguchi, A., Watanabe, T., Yamamoto, H., Hosomi, T., and Miyata, S., *Polym. J.*, 23, 1209, 1991. With permission.)

L1 (A = NO$_2$, D = OCH$_3$)
L2 (A = NO$_2$, D = O-C$_6$H$_4$-OCH$_3$)
L3 (A = NO$_2$, D = NC$_5$H$_{10}$)
L4 (A = NO$_2$, D = NC$_6$H$_{12}$)

L5 (A = NO$_2$, D = OCH$_3$)
L6 (A = NO$_2$, D = NC$_5$H$_{10}$)

The polymer host used in this experiment was Ultem, an amorphous soluble polyetherimide (General Electric products). The pure Ulten had a T_g of 210°C, which decreased as the concentration of dye molecules increased in the guest–host systems. The T_g values of 9%, 16%, and 26% L1 in Ultem was 190, 180, and 150 °C while for 9% L2 in Ultem was 185 °C. The d_{33} of composites increased with dye concentration and reached 17 pm/V for 26% L1 in Ultem. Authors analyzed the temporal decay of SHG activities for guest–host systems. The temperature dependence of the SHG decay was found to be non-Arrhenius for these composites. The decay can generally be described by a stretched exponential (Kohlrausch-Williams-Watts) equation.

$$\frac{d_{\text{eff}}(t)}{d_{\text{eff}}(0)} = \exp(-t/\tau)^{\beta} \tag{59}$$

The width of decay time distribution characterized by β and peak of the distribution characterization τ. The decay of SHG activity could be strongly correlated with the glass transition temperature of the guest–host systems using an empirical relationship similar to the Williams-Landel-Ferry or Vogel-Tamann-Fulcher equation.

Ozaki et al.[441] reported high SHG efficiency in *p*-NA-doped amorphous polymers poly(methyl methacrylate) and polycarbonate composites without poling treatment. The *p*-NA/PMMA composites were prepared from anisol solutions, whereas the *p*-NA/Pc composites were obtained from methylene chloride solutions. For *p*-NA/PMMA composite, the SHG intensity varied strongly with dye concentrations, reaching largest at 25% of *p*-NA concentration, which is 40 times larger that of urea. The *p*-NA/PMMA composite showed spontaneous SHG at high concentrations indicating that the crystallinity in the matrix polymer may not be important in the noncentrosymmetric structure of the composite. The SHG activity did not decay and was retained for at least 2 months. The SHG efficiency also depends on the conditions of preparation of composite thin films. For example, *p*-NA single crystals obtained after evaporation of solvent at room temperature for 3 days were not SHG active. When a *p*-NA/PMMA composite incorporated with 31 mol% of *p*-NA was heated up to the T_g of 110 °C, the SHG intensity decreased at 60°C and completely disappeared above 80°C. The *p*-NA/Pc composite showed the similar behavior and maximum SHG intensity was 10 times that of urea. Thin films prepared from the *p*-NA/Pc composite were mechanically strong and stable.

Wang[442a] reported the efficient SHG from low-dimensional thiapyrylium (TPY) dye with PF_6 and BF_4 as the counterion that aggregates spontaneously in a polymer matrix. Thin films of TPY/polycarbonate on a glass substrate were prepared by spin coating of a methylene chloride/methanol solution. JAP/CALTECH group[442b] investigated EO properties of guest-host systems having DR1 and two new dyes (JC-Dye-1 and JC-Dye-2) in PMMA, polysufone, polycarbonate and cured epoxy resins as the hosts. We have summarized second-order NLO properties of guest-host systems in Table 64.

(JC-Dye1)   (JC-Dye 2)

The guest–host systems were the first organic materials in which SHG was induced by noncentrosymmetric ordering of NLO chromophores under electrical poling. A guest–host system almost virtually guarantees creation of the noncentrosymmetric crystal structures and a right combination of dye–polymer system could display sufficiently large SHG after electrical poling. One of the main problems encountered with these systems was the low concentration of NLO chromophores dissolved into a hot polymer matrix and the instability of SHG properties as a function of time. Several chemical strategies have been implemented to attach NLO dye molecules into a polymeric backbone. A wide variety of such examples dealing with the molecular designs will be provided in the next sections.

1. *NLO-chromophore functionalized side-chain polymers.* This class presents linear polymers with covalently attached NLO-active side groups. A NLO dye is covalently bound as a pendant group to the polymer backbone via a spacer.
2. *NLO chromophoric main-chain polymers.* In this class, the polymer main chain comprises NLO chromophores in a head-to-tail or tail-to-head configuration.
3. *Chromophoric main-chain polymers with NLO-active side chains.* The NLO chromophores are directly attached as a side chain to the chromophoric main-chain polymer backbone, allowing a much higher density of NLO moieties.
4. *Cross-linked NLO polymers.* Polymers either with side-chain NLO chromophores or chromophoric main-chain polymers are transformed into a cross-linked network polymer via cross-linking agents.
5. *Ferroelectric polymers.* A necessary condition for a piezoelectric polymer is to have a noncentrosymmetric structure, therefore, this class of polymers is inherently active to second-order NLO effects. Poly(vinylidene fluoride) PVDF, copolymers of PVDF, cyano polymers, ferroelectric liquid crystal polymers, odd-numbered nylons, and polyureas are examples of such polymers.

Table 64 Second-Order NLO Properties (d_{33} and r_{33} coefficients) of Guest–Host Systems

NLO dye	Host Polymer	Refractive Index at λ	NLO Coefficient (pm/V)	Ref.
DANS	LCP		$d_{33} = 0.63$ (1.06 μm)	413
	PMMA	–	$d_{33} = 0.84$ (1.06 μm)	417
DR1	PMMA	1.52 (0.6328 μm)	$d_{33} = 2.5$ (1.58 μm)	418
		1.52	$d_{33} = 8.4$ (1.06 μm)	419
			$d_{33} = 8.37$ (1.58 μm)	420
			$r_{33} = 0.8$ (820 nm)	442b
	Epoxy (E828)		$r_{33} = 0.184$ (820 nm)	442b
	Epoxy (Epofix)		$r_{33} = 0.091$ (820 nm)	442b
DCV	PMMA	1.58 (0.8 μm)	$d_{33} = 31$ (1.58 μm)	420
			$d_{31} = 11$	
			$d_{33} = 42$ (1.36 μm)	421
TCV	PMMA		$d_{33} = 84$ (1.36 μm)	421
p-NA	POE	1.621 (0.532 μm)	$d_{33} = 22$ (1.06 μm)	422
		1.535 (1.064 μm)		
p-NMDA	PVP	1.633 (0.532 μm)	$d_{33} = 2.4$ (1.06 μm)	423
		1.601 (1.064 μm)		
p-NMDA	PMMA	1.515 (0.532 μm)	$d_{33} = 1.5$ (1.06 μm)	424
		1.489 (1.064 μm)	$d_{31} = 0.7$	
TCSF	PMMA		$d_{33} = 15$	425
DPMP	PMMA		$d_{33} = 27$ (1.32 μm)	426
DDDSC	PMMA		$d_{33} = 26$ (1.32 μm)	426
DDDTC	PMMA		$d_{33} = 38$ (1.32 μm)	426
DR1	PSP	1.577 (0.532 μm)	$d_{33} = 1.54$ (1.06 μm)	415
		1.488 (1.064 μm)	$d_{31} = 0.47$ (1.06 μm)	
CPABMCA	P(VDF-TrFE)	1.647 (0.589 μm)	$d_{33} = 2.6$ (1.06 μm)	427
Calixarenes	PPMA		$d_{33} = 51$ (resonance)	428
9% L1	Ultem		$d_{33} = 6.0$ (1.06 μm)	440
			$d_{31} = 2.0$	
16% L1	Ultem		$d_{33} = 11$ (1.06 μm)	440
			$d_{31} = 3.2$	
26% L1	Ultem		$d_{33} = 17$ (1.06 μm)	440
			$d_{31} = 6.8$	
9% L2	Ultem		$d_{33} = 3.4$ (1.06 μm)	440
			$d_{31} = 1.2$	
10% L3	Ultem		$d_{33} = 7.0$ (1.06 μm)	440b
20% L3	Ultem		$d_{33} = 15.6$ (1.06 μm)	440b
			$d_{31} = 5.2$	
10% L4	Ultem		$d_{33} = 15.2$ (1.06 μm)	440b
			$d_{31} = 5.4$	
15% L5	Ultem		$d_{33} = 4.5$ (1.06 μm)	440b
15% L6	Ultem		$d_{33} = 11$ (1.06 μm)	440b
JC-Dye-1	PMMA		$r_{33} = 5.0$ (820 nm)	422b
	Epoxy (E828)		$r_{33} = 0.044$ (820 nm)	442b
JC-Dye-2	PMMA		$r_{33} = 6.0$ (820 nm)	442b
	Polysulfone		$r_{33} = 4.2$ (820 nm)	442b
	Polycarbonate		$r_{33} = 2.8$ (820 nm)	442b

3. NLO Chromophore-Functionalized Polymers

The NLO chromophore functionalized polymers were found to be more effective because they offer several advantages over the guest–host systems. The advantages of polymers containing NLO chromophores are as follows:

1. A very high concentration of the NLO chromophores can be introduced by covalently functionalized polymers. Tailoring of NLO properties and a whole material performance is possible via chemical modifications.

2. Poling-induced SHG activity of guest–host systems decreases over a prolonged period because of the orientational relaxation process, but polymers having covalently attached NLO chromophores show increased stability.
3. The NLO-dye dispersed in the host polymer matrix is distributed inhomogeneously causing significant scattering losses, whereas in NLO chromophore-functionalized polymer systems, a very high film homogeneity can be achieved because the NLO chromophores are covalently bonded, hence, absence of phase separation lessens the scattering losses. Fresnel reflection losses can be reduced because of the low indices easily matching to the index of glass fibers.
4. NLO chromophores can be processed in thin films onto a variety of substrates with large dimensions.
5. Quasi-phase matching is possible, which facilitates the design of high-frequency components.
6. Several techniques such as plasma etching, optically induced index changes, laser ablation, electrical poling, or photoresist formation can be applied.
7. Multilayers formation assists in easy integration with electronic and optical components.

These advantages provided by a NLO chromophore-functionalized polymer system make it more interesting than the guest–host system. In dye-functionalized polymers, the dye molecules are covalently bonded to the polymer backbone; hence, two chemical entities, which are joined together on a molecular level, offer greater stability than the guest–host polymer systems. The NLO dye molecules can be covalently attached to the linear chain of a polymer by the simple chemical processes. Although a wide variety of polymer backbone have been used for grafting NLO-active side groups, the most promising are the conventional organic polymers of poly(styrene) and poly(methyl methacrylate) because both polystyrene and PMMA backbones not only provide high optical transparency over a broad frequency range but also high glass transition temperature, low dielectric constant, and good processability. A variety of NLO chromophore-functionalized polymers discussed in this section are based on the PMMA backbone.

A very wide variety of polymers with NLO chromophores using different polymer backbones have been synthesized for second-order nonlinear optics. Ye et al.[443] prepared functionalized polystyrene with azo dye Disperse Red 1 (DR1). The functionalized systems include polystyrene derived from chloromethylated polystyrene (PS) and functionalized poly(p-hydroxystyrene) (PHS).

(PS)-CH$_2$-DR1

Using the polystyrene backbones containing 4.5% functionalized benzene rings with DR1, (PS)-CH$_2$-I was obtained; and using polystyrene backbones containing 12.5% functionalized benzene rings with pyridinium chromophores, (PS)-CH$_2$-II was prepared. The (PS)-CH$_2$-I and (PS)-CH$_2$-II exhibit d_{33} coefficients of 1.1 pm/V and 0.05 pm/V, respectively, at a poling field of 0 3 MV/cm. The d_{33} coefficient of (PS)-CH$_2$-I is linearly proportional to the applied poling field up to 0.50 MV/cm. The SHG efficiency of (PS)-CH$_2$-I does not decrease on storage for several days at ambient conditions, but heating to 115 °C leads to a remarkable decrease. For (PS)-CH$_2$-II, the d_{33} coefficient is only linearly dependent on the poling field up to 0.30 MV/cm, and these effects are associated with the ionic nature of the (PS)-CH$_2$-II films. The polystyrene backbone functionalized by NPP shows a d_{33} value of 1.3 pm/V at 0.30 MV/cm.

The d_{33} coefficient also increased as the level of functionalization increases and a similar situation exists with the poling field.[444–446] For example, the d_{33} value of a 48% chromophore-functionalized (PS)-O-NPP film increased to 7.6 pm/V at a poling field of 1.6 MV/cm. The d_{33} of (S)-NPP-PHS increased with the increase of functionalization level (% phenyl rings). The d_{33} values of (S)-NPP-PHS with 16, 36, 48, 62, 75, and 90% functionalization level were 13, 45, 58, 66, 72, and 80 $\times$ 10^{-9} esu, respectively. The d_{33} values of a 42 and 77% chromophore-functionalized (S)-NSP-PHS films were 50 and 160 $\times$ 10^{-9} esu at 4 MV/cm, respectively. The SHG activity shows good stability over a long time.

Eich et al.[447] covalently bonded the *p*-NA moieties to a polyethylene-type backbone. The yellow poly-*p*-nitroaniline (PPNA) polymer showed good solubility in DMSO and *N*-methylpyrrolidinone (NMP).

(PPNA)

PPNA-polymer has a glass transition temperature at 125 °C, and thermogravimetry analysis indicated a sharp weight loss process above 260 °C due to the thermal decomposition. The polymer showed absorption maxima at about 400 nm. Spin-coated thin films with a concentration of 30% by weight of the PPNA can be prepared from a NMP solution. The refractive indexes were 1.824 and 1.690 at 0.532 and 1.064 μm, respectively, and showed no noticeable anisotropy between in-plane and out-plane polarization. An appropriate working voltage was 20 kV at 140 °C; if the voltage exceeded 25 kV, the PPNA films were damaged. The SH activities leveled off at poling voltages above 15 kV for a 1.2 μm thick PPNA film. The poled PPNA films exhibited $d_{33} = 31$ pm/V and $d_{31} = 5$ pm/V. After 5 days, the d_{33} coefficient decreased from 31 to 19 pm/V (40% decrease) and stabilizes at this value. The estimated d_{33} value ranges from 21.8 to 50.8 pm/V and is in good agreement, compared with $d_{33} = 31$ pm/V obtained immediately after corona poling.

Shuto et al.[448] reported the noncolinearly phase-matched SHG in a thin film waveguide of 4-hexyloxy-4'-nitrostilbene (HNS) attached methyl methacrylate polymer (MMA). The random copolymer has refractive indexes n_1 and n_2 of 1.5537 and 1.6045 at 1.064 and 0.532 μm wavelengths. The corona-poled HNS-MMA film exhibits d_{33} coefficient 6.7 pm/V at 1.064 μm. A SHG efficiency of -5×10^{-5}% for fundamental peak power of about 130 W was estimated. The noncolinearly phase-matched SH power showed a very broad angular dependence, indicating that the interaction length for the noncolinear SHG was very short (0.1–1 mm) order compared to that of normal colinear SHG.

Amano and Kaino[449] reported the SHG of novel diazo-dye-attached polymers.

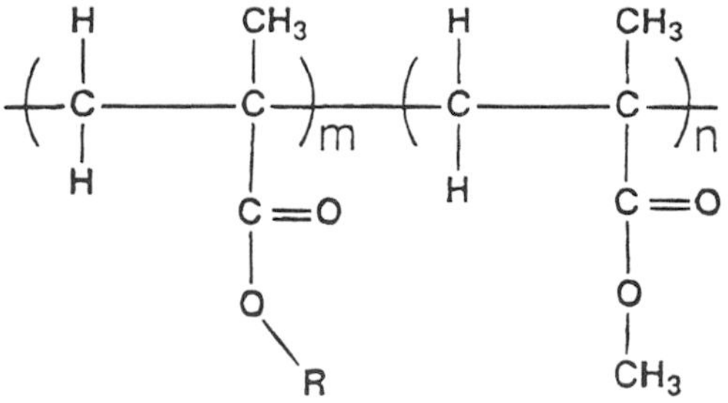

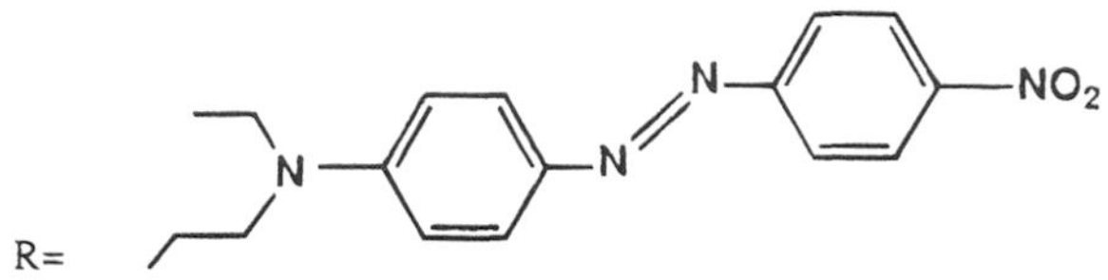

$$R=$$

(2R)

$$R=$$

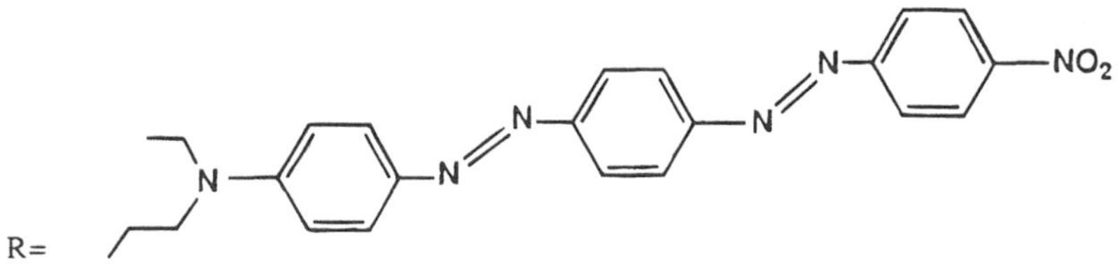

(3R)

The second-order NLO properties were measured at 1064 nm and 1500 to 1700 nm. The NLO coefficients of 2R polymer and 3R polymer were found to be 42 pm/V and 73 pm/V at 1064 nm, respectively. The off-resonant value of d_{33} of 2R polymer and 3R polymer were found to be 17 pm/V and 31 pm/V, respectively.

Amano et al.[450] reported the SHG in a homopolymer and copolymer of methyl methacrylate containing stilbene moiety as the side chain.

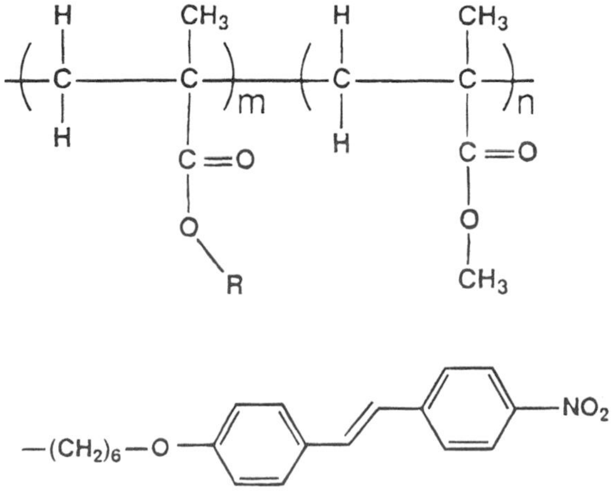

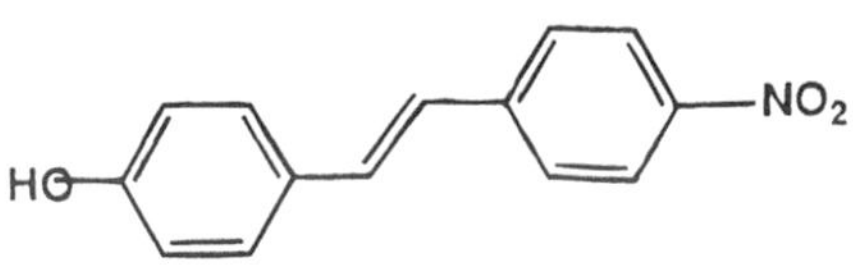

$$R = \quad -(CH_2)_6-O-$$

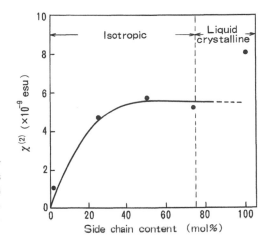

Figure 26 Variation of second-order NLO susceptibility as a function of side-chain content. The liquid phase is obtained above the side-chain content of 75 mol% which shows enhanced in optical nonlinearity. (From Amano, M., Kaino, T., Yamamoto, F., and Takeuchi, Y., *Mol. Cryst. Liq. Cryst.*, 182A, 81, 1990. With permission.)

The d_{33} values of the poled homopolymer and copolymer were 1.7 pm/V and 1.1 pm/V, respectively. The d_{33} values increased with the increase of stilbene moiety and leveled off around 25 mol%. Also, the d_{33} value increased proportionally with the strength of the poling field. An electric field of 15 kV/μm raised the d_{33} value as high as 2.7 pm/V. The dye-functionalized polymer showed a liquid crystalline phase when the stilbene moieties are higher than 75 mol%, and the d_{33} values are larger in anisotropic media. Figure 26 shows the $\chi^{(2)}$ values as a function of side-chain content. When the homopolymer is doped with 4-hydroxy-4′-nitrostilbene, the d_{33} value showed an increase up to a doping level of 20 mol%, and a d_{33} value 1.6 times larger than the undoped homopolymer is obtained.

The copolymers of MMA and azo dye-substituted methacrylate were prepared by Sohn et al.[451] A random copolymer containing dicyanovinyl azo dye, immediately after corona poling showed d_{31} and d_{33} values of 21.4 pm/V and 7.1 pm/V, respectively. The copolymer system shows an increased stability compared with that of corona-poled films of the guest–host system prepared from the same materials. The stilbene-functionalized PMMA was reported by the Hoechst Celanese research group.

$n = 6, 11, 12$ (HCC-1232)

The d_{33} value of a brand name material abbreviated as HCC-1232 polymer was 29.8 pm/V. The polymer had a d_{33} value of 65.1 pm/V as reported by Buckley et al.[452] A d_{33} value of 150 pm/V at 1.3 μm for the HCC-1232 polymer was reported by Hass et al.[453] This is a large NLO coefficient for a poled polymer system.

Ore et al.[454] synthesized the side-chain copolymers of benzylidene aniline and azobenzene chromophores by chemically attaching to the copolymers of styrene, acrylic acid, methyl methacrylate, and methacrylic acid.

$$X = CH \qquad Poly(St\text{-}co\text{-}MA\text{-}BA)$$
$$X = N \qquad Poly(St\text{-}co\text{-}MA\text{-}AB)$$

The d_{33} coefficient of poly(St-co-MA-BA) was found to be 30 pm/V; the d_{33} coefficient of poly(St-co-MA-AB) was 41 pm/V. These materials exhibit greater stability than that of guest–host mixtures.

Dai et al.[455,456] reported the NPP-functionalized poly(2,6-dimethyl-1,4-phenylene oxide) (PPO).

The analysis showed 1.43 to 1.63 NNPO groups per PPO repeat unit and their solution processible PPO-NPP films have good transparency. A corona-poled film exhibited a d_{33} coefficient of 27.3 pm/V at a functionalization level of 1.43. The d_{33} values were 5.5 and 27 pm/V for a functionalization level of 1.4 and 23 pm/V for a functionalization level of 1.6. These polymers show good thermal stability.

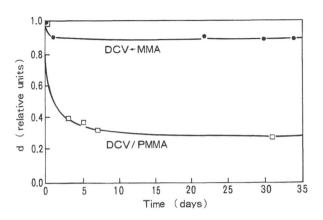

Figure 27 Stability of NLO coefficient of a guest–host system where dicyanovinyl dye is dispersed in PMMA matrix (DCV/PMMA) and DCV-MMA is a copolymer in which dye is grafted as a side chains. The dye-grafted polymer shows enhanced stability of the second-order coefficient over the guest–host system. (From Singer, K.D., Kuzyk, M., Holland, W., Sohn, J., Lalama, S., Comizzoli, R., Katz, H., and Schilling, M., *Appl. Phys. Lett.*, 53, 1800, 1988. With permission.)

Singer et al.[457] measured optical SHG in thin films of a copolymer containing a dicyanovinyl (DCV)-terminated azo dye side chain.

(DCV-MMA)

The corona-poled DCV-MMA system showed d_{33} and d_{31} coefficients of 21.4 pm/V and 7.1 pm/V, respectively, before initial decay. The d_{33} and d_{31} coefficients after decay were 8.0 pm/V and 2.5 pm/V, respectively. The phase matching using Cerenkov type was proposed using the abnormal dispersion of this polymer. Figure 27 shows a comparison of the SHG stability for a guest–host system (DCV/PMMA) and DCV chromophore grafted to PMMA backbone (DCV-MMA). The side-chain polymer was made by random copolymerization of DCV with methyl methacrylate. In both cases, the NLO chromophore was the same. It is apparent that the DCV-grafted copolymer retains SHG activity over an extended period due to the covalent bonding of the chromophore to the polymer backbone.

Stenger-Smith et al.[458] reported the main-chain type polymers containing the NLO chromophore. The T_g of this polymer was about 96 °C. The molecular weight of M_n and M_w are 2200 and 5200, respectively. Preliminary investigation of the second-order NLO coefficients yielded a NLO coefficient of 7 pm/V at 1064 nm.

Allcock et al.[459] reported the second-order NLO properties of polyphosphazene in which a nitrostilbene moiety was covalently linked to the polymer chain through a tris(ethylene oxide) spacer group. NMR studies showed a 36% content of the nitrostilbene side chains.

$$36 \% \text{ OR } = \text{O(CH}_2\text{CH}_2\text{O)n}$$

$$64 \% \text{ OR } = \text{OCH}_2\text{CF}_3$$

Many kinds of NLO dye were introduced into the phosphazene polymers. The d_{33} value ranged between 4.1 and 34 pm/V at 1.064 μm depending on the chromophore. For $n = 1$, the d_{33} values were 4.7 pm/V while for both $n = 2$ and 3, the d_{33} was 5.0 pm/V. The following chromophore shows a d_{33} of 34 pm/V at 1.064 μm at a functionalization level of 33%. These polymers have low T_g below room temperature therefore these polymers cannot be corona-poled due to the ionic conductivity.

Mortazavi et al.[460] reported the NLO properties of copolymers containing coumaromethacrylate. The NLO coefficient of this polymer was found to be 13 pm/V at 1064 nm. The electro-optic coefficients exhibited strong dispersion from 2 to 12 pm/V.

The side-chain polymers containing the 4-alkoxy-4′-alkylsulfone stilbene for frequency doubling were reported by Philips Research Laboratories.[461–464] The cutoff wavelength of this polymer was about 400 nm. The dispersion of refractive index at 820 nm and 410 nm was approximately 0.04. The homopolymer and copolymers where $R = C_6H_{13}$ tend to crystallize and have high scattering losses. For $R = CH_3$ clear and amorphous copolymers were obtained. The highest d_{33} of this copolymer was 9 pm/V at 820 nm for the 45% (w/w) copolymer. The temporal stability of SHG was good; however, storage at ambient conditions led to a rapid decrease in d_{33} after a few days due to water diffusing into the copolymer. Copolymer was found to contain 2 wt% water after 2 weeks. Phase-matching by spatially periodic photobleaching was also demonstrated.

Chen et al.[465] reported linear and nonlinear optical properties of a comb-like polymer comprising MMA and 4-[4-(methylacryloxy)octyloxy-2-methylphenyliminomethyl] cyanobenzene. The copolymer has a T_g of 89 °C. The refractive index of unpoled polymer was 1.525, while for the poled sample, the refractive indexes were 1.523 for TE mode and 1.534 for TM mode at 633 nm. The d_{33} value of 3 pm/V was obtained for corona poled films. The waveguide loss was 10 dB/cm at 633 nm.

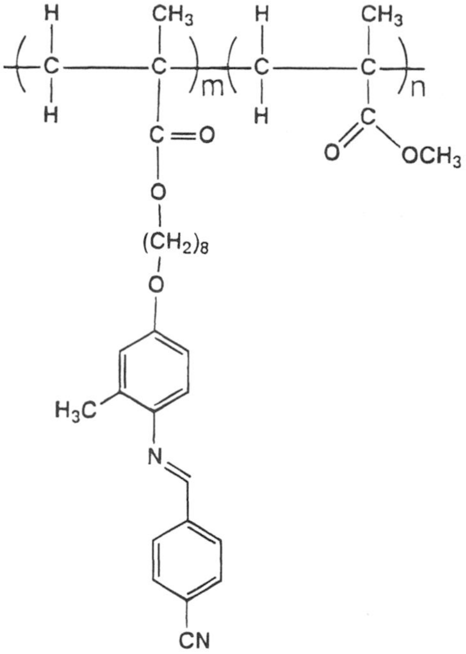

The side chain poled polymers of vinyl chromophore monomers, styrene, methacrylate, and vinyl benzoate derivatives having one-aromatic-ring push-pull chromophores were reported by Hayashi et al.[466] The d_{33} values were 22.0, 10.4, 1.0, and 0.5 pm/V and d_{31} values were 7.0, 4.2, 0.4, 0.2 pm/V for polymers (1), (2), (3), and (4) respectively. Polymer M1 showed a d_{33} of 30 pm/V and a d_{31} of 8.8 pm/V. The polymers show relatively slow decay and 90% of SHG activity, compared with the initial value, was retained after 1000 h.

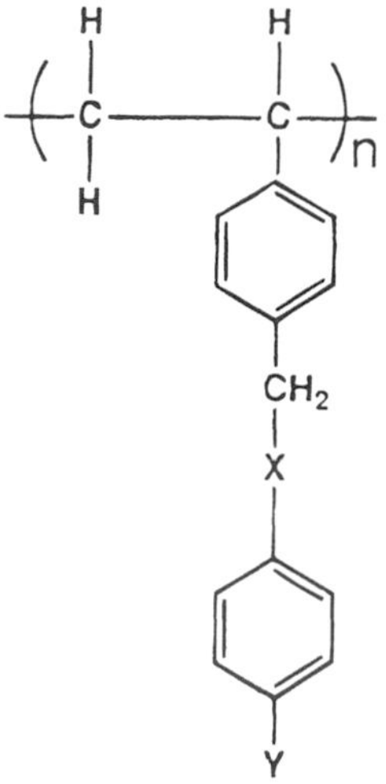

(1) $X = NCH_3$, $Y = NO_2$
(2) $X = NH$, $Y = NO_2$
(3) $X = NH$, $Y = CN$
(4) $X = O$, $Y = CN$

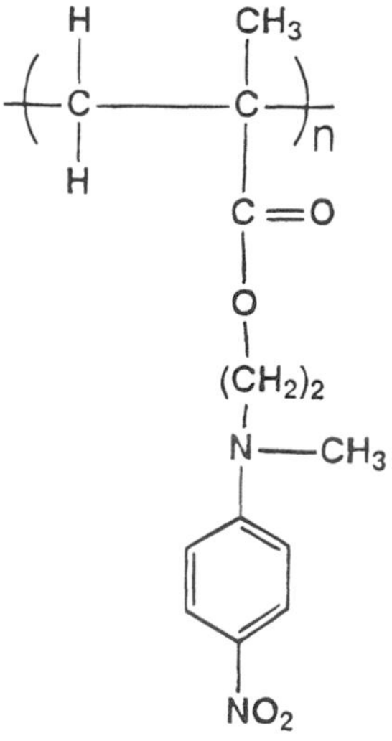

(M1)

A main chain polymer was reported by Xu et al.[467] The chromophore monomer was prepared by the diazonium-coupling reaction of 4-aminophenyl 6-hydroxyhexylsulfone and N-(2-hydroxyethyl)-N-methylaniline. Polymer was obtained by reacting monomer with 1,6-diisocyanatohexane in dioxane. The polymer films show good optical quality and exhibited a d_{33} of 40 pm/V at 1064 nm. The cross-

Table 65 T_g, Initial Decomposition Temperature (T_d), Absorption Maxima (λ_{max}), Refractive Indexes, and NLO Coefficients of Aminosulfone Main-chain and Other Side-Chain Polymers

Polymer	NLO (wt%)	T_g (°C)	T_d (°C)	λ_{max} (nm)	Refractive Indexes at λ (μm)		d_{33} (pm/V)
					0.532	1.064	
MC1	52	62	275	439	1.729	1.619	60
MC2	55	114	280	436	1.854	1.652	125
MC3	62	132	240	436	1.824	1.730	125
MC4	53	108	298	443	1.899	1.684	150
SC1	36	124	245	436	1.717	1.571	100
SC2	48	92	120	460	–	–	108
SC3	68	120	190	470	–	–	250
SC4	47	122	–	455	–	–	119
SC5	52	140	–	463	–	–	223

Polymers MC1 to MC4 and SC1 are from Xu, C., et al., *Macromolecules*, 25, 6714, 1992, and SC2 to SC5 are from Chen, M., et al., *Macromolecules*, 24, 5421, 1991. With permission.

linked polymer was obtained by heating and poling the polymer with 3,3′-dimethoxy-4,4′-biphenylene diisocyanate at 120 °C. The uncross-linked polymer showed a fast decay whereas the NLO stability was significantly enhanced by cross-linking the main chain.

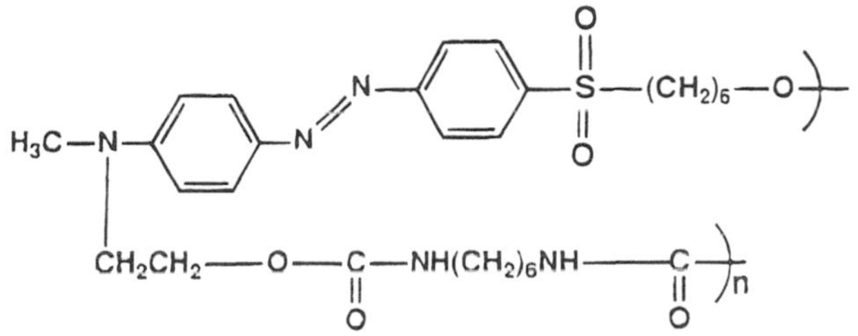

Xu et al.[468] reported a novel double-end, cross-linkable (DEC) chromophore for second-order NLO materials. DEC was polymerized with terephthaloyl chloride by refluxing in dioxane and pyridine to get a cross-linkable polymer I cross-linked via donor end. DEC was also polymerized with MMA at a 1:3 molar ratio in DMSO at 70 °C to prepare polymer II cross-linked via acceptor end. Both of these polymers were orange powders. The corona poled films of polymers I and II showed d_{33} values of 30 and 50 pm/V at 1.064 μm, respectively. Uncross-linked film showed continuing decay of SHG at room temperature while the cross-linked polymer had no temporal decay of SHG for 400 h. For cross-linked polymer, no SHG decay was observed for more than 2000 h at room temperature and 95% after 100 h of annealing at 90 °C.

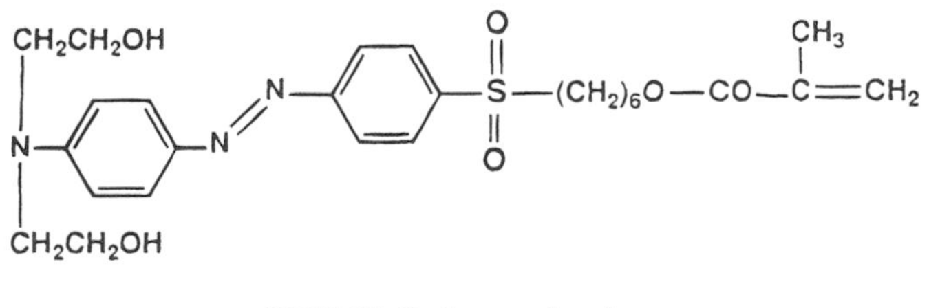

(DEC NLO chromophore)

Xu et al.[469] reported main-chain NLO polymers with aminosulfone azobenzene chromophores randomly incorporated into the polymer backbone. Polymer structures are shown below; it can be seen that polymer MC1 is a polyurethane, MC2 and MC3 are polyesters, and SC5 is an acrylate polymer. Table 65 lists physical as well as NLO properties of these polymers. The main-chain (MC) polymers have high chromophore density up to 62%. The T_g of main-chain polymers can be tailored by tuning the polymer structure which ranges from 62 to 132 °C. Comparatively, the side-chain polymer SC5 has a loading density of 36% and a T_g of 124 °C, although its molecular weight is quite high.

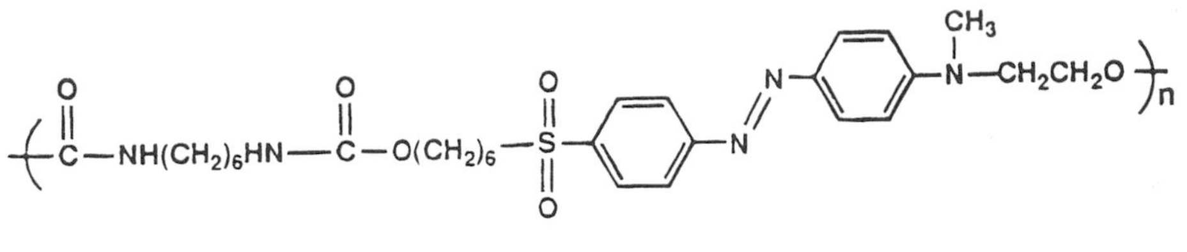

(Polymer MC1)

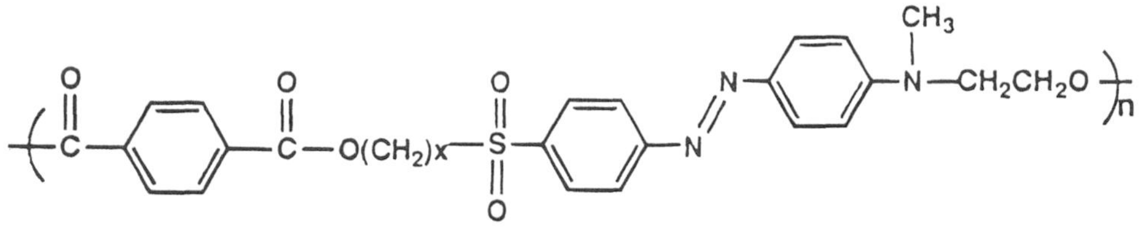

(Polymer MC2; × = 6; Polymer MC3, × = 2)

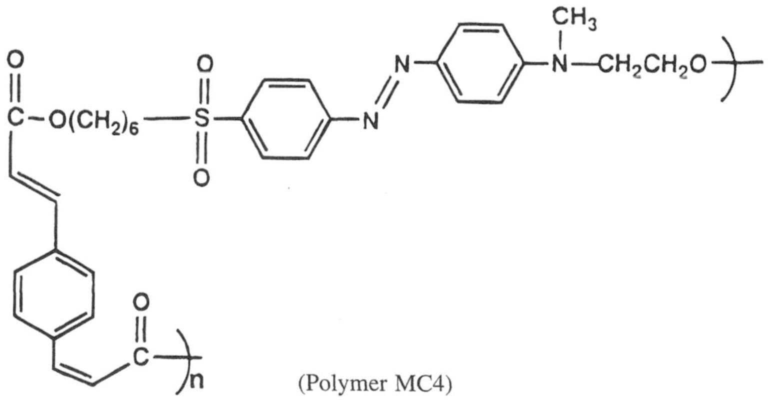

(Polymer MC4)

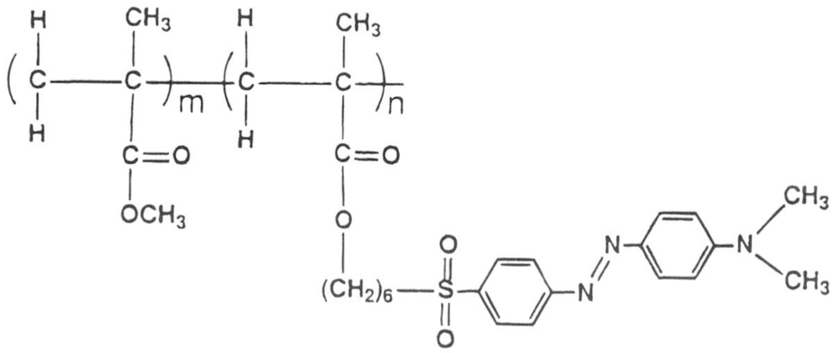

(Polymer SC1)

The MC polymers show very large d_{33} values as high as 150 pm/V which support efficient poling. The large d_{33} values are also related to the resonance contribution because of the absorption tail at SH wavelength (532 nm). Figure 28 shows the temporal decay of SHG at room temperature. Polymer MC1 shows a more rapid decay than those of polymers MC2 and MC4, which are associated with very long flexible segments between the rigid NLO chromophore of MC1. Polymers MC1 and MC4 are cross-linkable. The NH groups in polymer MC1 can easily react with the isocynate groups of 4,4'-diisocyanato-3,3'-dimethoxydiphenyl to form a 3-D polymer network to increase temporal SHG activity. The polymer MC1 was cross-linked and a comparison of the temporal NLO stability between the uncross-linked and

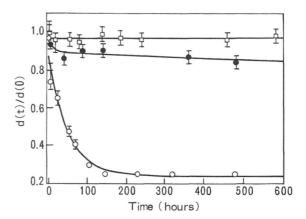

Figure 28 Normalized SHG as a function of time at room temperature for main-chain NLO polymers having aminosulfone chromophores. Square, polymer MC1; closed circle, polymer MC2; open circle; polymer MC4. (Reprinted with permission from Xu, C., Wu, B., Becker, M. W., Dalton, L. R., Ranon, P. M., Shi, Y., and Steier, W. H., *Chem. Mater.*, 5, 1439, 1993, Copyright 1993 American Chemical Society.)

cross-linked samples is shown in Figure 29. The uncross-linked film showed 25% decrease of the initial d_{33} value rapidly, whereas no decay was noticed for the cross-linked film. Cross-linking proved to be an important step in preventing the temporal NLO stability of poled polymer.

Chen et al.[470] reported new side-chain polymers containing Disperse Red (DR1) dye as a NLO chromophore in conjunction with photo-cross-linkable groups. Table 65 lists physical as well as NLO properties of side-chain polymers SC2 to SC5. Polymers SC2 to SC4 show photo-bleaching on exposure to the UV light. The d_{33} values decreased after exposure, and refractive index changed significantly due to photo-induced *cis-trans* isomerization in the azo unit in NLO chromophores. Both refractive index and d_{33} were partly recovered after thermal annealing of SC2 at 80 °C and SC3 to SC5 at 140 °C because the cross-linked polymer matrix assisted in preventing *cis-trans* isomerization.

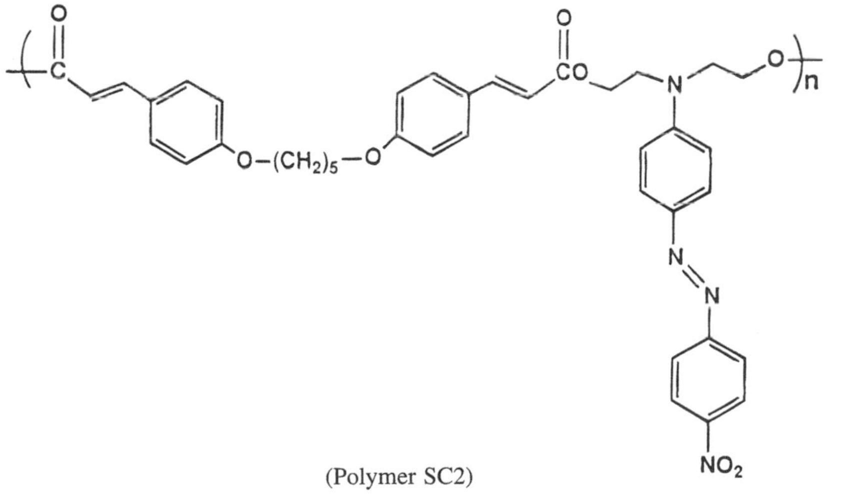

(Polymer SC2)

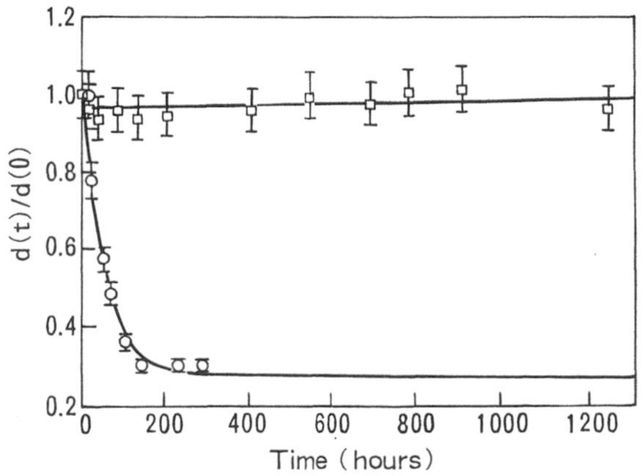

Figure 29 A comparison of temporal NLO stability between the uncross-linked (circle) and cross-linked (square) MC1. (Reprinted with permission from Xu, C., Wu, B., Becker, M. W., Dalton, L. R., Ranon, P. M., Shi, Y., and Steier, W. H., *Chem. Matter.*, 5, 1439, 1993, Copyright 1993 American Chemical Society.)

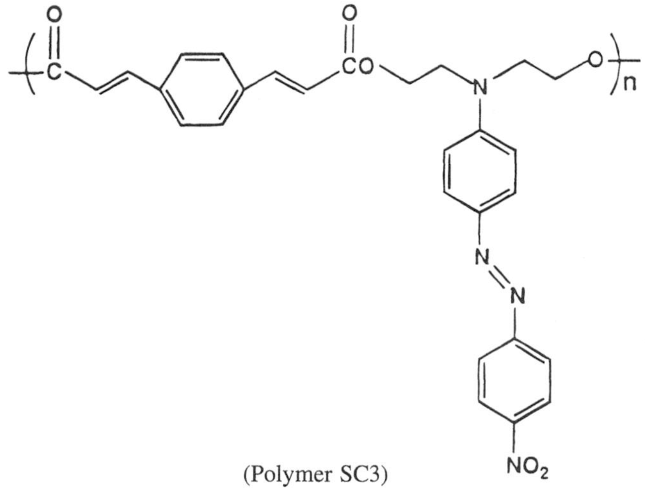

(Polymer SC3)

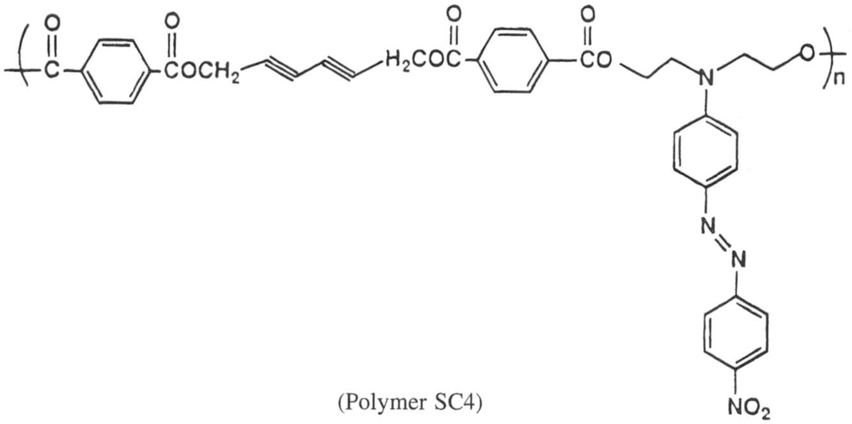

(Polymer SC4)

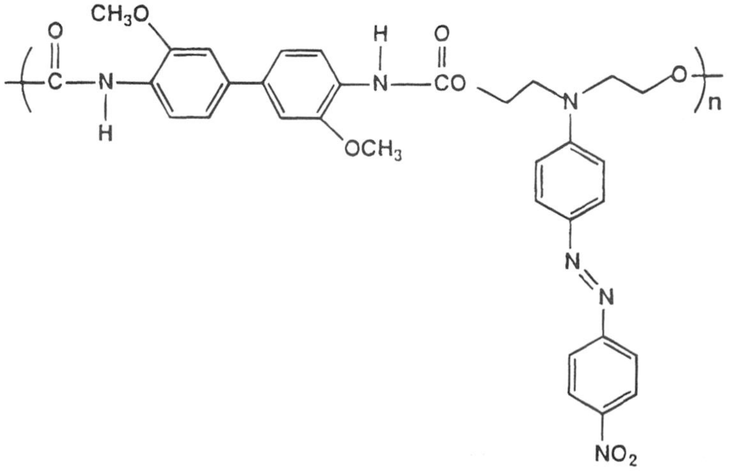

(Polymer SC5)

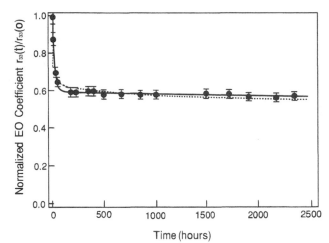

Figure 30 Temporal stability of electro-optic (EO) coefficient r_{33} at 80 °C for a thermally stable guest–host system consisting of derivatized cinamylthiophene chromophore and a high T_g polyquinoline host polymer. (From Cai, Y. M., and Jen, A. K.-Y, *Appl. Phys. Lett.*, 67, 299, 1995, American Institute of Physics. With permission.)

The d_{33} values of these polymers were very large due to resonance enhancement and showed high stability without exposure to UV light. The NLO coefficient of SC2 was found to decrease 40% within 100 h and then stabilized at 40% of the initial value which is associated with the low T_g, 92 °C. On the other hand, SC3 and SC4 showed 80 to 90% stabilized d_{33} values. In polymer SC5, hydrogen bonding helped in cross-linking which leads to stabilized dipole orientation. The largest d_{33} value of SC3 appears because of it highest chromophore density and resonant enhancement. The r_{33} coefficients were calculated from the d_{33} value using a two-level model. The estimated r_{33} values were 23, 26, 24, and 22 pm/V for SC2, SC3, SC4, and SC5 polymers, respectively. These photocross-linkable polymers are advantagous because the large photo-induced refractive index change can be useful for fabricating channel waveguide structures.

Cai and Jen[471a] reported a very thermally stable guest–host system comprising polyquinoline (PQ-100) and a heteroaromatic chromophore diethyl-amino-tricyanovinyl-substituted cinamyl thiophene (RT-9800). The host polyquinoline has thermal stability up to 500 to 600 °C and a T_g of 265 °C. The T_g of the guest–host system with 20% NLO chromophore was around 180 °C. This composite system poled at 0.8 MV/cm showed a nonresonant r_{33} coefficient of 45 pm/V at 1.3 μm, which is remarkably large. Thermal stability of the poled composite films was studied by heating isothermally at 80 °C over a period of 2000 h. Figure 30 shows the temporal decay of electro-optic coefficient at 80 °C. An initial decay of r_{33} about 40% was observed for 100 h, electro-optic coefficient r_{33} of 26 pm/V was remained over a period of more than 2000 h. This study is of great signifcance because very large electro-optic coefficient and long-term thermal stability was achieved in a guest–host system.

(PQ-100)

(RT-9800)

Chen et al.[471b] covalently linked side-chains of *p*-(diethylamino)tricyanovinylbenzene chromophore to polyquinoline backbone to get large second-order optical nonlinearity and high thermal stability EO polymers (structure shown below). The PQL-1 and PQL-2 show λ_{max} at 519 nm and 530 nm due to the $\pi-\pi^*$ charge-transfer band of the NLO chromophore of the *p*-(diethylamino)tricyanovinyl-benzene, respectively. The PQL-1 having 27% wt of NLO chromophore loading showed r_{33} value of 21 pm/V at 0.63 μm and 9 pm/V at 0.83 μm. On the other hand, PQL-2 containing 29% wt NLO chromophore loading showed r_{33} value of 29 pm/V at 0.63 μm and 16 pm/V at 0.83 μm with a poling field of 1.0 MV/cm. The PQL-1 retained 95% of the r_{33} value at 110 °C for 500 h while PQL-2 retained about 80% of the original r_{33} value at 80 °C for more than 500 h. The dielectric constants of 3.89 for PQL-1 and 3.74 for PQL-2 were measured.

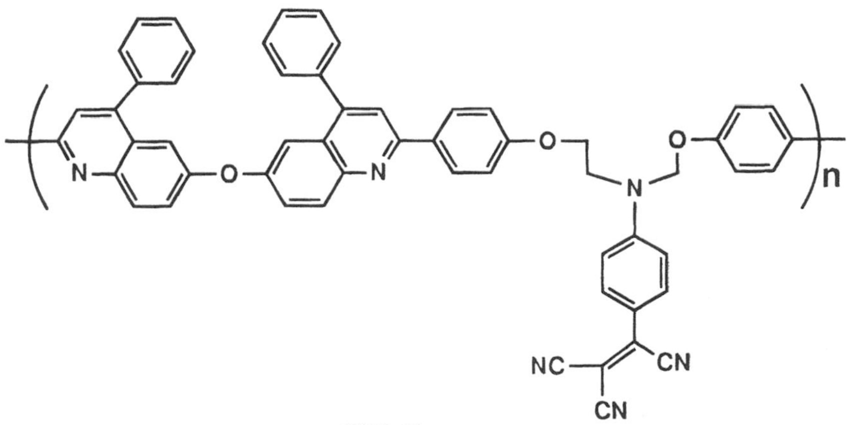

(PQL-1)

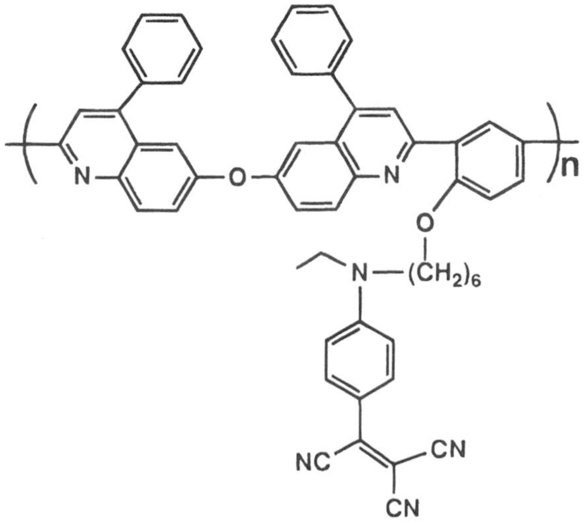

(PQL-2)

Shi et al.[471b] reported high thermal stability for fused-ring guest electro-optic chromophores, 1,8-naphthoylene-benzimidazoles for incorporation into polyimide host. The fused-ring chromophores and polyimides have structural similarities that are easy to incorporate into polyimides. Four different types of chromophores were prepared; (1) $R_1 = R_2 = H$ named as SY156, (2) $R_1 = H$, $R_2 = OCH_3$ named as SY165, (3) $R_1 = NH_2$, $R_2 = H$ named as SY177 and (4) $R_1 = $ pyrrolidine, $R_2 = H$ named as SY215. Thermal stability of chromophores increases upon substitution with donor groups of increased strength; for example, SY215 is stable up to 400 °C. The $\mu\beta$ value determined from DCSHG measurements at 1907 nm was 1500×10^{-48} esu for SY215. The chromophore SY177 showed thermal stability $>350\,°C$ and a β value of 93×10^{-30} esu at 1097 nm. SY177 is soluble in 1,4-dioxane and its absorption maximum occurs at 562 nm. The r_{33} coefficient for SY177-polymide composite was evaluated as 8 pm/V with 10% chromophore loading.

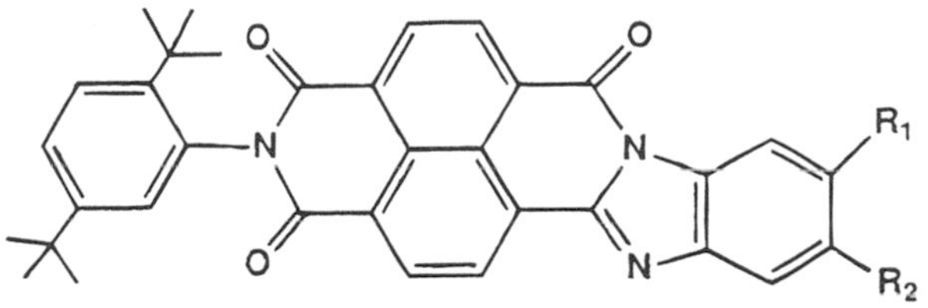

(1) $R_1 = R_2 = H$ SY165
(2) $R_1 = H$, $R_2 = OCH_3$ SY165
(3) $R_1 = NH_2$, $R_2 = H$ SY177

Kitipichai et al.[472] reported SHG properties of polyurethanes prepared from six diol and one triol monomers bearing donor–acceptor chromophores with 2,4-toluenediisocyanate (TDI). The chemical structures of NLO chromophore-bearing polyurethanes are shown below. The monomers contain the N,N'-bis(2-hydroxyethyl)aniline unit with nitro-, tricyanovinyl, cyclobutene-1,2-dione, and imidazolidine-2,4-dione acceptor groups attached via azo or olefin linkages. Polyurethanes with high T_g values show better temporal stability, and percent decay of SHG depends on the T_g and size of the NLO chromophore. The SHG activity of poled polyurethanes (3), (4) and (6) retained about 60 ~ 70% of their initial values after 180 days, whereas the SHG activity of poled polyurethanes (1) and (5) remained 25 ~ 35% over the same period. In particular, the polyurethane (6) prepared from the diol having an imidazolidine-2,4-dione acceptor group showed only 30% decay in SHG activity over 200 days at room temperature although it is not cross-linked, which results from the hydrogen bonding between the urethane and imidazolidine-2,4-dione groups. No decay in the SHG activity was observed at 100 °C for a few hours from a cross-linked film prepared from the triol and TDI; this film has remarkably stable SHG at room temperature. Repoling of polyurethane films after dipole relaxation does not restore the initial SHG values. The NLO coefficients of NLO chromophore-bearing polyurethanes were 4.62, 21.3, 22.75, 23.77, 6.16, and 8.74×10^{-9} esu at 15 kV for (1), (2), (3), (4), (5), and (6), respectively. The d_{33} showed dependence on poling field.

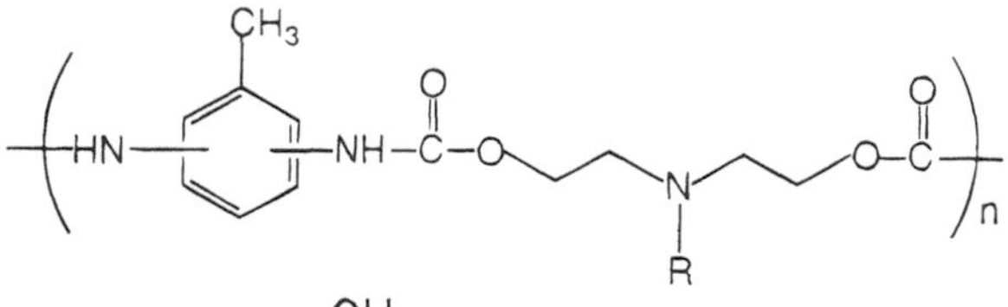

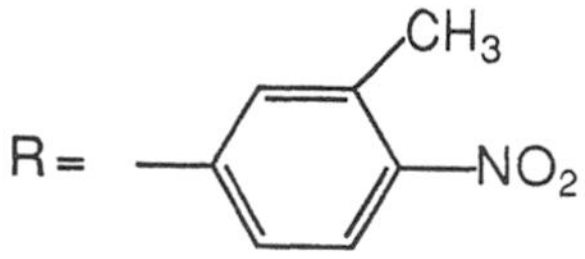

$R = $ (1)

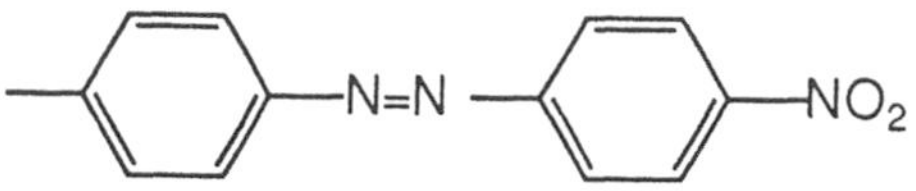

(2)

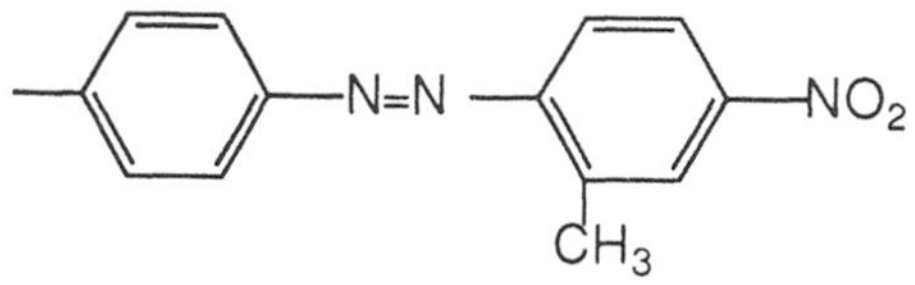

(3)

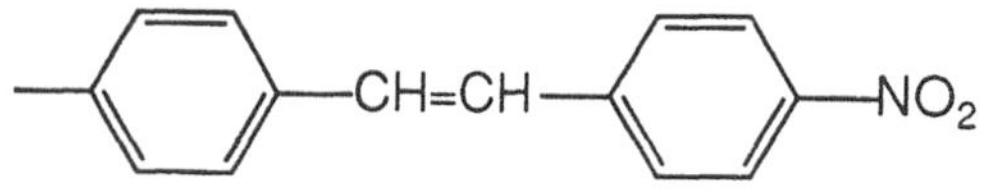

(4)

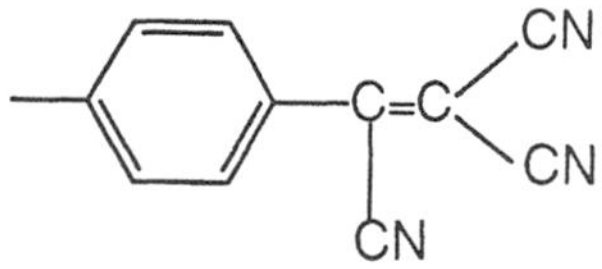

(5)

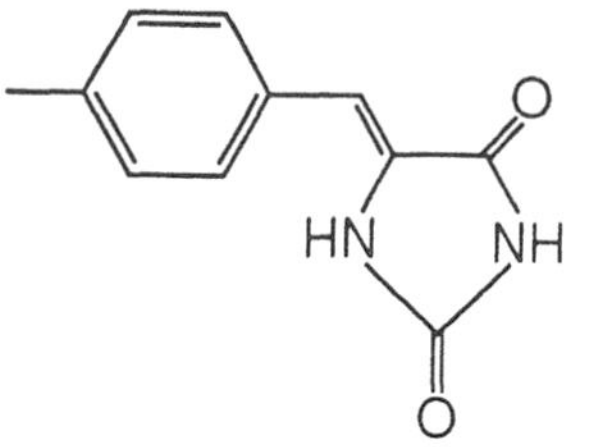

(6)

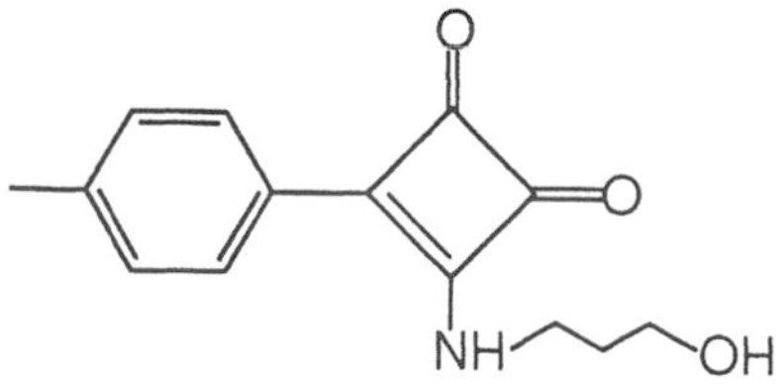

(7)

Gulotty et al.[473] reported SHG and electro-optic coefficients of 4-nitrophenylhydrazone-based phenoxy polymers. The TP7 had the largest d_{33} and r_{33} coefficients and refractive indexes (TE mode) of 1.718 at 633 nm and 1.625 at 1.320 μm. The polymers showed losses of <1 dB/cm in slab waveguides and losses of 2 to 3 db/cm channel waveguides at 1.320 μm. Comparisons were made of r_{33} coefficients obtained from EFISH data from μβ at 1.064 μm and d_{33} data at 1.579 and 1.064 μm from film SHG and electro-optic measurements at 1.320 μm. The d_{33} values of TP7, TP4, and TP31 were 11.3, 10.05, and 8.37 pm/V at 1.064 μm and 4.6, 3.18, and 3.05 pm/V at 1.579 μm, respectively. The r_{33} values of TP7, TP4, and TP31 were 33, 13, and 26 pm/V at 1.064 μm from EFISH μβ data and 15, 9.2, and 10 pm/V from electro-optic measurements at 1.32 μm, respectively. The r_{33} values of TP7, TP4, and TP31 calculated from SHG measurements were 11, 9.7, and 7.8 pm/V at 1.064 μm and 3.9, 3.1, and 2.9 pm/V at 1.579 μm, respectively. The SHG activity of parallel plate poled films of TP31 initially drops about 20% over a period of 220 days at 100 °C in air. The electro-optic activity of a TP31-based Mach-Zehnder modulator (9 pm/V at 1.320 μm) is stable until 140 °C.

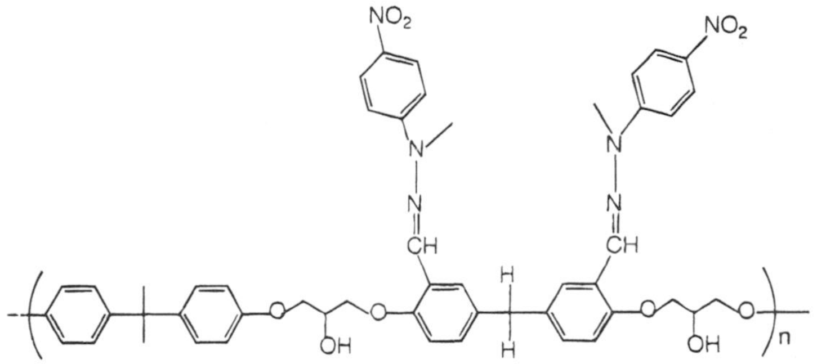

(TP7)

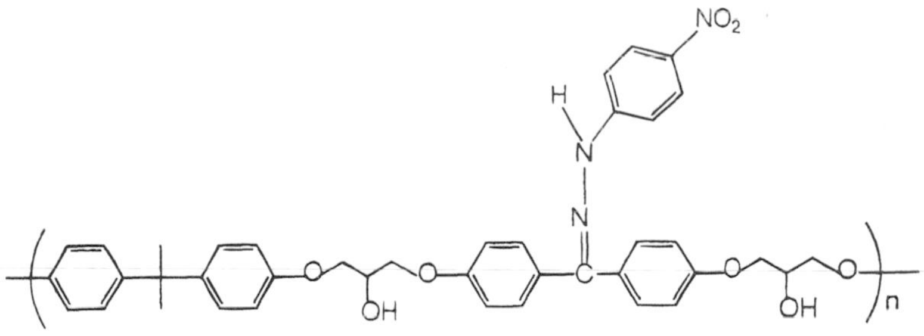

(TP4)

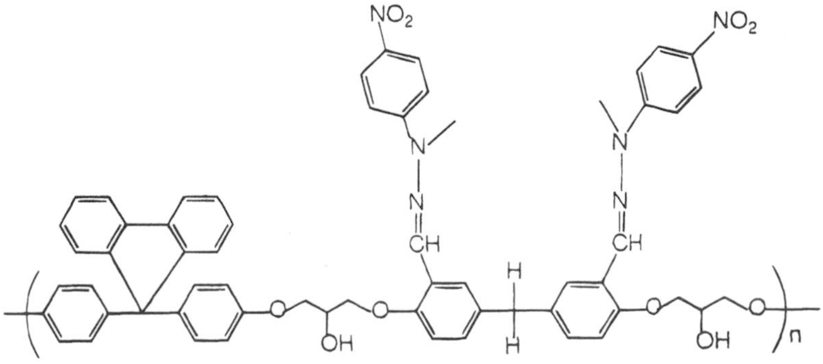

(TP31)

Ching et al.[474] reported the NLO properties of polymers-bearing pendant phosphine oxide chromophores which were derived from azobenzene dye and have an amino group as donor and a phosphoryl group as acceptor. Two types of polymers were prepared, one as a side chain of MMA backbone (PO-MMA) and another with a polyurethane backbone from isophorone di-isocyanate (PO-IDI) and hexamethylene diisocyanate (PO-HDI). PO-MMA, PO-IDI, and PO-HDI showed T_g values of 157 (chromophore density 75%), 163, and 120 °C, respectively. The d_{33} value of PO-MMA increased linearly up to 20% chromophore level and plateaued at approximately 45% where d_{33} values were of the same order. The d_{33} values were 85, 90, and 70 pm/V for 10%, 20%, and 75% chromophore levels. The d_{33} values for polyurethane backbones PO-IDI and PO-HDI were 100 and 135 pm/V, respectively, larger than PO-MMA. The UV-visible and SHG measurements showed that all three polymers retain 60 to 80% of their initial d_{33} value after 46 days at room temperature. PO-IDI having high T_g values retained 60% of the initial d_{33} value after 3 h at 80 °C, or 20 min at 100 °C and 50% after 20 min at 120 °C. The bulky diphenylphosphoryl acceptor groups were found to be useful for high T_g and good temporal stability.

(PO-MMA)

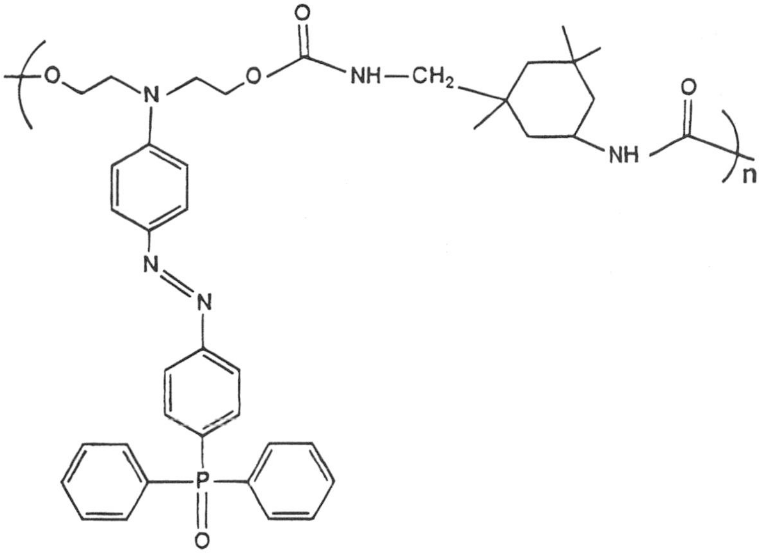

(PO-IDI)

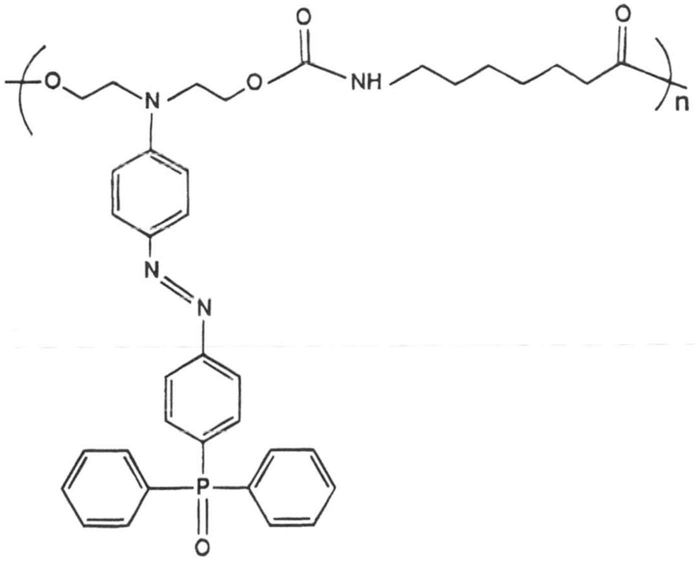

(PO-HDI)

Nahata et al.[475] reported NLO properties of a covalently functionalized copolymer of DR1 with MMA backbone. The copolymer had an absorption peak at approximately 470 nm and a T_g of 132 °C. The 16.5 mol% DR1-MMA copolymer showed r_{33} coefficients of 8.6, 54, 3.6, and 2.1 pm/V at 632.8, 670, 810, and 1305 nm, respectively. The $\chi^{(2)}$ ranged between 4 and 6.5 pm/V for the copolymer at 0.5 MV/cm. Levenson et al.[476] reported three types of NLO polymers: side-chain polymers bearing aminocyano azobenzene (Orange1), DR1 dyes, and cross-linked DR1-MMA polymer (CL-DR1-MMA). The refractive indexes of these polymers depends on dye concentrations. The d_{33} values were 44 pm/V for Orange 1-MMA (9.3%), 40 pm/V for DR1-MMA (40%), and 20 pm/V for C1-DR1-MMA (30%). The r_{33} coefficients were 12.4, 5.5, and 12 pm/V at 1324 nm, and 13.5, 4.4, and 12.6 pm/V at 1064 nm for these polymers, respectively. The r_{33} values at 860 nm were slightly higher. The replacement of the nitro group by a cyano group in Orange1 increased thermal stability although Orange 1-MMA polymer had a lower SHG activity. Thermally cross-linked CL-DR1-MMA showed improved stability when the poling and cross-linking processes are optimized and its r_{33} coefficient stabilized during a week at 85 °C. Swalen et al.[477] reported three different polymers having side chains of DR1, p-nitroaminotolane (NAT), and trifluoromethyl sulfonylaminotolane (1FSAT) chromophores attached to a PMMA backbone. The λ_{max} values were at 491 nm for DR1-MMA, 402 nm for NAT-MMA, and 388 nm for 1FSAT-MMA. The measured d_{33} values were 47, 18, and 9 pm/V for DR1-MMA, NAT-MMA, and 1FSAT-MMA polymers, respectively. The r_{33} values were 13 pm/V for NAT-MMA and 9 pm/V for 1FSAT-MMA polymers. The optical losses measured at 830 nm ranged between 1 and 9 dB/cm.

The second-order NLO properties of side-chain polymer containing the organic salt dye were reported by Choi et al.[478] The polymers were synthesized from the following monomers, which is a molecular ionic (N-methylpyridinium) NLO chromophore.

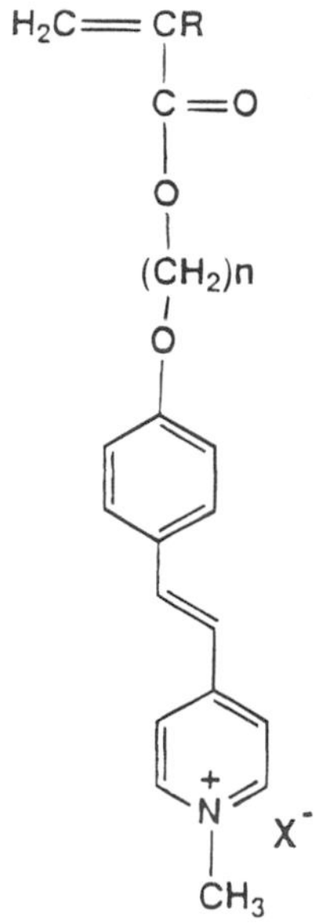

The abbreviations used for polymers are as follows:

1. HPBr15; Poly{1-hexadecyl-4-[2-(4-methacryloxyphenyl)vinylpyridinium bromide, $X = $ Br}
2. HPBr21; Poly{1-docosyl-4-[2-(4-methacryloxyphenyl)vinyl]pyrodonium bromide, $X = $ Br}
3. CP6I-HEMA; Poly(1-methyl-4-(2-[4-(6-methacryloxyhexyloxy)phenyl]vinyl} pyrodonium iodide-co-2-hydroxyethyl methacrylate)
4. HP6I; Poly(1-methyl-4-{2-[4-(6-methacryloxyhexyloxy) phenyl]vinyl}pyridinium iodide, $X = $ I).

The T_g and d_{33} values of HPBr15, HPBr21, CP6I-HEMA, and HP6I were 142, 143, 93, and 138 °C and 5.9, 5.4, 10, and 15.9 pm/V, respectively. It is generally believed that organic salts are difficult to pole because of their high conductivity. In these polymers, poling efficiency and temporal stability of SHG activity was improved by varying the counterion in the polymer. A bulky tetraphenylborate (TPB) counterion was also substituted to reduce the migration of the counterion during poling. By the use of TPB, the poling-induced chromophore alignment was found to increase to that of the iodide- or bromide-containing polymers. Furthermore, SHG of TPB-containing polymer was five times larger than its iodide analog and temporal stability was improved.

Yokoh et al.[479] reported the acetalized PVA containing NLO chromophore that had the following structure. The cutoff wavelength of this polymer was about 480 nm. The d_{33} coefficient was estimated to be 41 pm/V. The sample was corona poled at 150 °C for 15 min. An experiment of blue SHG was carried out using a mode-locked Ti:sapphire laser operating at 887 nm. Second harmonic power of 0.39 W was obtained at 14.4 kW fundamental input.

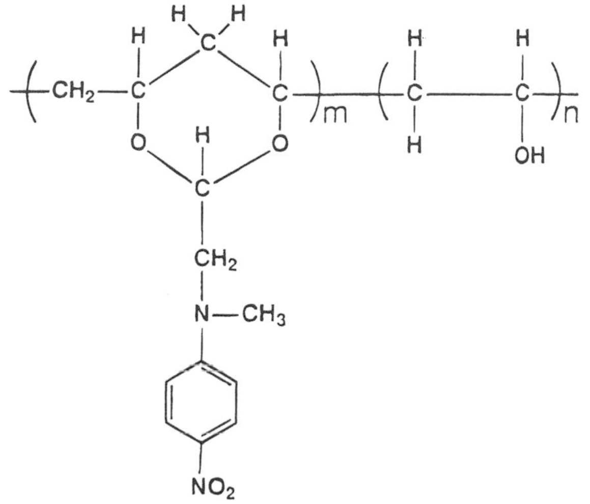

(pNAn-PVA)

Lindsay et al.[480] reported chromophoric polymers with self-ordering natures to prepare high-T_g polymers based on cinnamamide chromophore linked with 1,2-diamidocyclohexyl- and o-xylyl-bridging groups. The folded syndioregic structure connecting α-cyanocinnamoyl chromophore and containing N-methyl and ethyl groups was prepared. The N-methyl-containing polymer had higher T_g values, higher chromophore density, better solubility and thermal stability than N-ethyl polymers. The T_g of higher molecular weight polymers having N-ethyl was approximately 208 °C and that having N-methyl was about 220 °C. The d_{33} value for the N-ethyl group containing polymer was measured at 10 pm/V. Wang et al.[481] also investigated these accordion-type polymers. The d_{33} value of several pm/V, were estimated for a BCSC polymer, 1,2-benzyl-1,2-cyclohecyl-bridged syndioregic α-cyanocinnamamide.

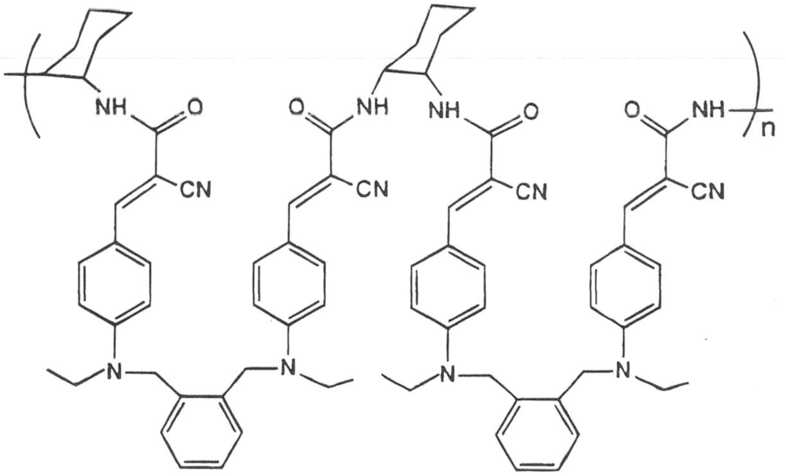

(BCSC)

Tao et al.[482] synthesized Λ-type main-chain polymers with two-directional charge-transfer chromophores by nucleophilic displacement polymerization using bis(4-fluoro-3-nitrophenyl)sulfone with both aliphatic and aromatic diamines in aprotical solvents. Figure 31 shows the UV spectrum of M1 before and after poling. Figure 32 shows the Maker fringes of M4 polymer for P polarization at 1.064 μm with its temporal stability at 100 °C. The advantage of this polymer is the NLO chromophore embedding in the main chain which is thermally stable because of the relatively small β relaxation. In addition

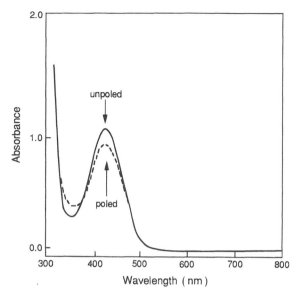

Figure 31 UV-visible absorption spectra of the M1 polymer before and after poling. (From Tao, X. T., Watanabe, T., Shimoda, S., Zou, D. C., Sato, H., and Miyata, S., *Chem. Mater.*, 6, 1961, 1994, American Chemical Society. With permission.)

Table 66 T_g, Initial Decomposition Temperature (T_d), Cutoff Wavelength (λ_{cutoff}), Refractive Indexes, and NLO Coefficients of Λ-Type Main-Chain Polymers

Polymer	T_g (°C)	T_d (°C)	λ_{cutoff} (nm)	Refractive Indexes at λ (μm) 0.532	Refractive Indexes at λ (μm) 1.064	NLO Coefficients (pm/V) d_{33}	NLO Coefficients (pm/V) d_{33}
M1	130	270	500	1.7427	1.6474 (nTE)	12	4.2
				1.7485	1.6483 (nTM)		
M2	78	275	510	1.7073	1.6381 (nTE)	7.6	2.5
				1.7103	1.6385 (nTM)		
M3	105	265	500	1.7027	1.6386 (nTE)	17.6	5.4
				1.7038	1.6340 (nTM)		
M4	192	260	540	1.6654	– (nTE)	9.2	3.0
				1.6541	– (nTM)		
M5	148	253	510	1.6966	1.6268 (nTE)	8.0	2.4
				1.6868	1.6164 (nTM)		

From Tao, X. T., et al., *Chem. Mater.*, 6, 1961, 1994, American Chemical Society. With permission.

the dipole moment of this polymer is located perpendicular to the molecular chain which is easy to pole the chromophore. The SHG intensity remained at approximately 60% of the original value. Most of the decay was ascribed to the diffusion of surface charges on the films after poling.

M1 $X = H$ $R = (CH_2)_3\text{-}CH\,(CH_3)\text{-}CH_2$
M2 $X = H$ $R = (CH_2)_{10}$
M3 $X = CH_3$ $R = (CH_2)_{10}$

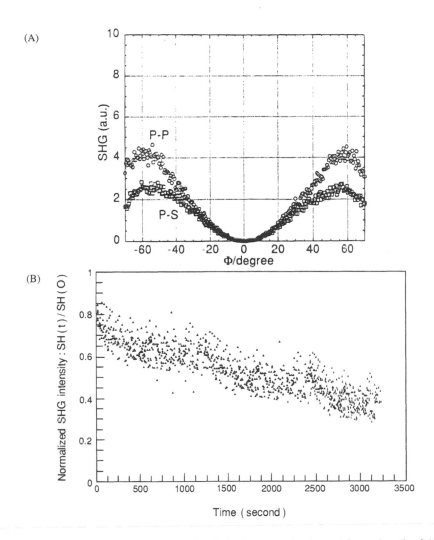

Figure 32 (A) Maker fringes of M4 for P and S polarizations at a fundamental wavelength of 1064 nm, (B) SHG temporal stability of M4 at 100 °C. (Reprinted with permission from Tao, X. T., Watanabe, T., Shimoda, S., Zou, D. C., Sato, H., and Miyata, S., *Chem. Mater.*, 6, 1961, 1994. Copyright 1994 American Chemical Society.)

M4 $X = H$

M5   diamine   HN ⬡ ⬡ NH

Watanabe et al.[483] reported SHG properties of polymers having 2-D and 1-D charge-transfer molecules. NLO chromophores of P-1 side-chain polymer have an electron-donating amino group located *para* to the nitro group. This polymer has the largest β_{xxx} component, but the smaller β_{xyy} component. In polymer D-1, electron-donating amino groups are located *para* as well as *ortho* to nitro groups, causing a 2-D nonlinearity. The β_{xyy} component of NLO chromophore is three to five times larger than that of the diagonal β_{xxx} component, whereas β_{xzz} is negligible. The d_{33} value was 6.2 pm/V for P-1 and 8.2 pm/V for D-1 polymer. The d_{31} values were 2.0 and 3.5 pm/V for P-1 and D-1, respectively. The larger β_{xyy} component led to larger d_{31} values of main-chain polymers by drawing and poling than d_{33}.

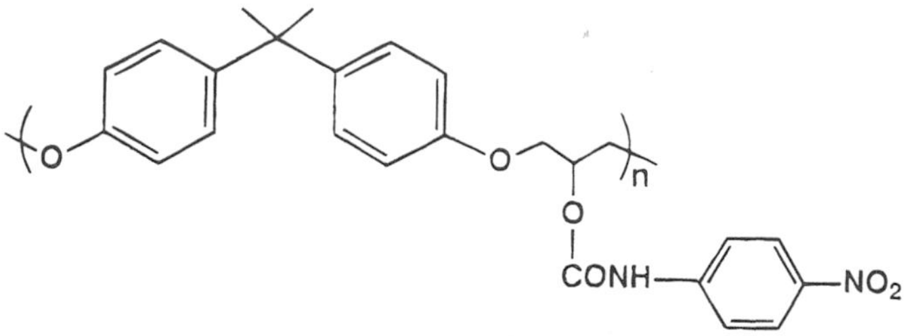

(P-1)

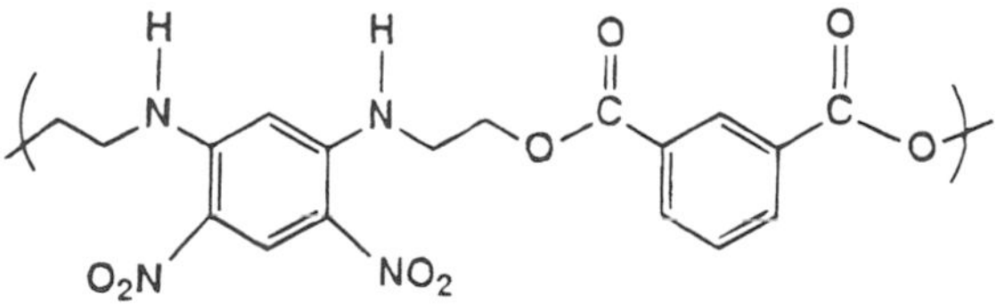

(D-1)

The SHG properties of polyureas (abbreviated as PU) was reported by Nalwa et al.[484–487] PU1 showed that the d_{33} coefficient of 5.5 pm/V and the estimated electro-optical coefficient r_{33} was 3.56 pm/V. The d_{33} of PU2 was evaluated as 8.37 pm/V and d_{31} coefficient was 2.9 pm/V. The introduction of a methyl group to the phenyl ring increased the d_{33} to 12 pm/V. The aromatic polyurea PU5 exhibited the largest d_{33} coefficient, 11.7 pm/V. The aromatic polyurea having a pyridine ring with a nitro group showed optical transparency up to 460 nm and a d_{33} coefficient of 7.5 pm/V. The SHG efficiency of PUs was many times larger than that of urea single crystal. Kajiwara et al.[488] also reported a d_{33} coefficient of 1.65 pm/V for polyurea films formed by vapor deposition polymerization, which showed indefinite stability at room temperature after an initial decay of a few percent. Aliphatic and aromatic polyureas with large positive birefringence have been reported by Tao et al.[489] The refractive indexes of PU7 ~ PU10 in TE modes were found to be higher than those of TM modes before poling. The refractive indexes of TE and TM modes of PU7 and PU8 were increased by corona poling. For example, the refractive index of PU7 at 532 nm in TE modes before poling was 1.6052, which increased to 1.6363 after poling. Polyureas were the first example of the poling-induced refractive indexes increase of TE modes. The unusual refractive indexes change may be associated with the conformation of the cyclohexane group. PU9 and PU10 have no such increase in refractive indexes because no conformation change occurred from corona poling. The d_{33} coefficients of PU7 ~ PU10 were found to range from 2 to 4 pm/V at 1.064 μm, smaller than the d_{33} coefficients of PU1 ~ PU6. The d_{33} coefficients of PU7, PU8, PU9, and PU10 were 3.1, 2.4, 2.8, and 3.4 pm/V, respectively, The estimated d_{31} coefficients of PU7, PU8, PU9, and PU10 were 1, 0.8, 0.9, and 1.1 pm/V, respectively, about one third that of d_{33} coefficients. PU7 had an optical loss of 1.26 dB/cm at 632.8 nm. Figure 33 shows the temporal decay of SHG intensity at 25, 80, and 140 °C of PU7. Initial, rapid temperature-dependent decay at short time was followed by a gradual decay process. The SHG activity remained at 50% even at 140 °C. This good stability was mainly due to the hydrogen bond between urea groups. Polyureas prepared from aliphatic and aromatic diamines showed the shortest cutoff wavelength of about 300 nm. Figures 34 and 35 shows the optical spectra of PU7 and PU1 films, respectively.

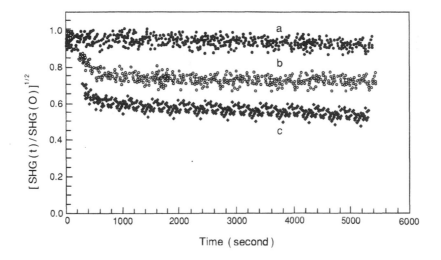

Figure 33 SHG temporal stability of PU7 polymer at different temperatures after poling field was removed. (a) at 25 °C; (b) at 80 °C₃; (c) at 140 °C. (Reprinted with permission from Tao, X.T., Watanabe, T., Zou, D. C., Shimoda, S., Sato, H., and Miyata, S., *Macromolecules,* 28, 2637, 1995. Copyright 1995 American Chemical Society.)

Figure 34 Optical transmission spectrum of PU7 film. (Reprinted with permission from Tao, X. T., Watanabe, T., Zou, D. C., Shimoda, S., Sato, H., and Miyata, S., *Macromolecules,* 28, 2637, 1995. Copyright 1995 American Chemical Society.)

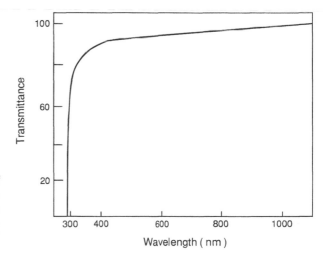

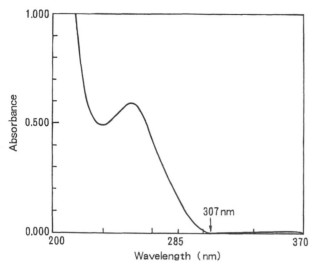

Figure 35 Optical transmission spectrum of aromatic polyurea PU1 having no pendant chromophore. Polyurea has transparency up to 300 nm. (From Nalwa, H. S., Watanabe, T., Kakuta, A., Mukoh, A., and Miyata, S., *Electron. Lett.,* 28, 1409, 1992. With permission.)

Polyureas	A	Y	X
PU1	CH₂–⟨⟩–CH₂	H	H
PU2	CH₂–⟨⟩–CH₂	H	NO₂
PU3	CH₂–⟨⟩–CH₂	CH₃	NO₂
PU4	CH₂	H	NO₂
PU5	CH₂	CH₃	NO₂
PU6	CH₂	H	COOCH₃

| PU7 | $X =$ ⟨⟩–CH₂–⟨⟩ | PU8 | $X=$ ⟨⟩ |
| PU9 | $X = H_2C-\overset{CH_3}{\underset{CH_3}{C}}-CH_2$ | PU10 | $X=$ ⟨⟩–SO₂–⟨⟩ |

A new side-chain polymer was reported by Watanabe et al.[490] based on the thiazole ring. The NLO properties of TR polymer were determined as $d_{33} = 44$ pm/V and $r_{33} = 12$ pm/V at 1300 nm. The propagation loss of the channel waveguide was 6.5 dB/cm, and the intrinsic loss of polymer was estimated as 1.2 dB/cm.

Nahata et al.[491] reported electro-optic properties of novel thermostable fluorene-based polyester polymer. The polymer was synthesized from 9,9-bis-4-hydroxyphenyl-2-diamino-7-nitrofluorene and the BISMAN polyester showed a T_g of 205 °C. The r_{33} coefficient was found to be 2.6 pm/V at 810 nm for a poling field of 0.5 MV/cm. The r_{33} value increased as a function of the poling field and, for a poling field of 2.5 MV/cm, the r_{33} was about 11 pm/V.

(BISMAN polyester)

For applications, it has been realized that poled NLO polymers should have good thermal stability at elevated temperatures; for example, at 80 °C for a few years and around 250 °C for a short time. Such temporal stability is highly desirable not only for the device's performance but also during the integration process. Polyimides are well known in packaging industries for their exceptionally high thermal stability and T_g. With this view, researchers have attempted to load NLO chromophores into polyimides either as guest molecules or as side chains into the backbone. Significant developments have been made in this direction which are important for electro-optic applications. Meinhardt et al.[492] reported electro-optical properties of Ultradel 9000D photosensitive polyimides doped with NLO chromophores [4-(dicyanomethylene)-2-methyl-6-(p-dimethylaminostyryl)-4H-pyran] (DCM) and [2,6-bis(2-(3-(9-(ethyl)carbazolyl)ethenyl)-4H-pyran-4-ylidene) propanedinitrile (DADC). The chemical structure of Ultradel 9000D photosensitive polyimides is shown where X is an alkylated photocross-linking group. The 9% DADC doped polyimide poled from 175 to 225 °C had an initial r_{13} coefficient of 0.26 pm/V at all poling temperatures. No decay of r_{13} was observed after thermal aging at 125 °C for 53 h. DADC doped film showed good electro-optic stability after 2 days at 120 °C. On the other hand, a significant drop in r_{13} was observed for 17% DCM doped polyimide aged at 125 °C for 42 h in air because of the slow decomposition of DCM and the incomplete cross-linking at poling temperatures below 200 °C.

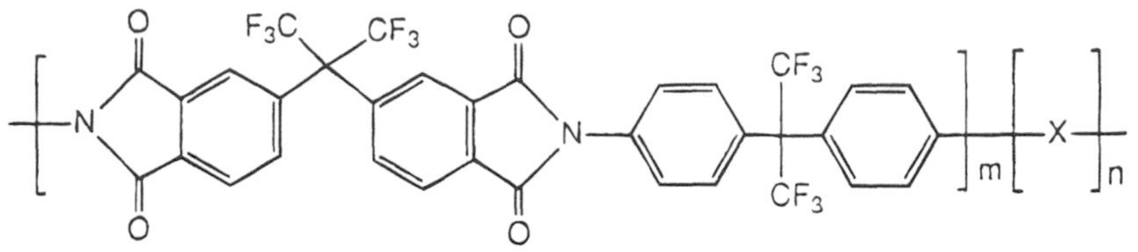

(Ultradel 9000D)

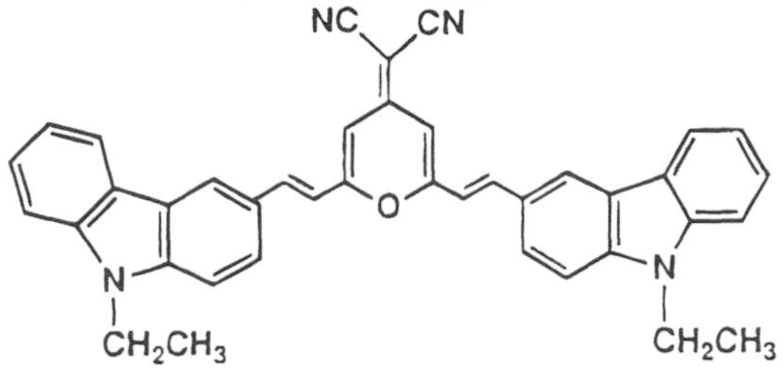

(DADC)

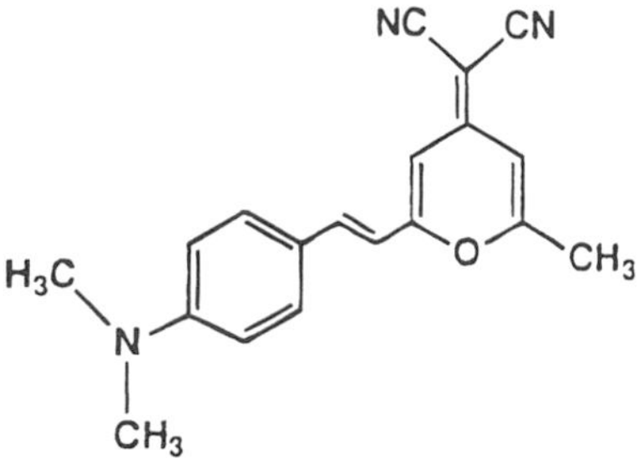

(DCM)

Pretre et al.[493] reported SHG and electro-optic properties of modified polyimides (structures not specified). The refractive indexes of A-095.11, A-097.07, and A-148.02 polymers were 1.65, 1.76, and 1.66 at 1.3 μm, respectively. The d_{33} values were 19–38, 30–40, 13–28 pm/V at 1542 nm for A-095.11, A-097.07, and A-148.02 polymers, respectively. The r_{33} values measured at 1313 nm were 7–13, 10–13, and 5–10 pm/V, respectively. The relaxation time of 5 years for temperatures up to 70 and 100 °C for A-095.11 and A-148.02 polyimides, respectively, were estimated from the SHG measurements. Becker et al.[494] reported NLO properties of a polyimide containing DR1 chromophore. The poled and cured film showed a d_{33} coefficient of 117.3 pm/V at 1064 nm. The cured polyimide films showed no decay of d_{33} values over hundreds of hours at 90 °C in air. This polymer was also investigated by Sotoyama et al.[495] The r_{33} coefficient of 4.3 pm/V by corona poling and 10.8 pm/V by contact poling was measured.

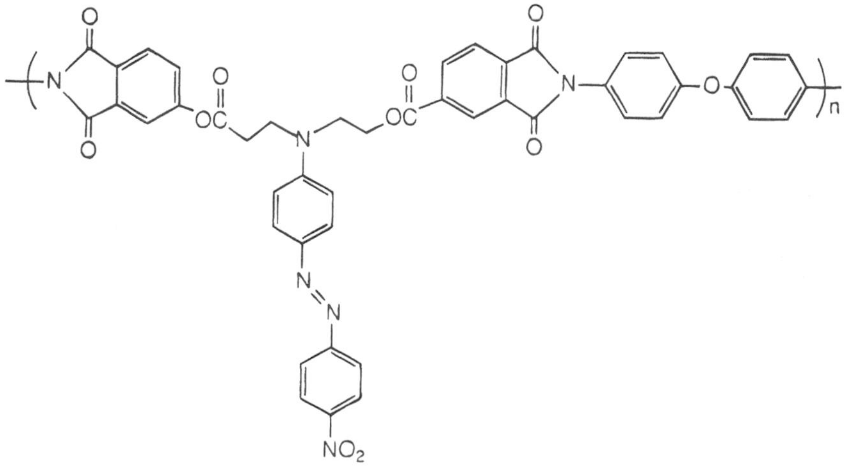

The new polyimide having NLO chromophore embedded in the main chain was reported by Weder et al.[496] The λ_{max} was at 471 nm for P1 and 473 nm for P2 polymer. The refractive indexes were 1.758 for P1 and 1.832 for P2 at 633 nm and 1.627 for P1 and 1.663 for P2 at 1295 nm. The d_{33} coefficient was 16 pm/V for P1 and 27 pm/V for P2, whereas the r_{33} values were 40 pm/V for P1 and 10 pm/V for P2. Preliminary results indicated that the significant dipolar relaxation did not occur in 120 days after poling.

P1: $x = 12$, P2: $x = 10$, P3: $x = 8$, P4: $x = 6$, P5: $x = 4$

A high T_g polymer was reported by Lin et al.[497] based on the polyimide structures. The T_g of this polymer was approximately 236 °C. Corona poling and final curing of this polymer were performed at various temperatures. The d_{33} values of polymers ranged from 4.6 to 5.5 pm/V. The estimated chromophore density was about 7×10^{20} molecules/cm^3. The minor decay of SHG activity of less than 10–15% was observed during 24 h at 85 °C.

The novel cross-linking polyimides were synthesized by Yu et al.[498] The glass transition of these polymers was 250 °C. The d_{33} coefficient was found to be 45.7 pm/V for polyimide A and 60.7 pm/V for polyimide B at 1064 nm. The off-resonant d_{33} was estimated to be 15 pm/V by a two-level model. These polymers are stable at more than 300 °C. Only 15% of decay from the initial value was observed even at 150 °C. The λ_{max} of the polymers was about 385 nm and the cutoff wavelength was 580 nm. The r_{33} were 6.1 and 8.0 pm/V at 632 nm and 4.3 and 4.0 pm/V at 780 nm for polyimide A and B, respectively.

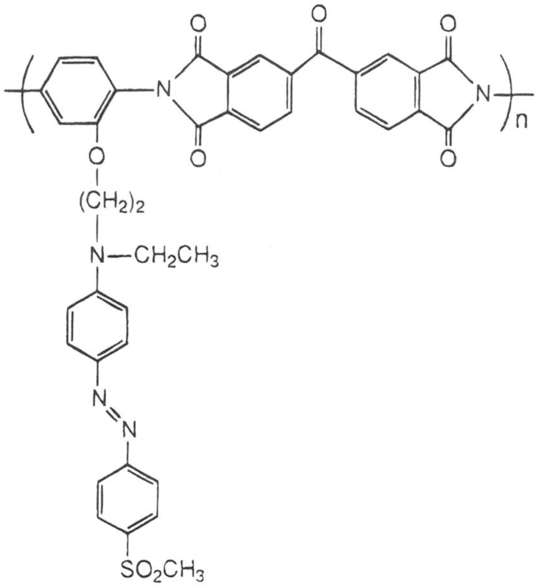

(Polyimide A)

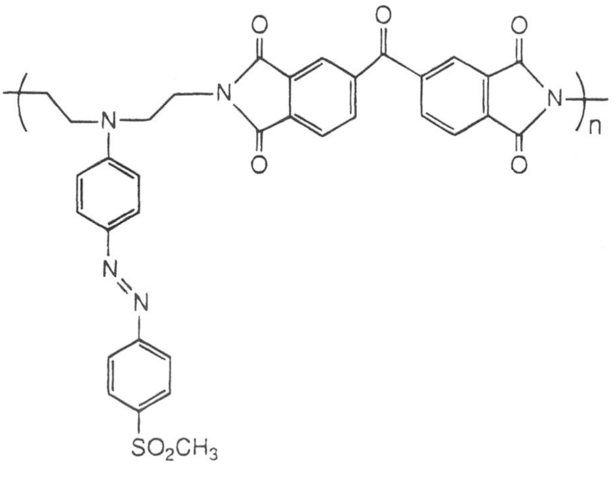

(Polyimide B)

The novel side-chain polymer was reported by Verbiest et al.[499] The polymer contains the following structures. The T_g values of DR1-ODPA and DPDR1-ODPA polyimide are 235 and 227 °C, respectively. The electro-optic coefficients of these polymers were found to be 8.2 to 10 pm/V. Polymer DPDR1-ODPA was stable even at 300 °C. The decay time was extrapolated from the aging experiments and found to be more than 1000 days at 85 °C. This was remarkably good stability for thermoplastic polymers without cross-linking systems.

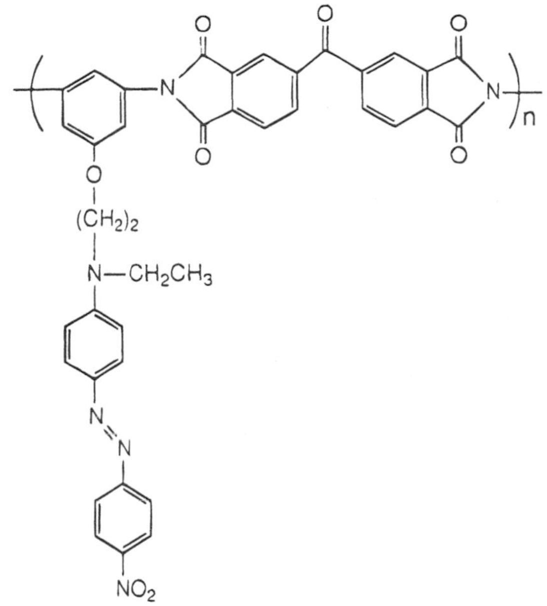

(DR1-ODPA)

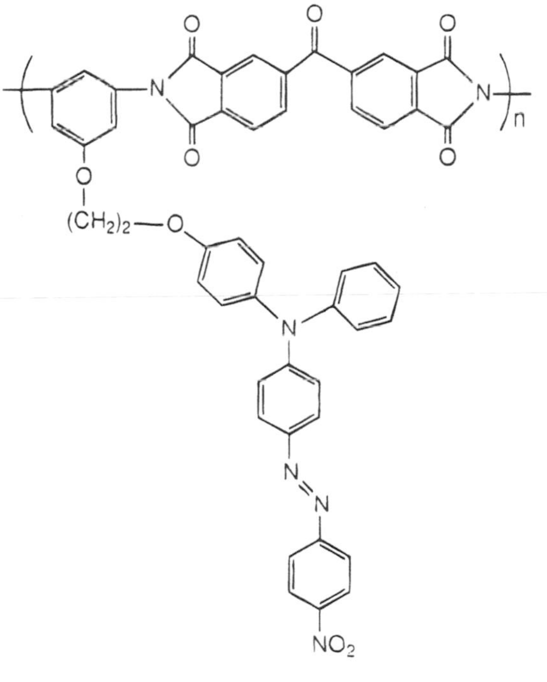

(DPDR1-ODPA)

Miller et al.[500] further modified polyimides using (hexafluoroisopropylidene)diphthalic anhydride as a starting material. The polyimide backbone had DR1 chromophore and fully imidized polymer was soluble in common organic solvents. The T_g of the polymer was very high at approximately 350 °C, which supported the increased rigidity of the donor-embedded structure. TGA showed only 5% weight loss above 500 °C and onset decomposition temperature higher than 400 °C. The r_{33} coefficient of the polymer was 1.7 pm/V at 1.3 μm for 75 V/μm poling field. From the extrapolation of the poling field to higher values of 250–300 V/μm, the r_{33} value of 6–7 pm/V at 1.3 μm was estimated. Using β value of the chromophore and loading level, a maximum r_{33} value of 10.5 pm/V at 310 °C and at a poling field of 250 V/μm was calculated. This NLO chromophore functionalized polyimide showed excellent temporal stability because of its very high T_g. The calculated relaxation time was 75 days at 250 °C. Figure 36 shows the SHG stability for samples prepared from the poly(amic acid) and from the chemically imidized material.

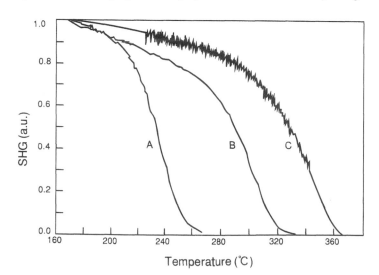

Nemoto et al.[501] studied polyesters containing NLO chromophores as side chains in high density. These NLO chromophore-grafted polymers were amorphous. The T_g recorded from DSC measurements were 118, 139, 87, 109, and 83 °C and λ_{cutoff} of spin-coated films was 606, 577, 492, 577, 617, and 390 nm for PE-1, PE-2, PE-3, PE-4, and PE-5, respectively. The d_{33} coefficient measured from SHG of the spin-coated films was 110 pm/V for PE-1, 24 pm/V for PE-2, 29 pm/V for PE-3, 202 pm/V for PE-4, and 3 pm/V for PE-5 after corona poling. PE-5 polymer showed transparency up to 390 nm but it had a rather low d_{33} value. On the other hand, PE-4 polymer showed λ_{max} at 407 nm and λ_{cutoff} at 617 nm, but its NLO coefficient was the largest among all polyesters. This study showed that both optical and NLO properties of polyesters can be tailored by using different NLO chromophores.

Figure 36 SHG stability of NLO side-chain polyamide (A) film prepared from the poly(amic acid) heated to a maximum temperature of 250 °C in a corona field, (B) film prepared from the poly(amic acid) heated to a maximum temperature of 280 °C in a corona field, (C) chemically imidized film corona poled at 310 °C. (Reprinted with permission from Miller, R. D., Burland, D. M., Jurich, M., Lee, V. Y., Moylan, C. R., Tackara, J. I., Twieg, R. J., Verbiest, T., and Volksen, W., *Macromolecules*, 28, 4970, 1995. Copyright 1995 American Chemical Society.)

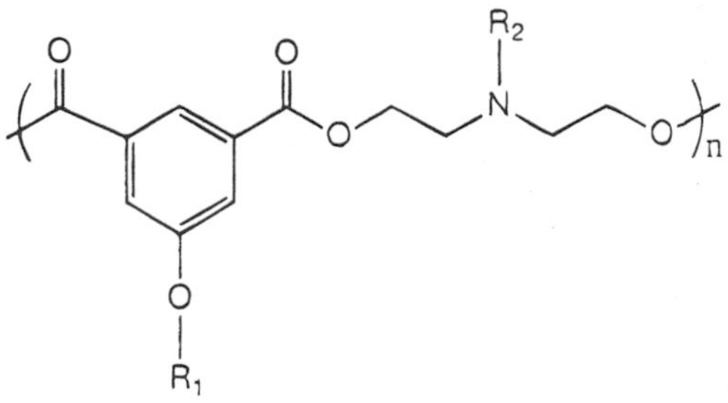

PE-1 R_1 R_2

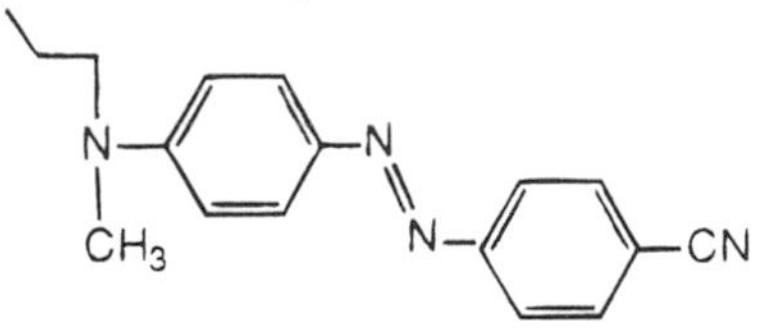

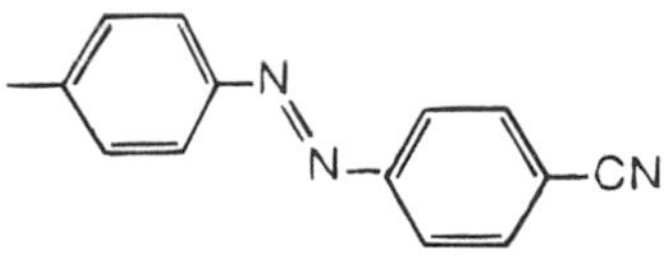

PE-2

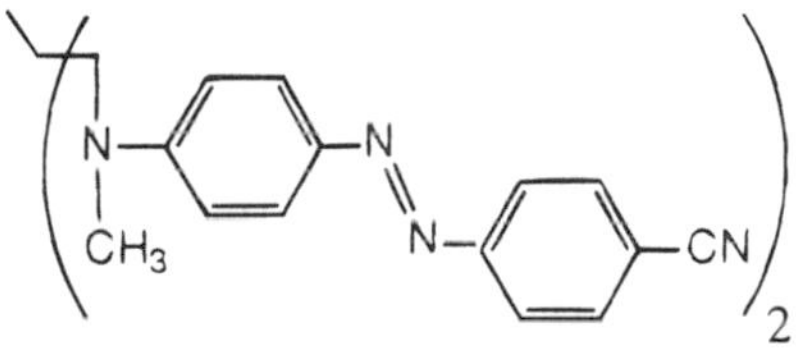

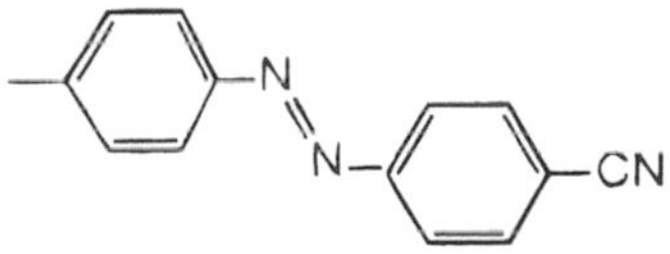

PE-3

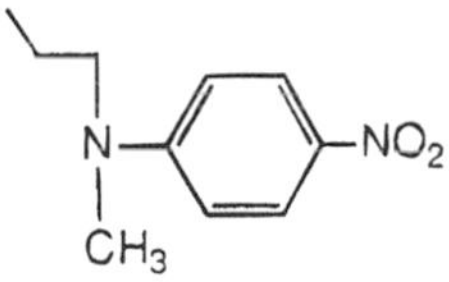

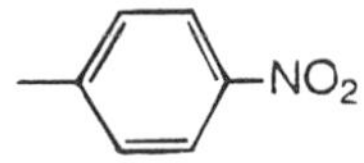

PE-4

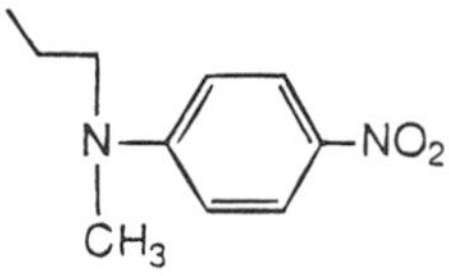

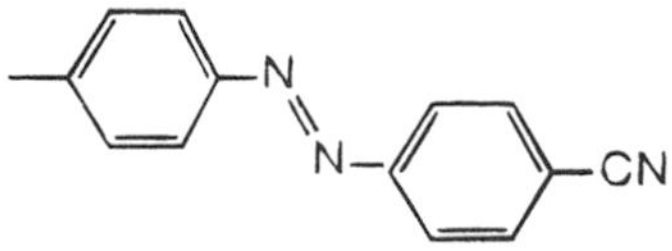

PE-5

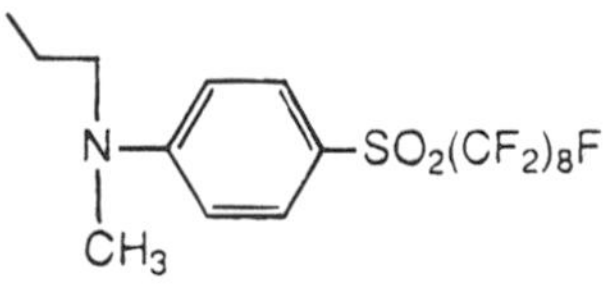

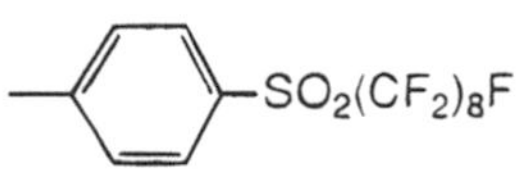

Suen et al.[502] reported NLO properties of homopolymers and copolymers based on tolane chromophores. The chemical structure and their λ_{max} values and d_{33} coefficients are shown below. The optical transparency of tolane chromophores having a trifluoromethyl terminal group are as low as 340 nm. For homopolymers with cyano and trifluoromethyl terminal groups, d_{33} value increases with decreasing spacer length due to higher chromophore density.

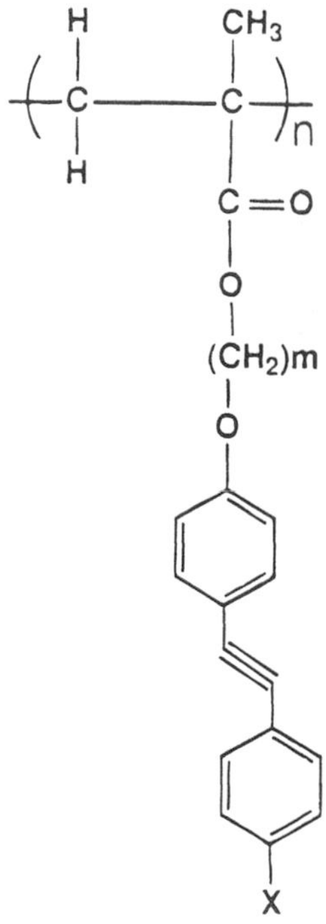

Polymer	X	m	λ_{max} (nm)	d_{33} (pm/V)
P3NO2	NO2	3	357	15.5
P6NO2	NO2	6	357	22.6
P3CN	CN	3	319	11.8
P6CN	CN	6	319	8.5
P3CF3	CF3	3	300	5.4
P6CF3	CF3	6	300	5.3
P11CF3	CF3	11	300	3.1

The d_{33} value was 4.1 pm/V for CP27, 3.2 pm/V for CP28, 1.1 pm/V for CP29, 10.1 pm/V for CP31, and 18.8 pm/V for CP32. The CPs with higher d_{33} values also showed λ_{max} values at longer wavelengths, for example, it was 410 nm for CP32 and 316 nm for CP29. The polymers 5% P6NO2, 20% P6NO2, 48% P6NO2, and 5% P6FC3 showed d_{33} values of 6.7, 10.4, 14.6, 1.4 pm/V, respectively. This indcates that NLO activity of P6NO2 polymer increases with the increase of chromophore density.

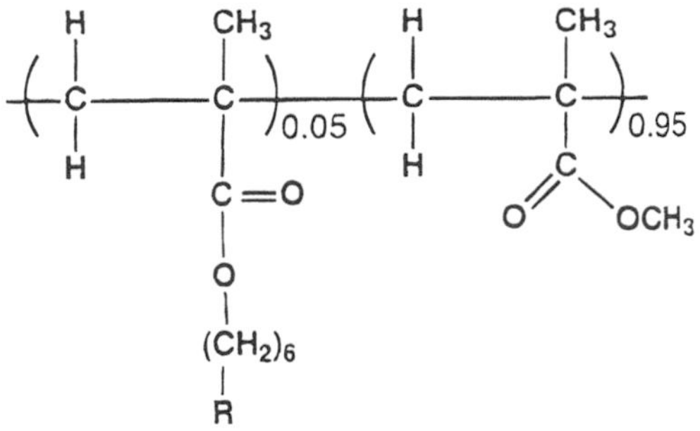

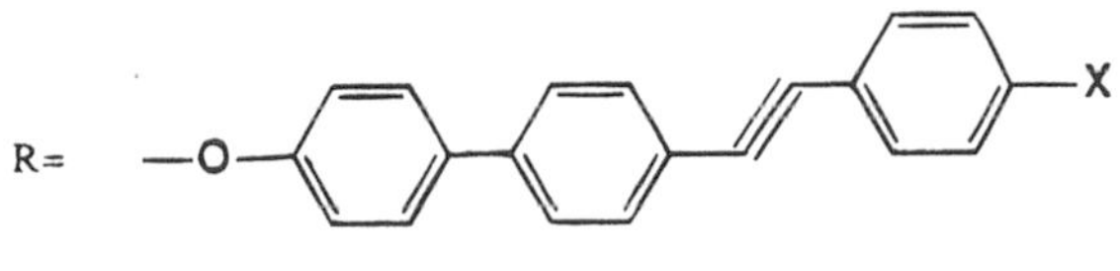

$X = NO_2$ (CP27) $X = CN$ (CP28) $X = CF_3$ (CP29)

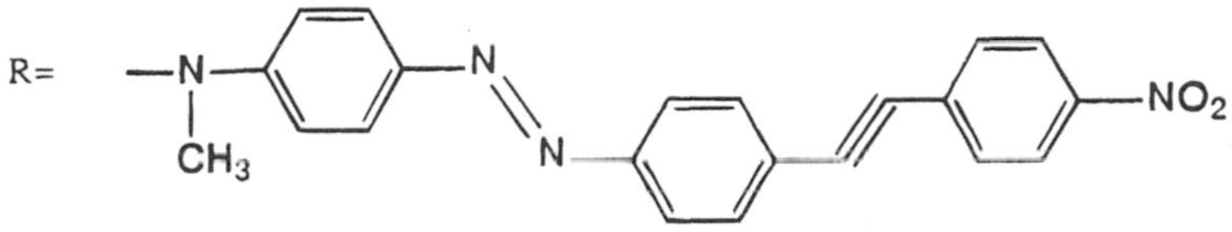

(CP31)

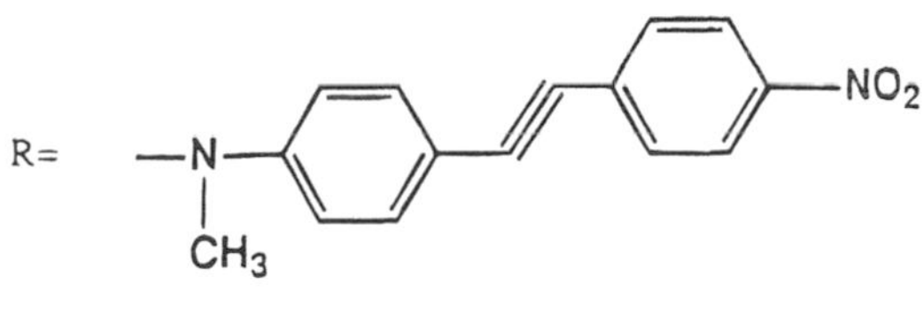

(CP32)

Tapolsky et al.[503] reported a maleimide-based, cross-linkable electro-optic polymer. The DR17 was reacted with maleimideobenzoyl chloride to obtain a dimaleimide-functionalized chromophore and the polymer was obtained using piperazine. The polymer showed a T_g of 140 °C. The r_{33} coefficient measured at 830 nm was 8 pm/V, whereas the r_{33} value calculated by the two-level model was 9 pm/V. The polymer showed good electro-optic stability at 150 °C and up to 200 °C. The polyurethane-based NLO polymers showed refractive indexes of 1.733 at 815 nm and 1.685 at 1300 nm.[504] The measured r_{33} value was 4.45 pm/V at 30 V/μm, 9.6 pm/V at 60 V/μm and 23 pm/V at 120 V/μm. The electro-optic stability was found to decreased by 8% in 17 h at 100 °C. The decay of electro-optic activity at 125 °C over 4 h was 35%. Piezoelectric contribution on r_{33} was small though reached 16% on r_{13}.

(maleimide based NLO polymer)

(polyurethane-based NLO polymer)

Moylan et al.[505] reported a NLO polymer having diethylaminonitrotolane tethered as side chain to PMMA backbone. Electro-optic coefficient r_{33} was determined at four wavelengths. The measured r_{33} values were 10.84 pm/V at 1300 nm, 11.96 pm/V at 1064 nm, 13.94 pm/V at 884 nm, and 14.48 pm/V at 832 nm. The calculated r_{33} values from β and concentration of chromophore matched well with measured values where the poling field was adjusted to 278 V/μm at 1300 nm.

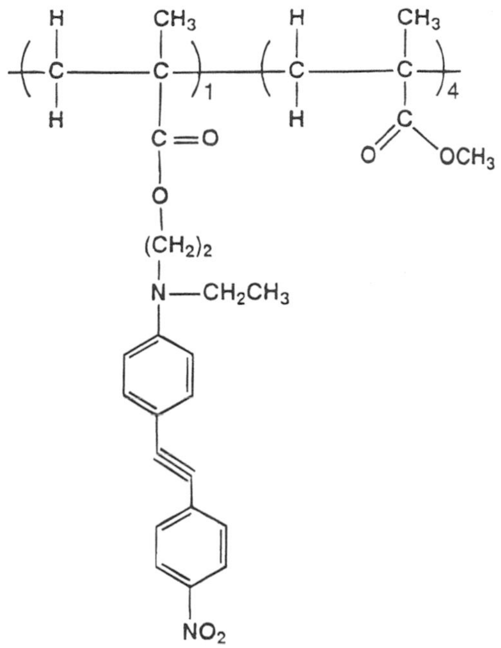

We have summarized a wide variety of NLO chromophores containing polymers. NLO chromophores are tethered as side chains to conventional polymer backbones, generally to PMMA. NLO chromophores with two reactive groups can be imparted into main chain leading to main-chain NLO polymers. Although electrical poling of NLO polymers with side chains are easy, similar polar orientation can not be achieved in main chain NLO polymers which was considered a drawback.

4. Cross-linked Polymers and Sol-Gel Systems

NLO polymers suffer from orientational relaxation problem because their second-order optical nonlinearity decreases with time at room temperature as well as elevated temperatures. To overcome this problem, NLO chromophores are incorporated in a polymer backbone with a high glass transition temperature (T_g). NLO polymers that have high T_g showed improved temporal stability of SHG and EO activity because molecular motions of NLO chromophores are related to the T_g of an individual polymer. In this view, polyimides and heteroaromatic ladder polymers proved to be the most promising. Another appropriate approach to prevent temporal decay of SHG was the use of cross-linkable polymeric systems. As discussed earlier the cross-linking process gives rise to stable SHG.[468-470] A wide variety of the cross-linked epoxy NLO polymers have been prepared by reacting bifunctional epoxy monomers with tetrafunctional nitro-substituted diamines by the IBM group.[506,507] The number average molecular weights (M_n) of the prepolymers prepared from 4-nitro-1,2-phenylenediamine and bisphenol-A diglycidylether was about 1000. The cross-linked polymers (bis-A-NPDA) have the following structures:

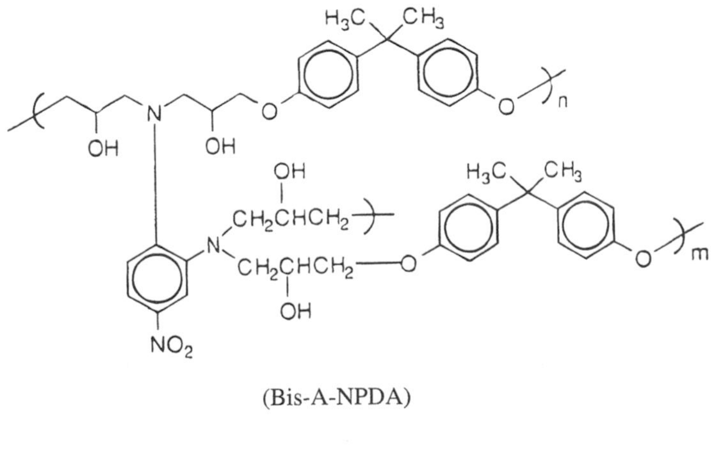

(Bis-A-NPDA)

The curing of these prepolymers above 100 °C led to the formation of a highly cross-linked polymer network in which the mobility of the NLO moieties was restricted, helping to stabilize the noncentrosymmetric alignment.[508] A cross-linked polymer (bis-A-NPDA) prepared from 1,2-diamino-4-nitrobenzene and bisepoxide glycidylether after curing at 140 °C showed d_{33} and d_{31} coefficients of 13.5 pm/V and 3.0 pm/V, respectively. This cross-linked system contained 20 wt% of NLO moieties and exhibited no decay after 3 weeks at room temperature. Furthermore, no tendency to decay was observed in the SHG signal even at 85 °C. A large d_{33} value of 8.4 pm/V was obtained in another cross-linked epoxy polymer having 63 wt% NLO moieties. The UV-visible spectra showed the π-π* transition at about 410 nm and cutoff wavelength at 500 nm.

Teraoka et al.[509] reported the stability of cross-linking polymers based on the epoxy resin. Two polymers were prepared by the reaction of diglycidylbisphenol A with 4-nitroaniline (Bis-A-NA) and N,N'-dimethyl(4-nitro-1, 2-phenylenediamine) (Bis-A-DMNPDA). The Bis-A-NA polymer was synthesized by melting the corresponding monomers under the nitrogen atmosphere for 6 to 8 at 150 °C. The Bis-A-DMNPDA was also synthesized by a similar method. The λ_{max} values were at 392 and 412 nm for Bis-A-NA and Bis-A-DMNPDA, respectively. Large anisotropy of refractive index was observed in TE and TM modes. The initial d_{33} and d_{31} coefficients of 31 and 7.5 pm/V and 22 and 5.3 pm/V after 28 days were measured for Bis-A-NA polymer. For the Bis-A-DMNPDA system, initial d_{33} and d_{31} coefficients were 11.7 and 3.9 pm/V and 10.6 and 3.5 pm/V after 32 h, respectively. In the dielectric relaxation measurements, both poled and unpoled samples showed the α relaxation related to the glass transition and β relaxation attributed to the local segmental motion. The critical temperature determined from the dielectric constant using the William-Landel-Ferry (WLF) equation showed a significant decrease with the increase of the poling field and reached 40 °C.

(Bis-A-NA)

(Bis-A-DMNPDA)

Jungbauer et al.[510] reported the cross-linking polymer containing tolane chromophores. The polymer was synthesized by a reaction of 4-amino-4′-nitrotolane (ANT) with diglycidylbisphenol A at 150 °C. The absorption peak of tolane monomer at 418 nm was red-shifted to 444 nm by reaction, and the dispersion of refractive indexes of Bis-A-ANT was studied. The birefringence of poled Bis-A-ANT was approximately 0.075, about 50% larger than that of Bis-A-NA. The d_{33} of Bis-A-ANT poled using the corona discharge was found to be as large as 89 pm/V. The decay of d_{33} constants was 16% after 20 h at 100 °C. The dielectric α relaxation temperature of poled polymer films was larger than that of unpoled films, this was ascribed to the thermotropic interaction between tolane groups.

(Bis-A-ANT)

Park et al.[511] reported the cross-linking (CL) polymer based on the NPP functionalized poly(*p*-hydroxystyrene). Without using any cross-linking agent, the polymer showed a d_{33} value of 3.7 pm/V. When cross-linking agents were applied to the polymer, the d_{33} value was 2.9 pm/V for CL agent 1 and 2.3 pm/V for CL agent 2 for the 0.5 stoichiometry of diepoxide/OH. The d_{33} value was dependent on the stoichiometry of diepoxide/OH. The τ_2 was 30 days for uncross-linked polymer, which increased to 100 days by applying CL agent 1 and to 74 days by applying CL agent 2. These results indicate that the polar stability of poling induced chromophore-functionalized NLO polymers can be improved by chemical cross-linking.

Cross-linking Agent 1   Cross-linking Agent 2

Mandel et al.[512] reported the cross-linked polymers based on a poly(vinyl cinnamate) (PVCN). Photochemical reaction was performed using PVCN with 3-cinnamoyloxy-4-[4-(*N,N*-diethylamino)-2-cinnamoyloxyphenylazo]nitrobenzene (CNNB-R). Thin films were corona poled with UV irradiation after spin coating. The poling temperature was 70 °C and the power of UV light was 2 mW/cm^2 at 254 nm. The irradiation time was within 3 and 10 min otherwise sample were decomposed. The d_{33} coefficient was measured as 11.5 pm/V for PVCN/CNNB-R (10%) and 21.5 pm/V for PVCN/CNNB-R (20%) at 1064 nm and 3.7 and 5.1 pm/V at 1540 nm, respectively. Polymers had λ_{max} at 520 nm and T_g between 81 and 85 °C.

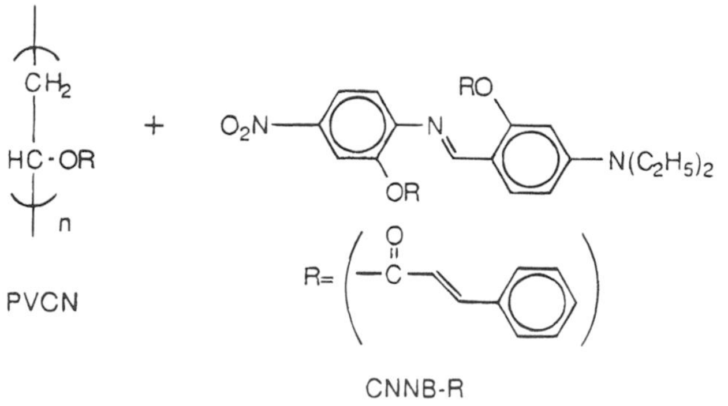

PVCN + CNNB-R

The novel photoactive NLO polymers were reported by Hayashi et al.[513] The absorption peak at 340 nm was found to decrease by poling. The change of the spectra was caused by the molecular alignment. In the case of AZ-2 polymers new absorption appeared after irradiation of UV light and poling around 300 and 400 nm, which is ascribed to the aminobenzoate and azobenzoate groups. The refractive indexes of AZ-1, AZ-2, and DA were 1.573, 1.574, and 1.570 at 1064 nm and 1.605, 1.602, and 1.613 at 532 nm, respectively. The measured d_{33} values were 0.31, 3.6, and 7.2 pm/V for AZ-1, AZ-2, and DA, respectively.

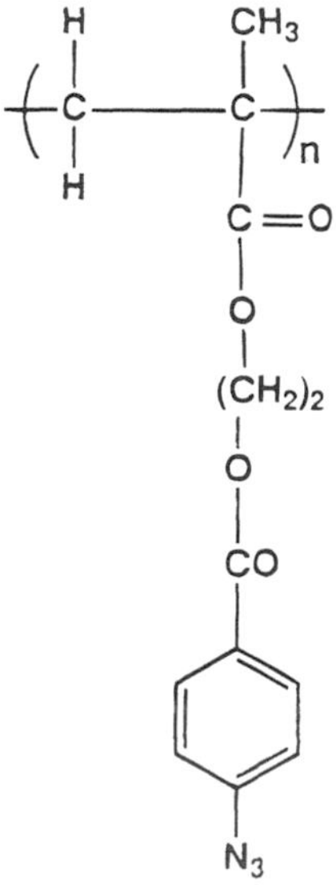

(AZI) 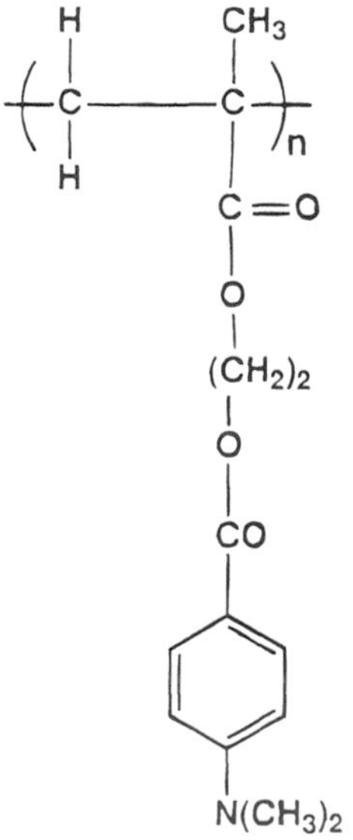   (DA)

Zhu et al.[514] reported the photocross-linkable polymers with stable second-order nonlinearity. P1 and P2 were functionalized with the cinnamoyl group which is used in a traditional photoreactive polymer such as polyvinylcinnamate. P3 and P4 contain the styrylacrylol group as the photocross-linking group which is sensitive to violet to UV radiations. The d_{33} coefficients measured at 1064 nm were 3.2, 7.0, 5.0, and 8.8 pm/V for P1, P2, P3, and P4 systems, respectively. The r_{33} values were 2.4, 4.8, 2.9, and 5.0 pm/V at 633 nm for P1, P2, P3, and P4 systems, respectively. The d_{33} value remained 95% of initial value after 1500 h aging and very stable compared with uncross-linked polymers.

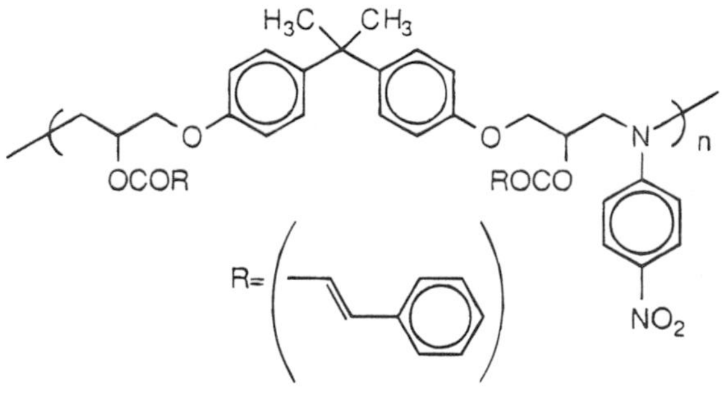

(P1)

(P3)

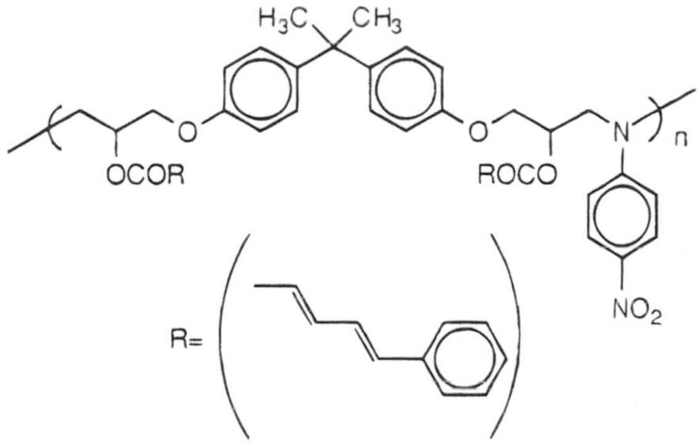

(P2)

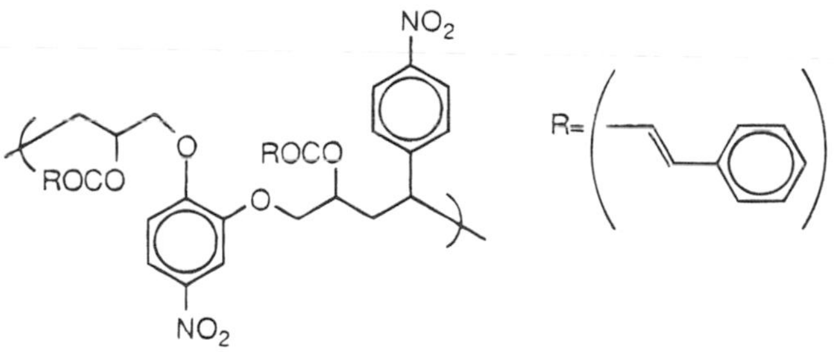

(P4)

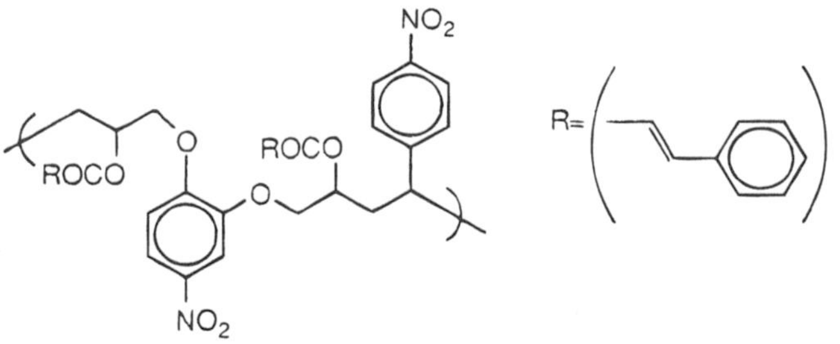

Dalton et al.[515] reported the cross-linking polymer containing the isocyanate groups. The sample was heated above T_g and poled. The cross-linking was thermally carried out between two monomers and exhibited greater than 95% retention of SHG for a period longer than 5000 h. The d_{33} and r_{33} of the polymers were found to be 150–200 pm/V and 40–60 pm/V, respectively.

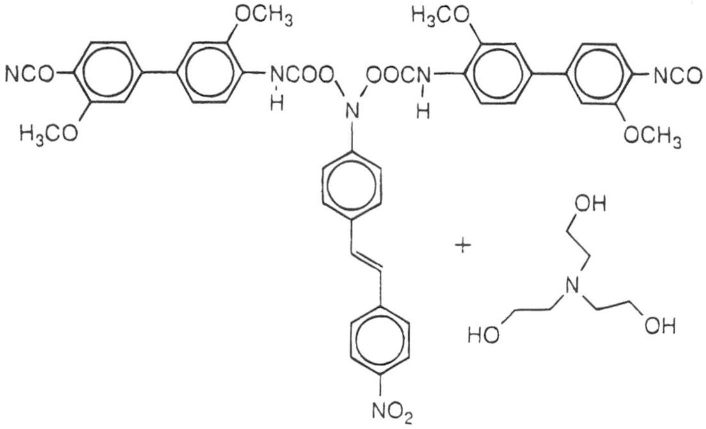

Muller et al.[516] reported the thermally cross-linkable polymers. The compositions of each system (COPO1/TETA, A/TETA, B/TETA) were optimized to provide the good optical quality films with thickness ranging from 1 to 1.5 μm. Chloroform was used as a solvent. The initial r_{33} values were 1.55, 2.0, 0.9, and 3.5 pm/V for COPO1/TETA, COPO1, A/TETA, and B/TETA, respectively. The relatively small electro-optic coefficients were probably due to the unoptimized poling conditions.

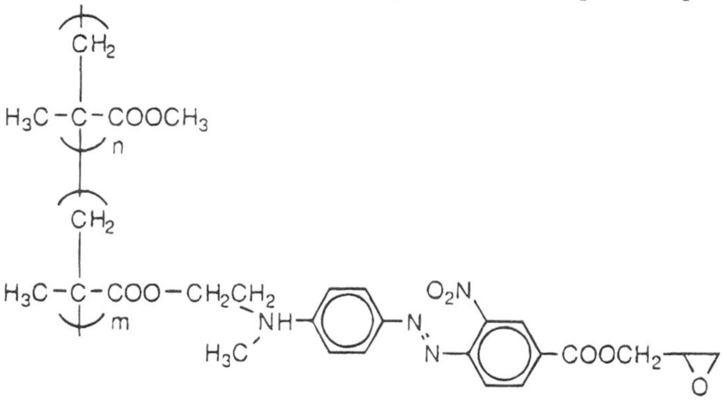

(COPO1)

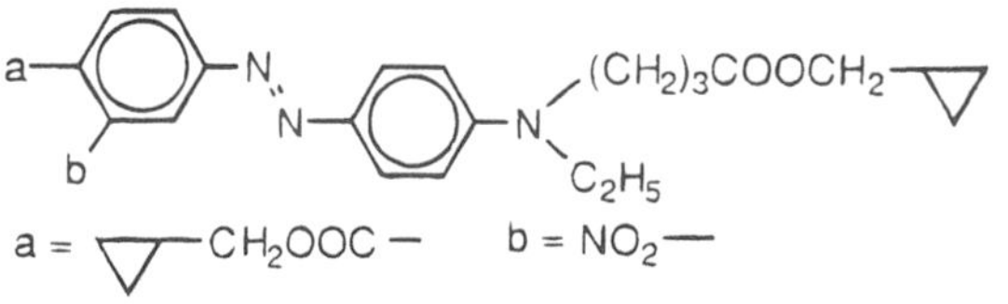

$a = \triangledown-CH_2OOC-$ $b = NO_2-$

Chromophore A :
$a = NO_2-$ $b = \triangledown-CH_2OOC-$

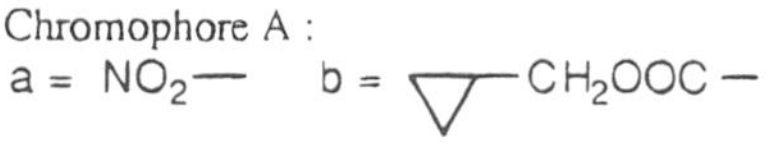

Chromophore B:

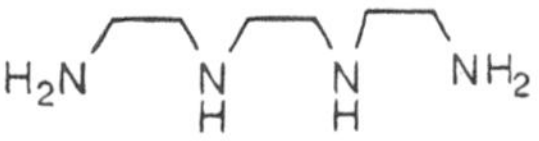

(TETA)

Kalluri et al.[517] reported the sol-gel polymers. An aminosulfone dye was covalently incorporated into a sol-gel network. The NLO chromophore loading density was more than 35%. Both ends of dye were locked by cross-linking; thus, the thermal stability of NLO properties was improved. The long-term stability at 100 °C was more than 500 h. The refractive index of cured sample was 1.663 at 850 nm. NLO coefficients of the system were found to be 27 pm/V when Nd:YAG laser was illuminated.

Gibbons et al.[518,519] reported the novel thermoset polymers containing the tetramethylcyclotetrasiloxane. The polymers were synthesized from three compounds, such as tetraethylcyclotetrasiloxane (TMSTS), NLO chromophore, and dicyclopentadiene. The advantage of these systems is in the two or three covalent bonds participating in the cross-linking networks from different sites of the NLO chromophore, which should be effective to lock the orientation of NLO chromophore in the matrix. The d_{31} values measured at 1300 nm were 5, 30, 28, 5, and 3 pm/V for dye 2:TMCTS, dye 4:TMCTS, dye 5:TMCTS, dye 5:DCPD:TMCTS, and dye 7:TMCTS, respectively. The d_{33} was about three times larger than that of d_{31}. Dye 7 and TMCTS systems exhibited thermal stability exceeding of 190 °C. The improvement of the curing and poling procedures of the system should exhibit good NLO properties.

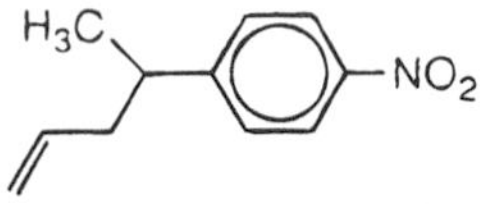

dye 1

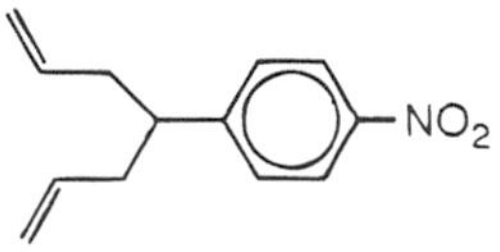

dye 2

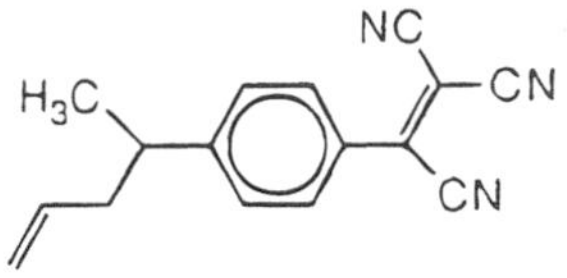

dye 3

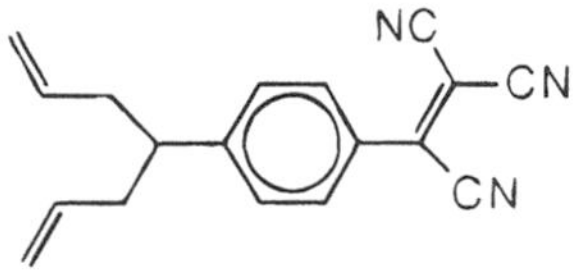

dye 4

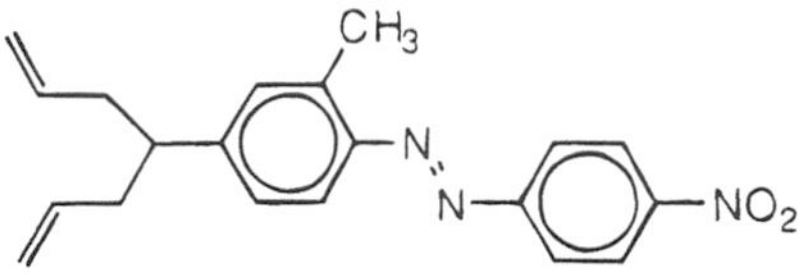

dye 5

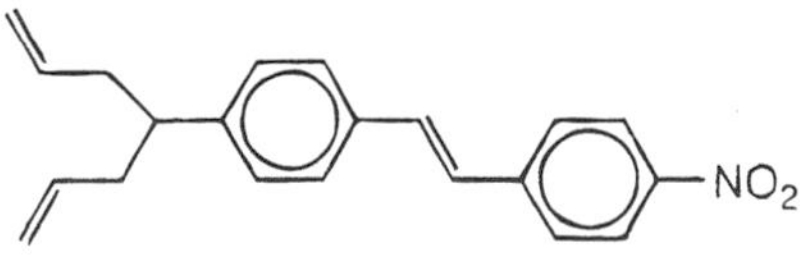

dye 6

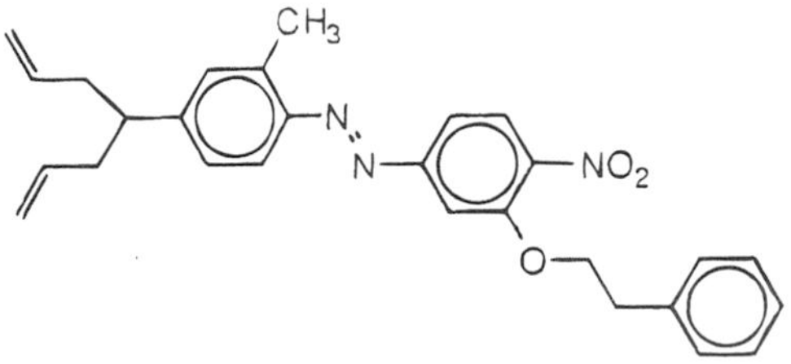

dye 7

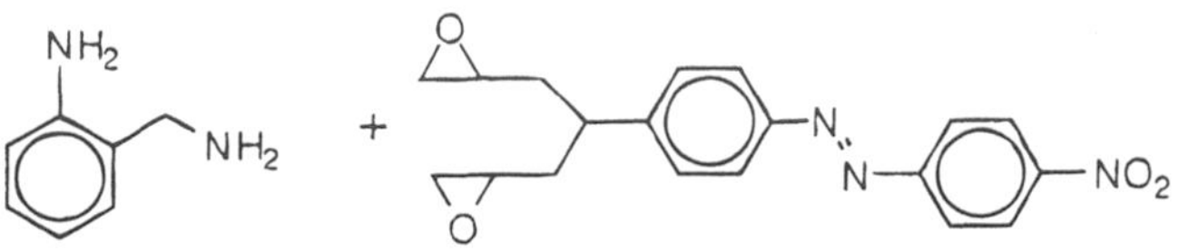

Boogers et al.[520] reported the photocross-linking polyurethanes. Two different acrylic monomers were converted by UV curing at modest temperatures into the highly stable NLO materials with high loading of NLO chromophores more than 50 w/w%. The degradation of polymers by UV light was observed. This degradation can be prevented by using the UV cut filters. The monomer UV1 (40/56/4 ITX) showed a d_{33} value of 34 pm/V, whereas monomer UV2 (49/49/2 Irg) showed a d_{33} value of 13 pm/V and monomer UV2 (96/-/4 Irg) showed a d_{33} value of 43 pm/V. The d_{33} relaxed values were 23, 13, and 8 pm/V, respectively. The components (w/w%) described for the monomer/P3016/PI system were isopropylthioxanthone (ITX) and Irgacure 907 (Irg). Although the azobenzene was partially degradated by UV light the stable SHG activity was observed. NLO chromophore 2 exhibits the relatively large NLO coefficients as much as 13 pm/V with long-term stability in excess of one year at room temperature and no relaxation for a week at 80 °C.

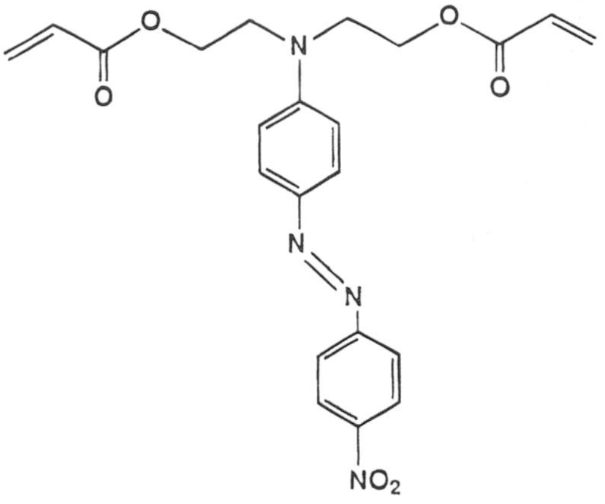

(UV1)

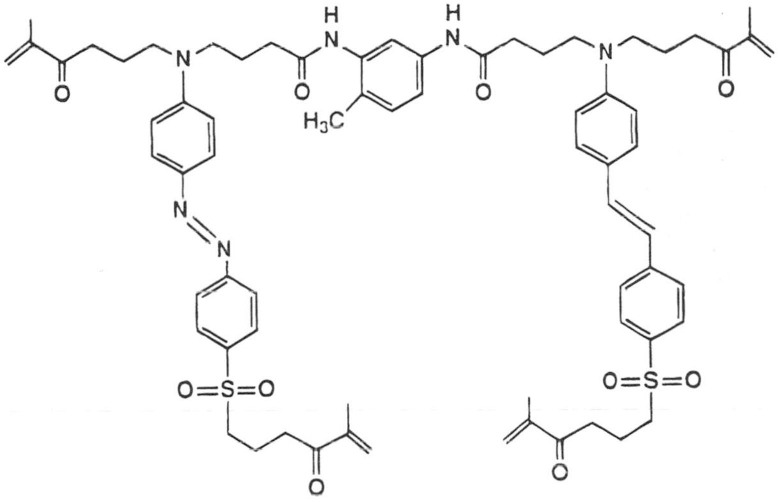

(UV2)

Boogers et al.[521] reported the cross-linking polyurethanes. NLO coefficient d_{33} was found to be 60 pm/V at 1064 nm. This d_{33} large value is associated with the resonance effect of chromophore because there is a strong absorption around 450 nm. The stability of SHG signal was followed at 70 °C. At this temperature, the d_{33} decreased to 40 pm/V after approximately 800 h.

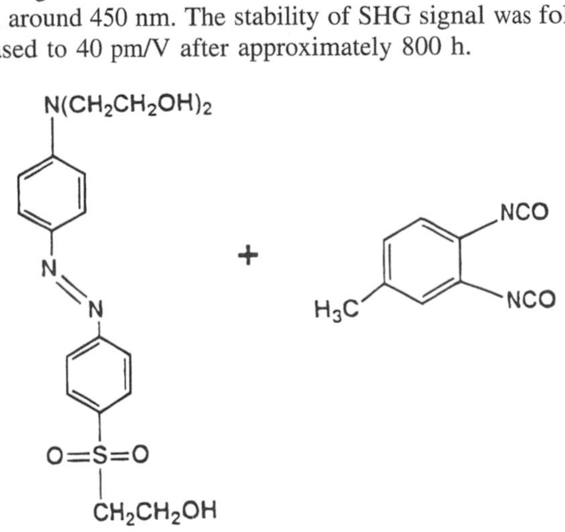

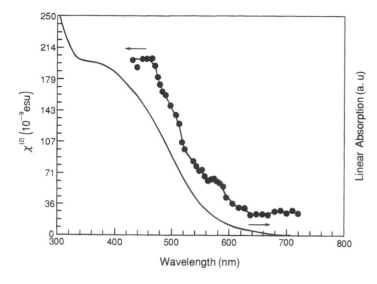

Figure 37 SHG wavelength dependence of $\chi^{(2)}$ and linear absorption spectrum (solid line) for a poled and cross-linked (S)-(+)-NPS-PHS-TPGX films. The $\chi^{(2)}$ was plotted as a function of output wavelength. (Reprinted with permission from Crumpler, E. T., Reznichenko, J. L., Li, D., Marks, T. J., Lin, W., Lundquist, P. M., and Wong, G. K., *Chem. Mater.*, 7, 596, 1995. Copyright 1995 American Chemical Society.)

The cross-linking polymers based on the hexa(methoxymethy)melamine (HMM) were developed by Jeng et al.[522] Two NLO-active dyes 4-(4′-nitrophenyl)aniline (DO3) and 4-amino-4′-nitrobiphenyl (ANB), were embedded into the melamine matrices. The T_g of cured HMM/DO3 was about 100 °C. The poled films of HMM/DO3 and HMM/ANB exhibited d_{33} values of 10.7 pm/V and 1.8 pm/V, respectively. When the sample was heated to 100 °C for 40 h, a reduction of 43% in d_{33} was observed in the case of HMM/DO3. In the case of HMM/ANB a reduction of 55% in d_{33} was observed under similar experimental conditions. Immediately after poling, the absorbance was reduced. This phenomenon is due to the dichroism of the NLO chromophore and the sublimation of dye. Waveguide measurements reveal the low optical loss of these systems.

(HMM)

Crumpler et al.[523] reported the high T_g polymers based on the polyfunctional epoxide systems. Cross-linking experiments were carried out with the difunctional diisocyanate reagent 4,4′-diisocyanato-3,3′-dimethoxybiphenyl (DIISO) and tetrafunctional tetraphenylglycidyl ether of *p*-xylene (TPGX). Poly(*p*-hydoxy styrene-co (S)+NSP), Poly(*p*-hydroxy styrene-co (S)+NSP) with cross-linking agent DIISO and poly(*p*-hydroxy styrene-co (S)+NSP) with cross-linking agent TPGX were spin coated and corona poled. The poling sample temperature was ramped in step and kept for 30 min at maximum temperature. Initial d_{33} was found to be 50 pm/V, 17.5 pm/V, and 18 pm/V where the chromophore ratio is 81 molar ratio. The d_{33} measured as a function of wavelength as shown in Figure 37. The dispersion curves is similar compared with the absorption loss. The maximum d_{33} was obtained at 41.5 pm/V at 840 nm and reduced to 6 pm/V at 1440 nm. The 90% of SHG activity was retained after 1600 h passed at room temperature.

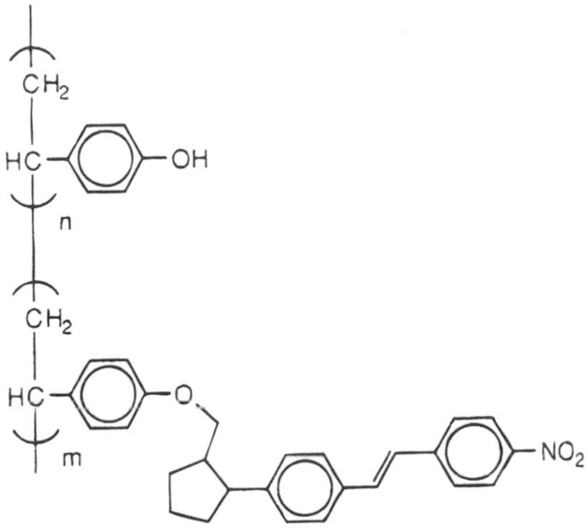

Poly(*p*hydroxy styrene-co (S)+NSP)

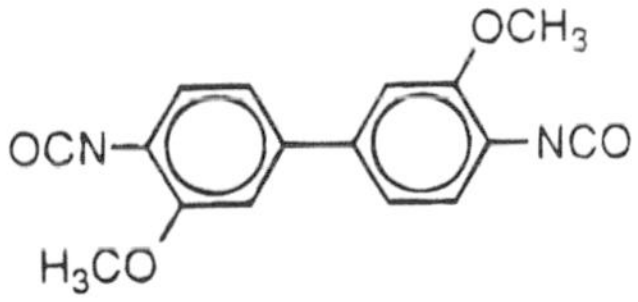

(DIISO)

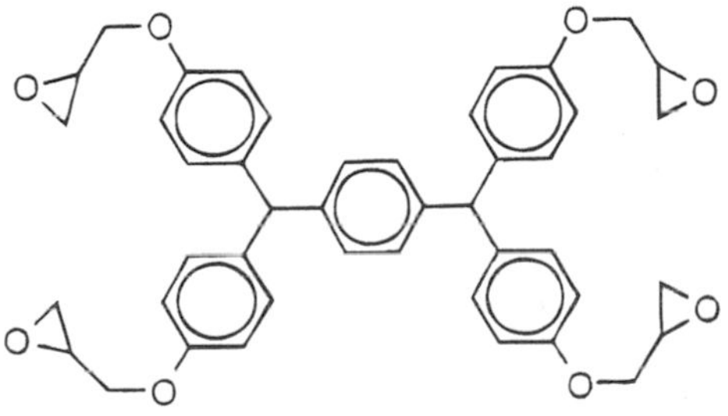

(TPGX)

The new cross-linked polyimide system was reported by Liang et al.[524] The T_g of polyamic acid was approximately 195 °C. The UV-visible near IR spectra of these polyimide without poling and with poling were recorded. The SHG exhibit the dynamic stability up to 160 °C and the a d_{33} coefficient of 70 pm/V.

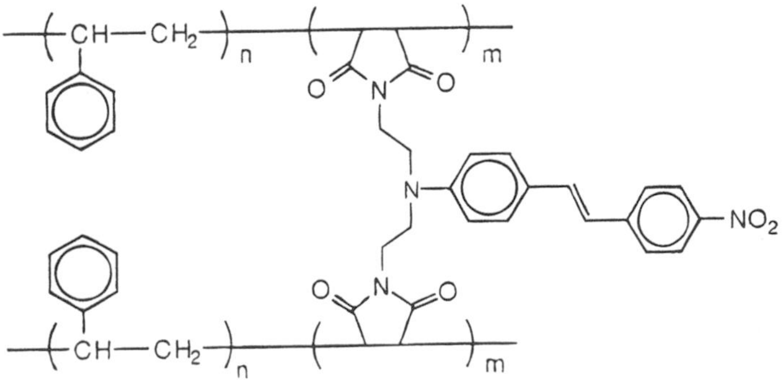

5. Ferroelectric Polymers

Ferroelectric polymers are highly polar materials which exhibit interesting pyroelectric, piezoelectric, and ferroelectric properties.[525,526] The strong piezoelectric effect in PVDF films arises from permanently oriented dipoles and noncentrosymmetric structure. Being noncentrosymmetric PVDF films exhibit both piezoelectricity and SHG. Second-order NLO properties of ferroelectric polymers have been documented recently by Watanabe et al.[527] SHG in PVDF thin film was reported as early as 1971 by Bergman et al.[528] and McFee et al.[529] The PVDF shows birefringence, and the refractive index of 1.425 was measured by Abbe refractometer at 632.8 nm. The measured NLO coefficients of PVDF were d_{33}(PVDF) = $2d_{31}$(PVDF) = d_{11}(quartz), and d_{32}(PVDF) = 0.

(PVDF)　　　　　　　　　　　　　P(VDF/TrFE)

Broussoux and Micheron[530] reported refractive indexes of n_1 = 1.444, n_2 = 1.436, and n_3 = 1.425 and electro-optic coefficients r_{51} = 0.10 pm/V and r_{42} = 0.21 pm/V. Robin et al.[531] reported SHG properties of vinylidene fluoride/trifluoroethylene P(VDF/TrFE) copolymer for the composition of VDF:TrFE = 65:35 mol%. Unpoled films showed no SHG, although a d_{33} coefficient of 0.6 pm/V was obtained after electrical poling. The d_{33} coefficient decreased with increasing temperature and vanished at the Curie temperature, which suggests that SHG was related to the remaining polarization. The d_{33} coefficient also showed a strong poling field dependence. Legrand et al.[532] observed no SHG in the paraelectric phase of a VDF/TrFE copolymer for a composition of 70:30 mol%. The SHG intensity was found to depend on the poling field. Sato and Gamo[533] reported SHG properties of P(VDF/TrFE) copolymer for a composition (VDF:TrFE = 78:22 in mol%). where the SHG signal showed a quadratic dependence as a function of input power. A d_{33} coefficient of 1.71 pm/V for P(VDF/TrFE) was reported by Pantelis et al.[427] Broussoux et al.[425] reported a d_{33} coefficient of 1 ± 0.5 pm/V for P(VDF/TrFE) with a composition of 75:25 mol%. The average electro-optic coefficient $\langle r \rangle$ measured by the surface–plasmon method was 16 pm/V and the electro-optic part of this coefficient was between 1 and 3 pm/V, similar to the value obtained from the d_{33} value. Berry et al.[534] reported an electro-optic coefficient of 15 pm/ V, refractive index = 1.4, and optical loss of 4 dB/cm at 633 nm for P(VDF-TrFE) copolymer. As can be seen from discussion above, that both SHG and electro-optic coefficients of copolymers are governed by the copolymer composition and measurement conditions.

Tsutsumi et al.[535] reported SHG from the blends P(VDF/TrFE) with 75:25 mol% ratio in PMMA. The blends of P(VDF/TrFE)/PMMA and P(VDF/TrFE)/PMMA/P(MMA-co-MMA-DR1) with low DR1 contents were optically transparent. The d_{33} coefficients 1.38, 0.96, and 0.38 pm/V were recorded for P(VDF-TrFE), P(VDF-TrFE)/PMMA blend, and P(VDF/TrFE)/PMMA/P(MMA-co-MMA-DR1), respectively. The temperature dependent studies showed a decrease in d_{33} for 90:10 P(VDF-TrFE)/ PMMA blend between 40 and 60 °C due to the molecular relaxation. The d_{33} of P(VDF/TrFE)/PMMA/ P(MMA-co-MMA-DR1) was thermally stable at T_g. The guest–host systems of ferroelectric polymers seem promising for stabilizing the SHG efficiency.

Cyano-polymers are interesting because of the large C≡N dipole moments in the molecules. Azumai et al.[536] reported frequency doubling in vinylidene cyanide/vinyl acetate P(VDCN/VAc) copolymer thin films. SHG was evaluated by using the Er:YAG 2.94 μm laser radiation. The SHG intensity increased significantly with increasing poling field and was enhanced both by poling and stretching (Fig. 38). The P(VDCN/VAc) copolymer showed an effective $\chi^{(2)}$ of 234 pm/V which is larger by a factor of 2.8 compared with lithium niobate crystal. Effective $\chi^{(2)}$ values of 41.4 and 234 pm/V were measured for the P(VDCN/VAc) thin films at 1.064 and 2.94 μm, respectively, by Azumai et al.[537] with slab waveguides in the Cerenkov radiation scheme. At 1.064 μm, 1.5 times enhancement was obtained from Cerenkov radiative scheme compared with that obtained from a uniform scheme.[538] Dumont et al.[539] measured electro-optic coefficient of P(VDCN/VAc) copolymer as 1.2 pm/V. Broussoux et al.[540] reported the d_{33} coefficient of 0.4 pm/V at 1.064 μm after poling 160 °C.

P(VDCN/VAc)

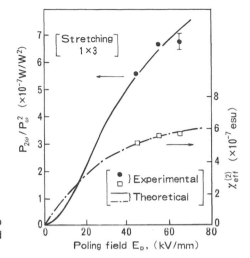

Figure 38 The SHG activity as a function of drawing ratio for P(VDCN/VAc) copolymer. (From Azumai, Y., Sato, H., and Seo, I., *Opt. Lett.*, 15, 932, 1990. With permission.)

Kishimoto et al.[541] measured SHG properties of VDCN and *p*-hydroxyvinyl benzoate copolymers having 4-nitro-4'-oxobiphenyl (BP) and 4-nitro-4'-oxostilbene (ST), referred to as P(VDCN/VBZ-ST) and P(VDCN/VBZ-BP), at 1.064 μm. The P(VDF/VAc) copolymer which has no pendant NLO chromophore showed a d_{33} value of 8 pm/V. The d_{33} value was increased significantly for copolymers having side-chain NLO chromophores. The d_{33} values of P(VDCN/VBZ-ST) and P(VDCN/VBZ-BP) were 15 and 7.5 times larger than that of P(VDF/VAc) copolymer due to the large dipole moment of the C-CN in the vinylidene cyanide segment and larger space provided by the vinyl acetate segment for easy rotation of CN dipoles under high electric fields.

P(VDCN/VBZ-BP)

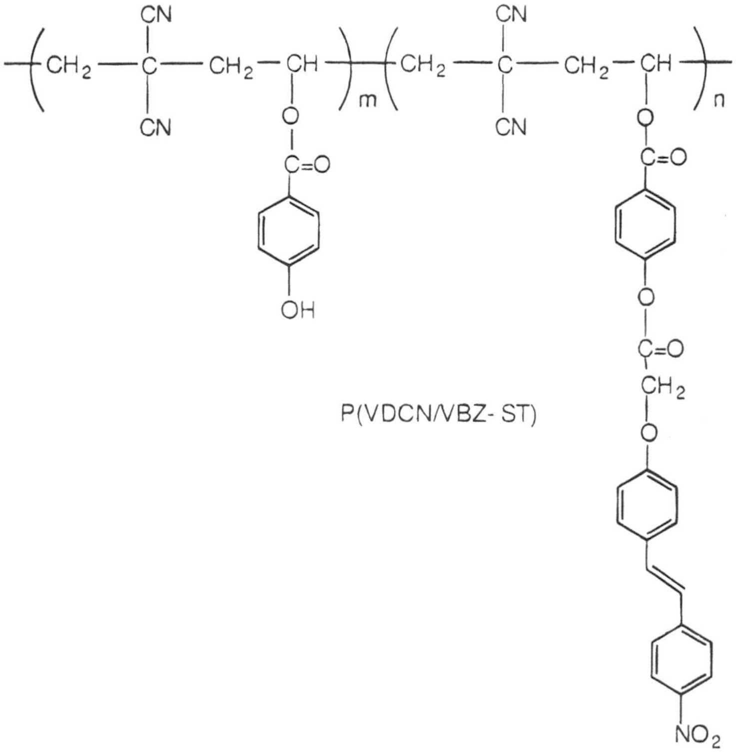

P(VDCN/VBZ- ST)

The SHG properties of aromatic polyureas with and without NLO chromophore were reported by Nalwa et al.[484–487,542,543] The PU1, which is a ferroelectric polymer, showed a d_{33} coefficient of 5.5 pm/V and d_{31} coefficient of 1.67 pm/V. Temporal studies showed stable SHG for PU1 at room temperature. Kajikawa et al.[488] reported a d_{33} value of 1.65 pm/V for polyurea (PU2) thin films formed by vapor deposition polymerization. The SHG was stable indefinitely at room temperature after a slight initial decay in the first week.

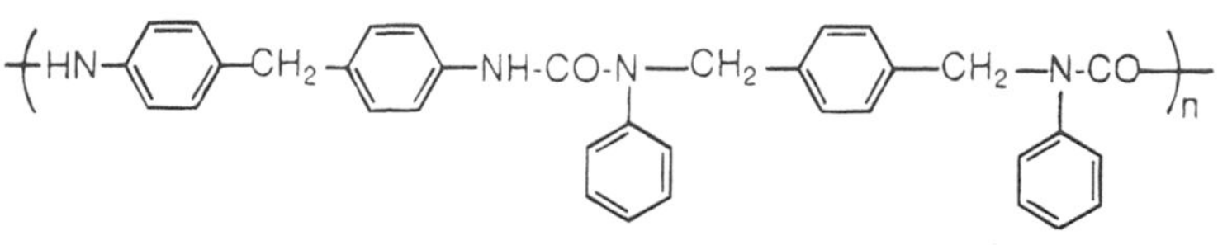

(PU1)

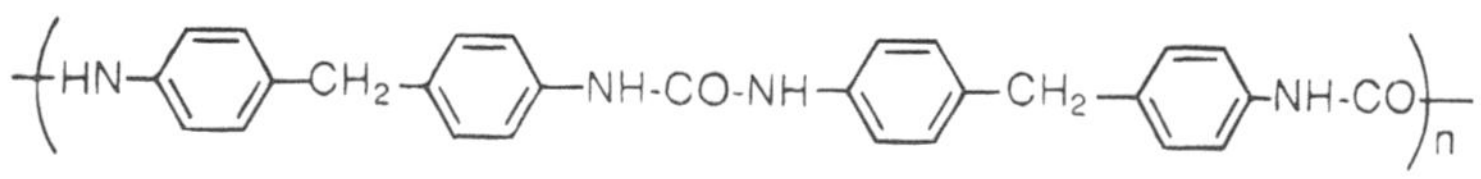

(PU2)

There are only a few reports on the SHG of ferroelectric liquid crystalline (FLC) polymers.[527,544,545] Ozaki et al.[546] reported the SHG properties of a FLC polymer having siloxane backbone with the chiral mesogen in the side chain. The SM A phase showed no SH signal being centrosymmetric, and the chiral smectic C(Sm C*) phase was SHG inactive. Under an electric field, the SHG intensity was found to increase in the Sm C* phase depending on the rotating angle. The FLC polymer attained a noncentrosymmetric structure and dipole moments were oriented in the direction of the 2-fold axis normal to the molecular long axis and parallel to the smectic layer. The effective NLO coefficient of 0.02 pm/V under the phase-matching conditions was evaluated. Type I phase matching was observed. The low concentration of the side-chain mesogen and poor homeotropic alignment may have been responsible for lower SHG activity.

Coles et al.[547] reported SHG in a FLC side-chain copolymer that has polysiloxane backbone with 30% of the sites functionalized with electroactive side groups. The SHG of the copolymer depended on temperature that coincides with the phase behavior, and the SHG efficiency of the copolymer was equal to that of quartz. Ferroelectric and NLO properties were investigated by varying the ratio of mesogen to chromophore content. Ferroelectric properties were found to be suppressed by increasing chromophore contents and no switching was observed for more than 50% contents.

The SHG efficiency of ferroelectric polymers with and without NLO chromophores range between 0.5 and 234 pm/V, depending on the chemical structures and measurement wavelength. Cyano copolymers seem interesting because they offer excellent optical transparency, high T_g, and large SHG efficiency. Likely, materials performance and SHG efficiency can be tailored for polyureas using NLO chromophores with large β values. Ferroelectric polymers are advantageous because they show stable SHG efficiency over an extended period at room temperature.

A wide variety of poled polymers have been discussed above. Poled polymers are promising NLO materials because they have large NLO coefficients ranging from tens to a few hundred pm/V. In past years, there has been considerable progress in applying various chemical and physical means to cover the SHG temporal stability problem. In this direction, chemical strategies using high T_g polymers, cross-linking, and composite formation proved quite effective in preventing molecular motion of NLO chromophores.

VIII. SHG EFFICIENCY VS. TRANSPARENCY TRADE-OFF

The trade-off between optical transparency and SHG efficiency is an important issue while designing devices using NLO materials. Zyss[548] has dealt with this in great detail. As discussed earlier, hyperpolarizability of an organic molecule is related to the charge-transfer characteristics eventually governed by the increasing conjugation length and the strength of donor and acceptor groups. Therefore there should exist a trade-off between hyperpolarizability and optical transparency, such as large β values can be obtained from highly polar molecules, but that leads to a loss in optical transparency. Both β and optical transparency are affected by the nature of the conjugated bonds, length of π-conjugation, strength of electron donor and acceptor substituents, and conformation.

Cheng et al.[549,550] demonstrated that electron-accepting groups such as the carbonyl, fluoromethyl, perfluoroalkyl, sulfonyl, and perfluoroalkyl sulfonylsulfimide are favorable for trade-off between hyperpolarizability β and optical transparency. The fluorinated sulfonyl and sulfonylsulfimide groups show better transparency trade-off because low-lying transitions are not required. Table 67 compares the absorption maxima, dipole moment, and β values for both fluorinated and unfluorinated compounds with nitro group-substituted compounds. If a comparison is made between nitrobenzenes and fluorinated sulfone then sulfone compounds show a lower λ_{max} value and also slightly lower β values, but larger dipole moments showing interesting μβ and transparency trade-off.

Duan et al.[551] reported that the |β| values of stilbazolium p-toluenesulfonates were 3 to 10 times larger than the |β| value of p-NA as measured by the HRS method.

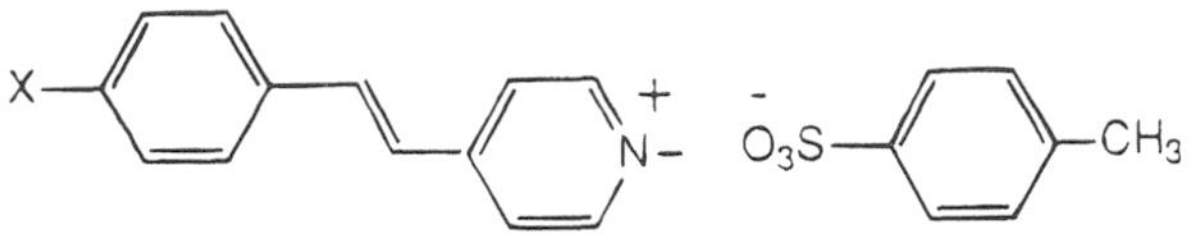

The |β| values of stilbazolium p-toluenesulfonates where X is varied from CN, Br, Cl, H, CH₃, OCH₃, to OH were $105 \pm 20 \times 10^{-30}$, $305 \pm 60 \times 10^{-30}$, $290 \pm 60 \times 10^{-30}$, $190 \pm 40 \times 10^{-30}$, $280 \pm 60 \times 10^{-30}$, $360 \pm 80 \times 10^{-30}$, and $390 \pm 80 \times 10^{-30}$ esu, respectively. These compounds where X is CN, Br, Cl, H, CH₃, OCH₃, and OH showed λ_{max} of 335, 348, 346, 347, 361, 381, and 391 nm, respectively. As seen here, the halogenated stilbazolium compounds show larger |β| values than unsubstituted compounds, although they have almost the same optical transparency.

Many examples illustrating SHG efficiency versus transparency trade-off can be found in this chapter. For more details, single crystals and poled polymers are referenced. Organic materials can be categorized into three main-target families depending on the transparency regions and their quadratic nonlinear optics as described by Zyss: **(1) transparent materials** having cutoff wavelength below the 450 nm region (e.g., urea-like molecules, polyureas); **(2) yellow materials** having cutoff wavelengths in the 500 nm region. The yellow materials are usually p-NA derivatives, MNA, NPP, POM, NPAN, MMONS, etc., and some NLO polymers; **(3) colored materials** (red, blue, and dark colors) having solid-state transparency cutoff wavelengths above the 550 nm region. The well-known colored systems are stilbene derivatives, diazo-molecules, phenylhydrazones, azulene derivatives, dicyano-divinylidene stilbenes, push-pull polyenes, organometallics, and poled polymers having these NLO chromophores.

Table 67 Absorption Maximum, Dipole Moment, and β Values of *para*-Disubstituted Benzenes

Donor	Acceptor	λ_{max} (nm)	μ (10^{-18} esu)	β (10^{-36} esu)
Donor—⟨benzene⟩—Acceptor				
F	$SO_2C_{10}F_{21}$	225	3.6	1.5
Br	$SO_2C_{10}F_{21}$	245	3.5	2.0
$N(C_2H_5)_2$	$SO_2C_{10}F_{21}$	314	7.3	9.0
NC_5H_{10}	$SO_2C_{10}F_{21}$	314	7.3	9.1
OH	SO_2CH_3	290	3.4	1.3
NH_2	$SO_2C_{16}H_{33}$	262	5.5	2.1
OCH_3	SO_2CF_3	290	5.4	3.3
$N(CH_3)_2$	NO_2	376	6.4	12
OCH_3	NO_2	302	4.6	5.1
OCH_3	$COCF_3$	292	4.0	3.6
$N(CH_3)_2$	$COCF_3$	356	5.9	10
NC_4H_8	SSI	336	9.0	13
Donor—⟨benzene-vinyl⟩—Acceptor				
OCH_3	SO_2R_f	316	5.5	14
OCH_3	NO_2	352	4.6	17
$N(CH_3)_2$	SO_2R_f	376	7.4	35
$N(CH_3)_2$	NO_2	440	6.5	50
Donor—⟨biphenyl⟩—Acceptor				
OCH_3	$SO_2C_3F_7$	305	6.0	9.1
OCH_3	$SO_2C_6F_{13}$	305	5.9	11
OCH_3	$COCF_3$	328	4.2	10
OCH_3	$COCH_3$	304	3.4	4.9
OCH_3	NO_2	332	4.5	9.2
$N(CH_3)_2$	$SO_2C_6F_{13}$	362	8.0	25
$N(CH_3)_2$	NO_2	390	5.5	50
Donor—⟨benzene-vinyl-benzene⟩—Acceptor				
OCH_3	$SO_2C_5F_{11}$	336	6.5	10
OCH_3	$SO_2C_6F_{13}$	347	7.8	14
OCH_3	$COCF_3$	368	4.2	16
OCH_3	NO_2	368	4.5	28

From Cheng, L. T., et al., *SPIE Proc.*, 1337, 203, 1990; Cheng, L. T., et al., *Nonlinear Optics*, 3, 69, 1992. With permission.

The solid-state cutoff wavelength results from the intermolecular as well as intramolecular interactions. The solid-state transparency-efficiency trade-off is an important factor in deciding the potential application of an individual NLO material. Figure 39 shows the three sets of efficiency-transparency materials with respect to their molecular hyperpolarizability, SHG, electro-optic coefficients. Therefore, molecular engineering of organic NLO materials should be considered in terms of efficiency-transparency basis. The colors of NLO chromophores can be tailored either by manipulating π-conjugated system or by donor–acceptor groups. For example, electron acceptor carbonyl groups would provide transparent materials compared to a nitro group. Likely, the cutoff wavelength would shift tremendously toward longer wavelengths with increasing π-conjugation lengths. One of the applications of NLO chromophores

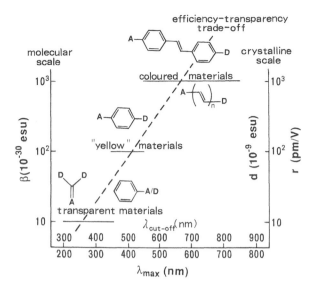

Figure 39 Transparency-efficiency trade-off scale for SHG active materials. Materials can be divided into three classes; transparent, yellow, and colored. Plots show the relationship between the wavelength, molecular hyperpolarizability, and second-order NLO coefficients. (From Zyss, J., in *Conjugated Polymeric Materials: Opportunities in Electronics, Optoelectronics and Molecular Electronics,* Bredas, J. L., and Chance, R. R., eds., NATO ASI Series E, Appl. Sci., Vol. 182, Kluwer Academic Press, Dordrecht, 1990, p. 545. Reprinted by permission of Kluwer Academic Publishers.)

is as semiconductor lasers. Regions of color spectrum of organic NLO chromophores corresponding to semiconductors laser systems are ultraviolet (366 nm), violet (400 nm, GaN), blue (450 nm, ZnS), green (500 nm, AlP), yellow (580 nm, $GaAS_{1-x}P_x$), orange (600 nm, $InxGa_{1-x}P$), and red (650 nm, GaP:N). Mostly semiconductor lasers are operated outside the UV-visible region. Several single crystals have shown potential in blue-green region. The nonlinear and transparency tradeoff is very important in designing organic molecules for efficient frequency doubling in the blue-green spectral region. On the other hand, colored NLO materials can be used for electro-optical applications. The desired wavelength for optical telecommunication system is 1.3 μm.

IX. APPLICATIONS IN PHOTONICS

The potential applications of second-order NLO materials are as great as those of electronic materials in which photons can do a better job in providing devices that are much faster and cleaner. The production in the optoelectronic industries is growing at a rapid rate. The examples of optoelectronic components are lasers, displays, optical fibers, light-emitting diodes (LED), photodectors; optoelectronic equipment includes optical sensors, optical disks (audio and video players), optical telecommunication devices, printers, laser processing equipments and optical telecommunication systems. NTT has developed a super-high-speed, large-capacity optical mass-storage system (OMSS) that can store a maximum of 1 Tbit of information, equivalent to 5 million newspaper pages. Recently, new products such as visible and high-power diode lasers, optical memory systems, and fiber optic systems have also emerged. The importance of electro-optic technologies is demonstrated by the fact that the gross value of production of photonic components, equipments, and systems reaches several billion dollars annually in the Asia-Pacific region alone.[552] The photonic technologies being developed may replace many of the electro-optical devices mentioned above because same type of setup could be applied. The potential applications are in integrated laser diodes, spatial light modulators, holography, telecommunications networks, optical computing, frequency up-and-down conversions, and all-optical processing. A simple example of an application of NLO materials is to change the frequency of laser light via generation of second and third harmonics. For simplicity, SHG materials can convert a 1064 nm Nd-YAG laser beam into a 532 nm intense green laser beam, whereas the THG materials can convert the same 1064 nm near-infrared laser beam into a 355 nm ultraviolet laser beam; hence, NLO materials can be used as frequency doublers and frequency triplers, respectively. In addition, the NLO materials can be exploited in generating one or more new laser frequencies either by interacting two laser beams of the same frequency or by differing frequencies. For example, a blue light at 473 nm can be produced by sum-frequency generation using a Nd-YAG laser and an 946 nm laser. The NLO processes are of both scientific and practical interest. Second-order NLO materials are receiving a great deal of attention as the most promising components of optical communication systems, frequency doubling, modulation, switching, directional couplors, voltage sensors, parametric oscillator, electromagnetic radiation detector, and other photonic devices.

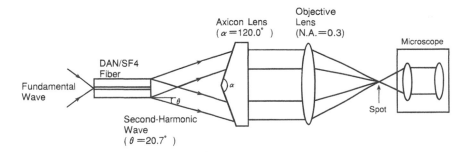

Figure 40 Experimental set-up for focusing the SH light. (From Harada, A., Okazaki, Y., Kamiyama, K., and Umegaki, S., *Appl. Phys. Lett.,* 59, 1535, 1991. With permission.)

A. FREQUENCY DOUBLING
1. High-Density Recording

Frequency doubling of a well-developed infrared semiconductor laser is required to produce the compact short-wavelength laser sources for high-density optical recording. Organic materials have great potential due to their large optical nonlinearities. The frequency doubling can be performed in a variety of ways including: (1) modal dispersion of waveguide, (2) quasi-phase matching, (3) anomalous dispersion, (4) Cerenkov radiation, and (5) counter-propagated waves. All techniques have been mentioned in chapter 5; here we will mention only the application of frequency doubler. Harada et al.[553] demonstrated the efficient frequency using Cerenkov radiation from organic single-crystal-cored fiber (OSCCF). The OSCCF has the following advantages: (1) greater amplitude of the nonlinear polarization can be obtained due to confinement of fundamental laser in a small region without the reduction of interaction length; and (2) the phase-matching condition can be fulfilled without birefringence. In addition the crystals are protected by a glass wall from the chemical and physical damage. Many kinds of commercially available glass were used as cladding materials. The DAN fiber generated the SH power of 29.4 mW when 124 mW Nd:YAG laser was illuminated. The highly efficient doubling of a semiconductor laser was performed using DMNP fiber. The largest NLO d_{32} coefficients of DMNP can be used for the linearly polarized LP_{01} fundamental mode. The diameter of the core and the cladding were 1.4 μm and 1.5 mm, respectively, and the interaction length was 5 mm, 17 mW blue light was observed. When 65 mW of the semiconductor was illuminated, SH light of Cerenkov configuration was radiated with an external angle θ defined as:

$$\cos \theta = N(\omega)/N_{\text{clad}}(2\omega) \tag{58}$$

The radiated SH pattern looks like a single ring and the wavefront of the generated SH wave was conical. Although the usual spherical lens cannot collimate the harmonic light, the SH wave can be collimated by using an Axicon lens, which is known as conical prism (α = 120°) and focused by an objective lens (N.A. = 0.3).[553b] The experimental set-up is shown in Figure 40. The spot size of main peak due to the Fraunhofer diffraction was 1.8 μm in a diameter. For the light beam with a circular aperture, the spot size D is given by:

$$D = 1.22 \ \lambda/\text{N.A.} \tag{59}$$

This gives the D value of 2.2 μm for the wavelength λ of 532 nm and for N.A. of 0.3. The experimental value agrees well with the theoretical calculation. This small spot size can increase the recording density of compact disc (CD) or improve the resolution of laser printer.

Frequency conversions in nonlinear waveguides have gained much attention recently. Various methods of phase matching have been reported in polymer waveguides that include quasi-phase matching (QPM), Cerenkov, modal dispersion, and noncollinear phase matching. Norwood et al.[554] have provided a comparison of quasi-phase-matched frequency doublers in organic and inorganic NLO polymers. In

Table 68 Phase-Matched Frequency Doubler from Organic NLO Materials

Material	Waveguide Format	Phase-matching Length (mm)	Efficiency (%/W)	Method of Phase-matching	Ref.
KTP	Channel	5	100	QPM	555
LiNbO$_3$	Channel	1	0.24	QPM	556
LiTaO$_3$	Channel	5	20	QPM	557
DCANP (LB film)	Channel		2×10^{-3}	Cerenkov	558
MBANP	Channel	2		Cerenkov	559
m-NA	Channel	5		Cerenkov	560
	Channel	5	0.04	QPM	561
Oxynitrostilbene copolymer	Slab	5	0.01	QPM	562
Oxysulfonestilbene MMA copolymer	Slab	5	0.4	MDPM	563
Oxynitrostilbene/ MMA copolymer	Slab	0.1 ~ 1	3×10^{-7}	Non-collinar	564
PMMA-co- DR1MA	Slab		1×10^{-5}	Cerenkov	565
MNA/PMMA	Slab		1.7×10^{-4}	MDPM	566
VDCN/VAC	Slab	2.0		Cerenkov	567

QPM = quasi-phase matching; MDPM = modal dispersion phase-matching; DCANP = 2-docosylamino-5-nitropyridine; MBANP = (−)2-(α-Methylbenzylamino)-5-nitropyridine.

Table 68, we have summarized single crystals, LB films, and poled polymers where frequency doubling have been demonstrated.

2. Compact Coherent Laser Source

Application of new inorganic as well as organic NLO materials are expected in the development of novel laser systems; for example, in optical recording systems. The semiconductor diode lasers are used in the digital audio disk or compact disk system in the 780 to 830 nm wavelength region. Diode lasers that weigh less than a gram show much better electrical-to-optical power conversion efficiency than gas lasers. Most of the recording media now are based on the semiconductor-diode lasers.[568] Recent developments in applications of NLO inorganic materials are the fabrication of semiconductor diode lasers. For high-density optical data storage systems, shorter wavelength lasers are desirable. Currently much attention has also been focused on the II-IV semiconductors as the new promising materials for the fabrication of blue (430 nm) and green (532 nm) laser diodes. Haase et al.[569] produced a blue-green semiconductor laser using a separate-confinement heterostructure (SCH) configuration. The research team at Sony and Philips Laboratories reported the blue-green injection lasers containing pseudomorphic $Zn_{1-x}Mg_xS_ySe_{1-y}$ cladding layers.[570] The II-VI semiconductors such as ZnCdSSe and MgZnCdSSe are potential candidates for compact and inexpensive green lasers (490 nm), but presently they have short life times. In that respect, single crystals of KTP, LiNbO$_3$, and KNbO$_3$ are well established for high-power lasers. Organic materials can have similar applications in blue- and green-light-emitting lasers because they show high SHG efficiency and can be tailored for high performance.

The microtip lasers have been investigated extensively because of their high conversion efficiency. Intracavity SHG using microtip solid-state lasers pumped by a laser diode (LD) combined with a NLO material offers a very simple, compact method to obtain shorter wavelength light with high conversion efficiency. If the organic material with a NLO coefficient of 50 pm/V is used for SHG, the interaction length of the crystal can be below 1 mm to get the same output energy from an inorganic crystal. The enhancement of SHG can be performed by using optical resonance. First let consider the external cavity SHG, which is shown in Figure 41, where the output power of the oscillator into the resonator is I_0,

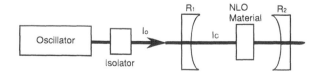

Figure 41 Schematic diagram of extracavity SHG. (From Sasaki, T., *Nonlinear Optics*, Miyata, S., ed., North-Holland, 1992, p. 445. With permission.)

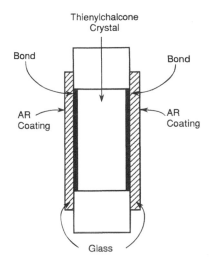

Figure 42 Schematic diagram of SHG device using thienylchalcone crystal. (From Sasaki, T., *Nonlinear Optics*, Miyata, S., ed., North-Holland, 1992, p. 445. With permission.)

the reflectivities of the input side and the other side mirrors are R_1 and R_2, and the circulation power of oscillator is I_c.[571] As it is well known, the highest enhancement of the circulation power by optical resonance can be obtained when $R_2 = 1$, $R_1 = 1 - \alpha$, that is the impedance matching is achieved and I_c is expressed as follows

$$I_c = I_0/(1 - \alpha) \tag{60}$$

where α is the total round trip loss in the cavity. If α is 2%, I_c becomes 50 times of I_0. When the conversion efficiency is small, the enhancement of SHG becomes 2500, as SHG is proportional to the square of I_0. Therefore, large enhancements are possible for resonator with low loss. To get high conversion efficiency by using an extracavity SHG technique, the following conditions are necessary:

1. Low loss in resonator
2. Narrower input spectrum than full width of half-maximum (FWHM) of transmission
3. Feedback system to coincide the spectrum of the oscillator with that of the resonator
4. Isolator between oscillator and resonator
5. Special mode matching between oscillator and resonator
6. Impedance matching to enhance fundamental power of resonator

On the other hand, in the intracavity SHG device an arbitrary frequency cannot be chosen like a laser diode. Interactive SHG needs a solid-state laser and the oscillation wavelength is determined by the ion doped in a solid-state laser. A feedback system for frequency doubling is not necessary. The intracavity SHG method can provide a high conversion efficiency of more than few percent by using a microtip laser; its construction is more compact and simpler than external SHG method, which makes the internal SHG method more practical. Sasaki et al.[571] developed the intracavity SHG laser using TC-28 and TC-121 crystals. Because organic crystals are very soft, it is difficult to polish the surfaces finely. Therefore the crystal was placed between two BK-7 glass plates, adhering by the optical bond as shown in Figure 42. A 3-mm-thick crystal cut for Type I phase matching, was buff-polished with 1-mm diamond powder. The substrates were antireflection coated at 1064 nm. The maximum transmittance of 96% was obtained at 1064 nm. This corresponds to the loss coefficient of 0.6 dB/cm in the crystal. The maximum power of 0.5 mW was obtained at the absorbed power 750 mW in Nd:YVO$_4$, but the higher power could not be obtained in spite of the straight increase of the cavity power at the fundamental wavelength. These phenomena ascribe to the phase mismatch induced by heating. These microtip lasers will be used instead of Ar ion and He-Cd lasers, not only in the scientific field but also in industrial fields.

B. MODULATORS

The modulation of propagated light in a medium is required for telecommunication. This modulation can be performed by use of electro-optic polymers. Here one must consider the propagation time τ of

320

light in a modulator, because constant voltage can not be applied if $\omega_m\tau$ is smaller than $\pi/2$. ω_m is the angular frequency of modulation. When $\omega_m\tau$ is smaller than $\pi/2$, the modulator can be treated as lumped circuit. The simple birefringent modulator was prepared using poled polymer.[572] The modulator structure is given in Figure 43. The active NLO layer is sandwiched between two buffer layers. The glass substrate sustains the first electrode. The first buffer layer is a photo-cross-linked polymer with a refractive index $n = 1.51$; the active polymer is spin coated on this layer. The second buffer layer is a fluorinated polymer of refractive index $n = 1.42$. Electrode 2 is evaporated aluminum. The laser beam is coupled into an active NLO layer via prism. The light intensity is represented by the following equation.

$$I = \sin^2 \frac{\varphi}{2} \tag{61}$$

where φ is the birefringence, with $\varphi = \varphi_0 + \varphi_{E0}$, φ_0 is the static birefringence. By variation of the modulating electric field, one can obtain the interference as a function of voltage. The difference of $r_{33} - r_{13}$ works as the effective electro-optic coefficient.

Most modulators have been fabricated in a Mach-Zehnder configuration. The fabrication of modulator is shown in Figure 44.[573,574] The phase difference between the two arms of the interferometer induce the modulation of propagated light. The largest electro-optic coefficients of poled polymer can be used in this configuration. The poled polymer was spun into high-quality thin film on the substrate. As spun, the film was isotropic and exhibited no SHG. In addition to inducing a noncentrosymmetric structure to achieve macroscopic electro-optic effects, the channel waveguide can be fabricated by electrical poling as shown in Figure 44. An electrode pattern defining the channel waveguide is first deposited and a planar buffer layer is applied to isolate optically the active waveguide layer from the metal electrodes. The nonlinear layer is then poled by applying an electric field above the polymer glass transition and cooling the sample to room temperature under an electric field. Because most side-chain polymers possess the anisotropic microscopic linear polarizability, the poled region becomes birefringent. The refractive index of poled region becomes larger than that of unpoled area so that TM waves can be confined in the lateral dimension. After poling, the electrode was removed and applied to an upper buffer layer and deposited on the pattern-switched electrode as shown in Figure 45. A prototype Y-branch interferometer was fabricated by first defining the waveguide pattern in an aluminum electrode on a glass substrate. The guide was 7 μm wide. The NLO-active polymer PC6S was poled at 90 °C for 5 min with an electric field of 100 V/μm, yielding an electro-optic coefficient of $r_{33} = 11$ pm/V. Modulation frequencies of up to 85 MHz were achieved. The modulation speed was restricted by both resistance and capacitance of modulator. To achieve high-frequency modulation, phase matching between the propagated light and traveling microwave is required. The modulation frequency band width is given by

$$\Delta f = \frac{1.4c}{\pi l |\sqrt{\epsilon_{\text{eff}}} - n|} \tag{62}$$

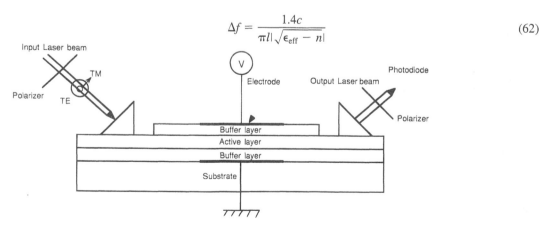

Figure 43 Schematic diagram of birefringent modulator. (From Dubois, J. C., Organic conjugated materials for optoelectronic and applications, in *Conjugated Polymeric Materials*, Bredas, J. L., and Chance, R. R., eds., Kluwer Academic Publishers, Netherlands, 1990, p. 321. Reprinted by permission of Kluwer Academic Publishers.)

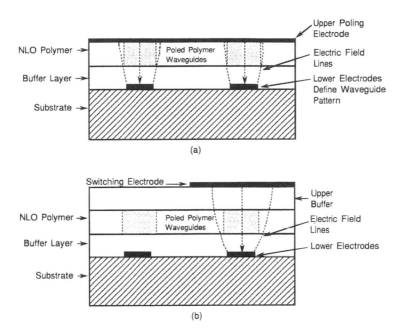

Figure 44 (a) Fabrication of channel waveguide by electric field poling. (b) Electro-optic modulation in a poled polymer electro-optic waveguide. (From Thackara, J. I., et al., *Nonlinear Optical Properties of Polymers,* Heeger, A. J., Orenstein, J., and Ulrich, D. R., eds., Materials Research Society, Pittsburgh, 109, 19, 1988. With permission.)

where c is the velocity of light and l is the propagation length of light. The advantage of the poled polymer is the small dielectric constant which can reduce the capacitance of modulator, since $\sqrt{\epsilon_{eff}} - n$ is almost zero. Figure 46 shows the traveling wave waveguide[575] The device electrode was designed in microstrip which offers greater flexibility in high-speed device and performance from the standard viewpoints of characteristic impedance, input/output interface, microwave losses. The characteristic impedance Z for microstrip lines is a function of the microstrip dimensions, and the dielectric constants of the line and surrounding media. The device was poled at 165 V/μm yielding electro-optic coefficients of $r = 21$ pm/V. The measured half-wave voltage was 6 V with a single-arm operation and 3 V in push-pull operation at 1.3 μm. The measured characteristic impedance of the microstrip and taper lines was vary close to 50 Ω. The frequency response was measured with a

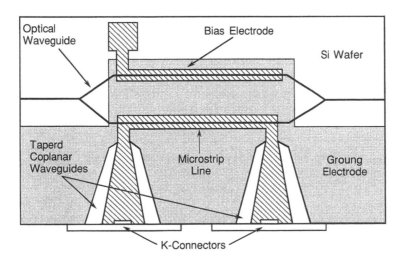

Figure 45 Schematic layout of the optical and microwave circuits of the modulator. (From Findakly, T. and Teng, C. C., *SPIE Proc.,* 2025, 526, 1993. With permission.)

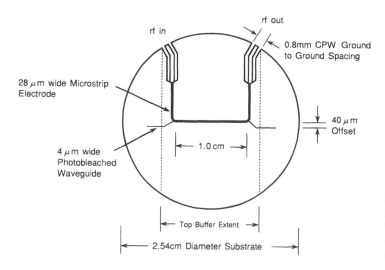

rf out

rf in

0.8mm CPW Ground
to Ground Spacing

28 μm wide Microstrip
Electrode

40 μm
Offset

4 μm wide
Photobleached
Waveguide

1.0 cm

Top Buffer Extent

2.54cm Diameter Substrate

Figure 46 Schematic layout of the traveling-wave phase modulator. (From Thackara, J. I., et al., *SPIE Proc.* 564, 2025, 1993. With permission.)

calibrated direct detection system. With the modulator biased at the center of its transfer function, there was a leveled output in the frequency range of 0.01–40 GHz. The signal was measured using a calibrated InGaAs high-speed detector. The 3-dB electrical bandwidth was evaluated to be above 40 GHz.

The 40-GHz modulation in LiNBO$_3$ single crystal was reported in 1990.[578] An 94-GHz phase modulation in MNA crystal has been reported by Wah et al.[579] The MNA crystal has dielectric constant of 2.2, loss tangent of 0.09, and electro-optic coefficient r_{11} of 30 pm/V at 94 GHz and showed the possibility of ultrahigh frequency modulations. Waveguide modulators such as a Mach-Zehnder type and a directional coupler type have been developed with electro-optical poled polymers. A comparison of waveguide modulator of LiNbO$_3$ and poled polymers has been discussed by Lytel.[580]

1. Phase Modulator

Here we consider the situation in which the phase of propagated light is modulated. The incident beam is polarized parallel to the Z axis of polymer and the poling electric field is applied to normal to the film. The electric field in the film was given by

$$E(x, t) = E_i \cos(\omega t - \varphi) \tag{63}$$

where the φ is the phase-shift in the waveguide when electric field V_0 is applied. The φ is given by the following equation

$$\varphi_z = k_0 n_z x = k_{0x}\left(n_e - \frac{1}{2} n_e^3 r_{33} \frac{V_0}{d}\right) \tag{64}$$

If the bias field is sinusoidal and is written as

$$V_0 = V_m \sin \omega_m t \tag{65}$$

At the output face ($x = 1$) of the waveguide of the electric field is given:

$$E_z(t, l) = E_i \cos(\omega t - \varphi_{0z} - \delta_z \sin \omega_m t) \tag{66}$$

where φ_z is the phase shift then $\varphi_{0z} = k_0 n_e l \delta_z$ is the phase shift induced by applying the electric field and represented by the following equation:

$$\delta_z = \left(\frac{\pi}{\lambda}\right) n_e^3 r_{33} \frac{l}{d} V_m \tag{67}$$

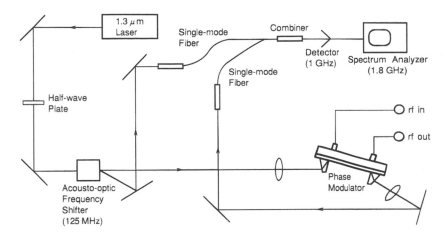

Figure 47 Schematic setup of the heterodyne measurement. (From Thackara, J. I., et al., *SPIE Proc.* 564, 2025, 1993. With permission.)

d is referred to as the phase-modulation index. The optical field is thus phase-modulated with a modulation index δ, if we use the Bessel function identities.[8]

$$\cos(\delta \sin \omega_m t) = J_0(\delta) + 2J_2(\delta)\cos 2\omega_m t + 2J_4(\delta)\cos 4\omega_m t \tag{68}$$

and

$$\sin(\delta \sin \omega_m t) = 2J_1(\delta)\sin \omega_m t + 2J_3(\delta)\sin 3\omega_m t \tag{69}$$

We can rewrite Equation 66 as:

$$E_{\text{out}} = A \begin{aligned}[t] &[J_0(\delta)\cos \omega_m t - J_1(\delta)\cos(\omega + \omega_m)t \\ &+J_1(\delta)\cos(\omega - \omega_m)t + J_2(\delta)\cos(\omega + 2\omega_m)t \\ &+J_2(\delta)\cos(\omega - 2\omega_m)t - J_3(\delta)\cos(\omega + 3\omega_m)t \\ &+J_3(\delta)\cos(\omega - 3\omega_m)t + J_4(\delta)\cos(\omega + 4\omega_m)t \\ &+J_4(\delta)\cos(\omega - 4\omega_m)t - \cdots\cdots] \end{aligned} \tag{70}$$

This form gives the distribution of energy in the sidebands as a function of modulation index, δ, and shows that the input energy was divided into a lot of sidebands according to the J_i value. The frequency difference of each sideband is ω_m. Thackara et al.[581] fabricated the phase modulator based on the PMMA-DR1 polymers as shown in Figure 46. The film was deposited on the fused quartz wafers, 2.54 cm in diameter and 500 μm thick. To achieve the high electrical efficiency and the 50 Ω electrical impedance needed for broadband operation., a 28-μm-wide microstrip electrode was used over the active portion of the devices. The optical waveguide was normally 4 μm and wide single mode in the lateral dimension. The waveguide consisted of two 2-mm-long s-bend sections and a 1-cm-long straight section, which was poled in the direction normal to the plane. The optical response of the phase modulators was measured using the heterodyne test bed shown in Figure 47. The measurement system stimulated the operation of a coherent communication system by generating the local oscillator with an acousto-optic coefficient of active layers in these devices. The system was operated at 1.32 μm and used a fiber optic coupler to recombine the reference and phase-modulated beams. The measurement bandwidth was limited by the photodetector. The sidebands impressed on the 1.32-μm optical carrier, generated by rf signals at υ_{rf} (5 MHz to 1 GHz) applied to the microstrip electrode, were downshifted by the heterodyne detection. The detected signal had a center beat at $\upsilon_a = 125$ MHz because the frequency shifter was driven at 125 MHz. The ratio of the detected electric powers in the central beat

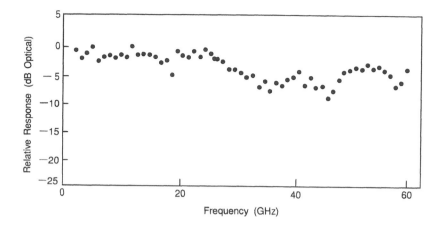

Figure 48 Optical frequency response of phase modulator, which had a 10 dB bandwidth of 60 GHz. (Reprinted with permission Dalton, L. R., et al., *Chem. Mater.*, 7, 1060, 1995. Copyright 1995 American Chemical Society.)

at υ_a, and the side peaks at $V_a \pm V_{rf}$ is a direct measurement of the modulation index M, which depends only on the device parameters and the experimental condition. M can be given by:

$$M = 2 \times 10^{-\Delta/20} \tag{71}$$

where Δ is the difference between the central and side peaks measured in decibels. This phase modulator developed by the IBM group at Almaden Research Center used PMMA-DR1 polymer. The electro-optical polymer modulator was able to transmit optically and receive six television signals in color pictures with audio.

In general, a 1-GHz bandwidth is large enough to transmit over 100 color TV channels. As the preliminary demonstration, the signal from a single television channel, with a carrier frequency of 211.25 MHz was applied to a modulator. The total insertion loss of modulator is about 30 dB/cm, the color signals are clearly visible in the received signal with the carrier peak being nearly 40 dB above the noise level. Dalton et al.[582] developed a novel thermo-setting polymer and it was applied to a phase modulator. The bandwidth response of this modulator reached up to 60 GHz as shown in Figure 48.

Figure 49 shows the schematic diagram of an array of polymeric integrated phase modulators reported by Akzo Research Laboratory.[583] An electro-optic polymer containing DANS chromophores was used. This polymer had a T_g of 140°C. The modulator consists of glass substrate, a gold-layer electrode, 2.4-μm-thick passive polymer cladding, 1.4-μm-thick DANS-containing polymer layer followed by 2.4-μm-thick upper cladding layer, and a top electrode. After a workout, the sandwich structure was poled at 130°C with an electric field of 100 V/μm. The phase modulator showed a r_{33} value of 8 pm/V in

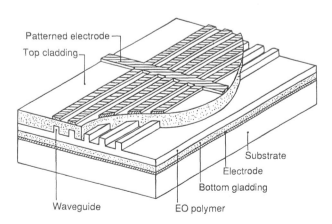

Figure 49 Schematic diagram of an array of polymeric integrated phase modulator. (From Mohlmann, G. R., et al., *SPIE Proc.*, 1337, 215, 1990. With permission.)

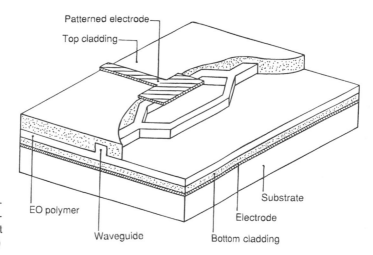

Figure 50 Sructure of an integrated Mach-Zahnder interferometer. (From Mohlmann, G. R., et al., *SPIE Proc.,* 1337, 215, 1990.)

TM mode and 2.3 pm/V in TE mode. Another polymeric phase modulator was also developed by this research team using the same DANS-polymer which showed a r_{33} value of 34 pm/V.

2. Intensity Modulator: Mach-Zahnder Interferometer

Akzo research group[583] also developed a Mach-Zahnder interferometer using the same technique and procedure applied for phase modulator. Figure 50 shows the integrated polymer Mach-Zahnder interferometer. It consists of a silicon wafer, 2-μm-thick quartz layer, Ti/Au bottom electrode, and a DANS-polymer waveguide core. A r_{33} value of 2 pm/V was obtained from 300 V of modulating potential at 1.3 μm for TE light over 4 mm active length. Mach-Zahnder interferometer structure was improved further and intensity modulation over 20 dB was obtained at 1.3 μm from 8 V over 14 mm long electrode covering the device. The r_{33} of 18.5 pm/V was evaluated and even dc-r_{33} of 32 pm/V was achieved and V_π value increased to 9 V for light modulation frequencies (100 Hz).

C. DIRECTIONAL MODE COUPLERS

Akzo research team[583] also developed a 2*2 mode coupling space switch in the form of directional mode couplers from electro-optic polymers (Fig. 51). Waveguiding channels, each a few microns wide and parallel to each other over a distance of centimeters separated by a few microns were fabricated. The technique and procedure was the same as used for phase modulator. Poling was carried out at 130 °C

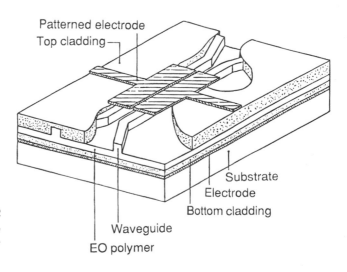

Figure 51 Schematic diagram of 2*2 mode coupler. (From Mohlmann, G. R., et al., *SPIE Proc.,* 1337, 215, 1990. With permission.)

for 10 min with an electric field of 100 V/μm. At a switching potential difference of 7.5 V over 14 mm electrode, light was switched from the cross to bar state with modulation depth of −17 dB.

D. POLYMER WAVEGUIDE ARRAY

Polymer waveguide components are attracting much attention in optical interconnections and optical data transmission because of their large optical nonlinearity, low losses, low cost production, simple fabrication, and compatibility with other substrates. The applications of polymeric NLO materials include optical modulators, frequency doublers, switch, and passive optical components such as combiners, interconnections, splitters, multiplexers, and demultiplexers. De Dobbelaere et al.[584] reported the first quasi-monolithic integration of an electrically pumped III-V semiconductor laser array (16 elements) with an array of NLO polymeric channel waveguides on a supporting Si substrate using the epitaxial lift-off (ELO) processes. The waveguide structure consists of a Si substrate, a polymer undercladding layer (3.2 μm), 4-dimethylamino-4′-nitrostilbene (DANS) side-chain polymer core layer (1.83 μm), and a polymer upper cladding layer (3.2 μm). The diode laser (GRINSCH InGaAs SQW) was developed by MOCVD method while purposely keeping both n- and p-type contacts on the topside of the device. A thin semiconductor film was placed in front of the etched polymer waveguide facet and bonded on the Si surface, which provides good transverse alignment between the polymer waveguide and the light source by photolithography. The coupling of light into the polymeric channel waveguide provided emission at 966 nm. The ELO-integration process was considered advantageous for polymeric waveguides because it offered operation at <100 °C, high alignment between the polymer waveguide and the light source, good thermal properties, integration of II-V components on large substrates, and versatile interconnections. Therefore quasi-monolithic integration provides CW operation of semiconductor lasers transplanted by ELO and coupling between laser and the polymer waveguides.

E. PHOTONIC DETECTION OF ELECTROMAGNETIC RADIATION

Most conventional techniques to detect the electromagnetic field require metal electrodes which have the disadvantage of deforming the field to be measured. Polymeric electro-optic materials have been attractive to detect the fields. The sensing was performed using a laser beam. Distorting the fields by inserting of electro-optic polymers is less than metal electrode because the dielectric constant of electro-optic polymer is relatively small. In general the sensor consists of the metallic antenna and electro-optic modulator to improve the sensitivity of detection as shown in Figure 52(a), although the metallic part distorts the fields. Hilliard and co-workers[585] developed completely low dielectric electromagnetic field sensors with Luneburg lens to improve the field gain as shown in Figure 52(b). The lens where the electro-optic modulator was placed allows field to penetrate into the lens and focused onto the modulator on the other side. Maximum energy coupling into the electro-optical modulator can be achieved by minimizing the reflection of fields at the surface of modulator. The reflection of energy to the source from the polymeric modulator is one seventh of inorganic modulator, such as $LiNbO_3$, which can improve the gain and resolution of detecting the electromagnetic radiation.

Some of the applications of second-order NLO materials in photonics devices have been discussed above in detail. Many other similar and related applications of electro-optic materials in photonics devices have been reported.[586-603] Lipscomb and Lytel[604,605] suggested that EO polymers may be used in photonic large-scale integration (PLSI) by hybrid integration of electronic and photonic devices combining the processing power of VLSI with a dense, high bandwidth, photonic interconnection and switching network in a single large format package. The applications of optical interconnections include in telecommunication, box-to-box, board-to-board, and chip-to-chip interconnects. Other potential applications of electro-optic polymers were indicated in optical multichip modules, reconfigurable optical connectors, reconfigurable optical backplanes, high-speed multiplexers and demultiplexers, high-speed switching networks, high-speed digital and analog-to-digital modulator arrays, high-speed analog-to-digital convertors, two-dimensional optical source arrays, pigtailed integrated optic devices, and linear lasers and packages.

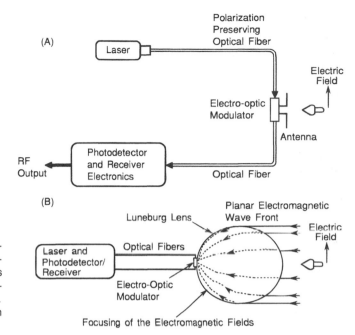

Figure 52 (A) Block diagram of a photonic sensor system. (B) Schematic representation of the use of a Luneburg lens with a single photonic sensor. (From Hilland, D. and Mensa, D., Proc. *IEEE Ant. Prop. Soc. Symp.*, 2, 720, 1992. With permission.)

X. PROGNOSIS

This review chapter on the developments of fundamental chemistry and physics of organic materials for second-order nonlinear optics demonstrates the remarkable progress that has been made in the molecular engineering of NLO materials. Second-order optical nonlinearity of organic materials is governed by conjugation length, nature of π-bonding sequence, strength of donor–acceptor groups, and location of substitution. In the past decade, SHG-active organic materials have been developed and created by using various chemical and physical approaches. Strategies of chirality, hydrogen bonding, steric hindrance, molecular asymmetry, reduced dipole–dipole interaction, Λ-type conformation, organometallic moieties, LB technique, co-crystallization, and electrical poling have been used for generating noncentrosymmetric structures required for the appearance of SHG. One-dimensional charge-transfer molecules seem to be the most promising second-order NLO materials. On the other hand, two-dimensional charge-transfer molecules and octupolar chemical species are also unique and interesting NLO materials. At present, photonics technologies have been dominated by inorganic, bulk, single-crystalline materials; $LiNbO_3$, $KTiOPO_4$ (KTP), and semiconductors but competitive organic materials have emerged. A comparison of the SHG described by a figure of merit of inorganic and organic materials substantiates the importance of organic materials in designing and fabricating photonics devices. Attractive features of organic materials include high optical damage thresholds (GW/cm^2), fast switching time (subpicoseconds), architectural flexibility, optical transparency, phase matching, chemical and thermal stability, ease of mechanical integration and processibility. For an ideal SHG material, these properties should coexist; this is a challenging task and requires understanding of materials design as well as fabrication for a particular end use. Let us take a case of frequency doubling, the material requirements include a large NLO coefficient, low optical losses at ω and 2ω, high thermal, mechanical, and environmental stability, easy processing, and low cost.

Material requirements for applications of SHG and electro-optic materials differ depending on the end use. For electro-optic applications, polymeric materials are of great interest; the requirements include: (1) high electro-optic coefficient ($r > 35$ pm/V); (2) no orientational relaxation during operation at 100 °C for the lifetime of the device and 250 °C for months (for the useful material, the T_g must be about 300 °C and the decomposition temperature higher than 450 °C); (3) no degradation either chemically or physically at temperatures exceeding 350 °C, which is needed in the device fabrication process; (4) optical loss of less than 2 dB/cm; (5) compatibility with other substrates; (6) solution processability; and (7) no scattering or absorption of light.

Although organic conjugated molecules having strong electron acceptor and electron donor groups at the opposite ends exhibit very large SHG, their fitting into a transparency–efficiency trade-off scale is important from an application viewpoint. For example, organic single crystals are best candidates for frequency doubling. Organic single crystals exhibit NLO coefficients from tens to several hundreds pm/V. Single crystals of N,N-dimethylamino-N'-methylstilbazolium p-toluenesulfonate (DAST) show d_{11} coefficients as high as 1900 ± 500 pm/V at 1.2 μm. The difficulties lie in growing large-size organic single crystals of interest. Moreover, organic single crystals are mechanically soft, which causes problems in the cutting and polishing. At present, this is a disadvantage, but developments of new techniques and materials may overcome this difficulty.

Langmuir-Blodgett (LB) technique is a molecular engineering tool to introduce noncentrosymmetric structures that display SHG activity. Many charge-transfer LB molecules exhibiting large SHG have been reported. For example, LB films of hemicyanine dyes show $\chi^{(2)}$ values in the range $500 \sim 1500$ pm/V. The currently available LB materials are soft and rather instable, hence their application depends on the development of new robust materials of interest.

In the 1990s, three main problems were identified for the use of organic SHG or electro-optic polymeric materials in photonics devices: (1) orientational relaxation of poling induced polar order in poled polymeric systems which caused remarkable loss of second-order optical nonlinearity; (2) poor NLO chromophore loading either into the polymer host matrix or backbones; (3) thermal stability of SHG or electro-optic coefficients at elevated temperatures. To meet this challenge, considerable effort and progress have been made on the materials front using high T_g polymers of polyimides either as host matrix or backbone for incorporating NLO chromophores. NLO dyes were covalently attached to many types of polymer backbones; PMMA, polycarbonate, polyurethane, polystyrene, polyurea, polyethylene, polyphosphazene, polyimides, epoxy resins, and heteroaromatic polymers. Guest–host systems are of particular interest because they are formed simply by dissolving NLO dyes in a conventional polymer matrix. These dye–polymer composites can be processed easily and quickly into thin films either by spin coating or solvent-evaporation techniques. The electrically poled composites guarantee the appearance of SHG or electro-optic activity by organizing molecules into noncentrosymmetric fashion. The disadvantages of guest–host systems are their low NLO coefficients and instability over prolonged periods because orientational relaxation causes a decrease of the NLO coefficient as low as 50% or more. Thiophene-based NLO chromophores in polyquinoline matrix showed the largest r_{33} coefficient of 45 pm/V which decreased to 26 pm/V over a period of more than 2000 h. This is a remarkable development in electro-optic polymers. The problem of NLO instability has been overcome by covalently attaching NLO dye either as side chains to conventional polymer backbones or imparting them in the main chain of a polymeric system. This molecular engineering approach provides materials with increased stability over guest–host systems by minimizing the NLO decay. The d_{33} value as high as 250 pm/V has been obtained for NLO side-chain polymers. Stability of NLO activity has been achieved through cross-linking process using either cross-linking agents or thermal or photochemical techniques. The cross-linked polymers in which NLO-active moieties are incorporated covalently into the polymer network also show very large and stable NLO coefficients. Even at elevated temperatures these cross-linked polymers show no detectable relaxations. Cross-linked polymers can help in decreasing orientational relaxation but they are plagued with several problems such as insolubility, high scattering losses, shrinkage, adhesion, and mechanical stress.

Applications of second-order NLO materials in frequency doubling, Mach-Zahnder modulators, voltage sensors, directional couplers, and electromagnetic radiation detection have been discussed. The development of NLO devices need organic materials that display not only large SHG or electro-optic coefficients but those that can meet the overall challenge of a target material. The emerging photonics technologies demands new high-performance NLO materials. The unlimited synthetic capabilities and architectural flexibility of organic structures promise that a collaborative effort of chemists, material scientists, and electronic and photonics engineers will be able to develop novel materials of interest for photonics applications.

ACKNOWLEDGMENT

The author expresses his thanks to Dr. Yasuo Imanishi of HRL, Hitachi Ltd, for his generosity in translating many articles from Japanese to English from conference proceedings.

REFERENCES

1. **Maiman, T. H.,** Stimulated optical radiation in ruby, *Nature,* 187, 493, 1960.
2. **Javan, A., Bannett, W. R, and Herriott, D. R.,** Population inversion and continuous optical maser oscillation in a gas discharge containing a He-Ne mixture, *Phys. Rev. Lett.,* 6, 106, 1961.
3. *Laser Focus Buyer's Guide,* Advanced Technology Publications, Newton, MA, 1979.
4. **Ready, J. F.,** *Industrial Applications of Lasers,* Academic Press, Inc., New York, 1978.
5. **Franken, P. A., Hill, A. E., Peters, C. W., and Weinreich, W.,** *Phys. Rev. Lett.,* 7, 118, 1961.
6. **Bloembergen, N.,** *Nonlinear Optics,* W. A. Benjamin, Inc., London, 1965.
7. **Shen, Y. R.,** *The Principle of Nonlinear Optics,* John Wiley & Sons, New York, 1984.
8. **Yariv, A.,** *Quantum Electronics,* John Wiley & Sons, New York, 1984.
9. **Reintjes, J. F.,** *Nonlinear Optical Parametric Processes in Liquids and Gases,* Academic Press, Inc., New York, 1984.
10. **Hanna, D. C., Yuratich, M. A., and Cotter, D.,** *Nonlinear Optics of Free Atoms and Molecules,* Springer-Verlag, Berlin, 1979.
11. **Flytzanis, C., Rabin, H., and Tang, C. L., eds.,** *Quantum Electronics: A Treatise,* Vol. 1, Academic Press, Inc., New York, 1975.
12. **Zernike, F. and Midwinter, J. E.,** *Applied Nonlinear Optics,* John Wiley & Sons, New York, 1973.
13. **Byer, R. L.,** *Nonlinear optical phenomena and materials, Annu. Rev. Mater. Sci.,* 4, 147, 1974.
14. **Zernike, F. and Midwinter, J.,** *Applied Nonlinear Optics,* John Wiley & Sons, New York, 1973.
15. **Davydov, B. L., Derkacheva, L. D., Dunina, V. V., Zhabotinskii, M. E., Zolin, V. K., Kreneva, L. G., and Samokhina, M. A.,** Connection between charge transfer and laser second harmonic generation, *JEPT Lett.,* 12, 16, 1970.
16. **Zyss, J.,** Nonlinear organic materials for integrated optics: a review, *J. Mol. Electron.,* 1, 25, 1985.
17. **Williams, D. J.,** Organic polymeric and non-polymeric materials with large optical nonlinearities, *Angew. Chem. Int. Ed. Engl.,* 23, 690, 1984.
18. **Burland, D. M., Miller, R. D., and Walsh, C. A.,** Second-order nonlinearity in poled-polymer systems, *Chem. Rev.,* 94, 31, 1994.
19. **Williams, D. J., ed.,** *Nonlinear Optical Properties of Organic and Polymeric Materials,* ACS Symposium Series 233, American Chemical Society, Washington, DC, 1983.
20. **Kobayashi, T., ed.,** Nonlinear optics of organic and semiconductors, *Springer Proceedings of Physics,* Springer-Verlag, Berlin, 1989.
21. **Hahn, R. A. and Bloor, D., eds.,** *Organic Materials for Nonlinear Optics,* Royal Society of Chemistry, London, 1990.
22. **Bredas, J. L. and Chance, R. R.,** Conjugated polymeric materials: opportunities in electronics, opto-electronics and molecular electronics, *NATO ASI Series E,* Applied Sciences, Vol. 182, Kluwer Academic Publishers, Dordrech, 1990.
23. **Chemla, D. S. and Zyss, J., eds.,** *Nonlinear Optical Properties of Organic Molecules and Crystals,* Vols. 1 and 2, Academic Press, New York, 1987.
24. **Kanis, D. R., Ratner, M. A., and Marks, T. J.,** Design and construction of molecular assemblies with large second-order optical nonlinearities. Quantum chemical aspects, *Chem. Rev.,* 94, 195, 1994.
25. **Armstrong, J. A., Bloembergen, N., Ducuing, J., Pershan, P. S.,** Interaction between light wave in a nonlinear dielectric *Phys. Rev.,* 127, 1918, 1962.
26. **Bloembergen, N. and Shen, Y. R.,** *Phys. Rev.,* 133, A37, 1964.
27. **Ward, J. F.,** Calculation of nonlinear optical susceptibility using diagrammatic perturbation theory, *Phys. Rev.,* 37, 1, 1965.
28. **Orr, J. B. and Ward, J. F.,** Perturbation theory of the nonlinear optical polarization of an isolated system, *Mol. Phys.,* 20, 513, 1971.
29. **Yariv, A. and Yeh, P.** *Optical Waves on Crystals,* John Wiley & Sons, New York, 1983.
30. **Zyss, J. and Oudar, J. L.,** Relation between microscopic and macroscopic lowest-order optical nonlinearities of molecular crystals with one- or two-dimensional units, *Phys. Rev.* A26, 2028, 1982.
31. **Oudar, J. L.,** Optical nonlinearities of conjugated molecules. Stilbene derivatives and highly polar aromatic compounds, *J. Chem. Phys.,* 67, 446, 1977.
32. **Oudar, J. L. and Chemla, D. S.,** Hyperpolarizabilities of the nitroanilines and their relations to the excited state dipole moment, *J. Chem. Phys.,* 66, 2664, 1977.
33. **Singer, K. D., Kuzyk, M. G., and Sohn, J. E.,** Second-order nonlinear-optical processes in orientationally ordered materials: relationship between molecular and macroscopic properties, *J. Opt. Soc. Am.,* B4, 968, 1987.
34. **Nye, J. F.** *Physical Properties of Crystals,* Oxford University Press, London, 1967.
35. **Bosshard, Ch., Sutter, K., Schlesser, R., and Gunter, P.,** Electro-optic effects in molecular crystals, *J. Opt. Soc. Am.,* B10, 867, 1993.
36. **Boyd, G. D. and Kleinman, D.,** Parametric interaction of focused Gaussian light beams, *J. Appl. Phys.,* 39, 3597, 1968.
37. **Zyss, J.,** Hyperpolarizabilities of substituted conjugated molecules. I. Perturbated INDO approach to monosubstituted benzene, *J. Chem. Phys.,* 70, 3333, 1979.

38. **Zyss, J.,** Hyperpolarizabilities of substituted conjugated molecules. II. Substituent effects and respective σ-π contributions, *J. Chem. Phys.,* 70, 3341, 1979.

39. **Zyss, J.,** Hyperpolarizabilities of substituted conjugated molecules. III. Study of a family of donor-acceptor disubstituted phenyl-polyenes, *J. Chem. Phys.,* 71, 909, 1979.

40. **Chemla, D. S., Oudar, J. L., and Zyss, J.,** Molecular engineering for modern optics, *L'echo des Recherches,* 47, 1981.

41. **Lalama, S. J. and Garito, A. F.,** Origin of the nonlinear second-order optical susceptibilities of organic systems, *Phys. Rev. A,* 20, 1179, 1979.

42. **Levine, B. F.,** Donor-acceptor charge transfer contributions to the second order hyperpolarizability, *Chem. Phys. Lett.,* 37, 516, 1976.

43. **Oudar, J. L. and Chemla, D. S.,** Hyperpolarizability of the nitroanilines and their relation to the excited state dipole moment, *J. Chem. Phys.,* 66, 2664, 1977.

44. **Stahelin, M., Burland, D. M., and Rice, J. E.,** Solvent dependence of the second-order hyperpolarizability in p-nitroaniline, *Chem. Phys. Lett.,* 191, 245, 1992.

45. **Cady, H. H. and Larson, A. C.,** The crystal structure of 1,3,5-Triamino-2,4,6-trinitrobenzene, *Acta. Crystallogr.,* 18, 485, 1965.

46. **Zyss, J.,** Octupolar organic systems in quadratic nonlinear optics: molecules and materials, *Nonlinear Opt.,* 1, 3, 1991.

47. **Nalwa, H. S., Nakajima, K., Watanabe, T., Nakamura, K., Yamada, A., and Miyata, S.,** Second harmonic generation in Langmuir-Blodgett monolayer of a two-dimensional charge transfer molecule: N, N'-dioctadecyl-4,6-dinitro-1,3-diaminobenzene, *Jpn. J. Appl. Phys.,* 30, 983, 1991.

48. **Nalwa, H. S., Watanabe, T., and Miyata, S.,** A comparative study of 4-nitroaniline, 1,5-diamino-2,4-dinitrobenzene, and 1,3,5-triamino-2,4,6-trinitrobenzene, *Opt. Mater.,* 2, 73, 1993.

49. **Bredas, J. L., Dehu, C., Meyers, F., and Zyss, J.,** *SPIE Proc.,* 1560, 98, 1991.

50. **Bredas, J. L., Meyers, F., Pierce, B. M., and Zyss, J.,** On the second-order polarizability of conjugated π-electron molecules with octupolar symmetry: the case of triaminotrinitrobenzene, *J. Am. Chem. Soc.,* 114, 4928, 1992.

51. **Ledoux, I., Zyss, J., Siegel, J. S., Brienne, J., and Lehn, J. M.,** Second-harmonic generation from non-dipolar non-centrosymmetric aromatic charge-transfer molecules, *Chem. Phys. Lett.,* 172, 440, 1990.

52. **Nalwa, H. S., Ogino, K., Watanabe, T., Sato, H., and Miyata, S.,** unpublished results.

53. **Zyss, J.,** Molecular engineering implications of rotational invariance in quadratic nonlinear optics: from dipolar to octapolar molecules and materials, *J. Chem. Phys.,* 98, 6583, 1993.

54. **Zyss, J.,** Molecular engineering implications of rotational invariance in nonlinear optics: octapolar systems for quadratic processes, *Proceedings of the 5th Toyota Conference on Nonlinear Optics,* S. Miyata, ed., Elsevier, Amsterdam, 1992, p. 33.

55. **Watanabe, T., Kagami, M., Miyamoto, H., Kidoguchi, A., and Miyata, S.,** *Proceedings of 5th Toyota Conference Nonlinear Optics,* S. Miyata, ed., Elsevier, Amsterdam, 1991, p. 201.

56. **Nalwa, H. S., Watanabe, T., and Miyata, S.,** 2-D charge transfer molecules for second-order NLO: role of off-diagonal orientation, *Adv. Mater.,* 7, 1001, 1995.

57. **Zyss, J., Nicoud, J. F., and Coquillay, M.,** Chirality and hydrogen-bonding in molecular crystals for phase-matched second harmonic generation. N-(4-nitrophenyl)-(L)-prolinol (NPP), *J. Chem. Phys.,* 81, 4160, 1984.

58. **Barzoukas, M., Josse, D., Fremaux, P., Zyss, J., Nicoud, J. F., and Morley, J. O.,** Quadratic nonlinear properties of N-(4-nitrophenyl)-(L)-prolinol and of a newly engineered molecular compound N-(4-nitrophenyl)-N-methyl-aminoacetonitrile: a comparative study, *J. Opt. Soc. Am. B,* 4, 977, 1987.

59. **Oudar, J. L. and Hierle, R.,** An efficient organic crystal for nonlinear optics: methyl-(2,4-dinitrophenyl)-aminopropanoate, *J. Appl. Phys.,* 48, 2699, 1977.

60. **Zyss, J. and Ledoux, I.,** Nonlinear optics in multipolar media: theory and experiments, *Chem. Rev.,* 94, 77, 1994.

61. **Nalwa, H. S., Watanabe, T., Nakajima, K., and Miyata, S.,** Two-dimensional charge-transfer molecules for optimizing second-harmonic generation in Langmuir-Blodgett films, *Proceedings of 5th Toyota Conference Nonlinear Optics,* S. Miyata, ed., North-Holland, Amsterdam, 1992, p. 271.

62. **Suslick, K. S., Chen, C. T., Meredith, and Cheng, L. T.,** Push-pull porphyrins as nonlinear optical materials, *J. Am. Chem. Soc.,* 114, 6928, 1992.

63. **Moylan, C. R., Miller, R. D., Twieg, R. J., Betterton, K. M., Lee, V. Y., Matray, T. J., and Nguyen, C.,** Synthesis and nonlinear optical properties of donor-acceptor substituted triaryl azole derivatives, *Chem. Mater.,* 5, 1499, 1993.

64. **Allen, S., McLean, T. D., Gordon, P. F., Bothwell, B. D., Hursthouse, M. B., and Karaulov, S. A.,** A novel organic electro-optical crystal: 3-(1,1-dicyanoethenyl)-1-phenyl-4,5-dihydro-1H-pyrazole, *J. Appl. Phys.,* 64, 2583, 1988.

65. **Nicoud, J. F. and Twieg, R. J.,** Design and synthesis of organic molecular compounds for efficient second-harmonic generation, in *Nonlinear Optical Properties of Organic Molecules and Crystals,* D. S. Chemla and J. Zyss, eds., Academic, New York, 1987, vol. 1, chap. II-3, p. 227.

66. **Rao, V. P., Jen, A. K., Wong, K. Y., Drost, K. J., and Minini, R. M.,** *SPIE Proc.,* 1775, 32, 1992.

67. **Wong, K. Y., Jen, A. K., Rao, V. P., Drost, K. J., and Minini, R. M.,** *SPIE Proc.,* 1775, 74, 1992.

68. **Jen, A. K., Rao, V. P., Wong, K. Y., and Drost, K. J.,** Functionalized thiophenes: second-order nonlinear optical materials, *J. Chem. Soc. Chem. Commun.,* 90, 1993.

69. **Jen, A. K., Rao, V. P., Drost, K. J., Cai, Y. M., Mininni, R. M., Kenney, J. T., Binkley, E. S., Dalton, L. R., and Marder, S. R.,** Progress on heteroaromatic chromophores in high temperature polymers for electro-optic applications, *SPIE Proc.,* 2143, 30, 1994.

70. **Huijts, R. A. and Hesselink, G. L. J.,** *Chem. Phys. Lett.,* 156, 209, 1989.

71. **Barzoukas, M., Blanchard-Desce, M., Josse, D., Lehn, J. M., and Zyss, J.,** Very large quadratic nonlinearities in solutions of two push-pull polyene series: effect of the conjugation length and of the end groups, *Chem. Phys.,* 133, 323, 1989.

72. **Blanchard-Desce, M.,** Functionalized polyenic oligomers and carotenoids with enhanced optical nonlinear responses, *Condensed Matter News,* 2, 12, 1993.

73. **Blanchard-Desce, M., Bloy, V., Lehn, J. M., Runser, C., Barzoukas, M., Fort, A., and Zyss, J.,** Push-pull polyenes and carotenoids with enhanced quadratic nonlinear optical susceptibilities, *SPIE Proc.,* 2143, 20, 1994.

74. **Cheng, L. T., Tam, W., Marder, S. R., Stiegman, A. E., Rikken, G., and Spangler, C. W.,** *J. Phys. Chem.,* 95, 10643, 1991.

75. **Morley, J. O., Docherty, V. J., and Pugh, D.,** Nonlinear optical properties of organic molecules. Part 2. Effect of conjugation length and molecular volume on the calculated hyperpolarizabilities of polyphenyls and polyenes, *J. Chem. Soc. Perkin Trans.* II, 1351, 1987.

76. **Cheng, L. T., Tam, W., Stevenson, S. H., Meredith, G. R., Rikken, G., and Marder, S. R.,** Experimental investigation of organic molecular nonlinear optical polarizabilities. 1. Methods and results on benzene and stilbene derivatives, *J. Phys. Chem.,* 95, 10631, 1991.

77. **Stiegman, A. E., Graham, E., Perry, K. J., Khundkar, L. R., Cheng, L. T. and Perry, J. W.,** The electronic structure and second-order nonlinear optical properties of donor-acceptor acetylenes: a detailed investigation of structure-property relationships, *J. Am. Chem. Soc.,* 113, 7658, 1991.

78. **Burland, D. M., Miller, R. D., Reiser, O., Twieg, R. J., and Walsh, C. A.,** *J. Appl. Phys.,* 71, 410, 1992.

79. **Kondo, K., Ohnishi, N., Tajemoto, K, Yoshida, H., and Yoshida, K.,** Synthesis of optically quadratic nonlinear phenylpyridylacetylenes, *J. Org. Chem.,* 57, 1623, 1992.

80. **Cheng, L. T., Tam, W., Meredith, G. R., Rikken G. L. J. A., and Meijer, E. W.,** *SPIE Proc.,* 1147, 61, 1989.

81. **Twieg, R. J. and Jain, K.,** Organic materials for optical second harmonic generation, *Proc. Am. Chem. Soc.,* 1983, p. 57.

82. **Zyss, J. and Berthier, G.,** *J. Chem. Phys.,* 77, 3635, 1982.

83. **Bierlein, J. D., Cheng, L. K., Wang, Y., and Tam., W.,** Linear and nonlinear optical properties of 3-methyl-4-methoxy-4'-nitrostilbene single crystals, *Appl. Phys. Lett.,* 56, 423, 1990.

84. **Levine, B. F., Bethea, C. G., Thurmound, C. D., Lynch, R. T., and Berstein, J. L.,** An organic crystal with an exceptionally large optical second harmonic cofficient: 2-methyl-4-nitroaniline, *J. Appl. Phys.,* 50, 2523, 1979.

85. **Tieke, B.,** *Adv. Mater.,* 2, 222, 1990.

86. **Stroeve, P. and Franses, E.,** *Thin Solid Films,* 152, 405, 1987.

87. **Peterson, I. R.,** *J. Phys. D: Appl. Phys.,* 23, 379, 1987.

88. **Tomaru, S., Zembutsu, S., Kawachi, M., and Kobayashi, M.,** *J. Chem. Soc. Chem. Commun.,* 1207, 1984.

89. **Wang, Y. and Eaton, D. F.,** *Chem. Phys. Lett.,* 120, 441, 1985.

90. **Miyazaki, T., Watanabe, T., and Miyata, S.,** Highly efficient second harmonic generation in p-nitroaniline-poly(lactone) systems, *Jpn. J. Appl. Phys.,* 27, L1724, 1988.

91. **Meredith, G. R., VanDusen, T. G., and Williams, D. J.,** Optical and nonlinear optical characterization of molecularly doped thermotropic liquid crystalline polymers, Macromolecules, 15, 1385, 1982.

92. **Sohn, J. E., Singer, K. D., Lalama, S. J., and Kuzyk, M.,** *Polym. Mater. Sci. Eng.,* 55, 532, 1986.

93. **Okamoto, N., Abe, T., Chen, D., Fujimura, H., and Matsushima, R.,** Large second harmonic generation from mixtures of para-nitroaniline and its N-alkyl derivatives, *Opt. Commun.,* 74, 421, 1990.

94. **Addadi, L., Berkovvitch-Yellin, Z., van Mill, J., Simon, L. J. W., Lahav, M., and Leiserowitz, L.,** *Angew. Chem. Int. Ed. Engl.,* 24, 466, 1985.

95. **Tabei, H., Kurihara, T., and Kaino, T.,** *Appl. Phys. Lett.,* 50, 1855, 1987.

96. **Nogami, T., Nakano, H., Shirota, Y., Umegaki, S., Shimizu, Y., Uemiya, T., and Yasuda, N.,** *Chem. Phys. Lett.,* 155, 338, 1989.

97. (a) **Yamamoto, H., Hosomi, T., Watanabe, T., and Miyata, S.,** Structure and nonlinear optical properties of methanediamines synthesized according to a novel molecular design method (lambda-type conformation), *J. Chem. Soc. Jpn. Chem. Indust. Chem.,* 7, 789, 1990.

    (b) **Yamamoto, H., Katogi, S., Watanabe, T., Sato, H., Miyata, S., and Hosomi, T.,** New molecular design method for noncentrosymmetric crystal structure: Lambda (Λ) shape molecules for frequency duobling, *Appl. Phys. Lett.,* 60, 935, 1992.

    (c) **Miyata, S., Watanabe, T., and Nalwa, H. S.,** Molecular design of nonlinear optical molecules and polymers by lambda (Λ)-type conformation, *Proceedings of the International School of Physics "Enrico Fermi", Nonlinear Optical Materials: Principles and Applications,* Degiorgio, V., and Flytzanis, C., eds., IOS Press, Amsterdam, (1995), pp. 225–241.

98. **Tasaka, S., Abe, T., Matsushima, R., Suzuki, M., Chen, D. Y., and Okamoto,** SHG-active polymorphism of p-nitroaniline mixed with its N-alkyl derivatives, *Jpn J. Appl. Phys.*, 30, 296, 1991.

99. **Nalwa, H. S., Watanabe, T., and Miyata, S.,** unpublished results

100. **Zyss, J. and Chemla, D. S.,** Demonstration of efficient nonlinear optical crystals with vanishing molecules dipole moment: Second-harmonic generation in 3-methyl-4-nitropyridine-1-oxide, *J. Chem. Phys.*, 74, 4800, 1981.

101. **Nalwa, H. S.,** Organometallic materials for nonlinear optics, *Appl. Organometal. Chem.*, 5, 349, 1991.

102. **Fan, Y. X., Eckardt, R. C., R. L. Byer, R. K. Route, and R. S. Fiegelson,** *App. Phys. Lett.*, 45, 313, 1984.

103. **Bierlein, J. D. and Vanherzeele, H.,** *J. Opt. Soc. Am.*, B6, 622, 1989.

104. **Mei, L., Wang, Y., Chen, C., and Wu, B.,** Nonlinear optical materials based on $MBe_2BO_3F_2$(M=Ma,K), *J. Appl. Phys.*, 74, 7014, 1993.

105. **Rentzepis, P. M. and Pao, Y. H.,** *Appl. Phys. Lett.*, 5, 156, 1964.

106. **Heilmer, G. H., Ockman, N., Braunstein, R., and Karmer, D. A.,** Relationship between second harmonic generation and electro-optic effects of the molecular crystal hexamine, *Appl. Phys. Lett.*, 5, 229, 1964.

107. **Orlov, R.,** *Sov. Phys. Crystallogr.*, 11, 410, 1966.

108. **Badan, J., Hierle, R., Perigaurd, A., and Vidakovic,** Growth and characterization of molecular crystals, in *Nonlinear Optical Properties of Organic Molecules and Crystals,* Vol. 1, Chemla, D. S., and Zyss, J., eds., Academic Press, New York, 1987, Chapter II-4, p. 297.

109. **Penn, B. G., Cardelino, B. H., Moore, C. E., Shields, A. W., and Frazier, D. O.,** Growth of bulk single crystals of organic materials for nonlinear optical devices: an overview, *Prog. Crystal Growth Charact.*, 22, 19, 1991.

110. **Karl, N.** *Crystals: Growth, Properties and Applications,* Freyhardt, H. C., ed., Springer-Verlag, Berlin, 1980.

111. **Gunter, P., Bosshard, Ch., Sutter, K., Arend, H., Chapuis, G., Twieg, R. J., and Dobrowolski, D.,** 2-Cyclooctylamino-5-nitropyridine, a new nonlinear optical crystal with orthorhombic symmetry, *Appl. Phys. Lett.*, 50, 486, 1987.

112. **Baumert, J. C., Twieg, R. J., Bjorklund, G. C., Logan, L. A., and Dirk, C. W.,** Crystal growth and characterization of 4-(N,N-dimethylamino)-3-acetamidonitrobenzene, a new organic material for nonlinear optics, *Appl. Phys. Lett.*, 51, 1484, 1987.

113. **Twieg, R. J., Jain, K., Cheng, Y. Y., Crowley, J. I., and Azema,** *Polym. Preprints*, 23, 47, 1982.

114. **Hierle, R., Badan, J., and Zyss, J.,** Growth and characterization of a new material for nonlinear optics: methyl-3-nitro-4-pyridine-1-oxide (POM), *J. Cryst. Growth*, 69, 545, 1984.

115. **Stevenson, J. L., and Skaski, A. C.,** *J. Phys. C., Solid State Phys.*, 5, L233, 1972.

116. **Garside, J., and Matsuoka, M.,** The role of melt composition on the crystal morphology of m-chloronitrobenzene, in *Proceedings of First International Workshop on Crystal Growth of Organic Materials (CGOM),* 1989, p. 171.

117. **Sherwood, J. N.,** The growth, perfection and properties of crystals of organic nonlinear optical materials, in *Proceedings of First International Workshop on Crystal Growth of Organic Materials (CGOM),* 1989, p. 180.

118. **Dmitriev, V. G., Gurzadyan, G. G., and Nikogosyan, D. N.,** *Handbook of Nonlinear Optical Crystals,* Springer-Verlag, Berlin, 1991.

119. **Halbout, J., Blit, S., Donaldson, W., and Tang, C. L.,** Efficient phase-matched second-harmonic generation and sum-frequency mixing in urea, *IEEE J. Quantum Electron.*, QE-15, 1176, 1979.

120. **Betzler, K., Hesse, H., and Loose, P.,** Optical second harmonic generation in organic single crystals: urea and ammonium-malate, *J. Mol. Struct.*, 47, 393, 1978.

121. **Morrell, J. A., Albrecht, A. C., Levin, K. H., and C. L. Tang,** *J. Chem. Phys.*, 71, 5063, 1979.

122. **Bauerle, D., Betzler, K., Hesse, H., Kappan, S., and Loose, P.,** Phase-matched second harmonic generation in urea, *Phys. Status Solidi* (A), 42, K199, 1977.

123. **Donaldson, W. R. and Tang, C. L.,** *Appl. Phys. Lett.*, 44, 25, 1984.

124. **Cassidy, C., Halbout, J. M., Donaldson, W., Tang, C. L.,** Nonlinear optical properties of urea, *Opt. Commun.*, 29, 243, 1979.

125. **Kato, K.,** *IEEE J. Quantum Electronic,* QE-16, 810, 1980.

126. **Ledoux, I. and Zyss, J.,** Influence of the molecular environment in solution measurements of the second-order optical susceptibility for urea and derivatives, *Chem. Phys.*, 73, 203, 1982.

127. **Matsushima, R., Hiramatsu, K., and Shimamoto, K.,** Second harmonic generation from single crystals and mixed crystals of p-nitroaniline and its derivatives, *Bull. Chem. Soc. Jpn.*, 67, 1479, 1994.

128. **Okamoto, N., Unoh, H., Sugihara, O., and Matsushima,** Large second harmonic generation in composite film of p-nitroaniline and its derivatives, Proceedings of the 4th Iketani Conference, Hawaii, May 17–20, 1994, p. 185.

129. **Kato, K.,** *IEEE J. Quantum Electronic,* QE-16, 1288, 1980.

130. **Davydov, B. L., Zolin, V. F., Korenneva, L. G., and Lavrovsky, E. A.,** *Opt. Spectros.* USSR, 39, 403, 1975.

131. **Koreneva, L. G., Zolin, V. F., and Davydov, B. L.,** Nonlinear Optics of Molecular Crystals, Nauka, Moscow, 1985 (in Russian).

132. **Davydov, B. L., Korenneva, L. G., and Lavrovsky, E. A.,** *Radio Eng. Electron. Phys.*, 19, 130, 1974.

133. **Carenco, A., Jerphagnon, J., and Perigaud, A.,** Nonlinear optical properties of some m-disubstituted benzene derivatives, *J. Chem. Phys.*, 66, 3806, 1977.

134. **Stevenson, J. L.,** The linear electro-optic coefficient of meta-nitroaniline, *J. Phys. D.*, 6, L13, 1973.

135. **Belikova, G. S., Golovey, M. P., Shigorin, V. D., Shipulo, G. P.,** *Opt. Spectros.* USSR, 38, 441, 1975.

136. **Levine, B. F., Bethea, C. G., Thurmound, C. D., Lynch, R. T., and Berstein, J. L.,** An organic crystal with an exceptionally large optical second harmonic coefficient: 2-methyl-4-nitroaniline, *J. Appl. Phys.,* 50, 2523, 1979.

137. **Lipscomb, G. F., Garito, A. F., and Narang, R. S.,** An exceptionally large linear electro-optic effect in the organic solid MNA, *J. Chem. Phys.,* 75, 1509, 1981.

138. **Grossman, C. H.,** Ph.D. thesis, University of Pennsylvania, 1984.

139. **Bosshard, C., Knopfle, G., Pretre, P., and Gunter, P.,** Dispersion of the molecular second-order nonlinear optical susceptibilities of nitropyridine derivatives, SPIE, 1775, 213, 1992.

140. **Singh, N. B., Henningsen, T., Hopkins, R. H., Mazelsky, R., Hamacher, R. D., Supertzi, E. P., Hopkins, F. K., Zelmon, D. E., and Singh, O. P.,** Nonlinear optical characteristics of binary organic system, *J. Cryst. Growth,* 128, 976, 1993.

141. **Gerbi, D. J., Boyd, G. T., Ender, D. A., and Leung, P. C. W.,** Nonlinear optical properties of chloronitroanilines, *Mat. Res. Soc. Symp. Proc.,* 109, 331, 1989.

142. **Ledoux, I., Lepers, C. Periguad, A., Badan, J., and Zyss, J.,** Linear and nonlinear optical properties of N-4-nitrpphenyl-L-prolinol single crystals, *Opt. Commun.,* 80, 149, 1990.

143. **Ledoux, I., Josse, D., Vidakovic, P., and Zyss, J.,** Highly efficient single-crystalline organic thin films for quadratic nonlinear optics, *Opt. Eng.,* 25, 202, 1986.

144. **Ledoux, I., Zyss, J., Migus, A., Etchepare, J., Grillon, G., and Antonetti, A.,** Generation of high peak power subpicosecond pulses in the 1.0–1.6 μm range by parametric amplification in an organic crystal, *Appl. Phys. Lett.,* 48, 1564, 1986.

145. **Davis, P. J., Hall, S. R., Jones, R. J., Kolinsky, P. V., Weinel, A. J., Hursthouse, M. B., and Karaulov, S. A.,** The examination of novel organic materials for second harmonic generation, in Organic Materials for Nonlinear Optics, Hann, R. A., and Bloor, D., eds., Special Publication No. 69, Royal Society of Chemistry, London, 1989, p. 163.

146. **Lequan, M., Lequan, R. M., Ching, K. C., Barzoukas, M., Fort, A., Bravic, G., Chasseau, D., Barrans, Y., and Huche, M.,** Quadratic nonlinear optical properties of N-(4-nitrostilbenyl) (s) prolinol, *Chem. Phys. Lett.,* 213, 71, 1993.

147. **Twieg, R. J. and Dirk, C. W.,** Molecular and crystal structure of the nonlinear optical material: 2-(N-prolinol)-5-nitropyridine (PNP), *J. Chem. Phys.,* 85, 3537, 1986.

148. **Bosshard, C., Sutter, K., and Gunter, P.,** Linear and nonlinear optical properties of 2-(N-prolinol)-5-nitropyridine, Ferroelectrics, 92, 395, 1989.

149. **Bosshard, C., Sutter, K., Schlesser, R., and Gunter, P.,** Electro-optic effects in organic single crystals, *SPIE,* 1775, 271, 1992.

150. **Bailey, R. T., Cruickshank, F. R., Guthrie, S. M. G., McAndre, B. J., McGillivray, G. W., Morrison, H., Pugh, D., Shepherd, E. A., Sherwood, J. N., Yoon, C. S., Kashyap, R., Nayar, B. K., and White, K. I.,** A full optical characterization of the organic nonlinear optical material (−)-2-(a-Methylbenzyamino)-5-nitropyridine (MBA-NP), in *Organic Materials for Nonlinear Optics,* Hann, R. A. and Bloor, D., eds., Special Publication No. 69, Royal Society of Chemistry, 1989, p. 129.

151. **Kondo, T., Morita, R., Ogasawara, N., Umegaki, S., and Ito, R.,** A nonlinear optical organic crystal for waveguiding SHG devices: (−)-2-(a-Methylbenzyamino)-5-nitropyridine (MBANP), *Jpn. J. Appl. Phys.,* 28, 1622, 1989.

152. **Vidakovic, P. V., Coquillay, M., Salin, F.,** N-(4-nitrophenyl)-N-methyl-amino-aceto-nitrile: a new organic material for efficient second harmonic generation in bulk and waveguide configurations. I. Growth, crystal structure, and characterization of organic crystal-cored fibers, *J. Opt. Soc. Am.,* B4, 988,1987.

153. **(a) Morita, R. and Vidakovic, P. V.,** Angle and temperature tuning of phase-matched second-harmonic generation in N-(4-nitrophenyl)-N-methylaminoacetonitrile, Appl. Phys. Lett., 61, 2854, 1992.
(b) Vidakovic, P. V., Second-harmonic generation in bulk crystals and crystal cored fibers of N-(4-nitrophenyl)-N-methylaminoacetonitrile, *Nonlinear Opt.,* 4, 253, 1993.

154. **Nicoud, J. F.,** Organic crystals for nonlinear optics: molecular engineering of noncentrosymmetric crystal structures, in Organic Materials for Nonlinear Optics, Hann, R. A., and Bloor, D., eds., Special Publication No. 69, Royal Society of Chemistry, London, 1989.

155. **Bosshard, Ch., Sutter, K., and Gunter,** Linear and nonlinear-optical properties of 2-cyclooctylamino-5-nitropyridine, *Ferroelectrics,* 92, 387, 1989.

156. **Bosshard, Ch., Sutter, K., Gunter, P., and Chapuis, G.,** Linear and nonlinear optical properties of 2-cyclooctylamino-5-nitropyridine, *J. Opt. Soc. Am.,* B6, 721, 1989.

157. **Kerkoc, P., Zgonik, M., Shutter, K., Bosshard, Ch. and Gunter, P.,** 4-(N,N-dimethylamino)-3-acetamidonitrobenzene single crystals for nonlinear-optical applications, *J. Opt. Soc., Am.* B, 7, 313, 1990.

158. **Norman, P. A., Bloor, D., Obhi, J. S., Karaulov, S. A., Hursthouse, M. B., Kolinsky, P. V., Jones, R. J., and Hall, S. R.,** Efficient second-harmonic generation in single crystals of 2-(N,N-dimethylamino)-5-nitroacetanilide, *J. Opt. Soc. Am.* B, 4, 1013, 1987.

159. **Bailey, R. T., Cruickshank, F. R., Guthrie, S. M. G., McArdle, B. J., Morrison, H., Pugh, D., Shepherd, E., Sherwood, J. N., and Yoon, C. S.,** Second order optical non-linearity and phase matching in 4-nitro-4'-methylbenzylidene aniline (NMBA), *Mol. Cryst. Liq. Cryst.,* 166, 267, 1989.

160. **Pugh, D.,** Linear and nonlinear optical properties of monoclinic organic crystals, in *Crystal Growth of Organic Materials,* Tokyo University of Agriculture and Technology, 1989, p. 238.

161. **Gotoh, T., Tsunekawa, T., Kondoh, T., Fukuda, S., Mataki, H., Iwamoto, M., and Maeda, Y.,** 4'-Nitrobenzylidene-3-acetamino-4-methoxyaniline (MNBA) crystal as a novel and excellent second-order nonlinear optical material, in *Crystal Growth of Organic Materials,* Tokyo University of Agriculture and Technology, 1989, p.234.

162. **Bailey, R. T., Bourhill, G., Cruickshank, F. R., Pugh, D., Sherwood, J. N., and Simpson, G. S.,** The linear and nonlinear optical properties of the organic nonlinear materials: 4-nitro-4'-methylbenzylidene aniline, *J. Appl. Phys.,* 73, 1591, 1993.

163. **Halfpenny, P. J., Shepherd, E. E. A., Sherwood, J. N., and Simpson, G. S.,** Organic crystals for optoelectronic devices, *SPIE Proc.,* 2025, 171, 1993.

164. **Tam, W., Guerin, B., Calabrese, J. C., and Stevenson, S. H.,** 3-methyl-4-methoxy-4'-nitrostilbene (MMONS): crystal structure of a highly efficient material for second-harmonic generation, *Chem. Phys. Lett.,* 154, 93, 1989.

165. **Bierlein, J. D., Cheng, L. K., Wang, Y., and Tam., W.,** Linear and nonlinear optical properties of 3-methyl-4-methoxy-4'-nitrostilbene single crystals, *Appl. Phys. Lett.,* 56, 423, 1990.

166. **Hosomi, T., Suzuki, T., Yamamoto, H., Watanabe, T., Sato, H., and Miyata, S.,** Efficient second harmonic generation in single crystal of N-methoxymethyl-4-nitroaniline, *SPIE Proc.,* 1147, 124, 1989.

167. **Sutter K., Bosshard, Ch., Ehrensperger, M., Gunter, P., and Twieg, R. J.,** Nonlinear optical and electrooptical effects in 2-methyl-4-nitro-N-methylaniline (MNMA) crystals, *IEEE J. Quantum Electron.,* QE-24, 2362,1988.

168. **Vizgert, R. V., Davydov, B. L., Kotovschchikov, S. G., and Starodubtseva, M. P.,** Generation of the second harmonic of a neodymium laser in powders of noncentrosymmetric organic compounds, *Sov. J. Quantum Electron.,* 12, 214, 1982.

169. **Rao, S. M., Batra, A. K., Lal, R. B., Evans, R. A., Metzger, R. M., and Lee, W. J.,** Mixed methyl-(2,4-dinitrophenyl)-aminopropranoate:2-methyl-4-nitroaniline crystal-a new nonlinear optical material, *J. Appl. Phys.,* 70, 6674, 1991.

170. **Zyss, J., Ledoux, I., Hierle, R., Raj, R. M., and Oudar, J. L.,** Optical parametric interactions in 3-methyl-4-nitropyridine-1-oxide (POM) single crystals, *IEEE J. Quantum. Electron.,* QE-21, 1286, 1985.

171. **Sigelle, K. and Hierle, R.,** *J. Appl. Phys.,* 52, 4199, 1981.

172. **Nahata, A., Horn, A., Yardley, J. T.,** A new organic electrooptic crystals 2,6-dibromo-N-methyl-4-nitroaniline (DBNMNA), *IEEE J. Quantum Electron.,* QE-26, 1521, 1990.

173. **Horn, K. A. and Nahata, A.,** Structure of 2,6-dibromo-N-methyl-4-nitroaniline (DBNMNA); a new electro-optic organic crystal, *Acta Cryst.,* C47, 1283, 1991.

174. **Tao, X. Z., Zhang, N., Yuan, D., Xu, D., Yu, W., and Jiang, M.,** New organometallic complex nonlinear optical crystal-thiosemicarbazide cadmium thiocyanate, *SPIE Proc.,* 1337, 385, 1990.

175. **Zhang, N., Jiang, M. H., Yuan, D. R., Xu, D., Shao, Z. S., and Tao, X. T.,** Study on the quality and performance of the organometallic complex triallylthiourea cadmium chloride, *SPIE Proc.,* 1337, 390, 1990.

176. **Yuan, D., Zhang, N., Yu, W., Xu, D., Tao, X. T., and Jiang, M.,** A new nonlinear optical crystal-triallylthiourea cadmium bromide, *SPIE Proc.,* 1337, 394, 1990.

177. **Marcy, H. O., Warren, L. F., Webb, M. S., Ebbers, C. A., Velsko, S. P., Kennedy, G. C., and Catella, G. C.,** Second-harmonic generation in zinc tris(thiourea) sulfate, *Appl. Opt.,* 31, 5051, 1992.

178. **Tomaru, S., Matsumoto, S., Kurihara, T., Suzuki, H., Ooba, N., and Kaino, T.,** Nonlinear optical properties of 2-Adamantylamino-5-nitropyridine crystals, Appl. Phys. Lett., 58, 2583, 1991.

179. **Wada, T., Yamada, A., and Sasabe, H.,** Crystal growth and phase matched second harmonic generation (PMSHG) of dicyanovinyl anisole (DIVA), in Proceedings of the First International Workshop on Crystal Growth of Organic Materials (CGOM) 1989, p. 229.

180. **Sutter, K., Knopfle, G., Saupper, N., Hulliger, J., Gunter, P., and Petter, J. W.,** Nonlinear-optical, optical, and crystallographic properties of N-(4-Nitro-2-pyridinyl)-phenylalaninol, *J. Opt. Soc. Am* B, 8, 1483, 1991.

181. **Goto, Y., Hayashi, A., Nakayama, M., Kitaoka, Y., Sasaki, T., Watanabe, T., Miyata, S., Honda, K., and Goto, M.,** Second harmonic generation and crystal growth of chalcone derivatives, Kobunshi Ronbunshu, 47, 791, 1990 (in Japanese).

182. **Goto, Y., Hayashi, A., Nakayama, M., Watanabe, T., Miyata, S., and Goto, M.,** Second harmonic generation and crystal structure of alkylthio group substituted chalcones, Nippon Kakingaku Kai Shi, 968, 1990, (in Japanese).

183. **Kitaoka, Y., Sasaki, T., Nakai, S., Yokotani, A., Goto, Y., and Nakayama, M.,** Laser properties of new organic nonlinear optical crystal chalcone derivatives, *Appl. Phys. Lett.,* 56, 2074, 1990.

184. (a) **Zhang, G. J., Kinoshita, T., Sasaki, K., Goto, Y., and Nakayama, M.,** Second harmonic generation of a new chalcone-type crystal, *Appl. Phys. Lett.,* 57, 221, 1990.
    (b) **Zhang, G. J., Kinoshita, T., Sasaki, K., Goto, Y., and Nakayama, M.,** Crystal growth of 4-Br-4'-methoxychalcone and its characteristics, *J. Cryst. Growth,* 100, 411, 1990.

185. **Kitaoka, Y., Sasaki, T., Nakai, S., and Goto, Y.,** New nonlinear optical thienylchalcone and its harmonic generation properties, *Appl. Phys. Lett.,* 59, 19, 1991.

186. **Goto, Y., Hayashi, A., Kimura, Y., and Nakayama, H.,** Second harmonic generation and crystal growth of substituted thienyl chalcone, *J. Cryst. Growth,* 108, 688, 1991.

187. **(a) Zhang, N., Yuan, D., Tao, Z., Shao, Z., Dou, S., Jiang, M., and Xu, D.,** New nonlinear optical crystal 3-methoxy-4-hydroxy-benzaldehyde and its phase-matched properties, *J. Cryst. Growth,* 123, 255, 1992.
    (b) Zhang, N., Tao, Z., Yuan, D., Zhnag, N., and Shao, Z., A new organic crystal-MHBA for potential dioed laser second harmonic generation, Mater, *Res. Soc. Symp. Proc.,* 247, 265, 1992.

188. **Wang, Y., Tam., W., Stevenson, S. H., Clement, R. A., and Calabrese, J.,** New organic nonlinear optical materials of stilbene and diphenylacetylene derivatives, *Chem. Phys. Lett.,* 148, 136, 1988.

189. **Oliver, S. N., Pantelis, P., and Dunn, P. L.,** Polymorphism and crystal-crystal transformations of the highly optically nonlinear organic compound a-[4'-Methoxyphenyl)methylene]-4-nitrobenzen-acetonitrile, *Appl. Phys. Lett.,* 56, 307, 1990.

190. **Sugiyama, Y., Suzuki, Y., Mitamura, S., and Nishiyama, T.,** Second-order optical nonlinearities of stilbene derivatives, Sen-i-Gakkai Symp. Preprints, p. B88, 1992.

191. **Sarma, J. A. R. P., Dhurjati, M. S. K., Ravikumar, K., and Bhanuprakash, K.,** Molecular and crystal enegineering studies of tow 2,4-dinitroalkoxystilbenes: an endeavor to generate efficient SHG crystals, *Chem. Mater.,* 6, 1369, 1994.

192. **Lee, R. W.,** *J. Opt. Soc. Am.,* 59, 1574, 1969.

193. **Halbout, J. M. and Tang, C. L.,** Phase-matched second harmonic generation in sucrose, *IEEE J. Quantum Electron.,* QE-18, 410, 1982.

194. **(a) Pu, L. S. and Tomono, T.,** A new electron acceptor: cyclobutenedione for control of polarizable molecular orientation in crystal to generate high SHG, Sen-i-Gakkai Symp. Preprints, B76, 1992.
    (b) Pu, L. S., Observing high second harmonic generation and control of molecular alignment in one dimension: cyclobutanediones as a promising new acceptor for nonlinear optical materials, in Materials for Nonlinear Optics: Chemical Perspectives, Marder, S. R., Sohn, J. E., and Stucky, G. D., eds., ACS Symposium Series 455, American Chemical Society, Wasington, DC, 1991, Chap. 22, p. 331.

195. **Pu, L. S.,** New materials: cuclobutenediones for second-order nonlinear optics, *Nonlinear Opt.* 3, 233, 1992.

196. **Asano, K., Morita, M., Nagasawa, Y., and Ninomiya, H.,** Crystallographic structure and optical characteristics of nonlinear optical materials MNP, Extended abstracts, No. 30a-Q-5, p. 1028, The Autumn Meeting, 1992, The Japanese Society of Applied Physics (in Japanese).

197. **Nagasawa, Y., Takai, K., Katoh, T., Ninomiya, H., and Takeuchi, Y.,** measurement of the dispersion of second-order nonlinear optical coefficient of MNP, in Extended abstracts, No. 16a-Z-6, p. 997, The 53rd Autumn Meeting, 1992, The Japanese Society of Applied Physics (in Japanese).

198. **Morita, M., Asano, K., Nagasawa, Y., and Ninomiya, H.,** Crystallographic structure and nonlinear optical properties of 4-hydroxy-3-methoxybenzonitrile, Extended abstracts, No. 30a-Q-6, p. 1028, The Autumn Meeting, 1992, The Japanese Society of Applied Physics (in Japanese).

199. **Katoh, T., Harada, A., Ishihara, M., Okazaki, K., Tonogaki, M., Ohba, S., and Saito, Y.,** Crystal structure and nonlinear optical properties in 1-(4-nitrophenyl)azole), Extended abstracts, No. 30a-Q-2, p. 1027, The Autumn Meeting, 1992, The Japanese Society of Applied Physics (in Japanese).

200. **Okazaki, M., Ishihara, Ogawa, K., Harada, A., Okazaki, Y., Kato, T., and Kamiyama, K.,** Nonlinear optical properties of N-(4-nitrophenyl)azoles, Nippon Kakingaku Kai Shi, No. 10, p. 1237, 1992 (in Japanese).

201. **Okazaki, Y., Harada, A., Kamiyama, K., Okazaki, M., Kudobera, S., Fukunaga, H., and Umegaki, S.,** Crystal structure and nonlinear optical property, in 1-(4-itrophenyl)azole) derivatives nonlinear optical material, Extended abstracts, No. 2a-G-7, p. 197, The Autumn Meeting, 1992, The Japanese Society of Applied Physics (in Japanese).

202. **Okazaki, M., Fukunaga, H., Ishihara, M., and Kubo, K.,** 2-(1-Pyrrolyl, diazolyl, and triazolyl)-substituted 5-nitropyridines for nonlinear optical materials, *Bull. Chem. Soc. Jpn.,* 67, 1936, 1994.

203. **Nakazumi, H., Maeda, K., Yagi, S., Kitao, T., and Ogawa, T.,** Synthesis and SHG efficiency of 2-(aminomethylene)-benzo[b]thiophe-3(2H)-ones derived by ring contraction of 1-benzothiopyran derivatives, Nippon Kakingaku Kai Shi, No. 10, p. 1223, 1992 (in Japanese).

204. **Francis, C. V. and Tiers, G. V. D.,** Straight-chain carbamyl compounds for second harmonic generation, *Chem. Mater.,* 4, 353, 1992.

205. **(a) Takashima, M., Katogi, S., Ukuda, H., Watanabe, T., Miyata, S., and Hosomi, T.,** Nonlinear optical properties of urea derivatives, in Extended abstracts, No. 16a-Z-3, p. 996, The 53rd Autumn Meeting, 1992, The Japanese Society of Applied Physics (in Japanese).
    (b) Katogi, S., Studies of organic SHG materials: improvement of transparency in the visible region, Master thesis, Tokyo University of Agriculture and Technology, Tokyo, 1991 (in Japanese).
    (c) Yamamoto, H., Funato, S., Kang, W. B., Okinawa, K., Sugiyama, T., Kinoshita, T., and Sasaki, K., Waveguide SHG device using organic nonlinear optical material IAPU (1), Extended Abstract of 41st Spring meeting of the Japan Society of Applied Physics and Related Societies, 1994, p. 1097 (in Japanese).

206. **Okunaka, M., Kato, M., Kanetake, T., Kumegawa, N., Kiguchi, M., Taniguchi, Y., and Kobayashi, K.,** Crystal structures and phase transitions of second-order optically nonlinear materials, hydrogen-bonding tolane derivatives, in Extended abstracts, No. 16a-Z-2, p. 999, The 53rd Autumn Meeting, 1992, The Japanese Society of Applied Physics (in Japanese).

207. **Watanabe, O., Noritake, T., Tsuchimori, M., Hirose, Y., Okada, A., and Kurauchi, T.,** SHG properties of 2-

amino-5-nitropyridine-(+)-tartarate, in Extended abstracts, No. 16a-Z-10, p. 998, The 53rd Autumn Meeting, 1992, The Japanese Society of Applied Physics (in Japanese).

208. **Marder, S. R., Perry, J. W., Schaefer, W. P., Tiemann, B. G., Groves, P. C., and Perry, K. J.,** New organic and organometallic salts for second-order nonlinear optics, *SPIE Proc.,* 1147, 108, 1989.

209. **Marder, S. R., Perry, J. W., and Schaefer, W. P.,** *Science,* 245, 626, 1989. Marder, S. R., Perry, J. W., and Yakymyshyn, C. P., Organic salts with large second-order optical nonlinearities, *Chem. Mater.,* 6, 1137, 1994, and references therein.

210. **(a) Lawrence, B.,** Master thesis, Massachusetts Institute of Technology, 1992.

   (b) Pan, F., Knopfle, G., Bosshard, Ch., Follonier, S., Spreiter, R., Wong, M. S., and Günter, P., Electro-optic properties of the organic salt 4-N,N-dimethylamino-4'-N'-methyl-stilbazolium tosylate, *Appl. Phys. Lett.,* 69, 13, 1996.

211. **Okada, S., Masaki, A., Sakai, K., Ohmi, T., Koike, T., Anzai, E., Umegaki, S., Matsuda, H., and Nakanishi, H.,** Crystal of stilbazolium p-toluenesulfonate for second-order nonlinear optics, in *Nonlinear Optics,* Miyata, S., ed., p. 237, 1992.

212. **Okada, S., Masaki, A., Matsuda, H., Nakanishi, H., Koike, T., Ohmi, T., Yishikawa, N., and Umegaki, S.,** Merocyaninep-ptoluenesulfonate acid complex with large second order nonlinearity, *SPIE Proc.,* 1337, 178, 1990.

213. **Sakai, K., Yoshikawa, N., Ohmi, T., Koike, T., Umegaki, S., Okada, S., Masaki, A., Matsuda, H., and Nakanishi, H.,** Crystal structure and nonlinear optical properties of stilbazolium derivatives having shortened cut-off wavelength, *SPIE Proc.,* 1337, 307, 1990.

214. **Sebutoviez, C., Nicoud, J. F., Fischer, J., Ledoux, I., and Zyss, J.,** Crystalline zwitterionic stilbazolium derivatives with large quadratic optical nonlinearities, *Chem. Mater.,* 6, 1358, 1994.

215. **Lacroix, P. G., Clement, R., Nakatani, K., Ledoux, I., and Zyss, J.,** Nonlinear optical properties of substituted stilbazolium-PMS$_3$ intercalation compounds, *Mater. Res. Soc. Symp.,* 328, 613, 1994.

216. **Ledoux, I., Zyss, J., Jutand, A., and Amatore, C.,** Nonlinear optical properties of asymmetric polyphenyls: efficiency versus transparency trade-off, *Chem. Phys.,* 150, 117, 1991.

217. **Zyss, J., Ledoux, I., Bertault, M., and Toupet, E.,** Dimethylaminocyanobiphenyl (DMACB): a new optimized molecular crystal for quadratic nonlinear optics in visible, *Chem. Phys.,* 150, 125, 1991.

218. **Takagi, K., Mizuno, A., Nakatsu, K., Abe, S., and Matsuoka, M.,** Crystal engineering; substituent effect of biphenyls for SHG materials, *Nonlinear Optics,* 1, 31, 1990.

219. **Ogawa, K., Yoshimura, S., Kaji, M., Kagawa, H., and Kakuta, A.,** 16-(p-Butoxybenzylidene)androsta-1,4-diene-3,17-dione, *Acta Cryst.,* C48, 1359, 1992.

220. **Kinoshita, T., Horinouchi, S., Sasaki, K., Okamoto, H., Tanaka, N., Fukaya, T., and Goto, M.,** Noncollinear phase-matched second harmonic generation in a 2-furyl methacrylic anhydride crystal, in Proceedings of the Third Optoelectronic Industry and Technology Development Association (OITDA) International Forum, Sasaki, K., ed., pp. 7–10, 1993.

221. **Kinoshita, T., Horinouchi, S., Sasaki, K., Okamoto, H., Tanaka, N., Fukaya, T., and Goto, M.,** Nonlinear optical properties of a novel organic crystal: 2-furyl methacrylic anhydride, *J. Opt. Soc. Am.,* A11, 986, 1994.

222. **Senoh, T., Koike, T., Hama, H., and Yamanaka, T.,** Nonlinear optical properties of carbohydrazide derivatives, in Proceedings of the Third Optoelectronic Industry and Technology Development Association (OITDA) International Forum, Sasaki, K., ed., p.15, 1993.

223. **Hama, H., Koike, K., Senoh, R., and Yamanaka, T.,** Extended abstracts 53th Autumn Meeting of the Japan Society of Applied Physics, 1992, p. 998.

224. **Yakovlev, Y. O., and Poezzhalov, V. M.,** Nonlinear optical properties of p,p'-dihydroxydiphenyl sulfone crystal (DHDPS), *Sov. J. Quantum Electron.,* 20, 694, 1990.

225. **Tsunekawa, T., Gotoh, T., Mataki, H., and Iwamoto, M.,** Dihedral rigid molecule in crystal engineering for optimized phase-matched SHG: 3,9-Dinitro-5a,6,11a,12-tetrahydro[1,4]-benzoxadino[3,2b] benzoxadine (DNBB), *SPIE Proc.,* 1337, 285, 1990.

226. **Terao, H., Itoh, Y., Ohno, K., Isogai, M., Kakuta, A., and Mukoh, A.,** Second-order nonlinear optical properties and polymorphism of benzophenone derivatives, *Opt. Commun.,* 75, 451, 1990.

227. **Guha, S., Frazier, C. C., Chauchard, and Chen, W. P,** Phase-matched second-harmonic generation in 4-(amino-benzophenone crystal, *SPIE Proc.,* 971, 89, 1988.

228. **Imanishi, Y., Itoh, Y., Kakuta, A., and Mukoh, A.,** Nonlinear optical properties of xanthone derivatives, *SPIE Proc.,* 1361, 570, 1990.

229. **Wada, T., Zhang, Y., Choi, Y. S., and Sasabe, H.,** Nonlinear optical properties of photoconductive materials, *SPIE,* 1775, 289, 1992.

230. **(a) Katogi, S., Yamamoto, H., Takashima, M., Watanabe, T., Miyata, S., and Hosomi, T.,** Studies on novel organic nonlinear optical materials with Λ-shaped conformation, Annual Meeting of Japanese Chemical Society, 1991, p. 2G547. (In Japanese)

   (b) Ukuda, H., Takashima, M., Watanabe, T., Miyata, S., and Hosomi, T., 4-Nitrophenylcarbamic acid ethylester, Extended Abstract of 40th Spring Meeting, The Japan Society of Applied Physics and Related Societies, 1993, No. 30a-ZG-4, p. 1123. (In Japanese)

231. **Bergman, J. G., Crane, G. R., Levine, B. F., and Bethea, C. G.,** Nonlinear optical susceptibility of 5-nitrouracil, *Appl. Phys. Lett.,* 20, 21, 1972.

232. **Puccetti, G., Perigaud, A., Badan, J., Ledoux, I., and Zyss, J.,** 5-Nitrouracil: a transparent and efficient nonlinear organic crystal, *J. Opt. Soc. Am.* B, 10, 733, 1993.

233. **Kotler, Z., Hierle, R., Josse, D., Zyss, J., and Masse, R.,** Quadratic nonlinear optical properties of anew transparent and highly efficient organic-inorganic crystal: -Amino-5-nitropyridinium dihydrogen phosphate (2A5NPDP), *J. Opt. Soc. Am.,* B9, 534, 1992.

234. **Samuel, I. D. W., Villacampa, B., Josse, D., Khodja, S., and Zyss, J.,** Efficient optical parametric generation in an organomineral crystal, *Appl. Phys. Lett.,* 66, 2019, 1995.

235. **Horiuchi, N., Lefaucheux, F., Robert, M. C., Josse, D., Khodja, S., and Zyss, J.,** Gel growth of 2-amino-5-nitropyridinium dihydrogen phosphate organomineral crystals: X-ray and nonlinear optics characterization, *J. Cryst. Growth,* 147, 361, 1995.

236. (a) **Masse, R., Beucher, M. B., Pecaut, J., Levy, J. P., and Zyss, J.,** Design of organic-inorganic polar crystals for quadratic nonlinear optics, *Nonlinear Opt.* 5, 413, 1993.
    (b) **Fur, Y. L., Beucher, M. B., Masse, R., Nicoud, J. F., and Levy, J. P.,** Crystal engineering of noncentrosymmetric structures based on 2-amino-5-nitropyridine and n-chloroacetic acid assemblies, *Chem. Mater.,* 8, 68, 1996.

237. **Fukaya, T., Nagawa, Y., Kodaka, M., Goto, M., Honda, K., and Nakanishi, H.,** Crystal structure of 1,1′-(1,8-naphthylene)-bis(1H-1,2,3-triazole and its optical nonlinearity, in Proceedings of First International Workshop on Crystal Growth of Organic Materials (CGOM), 1989, p. 221.

238. **Eimerl, D., Velsko, E., Davis, L., Wang, F., Laiacono, G., and Kennedy, G.,** *IEEE J. Quantum Electron.,* QE-25, 179, 1989.

239. **Yokotani, A., Sasaki, T., Yoshida, K., and Nakai, S.,** *Appl. Phys. Lett.,* 55, 2692, 1989.

240. **Sasaki, T., Yokotani, A., Fujioka, Y., and Nakai, S.,** Growth and laser properties of inorganic and organic nonlinear materials, in Proceedings of the First International Workshop on Crystal Growth of Organic Materials (CGOM), 1989, p. 2216.

241. **Haussühl, S., Chrosch, J., Gnanam, F., Fiorentini, E., Recker, K. J., and Wallrafen, F.,** Crystal growth and physical properties of monoclinic L-arginine hydrochloride monohydrae, $C_6H^{14}O_2N_4HCl \cdot H_2O$ and L-arginine hydrobromide monohydrate, $C_6H^{14}O_2N_4HBr \cdot H_2O$, *Cryst. Res. Technol,* 25, 617, 1990.

242. **Schuler, L., Betzler, K., Hesse, H., and Kapphan, S.,** *Opt. Commun.,* 43, 157, 1982.

243. **Kondo, K., Ochiai, S., Takemoto, K., and Irie, M.,** Nonlinear optical properties of p-substituted benzalbarbituric acids, *Appl. Phys. Lett.,* 56, 718, 1990.

244. **Miyata, S., Watanabe, T., and Nalwa, H. S.,** Molecular design of nonlinear optical molecules and polymers by lambda (A)-type conformation, *Nonlinear Opt.,* 1994

245. **Sano, K., Mori, Y., Todori, K., Pellet, C., Touri-iwa, N., and Kobayashi, K.,** Second harmonic generation of 2-cyano-2-propenoic acid derivatives, *Nonlinear Op.,* 2, 89, 1992.

246. **Nakatani, H., Hayashi, H., and Hidaka, T.,** Linear and nonlinear optical properties of 2-cyano-3-(2-methoxyphenyl)-2-propenoic acid methyl ester, *Jpn. J. Appl. Phys.,* 31, 1802, 1992.

247. (a) **Abdel-Halim, A., Cowan, D. O., Robinson, D. W., Wiygul, F. M., and Kimura, M.,** Preliminary study of the nonlinear optical properties of 4-amino-4′-nitrodiphenyl sulfide, *J. Phys. Chem.,* 90, 5654, 1986.
    (b) **Bakzoukas, M., Josse, D., Zyss, J., Gordon, P., and Morley, J.,** *Chem. Phys.,* 139, 359, 1989.

248. (a) **Verbiest, T., Clays, K., Persoons, A., Samyn, C., Meyers, F., and Bredas, J. L.,** Investigation of the β-response in organic molecules from dipolar to octapolar systems, *SPIE Proc.,* 2025, 31, 1993.
    (b) **Stadler, S., Bräuchle, Ch., Brandl, S., and Gompper, R.,** Determination of the first hyperpolarizability of octopolar molecular ions made from symmetric cyanine dyes, *Chem. Mater.,* 8, 676, 1996.

249. **Kagawa, H., Sagawa, M., Kawabata, Y., Hamada, T., and Kakuta, A.,** Synthesis and nonlinear optical properties of tribenzamide derivatives, *J. Chem. Soc. Jpn., Chem. Indust. Chem.,* 1261, 1992, (in Japanese).

250. **Bacquet, G., Bassoul, P., Combellas, C., Simon, J., Thiebault, A., and Tournilhac, F.,** Highly polarizable zwitterions for nonlinear optics: synthesis and properties of phenoxide-pyridium derivatives, *Adv. Mater.,* 2, 311, 1990.

251. **Uemiya, T. and Nogami, T.,** Organic second-order materials, in Advanced Nonlinear Optical Organic Materials I, Nakanishi, H., et al., eds., CMC Press, Tokyo, 1991, Chap. 2, p. 9 (in Japanese).

252. **Youping, H., Genbo, S., Jianqiu, S., Guiming, Y., and Rihong, J.,** New organic nonlinear optical crystals of indole-3-aldehyde, *J. Cryst. Growth,* 130, 444, 1993.

253. **Kagawa, H., Sagawa M and Kakuta, A.,** 3-(4-tert-Butylphenyl)-3-hydroxy-1-(4-methoxyphenyl)-2-propen-1-one: an SHG active b-diketone, *Acta Crystallogr.,* C49, 2181, 1993.

254. **Ogawa, K., Kaji, M., Kagawa, H., Sagawa, M., and Kakuta, A.,** 8-(4′-Acetylphenyl)-1,4-dioxa-8-azaspiro[4,5]decane: a new potential nonlinear optical material, *Acta Crystallogr.,* C50, 95, 1994.

255. **Kagawa, K. Sagawa, M., Kakuta, A., Kaji, M., Nakayama, H and Ishii, K.,** Single crystal growth and characterization of a new organic nonlinear optical materials:8-(4′-acetylphenyl)-1,4′-dioxa-8-azaspiro[4,5]decane (APDA), *J. Cryst. Growth,* 139, 309, 1994.

256. **Sagawa, M., Kagawa, H., Kakuta, A., and Kaji., M.,** Blue second-harmonic light generation from a new organic

single crystal: 8-(4'-acetylphenyl)-1,4'-dioxa-8-azaspiro[4,5]decane (APDA), *Mater. Res. Soc. Symp. Proc.,* 328, 601, 1994.

257. **Yamamoto, H., Funato, S., Sugiyama, T., Kinoshita, T., and Sasaki, K.,** Linear and nonlinear optical properties of organic crystal ASBA, Sen-i-Gakkai Preprints, 1994, p. G-130.

258. **Yamamoto, H., Funato, S., Sugiyama, T., Johnson, R. E., Norwood, R. A., Jung, J., Kinoshita, T., and Sasaki, K.,** Linear and nonlinear optical properties of a new organic crystal N-(4-Aminobenzenesulfonyl)acetamide *J. Opt. Soc. Am. B.,* 13, 837, 1996.

259. **(a) Kitazawa, M., Higuchi, R., Takahashi, M., Wada, T., and Sasabe, H.,** Ultraviolet generation at 266 nm in a novel organic nonlinear optical crystal: L-pyrrolidone-2-carboxylic acid, *Appl. Phys. Lett.,* 64, 2477, 1994.
   (b) Kitazawa, Takahashi, M., and Matsuoka, M., Growth behavior of ʟ-pyrrolidone-2-carboxylic acid (L-PCA) single crystals, *J. Cryst. Growth,* 141, 425, 1994.

260. **Yonehara, H., Tachikawa, Y., Pac, C., and Maki, H.,** Molecular orbital calculations of nonlinear optical properties of 4-nitrophenylhydrazones, *Nonlinear Opt.,* 6, 51, 1993.

261. **Yonehara, H., Kang, W. B., and Pac, C.,** Synthesis of 4-nitrophenylhydrazone derivatives and their second-harmonic generation properties, *Nonlinear Opt.,* 4, 357, 1993.

262. **Serbutoviez, C., Bosshard, Ch., Knopfle, G., Wyss, P., Pretre, P., Gunter, P., Schenk, K., Solari, E., and Chapuis, G.,** Hydrazone derivatives, and efficient class of crystalline materials for nonlinear optics, *Chem. Mater.,* 7, 1198, 1995.

263. **Yonehara, H., Kang, W., Kawara, T., and Pac, C.,** Synthesis and second harmonic generation properties of 2-(4-nitroanilino)-1,3,5-triazine derivatives, *J. Mater. Chem.,* 4, 1571, 1994.

264. **Matsushima, R., Tatemura, M., and Okamoto, N.,** Second-harmonic generation from 2-Arylideneindan-1,3-diones studied by the powder method, *J. Mater. Chem.,* 2, 507, 1992.

265. **Sugiyama, T., Shigemoto, T., Komatsu, H., Sakaguchi, Y., and Ukachi, T.,** Linear and nonlinear optical properties of new organic materials, *Mol. Cryst. Liq. Cryst.,* 224, 45, 1993.

266. **Ukachi, T., Shigemoto, T., Komatsu, H., and Sugiyama, T.,** Crystal growth and characterization of a new organic nonlinear optical material: L-N-(5-Nitro-2-pyridyl)leucinol, *J. Opt. Soc. Am. B,* 10, 1372, 1993.

267. **Osaka, H., Ishida, T., Nogami, T., Yamazaki, R., Yasui, M., Iwasaki, F., Mizoguchi, A., Kubata, M., Uemiya, T., and Nishimura, A.,** Organic second-order nonlinear optical materials with transparency in the blue wavelength region: ethylene derivatives possessing electron-donor and acceptor groups, *Bull. Chem. Soc. Jpn.,* 67, 918, 1994.

268. **Magnani, G., Kramer, A., Puccetti, G., Ledoux, I., Soula, G., Zyss, J., and Meyrueix, R.,** *Organometallics,* 9, 2640, 1990.

269. **Magnani, G., Kramer, A., Puccetti, G., Ledoux, I., Soula, G., and Zyss, J.,** *Mol. Eng.,* 1, 11, 1991.

270. **Hissink, D., Bolink, H. J., Eshuis, J. W., van Hutten, P. F., Malliaras, G. G., and Hadziioannou, G.,** Silicon based donor-acceptor compounds: synthesis and nonlinear optical chracterization, SPIE Proc., 2025, 37, 1993.

271. **Kelderman, E., Verboom, W., Engbersen, J. F. J., Harkema, S., Heesink, G. J. T., Lehmusvaara, E., Hulst, N. F., Reinhoudt, D. N., Derhaeg, L., and Persoons, A.,** Chiral pyrrolo[1,2-a]quinolines as second-order nonlinear optical materials, *Chem. Mater.,* 4, 626, 1992.

272. **Dastidar, P., Row, T. N. G., and Venkatesan,** Studies of nonlinear optical organic materials: crystal and molecular structure of 2-dicyanomethylene-1,3-dioxolane, *J. Mater. Chem.,* 1, 1057, 1991.

273. **Begum, N. S. and Venkatesan, K.,** Studies of nonlinear optical materials: structure of diphenylmethyl(Z)-1-(1-methylthio-2-nitrovinyl)tetrahydropyrrole-2-carboxylate, *Acta Cryst.,* C48, 902, 1992.

274. **(a) Kohler. D., Enderle, T., Stahelin, M., and Granacher, I. Z.,** Experimental characterization of novel materials for nonlinear optics represented by iodoform complexes, in Proceedings of the 5th International Conference on Unconventional Photoactive Solids, Okazaki, Japan, October 13–17, 1991, p. 289.
   (b) Samoc, A., Samoc, M., Kohler. D., Stahelin, M., Funfschilling, J., and Granacher, I. Z., Linear and second order optical properties of the trigonal adducts of triiodomethane with sulfur and quinoline, *Nonlinear Opt.,* 2, 13, 1992.

275. **Loos-Wildenauer, M., Kunz, S., Voigt-Martin, I. G., Yakimanski, A., Wischerhoff, E., Zentel, R., Tschierske, C., and Muller, M.,** Second harmonic generation in ferroelectric liquid crystalline thiadiazole derivatives, *Adv. Mater.,* 7, 170, 1995.

276. **Lambert, C., Machell, J. C., Mingos, M. P., and Stolberg, T. L.,** Preparation and second-harmonic generation properties of tris(pyrocatecholato)stannate (IV) compounds, *J. Mater. Chem.,* 1, 775, 1991.

277. **Calabrese, J. C. and Tam, W.,** Organometallics for nonlinear optics: metal-pyridine and bipyridine complexes, *Chem. Phys. Lett.,* 133, 244; 1987.

278. **Tam, W. and Calabrese, J. C.,** Oxidative addition of Pd(0) and Pt(0) with aromatic halides: materials for second harmonic generation, *Chem Phys. Lett.,* 144, 79, 1988.

279. **Green, M. L. H., Marder, S. R., Thompson, M. E., Bandy, J. A., Bloor, D., Kolinsky, P. V., and Jones, R. J.,** Synthesis and structure of (cis)-[1-Ferrocenyl-2-(4-nitrophenyl)ethylene], an organotransition metal compound with a large second-order optical nonlinearity, *Nature,* 330, 360, 1987.

280. **Bandy, J. A., Bunting, H. E., Garcia, M. H., Green, M. L. H., Marder, S. R., Thompson, M. E., Bloor, D.,**

Kolinsky, P. V., and Jones, R. J., The synthesis of organometallic compounds with second-order optical nonlinearities, in Organic Materials for Nonlinear Oprics, Hann, R. A., and Bloor, D., eds., Special Publication No. 69, Royal Society of Chemistry, London, 1989, p. 225.

281. **Coe, B. J., Foulon, J. D., Hamor, T. A., Jones, C. J., McCleverty, J. A., Bloor, D., Cross, G. H., and Axon, T. L.,** Nonlinear optical materials containing molybdenum or tungsten mononitrosyl redox centres: diaromatic azo derivatives, *J. Chem. Soc. Dalton Trans.* 3427, 1994.

282. **Coe, B. J., Jones, C. J., McCleverty, J. A., Bloor, D., Kolinsky, P. V., and Jones, R. J.,** Second harmonic generation by bimetallic complexes containing ferrocenyl and molybdenum or tungsten mononitrosyl redox centres, *J. Chem. Soc. Chem. Commun.,* 1485, 1989.

283. **Coe, B. J., Kurek, S. S., Rowley, N. M., Foulon, J. M., Hamor, T. A., Harman, M. E., Hursthouse, M. B., Jones, C. J., McCleverty, J. A., and Bloor, D.,** Structural studies of some metal organic complexes which exhibit nonlinear optical properties, *Chemtronics,* 5, 23, 1991.

284. **Marder, T. B., Lesley, G., Yuan, Z., Fyfe, H. B., Chow, P., Stringer, G., Jobe, I. R., Taylor, N. J., Williams, I. D., and Kurtz, S. K.,** Transition metal acetylides for nonlinear optical properties, in Materials for Nonlinear Optics: Chemical Perspectives, ACS Symposium Series, Marder, S. R., Sohn, J. E., and Stucky, G. D., eds., American Chemical Society, Washington, D. C., 1991, Chap. 40, p. 605.

285. **Kawamata, J. and Inoue, K.,** Second-order nonlinear optical coefficients of bis(benzylidene)cycloalkanone derivatives, Extended abstract of the Japanese Society of Applied Physics and Related Sciences, 42nd Spring Meeting, March 28–31, 1995, p. 1164, (in Japanese).

286. **Kawamata, J. and Inoue, K.,** New second-order nonlinear optical materials with a cutoff wavelength of 350 nm. 3-benzylidene-D-camphor derivatives, *Chem. Lett.,* 921, 1993.

287. **Hendricks, E., Clays, K., Persoons, A., Dehu, C., and Bredas, J. L.,** The bacteriorhodopsin chromophore retinal and derivatives: an experimental and theoretical investigation of the second-order optical properties, *J. Am. Chem. Soc.,* 117, 3547, 1995.

288. **Bhattacharya, S., Dastidar, P., and Guru Row, T. N.,** Hydrogen-bond-directed self-assembly of D-(+)-dibenzoyl tartaric acid and 4-aminopyridine: optical nonlinearities and stoichiometry-dependent novel structural features, *Chem. Mater.,* 6, 531, 1994.

289. **Tam, W., Eaton, D. F., Calabrese, J. C., Williams, I. D., Wang, and Anderson, A. G.,** Channel inclusion complexation of organometallics: dipolar alignment for second harmonic generation, *Chem. Mater.,* 1, 128, 1989.

290. (a) **Qi, F., Hua, J., Zheng, Q., Hua, C. J., Hong, L., Tao, Y. W., and Zhuang, Z.,** Nonlinear optical properties of DMIT derivatives, *J. Mater. Chem.,* 4, 1041, 1994.
     (b) **Qi, F., Hua, J. and Dong, X.,** Synthesis, structure, and SHG properties of multi-sulphur type organic crystal α-BNPT-DTT, β-BNPT-DTT, α-BNPT-DTO, β-BNPT-DTO, Extended Abstract of Second International Conference on Organic Nonlinear Optics (ICONO′2), Kusatsu, Japan, July 23–26, 1995, p. 125.

291. **Youping, H., Genbo, S., Guiming, Y., Xiangjin, H., and Rihong, J.,** Growth and characterization of new organic nonlinear optical crystal: 1-(3-thienyl)-3-(4-chlorophenyl)-propen-1-one, *J. Cryst. Growth,* 141, 389, 1994.

292. **Donald, D. S., Cheng, L. T., Desiraju, G., Meredith, G. R., and Zumsteg, F. C.,** New second-order nonlinear optical organic crystals, *Mater. Res. Soc. Symp. Proc.,* 173, 487, 1990.

293. **Marcy, H. O., Rosker, M. J., Warren, L. F., Cunningham, P. H., Thomas, C. A., Deloach, L. A., Velsko, S. P., Ebbers, C. A., Liao, J. -H., and kanatzidis, M. G.,** l-Histidine tetrafluoroborate: a solution-grown semiorganic crystal for nonlinear frequency conversion, *Opt. Lett.,* 20, 252, 1995.

294. **Umegaki, S.,** Organic Nonlinear Optical Materials, Bunshin Press, Tokyo, 1990 (in Japanese).

295. **Umegaki, S.,** Proceedings of the 5th Toyota Conference, Nonlinear Optics, S. Miyata, ed., Elsevier, Amsterdam, 1992, p. 431.

296. **Cheng, C. T., Tam, W., Meredith, G. R., and Marder, S. R.,** Quadratic hyperpolarizabilities of some organometallic compounds, *Mol Cryst. Liq. Cryst.,* 198, 137, 1990.

297. **Calabrese, J. C., Cheng, L. T., Green, J. C., Marder, S. R., and Tam, W.,** Molecular second-order optical nonlinearities of metallocenes, *J. Am. Chem. Soc.,* 113, 7227, 1991.

298. **Marder, S. R., Perry, J. W., Schaefer, W. P., Tiemann, B. G., Groves, P. C., and Perry, K. J.,** New organic and organometallic salts for second-order nonlinear optics, *SPIE Proc.,* 1147, 108, 1989.

299. **Doisneau, G., Balavoine, G., Khan, T. F., Clinet, J. C., Delaire, J., Ledoux, I., Loucif, R., and Puccetti, G.,** synthesis and nonlinear optical properties of new bimetallic iron/palladium complexes, *J. Organometal. Chem.,* 421, 299, 1991.

300. **Loucif, R., Delaire, J., Bonazzola, L., Doisneau, G., Balavoine, G., Khan, T. F., and Ledoux, I.,** Bimetallic ferrocenyl derivatives: molecular second order hyperpolarizabilities and second harmonic generation, *Mol. Eng.,* 2, 221, 1992.

301. **Wright, M. E., Topikar, E. G., Kubin, R. F., and Seltzer, M. D.,** Organometallic nonlinear optical (NLO) polymers. 1. Pendant ferroecene NLO-phores in a poly(methylmethacrylate) copolymer. The first $\chi^{(2)}$ oganometallic NLO polymer, Macromolecules, 25, 1838, 1992.

302. **Qin, J., Liu, D., Dai, C., Gong, X., Liao, J., Wu, X., Chen, C., and Wu, B.,** Organometallics for nonlinear optics, Sen-i-Gakkai Symp. Preprints, Japan, A46, 1991.

303. **Pollagi, T. P., Stoner, T. C., Dallinger, R. F., Gilbert, T. M., and Hofkins, M. D.,** Nonlinear optical and excited state properties of conjugated one-dimensional [N≧M(OR)$_3$]$_n$ polymers, *J. Am. Chem. Soc.,* 113, 703, 1991.

304. **Frazier, C. C., Harvey, M. A., Cockerham, M. P., Hand, H. M., Chauchard, E. A., and Lee, C. H.,** Second harmonic generation in transition-metal-organic compounds, *J. Phys. Chem.,* 90, 5703, 1986.

305. **Kanis, D. R., Ratner, M. A., and Marks, T. J.,** Design and construction of molecular assemblies with large second-order optical nonlinearities. Quantum chemical aspects, *Chem. Rev.,* 94, 195, 1994.

306. **Kanis, D. R., Lacroix, P. G., Ratner, M. A., and Marks, T. J.,** Electronic structure and quadratic hyperpolarizabilities in organotransition-metal chromophores having weakly coupled π-networks. unusual mechanisms for second-order response, *J. Am. Chem. Soc.,* 116, 10089, 1994.

307. **Chollet, P. A., Kajzar, F., and Moigne, J. L.,** Structure and nonlinear optical properties of copper phthalocyanine thin films, *SPIE Proc.,* 1273, 87, 1990.

307. **Kumagai, K., Mizutani, G., Tsukioka, H., Yamauchi, T., and Ushioda, S.,** Second-harmonic generation in thin films of copper phthalocyanine, *Phys. Rev.,* B, 48, 14488, 1993.

308. **Yamada, T., Hoshi, H., Ishikawa, K., Takezoe, H., and Fukuda, A.,** Origin of second-harmonic generation in vacuum-evaporated copper phthalocyanine film, *Jpn. J. Appl. Phys.,* 34, L299, 1995.

309. **Hoshi, H., Hamamoto, K., Yamada, T., Ishikawa, K., Takezoe, H., Fukuda, A., Fang, S., Kohama, K., and Maruyama, Y.,** Thickness dependence of the epitaxial structure of vanadyl phthalocynine film, *Jpn. J. Appl. Phys.,* 33, L1555, 1994.

310. **Liu, Y., Xu, Y., Zhu, D., Wada, T., Sasabe, H., Liu, L., and Wang, W.,** Langmuir-Blodgett films of an asymmetrically substituted metal-free phthalocyanine and the second-order nonlinear optical properties, Thin Solid Films, 244, 943, 1994.

311. **Liu, Y., Xu, Y., Zhu, D., Wada, T., Sasabe, H., Zhao, X., and Xie, X.,** Optical second-harmonic generation from Langmuir-Blodgett films of an asymmetrically substituted phthalocyanine, *J. Phys. Chem.,* 99, 6957, 1995.

312. **Neuman, R. D., Shah, P., and Akki, U.,** Nonlinear grating behavior of phthalocyanine Langmuir-Blodgett films, *Opt. Lett.,* 17, 798, 1992.

313. **Hoshi, H., Nakamura, N., and Maruyama, Y.,** Second- and third-harmonic generations in ultrathin phthalocyanine films prepared by the molecular-beam epitaxy technique, *J. Appl. Phys.,* 70, 7244, 1991.

314. **Lequan, M., Branger, C., Simon, J., Thami, T., Chauchard, E., and Persoons, A.,** First hyperpolarizability of organotin compounds with T$_d$ symmetry, *Adv. Mater.,* 6, 851, 1994.

315. **Chiang, W., Thompson, M. E., and Engen, D. V.,** Synthesis and nonlinear optical properties of inorganic coordination polymers, in Organic Materials for Nonlinear Optics II, Hahn, R. A., and Bloor, D., eds., Royal Society of Chemistry, London, 1991, p. 211.

316. **(a) Bella, S. D., Fragala, I., Ledoux, I., Diaz-Garcia, M. A., Lacroix, P. G., and Marks, T. J.,** Sizable second-order nonlinear optical response of donor-acceptor bis(salicylaldiminato)nickel(II) Schiff base complexes, *Chem. Mater.,* 6, 881, 1994.
    (b) Lacroix, P. G., Bella, S. D., and Ledoux, J., Synthesis and second-order nonlinear optical properties of new copper(II), nickel(II) and zinc(II) Schiff-base complexes. Towards a role of inorganic chromophores for second-harmonic generation, *Chem. Mat.,* 8, 541, 1996.

317. **(a) Laidlaw, W. M., Denning, R. G., Verbiest, T., Chauchard, E., and Persoons, A.,** Second-order nonlinearity in mixed-valenced metal chromophores, *SPIE Proc.,* 2143, 14, 1994.
    (b) Laidlaw, W. M., Denning, R. G., Verbiest, T., Chauchard, E., and Persoons, A., Large second-order optical polarizabilities in mixed-valency metal complexes, *Nature,* 363, 58, 1993.

318. **Nalwa, H. S.,** Recent developments in ferroelectric polymers, *J. Macromol. Sci. Rev. Macromol. Chem. Phys.,* C31, 341, 1991.

319. **Pillai, P. K. C.,** Polymeric electrets, in *Ferroelectric Polymers.* Nalwa, H. S., ed. Marcel Dekker, New York, 1995, Chapter 1, p.1

320. **Nalwa, H. S. and Kakuta, A.,** Organometallic Langmuir-Blodgett films for electronics and photonics, *Appl. Organometal. Chem.,* 6, 645, 1992.

321. **Aktsipetrov, O. A., Akhmediev, N. N., Mishina, E. D., and Novak, V. R.,** Second-harmonic generation of reflection from a monomolecular Langmuir layer, *JEPT Lett.,* 37, 207, 1983.

322. **Girling, I. R., Cade, N. A., Kolinsky, P. V., and Montgomery, C. M.,** Observation of second-harmonic generation from a Langmuir-Blodgett monolayer of a merocyanine dye, *Electron. Lett.,* 21, 169, 1985.

323. **Girling, I. R., Kolinsky, P. V., Cade, N. A., Earls, J. D., and Peterson, I. R.,** Second harmonic generation from alternating Langmuir-Blodgett films, *Opt. Commun.,* 35, 289, 1985.

324. **Stroeve, P., Srinivasan, M. P., Higgins, B. G., and Kowel, S. T.,** Langmuir-Blodegtt multilayers of polymer-merocyanine dye mixtures, Thin Solid Films, 146, 209, 1987.

325. **Cross, G. H., Peterson, I. R., Girling, I. R., Cade, N. A., Goodwin, M. J., Carr, N., Sethi, R. S., Marsden, R., Gray, G. W., Lacey, D., McRoberts, A. M., Scrowston, R. M., and Toyne, K. J.,** A comparison of second harmonic generation and electro-optic studies of Langmuir-Blodgett monolayers of new hemicyanine dyes, Thin Solid Films, 156, 39, 1988.

326. **Hayden, L. M., Kowel, S. T., and Srinivasan, M. P.,** Enhanced second harmonic generation from multilayered Langmuir/Blodgett films of dye, *Opt. Commun.,* 61, 5, 1987.

327. **Hayden, L. M., Anderson, B. L., Lam, J. Y. S., Higgins, B. G., Stroeve, P., and Kowel, S. T.,** Second-harmonic generation in Langmuir Blodegtt films of hemicyanine-PODMA and hemicyanine-behenic acid, Thin Solid Films, 160, 1988.

328. **Neal, D. B., Petty, M. C., Roberts, G. G., Ahmad, M. M., Feast, W. J., Girling, I. R., Cade, N. A., Kolinsky, P. V., and Peterson, I. R.,** Paper presented at the IEEE International Symposia on the Application of Ferroelectrics, 1986

329. **Anderson, B. L., Hall, R. C., Higgins, B. G., Lindsay, G., Stroeve, P., and Kowel, S. T.,** Quadratically enhanced second harmonic generation in polymer-dye Langmuir-Blodgett films: A new bilayer architecture, *Synth. Metals,* 28, D683, 1989.

330. **Lupo, D., Prass, W., Scheunemann, U., Laschewsky, A., Ringsdorf, H., and Ledoux, I.,** Second-harmonic generation in Langmuir-Blodgett monolayers of stilbazium salt and phenylhydrazone dyes, *J. Opt. Soc. Am. B,* 5, 300, 1988.

331. **Ledoux, I., Josse, D., Vidakovic, P., Zyss, J., Hann, R. A., Gordon, P. F., Bothwell, B. D., Gupta, S. K., Allen, A., Robin, P., Chastaing, E., and Dubois, J. C.,** Second harmonic generation by Langmuir-Blodgett multilayers on an organic azo dye, *Europhys. Lett.,* 3, 803, 1987.

332. **Decher, G., Tieke, B., Bosshard, C., and Gunter, P.,** Optical second-harmonic generation in Langmuir-Blodgett films of 2-docosylamino-5-nitropyridine, *J. Chem. Soc. Chem. Commun.,* 933, 1988.

333. **Decher, G., Tieke, B., Bosshard, G., and Gunter, P.,** Optical second-harmonic generation in Langmuir-Blodgett films of novel donor acceptor substituted pyridine and benzene derivatives, *Ferroelectrics,* 91, 193,1989.

334. **Decher, G., Klinkhammer, F., Peterson, I. R., and Steitz, R.,** Structural investigations of Langmuir-Blodgett films of 2-docosylamino-5-nitropyridine, a new type of noncentrosymmetric multilayer for use in nonlinear optics, Thin Solid Films, 178, 445, 1989.

335. **Richardson, T., Roberts, G. G., Polywka, M. E. C., and Davies, S. G.,** The characterization of organoruthenium complexes, Thin Solid Films, 179, 405, 1989.

336. **Davies, S. G., Richardson, T., Roberts, G. G., and Polywka, M. E. C.,** *Eur. Patent,* EP-88-306310 (July 11, 1988).

337. **Sakaguchi, H., Nagamura, T., and Matsuo, T.,** Jpn. *J. Appl. Phys.,* 30, L377, 1991.

338. **Carr, N. and Goodwin, M. J.,** Second harmonic generation in a monomolecular Langmuir-Blodgett film of a performed polymer, *Makromol. Chem. Rapid Commun.,* 8, 487, 1987.

339. **Wang, X. K., Zhang, T. G., Lin, W. P., Liu, S. Z., Wong, G. K., Kappes, M. M., Chang, R. P. H., and Ketterson, J. B.,** Large second-harmonic response of C60 thin films, *Appl. Phys. Lett.,* 6, 810, 1992.

340. **Kajzar, F., Taliani, C., Zamboni, R., Rossini, S., and Danieli, R.,** Nonlinear optical properties of sublimed C60 thin films, *Synth. Metals,* 54, 21, 1993.

341. **Gan, L. B., Zhou, D. J., Luo, C. P., Huang, C. H., Li, T. K., Bai, J., Zhao, X. S., and Xia, X. H.,** Langmuir-Blodgett film and second harmonic generation of $C_{60}(C_4H_8N_2)$, *J. Phys. Chem.,* 98, 12459, 1994.

342. **Zhou, D., Gan, L., Luo, C. Tan, H., Huang, C., Liu, Z., Wu, Z., Zhao, X., Xia, X., Zhang, S., Sun, F., and Zou, Y.,** Langmuir-Blodgett film and nonlinear optical property of C^{60} glycine ester derivatives, *Chem. Phys. Lett.,* 235, 548, 1995.

343. **Leigh, D. A., Moody, A. E., Wade, F. A., King, T. A., West, D., and Bahra, G. S.,** Second harmonic generation from Langmuir-Blodgett films of fullerene=aza-crown ethers and their potassium ion complexes, Langmuir, 11, 2334, 1995.

344. **Cresswell, J. P., Tsibouklis, J., Petty, M. C., Feast, W. J., Carr, N., Goodwin, M., and Lvov, Y. M.,** Langgmuir-Blodgett alternate layer structures for second-order nonlinear optics, *SPIE Proc.,* 1337, 358, 1990.

345. **Bosshard, Ch., Kupler, M., Florscheimer, M., Borer, T., Gunter, P., Tang, Q., and Zahir, S.,** Investigation of chromophore orientation of 2-docosylamino-5-pyridine and derivatives by nonlinear optical techniques, Thin Solid Films, 210, 198, 1992.

346. **Bosshard, Ch., Florscheimer, M., Kupler, M., and Gunter, P.,** *Opt. Commun.,* 85, 247, 1991.

347. **Tang, Q., Zahir, S., Bosshard, Ch., Florscheimer, M., Kupler, M., and Gunter, P.,** Polymerized nonlinear optical Langmuir-Blodgett films based on 2-(21-docosenyl)amino-5-nitropyridine, Thin Solid Films, 210, 195, 1992.

348. **Kupfer, M., Florscheimer, M., Baumann, W., Bosshard, S., Gunter, P., Tang, Q., and Zahir, S.,** Optical second harmonic generation from polymerized Langmuir-Blodgett films of 2-(21-docosenyl)amino-5-nitropyridine, Thin Solid Films, 226, 270, 1993.

349. **Wang, K. Z., Huang, C. H., Xu, G. X., Xu, Y., Liu, Q., Zhu, D. B., Zhao, X. S., Xie, X. M., and Wu, N. Z.,** Preparation, characterization, and second harmonic generation of a Langmuir-Blodgett film based on a rare-earth coordination compound, *Chem. Mater.,* 6, 1986, 1994.

350. **Dentan, V., Blanchard-Desce, M., Palacin, S., Ledoux, I., Barraud, A., Lehn, J. M., and Zyss, J.,** Second harmonic generation in mixed carotenoid-fatty acid and carotenoid-cyclodextrin Langmuir-Blodgett films, Thin Solid Films, 210, 221, 1992.

351. **Ashwell, G. J., Jackson, P. D., Lochun, D., Crossland, W. A., Thompson, P. A., Bahra, G. S., Brown, C. R.,**

and Jasper, C., Second harmonic generation from alternate-layer LB films: quadratic enhancement with film thickness, *Proc. Roy. Soc. Lond A.,* in press, 1993.

352. **Ashwell, G. J., Crossland, W. A., and Jackson, P. D.,** Z-type Langmuir-Blodgett films for second harmonic generation, *Nature,* 1993.

353. *Ashwell, G. J., Hargeaves, R. C., Baldwin, C. E., Bahra, G. S., and Brown, C. R.,* Improved second harmonic generation from Langmuir-Blodgett films of hemicyanine dyes, *Nature,* 357, 393, 1992.

354. **Ashwell, G. J., Jackson, P. D., and Crossland, W. A.,** Noncentrosymmetry and second harmonic generation in Z-type Langmuir-Blodgett films, *Nature,* 368, 438, 1994.

355. **Marowsky, G., Chi, L. F., Mobius, D., Steinhoff, R., Shen, Y. R., Dorsch, D., and Rieger, B.,** Nonlinear optical properties of hemicyanine monolayers and the protonation effect, *Chem. Phys. Lett.,* 147, 420, 1988.

356. **Howarth, V. A., Asai, N., Kishi, N., and Fujiwara, I.,** Optical second harmonic generation in Langmuir-Blodgett films of n-docosyl-2-methyl-4-nitroaniline, *Appl. Phys. Lett.,* 61, 1616, 1992.

357. **Nakamura, K., Era, M., Tsutsui, T., and Saito, S.,** Enhancement of second harmonic generation in Langmuir-Blodgett monolayer films by molecular mixing of homologous amphiphiles, *Jpn. J. Appl. Phys.,* 29, L628, 1990.

358. **Era, M., Kawafuji, H., Tsutsui, T., Saito, S., Takehara, K., Takehara, K., Isomura, K., and Taniguchi, H.,** Second-order nonlinear optical Langmuir-Blodgett films of pyrazine derivatives, *Thin Solid Films,* 210, 163, 1992.

359. **Yao, Z. Q., Liu, P., Yan, R. Z., Liu, L. Y., Liu, X. H., and Wang, W. C.,** Studies on nonlinear optical properties of Langmuir-Blodgett films formed from azobenocrown ether derivatives, *Thin Solid Films,* 210, 208, 1992.

360. **Kajzar, F. and Ledoux, I.,** Quadratic nonlinear spectroscopy in Langmuir-Blodgett films of charge-transfer diazostilbenes and polyenes, *Thin Solid Films,* 179, 359, 1989.

361. **Allen, S., McLean, T. D., Gordon, P. F., Bothwell, B. D., Robin, P., and Ledoux, I.,** Properties of polyenic Langmuir-Blodgett films, *SPIE Proc.,* 971, 206, 1988.

362. **Ledoux, I., Josse, D., Fremaux, P., Piel, J. P., Post, G., Zyss, J., MacLean, T., Hann, R. A., Gordon, P. F., and Allen, S.,** Second harmonic generation in alternate non-linear Langmuir-Blodgett films, *Thin Solid Films,* 160, 217, 1988.

363. **Huang, J., Lewis, A., and Rasing T.,** Second harmonic generation from Langmuir-Blodgett films of retinal and retinal Schiff bases, *J. Phys. Chem.,* 92, 1756, 1988.

364. **Nalwa, H. S., Watanabe, T., Nakajima, K., and Miyata, S.,** Optical second harmonic generation in Langmuir-Blodgett monolayers of N-octadecyl-2,4-dinitro-5-fluoroaniline, *Thin Solid Films,* 227, 205, 1993.

365. **Nalwa, H. S., Watanabe, T., Nakajima, K., and Miyata, S.,** in Nonlinear Optics, Miyata, S., ed., Elsevier, Amsterdam, 1992, p. 271.

366. **Kalina, D. W. and Grubb, S. G.,** Langmuir-Blodgett films of noncentrosymmetric azobenzene dyes for non-linear optical applications, *Thin Solid Films,* 160, 363, 1988.

367. **Laschewsky, A., Paulus, W., Ringsdorf, H., Schuster, A., Frick, G., and Mathy, A.,** Mixed polymeric monolayers and Langmuir-Blodgett multilayers with low molecular weight guest compounds, *Thin Solid Films,* 210, 191, 1992.

368. **Bubeck, C., Laschewsky, A., Lupo, D., Neher, D., Ottenbreit, P., Paulus, W., Prass, W., Ringsdorf, H., and Wegner, G.,** Amphiphilic dyes for nonlinear optics: dependence of second harmonic generation on functional group substitution, *Adv. Mater.,* 3, 54, 1991.

369. **Lupo, D., Ringsdorf, H., Schuster, A., and Seitz, M.,** Amphiphilic nonlinear optical bis-chromophores and their mixtures with amphotropic copolymers: preparation of monolayers and Langmuir-Blodgett multilayers, *J. Am. Chem. Soc.,* 116, 10498, 1994.

370. **Heinz, T. F., Tom, H. W. K., and Shen, Y. R.,** Determination of molecular orientation of monolayer adsorbates by optical second-harmonic generation, *Phys. Rev. A,* 28, 1883, 1983.

371. **Heinz, T. F., Chen, C. K., Ricard, D., and Shen, Y. R.,** Spectroscopy of molecular monolayers by resonant second-harmonic generation, *Phys. Rev. Lett.,* 48, 478, 1982.

372. **Berkovic, G., Rasing, Th., and Shen, Y. R.,** Second-order nonlinear polarizability of various biphenyl derivatives, *J. Opt. Soc. Am. B,* 4, 945, 1987.

373. **Miyamoto, Y., Kaifu, K., Koyano, T., Saito, M., and Kato, M.,** *Jpn. J. Appl. Phys.,* 30, L1647, 1991.

374. **Kajiwara, K., Anzai, T., Takezoe, H., Fukuda, A., Okada, S., Matsuda, H., Nakanishi, H., Abe, T., and Ito, H.,** Second harmonic generation in dispersed-red-labeled poly(methyl methacrylate) Langmuir-Blodgett film, *Appl. Phys. Lett.,* 62, 2161, 1993.

375. **Ou, S. H., Mann, J. A., Iando, J. B., Zhou, L., and Singer, K. D.,** Nonlinear optical studies of polar polymeric Langmuir-Schaefer films, *Appl. Phys. Lett.,* 61, 2284, 1992.

376. **Penner, T. L., Armstrong, N. J., Willand, C. S., Schidkraut, J. S., and Robello, D. R.,** SPIE Proc., 1560, 377, 1991.

377. **Verbiest, T., Samyn, C., and Persoons, A.,** Second harmonic generation in Langmuir-Blodgett films of preformed polymers, *Thin Solid Films,* 210, 188, 1992.

378. **Bjornholm, T., Geisler, T., Larsen, J., and Jorgensen, M.,** Nonlinear optical phenomena due to donor-acceptor interfaces created in Langmuir-Blodgett films, *J. Chem. Soc. Chem. Commun.,* 815, 1992.

379. **Armand, F., Sakuragi, H., Takumaru, K., Yamada, T., Kajikawa, K., Takezoe, T., Okada, S., Matsuda, H., and Nakanishi, H.,** Second harmonic generation in monolayer of coordination compound of pentacyanoferrate (III),

Extended Abstract of the 40th Spring Meeting, 1993, The Japan Society of Applied Physics and Related Societies, No. 30a-ZG-9, p. 1125 (in Japanese).

380. **Marks, T. J. and Ratner, M. A.,** Design, synthesis, and properties of molecule-based assemblies with large second-order optical nonlinearities, *Angew. Chem. Int. Ed. Engl.,* 34, 155, 1995.

381. **Li, D., Ratner, M. A., Marks, T. J., Zhang, C., Yang, J., and Wong, G. K.,** Chromophoric self-assembled multilayers. Organic superlattice approaches to thin-film nonlinear optical materials, *J. Am. Chem. Soc.,* 112, 7389, 1990.

382. **Yitzchaik, S., Roscoe, S. B., Kakkar, A. K., Allan, D. S., Marks, T. J., Xu, Z., Zhang, T., Lin, W., and Wong, G. K.,** Chromophoric self-assembled NLO multilayer materials: real time observation of monolayer growth and microstructural evolution by in situ second harmonic generation techniques, *J. Phys. Chem.,* 97, 6958, 1993.

383. **Kakkar, A. K., Yitzchaik, S., Roscoe, S. B., Kubota, F., A. K., Allan, D. S., Marks, T. J., Lin, W., and Wong, G. K.,** Chromophoric self-assembled NLO multilayer materials: synthesis, properties and structural interconversions of assemblies with rod-like alkynyl chromophores, Langmuir, 9, 388, 1993.

384. **Lundquist, P. M., Yitzchaik, S., Zhang, T., Kanis, D. R., Ratner, M. A., Marks, T. J., and Wong, G. K.,** Dispersion of second-order optical nonlinearity in chromophoric self-assembled films by optical paramteric amplification: experimental and theory, *Appl. Phys. Lett.,* 64, 2194, 1994.

385. **Yitzchaik, S., Lundquist, P. M., Lin, W., Marks, T. J., and Wong, G. K.,** $\chi^{(2)}$ dispersion and waveguiding measurements in acentric chromophoric self-assembled NLO materials, *SPIE Proc.,* 2285, 282, 1994.

386. **Roscoe, S. B., Yitzchaik, S., Kakkar, A. K., Marks, Lin, W., and Wong, G. K.,** Ion exchange processes and environmental effects in chromophoric self-assembled superlattices. manipulation of microstructure and large enhancement in nonlinear optical response, Langmuir, 10, 1337, 1994.

387. **Lin, W., Yitzxchaik, S., Lin, W., Malik, A., Wong, G. K., Dutta, P., and Marks, T. J.,** New nonlinear optical materials: expedient topotatic self-assembly of acentric chromophoric superlattices, *Angew. Chem. Int. Ed. Eng.,* 34, 1497, 1995.

388. **Katz, H. E., Scheller, G., Putvinski, T. M., Schilling, M. L., Wilson, W. L., and Chidsey, C. E. D.,** *Science,* 254, 1485, 1991.

389. **Katz, H. E., Wilson, W. L., and Scheller, G.,** Chromophore structure, second harmonic generation and orientational order in zirconium phosphate/phosphate self-assembled multilayers, *J. Am. Chem. Soc.,* 116, 6636, 1994.

390. **Schmitt, K., Herr, R. P., Schadt, M., Funfschilling, J., Buchecker, R., Chen, X. H., and Benecke, C.,** Strongly nonlinear optical ferroelectric liquid crystals for frequency doubling, *Liq. Cryst.,* 14, 1735, 1993.

391. **Yoshino, K., Utsumi, M., Morita, Y., Sadohara, Y., and Ozaki, M.,** Second harmonic generation in ferroelectric liquid crystals and their mixtures, *Liq. Cryst.,* 14, 1021, 1993.

392. **Shtykov, N. M., Barnik, M. I., Beresnev, L. A., Blinov, L. M.,** *Mol. Cryst. Liq. Cryst.,* 124, 379, 1985.

393. **Ozaki, M., Utsumi, M., Gotou, T., Morita, Y., Daido, I. K., Sadohara, Y., and Yoshino, K.,** *Ferroelectrics,* 121, 259, 1991.

394. **Walba, D. M., Ros, M. B., Clark, N. A., Shao, R., Robinson, M. G., Liu, J. Y., Johnson, K. M., and Doroski, D.,** Design and synthesis of new ferroelectric liquid crystals. 14. An approach to the stereocontrolled synthesis of polar organic thin films for nonlinear optical applications, *J. Am. Chem. Soc.,* 113, 5471, 1991.

395. **Liu, J. Y., Robinson, M. G., Johnson, K. M., and Doroski, D.,** Second harmonic generation in ferroelectric liquid crystals, *Opt. Lett.,* 15, 267, 1990.

396. **Schadt, M.,** Linear and nonlinear liquid crystal materials, electro-optical effects and surface interactions. Their applications in present and future devices, *Liq. Cryst.,* 14, 73, 1993.

397. **Guyot-Sionnest, P., Hsuing, P., and Shen, Y. R.,** Surface polar ordering in a liquid crystal observed by optical second harmonic generation, *Phys. Rev. Lett.,* 57, 2963, 1986.

398. **Chen, W., Feller, M. B., and Shen, Y. R.,** Investigation of anisotropic molecular orientational distributions of liquid crystal monolayers by by optical second harmonic generation, *Phys. Rev. Lett.,* 63, 2665, 1989.

399. **Smith, D. A. and Coles, H. J.,** The second and third-order nonlinear optical properties of liquid crystalline polymers, *Liq. Cryst.,* 14, 937, 1993.

400. **Yitzchaik, S., Berkovic, G., and Krongauz, V.,** Polymers and blends exhibiting two-dimensional asymmetry and optical nonlinearity upon poling by an electric field, *Macromolecules,* 23, 3539, 1990.

401. **Singer, K. D., Lalama, S. L., Sohn, J. E., and Small, R. D.,** Electro-optic organic materials, in Nonlinear Optical Properties of Organic Molecules and Crystals, Chemla, D. S., and Zyss, J., eds., Academic Press, Orlando, 1987.

402. **Sessler, G. M.,** Electrets-Topics in Appled Physics, Springer-Verlag, Berlin, 1987.

403. **Mortazavi, M. A., Knoesen, A., Kowel, S. T., Higgins, B. G., and Dienes, A.,** Second-harmonic generation and absorption studies of polymer-dye films oriented by corona-onset poling at elevated temperatures, *J. Opt. Soc. Am. B,* 6, 733, 1989.

404. **Singer, K. D., Sohn, J. E., and Lalama, S. J.,** Second harmonic generation on poled polymer films, *Appl. Phys. Lett.,* 49, 248, 1986.

405. **Sekkat, Z. and Dumont, M.,** *Appl. Phys. B,* 54, 733, 1992.

406. **Blanchard, P. M. and Mitchell, G. R.,** *J. Phys. D: Appl. Phys.,* 26, 500, 1993.

407. **Dao, P. T. and Williams, D. J.,** Constant current corona charging as a technique for poling organic nonlinear optical films and the effect of ambient gas, *J. Appl. Phys.,* 73, 2043, 1993.

408. **Yang, G., Bauer-Gogonea, S., and Sessler, G. M.,** Selective poling of nonlinear optical polymer films by means of a monoenergetic electron beam, *Appl. Phys. Lett.,* 64, 22, 1994.

409. **Otomo, A., Stegeman, G. I., Horsthuis, W. H. G., and Mohlmann, G. R.,** Strong field, in-plane poling for nonlinear optical devices in highly nonlinear side chain polymers, *Appl. Phys. Lett.,* 65, 2389, 1994.

410. **Hill, R. A. and Knoesen, A.,** Corona poling of nonlinear polymer thin films for electro-optic modulators, *Appl. Phys. Lett.,* 65, 1733, 1994.

411. **Meredith, G. R., Vandusen, J. G., and Williams, D. J.,** Characterization of liquid crystalline polymers for electro-optic applications, in Nonlinear Optical Properties of Organic and Polymeric Materials, Williams, D. J., ed., American Chemical Society, Washington, D. C., 109, 233, 1983.

412. **Kielich, S.,** Optical second-harmonic generation by electrically polarized isotropic media, *IEEE J. Quantum Electron.,* QE-5, 562, 1969.

413. **Meredith, G. R., VanDusen, J. G., and Williams, D. J.,** Optical and nonlinear optical characterization of molecularly doped thermotropic liquid crystalline polymers, *Macromolecules,* 15, 1385, 1982.

414. **Singer, K. D., Lalama, J., and Sohn, J. E.,** *SPIE Proc.,* 578, 130, 1985.

415. **Jeng, R. J., Chen, Y. M., Jain, A. K., Tripathy, S. K., and Kumar, J.,** *Opt. Commun.,* 89, 212, 1992.

416. **Miyazaki, T., Watanabe, T., and Miyata, S.,** Highly efficient second harmonic generation in p-nitroaniline/poly(e-caprolactone) systems, *Jpn. J. Appl. Phys.,* 27, L1724, 1988.

417. **Hampsch, H. L., Yang, J., Wong, G. K., and Torkelson, J. M.,** *Macromolecules,* 21, 528, 1988.

418. **Singer, K. D., Sohn, J. E., and Lalama, S. J.,** *J. Appl. Phys.,* 49, 248, 1986.

419. **Mortazavi, M. A., Knoesen, A., Kowel, S. T., Higgins, B. G., and Dienes, A.,** *J. Opt. Soc. Am.,* B6, 733, 1989.

420. **Sohn, J. E., Singer, K. D., Kuzyk, M. G., Holland, W. R., Katz, H. E., Dirk, C. W., Schilling, M. L., and Comizzoli, R. B.,** Nonlinear Optical Effects in Organic Polymers, Messier, J., ed., Kluwer Academic Publishers, Dordrecht, 1989, p. 291.

421. **Katz, H. E., Dirk, C. W., Schilling, M. L., Singer, K. D., and Sohn, J. E.,** *Mater. Res. Soc. Symp. Proc.,* 109, 127, 1988.

422. **Watanabe, T., Yoshinaga, K., Fichou, D., and Miyata, S.,** *J. Chem. Soc. Chem. Commun.,* 250, 1988.

423. **Lee, S. C., Kidoguchi, A., Watanabe, T., Yamamoto, H., Hosomi, T., and Miyata, S.,** *Polym. J.,* 23, 1209, 1991.

424. **Watanabe, T., Kagami, M., Yamamoto, H., Kidoguchi, A., Hayashi, A., Sato, H., and Miyata, S.,** *Polym. Preprints,* 32, 88, 1991.

425. **Broussoux, D., Esselin, S., Le Barny, P., Pocholle, J. P., and Robin, P.,** Nonlinear Optics of Organics and Semiconductors, Kobayashi, T., ed., Springer-Verlag, Berlin, 1989, p. 126.

426. **Gadret, G., Kajzar, F., and Raimond, P.,** *SPIE Proc.,* 1560, 226, 1991.

427. **(a) Hill, J. R., Dunn, P. L., Davis, G. J., Oliver, S. N., Pantelis, P., and Rush, J. D.,** Efficient frequency doubling in a poled PVDF copolymer guest/host composite, *Electron. Lett.,* 23, 700, 1987.
    (b) Pantelis, P., Hill, J., and Davies, P. L., Poled copoly(vinylidene fluoride-trifluoroethylene) as a host for nonlinear optical molecules, in Nonlinear Optical and Electroactive Polymers, American Chemical Society, Washington, D.C., 1987.

428. **Heesink, G. J., van Hulst, N. F., Bolger, B., Kelderman, E., Engbersen, J. F. J., Verboom, W., and Reinhoudt, D. N.,** Novel calixarenes in thin films for efficient second harmonic generation, *Appl. Phys. Lett.,* 62, 2015, 1993.

429. **Hirota, K., Hosoda, M., Joglekar, B., Matsui, M., and Muramatsu, H.,** *Jpn. J. Appl. Phys.,* 32, L1811, 1993.

430. **Gadret, G., Kajzar, F., Raimond, P.,** Nonlinear optical properties of poled polymers, *SPIE Proc.,* 1560, 226, 1991.

431. **Levenson, R., Linage, J., Toussaere, E., Carenco, A., Zyss, J.,** A quasi-relaxation free guest-host poled polymer waveguide modulator: material, technology and characterization, *SPIE Proc.,* 1560, 251, 1991.

432. **Nalwa, H. S., Watanabe, T., and Miyata, S.,** unpublished results.

433. **Watanabe, T., Yoshinaga, K., Fichou, D., and Miyata, S.,** Large second harmonic generation in electrically ordered p-nitroaniline-poly(oxyethylene) 'guset-host' systems, *J. Chem. Soc. Chem. Commun.,* 250, 1988.

434. **Watanabe, T., Miyazaki, T. and Miyata, S.,** Nonlinear optics in host guest systems-crystalline polymer-dye binary mixtures, MRS *Int. Mtg. Adv. Mater,* 23, 1989.

435. **Watanabe, T. and Miyata, S.,** Effect of crystallization process on the second harmonic generation of poly(oxyethylene)/p-nitroaniline systems, *SPIE Proc.,* 1147, 101, 1989.

436. **Nakanishi, H., Kagami, M., Hamazaki, N., Watanabe, T., Sato, H., and Miyata, S.,** Nonlinear optical properties of polymer dye systems, *SPIE Proc.,* 1147, 84, 1989.

437. **Kamino, Y., Watanabe, T., Kajikawa, K., Takezoe, H., Fukuda, A., Toyooka, T., and Ishii, T.,** Second-harmonic generation in noncentrosymmetric p-nitroaniline spontaneously crystallized by quenching from lyotropic liquid crystalline poly(g-benzyl-L-glutamate), *Jpn. J. Appl. Phys.,* 30, 1710, 1991.

438. **Boyd, G. T., Francis, C. V., Trend, J. E., and Ender, D. A.,** Second-harmonic generation as a probe of rotational mobility in poled polymers, *J. Opt. Soc. Am.* B, 8, 887, 1991.

439. **Lee, S. C., Kidoguchi, A., Watanabe, T., Yamamoto, H., Hosomi, T., and Miyata, S.,** The effect of physical aging on the tine-decay behavior of second harmonic generation activity in a poled polymer film, *Polym. J.,* 23, 1209, 1991.

440. **(a) Stahelin, M., Walsh, C. A., Burland, D. M., Miller, R. D., Twieg, R. J., and Volksen, W.,** Orientational decay in poled second-order nonlinear optical guest-host polymers: Temperature dependence and effects of poling geometry, *J. Appl. Phys.,* 73, 8471, 1993.

      (b) Twieg, R. J., Betterton, K. M., Burland, D. M., Lee, V. Y., Miller, R. d., Moylan, C. R., Volksen, W., and Walsh, C. A., Progress in nonlinear optical chromophores and polymers for practical electro-optic waveguide applications, *SPIE Proc.,* 2025, 94, 1993.

441. **Ozaki, M., Daiodo, K., and K. Yoshino,** Optical second harmonic generation in doped polymer composites, *Technol. Rep. Osaka Univ.,* 40, 273, 1990.

442. **(a) Wang, Y.,** Efficient second harmonic generation from low dimensional dye aggregates in thin polymer film, *Chem. Phys. Lett.,* 126, 209, 1986.

      (b) Skindhoj, J., Perry, J. W., and Marder, S. R., Development of electrooptic polymers for high voltage instrument transformers, *SPIE Proc.,* 2285, 116, 1994.

443. **Ye, C., Marks, T. J., Yang, J., and Wong, G. K.,** Synthesis of molecular arrays with nonlinear optical properties. Second-harmonic generation by covalently functionalized glassy polymers, *Macromolecules,* 20, 2322, 1987.

444. **Ye, C., Minami, N., Marks, T. J., Yang, J., and Wong, G. K.,** Persistent, efficient frequency doubling by poled annealed films of a chromophore functionalized poly(p-hydroxystryrene), *Macromolecules,* 20, 2899, 1988.

445. **Ye, C., Minami, N., Marks, T. J., Yang, J., and Wong, G. K.,** Synthetic approaches to stable and efficient polymeric frequency doubling materials. Second-order nonlinear optical properties of poled, chromophore-functionalized glassy polymers, in Nonlinear Optical Effects in Organic Polymers, Messier, J., Kajar, F., Prasad, P., and Ulrich, D., eds., Kluwer Academic Publishers, Dordrecht, 173, 162, 1989.

446. **Marks, T. J. and Ratner, M. A.,** Design, synthesis, and properties of molecule-based assemblies with large second-order optical nonlinearities, Angew. *Chem. Int. Ed. Engl.,* 34, 155, 1995.

447. **Eich, M., Sen, A., Looser, H., Bijorklund, G. C., Swalen, J. D., Twieg, R., and Yoon, D. Y.,** Corona poling and real-time second-harmonic generation study of a novel covalently functionalized amorphous optical polymer, *J. Appl. Phys.,* 66, 2559, 1989.

448. **Shuto, Y., Takara, H., Amano, M., and Kaino, T.,** Noncollinearly phase-matched second harmonic generation in stilbene-dye-attached polymer thin films, *Jpn. J. Appl. Phys.,* 28, 2508, 1989.

449. **Amano, M. and Kaino, T.,** Second-order nonlinearity of a novel diazo-dye-attached polymer, *J. Appl. Phys.,* 68, 6024, 1990.

450. **Amano, M., Kaino, T., Yamamoto, F., and Takeuchi, Y.,** Second-order nonlinear optical properties of polymers containing mesogenic side chains, *Mol. Cryst. Liq. Cryst.,* 182A, 81, 1990.

451. **Sohn, J. E., Singer, K. D., Kuzyk, M. R., Holland, W. R., Katz, H. E., Dirk, C. W., Schilling, M. L., and Comizzoli, R. B.,** Orientational ordered nonlinear optical polymer films, in Nonlinear Optical Effects in Organic Polymers, Messier, J., ed., Kluwer Academic, Dordrecht, 291, 162, 1989.

452. **Buckley, A. and Stamatoff, J. B.,** Nonlinear optical polymers for active optical devices, in Nonlinear Optical Effects in Organic Polymers, Messier, J., ed., Kluwer Academic, Dordrecgt, 327, 162, 1989.

453. **Hass, D., Yoon, H., Man, H. T., Cross, G., Man, S., and Persons, N.,** Polymeric electro-optic waveguide modulator, materials and fabrication, *SPIE Proc.,* 1147, 222, 1989.

454. **Ore, F. R. J., Hayden, L. M., Sauter, G. F., Pasillas, P. L., Hoover, J. M., Henry, R. A., and Lindsay, G. A.,** Electro-optic properties of new nonlinear side-chain polymers, *SPIE Proc.,* 1147, 26, 1989.

455. **Dai, D., Marks, T. J., Yang, J., Lundquist, P. M., and Wong, G. K.,** Polyphenylene ether based thin film nonlinear optical materials having high chromophore densities and alignment stability, *Macromolecules,* 23, 1891, 1990.

456. **Dai, D. R., Hubbard, M. A., Park, J., Marks, T. J., Wang, J., and Wong, G. K.,** Rational design and construction of polymers with large second-order optical nonlinearities. Synthetic strategies for enhanced chromophore number densities and frequency doubling temporal stabilities, *Mol. Cryst. Liq. Cryst.,* 189, 93, 1990.

457. **Singer, K. D., Kuzyk, M., Holland, W., Sohn, J., Lalama, S., Comizzoli, R., Katz, H., and Schilling, M.,** Electro-optic phase modulation and optical second-harmonic generation in corona-poled polymer films, *Appl. Phys. Lett.,* 53, 1800, 1988.

458. **Stenger-Smith, J. D., Fischer, J. W., Henry, R. A., Hoover, J. M., Nadler, M. P., Nissan, P. A., and Lindsay, G. A.,** Poly[(4-N-ethylene-N-ethylamino)-a-cyanocinnamate]: A nonlinear optical polymer with a chromophoric mainchain. I. Synthesis and spectral characterization, *J. Polym. Sci. Part A. Polym. Chem.,* 29, 1623, 1991.

459. **Allcook, H., Dembek, A., Kim, C., Devine, R., Shi, Y., Steier, W., and Spangler, C.,** Second-order nonlinear optical poly(organophosphazenes): synthesis and nonlinear optical characterization, *Macromolecules,* 24, 1000, 1991.

460. **Mortazavi, M. A., Knoesen, A., Kowel, T., Henry, R. A., Hoover, J. M., and Lindsay, G. A.,** Second-order nonlinear optical properties of poled coumaromethacrylate copolymers, *Appl. Phys.* B, 53, 287, 1991.

461. **Seppen, C. J. E., Rikken, G. L. J. A., Staring, E. G. J., Nijhuis, S., and Venhuzen, A. H. J.,** Linear optical properties of frequency doubling polymers, *Appl. Phys.* B, 53, 282, 1991.

462. **Rikken, G. L. J. A., Seppen, C. J. E., Nijhuis, S., and Staring, E. G. J.,** Poled polymers for frequency doubling of diode lasers, *SPIE Proc.,* 1337, 35, 1990.

463. **Staring, E. G. J., Rikken, G. L. J. A., Seppen, C. J. E., Nijhuis, S., and Venhuzen, A. H. J.,** Organic materials for frequency doubling, *Adv. Mater.,* 3, 401, 1991.

464. **Rikken, G. L. J. A., Seppen, C. J. E., Nijhuis, S., and Meijer, E. W.,** Poled polymers for frequency doubling of diode lasers, *Appl. Phys. Lett.,* 58, 435, 1991.

465. **Chen, Y., Rahman, M., Takahashi, T., Mandal, B., Lee, J., Kumar, J., and Tripathy, S.,** Linear and nonlinear optical properties of a comb-like polymer, *Jpn. J. Appl. Phys.,* 30, 672, 1991.

466. **Hayashi, A., Goto, Y. M. N., Kaluzynski, K., Sato, H., Kato, K., Kondo, K., Watanabe, T., and Miyata, S.,** Second-order nonlinear optical properties of poled polymers of vinyl chromophore monomers: styrene, methacrylate, and vinyl benzoate derivatives having one-aromatic-ring push-pull chromophores, *Chem. Mater.,* 4, 555, 1992.

467. **Xu, C., Wu, B., Dalton, L. R., Ranon, P. M., Shi, Y., and Steier, W. H.,** New random main-chain, second-order nonlinear optical polymers, *Macromolecules,* 25, 6716, 1992.

468. **Xu, C., Wu, B., Dalton, L. R., Shi, Y., Ranon, P. M., and Steier, W. H.,** Novel double-end cross-linkable chromophore for second-order nonlinear optical materials, *Macromolecules,* 25, 6714, 1992.

469. **Xu, C., Wu, B., Becker, M. W., Dalton, L. R., Ranon, P. M., Shi, Y., and Steier, W. H.,** Main-chain second-order nonlinear optical polymers: random Incorporation of amino-sulfone chromophores, *Chem. Mater.,* 5, 1439, 1993.

470. **Chen, M., Yu, L., Dalton, L. R., Shi, Y., and Steier, W. H.,** New polymers with large and stable second-order nonlinear optical effects, *Macromolecules,* 24, 5421, 1991.

471. (a) **Cai, Y. M. and Jen, A. K.-Y,** Thermally stable poled polyquinoloine thin film with very large electro-optic response, *Appl. Phys. Lett.,* 67, 299, 1995.
(b) **Chen, T. A., Jen, A. K.-Y., and Cai, Y. M.,** A novel class of nonlinear optical side chain polymers: polyguinolines with large second-order nonlinearity and thermal stability, *Chem. Mater.,* 8, 607, 1996.
(c) **Shi, R. F., Wu, M. H., Yamada, S., Cai, Y. M., and Garito, A. F.,** Dispersion of second-order optical properties of new high thermal stability guest chromophores, *Appl. Phys. Lett.,* 63, 1173, 1993 (and private communications).

472. **Kitipichai, P., La Peruta, L., Korenowski, G. M., and Wnek, G. E.,** *J. Polym. Sci. Polym. Chem.,* 31, 1365, 1993.

473. **Gulotty, R. J., Brennan, D. J., Chartier, M. A., Gille, J. K., Haag, A. P., Hazard, K. A., Inbasekaran, M. N., Ashley, P. R., and Tumolillo, T. A., Jr.,** in Organic Thin Films for Photonic Applications, ACS/OSA Joint Conference, Toronto, October 5–7, 1993, Technical Digest Series, vol. 17, p. 6.

474. **Ching, K. C., Lequan, M., Lequan, R. M., and Kajzar,** Nonlinear optical properties of polymers bearing pendant phosphine oxide chromophores, *Chem. Phys. Lett.,* 242, 598, 1995.

475. **Nahata, A., Shan, J., Yardley, J. T., and Wu, C.,** Electro-optic determination of the nonlinear optical properties of a covalenltly functionalized dispersed red 1 copolymer, *J. Opt. Soc. Am.* B, 10, 1553, 1993.

476. **Levenson, R., Liang, J., Beylen, M. V., Samyn, C. Foll, F., Rousseau, and Zyss, J.,** in Organic Thin Films for Photonic Applications, ACS/OSA Joint Conference, Toronto, October 5–7, 1993, Technical Digest Series, vol. 17, p. 81.

477. **Swalen, J. D., Bjorklund, G. C., Fleming, W., Hung, R., Jurich, M., Lee, V. Y., Miller, R. D., Moerner, W. E., Morichere, D. Y., Skumanich, A., and Smith, B. A.,** *SPIE Proc.,* 1775, 369, 1992.

478. (a) **Choi, D. H., Kim, H. M., Wijekoon, W. M. K. P., and Prasad, P. N.,** Synthesis and second-order nonlinear optical properties of polymethacrylates containing organic salt dye chromophore, *Chem. Mater.,* 4, 1253, 1992.
(b) **Choi, D. H., Wijekoon, W. M. K. P., Kim, H. M., and Prasad, P. N.,** Second-order nonlinear optical effects in novel polymethacrylate containing a molecular-ionic chromophore in the side chain, *Chem. Mater.,* 6, 234, 1994.

479. **Yokoh, Y. and Ogata, N.,** Thin films of acetalized poly(vinyl alcohol) as nonlinear optical materials, *Polym. J.,* 24, 63, 1992.

480. **Lindsay, G. A., Stenger-Smith, J. D., Henry, R. A., Nissan, R. A., Merwin, L. H., Cafin, A. P., and Yee, R. Y.,** High temperature sierrulate nonlinear optical polymers, in Organic Thin Films for Photonic Applications, ACS/ OSA Joint Conference, Toronto, October 5–7, 1993, Technical Digest Series, vol. 17, p. 14.

481. **Wang, C. H. and Guan, H. W.,** Second-harmonic generation and optical anisotropy of a spin-cast polymer film, *J. Polym. Sci. Polym. Phys., 31, 1983, 1992.*

482. **Tao, X. T., Watanabe, T., Shimoda, S., Zou, D. C., Sato, H., and Miyata, S.,** Λ-type main chain polymer for second harmonic generation, *Chem. Mater.,* 6, 1961, 1994.

483. **Watanabe, T., Tao, X. T., Kim, J., Miyata, S., Nalwa, H. S., and Lee, S. C.,** Molecular design of nonlinear optical molecules and polymers based on two-dimensional charge transfer, Extended Abstract of Second International Conference on Organic Nonlinear Optics (ICONO-2), Kusatsu, Gumma, July 23–26, 1995, p. 96.

484. **Nalwa, H. S., Watanabe, T., Kakuta, A., and Miyata, S.,** Aromatic polyureas: a new class of chromophoric main-chain polymers for second-order nonlinear optics, *Nonlinear Opt.,* 8, 157, 1994.

485. **Nalwa, H. S., Watanabe, T., Kakuta, A., Mukoh, A., and Miyata, S.,** Aromatic polyurea exhibiting large second harmonic generation and UV transparency, *Synth. Metals,* 57, 3895, 1993.

486. **Nalwa, H. S., Watanabe, T., Kakuta, A., Mukoh, A., and Miyata, S.,** Aromatic polyurea exhibiting large second harmonic generation and optical transparency down to 300 nm, *Appl. Phys. Lett.,* 62, 3223, 1993.

487. **Nalwa, H. S., Watanabe, T., Kakuta, A., Mukoh, A., and Miyata, S.,** Aromatic Polyureas: a new class of nonlinear optical materials with UV transparency, *Electron. Lett.,* 28, 1409, 1992.

488. **Kajiwara, K., Nagamori, H., Takezoe, H., Fukada, A., Ukishima, S., Takahashi, Y., Iijima, M., and Fukada, E.,** *Jpn. J. Appl. Phys.,* 30, 1737 1991.

489. **Tao, X. T., Watanabe, T., Zou, D. C., Shimoda, S., Sato, H., and Miyata, S.,** Polyurea with large positive birefringence for second harmonic generation, *Macromolecules,* 28, 2637, 1995.

490. **Watanabe, T., Amano, M., and Tomaru, S.,** Second-order optical nonlinearities of an electrooptic polymer containing a thiazole ring and its use in single-mode waveguide fabrication, *Jpn. J. Appl. Phys.,* 33, L1683, 1994.

491. **Nahata, A., Wu, C., Knapp, C., Lu, V., Shan, J., and Yardley, J. T.,** Thermally stable polyester polymers for second-order nonlinear optics, *Appl. Phys. Lett.,* 64, 3371, 1994.

492. **Meinhardt, M. B., Cahil, P. A., Seager, C. H., Beuhler, A. J., Wargowski, D. A., Singer, K. D., Kowalczyk, T. C., Kosc, T. Z., and Ermer, S.,** Characterization of cross-linked electro-optic polyimides, *SPIE Proc.,* 2143, 110, 1994.

493. **Pretre, Ph., Kaatz, P. G., Bohren, A., Gunter, P., Zysset, B., Ahlheim, M., and Lehr, F.,** in Organic Thin Films for Photonic Applications, ACS/OSA Joint Conference, Toronto, October 5–7, 1993, Technical Digest Series, vol. 17, p. 26.

494. **Becker, M. W., Sapochak, L. S., Ghosen, R., Xu, C., Dalton, L. R., Shi, Y., Steir, W. H., and Jen, A. K.-Y.,** Lrage and stable nonlinear optical effects observed from a polyimide covalently incorporating a nonlinear optical chromophore, *Chem. Mater.,* 6, 104, 1994.

495. **Sotoyama, W., Tatsuura, S., and Yoshimura, T.,** Electoro-optic side-chain polyinide system with large optical nonlinearity and high thermal stability, *Appl. Phys. Lett.,* 17, 2197, 1994.

496. **Weder, C., Neuenschwander, P., and W., S. U.,** New polyamides with large second-order nonlinear optical properties, *Macromolecules,* 27, 2181, 1994.

497. **Lin, J. T., Hubbard, M. A., and Marks, T. J.,** Poled polymeric nonlinear optical materials. Exceptional second harmonic generation temporal stability of a chromophore-functionalized polimide, *Chem. Mater.,* 4, 1148, 1992.

498. **Yu, D., Gharavi, A., and Yu, L.,** Novel second-order nonlinear optical aromatic, and aliphatic polyimides exhibiting high-temerature stability, *Appl. Phys. Lett.,* 66, 1050, 1995.

499. **Verbiest, T., Burland, D. M., Jurich, M. C., Lee, V. Y., Miller, R. D., and Volksen, W.,** Electrooptic properties of side-chain polyimides with exceptional thermal stabilities, *Macromolecules,* 28, 3005, 1995.

500. **Miller, R. D., Burland, D. M., Jurich, M., Lee, V. Y., Moylan, C. R., Tackara, J. I., Twieg, R. J., Verbiest, T., and Volksen, W.,** Donort-embedded nonlinear optical side chain polyimides containing no flexible tether: materials of exceptional thermal stability for electrooptical applications, *Macromolecules,* 28, 4970, 1995.

501. **Nemoto, N., Miyata, F., Nagase, Y., Abe, J., Hasegawa, M., and Shirai, Y.,** Novel types of polyesters containing second-order NLo-active chromophores with high density, Extended Abstract of Second International Conference on Organic Nonlinear Optics, (ICONO-2), Kusatsu, Gumma, July 23–26, 1995, p. 155.

502. **Suen, T. H., Lee, R. R., Shy, J. T., Hsue, G. H., Wu, L. H., Kuo, W. C., and Jeng, R. J.,** Nonlinear optically active liquid crystalline polymers and their copolymers based on tolane chromophores, Extended Abstract of Second International Conference on Organic Nonlinear Optics, (ICONO-2), Kusatsu, Gumma, July 23–26, 1995, p. 169.

503. **Tapolsky, G., Lecomte, J. P., and Meyrueix, R.,** Maleimide-based crosslinkable electrooptic polymers with excellent thermal stability characteristics, *Macromolecules,* 26, 7383, 1993.

504. **Meyrueix, R., Tapolsky, G., Dickens, M., and Lecomte, J. P.,** NLO polyurethane and crosslinked bismaleimide network with exceptional thermal stability, *SPIE Proc.,* 2025, 117, 1993.

505. **Moylan, C. R., Swanson, S. A., Walsh, C. A., Tackara, J. I., Twieg, R. J., Miller, R. D., and Lee, V. Y.,** From EFISH to electrooptic measurements of nonlinear chromophores, *SPIE Proc.,* 2025, 192, 1993.

506. **Reck, B., Eich, M., Jungbauer, D., Twieg, R. J., Willson, C. G., Y., Y. D. and Bijorklund, G. C.,** Crosslinked epoxy polymers with large and stable nonlinear optical susceptibilities, *SPIE Proc.,* 1147, 74, 1989.

507. **Twieg, R. J., Bijorklund, G. C., Lee, V., Baumert, C., Looser, H., Ducharme, S., Moerner, W. E., Willson, G. C., Reck, B., Swalen, J. D., Eich, M., Jungbauer, D, and Yoon, D. Y.,** Selection fabrication and applications of organic nonlinear optical materials, Proceedings of First International Conference on Crystal Growth of Organic Materials, Tokyo University of Agriculture and Technology, Tokyo, 1989, p. 262.

508. **Eich, M., Rock, B., Yoon, D. Y., Willson, C. G., and Bjorklund, G. C.,** Novel second-order nonlinear optical polymers via chemical cross-linking-induced vitrification under electric field, *J. Appl. Phys.,* 66, 3241, 1989.

509. **Teraoka, I., Jungbauer, D., Reck, B., Y., Y. D., Twieg, R. and Willson, C. G.,** Stability of nonlinear optical characteristics and dielectric relaxations of poled amorphous polymers with main-chain chromophores, *J. Appl. Phys.,* 69, 2568, 1991.

510. **Jungbauer, D., Teraoka, I., Yoon, D. Y., Reck, B., Swalen, J. D., Tweig, R., and Willson, C. G.,** Second-order nonlinear optical properties and relaxation characteristic of poled linear epoxy polymers with tolane chromophores, *J. Appl. Phys.,* 69, 8011, 1991.

511. **Park, J., Marks, T. J., Yang, J., and Wong, G. K.,** Chromophore-functionalized polymeric thin-film nonlinear optical materials. Effect of in-site cross-linking on second harmonic generation temporal characteristics, *Chem. Mater.,* 2, 229, 1990.

512. **Mandel, B. K., Chen, Y. M., Lee, J. Y., Kumar, J., and Tripathy, S.,** Cross-linked stable second-order nonlinear optical polymer by photochemical reaction, *Appl. Phys. Lett.,* 58, 2459, 1991.

513. **Hayashi, A., Goto, Y., Nakayama, M., Sato, H., Watanabe, T., and Miyata, S.,** Novel photoactivatable nonlinear optical polymers: poly[((4-azidophenyl)carboxy)ethyl methacrylate], *Macromolecules,* 25, 5094, 1992.

514. **Zhu, Z., Chen, Y. M., Li, L., Jeng, R. J., Mandel, B. K., Kumar, J., and Tripathy, S. K.,** Photocrosslinkable polymers with stable second-order optical nonlinearity, *Opt. Commun.,* 88, 77, 1992.

515. **Dalton, L. R., Yu, L. P., Chen, M., Sapochak, L. S., and Xu, C.,** Recent advances and characterization of nonlinear optical materials second-order materials, *Synth. Metals,* 54, 155, 1993.

516. **Muller, S., Chastaing, E., Le., B. P., and Robin, P.,** Quadratic nonlinear optical properties of thermally crosslinkable polymers, *Synth. Metals,* 54, 139, 1993.

517. **Kalluri, S., Shi, Y., and Steier, W. H.,** Improved poling and thermal stability of sol-gel nonlinear optical polymers, *Appl. Phys. Lett.,* 65, 2651, 1993.

518. **Gibbons, W. M., Grasso, R. P., O'Brien, M. K., Shannon, P. J., and Sun, S. T.,** Thermoset nonlinear optical polumers via hydrosilation of chromophores with tetramethylcyclotetrasiloxane, *Macromolecules,* 27, 771, 1994.

519. **Gibbson, W. M., Grasso, R. P., O'Brien, M. K., Shannon, P. J., and Sun, T.,** Nonlinear optical characterization of new thermoset hydrosilation polymers, *Appl. Phys. Lett.,* 64, 2628, 1994.

520. **Boogers, J. A. F., Klaase, P. T. A., Vlieger, J. J., Alkema, D. P. W., and Tinnemans, A. H. A.,** Cross-linked polymer materials for nonlinear optics. 1. UV-cured acrylic monomers bearing azobenzene dyes, *Macromolecules,* 27, 197, 1994.

521. **Boogers, J. A. F., Klaase, P. T., Vlieger, J. J. D., and Tinnemans, A. H. A.,** Crosslinked polymer materials for nonlinear optics. 2. Polyurethanes bearing azobenzene dyes, *Macromolecules,* 27, 205, 1994.

522. **Jeng, R. J., Hsiue, G. H., Chen, J. I., Marturunkakul, S., Li, L., Jiang, X. L., Moody, R. A., Masse, C. E., Kumar, J., and Tripathy, S. K.,** Low loss second-order nonlinear optical polymers based on all organic sol-gel materials, *J. Appl. Polym. Sci.,* 55, 209, 1995.

523. **Crumpler, E. T., Reznichenko, J. L., Li, D., Marks, T. J., Lin, W., Lundquist, P. M., and Wong, G. K.,** Modular high-Tg second-order polymeric nonlinear optical materials: polyfunctional epoxide and diisocyanate cross-linked chromophoric polyhydroxystyrene, *Chem. Mater.,* 7, 596, 1995.

524. **Liang, Z., Dalton, L. R., Garner, S. M., Kalluri, S., Chen, A., and Steier, W. H.,** A cross-linkable polyimide for second-order optical nonlinearities, *Chem. Mater.,* 7, 941, 1995.

525. **Nalwa, H. S., ed.,** Ferroelectric Polymers, Marcel Dekker, New York, 1995.

526. **Nalwa, H. S.,** Recent developments in ferroelectric polymers, *J. Macromol. Sci. Rev. Macromol. Chem. Phys.,* 31, 341, 1991.

527. **Watanabe, T., Miyata, S., and Nalwa, H. S.,** Nonlinear optical properties of ferroelectric polymers, in Ferroelectric Polymers, Nalwa, H. S., ed., Marcel Dekker, New York, 1995, chap. 12, p. 611.

528. **Bergman, Jr. J. G., McFee, J. H., and Crane, G. R.,** Pyroelectricity and optical second harmonic generation in polyvinylidene fluoride films, *Appl. Phys. Lett.,* 18, 203, 1971.

529. **McFee, J. H., Bergman, Jr. J. G., and Crane, G. R.,** Pyroelectric and nonlinear optical properties of poled polyvinylidene fluoride films, *Ferroelectrics,* 3, 305, 1972.

530. **Broussoux, D. and Micheron, F.,** Electro-optic and elasto-optic effects in polyvinylidene fluoride, *J. Appl. Phys.,* 51, 2020, 1980.

531. **Rabin, P., Chastaing, E., Broussour, D., Raffy, J., and Pocholle, J. P.,** Optical second harmonic generation in copolymer P(VDF/TrFE), *Ferroelectrics,* 94, 133, 1989.

532. **Legrand, J. F., Lajzerowicz, J., Berge, B., Delzenne, P., Macchi, E., Leonard, C. B., Wicker, A., and Kruger, J. K.,** Ferroelectricity in PVF_2 based copolymer, *Ferroelectrics,* 92, 267, 1984.

533. **Sato, H. and Gamo. H.,** New SHG observation in vinylidene fluoride/trifluoroethylene copolymer film using a pulsed YAG laser, *Jpn. J. Appl. Phys.,* 25, L990, 1986.

534. **Berry, M. H., Gookin, D. M., and Jacobs, E. W.,** Nonlinear optical properties of materials, *Tech. Digest Ser.,* 9, 121, 1988.

535. **Tsutsumi, N., Ono, T., and Kiyotsukuri, T.,** Internal electric field and second harmonic generation in the blends of vinylidene fluoride-trifluoroethylene copolymer and poly(methyl methacrylate) with a pendant nonlinear optical dye, *Macromolecules,* 26, 5447, 1993.

536. **Azumai, Y., Sato, H., and Seo, I.,** Second-harmonic generation of Er:YAG 2.94 mm laser radiation using an organic vinylidene cyanide/vinyl acetate thin film, *Opt. Lett.,* 15, 932, 1990.

537. **Azumai, Y., Seo, I., and Sato, H.,** Enhanced second harmonic generation with Cerenkov radiation scheme in organic film slab-guide at IR line, *IEEE J. Quantum Electron.,* 28, 231, 1992.

538. **Azumai Y. and Sato, H.,** Improvement of the Cerenkov radiative second harmonic generation in the slab waveguide with a periodic nonlinear optical susceptibility, *Jpn. J. Appl. Phys.,* 32, 800, 1993.

539. **Dumont, M., Levy, Y., and Morichere, D.,** Electro-optic organic waveguides: optical characterization, in Organic Molecules for Nonlinear Optics and Photonics, Messier, J., et al., eds., Kluwer Academic Publishers, Netherland, 1991, p. 461.

540. **Broussoux, D. E., Chastaing, S., Esselin, O., Le Barny, R., Robin, Y., Pocholle, J. P., and Raffy, J.,** Organic materials for nonlinear optics, *Rev. Tech. Thomson-CSF,* 20–12, 151, 1989.

541. **Kishimoto, M., Zou, D., and Seo, I.,** New polymers for nonlinear optics, nonlinear guided-wave phenomena 1991, Technical Digest Series, Vol. 15, p. 82, 1991.

542. **Nalwa, H. S., Watanabe, T., Kakuta, A., Mukoh, A., and Miyata, S.,** N-Phenylated aromatic polyureas: a new class of nonlinear optical materials exhibiting large second harmonic generation and u.v. transparency, *Polymer,* 34, 657, 1993.

543. **Nalwa, H. S., Watanabe, T., Kakuta, A., Mukoh, A., and Miyata, S.,** Aromatic Polyurea for second-order nonlinear

optics, in Proceedings of ACS/OSA Topical Meeting on Organic Thin Films for Photonic Applications, 1993 Technical Digest Series, Vol. 17, pp. 85–88, 1993.

544. **Scherowsky, G.,** Ferroelectric liquid crystal (FLC) polymers, in Ferroelectric Polymers, Nalwa, H. S., ed., Marcel Dekker, New York, 1995, chap. 10, p. 435.

545. **Kiefer, R.,** Applications of ferroelectric liquid crystalline polymers, in Ferroelectric Polymers, Nalwa, H. S., ed., Marcel Dekker, New York, 1995, chap. 19, p. 815.

546. **Ozaki, M., Utsumi, M., Yoshino, K., and Skarp, K.,** Second harmonic generation in ferroelectric liquid crystalline polymer, *Jpn. J. Appl. Phys.,* 32, L852, 1993.

547. **Coles, H. J., Redmond, M., Mondain-Monval, O., Wischerhoff, E., and Zentel, R.,** A study of second harmonic generation in side-chain ferroelectric liquid crystalline copolymers, in Proceedings of the Fourth International Conference on Ferroelectric Liquid Crystals, p. 83, Tokyo, Japan, Septemebr 28–October 1, 1993.

548. **Zyss, J.,** in Conjugated Polymeric Materials: Opportunities in Electronics, Optoelectronics and Molecular Electronics, Bredas, J. L., and Chance, R. R., eds., NATO ASI Series E, Appl. Sci. vol. 182, Kluwer Academic Press, Dordrecht, 1990, p. 545.

549. **Cheng, L. T., Tam, W., Feiring, A., and Rikken, G. L. J. A.,** Quadratic hyperpolarizabilities of fluorinated sulfonyl and carbonyl aromatics: optimization of nonlinearity and transparency trade-off, *SPIE Proc.,* 1337, 203, 1990.

550. **Cheng, L. T., Tam, W., and Feiring, A.,** Molecular hyperpolarizabilities of fluorinated sulfonyl aromatics, *Nonlinear Opt.,* 3, 69, 1992.

551. **Duan, X. M., Okada, S., Nakanishi, H., Clays, K., Persoons, A., and Matsuda, H.,** Evaluation of β of stilbazolium p-toluenesulfonates by the hyper Rayleigh scattering method, *SPIE Proc.,* 2143, 41, 1994.

552. Laser Focus World, September 1991.

553. **(a) Harada, A., Okazaki, Y., Kamiyama, K., and Umegaki, S.,** Generation of blue coherent light from a continuous-wave semiconductor laser using an organic crystal-cored fiber, *Appl. Phys. Lett.,* 59, 1535, 1991.
(b) Uemiya, T., Uenishi, N., Shimizu, Y., Okamoto, S., Chikuma, K., Tohma, T., and Umegaki, S. Crystal cored fiber using organic material and focusing properties of generated second-harmonic wave, SPIE, San Diego, 1989, p. 207.

554. **Norwood, R. A., Findakly, T., Goldberg, H. A., Khanarian, G., Stamatoff, J. B., and Yoon, H. N.,** in Polymers for Lightwave and Integrated Optics, Hornak, L. A., ed., Marcel Dekker, New York, 1992, chap. 11, p. 287.

555. **van der Poel, C. J., Bierlein, J. D., Brown, J. B., and Colak, S.,** *Appl. Phys. Lett.,* 57, 2074, 1990.

556. **Lim, E. J., Fejer, M. M., Beyer, R. L., and Kozlovsky, W. J.,** *Electron. Lett.,* 25, 731, 1989.

557. **Yamamoto, K., Mizuuchi, K., and Taniuchi, T.,** *Opt. Lett.,* 16, 1156, 1991.

558. **Florsheimer, M., Kupfer, M., Bosshard, Ch., and Gunter, P.,** in *Nonlinear Optics,* Miyata, S., ed., Elsevier, Amsterdam, 1991, p. 255.

559. **Kondo, T., Hashizume, N., Miyoshi, S., Morita, R., Ogasawara, N., Umegaki, S., and Ito, R.,** *SPIE Proc.,* 1337, 53, 1990.

560. **Suhara T. and Nishihara, H.,** in *Nonlinear Optics,* S. Miyata, S., ed., Elsevier, Amsterdam, 1991, p. 501.

561. **Suhara, T., Morimoto, T., and Nishihara, H.,** The Fourth Micro-optics Conference and the Eleventh Topical Meeting of Gradient-Index Optical Systems (MOC/GRIN'93), Technical Digest, The Japan Society of Applied Physics, October 20 ~ 22, 1993, Kawasaki, Japan, p. 76.

562. **Norwood R. A. and Khanarian, G.,** *Electron. Lett.,* 26, 2105, 1990.

563. **(a) Rikken, G. L. J. A., Seppen, C. J. E., Nijhuis, S., and Meijer, E. W.,** *Appl. Phys. Lett.,* 58, 435, 1991.
(b) Rikken, G. L. J. A., Seppen, C. J. E., Staring, E. G. J., and Venhuizen, A. H. J., *Appl. Phys. Lett.,* 62, 2483, 1993.

564. **Shuto, Y., Takara, H., Amano, M., and Kaino, T.,** *Jpn. J. Appl. Phys.,* 28, 2508, 1989.

565. **Kinoshita, T., Nonaka, Y., Nihei, E., Koike, Y., and Sasaki, K.,** in *Nonlinear Optics,* Miyata, S., ed., Elsevier, Amsterdam, 1991, p. 479.

566. **Sugihara, O., Kinoshita, T., Okabe, M., Kunioka, S., Nonaka, Y., and Sasaki, K.,** *Appl. Opt.,* 30, 2957, 1991.

567. **Sato, H., Nozawa, H., Azumi, Y., and Seo, I.,** Demonstration of enhanced Cerenkov-radiative SHG with chirped nonlinear optical susceptibility in organic polymer waveguide, *Nonlinear Optics,* 10, 319, 1995, and references therein.

568. *Laser Focus World Digest,* 15, 1992.

569. **Haase, M. A., Qui, J., DePuydt, M., and Cheng, H.,** *Appl. Phys. Lett.,* 59, 1272, 1991.

570. **(a) Gaines, J. M., Drenten, R. R., Haberern, K. W., Marshall, T., Mensz, P., and Petruzello, J.,** Blue-green injection lasers containing pseudomorphic $Zn_{1-x}Mg_xS_ySe_{1-y}$ cladding layers and operating up to 394 K, *Appl. Phys. Lett.,* 62, 2462, 1993.
(b) Itoh, S., Okuyama, H., Matsumoto, S., Nakayama, N., Ohata, T., Miyajima, T., Ishibashi, A., and Akimoto, K., Room temperature pulsed operation of 498 nm laser with ZnMgSSe cladding layers, *Electron. Lett.,* 29, 766, 1993.

571. **Sasaki, T.** Intracavity Second Harmonic Generation by Organic and Inorganic Materials, *Proceedings of 5th Toyota Conference, Nonlinear Optics,* Miyata, S., ed., North-Holland, 1992, p. 445.

572. **Dubois, J. C.,** Organic conjugated materials for optoelectronic and applications, in Conjugated Polymeric Materials, Bredas, J. L., and Chance, R. R., eds., Kluwer Academic Publishers, Netherlands, 321, 1990.

573. **Thackara, J. I., Lipscomb, G. F., Stiller, M. A., Ticknor, A. J., and Lytel, R.,** Poled electro-optic waveguide formation in thin-film organic media, *Appl. Phys. Lett.,* 52, 1031, 1988.

574. **Thackara, J. I., Lipscomb, G. F., Lytel, R. S., and Ticknor, A. J.,** Advances in organic electro-optic devices, in *Nonlinear Optical Properties of Polymers,* Heeger, A. J., Orenstein, J., and Ulrich, D. R., eds., Materials Research Society, Pittsburgh, 19, 109, 1988.

575. **Findakly, T, and Teng, C. C.,** Wideband NLO polymer modulatol, *SPIE Proc.,* 2025, 526, 1993.

578. **Sueta, T., and Izutsu, M.,** *IEEE Trans. Microwave Theory Tech.,* 38, 477, 1990.

579. **Wah, C. K. L., Iizuka, K., and Freundorfer, A. P.,** *Appl. Phys. Lett.,* 63, 3110, 1993.

580. **(a) Lytel, R.,** *SPIE Proc.,* 1216, 30, 1990.
    (b) Lytel, R., Limscomb, G. F., Kenney, J. T., and Binkley, E. S., in Polymers for Lightwave and Integrated Optics, Hornak, L. A., ed., Marcel Dekker, New York, 1992, chap. 16, p. 433.

581. **Thackara, J. I., Bjorklund, G. C., Fleming, W., Jurich, M., Smith, B. A., and Swalen, J. D.,** A polymeric electro-optic phase-modulator for broadband data transmission, *SPIE Proc.,* 2025, 564, 1993.

582. **Dalton, L. R., Harper, A. W., Ghosn, R., Steier, W. H., Ziari, M., Fetterman, H., Shi, Y., Mustacich, R. V., Jen, A. K.-Y., and Shea, K. J.,** Synthesis and processing of improved organic second-order nonlinear optical materials for applications in photonics, *Chem. Mater.,* 7, 1060, 1995.

583. **Mohlman, G. R., et al.,** Optically nonlinear polymeric switches and modulators, *SPIE Proc.,* 1337, 215, 1990.

584. **P. De Dobbelaere, F. Vermaerke, G. Vermeire, P. Demester, P. Van Daele, G. R. Mohlmann, and W. H. G. Horsthuis,** The Fourth Micro-optics Conference and the Eleventh Topical Meeting of Gradient-Index Optical Systems (MOC/GRIN'93), Technical Digest, The Japan Society of Applied Physics, October 20 ~ 22, 1993, Kawasaki, Japan, p. 88.

585. **Hilland, D., and Mensa, D.,** *Proc. IEEE Ant. Prop. Soc. Symp.,* 2, 720, 1992.

586. **Van Eck, T. E., Ticknor, A. J., Lytel, R. S., and Lipscomb,** Complementary optical tap fabricated in an electro-optic polymer waveguide, *Appl. Phys. Lett.,* 58, 1588, 1991.

587. **De Dobbelaere, P., Vermaerke, F., Vermeire, G., Demester, P. Van Daele, P., Mohlmann, G. R., Heidemann, J. L. P., and Horsthuis, W. H. G.,** Integration technology for light source arrays with polymeric optical waveguide arrays, *SPIE Proc.,* 2285, 352, 1994.

588. **Girton, D. G., Kwiatkowski, S., Lipscomb, G. F., and Lytel, R.,** 20 Ghz electro-optic polymer Mach-Zahnder modulator, *Appl. Phys. Lett.,* 58, 1730, 1991.

589. **Van Tomme, E., Van Daele, P. P., Baets, R. G., and Lagasse, P. E.,** Integrated optic devices based on nonlinear optical polymers, *IEEE J. Quantum Electron.* QE-27, 778, 1991.

590. **Teng, C. C.,** Travelling-wave polymeric optical intensity modulator with more than 40 GHz of 3 dB electrical bandwidth, *Appl. Phys. Lett.,* 60, 1538, 1992.

591. **Fawlett, G., Johnstone, W., Andonvic, I., Bone, D. J., Harvey, T. G., Carter, N., and Ryan, T. G.,** In-line fibre-optic intensity modulator using electro-optic polymer, *Electron. Lett.,* 28, 985, 1992.

592. **Schaefer, S. R., and Swafford, W. J.,** Polymer modulators for broadband communications, *SPIE Proc.,* 2025, 460, 1993.

593. **Suyten, F. M. M., Diemeer, M. B. J., and Henriksen, B.,** A digital thermo-optic 2 × 2 switch based on nonlinear optic polymer, *SPIE Proc.,* 2025, 479, 1993.

594. **Tumolillo, Jr., T. A., and Ashley, P. R.,** Fabrication and design considerations for multilevel active polymeric devices, *SPIE Proc.,* 2025, 507, 1993.

595. **Ticknor, A. J., Lipscomb, G. F., and Lytel, R.,** Practical considerations for polymer photonics devices, *SPIE Proc.,* 2285, 386, 1994.

596. **Yardley, J. T., Eldada, L., Stengel, K. M. T., Shacklette, L. W., Wu, C., and Xu, C.,** Toward the practical applications of polymeric optical interconnection technology, Extended Abstract of Second International Conference on Organic Nonlinear Optics (ICONO'2), Kusatsu, Japan, July 23–26, 1995, p. 72.

597. **Stegeman, G. I.,** Lightwave manipulation in guided wave geometries, Extended Abstract of Second International Conference on Organic Nonlinear Optics (ICONO'2), Kusatsu, Japan, July 23–26, 1995, p. 6.

598. **Liang, J., Levenson, R., Zyss, J., Bosc, D., and Foll, F.,** Design and fabrication of electro-optic polymer waveguides, *Nonlinear Opt.,* 10, 431, 1995.

599. **Dubois, J.,** Polymers in new electro-optic components, *Nonlinear Opt.,* 10, 439, 1995.

600. **Dalton, L. R., Xu, C., Harper, A. W., Ghosn, R., Wu, B., Liang, Z., Montegomery, R., and Jen, A. K.-Y.,** Development and applications of organic electro-optic modulators, *Nonlinear Opt.,* 10, 383, 1995.

601. **Wang, C. H., Wherrett, B. S., Cresswell, J. P., Petty, M. C., Ryan, T., Allen, S., Ferguson, I., Hutchings, M. G., and Devonald, D. P.,** Observation of electro-optic and electro-absorption modulations in a Langmuir-Blodegtt film Fabry-Pert-etalon, *Opt. Lett.,* 20, 1533, 1995.

602. **Dalton, L. R.,** High frequency electro-optic modulation and integration of photonic and VLSI electronic circuitry demonstrated, *Photonics Science News,* 1, 7, 1995.

603. **Dagani, R.,** *Chem. Eng. News,* March 4, 1996, pp. 22–27.

604. **Lipscomb, G. F., and Lytel, R.,** Photonic large scale integration, *Nonlinear Opt.,* 3, 41, 1992.

605. **Lipscomb, G. F., Lytel, R., and Ticknor, A. J.,** Electro-optic polymer waveguide arrays in digital systems, *Nonlinear Opt.,* 10, 421, 1995.

Chapter 5

Phase-Matched Second Harmonic Generation in Organic Materials

Toshiyuki Watanabe, Hari Singh Nalwa, and Seizo Miyata

CONTENTS

I. INTRODUCTION

A wide variety of organic second harmonic generation (SHG) materials are summarized in Chapter 4, including single crystals, poled polymers, Langmuir-Blodgett films, liquid crystals, and guest–host systems. In this chapter, we briefly describe the phase-matching conditions in bulk medium as well as in waveguides of SHG materials. Phase matching of the fundamental and harmonic waves in the second harmonic generation is of great concern. A number of techniques have been demonstrated in organic crystals, Langmuir-Blodgett films, and poled polymers. In order to utilize materials in photonic applications, the effectively induced nonlinear optical polarizations and the phase-matching techniques are of great significance to achieve the efficient nonlinear optical interactions. The second harmonic power is expressed by:[1]

$$P(2\omega) = \frac{2\omega^2}{\varepsilon_0 c^3} \frac{d^2}{n(\omega)^2 n(2\omega)} \frac{P(\omega)^2}{A} L^2 \times \frac{\sin^2(\Delta k L/2)}{(\Delta k L/2)} \tag{1}$$

where ω = angular frequency of the fundamental wave; A = fundamental beam area; d = nonlinear optical coefficient; $n(\omega)$ = refractive index of material at the fundamental wavelength (ω); $n(2\omega)$ = refractive index of material at the second harmonic wavelength (2ω); c = velocity of light in vacuum; $\Delta k = 2k(\omega) - k(2\omega)$ wave vector mismatch; and L = optical pass length of material.

If Δk is not zero, the second harmonic power is oscillated as a function of propagation length and the highly efficient conversion efficiency cannot be obtained. If $\Delta k = 0$, the conversion efficiency increases with the square of the propagation length and enables us to provide a highly efficient conversion efficiency of more than 10% in an ideal case. In general, phase-matching conditions are described by

352

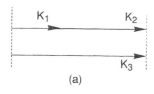

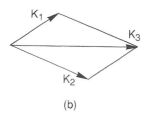

Figure 1 The collinear (a) and noncollinear (b) phase matching for three-wave interactions. (From Dmitriev, V. G., et al., *Handbook of Nonlinear Optical Crystals,* Springer-Verlag, Berlin, 1991. With permission.)

wave number as follows:

$$k_3 = k_2 + k_1 \tag{2}$$

where k_i are the wave vectors corresponding to the wavelength. An excellent description of uniaxial and biaxial crystals and their phase-matching conditions can be found in reference 2. There are two configurations of wave vector under phase-matching conditions as shown by Figure 1. One is called collinear phase matching, while the other is called noncollinear phase matching.[2] Under collinear phase matching for frequency doubling, the relation between fundamental beam and output beam is represented by the following equation:

$$k_3 = 2k_1 \tag{3}$$

This relationship between fundamental beam and harmonic beam can never be accomplished in an isotropic medium (and also in anisotropic medium for waves of identical polarization, which means diagonal tensor components such as d_{33} cannot be utilized in bulk state for phase matching) in the optical transparent region due to the normal dispersion of the refractive index. Figure 2 shows the normal dispersion of the refractive index in a biaxial medium. The phase-matching condition using birefringence is accomplished only in anisotropic crystals with interaction of different polarized waves.

Organic materials possess the large diagonal tensor such as d_{33} because of highly aligned nonlinear optical (NLO) chromophores. For example, the d_{33} of 2-methyl-4-nitroaniline (MNA) is one order of magnitude larger than that of $LiNbO_3$. Consequently, the figure of merits represented by d^2/n^3, which is directly related to the conversion efficiency, is three orders of magnitude larger than that of $LiNbO_3$ crystals. But this tensor cannot be used in a bulk material. In order to utilize the diagonal tensor components, a waveguide geometry is proposed. Phase-matching techniques using waveguide geometry are described in Section 3.

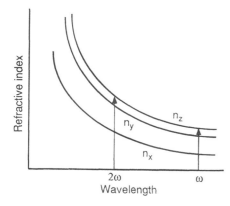

Figure 2 Dispersion of the refractive index of a biaxial crystal. (From Dmitriev, V. G., et al., *Handbook of Nonlinear Optical Crystals,* Springer-Verlag, Berlin, 1991. With permission.)

II. OPTICAL CRYSTALS AND BIREFRINGENCE PHASE MATCHING

A. UNIAXIAL CRYSTALS

The merit of phase matching from birefringence is that it allows the use of collinear mixing configuration and is more readily adaptable to find a phase-matching direction just by rotating the crystal.[3] Furthermore, birefringence phase matching is easy to couple with the fundamental beam into the materials and it possesses relatively wide acceptance for temperature fluctuation. Let us introduce the optical dielectric axis of a medium before discussing phase matching. The refractive index of any material can be represented using an ellipsoid (normal surface). This means that we can define a three-orthogonal principle axis in any medium. In the case of isotropic media, the refractive index is the same for any polarization and any given propagation direction. If the material has an anisotropy, the normal surface is given by

$$\frac{n^2 s_x^2}{n^2 - n_x^2} + \frac{n^2 s_y^2}{n^2 - n_y^2} + \frac{n^2 s_z^2}{n^2 - n_z^2} = 1 \tag{4}$$

$$k = \frac{2\pi n}{\lambda_0} s = \frac{\omega n}{c} s \tag{5}$$

where k represents the wave vector, s is the unit vector of k, λ_0 is the wavelength of light in vacuum, c is the velocity of light in vacuum, and ω is the angular frequency. The projection of s is given by

$$s_x = \sin \theta \cos \varphi \tag{6}$$

$$s_y = \sin \theta \sin \varphi \tag{7}$$

$$s_z = \cos \theta \tag{8}$$

$$s_x^2 + s_y^2 + s_z^2 = 1 \tag{9}$$

If s_x and s_y are same, this medium is classified as uniaxial. If s_x, s_y, and s_z are not equal, the medium belongs to biaxial materials. There is a special direction in the ellipsoid called the optic axis. The refractive index of materials does not depend on the polarization when the propagation of light is parallel to the optic axis. A uniaxial crystal has only one optic axis, whereas a biaxial crystal possesses two. The plane containing the z axis and the wave vector k of the light is termed the principal plane. In the case of a uniaxial crystal, the polarized light is normal to the principal plane, which is called an ordinary beam (o beam). The beam polarized in the principal plane is known as the extraordinary beam (e beam). The refractive index of the o beam does not depend on the propagation direction, but for the e beam, it does. The refractive index for the e beam and o beam are represented as n_e and n_o, respectively. The difference between the refractive index of n_o and n_e is known as birefringence (Δn). The Δn is zero when a light beam solely propagates along the optic axis and is maximum in the direction normal to this axis. The refractive index of the extraordinary wave is a function of the polar angle φ and θ as shown in Figure 3.

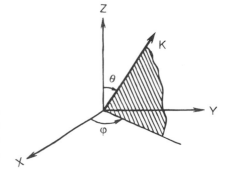

Figure 3 Polar coordinate systems for description of the refractive index ellipsoid of a uniaxial crystal. Z is the optic axis; k is the wave vector and represents the propagation direction of light. (Dmitriev, V. G., et al., *Handbook of Nonlinear Optical Crystals*, Springer-Verlag, Berlin, 1991. With permission.)

354

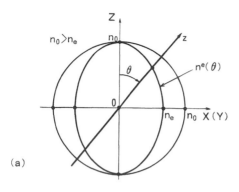

(a)

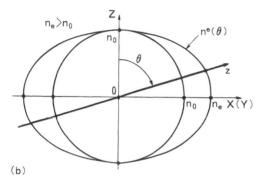

(b)

Figure 4 The indicatrices of the refractive indices for negative (a) and positive (b) crystals. (Dmitriev, V. G., et al., *Handbook of Nonlinear Optical Crystals,* Springer-Verlag, Berlin, 1991. With permission.)

In the case of uniaxial material, the following relation is derived:

$$n_e(\theta) = n_o[(1 + \tan^2\theta)/(1 + (n_o/n_e)^2\tan^2\theta)]^{1/2} \tag{10}$$

where

$$n_o(\theta) = n_o$$
$$n_e(0) = n_o$$
$$n_e(90) = n_e$$

Figure 4 gives the indicatrices of the refractive indices for ordinary and extraordinary waves in negative and positive uniaxial crystals. If $n_o > n_e$, the crystal is called negative; if $n_o < n_e$, it is called positive. For the negative crystal, the ellipsoid is in the sphere. On the other hand, for the positive crystal, the sphere is in the ellipsoid. The quantity of n_e does not depend on the azimuthal angle φ. The phase-matching condition in a uniaxial crystal is given by the wave vector. In a negative crystal[4]

$$k_1^o + k_2^o = k_3^e(\theta) \tag{11}$$

This configuration is defined as the *ooe* interaction. In a positive crystal

$$k_1^e + k_2^e = k_3^o(\theta) \tag{12}$$

This configuration is called the *eeo* interaction. Type I phase matching requires the mixing waves to be in parallel polarization, whereas type II phase matching requires the mixing wave to be in orthogonal polarization. In a negative crystal

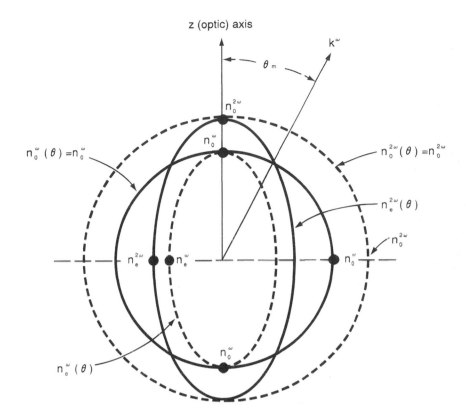

Figure 5 Phase-matching direction of a uniaxial crystal. (From Yariv, A., *Quantum Electronics*, John Wiley & Sons, New York, 1975. With permission.)

$$k_1^o + k_2^e = k_3^e(\theta) \ (oee \text{ interaction}) \tag{13}$$

$$k_1^e + k_2^o = k_3^e(\theta) \ (eoe \text{ interaction}) \tag{14}$$

In a positive crystal

$$k_1^o + k_2^e = k_3^o(\theta) \ (oeo \text{ interaction}) \tag{15}$$

$$k_1^e + k_2^o = k_3^o(\theta) \ (eoo \text{ interaction}) \tag{16}$$

Figure 5 shows that how one can find out the phase-matching direction of collinear phase matching for type I SHG in uniaxial negative crystals.[5] For the *ooe* interaction $n_\omega^o = n_{2\omega}^e(\theta)$ must be satisfied at $\theta = \theta m$. The cross-sectional points between n_ω^o and $n_{2\omega}^e(\theta)$ curves give the phase-matching condition. The angle θm is obtained from the intersection of the sphere.

B. BIAXIAL CRYSTALS

The refractive index of a biaxial crystal depends on the propagating direction.[2] Therefore, all refractive indices are similar to the extraordinary one of a uniaxial crystal. In general, the principal dielectric axes of a biaxial crystal are defined as $n_z > n_y > n_x$. Indicatrices of the refractive index in a biaxial crystal are shown in Figure 6. The optic axis can be defined as follows:

$$\sin V_z = \frac{n_z}{n_y} \left(\frac{n_y^2 - n_x^2}{n_z^2 - n_x^2} \right)^{1/2} \tag{17}$$

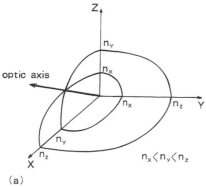

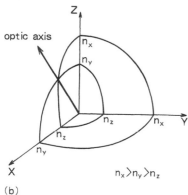

Figure 6 The indicatrices of a biaxial crystal. (Dmitriev, V. G., et al., *Handbook of Nonlinear Optical Crystals,* Springer-Verlag, Berlin, 1991. With permission.)

In the *XY* plane, the refractive index of a wave polarized normally to this plane is constant and equals n_z, and that of a wave polarized in this plane changes from n_y to n_x.

$$n_{XY}(\theta) = n_y\left(\frac{1 + \tan^2\theta}{1 + (ny/nx)^2\tan^2\theta}\right)^{1/2} \tag{18}$$

Analytically, the index surface in a biaxial crystal for any given frequency is represented by the following equations:

$$A^{(\omega)}n^4 - B^{(\omega)}n^2 + C^{(\omega)} = 0 \tag{19}$$

$$A^{(2\omega)}n^4 - B^{(2\omega)}n^2 + C^{(2\omega)} = 0 \tag{20}$$

where $A^{(\omega)}$, $B^{(\omega)}$, and $C^{(\omega)}$ are coefficients for the fundamental beam and are defined as follows:

$$A^{(\omega)} = (n_x^{\omega})^2 s_x^2 + (n_y^{\omega})^2 s_y^2 + (n_z^{\omega})^2 s_z^2 \tag{21}$$

$$B^{(\omega)} = (n_y^{\omega})^2(n_z^{\omega})^2(s_y^2 + s_z^2) + (n_z^{\omega})^2(n_x^{\omega})^2(s_z^2 + s_x^2) + (n_x^{\omega})^2(n_y^{\omega})^2(s_x^2 + s_y^2) \tag{22}$$

$$C^{(\omega)} = (n_x^{\omega})^2(n_y^{\omega})^2(n_z^{\omega})^2 \tag{23}$$

$A^{(2\omega)}$, $B^{(2\omega)}$, and $C^{(2\omega)}$ are coefficients for the harmonic light and are given by

$$A^{(2\omega)} = (n_x^{2\omega})^2 s_x^2 + (n_y^{2\omega})^2 s_y^2 + (n_z^{2\omega})^2 s_z^2 \tag{24}$$

$$B^{(2\omega)} = (n_y^{2\omega})^2(n_z^{2\omega})^2(s_y^2 + s_z^2) + (n_z^{2\omega})^2(n_x^{2\omega})^2(s_z^2 + s_x^2) + (n_x^{2\omega})^2(n_y^{2\omega})^2(s_x^2 + s_y^2) \tag{25}$$

$$C^{(2\omega)} = (n_x^{2\omega})^2(n_y^{2\omega})^2(n_z^{2\omega})^2 \tag{26}$$

From Equations 19 through 26, one can obtain the two refractive indices for the different polarizations.

$$n_i^{(\omega)} = \left\{ \frac{B^{(\omega)} \pm [(B^{(\omega)})^2 - 4A^{(\omega)}C^{(\omega)}]^{1/2}}{2A^{(\omega)}} \right\}^{1/2} \tag{27}$$

$$n_i^{(2\omega)} = \left\{ \frac{B^{(2\omega)} \pm [(B^{(2\omega)})^2 - 4A^{(2\omega)}C^{(2\omega)}]^{1/2}}{2A^{(2\omega)}} \right\}^{1/2} \tag{28}$$

Therefore the type I phase-matching angle is obtained to find the combination of θ and φ to satisfy Equation 29.

$$\left\{ \frac{B^{(2\omega)} - [(B^{(2\omega)})^2 - 4A^{(2\omega)}C^{(2\omega)}]^{1/2}}{2A^{(2\omega)}} \right\}^{1/2} = \left\{ \frac{B^{(\omega)} + [(B^{(\omega)})^2 - 4A^{(\omega)}C^{(\omega)}]^{1/2}}{2A^{(\omega)}} \right\}^{1/2} \tag{29}$$

The type II phase-matching angle is also obtained by a similar method.

$$2 \cdot \left\{ \frac{B^{(2\omega)} - [(B^{(2\omega)})^2 - 4A^{(2\omega)}C^{(2\omega)}]^{1/2}}{2A^{(2\omega)}} \right\}^{1/2} = \left\{ \frac{B^{(\omega)} - [(B^{(\omega)})^2 - 4A^{(\omega)}C^{(\omega)}]^{1/2}}{2A^{(\omega)}} \right\}^{1/2}$$

$$+ \left\{ \frac{B^{(\omega)} + [(B^{(\omega)})^2 - 4A^{(\omega)}C^{(\omega)}]^{1/2}}{2A^{(\omega)}} \right\}^{1/2} \tag{30}$$

C. EFFECTIVE NONLINEAR OPTICAL COEFFICIENTS

Phase-matching direction does not coincide with the principal dielectric axes except in noncritical phase matching. Therefore, we must consider the projection of both the fundamental electric field and the nonlinear polarization to the each tensor. Any linearly polarized light in an uniaxial crystal can be represented by superposition of two waves with ordinary and extraordinary polarizations. Unit vector p is introduced which is given by polar coordinates θ and ϕ. Z is the optic axis. The projection of polarizations is defined as

$$P_x^o = -\sin \varphi, \qquad P_x^e = \cos \theta \cos \varphi \tag{31}$$

$$P_y^o = \cos \varphi, \qquad p_y^e = \cos \theta \sin \varphi \tag{32}$$

$$P_z^o = 0, \qquad P_z^e = -\sin \theta \tag{33}$$

The equations for calculating the conversion efficiency employ the effective nonlinear coefficients, which consider the projection of all summation operations along the polarization direction of interacting waves. The effective nonlinear optical coefficients[2] for uniaxial crystals of 11 point groups and for biaxial crystals of 4 point groups are listed in Table 1.

D. WALK OFF

When harmonic wave (vector **k**) propagates in anisotoropic crystal, the direction of propagation of the wave phase generally does not coincide with that of the wave energy (vector **S**) except in noncritical phase matching. The direction of **S** can be defined as the normal to the tangent drawn at the point of intersection of vector **k**. In the case of a uniaxial crystal, the **k** and **s** propagate in different directions as defined by angle ρ.

$$\cos \rho = \frac{\mathbf{S}_2^\omega \cdot \mathbf{S}_1^{2\omega}}{|\mathbf{S}_2^\omega||\mathbf{S}_1^{2\omega}|} \tag{34}$$

Table 1 Expressions for d_{eff} in Nonlinear Crystals of Different Point Groups When Kleinman Symmetry Relations Are Valid

Point Group	Principal Plane	Type of Interaction	
		ooe, oeo, eoo	*eeo, eoe, oee*
2	XY	$d_{23}\cos\varphi$	$d_{36}\sin 2\varphi$
	YZ	$d_{21}\cos\theta$	$d_{36}\sin2\theta$
	XZ	0	$d_{21}\cos^2\theta + d_{23}\sin^2\theta - d_{36}\sin2\theta$
m	XY	$d_{13}\sin\varphi$	$d_{31}\sin^2\varphi + d_{32}\cos^2\varphi$
	YZ	$d_{31}\sin\theta$	$d_{13}\sin^2\theta + d_{12}\cos^2\theta$
	XZ	$d_{12}\cos\theta - d_{32}\sin\theta$	0
mm2	XY	0	$d_{31}\sin^2\varphi + d_{32}\cos^2\varphi$
	YZ	$d_{31}\sin\theta$	0
	XZ	$d_{32}\sin\theta$	0
222	XY	0	$d_{36}\sin2\varphi$
	YZ	0	$d_{36}\sin2\theta$
	XZ	0	$d_{36}\sin2\theta$
4,4mm 6,6mm		$d_{15}\sin\theta$	0
$\bar{6}m2$		$d_{22}\cos\theta\sin3\varphi$	$d_{22}\cos^2\theta\cos\varphi$
3m		$d_{15}\sin\theta - d_{22}\cos\theta\sin3\varphi$	$d_{22}\cos^2\theta\cos3\varphi$
$\bar{6}$		$(d_{11}\cos3\varphi - d_{22}\sin3\varphi)\cos\theta$	$(d_{11}\sin3\varphi + d_{22}\cos3\varphi)\cos^2\theta$
3		$(d_{11}\cos3\varphi - d_{22}\sin3\varphi)\cos\theta + d_{15}\sin\theta$	$(d_{11}\sin3\varphi + d_{22}\cos3\varphi)\cos^2\theta$
32		$d_{11}\cos\theta\cos3\varphi$	$d_{11}\cos^2\theta\sin3\varphi$
$\bar{4}$		$(d_{14}\sin2\varphi + d_{15}\cos2\varphi)\sin\theta$	$(d_{14}\cos2\varphi - d_{15}\sin2\varphi)\sin2\theta$
$\bar{4}2m$		$d_{36}\sin\theta\sin2\varphi$	$d_{36}\sin2\theta\cos2\varphi$

The following assignment of the crystallophysical and crystallographic axes was assumed in deriving the equations for d_{eff} for biaxial crystals: for point groups 2 and m, $Y \parallel b$; for point groups mm2 and 222, $X \parallel a$, $Y \parallel b$, $Z \parallel c$.
Adapted from Dmitriev, V. G., et al. *Handbook of Nonlinear Optical Crystals*, Springer-Verlag, Berlin, 1991. With permission.

or

$$\frac{\mathbf{S}_1^{\omega} \cdot \mathbf{S}_2^{2\omega}}{|\mathbf{S}_1^{\omega}||\mathbf{S}_2^{2\omega}|}$$

where $\mathbf{S}_1 > \mathbf{S}_2$. The subscript 1 represents that the electric field is in the $x - y$ plane while subscript 2 represents that it is parallel to Z. When ρ is not zero, the energy of the nonlinear polarization cannot convert to the harmonic beam. In this case, the conversion efficiency cannot be increased with the square of the crystal length. This condition is called critical phase matching. To avoid critical phase matching, noncritical phase matching where $\theta = 90°$ is used. But noncritical phase matching is difficult to achieve in any wavelength. In general, noncritical phase matching is performed by changing the temperature of the crystals. One exception is with polymeric material, since the refractive index of the main-chain poled polymer can be controlled by mechanical drawing and electrical poling. This means that main-chain poled polymers can satisfy the noncritical phase-matching condition at any wavelength by drawing and poling. The detail of birefringence phase matching of poled polymers is described in Section II.F.

E. BIREFRINGENCE PHASE MATCHING IN SINGLE CRYSTALS

Several reports are available on phase matching in single crystals. Therefore, we provide an example for phase matching of organic single crystals. *N*-methoxymethyl-4-nitroaniline (MMNA), which belongs to the point group 222 and is a biaxial crystal, was synthesized by Hosomi et al.[6] Therefore d_{15}, d_{24}, and d_{36} can be utilized for birefringence phase-matchable components. The crystal was grown to $1 \times 2 \times 20$ mm from tetrahydrofuran solution. Phase matching can be performed using the natural grown surface of MMNA. When a 370-kW Nd:YAG laser (1064 nm) was illuminated to the [110] plane of a 1.9-mm-thick MMNA crystal, 12% conversion efficiency was obtained. The θ and φ for phase

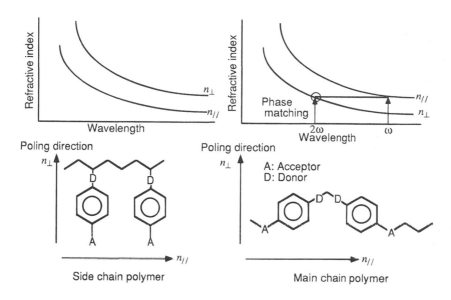

Figure 7 The dispersion of the refractive index of side-chain and main-chain polymers. There is no phase-matching point for side-chain polymers, whereas a phase-matching point can be found in main-chain polymers with appropriate birefringence.

matching are 56.7° and 21°, respectively. The effective nonlinear optical coefficient is 1.65 times larger than that of the KTP crystal.

$$O_2N-\!\!\!\bigcirc\!\!\!-NHCH_2OCH_3 \quad (MMNA)$$

F. BIREFRINGENCE PHASE MATCHING IN POLED POLYMERS

Many materials have been developed for SHG applications such as bulk single crystals and thin films of poled polymers. In single crystals, phase matching can be realized by using birefringence. The refractive index dispersion can be offset by natural birefringence in certain crystals. This method has been generally used in practical SHG single-crystal devices. However, it is difficult to design and synthesize such non-centrosymmetric crystal materials; moreover, the crystal growth, cutting, and polishing are time consuming and it is not easy to get large crystals of good optical quality, especially for organic materials. Another attractive technique of generating noncentrosymmetric material into a medium for NLO application is to align the NLO moieties by an external electric field. The polymeric thin film is heated to increase the mobility of the nonlinear optical active chromophores. When an electric field is applied, due to the electrostatic interaction with the chromophore dipoles, the nonlinear molecules are aligned along the electric field. Such film posseses a noncentrosymmetry, a requirement for the appearance of macroscopic second-order nonlinear optical properties. Birefringence phase matching in poled polymers are discussed below.[4]

The poled polymers belong to ∞mm symmetry and only three of the tensor elements are nonzero: d_{33}, d_{31}, and d_{15}. In accordance with the Kleinman symmetry, there are only two independent nonzero tensor elements for the poled polymer system. Phase-matchable NLO coefficients are d_{31} and d_{15}. Therefore, to realize bulk phase matching in poled polymers, the refractive indices of $n//$ (in-plane refractive index) must be larger than those of $n\perp$ (out-of-plane refractive index; this is the direction of poling) and the refractive indices' dispersion must be offset by positive birefringence. It is assumed that the poled polymer systems cannot be phase matched because $n_\omega \neq n_{2\omega}$. However, some main-chain polymers have been found with large positive birefringence even after corona poling. Figure 7 shows the dispersion of the refractive index of a side-chain polymer and a main-chain polymer. There is no phase-matching point for the side-chain polymers since the refractive index of the poling direction is larger than that of the in-plane. As a result, the d_{31} of the side-chain polymer cannot be utilized. Due to this reason, there is no report on birefringence phase matching in poled polymers. Whereas the refractive index in the poling direction for the main-chain polymer is smaller than that of the in-plane due to the in-plane orientation of benzene rings embedded into polymer chains.

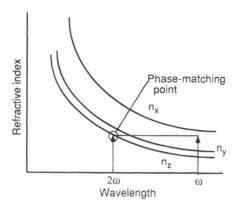

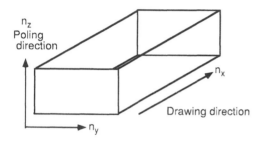

Figure 8 The geometry of the drawn and poled polymer film.

The main-chain polymer, polyurea (PU1) depicted below, was used for the birefringence phase matching experiment.[7]

(Polyurea)

Films were prepared by spin coating or casting from the N-methylpyrrolidone (NMP) solution on a slide glass. The films were dried at 140 °C in a vacuum oven to remove all the solvent. The polymer films can be easily taken off from the substrate with tapes to get freestanding films. The films were characterized by an initial positive birefringence ($\Delta n = n_{\parallel} - n_{\perp}$ and molecular weights of $M_n = 3.6 \times 10^4$ and $M_w = 12.0 \times 10^4$ were determined by gel permeation chromatography with polystyrene as a reference and NMP as a solvent. The required noncentrosymmetric structure of polymer films for SHG was induced by corona discharge poling. The poling temperature was 165 °C, which was selected according to the in-situ SHG experiments. To increase the positive birefringence, a mechanical drawing experiment was performed. The freestanding films, 5 to 15 μm thick, were mounted in a tensile testing machine, so that an area 30 × 6 mm was free between the clamps. They were drawn parallel to the length direction at 170 °C. The speed of cross-head was 0.5 mm/min (a speed low enough to ensure isothermal conditions). After drawing, the temperature of the oven was rapidly decreased while keeping the polymer films under the stress between clamps. The geometry of the drawn and poled polymer film is shown in Figure 8. The refractive indices are defined as n_z, n_y, and n_x for the polarization vector in the film normal direction (poling direction) (3-axis), the transverse direction in the plane of film (2-axis), and parallel to the draw direction (1-axis), respectively. The θ and φ are polar coordinates (referred to as z and x, respectively as shown in Figure 3). The refractive indices and its wavelength dispersion for all samples were measured by the m-line method. By drawing, the polymer chain would lie along the draw direction and hence a wave polarized parallel to this direction has larger refractive indices

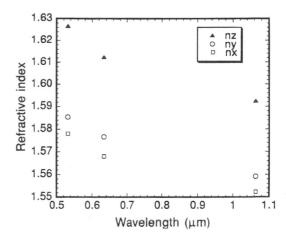

Figure 9 The refractive indices wavelength dispersion of poled polyurea with a draw ratio of 1.4. (From Tao, X. T., et al., *J. Opt. Soc. Am. B,* 12, 1581, 1995. With permission.)

(n_x). On the other hand, a wave polarized perpendicular to the molecular plane will have relatively small refractive indices n_z. The refractive indices along the transverse direction are expected to lie between those of n_x and n_z ($n_x > n_y > n_z$). The refractive indices as a function of wavelength for poled polyurea film with draw ratio of 1.4 are shown in Figure 9.

By corona poling, the dipoles of chromophores will be aligned along the poling direction which is perpendicular to the film plane. If we take this direction as 3-axis, then the refractive indices in this direction are larger than those of in-film plane and the system is isotropic in the 1–2 plane. The appropriate point group of poled films is ∞mm, whose symmetry operations include an infinite-fold rotation about the poling field direction and an infinity of mirror planes including this direction. In the present study, the electric poling field and the uniaxial drawing were perpendicular to each other and the plane, becoming anisotropic. The symmetry characters are apparently different from the reported poled polymers. It was reported that the stretched and poled polyvinylidene fluoride (PVDF) and other polymeric films possess symmetry $mm2$. The drawn and poled polyureas also belong to point group $mm2$. The symmetry operations include a two-fold axis along the film normal direction and two mirror planes perpendiculars to each other and includes the 2-fold axis. For a system with an $mm2$ point group, the nonlinear polarization P_{NL} is given by

$$P_{NL} = \begin{pmatrix} 0 & 0 & 0 & 0 & d_{15} & 0 \\ 0 & 0 & 0 & d_{24} & 0 & 0 \\ d_{31} & d_{32} & d_{33} & 0 & 0 & 0 \end{pmatrix} \begin{pmatrix} E_1^2 \\ E_2^2 \\ E_3^2 \\ 2E_2 E_3 \\ 2E_1 E_3 \\ 2E_1 E_2 \end{pmatrix} \tag{35}$$

where five nonzero d coefficients exist. Under the Kleinman symmetry restriction, $d_{15} = d_{31}$ and $d_{24} = d_{32}$. The second-order NLO properties are finally governed by the three independent nonzero coefficients.

The NLO coefficients were determined by the Maker fringe method. In the experiment, a 6 μm poled polyurea film with draw ratio of 1.4 was used. No fringes were observed but only an increase in the SH signal intensity with an increasing angle of incidence because the film thickness was small compared with the coherence length. The nonlinear optical coefficients of the drawn and poled polyurea films were calculated using a Y-cut quartz (d_{11} = 0.4 pm/V) plate as reference. The d_{33} = 1.4 pm/V, d_{32} = 0.2 pm/V, and d_{31} = 0.6 pm/V were estimated. To obtain high SH conversion efficiency, phase matching is of critical importance because in SHG, a fundamental wave with frequency ω_1 and wavelength λ interacts with the second-order nonlinear susceptibility of a material to produce a polarization wave at the second harmonic frequency $\omega_2 = 2\omega_1$. Since the polarization wave is forced by the fundamental wave, it travels with the same velocity, determined by $n(\omega)$, the refractive index at the fundamental wavelength. The polarization wave radiates a free second harmonic wave which travels at a velocity

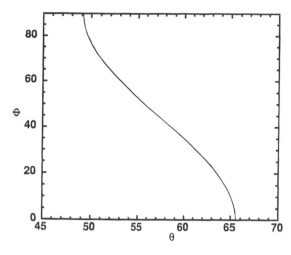

Figure 10 The calculated type I phase-matching locus of polyurea with a draw ratio of 1.4. (From Tao, X. T., et al., *J. Opt. Soc. Am. B*, 12, 1581, 1995. With permission.)

determined by $n(2\omega)$, the refractive index at the harmonic wavelength. In normal dispersion materials, $n(2\omega) > n(\omega)$, so that the fundamental and second harmonic waves travel at different phase velocities. The sign of power flow from one wave to the other is determined by the relative phase between the waves. The continuous phase slip between these waves causes a repetitive growth and decay of the second harmonic intensity along the length of the interaction. This can be clearly seen from the SHG conversion efficiency equation for bulk phase matching. To get high conversion efficiency, Δk should be zero. In this case, the SHG conversion efficiency is proportional to L^2. This condition is called phase matching and can be realized by using birefringence in an anisotropic material.

Analytically, the index surface in a biaxial crystal for any given frequency is given by the two solutions of Equations 27 and 28. After drawing, polyurea has three principal refractive indices and behaves as a biaxial crystal. Using the refractive indices of polyurea with a draw ratio of 1.4, the type I phase-matching locus was calculated. The results are shown in Figure 10. It belongs to type VIII phase matching as classified by Hobden et al.[8] According to the calculated phase-matching direction, a film with $\varphi = 90°$ and $\theta = 49.2°$ for the phase-matching experiment was cut. The source used in this study was a Quantel Q-switched Nd:YAG laser emitting 1064 nm. The energy of the emerging pulses was less than 0.2 mJ and their duration was 10 ns in a Q-switched regime at a repetition rate of 10 Hz. The fundamental beam is *s*-polarized and the second harmonic beam is *p*-polarized. The beams were guided into and coupled out of the film by two prisms spaced 1 mm apart. The sample holder can be rotated by a stepping motor. The harmonic beam emerging from the sample is detected after filtering by a photomultiplier and subsequently displayed. The results of angle turning type I phase-matched SHG at room temperature are shown in Figure 11, which clearly shows that phase-

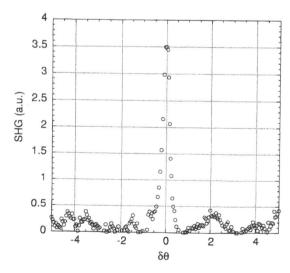

Figure 11 The angle turning type I phase-matched SHG at room temperature of polyurea. (From Tao, X. T., et al., *J. Opt. Soc. Am. B*, 12, 1581, 1995. With permission.)

matched SHG was obtained. The calculated phase-matching direction at the y-z plane was confirmed by the experiment, indicating that the refractive indices measurements have good accuracy. This phase matching is critical. In the case of main-chain-type polymers, the noncritical phase matching is also possible by controlling the drawing ratio.

III. PHASE MATCHING IN WAVEGUIDES

A. Modal Dispersion Phase Matching

1. Three-Layer Waveguides

When the refractive index of the guiding layer is higher than that of both substrate and overlayer as shown in Figure 12, the light can propagate in a medium with high power density. This phenomenon is known as waveguide. The waveguide geometry has advantages for nonlinear optical applications, because of the high power density that can be achieved over a long propagation length and utilization of diagonal tensor components.[9] Diagonal tensor components cannot be utilized due to the normal dispersion of the refractive index. By carefully adjusting the waveguide thickness, the dispersion of the medium allows different modes at the two wavelengths to have the same effective refractive index n_{eff}, which is called modal dispersion phase matching (PM).[10] In the waveguide geometry even diagonal tensor components can be utilized. The effective refractive index of waveguide can be determined.[11] The propagation of light in the waveguide can be explained by a dispersion equation. In the case of a transverse electric field (TE), the dispersion is given by:

$$k_x W = (m + 1)\pi - \tan^{-1}(k_x/\gamma_s) - \tan^{-1}(k_x/\gamma_c) \tag{36}$$

In the case of a transverse magnetic field (TM), the dispersion equation is given by Equation 37. The direction of the electric field is shown in Figure 12.

$$k_x W = (m + 1)\pi - \tan^{-1}[(n_0/n)^2((k_x/\gamma_s))] - \tan^{-1}[(n_2/n)^2((k_x/\gamma_c))] \tag{37}$$

where the propagation constants, γ_c, γ_s, and k_x are given by Equations 38, 39, and 40, respectively.

$$\gamma_c = k_0(n_{eff}^2 - n_2^2)^{1/2} \tag{38}$$

$$\gamma_s = k_0(n_{eff}^2 - n_0^2)^{1/2} \tag{39}$$

$$k_x = k_0(n^2 - n_{eff}^2)^{1/2} \tag{40}$$

where n_2 and n_0 are the refractive index of the overlayer and the substrate of the waveguide, respectively, and n is the refractive index of the material, W represents the waveguide dimension. From Equations 36 and 37, one can find that a different mode gives a different effective refractive index. Therefore, by changing the waveguide thickness, we can find the phase-matching point in a waveguide geometry.

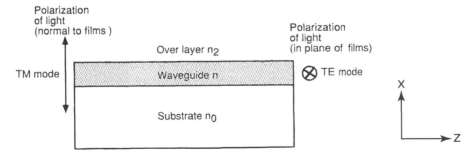

Figure 12 The waveguide geometry.

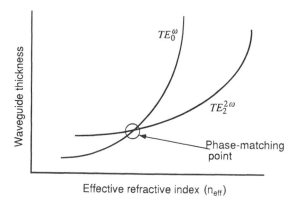

Figure 13 The mode dispersion curves of the waveguide. The polarization of both the fundamental and harmonic waves is in the TE mode.

Figure 13 shows the mode dispersion curves of the waveguide. Each crossing point of the fundamental and SH curves corresponds to the phase-matching conditions. The advantage of the waveguide is the high density of the fundamental beam compared with birefringence phase matching.[9] The second harmonic power can be expressed as follows:

$$P^{2\omega} = \frac{2\omega^2}{\varepsilon_0 c^3} \frac{d^2}{n_{\text{eff}}^{2\omega}(n_{\text{eff}}^\omega)^2} \frac{(P^\omega)^2}{WD} S_{ij}^2 L^2 \frac{\sin^2(\Delta\beta L/2)}{(\Delta\beta L/2)^2} \tag{41}$$

where L is the propagation length, d is the nonlinear optical coefficient, and n_{eff} is the effective refractive index. c represents the speed of light, W is the beam width, D is the effective waveguide thickness, $\Delta\beta = \beta^{2\omega} - 2\beta^\omega k_{2\omega} n_{\text{eff}}^{2\omega} - 2k_\omega n_{\text{eff}}^\omega$ is the phase mismatch, ω is the fundamental frequency, and S_{ij} is the overlap integral defined as

$$S_{ij} = \frac{\displaystyle\int_0^t d_{\text{eff}}(F_\omega^{(j)})^2 (F_{2\omega}^{(i)}) dx}{\displaystyle\int_{-\infty}^\infty (F_\omega^{(j)})^2 dx \sqrt{\int_{-\infty}^\infty (F_{2\omega}^{(i)}) dx}} \tag{42}$$

where d_{eff} represents the normalized nonlinear optical coefficient, F is the normalized electric field, and t is the thickness of the NLO waveguide.

It is difficult to satisfy the phase-matching condition using modal dispersion of the waveguide, since the effective refractive index of the waveguide critically depends on the waveguide dimensions. Figure 14 shows the fabrication tolerance of an α-quartz waveguide. Even using α-quartz with relatively small dispersion of the refractive index, fabrication tolerance is less than 5 nm. In the case of organic material with a large diagonal tensor component, this fabrication tolerance becomes more critical, less than a few nanometers.

Therefore, a tapered slab waveguide is used to adjust the film thickness in both the single crystal and polymer films. Sasaki et al.[12] demonstrated the phase matching using a tapered MNA slab waveguide.

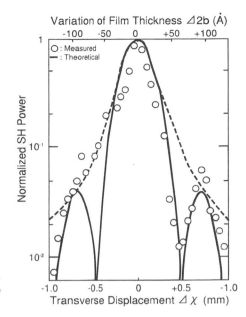

Figure 14 The fabrication tolerance of slab waveguide prepared from α-quartz. (From Suematsu, Y., et al., *IEEE J. Quantum Electron.*, QE-10, 222, 1974. With permission.)

The crystal was prepared by vapor phase growth using a thermocontrolled oil bath with an accuracy of ± 01°C. Crystals were grown at 120°C for 10 days in a sealed glass tube. A fundamental beam (1064 nm) was coupled into the waveguide by a prism. The phase-match condition between the fundamental wave and the second harmonic wave is realized by coincidence of both propagation constants with transitional adjustment of the thickness of the waveguide. The conversion efficiency was very small due to the small overlap integral and fluctuation of the film thickness.

The coupling of the fundamental beam into the waveguide using a prism is very difficult since the prism damages the crystal under strong pressure. A grating coupler was used for SHG experiments of MNA single crystals by Sugihara and Sasaki.[13] The advantages of using a grating coupler are as follows:

1. It is possible to avoid the mechanical damage to single crystals.
2. Coupling is easier, and its efficiency is higher.

The input and output coupling efficiency of the grating and propagation loss were measured to be 11.4%, 43.2%, and 11.3 dB/cm respectively. First, a three-layer waveguide was used but light could not propagate as far as 5 mm. Therefore, a four-layer waveguide was used in this experiment. Figure 15 shows the schematic structures of the sample waveguide. The four-layer structure consists of a Pyrex substrate with two gratings, a Corning 7059 glass waveguide, a tapered MNA single crystal, and a Pyrex substrate without gratings. The typical thickness of the Corning 7059 was 0.88 μm. The taper gradient of MNA was 0.05 μm/mm. The phase-matching condition is satisfied between the first fundamental mode and the third, second harmonic mode. The SHG conversion efficiency was estimated to be $7.04 \times 10^{-2}\%$ overall and 1.7% in a waveguide.

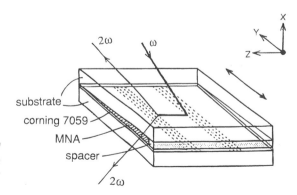

Figure 15 The schematic structures of the sample waveguide with a grating coupler. (From Sugihara, O. and Sasaki, K., *J. Opt. Soc. Am. B*, 9, 104, 1992. With permission.)

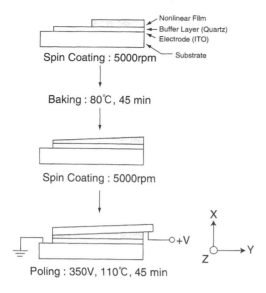

Spin Coating : 5000rpm

Baking : 80°C, 45 min

Spin Coating : 5000rpm

Poling : 350V, 110°C, 45 min

Figure 16 Sample preparation process of a tapered polymer waveguide. (From Sugihara, H., et al., *Appl. Opt.*, 30, 2957, 1991. With permission.)

The modal dispersion phase matching of poled polymer films was demonstrated by Sugihara et al.[14] In order to control the waveguide thickness, a two-step spin coating was carried out to prepare the tapered waveguide. Figure 16 shows the sample preparation procedure. Poly(methylmethacrylate) PMMA and MNA (20 wt% ratio) were dissolved in a chloroform solvent. The indium tin oxide glass with sputtered fused silica buffer layer was used as a substrate. First, the half-part of the substrate was masked, and the solution was spin coated onto the other parts. After drying the substrate at 80°C for 45 min, the mask was removed and the solution was spin coated again on the whole part of the substrate. The films were poled by a parallel plate electrode with 350 V DC for 45 min at 110°C. A Nd:YAG laser was used as a laser source. The phase matching was fulfilled by translating the substrate to the laser beam. The obtained conversion efficiency was estimated to be 7.4×10^{-3} %. In order to improve the conversion efficiency, optical field overlap must be large between the propagating waveguide modes. Four-layer waveguides and five-layer waveguides were prepared to increase the overlap integral.

2. Four-Layer Waveguides

Florsheimer et al.[15] demonstrated the four-layer waveguide prepared by the Langmuir-Blodgett technique. The sample used in this experiment was 2-docosylamino-5-nitropyridine (DCANP).

$$NH(CH_2)_{21}CH_3$$

(DCANP)

The DCANP was spread on the ultrapure water from the hexane solution. The DCANP film was deposited at 18.0 mN/m and dipping velocity was 1.0 cm/min. The film was deposited on TiO_2/SiO_2 waveguides which were hydrophobized using freshly distilled octadecyltrichlorosilane. In the case of the three-layer waveguide, the overlap integral is small unless all the interacting modes have the same mode number. Ito and Inaba[16] introduced a method to optimize the overlap integral using the four-layer waveguide with both linear and nonlinear materials. The cancellation of positive and negative parts of the overlap integral is avoided since there is only contribution from the nonlinear optical region of the waveguide as shown in Figure 17. The optical loss of the waveguide was 5.5 dB/cm.[17] The conversion efficiency of this waveguide was 0.6% when a 30 W fundamental beam was illuminated. The conversion efficiencies of three-layer and four-layer waveguides are listed in Table 2. The three-layer waveguide does not exhibit any SHG due to the small overlap integral. The four-layer waveguide is a useful technique to increase the overlap integral.

The similar kind of four-layer waveguide structure was demonstrated by Clays et al.[18] using a stilbene polymer (Poly-co-{[4-carbo-{2-ethyleneoxy-(4-oxy-4′-methyl-sulfonylstilbene)}styrene] [4-carbo-{2-hexyloxy-(4-oxy-4′-methylsulfonylstilbene)}styrene]} and a phenyl polymer (Poly[4-carbo-{4-oxy-1-(4-nonafluorobutylsulfonylphenyl)piperidine} styrene].

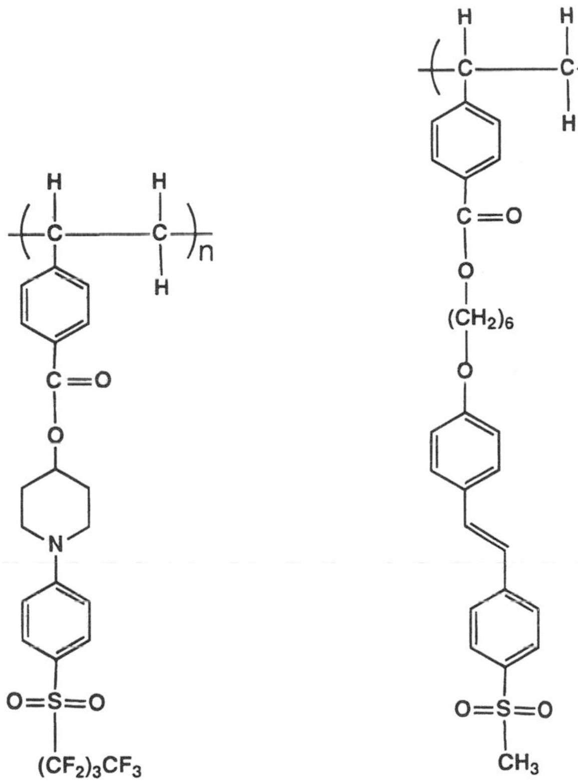

(Phenyl polymer)

(Stilbene polymer)

The waveguide consists of a glass substrate, a nonlinear-optical polymer with a passive polymer such as polystyrene and an air cover. The refractive indices of the polymer were measured by the m-line method (Table 3). The electro-optic coefficients are 0.69 and 0.75 pm/V for the stilbene and phenyl polymers, respectively. The optical loss of the four-layer waveguide was 19.8 dB/cm at 425 nm and 0.2 dB/cm at 850 nm. The phase matching was carried out by using different kinds of a dye-laser pumped by a Nd:YAG laser. The obtained conversion efficiency was less than 0.001%. Such a low conversion efficiency may be related to the fluctuation of the film thickness. The nonlinear optical susceptibility inversion structure (NIST) across the film structure was prepared using polymeric films. Penner et al.[19] demonstrated the blue phase-matched SHG in a guided waveguide prepared by alternate deposition of Langmuir-Blodgett films. The overlap integral was drastically increased due to the inversion of nonlinear polarization in the four-layer waveguide. The

Table 2 Comparison of Frequency Doubling Conversion Efficiencies in DCANP Langmuir-Blodgett Waveguides

Layers	Configuration	λ (nm)	$P\omega$ (W)	L	ηmeas (%/W)
Three-layer	Mode-conversion	926	30	2	—
Four-layer	Mode-conversion	926	30	2	2×10^{-2}

Adapted from Kupfer, M., et al. *SPIE Proc.*, 1775, 340, 1992. With permission.

NIST can be prepared by rotating the substrate 180° during the deposition. The chemical structure of the NLO active polymer and NLO passive poly(tert-butyl methacrylate) used in this experiment are as follows:

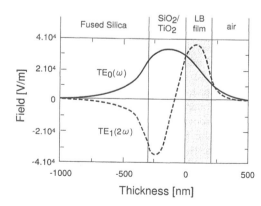

The molar ratio of n and m for a NLO active polymer was 1:3. The optical loss of this waveguide was 1.5 and 2.0 dB/cm at 457.9 nm for both TE and TM modes, respectively, which is superior to that of any previously reported LB films. Modal dispersion phase matching was performed using a continuous wave Ar^+ ion-pumped Ti-sapphire laser tunable over 800–920 nm. Phase matching was confirmed for both $TE_0(\omega) \rightarrow TM_1(2\omega)$ and $TM_0(\omega) \rightarrow TM_1(2\omega)$ polarizations. An SHG intensity of 1 μW at 410 nm was observed for 50 mW of fundamental power at 819 nm. This yielded a power-corrected conversion

Figure 17 Electric field distributions of fundamental TE_0 mode (λ = 926 nm) and a second harmonic TE_1 mode in a four-layer waveguide. The two modes are phase matchable. Only the shaded area contributes to the overlap integral. (From Kupfer, M. et al., *SPIE Proc.*, 1775, 340, 1992. With permission.)

Table 3 Refractive Indices for Polystyene and NLO Polymers Three-layered Waveguides

Wavelength (nm)	Polystyrene	Stilbene Polymer	Phenyl Polymer
457.9	1.6046	1.6789	1.5756
488.0	1.5942	1.6553	1.5599
632.8	1.5820	1.6306	1.5434
830.0	1.5720	1.6128	1.5318

Adapted from Clays, K., et al. *J. Opt. Soc. Am. B,* **11,** 655, 1994. With permission.

efficiency of 0.04%/W, which is a high value, considering that there is no lateral confinement of the light in the waveguide.

3. Five-Layer Waveguides

Rikken et al.[20] reported that five-layer waveguides improve the overlap integral. By introducing a very thin and high refractive index satellite layer, the phase-match condition can be fulfilled with a large overlap integral between the two waves, resulting in efficient doubling. The Si_3N_4/SiO_2 materials were used as the satellite and substrate layers. These layers can be deposited with great accuracy by standard silicon processing facilities onto silicon wafers. A 3-μm SiO_2 substrate layer was used as a buffer layer prepared by thermal oxidation of a silicon wafer. The structure and field distribution of the NLO waveguide is shown in Figure 18. The 75 nm-thick Si_3N_4 was prepared by the chemical vapor deposition (CVD) technique. The methylmethacrylate copolymer having 4-alkoxy-4′-alkylsulfone stilbene side chains as NLO moieties (MSMA) was deposited by spin coating from chlorobenzene with a typical thickness of 0.7 μm.

(MSMA)

Poling was carried out by using a vapor-deposited gold electrode on top of the cladding polymer and the silicon wafer as a ground electrode at 100°C. Figure 19 shows the NLO active layer thickness required to obtain $TM_0^\omega \rightarrow TM_1^{2\omega}$ at a 860 nm wavelength, as a function of the Si_3N_4 thickness. Also the overlap integral S_{01} is shown in Figure 19. The maximum conversion efficiency obtained by using modal dispersion phase matching was 0.4%/W, measured at the channel output. By improving the electrical properties of the polymer and the oxide/nitride layers as well as new NLO polymers with large nonlinearities, this efficiency could be increased up to 50%/W. The increase of the overlap integral and the fabrication tolerance for mode conversion for the five-layer waveguide were proposed and

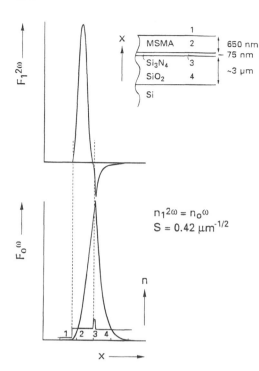

Figure 18 Refractive index profile and normalized electric field profiles of the TM_0^ω and $TM_1^{2\omega}$ modes of a four-layer slab waveguide with a high index satellite layer. The inset shows an implementation of such a waveguide structure. (From Rikken, G. L. J. A., *Appl. Phys. Lett.*, 62, 2483, 1993. With permission.)

demonstrated by Watanabe et al.[21] Theoretical results show a drastic increase not only in dimension tolerance but also in the overlap integral. Let us introduce the theoretical description to calculate the phase-matching thickness for a five-layer waveguide considering the boundary conditions of the electromagnetic wave. In general, the waveguides could usually support several modes m with different propagation constants. The phase-matching condition is fulfilled if the fundamental propagates in a lower mode than the second harmonic wave. The overlap integral of the five-layer structure can be modified from Equation 42, subscripts i and j represent the different modes.

$$S_{ij} = \frac{\int_0^1 \frac{d_{eff}}{|d_{eff}|}(F_\omega^{(j)})^2(F_{2\omega}^{(i)})dx + \int_2^3 \frac{d_{eff}}{|d_{eff}|}(F_\omega^{(j)})^2(F_{2\omega}^{(i)})dx}{\int_{-\infty}^\infty (F_\omega^{(j)})^2 dx \sqrt{\int_{-\infty}^\infty (F_{2\omega}^{(i)})dx}} \tag{43}$$

where T_1, T_2, and T_3 are absolute thickness (see Figure 20). $F_\omega^{(m)}$ is the normalized field distribution for the mth mode at frequency ω. The integral runs only over the range of the nonlinear optical layers.

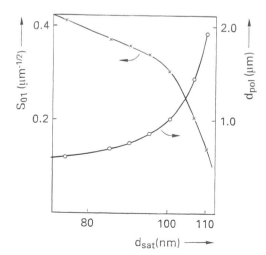

Figure 19 Polymer thickness required for $TM_0^\omega \rightarrow TM_1^{2\omega}$ phase matching at 860 nm wavelength as a function of the Si_3N_4 satellite layer thickness with overlap integral. (From Rikken, G. L. J. A., *Appl. Phys. Lett.*, 62, 2483, 1993. With permission.)

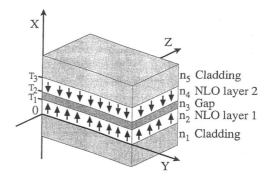

Figure 20 The geometry of a five-layer waveguide. (From Watanabe, T., et al., *Opt. Commun.*, 123, 76, 1996. With permission.)

The S should have a value which significantly differs from zero to achieve a high second harmonic power. For simple three-layer structures, the overlap integral value is rather small due to the nearly orthogonal solutions of Maxwell's equations. This structure avoids the cancellation of the overlap integral between the two regions divided by the node of the second harmonic field. Five-layer waveguide structures are known, but experimental data are not available because of the delicate preparation conditions for inorganic materials. When $\chi^{(2)}$ is used for the NLO layer introducing the inversion of nonlinear susceptibilities, the overlap integral is drastically increased.

A complete description of the guiding modes of a five-layer slab waveguide requires that the Maxwell's wave equations be solved:

$$\nabla \times \mathbf{H} = \varepsilon_o n^2 \frac{\partial \mathbf{E}}{\partial t} \tag{44}$$

$$\nabla \times \mathbf{E} = -\mu_0 \frac{\partial \mathbf{H}}{\partial t} \tag{45}$$

The light travels only in the z direction. The xy plane is assumed to be infinite. For the two-dimensional analysis, $\partial/\partial y = 0$, the field components for the TE polarization consist of E_Y, H_X, and H_Z, whereas for the TM polarization they are H_Y, E_X, and E_Z only.

Some straightforward calculations give the one-dimensional wave equation

$$\frac{\partial^2 E_y}{\partial x^2} + k_0^2(n^2 - n_{\text{eff}}^2)E_y = 0 \tag{46}$$

for the y components of the electric field in the case of TE polarization. In the case of TM-guided modes, the magnetic field component of the wave equation is

$$\frac{\partial^2 H_y}{\partial x^2} + k_0^2(n^2 - n_{\text{eff}}^2)H_y = 0 \tag{47}$$

$$E_z = \left(\frac{-i}{n^2\omega\varepsilon_0}\right) \frac{\partial H_y}{\partial x} \tag{48}$$

The propagation constant for the incident beam is given by

$$k_o = \frac{2\pi}{\lambda} = \frac{\omega}{\sqrt{\varepsilon_0\mu_0}} \tag{49}$$

and $n_{\text{eff}} = n \cos\theta$, describing the propagation of the wave inside the waveguide as a function of the internal reflection angle θ. The solution of the wave equations (Equations 46 and 47) requires the boundary conditions that the fields of the guided modes must vanish at $x = \pm \infty$ and that the tangential H and E fields must be continuous at the discontinuity between the layers.

Throughout these equations, the subscripts 1 to 5 are used as shown for the waveguide in Figure 20. We outline the solutions for the TE and TM polarization with the guiding layers 2 and 4 (n_2, $n_4 > n_1$, n_3, n_5). Now defining an abbreviation for the effective wave vectors in each layer,

$$k_i = \begin{cases} k_0\sqrt{n_i^2 - n_{\text{eff}}^2}, & \text{if} \quad n_i^2 > n_{\text{eff}}^2 \\ k_0\sqrt{n_{\text{eff}}^2 - n_i^2}, & \text{if} \quad n_{\text{eff}}^2 > n_i^2, \end{cases} \quad i = 1, 2, 3, 4, 5 \tag{50}$$

For convenience, we introduce q_{ij} as a polarization dependent coefficient which is

$$q_{ij} = \begin{cases} 1 & \text{for TE polarization} \\ n_j^2/n_i^2 & \text{for TM polarization} \end{cases} \tag{51}$$

The calculations carried out from equations 53–76 give the second NLO layer thickness

$$T_3 - T_2 = \frac{1}{k_4} \arctan\left(\frac{q_{45}\dfrac{k_4}{k_5} B_4 + A_4}{q_{45}\dfrac{k_4}{k_5} A_4 - B_4} \right) \tag{52}$$

which is now only a function of the effective refractive index n_{eff}, the thickness of the first guiding layer T_1, and the gap width $T_2 - T_1$ as well as the amplitude A_1 (for definitions of amplitudes A_i and B_i, see below). For our calculations of the overlap integral value, the value of A_i has no relevance because of using normalized fields in Equation 43. In general, it is difficult to obtain large dimension tolerance, together with high overlap integral for modal dispersion phase matching. By choosing the proper thickness T_1, T_2, and T_3, the mode dispersion curves at different wavelengths can cross in the wide range of the waveguide dimensions over 60 nm. The calculation is shown below from Equations 53 through 76. In the case of TM modes, the amplitude of the magnetic field component for all layers can be given by solving Equations 46 and 47.

$$H_y = A_1 \exp(k_1 x) \qquad\qquad (x \leq 0) \tag{53}$$

$$H_y = A_2 \cos(k_2 x) + B_2 \sin(k_2 x) \qquad\qquad (0 < x \leq T_1) \tag{54}$$

$$H_y = A_3 \exp[k_3(x - T_1)] + B_3 \exp[-k_3(x - T_1)] \qquad (T_1 \leq x \leq T_2) \tag{55}$$

$$H_y = A_4 \cos[k_4(x - T_2)] + B_4 \sin[k_4(x - T_2)] \qquad (T_2 \leq x \leq T_3) \tag{56}$$

$$H_y = A_5 \exp[-k_5(x - T_3)] \qquad\qquad (x \geq T_3) \tag{57}$$

Calculating from these equations, the E_z components give

$$E_z = \frac{-i}{n_1^2 \omega \varepsilon_0} k_1 A_1 \exp(k_1 x), \qquad\qquad (x \leq 0) \tag{58}$$

$$E_z = \frac{-i}{n_2^2 \omega \varepsilon_0} k_2 [B_2 \cos(k_2 x) - A_2 \sin(k_2 x)], \qquad\qquad (0 \leq x \leq T_1) \tag{59}$$

$$E_z = \frac{-i}{n_3^2 \omega \varepsilon_0} k_3 \{A_3 \exp[k_3(x - T_1)] - B_3 \exp[-k_3(x - T_1)]\}, \qquad (T_1 \leq x \leq T_2) \tag{60}$$

$$E_z = \frac{-i}{n_4^2 \omega \varepsilon_0} k_4 \{B_4 \cos[k_4(x - T_2)] - A_4 \sin(k_4[x - T_2])]\}, \qquad (T_2 \leq x \leq T_3) \quad (61)$$

$$E_z = \frac{-i}{n_5^2 \omega \varepsilon_0} k_5 A_5 \exp[-k_5(x - T_3)]. \qquad (x \geq T_3) \quad (62)$$

where k_i is represented as follows:

$$k_1 = k_0 \sqrt{n_{\text{eff}}^2 - n_1^2} \qquad (63)$$

$$k_2 = k_0 \sqrt{n_2^2 - n_{\text{eff}}^2} \qquad (64)$$

$$k_3 = k_0 \sqrt{n_{\text{eff}}^2 - n_3^2} \qquad (65)$$

$$k_4 = k_0 \sqrt{n_4^2 - n_{\text{eff}}^2} \qquad (66)$$

$$k_5 = k_0 \sqrt{n_{\text{eff}}^2 - n_5^2} \qquad (67)$$

Solving $H_{z,1} = H_{z,2}$ and $E_{z,1} = E_{z,2}$ simultaneously for $x = 0$:

$$A_2 = A_1 \qquad (68)$$

$$B_2 = q_{12} \frac{k_1}{k_2} A_1 \qquad (69)$$

for $x = T_1$:

$$A_3 = \frac{1}{2} \left\{ \left[A_2 + q_{23} \frac{k_2}{k_3} B_2 \right] \cos(k_2 T_1) + \left[B_2 - q_{23} \frac{k_2}{k_3} A_2 \right] \sin(k_2 T_1) \right\} \qquad (70)$$

$$B_3 = \frac{1}{2} \left\{ \left[A_2 - q_{23} \frac{k_2}{k_3} B_2 \right] \cos(k_2 T_1) + \left[B_2 + q_{23} \frac{k_2}{k_3} A_2 \right] \sin(k_2 T_1) \right\} \qquad (71)$$

for $x = T_2$:

$$A_4 = A_3 \exp[k_3(T_2 - T_1)] + B_3 \exp[-k_3(T_2 - T_1)] \qquad (72)$$

$$B_4 = q_{34} \frac{k_3}{k_4} \{A_3 \exp[k_3(T_2 - T_1)] - B_3 \exp[-k_3(T_2 - T_1)]\} \qquad (73)$$

for $x = T_3$:

$$A_5 = A_4 \cos[k_4(T_3 - T_2)] + B_4 \sin[k_4(T_3 - T_2)] \qquad (74)$$

$$B_5 = q_{45} \frac{k_4}{k_5} \{-B_4 \cos[k_4(T_3 - T_2)] + A_4 \sin[k_4(T_3 - T_2)]\} \qquad (75)$$

374

Elimination of A_5 in the last two equations gives

$$T_3 = \frac{1}{k_4} \arctan\left(\frac{q_{45}\dfrac{k_4}{k_5}B_4 + A_4}{q_{45}\dfrac{k_4}{k_5}A_4 - B_4}\right) + T_2 \tag{76}$$

In order to confirm this theory, Watanabe et al.[22] reported five-layer organic waveguides using Langmuir-Blodgett techniques. Langmuir-Blodgett (LB) films are good candidate to validate the waveguide theory of the five-layer NLO waveguide because film thickness can be precisely controlled by numbers of dipping cycle within a few nanometers. Materials used in this experiment are shown below.

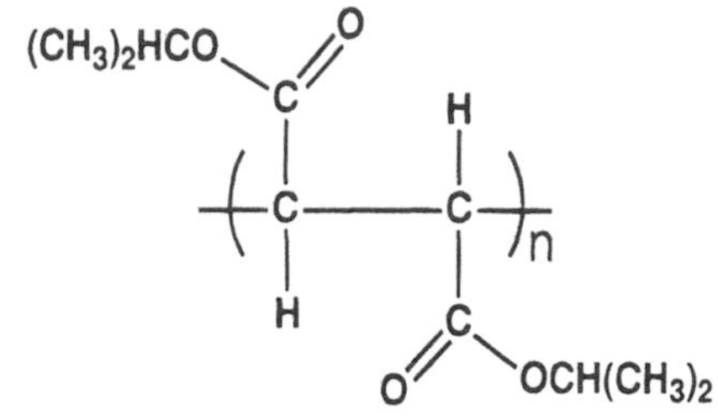

(PDiPF)

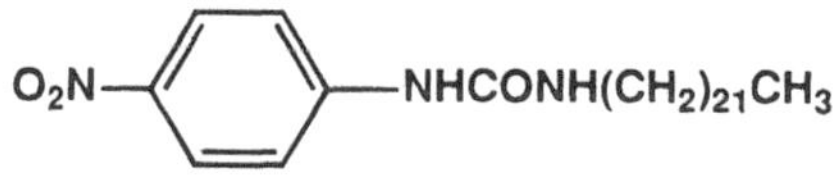

(DONPU)

The three-layer waveguide and four-layer waveguide with structural inversion were also fabricated for a comparison with the five-layer waveguide. The five-layer Langmuir-Blodgett films were prepared from a combination of low molecular weight amphiphilic NLO dye and SHG inactive polymer. n-Docosyl-4-nitrophenylurea (DONPU) was deposited as Y-type LB films. DONPU can be fabricated without introducing centrosymmety in the films, similar to 2-docosylamino-5-nitropyridine (DCANP).[23] The short cutoff wavelength of the nonlinear chromophore (400 nm) ensures low absorption losses. DONPU films were deposited on fused silica substrates with 1,1,1,3,3,3-hexamethyldisilazane prior to the transfer process. The noncentrosymmetry was introduced by a herringbone-type structure which shows strong in-plane anisotropy.[24] The SHG efficiency of this film was dominated by the contribution from in-plane components such as d_{33} (3 is dipping direction). Other tensor components were almost negligible. The DONPU monolayer was deposited as the first NLO layer at 20 mN/m. The Y-type multilayer films show a quadratic increase in SHG along the dipping direction as the number of layers increases. SHG inactive poly(diisopropyl-fumalate) (PDiPF) film was deposited as Y-type LB film onto the DONPU layer at 20 mN/m. This second layer increases the thickness tolerance for phase matching. PDiPF also can form a good monolayer and can be easily deposited to multilayers. The second NLO layer was deposited on PDiPF films after reversing the dipping direction to introduce the inversion of the nonlinear susceptibilities across the film. The refractive indices of DONPU films were measured by the prism coupling method. The optical loss of DONPU LB films was measured by monitoring the light scattered out of the waveguide and found to be 12 dB/cm. This optical loss is relatively smaller than that of DCANP and not so large compared with other low molecular weight LB film.[25] The accuracy of measurements was limited by the scattering loss of DONPU LB films resulting in some uncertainty in calculating the exact phase-matching thickness. The refractive indices of PDiPF were measured using an ellipsometry technique because the refractive index of PDiPF was smaller than that of fused silica substrates.

Preliminary nonlinear optical characterization of the LB films was carried out by the Maker fringe experiments using multilayer films. The d_{33} of DONPU was found to be 0.8 pm/V, whereas PDiPF film does not exhibit any SHG activity. Although DONPU possesses a small $\chi^{(2)}$ value, it can be deposited up to several hundred layers without deforming the structures. Therefore, DONPU is a good candidate to confirm the large dimension tolerance of the five-layer waveguide. In this section, the full width of half-maximum (FWHM) is defined as a fabrication tolerance of waveguide dimensions which one can obtain at 50% of maximum SHG intensity. In general, the fabrication tolerance of the modal dispersion PM is very small, such as a few nanometers. A five-layer waveguide with inversion susceptibil-

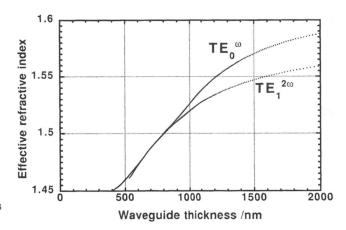

Figure 21 The mode dispersion curves of a five-layer waveguide.

ities is mainly studied to increase the overlap integral by theoretical analysis due to the difficulty of fabrication.[26,27] The overlap integral S_{01} of the five-layer waveguide was about 74% of the S_{01} of the four-layer waveguide. This value is about two orders of magnitude larger than the S_{02} of the three-layer waveguide, since nonlinear polarization in the five-layer waveguide does not cancel out due to the inversion of nonlinear susceptibilities across the film. The small reduction of the overlap integral is due to no nonlinear polarization in the SHG inert layer. To increase the overlap integral, LB films with small refractive indices are required since the thickness of the SHG inert layer can be reduced. The refractive index of PDiPF is 1.4306 and 1.4552 at 1064 and 532 nm, respectively, suitable for this purpose.

The FWHM of the four-layer waveguide is reported as 6 nm for a phase-matched interaction length of 1.5 mm.[19] This small allowance ascribed to the different slopes of mode dispersion curves at fundamental and second harmonic wavelengths. By selecting the proper combination of the SHG inert layer and SHG active layers, the mode dispersion curves were compensated to have the same slopes. The refractive indices used in the computer simulation were $n_s(\omega) = 1.44963$, $n_f(\omega) = 1.5774$ at the fundamental wavelength and $n_s(2\omega) = 1.46071$, $n_f(2\omega) = 1.6109$ at the harmonic wavelength, where n_s and n_f represent the refractive index of substrate and DONPU, respectively. There are three intersections of the mode dispersion curves, and a very good thickness tolerance for phase matching can be expected. One can find a range (Figure 21) where both dispersion curves run nearly with the same slope, making the small phase mismatch. At the crossing points at higher order where Δk becomes also zero, the overlap integral decreases. Computer simulations show that FWHM is more than 85 nm for a phase-matched interaction length of 2 mm. This large FWHM for a five-layer waveguide is at least 14 times larger than that of a three- or four-layer waveguide with inversion susceptibilities.

To demonstrate phase matching, the second harmonic (2ω) output from the waveguide was monitored as a function of the waveguide thickness. The experimental SH intensity in Figure 22 was normalized to fit the theoretical calculation. In the case of the five-layer waveguide, the number of first DONPU layers and PDiPF layers was fixed and second DONPU layers were changed. The SHG peak was

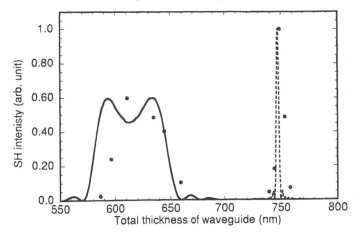

Figure 22 The SH intensity of five-layer and four-layer waveguides as a function of thickness.

observed on 610 nm-thick film for $TE_0(\omega)$ and $TE_1(2\omega)$ conversion. In the case of the four-layer waveguide, first DONPU layers were fixed and second DONPU layers were changed. The SHG peak in the four-layer waveguide was observed on 715 nm-thick film. The FWHM of the five-layer waveguide was found to be 56 nm from this experiment, which is 8.6 times larger than that of the four-layer waveguide with inversion susceptibilities. Only Langmuir-Blodgett film can sustain absolute thickness within 1 nm. This is the first example to demonstrate the precise effects of waveguide dimension for modal dispersion PM on conversion efficiency. The conversion efficiency of the four-layer waveguide drastically decreases when the difference between phase-matched thickness and actual waveguide thickness is only one bilayer. The small difference between theory and experimental data may arise from the inaccuracy of the refractive index of PDiPF LB films, since we assumed that PDiPF LB films are isotropic for the calculation of refractive indices. The conversion efficiency of the five-layer waveguide can be improved by introducing a fluorinated polydialkyl-fumalate as an SHG inert layer for increasing the overlap integral.

A three-layer waveguide was fabricated to confirm the inversion of nonlinear susceptibilities. No detectable SHG was observed from the three-layer waveguide because of the small overlap integral. This means that the sign of nonlinear susceptibilities of four- and five-layer waveguides is actually reversed. SHG intensity at 532 nm of 13 mW was measured by a photomultiplier tube calibrated by a power meter. The input energy was 82 W at 1064 nm. Conversion efficiency of $\eta = 0.016\%$ was obtained. With the corresponding propagation length of $L = 2$ mm and a beam diameter of $W = 0.4$ mm, a conversion efficiency of $\eta = 0.04\%$ was calculated. The calculated and measured values agree reasonably well if one considers the inaccuracy of the coupling efficiency of the fundamental beam. The measured conversion efficiency was normalized and compared with other experiments. The efficiency of modal conversion for DCANP $(2 \times 10^{-2}\%/W)$[15] is higher than DONPU data $(1.9 \times 10^{-4}\%/W)$. The relatively low efficiency of device is due to the small d_{33} of DONPU. By replacing the DONPU layers by large $\chi^{(2)}$ film, such as polymeric LB films developed by Clays et al.,[28] DCANP conversion efficiency will be drastically improved. Polymeric LB films are especially useful for the SHG device because of their low optical loss and good thermal stability.

B. ELECTRIC FIELD-INDUCED PHASE MATCHING

In this section, we discuss electric field-assisted dynamic phase matching in a guided mode using a soft-gel NLO material. The effective phase-matching thickness can be controlled by an applied electric field.[29] A low Tg material was used in this experiment which consists of p-nitroaniline (p-NA) units coupled to an ethylene glycol chain (PEG) to form a ω,ω'-di(p-nitroaniline)-ethylene glycol.

(PEG-p-NA)

A PEG-p-NA waveguide was prepared between two indium tin oxide (ITO) glass plates covered by a fused silica buffer layer. The thickness of the buffer layer was 0.95 μm and the taper gradient of PEG-p-NA was 0.05 μm/mm. The d_{33} of PEG-p-NA films was determined to be 1–5 pm/V. The Nd:YAG laser was TM polarized in order to use the d_{33} of PEG-p-NA efficiently. The fundamental beam was focused by 40 times magnified microscope objective lens and coupled into the waveguide from the edge of the substrate. The generated phase-matched SH beam was detected by a pin-photodiode behind the YAG-cut filter and interference filter.

Experimentally obtained refractive indices of PEG-p-NA and calculated phase-matching thickness as a function of poling voltage are listed in Table 4. The film thickness of the unpoled sample corresponds to the hypothesis phase-matching thickness because of its SHG inactivity. The phase-matched conversion between the fundamental zero-order mode and SH second-order mode was selected. The refractive indices for the TM mode increased with an increasing poling electric field, whereas the phase-matching thickness decreased. The largest polarizability α_{xx} of PEG-p-NA aligns along the poling electric field,

Table 4 Phase-Matching Thickness of PEG-*p*-NA Waveguide

Poling Voltage Thickness (V/mm)	n (1.064 μm)	n (0.532 μm)	Phase Matching (mm)
0	1.500	1.558	1.2719
15	1.501	1.559	1.2682
30	1.502	1.560	1.2645
45	1.503	1.561	1.2609

Adapted from Watanabe, T., et al., *Mol. Cryst. Liq. Cryst.,* 255, 95, 1994. With permission.

which induces the large refractive index change by poling. The dispersion of the refractive indices of poled material is larger than that of unpoled material due to the red-shift of absorption maxima induced by poling. It is clear that the effective phase-matching thickness can be controlled from 1.272 μm to 1.260 μm using an external electric field. This effective phase-matching thickness variation induced by poling is very helpful for achieving a phase-matching condition of taperless waveguide. Figure 23 shows the mode dispersion curves of the PEG-*p*-NA waveguide. The phase-matched points are represented by the intersection points between the fundamental zero-order mode curve and the SH second-order mode curve. Unless the red-shift of cutoff wavelength is induced by poling, a wide range of effective phase-matching conditions cannot be expected. Fabrication tolerance for guided mode phase matching using electric field-induced phase matching was increased about 6 times that of a conventional three-layer waveguide. Figure 24 shows the conversion efficiency of the PEG-*p*-NA waveguide. The phase matching was satisfied at 30 V/μm. Nonlinear optical coefficients were increased by increasing the poling electric field. SH intensity decreases due to the phase mismatch in the waveguide. The

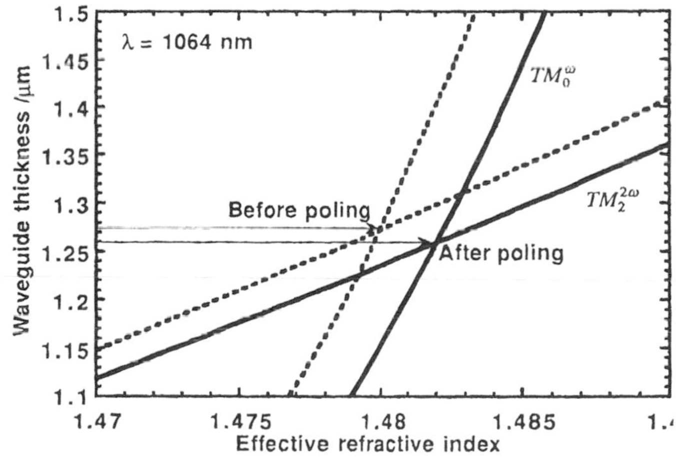

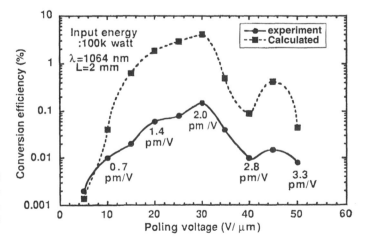

Figure 23 The mode dispersion curves of a low Tg NLO waveguide. (From Watanabe, T., et al., *Mol. Cryst. Liq. Cryst.,* 255, 95, 1994. With permission.)

Figure 24 SH intensity of low Tg NLO waveguide as a function of temperature. (From Watanabe, T., et al., *Mol. Cryst. Liq. Cryst.,* 255, 95, 1994. With permission.)

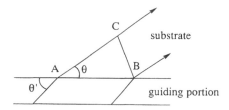

Figure 25 Cerenkov radiation-type phase matching. (From Chikuma, K. and Umegaki, S., *J. Opt. Soc. Am. B, 7,* 768, 1990. With permission.)

difference between experimentally obtained SH intensity and calculated SH intensity was ascribed to the relatively large area of the fundamental beam compared with the phase-matching area.

C. CERENKOV PHASE MATCHING

Cerenkov phase matching can be applied to any dimension of a waveguide structure. The principle of Cerenkov phase matching can be expressed by considering the velocity of the light in a waveguide.[30] Figure 25 shows the waveguide geometry. In type I phase matching, the velocity of the nonlinear polarization wave coincides with that of the fundamental wave. Consider a nonlinear polarization wave radiating an optical second harmonic light with angle θ at point A in Figure 25. After point A, the polarization wave again radiates the optical second harmonics with angle θ at point B. If AC $\perp$ BC, the front surface of the optical second harmonic wave represented by CB would be constructed during the propagation of the polarization wave from A to B. This condition is the same as phase matching. This condition for AC $\perp$ BC is given by

$$V^{\omega} \cos \theta = V_{2\omega} \tag{77}$$

where V_{ω} and $V_{2\omega}$ are the velocity of the fundamental mode and optical second harmonic wave in the substrate. This condition can be written using refractive index.

$$N(\omega) = n_s(2\omega)\cos \theta \tag{78}$$

where $N(\omega)$ is the effective refractive index, and $n_s(2\omega)$ is the refractive index of substrate. If $N(\omega) < n_s(2\omega)$, then the phase-matching condition is automatically satisfied.

In the case of a crystal cored fiber, a ring-shaped pattern of the second harmonic wave can be observed from the end surface of the fiber as shown in Figure 26. The second harmonic power of Cerenkov radiation is given by

$$P^{2\omega} = \frac{2\omega^2}{\varepsilon_0 c^3} \frac{d^2}{n_{\text{eff}}^{2\omega}(n_{\text{eff}}^{\omega})^2} \frac{(P^{\omega})^2}{WD} S_{ij}^2 L \frac{\sin^2(\Delta\beta L/2)}{(\Delta\beta L/2)^2} \tag{79}$$

The difference between modal dispersion and Cerenkov radiation is the dependence of propagation length on power. The former is proportional to the square of propagation length L, whereas the latter is proportional to L.

In the case of a crystal cored fiber, one has to consider the diffraction of the radiative harmonic light at a far field. Especially the phase-matching direction θ given by Equation 78 must be in the zero-order diffraction mode, otherwise conversion efficiency would drastically decrease. Thus one cannot

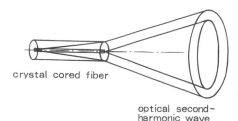

Figure 26 Ring-shaped pattern of the optical second harmonic wave from the crystal cored fiber. (From Chikuma, K. and Umegaki, S., *J. Opt. Soc. Am. B, 7,* 768, 1990. With permission.)

use the crystals with large diagonal tensor components such as MNA due to the large dispersion of the refractive index.

If the materials satisfy the following relations, one can obtain a good conversion efficiency using crystal cored fiber.[31]

$$n_f(2\omega)^2 - n_f(\omega)^2 < 0.093(\lambda_\omega/a)^2 \tag{80}$$

where $n_f(2\omega)$ and $n_f(\omega)$ are the refractive indexes of crystal at harmonic and fundamental waves, respectively. In general, the material that can fulfill the above condition is limited in crystals with proper birefringence. Cerenkov radiation can use off-diagonal tensor components such as d_{31}.

Harada et al.[32] demonstrated an efficient SHG in a crystal cored fiber prepared from 3,5-dimethyl-1-(4-nitrophenyl)pyrazole (DMNP), as shown in the following structure.

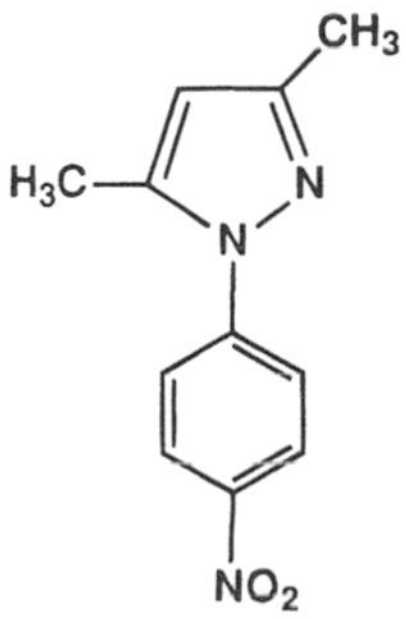

(DMNP)

The d_{32} of DMNP crystal is 90 pm/V which has good advantages for frequency doubling in a crystal cored fiber. The refractive indices n_x, n_y, and n_z of DMNP are 1.567, 1.997, and 1.822 at 442 nm and 1.513, 1.793, and 1.696 at 884 nm respectively. The cutoff wavelength of an axis was around 450 nm. The crystal cored fiber was prepared using the Bridgman-Stockbarger method. The x axis of crystal is always parallel to the fiber axis, whereas y and z axes are perpendicular to the fiber axes. This means that the largest tensor component of DMNP can be utilized for phase matching. The diameter of the core was 1.4 μm and the length as 5 mm. When a 16.6 mW laser-diode at 884 nm was coupled into the fiber, 64.5 μW SH power was obtained. SH power reached 0.16 mW by increasing the cladding layer.

Cerenkov radiation was demonstrated using Langmuir-Blodgett films by Florsheimer et al.[33] DCANP was used in this experiment. The conversion efficiency was measured at various wavelengths as listed in Table 5. Asai et al.[34] showed high overlap integral of four-layer waveguide with nonlinear optical susceptibility inversion structures across the films. This structure can be prepared using Langmuir-Blodgett films to construct the waveguide with NIST structure. The spontaneous polarization of DCANP exists in the plane of films. Therefore, the domain inversion across the films was carried out by inverting the dipping direction. The SHG intensity from the NIST waveguide was found to be 20–30 times larger than that of a conventional waveguide.

Table 5 Conversion Efficiency of DCANP Waveguides

λ(nm)	$d_{33(pm/V)}$	$\eta_{calc.}$	$\eta_{meas.}$	$\eta_{meas}(W^{-1}cm^{-1})$
1064	7.8	1.5×10^{-6}	3×10^{-7}	1.4×10^{-9}
910	12	6×10^{-3}	2×10^{-3}	4×10^{-6}
860	18	6×10^{-3}	2×10^{-3}	4×10^{-6}
820	27	8×10^{-3}	2×10^{-3}	4×10^{-6}

Adapted from Florsheimer, M., et al., *Nonlinear Optics*, North-Holland, Amsterdam, 1992. With permission.

Sugihara et al.[35] demonstrated Cerenkov phase matching using poled copolymer MMA-co-DR1-MA as shown below.

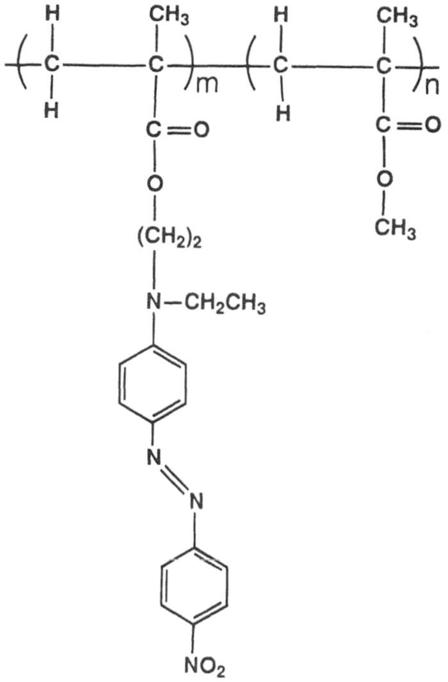

The molar ratio of MMA/DR1 is 70:1. The sample was poled by the corona poling technique. The d_{33} of this polymer was 6.64 pm/V. Equation 78 can be satisfied easily by spin-coated films without precise control of the film thickness. Cerenkov phase matching was obtained within the guide thickness from 1.0 to 3.3 μm. The maximum SH peak power was 3.38 W at a fundamental power of 196 kW (1064 nm). The total conversion efficiency was $1.72 \times 10^{-3}\%$. By considering the coupling loss, SHG efficiency was estimated as 0.21%. A conversion efficiency of about 0.3% at 5.3 W in the waveguide of 4-[-ethyl-N-(2-hydroxye-thyl)]-amino-4′-nitrobenzene was obatined at doubled wavelength of 812 nm. Frequency doubling for green and purple-blue wavelengths was demonstrated.

Tomono et al.[36] reported that the Cerenkov phase matching in poled polymer films based on poly(methyl-methacrylate) containing p-aminophenyl-cyclobutenedione moieties as the side chains, having the following structure.

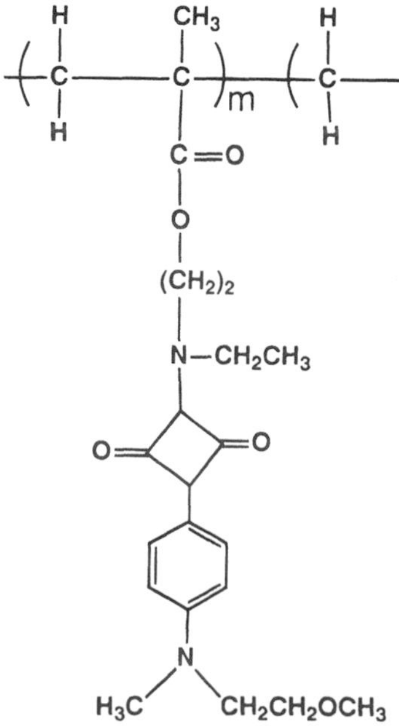

The glass transition temperature of this polymer was in the range of 120–140°C depending upon chromophore contents. The d_{33} of this polymer with 10 wt% chromophore concentration is 8.1 pm/V after poling and 3.6 pm/V after decay. The refractive indices of polymer are 1.495 and 1.512 for TM-polarized light at the wavelengths of 1064 nm and 532 nm, respectively. The conversion efficiency was found to be 3.3×10^{-8}%/W by experiment.

Clays et al.[37] demonstrated the blue and green phase-matched SHG in a Langmuir-Blodgett waveguide using Cerenkov-type geometry. The polymer used was a random copolymer of the monomer containing the NLO chromophore 4-(dimethylamino)-4′-(methylsulfonyl)azobenzene with hydroxy-ethylacrylate in a 1:6 molar ratio. Poly(tert-butyl methacrylate) was used as NLO passive polymer to preserve the noncentrosymmetry in the LB film. The refractive indices of the polymer films are given in Table 6. The intrinsic loss in the LB film was 1.0 dB/cm.

The optical loss of this waveguide was 0.5 dB/cm at 633 nm. When the sample was illuminated with a Nd:YAG laser, green SHG was observed. Blue phase-matched SHG was also observed using the pulsed dye-laser. Cerenkov radiation of the second harmonic wave in the planar waveguide was reported by Kinoshita et al.[38] from a polymer having the following structure.

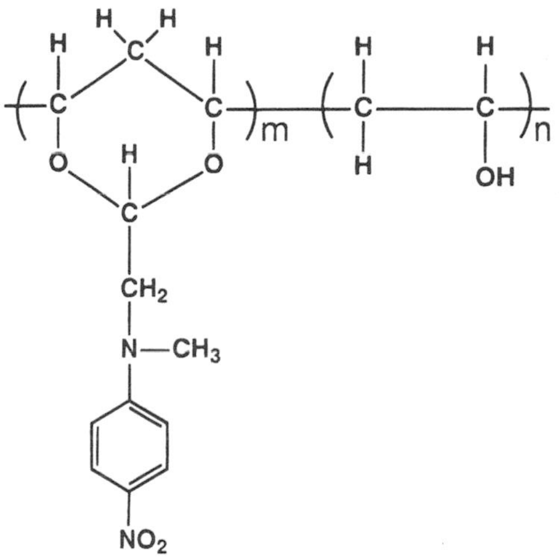

The sample was corona poled at 150°C for 15 min. The d_{33} of this film was found to be 10 pm/V. An experiment of blue SHG was carried out using a mode-locked Ti: sapphire laser operating at 887 nm. Second harmonic power of 0.39 W was obtained at 14.4 kW fundamental input. The conversion efficiency of this film was 0.0027%. The generated second harmonic light was shorter than the cutoff wavelength of polymer.

Izawa et al.[39] reported the phase-matched SHG from the silica films doped with Disperse Red 1 (DR1). DR1, tetramethoxysilane, ethanol, and N,N-dimethylformamide were mixed and hydrochloric acid was added to the solution. The sol-gel film was prepared by spin coating from the solution. After spin coating, the film was aged for 10 min at room temperature and heated up to 150°C. During the

Table 6 Refractive Indices of Substrate, Obtained Film, and Effective Refractive Indices for a 0.504-μm LB Film

Wavelength (nm)	n(Pyrex)	n(ord)	n (extra)	n_{eff} (TE)	N_{eff}(TM)
1064	1.462	1.524	1.547	1.462	1.463
860	1.4635	1.529	1.5545	1.471	1.475
830	1.464	1.5295	1.555	1.473	1.478
633	1.474	1.5365	1.566	1.493	1.507

Adapted from Clays, K., et al., *J. Opt. Soc. Am. B*, 10, 886, 1993. With permission.

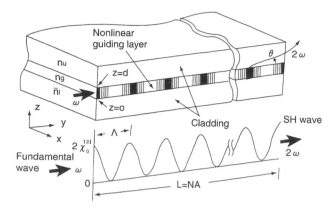

Figure 27 Theoretical model of an asymmetric waveguide with a periodically corrugated NLO polymer. (From Azumai, Y. and Sato, H., *Jpn. J. Appl. Phys.*, 32, 800, 1993. With permission.)

curing, the sample was corona poled. The final concentration of DR1 was 21.7 wt%. The largest nonlinear optical coefficients d_{33} was found to be 75 pm/V. The phase-matched SHG was observed from films when a Nd:YAG laser was illuminated. Cerenkov radiation can match the velocity of light automatically, but the overlap integral must be optimized in the planar waveguide. The overlap integral of Cerenkov radiation can be improved by introducing the NIST structure across the film, otherwise the generated harmonic light is reduced during the propagation because of the multiple reflection. If the materials possess the proper birefringence to fulfill the $N^\omega - N^{2\omega} = 0$, one can also obtain the high conversion efficiency without NIST structure. Therefore, a single crystal with proper birefringence to satisfy the phase-matching condition in a bulk medium can generate the highly efficient SHG in Cerenkov-type geometry. That is why DMNP yields high conversion efficiency while 2-(α-methylbenzylamino)-5-nitropyridine (MBA-NP) does not generate the efficient SHG.

In order to improve the efficiency of Cerenkov phase matching for poled polymers, a periodic structure was proposed by Azumai and Sato.[40] Figure 27 shows the waveguide with a periodically poled polymer. In the experiment, the 60 μm periodic structure was prepared using poly(vinylidene cyanide-vinyl acetate) P(VDCN/VAc); the chemical structure of the polymer is shown below.

By introducing the periodic structure into the slab waveguide, the conversion efficiency was enhanced by a factor of 1.5 in comparison with the uniform structure.

D. NONCOLLINEAR PHASE MATCHING

Noncollinear phase matching can be realized by two fundamental beams to match the velocity between fundamental and harmonic waves. The phase-matching condition is given by[41]:

$$N^\omega \cos \theta_M = N^{2\omega} \tag{81}$$

where N^ω and $N^{2\omega}$ are the effective refractive indices. θ_M is the angle between the propagation direction of the fundamental wave and its harmonic wave and can be assumed to be the phase-matching angle. The experimental setup is shown in Figure 28. The two separated fundamental beams are focused by the lens and coupled by the prism into the waveguide. By changing the angle θ, the phase-matching condition is observed.

The noncollinear phase matching of a MNA crystal was demonstrated by Sugihara et al.[42] using the grating coupler. A MNA crystal with tapered structure was prepared on the grating coupler by the zone-

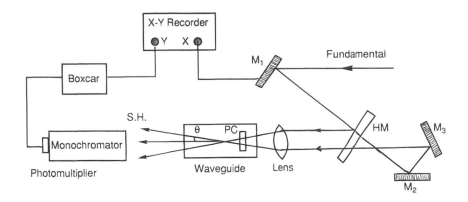

Figure 28   The experimental setup for noncollinear phase matching. M_1, M_2, M_3 are the mirrors; HM is the half-waveplate; PC is the prism coupler. (From Umegaki, S., et al., *Opt. Commun.*, 45, 80, 1983. With permission.)

melting technique. The two separated beams were focused by a lens and coupled into a waveguide. By fixing the angle of the fundamental beam, the output SH was measured as a function of lateral position. The obtained conversion efficiency was estimated as 0.05% when a 49 kW laser was illuminated. Noncollinear phase matching was also demonstrated by Tomono et al.[36] The conversion efficiency of $4.3 \times 10^{-8}\%/W$ was obtained at 1064 nm. The small conversion efficiency is due to the small interaction length of noncollinear phase matching.

E. PERIODIC PHASE MATCHING (QUASI PHASE MATCHING)

Quasi phase matching using a periodic structure of $\chi^{(2)}$ grating was proposed by Bloembergen and Sievers.[43] The quasi phase matching was analyzed by Somekh and Yariv.[44] The waveguide structure with a periodic modulation is shown in Figure 29. The nonlinear optical coefficient is modulated periodically with a period Λ, while the refractive index of materials does not change. The modulation period is equal to $2\pi/\Lambda = \Delta\beta$ where $\Delta\beta$ is the phase mismatch of the wavenumber between the fundamental and the harmonic wave. The conversion efficiency of quasi phase matching is given by

$$P(2\omega) = \frac{2\omega^2}{\varepsilon_0 c^3} \frac{d_{\text{eff}}^2}{n(\omega)^2 n(2\omega)} \frac{P(\omega)^2}{A} L^2 \times \frac{\sin^2(\Delta\beta L/2)}{(\Delta\beta L/2)} \tag{82}$$

where $d_{\text{eff}} = d/m\pi$ (m is an order of the modulation). Figure 29 also shows the SH intensity as a function of propagation length. The SH intensity of quasi phase matching is also proportional to the square of the propagation length, whereas the NLO constant is reduced by a factor of $1/\pi$. By using this technique, one can utilize the largest diagonal tensor components of a single crystal. However, the fabrication of the periodic structures is rather difficult in the waveguide geometry. Quasi phase matching can be applied to bulk materials. In this case, phase matching is achieved by rotating the sample with periodic structures. This is relatively convenient since phase matching can be satisfied for different wavelengths by rotating the crystals. Quasi phase matching has been demonstrated in periodic waveguides of inorganic crystals KTP, $LiTaO_3$, and $LiNbO_3$.

The first experiment of quasi phase matching was performed by Levine et al.[45] using organic liquid-filled waveguides. The phase matching was satisfied by rotating the liquid cell. The structure of the liquid cell is shown in Figure 30. By rotating the gratings relative to the propagation direction, phase matching can be achieved. Khanarian et al.[46] demonstrated the quasi phase matching using a poled polymer. The periodic structure was prepared in the waveguide geometry. The refractive indexes of the waveguide were: for P6CS/MNA (guide layer), 1.5871 at ω and 1.6209 at 2ω; for polysiloxane (upper layer), 1.404 at ω and 1.412 at 2ω; and for polyvinyl alcohol (lower cladding), 1.4925 at ω and 1.5006 at 2ω. The d constant of the polymer was 2.39 pm/V poled at 70 V/μm. The polymer used in this

384

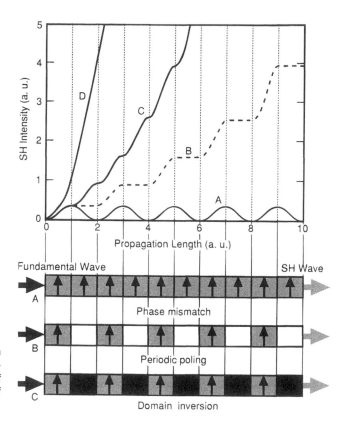

Figure 29 The waveguide geometry with a periodic modulation of the nonlinear optical susceptibilities. The SH intensity of quasi phase matching as a function of propagation length is also presented.

experiment was a copolymer of methylmethacrylate and a monomer containing 4-oxy-4′-nitrostilbene, termed P6CS/MMA/50/50.

In order to introduce the periodic modulation, the aluminum finger electrodes were deposited on the polymer with a 12 μm periodicity. The sample was poled at 90°C near Tg and immediately tested

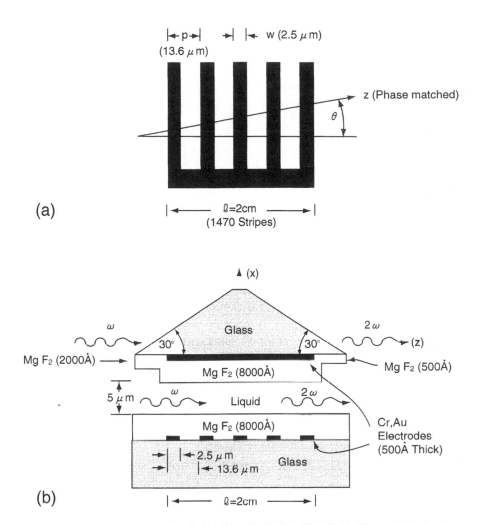

Figure 30 (a) The periodic electrode geometry for quasi phase matching. The period is 13.6 μm; the strip width is 2.5 μm; and the total propagation length is 2 cm. (b) Cross-sectional view of the waveguide structure. (From Levine, B. F., et al., *Appl. Phys. Lett.*, 26, 375, 1975. With permission.)

for phase matching after removal of the electrode by etching. The difference in the refractive index between poled and unpoled areas is about 0.0006, confirmed by a Mach-Zehnder interference microscope. The advantage of quasi phase matching is that a guided mode can use the same mode number as both the fundamental and harmonic waves. This means that the overlap integral of a guided mode for quasi phase matching is quite high. The waveguide is rotated to fulfill the phase matching conditions and phase-matched SHG was observed at an incident angle of 24°. The efficiency of this phase matching was estimated to be $1.2 \times 10^{-4}\%$.

Khanarian et al.[47] reported quasi phase matching from freestanding films. The poled polymer used in this experiments was a 10/90 copolymer of 4-oxy-4′-nitrostilbene and methylmethacrylate (P6CS/MMA). Freestanding film was prepared by spin coating on a silicon wafer. At first the water-soluble polymer poly(acrylic acid) was coated and then P6CS/MMA was coated. The sample was poled at 110°C. The sample was cut into a small size using an excimer laser ($\lambda = 193$ nm). Then the silicon wafer was kept in the water to dissolve the poly(acrylic acid). The film that floated on the water surface was deposited on the glass substrate to make sure that the poling direction was alternately changed to give the periodical modulation of $\chi^{(2)}$. A sample with 1 to 52 layers was prepared. A Q-switched Nd:YAG laser was illuminated into the films. The peak power of the fundamental wave was 225 W. Figure 31 shows the SH intensity of the freestanding film with periodical modulation of $\chi^{(2)}$. The $N = 52$

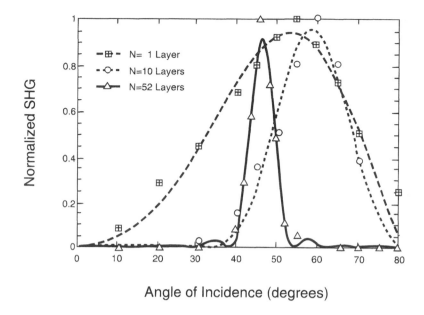

Figure 31 Quasi-phase-matched SHG from N = 1, 10, and 52 periodic films. (From Khanarian, G., et al., *Appl. Phys. Lett.*, 63, 1462, 1993. With permission.)

shows the very wide angular tuning curves for phase matching. A significant improvement can be made in frequency doublers of poled polymers to obtain high efficiencies comparable to those of inorganic crystal-based doublers.

F. COUNTER-PROPAGATING PHASE MATCHING

Counter-propagating phase matching was demonstrated by Bosshard et al.[48] The phase-matched SHG was generated from the surface of the films. The DCANP used in this experiment was deposited onto the TiO_2/SiO_2 substrate using the Langmuir-Blodgett technique. The thickness of TiO_2/SiO_2 and LB films was 145 nm and 230 nm, respectively. The fundamental beam was split in half and coupled into both sides of the waveguide to excite the TE_0 mode. The SH power from counter propagation is given by

$$I(2\omega) = A^{NL}I_+(\omega)I_-(\omega) \tag{83}$$

where $I_\pm(\omega)$ is the line intensity,

$$A^{NL} = \varepsilon_o\omega k_0(2\omega)n_c(2\omega)|C^4||S^2| \tag{84}$$

where A^{NL} is the nonlinear cross section, S is the overlap integral, and C is the power normalized constant of the guided mode. A^{NL} depends on the overlap integral. The thickness of LB films was determined to optimize A^{NL}. FWHM of counter-propagating phase matching is about 100 nm. This condition is very moderate compared with modal dispersion phase matching. Figure 32 shows the waveguide configuration of this experiment. The SHG light was emitted from the surface of the films. The second harmonic power of 6 nW was obtained when an 12.6 mW laser was coupled.

IV. CONCLUDING REMARKS

Waveguide geometry has a big advantage, based on the high optical power density in producing the highly efficient SHG. However, phase-matching conditions using the waveguide geometry are very severe. Therefore, many approaches have been demonstrated to improve both the fabrication tolerance and the overlap integral. These streams start from modal dispersion phase matching and reached to counter-propagation phase matching. The problem of trade-off between high conversion efficiency and wide fabrication toler-

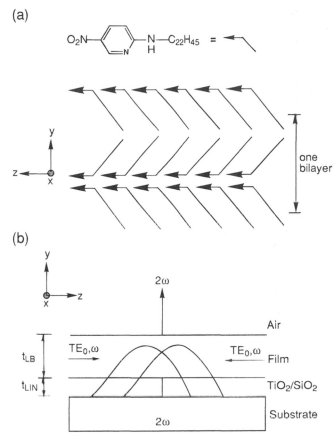

Figure 32 (a) Orientation of DCANP in the LB films. The arrows in the figure show the dipoles. (b) Waveguide geometry and schematic diagram of the counter-propagating experiments. (From Bosshard, C., et al., *Appl. Phys. Lett.*, 64, 2076, 1993. With permission.)

ance still remains. Another serious problem encountered for practical applications is photo bleaching. For frequency doubling, continuous wave operation is required. If the SHG is near to the cutoff wavelength of the waveguide, photo bleaching can cause serious problems. In order to overcome this problem, efforts are required from molecular engineering to optimize the relation between cutoff wavelength and second-order optical nonlinearity.

Several techniques have been demonstrated in poled polymers, including quasi phase matching, modal dispersion, Cerenkov, and noncollinear phase matching. Poled polymer waveguides seem quite promising for frequency doublers comparable to that of well-known conventional inorganic single crystals, such as KTP and LiNbO$_3$. The birefringence phase matching using a poled polymer shows great potential for frequency doubling. The poled polymers with cavity mirror may drastically improve the conversion efficiency coupled with intra- or intercavity geometry. Further studies on the fluctuation of the refractive index induced by poling, mechanical drawing, and thermal conductivity are necessary. Efficient conversion can be obtained by selecting appropriate SHG materials and optimal design of waveguide structures. We hope that practical frequency doublers will be realized in the near future using organic molecular and polymeric materials.

REFERENCES

1. **Maker, P. D., Terhune, R. W., Nisenoff, M., and Savage, C. M.,** Effects of dispersion and focusing on the production of optical harmonics, *Phys. Rev. Lett.,* 8, 21, 1962.
2. **Dmitriev, V. G., Gurzadyan, G. G., and Nikogosyan, D. N.,** *Handbook of Nonlinear Optical Crystals,* Springer-Verlag, Berlin, 1991.
3. **Nye, J. F.,** *Physical Properties of Crystals,* Oxford University Press, London, 1957.
4. **Tao, X. T., Watanabe, T., Zou, D. C., Ukuda, H., and Miyata, S.,** Phase-matched second-harmonic generation in poled polymers by the use of birefringence, *J. Opt. Soc. Am. B,* 12, 1581, 1995.
5. **Yariv, A.,** *Quantum Electronics,* John Wiley & Sons, New York, 1975.

6. **Hosomi, T., Suzuki, T., Yamamoto, H., Watanabe, T., Sato, H., and Miyata, S.,** Efficient second-harmonicgeneration on single crystal of *N*-methoxymethyl-4-nitroaniline, *SPIE Proc.,* 1147, 124, 1989.

7. **Tao, X. T., Watanabe, T., Zou, D. C., Shimoda, S., Sato, H., and Miyata, S.,** Polyurea with large positive birefringence for second harmonic generation, *Macromolecules,* 12, 2637, 1995.

8. **Hobden, M. V.,** Phase-matched second-harmonic generation in biaxial crystals, *J. Appl. Phys.,* 38, 4365, 1967.

9. **Stegeman, G. I. and Stolen, R. H.,** Waveguides and fibers for nonlinear optics, *J. Opt. Soc. Am. B,* 6, 652, 1989.

10. **Suematsu, Y., Sasaki, Y., Furuya, K., Shibata, K., and Ibukuro, S.,** Optical second-harmonic generation due to guided-wave structure consisting of quartz and glass film, *IEEE J. Quantum Electron.,* QE-10, 222, 1974.

11. **Ulrich, R. and Torge, R.,** Measurement of thin film parameters with a prism coupler, *Appl. Opt.,* 12, 2901, 1973.

12. **Sasaki, K., Kinoshita, T., and Karasawa, N.,** Second harmonic generation of 2-methyl-4-nitroaniline by a neodymium: yttrium aluminum garnet laser with a tapered slab-type optical waveguide., *Appl. Phys. Lett.,* 45, 333, 1984.

13. **Sugihara, O. and Sasaki, K.,** Phase-matched second-harmonic generation in a 2-methyl-4-nitroaniline single-crystal waveguide: combined structure of grating couplers and four-layer waveguide, *J. Opt. Soc. Am. B,* 9, 104, 1992.

14. **Sugihara, H., Kinoshita, T., Okabe, M., Kunioka, S., Nonaka, Y., and Sasaki, K.,** Phase-matched second harmonic generation in poled dye/polymer waveguide, *Appl. Opt.,* 30, 2957, 1991.

15. **Florsheimer, M., Kupfer, M., Bosshard, C., Looser, H., and Gunter, P.,** Phase-matched optical second-harmonic generation in Langmuir-Blodgett film waveguides by mode conversion, *Adv. Mater.,* 4, 795, 1992.

16. **Ito, H. and Inaba, H.,** Efficient phase-matched second-harmonic generation method in four-layered optical-waveguide structure, *Opt. Lett.,* 2, 139, 1978.

17. **Kupfer, M., Florsheimer, M., Bosshard, C., Looser, H., and Gunter, P.,** Phase-matched second-harmonic blue light generation in 2-docosylamino-5-nitropyridine Langmuir-Blodgett waveguide, *SPIE Proc.,* 1775, 340, 1992.

18. **Clays, K., Schildkraut, J. S., and Williams, D. J.,** Phase-matched second-harmonic generation in a four-layered polymeric waveguide, *J. Opt. Soc. Am. B,* 11, 655, 1994.

19. **Penner, T. L., Motschmann, H. R., Armstrong, N. J., Ezenyilimba, M. C., and Williams, D. J.,** Efficient phase-matched second-harmonic generation of blue light in an organic waveguide, *Nature,* 367, 49, 1994.

20. **Rikken, G. L. J. A., Seppen, C. J. E., Staring, E. G. J., and Venhuizen, A. H. J.,** Efficient modal dispersion phase-matched frequency doubling in poled polymer waveguides, *Appl. Phys. Lett.,* 62, 2483, 1993.

21. **Watanabe, T., Edel, V., Tao, X. T., Shimoda, S., and Miyata, S.,** A polymeric five-layer nonlinear optical waveguide with a large dimension tolerance and large overlap integral for mode conversion phase-matching, *Opt. Commun.,* 123, 76, 1996.

22. **Watanabe, T., Ogasawara, H., Edel, V., Nakajima, K., Shigehara, A., Kaiya, N., and N., A.,** Phase-matched second-harmonic generation from five-layer Langmuire-Blodgett film waveguides with large fabrication tolerance and overlap integral for mode conversion, *Opt. Rev.,* submitted, 1996.

23. **Decher, G., Tieke, B., Bosshard, C., and Gunter, P.,** Optical second-harmonic generation in Langmuir-Blodgett films of 2-docosylamino-5-nitoropyridine, *Chem. Soc. Chem. Commun.,* 933, 1988.

24. **Wijeloon, W. M. K. P., Karana, S. P., Talapatra, G. B., and Prasad, P. N.,** Second-harmonic generation studies of differences in molecular orientation of Langmuir-Blodgett films fabricated by vertical and horizontal dipping techniques, *J. Opt. Soc. Am. B,* 10, 213, 1993.

25. **Bosshard, C., Kupfer, M., and Gunter, P.,** Optical waveguiding and nonlinear optics in high quality 2-docosylamino-5-nitropyridine Langmuir-Blodgett films, *Appl. Phys. Lett.,* 56, 1204, 1990.

26. **Lifante, G. and Townsend, P. D.,** Analysis of SHG in a planar double waveguide structures, *J. Mod. Opt.,* 39, 1353, 1992.

27. **Duguay, M. A. and Weiner, J. S.,** Five-layer nonlinear waveguide for second harmonic generation, *Appl. Phys. Lett.,* 47, 547, 1985.

28. **Clays, K., Armstrong, N. J., Ezenyilimba, M. C., and Penner, T. L.,** Blue light guiding in a polymeric nonlinear optical Langmuir-Blodgett waveguide, *Chem. Mater.,* 5, 1032, 1993.

29. **Watanabe, T., Zou, D., Shimoda, S., Tao, X., Usui, H., Miyata, S., Claude, C., and Okamoto, Y.,** A novel phase-matching technique for a poled polymer waveguide, *Mol. Cryst. Liq. Cryst.,* 255, 95, 1994.

30. **Chikuma, K. and Umegaki, S.,** Characteristics of optical second-harmonic generation due to Cerenkov-radiation type phase-matching, *J. Opt. Soc. Am. B,* 7, 768, 1990.

31. **Chikuma, K. and Umegaki, S.,** Theory of optical second-harmonic generation in crystal-cored fibers based on phase-matching of Cerenkov-type radiation, *J. Opt. Soc. Am. B,* 9, 1083, 1992.

32. **Harada, A., Okazaki, Y., Kamiyama, K., and Umegaki, S.,** Generation of blue coherent light from a continuous-wave semiconductor laser using an organic crystal-cored fiber, *Appl. Phys. Lett.,* 59, 1535, 1991.

33. **Florsheimer, M., Kupfer, M., Bosshard, C., and Gunter, P.,** Phase-matched frequency doubling in Langmuir-Blodgett film waveguide using the Cerenkov-type configuration, in *Nonlinear Optics,* Miyata, S., Ed., North-Holland, Amsterdam, 1992, 255.

34. **Asai, N., Tamada, H., Fujiwara, I., and Seto, J.,** An optical waveguide with a nonlinear optical susceptibility inversion structure in the thickness direction, *J. Appl. Phys.,* 72, 4521, 1992.

35. **Sugihara, O., Kunioka, S., Nonaka, Y., Aizawa, R., Koike, Y., and Kinoshita, T.,** Second-harmonic generation by Cerenkov-type phase matching in a poled polymer waveguide, *J. Appl. Phys.,* 70, 7249, 1991.

36. **Tomono, T., Nishikata, Y., Pu, L. S., Sassa, T., Kinoshita, T., and Sasaki, K.,** Phase-matched second harmonic generations in poled polymer films based on poly(methyl methacrylate) containing a *p*-aminophenyl-cyclobutenedione moiety as the side chains, *Mol. Cryst. Liq. Cryst.,* 227, 113, 1993.

37. **Clays, K., Armstrong, N. J., and Penner, T. J.,** Blue and green Cerenkov-type second-harmonic generation in a polymeric Langmuir-Blodgett waveguide, *J. Opt. Soc. Am. B,* 10, 886, 1993.

38. **Kinoshita, T., Tsuchiya, K., Sasaki, K., Yokoh, Y., Ashitaka, H., and Ogata, N.,** Cerenkov radiation of second harmonic wave by poled polymer planar waveguide of pNAn-PVA, *IEICE. Trans. Elctron.,* 77, 679, 1994.

39. **Izawa, K., Okamaoto, N., and Sugihara, O.,** Stable and large second harmonic generation in sol-gel-processed poled silica waveguide doped with organic azo dye, *Jpn. J. Appl. Phys.,* 32, 807, 1993.

40. **Azumai, Y. and Sato, H.,** Improvement of the Cerenkov radiative second harmonic generation in the slab waveguide with a periodic nonlinear optical susceptibility, *Jpn. J. Appl. Phys.,* 32, 800, 1993.

41. **Umegaki, S., Anabuki, Y., Ohta, K., Inoue, K., and Tanaka, S.,** Optical second-harmonic generation in a thin-film waveguide without control of film thickness, *Opt. Commun.,* 45, 80, 1983.

42. **Sugihara, O., Kai, S., Uwatoko, K., Kinoshita, T., and Sasaki, K.,** Phase-matched noncollinear second harmonic generation in 2-methyl-4-nitroaniline single-crystal film waveguide, *J. Appl. Phys.,* 68, 4990, 1990.

43. **Bloembergen, N. and Sievers, A. J.,** Nonlinear optical properties of periodic laminar structures, *Appl. Phys. Lett.,* 17, 483, 1970.

44. **Somekh, S. and Yariv, A.,** Phase matching by periodic modulation of the nonlinear optical properties, *Opt. Commun.,* 6, 301, 1972.

45. **Levine, B. F., Bethea, C. G., and Logan, R. A.,** Phase-matched second harmonic generation in a liquid-filled waveguide, *Appl. Phys. Lett.,* 26, 375, 1975.

46. **Khanarian, G., Norwood, R. A., Haas, D., Feuer, B., and Karim, D.,** Phase-matched second-harmonic generation in a polymer waveguide, *Appl. Phys. Lett.,* 57, 977, 1990.

47. **Khanarian, G., Mortazavi, M. A., and East, A. J.,** Phase-matched second-harmonic generation from free-standing periodically stacked polymer films, *Appl. Phys. Lett.,* 63, 1462, 1993.

48. **Bosshard, C., Otomo, A., and Stegeman, G. I.,** Surface-emitted green light generated in Langmuir-Blodgett film waveguide, *Appl. Phys. Lett.,* 64, 2076, 1993.

Chapter 6

Electro-optic Effects in Molecular Crystals and Polymers

Christian Bosshard and Peter Günter

CONTENTS

0-8493-8923-2/97/$0.00+$.50

I. INTRODUCTION

Materials in which the optical properties can be changed by electric fields are of basic importance for optoelectronics. In contrast to pure electronics or photonics where the unit information carriers are electrons and photons, respectively, in optoelectronics one has both electrons and photons as information vehicles and they can interact with each other within the electro-optic material. Optoelectronic devices can replace well-known purely electronic functions in communication and measurement systems: propagation, deflection, modulation, amplification, analog-digital conversion, sampling, and storage.

Different effects can give rise to a material's electro-optic response. The polarizability of a crystal is generally composed of contributions from, first, the lattice components (atoms, molecules) and, second, the interaction between these components. Whereas the second effect is dominant in inorganic materials with their weakly polarizable atoms and complexes, the first contribution is dominant in organic materials because of the weak intermolecular bonding (Van der Waals, dipole–dipole interactions, hydrogen bonds). Nonlinear optical effects that mainly depend on the change of polarizability of the electrons in the π-bonding orbitals are much faster than effects due to lattice vibrations (occurring in a frequency range of typically 1 MHz up to 100 MHz). Therefore organic materials are well suited for high-speed applications, e.g., nonlinear optics with ultrashort pulses or high data rate electro-optic modulation.

Organic electro-optic materials based on extended π-electron systems were proposed for electro-optic applications several years ago. Since the first detailed electro-optic characterization of an organic material, hexamethylenetetramine,[1] only a few organic substances have been thoroughly investigated for their electro-optic response.

In this review we first describe the electro-optic effect in terms of its electro-optic and polarization optic coefficients. The dispersion of the effect with regard to the wavelength of the incoming light and the frequency of the applied electric fields is discussed. The most relevant relations between microscopic and macroscopic nonlinear parameters as well as between electro-optic and nonlinear optical coefficients needed for further insight are given. A rough estimation of electro-optic coefficients in molecular crystals based on measured parameters is presented. In Section III several measurement techniques for electro-optic coefficients are described. Section IV is devoted to electro-optic materials. We present a detailed list of material parameters relevant for electro-optics for inorganics, semiconductors, molecular crystals, polymers, and Langmuir-Blodgett films. In Section V we discuss some electro-optic devices with special focus on relevant figures-of-merit. Finally, some concluding remarks concerning the potential of organic materials for nonlinear and electro-optics are made.

II. ELECTRO-OPTIC EFFECTS

In this work, upper case indices are used to describe coefficients of macroscopic tensorial quantities with respect to the dielectric coordinate system, whereas lower case indices denote molecular parameters. In equations involving tensors, summation over common indices is assumed.

A. ELECTRO-OPTIC EFFECTS IN SOLIDS

Changes (deformation, rotation) of the optical indicatrix in noncentrosymmetric substances can be induced by application of an electric field.[2] These electro-optic effects are most commonly described by the tensors r_{IJK} and R_{IJKL} for the linear (Pockels electro-optic coefficient) and quadratic case:

$$\Delta\left(\frac{1}{n^2}\right)_{IJ} = \Delta\left(\frac{1}{\varepsilon}\right)_{IJ} = r_{IJK}E_K + R_{IJKL}E_KE_L \tag{1}$$

The change in the refractive index Δn resulting from the linear effect can be approximated as follows (for $r \cdot E \ll 1/n^2$)[3]:

$$\Delta n_I = -\frac{n_I^3 r_{IIK} E_K}{2} \qquad (\text{for } I = J) \tag{2}$$

An important parameter for the characterization of electro-optic materials for modulator applications is the half-wave voltage, V_π, which is the voltage required to shift the phase of the transmitted beam by π in an appropriate modulator configuration[4]

$$V_\pi = \frac{\lambda}{(n^3 r)_{\text{eff}}} \times \frac{d}{L} = v_\pi \frac{d}{L} \tag{3}$$

where $(n^3 r)_{\text{eff}}$ is the effective electro-optic coefficient. Low modulation voltages require a high value of $(n^3 r)_{\text{eff}}$ as well as large geometry factors L/d (L = sample thickness, d = electrode distance) achieved using the transverse modulator configuration where the light propagation direction is perpendicular to the electric field. The reduced half-wave voltage $v_\pi = \lambda/(n^3 r)_{\text{eff}}$ is often used for characterizing an electro-optic material as a sample independent quantity.

An appropriate figure-of-merit (FM_{EO}) for linear electro-optic effects to be used for amplitude modulators is[5]

$$FM_{EO} = n^3 r_{IJK} \qquad \text{with } n = n_I \qquad\qquad \text{for } I = J$$

$$n = \sqrt{2}\,\frac{n_I n_J}{\sqrt{n_I^2 + n_J^2}} \qquad \text{for } I \neq J \tag{4}$$

This FM_{EO} is discussed in more detail in Sections II. G. and V.

B. POLARIZATION OPTIC COEFFICIENTS

Using the field-induced polarization P_K, the electro-optic effect is frequently expressed as[2]

$$\Delta\left(\frac{1}{n^2}\right)_{IJ} = f_{IJK} P_K + g_{IJKL} P_K P_L \tag{5}$$

where f_{IJK} and g_{IJKL} are the linear and quadratic polarization optic coefficients. The connection between these coefficients and the electro-optic coefficients is given by

$$r_{IJK} = \varepsilon_o f_{IJM}(\varepsilon_{MK} - \delta_{MK})$$

$$R_{IJKL} = \varepsilon_o^2 g_{IJMN}(\varepsilon_{MK} - \delta_{MK})(\varepsilon_{NL} - \delta_{NL}) \tag{6}$$

$$\delta_{KL} = \begin{cases} 1, & \text{for } K = M \\ 0, & \text{otherwise} \end{cases}$$

where ε_{KL} is the dielectric constant at the frequency of the modulating field. Polarization optic coefficients are of special importance in the characterization of oxygen-octahedra ferroelectric materials since they can be used to describe the linear electro-optic effect as well as the birefringence in these materials. In this work we focus on the linear electro-optic coefficients and, therefore, quadratic effects are neglected.

C. FREQUENCY DEPENDENCE OF ELECTRO-OPTIC COEFFICIENTS: RESONANCE EFFECTS

In strongly polar inorganic materials such as $KNbO_3$[5], lattice vibrational effects due to acoustic and optic phonons can significantly contribute to the linear electro-optic effect and are often similar in size to the electronic contributions. In organic materials, electro-optic effects are assumed to be of purely electronic origin in the visible spectral range.

The free electro-optic coefficient r^T contains three contributions: the electronic part (r^e) and the effects from optic (r^o) and acoustic (r^a) phonons[6]:

394

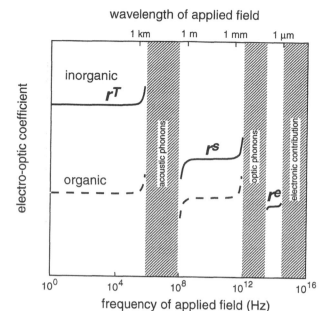

wavelength of applied field

Figure 1 Sketch of the dependence of the electro-optic coefficient on the frequency of the applied electric field. A different behavior is expected as well as experimentally confirmed for inorganic and organic materials.

$$r^T = r^e + r^o + r^a = r^s + r^a \tag{7}$$

r^s is called the clamped electro-optic coefficient and is defined through Equation 7. The various physical processes contributing to the electro-optic effect have been explained by considering field-induced changes of the optical dielectric constant with contributions from optic (Raman scattering processes) and acoustic vibrational modes (Brillouin scattering processes).[7-9]

We first have a contribution that is purely electronic in origin. The second contribution is due to the optic modes and is nonzero only for vibrational modes being both Raman and infrared active. The third contribution arises from the photoelastic effect, i.e., from lattice strains induced by the inverse piezoelectric effect driven by the applied electric field. It should be mentioned that all three contributions can have either sign.

The acoustic mode contribution can be described by considering elasto-optic effects due to strains $s_{JK} = a_{IJK}E_I$ induced by the inverse piezoelectric effect for electric fields E_I.[10] In measurements of electro-optic coefficients the applied electric field first induces a change of the refractive index via the electro-optic coefficient r^s (clamped electro-optic effect). In addition, the applied field can also induce a strain (inverse piezoelectric effect) which leads to a change in the refractive index via the elasto-optic coefficients p_{IJMN} defined by

$$\Delta\left(\frac{1}{n^2}\right)_{IJ} = p_{IJMN}s_{MN} \tag{8}$$

This leads to

$$\Delta\left(\frac{1}{n^2}\right)_{IJ} = (r^s_{IJK} + p_{IJMN}a_{KMN})E_K = r^T_{IJK}E_K \tag{9}$$

with

$$r^a_{IJK} = p_{IJMN}a_{KMN} \tag{10}$$

At low frequencies acoustic and optic phonons can be excited (already below 1 MHz). Above the fundamental acoustic resonances the acoustic modes are clamped and above the "Reststrahlen" region (around 10^{12} Hz) also the optic mode contributions disappear. The electro-optic coefficients thus show a significant dependence on the frequency of the applied electric field (Figure 1). Since the phonon contributions are small in organic materials, this frequency dependence is expected to be small for organic materials.

The applied electric field (e.g., along the optical wave vector direction K) can also cause a piezoelectrically induced change of the light path ΔL_J along the propagation direction[11]:

$$\left(\frac{\Delta L}{L}\right)_{JJ} = a_{KJJ}E_K \tag{11}$$

It has been shown that this change can considerably contribute to the electrically induced phase shift in an interferometric measurement of electro-optic coefficients also away from any acoustic resonances.[11]

D. MOLECULAR HYPERPOLARIZABILITIES

The nonlinear optical and electro-optic response in organic materials arises from the nonlinearity of the basic molecular units. Many properties of the solid can therefore be explained by the properties of the molecules (Section II.E). We first define the relevant physical quantities in the following. The origin of the nonlinear optical effects lies in the nonlinear response of a material to an electric field E. The polarization p of a molecule in the presence of an external field E can be described by

$$p_i = \mu_{g,i} + \varepsilon_o(\alpha_{ij}E_j + \beta_{ijk}E_jE_k + \gamma_{ijkl}E_jE_kE_l + \cdots), \qquad i, j, k = x, y, z$$

$$(x,y,z) = \text{molecular coordinate system} \tag{12}$$

where $\mu_{g,i}$ is the ground state dipole moment, α_{ij} is the polarizability tensor, β_{ijk} is the second-order polarizability or first-order hyperpolarizability tensor, and γ_{ijkl} is the second-order hyperpolarizability tensor. E is the electric field strength at the location of the molecule. For molecules in a solid or in solution, this electric field has to be calculated from the external field by taking into account appropriate local field corrections.

Assuming a two-level model[12] where only terms involving either the ground or the first excited state of the molecule are considered, the microscopic second-order polarizability (or first-order hyperpolarizability) β can be calculated. This model holds quite well for many organic compounds since there often exists one state with a combination of low energy, high oscillator strength and a large change in dipole moment on excitation.

For molecules with strong nonlinearities along a single charge-transfer axis (donor–acceptor groups at the end of π-electron systems), it is often assumed that one component β_{zzz} of the hyperpolarizability tensor along this axis is sufficient to describe the nonlinearity in first approximation. Using the simple two-level model for the molecular hyperpolarizability with one ground state g and one excited state e, β_{zzz} can be written as (for the case of sum-frequency generation)[13]

$$\beta_{zzz}^{(-\omega_3,\omega_1,\omega_2)} = \frac{1}{2\varepsilon_o\hbar^2} \frac{\omega_{eg}^2(3\omega_{eg}^2 + \omega_1\omega_2 - \omega_3^2)}{(\omega_{eg}^2 - \omega_1^2)(\omega_{eg}^2 - \omega_2^2)(\omega_{eg}^2 - \omega_3^2)} \Delta\mu\mu_{eg}^2 \tag{13}$$

where ω_{eg} denotes the resonance frequency of the transition. Note that ω_{eg} of the nonlinear optical molecules is red-shifted in a dielectric medium with $\varepsilon > 1$ due to local field effects. $\Delta\mu = \mu_e - \mu_g$ is the difference between excited and ground-state electric dipole moments, and μ_{eg} is the transition dipole moment between excited and ground state. We can derive the appropriate hyperpolarizability describing the electro-optic effect by setting $\omega_3 = \omega$, $\omega_1 = \omega$ and $\omega_2 = 0$.

We can now introduce the dispersion-free hyperpolarizability β_o extrapolated to infinite optical wavelengths away from the electronic resonances

$$\beta_o = \frac{3}{2\varepsilon_o\hbar^2} \frac{\Delta\mu\mu_{eg}^2}{\omega_{eg}^2} \tag{14}$$

From this equation we obtain for the linear electro-optic effect:

$$\beta^{(-\omega,\omega,0)} = \frac{\omega_{eg}^2(3\omega_{eg}^2 - \omega^2)}{3(\omega_{eg}^2 - \omega^2)^2} \beta_o \qquad (15)$$

E. RELATION BETWEEN MICROSCOPIC AND MACROSCOPIC ELECTRO-OPTIC AND NONLINEAR OPTICAL COEFFICIENTS

The macroscopic polarization P of a nonlinear medium under the influence of an external field E can be described by the following equation:

$$P_I = P_{0I} + \varepsilon_o(\chi_{IJ}^{(1)}E_J + \chi_{IJK}^{(2)}E_JE_K + \chi_{IJKL}^{(3)}E_JE_KE_L + \cdots) \qquad I, J, K, L = 1, 2, 3 \qquad (16)$$

where P_{0I} is the spontaneous polarization and $\chi^{(n)}$ is the nonlinear optical susceptibility of nth order. In the case of frequency doubling we have

$$P_I^{(2\omega)} = \varepsilon_o d_{IJK}^{(-2\omega,\omega,\omega)}E_J^{(\omega)}E_K^{(\omega)} \qquad (17)$$

where d_{IJK} is the nonlinear optical coefficient related to $\chi^{(2)}$ through $d_{IJK}^{(-2\omega,\omega,\omega)} = 1/2\,\chi_{IJK}^{(-2\omega,\omega,\omega)}$.

Based on bond hyperpolarizabilities in inorganic materials, microscopic and macroscopic nonlinear optical properties were first related by Bergman and Crane.[14] Zyss and Oudar later applied their model to organic crystals.[15] Their so-called oriented gas model neglects all intermolecular interactions contributing to the optical nonlinearity except for local field corrections. Therefore, only intramolecular interactions are taken into account. This leads to a simple dependence of the the nonlinear optical susceptibilities for frequency doubling on structural parameters and molecular hyperpolarizabilities

$$d_{IJK} = N\frac{1}{n(g)}f_I^{(2\omega)}f_J^{(\omega)}f_K^{(\omega)}\sum_s^{n(g)}\sum_{ijk}^3 \cos(\theta_{Ii}^s)\cos(\theta_{Jj}^s)\cos(\theta_{Kk}^s)\beta_{ijk}^{(-2\omega,\omega,\omega)} \qquad (18)$$

A similar relation can be obtained for electro-optics:

$$r_{IJK} = -N\frac{1}{n(g)}\frac{4}{n_I^2(\omega)n_J^2(\omega)}f_I^{(\omega)}f_J^{(\omega)}f_K^{(0)}\sum_s^{n(g)}\sum_{ijk}^3 \cos(\theta_{Ii}^s)\cos(\theta_{Jj}^s)\cos(\theta_{Kk}^s)\beta_{ijk}^{(-\omega,\omega,0)} \qquad (19)$$

where θ_{Ii}^s are the angles between dielectric and molecular axes I and i, N is the number of molecules per unit volume, $n(g)$ is the number of symmetry equivalent positions in the unit cell, s denotes a site in the unit cell, $f_I^{(\omega)}$ are local field corrections at the different frequencies, and β_{ijk} is the molecular hyperpolarizability. As mentioned above, it is appropriate for molecules with strong nonlinearities along a single charge-transfer axis to assume a one-dimensional hyperpolarizability β_{zzz} along such an axis. One should note the factor-of-two difference in the values of β used in Equation 12 and Equations 18 and 19: β values from electric field-induced second harmonic generation measurements,[16] normally used for an estimation of d_{IJK} and r_{IJK}, are usually calibrated with some crystal d values (e.g., quartz). Therefore, we have $1/2\,\beta$ (Equation 12) = β (Equations 18 and 19) since $d = 1/2\,\chi$.

Using the oriented gas model, measured values of β_{zzz} could be compared reasonably well with the measured d_{IJK} and r_{IJK}.[17,18] When the intermolecular interactions do not significantly contribute to the optical nonlinearity, one generally only has to take into account a red shift of the resonance frequency when describing macroscopic electro-optic or nonlinear optical coefficients with microscopic hyperpolarizabilities. Such red shifts can be deduced from, for example, a comparison of Sellmeier coefficients of bulk crystals and absorption measurements of molecules in a solution (see below). The local field correction factors f^ω are most often calculated within the simple approximation of the Lorentz model[19]:

$$f^{(\omega)} = \frac{(n^{(\omega)})^2 + 2}{3} \tag{20}$$

F. WAVELENGTH DEPENDENCE OF THE LINEAR ELECTRO-OPTIC EFFECT

Combining the results from Sections II.D and II.E and assuming a one-dimensional charge-transfer axis along z, we can calculate the wavelength dependence of the electro-optic coefficients of polar organic materials:

$$r_{IJK}^{(\omega)} = -\frac{\omega_{eg}^2}{3} \beta_o \frac{4N}{n_I^2(\omega)n_J^2(\omega)} \frac{n_I^2(\omega) + 2}{3} \frac{n_J^2(\omega) + 2}{3} \frac{\varepsilon_K + 2}{3}$$

$$\cdot \cos(\theta_{Iz})\cos(\theta_{Jz})\cos(\theta_{Kz}) \frac{3\omega_{eg}^2 - \omega^2}{(\omega_{eg}^2 - \omega^2)^2}$$

$$= K \frac{1}{n_I^2(\omega)n_J^2(\omega)} \frac{n_I^2(\omega) + 2}{3} \frac{n_J^2(\omega) + 2}{3} \frac{3\omega_{eg}^2 - \omega^2}{(\omega_{eg}^2 - \omega^2)^2} \tag{21}$$

introducing the parameter K.

G. DISPERSION OF THE FIGURE-OF-MERIT FOR ELECTRO-OPTIC MODULATION

We will now give a description of the wavelength dependence of the figure-of-merit for electro-optic modulation for organic and inorganic materials based on simple arguments.

1. Organic Materials

For organic materials, $n_I^3 r_{IIK}^{(\omega)}$ is given by (for $I = J$)[9]

$$n_I^3 r_{IIK}^{(\omega)} = K' g(\omega) \frac{\omega_{eg}^2 (3\omega_{eg}^2 - \omega^2)}{(\omega_{eg}^2 - \omega^2)^2} \tag{22}$$

with

$$g(\omega) = \frac{-4}{n_I(\omega)} \left(\frac{n_I^2(\omega) + 2}{3} \right)^2$$

where K' can be obtained from K and Equation 21. If we look at $g(\omega)$ in more detail, we see that this factor only contains refractive indices. The wavelength dependence of the refractive indices is generally much smaller than the one of the last term in Equation 22. This can be illustrated with an example: Typical values for a refractive index of an organic material optimized for electro-optics (that is, parallel alignment of all chromophores along the polar axis I) are of the order of $n_I = 1.8$ in the near infrared and about $n_I = 2.4$ near the absorption edge of the material. This gives values for $g(\omega)$ of -6.8 in the near infrared and $g(\omega) = -11.2$ near the absorption edge. For the same wavelength range, the first-order hyperpolarizability β increases by several hundred percent. Therefore, in first approximation we can assume $g(\omega)$ to be constant when compared to the last term in Equation 22. This leads to a rough approximation of the dispersion of the electro-optic figure-of-merit

$$n_I^3 r_{IIK}^{(\omega)} \cong F \frac{\omega_{eg}^2 (3\omega_{eg}^2 - \omega^2)}{(\omega_{eg}^2 - \omega^2)^2} \tag{23}$$

with $F = K' \times g(\omega)$. Equation 23 contains only two parameters, F and ω_{eg}. ω_{eg} (the average oscillator frequency of the crystals) is reasonably well known for most of the nonlinear optical organic compounds from absorption measurements or from the coefficients describing the dispersion of the refractive indices (Sellmeier parameters). F can be found once we know ω_{eg} and $n^3 r$ at one wavelength. Thus Equation 23 allows us to plot an approximation of the figure-of-merit for electro-optic modulation for many

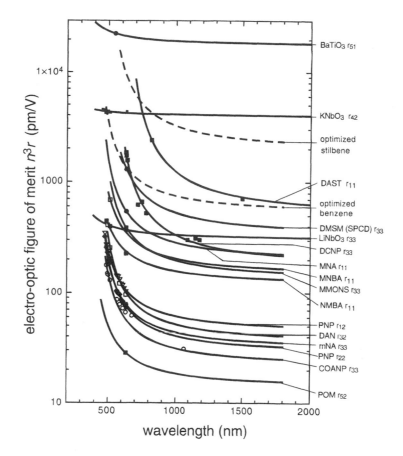

Figure 2 The wavelength dependence of the electro-optic figure-of-merit for different organic and inorganic crystals. The curves are drawn over the transmission ranges of the respective materials.

organic electro-optic materials if we assume that the electro-optic response is dominated by electronic contributions.

Figure 2 shows the electro-optic figure-of-merit as a function of wavelength for the most important organic and inorganic crystals. The parameters used for the calculation of the dispersion of different molecular crystals are tabulated in Section IV.C.

2. Inorganic Materials

For the inorganic ferroelectric materials shown in Figure 2 the dispersion was estimated using a formula based on[6]

$$n^3r = n^3K'\left(1 - \frac{1}{n^2}\right)^2\left[(1 - K) + (1 + K)\left(\frac{\lambda_{eg}}{\lambda}\right)^2\right] \quad \text{with} \quad K' = \frac{\varphi}{E_d}(\varepsilon - 1) \qquad (24)$$

where K is the dispersion constant, ϕ is the linear polarization potential, and E_d is the dispersion energy. The parameters used for the calculation in Equation 24 can be found in reference 9. Note that unclamped values of the electro-optic coefficients were used.

It can be seen from Figure 2 that n^3r varies over several orders of magnitude (some 10 pm/V up to several times 1000 pm/V). The FM_{EO} is increased by one order for wavelengths close to the absorption edge where resonance effects become dominant. Different electro-optic materials are compared in Section IV.

H. RELATION BETWEEN ELECTRO-OPTIC AND NONLINEAR OPTICAL COEFFICIENTS

The electronic contributions to the electro-optic effect r_{IJK}^e can be directly calculated from the nonlinear optical coefficients d_{IJK}^{EO} using[20]

$$r_{IJK}^e = -\frac{4}{n_I^2(\omega)n_J^2(\omega)} d_{IJK}^{EO} \tag{25}$$

Again assuming a one-dimensional two-level model for the charge transfer, d_{IJK}^{EO} is given by

$$d_{IJK}^{EO} = d_{IJK}^{(-\omega,\omega,0)} = \frac{f_I^\omega f_J^\omega f_K^0}{f_K^{2\omega'} f_I^{\omega'} f_J^{\omega'}} \frac{(3\omega_{eg}^2 - \omega^2)(\omega_{eg}^2 - \omega'^2)(\omega_{eg}^2 - 4\omega'^2)}{3(\omega_{eg}^2 - \omega^2)^2\omega_{eg}^2} d_{KIJ}^{(-2\omega',\omega',\omega')} \tag{26}$$

Please note the typing error in the same formula in Equation 16 of Bosshard et al.[9] With the above two equations, nonlinear optical coefficients measured at frequency ω' can be compared to electro-optic coefficients measured at frequency (of light) ω (see Section III.F).

I. ESTIMATION OF UPPER LIMITS OF ELECTRO-OPTIC COEFFICIENTS IN ORGANIC MATERIALS

Equations 13 and 19 describe the electro-optic coefficient r in terms of molecular and crystallographic parameters and refractive indices. They also describe its dispersion by the frequency dependence of the second-order polarizability β and the local field factors.

In order to estimate the maximum electro-optic coefficient of a material based on charge-transfer molecules, the limits of the following parameters must be known[9]:

• *The molecular second-order polarizability* β (Equation 13). This is the parameter that will have the strongest influence on the electro-optic coefficient. It primarily depends on the shape and size of the molecule and the nature of its donor and acceptor substituents. It also depends strongly on the frequency of the optical fields and shows resonance enhancement near the charge transfer transition.
• *The structural parameters* N (number of molecules per unit volume) and θ_{ji}^z (angle between molecular charge-transfer axis and reference system) or θ_p (angle between molecular charge-transfer axis and polar crystal axis).
• *The refractive indices* n_I, which are used in Equations 19 and 25 and which give an estimate for the local field factors f_I. The refractive indices are estimated to about 2.2–2.4 for stilbenes and thiophenes and around 2.0 for benzenes in spectral regions of moderate absorption.

The first two parameters in this list, the hyperpolarizability and the molecular density, are discussed in more detail in the following two sections.

1. Maximum Second-Order Polarizabilities

Maximum second-order polarizabilities can be estimated from measured values. For a semiempirical estimation of β we used experimental results published by L.-T. Cheng et al.[21] and Rao et al.[22–24] The measurements were performed at $\lambda = 1.9$ μm to be truly nonresonant. These authors have measured the hyperpolarizabilities of a large number of benzene, stilbene (Figure 3a), and thiophene derivatives by electric field-induced second harmonic generation (EFISH). Figure 3b shows a plot of the hyperpolarizability extrapolated to infinite wavelengths β_o (see Equation 15) as a function of the maximum absorption wavelength λ_{eg}. Donor and acceptor influence both ω_{eg} and β so that a direct relationship between β and ω_{eg} can be deduced.

From the two parameters λ_{eg} and β_o, it is possible to calculate the hyperpolarizabilities $\beta^{(-\omega,\omega,0)}$ at the desired wavelengths using the relation in Equation 15.[9]

Note that when transferring the data in Figure 3b to solids a shift $\Delta\lambda$ of λ_{eg} (= $2\pi c/\omega_{eg}$) towards the red is observed due to higher dipole–dipole interactions: $\lambda_{eg}^{solid} = \lambda_{eg}^{solution} + \Delta\lambda$. For benzenes we have $\Delta\lambda \approx 40$ nm, and for stilbenes $\Delta\lambda \approx 60$ nm (for solutions in 1,4-dioxane). Such red shifts can be deduced from a comparison of Sellmeier coefficients and absorption measurements in solutions of various compounds in nonpolar solvents.[9] To take full advantage of the resonance enhancement, the operating wavelength should lie close to the absorption edge. As a crude rule the absorption edge for

(a)

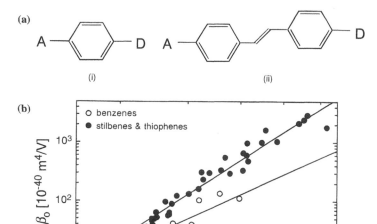

(b)

Figure 3 (a) Representatives of intramolecular charge-transfer molecules (A, acceptor substituent; D, donor substituent). i, benzene; ii, stilbene. (b) Second-order polarizability β_0 extrapolated to infinite wavelengths vs. wavelength of maximum absorption. Data points originate from Cheng et al.,[21] Jen et al.,[22] and Rao et al.[23,24]

the molecules considered here lies 100 nm from its solid-state peak absorption wavelength λ_{eg}^{solid}, i.e. $\lambda_{eg}^{solid} = \lambda_c - 100$ nm. Therefore, we will calculate the electro-optic coefficients at $\lambda_{eg}^{solution} + 100$ nm but plot them vs. $\lambda_c = \lambda_{eg}^{solution} + 100$ nm $+ \Delta\lambda$ (see Section II.I.3).

2. Limitations due to Molecular Packing

The relation between the microscopic and macroscopic electro-optic response (Equation 19) depends on the structural parameters N and θ_p. N is the number of molecules per unit volume and θ_p the angle between the molecular charge-transfer axis and the polar crystalline axis. In order to maximize a coefficient of r_{IJK} it is easy to see from Equation 19 that θ_{ii}^s (or θ_p) should be close to zero. Such a configuration optimizes the diagonal coefficient r_{III}. In other words, the charge-transfer axes of all molecules should be parallel.

For quasi phase-matched frequency conversion in bulk crystals and waveguides or phase matching using modal conversion in waveguides, the highest conversion efficiencies are obtained for the same orientation of the molecules as for electro-optics.

The molecular density N can be estimated from typical densities and molecular weights for benzene and stilbene derivatives. A typical benzene derivative with a nearly optimized structure is 2-methyl-4-nitroaniline (MNA)[25]; a typical stilbene is 3-methyl-4-methoxy-4′-nitrostilbene (MMONS).[26,27]

The molecular density for MNA is $N \approx 6 \times 10^{27}$ m^{-3}; for MMONS it is $N \approx 3 \times 10^{27}$ m^{-3}. Since the packing in most molecular crystals is very tight, these numbers do not vary much for other molecules. Typical chromophore densities of electro-optic polymers are around $N \approx 1.5 \times 10^{27}$ m^{-3}.

3. Maximum Electro-optic Coefficients

Using the optimized hyperpolarizabilities β_{max} as explained in Section II.I.1, we can now find the upper limits of the electro-optic coefficients r_{IJK}.[9] If all intermolecular contributions (except for local field corrections) to the nonlinearity are neglected, optimized structures for electro-optic applications can easily be obtained (see Equation 19). For a series of compounds, such as, para-disubstituted benzene derivatives for example, we chose a small molecule to use the corresponding large number of molecules per unit volume and refractive index in order to get an approximation for the upper limit for electro-optic and nonlinear optical effects. These values were then used for all compounds in this series. Since MNA and MMONS are well characterized, we chose the values from MNA[25,28] ($N \approx 6 \times 10^{27}$ m^{-3}, $n^{2\omega} = 2.2$, $n^{\omega} = 2.2 - 0.3$) for the benzene derivatives and MMONS[26] ($N \approx 3 \times 10^{27}$ m^{-3}, $n^{2\omega} = 2.4$, $n^{\omega} = 2.4 - 0.3$) for the stilbene and thiophene derivatives as was already mentioned in Section II.I.2.

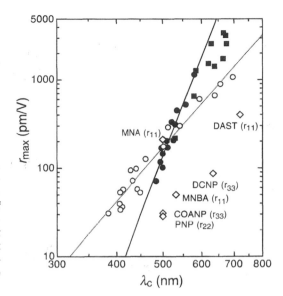

Figure 4 Upper limits of electro-optic coefficients. The graph shows maximum values of r_{IJ} vs. cutoff wavelength calculated from measured values of hyperpolarizabilities assuming an optimum parallel alignment of all molecules. ○, benzene derivatives; ●, stilbene derivatives; ■, thiophene derivatives. The *solid line* is a fit for the stilbene derivatives and the *dashed line* is a fit for benzene derivatives. Measured values of electro-optic coefficients of some molecular crystals are also plotted.

Taking these conditions into account and using Equation 19, one obtains the upper limits for the electro-optic coefficients r_{max} as shown in Figure 4. Note that the values are again calculated at $\lambda_{eg}^{solution} + 100$ nm but plotted vs. $\lambda_c = \lambda_{eg}^{solution} + 100$ nm $+ \Delta\lambda$ ($\Delta\lambda = 40$ nm for benzene derivatives and $\Delta\lambda = 60$ nm for the others) to account for the red shift as explained in Section II.I.1.

It can be seen from Figure 4 that the electro-optic coefficients of stilbene and benzene derivatives can be expressed by $r_{max} = A \times \lambda_c^n$, where $n \approx 14$ for stilbenes and $n \approx 6.3$ for benzenes and where λ_c denotes the absorption edge.[9] Shifting the absorption edge to longer wavelengths by adding stronger donors and acceptors will consequently increase the maximum value of the electro-optic coefficient. However, it should be noted that the absorption edge can only be shifted up to values around 700–800 nm for very strong donors and acceptors. Also, the addition of strong donors and acceptors increases the ground-state dipole moment of a molecule, which often leads to centric crystal structures.

For poled electro-optic polymers the situation is slightly different. The refractive indices of common electro-optic polymers are smaller than the ones of molecular crystals (typically around 1.6–1.9). In comparison with stilbene-type crystals the chromophore density is about half as big (typically around $N \approx 1.5 \times 10^{27}$ m^{-3}). Taking these considerations into account, we obtain upper limits for electro-optic coefficients r_{max} for poled polymers that are about half as large as the ones for the corresponding stilbene and thiophene-type molecular crystals.

For the estimation of the maximum electro-optic response at longer wavelengths, away from the absorption edge, a discussion of the dispersion of the electro-optic coefficient can be found in Section II.F.

Figure 4 also displays measured values of electro-optic coefficients of different organic single crystals. We see that most materials known up to now are far from being optimized. The major problem is not the molecular but the crystal engineering. There are already many new molecules with exceptionally large nonlinearities. Few of them could, however, be used to grow optimized noncentrosymmetric crystals. Therefore, a large potential in the development of better electro-optic molecular crystals remains to be exploited.

J. STABILITY OF ELECTRO-OPTIC PROPERTIES IN ORGANIC MATERIALS

An important issue is the stability of electro-optic materials. Generally organic materials are more fragile than semiconductors or oxides crystals, for example. Especially light and chemicals (e.g., solvents) can lead to damage in organic materials. In addition, mechanical stability may also be a problem. In this section we will discuss the problem of temporal stability of the electro-optic effect due to orientational relaxation.

For molecular crystals no such relaxation is known or expected because the molecules are fixed in a crystal lattice.

For polymers the situation is different: to fully exploit the electro-optic properties one has to break the isotropic structure of polymer films. Poling under an electric field leads to the desired noncentrosymmetry through the alignment of the nonlinear optical active moieties. Unfortunately, this electric field-induced order tends to decay when the poling field is turned off, especially at the higher temperatures.

In the past there have been different attempts to overcome the stability problem: densification of the polymer matrix using thermal or photo-cross-linkable network polymers,[29–33] reduction of the mobility of the nonlinear optical chromophore by incorporating it into the polymer main-chain backbone[34–36] or preparation of guest–host systems with polyimide polymers which are known for their excellent thermal stability.[37–40]

The stability of the poling-induced order is to a large extent most strongly coupled with aspects of the glass transition as the prominent property of polymeric systems. A variety of theoretical expressions exist to describe the temperature dependence of the relaxation time based on free volume, entropy, or other phenomenological expressions.[41] In the Adam-Gibbs entropic formulation of the glass transition, the structural relaxation of polymeric liquids involves the cooperative rearrangement of increasingly larger numbers of molecular segments as the temperature decreases.[42,43] The structural relaxation time τ can be described by an Adam-Gibbs expression which reproduces the Vogel-Fulcher/Arrhenius behavior above/below the glass transition T_g.[44] It was found that the temperature dependence of the relaxation processes associated with the large-scale polymer chain motion at temperatures below the glass transition should scale in the following way[45]:

$$\tau(T) \propto \frac{T_g - T}{T} \tag{27}$$

Equation 27 immediately implies that the stability can be increased by increasing the glass transition temperature T_g.

Relaxation of the chromophores in electro-optic polymers below the glass transition is often investigated by the decay of the second harmonic signal from corona poled films. The time dependence of the second harmonic decay is usually found to be well represented by a Kohlrausch-Williams-Watts stretched exponential function,

$$\mathbf{d}(t) = \mathbf{d}_0 \exp - (t/\tau)^b \tag{28}$$

where $0 < b \leq 1$ is a measure of the breadth of the distribution function and the extent of deviation from a single exponential behavior. The characteristic relaxation time τ is the time that is required for the nonlinear optical coefficient to decay to $1/e$ of its initial value. Experimental results and estimated lifetimes of electro-optic polymers are given in Section IV.D3.

III. MEASUREMENT OF ELECTRO-OPTIC EFFECTS

A. INTERFEROMETRIC TECHNIQUE

1. Determination of Absolute Values

One of the most suitable methods for the measurement of electro-optic coefficients is the phase modulation technique using Michelson and Mach Zehnder interferometers (Figure 5, see, e.g., reference 46). In this experiment the laser beam passes through the crystal once (Mach Zehnder) or twice (Michelson interferometer).

Two monochromatic plane waves of the same polarization with intensities I_1 and I_2 are combined at the output of a Michelson interferometer where they interfere (Figure 5) (externally applied electric field along J, polarization along I). If one of the beams passes through the crystal, the resulting intensity of the interference pattern at the output is given by

$$I = I_1 + I_2 + 2\sqrt{I_1 I_2} \cos(\Delta\phi) \tag{29}$$

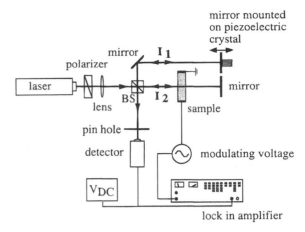

Figure 5 Setup for the determination of electro-optic coefficients using a Michelson interferometer.

lock in amplifier

where

$$\Delta\phi = \Delta\phi_o + \Delta\phi(E) \qquad (30)$$

$$= \frac{2\pi}{\lambda}[\Delta L + (n_I - 1)L] - \frac{\pi}{\lambda}n_I^3 r_{IJ}LE_J$$

where n_I is the appropriate refractive index, L is the length of the beam path in the crystal (equal to one or two times the crystal thickness for perpendicular incidence), and ΔL is the difference in length between the two interferometer arms. ΔL can be adjusted with a piezoelectric mirror (Figure 5). For a preset ΔL with $2\pi/\lambda(\Delta L + (n_I - 1) L) = (2m - 1/2)\pi$, $m = 0, \pm 1, \pm 2, \ldots$, the application of a weak sinusoidal modulation voltage U

$$U = U_o \sin(2\pi vt) \qquad \left(U_o << \frac{d\lambda}{\pi n_I^3 r_{IJ}L} = \frac{V_\pi}{\pi}\right)$$

(d = electrode distance, V_π = half-wave voltage) leads to

$$I = I_1 + I_2 - 2\sqrt{I_1 I_2}\,\frac{\pi}{\lambda}\,n_I^3 r_{IJ}\left(\frac{L}{d}\right)U_o \sin(2\pi vt) \qquad (31)$$

With the two definitions (see Figure 6)

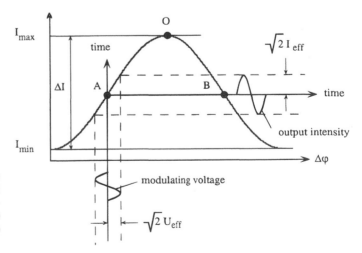

Figure 6 Intensity at the output of the interferometer. The modulator is biased to the point $I_1 + I_2$ with, e.g., a piezoelectric mirror. A small sinusoidal voltage applied to the sample then modulates the light intensity about the bias point A.

$$\Delta I \equiv I_{\max} - I_{\min} = 4\sqrt{I_1 I_2}, \qquad \text{intensity contrast}$$

$$\left.\begin{aligned} I_{\text{eff}} &= \left(\frac{\Delta I}{2}\right)\left(\frac{\pi}{\lambda}\right) n_I^3 \, | \, r_{IJ} \, | \left(\frac{L}{d}\right) U_{\text{eff}} \\[1em] U_{\text{eff}} &\equiv \sqrt{\langle U^2 \rangle_t} = \frac{U_o}{\sqrt{2}} \end{aligned}\right\} \qquad \text{modulation signal}$$

and the measurement of the contrast ΔI and the effective modulation signal I_{eff} (Figure 6) the electro-optic coefficients can be determined[47]:

$$| \, r_{IJ} \, | = \frac{2\lambda}{\pi} \frac{I_{\text{eff}}}{\Delta I} \frac{d}{L} \frac{1}{n_I^3} \frac{1}{U_{\text{eff}}} \tag{32}$$

Typically the two beams have approximately equal intensities $I_1 \approx I_2$. The electric field is applied along crystallographic axes. The polarization of the light beam propagating in the crystal is chosen to select a particular electro-optic coefficient. It is desirable to have crystal dimensions such that $d/L \ll 1$.

For the measurements, photomultipliers and photodiodes are suitable for the detection of intensity changes in the interference pattern. A lock-in amplifier is usually used for phase-sensitive detection of the field-induced phase shift. It was shown that with electrodes either made of silver paste or with the sample mounted between two parallel metal plates with areas exceeding those of the samples, the same results can be obtained.[9] Modulation voltages usually vary between several to some hundred volts, and modulation frequencies are in the kHz range.

For thin films such as electro-optic polymers, where the longitudinal electro-optic effect is characterized, Equation 32 looks slightly different:

$$r_{13} = \frac{\lambda}{\pi} \frac{I_{\text{eff}}}{\Delta I} \frac{1}{n^3} \frac{1}{U_{\text{eff}}} \tag{33}$$

The 3-axis indicates the polar direction perpendicular to the substrate, whereas the 1- and 2-axes are within the substrate plane. Due to the C_∞ symmetry of poled polymers, we have $r_{13} = r_{23}$. In the case of poled polymers, the thickness of the film does not have to be known since the optical path and the electrode spacing are the same.

2. Sign of Electro-optic Coefficients

For inorganic crystals the determination of the absolute sign of the electro-optic coefficients with respect to the direction of the polar axis (equal to the spontaneous polarization in ferroelectric materials) is not difficult since this direction is often known. In contrast, this is often not the case in molecular crystals. Once the direction of the polar axis is known, the absolute sign determination is then the same for all materials.

A positive field direction in the crystals can be defined using a pyroelectric measurement.[9] At room temperature the voltage across the crystal is 0 (due to a compensation by surface charges). By heating up the crystal, pyroelectric charges are generated leading to positive and negative surface charges S_+ and S_- (Figure 7). One then defines a positive sign for an electric field E with a field vector pointing from S_+ to S_-.

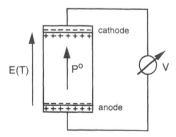

Figure 7   Determination of the direction of the spontaneous polarization by the pyroelectric method.

This measurement allows us to define a positive field direction along the polar axis of crystals. Note that such experiments with crystals of 2-cyclooctylamino-5-nitropyridine (COANP) grown from undercooled melt and crystals of 2-(N-prolinol)-5-nitropyridine (PNP) obtained from solution showed that these crystals always grow in the direction along which negative pyroelectric charges appear upon heating (along the c axis (COANP) or b axis (PNP) of the crystal).[9]

Under the assumption that the dielectric polarization P_I^o decreases with increasing temperature due to increased thermal motions of the molecules, the sign of the pyroelectric coefficient p_I defined by[10]

$$p_I = \frac{dP_I^o}{dT} \tag{34}$$

can also be determined. For the case of COANP and PNP, we obtained $p_I < 0$. However, the above assumption of decreasing polarization with increasing temperature has first to be verified.

In the second step, which is the same for all material classes, a thin glass plate can be mounted in the path of the reference beam of the Michelson interferometer and then be rotated in such a way that the applied voltage and the signal I are in phase. Then an increase of U will also increase the intensity measured by the detector. If a further rotation of the glass plate leads to an increase of the optical path length in this arm of the interferometer and to an increase in average output intensity, this corresponds to a decrease of the optical path of the probe beam and therefore $\Delta n < 0$ for an increase of U. Using Equation 2, $\Delta n < 0$ leads to $r > 0$. The other way around we would get $r > 0$.

A comprehensive list of measured electro-optic coefficients can be found in Section IV.

B. AMPLITUDE MODULATION TECHNIQUE

In addition to the interferometric measurements, experiments using field-induced birefringence with crossed polarizers ($\pm 45°$) are also often performed.[3] A typical experimental setup is illustrated in Figure 8. In the setup shown, a $LiNbO_3$ crystal is used as compensator (via the linear electro-optic effect). This method is much simpler than the interferometric technique described above. However, it often only allows the determination of linear combinations of electro-optic coefficients $|n_I^3 r_{II} - n_J^3 r_{JI}|$ when working in the transversal configuration. Single coefficients can only be obtained if one tensor element dominates all the others.[48] Due to its simplicity this technique was chosen to investigate the dependence of $|n_I^3 r_{II} - n_J^3 r_{JI}|$ on the modulation frequency for several molecular crystals (from 1 kHz to 1 MHz, see Section IV.C.2).[9,48] In analogy to the discussion above, $|n_I^3 r_{II} - n_J^3 r_{JI}|$ can be determined using[5,47]

$$|n_I^3 r_{II} - n_J^3 r_{JI}| = \frac{2\lambda}{\pi} \frac{I_{\text{eff}}}{I_{\text{max}} - I_{\text{min}}} \frac{d}{L} \frac{1}{U_{\text{eff}}} \tag{35}$$

I_{max} and I_{min} are the beam intensities for field-dependent maximum and minimum transmission. The other parameters have been defined above.

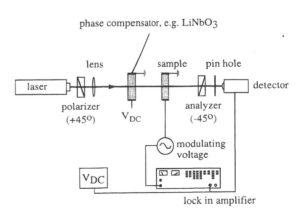

Figure 8 Setup for the determination of electro-optic coefficients using field-induced birefringence.

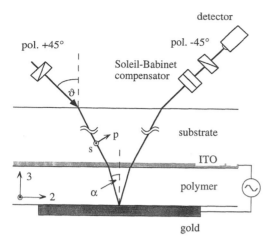

Figure 9 Ellipsometric setup for electro-optic measurements on thin films.

C. REFLECTION TECHNIQUE

The reflection technique is most often applied in polymeric thin films.[49] For this method an ellipsometric configuration (Figure 9), where the input laser beam is polarized 45° with respect to the plane of incidence, is used. The output intensity is given by

$$I = \Delta I \cdot \sin^2[(\Delta\varphi_0 + \delta\varphi_{sp}(E))/2] \tag{36}$$

where $\Delta\varphi_0$ can be set to $\pi/2$ (half-intensity output) with a Soleil-Babinet compensator. The intensity contrast is again ΔI. A phase difference $\delta\varphi_{sp}(E)$ is acquired between the parallel (p) and perpendicular (s) components of the beam as they pass through the polymer film when the modulating field E is applied. The phase difference is due to the different effective electro-optic coefficients experienced by the s- and p-polarized waves and the resulting difference in the optical path length L between the two waves in the polymer film:

$$\delta\varphi_{sp} = \frac{2\pi}{\lambda} (L \cdot \delta n_{sp} + n \cdot \delta L_{sp}) \tag{37}$$

For small modulation voltages, the modulated signal I_{eff} is proportional to this phase difference:

$$I_{eff} \approx 1/(2 \cdot \sqrt{2}) \cdot \Delta I \cdot \delta\varphi_{sp} \tag{38}$$

With the approximation $n_s \approx n_p \approx n$ and the assumption $r_{33} = 3 \cdot r_{13}$[50] one gets[49]:

$$r_{13} = \frac{\lambda}{2\pi} \frac{I_{eff}}{\Delta I} \cdot \frac{(n^2 - \sin^2\vartheta)^{3/2}}{n^2(n^2 - 2\sin^2\vartheta)} \cdot \frac{1}{\sin^2\vartheta} \frac{1}{U_{eff}} \tag{39}$$

where ϑ is the external angle of incidence. As in the case of the interferometric method for thin polymeric films, knowledge of the film thickness is not needed.

Figure 6 shows the representative output characteristics for the interferometric case where the intensity function is proportional to $\cos(\Delta\varphi)$ and for the ellipsometric experiment where it is proportional to $\sin^2(\Delta\varphi/2)$. For a purely linear electro-optic effect, one gets the same modulated signal at the bias points A and B except for a sign reversal. At point O there should not be any modulation at the same frequency. However, due to multiple reflection and absorption effects in the different layers, the signals in A and B are often not symmetric and one can even measure a modulation at point O. In this case the results have to be interpreted very carefully.[51]

D. ATTENUATED TOTAL REFLECTION TECHNIQUE

Another widely used method for the determination of electro-optic coefficients in thin films is attenuated total reflection. The method is based on the Kretschmann method for the observation of surface plasmon

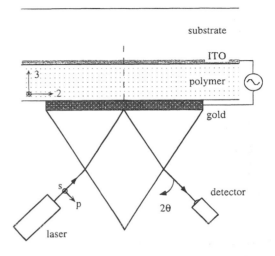

Figure 10 Setup for electro-optic measurements on thin films using attenuated total reflection.

waves at metallic interfaces.[52] It is often used to measure the refractive index and the thickness of thin dielectric films on metallic layers as well as electro-optic coefficients.[52–55] A typical experimental setup for electro-optic measurements is sketched in Figure 10.[54] A configuration of prism/electrode (often gold)/nonlinear optical film/electrode (semitransparent, e.g., ITO)/substrate is shown. A collimated monochromatic light beam is reflected at the multilayered structure. The reflectivity $R(\theta)$ is recorded as a function of angle of incidence θ ($\theta = 0$ corresponds to the prism normal). Various waveguide modes can be excited as a function of θ. The extracted energy from the beam leads to sharp dips in the reflected signal (Figure 11). Fitting of such experimental curves yields refractive indices as well as the corresponding film thicknesses. Since electro-optic thin films (in general polymers) are uniaxial, polarization-dependent measurements yield n_o and n_e.

Electro-optic coefficients can now be determined by measuring the shift in coupling angles of the

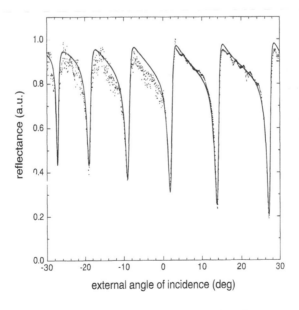

Figure 11 Measured reflectivity as a function of the external angle θ ($\theta = 0$ corresponds to the prism normal). Example of a 2.33-μm-thick polystyrene film on a silver-coated medium flint prism.

waveguide modes with applied voltage. For s-polarized light and for each mode of order m the change in reflectivity can be related to the change in refractive index by

$$\Delta R_s = \frac{\partial R_s}{\partial n_o} \Delta n_o \tag{40}$$

For p-polarized light the corresponding relation is

$$\Delta R_p = \frac{\partial R_p}{\partial n_o} \Delta n_o + \frac{\partial R_p}{\partial n_e} \Delta n_e \tag{41}$$

The partial derivatives of R with respect to n_o and n_e can be calculated numerically from the Fresnel equations. Δn_o and Δn_e can then be determined from the experimental modulation curves with the above two equations. With known applied electric field and a theoretical fit of the modulation curves for all modes the electro-optic coefficients r_{33} and r_{13} can be obtained using

$$r = \frac{2}{n^3} \times \frac{\Delta n}{\Delta E} \tag{42}$$

E. PULSED ELECTRO-OPTIC MEASUREMENTS

It is well established that the large linear electro-optic effect in perovskite-type crystals is mainly due to acoustic and optic phonon contributions. Therefore, the electro-optic response of these materials is strongly dependent on the frequency of the applied electric field. Typically the clamped coefficients $r^s = r^b + r^e$ are smaller by about a factor of 2 in comparison with the unclamped r^T. On the other hand, acoustic phonon contributions are expected to be small for organic materials.

The value of r^a can be determined using an experimental technique based on a time-resolved interferometric measurement of the applied phase-shifts $\Delta\phi(t)$ induced by the application of rectangular-shaped voltage pulses to the samples.[56,57] It is essentially the same experimental setup using a Michelson interferometer as described in Section III.A. Square wave voltage pulses were applied to the samples (electric field amplitude applied to the samples: 2–20 V/cm, frequency 500 Hz, short rise time ≤ 10 ns).[57] A broadband detection system consisting of a silicon photodiode and amplifier operating up to 100 MHz was used. The measurement of I_{eff} was not taken from a lock-in amplifier but was directly read with a digital oscilloscope typically averaging over 1000 pulses. As in the normal interferometric setup (Section III.A), the output is first set to 50% of the total possible output before the pulses are applied.

When a step-like electric field E_K is applied to the crystal, the electro-optic response of the bound electrons and optical phonons shows up instantaneously on a nanosecond time scale. The contribution from acoustic phonons, however, is delayed. They dynamically influence the optical signal as the crystal starts ringing in mechanical resonances. Thus a time-resolved measurement of the electro-optic response $r_{IIK}^*(t)$, defined ad hoc as

$$r_{IIK}^*(t) = -\Delta\phi(t) \frac{\lambda}{\pi n_I^3 L E_K} \tag{43}$$

allows us to separate the quasi-instantaneous reaction of the bound electrons (still containing the optic phonon contributions) from the unclamped electro-optic effect described by $r^T r^T$ can be measured after the relaxation of the acoustic resonances. This pulsed experiment thus allows a precise measurement of the ratio $r^s/r^T = (r^e + r^o)/r^T$. The first precise determination of r^T with the setup in Section III.A in combination with the pulsed method yields a reliable value of r^s (see Section IV for experimental results).

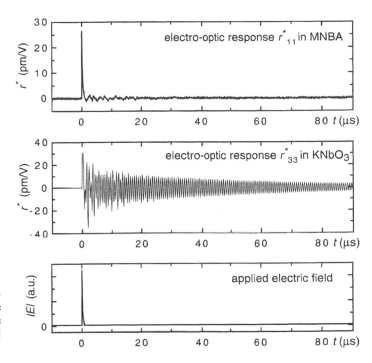

Figure 12 Demonstration of the different time-dependent electro-optic response of ferroelectric and organic crystals on application of a pulsed electric field.

Figure 12 illustrates the difference between a ferroelectric material and an organic compound in the time dependence of the electro-optic response when a pulsed electric field is applied. For the ferroelectric material (here $KNbO_3$), the crystal starts ringing in its mechanical resonances which are damped after typically 0.1 ms. For the organic compound (here MNBA), almost no ringing is observed.

F. SECOND HARMONIC GENERATION

The purely electronic contribution to the electro-optic coefficient r^e can be calculated from the nonlinear optical coefficients determined through second-order frequency-mixing experiments such as second harmonic generation and sum-frequency generation. Second harmonic generation (using the nonlinear optical coefficient d_{IJK}) is most often used. r^e_{IJK} is related to the nonlinear optical coefficient at frequencies $(\omega, \omega, 0)$, d^{EO}_{IJK}, by Equation 25.

For organic materials, the calculation of d^{EO}_{IJK} from a measured d_{IJK} at frequencies $(2\omega', \omega', \omega')$ is most often performed using Equation 26. For inorganics, this calculation is usually done with the Miller δ-coefficient[58]:

$$d^{EO}_{IJK} = d^{(-\omega,\omega,0)}_{IJK} = \frac{\chi^{\omega}_I \chi^{\omega}_J \chi^0_K}{\chi^{2\omega'}_K \chi^{\omega'}_I \chi^{\omega'}_J} d^{(-2\omega',\omega',\omega')}_{KIJ}$$

$$= \varepsilon_o \delta_{KIJ} \chi^{\omega}_I \chi^{\omega}_J \chi^0_K \tag{44}$$

If the nonlinear optical coefficient d_{IJK} at frequency ω' as well as the linear susceptibilities $\chi^{(1)}$ at the different wavelengths are known, d^{EO} can be calculated from Equation 44.

A comparison between measured r^s and calculated r^e also allows an estimation of the optic phonon contribution to the linear electro-optic effect, r^o. Values of r^e determined from second harmonic generation measurements are tabulated in Section IV.

G. OTHER EXPERIMENTAL METHODS

Besides the selection of experimental techniques for the characterization of electro-optic effects described here, there exists a variety of other possibilities. One interesting example was presented by Nahata et al., who developed an apparatus for the measurement of electro-optic coefficients for a coplanar electrode configuration especially suited for poled polymers.[59] In another aproach Yoshimura used a special ac modulation technique to characterize thin crystalline films.[60]

IV. ELECTRO-OPTIC MATERIALS

A. INORGANIC MATERIALS

1. Wavelength Dependence of the Linear Electro-optic Effect

Not too many measurements on the wavelength dependence of the linear electro-optic effect of inorganic materials are known. We will summarize some results of $KNbO_3$ only (Figure 13).[5,47] As expected, the wavelength dependence of both unclamped values $r_{33}^T - (n_2/n_3)^3 r_{23}^T$ and r_{42}^T of $KNbO_3$ is small. This is in agreement with the Miller δ-rule as well as with the fact that both acoustic and optic phonon contributions, which are independent of the wavelength of light, dominate the electro-optic effect in $KNbO_3$.

2. Linear Electro-optic Coefficients and Dielectric Constants of $KNbO_3$, $BaTiO_3$, and $LiNbO_3$

Table 1 lists experimental results of the clamped (r^s) and unclamped (r^T) electro-optic coefficients of $KNbO_3$,[61] $BaTiO_3$,[62] and $LiNbO_3$.[63] In addition, calculated values of r^e from measured nonlinear optical coefficients are also given (see Section III.F). The data required for these calculations were gathered from several publications.[63–68] For $KNbO_3$ and $BaTiO_3$, r^T is much larger than r^s, showing the important influence of acoustic phonons. In $LiNbO_3$ this is much less the case, especially for r_{33} for which there is only a small difference between clamped and unclamped values. However, for all the cases listed, r^e is several times smaller than r^s. This shows that indeed the main contribution to the electro-optic effect at high frequencies (above the acoustic phonon contributions) in the materials mentioned here arises from optic phonons. Note that currently there is an ongoing discussion on absolute values of the nonlinear optical coefficients (see, e.g., reference 69). The values of the nonlinear optical coefficients used for the calculation of r^e ($KNbO_3$,[64] $BaTiO_3$,[67] and $LiNbO_3$[66]) may therefore not be the best ones. Nevertheless, the main statement of $r^e \ll r^s$ remains valid.

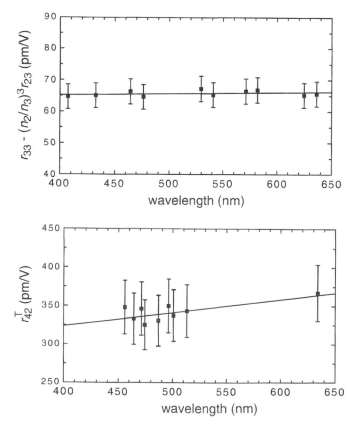

Figure 13 The wavelength dependence of the unclamped values $r_{33} - (n_2/n_3)^3 r_{23}$ (a) and r_{42}^T (b) of $KNbO_3$.

Table 1 Electro-optic Coefficients r_{IJK} (at $\lambda = 632.8$ nm, in pm/V), Nonlinear Optical Coefficients d_{IJK} (in pm/V), and Dielectric Constants ϵ_I of KNbO$_3$ (Point Group mm2), BaTiO$_3$ (Point Group 4mm), and LiNbO$_3$ (Point Group 3m) at Room Temperature

IJK	r^T	r^s	r^e	d^a	ϵ^T	ϵ^s	ϵ^e $(=n^2)$
			KNbO$_3$				
113	34 ± 2	20.1 ± 2	$+2.4$	-15.8			
223	6 ± 1	7.1 ± 0.5	$+2.6$	-18.3			
333	63.4 ± 1	34.4 ± 2	$+5.0$	-27.4			
131	120 ± 10	27.8 ± 3	$+2.6$	-16.5			
232	450 ± 30	360 ± 30	$+2.6$	-17.1			
1					150 ± 5	37 ± 2	4.8338
2					985 ± 20	780 ± 50	4.9845
3					44 ± 2	24 ± 2	4.4192
			BaTiO$_3$				
113	8 ± 2	10.2 ± 0.6	1.7	14.4			
333	105 ± 10	40.6 ± 2.5	0.7	5.5			
131	1300 ± 100	730 ± 100	1.6	13.6			
1					4400 ± 400	2200 ± 200	5.292
3					129 ± 5	56 ± 3	5.131
			LiNbO$_3$				
113	10.0	7.7	$+0.87$	-5.95			
333	32.2	28.8	$+5.4$	-34.4			
3					32	28	4.482

Note: The nonlinear optical coefficients were measured at $\lambda = 1064$ nm (KNbO$_3$ and LiNbO$_3$) and $\lambda = 1058$ nm (BaTiO$_3$). ϵ^e is assumed to be equal to n^2, extrapolated to infinite wavelengths. r_{IIJ} corresponds to d_{JII}.

The same observation as for the electro-optic coefficients can be made for the dielectric constants: Table 1 shows the distinct differences between ϵ^T, ϵ^s, and ϵ^e due to contributions from acoustic and optic phonons.

B. SEMICONDUCTORS

Table 2 shows electro-optic coefficients, dielectric constants, and refractive indices for three semiconductors: GaAs, InP (both III-V), and CdTe (II-VI). All data are from the *CRC Handbook of Laser Science and Technology*.[70] The electro-optic coefficients are smaller than the ones of the materials in Section IV.A. On the other hand, a comparison of dielectric constants and refractive indices shows that $\epsilon \approx \sqrt{n^2}$, which leads to large electro-optic modulation bandwidths (see Section V.A). In addition, we can conclude that the measured values of the electro-optic coefficients and the dielectric constants are mainly of electronic origin.

Table 2 Electro-optic Coefficients, Dielectric Constants, and Refractive Indices of Selected Semiconductors (GaAs, InP, CdTe)

Material	r^T_{41}	r^s_{41}	ϵ^T	ϵ^s	n
GaAs	—	1.2 ($\lambda = 1.02$ μm)	12.5	13.2	3.5 ($\lambda = 1.02$ μm)
	1.51 ($\lambda = 10.6$ μm)	—			—
InP	—	-1.34 ($\lambda = 1.06$ μm)	—	12.6	3.29 ($\lambda = 1.06$ μm)
CdTe	6.8 ($\lambda = 3.39$ μm)	—	—	9.4	2.82 ($\lambda = 1.3$ μm)
	6.8 ($\lambda = 10.6$ μm)	5.3 ($\lambda = 10.6$ μm)			2.67 ($\lambda = 10.6$ μm)
					2.60 ($\lambda = 10.6$ μm)

Note: All of them have point group $\overline{4}3$ m.

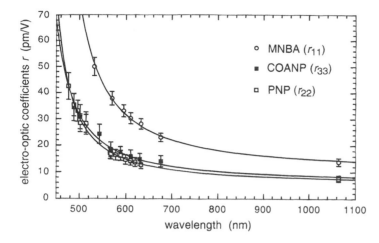

Figure 14 The wavelength dependence of some electro-optic coefficients of COANP, PNP, and MNBA.

C. MOLECULAR CRYSTALS

1. Wavelength Dependence of the Linear Electro-optic Effect

As in the case of inorganics, only relatively few wavelength-dependent measurements of electro-optic coefficients in molecular crystals have been reported. Figure 14 shows examples of 2-cyclooctylamino-5-nitropyridine (COANP), 2-(N-prolinol)-5-nitropyridine (PNP), and 4'-nitrobenzylidene-3-acetamino-4-methoxyaniline (MNBA). Note that for all cases in Figure 14 the two-level model describes the dispersion of r reasonably well (see Equation 21). The resonance frequencies ω_{eg} were determined from the values obtained from the measurement of the dispersion of the refractive indices.[9,48]

2. Dependence of the Linear Electro-optic Effect on the Frequency of the Applied Electric Field

Figure 15 shows the dependence of the effective electro-optic coefficient $|n_3^3 r_{33} - n_2^3 r_{23}|$ on the modulation frequency ν measured at $\lambda = 514.5$ nm for COANP.[9] The dotted line in the main graph

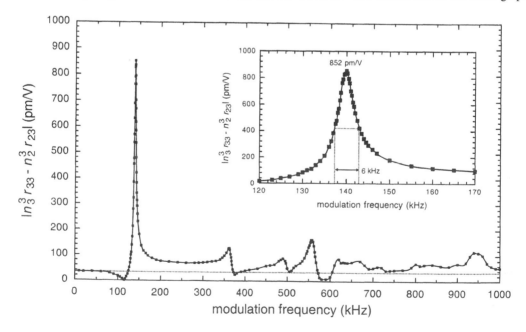

Figure 15 Dependence of the effective electro-optic coefficient $|n_3^3 r_{33} - n_2^3 r_{23}|$ on the modulation frequency ν measured at $\lambda = 514.5$ nm (applied voltage: 7 V (rms), electric field $|E|_{rms} = 4.1$ kV/m) for COANP. The *dashed line* shows the value of $|n_3^3 r_{33} - n_2^3 r_{23}|$ calculated from r_{33} and r_{23} (interferometric measurement at 1 kHz). The crystal was mounted between metal plates. *Inset:* First resonance with a better resolution.

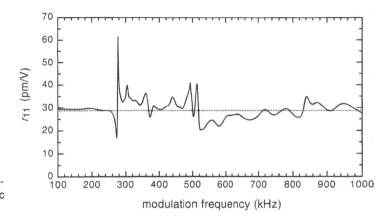

Figure 16 Frequency dependence of the largest electro-optic coefficient r_{11} of MNBA.

shows the value of $|n_3^3\, r_{33} - n_2^3\, r_{23}|$ calculated from r_{33} and r_{23} (interferometric measurement at 1 kHz). The crystal was fixed between two parallel metal plates where care was taken not to clamp the crystal too strongly. We see that there are several resonances starting at about 150 kHz with enhancements of $|n_3^3\, r_{33} - n_2^3\, r_{23}|$ up to a factor of 25. In order to identify these resonances the measurements were repeated for crystals with different geometries. It could be clearly observed that the peak at $v_l = 140$ kHz was due to a standing longitudinal wave in the b direction. A rough estimation of the velocity of sound for longitudinal waves along the b direction yielded $c_s = 2100$ m/s. From c_s also an approximate value for the elastic constant $c_{33} = c_s^2\, \rho$ could be deduced, $c_{33} = 5.5 \cdot 10^9$ Nm^{-2}. Models explaining resonance frequencies for different boundary conditions can be found in reference 9.

The method also allows the separation of the unclamped (r^T) from the clamped (r^s) electro-optic coefficients if the frequency of the applied electric field is extended above all acoustic resonances. Figure 16 shows another example of MNBA.[18] We see that for MNBA almost no influence of acoustic phonons on the linear electro-optic effect can be seen confirming the almost exclusively electronic origin of the effect in MNBA. This result has also been confirmed with a pulsed electro-optic measurement (see below).[57]

Figures 17 and 18 show the time evolution of the electro-optic response for mechanically unclamped crystals of KNbO$_3$ and MNBA on application of a square-wave voltage (see Section III.E). A comparison between the graphs in Figure 18 shows that indeed acoustic-phonon contributions are of minor importance in the organic MNBA, where 95% of the unclamped electro-optic effect can be used for high-frequency modulation purposes. This is in contrast to KNBO$_3$, where the ratio of r_{33}^s/r_{33}^T gives only 54% (Figure 17). Such experiments have confirmed that organic materials show greatly improved response characteristics as compared to inorganic perovskite crystals. Further results of such measurements are given in Sections IV.A.2 and IV.C.3.

3. Compilation of Data on the Linear Electro-optic Effect

Table 3 lists experimental results of electro-optic coefficients of various molecular crystals. The wavelength of maximum absorption of the bulk is also given. For diagonal tensor elements the factors θ_p and $\cos^3\theta_p$ (θ_p is the angle between the molecular charge-transfer axis and the polar crystalline axis) are included. For an optimized structure we have $\theta_p = 0$. Therefore, $\cos^3\theta_p$ indicates the decrease in the electro-optic coefficient compared to an ideal arrangement of the molecules in the crystal lattice (neglecting possible local field effects). Dielectric constants of some of the materials are tabulated below (Sections V.A and V.B). A few of the materials of Table 3 are discussed in more detail in the following paragraphs.

The largest electro-optic coefficients published so far for an organic single crystal are found in DAST (4'-dimethylamino-N-methyl-4-stilbazolium tosylate).[82] The molecular arrangement ($\theta_p = 20°$) is almost ideal for electro-optics. Perry et al.[82] measured an electro-optic coefficient $r_{11} = 400$ pm/V at 820 nm leading to $n_1^3 r_{11} = 4200$ pm/V which is a value comparable to that found in the most efficient inorganic electro-optic and photorefractive materials.[61,62] Very recent measurements, as well as estimations based on measured nonlinear optical coefficients, have shown that a more reliable value of r_{11} is around 160–170 pm/V.[71,72] This value is still huge when considering that it is purely electronic in origin. It is also huge in comparison with polymers (see Table 4). Our estimation based on measured molecular hyperpolarizabilities show, however, that bigger values should be feasible in optimized crystalline structures.

414

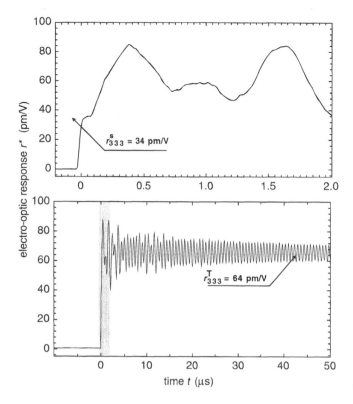

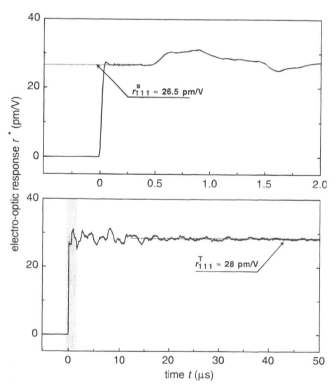

Figure 17 Electro-optic response of a mechanically unclamped inorganic ferroelectric crystal (KNbO$_3$).

Figure 18 Electro-optic response of a mechanically unclamped organic crystal (MNBA).

Table 3 Examples of Molecular Crystals that Have Been Investigated for Their Electro-optic Response

Material	Point Group	r(pm/V), n	θ_p (deg), $\cos^3\theta_p$	$\lambda_{eg} = 2\pi c/\omega_{eg}$ of bulk (nm)	Ref.
DAST 4'-Dimethylamino-N-methyl-4-stilbazolium tosylate	m	r_{11}(820 nm) = 160 n_1 (820 nm) = 2.455 r_{11} (1500 nm) = 70 n_1 (1500 nm) = 2.162	$\theta_p = 20$ $\cos^3\theta_p \approx 0.83$	600 Assumed: 100 nm below cutoff wavelength at 700 nm	71, 72
DMSM (SPCD) 4'-Dimethylamino-N-methyl-4-stilbazolium methylsulfate	mm2	r_{33} (632.8 nm) = 430 $n_3^3 r_{33}$(632.8 nm) = 1300 n_3 (632.8 nm) = 1.55	$\theta_p = 34$ $\cos^3\theta_p \approx 0.57$	460 Estimated from absorption peak at 2.7 eV	60, 73
DCNP 3-(1,1-Dicyanoethenyl)-1-phenyl-4,5-dihydro-1H-pyrazole	m	r_{33} (632.8 nm) = 87 n_3 (632.8 nm) = 2.7	—	531 From the fit of the measured values of $n^3 r$ using Equation 23	74
MMONS 3-Methyl-4-methoxy-4'-nitrostibene	mm2	r_{33}(632.8 nm) = 39.9 n_3 (632.8 nm) = 2.129	$\theta_p = 34$ $\cos^3\theta_p \approx 0.57$	421 From the Sellmeier coefficients of n_3	26

Continued

Table 3 *Continued*

Material	Point Group	r(pm/V), n	θ_p (deg), $\cos^3\theta_p$	$\lambda_{eg} = 2\pi c/\omega_{eg}$ of bulk (nm)	Ref.
MNBA 4'-Nitrobenzylidene-3-acetamino-4-methoxyaniline	m	$n_1^3 r_{11}$ = 685 pm/V at 514.5 nm r_{11} (532 nm) = 50 ± 4 r_{11} (632.8 nm) = 29 ± 2 n_1 (532 nm) = 2.136 n_1 (632.8 nm) = 2.024	θ_p = 18.7 $\cos^3\theta_p \approx 0.85$	390 From the Sellmeier coefficients	48
NMBA 4-Nitro-4'-methylbenzylidene aniline	m	r_{11} (488 nm) = 37.2 r_{11} (514.5 nm) = 36.4 r_{11} (632.8 nm) = 25.2 n_1 (488 nm) = 2.283 n_1 (514.5 nm) = 2.216 n_1 (632.8 nm) = 2.078	$\theta_p \approx 18$ $\cos^3\theta_p \approx 0.86$	353 From the fit of the measured values of $n^3 r$ using Equation 23	75
MNA 2-Methyl-4-nitroaniline	m	r_{11} (632.8 nm) = 67 n_1 (632.8 nm) = 2.00	θ_p = 21 $\cos^3\theta_p \approx 0.81$	414 From the Sellmeier coefficients of n_1	25, 28
DBNMNA 2,6-Dibromo-N-methyl-4-nitroaniline	mm2	$n^3 r(514.5\text{ nm})$ = 367 $n^3 r(632.8\text{ nm})$ = 79 $n^3 r(810\text{ nm})$ = 40	—	458 From the fit of the measured values of $n^3 r$ using Equation 23	59

Compound	Point group	Electro-optic and refractive index data	Phase-matching	Cutoff (nm)	Notes	Ref.
PNP 2-(N-prolinol)-5-nitropyridine	2	r_{22} (514.5 nm) = 28.3 r_{22} (632.8 nm) = 12.8 r_{12} (514.5 nm) = 20.2 r_{12} (632.8 nm) = 13.1 n_1 (514.5 nm) = 2.164 n_1 (632.8 nm) = 1.994 n_2 (514.5 nm) = 1.873 n_2 (632.8 nm) = 1.788	$\theta_p = 59.6$ $\cos^3\theta_p \approx 0.13$	404	From the Sellmeier coefficients of the refractive indices	9, 76
COANP 2-Cyclooctylamino-5-nitropyridine	mm2	r_{33} (514.5 nm) = 28 r_{33} (632.8 nm) = 15 n_3 (514.5 nm) = 1.715 n_3 (632.8 nm) = 1.647	$\theta_p = 61.8$ $\cos^3\theta_p \approx 0.11$	410	From the Sellmeier coefficients of n_3	9, 77
MBANP (−)2-(α-Methylbenzylamino)-5-nitropyridine	2	r_{eff} (488 nm) = 31.4 ± 0.1 r_{eff} (514.5 nm) = 26.6 ± 0.1 r_{eff} (632.8 nm) = 18.2 ± 0.2	$\theta_p = 33.2$ $\cos^3\theta_p \approx 0.59$	≈ 350	Assumed: 100 nm below cutoff at 450 nm	78
DAN 4-(N,N-dimethylamino)-3-acetamidonitrobenzene	2	r_{32} (632.8 nm) = 13 n_3 (632.8 nm) = 1.949	—	419	From the Sellmeier coefficients of n_3	79, 80
mNA meta-Nitroaniline	mm2	r_{33} (632.8 nm) = 16.7 n_3 (632.8 nm) = 1.675	—	400	Estimated	81

Continued

Table 3 *Continued*

Material	Point Group	r(pm/V), n	θ_p (deg), $\cos^3\theta_p$	$\lambda_{eg} = 2\pi c/\omega_{eg}$ of bulk (nm)	Ref.
POM					
3-Methyl-4-nitropyridine-1-oxide	222	r_{52}(632.8 nm) = 5.2 n_2 (632.8 nm) = 1.92 n_3 (632.8 nm) = 1.64	—	358 From the Sellmeier coefficients of the refractive indices	46
Optimized benzene (λ_{eg} (solid state) = 401 nm)	—	r (501 nm) = 210 n(501 nm) = 2.2	$\theta_p = 0$ $\cos^3\theta_p = 1$	401	9
Optimized stilbene (λ_{eg} (solid state) = 479 nm)	—	r(579 nm) = 1150 n(579 nm) = 2.4	$\theta_p = 0$ $\cos^3\theta_p = 1$	479	9

Table 4 Electro-optic and Nonlinear Optical Properties of Selected Polymer Systems

Type[a]	Matrix and Nonlinear Optical Moiety	E_p (V/μm)	Chromophore Concentration (weight %)[b]	d (pm/V) (@λ, nm)	r (pm/V) (@λ, nm)	T_g (°C)	Ref.
GH	Poly(methylmethacrylate) (PMMA)	37	(2.74)	—	$r_{33} = 2.5$ (633)	—	50
		62	(2.74)	$d_{33} = 2.5$ (1580)	—	—	50
	4[N-Ethyl-N-(2 hydroxyethyl)] amino-4′-nitroazobenze (Disperse Red 1) (DRI)	240	(2.0)	—	$r_{33} = 11$ (633)	≈90	97
		—	(2.3)	$d_{33} = 8.4$ (1580)	—	—	96
GH	PMMA + 4-dicyanovinyl-4′-(dimethylamino) azobenzene (DCV)	—	10	$d_{33} = 15.1$ (1356)	—	—	98
		—	(2.3)	$d_{33} = 31$ (1580)	—	127	96
GH	PMMA + 4-tricyanovinyl-4′-(diethylamino) azobenzene (TCV)	—	10	$d_{33} = 84$ (1356)	—	—	98

Continued

Table 4 *Continued*

420

Type[a]	Matrix and Nonlinear Optical Moiety	E_p (V/μm)	Chromophore Concentration (weight %)[b]	d (pm/V) (@λ, nm)	r (pm/V) (@λ, nm)	T_g (°C)	Ref.
GH	Pyralin 2611D (polyimide) 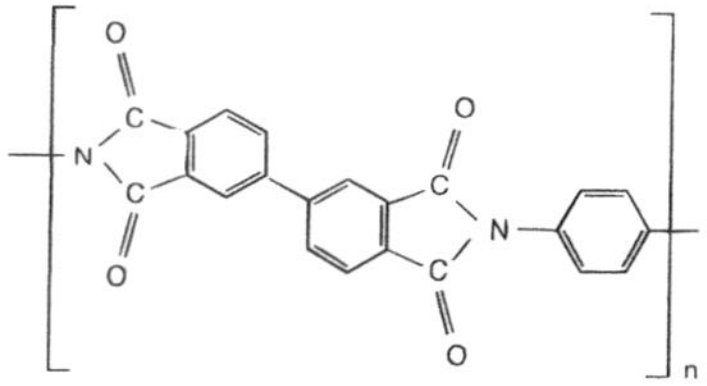 + Eriochrome Black T 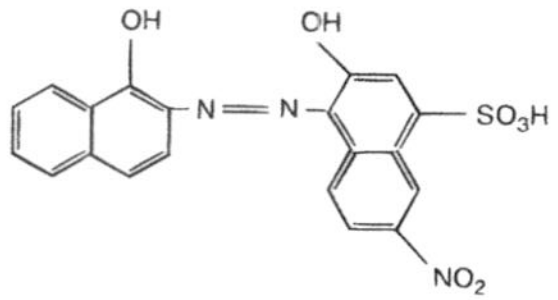	50	10	—	a few pm/V (633)	≈320	39
GH	Polyamic acid PIQ-2200 + N,N'-diethylaminotricyanovinyl substituted thiophene stilbene 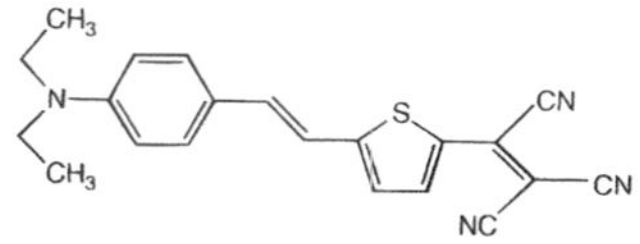	100	12	—	$r_{33} = 10.8$ (1520)	≈220	99
CL	N,N-Diglycidyl-4 nitroaniline (NNDN) + N-(2-aminophenyl)-4-nitroaniline (NAN) 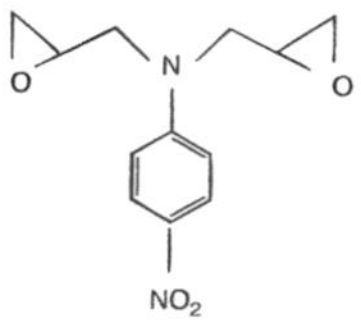 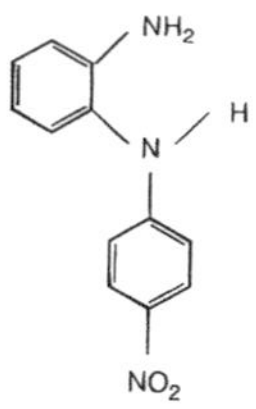	—	63	$d_{33} = 50$ $d_{31} = 16$ (1064)	$r_{13} = 6.5$ (530.9)	≈110	100

421

CL	PMMA-based + di-[2-hydroxyethyl]aminohexanylsulfonyl azostilbene	—	32	$d_{33} = 60$ (1064)	—	—	101
	2,2′-Methoxy-4,4′-biphenyl diisocyanate (cross-linker)						
MC	Polyamide main-chain polymer + 2′,5′-diamino-4-(dimethylamino)-4′nitrostilbene (DDANS) 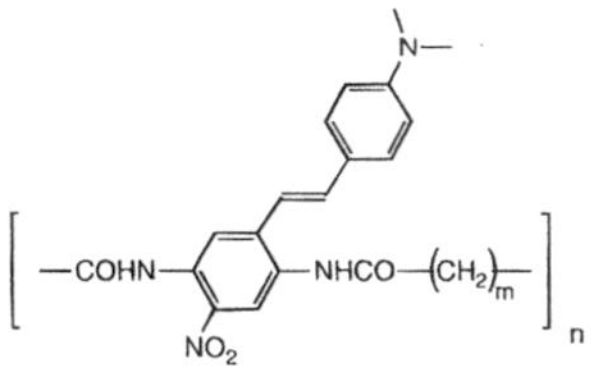	—	67.7	—	$r_{33} = 16$ (1300)	125 ($m = 12$)	34
			75.8	—	—	159 ($m = 8$)	
			86.3	—	—	206 ($m = 4$)	
SC	PMMA + 4-dicyanovinyl-4′-[N-ethyl-N-(2-hydroxyethyl)] azobenzene DCV 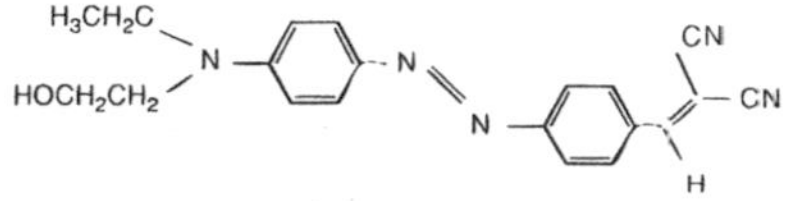	—	(8.0)	$d_{33} = 21$ (1580)	$r_{33} = 18$ (799)	127	96

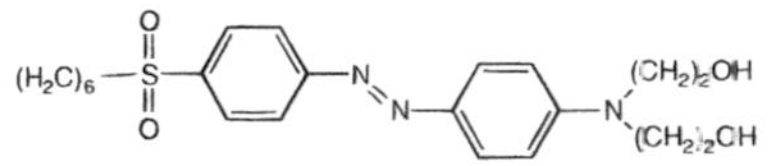

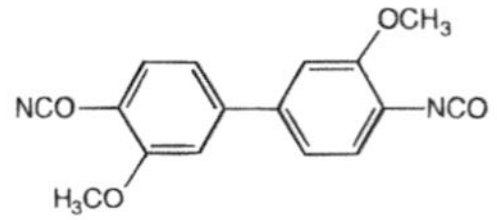

Continued

Table 4 *Continued*

Type[a]	Matrix and Nonlinear Optical Moiety	E_p (V/μm)	Chromophore Concentration (weight %)[b]	d (pm/V) (@λ, nm)	r (pm/V) (@λ, nm)	T_g (°C)	Ref.
SC	MMA + 4-dicyanovinyl-4'-(diethylamino) diazobenzene 3RDCVXY	≈200	(4.0)	d_{33} = 420 (1064)	r_{33} = 40 (633)	<140	102
SC	PMMA-based + DR1-based	50	—	—	r_{33} = 2.1 (1305)	132	86
SC	PMMA + N-ethylamino-N'-[6-hydroxyhexyl]-tricyanovinyl substituted thiophene stilbene	50	35	—	r_{33} = 23 (1520)	≈115	104
SC	+ 4-Dimethylamino-4'-nitrostilbene (DANS)[c]	170	35	—	r_{33} = 34 (1300)	142	105
SC		300	35	d_{22} = 153 (1064)	—	140	90
SC	+DANS based[c]	≈200	—	—	r_{33} = 38 (1300) r_{33} = 98 (633)	—	106

	Structure / Name						
SC	polyimide structure; with 4-(*p*-ethoxydiphenylamino-4′-nitro)-azostilbene; R=	—	47	—	$r_{33} = 10.5$ (1300)	227	107
SC	Polyurethane polymers with naphthoquinone methide dye	50	23	—	$r_{33} = 11$ (1300)	—	108
SC	Fluorene-based cardo-type polymer (BISMAN polyester)	50	14.4	—	$r_{33} = 1.7$ (1305)	205	103

Continued

Table 4 *Continued*

Type[a]	Matrix and Nonlinear Optical Moiety	Chromophore Concentration (weight %)[b]	E_p (V/μm)	d (pm/V) (@λ, nm)	r (pm/V) (@λ, nm)	T_g (°C)	Ref.
SC	Alternating styrene-maleic-anhydrid copolymer (polyimide) + DR1	56	150	$d_{31} = 19$ (1338)	$r_{13} = 6.5$ (1313)	172	45
SC	Polyimide + DR 1	—	170	—	$r_{33} = 10.8$ (1300)	—	109

Note: The poling process induces a symmetry C_∞ leading to $r_{33} = 3 \times r_{13}$.

[a] GH, guest–host system; CI, cross-linked polymer; SC, side-chain polymer; MC, main-chain polymer.

[b] Values in parentheses refer to the number density $N(10^{26}\ m^{-3})$.

[c] Backbone polymer not known.

Large electro-optic coefficients have also been found in SPCD (styrilpyridinium cyanine dye) also abbreviated as DMSM (4'-dimethylamino-N-methyl-4-stilbazolium methylsulfate).[60,73] Crystals of SPCD have point group symmetry mm2 and the charge-transfer axis of the cations is inclined by $\theta_p = 34°$ towards the symmetry axis. Yoshimura[60] measured an electro-optic figure-of-merit $n_3^3 r_{33} = 1300$ pm/V at 632.8 nm. DCNP (3-(1,1-dicyanoethenyl)-1-phenyl-4,5-dihydro-1H-pyrazole) is another material with very strong electro-optic response.[74,83] Its electro-optic figure-of-merit $n^3 r$ reaches values around 1720 pm/V at 632.8 nm. However, this is a wavelength very close to the absorption edge and measurements at longer wavelength show a drastic decrease by nearly a factor of 5 over a wavelength range of 300 nm.[83] The molecular charge-transfer axis of DCNP lies nearly parallel to the polar axis of the crystal which is an ideal orientation for electro-optic applications. MMONS (3-methyl-4-methoxy-4'-nitrostilbene) has a molecular arrangement ($\theta_p = 34°$) which is not ideal for electro-optics but comparable to DMSM.[26] However, its electro-optic figure-of-merit is less than half of that of DMSM.

MNBA (4'-nitrobenzylidene-3-acetamino-4-methoxyaniline) and NMBA (4-nitro-4'-methylbenzylidene aniline) are very similar two-ring systems. Both have an aza substitution in the 1-position that should lower the second-order polarizability by about a factor of 2 as compared to the corresponding stilbene derivatives.[21] Both have a very similar angle between the molecular charge-transfer axis and the polar crystal axis ($\theta_p = 18.7°$, MNBA,[48] $\theta_p \sim 18°$, NMBA[75]). Therefore, the values for r_{11} of 29 pm/V for MNBA[48] and 25.2 pm/V for NMBA[75] are reasonable when comparing them with measured hyperpolarizabilities β. MNA (2-methyl-4-nitroaniline) is the best electro-optic material among the group of compounds with benzene and pyridine rings.[25] Its molecular alignment in the crystal lattice is nearly perfect for electro-optics ($\theta_p = 21°$).

The last two materials are optimized with regard to their electro-optic response as discussed in Section II.I. All other compounds in Table 3 and displayed in Figure 2 are single π-electron ring systems with molecular hyperpolarizabilities comparable to MNA. However, the molecular arrangement is far from ideal for electro-optics.

D. POLYMERS

1. Dependence of the Linear Electro-optic Effect on the Poling Field

Noncentrosymmetry in polymers is induced by alignment of the chromophores through the coupling of their dipole moment to an external poling field. Four statistical models describing the molecular distribution in the polymer are treated in the literature: the isotropic and the Ising model,[84] the model of Singer et al. (SKS),[50] and the model of van der Vorst and Picken (VVP).[85] The most often used expressions for the macroscopic susceptibilities are the linear approximations of the isotropic model in the weak field regime. In this case, the electro-optic coefficient r_{33} is linearly proportional to the poling field. Figure 19 illustrates the dependence of the electro-optic coefficient on the poling field for a covalently functionalized Disperse Red 1 copolymer at 810 nm (Table 4[86]). The dashed line is a linear extrapolation, and the solid line is a nonlinear least-square fit to Langevin functions.[86] Above about 100 V/μm the data points start to deviate from the linear extrapolation: The weak field limit is no longer valid.

Besides poling perpendicularly to the film surface, in-plane poling of electro-optic polymers can

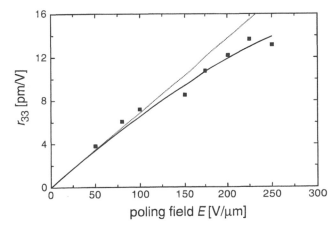

Figure 19 Electro-optic coefficient r_{33} vs. poling field for a covalently functionalized Disperse Red 1 copolymer at 810 nm. (Adapted from Nahata, A., et al. *J. Opt. Soc. Am.*, B10, 1553, 1993. With permission.) The *dashed line* is a linear extrapolation; the *solid line* is a nonlinear least-square fit to Langevin functions.[86] Above about 100V/μm the data points start to deviate from the linear extrapolation: the weak field limit is no longer valid in these experiments.

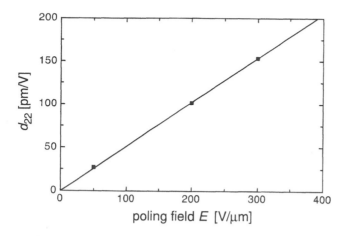

Figure 20 Nonlinear optical coefficient as a function of the poling field for in-plane poled DANS side-chain polymers. A remarkable linear response was observed up to a poling field of 300 V/μm. (Adapted from Otomo, A., et al., *Appl. Phys. Lett.*, 65, 2389, 1994. With permission.)

lead to many interesting applications using counterpropagating beams. Among them are high-speed signal processing such as pulse convolution and wavelength division multiplexing.[87] In addition, a large potential for efficient frequency doubling due to large resonant nonlinearities exists.[88,89] Poling fields as high as 370 V/μm in an in-plane geometry were achieved in 4-dimethylamino-4'-nitrostilbene (DANS) side-chain polymers before breakdown occurred (Table 4[90]). A linear dependence of the nonlinear optical coefficient d_{22} on the poling field up to 300 V/μm was observed in these experiments (Figure 20).

2. Dependence of the Linear Electro-optic Effect on the Chromophore Concentration

At low concentrations conventional theory predicts a linear dependence of the electro-optic response on the chromophore concentration for glassy polymers. At very high concentrations a deviation from this linear dependence might occur. Possible reasons can be chromophore aggregation or increased dipole interactions leading to a reduced orientation of the chromophores. Figure 21 shows the linearity of the electro-optic coefficient with number density for a covalently functionalized Disperse Red 1 copolymer (Table 4[86]). Samples with molar fractions from 0 mol% up to 100 mol% of the chromophore were prepared. A poling field of 50 V/μm was applied to all samples. These experiments confirmed that the nonlinear optical chromophores respond to the poling field as independent molecules up to concentrations of 1.7×10^{27} m^{-3}. Note that larger chromophore concentrations usually lead to a decrease of the glass transition temperature.[45]

3. Temporal Stability of the Linear Electro-optic Effect

Insufficient temporal stability of poled polymers at elevated temperatures is one factor that precludes the present use of these materials for many potential device applications. According to the description in Section II.J, relaxation times of electro-optic polymers at different temperatures can be deduced from second harmonic generation (SHG) experiments, for example.

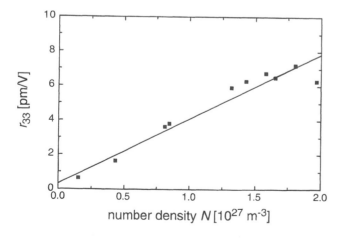

Figure 21 Concentration dependence of the electro-optic coefficient r_{33} (poling field of 50 V/μm) of a covalently functionalized Disperse Red 1 copolymer at 810 nm. (Adapted from Nahata, A., et al., *J. Opt. Soc. Am.*, B10, 1553, 1993. With permission.) The *solid line* is a least-square fit to the data.

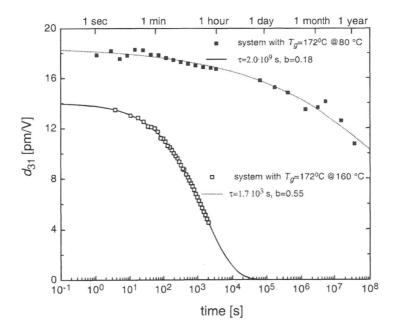

Figure 22 Orientational relaxation of the nonlinear optical susceptibility of a polyimide side-chain polymer (T_g = 172°C) as measured with second harmonic generation at two different temperatures.[91]

Figure 22 shows examples of the decay curves for an alternating styrene-maleic-anhydrid copolymer with Disperse Red 1 side chains at two different decay temperatures in a semilogarithmic plot which is the most appropriate representation for this kind of relaxation (Table 4[45,91]). All of the decay experiments exhibit a fast initial decay amounting to about 20–30% of the total initial SHG signal during the first 1–10 min. This decay is attributed to the release of charge carriers trapped in the film after the corona voltage has been turned off. The contribution from this fast decay has to be neglected when analyzing the decay using Equation 28. The fits in Figure 22 nicely represent the data points corrected for their fast decay.

From Figure 22 we expect the side-chain polymer with T_g = 172°C to be stable up to 100 years at temperatures up to 80°C.[91] New chromophores with improved nonlinear optical properties would allow one to reduce the chromophore content without loosing any of the electro-optic efficiency but with higher glass transition temperatures and therefore even better long-term stability.

The results of second harmonic generation experiments allow further predictions of relaxation times. Figure 23 shows the scaling behavior below the glass transition for two types of polyimide polymer systems.[45] In first approximation the relaxation time at any temperature can be deduced from that figure. For example, a side-chain polymer with T_g = 172°C is predicted to have a relaxation time of about 3 years at an operating temperature of 100°C. The data are plotted along with the results of Stähelin et al. for guest–host polymer matrices with glass transition temperatures comparable to our systems.[38] The second harmonic generation results indicate that the relaxation time of the nonlinear optical chromophores of side-chain polymers correlate better with this main-chain segmental motion than do guest–host nonlinear optical polymer matrices.

4. Wavelength Dependence of the Linear Electro-optic Effect

As in the case of molecular crystals, the dispersion of the electro-optic coefficient of electro-optic polymers can be well described by the two-level model for wavelengths away from linear absorption. An example of the wavelength dependence of the electro-optic coefficient r_{33} is shown in Figure 24[91] for the same polyimide polymer as discussed in Section IV.D.3. Values for poling fields of 100 V/μm and 220 V/μm are shown. The almost linear dependence of the electro-optic coefficient on the poling field can be confirmed for this polymer. The wavelength of maximum absorption as deduced from absorption measurements of this material is λ_{eg} = 470 nm. Measurements of the electro-optic coefficients

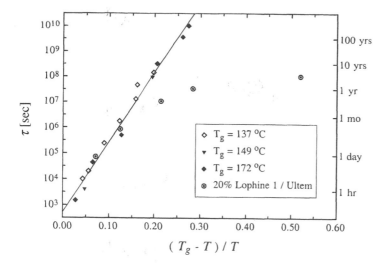

Figure 23 Second harmonic decay times below the glass transition with $(T_g - T)/T$ as the relevant scaling parameter for polyimide side-chain polymers.[45] For comparison, data points (*circles*) for guest–host polymer matrices with glass transition temperatures comparable to the polyimide systems are plotted.[38] The second harmonic generation results indicate that the relaxation time of the nonlinear optical chromophores of side-chain polymers correlate better with this main-chain segmental motion than do guest–host nonlinear optical polymer matrices.

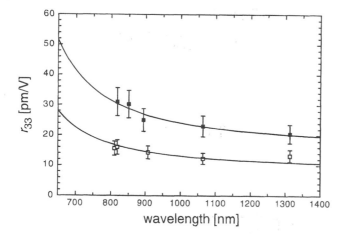

Figure 24 The wavelength dependence of a polyimide side-chain polymer (*dashed line*, poling field of 100V/μm; *solid line*, poling field of 220 V/μm).[91] Theoretical curves using the two-level model ($\lambda_{eg} = 470$ nm) are also shown.

of other materials inside the absorption range have also been carried out by other groups.[92] The two-level, however, does not properly describe the dispersion in this case.

5. Compilation of Data on the Linear Electro-optic Effect

In recent years a considerable amount of research has been directed towards the development of nonlinear optical and electro-optic polymers. These polymers are thought to have considerable potential for efficient, ultrafast, and low-voltage electro-optic modulators and related devices.[93] Potential advantages of polymeric materials are high nonlinear optical coefficients, ease of processing, and device fabrication. Low production costs are envisioned for these materials. However, additional requirements must be satisfied before polymeric materials can demonstrate their potential as actual device materials. Some of these include good long-term stability of the nonlinear optical or electro-optic effect, good structural and chemical stability, and low optical propagation losses. Recent reports in the literature have demonstrated that all of these properties can be achieved with polymeric materials, but presently no single polymer can satisfy all the requirements of current device specifications.[94,95]

Experimental results on electro-optic polymers can be found in Table 4. The list only contains selected polymers and is by no means complete. It lists guest–host, crossed-linked, side-chain, and main-chain polymers with corresponding poling fields, chromophore densities, nonlinear and electro-optic coefficients, and glass transition temperatures. Initial research using relatively simple guest–host systems demonstrated the advantages of this new class of organic materials and were later used to test new chromophores for their electro-optic potential.[39,50,96–99] Work then continued with the search for better nonlinear optical dyes on one side and the amelioration of the long-time stability through high

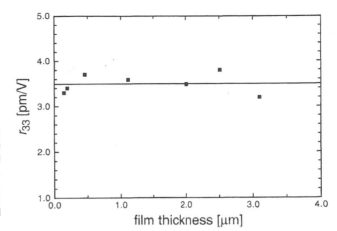

Figure 25 Dependence of the electro-optic coefficient r_{33} (poling field of 50 V/μm) on the film thickness for a covalently functionalized Disperse Red 1 copolymer at 810 nm. (Adapted from Nahata, A., et al. *J. Opt. Soc. Am.*, B10, 1553, 1993. With permission.) Very reproducible values were measured from 0.1 to 3 μm.

glass transition temperatures (T_g) or cross-linked systems on the other hand. Examples of improving the stability using guest–host systems with polyimide polymers which are known for their excellent thermal stability and thermal or photo-cross-linkable network polymers are listed in Table 4.[39,100,101]

Most work has concentrated on side-chain polymers (see Table 4[45,86,90,96,102–109]). Large electro-optic coefficients (e.g., r_{33} = 34 pm/V at 1300 nm) were reported.[105] Electro-optic polymers based on polyimides showed large electro-optic coefficients with very good long-term stability at 80°C.[45,110]

Polymers with even better long-term stability (glass transition temperature above 200°C) were developped.[103,107] Note that for high-temperature electro-optic polymers (e.g., stability at 150°C for long periods) new chromophores with increased thermal stability have to be developed as well.[107,111]

The preparation of polymeric electro-optic thin films is of good reproducibility. Figure 25 shows an example of measurements of the electro-optic coefficient as a function of film thickness ($λ$ = 810 nm, poling field of 50 V/μm).[86] The electro-optic coefficient r_{33} stays constant over a thickness range from 0.1 to 3 μm.

In terms of electro-optic coefficients, polymers are as good as the inorganic materials, but they cannot compete with molecular crystals. On the other hand, electro-optic polymers are ideal for waveguide applications with high-speed modulation. A more detailed discussion on device issues can be found in Section V and especially in other chapters of this book.

More information on electro-optic polymers can be found in the recent proceedings of the topical meetings on "Organic Thin Films for Photonic Applications"[94,95] and "Nonlinear Optical Properties of Organic Materials VII."[112] In addition, an extensive review was presented by Burland et al.[113]

E. LANGMUIR-BLODGETT FILMS

Experimental results of electro-optic measurements are given in Table 5. Note that not much work has been done with regard to electro-optic Langmuir-Blodgett (LB) films. Therefore, Table 5 has been added mainly for completeness. Most values listed are quite large. This is mainly due to a resonance enhancement, as can also be seen from the large imaginary part of the electro-optic coefficient. Only monolayers have been investigated for their electro-optic response. It is not guaranteed that molecules which form LB monolayers also form well-ordered multilayers.

In general, nonlinear optical data on monolayers can be valuable for a comparison between monolayers of different molecules. In addition, such measurements can provide possible guidelines for further research. However, an extrapolation to multilayers (e.g., 100–200 monolayers), required for applications, is not straightforward.

A detailed description of various aspects of Langmuir-Blodgett films can be found in Bosshard et al.[18] Preparation of different types of multilayers, thermodynamic aspects, short- and long-range order, defects, nonlinear optical properties, waveguiding properties, and more are covered in that work. Very recently, nonlinear optical effects of LB films with special emphasis on frequency doubling and guided-wave nonlinear optics have been summarized and discussed by Bosshard and Küpfer.[117]

Table 5   Data on Electro-optic Coefficients of Langmuir-Blodgett Films

Molecule	r_{33} (pm/V)	Ref.
$C_{18}H_{37}$—NH—⟨⟩—CH=CH—⟨N⟩$^+$—CH$_3$	$-317.7 - i\ 398.7$	114
$C_{18}H_{37}$—O—⟨⟩—CH=CH—⟨N⟩$^+$—CH$_3$	$-37.4 + i\ 32$	114
$(C_{10}H_{21})_2$N—⟨⟩—CH=CH—⟨N⟩$^+$—C$_2$H$_4$—COO$^-$	$-279 + i\ 13.98$	115
$(C_{14}H_{29})_2$N—⟨⟩—CH=CH—⟨N⟩$^+$—C$_2$H$_4$—COO$^-$	$-169 - i\ 48.6$	115
$(C_{18}H_{37})_2$N—⟨⟩—CH=CH—⟨N⟩$^+$—C$_2$H$_4$—COO$^-$	$-156.1 - i\ 5.82$	115
$C_{22}H_{45}$—⟨N⟩$^+$—CH=CH—⟨⟩—N(CH$_3$)$_2$ Br$^-$	$223.7 + i\ 106$	116
		114

Note: All experiments were performed at 632.8 nm. All films were deposited on silver. Only monolayers were measured.

V. ELECTRO-OPTIC DEVICES

There already exist various electro-optic waveguide applications using LiNbO$_3$. Typical devices are the Mach-Zehnder interferometer and the directional coupler (Figure 26). Figure 27 shows an example of constructive and destructive interference at the output of a Mach-Zehnder interferometer. Relevant properties for such applications are large, nonresonant optical nonlinearities, low dc and microwave dielectric constants, and optical transparency in a wide wavelength range. In this section we discuss some important parameters and figures-of-merit related to electro-optic modulators and photorefractive applications. Other applications such as electro-optic deflectors are not discussed in this work. A detailed description of the principles and applications of electro-optic deflectors can be found in the book by Gottlieb et al.[118]

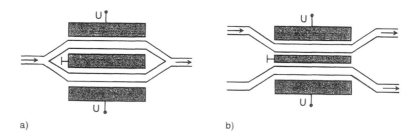

a)   b)

Figure 26   Configuration for (a) a Mach-Zehnder interferometer and (b) a directional coupler.

TE$_{00}$ TE$_{01}$

$\Delta\phi = 0°$ $\Delta\phi = 180°$

Figure 27 Amplitude modulation through electro-optically induced interference in a Mach-Zehnder interferometer.

A. ELECTRO-OPTIC MODULATION

1. Electro-optic Modulation Bandwidth

One of the big problems arising in the context of electro-optic amplitude modulation in waveguides is the bandwidth. The wavelength of the modulating electric field becomes shorter than the modulator length at high frequencies. In this case, the modulation of the optical beam is achieved with traveling microwaves. There are limits to the modulation bandwidth due to a refractive index mismatch between microwaves and optical waves. The 3-dB bandwidth of the modulator (frequency at which the power in the optical side bands is reduced by one-half) is given by (see reference 119)

$$\Delta f_{3\,\mathrm{dB}} = \frac{1.4c}{\pi \, | n_o - n_m | L} \tag{45}$$

where n_o and n_m ($= \sqrt{\varepsilon_m}$) are the refractive indices at optical and microwave frequencies, respectively, c is the speed of light, and L is the waveguide length. Table 6 lists $\Delta f_{3\,\mathrm{dB}}$ for some selected materials.

For $\Delta f_{3\mathrm{dB}}$ the advantage of semiconductor (e.g., GaAs) and organic materials comes into play. Both have a small dispersion in ε and, therefore, a small phase mismatch in waveguide modulators. In ionic crystals, lattice vibrations contribute significantly to the electro-optic coefficients and the dielectric constant leading to a strong frequency dependence of both. These ionic contributions increase the electro-optic effects considerably but, due to the increase of the dielectric constants, they also limit the bandwidth (see Equation 45). Nevertheless, inorganic materials can be used for fast electro-optic modulators: By using special electrode-waveguide geometries, the microwave speed can be increased to a certain degree and a 40-GHz Ti:LiNbO$_3$ modulator has been demonstrated.[120] On the other hand, in organic substances the contributions are mainly of electronic origin yielding smaller electro-optic effects, but also a small dependence on the frequency of the applied electric field. Recently, a traveling-wave polymeric optical intensity modulator was reported with more than 40 GHz of 3-dB electrical bandwidth.[121]

2. Power Requirements in Electro-optic Modulators

There are other limitations imposed on wideband optical modulators for communication purposes based on fundamental physics. One of them is the power requirement in electro-optic amplitude modulators. For the simple case of a bulk crystal modulator employing the transverse electro-optic effect, the power requirement per unit bandwidth is given by[6]

$$\left\langle \frac{P}{\Delta f} \right\rangle_t = \varepsilon_o (S^2 \lambda^3) \frac{\varepsilon}{n^7 r_{\mathrm{eff}}^2} \tag{46}$$

Table 6   Calculated Approximate 3-dB Bandwidths for a Selection of Materials

Material	n_o	n_m	$\Delta f_{3\,\mathrm{dB}}$ (GHz) for $L = 1$ cm
LiNbO$_3$	2.2 (@632.8 nm)	4.2	6.7
KNbO$_3$	2.1687 (@632.8 nm)	4.9	4.9
GaAs	3.5 (@1020 nm)	3.63	103
DAST	2.2 (@1300 nm)	2.7	27
MNBA	2.02 (@632.8 nm)	2.0	670
MNBA	1.90 (@1300 nm)	2.0	134
COANP	1.65 (@632.8 nm)	1.61	330
mNA	1.7 (@632.8 nm)	1.8	130
Alternating styrene-maleic-anhydrid copolymer (see Table 4)	1.66 (@1300 nm)	~1.9	56

where S is a safety factor to account for diffraction of focused beams. In practice, diffraction losses are negligible for $S \geq 3.$[6] Note that the above expression is only a simple approximation since, for example, any circuit capacitance and conductance in parallel with the crystal was neglected. $n^7 r_{eff}^2 / \varepsilon$ can be regarded as a figure-of-merit for this configuration if a minimum drive power is required. Table 7 lists some examples of this figure-of-merit together with reduced half-wave voltages v_π (Equation 3).

Clearly, both v_π and $n^7 r_{eff}^2 / \varepsilon$ are much more advantageous for organics than inorganic materials. This is, of course, due to the larger electro-optic coefficients as well as the lower dielectric constants of organics. For most material classes (except polymers) the refractive indices are large and do not play a decisive role here. For $S = 3$ we get drive powers per unit bandwidth of typically 3 mW/MHz for inorganics at $\lambda = 632.8$ nm and 0.1 mW/MHz for, e.g., the optimized stilbene compound at $\lambda = 1500$ nm.

B. PHOTOREFRACTIVE APPLICATION: FIGURE-OF-MERIT

Electro-optic materials with the possibility of photoexcitation of carriers can change their index of refraction or absorption constant when they are irradiated with light. Among the resulting effects are photochromism, photoinduced orientational gratings in liquid crystals, thermal gratings, electron population density gratings, and the space charge-induced photorefractive effect.[122] Several of these mechanisms can lead to permanent as well as reversible refractive index or absorption changes.

Here we focus on the space charge-induced photorefractive effect. This effect is often simply called *the photorefractive effect*, a convention that will be used here also. The photorefractive effect is a consequence of the interaction of different physical processes (photoinduced charge carrier generation, charge transport and trapping, and the electro-optic effect).

The experimental situation for a standard photorefractive experiment is shown in Figure 28. A photorefractive crystal is illuminated by two interfering beams. The direction of the grating vector **K** is denoted by x, the surface normal by z, y is perpendicular to x and z. The (internal) angle of incidence of the writing beams is given by θ (and is equal for both beams). The axes of the dielectric reference system of the crystal are assumed to be parallel to x, y, and z.

We will use two figures-of-merit for characterizing the photorefractive sensitivity of a material. The first one, S_n, describes how much optical energy absorbed per unit volume is needed to produce a given refractive index change; the second one, S_η, gives the energy needed for a diffraction efficiency of 1% for a 1-mm-thick storage material. The first one is given by[123]

Table 7 Calculated Reduced Half-wave Voltages and Figures-of-merit for Minimum Drive Power in Electro-optic Modulators Employing Bulk Crystals in the Transverse Configuration

Material	Wavelength (nm)	n	ε^s	r^s (pm/V)	v_π (kV)	$n^7 r_{eff}^2/\varepsilon$ ($\times 10^3$ (pm/V)^2)
KNbO$_3$ (r_{33})	632.8	2.167	24	34	1.8	11
KNbO$_3$ (r_{33})	1300	2.109	24	34	4.1	8.9
LiNbO$_3$ (r_{33})	632.8	2.196	28	28.8	2.1	7.3
LiNbO$_3$ (r_{33})	1300	2.134	28	28.8	4.6	6.0
GaAs (r_{41})	1020	3.50	13.2	1.2	20	0.7
MNA (r_{11})	632.8	2.0	3.5	67	1.2	160
MNBA (r_{11})	632.8	2.024	~4	30	2.5	31
DAST (r_{11})	820	2.45	~7–8	~160	0.34	4100
DAST (r_{11})	1300	2.19	~7–8	~75	1.65	180
DAST (r_{11})	1500	2.16	~7–8	~70	2.1	140
Optimized stilbene (r_{33})	1300	~2	~3–7	~330	~0.5	~1990–4650
Optimized stilbene (r_{33})	1500	~2	~3–7	~310	~0.6	~1760–4100
Alternating styrene-maleic-anhydrid copolymer (see Table 4) (r_{33})	1300	1.66	~3.6	~20	~14.2	~3.9

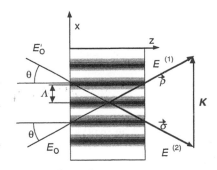

Figure 28 Experimental setup for the formation of a photorefractive grating.

$$S_n = \frac{n^3 r_{\text{eff}}}{2\varepsilon\epsilon_o} \times \frac{\phi}{h\nu} \times eL_{\text{eff}}m \qquad (47)$$

where L_{eff} is the effective charge transport length, m is the modulation factor, and ϕ is the quantum efficiency of carrier generation. The second figure-of-merit, S_η, is given by[123]

$$S_\eta = \frac{\pi}{\lambda \cos\theta} S_n \qquad (48)$$

where 2θ is the internal angle between the two writing beams in a photorefractive experiment. As expected, one of the material parameters, $n^3 r_{\text{eff}}/\varepsilon$, is related to electro-optics. Table 8 gives examples of the magnitude of $n^3 r_{\text{eff}}/\varepsilon$ for different material classes. For KNbO$_3$, BaTiO$_3$, and LiNbO$_3$, the unclamped coefficients were used. Note that the use of the clamped values only slightly changes $n^3 r_{\text{eff}}/\varepsilon$. The table also contains limits of $n^3 r_{\text{eff}}/\varepsilon$ for organic single crystals and polymers as derived in Bosshard et al.[18] assuming a space charge-induced photorefractive effect and neglecting any orientational contributions.[124]

Table 8 Material Parameter $n^3 r_{\text{eff}}/\varepsilon$ Relevant for Photorefractive Applications

Material	Wavelength (nm)	n	ε^T	r^T (pm/V)	$\lvert n^3 r_{\text{eff}}/\varepsilon\rvert$ (pm/V)
KNbO$_3$ (r_{33})	632.8	2.167	44	63.4	15
KNbO$_3$ (r_{42})	632.8	2.244	985	450	5
BaTiO$_3$ (r_{33})	632.8	2.359	129	105	11
BaTiO$_3$ (r_{51})	632.8	2.385	4400	1300	4
LiNbO$_3$ (r_{33})	632.8	2.196	32	32.2	11
MNBA (r_{11})	632.8	2.024	~4	30	62
DAST (r_{11})	820	2.45	~7–8	160	314
DAST (r_{11})	1300	2.19	~7–8	75	105
Limit for organic single crystals (optimized stilbene)[3]	580	~2.4	~3–7	~1200	5300
Alternating styrene-maleic-anhydrid copolymer[a] (see Table 4) (r_{33})	700	1.75	~3.6	~38	57
Limit for unordered saturated polymer (with applied electric field)	580	1.6–2.4	~3–7	550	2500

[a] Note that this polymer is not photorefractive and is only used to provide a typical value for $n^3 r_{\text{eff}}/\varepsilon$.

Again the advantage of organic materials with regard to this figure-of-merit is obvious. It has also been estimated that S_n and S_η could be of the same order of magnitude or even larger for organics in comparison with inorganics.[18]

VI. CONCLUDING REMARKS AND OUTLOOK

We have given an overview of the field of electro-optic effects in organic materials. In the first part of this review our work focuses mainly on molecular crystals. The experimental results are compared with theoretical calculations of limiting values of nonlinear optical coefficients. From this estimation we can conclude that still quite some improvement in the electro-optic coefficients of molecular crystals is possible. Our simple equation describing the wavelength dependence of the electro-optic figure-of-merit $n^3 r$ allows an easy comparison of different materials. This equation obviously also holds for polymers.

Our simple expression for the estimation of the upper limits for electro-optic coefficients shows that the current organic materials are not optimized. Optimized benzene derivatives with a cutoff at $\lambda = 500$ nm such as, e.g., MNA ($r_{11} = 67$ pm/V at $\lambda = 632.8$ nm) could have electro-optic coefficients around 210 pm/V ($\lambda = 500$ nm) or 83 pm/V ($\lambda = 632.8$ nm) for $\theta_p = 0°$. Materials with a cutoff at $\lambda = 580$ nm such as stilbene derivatives, e.g., DMSM ($r_{33} = 430$ pm/V at $\lambda = 632.8$ nm; DMSM is a salt which means that strong coulombic interactions are present!) could show electro-optic coefficients around 1150 pm/V.

We have reviewed the recent developments made in the field of nonlinear optical polymers. Initial research using relatively simple guest–host systems demonstrated the advantages of this new class of organic materials. Work then continued with the search for better nonlinear optical dyes on one side and the amelioration of the long-time stability through high glass transition temperatures (T_g) or cross-linked systems on the other hand.

For applications using electro-optic polymers as optical interconnects in integrated circuits the polymer must have an extremely high thermal stability (for the survival of soldering steps during processing and high operational temperatures), but only modest nonlinearity. On the other hand, for applications requiring high sensitivity and low power consumption, moderate thermal stability is sufficient, but large nonlinearities are essential. Therefore, the development of new materials for poled polymers can be categorized by the following: (1) synthesis of polymers with moderate nonlinearities that are stable at high temperatures ($\geq 300°C$) for short periods and at 150°C for long periods, and (2) polymers with large nonlinearities (electro-optic coefficient greater than 30 pm/V at the telecommunication wavelengths of 1.3 μm and 1.5 μm) with excellent long-term stability at around 80°C.

The current electro-optic modulators use mainly the inorganic crystal LiNbO$_3$ with an electro-optic coefficient of 30 pm/V. On the other hand, up to now there are few reports of thermally stable polymers with electro-optic coefficients larger than about 20 pm/V (at, e.g., 1.3 μm). Higher nonlinearities in polymers will lead to lower switching voltages in electro-optic devices.

The requirements for electro-optic polymers to be used in modulators can be summarized as follows:

- operation at the telecommunication wavelengths of 1.3 and 1.5 μm
- electro-optic coefficients as high as 30 pm/V
- flat electro-optic response up to 60 GHz
- thermal stability at 80°C (maximum decrease of the electro-optic coefficient of less than 5% over 10 years at 80°C)
- large refractive indices (leading to low half-wave voltages (see also Table 7))
- low losses in waveguides (<1 dB/cm)

An important point is the large refractive index. As can be seen from Table 7, the lower reduced half-wave voltage v_π of LiNbO$_3$ ($v_\pi = 4.6$ kV, $n_3 = 2.134$) compared to the alternating styrene-maleic-anhydrid copolymer ($v_\pi \sim 14.2$ kV, $n = 1.66$) is to a considerable extent due to the larger refractive index of LiNbO$_3$.

Current research is being directed towards integrated electro-optic devices with a view to a rapid introduction of such products into the market, although much work remains to be done in fully characterizing the electro-optic properties of, for example, Langmuir-Bodgett thin films (summarized in Table 5).

It can be concluded from the work presented in this chapter that, in terms of basic physical properties, electro-optic coefficients, and relevant figures-of-merit, organic substances present themselves as very promising materials for electro-optic device applications.

ACKNOWLEDGMENTS

Many discussions with Philippe Prêtre, Graeme Ross, Raoul Schlesser, Georg Knöpfle, and Marko Zgonik are acknowledged.

REFERENCES

1. **McQuaid, R. W.,** The Pockels effect of hexamethylenetetramine, *Appl. Opt.,* 2, 310, 1963.
2. **Wemple, S. H. and DiDomenico, M. J.,** Oxygen-octahedra ferroelectrics. I. Theory of electro-optical and nonlinear optical effects, *J. Appl. Phys.,* 40, 720, 1969.
3. **Yariv, A.,** *Quantum Electronics,* John Wiley & Sons, New York, 1975.
4. **Chen, F. S.,** *Modulators for Optical Communications,* Proceedings of the IEEE, 58, 1440, 1970.
5. **Günter, P.,** Electro-optical properties of KNbO₃, in *Proceedings of Electro-optics/Laser International '76,* IPC Science and Technology, Surrey, England, 1976, 121.
6. **Wemple, S. H. and DiDomenico, M., Jr.,** Electrooptical and nonlinear optical properties of crystals, in *Advances in Materials and Device Research,* Academic Press, New York, 1972, 263.
7. **Fousek, J.,** Refractive indices and electro-optics at ferroelectric and structural phase transitions, *Ferroelectrics,* 20, 11, 1978.
8. **Günter, P.,** Electro-optical and nonlinear optical materials, *Ferroelectrics,* 24, 35, 1980.
9. **Bosshard, C., Sutter, K., Schlesser, R., and Günter, P.,** Electro-optic effects in molecular crystals, *J. Opt. Soc. Am.,* B10, 867, 1993.
10. **Nye, J. F.,** *Physical Properties of Crystals,* Clarendon, Oxford, 1967.
11. **Ducharme, S., Feinberg, J., and Neurgaonkar, R.,** Electrooptic and piezoelectric measurements in photorefractive Barium Titanate and Strontium Barium Niobate, *IEEE J. Quantum Electron.,* QE-23, 2116, 1987.
12. **Oudar, J. L.,** Optical nonlinearities of conjugated molecules. Stilbene derivatives and highly polar aromatics., *J. Chem. Phys,* 67, 446, 1977.
13. **Teng, C. C. and Garito, A. F.,** Dispersion of the nonlinear second-order optical susceptibility of organic systems, *Phys. Rev.,* B28, 6766, 1983.
14. **Bergman, J. G. and Crane, G. R.,** Structural aspects in nonlinear optics: optical properties of KIO₂F₂ and its related iodates, *J. Chem. Phys.,* 60, 2470, 1974.
15. **Zyss, J. and Oudar, J. L.,** Relation between microscopic and macroscopic low-order optical nonlinearities of molecular crystals with one- or two-dimensional units, *Phys. Rev.,* A26, 2028, 1982.
16. **Levine, B. F. and Bethea, C. G.,** Second and third order hyperpolarizabilities of organic molecules, *J. Chem. Phys.,* 63, 2666, 1975.
17. **Zyss, J. and Chemla, D. S.,** Quadratic nonlinear optics and optimization of the second-order nonlinear optical response of molecular crystals, in *Nonlinear Optical Properties of Organic Molecules and Crystals,* Academic Press, Orlando, 1987, 3.
18. **Bosshard, C., Sutter, K., Prêtre, P., Hulliger, J., Flösheimer, M., Kaatz, P., and Günter, P.,** Organic nonlinear optical materials, in *Advances in Nonlinear Optics I,* Gordon & Breach, Amsterdam, 1995.
19. **Boettcher, C. J.,** *Theory of Electric Polarization,* Elsevier, Amsterdam, 1952.
20. **Boyd, G. D. and Kleinman, D. A.,** Parametric interaction of focused Gaussian light beams, *J. Appl. Phys.,* 39, 3597, 1968.
21. **Cheng, L. T., Tam, W., Stevenson, S. H., Meredith, G. R., Rikken, G., and Marder, S.,** Experimental investigations of organic molecular nonlinear optical polarizabilities. 1. Methods and results on benzene and stilbene derivatives, *J. Phys. Chem.,* 95, 10631, 1991.
22. **Jen, A. K.-Y., Rao, V. P., Wong, K. Y., and Drost, K. J.,** Functionalized thiophenes: second-order nonlinear optical materials, *J. Chem. Soc., Chem. Commun.,* 1993, 90, 1993.
23. **Rao, V. P., Jen, A. K.-Y., Wong, K. Y., and Drost, K. J.,** Novel push-pull thiophenes for second-order nonlinear optical applications, *Tetrahedron Lett.,* 34, 1747, 1993.
24. **Rao, V. P., Jen, A. K.-Y., Wong, K. Y., and Drost, K. J.,** Dramatically enhanced second-order nonlinear optical susceptibilities in tricyanovinylthiophene derivatives, *J. Chem. Soc., Chem. Commun.,* 1993, 1118, 1993.
25. **Lipscomb, G. F., Garito, A. F., and Narang, R. S.,** A large linear electro-optic effect in a polar organic crystal 2-methyl-4-nitroaniline (MNA), *Appl. Phys. Lett.,* 38, 663, 1981.
26. **Bierlein, J. D., Cheng, L. K., Wang, Y., and Tam, W.,** Linear and nonlinear optical properties of 3-methyl-4-methoxy-4′-nitrostilbene single crystals, *Appl. Phys. Lett.,* 56, 423, 1990.

27. **Tam, W., Guerin, B., Calabrese, J. C., and Stevenson, S. H.,** 3-Methyl-4-methoxy-4'-nitrostilbene (MMONS): crystal structure of a highly efficient material fo second-harmonic generation, *Chem. Phys. Lett.,* 154, 93, 1989.

28. **Morita, R., Ogasawara, N., Umegaki, S., and Ito, R.,** Refractive indices of 2-methyl-4-nitroaniline (MNA), *Jpn. J. Appl. Phys.,* 26, L1711, 1987.

29. **Eich, M., Reck, B., Yoon, D. Y., Willson, C. G., and Bjorklund, C. G.,** Novel second-order nonlinear optical polymers via chemical crosslinking-induced vitrification under electric fields, *J. Appl. Phys.,* 66, 3241, 1989.

30. **Mandal, B. K., Chen, Y. M., Lee, J. Y., and Tripathy, J. K.,** Cross-linked stable second-order nonlinear optical polymer by photochemical reaction, *Appl. Phys. Lett.,* 58, 2459, 1991.

31. **Müller, H., Müller, I., Nuyken, O., and Strohriegl, P.,** Novel nonlinear optical polymers with photocrosslinkable cinnamoyl group, *Makromol. Chem. Rapid Commun.,* 13, 289, 1992.

32. **Zhu, X., Chen, Y. M., Li, L., Jeng, R. J., Mandal, B. K., Kumar, J., and Tripathy, S. K.,** Photocrosslinkable polymers with stable second-order optical nonlinearity, *Opt. Comm.,* 88, 77, 1992.

33. **Chen, M., Yu, L., Dalton, L. R., Shi, Y., and Steier, W. H.,** New polymers with large and stable second-order nonlinear optical effects, *Macromolecules,* 24, 5421, 1991.

34. **Weder, C., Neuenschwander, P., Suter, U. W., Prêtre, P., Kaatz, P., and Günter, P.,** New polyamides with large second-order nonlinear optical properties, *Macromolecules,* 27, 2181, 1994.

35. **Köhler, W., Robello, D. R., Dao, P. T., Willand, C. S., and Williams, D. J.,** Electric-field induced orientation and relaxation studies in polymers for second-order nonlinear optics, *Mol. Cryst. Liq. Cryst. Sci. Tech. B,* 3, 83, 1992.

36. **Lin, J. T., Hubbard, M. A., Marks, T. J., Lin, W., and Wong, G. K.,** Poled polymeric nonlinear optical materials. Exceptional second harmonic generation temporal stability of a chromophore-functionalized polyimide, *Chem. Mater.,* 4, 1148, 1992.

37. **Jen, A. K. Y., Wong, K. Y., Drost, K., Rao, V. P., Caldwell, B., and Mininni, R. M.,** Thermally stable poled polymers: highly efficient heteroaromatic chromophores in high temperature polymers, in *Organic Thin Films for Photonic Applications,* paper FB3, Optical Society of America, Toronto, Canada, 1993, 286.

38. **Stähelin, M., Burland, D. M., Ebert, M., Miller., R. D., Smith, B. A., Twieg, R. J., Volksen, W., and Walsh, C. A.,** Re-evaluation of the thermal stability of optically nonlinear polymeric guest-host systems, *Appl. Phys. Lett.,* 61, 1626, 1992.

39. **Wu, J., Valley, J. F., Ermer, S., Binkley, E. S., Kenney, J. T., Lipscomb, G. F., and Lytel, R.,** Thermal stability of electro-optic response of poled polyimide systems, *Appl. Phys. Lett.,* 58, 225, 1991.

40. **Wu, J. W., Binkley, E. S., Kenney, J. T., Lytel, R., and Garito, A. F.,** Highly thermally stable electro-optic response in poled guest-host polyimide systems cured at 360°C, *J. Appl. Phys.,* 69, 7366, 1991.

41. **Kovacs, A. J., Hutchinson, J. M., and Aklonis, J. J.,** Isobaric volume and enthalpy recovery of glasses. I. A critical survey of recent phenomenological approaches, in *The Structure of Non-Cristalline Materials,* Taylor & Francis, London, 1977, 153.

42. **Gibbs, J. H. and DiMarzio, E. A.,** Nature of the glass transition and the glassy state, *J. Chem. Phys.,* 28, 373, 1958.

43. **Adam, G. and Gibbs, J. H.,** On the temperature dependence of cooperative relaxation properties in glass-forming liquids, *J. Chem. Phys.,* 43, 139, 1965.

44. **Hodge, I. M.,** Adam-Gibbs formulation of nonlinearity in glassy-state relaxations, *Macromolecules,* 19, 936, 1986.

45. **Prêtre, P., Kaatz, P., Bohren, A., Günter, P., Zysset, B., Ahlheim, M., Stähelin, M., and Lehr, F.,** Modified polyimide side-chain polymers for electro-optics, *Macromolecules,* 27, 5476, 1994.

46. **Sigelle, M. and Hierle, R.,** Determination of the electrooptic coefficients of 3-methyl 4-nitropyridine 1-oxide by an interferometric phase modulation technique, *J. Appl. Phys.,* 52, 4199, 1981.

47. **Günter, P.,** Electro-optical properties of KNbO3, *Opt. Commun.,* 11, 285, 1974.

48. **Knöpfle, G., Bosshard, C., Schlesser, R., and Günter, P.,** Optical, nonlinear optical, and electrooptical properties of 4'-nitrobenzylidene-3-acetamino-4-methoxyaniline (MNBA) crystals, *IEEE J. Quantum. Electron.,* 30, 1303, 1994.

49. **Teng, C. C. and Man, T. H.,** Simple reflexion technique for measuring the electro-optic coefficient of poled polymers, *Appl. Phys. Lett.,* 56, 1734, 1990.

50. **Singer, K. D., Kuzyk, M. G., and Sohn, J. E.,** Second-order nonlinear-optical processes in orientationally ordered materials: relationship between molecular and macroscopic properties, *J. Opt. Soc. Am. B,* 4, 968, 1987.

51. **Levy, Y., Dumont, M., Chastaing, E., Robin, P., Chollet, P. A., Gadret, G., and Kajzar, F.,** Reflection method for electro-optical coefficient determination in stratified thin film structures, *Mol. Cryst. Liq. Cryst. Sci. Technol. B,* 4, 1, 1993.

52. **Kretschmann, E.,** Die Bestimmung optischer Konstanten von Metallen durch Anregung von Oberflächenplasmaschwingungen, *Z. Phys.,* 241, 313, 1971.

53. **Herminghaus, S., Smith, B. A., and Swalen, J. D.,** Electro-optic coefficients in electric-field-poled polymer waveguides, *J. Opt. Soc. Am. B,* 8, 2311, 1991.

54. **Morichere, D., Chollet, P.-A., Fleming, W., Jurich, M., Smith, B. A., and Swalen, J. D.,** Electro-optic effects in two tolane side-chain nonlinear-optical polymers: comparison between measured coefficients and second-harmonic generation, *J. Opt. Soc. Am. B,* 10, 1894, 1993.

55. **Dumont, M., Levy, Y., and Morichere, D.,** Electro-optic organic waveguides: optical characterization, in *Organic Molecules for Nonlinear Optics and Photonics,* Kluwer Academic, Dordrecht, 1991, 461.

56. **Johnston, A. R.,** The strain-free electro-optic effect in single-crystal barium titanate, *Appl. Phys. Lett.,* 7, 195, 1965.
57. **Schlesser, R., Bernasconi, P., Bosshard, C., Zgonik, M., and Günter, P.,** A comparative study on the clamped linear electro-optic effect in organic and inorganic crystals, in *Conference on Lasers and Electro-Optics '94 (CLEO/Europe '94),* Amsterdam, The Netherlands, 1994, 343.
58. **Miller, R. C.,** Optical second-harmonic generation in piezoelectric crystals, *Appl. Phys. Lett.,* 5, 17, 1964.
59. **Nahata, A., Wu, C., and Yardley, J. T.,** Electrooptic characterization of organic media, *IEEE Trans. Instrum. Meas.,* 41, 128, 1992.
60. **Yoshimura, T.,** Characterization of the electro-optic effect in styrylpyridium cyanine dye thin-film crystals by an ac modulation method, *J. Appl. Phys.,* 62, 2028, 1987.
61. **Zgonik, M., Schlesser, R., Biaggio, I., Voit, E., Tscherry, J., and Günter, P.,** Materials constant of $KNbO_3$ relevant for electro- and acousto-optics, *J. Appl. Phys.,* 74, 1287, 1993.
62. **Zgonik, M., Bernasconi, P., Duelli, M., Schlesser, R., Günter, P., Garrett, M. H., Rytz, D., Zhu, Y., and Wu, X.,** Dielectric, elastic, piezoelectric, electro-optic, and elasto-optic tensors in $BaTiO_3$ crystals, *Phys. Rev. B,* 50, 5941, 1994.
63. **Weis, R. S. and Gaylord, T. K.,** Lithium niobate: summary of physical properties and crystal structure, *Appl. Phys. Lett.,* A37, 191, 1985.
64. **Baumert, J.-C., Hoffnagle, J., and Günter, P.,** Nonlinear optical effects in $KNbO_3$ crystals at $Al_xGa_{1-x}As$, dye, ruby and Nd:YAG laser wavelengths, in *1984 European Conference on Optics, Optical Systems, and Applications,* SPIE–The International Society for Optical Engineering, Bellingham, WA, 1984, 374.
65. **Buse, K., Riehemann, S., Loheide, S., Hesse, H., Mersch, F., and Krätzig, E.,** Refractive indices of single domain $BaTiO_3$ for different wavelengths and temperatures, *Phys. Stat. Sol. A,* 135, K87, 1993.
66. **Choy, M. M. and Byer, R. L.,** Accurate second-order susceptibility measurements of visible and infrared nonlinear crystals, *Phys. Rev. B,* 14, 1693, 1976.
67. **Landolt-Börnstein,** *Ferroelektrika und verwandte Substanzen,* Springer Verlag, Berlin, 1981.
68. **Zysset, B., Biaggio, I., and Günter, P.,** Refractive indices of orthorhombic $KNbO_3$. I. Dispersion and temperature dependence, *J. Opt. Soc. Am. B,* 19, 380, 1992.
69. **Roberts, D. A.,** Simplified Characterization of uniaxial and biaxial nonlinear optical crystals: a plea for standardization of nomenclature and conventions, *IEEE J. Quantum Electron.,* 28, 2057, 1992.
70. **Kaminov, I. P.,** Linear electrooptic materials, in *CRC Handbook of Laser Science and Technology, Vol. IV, Optical Materials, Part 2: Properties,* CRC Press, Boca Raton, FL, 1984, 253.
71. **Knöpfle, G., Schlesser, R., Ducret, R., and Günter, P.,** Optical and nonlinear optical properties of 4′-dimethylamino-N-methyl-4-stilbazolium tosylate (DAST) crystals, *Nonlinear Opt.,* 9, 143, 1995.
72. **Marder, S. R., Perry, J. W., and Yakymyshyn, C. P.,** Organic salts with large second-order optical nonlinearities, *Chem. Mater.,* 6, 1137, 1994.
73. **Yoshimura, T. and Kubota, Y.,** Pockels effect in organic thin films, in *Nonlinear Optics of Organics and Semiconductors,* Springer-Verlag, Berlin, 1989, 222.
74. **Allen, S., McLean, T. D., Gordon, P. F., Bothwell, B. D., Hursthouse, M. B., and Karaulov, S. A.,** A novel organic electro-optic crystal: 3-(1,1-dicyanoethenyl)-1-phenyl-4,5-dihydro-1H-pyrazole, *J. Appl. Phys.,* 64, 2583, 1988.
75. **Bailey, R. T., Bourhill, G. H., Cruickshank, F. R., Pugh, D., Sherwood, J. N., Simpson, G. S., and Varma, K. B. R.,** Linear electro-optic effect and temperature coefficient of birefringence in 4-nitro-4′-methylbenzylidene aniline single crystals, *J. Appl. Phys.,* 71, 2012, 1992.
76. **Sutter, K., Bosshard, C., Wang, W. S., Surmely, G., and Günter, P.,** Linear and nonlinear optical properties of 2-(N-prolinol)-5-nitropyridine, *Appl. Phys. Lett.,* 53, 1779, 1988.
77. **Bosshard, C., Sutter, K., Günter, P., and G. Chapuis,** Linear and nonlinear optical properties of 2-cyclooctylamino-5-nitropyridine, *J. Opt. Soc. Am. B,* 6, 721, 1989.
78. **Bailey, R. T., Bourhill, G. H., Cruickshank, F. R., Pugh, D., Sherwood, J. N., Simpson, G. S., and Varma, K. B. R.,** Linear electro-optic dispersion in (−)-2-(α-methylbenzylamino)-nitropyridine single crystals, *J. Appl. Phys.,* 75, 489, 1994.
79. **Kerkoc, P.,** Growth of Bulk Crystals and Single Crystal Cored Fibers of 4-(N,N-Dimethylamino)-3-acetamidonitrobenzene (DAN) and Their Nonlinear Optical Properties, Ph.D. thesis 9413, ETH Zürich, Switzerland, 1991.
80. **Kerkoc, P., Zgonik, M., Bosshard, C., Sutter, K., and Günter, P.,** Optical and nonlinear optical properties of 4-(N,N-dimethylamino)-3-acetamidonitrobenzene single crystals, *Appl. Phys. Lett.,* 54, 2062, 1989.
81. **Stevenson, J. L.,** The linear electro-optic coefficient of meta-nitroaniline, *J. Phys. D* 6, L13, 1973.
82. **Perry, J. W., Marder, S. R., Perry, K. J., Sleva, E. T., Yakymyshyn, C., Stewart, K. R., and Boden, E. P.,** Organic salts with large electro-optic coefficients, in *Nonlinear Optical Properties of Organic Materials IV,* SPIE–International Society for Optical Engineering, Bellingham, WA 1991, 302.
83. **Allen, S.,** Electro-optic properties of a new organic pyrazole crystal, in *Organic Materials for Non-linear Optics,* The Royal Society of Chemistry, London, 1989, 137.

438

84. **Meredith, G. R., Vandusen, J. G., and Williams, D. J.,** Characterization of liquid crystalline polymers for electro-optic applications, in *Nonlinear Optical Properties of Organic and Polymeric Materials,* Washington, DC, 1983, 109.

85. **van der Vorst, C. P. J. M. and Picken, S. J.,** Electric field poling of acceptor-donor molecules, *J. Opt. Soc. Am.,* B7, 320, 1990.

86. **Nahata, A., Shan, J., Yardley, J. T., and Wu, C.,** Electro-optic determination of the nonlinear optical properties of a covalently functionalized Disperse Red 1 copolymer, *J. Opt. Soc. Am.,* B10, 1553, 1993.

87. **Otomo, A., Mittler-Neher, S., Bosshard, C., Stegeman, G. I., Horsthuis, W. H. G., and Möhlmann, G. R.,** Second harmonic generation by counter propagating beams in 4-dimethylamino-4'-nitrostilbene side-chain polymer channel waveguides, *Appl. Phys. Lett.,* 63, 3405, 1993.

88. **Otomo, A., Bosshard, C., Mittler-Neher, S., Stegeman, G. I., Küpfer, M., Flörsheimer, M., Günter, P., Horsthuis, W. H. G., and Möhlmann, G. R.,** Second-harmonic generation by counterpropagating beams in organic thin films, *Nonlinear Opt.,* 10, 331, 1995.

89. **Bosshard, C., Otomo, A., Stegeman, G. I., Küpfer, M., Flörsheimer, M., and Günter, P.,** Surface emitted green light generated in Langmuir-Blodgett film waveguides, *Appl. Phys. Lett.,* 64, 2076, 1994.

90. **Otomo, A., Stegeman, G. I., Horsthuis, W. H. G., and Möhlmann, G. R.,** Strong field, in-plane poling for nonlinear optical devices in highly nonlinear side chain polymers, *Appl. Phys. Lett.,* 65, 2389, 1994.

91. **Prêtre, P.,** private communication, 1994.

92. **Fiorini, C., Chollet, P.-A., Charra, F., Kajzar, F., and Nunzi, J.-M.,** Quadratic optical polarizabilities in polymer films obtained by dc-electric field and pure optical poling, in *Nonlinear Optical Properties of Organic Materials VII,* SPIE–The International Society for Optical Engineering, Bellingham, WA, 1994, 92.

93. **Lytel, R., Lipscomb, G. F., Kenney, J. T., and Binkley, E. S.,** Large-scale integration of electro-optic polymer waveguides, in *Polymers for Lightwave and Integrated Optics,* Marcel Dekker, New York, 1992, 433.

94. **Organic Thin Films for Photonic Applications, Optical Society of America, Washington, DC, 1993.**

95. **Organic Thin Films for Photonic Applications, Optical Society of America, Washington, DC, 1994.**

96. **Singer, K. D., Kuzyk, M. G., Holland, W. R., Sohn, J. E., and Lalama, S. J.,** Electro-optic phase modulation and optical second-harmonic generation in corona-poled polymer films, *Appl. Phys. Lett.,* 53, 1800, 1988.

97. **Page, R. H., Jurich, M. C., Reck, B., Sen, A., Twieg, R. J., Swalen, J. D., Bjorklund, G. C., and Willson, C. G.,** Electrochromic and optical waveguide studies of corona-poled electro-optic polymer films, *J. Opt. Soc. Am. B,* 7, 1239, 1990.

98. **Katz, H. E., Dirk, C. W., Schilling, M. L., Singer, K. D., and Sohn, J. E.,** Optimization of second order nonlinear optical susceptibilities in organic materials, in *Nonlinear Optical Properties of Polymers,* Materials Research Society, Pittsburgh, PA, 1988, 127.

99. **Wong, K. Y. and Jen, A. K.-Y.,** Thermally stable poled polyimides using heteroaromatic chromophores, *J. Appl. Phys.,* 75, 3308, 1994.

100. **Jungbauer, D., Reck, B., Twieg, R., Yoon, D. Y., Willson, C. G., and Swalen, J. D.,** Highly efficient and stable nonlinear optical polymers via chemical cross-linking under electric field, *Appl. Phys. Lett.,* 56, 2610, 1990.

101. **Shi, Y., Ranon, P. M., Steier, W. H., Xu, C., Wu, B., and Dalton, L. R.,** Improving the thermal stability by anchoring both ends of chromophores in the side-chain nonlinear optical polymers, *Appl. Phys. Lett.,* 63, 2168, 1993.

102. **Shuto, Y., Amano, M., and Kaino, T.,** Electrooptic light modulation and second-harmonic generation in novel diazo-dye-substituted poled polymers, *IEEE Trans. Photon. Techn. Lett.,* 3, 1003, 1991.

103. **Nahata, A., Wu, C., Knapp, C., Lu, V., Shan, J., and Yardley, J. T.,** Thermally stable polyester polymers for second-order nonlinear optics, *Appl. Phys. Lett.,* 64, 3371, 1994.

104. **Drost, K. J., Rao, V. P., and Jen, A. K.-Y.,** A new synthetic approach for the incorporation of highly efficient second-order nonlinear optical chromophores containing tricyanovinyl electron acceptors into methycrylate polymers, *J. Chem. Soc., Chem. Commun.,* 1994, 369, 1994.

105. **Möhlmann, G. R., Horsthuis, W. H. G., McDonach, A., Copeland, M. J., Duchet, C., Fabre, P., Diemeer, M. B. J., Trommel, E. S., Suyten, F. M. M., Van Tomme, E., Baquero, P., and Van Daele, P.,** Optically nonlinear polymeric switches and modulators, in *Nonlinear Optical Properties of Organic Materials III,* SPIE–The International Society for Optical Engineering, Bellingham, WA, 1990, 215.

106. **Haas, D., Yoon, H., and Man, H.-T.,** Polymeric electro-optic waveguide modulator: materials and fabrication, in *Nonlinear Optical Properties of Organic Materials II,* SPIE–The International Society for Optical Engineering, Bellingham, WA, 1989, 222.

107. **Moylan, C. R., Twieg, R. J., Lee, V. Y., Miller, R. D., Volksen, W., Thackara, J. I., and Walsh, C. A.,** Synthesis and characterization of thermally robust electro-optic polymers, in *Nonlinear Optical Properties of Organic Materials VII,* SPIE–The International Society for Optical Engineering, Bellingham, WA, 1994, 17.

108. **Aramaki, S., Okamoto, Y., Murayama, T., and Kubo, Y.,** Novel naphthoquinone methide dyes for poled polymers, in *Nonlinear Optical Properties of Organic Materials VII,* SPIE–The International Society for Optical Engineering, Bellingham, WA, 1994, 58.

109. **Sotoyama, W., Tatsuura, S., and Yoshimura, T.,** Electro-optic side-chain polyimide system with large optical nonlinearity and high thermal stability, *Appl. Phys. Lett.,* 64, 2197, 1994.

110. **Ahlheim, M. and Lehr, F.,** Electrooptically active polymers. Non-linear optical polymers prepared from maleic anhydride copolymers by polymer analogous reactions, *Makromol. Chem,* 195, 361, 1994.

111. **Rao, V. P., Wong, K. Y., Jen, A. K.-Y., and Drost, K. J.,** Functionalized fused thiophenes: a new class of thermally stable and efficient second-order nonlinear optical chromophores, *Chem. Mater.,* 6, 2210, 1994.

112. **Möhlmann, G. R., Ed.,** *Nonlinear Optical Properties of Organic Materials VII,* SPIE–The International Society for Optical Engineering, Bellingham, WA, 1994.

113. **Burland, D. M., Miller, R. D., and Walsh, C. A.,** Second-order nonlinearity in poled-polymer systems, *Chem. Rev.,* 94, 31, 1994.

114. **Cross, G. H., Peterson, I. R., Girling, I. R., Cade, N. A., Goodwin, M. J., Carr, N., Sethi, R. S., Marsden, R., Gray, G. W., Lacey, D., McRoberts, A. M., Scrowston, R. M., and Toyne, K. J.,** A comparison of second harmonic generation and electro-optic studies of Langmuir-Blodgett monolayers of new hemicyanine dyes, *Thin Solid Films,* 156, 39, 1988.

115. **Cross, G. H., Girling, I. R., Peterson, I. R., Cade, N. A., and Earls, J. D.,** Optically nonlinear Langmuir-Blodgett films: linear electro-optic properties of monolayers, *J. Opt. Soc. Am. B,* 4, 962, 1987.

116. **Schildkraut, J. S., Penner, T. L., Willand, C. S., and Ulman, A.,** Absorption and second-harmonic generation of monomer and aggregate hemicyanine dye Langmuir-Blodgett films, *Opt. Lett.,* 13, 134, 1988.

117. **Bosshard, C. and Küpfer, M.,** Oriented molecular systems, in *Science and Technology of Organic Thin Films for Waveguiding Nonlinear Optics,* F. Kajzar, F. and Swalen, J. D., Eds., Gordon & Breach, New York, in press.

118. **Gottlieb, M., Ireland, C. L. M., and Ley, J. M.,** *Electro-Optic and Acousto-Optic Scanning and Deflection,* Marcel Dekker, New York, 1983.

119. **Voges, E.,** Integrated electro-optic devices, in *Electro-optic and Photorefractive Materials,* Springer-Verlag, Berlin, 1987, 150.

120. **Noguchi, K., Mitomi, O., Kawano, K., and Yanagibashi, M.,** Highly efficient 40-GHz bandwidth Ti:LiNbO$_3$ optical modulator employing ridge structure, *IEEE Photon. Techn. Lett.,* 5, 52, 1993.

121. **Teng, C. C.,** Traveling-wave polymeric optical intensity modulator with more than 40 GHz of 3-dB electrical bandwidth, *Appl. Phys. Lett.,* 60, 1538, 1992.

122. **Eichler, H. J., Günter, P., and Pohl, D. W.,** *Laser-Induced Dynamic Gratings,* Springer-Verlag, Berlin, 1986.

123. **Günter, P.,** Holography, coherent light amplification and optical phase conjugation with photorefractive materials, *Phys. Rep.,* 93, 199, 1982.

124. **Moerner, W. E., Silence, S. M., Hache, F., and Bjorklund, G. C.,** Orientationally enhanced photorefractive effect in polymers, *J. Opt. Soc. Am. B,* 11, 320, 1994.

Chapter 7

High-Speed Electro-optic Modulators from Nonlinear Optical Polymers

Chia-Chi Teng

CONTENTS

I. INTRODUCTION

A demand for high bandwidth and low drive voltage electro-optic modulating devices has been stimulated by the advancing technology of high-capacity fiber-optic communication networks where a very high data bit rate is applied. Organic electro-optic (EO) materials[1,2] offer potential advantages over inorganic materials such as $LiNbO_3$ and GaAs crystals by virtue of their low dielectric constants and potentially high electro-optic coefficients. The characteristic of having a low dielectric constant in organic systems is derived from their electronic origin[3] of EO responses. The overall electro-optic effect is an addition of virtual electronic excitations of individual nonlinear optical (NLO) molecules. The dispersion of electronic polarizability from DC to optical frequencies is low. In the case of inorganic crystals, the EO effect is based on the ionic displacement in a noncentrosymmetric potential. The polarizability from the ionic displacement is quite large when driven by a low-frequency electric field. As a result, the velocity mismatch between the ratio frequency (rf) and optical waves is large in inorganic crystals and almost negligible in organic materials. This is a key factor affecting the bandwidth and efficiency of an EO modulator. The recent success[4-7] in molecular designs of highly conjugated asymmetric π—electron systems results in big improvements of the molecular second-order hyperpolarizability. Potentially very high electro-optic organic materials can be developed and subsequently very low drive voltage and very high bandwidth EO modulators are within hope. Due to the difficulty of growing defect-free noncentrosymmetric crystals out of NLO molecules, nonlinear optical polymer[8-10] offers an alternative way to incorporate these molecules into an optically transparent medium for electro-optic applications. The excellent velocity match between rf and optical waves of these materials has been demonstrated[11-13] experimentally and resulted in a very high bandwidth and low drive voltage EO modulator. In Section II, an introduction to NLO polymers, including their microscopic origin of EO effects, is given. In Section III, a technical review of developments of polymeric high-speed EO modulators is presented.

II. NONLINEAR OPTICAL POLYMERS

A. NLO MOLECULES AND ELECTRO-OPTIC EFFECTS

Conjugated molecules with asymmetric delocalized charge exhibit unusually large[14] molecular second-order susceptibilities β_{uvw} (the indices u, v, w refer to the molecular coordinate system). Due to the weak intermolecular interactions, the macroscopic second-order nonlinearity $\chi_{ijk}^{(2)}$ is an ensemble addition of all contributions from the molecular constituents in the medium:

$$\chi_{ijk}^{(2)}(\omega'; \omega_1, \omega_2) = Nf_i^{\omega'}f_j^{\omega_1}f_k^{\omega_2}\langle\beta_{ijk}(\omega'; \omega_1, \omega_2)\rangle \tag{1}$$

where N is the number of molecules per unit volume, f are the local field factors, $\langle\ \rangle$ represents the orientational ensemble average of the molecules. For a fix orientational distribution, $\chi_{ijk}^{(2)}$ is proportional to β_{uvw}.

Under the presence of a low-frequency (DC to rf) electric field $\mathbf{E}^0$, the first-order optical susceptibility $\chi^{(1)}$ subjects to a change according to:

$$P_i^\omega = \chi_{ij}^{(1)}(\omega)E_j^\omega + \chi_{ijk}^{(2)}(\omega; \omega, 0)E_j^\omega E_k^0 = [\chi_{ij}^{(1)} + \Delta\chi_{ij}^{(1)}]\cdot E_j^\omega \tag{2}$$

where $\Delta\chi_{ij}^{(1)} = \chi_{ijk}^{(2)}(\omega; \omega, 0)E_k^0$

The linear electro-optic constant $\mathbf{r}$ is defined[15] by:

$$\Delta(n^{-2})_i = r_{ij}E_j^0 \tag{3}$$

where $(n^{-2})_i$ are the coefficients of the index ellipsoid with $i = 1$–6.

For $i = 1, 2, 3$, $(n^{-2})_i|_{E=0} = n_i^{-2}$, where n_i are the indices of refraction of the principal dielectric axes with $1 = x$, $2 = y$, $3 = z$. One can derive that:

$$r_{ij} = -\frac{2\cdot\chi_{iij}^{(2)}(\omega; \omega, 0)}{n_i^4(\omega)} \tag{4}$$

where $\chi_{iij}^{(2)}(\omega; \omega, 0)$ is related to $\beta_{uvw}(\omega; \omega, 0)$ by Equation 1.

In one-dimensional molecular structures, $\chi_{iij}^{(2)}(\omega'; \omega_1, \omega_2)$ is primarily contributed from the component β_{zzz} of β_{uvw} along the molecular axis, and usually there is only one major excitation energy level which contributes to the second-order processes. Under a two-level approximation[5]:

$$\beta_{zzz}(\omega'; \omega_1, \omega_2) \cong a\cdot F(\omega_1, \omega_2) \tag{5}$$

Table 1 $\beta_0\mu$ Relative to DANS and λ_{max} (nm) of Highly Conjugated Molecules

molecule	λ_{max}	$\beta_0\mu$	molecule	λ_{max}	$\beta_0\mu$
DANS	450 [b]	1 [d]		480 [c]	1.2 [f]
DCVHT	518 [b]	1.9 [d]		579 [c]	2.3 [f]
	550 [a]	2.4 [d]		513 [c]	2.6 [f]
	594 [a]	3.9 [e]		582 [c]	4.3 [g]
	640 [a]	7.9 [e]		645 [c]	3.9 [f]
	718 [a]	6.7 [e]		662 [a]	11 [e]

a In dioxane.

b In DMSO.

c In methylene chloride.

d Ref. 7.

e Ref. 6.

f Ref. 5.

g Ref. 4.

where $a = e^3|\mu_{01}|^2(\mu_{11} - \mu_{00})/\hbar^2$ is a dispersionless quantity and

$$F(\omega_1, \omega_2) = (\omega_0 - \omega_1 - \omega_2)^{-1}(\omega_0 - \omega_1)^{-1} + (\omega_0 - \omega_1 - \omega_2)^{-1}(\omega_0 - \omega_2)^{-1}$$
$$+ (\omega_0 + \omega_1 + \omega_2)^{-1}(\omega_0 + \omega_1)^{-1} + (\omega_0 + \omega_1 + \omega_2)^{-1}(\omega_0 + \omega_2)^{-1}$$
$$+ (\omega_0 + \omega_2)^{-1}(\omega_0 - \omega_1)^{-1} + (\omega_0 + \omega_1)^{-1}(\omega_0 - \omega_2)^{-1}$$

is a dispersion factor, μ_{01} is a transition moment, μ_{00} and μ_{11} are ground and excited state dipole moments, and $\hbar\omega_0 = 2\pi\hbar c/\lambda_0$ is the excitation energy of the molecule.

Research on enhancing β_{uvw} of organic molecular structures has been carried on extensively during last decade.[4-7] Table 1 lists some of the discovered nIO molecules with large second-order nonlinearities β_{zzz}, which were measured from DC-induced second harmonic generation[17-18] at various laser wavelengths and solvent media. The second-order molecular susceptibility β_{zzz} $(2\omega; \omega, \omega)$ for second harmonic generation and β_{zzz} $(\omega; \omega, 0)$ for electro-optic effect are interrelated through the dispersionless parameter a and dispersion factor $F(\omega_1, \omega_2)$.

From Equation 5:

$$\beta_{zzz}(2\omega; \omega, \omega) = a \cdot \frac{3\omega_0^2}{(\omega_0^2 - \omega^2)(\omega_0^2 - 4\omega^2)} \qquad (6)$$

$$\beta_{zzz}(\omega; \omega, 0) = a \cdot \frac{(3\omega_0^2 - \omega^2)}{(\omega_0^2 - \omega^2)^2} \qquad (7)$$

When the optical frequencies are far off resonance, i.e., $\omega \ll \omega_0$ and $2\omega \ll \omega_0$, both $\beta_{zzz}(2\omega; \omega, \omega)$ and $\beta_{zzz}(\omega; \omega, 0)$ approach to a dispersionless quantity $\beta_0 = 3a/\omega_0^2$. In Table 1, the quantity $\beta_0\mu$ of molecular structures relative to $(\beta_0\mu)_{DANS}$ of dimethylaminonitrostilbene (DANS), where $\mu \equiv \mu_{00}$, is listed for comparison. In actual applications of these molecules for electro-optic media, the figure-of-merit may be slightly different from $\beta_0\mu$ due to solvent shift effect[18] which lowers the resonance frequency ω_0.

B. POLING FOR ELECTRO-OPTIC EFFECTS

Polar molecules tend to form pairs in their crystalline structure. It is very difficult to grow defect-free noncentrosymmetric single crystal out of highly polar NLO molecules. Embedding them in an amorphous polymeric matrix provides an alternative way to utilize the large second-order nonlinear optical property of these molecules. For second-order nonlinear optical effects, a poling process is used to orient the NLO moieties to a preferred direction. The polymer is heated to a temperature T which is close to the glass transition temperature of the material. A strong DC electric field $\mathbf{E}_p$ (along z axis) is applied to the medium. The local poling field $\mathbf{e}_p$ aligns the permanent ground-state dipole moment $\boldsymbol{\mu}$ of the NLO moieties according to the orientational distribution function[19]:

$$G(\theta) = \exp\left(-\frac{U_0(\theta) + U_1(\theta) + U_2(\theta)}{kT}\right) \cdot \left[\int_{-1}^{1} d[\cos(\theta)] \cdot \exp\left(-\frac{U_0(\theta) + U_1(\theta) + U_2(\theta)}{kT}\right)\right]^{-1} \quad (8)$$

where $e_p = f_p E_p$, f_p is a local field factor, and θ is the angle between the z axis and $\boldsymbol{\mu}$. $U_0(\theta)$ is a mean field potential accounting for the hindered molecular rotation from the polymer chains and the mutual interaction between the NLO moieties; $U_1(\theta)$ is the potential of the dipole moment and the poling field:

$$U_1(\theta) = -\mu f_p E_p \cos(\theta) \quad (9)$$

$U_2(\theta)$ is the energy of the induced dipole moment of the NLO moieties and the poling field and is proportional to E_p^2. $U_0(\theta)$ does not depend on E_p directly. If the poling field is not too strong, $U_2(\theta)$ is negligible. We can further neglect $U_0(\theta)$ under the condition that the interaction between the NLO moieties are weak (nonliquid crystalline) and the hindered rotation is not significant. (This would be true if the poling temperature is above the glass transition temperature of the polymer so that the polymer chains have enough freedom of movement.) We will use these approximations for the rest of the chapter.

The orientation of $\boldsymbol{\mu}$ is retained after cooling down the sample with the poling field on. In the presence of a modulating electrical field $\mathbf{E}^0$ and an optical field $\mathbf{E}^\omega$, the polarization $\mathbf{P}^\omega$ of the poled media is given by:

$$P_i^\omega = (\chi_0^{(1)} + \chi_{ij}^{(1)})E_j^\omega + \chi_{ijk}^{(2)}E_j^\omega E_k^0 \quad (10)$$

where $\chi_0^{(1)}$ is the first-order susceptibility of the background polymer, $\chi_{ij}^{(1)}$, $\chi_{ijk}^{(2)}$ are the first- and second-order susceptibilities of the NLO moieties: $\chi_{ij}^{(1)} = N f_j^\omega \langle \alpha_{ij} \rangle$, $\chi_{ijk}^{(2)} = N f_i^\omega f_j^\omega f_k^0 \langle \beta_{ijk} \rangle$ with $\langle \alpha_{ij} \rangle = \langle b_{ui} b_{vj} \rangle \alpha_{uv}$, $\langle \beta_{ijk} \rangle = \langle b_{ui} b_{vj} b_{wk} \rangle \beta_{uvw}$, where b_{ui} are the transformation matrix elements from the molecular frame (X, Y, Z with $\boldsymbol{\mu}$ along Z axis) to the laboratory frame (x, y, z). By keeping only the major components of α_{uv} and β_{uvw}, one can express the indices of refraction of a poled medium in terms of $a = \mu E_p/kT$:

$$n_i^2 = 1 + 4\pi(\chi_0^{(1)} + \chi_{ii}^{(1)}) = n_{0i}^2 + A_i f_i^\omega\left[L_2(f_p a) - \frac{1}{3}\right] = n_{0i}^2 + A_i f_i^\omega\left[\frac{2f_p^2 a^2}{45} + \cdots\right] \quad (11)$$

where $A_z = 4\pi N\alpha$, $A_x = A_y = -A_z/2$, $n_{0i}^2 = 1 + 4\pi[\chi_0^{(1)} + N f_i^\omega(\alpha_0 + \frac{\alpha}{3})]$, $\alpha_{XX} \approx \alpha_{YY} \equiv \alpha_0$, $\alpha_{ZZ} - \alpha_0 \equiv \alpha$, and the electro-optic constants ($\beta \equiv \beta_{ZZZ}$):

$$r_{33} = -\frac{2\chi_{zzz}^{(2)}}{n_z^4} = B_z L_3(f_p a) = B_z\left(\frac{f_p a}{5} - \frac{f_p^2 a^2}{105} + \cdots\right) \quad (12)$$

$$r_{13} = r_{23} = -\frac{2\chi_{xxz}^{(2)}}{n_x^4} = \frac{B_x}{2} [L_1(f_p a) - L_3(f_p a)] = B_x \left(\frac{f_p a}{15} - \frac{2f_p^2 a^2}{315} + \cdots \right) \qquad (13)$$

with

$$B_i = -\frac{2N(f_i^\omega)^2 f_z^0 \beta}{n_i^4}$$

and

$$L_n(x) \equiv \frac{\int_{-1}^{1} d[\cos(\theta)] \cdot e^{-x\cos(\theta)} \cdot \cos^n(\theta)}{\int_{-1}^{1} d[\cos(\theta)] \cdot e^{-x\cos(\theta)}}$$

is the nth order Langevin function.

From Equations 12 and 13, we have:

$$R \equiv \frac{r_{33}}{r_{13}} \cong \frac{B_z}{B_x} \left[3 + \frac{f_p^2 a^2}{7} - \cdots \right] \qquad (14)$$

Onsager derived the local field factors by considering the contribution of the reaction field from the induced dipole moment of the molecule in the cavity to the cavity field[20]:

$$f_i = \frac{n_i^2(n_a^2 + 2)}{2n_i^2 + n_a^2}, \qquad f_p = \frac{\varepsilon(n_a^2 + 2)}{2\varepsilon + n_a^2} \qquad (15)$$

with

$$\frac{n_a^2 - 1}{n_a^2 + 2} \equiv \frac{\alpha_{ZZ}}{a^3}$$

where n_a represents the index of refraction of a medium composed of molecules with polarizability α_{zz} and packed together with cavity radius a. Notice that when $n_a \approx n_i$ and $n_a^2 \approx \varepsilon$, the Onsager local field factors default back to the Lorenz-Lorentz model where $f = (\varepsilon + 2)/3$. For a material containing molecules with a large molecular dipole moment and a different polarizability than the ambient medium, n_a is quite different from n_i and ε. The Onsager model is more suitable to describe the actual local fields inside the cavity. n_a can be calculated from extrapolation of the experimental data of refractive indices versus moieties concentration and poling fields using Equation 11. For determining local poling field, the use of Onsager model is essentially important. At the poling temperature, the dipole moments of the NLO moieties are free to rotate. The dielectric constant ε of the medium is significantly larger than n_a^2 and can be estimated using the Onsager equation:

$$\mu^2 = \frac{9kT}{4\pi N} \frac{(\varepsilon - n_a^2)(2\varepsilon + n_a^2)}{\varepsilon \cdot (n_a^2 + 2)^2} \qquad (16)$$

As an example, n_a and ε of P2ANS/MMA 50/50 which contains 65% of DANS (see Section II.C.2) are about 2.3 and 50, respectively, at the poling temperature. For $\varepsilon \gg n_a^2$, $f_p \approx (n_a^2 + 2)/2$ and the local field factor for poling field is basically determined by n_a.

C. GUEST–HOST AND SIDE-CHAIN POLYMERIC SYSTEMS

NLO molecules can be embedded into a polymeric matrix in various ways. Guest–host[21] and side-chain[19] NLO polymers are the two major systems studied by researchers. In the guest–host system, the dye molecules are dissolved into a polymeric medium and in the side-chain system, the molecules are chemically linked to the polymer chain. We shall discuss the advantages and disadvantages of these two systems.

1. Guest–Host Polymers

A mixture of an optically transparent polymer and a NLO dye is dissolved into a common solvent of the materials. The solution is used to form a film on a substrate by spin or dip coatings. After baking out the solvent, a guest–host NLO polymeric film is formed. Because of the sublimation of the dye molecules during baking, high boiling-point solvents should be avoided for this process. One of the major disadvantage of the guest–host system is that there is an upper limit of the concentration of the dye molecules inside the polymeric matrix. Beyond this limit, aggregation into dimers or crystallites of the dye molecules occurs. Depending on the dye and polymeric systems, the maximum dye concentration in a guest–host polymer is from 10% to 20% by weight. Furthermore, the stability of orientation in a poled guest–host polymer is usually poor.[22] The electro-optic constant suffers a large initial decay during the first few days for the polymethylmethacrylate (PMMA) guest–host system. Recent research focuses on the use of polyimides as host media[23] for NLO molecules. The process begins with spin coating of a solution of polyamic acid resin and NLO dyes mixture and followed by an imidization and poling process[24,25] at an elevated temperature from 210°C to 250°C. During this process, the polyamic acid condenses to imide ring linkages and results in a shrinkage of film thickness in the order of 40%. The nature of high T_g (over 300°C, which is compared with 90°C of PMMA) and the reduction of free volumes in the polymeric matrix during the imidization process accounts for the thermal stability of a poled polyimide guest–host system. Further densification[26] occurs at a temperature over 340°C and results in a highly thermally stable poled film. Furthermore, the thermal expansion coefficients of polyimide guest–host systems are a few times smaller than those of conventional polymeric systems such as PMMA. This is a big advantage over the conventional systems for the applications of EO devices which require a low sensitivity $\partial n/\partial T$ of the refractive indices versus temperature, such as the Mach-Zehnder type of EO intensity modulator. The NLO dyes qualified for this system have a requirement of chemical stability of over 200°C. This requirement prohibits the use of most strong NLO dyes. Research on developing highly thermally stable NLO molecular structures have been initiated.[24] Another disadvantage of this system is that the maximum poling field is limited to the range of 100 V/μm due to the electrical conductivity of the material at the poling temperature which results from the mobility of undesirable ionic impurities in the system.

2. Side-Chain Polymers

The idea of chemically attaching NLO molecules onto polymeric chains for an improvement of orientational stability and aggregation of dye molecules at a high volume concentration was proven valid.[2,22] For PMMA backbone, a weight fraction as high as 75% of NLO dyes is achievable without running into the problem of dye crystallites formation. The thermal stability of a poled side chain polymeric film is significantly improved over the guest–host systems.[22] Figure 1 illustrates a two-step synthesis[27] of a PMMA backbone NLO polymer from a hydroxyethyl-dye molecule. The first step is the preparation of NLO monomer by reacting methacrylate anhydride with the hydroxyethyl-dye under a dimethylamino-pyridine (DMAP) catalyst. The second step is the radical polymerization[9] by adding initiator azobis(isobutyronitrile) (AIBN) to a prewarmed solution containing a mixture of x moles of NLO monomer and $100 - x$ moles of PMMA monomer. The final product is a copolymer of Dye/MMA $x/(100 - x)$. The glass transition temperature T_g of the side-chain polymer is higher compared with that of the intrinsic polymer and increases with the molar ratio $x/(100 - x)$. T_g also depends on the structure of the dye molecule and the spacer which links the dye molecule to the polymer main chain. For example, T_g of P2ANS/MMA 50/50 (with side-group DANS and ethyl spacer) is 135°C and of P2DCVHT/MMA 50/50 (with side-group DCVHT) is 160°C. The highest usable molar ratio $x/(100 - x)$ is about 70/30. Beyond this limit, the solubility of the polymer decreases rapidly. Due to a large difference between the molecular weight of the NLO dye and the MMA monomer, the actual weight fraction F_w of the NLO moiety inside the polymer is higher than $x\%$. For P2ANS/MMA 50/50, F_w is 65%. The attainable poling field of this polymer is over 200 V/μm before an electrical breakdown occurs. At such high concentration of the dyes, mutual interactions between the dipoles are no longer negligible, i.e., $U_0(\theta)$ in Equation 8 should be taken into account when performing ensemble average of a poled film. Experimental results showed that the ratio $R = r_{33}/r_{13}$ of a poled P2ANS/MMA 50/50 film increases

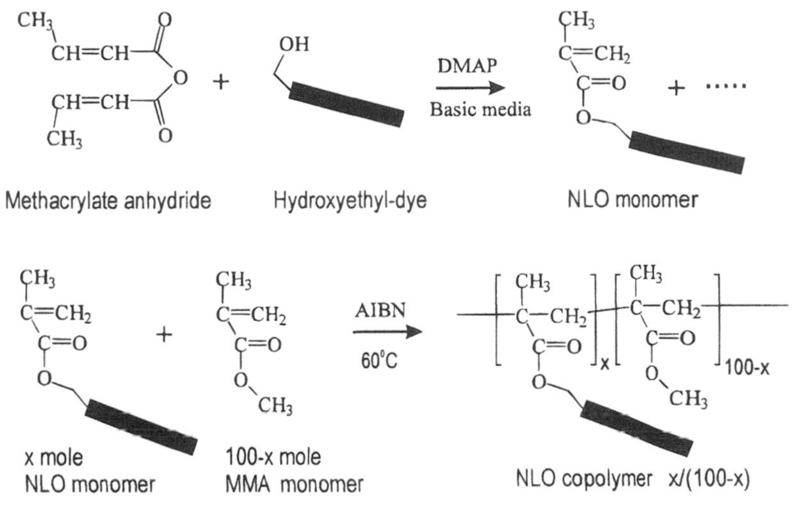

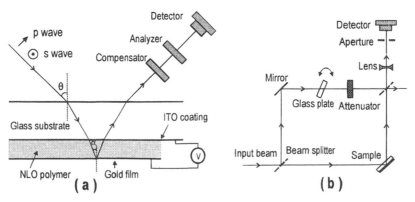

Figure 1 Two-step synthesis of NLO polymers with PMMA backbones.

from 3 to 4.9 when increasing the poling field from 0 to 170 V/μm, where $f_p a \approx 2.6$. The dependence of R on the poling field is stronger than what Equation 14 predicted, where we have omitted the term $U_0(\theta)$ and $U_2(\theta)$ during ensemble average.

D. MEASURING ELECTRO-OPTIC CONSTANTS OF A POLED FILM

1. s and p Waves Method

A simple way to measure the EO constant of a poled film is to utilize the interference between the s and p waves of a laser beam incident at an angle relative to the film surfaces.[28-30] Figure 2a illustrates the optical configuration of the technique, where a reflection scheme is used to double the path length of the beam inside the poled medium. A thin film of polymer, of the order of a few microns, is spin coated onto a glass substrate which is precoated with a thin transparent conducting layer of indium tin oxide (ITO). The ITO serves as one of the two electrodes used for the poling field as well as the modulating field in the EO measurement. The second electrode is a metallic layer thermally evaporated on top of the film. A laser beam is incident from the back of the substrate at an angle θ. It propagates through the substrate, the ITO, the polymer layer, and then it is reflected back out into the air by the top electrode. The polarization of the input beam is set at an angle relative to the plane of incidence so that the parallel (p wave) and perpendicular (s wave) components of the optical field of the reflected beam are equal in amplitude. The reflected beam propagates through a Soleil-Babinet compensator, an analyzer, into a detector. A modulating voltage $V = V_m \sin \omega_m t$ is applied across the electrodes. The modulated signal from the detector is measured using a lock-in amplifier.

Figure 2 Measurement of EO coefficients of poled films using reflection scheme. (a) Interference between s and p waves. (b) Interference between arms of a split beam.

The phase shift of the s wave propagated through the film is given by:

$$\phi_s = \frac{4\pi d}{\lambda} n_{/\!/} \cos \alpha_s = \frac{4\pi d n_{/\!/}}{\lambda} \sqrt{1 - \frac{\sin^2\theta}{n_{/\!/}^2}} \tag{17}$$

and the p wave:

$$\phi_p = \frac{4\pi d}{\lambda} n_p \cos \alpha_p = \frac{4\pi d n_{/\!/}}{\lambda} \sqrt{1 - \frac{\sin^2\theta}{n_\perp^2}} \tag{18}$$

where d is the film thickness, λ is the wavelength, and $n_{/\!/}$ and $n_\perp$ are the refractive indices parallel and perpendicular to the film. α_s and α_p are the refractive angles of the s and p waves, with $n_p^{-2} = n_\perp^{-2} \sin^2 \alpha_p + n_{/\!/}^{-2} \cos^2 \alpha_p$.

The induced phase shifts from the applied modulating voltage is given by:

$$\delta\phi_s = -\frac{2\pi n_{/\!/}^4 r_{13} V_m \sin \omega_m t}{\lambda} (n_{/\!/}^2 - \sin^2\theta)^{-1/2} \tag{19}$$

$$\delta\phi_p = -\frac{2\pi n_{/\!/} V_m \sin \omega_m t}{n_\perp \lambda} [(r_{33} n_\perp^2 - r_{13} n_{/\!/}^2)\sin^2\theta + r_{13} n_{/\!/}^2 n_\perp^2](n_\perp^2 - \sin^2\theta)^{-1/2} \tag{20}$$

with $\delta n_{/\!/}^{-2} = r_{13}(V_m/d)\sin \omega_m t$, $\delta n_\perp^{-2} = r_{33}(V_m/d)\sin \omega_m t$.

The output intensity is a function of the total phase retardation Φ_{sp} between the s and p waves:

$$I = 2I_c \sin^2(\Phi_{sp}/2) \tag{21}$$

where I_c is the midpoint intensity, $\Phi_{sp} = \Phi_{sp}^0 + \delta\phi_s - \delta\phi_p$. By tuning the compensator so that $I \cong I_c$, which corresponds to $\Phi_{sp}^0 = (n + \frac{1}{2})\pi$, where n is an integer. We have:

$$\frac{I}{I_c} \cong 1 \pm (\delta\phi_s - \delta\phi_p) \tag{22}$$

For low poling field, $n_{/\!/} \cong n_\perp$, and from Equation 23, $r_{33} \cong 3r_{13}$. Under these approximations, we can calculate r_{33} from the measured amplitude of modulation I_m:

$$r_{33} \cong \frac{3\lambda I_m}{4\pi V_m I_c n_{/\!/}^2} \frac{\sqrt{n_{/\!/}^2 - \sin^2\theta}}{\sin^2\theta} \tag{23}$$

2. The Interferometer

The previous method offers a simple way to measure r_{33} at a poling field of below 100 V/μm without suffering too much inaccuracy. For higher poling fields, $R = r_{33}/r_{13}$ is significantly larger than 3, as mentioned in Section II.C.2. Furthermore, $n_{/\!/}$ is quite different from $n_\perp$. The following interferometer should be used for a reliable measurement of r_{33} and r_{13} independently. Figure 2b illustrates the optical configuration of this method. The preparation of the sample is the same as discussed in the previous section. The laser beam is split into two arms: one arm propagates through the film and is reflected by the top electrode of the sample and the other arm passes through a rotatable parallel plate to control the phase shift and a variable attenuator to adjust the intensity of the arm. The two arms, at equal intensities, are then combined together for interference. A lens and aperture system is placed in front of the detector to assure a uniform interference of the two arms in the area which pass through the aperture. This could be checked by rotating the parallel plate and monitoring the minimum beam intensity at the detector. During the measurement, the beam intensity is again biased at the midpoint

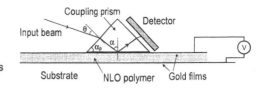

Figure 3 Measurement of EO coefficients of poled films using attenuated total reflection.

I_c between the maximum and minimum. For a vertically polarized input laser beam, r_{13} can be calculated from Equation 19, with $\theta = 45°$:

$$r_{13} \cong \frac{\lambda I_m}{\sqrt{2}\pi V_m I_c n_{//}^4}\sqrt{2n_{//}^2 - 1} \tag{24}$$

and for a horizontally polarized beam, from Equation 20:

$$r_{33} \cong \frac{\lambda I_m}{\sqrt{2}\pi V_m I_c n_{//}n_\perp}\sqrt{2n_\perp^2 - 1} - r_{13}\left(2n_{//}^2 - \frac{n_{//}^2}{n_\perp^2}\right) \tag{25}$$

3. Attenuated Total Reflection

The NLO polymer film is spin coated onto a substrate precoated with a metallic layer.[31] A semi-transparent thin layer of metal is evaporated or sputtered onto the polymer film. The metallic layers serve as poling and modulating electrodes. Highly attenuated guided modes exist in this metal–polymer–metal structure. Mode index n of a guided mode in a thin-film waveguide can be determined via prism coupling method, as illustrated in Figure 3, with:

$$n = n_p \sin \alpha = n_p \sin\left[\alpha_p + \sin^{-1}\left(\frac{\sin \theta}{n_p}\right)\right] \tag{26}$$

where n_p and α_p are the refractive index and angle of the coupling prism; θ and α are the incident and coupling angles. A scan of the reflected power R from the coupling point vs. n shows sharp valleys at positions corresponding to guided modes. Applying an AC voltage $V_m \sin \omega_m t$ to the electrodes, the modulated signal in the reflected beam is given by, for a TE guided mode:

$$\delta R = -\frac{\delta R}{\delta n}\frac{\delta n}{\delta n_{//}}n_{//}^3 r_{13}\frac{V_m}{d}\sin \omega_m t \tag{27}$$

and for a TM mode:

$$\delta R = -\frac{\delta R}{\delta n}\frac{\delta n}{\delta n_\perp}n_\perp^3 r_{33}\frac{V_m}{d}\sin \omega_m t \tag{28}$$

where $\delta R/\delta n$ is the slope of R vs. n and can be determined experimentally. $\delta n/\delta n_\parallel$ and $\delta n/\delta n_\perp$ are the modulating efficiencies of the waveguide and can be calculated by mode analysis. By measuring δR at the maximum slope of the scan curve of R vs. n, one can calculate r_{13} and r_{33}.

III. POLYMERIC HIGH-SPEED ELECTRO-OPTIC MODULATOR

A. GENERAL STRUCTURE AND FABRICATION PROCESS

A polymeric high-speed EO intensity modulator consists of a Mach-Zehnder interferometer which is in the form of a channel single-mode optical waveguide circuit[32] fabricated onto a three-layer structure of poled NLO polymeric thin films. Radio frequency waveguide circuits are formed onto the two arms

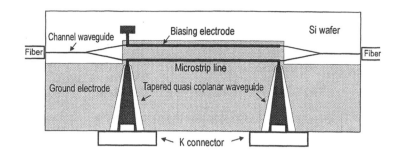

Figure 4 A schematic drawing of a traveling wave EO modulator with a single microstrip line.

of the interferometer for phase modulation of the guided light. Figure 4 illustrates a design of a high bandwidth EO modulator by Teng.[12] The input laser power is fiber-coupled into the channel optical waveguide. A Y junction splits the optical power into two arms. After propagating through the EO active area in between the electrodes of the rf waveguides, two arms are combined together through another Y junction and the guided optical power is coupled into an output single-mode fiber. Through the EO effect of the guiding layer, the applied electrical field on the electrodes induces a phase shift $\delta\Phi$ of the optical fields propagating in the two arms of the interferometer. The output power collected by the fiber is a function of $\delta\Phi$:

$$P = P_{\max} \sin^2 \frac{\delta\Phi + \Phi_0}{2} + P_{\min} \tag{29}$$

Figure 5 explains how this works. The optical field profile propagating along each branch of the Y junction can be considered as a superposition of a first-order mode and a second-order mode at the Y junction. The waveguide tapers down to a single-mode waveguide gradually. The second-order mode radiates out while the first-order mode continues to guide. When the optical fields of the two arms are in phase at the junction, the second-order modes cancel out each other and there are no radiating optical power. On the contrary, if the optical fields are 180° out of phase, the first-order modes cancel out each other and there is only a second-order mode at the junction. This mode radiates out while propagating toward the single-mode section of the Y junction, and subsequently no optical power will be collected by the output fiber.

Following is a description of the fabrication process of Teng's high bandwidth polymeric EO modulator. A Si wafer is used as a substrate for the polymeric films because of its cleavability. Cleaving the Si wafer along its crystal plane results in an end surface of the polymeric films with an acceptable quality for fiber coupling. A patterned gold layer of ground electrode (with a thickness of around 0.6 μm, which is slightly thicker than the skin depth at 40 GHz rf frequency) is coated onto the Si substrate and subsequently three layers of polymers are spin coated onto the ground electrode. The lower and the middle layers are made of the same NLO polymer with a lower NLO chromophore concentration for the lower layer which serves as the lower cladding layer for the channel waveguide. The upper layer is a very thin (in the order of 0.5 μm) transparent polymeric layer which isolates the top electrode from the guided optical wave. The index of refraction of this layer is very low compared to that of the guiding layer. The structure is poled locally for EO active areas after the deposition of a patterned poling electrode. The poling electrode is then removed and the channel waveguide circuit is formed

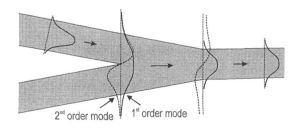

Figure 5 Transformation of a guided optical field profile at a Y junction of single-mode channel waveguides.

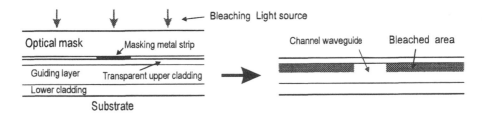

Figure 6 Fabrication of channel waveguides in NLO polymeric films via photobleaching.

by a photobleaching process (see Figure 6). In this process, an intense collimated light source (Xenon lamp) is used to irradiate the sample through an optical mask. The exposed regions of the film undergo a reduction of refractive index due to photochemistry processes which transform the structure of the NLO chromophores, thus creating the optical confinement along the patterned circuit. Single-mode channel waveguides are obtained through control of exposure time and intensity of the light source. The bleaching efficiency is the highest for a light source corresponding to the absorption peak of the NLO chromophores. However, for better penetration depth, an off-resonance light source is preferred. After photobleaching, The rf transmission lines (2-μm-thick gold strips) are formed by a combination of processes including thermal evaporation, masking with photoresist, electroplating, and a removal of the residual gold film, as illustrated in Figure 7.

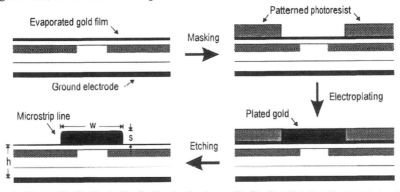

Figure 7 Formation of microstrip lines on polymeric films.

B. SPECIFICATIONS OF AN ELECTRO-OPTIC MODULATOR
1. Drive Voltage V_π
The drive voltage V_π of an EO modulator is defined as the minimum voltage applied to the electrodes to modulate the output optical power from P_{min} to P_{max}, as illustrated in Figure 8. The lower the drive voltage, the lower the rf power required for achieving certain modulated optical power.

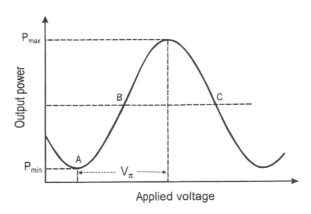

Figure 8 Transfer curve of an EO modulator.

2. Extinction Ratio

The extinction ratio η, in units of dB, can be defined as:

$$\eta \equiv 10 \cdot \log(P_{max}/P_{min}) \tag{30}$$

A high-performance $LiNbO_3$ EO modulator has an extinction ratio higher than 25 dB. For polymeric modulators, an extinction ratio of 20 dB has been demonstrated by Teng.[12]

3. Insertion Loss

The insertion loss is the optical transmission from fiber to fiber at P_{max}. Sources contribute to insertion loss are: fiber-to-channel waveguide coupling losses, optical waveguiding loss, Y-junction losses, and poling-induced losses,[33] in the case of polymeric EO modulators.

4. Frequency Response—dB Optical and dB Electrical

The frequency response of an EO modulator, in units of dB optical, is the relative modulated output optical power $P_{opt}(f)$ vs. modulating radio frequency f at a fixed applied rf power to the electrodes:

$$F_{opt}(f) = 10 \cdot \log(P_{opt}(f)/P_{opt}(0)) = 10 \cdot \log(a(f)/a(0)) \tag{31}$$

where $P_{opt}(f) = a(f)V(f)$ with the RMS applied voltage $V(f) = V(0)$. The $-3dB$ bandwidth BW_{opt} is defined as the frequency f at which $P_{opt}(f)$ is 3 dB smaller than $P_{opt}(0)$.

The frequency response, in units of dB electrical is defined as the relative input rf power $P_{elc}(f)$ applied to the electrodes vs. modulating radio frequency f at a fixed modulation depth of the output optical power:

$$F_{elc}(f) = -10 \cdot \log(P_{elc}(f)/P_{elc}(0)) = -20 \cdot \log(V(f)/V(0)) = 20 \cdot \log(a(f)/a(0)) \tag{32}$$

with $a(f)V(f) = a(0)V(0)$. The $-3dB$ bandwidth BW_{elc} is defined as the frequency f at which $P_{elc}(f)$ is 3 dB larger than $P_{elc}(0)$.

From Equation 31 and 32, we have:

$$F_{elc}(f) = 2 \cdot F_{opt}(f) \tag{33}$$

Therefore, the frequency response curve plotted in the unit of dB optical declines twice as slow as that plotted in dB electrical and the bandwidth BW_{opt} is much larger than BW_{elc}. The industry prefers to use the unit of dB electrical. Some authors use the unit of dB optical in their publications. Attention on the frequency response unit should be made on comparing the bandwidth of the modulators.

C. TRAVELING WAVE HIGH-SPEED ELECTRO-OPTIC MODULATION

Consider a rf field propagates along the electrodes of the modulator which is a planar rf transmission line with a characteristic impedance Z_0. The rf field $E_{rf}(x,t) = E_m e^{-\alpha x} \sin(k_m x - \omega_m t)$ induces a change of the refractive index of the guiding medium in between the electrodes from the EO effect:

$$\delta n(x, t) = -\frac{n^3 r}{2} E_m e^{-\alpha x} \sin(k_m x - \omega_m t) \tag{34}$$

where r is the EO coefficient, $f = \omega_m/2\pi$ is the frequency of the rf wave, $k_m = \sqrt{\varepsilon_{eff}} \cdot \omega_m/c$, where ε_{eff} is the effective dielectric constant of the transmission line, and α is the attenuation per unit length of the rf field propagating along the transmission line. $\delta n(x,t)$ is a function of space and time. A guided optical field traveling along the same direction as the rf field experiences different part of $\delta n(x,t)$ unless the speed of light c/n_{eff} in the guided medium is the same as the propagation speed of rf field $c/\sqrt{\varepsilon_{eff}}$, where n_{eff} is the effective index of refraction of the guided optical mode. Optimum phase shift occurs when the propagation of the rf and optical fields are synchronized. The bandwidth of the EO modulation under this circumstance is limited primarily by the attenuation of the rf power in the transmission line.

There are two major sources contributing to the attenuation of rf power in the transmission line, namely conductor loss and dielectric loss. The conductor loss is essentially an Ohmic loss in the electrodes accelerated by the skin depth effect from which the rf current concentrates on the electrode's surfaces and damps exponentially within a skin depth:

$$\gamma = \frac{c}{2\pi\sqrt{f\sigma_c}} \tag{35}$$

where σ_c is the conductivity of the electrodes. The frequency dependence of the conductor loss is approximately proportional to $\sqrt{f}$. The dielectric loss is derived from the loss tangent $\tan\delta$ of the dielectric medium. For NLO polymer, this loss is insignificant compared to the conductor loss. However, for a medium such as Si wafer, the dielectric loss is quite large due to the finite conductivity σ_d of the medium where $\tan\delta \propto \sigma_d$.

The total phase shift of an optical wave with wavelength λ, after propagating a distance L in between the electrodes of a transmission line with an input rf power P, is given by:

$$\delta\Phi = M\int_0^L e^{-\alpha x}\sin(Kx - \omega_m t)dx \tag{36}$$

where

$$M = \frac{\pi n^3 rg}{\lambda h}\sqrt{2PZ_0}, \qquad K = \frac{\omega_m}{c}(\sqrt{\varepsilon_{eff}} - n_{eff}),$$

h is the spacing between the electrodes, and g is an overlap integral between the rf and optical fields. $\delta\Phi$ can be expressed as:

$$\delta\Phi = M\sqrt{\frac{e^{-2\alpha L} - 2e^{-\alpha L}\cos KL + 1}{\alpha^2 + K^2}}\sin(\phi - \omega_m t) \tag{37}$$

where ϕ is the phase delay of the modulated signal:

$$\tan\phi = \frac{\alpha\sin KL + K\cos KL - Ke^{\alpha L}}{\alpha\cos KL - K\sin KL - \alpha e^{\alpha L}} \tag{38}$$

When biasing at the midpoint (point B or C in Figure 8) between P_{min} and P_{max}, the frequency response of the EO modulator in units of dB electrical, based on small signal approximation, is given by:

$$F_{elc}(f) = 10\cdot\log\left(\frac{e^{-2\alpha(f)L} - 2e^{-\alpha(f)L}\cos K(f)L + 1}{[\alpha^2(f) + K^2(f)]L^2}\right) \tag{39}$$

with $K(0) = 0$ and $\alpha(0) \cong 0$. If the synchronization between the waves is near perfect, i.e., $K \approx 0$, one can express $F_{elc}(f)$ by:

$$F_{elc}(f) \cong 20\cdot\log\left(\frac{1 - e^{-\alpha(f)L}}{\alpha(f)L}\right) \tag{40}$$

Assuming the rf power loss is primary due to conductor loss, the attenuation coefficient $\alpha(f)$ is given by:

$$\alpha(f) \cong a_0\frac{\ln 10}{20}\sqrt{f} \tag{41}$$

454

where a_0 is a constant represented the conductor loss in units of dB/unit length at unit frequency. The bandwidth BW_{elc} and BW_{opt} defined in Section III.B.4 is approximately equal to:

$$BW_{\text{elc}} \cong \frac{40.8}{(a_0 L)^2} \cong 0.21 \cdot BW_{\text{opt}} \tag{42}$$

As mentioned in Section III.B.4, the bandwidth defined in dB optical is much larger than that defined in dB electrical and is almost 5 times larger in this case. The conductor loss a_0 of a transmission line is in general inversely proportional to the electrode spacing h at a fixed characteristic impedance (see next section). Therefore, the bandwidth is proportional to $(h/L)^2$. At the mean time, the drive voltage V_π is proportional to h/L. To compare between performances of EO modulators, it is necessary to define a figure-of-merit:

$$M = \sqrt{BW_{\text{elc}}}/V_\pi \tag{43}$$

the higher the figure-of-merit M, the better the design performance.

D. MICROSTRIP LINE

Due to the requirement of a ground electrode below the three-layer structure of polymeric films for the poling process, microstrip lines are employed as the rf transmission lines for the traveling wave EO modulation. Figure 7 shows a cross section of a microstrip line and a channel polymeric optical waveguide. Under this configuration, the overlap integral g in Equation 37 is close to 1. This results in an optimum modulation efficiency.

1. Quasi-static Analysis of TEM Fundamental Mode

A static approximation for the quasi TEM fundamental mode of a microstrip line remains valid if the radio frequency is below[34]:

$$f_g \approx \frac{0.07c}{(w + 2h)\sqrt{\varepsilon + 1}} \tag{44}$$

where w and h are the width and height of the microstrip. w and h employed for the integrated EO modulator is in the order of 10 μm. This means the static analysis is good for frequencies as high as a few hundred GHz. The characteristic impedance and effective dielectric constant from the static analysis are given by, with $c = 1/\sqrt{L_a C_a}$:

$$Z_0 = \frac{1}{c \cdot C_a \sqrt{\varepsilon_{\text{eff}}}}, \qquad \varepsilon_{\text{eff}} = \frac{C}{D_a} \tag{45}$$

where C is the capacitance per unit length, and C_a and L_a are the capacitance and inductance per unit length, respectively, with the dielectric medium replaced by air. The closed form expressions for Z_0 and ε_{eff} are given by[35]:

$$Z_0 = \frac{120\pi}{\sqrt{\varepsilon_{\text{eff}}}} \left\{ \frac{1}{2\pi} F(1) \cdot \ln\left(\frac{8}{R_{we}} + 0.25 R_{we} \right) \right.$$

$$\left. + (1 - F(1)) \cdot (R_{we} + 1.393 + 0.667 \cdot \ln(R_{we} + 1.444))^{-1} \right\} \tag{46}$$

$$\varepsilon_{\text{eff}} = \frac{\varepsilon + 1}{2} + \frac{\varepsilon - 1}{2}\left(1 + \frac{12}{R_w}\right)^{-1/2} + 0.04 \cdot F(1) \cdot (1 - R_w)^2 - \frac{(\varepsilon - 1)R_s}{4.6\sqrt{R_w}} \tag{47}$$

where $R_w = w/h$, $R_s = s/h$, $R_{we} = R_w + 0.4R_s \,[1 + \ln(2/R_s) + F(1/2\pi) \cdot \ln(2\pi R_w)]$, and

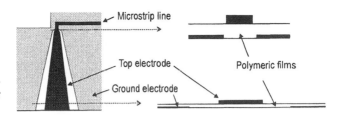

Figure 9 Coupling of rf power into microstrip line via a tapered quasi-coplanar waveguide.

$$F(a) \equiv \frac{1}{2}\left(1 - \frac{R_w - a}{|R_w - a|}\right)$$

is a step function of R_w.

The conductor loss and dielectric loss in units of dB/unit length can be approximated by:

$$a_c = \frac{1.21 \times 10^{-7}A}{h}\sqrt{\frac{f}{\sigma_c}}\left\{2.25 \times 10^4 \frac{F(1)}{Z_0}\left[\frac{32 - R_{we}^2}{32 + R_{we}^2}\right] + (1 - F(1))Z_0\varepsilon_{\text{eff}}\left(R_{we} + \frac{0.667R_{we}}{R_{we} + 1.444}\right)\right\}$$

(48)

where $A \equiv 1 + \dfrac{1}{R_{we}}(1 + 0.4 \cdot \ln(2/R_s) + 0.4 \cdot F(1/2\pi) \cdot \ln(2\pi R_w))$

$$a_d = \begin{cases} 27.3\dfrac{f}{c}\dfrac{\varepsilon}{\varepsilon - 1}\dfrac{\varepsilon_{\text{eff}} - 1}{\sqrt{\varepsilon_{\text{eff}}}}\tan\delta & \text{or} \\[3mm] 1.64 \times 10^3\sigma_d\dfrac{1}{\varepsilon - 1}\dfrac{\varepsilon_{\text{eff}} - 1}{\sqrt{\varepsilon_{\text{eff}}}} & \end{cases}$$

(49)

where σ_c And σ_d are the conductivities of the electrodes and dielectric medium in unit of $\Omega^{-1}\text{m}^{-1}$. One can relate the total loss $a = a_c + a_d$ with α in Equation 39 by: $\alpha = a \cdot \ln 10/20$.

2. Coupling to Microstrip Line

Wiltron's K and V connectors are two commercially available rf connectors with a bandwidth of 46 GHz and 60 GHz, respectively. The inner diameters of these connectors are in the order of 3 mm (smaller for the V connector). Coupling the rf power from these connectors to a 10-μm-size microstrip line with minimum reflection and radiation losses requires some engineering efforts. Figure 9 illustrates a design by Teng using a tapered quasi-coplanar waveguide as a transition element between the K connector and the microstrip line for low loss coupling. The structure is fabricated on a high-resistivity (2000 Ωcm) Si wafer to minimize the dielectric loss of the substrate in the quasi-coplanar section. Notice that this loss is independent of the rf frequency. The quasi-coplanar waveguide matches the dimension of the K connector at one end and of the microstrip line at the other end. The waveguide structure is similar to a coplanar waveguide at the K connector end and to a microstrip line at the other end. To reduce the rf power loss in this transition section, the length of the waveguide is minimized to 5 mm. After the transition section, the microstrip line makes a 90° sharp turn. The loss from this turn is negligible for the dimension of the microstrip line that we are considering (in the order of 10 μm). Figure 10, line, d shows the experimental coupling loss of this design from DC to 40 GHz. The loss increases from −0.3 dB at low frequency to −0.9 dB at 40 GHz. Among the 0.9 dB loss at 40 GHz, 0.3 dB could be attributed to the dielectric loss in the transition section due to the finite conductivity of the Si substrate. This loss could be reduced by using substrates with higher resistivities.

E. DESIGN OF HIGH-SPEED EO MODULATOR
1. Single-Arm Modulation

In the single-arm modulation design, a single microstrip line is fabricated onto one arm of the Mach-Zehnder interferometer. A DC electrode is placed onto the other arm for DC bias, as shown in Figure

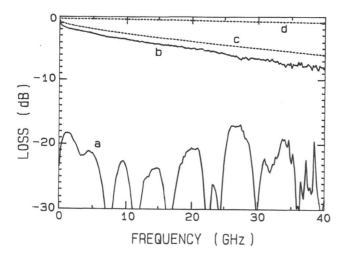

Figure 10 Radio frequency power losses vs. frequency from ref. 12. (a) Return loss; (b) total transmission loss; (c) loss from the 12-mm microstrip line; (d) input coupling loss to the microstrip line.

5. The following analysis is based on an EO modulator made by P2ANS/MMA 50/50. The dielectric constant ε and the refractive index $n_\perp$ of the polymer are 3.5 and 1.64, respectively. The characteristic impedance Z_0 and the effective dielectric constant ε_{eff} vs. $R_w = w/h$ are plotted in Figure 11. In Figure 12, the conductor loss of a microstrip made of pure gold ($\sigma_c^{-1} = 2.2 \times 10^{-8}$ Ωm) at 1 GHz radio frequency is plotted vs. w/h at $h = 6$, 12, and 18 μm. A value of $R_s = s/h = 0.3$ has been used for all the plots. For $Z_0 = 50$ Ω, $w/h = 2.4$, and the effective dielectric constant ε_{eff} is 2.65. $\Delta n = \sqrt{\varepsilon_{eff}} - n_{eff} \approx -0.02$ is close to zero. Notice that Δn is negative. Further improvement for very high modulating frequency can be done by overcoating a dielectric layer on top of the microstrip to increase the effective dielectric constant ε_{eff}. The bandwidth of the modulator is primarily limited by the conductor loss of the microstrip line. As expressed by Equation 48, the conductor loss is inversely proportional to the spacing h of the electrodes at a fix characteristic impedance. The larger the spacing, the lower the loss and the wider the bandwidth. However, increasing the electrodes spacing results in a higher drive voltage V_π. Similarly, from Equation 42 a trade-off between the drive voltage and the bandwidth of the modulator can be made by shortening the total electrode length L. To maintain a certain value of V_π, the ratio L/h is fixed, and from Equation 42 the bandwidth remains the same. Figure 13 plots the theoretical bandwidth BW_{eff} vs. h/L. This represents the maximum bandwidth that one could achieve

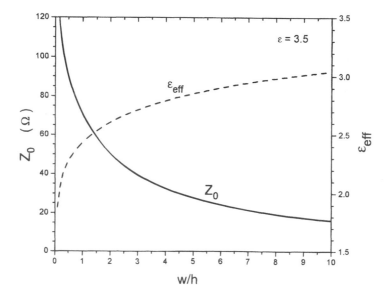

Figure 11 Characteristic impedance Z_0 and effective dielectric constant ε_{eff} vs. w/h of a microstrip line with $\varepsilon = 3.5$ and $s/h = 0.3$.

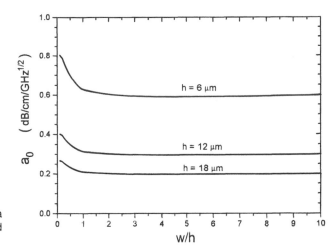

Figure 12 Conductor loss vs. *w/h* of a microstrip line made of gold with ε = 3.5 and *s/h* = 0.3.

theoretically at a particular ratio of *h/L*. In practice, the bandwidth of the modulator is somewhat lower than what has been predicted. The discrepancy is mainly from the underestimation of the conductor loss in the microstrip line and the coupling loss from the rf connector to the microstrip line. Figure 14 shows the experimental frequency response of the traveling wave modulator using P2ANS/MMA 50/50 designed by Teng[12] (refer to Figure 4 for a schematic drawing of the design). The electrode's length and spacing of the modulator are 1.2 cm and 6.5 μm, respectively. The modulation suffers a quick 1-dB (electrical) drop at very low frequencies due to the orientational effect of the NLO chromophore. The 3-dB electrical bandwidth BW_{elc} of the modulator from extrapolation of the data points is about 50 GHz, which is much smaller than the number of 91 GHz predicted in Figure 13. The primary reason for the discrepancy is due to an increase in conductor loss compared to theoretical calculations. This is mainly from the higher actual resistivity of the plated gold electrodes compared to pure gold. The measured average resistivity of the plated gold is 3×10^{-6} Ωcm, which is about 35% higher than that of the book value. It may be caused by the impurities and the porosity of the plated electrodes, especially after the final process which was used to remove the residual gold film. Furthermore, the frequency-dependent part of the input coupling rf power loss degrades the performance of the modulator in addition to the conductor loss. A drive voltage of 6 V at 1.3 μm input optical wavelength was achieved with an average poling field of 165 V/μm, yielding a $n_\perp$ of 1.75 and r_{33} of 21 pm/V. The figure-of-merit *M* defined in Equation 43 is 1.2 $\sqrt{GHz}$/V at 1.3 μm wavelength, which is compared with the number 1 $\sqrt{GHz}$/V of a representative work[36] by Noguchi et al. on LiNbO$_3$ crystal. Noguchi used a coplanar rf waveguide with a shielding plane to improve the large velocity mismatch found in LiNbO$_3$, with a trade-off in drive voltage. This design was based on a push–pull modulation, which doubles the modulating efficiency (see next section), and with a material figure-of-merit n^3r 3 times higher than

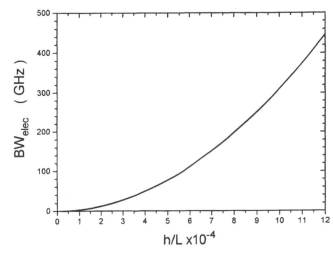

Figure 13 Theoretical bandwidth BW_{elec} vs. *h/L* of a traveling wave polymeric EO modulator.

458

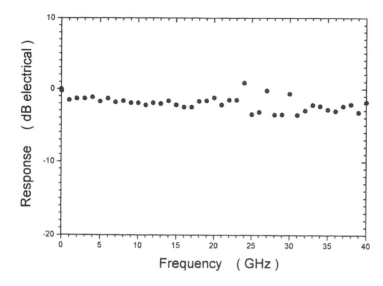

Figure 14 Experimental frequency response of a traveling wave polymeric EO modulator. (From Teng, CC., *Appl. Phys. Lett.*, 60, 1538, 1992. With permission.)

that of P2ANS/MMA 50/50 poled at 165 V/μm. Therefore, the merit of using polymeric material is about 7 times better than LiNbO₃ crystal.

2. Push–Pull Modulation

When modulating at the same time on both arms of the interferometer, the induced total phase shift $\delta\Phi$ in Equation 29 is given by:

$$\delta\Phi = \delta\Phi_1 - \delta\Phi_2 \propto r_1 E_1 - r_2 E_2 \qquad (50)$$

where r_1, r_2 and E_1, E_2 are the local EO coefficients and modulating fields at the first and second arms. To achieve efficient modulation, $r_1 E_1$ and $r_2 E_2$ should be in opposite sign. If $E_1 = E_2$ and $r_1 = -r_2$, $\delta\Phi$ is double and V_π is reduced to a half compared to the case of single-arm modulation. In theory, the figure-of-merit defined in Equation 43 should be double. This is only true for low-frequency modulations. For modulations at rf frequency, careful tailoring the characteristic impedance of the microstrip lines is necessary. For a total impedance of 50 Ω, the impedance of each individual microstrip line is 100 Ω. From Equation 46, a ratio of $w/h \approx 0.3$ is required. According to Equation 48, the conductor loss of the microstrip line is about 25% higher. Furthermore, the microstrip line is narrower than the optical waveguide and the overlapping integral g in Equation 37 is smaller than 1. The overall gain in the figure-of-merit is less than 1.5. The condition $r_1 = -r_2$ can be satisfied with a push–pull poling process. Figures 15 and 16 illustrate a push–pull modulation scheme designed by Teng.[37] The positive and negative poling electrodes are buried under the polymeric films and the top poling electrode serves as a ground plane to avoid air breakdown in between the positive and negative electrodes. The structure is fabricated onto a Si wafer isolated by a layer of SiO₂. The threshold of dielectric breakdown in SiO₂ is a few times higher than that of the NLO polymer.

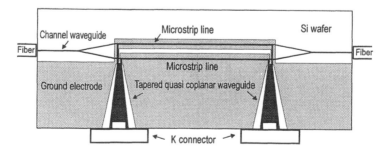

Figure 15 Schematic drawing of a double microstrip-line traveling wave polymeric EO modulator for push–pull modulation.

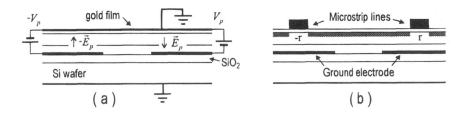

Figure 16 Cross section views of the double-microstrip-line design. (a) Push–pull poling configuration. (b) Push–pull RF modulation.

A layer of 2- to 3-μm-thick SiO_2 is sufficient for the isolation of the poling electrodes. After poling, the top electrode is removed and replaced by a double microstrip line.

F. METHODS OF MEASURING FREQUENCY RESPONSE
1. Direct Measurement
This method requires a calibrated high-speed optical detector and a rf synthesizer with a capability of amplitude modulation (AM) of the rf power with a low-frequency signal. Figure 17 illustrates a schematic drawing of the measurement. The EO modulator being evaluated is DC biased at the midpoint of its transfer curve (point B or C in Figure 8). The leveled output of the rf synthesizer is fed directly to the rf input connector of the modulator and the output is terminated with a 50-Ω load. The rf power is amplitude modulated with a sinusoidal signal at frequency $f_L = \omega_L/2\pi$ which is in the kHz range. The optical power is fiber coupled into the modulator and the output optical power from the modulator is fed into the high-speed optical detector which converts a modulated optical power into a rf power following its frequency response curve. If the input optical power and rf power remain constant while tuning the rf frequency, measuring the converted rf power with a correction of the frequency response of the detector gives the frequency response of the modulator in units of dB electrical. The converted rf power from the high-speed optical detector is usually very low. It is necessary to use a combination of a rf amplifier and a high-speed rf detector to convert the AM rf power (from the optical detector) into a signal at frequency f_L. This signal can be easily picked up by a lock-in amplifier. The high-speed rf detector is commercially available. It is composed of a low-barrier Schottky diode. The response time of the rf detector is below 1 μs. The power gain of the rf amplifier and the output of the rf detector is frequency dependent and usually nonlinear with input rf power. Careful calibration of the system is necessary. Following is a simple procedure of the frequency response measurement and calibration:

Step 1: Record the output signal from the rf detector as a function of rf frequency.
Step 2: Replace the optical detector with the calibrated rf power synthesizer and at each rf frequency, adjust the input AM rf power level until the output from the rf detector matches the output signal corresponding to the frequency response of the modulator which we recorded in the previous step. Record this rf power level as a function of rf frequency.
Step 3: Correct for the frequency roll-off of the optical detector.

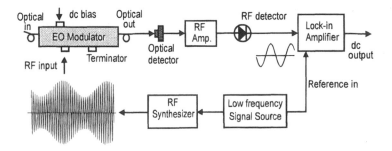

Figure 17 Measuring frequency response of a high-bandwidth EO modulator using high-speed optical and rf detectors.

2. Indirect Measurement

When biasing at the bottom of the transfer curve (point A in Figure 8), the modulated optical power has a quadratic dependence of a small phase shift $\delta\Phi$ induced by EO effect, according to Equation 29:

$$P \approx P_{\max}(\delta\Phi/2)^2 + P_{\min} \qquad (51)$$

An amplitude modulated rf voltage can be expressed as:

$$V = (a - b \sin \omega_L t) \cdot \sin(\omega_m t + \phi) \qquad (52)$$

where $2a = V_1 + V_2$ and $2b = V_2 - V_1$ with V_1 and V_2 are the minimum and maximum voltages.

Applying this voltage source to the modulator yields the following time-dependent terms of the modulated optical power:

$$(\delta\Phi)^2 \propto \left(a^2 + \frac{b^2}{2} - 2ab \sin \omega_L t - \frac{b^2}{2} \cos 2\omega_L t \right) \cdot (1 - \cos(2\omega_m t + 2\phi)) \qquad (53)$$

The existence of the term of modulated optical signal at frequency ω_L allows us to measure the frequency response of a high-speed modulator using a low-speed optical detector. The signal is proportional to ab/V_π^2 and is one to two orders of magnitude smaller than that of the direct measurement described earlier. However, the fact that a low-bandwidth optical detector is required for the measurement highly increases the feasibility of this method. A low-frequency optical detector with a sensitivity a few orders of magnitude higher compared to that of a very high-bandwidth detector is easily obtained using a high-gain trans-impedance amplifier. The overall signal amplitude in this case is larger than that of the previous method. Furthermore, biasing at the minimum output optical power results in a high signal-to-noise ratio if the noise in the optical source is an issue.

G. PRACTICAL PROBLEMS OF POLYMERIC EO MODULATORS

Despite the advantages of using NLO polymers for high-speed electro-optic modulators, there are quite a few issues need to be solved before employing polymeric EO modulators for practical applications. We shall discuss these issues and their possible solutions.

1. Intrinsic Optical Losses

A polymeric EO modulator suffers a significantly higher optical loss than that of an EO modulator made of $LiNbO_3$ crystal. For a channel waveguide fabricated in $LiNbO_3$ by proton exchange, the optical guiding loss is below 0.2 dB/cm at 1.3 μm optical wavelength. In the case of NLO polymer, a guiding loss of 1 dB/cm or higher is typically obtained for a polymer which is highly loaded with NLO chromophores. The guiding loss is mainly attributed to scattering over microdomain boundaries[33] in the material. These domains may be originated from the interaction between the dipole moments of the closely packed chromophores, forming microvolumes with preferred orientations of the chromophores. The molecular polarizability of the chromophores is highly anisotropic and is much larger along the primary molecular axis. This results in a difference in index of refraction (corresponding to an optical polarization along certain direction) across the domain boundaries. For P2ANS/MMA 50/50, the one-dimensional waveguiding loss γ_1 at $\lambda_1 = 1.3$ μm is about 0.64 dB/cm, among which 0.3 dB/cm can be attributed to C–H vibrational stretch overtone absorption.[38] At a shorter wavelength λ_2 of 0.83 μm, the guiding loss γ_2, with a measured value of 2.3 dB/cm, is significantly higher while the overtone absorption is negligible. The ratio $(\gamma_1 - 0.3)/\gamma_2$ is approximately equal to $(\delta n_1^2 / \delta n_2^2)(\lambda_2/\lambda_1)^4$, which confirms that the nature of the loss is scattering. A 5-μm-wide photobleached single-mode channel waveguide shows an additional guiding loss of 0.5 dB/cm at 1.3 μm wavelength. There are two sources for this additional loss. First, the confinement of optical power in a single-mode channel waveguide structure is more vulnerable to inhomogeneities inside the waveguide and more optical power is scattered into radiation modes. Second, the scattering loss of photobleached areas of a NLO polymeric film is relatively high, which could be attributed to bleaching inhomogeneously across the microdomains with different preferred orientations of the chromophores.

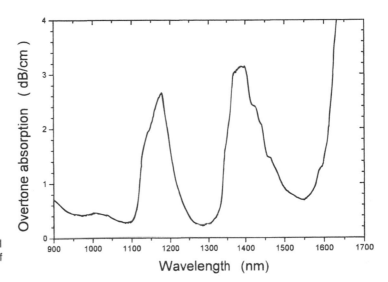

Figure 18 C–H vibrational stretch overtone absorption of P2ANS/MMA 50/50.

The C–H vibrational stretch overtone absorption becomes more eminent as optical communication wavelength moves toward 1.55 μm. Figure 18 illustrates the overtone absorption of P2ANS/MMA 50/50, which is contributed primarily from the aliphatic C–H stretch. Fortunately, both major fiber communication wavelengths, namely 1.3 and 1.55 μm, are located at valleys of the absorption bands. The absorption at 1.55 μm for this system is about 0.8 dB/cm. As demonstrated by McCulloch and Yoon,[39] the overtone absorption of a system containing primarily aromatic C–H bonds becomes less serious. The polyimide systems mentioned in Section II.C.1 have less overtone absorption compared with aliphatic polymers, since these systems contain primarily aromatic rings. Fluorination and deuteration offer alternative solutions to the problem.

2. Poling-Induced Loss

The electro-optically active area of a polymeric modulator suffers an excessive optical loss due to the poling process.[40,41] This excess loss is highly dependent on the poling field and, interestingly, the poling-induced optical loss of a TM guiding mode is much higher than that of a TE mode.[40] Teng[33] developed a theoretical explanation of the effect based on poling-enhanced inhomogeneities in the film. The theory considers volumes δV of inhomogeneities in the guiding medium which experience a different local poling field $\delta f_p E_p$. From Equation 11, the variation of indices of refraction $\delta n_\perp$ (TM mode) and $\delta n_{//}$ (TE mode) in these volumes from poling, under low poling field approximation, can be expressed as:

$$\delta n_i \approx A_i f_i^\omega \delta f_p^2 a^2 / 45 n_i \tag{54}$$

The poling-induced inhomogeneities serve as scattering sites of the guided optical power. The excess waveguiding loss induced by poling is proportional to the total power scattered into radiation modes of the channel waveguide:

$$\gamma_i \propto (\delta n_i)^2 n_i^2 (\delta V)^2 \lambda^{-4} \tag{55}$$

With $A_\perp = -2A_{//}$, the excess poling loss $\gamma_\perp$ of TM guide modes is about 4 times the loss $\gamma_{//}$ in TE modes:

$$\gamma_\perp \cong 4\gamma_{//} \tag{56}$$

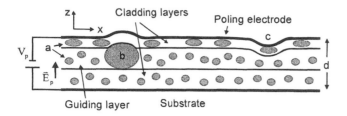

Figure 19 Inhomogeneities in NLO polymeric films causing poling-induced losses. a, b, c: types of inhomogeneities.

There are different types of inhomogeneities inside a spin-coated NLO polymeric film, as illustrated by Figure 19. Type a inhomogeneities are small domains distributed close to each other. These include the type of microdomains mentioned in last section and could also result from evaporation of solvent from the spin-coating process, which leave behind free volumes in the polymer and textures on the surface of the film. Because of their small volumes, the poling-induced loss generated from this type of inhomogeneities is relatively small. An excess loss of below 1 dB/cm is normally obtained for a poling field as high as 170 V/μm. Type b represents large volume defects which could result from aggregation of polymer in the solution, foreign particles, and air bubbles during spin coating. The excess poling loss from this type of inhomogeneity is large and unpredictable. It depends on the size of the volume and the frequency of occurrence. Type c defects result from the depression of the top electrode over weak spots in the films due to electrostatic force of the poling field.

The fundamental source for poling-induced optical loss and intrinsic scattering loss (mentioned in last section) of a NLO polymeric film is the large anisotropy in molecular polarizabilities of NLO chromophores. To reduce these losses, it is necessary to design NLO chromophores with a small anisotropy in their molecular polarizability, while maintaining the one-dimensional asymmetry of the molecular structure for a large second-order optical susceptibility. One possible solution is to attach symmetric one-dimensional molecules transversely onto the asymmetric NLO chromophores. The anisotropy in polarizabilities of both types of molecules balances out each other, while the symmetric molecules do not affect the large second-order optical susceptibility of the NLO chromophore.

3. Optical Degradation

One of the major issues of using NLO polymer as an electro-optically active material for integrated optical devices is the optical degradation of NLO chromophores. It was found[42] that under very high optical intensity, the NLO chromophores bleach away over time even at an optical frequency far away from the excitation energy of the molecule. The bleaching of NLO chromophores is accompanied by a decrease in index of refraction of the guiding medium and subsequently destroys the confinement condition of the channel waveguide. The degradation time was found inversely proportional to the optical intensity inside the waveguide. This rules out the possibility of molecular excitation through two-photon absorption. At an input optical power of 10 mW at 1.3 μm, a 4 $\times$ 5 μm photobleached single-mode channel waveguide made of P2ANS/MMA 50/50 begins losing confinement after a few days of continuously guiding. The same waveguide lasts up to a few weeks longer if operating in an oxygen-free environment. The degradation time is much shorter for waveguides made by polymers composed of DCVHT. The effect is very likely originated from the excitation of the chromophore at the broadening tail of the absorption spectrum of the molecule. For 10-mW input power, the optical intensity inside the 4 $\times$ 5 μm guide is about 5 $\times$ 10^4 W/cm^2, which is about 5 orders of magnitude higher than the intensity of the light source used for photobleaching to form the waveguide circuit (the bleaching time is in the order of hours for a bleaching wavelength near the absorption peak). Although the input wavelength at 1.3 μm is far away from the absorption peak of 426 nm of DANS, a tiny residual absorption at this wavelength could lead to the optical degradation at a very high optical intensity and a very long period of time. The chromophores are unstable at the excited state; they either undergo *trans-cis* transformation, or chemical decomposition, or oxidation when oxygen is available. Until NLO chromophores with a strong chemical stability at their excitational states are found, polymeric integrated devices are only suitable for short-term applications or perhaps used in area where disposable components are applicable.

4. Temperature Stability

There are two kinds of temperature stability involved, namely the stability of the orientational distribution of a poled film at elevated temperature and the temperature dependence of the refractive index from thermal expansion of the polymeric film. A poled NLO polymeric film is a thermodynamically, unstable system. At elevated temperature, mobility of the polymeric chain increases and, subsequently, relaxation of the aligned NLO chromophores occurs. Side-chain NLO polymer is a one-step advance in improving the thermal stability of the poling. Further efforts include cross-linking[22] the polymer during poling and using polyimide class[24,25] of polymers for higher glass transition temperature and lower free volume. The second issue is crucial to interference type of devices such as the Mach-Zehnder type modulator. The bias point of the interferometer fluctuates with temperature due to the local change in index of refraction of the film from thermal expansion. The problem is less serious for an interferometer with well-balanced arms. This means the shape of the electrodes and the rf power applied to the electrodes should be identical on both arms. By nature of their large thermal expansion coefficient, NLO polymers perform poorly in this regard. The polyimide class of NLO polymer offers a possible solution to this problem. The coefficient of thermal expansion of these polymers is a few times smaller than that of conventional polymeric systems.

IV. CONCLUSIONS

By nature of their low dielectric constant, NLO polymers offer significant advantages for high-bandwidth electro-optic modulator over inorganic crystals such as $LiNbO_3$. While researchers have basically maximized the performance of inorganic crystals on the issues of high bandwidth and low drive voltage, NLO polymers still have plenty of room for improvements. The figure-of-merit $\sqrt{BW_{elec}}/V_\pi$ of a polymeric EO modulator, using the first-generation NLO polymer P2ANS/MMA 50/50, is already higher than that of an EO modulator by $LiNbO_3$. The material figure-of-merit n^3r of the NLO polymers could be improved significantly by incorporation of some of the stronger NLO chromophores recently developed. As a result, a very high-bandwidth and low-drive voltage EO modulator is possible with NLO polymers. For a polymeric base of EO modulator employed in practical applications, several issues have to be considered and provided with a practical solution. In Section III.G, some of the practical problems associated with polymeric EO modulators have been described in details with suggested solutions. Hopefully in the near future, NLO polymers will emerge as a practical solution to the increasing demand for high-bandwidth and low-drive power EO modulating devices in high-speed fiber-optic communication networks.

REFERENCES

1. **Demartino, R., Haas, D., Khanarian, G., Leslie, T., Man, H. T., Riggs, J., Sansone, M., Stamatoff, J., Teng, C., and Yoon, H.** Nonlinear optical polymers for electrooptical devices, *Mater. Res. Soc. Symp. Proc.,* 109, 65, 1988.
2. **Man, H. T., Chiang, K., Haas, D., Teng, C. C., and Yoon, H. N.,** Polymeric materials for high speed electro-optic waveguide modulators, *Proc. SPIE,* 1213, 7, 1990.
3. **Lalama, S. J. and Garito, A. F.,** Origin of the nonlinear second-order optical susceptibilities of organic systems, *Phys. Rev. A,* 20, 1179, 1979.
4. **Singer, K. D., Sohn, J. E., King, L. E., Gordon, H. M., Katz, H. E., and Dirk, C. W.,** Second-order nonlinear-optical properties of donor- and acceptor-substitued aromatic compounds, *J. Opt. Soc. Am. B,* 6, 1339, 1989.
5. **Dirk, C. W., Katz, H. E., Schilling, M. L., and King, L. A.,** Use of thiazole rings to enhance molecular second-order nonlinear optical susceptibilities, *Chem. Mater.,* 2, 700, 1990.
6. **Jen, A., K.-Y., Wong, K. Y., Pushkara Rao, V., Drost, K., and Cai, Y. M.,** Thermally stable poled polymers highly efficient heteroaromatic chromophores in high temperature polyimides, *J. Electronics Mater.,* 23, 653, 1994.
7. **Man, H. T., Shu, C. F., Althoff, O., Mcculloch, I. A., Polis, D., and Yoon, H. N.,** Molecular and macroscopic NLO properties of organic polymers, *J. Appl. Polym., Sci.,* 53, 641, 1994.
8. **Prasad, P. N. and William, D. J.,** Bulk nonlinear optical susceptibility, in *Introduction to Nonlinear Optical Effects in Molecules and Polymers,* Wiley-Interscience, New York, 1991, chap. 4.
9. **Demartino, R., Choe, E. W., Khanarian, G., Haas, D., Leslie, T., Nelson, G., Stamatoff, J., Stuetz, D., Teng, C. C., Yoon, H.,** Development of polymeric nonlinear optical materials, in *Nonlinear Optical and Electroactive Polymers,* Prasad, R. N. and Ulrich, D. R., Eds., Plenum Publishing, New York, 1988.
10. **Garito, A. F., Wu, J., Lipscomb, G. F., and Lytel, R.,** Nonlinear optical polymer: challenges and opportunities in photonics, *Mater. Res. Soc. Symp. Proc.,* 173, 467, 1990.
11. **Girton, D. G., Kwiatkowski, S. L., Lipscomb, G. L., and Lytel, R. S.,** 20 GHz Electro-optic polymer Mach-Zehnder modulator, *Appl. Phys. Lett.,* 58, 1730, 1991.

12. **Teng, C. C.,** Traveling-wave polymeric optical intensity modulator with more than 40 GHz of 3-dB electrical bandwidth, *Appl. Phys. Lett.,* 60, 1538, 1992.

13. **Teng, C. C. and Findakly, T.,** Wideband NLO polymeric modulators, *Proc. SPIE,* 2025, 526, 1993.

14. **Garito, A. F., Singer, K. D., and Teng, C. C.,** Molecular optics: nonlinear optical properties of organic and polymeric crystals, *Am. Chem. Soc. Sym. Ser.,* 233, 1, 1983.

15. **Yariv, A. and Yeh, P.,** Electro-optics, in *Optical Waves in Crystals: Propagation and Control of Laser Radiation,* Wiley-Interscience, New York, 1984, chap. 7.

16. **Ward, J. F.,** Calculation of nonlinear optical susceptibilities using diagrammatic perturbation theory, *Rev. Mod. Phys.* 37, 1, 1965.

17. **Singer K. D. and Garito, A. F.,** Measurements of molecular second-order optical susceptibilities using dc-induced second harmonic generation, *J. Chem. Phys.* 75, 3572, 1981.

18. **Teng, C. C. and Garito, A. F.,** Dispersion of the nonlinear second-order optical susceptibility of organic systems, *Phys. Rev. B,* 28, 6766, 1983.

19. **van der Vorst, C. P. J. M., Horsthuis, W. H. G., and Mohlmann, G. R.,** Nonlinear optical side-chain polymers and electro-optic test devices, in *Polymers for Lightwave and Integrated Optics: Technology and Application,* Hornak, L. A., Ed., Marcel Dekker, New York, 1992, chap. 14.

20. **Bottcher, C. J. F.,** The dielectric constant in the continuum approach to the environment of the molecule, in *Theory of Electric Polarization,* Vol. 1, Elsevier, New York, 1973, chap. V.

21. **Pantelis, P. and Hill, J. R.,** Guest–host polymer systems for second-order optical nonlinearities, in *Polymers for Lightwave and Integrated Optics: Technology and Application,* Hornak, L. A., Ed., Marcel Dekker, New York, 1992, chap. 13.

22. **Singer, K.,** Molecular polymeric materials for nonlinear optics, in *Polymers for Lightwave and Integrated Optics: Technology and Application,* Hornak, L. A., Ed., Marcel Dekker, New York, 1992, chap. 12.

23. **Reuter, R., Franke, H., and Feger, C.,** Evaluating polyimides as lightguide materials, *Appl. Opt.,* 27, 4565, 1988.

24. **Wu, J. W., Valley, J. F., Ermer, S., Binkley, E. S., Kenney, J. T., Lipscomb, G. F., and Lytel, R.,** Thermal stability of eletro-optic response in poled polyimide systems, *Appl. Phys. Lett.,* 58, 225, 1991.

25. **Wong, K. Y. and Jen, A. K.-Y.,** Thermally stable poled polyimides using heteroaromatic chromophores, *J. Appl. Phys.,* 75, 3308, 1994.

26. **Wu, J. W., Binkley, E. S., Kenney, J. T., Lytel, R., and Garito, A. F.,** Highly thermally stable eletro-optic response in poled guest-host polyimide systems cured at 360°C, *J. Appl. Phys.,* 69, 7366, 1991.

27. **McCulloch, I., Man, H. T., Marr, B., Teng, C. C., and Song, K.,** Synthesis and electrooptic characterization of a novel highly active indoline nitroazobenze methacrylate copolymer, *Chem. Mater.,* 6, 611, 1994.

28. **Khanarian, G., Che, T., Demartino, R. N., Haas, D., Leslie, T., Man, H. T., Sansone, M., Stamatoff, J., Teng, C. C., and Yoon, H.,** Characterization of polymeric nonlinear optical materials, *Proc. SPIE.,* 824, 72, 1987.

29. **Teng, C. C. and Man, H. T.,** Simple reflection technique for measuring the electro-optic coefficient of poled polymers, *Appl. Phys. Lett.,* 56, 1734, 1990.

30. **Schildkraut, J. S.,** Determination of the electrooptic coefficient of a poled polymer film, *Appl. Opt.,* 29, 2839, 1990.

31. **Cross, G. H., Girling, I. R., Peterson, I. R., Cade, N. A., and Earls, J. D.,** *J. Opt. Soc. Am. B,* 4, 962, 1987.

32. **Nishihara, H., Haruna, M., and Suhara, T.,** Optical waveguide theory, in *Optical Integrated Circuits,* McGraw-Hill, New York, 1985, chap. 2.

33. **Teng, C. C., Mortazavi, M. A., and Boughoughian, G. K.,** Origin of the poling-induced optical loss in a nonlinear optical polymeric waveguide, *Appl. Phys. Lett.,* 66, 667, 1995.

34. **Hoffmann, R. K.,** in *Handbook of Microwave Integrated Circuits,* Artech House, 1987, sect. 3.2.

35. **Gupta, K. C., Garg, R., and Bahl, I. J.,** Microstrip lines II: fullwave analyses and design considerations, in *Microstrip Lines and Slotlines,* Artech House, 1979.

36. **Noguchi, K., Kawano, K., Nozawa, T., and Suzuki, T.,** A Ti:LiNbO$_3$ optical intensity modulator with more than 20 GHz bandwidth and 5.2 V driving voltage, *IEEE Photon. Tech. Lett.,* 3, 333, 1991.

37. **Teng, C. C.,** unpublished.

38. **Groh, W.,** Overtone absorption in macromolecules for polymer optical fibers, *Macromol. Chem.,* 189, 2861, 1988.

39. **McCulloch, I. and H. N. Yoon,** Fluorinated NLO polymers with improved optical transparency in the near infrared, *J. Polym. Sci. A,* 33, 1177, 1995.

40. **Diemeer, M. B., Suyten, F. M. M., Trommel, E. S., McDonach, A., Copeland, J. M., Jenneskens, L. W., and Horsthuis, W. H. G.,** Photoinduced channel waveguide formation in nonlinear optical polymers, *Elect. Lett.,* 26, 379, 1990.

41. **Tumolillo, T. A., Jr. and Ashley, P. R.,** A novel pulse poling technique for EO polymer waveguide devices using device electrode poling, *IEEE Photon. Tech.,* 4, 142, 1992.

42. **Mortazavi, M. A., Yoon, H. N., and Teng, C. C.,** Optical power handling properties of polymeric nonlinear optical waveguides, *J. Appl. Phys.,* 74, 4871, 1993.

Chapter 8

An Introduction to Photorefractive Polymers

B. Kippelen, K. Meerholz, and N. Peyghambarian

CONTENTS

I. INTRODUCTION

As current communication systems are becoming more powerful and complex, optical technology is expected to contribute significantly to the future electronic information processing networks. The free-space nature of optics allowing highly parallel interconnects, the immunity to electromagnetic interference together with the possibility to develop high-bandwidth optical modulators and switching elements, make this new technology very attractive for the transport, processing, and storage of large amounts of information. To make this emerging optical technology viable by the simultaneous development of enabling technologies (spatial light modulators, laser sources arrays) and new architectures, the search for novel optical materials is of crucial importance. Advances in materials are focusing on the development of optical materials with high optical quality, high efficiency, high sensitivity, long lifetime, low cost, and good processing capabilities. With such requirements, it is challenging to meet all these constraints simultaneously in a single material.

Among the most sensitive materials for nonlinear optical applications, photorefractive (PR) materials have been the subject of active research during the last 25 years. Discovered in $LiNbO_3$ as undesirable optical damage,[1] the photorefractive effect has been identified[2-4] and studied mainly in inorganic crystalline materials such as ferroelectrics, sillenites, and semiconductors.[5] The photorefractive effect is based on a combination of photoconducting and electro-optic properties and can lead to high refractive index variations under the illumination of a low-power laser source. In contrast to many grating formation mechanisms with thermal, electronic, or chemical origins, for instance, the photorefractive effect has a unique feature related to the transport of carriers during PR grating formation which leads to a dephasing between the light distribution that induces the grating and the resulting refractive index modulation. This nonlocal response is a fingerprint of the photorefractive effect and enables energy transfer between two coherent laser beams.[6] Many applications such as optical correlation, real-time holography, optical memories, and reconfigurable interconnects have been demonstrated with photorefractive materials.[7]

More recently, the progress achieved in the field of molecular engineering has led to the development of a new generation of photorefractive materials that are organic.[8] The field of organic photorefractive materials has been initiated in 1990 with the first evidence of the photorefractive effect in an organic crystal[9,10] and has grown rapidly after the demonstration of photorefractivity in a polymeric system.[11,12] Since the first proof of principle of photorefractivity in a polymer, the progress achieved in this new class of materials has been tremendous. Numerous PR polymer composites have been synthesized by using different synthetic approaches,[13–16] guest matrices,[17] sensitizers,[18,19] transport agents,[20] and electro-optic chromophores. However, for a long time, the index modulations generated in these materials remained too small to enable large diffraction efficiencies or to observe net gain in two-beam coupling experiments. Significant improvement in the performance of PR polymers was obtained by using the photoconductive polymer poly(N-vinylcarbazole) (PVK) as the composite host and by doping with nonlinear molecules. Such materials doped with the nonlinear optical molecule F-DEANST showed diffraction efficiencies up to 1% in 125-μm-thin films.[21] The photorefractive origin of these efficient gratings was confirmed by asymmetric two-beam coupling experiments and, for the first time, PR gain exceeding the absorption of the sample could be observed in a polymer. In PVK-based polymers doped with the chromophore DMNPAA and an additional plasticizer, significantly higher efficiencies (6%) could be measured in 105-μm-thin samples.[22] Further improvement of this composite has led recently to the demonstration of diffraction efficiencies close to 100% and net gain coefficients exceeding 200 cm^{-1}.[23] Within the past 4 years, photorefractive polymers have reached performance levels comparable or in some respect even superior to the best known inorganic crystals.

The significance of the development of such polymers with extremely high photorefractive efficiencies can be illustrated by considering the promising properties of polymers for photonic applications in general. For a long time, organic materials have been regarded as inefficient, impure, and unstable materials for photonic applications and most of the attention was paid to inorganic materials. In the early developments of photoconductive materials for instance, most of the research effort was focused on crystalline and amorphous semiconductors such as silicon. However, within the past 15 years an intensive effort in the search of new organic photoconductors has led successfully to many industrial applications in the copier area and, at present, nearly 90% of the photoreceptors are made of organic materials.[24] Likewise, organic materials are increasingly being recognized as high-performing materials for a variety of applications[25] and the misconception about the fragility of organic materials is gradually being erased by the synthesis of new materials with better environmental and thermal stability. The processability and structural flexibility of polymers give them an important technological potential and have driven intensive research efforts in the development of functional and multifunctional polymers. Electro-optic poled polymers for instance have matured in the last few years and can exhibit electro-optic coefficients close to those of lithium niobate ($LiNbO_3$) which is used currently in the electro-optic industry. In contrast to organic photoconductors, nonlinear optical molecules and polymers have not found major technological applications, but the worldwide research effort under progress should ensure a gradual evolution from purely electronic devices to optoelectronic devices based on semiconductors and polymers.

In addition to the capacity to form thin films compatible with semiconductor multilayer processing for integrated optics, polymers have physical properties which are favorable for the photorefractive effect. These features, including a low dielectric constant and a high electro-optic coefficient, are common to organic materials and lead to a high figure-of-merit for the photorefractive effect. With the tremendous progress achieved recently in this new class of materials, photorefractive polymers look attractive for photonic devices and can become an alternative to existing inorganic PR materials, especially when processing and low cost are of primary importance. Further development is expected owing to the important progress that is achieved simultaneously in the fields of molecular nonlinear optical materials and organic photoconductive materials.

The development of photorefractive polymers is a truly multidisciplinary field associating chemists and physicists. The aim of this chapter is to provide state-of-the-art information on the physics and chemistry of photorefractive polymers. Section II begins with a background on the photorefractive effect as it is modeled and understood in inorganic crystals. At this stage, there exists no real model for the photorefractive effect in organic materials that takes into account the differences in the physics of organic materials with respect to inorganic materials such as the field dependence of both the photogeneration efficiency and the mobility of the carriers. However, the commonly used model for inorganic crystals, namely the Kukhtarev model,[26] still provides a solid framework for the description of the formation of a photorefractive grating in organic materials.

In Section III, the physical properties required for observing photorefractivity, including charge generation, transport, and nonlinear optical properties, are discussed for amorphous polymers on both the molecular and the bulk levels. Four-wave mixing experiments, in which an independent probe beam is diffracted from a grating written previously by two interfering pump beams, and self-diffraction experiments, in which two interfering pump beams are exchanging energy, are described. The formation of PR index gratings are discussed for two kinds of polymers: permanently prepoled polymers whose glass transition temperature is high compared with the operating temperature and polymers with low glass transition temperatures that can be poled at room temperature. In the latter case, the internal space-charge field can influence the orientation of the nonlinear molecules and lead to higher diffraction efficiencies.

Section IV presents different approaches that have been adopted for the design of photorefractive polymers. The required multifunctionality including photosensitivity, photogeneration, transport, trapping, and nonlinear optical properties can be provided in different ways: the so-called guest–host approach, where the different functionalities are implemented by doping a polymer with functionalized low molecular weight compounds, and fully functionalized polymers, where the functional groups are attached to the polymer backbone. Section IV gives also a brief survey of the young field of photorefractive polymers and discusses some examples of photorefractive polymer composites. This section concludes with a summary of some differences in the fundamental physical properties between inorganic crystalline and organic amorphous materials.

Experimental results on PVK-based photorefractive polymers which have shown the highest efficiencies to date are described in detail in Section V. In addition to wave-mixing experiments, independent characterization experiments are described, such as measurements of the photogeneration efficiency, photoconductivity, and electro-optic properties. They provide useful information for the modeling of the photorefractive effect in polymers. We demonstrate the use of PR polymers in some possible applications such as dynamic holographic storage.

The final section gives a summary of the requirements for a well-performing PR polymer for holographic applications. Future trends including material and device aspects are discussed. We also describe new opportunities for fundamental research that these new materials offer.

II. FUNDAMENTALS OF PHOTOREFRACTIVITY

The photorefractive effect is based on a redistribution of photogenerated charges under a nonuniform illumination leading to an internal electric field pattern that modulates the refractive index through the electro-optic effect.[27] Photorefractivity can, therefore, be observed in materials that are photoconductive and electro-optic simultaneously. A photoconductor is a material that is a good insulator in the dark and becomes conductive upon illumination. The electrical conductivity of the material is due to mobile charge carriers that are created by absorbed photons. Photorefractive materials are typically illuminated by two coherent laser beams with the same wavelength as shown in Figure 1. These beams interfere to produce a spatially modulated intensity distribution $I(x)$ given by[28]:

$$I(x) = I_0[1 + m \cos(2\pi x/\Lambda)] \tag{1}$$

where $I_0 = I_1 + I_2$ is the total incident intensity, i.e., the sum of the intensities of the two beams, $m = 2(I_1 I_2)^{1/2}/(I_1 + I_2)$ is the fringe visibility, and Λ is the grating spacing which depends on the wavelength and the angle between the two beams (the smaller the angle, the larger the grating spacing).

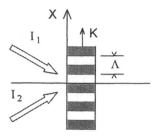

Figure 1 Schematic of the writing process of an hologram with two interfering light beams.

With visible light, the period of the sinusoidal light distribution, which is given by the value of the grating spacing, can vary generally from a fraction of a micrometer to a couple of tens of micrometers. Following a spatially periodic photoexcitation, more carriers are generated in the regions near the light maxima because the photogeneration rate can be assumed proportional to the local value of the optical intensity. In the simplest model, the majority carriers (electrons or holes) migrate from the positions of high intensity where they were generated and leave behind fixed charges of the opposite sign. In some materials, the minority carriers can migrate in the opposite direction but with a mobility that is much smaller than the mobility of the majority carriers. In traditional photorefractive crystals, the driving forces for carrier migration are diffusion due to concentration gradients and drift when a static electric field is applied.[29] The migration process is limited by traps present in the material under the form of impurities or more generally any structural modification of the local environment of a site that leads to a deformation of the local potential and to a lower energy state for the carriers. The transport process being optically activated, the carriers can to some extent move away from the brighter regions to the darker regions where the conductivity is much lower and where they get trapped. The characteristic distance over which the carriers migrate in an efficient photorefractive material is, therefore, in the micrometer range. The carriers trapped in the darker regions and the fixed charged ions left behind in the brighter regions give rise to an inhomogeneous space-charge distribution. The space-charge distribution is in phase with the light distribution if transport is governed by diffusion only, and can be phase shifted when drift in response to an applied field is present. If the light is removed, this space-charge distribution can remain in place for a period of time which can vary between nanoseconds and years depending on the photorefractive material class. For a one-dimensional light distribution described by Equation 1, a space-charge distribution $\rho(x)$, in a material with a dielectric constant ϵ, induces an internal space-charge field $E_{sc}(x)$ via Poisson's equation which writes $dE_{sc}/dx = 4\pi\rho/\epsilon$. The different steps of the buildup of the space-charge field are summarized in Figure 2. The last step in the buildup of a photorefractive grating is the modulation of the refractive index of the electro-optic material by this internal space-charge field via the electro-optic effect. The spatial derivative that appears in Poisson's equation is of paramount importance for the photorefractive effect since it is the origin of the nonlocal response of photorefractive materials, namely the dephasing between the refractive index modulation and the initial light distribution. This phase shift is unique to photorefractive materials and can lead to energy transfer between two incident beams. It shows that the underlying meaning of the term *photorefractivity* is more than a refractive index variation induced by light and makes the photorefractive effect very different from many other mechanisms that can lead to grating formation such as thermal, chemical, and electronic nonlinearities. In contrast to many processes studied in nonlinear optics and described in other chapters of this book, the photorefractive effect is very sensitive and can be observed using milliwatts of laser power. Other unique properties of photorefractive materials are the possibility to store gratings during a long period of time but with the possibility to erase the grating at any time with a uniform light source. All these properties make photorefractive materials important for dynamic holographic recording/retrieval and all the related applications.[7]

The photorefractive effect has been studied mainly in inorganic crystals,[29] including lithium niobate ($LiNbO_3$), barium titanate ($BaTiO_3$), and potassium niobate ($KNbO_3$), which belong to the class of ferroelectrics, bismuth silicon oxide ($Bi_{12}SiO_{20}$) and other sillenites, and semiconductors such as GaAs, InP, CdTe,[30] and semiconductor multiquantum wells (MQW).[31–37] The basic physical model for photorefractivity in inorganic materials is based on the Kukhtarev model.[26,38–41] It gives the first-order component

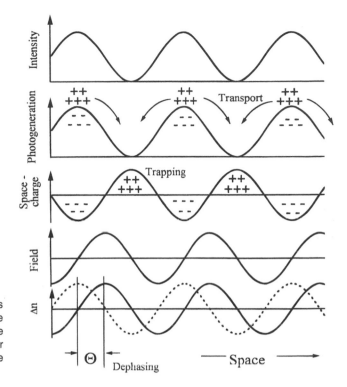

Figure 2 Schematic of the different steps of the buildup of a space charge, the resulting space-charge field, and finally the refractive index grating through the linear electro-optic effect in a photorefractive material.

of the steady-state space-charge field created by a sinusoidal light distribution. The amplitude of the space-charge field is E_{sc} given by:

$$E_{sc} = m\left(\frac{(E_0^2 + E_D^2)}{(1 + E_D/E_q)^2 + (E_0/E_q)^2}\right)^{1/2} \qquad (2)$$

where E_0 is the component of the applied field along the grating vector. The diffusion field E_D is defined as:

$$E_D = \frac{Kk_BT}{e} \qquad (3)$$

where $K = 2\pi/\Lambda$ is the grating vector, k_B is the Boltzmann constant, T the temperature, and e the elementary charge. In Equation 2, the trap-limited field E_q is given by:

$$E_q = \frac{eN_T}{K\epsilon\epsilon_0} \qquad (4)$$

where ϵ is the dielectric constant (dc), ϵ_0 the permittivity, and N_T the density of photorefractive traps. The phase shift Θ between the space-charge field and the light interference pattern resulting directly from transport is given by:

$$\Theta = arctg\left[\frac{E_D}{E_0}\left(1 + \frac{E_D}{E_q} + \frac{E_0^2}{E_DE_q}\right)\right] \qquad (5)$$

If transport takes place only by diffusion, the phase shift is $\pi/2$, otherwise it depends on the relative strength of the diffusion and drift processes.

The internal space-charge field creates a refractive index modulation $\Delta n(x)$ according to the linear electro-optic (Pockels) effect:

$$\Delta n(x) = -\frac{1}{2} n^3 r_{\text{eff}} E_{sc}(x) \qquad (6)$$

where n is the average refractive index of the material. The effective electro-optic coefficient r_{eff} is a combination of different electro-optic tensor elements and depends on the symmetry and the orientation of the material during the interaction process. The electro-optic tensor elements can be linked to the components of the second-order nonlinear susceptibility of the type $\chi^{(2)}_{IJK}(-\omega; 0, \omega)$ where an optical field interacts with a dc or low-frequency electric field. As for any material that exhibits second-order nonlinear properties, photorefractive materials must be noncentrosymmetric.

This model introduced by Kukhtarev and co-workers has provided a solid framework for the description of the formation of photorefractive gratings in inorganic crystals and agrees reasonably well with experimental results. However, at this early stage of the research in organic photorefractive materials, it is difficult to speculate on the validity of this model since it does not take into account some fundamental differences between inorganic and organic materials such as the field dependence of the photogeneration efficiency and mobility (see Section IV.B). Nevertheless, the following basic physical trends can be extracted from this model and applied to organic materials. Equation 4 shows that the maximum value for the space-charge field depends, for a given material, on the trap density N_T and on its low-frequency dielectric constant ϵ. In most of the inorganic crystals, these two parameters are not very favorable to the buildup of a strong space-charge field. High doping levels are difficult to achieve in crystals because crystals have the tendency to be repellent to impurities. In addition, since the electro-optic properties are often driven by ionic polarizability in inorganic crystals, high electro-optic properties are accompanied by a high dielectric constant which limits the value of the space-charge field.

As will be discussed in more detail in Section IV.B, in organics, the electronic origin of the nonlinearity leads to materials with high second-order nonlinear optical properties and simultaneously a low dielectric constant at low frequencies. Moreover, traps are not due to impurities but rather to structural defects or conformational disorder characteristic of noncrystalline materials. As a result, very high trap densities can be present and the space-charge field can reach saturation. Saturation occurs when the value of the internal space-charge field balances the external field. Polymers prepared in thin films can support very large electric fields, leading to significantly higher space-charge fields in polymers compared to inorganic crystals. These considerations make organic materials and especially polymers very attractive for photorefractivity and have been the driving forces in the early development of organic photorefractive materials.

III. PHYSICS OF PHOTOREFRACTIVE POLYMERS

In order to be photorefractive, a material has to combine photosensitivity to generate charge carriers and photoconductivity to separate them through transport, provide traps for them to store the space-charge field, and possess a field-dependent refractive index such as through the electro-optic effect. In the following, we will present the physical models currently used to describe charge generation and transport in amorphous organic photoconductors, and discuss the poling-induced linear and nonlinear (electro-optic) optical properties of materials containing molecular dipoles (the so-called poled polymers).[42] We will also describe the formation and testing of photorefractive gratings. As will be shown, these properties are different in organic materials in comparison with inorganic photorefractive materials.

A. CHARGE GENERATION

The space-charge buildup process can be divided into two steps: the electron-hole generation process followed by the transport of one carrier species. The creation of a bound electron-hole pair (or exciton) after absorption of a photon can be followed by recombination. This process limits the formation of free carriers that can participate in the transport process and is, therefore, a loss for the formation of the space charge. In organic materials, owing to a low dielectric constant, the screening of the Coulomb interaction between the electron and the hole is small, resulting in a low photogeneration efficiency unless an electric field is applied to counteract recombination. The quantum efficiency for carrier

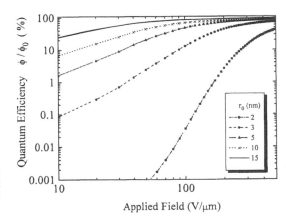

Figure 3 Normalized photogeneration efficiency vs. electric field according to Onsager's theory for different values of the thermalization radius r_0 (T = 300 K and ϵ = 3.5).

generation is, therefore, strongly field dependent and increases with the applied field. A theory developed by Onsager[43] for the dissociation of ion pairs in weak electrolytes under an applied field has been found to describe reasonably well the temperature and field dependence of the photogeneration efficiency in some of the organic photoconductors. A good numerical approximation to Onsager's quantum efficiency $\phi(E)$ was given by Mozumder[44] in terms of the infinite sum:

$$\phi(E) = \phi_0\left[1 - \zeta^{-1} \sum_{n=0}^{\infty} A_n(\kappa)A_n(\zeta)\right] \tag{7}$$

where $A_n(x)$ is a recursive formula given by:

$$A_n(x) = A_{n-1}(x) - \frac{x^n \exp(-x)}{n!} \tag{8}$$

and

$$A_0(x) = 1 - \exp(-x) \tag{9}$$

In Equation 7, ϕ_0 is the primary quantum yield, i.e., the fraction of absorbed photons that results in bound thermalized electron-hole pairs. It is considered independent of the applied field. $\kappa = r_c/r_0$ and $\zeta = er_0E/k_BT$, where $r_c = e^2/4\pi\epsilon_0\epsilon k_BT$ and r_0 is a parameter that describes the thermalization length between the bound electron and hole. For the infinite sum in Equation 7 good convergence is found for $n = 10$. Figure 3 shows the normalized photogeneration efficiency according to this model for different values of the thermalization length r_0 (T = 300 K and ϵ = 3.5). The dissociation of the carriers by the electric field is much easier when the thermalization radius is large, i.e., when the Coulomb interaction is weaker. As this model describes pretty well the field dependence of the photogeneration process in some organic photoconductors such as poly(N-vinylcarbazole),[45–47] more complex extrinsic photogeneration processes can occur involving reactions of the excited state with an electron donor–acceptor prior to the ionization of the electron-hole pair. In this case, the situation can be more complicated and the Onsager model is no longer sufficient to describe the photogeneration efficiency.

B. CHARGE TRANSPORT

After photogeneration of free carriers, the next step in the buildup of a space charge is their transport from brighter regions of the interference pattern, where they are generated, to the darker regions, where they get trapped. In contrast to inorganic photorefractive crystals with a periodic structure, photorefractive polymers have a nearly amorphous structure. The local energy level of each molecule/moiety is affected by its nonuniform environment. The disorder in amorphous photoconductors splits the conduction bands of molecular crystals into a distribution of localized electronic states. As a result, transport can no longer be described by band models but is attributed to intermolecular hopping of carriers between neighboring molecules or moieties. Pioneering work by Scher and Montroll[48] explained this hopping-

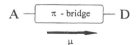

μ

Figure 4 Schematic structure of an intramolecular charge-transfer molecule (push–pull molecule) known for showing strong nonlinear optical properties.

type transport by a continuous time random walk model (CTRW) in which the charge carriers are hopping between equally spaced lattice sites. Other models based on multiple trapping/detrapping have also been proposed.[49] However, discrepancies from these simple models have been found experimentally[50] and more elaborate models that include the effects of disorder between the different sites are being developed.[51] In the latest models, a Gaussian shape is assumed for the density of states. The charge carriers are randomly generated within this distribution and relax during their hopping motion to the energetically lowest states. An activation energy E_A is then needed to continue the transport process. The activation energy is usually field dependent and decreases when the electric field is increased. This process has direct influence on the trap-limited mobility and leads to a field dependence of the mobility $\mu(E)$ of the type[52,53]:

$$\mu(E) \propto \exp[-(E_A(E) - \beta_T E^{1/2})/k_B T] \tag{10}$$

where E_A is the activation energy and β_T is a constant which is positive below the glass transition temperature of the polymer but can become negative above. The activation energy also decreases with decreasing intersite hopping distance and is expected to scale with the square root of the intermolecular distance. Such an activated or trap-limited transport process provides sufficient high carrier mobility in the brighter regions and low conductivity in the darker regions of the interference pattern. Structural anomalies such as impurities, defects of the monomeric units, or chain irregularities can act as traps. No specific extrinsic traps are needed as in inorganic crystals to ensure the storage of a space charge since trapping is due to the physical nature of the transport process itself. Photorefractive crystals are periodic structures and are, therefore, repellent to doping. As a result, the density of traps that can be reached by doping is often in the range 10^{16}–10^{17} cm^{-3}. In contrast, the density of traps in amorphous photoconductors is very high. It can be as high as the density of hopping sites. This property together with the low dielectric constant and high dielectric strength of organic photoconductors can lead to high space-charge field values and consequently to high photorefractive efficiencies. However, the magnitude of the activation energy can limit the storage time of the grating since the transport mechanism can also be activated thermally and not only optically. To get longer storage times, addition of molecules that form deep traps is needed, but this also reduces the mobility and, therefore, increases the response time.

C. PHYSICS OF POLED POLYMERS—ORIENTED GAS MODEL

In order to improve the performance of photorefractive polymers by molecular engineering, it is of primary importance to be able to relate their macroscopic properties to the individual molecular properties and to establish a structure–property relationship. Such a task is very challenging and is the subject of intensive research. However, simple descriptions based on the oriented gas model exist[54,55] and have proven to be in many cases a good approximation for the description of poled electro-optic polymers.[56] They can be applied to photorefractive polymers to describe their linear and second-order properties. Here, we will restrict ourselves to the description of the linear and second-order properties of photoconducting polymers doped with intramolecular charge-transfer molecules referred to as chromophores. In general, charge-transfer molecules consist of an electron-accepting group A connected to an electron-donating group D through a π-conjugated bridge as shown Figure 4. Such molecules are also called push–pull molecules. They have a strong permanent dipole moment and are highly polarizable, leading to a strong nonlinear optical response on the molecular level. The orientation of these chromophores by an external electric field leads to the electro-optic response needed for the photorefractive effect. Orientation is achieved by the torque exerted by an external electric field on the molecules. The field forces them to orient along its direction. This effect is counteracted by thermal energy and can be hindered by the interaction of the molecules with the surrounding host matrix.[57,58] Here, we deal exclusively with materials where the rotational mobility of the chromophores during poling can be assumed free and coupling to the host is neglected. This approximation is valid only when poling is achieved closely below or preferably above the glass transition temperature of the polymer.

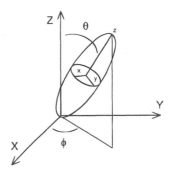

Figure 5 Schematic of the poling process of an individual rodlike molecule in the laboratory frame $\{X, Y, Z\}$ when an electric field is applied along the Z axis. $\{x, y, z\}$ is the frame attached to the molecule.

The main interest in order to model the experimental observations is to describe the nonlinear optical properties on a macroscopic level and to relate them to the microscopic properties of the chromophores which can also be characterized independently. This can be done using the oriented gas model where the macroscopic optical properties can be deduced from the orientational distribution of the molecules for a given field, the density of molecules, and their microscopic nonlinear optical properties. In this approximation, the degree of orientation can be described simply by assuming statistical Boltzmann distributions for the molecules. Push–pull molecules have typically a rodlike shape as shown Figure 5 and cylindrical symmetry around the main z axis of the molecule is assumed for simplicity. Due to the charge-transfer resonance, these molecules have a strong nonlinear optical response.[25] In the dipole approximation, the microscopic polarization components for a nonresonant excitation can be written on a molecular level as:

$$p_i = \mu_i + \alpha_{ij}E_j + R\beta_{ijk}E_jE_k + S\gamma_{ijkl}E_jE_kE_l + \cdots \tag{11}$$

where α_{ij}, β_{ijk}, γ_{ijkl} are tensor quantities and E_j, E_k, E_l are components of the electric field. The subscripts i, j, k, and l refer to components expressed in the molecular frame. R and S are degeneracy factors which depend on the nonlinear process and the frequencies of the electric fields involved in the interaction. When the second-order interaction is the linear Pockels effect, i.e., an interaction in which an external dc field is mixed with the optical field at frequency ω, the degeneracy factor R equals 2.[59]

On a macroscopic level in the frame (X, Y, Z) of the laboratory, the nonlinear polarization induced in the polymer can be described by:

$$P_I = \chi_{IJ}^{(1)}E_J + R\chi_{IJK}^{(2)}E_JE_K + S\chi_{IJKL}^{(3)}E_JE_KE_L + \cdots \tag{12}$$

where the indices I, J, K, and L refer to components expressed in the laboratory frame.

The orientation of the chromophores in the polymer matrix is of primary importance for the derivation of the relationship between the microscopic and macroscopic quantities for both first- and second-order properties. In order to obtain macroscopic second-order properties, an orientational order of the chromophores is needed to break the centrosymmetry of the bulk material where the molecules are initially randomly distributed. Due to their rodlike shape, the chromophores possess a linear polarizability that is very different for directions of the optical field parallel or perpendicular to the molecular axis. Orientation of the molecules by an external field, referred to as poling, leads, therefore, to birefringence and second-order nonlinear optical properties in the polymer film.[60] In purely electro-optic polymers where poling is initially achieved by a spatially uniform dc field, the poling induces a permanent birefringence that is spatially uniform and plays generally only a minor role during electro-optic modulation. In contrast, in photorefractive polymers with a glass transition temperature close to room temperature, spatially modulated birefringence[61] can be induced by the modulated internal space-charge field and can be of paramount importance since it can lead to strong refractive index gratings as will be shown later in Section V.D.

In the oriented gas model, the average orientation of the molecules is described by the Maxwell-Boltzmann distribution function $f(\Omega)$ given by:

$$f(\Omega)d\Omega = \exp[-U(\theta)/k_BT]\sin\theta\, d\theta \tag{13}$$

474

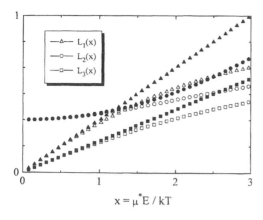

$$x = \mu^*E \,/\, kT$$

Figure 6 Plot of the Langevin functions $L_1(x)$, $L_2(x)$, and $L_3(x)$ (open symbols) and their low field limit approximations (corresponding solid symbols).

where k_BT is the thermal energy at the poling temperature and θ is the polar angle between the poling field direction and the dipole moment of the molecule (see Figure 5). The energy $U(\theta)$ describes the interaction between the poling field and the dipole:

$$U(\theta) = -\vec{\mu}^*.\vec{E}_{\hat{T}} - \frac{1}{2}\,\mathrm{p}.\mathrm{E}_T \approx -xk_BT\cos\theta \quad \text{with} \quad x = \frac{\mu^*E_T}{k_BT} \tag{14}$$

where E_T is the modulus of the total poling field and is pointing in the direction of the laboratory Z axis. μ^* is the effective dipole moment of the molecule taking into account the influence of the matrix. Push–pull molecules usually have a strong permanent dipole moment. It is, therefore, a good approximation to neglect the interaction of the poling field with the induced dipole moment **p**. As will be shown later in this section, orientational order can be described by averaged quantities in the form $\langle\cos^n\theta\rangle$ which with these approximations are given by:

$$\langle\cos^n\theta\rangle = \frac{\displaystyle\int \cos^n\theta\, f(\Omega)d\Omega}{\displaystyle\int f(\Omega)d\Omega} = \frac{\displaystyle\int_0^{\pi} \cos^n\theta\, \exp(x\cos\theta)\sin\theta\, d\theta}{\displaystyle\int_0^{\pi} \exp(x\cos\theta)\sin\theta\, d\theta} = L_n(x) \tag{15}$$

where x is the dimensionless parameter defined in Equation 14 and $L_n(x)$ are the Langevin functions. In the weak poling field limit, i.e., for values of the electric field such as $x \ll 1$, the Langevin functions can be approximated to:

$$L_1(x) \approx \frac{x}{3}, \qquad L_2(x) \approx \frac{1}{3} + \frac{2x^2}{45}, \qquad L_3(x) \approx \frac{x}{5} \tag{16}$$

Indeed, the expressions in Equation 16 are good approximations to the Langevin functions for values of x up to $x = 1$ as illustrated in Figure 6. The condition $x \leq 1$ is satisfied in most of the experimental conditions. For instance, for a molecule with $\mu^* = 10 \times 10^{-18}$ esu (or 10 Debye) at room temperature ($T = 300$ K), the approximation Equation 16 is valid for electric field values up to $E \leq 125$ V/μm.

1. Linear Optical Properties of Poled Polymers

The polarizability tensor of the molecule (see Equation 11) can be written in the molecular principal axes (x, y, z) as:

$$\tilde{\alpha} = \begin{pmatrix} \alpha_{xx} & 0 & 0 \\ 0 & \alpha_{yy} & 0 \\ 0 & 0 & \alpha_{zz} \end{pmatrix} = \begin{pmatrix} \alpha_{\perp} & 0 & 0 \\ 0 & \alpha_{\perp} & 0 \\ 0 & 0 & \alpha_{/\!/} \end{pmatrix} \tag{17}$$

where the subscripts ⊥ and ∥ refer to the direction perpendicular and parallel to the main z axis of the molecule. In the linear regime and on a microscopic level, an optical field at frequency ω interacting with the polymer film will induce the molecular polarization:

$$p_i = \alpha_{ii}E_i(\omega) \tag{18}$$

where the index i stands for the different components in the frame of the molecule ($i = x, y, z$).

The macroscopic polarization is given by the sum of the molecular polarizations. Since the molecules exhibit a strong anisotropic linear polarizability, the macroscopic polarization will depend on their average orientation. Since the average orientation can be changed by applying a poling field, the macroscopic polarizability can be continuously tuned by changing the value of the applied field. For an electric field applied along the laboratory Z axis, the macroscopic component of the polarization along the Z axis is given by:

$$P_Z = N \int (p_x \cos(\hat{x},\hat{Z}) + p_y \cos(\hat{y},\hat{Z}) + p_z \cos(\hat{z},\hat{Z}))f(\Omega)d\Omega \ / \int f(\Omega)d\Omega \tag{19}$$

where $\cos(\hat{i},\hat{Z})$ are the direction cosines which are given by:

$$\cos(\hat{x},\hat{Z}) = \sin\theta \cos\phi, \qquad \cos(\hat{y},\hat{Z}) = \sin\theta \sin\phi, \qquad \cos(\hat{z},\hat{Z}) = \cos\theta \tag{20}$$

Inserting Equations 18 and 20 into Equation 19 and taking the projections of the field components in the molecular frame (x, y, z) on the Z axis of the laboratory gives the following expression for the macroscopic polarization:

$$P_Z = N \int [\alpha_\perp \sin^2\theta \cos^2\phi + \alpha_\perp \sin^2\theta \sin^2\phi + \alpha_\parallel \cos^2\theta]E_Z(\omega)f(\Omega)d\Omega \ / \int f(\Omega)d\Omega$$

$$= N[\int (\alpha_\perp + (\alpha_\parallel - \alpha_\perp)\cos^2\theta)f(\Omega)d\Omega]E_Z(\omega) \ / \int f(\Omega)d\Omega \tag{21}$$

$$= \chi_{ZZ}^{(1)}(-\omega;\omega)E_Z(\omega)$$

For an unpoled film, the chromophores are randomly oriented and the distribution function $f_0(\Omega)$ is equal to unity. For a poling field applied along the Z axis, the change in linear susceptibility along the poling axis between an unpoled and poled film is given by:

$$\Delta\chi_{ZZ}^{(1)}(-\omega;\omega) = NF^{(1)}(\alpha_\parallel - \alpha_\perp)(\langle\cos^2\theta\rangle - 1/3) \tag{22}$$

where $F^{(1)}$ is a local field correction factor. Recalling the relationship between the refractive index and the linear susceptibility:

$$n^2(\omega) = \epsilon(\omega) = 1 + 4\pi\chi^{(1)}(-\omega;\omega) \tag{23}$$

the refractive index change induced by a poling field applied along Z, for an optical wave polarized along the Z axis is given (in CGS units) by:

$$\Delta n_Z^{(1)}(\omega) = \frac{2\pi NF^{(1)}}{n}(\alpha_\parallel - \alpha_\perp)\left(\langle\cos^2\theta\rangle - \frac{1}{3}\right) \tag{24}$$

The same derivation can be made for an optical field polarized along a direction perpendicular to the poling axis and the refractive index change is given by:

$$\Delta n_X^{(1)}(\omega) = \Delta n_Y^{(1)}(\omega) = -\frac{1}{2} \Delta n_Z^{(1)}(\omega) \tag{25}$$

2. Second-Order Nonlinear Optical Properties of Poled Polymers

As for the first-order properties, the electro-optic properties of a poled polymer film can be related to the microscopic hyperpolarizability of the chromophores through the oriented gas model. As axial order is sufficient to change the linear refractive index of the polymer composite, polar order is required in order to break the centrosymmetry and to induce electro-optic properties. Poled polymers belong to the ∞mm symmetry group. With this symmetry, the second-order susceptibility tensor has only three independent tensor elements which can be further reduced to two independent components when Kleinmann symmetry holds, i.e., away from any electronic resonance of the chromophore. For this symmetry, the third-rank tensor of the second-order susceptibility is:

$$\chi^{(2)} = \begin{pmatrix} 0 & 0 & 0 & 0 & \chi_{15}^{(2)} & 0 \\ 0 & 0 & 0 & \chi_{24}^{(2)} & 0 & 0 \\ \chi_{31}^{(2)} & \chi_{32}^{(2)} & \chi_{33}^{(2)} & 0 & 0 & 0 \end{pmatrix} \tag{26}$$

with $\chi_{31}^{(2)} = \chi_{32}^{(2)} = \chi_{24}^{(2)} = \chi_{15}^{(2)} = \chi_{ZXX}^{(2)}$ and $\chi_{33}^{(2)} = \chi_{ZZZ}^{(2)}$. Here the tensor elements are written in the contracted notation with $I = 1, 2, 3$ and $J = 1, 2 \ldots 6$. With a poling field applied along the Z axis and for molecules having their dipole moment oriented along the z axis, the two independent second-order susceptibility tensor elements $\chi_{IJk}^{(2)}$ can be approximated to:

$$\chi_{ZZZ}^{(2)} = NF^{(2)}\beta_{zzz}\langle\cos^3\theta\rangle \tag{27}$$

$$\chi_{ZXX}^{(2)} = NF^{(2)}\beta_{zzz}(\langle\cos\theta\rangle - \langle\cos^3\theta\rangle)/2 \tag{28}$$

where it was assumed that the only nonvanishing component of β_{ijk} was β_{zzz}, the component along the z axis of the molecule. This approximation is satisfactory for a rodlike molecule. In the following the first hyperpolarizability will be referred to as β and the indexes will be intentionally omitted. In Equations 27 and 28, $F^{(2)}$ is a local field correction factor that takes into account the effects of the surrounding matrix on the electric fields. For a dc or a low-frequency field, the local field factor can be approximated by the Onsager expression:

$$f_0 = \frac{\epsilon(\epsilon_\infty + 2)}{(2\epsilon + \epsilon_\infty)} \tag{29}$$

where ϵ_∞ is the dielectric function which is related to the refractive index n at frequency ω by $\epsilon_\infty = n^2$ and ϵ is the static (or low-frequency) dielectric constant. For an optical field, the local field factor can be approximated by a Lorentz-Lorentz type expression:

$$f_\infty = \frac{\epsilon_\infty + 2}{3} \tag{30}$$

The local field correction factor for the change in susceptibility in Equation 22 can be approximated to $F^{(1)} = f_\infty$ and the one for the electro-optic contribution in Equations 27 and 28 to $F^{(2)} = f_0 f_\infty f_\infty$. The second-order susceptibility tensor elements are related to the electro-optic tensor elements r_{IJ} by[62]:

$$r_{IJ} = -8\pi\chi_{IJ}^{(2)}(-\omega;0,\omega)/n^4 \tag{31}$$

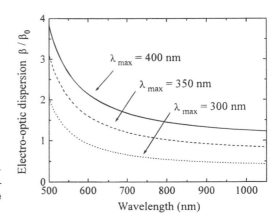

Figure 7 Normalized dispersion β/β_0 of the first hyperpolarizability as a function of the wavelength for molecules with a different absorption maximum λ_{max} (in the two-level model).

where r_{IJ} and $\chi_{IJ}^{(2)}$ are expressed in the contracted notation. Equation 31 is valid when r_{IJ} and $\chi_{IJ}^{(2)}$ are given in CGS units. When the tensor elements are expressed in MKS units, the right-hand side of Equation 31 has to be divided by 4π. The conversion between the two systems of units is given by $\chi_{IJ}^{(2)}$ (pm/V) $= 4.18 \times 10^8 \chi_{IJ}^{(2)}$ (esu).

The value of the first hyperpolarizability β of the chromophore can be deduced in solution from independent characterization experiments such as EFISH experiments (electric field-induced second harmonic generation). EFISH experiments have been performed on a large number of different chromophores and the literature is well documented on their hyperpolarizability values.[63–66] These experiments are usually performed off-resonance at a wavelength that is different from the one at which the photorefractive experiments are performed. Therefore, it is convenient to extrapolate the dispersion-free value β_0 from the β_{EFISH} value that is measured by EFISH experiments. This calculation can be done by applying the dispersion formula given by the two-level model[67]:

$$\beta_0 = \beta_{EFISH}(\lambda^2 - 4\lambda_{max}^2)(\lambda^2 - \lambda_{max}^2)/\lambda^4 \qquad (32)$$

where λ_{max} is the spectral position of the resonance of the molecule, i.e., the position of the maximum of the absorption spectrum, and λ is the wavelength at which the EFISH experiments were performed. Likewise, in the two-level model, the value of β for electro-optic experiments performed at the wavelength λ can be deduced from zero-frequency value β_0 according to:

$$\beta(-\omega,\omega,0) = \beta_0 \frac{(3 - \lambda_{max}^2/\lambda^2)}{3(1 - \lambda_{max}^2/\lambda^2)^2} \qquad (33)$$

This dispersion relation shows that the second-order nonlinearity is enhanced when the operating wavelength is close to the absorption maximum of the chromophore. An example of such a dispersion is shown Figure 7. This resonance enhancement has important consequences for the design of photorefractive polymers. In contrast to poled polymers designed for second harmonic generation, the transparency of the material at half of the wavelength of the exciting beam is no longer required and, as a result, the nonlinearity can be much higher if the operating wavelength is closer to the absorption maximum but still in a region of small absorption.

To summarize, the electro-optic properties of a photorefractive polymer are influenced by the magnitude of the nonlinearity on a molecular level, the degree of order of the chromophores after poling, their concentration, and also the value of the dielectric constants (static and optical) of the host which plays a role in the poling process through the local field factors given by Equations 29 and 30. According to the dispersion relations for the first hyperpolarizability (Equation 33), the detuning between the operating wavelength and the absorption maximum also plays an important role and has to be considered when the characteristics of different chromophores are compared.

D. DIFFRACTION THEORY OF THICK SLANTED GRATINGS

As shown in Sections III.A and III.B, an applied electric field plays an important role in both the photogeneration and the transport processes. Moreover, the electric field is also essential for the poling

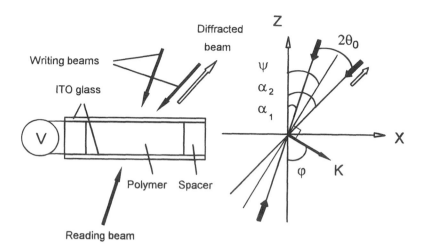

Figure 8 Schematic of the slanted four-wave mixing experiment used to study photorefractive polymer samples. The polymer is sandwiched between two transparent conducting electrodes.

and consequently for the electro-optic properties of the polymer film. A convenient way to apply electric field to photorefractive polymers is to sandwich them between two transparent electrodes such as ITO (indium tin oxide)-coated glass slides. The electric field is then applied in the direction perpendicular to the polymer film as shown Figure 8. With this geometry, the sample has to be slanted for two reasons. The first reason is related to the symmetry of polymers poled this way and the cancellation of the effective electro-optic coefficient in a four-wave mixing experiment when the sample is normal to the bisector of the writing beams, as is discussed in Section III.D.1. The second reason is due to the need of a component of the applied field along the grating vector in order to have a drift source for the transport of the carriers from the brighter regions to the darker regions of the interference pattern.

Although other poling configurations can be developed in the future, most of the diffraction experiments have been performed so far in the slanted geometry where the electric field is perpendicular to the polymer film. Therefore, in the following section, we review some elements of Kogelnik's coupled-wave model[68] for slanted thick phase holograms and apply this theory to describe the photorefractive gratings written in the slanted geometry in poled polymers.

In a four-wave mixing experiment, a reading beam is diffracted on the grating induced by two interfering writing beams. The photorefractive properties of the material are investigated by measuring the diffraction efficiency η defined as the ratio between the intensity of the diffracted beam measured after the sample and the intensity of the incident reading beam measured before the sample. In order to neglect the effects of the reading beam on the grating, its intensity is much weaker. It can have the same or a different wavelength. In both cases, the direction of the incident reading beam has to be adjusted in order to fulfill the Bragg condition. A convenient configuration is the backward degenerate four-wave mixing configuration shown in Figure 9 where all the beams have the same wavelength and where the reading beam is counterpropagating with one of the writing beams. In this case, the diffracted beam is counterpropagating with respect to the second writing beam and the phase matching imposed by the Bragg condition is automatically fulfilled. The slant of the sample is characterized by the slant

a)   b)

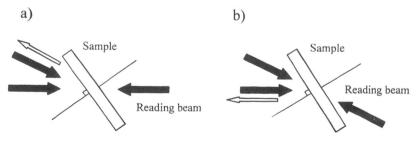

Figure 9 Schematic of different configurations of slanted backward degenerate four-wave mixing experiments.

angle φ defined as the angle between the grating vector from **K** and the laboratory Z axis which is perpendicular to the surface of the sample as shown Figure 9. This slant angle defined inside the material is given by:

$$\varphi = \frac{\pi}{2} - \frac{(\alpha_1 + \alpha_2)}{2} \tag{34}$$

where α_1 and α_2 are the angles of the writing beams with respect to the laboratory Z axis measured inside the sample. The grating spacing is given by:

$$\Lambda = \frac{\lambda}{2n \sin[(\alpha_2 - \alpha_1)/2]} \tag{35}$$

Note that in the slanted configuration for a fixed angle between the writing beams outside the sample, the grating spacing changes with the value of the slant angle since Descartes's (or Snell's) law for refraction is not a linear function and, therefore, the value of $(\alpha_2 - \alpha_1)$ is a function of the orientation of the sample.

The diffraction efficiency can be obtained from the coupled-wave theory for thick holograms developed by Kogelnik.[68] With the notation used in this chapter, the amplitude of the diffracted wave for Bragg incidence on a lossy thick phase grating in a material with low absorption and without absorption modulation is given by:

$$\eta = \exp\left(\frac{-\alpha d}{2}\left(\frac{1}{c_i} + \frac{1}{c_d}\right)\right) \frac{\sin^2(\nu^2 - \xi^2)^{1/2}}{(1 - \xi^2/\nu^2)} \tag{36}$$

with

$$\nu = \frac{\pi \Delta n d}{\lambda (c_i c_d)^{1/2}} \hat{e}_i \cdot \hat{e}_d, \quad \xi = \frac{\alpha d}{4}\left(\frac{1}{c_i} - \frac{1}{c_d}\right) \tag{37}$$

where $\hat{e}_i$, and $\hat{e}_d$ are the polarization vectors of the incident reading beam and the diffracted beam, respectively, and d is the thickness of the grating. In Equation 37, Δn is the refractive index modulation amplitude responsible for the phase grating formation. In the terminology of Kogelnik's paper, c_i and c_d are the obliquity factors of the incident reading beam and the diffracted beam, respectively. They are defined as:

$$c_i = \cos \alpha_i, \quad c_d = \cos \alpha_i - \frac{\lambda}{n\Lambda} \cos \varphi \tag{38}$$

where α_1 is the Bragg angle of the incident beam with respect to the Z axis. As shown in Figure 9, in the slanted backward degenerate four-wave mixing configuration, the reading beam can be counterpropagating with respect to either the writing beam that makes an angle α_1 with the Z axis (Figure 9a) or the other one that makes an angle α_2 ($\alpha_2 > \alpha_1$) (Figure 9b). In the first case, inserting the expressions for the grating spacing Λ (Equation 35) and the slant angle φ (Equation 34) in the expressions (38), the obliquity factors reduce to $c_i = \cos \alpha_1$ and $c_d = \cos \alpha_2$. In the second case, the same simple algebra leads to $c_i = \cos \alpha_2$ and $c_d = 2 \cos \alpha_2 - \cos \alpha_1$. In the following we will restrict our discussion to the geometry corresponding to the first case (Figure 9a). Equation 36 is valid for thick gratings, i.e., when the factor $Q' = \pi \lambda d/n\Lambda^2$ is larger than unity or preferably larger than 5. For 100-μm-thick samples for instance, this condition requires $\Lambda < 5$ μm. By applying Equation 36 to describe the diffraction efficiency in our polymers, we neglect the bending of the grating that occurs during the writing and which is due to the photorefractive coupling of the two writing beams[39] as is discussed later in Section III.E. Other approximations for the validity of Equation 36 include small absorption and sinusoidal modulation of the refractive index. These conditions can be met by choosing an operating wavelength at which the photorefractive material has high transmission.

To characterize a given slanted configuration, it is often more convenient to use the set of angles (ψ, $2\theta_0$) instead of the set of angles (α_1, α_2) where ψ is the angle between the bisector of the writing beams and the laboratory Z axis (tilt angle measured outside the sample) and $2\theta_0$ is the angle between the two writing beams measured outside the sample as shown Figure 8. The transformation between the two sets of angles is given by:

$$\alpha_1 = \arcsin\left(\frac{\sin(\psi - \theta_0)}{n}\right), \quad \alpha_2 = \arcsin\left(\frac{\sin(\psi + \theta_0)}{n}\right) \tag{39}$$

1. Purely Electro-optic Photorefractive Gratings in Prepoled Polymers

In this section, we apply Kogelnik's formula to the slanted geometry described in the previous section to the case where the index modulation amplitude Δn appearing in Equation 37 is purely due to the electro-optic effect. As photorefractive materials are noncentrosymmetric by definition, the electro-optic refractive index modulation given by Equation 6 is usually strongly anisotropic and depends on the configuration, the symmetry class of the sample, and the polarization of the reading beam. By analogy with a second harmonic generation process,[69] the effective electro-optic coefficient in a four-wave mixing process is given by the tensorial product[70,71]:

$$r_{\text{eff}} = \hat{e}_d^*.\hat{\epsilon}.[\hat{R}.\hat{k}.\hat{\epsilon}.\hat{e}_i] \tag{40}$$

where $\hat{\epsilon}$ and $\hat{R}$ are the second-rank dielectric tensor and the third-rank electro-optic tensor, respectively. The latter has a symmetry identical to the second-order susceptibility tensor defined in Equation 26 and with electro-optic tensor elements related to the second-order tensor elements by Equation 31. Here, the symmetry axis is the poling direction and is, therefore, the laboratory Z axis. For the geometry shown Figure 9a, the polarization vectors for the incident and diffracted beams write $\hat{e}_i = (\cos \alpha_1, 0, \sin \alpha_1)$ and $\hat{e}_d = (\cos \alpha_2, 0, \sin \alpha_2)$, respectively, for p-polarized light, and $\hat{e}_i = \hat{e}_d = (0, 1, 0)$ for s-polarized light. The grating unitary vector $\hat{k}$ is given by $\hat{k} = (\sin \varphi, 0, \cos \varphi)$. If the small contribution from the uniform anisotropy of the background refractive index in the sample is neglected, the effective electro-optic coefficients for s-polarized and p-polarized readout beams are:

$$r_{\text{eff}}^s = r_{13} \cos \varphi \tag{41}$$

$$r_{\text{eff}}^p = r_{13} \cos \alpha_1 \cos \alpha_2 \cos \varphi + r_{13} \sin(\alpha_1 + \alpha_2)\sin \varphi + r_{33} \sin \alpha_1 \sin \alpha_2 \cos \varphi \tag{42}$$

Equations 41 and 42 show that for an unslanted configuration ($\alpha_1 = -\alpha_2$) the effective electro-optic coefficient for both polarizations vanishes. For both polarizations, the effective electro-optic coefficient increases with the slant angle and when Kleinmann symmetry is assumed $r_{33} = 3r_{13}$. As a result, the effective electro-optic coefficient is stronger for p-polarized light than for s-polarized light. For a given geometry, the ratio in diffraction efficiency measured for p- and s-polarized reading beams can be deduced from the inherent polarization anisotropy of the expression of ν (through the polarization vectors in Equation 37) and, in addition, from the polarization anisotropy of the refractive index modulation related to the effective electro-optic coefficients through Equation 6.

2. Birefringence and Electro-optic Photorefractive Gratings in Low Glass Transition Temperature Polymers

In the preceding section it was assumed that the refractive index modulation amplitude was solely due to the electro-optic effect. This behavior is expected in polymers that are prepoled and that exhibit a long-term poling stability. As for purely electro-optic polymers, a quasistable poling can be achieved in photorefractive polymers if the polymer has a glass transition temperature T_g much higher than room temperature. In this case, poling is achieved by heating the polymer above T_g, by applying the electric

a)

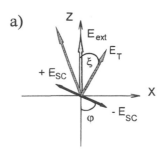

b)

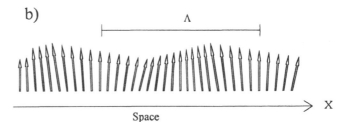

Figure 10 Schematic of the total poling field in a low glass transition temperature polymer (a) and the spatially modulated poling in the sample (b).

field, and by cooling the polymer down to room temperature with the electric field applied. The orientation of the chromophores is then frozen in the polymer and the electro-optic behavior can be observed without the presence of the poling field. The orientational stability is not infinite because the polymer is not in thermodynamical equilibrium, but it can last several years in some cases. The improvement of the poling stability in poled polymers is an active field of research.

In this section, we consider polymer composites with a T_g close to room temperature, meaning that poling can be achieved when an electric field is applied without heating the sample previously. So far, the most efficient photorefractive polymers that have been developed[22,23] are guest–host type polymers as discussed in Section V. These polymers usually have a glass transition temperature close to room temperature because the high concentration of guest chromophores acts as a plasticizer for the host. An additional plasticizer can be introduced to further lower the glass transition temperature. The electric-field orientation at room temperature has important consequences on the properties of such polymers since the poling field is no longer the external field but the total electric field namely the superposition of the external field and the internal photorefractive space-charge field as shown in Figure 10a. The total poling field is written:

$$\mathbf{E}_T(\mathbf{r}) = [E_{sc}(\mathbf{r})\sin \varphi]\hat{X} + [(E_{ext} + E_{sc}(\mathbf{r})\cos \varphi]\hat{Z} \qquad (43)$$

where the space-charge field is spatially modulated and is given by $E_{sc}(\mathbf{r}) = E_{sc} \exp(i\mathbf{Kr})$. The effect of the space-charge field on the poling is even more pronounced since it changes the magnitude of the total field and simultaneously its direction as shown in Figure 10b. [61] The total poling field makes an angle ξ with respect to the laboratory z axis given by:

$$\xi = \arctan\left[\frac{E_{sc}(\mathbf{r})\sin \varphi}{E_{ext} + E_{sc}(r)\cos \varphi}\right] \qquad (44)$$

This angle is periodically changing in space with the spatial periodicity of the space-charge field. In polymers poled previously by a spatially uniform electric field, the poling-induced birefringence discussed in Section III.C.1 plays generally a minor role on the photorefractive properties of the polymer. In contrast, in the low T_g polymers discussed here, the modulated internal space-charge field induces a spatially modulated birefringence. This effect is of paramount importance since it can lead to high refractive index modulations. In this case, the refractive index modulation has two contributions: the electro-optic and birefringence contributions. Moreover, the electro-optic effect is enhanced. Since the symmetry axis of the sample defined by the poling direction is changing in space, a local frame defined by the direction of the total poling field has to be introduced and differentiated from the laboratory frame defined by the sample geometry. This behavior makes the derivation of the diffraction efficiency more complex. The tensors for the linear polarizability and the first hyperpolarizability have a diagonal representation in the local frame and have to be derived in the frame of the laboratory. This frame transformation corresponds to a rotation ξ around the Y axis and can be achieved by applying the transformation matrix $T(\mathbf{r})$ to the different coordinate systems. This rotation matrix is written[72]:

$$T(\mathbf{r}) = \begin{pmatrix} \cos\xi & 0 & -\sin\xi \\ 0 & 1 & 0 \\ \sin\xi & 0 & \cos\xi \end{pmatrix} \tag{45}$$

The contribution of the birefringence to the refractive index modulated at the spatial frequency K for a given slanted four-wave mixing configuration can be written as:

$$\Delta n_{\text{eff}}^{(1)} = \frac{2\pi}{n}\, \hat{e}_d^*.T(\mathbf{r})\Delta\tilde{\chi}^{(1)}T^{-1}(\mathbf{r}).\hat{e}_i \tag{46}$$

where $\hat{e}_i$ and $\hat{e}_d$ are the unitary vectors along the polarization of the incident and diffracted beams, respectively as in the previous section (Equation 40). The diagonal matrix $\Delta\tilde{\chi}^{(1)}$ is given by:

$$\Delta\tilde{\chi}^{(1)} = \begin{pmatrix} \Delta\chi_{XX}^{(1)} & 0 & 0 \\ 0 & \Delta\chi_{YY}^{(1)} & 0 \\ 0 & 0 & \Delta\chi_{ZZ}^{(1)} \end{pmatrix} \tag{47}$$

The matrix elements $\Delta\chi_{JJ}^{(1)}$ are given by Equations 22 and 25. In the weak poling field limit ($x \ll 1$), they can be written according to Equations 14–16 and 22 as:

$$\Delta\chi_{ZZ}^{(1)} = \frac{2}{45}\, Nf_\infty(\alpha_{/\!/} - \alpha_\perp)\left(\frac{\mu^*}{k_BT}\right)^2 E_T^2 \tag{48}$$

$$\Delta\chi_{XX}^{(1)} = -\frac{1}{45}\, Nf_\infty(\alpha_{/\!/} - \alpha_\perp)\left(\frac{\mu^*}{K_BT}\right)^2 E_T^2 \tag{49}$$

Note the opposite signs in these expressions. Injecting these expressions into Equation 46, and keeping only the linear terms which are modulated at the spatial frequency K, we get for the two polarizations (CGS units):

$$\Delta n_{K,S}^{(1)} = -\frac{2\pi}{n}\, BE_{\text{ext}}E_{sc}\,\cos\varphi \tag{50}$$

$$\Delta n_{K,P}^{(1)} = \frac{2\pi}{n}\, BE_{\text{ext}}E_{sc}\, [2\cos\varphi\sin\alpha_1\sin\alpha_2 - \cos\varphi\cos\alpha_1\cos\alpha_2 + \frac{3}{2}\sin\varphi\sin(\alpha_1 + \alpha_2)] \tag{51}$$

with

$$B = \frac{2}{45}\, Nf_\infty(\alpha_{/\!/} - \alpha_\perp)\left(\frac{\mu^*}{k_BT}\right)^2 \tag{52}$$

Likewise, the contribution of the electro-optic nonlinearity to the refractive index modulation can be written as:

$$\Delta n^{(2)}_{\text{eff}} = \frac{4\pi}{n}\, \hat{e}^*_d . T(\mathbf{r})\tilde{\chi}^{(2)} E_T T^{-1}(\mathbf{r}).\hat{e}_i \tag{53}$$

when the tensorial nature of the second-order susceptibility is approximated to a diagonal matrix representation with the diagonal tensor components $\chi^{(2)}_{ZXX}$ and $\chi^{(2)}_{ZZZ}$.[61] By applying the same procedure as for the modulated birefringence effect, the expressions for the refractive index modulation with the spatial frequency K for the two polarizations can be derived from Equations 27, 28, and 53 and they are written:

$$\Delta n^{(2)}_{K,S} = \frac{8\pi}{n}\, CE_{\text{ext}}E_{sc}\cos\varphi \tag{54}$$

$$\Delta n^{(2)}_{K,P} = \frac{8\pi}{n}\, CE_{\text{ext}}E_{sc}\,[\cos\varphi\,\cos\alpha_1\,\cos\alpha_2 + 3\cos\varphi\,\sin\alpha_1\,\sin\alpha_2 + \sin\varphi\,\sin(\alpha_1+\alpha_2)] \tag{55}$$

with

$$C = \frac{Nf_0 f_\infty f_\infty \beta\mu^*}{15k_BT} \tag{56}$$

It can be easily shown that the electro-optic contribution is enhanced by a factor of 2 in low T_g polymers in contrast to Equations 41 and 42 derived in Section III.D.1 for high T_g polymers.

The total refractive index modulation Δn appearing in the expression of ν (Equation 37) is then given by the sum of the two contributions described by Equations 50, 51, 54, and 55, for each polarization of the reading beam. These two contributions are not exclusive and other contributions might also contribute to the overall diffraction efficiency. Higher order effects such as quadratic electro-optic effects (or Kerr effects)[73] and electronic third-order nonlinear effects have to be investigated in future experiments. The relative sign of the different contributions plays an important role in the overall efficiency.

E. SELF-DIFFRACTION THEORY

During the writing of a grating with two interfering beams in a nonlinear medium, the two writing beams diffract from the forming grating, resulting in a change of the intensity profile within the thickness of the crystal. This effect, referred to as self-diffraction, modifies simultaneously the modulation depth of the grating and the phase of the fringe system, leading to a bending of the grating lines. The writing of a grating in a nonlinear medium is, therefore, nontrivial and the dynamics of self-diffraction are rather complicated.[74]

An important application of self-diffraction is coherent energy coupling between the two writing beams, in which the intensity of one of the beams can be amplified at the expense of the loss for the intensity of the other beam. In stationary conditions, such an energy transfer is only allowed if the material has a nonlocal response, meaning that the light distribution inducing the grating and the grating induced by the interfering beams are phase shifted. Due to the transport-related nonlinearity involved in the photorefractive effect, the required nonlocal response condition is uniquely achieved in photorefractive materials. The coherent energy transfer in stationary conditions, also called two-beam coupling is, therefore, a direct fingerprint of the photorefractive effect. Maximum energy transfer occurs when the light distribution and the refractive index modulation are phase shifted by 90°. This optimum phase shift occurs naturally when the buildup of the space charge is only driven by diffusion, i.e., without any applied field. However, in the diffusion regime the typical space-charge fields that are generated are small especially in polymers where the photogeneration efficiency is strongly field dependent. Sensitivity can be drastically enhanced when an electric field is applied to assist the transport of the carriers through drift. Unfortunately, the resulting phase shift is often not optimized for energy exchange. Nevertheless, several techniques have been developed and tested with inorganic crystals to optimize this phase shift. The first technique is based on the tuning of the phase of one of the writing beams resulting in a moving grating with constant velocity.[75-77] For a particular velocity, the phase shift between the interference pattern and the grating can be adjusted to the desired value of 90°. The second technique is based on the writing of the photorefractive grating with alternating electric fields.[78] It requires the period of the applied field to be longer than the carrier lifetime and shorter than the grating formation time.

The coherent energy exchange in a photorefractive material is generally described by the gain coefficient Γ which is given by:

$$\Gamma = \frac{4\pi}{\lambda} (\hat{e}_1 . \hat{e}_2^*) \Delta n \sin \Theta \tag{57}$$

where Θ is the phase shift between the space-charge field and the interference pattern generated by the interacting beams. $\hat{e}_i$ are the polarization vectors of the two beams and Δn is the total refractive index modulation. When a strong pump beam with intensity I_2 is amplifying a weaker signal beam with intensity I_1 ($I_2 \gg I_1$), the gain in energy of the signal beam interacting over a distance l in a medium with a gain coefficient Γ is:

$$I_1(d) \approx I_1(0) \exp[(\Gamma - \alpha)l] \tag{58}$$

where α is the linear absorption coefficient. Equation 58 shows that in order to have net gain the gain coefficient must exceed the absorption coefficient.

Experimentally, the energy transfer can be characterized by the gain factor γ_0 defined as the signal beam intensity with the pump beam on divided by the signal intensity with the pump beam off ($\gamma_0 = I_1(I_2 \neq 0)/I_1(I_2 = 0)$). The gain factor γ_0 is derived from the coupled-wave equations[29] and is related to the gain coefficient Γ through the relation[29]:

$$\gamma_0 = \frac{(1 + b)\exp(\Gamma l)}{b + \exp(\Gamma l)} \quad \text{or} \quad \Gamma = \frac{1}{l}[\ln b\gamma_0 - \ln(1 + b - \gamma_0)] \tag{59}$$

where $b = I_2/I_1$ is the ratio between the intensities of the pump and the signal beams before the sample.

The direction of the energy transfer depends on the sign of the gain coefficient Γ which depends on the polarization of the beams and the sign of Δn, the refractive index modulation. In the configuration shown in Figure 9a, the sign of the effective electro-optic coefficients for s and p polarization is changing when the sample is rotated by 180°. This can be easily seen by replacing the values of the angles of α_1 and α_2 by the values $\alpha_1 + \pi$ and $\alpha_2 + \pi$ in Equations 41 and 42. According to Equation 34, the angle φ becomes $\varphi - \pi$. The change in sign of the effective electro-optic coefficient results in a change in sign of the gain coefficient [the sign of the polarization factor $(\hat{e}_1 . \hat{e}_2^*)$ remains unchanged] and the direction of the energy transfer is changed. This behavior, called asymmetric energy transfer, is the proof of the photorefractive nature of a grating. In low glass transition temperature polymers, the laboratory Z axis is pointing in the direction of the applied field. Thus changing the polarity of the poling field is equivalent to a rotation of the sample by 180°. As a result, the direction of the energy transfer is changing when the polarity of the applied field is reversed.

IV. CHEMISTRY OF PHOTOREFRACTIVE POLYMERS

A large number of PR polymers can be synthesized using different combinations of photosensitizers, photoconductor moieties, and electro-optic (EO) chromophores. In this section, different material classes are discussed together with a brief history of photorefractive polymers. We review, from a general point of view, some of the basic properties of some existing materials that are used in other applications and that can become possible future candidates for photorefractive polymers. The properties of PVK and other photoconductors are emphasized. Differences between inorganic crystalline materials and organic amorphous photorefractive materials are discussed also.

A. MATERIAL CLASSES—A SURVEY OF PHOTOREFRACTIVE POLYMERS

There are several possibilities to combine all functionalities to form a PR polymer. For example, a fully functionalized polymer can be synthesized that contains all functional groups either in the polymer backbone or attached to it as pendent side groups. Another possibility is to mix low molecular weight molecules with the desired properties into a polymer matrix. This so-called guest–host approach opens a lot of opportunities to obtain a PR polymer. The host polymer can, for instance, be an inert binder without any functionality; however, this choice is usually less favorable, because a major part of the material is inactive this way, diminishing the bulk photoconductivity and EO properties. Therefore, in most cases the host polymer is partially functionalized. Electro-optic polymers can be mixed with a low molecular weight photoconductor or photoconductive polymers can be mixed with different low molecular weight EO chromophores. Some of the components can be bifunctional. To get polymers with a high glass transition temperature and consequently a good poling stability, the side-chain approach is preferred to the guest–host approach.

Photorefractivity in an organic material has been reported first in the organic single crystal 2-cyclooctylamino-5-nitropyridine (COANP) (**I**) doped with the electron acceptor 7,7,8,8-tetracyanoquinodimethane (TCNQ) (**II**) to provide photoconductivity.[9,10] Photorefractive grating formation was investigated by two-beam coupling experiments and a phase shift between the refractive index modulation and the light pattern could be evidenced. Doping with TCNQ provided photosensitivity between 600 and 700 nm and increased the photoconductivity of the material. The electro-optic coefficient of the material was of the order of 10 pm/V. The recording time of a PR grating was very slow (several tens of minutes) and can probably be attributed to the weak photoconductivity. The maximum refractive index modulation achieved in this material was $\Delta n = 1.8 \times 10^{-6}$. The same year, J. S. Schildkraut[79] at Eastman Kodak reported on the photoconducting and electro-optic properties of thin polymer films consisting of an acrylic polymer containing a side-chain stilbene chromophore (**III**) doped with a perylene dye (**IV**) as photosensitizer and triarylamine molecules (**V**) as transport agent. However, no grating formation was demonstrated in this compound.

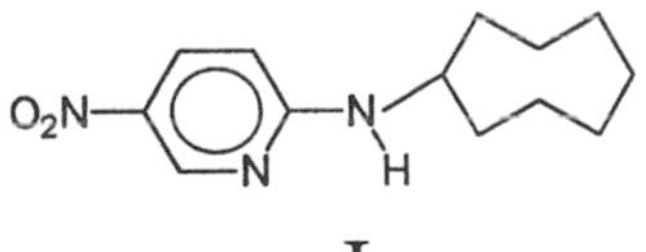

I

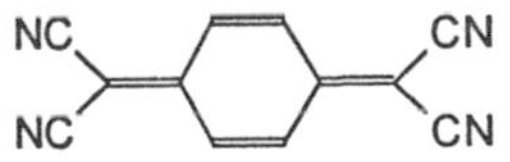

II

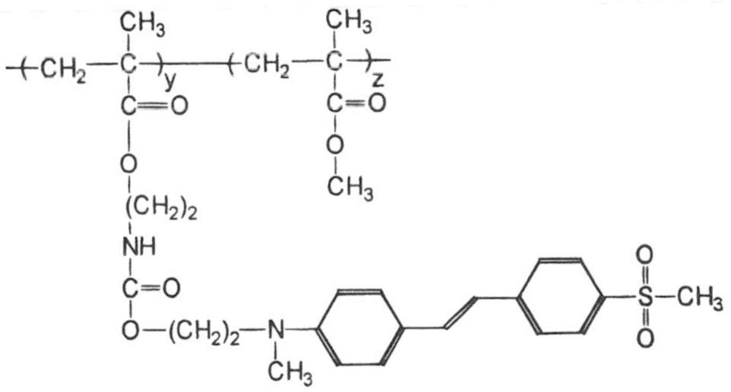

III

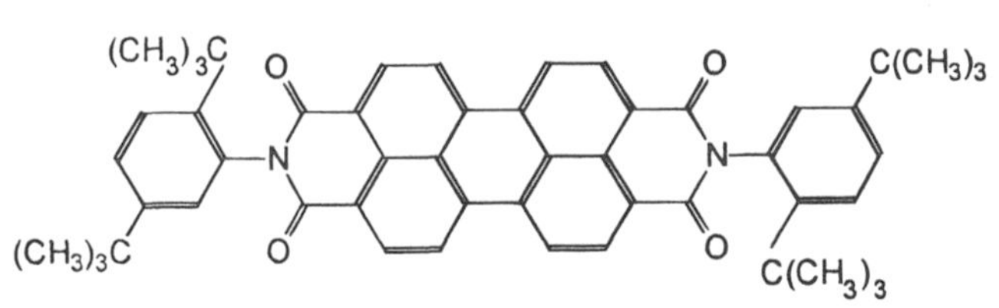

IV

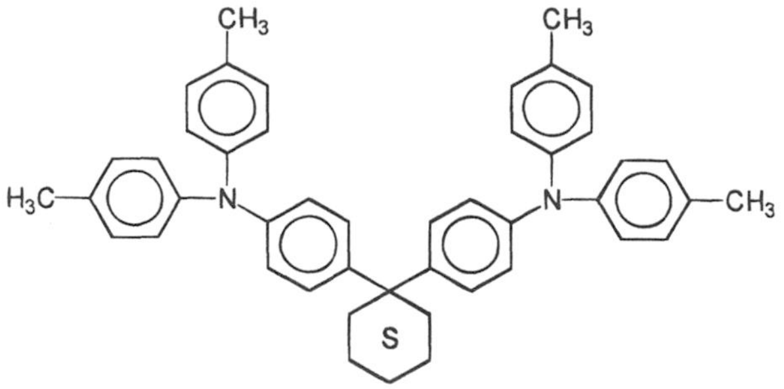

V

The first observation of the photorefractive effect in a polymer was reported shortly after at IBM Almaden[11,12] in a partially cross-linked form of an electro-optic polymer doped with the transport agent diethylamino-benzaldehyde diphenylhydrazone (DEH) (**VI**). The electro-optic polymer bisA-NPDA (**VII**) was composed of bis-phenol-A-diglycidylether (bis-A) and 4-nitro-1,2-phenylenediamine (NPDA) and NPDA provided photosensitivity also. As many other guest–host-type photorefractive polymers were obtained by doping, another approach was developed at the University of Arizona[13] which consisted of the synthesis of a fully functionalized side-chain polymer that showed electro-optic response, photosensitivity, and photoconductivity intrinsically. Carbazole moities, such as in PVK, a well-known photoconductor, were connected to a polyacrylate through alkylene spacers to enable poling (**VIII**). Some of them were modified with tricyanovinylcarbazole groups to obtain electro-optic activity and photosensitivity. This polymer had a glass transition temperature of $T_g = 95$–$100°C$ and showed a good poling stability (75% of the initial poling measured after 380 at room temperature). Photorefractive grating formation in these materials was studied by field-dependent four-wave mixing experiments[14,15] but the efficiency of the material was too small to perform two-beam coupling experiments. A similar side-chain approach was also developed at the University of Chicago.[16] Several new photorefractive polymer composites were synthesized by using different matrices as guest such as polymethyl methacrylate (PMMA)[17] or by using new sensitizers such as fullerene molecules[18,19] and borondiketonate (BDK) (**IX**) or new transport agents such as tri-p-tolylamine (TTA)[20] (**X**). As the description and characterization of the photorefractive effect in polymers was improved,[80,81] the performance of new polymers such as PMMA-PNA (**XI**) doped with DEH were gaining in speed (<1 s)[17] and in diffraction efficiency but the overall diffraction efficiencies were too small ($\eta < 10^{-4}$) to observe energy exchange in two-beam coupling experiments.[82–84]

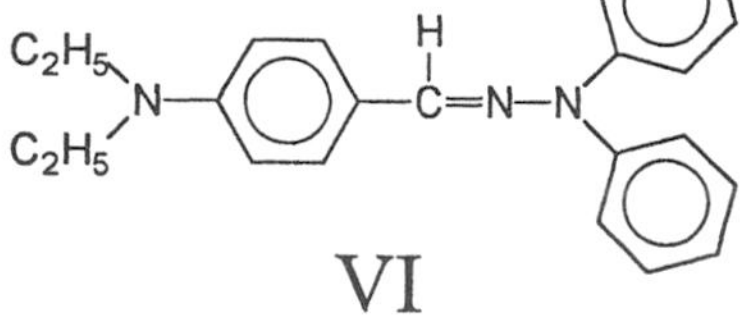

VI

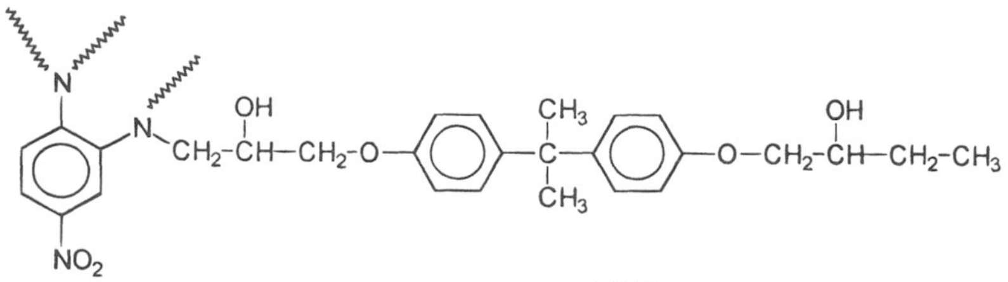

VII

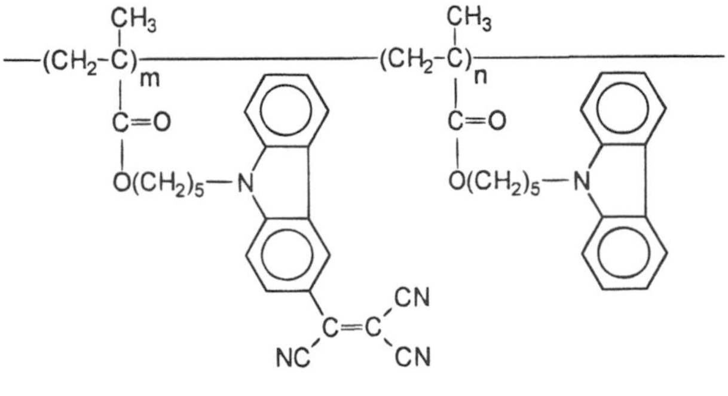

VIII

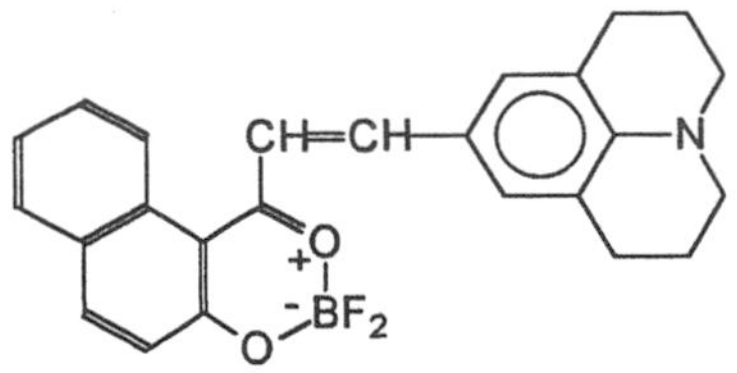

IX

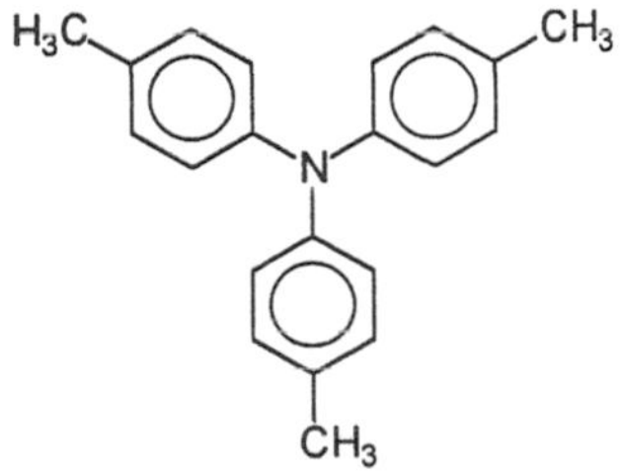

X

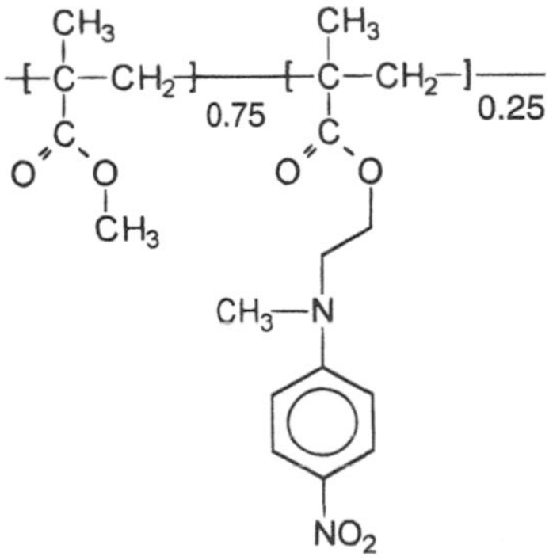

XI

Significant improvement in the performance of PR polymers was observed in the photoconductor matrix PVK (**XII**) doped with 2,4,7-trinitrofluorenone (TNF) (**XIII**) for charge generation and the nonlinear chromophore 3-fluoro-4-*N,N*-diethylamino-β-nitrostyrene (F-DEANST) (**XIV**).[21] In this polymer a diffraction efficiency of 1% could be measured in 125-μm-thick films at an applied field of 40 V/μm and a grating growth time of 100 ms for writing intensity of 1 W/cm². The improvement in diffraction efficiency compared to previous materials was two orders of magnitude. The photorefractive origin of these efficient gratings was confirmed by asymmetric two-beam coupling experiments and, for the first time, PR gain exceeding the absorption of the sample could be observed in a polymer. A few months later, a new polymer was reported which was based also on the PVK:TNF charge-transfer complex but doped with the chromophore 2,5-dimethyl-4-(*p*-phenylazo)anisole (DMNPAA) (**XV**) with higher doping level and also doped with a photoconducting plasticizer (*N*-ethylcarbazole) (**XVI**) in order to facilitate the poling process at room temperature.[22,85] In 105-μm-thick samples of this material, the diffraction efficiency was gradually improved from the value of 6% to 35% at 40 V/μm by increasing the dye concentration and the experimental configuration.[86,87] Further improvement in the processing of the samples enabled higher voltages to be applied, leading to a diffraction efficiency close to the maximum value of 100% at an applied field of 61 V/μm.[23] In these samples a net two-beam coupling gain coefficient of more than 200 cm⁻¹ was observed at 90 V/μm. This value has to be compared with the gain coefficient of BaTiO₃ which is of the order of 30 cm⁻¹.[88] These high values show the significance of the breakthrough that has been achieved within the past 3 years in these new materials and are very encouraging for the future.

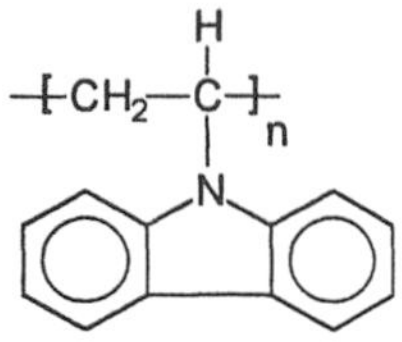

XII

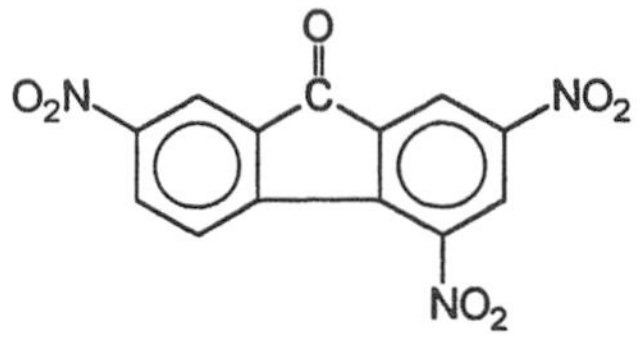

XIII

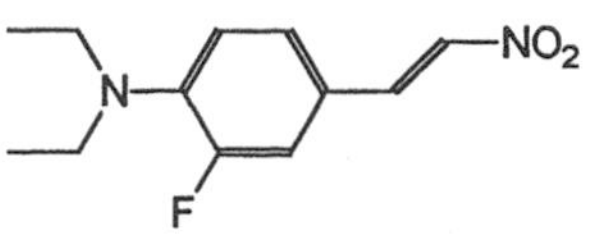

XIV

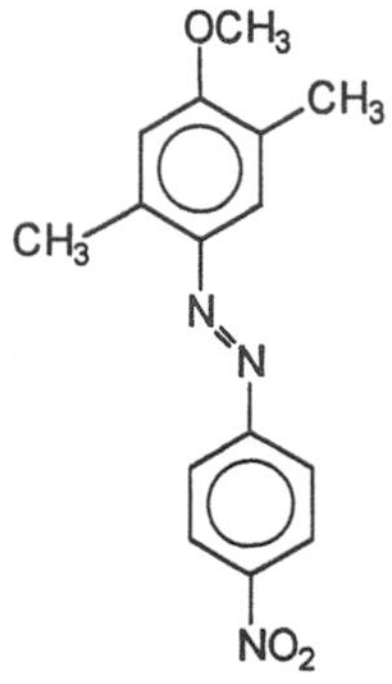

XV

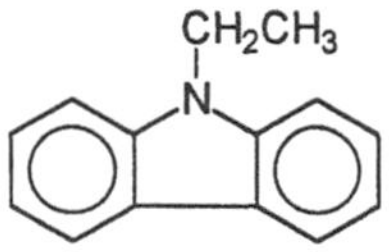

XVI

As high efficiencies have been demonstrated, the speed of these materials is still rather slow, on the order of couple of hundreds of milliseconds for a writing intensity of 1 W/cm^2. This can be attributed to the low hole mobility of PVK-based polymers. Attempts to increase the speed by using polysilane-based materials as photoconductor, which are known for their higher mobility compared to PVK, have revealed that the dynamics were also strongly limited by the photogeneration efficiency.[89] Simultaneously progress was also made in the field of organic photorefractive crystals and beam-coupling gain of 2.2 cm^{-1} was demonstrated in single crystals of 4′-nitrobenzylidene-3-acetamino-4-methoxyaniline (MNBA) (**XVII**).[90]

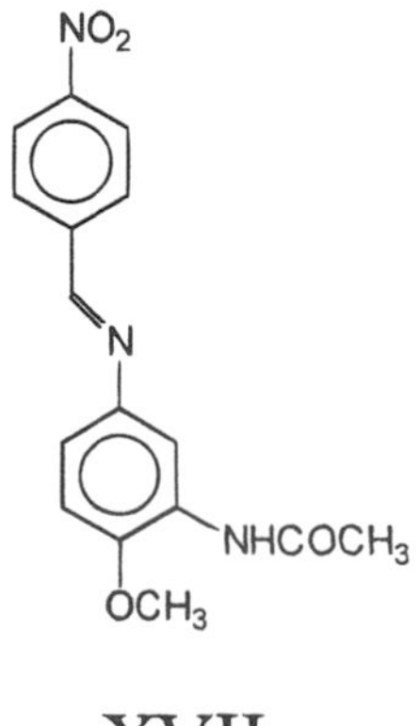

XVII

Owing to the rich flexibility of polymers, the number of new photorefractive polymers that will be synthesized in the future is expected to grow exponentially. In the following, we discuss from a chemical point of view, the general properties of existing photoconducting materials and emphasize on the properties of PVK which has been widely used so far in photorefractive polymers. The field of electro-optic polymers is a very active research area and several textbooks[25,62,91] and reviews[64–66,92] have been published recently.

a)   b)

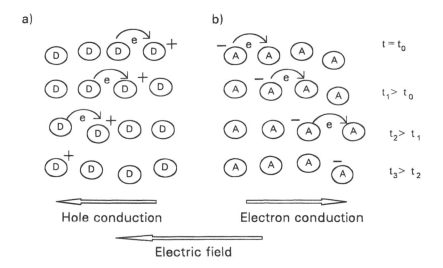

Hole conduction   Electron conduction

Electric field

Figure 11 Schematics of (a) hole transport and (b) electron transport through hopping in an organic photoconductor.

Photoconductors are compounds that become conductive upon absorption of a photon that leads to the generation of an electron-hole pair. Their spectral sensitivity is limited to their absorption band. However, the photosensitivity curve can be tuned to the desired wavelength by doping with an adapted sensitizer, or can be induced by a charge-transfer complex. Depending on the nature of the majority charge carrier in a given photoconductor, the photosensitizer needs to be either a good electron acceptor for hole transport or a good electron donor for electron transport. For hole (electron) transport, the conductive groups are donor (acceptor) type. Radical cations (anions) are formed upon photoexcitation and charge transport is due to a sequence of electron transfer steps between neighboring neutral and charged moieties as shown schematically in Figure 11. In other words, photoconductivity can be provided by redox active moieties that are easily oxidized or reduced. The transport proceeds via hopping between those redox sites which undergo a series of oxidation/reduction steps. Due to the tremendous development of organic photoconductors in the printer and photocopier industry, there exist a great number of photoconductors but detailed studies of their properties are not always available. A review of these materials has been presented recently by Law.[24] The most commonly used sensitizers are phthalocyanines, squaraines, azo dyes, perylene dyes, or thiapyrylium salts as shown Figure 12.

The polymeric photoconductors can be divided into two classes. One class consists of polymers where the photoconductive groups are built in the polymer structure itself as pendent groups or in-chain groups. The second class, referred to as molecularly doped polymers, are composites consisting of a neutral polymer binder which is chosen for its good optical and mechanical properties doped with small molecules which have the functionality required for photogeneration and transport. Although these two classes have structural differences, the description of their physical properties for photogeneration and transport have been identical so far. Most of the systematic studies on the photogeneration and transport properties have been carried out on PVK and more precisely on the charge-transfer complex that PVK forms with the electron acceptor TNF which provides photosensitivity in the visible[52] while pure PVK is sensitive only in the UV. Although the PVK:TNF photoconductor is no longer used in commercial systems because of the toxicity of TNF and its poor mechanical strength, it is a well-documented material and can, therefore, be used as a model photoconductor to describe and understand the physics of photorefractive polymers based on this charge-transfer-type photoconductor. PVK is a p-type photo-conductor, as are most of the known organic photoconductors, and hole transport is due to electron hopping between the carbazole moieties that are attached to the polymer backbone. The nitrogen gives a donor-like character to the carbazole group. In pure PVK conduction is, therefore, due to holes only. In PVK:TNF mixtures, at low TNF concentrations, transport is mainly due to holes, while at high concentrations of TNF, electron transport dominates. For mixtures of PVK:TNF close to 1:1, the conduction is then bipolar, meaning that both holes and electrons can be transported. In photorefractive materials equal hole and electron conduction is not desired since migration of each carrier species in

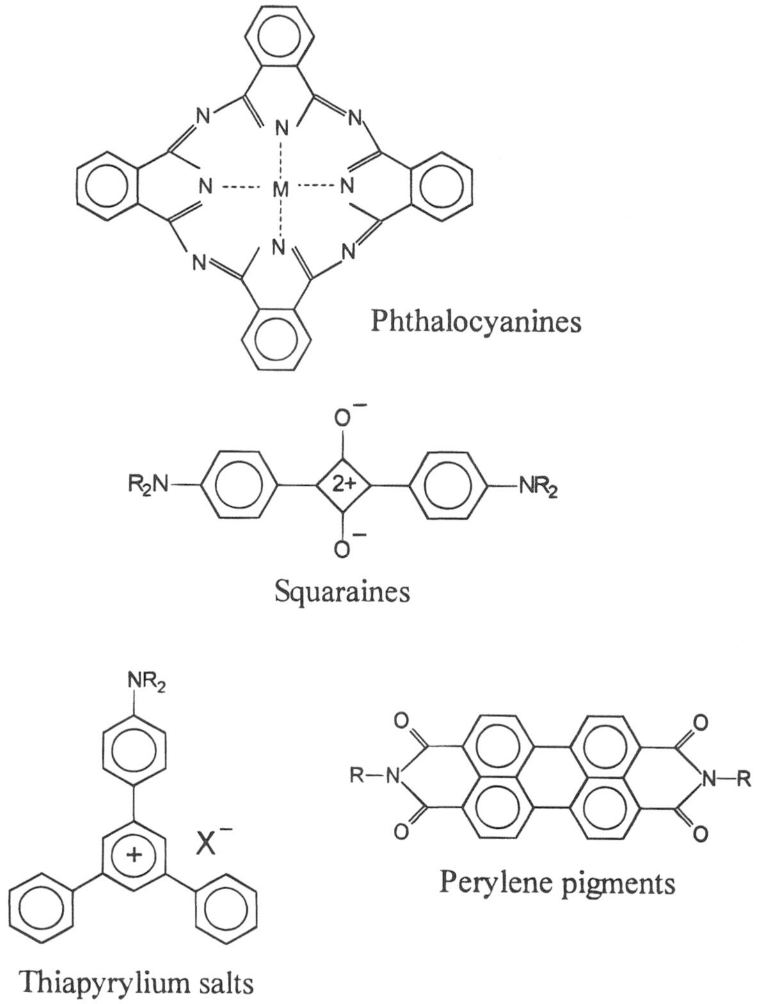

Phthalocyanines

Squaraines

Thiapyrylium salts

Perylene pigments

Figure 12 Examples of sensitizers used in organic photoconductors.

the opposite direction leads to a cancellation of the space charge. Therefore, the concentration of TNF in PVK has to be kept small (1%) such that conduction is due mainly to holes. The values reported in the literature for the activation energy for hole transport in PVK are ranging between 0.4 and 0.7 eV.[47,93] In addition, trapping in PVK is also attributed to the formation of carbazole sandwich pairs.[94] The mobility in this material remains small, even if high purity conditions are fulfilled, and is limited to the range of $\mu \approx 10^{-7}$ cm²/Vs at an applied field of 10^5 V/cm.

B. DIFFERENCES BETWEEN INORGANIC CRYSTALLINE AND ORGANIC AMORPHOUS MATERIALS

In addition to the capacity to form thin films with semiconductor compatible multilayer processes for integrated optics, polymers have physical properties which are more favorable for the photorefractive effect compared to inorganic crystals. These features, namely low dielectric constant and high electro-optic coefficients, are common to organic materials, including organic crystals. Although organic materials have generally a lower refractive index than inorganic materials, they have a high figure-of-merit and are, therefore, very promising. To illustrate this, one figure-of-merit often used to compare photorefractive materials, $Q = n^3 r/\epsilon$, is given for different material classes in Table 1. This figure-of-merit is not the absolute selection criterion for photorefractivity since there exist as many figures-of-merit as specific applications, but it gives a good idea about the potential of organics compared to inorganics.

In the following, we review the fundamentally different physical properties of conventional inorganic

Table 1 Properties of Selected Electro-optic Materials and Comparison of Their Figure-of-Merit for Photorefractivity

Materials	λ (μm)	r (pm/V)	n	ε	n^3r/ϵ (pm/V)	Response Time (ms)	Storage Time (s)
Inorganic							
LiNbO$_3$	0.6	$r_{33} = 31$	$n_e = 2.2$	32	10.3	1000	10^6
BaTiO$_3$	0.5	$r_{42} = 1640$	$n_e = 2.4$	3600	6.3	300	3×10^4
GaAs	1.06	$r_{12} = 1.43$	$n_e = 3.4$	12.3	4.6	<0.05	10^{-4}
BSO	0.6	$r_{41} = 5$	$n = 2.54$	56	1.5	2	10^{-2}
LiTaO$_3$	0.6	$r_{33} = 31$	$n_e = 2.2$	45	7.3	—	—
KNbO$_3$	0.6	$r_{42} = 380$	$n = 2.3$	240	19.3	—	—
InP	1.06	$r_{41} = -1.34$	$n = 3.29$	12.5	3.8	<0.05	10^{-4}
GaP	0.56	$r_{41} = 1.07$	$n = 3.45$	12	3.7	—	—
SBN	0.5	$r_{33} = 1340$	$n_e = 2.3$	3400	4.8	—	—
Organic							
MNA	0.6	$r_{11} = 65$	$n_X = 2$[a]	—	—	—	—
DCNP	0.6	$r_{33} = 82$	$n_X = 1.9$ $n_Z = 2.8$	—	—	—	—
MNBA	0.5	$r_{11} = 50$		4	217	5[b] $\times 10^3$	—
PPNA[c]	0.6	$r \approx 15$	$n = 1.7$	3.7	20		
PVK:TNF:FDEANST[d]	0.7	$r_{33} = 3.1$	$n = 1.7$	7	2.2	100	—
PVK:ECZ:TNF: DMNPAA	0.6	$r_{33} = 35$[e]	$n = 1.75$	5.5	34.1	100	—

[a] From Morita et al.[122] MNA, (2-methyl-4-nitroaniline); DCNP, 3-(1,1-dicyanoethenyl)-1-phenyl-4,5-dihydro-1H-pyrazole[123]; MNBA, 4'-nitrobenzylidene-3-acetamino-4-methoxyaniline.[90]

[b] At 130 mW/cm^2.

[c] From Table 4 in Burland et al.[56]

[d] From Donckers et al.[21]

[e] At 100 V/μm and 633 nm.[23]

PR crystals such as LiNbO$_3$ and amorphous PR polymers, including the electronic structure, the nature and number of PR traps, and the origin of the nonlinearity.

Inorganic PR crystals are characterized by electronic band structures which are principally defined over the entire crystal (delocalized charge carriers). The absorption of a photon transfers an electron into the conduction band of the crystal. The excited electron is free to migrate (i.e., diffuse) and the charge generation efficiency is generally close to one and is only slightly field dependent. Due to the electronic band structure, the conductivity of inorganic PR crystals is rather high. By contrast, in amorphous organic materials such as PR polymers, electrons are strongly localized on individual molecules due to the poor interaction (overlap of the wave functions) between different molecules. The absorption of a photon primarily creates a bound electron-hole pair within a sensitive moiety. The separation of this pair to generate free carriers (involving the transfer of an electron from the sensitizer to the photoconductor) competes with its recombination, leading to a strongly field-dependent charge generation efficiency (Onsager type) in the PR polymers (see Section III.A). Charge transport proceeds via hopping between spatially separated redox sites which undergo a sequence of oxidation and reduction steps (trapping/detrapping). The carrier mobility strongly depends on the spatial separation of the conducting redox sites, on the strength of an applied electric field, and on the temperature (see Section III.B).

In PR crystals mainly ionic impurities act as traps. Those are naturally at low concentration, limiting the maximum strength of the internal space charge field. By contrast, in PR polymers, basically every redox site involved in the conducting pathway may act as a trap as long as it is not photoreactivated. In other words, the number of possible traps in PR polymers equals basically the number of photoconducting moieties, enabling the buildup of strong space-charge fields comparable to the external field. The maximum value is given by the projection of external field along the grating vector.

The nonlinearity in inorganic crystals is a bulk property mainly driven by the ionic polarizability of the material; large nonlinearities are thus always accompanied by high bulk dielectric constants, resulting in small variations of the commonly used figure-of-merit. By contrast, in organic materials the nonlinearity is a molecular property, originating from the asymmetry of the electronic states of the EO chromophore; large nonlinearities and low dielectric constants can be thus obtained at the same time. Based on these considerations, a very high figure-of-merit can be obtained, therefore, with PR polymers as illustrated previously.

The periodic structure of inorganic PR crystals is unaffected by any electric field. Thus the PR response originates solely from the linear EO effect. By contrast, in PR polymers containing nonlinear chromophores, the second-order nonlinearity is produced by poling of the chromophores, and the bulk properties strongly depend on the poling conditions. In other words, the nonlinearity can be continuously adjusted with the poling field and can be switched "on" or "off."

V. EXPERIMENTAL RESULTS IN PVK-BASED POLYMERS

The aim of this section is to present experimental results on photorefractive polymers and to review some of the experimental techniques that can be used for their investigation. We intentionally restrict ourselves to the results we obtained in our laboratory on PVK-based polymers. A review of the results obtained by other teams can be found in Silence et al.[95] and Moerner and Silence.[8]

A. SAMPLE PREPARATION

Typically, the polymer mixtures were prepared by dissolving appropriate amounts of all components of the composite in chloroform or toluene. The solutions were then passed through a 1-μm filter before further processing. For the photoconductivity and electro-optic investigations, thin films (several micrometers) were spin-coated from these solutions on transparent (ITO) electrodes. For the holographic investigations, the solvent was removed and the remaining solid was ground into a fine powder. Some material was placed onto an ITO electrode and spacers were placed around it. The material was then heated to $\approx 150°C$, where it became a viscous liquid. Finally, a second ITO electrode was pressed on top. The result was optically clear films of given thickness (≈ 100 μm).

B. PHOTOCONDUCTIVITY CHARACTERIZATION

The study of the developed photorefractive polymer starts usually with the characterization of the basic properties for photorefractivity, namely photoconduction and electro-optic activity.

The characterization of the photoconductive properties of a given photorefractive polymer are important since they control the buildup of the internal space-charge field that is responsible for the refractive index modulation through the electro-optic effect. To check the photoconductive properties of a polymer, several techniques can be used. A first technique, often referred to as the discharge technique, is widely used in the field of xerography and is based on the measurement of the decay of the surface potential of a sample under illumination initially charged by corona discharge.[96] Typically, the sample is coated on a grounded electrode and mounted on a mobile sample holder. The free top surface of the sample can then be charged by moving the sample under a corona emitter. After charging, the sample is rapidly moved under an electrometer that measures the decay of the surface potential. The photodischarge is measured under dark conditions and under the illumination of monochromatic light of known intensity I. The thin sample can be described as a discharging capacitor and the rate of change of the surface charge density, dQ/dt, is given by:

$$\frac{dQ}{dt} = -C\left(\frac{dV}{dt}\right) = \phi(E)\delta e = \sigma(E)E \tag{60}$$

where C is the capacitance per unit area ($C = \epsilon\epsilon_0/d$ where d is the sample thickness), dV/dt is the rate of change of the surface potential, $\phi(E)$ is the photogeneration efficiency, δ is the number of absorbed photons per unit area and per second, $\sigma(E)$ is the photoconductivity, and E is the value of the electric field in the sample and is assumed uniform. The charged photoconductor shows usually a small dark discharge that is independent upon illumination. The photoconductivity $\sigma(E)$ is then given by:

$$\sigma(E) = \frac{-\epsilon\epsilon_0}{V}\left(\frac{dV}{dt}\bigg|_{Light} - \frac{dV}{dt}\bigg|_{Dark}\right) \tag{61}$$

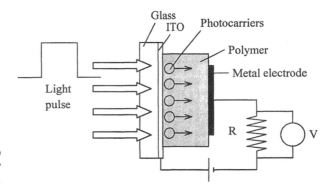

Figure 13 Schematic of a time-of-flight setup for the determination of the photoconductivity and the photogeneration quantum efficiency.

In the weak absorption regime ($\alpha d < 1$) and under steady-state conditions, the number of absorbed photons per unit area per second is given by:

$$\delta = \frac{I\alpha\tau d}{\hbar\omega} \tag{62}$$

where I is the power density of the illuminating source with photon energy $\hbar\omega$, α is the absorption coefficient, and τ is the lifetime of the photogenerated carriers which can be identified here with the transit time of carriers through the sample; $\tau = d/\mu E$ where μ is the drift mobility. The photogeneration efficiency (sometimes called xerographic gain) can be deduced from the measurement of the photoconductivity. According to Equations 60 and 62, the photogeneration efficiency can be expressed as:

$$\phi(E) = \hbar\omega\,\frac{\sigma(E)}{I}\,\frac{E}{\alpha de} \tag{63}$$

Another technique for the measurement of the photoconducting and transport properties of a material is the so-called time-of-flight technique shown in Figure 13. The polymer is coated on a transparent conducting electrode (typically ITO coated glass) and a metal electrode (Au, Ag, or Al) is evaporated on top of the polymer surface. Here, the sample can be identified with a capacitor having an optically transparent window. The capacitor is illuminated through this window and the absorbed light generates photocarriers which are collected by the metal electrode after transport through the sample. These collected carriers give rise to a photocurrent. This is detected by measuring the voltage across a resistor R placed after the sample in a circuit that contains also a voltage source. If the sample is excited with short light pulses with nanosecond duration, the mobility of the carriers can be deduced from the shape of the transient photocurrent. For photoconductivity studies, the sample is excited with light pulses whose duration (typically 100 ms) is much longer than the transit time of the carriers in the material. Thus, quasi-steady-state excitation conditions can be assumed during light exposure. The pulsed excitation is preferred to a cw excitation in order to reduce internal space-charge formation effects. For the same reasons, the light intensity is kept at its lowest level. The photocurrent i_{Ph} and the photoconductivity $\sigma(E)$ are given by:

$$i_{Ph} = \frac{V_{Ph}}{R} = \sigma(E)ES = n_D e\mu ES \tag{64}$$

where V_{Ph} is the voltage measured across the resistor R, S is the surface of the illuminated area of the sample, and n_D is the number of photocarriers per cubic centimeter given by:

$$n_D = \phi(E)\,\frac{I\alpha\tau}{\hbar\omega} \tag{65}$$

At equilibrium, the number of carriers collected at the metal electrode can be approximated by the number of carriers photogenerated at the transparent window, so that the lifetime of the carriers in the

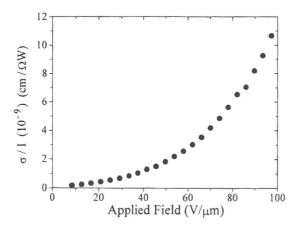

Figure 14 Normalized photoconductivity measured as a function of the applied field in a DR1:PVK:TNF thin film (see text for details).

sample can be identified with their transit time ($\tau = d/\mu E$). In this case, the photogeneration efficiency can be deduced from:

$$\phi(E) = \sigma(E) \frac{\hbar\omega}{e\alpha dI} E \qquad (66)$$

Figure 14 shows results of photoconductivity measurements performed using the time-of-flight configuration in a 1.2-μm-thick PVK polymer composite doped with TNF and with the azo dye Disperse Red 1 (DR1).[97] The proportions in the mixture of DR1:PVK:TNF were 24:75:1 %wt. The sample was then masked and coated with circular gold electrodes. These experiments were performed with 100-ms light pulses with 10-mW intensity originating from the 458-nm line of a cw Ar laser chopped at a frequency of 1 Hz with a mechanical shutter. At this wavelength, the absorption coefficient of the DR1:PVK:TNF sample, whose linear absorption spectrum is shown in Figure 15, is $\alpha = 2.5 \times 10^4$ cm^{-1}. As shown in Figure 14, the photoconductivity normalized to the power density of the incident light is strongly field dependent and reaches a maximum value of $\sigma/I = 1.07 \times 10^{-8}$ cm/ΩW at an applied field of 97 V/μm. The values of the quantum efficiency for photogeneration can be deduced from the values of the photoconductivity according to Equation 66. The experimental values as well as a theoretical fit using Onsager's model described in Section III.A are shown in Figure 16. The Onsager fit has been obtained with the following parameters: $r_0 = 1.29$ nm, $\phi_0 = 1$, $\epsilon = 3.4$. Figure 16 shows that the photogeneration efficiency at low fields is very small ($< 10^{-5}$) and that it can be increased by more than 3 orders of magnitude when the applied field is increased from 0 to 100 V/μm. This behavior demonstrates the necessity to apply an electric field to the polymeric photorefractive sample during the writing of a photorefractive grating. Since both photogeneration efficiency and mobility are field dependent, the dynamics of the buildup and erasure of a grating is field dependent as well. This is illustrated in Figure 17 where the erasure of a photorefractive grating is shown for different values of the applied field in a sample of DMNPAA:PVK:ECZ:TNF (40:40:19:1 %wt).

The dynamics of the buildup of a photorefractive grating can be described mainly by two time

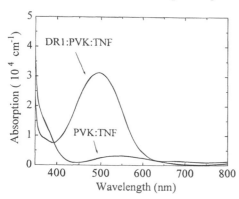

Figure 15 Linear absorption spectra of a PVK:TNF and a DR1:PVK:TNF thin film.

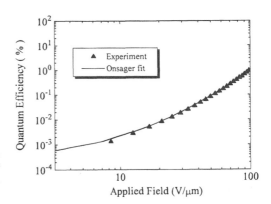

Figure 16 Photogeneration quantum efficiency measured as a function of the applied field in a thin film of DR1:PVK:TNF. The full line is a fit according to Onsager's model.

constants: a transit time constant $\tau_{TR}(E) = \Lambda/4\mu(E)E$ which is directly related to the time it takes for the carriers to be transported over a distance that is of the order of $\Lambda/4$ (corresponding to a 90° phase shift), and a generation time constant τ_G which is related to the time it takes to generate enough carriers in order to reach the saturated trap-limited space-charge field. The rate of photogeneration of carriers can be described by the following rate equation:

$$\frac{dn_D}{dt} = \phi(E)\frac{I\alpha}{\hbar\omega} - \frac{n_D}{\tau} \tag{67}$$

where τ is the lifetime of the photogenerated carriers. When the lifetime of the carriers is much longer than the transient time that is of interest, the solution of Equation 67 can be approximated to $n_D(t) = [\phi(E)I\alpha/\hbar\omega]t$. This is generally a good approximation when the storage time of a grating is much longer than its buildup time. The builtup time constant related to the photogeneration can then be identified with the time it takes to generate the density of carriers necessary to reach the steady-state space-charge field. The upper limit for the space-charge field is reached when it balances the external field E_{ext}. The combination of Equations 4 and 66 and the linear solution of Equation 67 give the following time constant:

$$\tau_G = \frac{\epsilon\epsilon_0}{\sigma} Kd \tag{68}$$

where $K = 2\pi/\Lambda$ is the grating vector and d is the thickness of the sample, and $\epsilon\epsilon_0/\sigma$ is the so-called dielectric response time τ_D. For a typical photorefractive polymer with 100 μm thickness having a photogeneration efficiency of $\phi(E) = 0.1\%$ at an applied field of 50 V/μm, a grating with a spacing of $\Lambda = 3$ μm written at 675 nm with 1 W/cm^2 writing power density, with an absorption coefficient of $\alpha = 10$ cm^{-1}, and with a dielectric constant of $\epsilon = 4$, the generation time constant defined in Equation 68 has a value of $\tau_D \approx 700$ ms. In contrast, for the same grating spacing, same applied field,

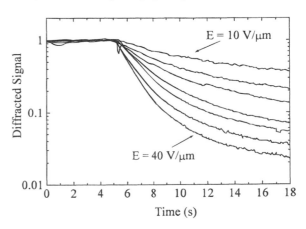

Figure 17 Erasure dynamics of a photorefractive grating as a function of the applied field in a 105-μm-thick sample of DMNPAA:PVK:ECZ:TNF.

496

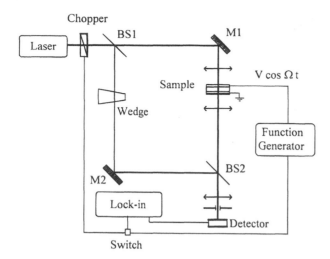

Figure 18 Experimental setup for Mach-Zehnder interferometric experiments.

and for a mobility of $\mu = 10^{-7}$ cm^2/Vs which is characteristic for a PVK-based material, the transit time is $\tau_{TR} \approx 1.5$ ms. These values indicate that at this stage of the research on photorefractive polymers, the buildup dynamics are mainly limited by the photogeneration of carriers rather than by the low carrier mobility in amorphous photoconductors. However, the dynamics of some photorefractive polymers also include rotational diffusion times of the chromophores and are, therefore, more complex and need to be analyzed in more detail in future studies.

C. ELECTRO-OPTIC CHARACTERIZATION

The second requirement for photorefractivity is electro-optic activity. The electro-optic coefficient can be determined by measuring the change in refractive index when a low-frequency (much smaller than optical frequencies) electric field is applied across the sample. Several techniques have been developed for the measurement of the components of the electro-optic tensor of a given material. Here, we review an interferometric technique based on a Mach-Zehnder interferometer[98] and an ellipsometric technique.[99,100] These techniques generally apply for prepoled samples and the effect of the field applied during the electro-optic characterization on the orientation of the chromophores can be neglected.

In the Mach-Zehnder technique shown Figure 18, the sample is placed in one arm of an interferometer and it changes the relative phase between the two arms of the interferometer, resulting in an intensity modulation of the light that is measured at the output of the interferometer which is given by:

$$I = \frac{1}{2}\left(E_1^2 + E_2^2 + 2E_1 E_2 \cos(\phi_2 - \phi_1)\right) \qquad (69)$$

where E_i are the field amplitudes of the light in each arm of the interferometer and $\phi_2 - \phi_1$ is their relative phase change. If an electric field with frequency Ω is applied to the electro-optic sample, the relative phase will be modulated as:

$$\phi_2 - \phi_1 = \phi_3 + A\cos(\Omega t) \qquad (70)$$

where ϕ_3 is the relative phase without electro-optic modulation. A is the modulation amplitude given by:

$$A = 2\pi \Delta n d/\lambda = -\pi r_{13} n^3 V/\lambda \qquad (71)$$

where V is the amplitude of the applied voltage. Note that the modulation amplitude is independent of the sample thickness. The refractive index modulation amplitude Δn is given by Equation 6. For the configuration shown in Figure 18 and with the ∞mm symmetry of poled polymers, the electro-optic coefficient that is responsible for the index modulation is r_{13}. Substituting Equation 70 into Equation 69 leads to the following modulated signal $\Phi(\Omega)$ at the output of the interferometer:

$$\Phi(\Omega) = E_1 E_2 \cos(\phi_3 + A\cos(\Omega t)) \tag{72}$$

For small modulation amplitudes A, the right-hand side of Equation 72 can be expanded in A and the modulation amplitude into the first order of A to:

$$\Phi(\Omega) = -E_1 E_2 A \sin(\phi_3)\cos(\Omega t) \tag{73}$$

The intensity at the output of the interferometer is modulated, therefore, at frequency Ω with an amplitude that depends on the product of the field amplitudes $E_1 E_2$, on the amplitude A and the relative phase ϕ_3 of the interferometer when no electro-optic modulation is applied to the sample. The product $E_1 E_2$ can be determined independently by measuring the intensity modulation of the interferometer when only the reference arm is modulated. This can be done by translating a wedge placed in the reference arm as shown in Figure 18 or by translating the mirror M2 in the reference arm by mounting it on a piezoelectric transducer. The maximum I_{max} and the minimum I_{min} intensities which are detected at the output of the interferometer are related to the product $E_1 E_2$ by:

$$E_1 E_2 = (I_{max} - I_{min})/2 \tag{74}$$

For this measurement, the s-polarized light is chopped and the output intensity is measured with the lock-in amplifier at the chopper frequency without applying a voltage to the sample. For the determination of the electro-optic coefficient, the phase of the reference arm is continuously changed with the translating wedge. The magnitude of the signal modulated at frequency Ω detected on the lock-in amplifier is oscillating periodically as shown by the $\sin(\phi_3)$ factor in Equation 73. According to Equations 71, 73, and 74, the electro-optic coefficient can then be deduced from the amplitude $I_s^{rms}(\Omega)$ of the oscillating signal:

$$r_{13} = \frac{\lambda}{\pi n^3 V^{rms}} \frac{I_s^{rms}(\Omega)}{I_{max} - I_{min}} \tag{75}$$

The analysis presented above stands for a purely linear electro-optic material. Other effects can also contribute to the refractive index change induced by the applied field. Piezoelectricity, for instance, leads to a thickness change of the sample and is, therefore, also inducing a phase change in the Mach-Zehnder apparatus. Since piezoelectricity is an effect that is linear in the electric field, its contribution cannot be discriminated from the linear electro-optic effect. Moreover, higher order effects such as the quadratic electro-optic effect or Kerr effect can also contribute to the modulated signal at frequency Ω, especially when a space-charge field introduced in the sample during the poling process leads to an internal DC field that is superimposed to the external AC field. Many other effects such as birefringence orientational effects,[60] higher order effects including electronic and orientational third-order effects, electrode attraction, electrostriction, or heating effects,[73] can possibly contribute to the overall refractive index changes. Higher order effects can lead to a signal modulated at frequency 2Ω which can yield valuable information on the third-order nonlinear optical properties of the polymer.[101] The relative strength of each of these possible contributions has not yet been clearly established in photorefractive polymers and more work needs to be done.

The ellipsometric technique is based on a change of polarization of a polarized light beam due to the electro-optic effect. This technique does not need very high mechanical stability in contrast to the interferometric technique. As shown in Figure 19, the sample is placed between a polarizer and a crossed analyzer. The incident light is linearly polarized at 45° with respect to the incidence plane of the sample so that the s component (perpendicular to the plane of incidence) and the p component (parallel to the plane of incidence) of the light field have equal amplitude. The relative phase between the s and p components can be continuously adjusted with the Soleil-Babinet compensator placed after the sample. When a modulated voltage is applied to the sample, the relative phase between the s and

498

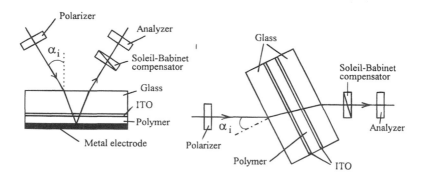

Figure 19 Schematic of the setup for ellipsometric measurements; (a) reflection geometry; (b) transmission geometry.

p components of the polarization of the light beam is changed, causing a change in the intensity transmitted through the analyzer. This intensity modulation is detected with a photodetector and a lock-in amplifier synchronized with the applied voltage. Figure 19 shows the apparatus for different geometries. The reflection geometry shown Figure 19a is convenient for the study of thin samples deposited on a transparent conducting electrode and coated with a reflecting metal contact. The transmission geometry shown in Figure 19b is adapted for the study of samples where the polymer is sandwiched between two transparent electrodes. Here we derive the theory for the transmission geometry. The equations for the reflection geometry differ only by a factor of 2, since the propagation length is 2 times higher in the reflection geometry. For simplicity we neglect the reflection losses at the different interfaces and assume that the sample is not absorbing. For a detailed analysis including these corrections see Levy et al.[102]

The output laser intensity after the analyzer can be expressed as:

$$I = I_i \sin^2[(\psi_{sp} + \psi_B)/2] \tag{76}$$

where ψ_{sp} is the phase difference between the s and p components after propagation in the polymer film, ψ_B is the adjustable phase mismatch between the s and p components that is introduced by the Soleil-Babinet compensator, and I_i is the incident light intensity. For the transmission geometry shown in Figure 19b, the phase difference ψ_{sp} is written:

$$\psi_{sp} = \frac{2\pi d}{\lambda}(n_s \cos\alpha_s - n_p \cos\alpha_p) \tag{77}$$

where the indexes s and p refer to the s and p components of the light field. Since the material is anisotropic, the refractive index sensed by each polarization is different and, consequently, the propagation angle inside the polymer is different for each component. The refractive indexes are given by[103]:

$$n_s = n_o \tag{78}$$

$$n_p = \left[n_o^2 + \sin^2\alpha_i\left(1 - \frac{n_o^2}{n_e^2}\right)\right]^{1/2} \tag{79}$$

where n_0 and n_e are the ordinary and extraordinary refractive indexes, respectively, and α_i, is the angle between the incident laser beam and the normal to the sample (before the sample). When a modulated voltage $V = V_m \sin \Omega t$ is applied to the polymer film, the refractive index change induced

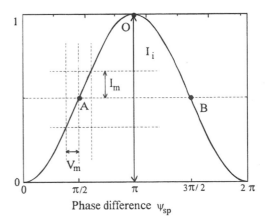

Figure 20 Measured intensity after analyzer as a function of the phase difference between the s and p components of a polarized laser beam.

through the electro-optic effect causes a change in the phase mismatch $\Delta\psi_{sp}$ between the s and p components given by:

$$\Delta\psi_{sp} = \frac{\partial\psi_{sp}}{\partial n_o}\Delta n_o + \frac{\partial\psi_{sp}}{\partial n_e}\Delta n_e \qquad (80)$$

where Δn_0 and Δn_e are the refractive index changes induced along the ordinary and the extraordinary axes, respectively. For a poled polymer when Kleinman symmetry is assumed, the extraordinary direction is along the poling direction and the ordinary one is in any perpendicular direction. For purely electro-optic modulation, in the weak poling limit, the refractive index modulations are given by:

$$\Delta n_e = -n_e^3 r_{33} V_m/2d \qquad (81)$$

$$\Delta n_o = -n_o^3 r_{13} V_m/2d \approx \Delta n_e/3 \qquad (82)$$

From Equations 77–79 and assuming a small anisotropy in the poled polymer film ($n_0 \approx n_e = n$), Equation 80 can be rewritten as:

$$|\Delta\psi_{sp}| = \frac{2\pi n^3 r_{33} V_m}{3\lambda G} \qquad (83)$$

where G is the geometrical factor given by:

$$G = \frac{n(n^2 - \sin^2\alpha_i)^{1/2}}{\sin^2\alpha_i} \qquad (84)$$

Since the phase difference ψ_{sp} scales with the thickness d of the sample (Equation 77) and since the electro-optic refractive index changes (Equations 81 and 82) scale with the inverse of the thickness, the phase mismatch $\Delta\psi_{sp}$ does not depend on the sample thickness. Indeed, the sensitivity of the experiment depends mainly on the value of amplitude of the modulated voltage that can be applied to the polymer film. Figure 20 shows the intensity measured after the analyzer as a function of the total phase difference between the s and p components introduced by the sample and the Soleil-Babinet compensator. For the determination of the electro-optic coefficient, the total phase is tuned with the compensator such that the output intensity is at half the incident intensity ($I_i/2$). At the points usually referred to as A and B, the sin^2x curve is at its most linear region and the slope is at its highest value,

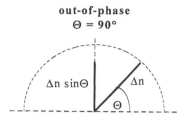

out-of-phase
$\Theta = 90°$

$\Delta n \sin\Theta$

Δn

Θ

in-phase
$\Theta = 0°$

Figure 21 Refractive index modulation amplitude tested in two-beam coupling and four-wave mixing experiments.

resulting in a maximum modulation of the output intensity. For small modulations, the modulated intensity amplitude detected at frequency Ω can be approximated to:

$$I_m^A(\Omega) = \Delta\psi_{sp}\frac{I_i}{2}$$

(85)

At point O, the first derivative of the curve is zero and no signal is observed for purely electro-optic modulation. Since the slope of the curve at point B is of opposite sign, the modulation intensity that is detected at this point has also opposite sign compared to the one detected in A. The final expression of the electro-optic coefficient r_{33} as a function of the modulated intensity $I_m(\Omega)$ is given by:

$$r_{33} = \frac{3\lambda G}{\pi n^3}\frac{I_m(\Omega)}{I_i V_m}$$

(86)

Note that the amplitude detected on the lock-in amplifier (peak to peak divided by 2) is the rms (root mean square) value and has to be converted.

The analysis described above applies only to prepoled samples. For polymer samples with a glass transition temperature close to the operating temperature (room temperature), the modulating field can also reorient the chromophores in the matrix leading to a strong contribution from the orientational birefringence as discussed for wave-mixing experiments in Section III.D.2. The contribution from the birefringence to the modulated signal in ellipsometric electro-optic experiments plays a role only if the frequency of the modulated field is compatible with the dynamics of the orientation of the molecules in the matrix. As a result, the modulated intensity is strongly frequency dependent. At low frequency, the modulated signal is due to the sum of the birefringence, second-order and higher order contributions, while at high frequency, the signal is only due to the second-order and higher order contributions. Frequency-dependent ellipsometric measurements give valuable information about the different contributions to the total refractive index changes.

D. FOUR-WAVE MIXING AND TWO-BEAM COUPLING EXPERIMENTS

Two standard holographic techniques are commonly used to characterize PR polymers, the four-wave mixing (4WM) and the two-beam coupling technique (2BC). The experimental geometry is identical for both techniques and has been discussed in Section III. D. Two coherent writing beams are overlapped in the sample to create a fringe pattern. They are both either s or p polarized. In the 4WM experiments, index gratings recorded in the PR material are probed by a weak beam, counterpropagating with the writing beam as shown Figure 8, and the intensity of the transmitted and the diffracted light is monitored and the diffraction efficiency is determined. In the 2BC experiments the counterpropagating beam is absent and the transmission of the two writing beams is measured, yielding the amplification factor γ_0 and the gain coefficient Γ as discussed in Section III.E.

The results in 4WM and 2BC experiments depend on experimental parameters such as the total light intensity incident on the sample, the fringe visibility (i.e., the intensity ratio of the writing beams), the polarizations of all beams, the grating spacing (mainly defined by the angle between the writing beams in the material and the wavelength), and the tilt angle of the sample.

Whereas in a 4WM experiment the total grating amplitude is probed independently of the phase shift between light pattern and index modulation, the beam-coupling experiment is sensitive only to the component of the grating which is 90° phase shifted with respect to the original light pattern as shown schematically in Figure 21. Therefore, 2BC experiments should be carried out together with

a)

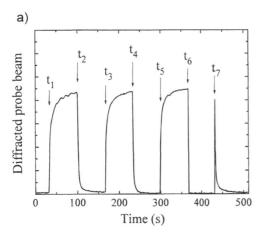

b)

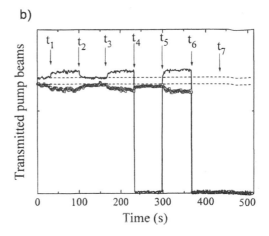

Figure 22 Diffracted probe beam in four-wave mixing experiments (a) and transmitted pump beams showing beam coupling (b) for different field and illumination conditions (see text for details).

4WM experiments in order to verify the nonlocal nature of the index grating and to prove its PR origin. Figure 22 shows the observations during 2BC and 4WM experiments. Originally there is no external field applied to the sample and no diffracted beam is observed. At $t = t_1$, the external electric field is switched on, and a photorefractive grating builds up. As a result, the probe beam is diffracted by this grating as shown in Figure 22a and asymmetric energy transfer is observed between the two pump beams as shown in Figure 22b: one pump beam is amplified, whereas the other is diminished. If the polarity of the applied field is changed, the direction of the energy transfer occurs in the opposite direction (not shown on Figure 22). When the field is switched off at $t = t_2$, the initial orientational distribution obtained with applied field is changed and the chromophores reorient solely under the influence of the internal space-charge field. Under these conditions, no diffracted signal can be observed in the first-order Bragg condition and beam coupling is no longer effective. When the electric field is switched back on at time t_3, the orientation of the chromophores by the total field is restored and some probe light is diffracted again. At time t_4, one of the pump beams is blocked, resulting in a spatially uniform illumination by the second pump beam leading to the erasure of the space-charge field and consequently the erasure of the grating. At time t_5, the second pump beam is restored and a new grating identical to the former grating is written. At time t_6, all beams are blocked and the electric field is switched off. At time t_7, the electric field is switched back on, and the probe beam is restored and is diffracted by the grating stored in the material. This demonstrates the storage capabilities of the material. After time t_7, the grating is erased by the probe beam.

The best steady-state performance of a PR polymer up to now was observed in the composite DMNPAA:PVK:ECZ:TNF (50:33:16:1 %wt).[23] These results will be presented here in more detail. These experiments were performed for the geometry $\psi_{ext} = 60°$ and $2\theta_{ext} = 22°$. Taking into account the refractive index of 1.75 measured by ellipsometry for this material, the internal angles are $\psi = 29°$ and $2\theta = 7°$, giving a grating spacing of 3.15 μm. The writing beam energy was 1 W/cm². These experiments were performed at 675 nm, where the 105-μm-thick samples show moderate absorption as shown in Figure 23 and its inset.

Figure 24 shows the field dependence of the amplification factor γ_0 defined by the intensity ratio of one of the writing beams with the other being present or absent, respectively, for three different intensity ratios b of p-polarized writing beams. In all cases, γ_0 increases monotonically with increasing field strength. For a beam ratio $b = 1.3$ the maximum amplification observed at the highest applied field is $\gamma_0^{(p)} = 2.1$ at 90 V/μm, i.e., very close to the maximum achievable value of $1 + b = 2.3$ when complete energy transfer would occur. By contrast, for a beam ratio of $b = 130$ the amplification factor at the identical voltage is much higher ($\gamma_0^{(p)} = 10.3$), but very far from the theoretical maximum value of 131. For the beam ratio $b = 13$ the behavior is between the two cases mentioned before. These observations are mainly related to the different modulation depth $m = 2\sqrt{b}/(1 + b)$ of the interfering writing beams.

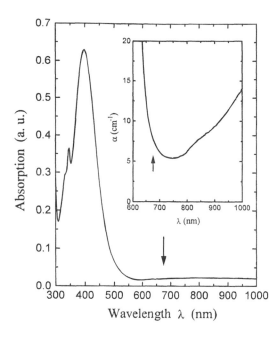

Figure 23 Linear absorption spectrum of a thin film of DMNPAA:PVK:ECZ:TNF and of a 105-μm-thick sample (inset).

Since the amplification factor depends on the intensity ratio of the writing beams, it is more convenient to characterize the material by its gain coefficient Γ defined in Equation 59. As shown in Figure 25, the gain coefficient for p-polarized beams Γ_p is essentially identical for the three beam ratios. Similar results were obtained for another PVK-based PR composite.[21] In the DMNPAA:PVK:ECZ:TNF, the maximum gain yields $\Gamma_p = 220$ cm^{-1} at an electric field of 90 V/μm. This value, by far, exceeds the absorption in the sample at this voltage ($\alpha = 13$ cm^{-1}), giving a net optical gain of $\Gamma_p - \alpha = 207$ cm^{-1}. The results of identical experiments performed with s-polarized beams are also shown Figure 25. In that case, the beam which gained energy when it was p-polarized, now loses energy for the identical field direction resulting in an opposite sign for Γ with this polarization. A monotonic increase of the absolute value of Γ_s with increasing field is observed, reaching $\Gamma_s = -40$ cm^{-1} at the maximum field of 90 V/μm. According to Equation 57, the change in the direction of the energy transfer for s- and p-polarized writing beams reflects opposite signs of the index modulations sensed by the two

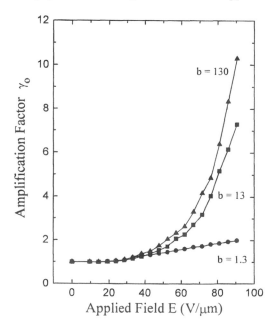

Figure 24 Two-beam-coupling experiment in DMNPAA:PVK:ECZ:TNF (the power density of each pump beam was 1 W/cm^2): external field dependence of the amplification factor γ_0 for p-polarized writing beams; variation of the intensity ratio of the writing beams: circles, $b = 1.3$; squares, $b = 13$; triangles, $b = 130$. The solid lines are guides to the eye.

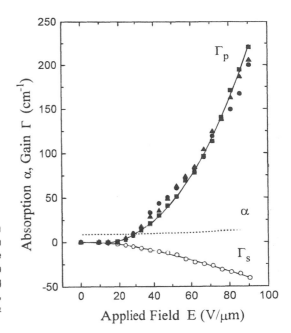

Figure 25 Two-beam-coupling experiment in DMNPAA:PVK:ECZ:TNF (the power density of each pump beam was 1 W/cm²): external field dependence of the gain coefficient Γ for s-polarized beams (open circles, beam ratio $b = 1.5$) and p-polarized writing beams (full circles, $b = 1.3$; squares, $b = 13$; triangles, $b = 130$); Also shown is the absorption coefficient α (dashed line). The solid lines are guides to the eye.

polarizations ($\Delta n_p / \Delta n_s < 0$) since the other factors in Equation 57 do not change sign. Note, that the absolute value of the gain for s-polarized beams is much smaller than for p-polarized beams. Therefore, 4WM experiments are usually performed with s-polarized writing beams in order to keep the beam coupling small and to avoid a nonuniform grating amplitude and phase throughout the sample in order to use Kogelnik's theory (Equations 36 and 37) for the description of the experimental data.

Figure 26 shows the 4WM results for p-polarized readout of a grating written with s-polarized beams in the identical material. The diffraction efficiency (squares) defined as the intensity ratio of the diffracted light and the incoming probe beam increases with the electric field and reaches a maximum value of 86% at $E = 61$ V/μm. At that field strength, the light is completely diffracted as demonstrated by the absence of any transmitted intensity (circles). Further increase of the field leads to periodic energy transfer between the diffracted and the transmitted beam; at $E = 81$ V/μm, all the light is again directed

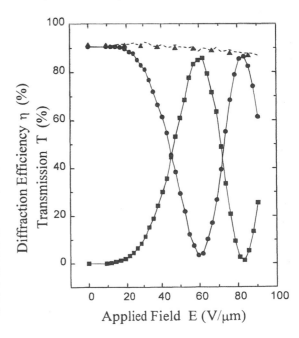

Figure 26 Degenerate four-wave-mixing experiment in DMNPAA:PVK:ECZ:TNF with s-polarized writing beams (power density ≈ 1 W/cm², beam intensity ratio $b = 1.3$) and a p-polarized reading beam (power density 0.35 mW/cm²): external field dependence of the diffraction efficiency η (squares) and the transmission in the presence (circles) and in the absence (triangles), respectively, of the writing beams. The dashed line is the sum of diffracted and transmitted intensity in the presence of the writing beams, indicating that the maximum achievable diffraction efficiency is limited by absorption and reflection losses (≈12%). The solid lines are guides to the eye.

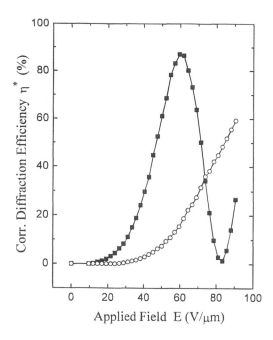

Figure 27 Degenerate four-wave-mixing experiment in DMNPAA:PVK:ECZ:TNF with *s*-polarized writing beams (power density ≈ 1 W/cm², beam intensity ratio *b* = 1.3): external field dependence of the corrected diffraction efficiency η* for *p*-polarized (squares) and *s*-polarized readout (power density 0.35 mW/cm²). The solid lines are guides to the eye.

into the original probe beam. The sum of diffracted and transmitted signals (dashed line) gradually decreases as the external field increases (an absorption change of 5 cm⁻¹ for field values ranging between 0 and 90 V/μm is observed). This is due to electric field-induced absorption changes in the sample[104] as we verified by an independent transmission measurement in the absence of the writing beams (triangles). Similar experiments were performed for *s*-polarized readout and no maximum was observed for fields up to 90 V/μm (not shown on Figure 26).

The periodic energy transfer between the transmitted and the diffracted beams observed in the 4WM experiments is in accordance with Kogelnik's coupled-wave model[68] described in Section III. E. Note, however, that the material shows a small increase in absorption when the field is varied from 0 to 90 V/μm, whereas the model assumes a constant absorption. In order to deduce the field dependence of the diffraction efficiency, we consider the effective diffraction efficiency defined as:

$$\eta^* = \frac{I_D}{(I_D + I_T)} \exp\left(\frac{-\alpha d}{2}\left(\frac{1}{\cos\alpha_1} + \frac{1}{\cos\alpha_2}\right)\right) \tag{87}$$

where I_D and I_T are the intensity of the diffracted beam and the intensity of the transmitted beam, respectively, measured after the sample. α is the value of the absorption coefficient at zero field. The other variables are defined in Section III.D. This way small field-dependent absorption changes can be removed from the experimental data and they can be compared directly to Kogelnik's formulas (Equations 36 and 37) described in Section III.D. This procedure is valid only for small values of the absorption and for small absorption changes. The field dependence of the effective diffraction efficiencies is shown in Figure 27 for both *s*- and *p*-polarized readout of a grating written by *s*-polarized beams. The absolute total index grating amplitude in the material $|\Delta n_{s,p}|$ for *s*- and *p*-polarized readout obtained following this procedure is shown in Figure 28. Note that due to the quadratic sine function in Equation 36 a priori only the absolute values of $\Delta n_{p,s}$ can be evaluated. However, the 2BC results described above indicate that Δn_s and Δn_p have opposite signs. Furthermore, as shown in Section III.C, Δn_p is positive for molecules with a positive hyperpolarizability value, whereas Δn_s can become negative when the birefringence contribution to the total index amplitude is dominating. Therefore, in Figure 28, Δn_p monotonically increases, whereas Δn_s decreases. At $E = 81$ V/μm, the field where all light is again transmitted rather than diffracted, the argument of the sine function in Equation 36 equals π, leading to index grating amplitudes of $\Delta n_s = -1.54 \times 10^{-3}$ and $\Delta n_p = +5.56 \times 10^{-3}$, respectively. Ellipsometric measurements performed on this material confirmed that such large field-induced refractive index

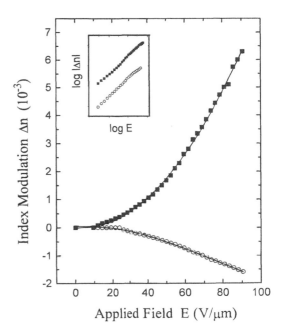

Figure 28 External field dependence of the index grating amplitudes Δn_s (circles) and Δn_p (squares) for *s*- and *p*-polarized readout, respectively, obtained from the coupled-wave model for slanted lossy dielectric transmission gratings, taking into account the change in absorption, but not the small phase and amplitude variations caused by energy exchange between the writing beams. The solid lines are guides to the eye. Inset: double logarithmic plot of the external field dependence of the index grating amplitudes $|\Delta n_s|$ (circles) and $|\Delta n_p|$ (squares) for *s*- and *p*-polarized readout, respectively.

changes can be achieved. The inset of Figure 28 shows the same results on a double-logarithmic plot. Two nearly parallel straight lines are obtained, indicating that the field dependence of the index modulations $\Delta n_{s,p}$ can nearly be described by a power law of the form $\Delta n_{s,p} \propto E^{\delta}$. The exponent ($\delta = 2.16$) is identical for both polarizations and the anisotropy ratio equals $\Delta n_p / \Delta n_s = -3.61$; it is nearly independent of the applied field.

The anisotropy ratio $\Delta n_p / \Delta n_s$ depends upon the microscopic properties of the EO chromophore ($\alpha_{/\!/} - \alpha_{\perp}$, β, and μ) and the properties of the host matrix such as its dielectric constant, i.e., local field corrections have to be taken into account.[55] The glass transition temperature of the DMNPAA:PVK:ECZ:TNF (50:36:13:1%wt) is below room temperature (15°C), allowing the orientational enhancement mechanism[61] to take place. Note that the ratio of the refractive index amplitudes for light polarizations parallel and perpendicular to the local poling-field direction is always positive for the EO effect, $(\Delta n_{/\!/}^{EO} / \Delta n_{\perp}^{EO}) > 0$, and negative for the electric field-induced BR, $(\Delta n_{/\!/}^{BR} / \Delta n_{\perp}^{BR}) < 0$. Assuming that the total index change is the sum of the individual changes induced by the BR and the EO effect, and for a positive first hyperpolarizability, the negative sign of $\Delta n_p / \Delta n_s$ states that the BR contribution to the total index modulation is dominant. Contributions from higher order phenomena can also be present and need to be investigated in future experiments. Diffraction from absorption gratings can be safely neglected since the maximum field-induced absorption change of 5 cm^{-1} gives an estimated diffraction efficiency of only $\eta \approx (\Delta \alpha d / 4)^2 \approx 1.5 \times 10^{-4}$.

The spacing of the gratings generated in PR materials can be easily varied by changing the opening angle $2\theta_0$ between the writing beams. Figure 29 shows the results of such variation in an 4WM experiment. The diffraction efficiency increases with decreasing opening angle $2\theta_0$ (i.e., increasing grating spacing). As discussed in Section III.D, for grating spacing values above 5 μm, the diffraction is no longer in the thick-grating Bragg regime and the diffraction efficiency is smaller than theoretically expected due the appearance of higher diffraction orders.

The dynamics of grating formation is complex. The speed depends upon the applied electric field, the light intensity, and the grating spacing. Under our experimental conditions (see above) and at 90 V/μm, the diffraction efficiency rises to ≈95% of the maximum value within approximately 100 ms, reaching a steady-state value after 10 s. The efficiency of the grating drops to 15% of the maximum value within 24 h after all beams and the electric field were switched off. The recorded index pattern can be erased by uniform illumination.

E. REAL-TIME HOLOGRAPHY

A holographic setup using a PR polymer as storage medium has been developed in order to demonstrate the ability to write and read two-dimensional holograms (Figure 30). The experiments were performed

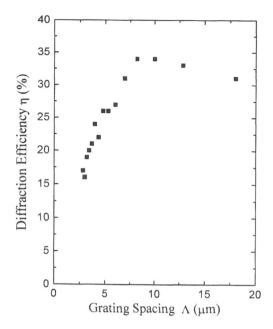

Figure 29 Degenerate four-wave-mixing experiment in DMNPAA:PVK:ECZ:TNF with s-polarized writing beams (power density ≈ 1 W/cm^2, beam intensity ratio $b = 1.3$) and a p-polarized reading beam (power density 0.35 mW/cm^2): grating spacing dependence of the diffraction efficiency η at external field of 50 V/μm.

with a semiconductor laser diode (LaserMax LAS-200-675-5) with 9 mW output power at 675 nm. The output beam was collimated, expanded, and then split into an object beam and a reference beam. The object beam was spatially intensity modulated after passing through a slide or an SLM (spatial light modulator). The s-polarized object and reference beams had equal pathlengths and were focused onto the PR polymer by Fourier transform lenses. The transmitted reference beam impinging on a perpendicular mirror placed after the sample was used to read-out the hologram. Its polarization was rotated by $\pi/2$, leading to a p-polarized beam. The images retrieved by the reading beam were inverse Fourier transformed by a lens, and finally imaged with a regular CCD camera. Bright and sharp images of the stored holograms could be reconstructed on the CCD camera. The resolution of the material was determined with a standard Air Force target to 300 1/mm corresponding to a resolution of approximately 3 μm in the material, i.e., comparable with the grating spacing. Figure 31 shows a retrieved phase hologram of a penny coin. For this experiment, the object beam was phase encoded after reflection on the penny coin and combined with the reference beam in the PR material. Encouraging preliminary studies have also demonstrated that photorefractive polymers can be used in optical correlators.[105]

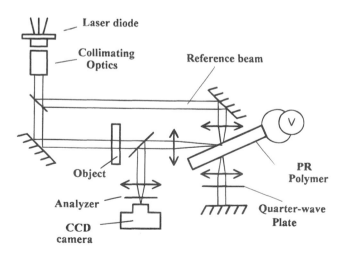

Figure 30 Schematic diagram of setup used for holographic storage and optical correlation.

Figure 31 Photograph of a retrieved hologram that was stored in a DMNPAA:PVK:ECZ:TNF sample.

VI. TRENDS AND OUTLOOK

A. REQUIREMENTS FOR A WELL-PERFORMING PHOTOREFRACTIVE POLYMER FOR HOLOGRAPHIC APPLICATIONS

With the invention of holography by Gabor in 1948,[106] a new area for information recording and processing has been initiated. The development of holography for photonic applications has been a fast-growing field of research, especially since the beginning of the 1960s with the development of highly coherent laser sources. At present, an intensive research effort is being undertaken to use holographic techniques for optical information storage. Holographic storage is technologically very appealing because the information storage capacity that can be reached with this technique, is considerably higher compared with other techniques. Figure 32 compares the storage capacity of different commercially available systems with the potential packing density of a holographic device. The predicted superiority of this technology is significant. The other powerful advantage of holographic storage is parallelism, meaning that all the information contained in an object is recorded and/or read at the same time, in contrast to many present systems that write and read the stored information bitwise sequentially. This can yield high data transmission rates.

The demand for better materials with higher capabilities for holographic storage applications is high. The speed of development of this advanced technology is strongly dependent on the discovery of new materials with better performance. With the increasing importance of the compactness of future commercial devices and available integrated laser sources, processing and cost are becoming the driving forces in the development of new materials. Therefore, polymeric materials, which are well known for their ease of processing and low cost, seem very promising. Early recording materials[107] such as silver halide emulsions and dichromated gelatins need wet processing steps, severely limiting their use in many applications. More recently, new polymeric materials have been investigated as a potential holographic

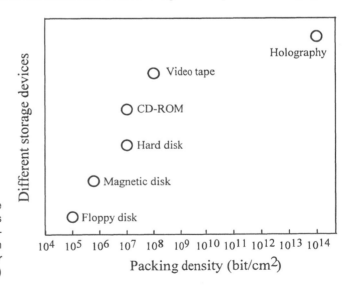

Figure 32 Comparison of the storage capacity of commercially available systems with the potential packing density of holographic storage devices. (Adapted from Manivannan, G., et al., *Trends in Polymer Science*, 2(8), 282, 1994. With permission.)

Table 2 Physical Properties of Several Polymer Materials for Holography

Materials	Thickness (μm)	Wavelength (nm)	Exposure Energy (J/cm²)	Resolution (lines/ mm)	η (%)	Reversibility
Photopolymerizable polymers[a]						
DuPont's polymer	1–200	350–550	0.1–0.4	3000	88–90	No
DuPont's omnidex (HRF series)	6–78	450–650	0.01–0.1	6000	99	No
DMP-128	1–30	442–647	0.005–0.03	5000	80–95	No
PMMA	2000	325	50–150	5000	70	No
Acrylamide	100	633	0.1	3000	80	No
Photocross-linking polymers[b]						
DCPVA	30–60	488	0.5	3000	68	No
DCPAA	40–60	488	4.0	3000	28	No
DCPAA-DMF	40–60	488	0.2	3000	70	No
FePVA	60	488	17	3000	80	No
PVK system	1.4–7.1	488	0.05–0.5	>3500	96	No
Photorefractive polymers						
PVK:FDEANST:TNF	125	676	0.1–1	—	1.3[c]	Yes
PVK:ECZ:DMNPAA:TNF	105	675	0.1–1	>300	86[d]	Yes

[a] Adapted from Table 1 in Manivannan and Lessard[108] and references cited therein.

[b] Adapted from Table 2 in Manivannan and Lessard[108] and references cited therein.

[c] At 40 V/μm.

[d] At 86 V/μm.

recording medium. These materials can be classified in two main categories: photopolymerizable and photocross-linking systems. In both types of systems, light exposure modifies permanently the structure of the material leading to a refractive index modulation and consequently to a hologram formation. Refractive index modulation is generally preferred to absorption modulation since absorption holograms can be retrieved only with a few percent efficiency according to diffraction theory, while an index hologram can be retrieved with an efficiency of 100%.[68] Several polymers cast into thin film have been studied recently for holographic recording applications.[108] Table 2 summarizes the performance of these materials. They show good processing capabilities, high optical clarity, high efficiency, good environmental stability, satisfactory resolution and sensitivity, but all lack reversibility. The ability to be erased rapidly is of paramount importance for a holographic material. Reversibility does not limit the field of applications to permanent storage only (WORM memories) but opens the use of such materials widely to real-time optical data processing, pushing the holographic technology one step closer to optical technology. Although advances in materials is of paramount importance for the development of a holographic technology, simultaneous progress is required in the development of enabling technologies (spatial light modulators, detector, and laser arrays) and novel architectures. The signal-to-noise ratios due to cross-talk effects have to be fully studied and compared with the performance of existing storage technologies. The emergence of new architectures is continuously redefining the requirements for the ideal holographic material. The most important requirements for a holographic material for real-time optical holographic recording and processing are: optical quality, sensitivity in the near IR, large refractive index modulation, self-processing, inert and long shelf lifetime, erasability, fixability, nondestructive read-out, and finally low cost. At this stage, it is still very challenging to meet all of these requirements together in a same material. Depending upon the application, the criteria listed above may vary but right now they give guidelines for the search of the ideal material. Nevertheless, the comparison of these properties with the performance of the photorefractive polymers investigated in Section V highlights the strength of these new materials as a holographic recording medium but shows also that a lot more progress is needed before they can be used in commercial storage devices.

B. TRENDS IN MATERIAL RESEARCH AND DEVICE DEVELOPMENT

Within the past 4 years, the progress achieved in the field of photorefractive polymers has been very stimulating. Starting from the demonstration of the principle,[12] a gradual and constant improvement of the efficiency of these new materials has been achieved[21,97,109] and has led to the demonstration of nearly 100% diffraction efficiency.[23] This initial investigation of fundamental principles has to be followed by an intensive search of new materials. Owing to the power of molecular engineering of organic materials, this field is expected to grow exponentially during the coming years. A synergistic effort between scientists working in the field of organic chemistry and optical sciences will contribute efficiently to this endeavor. The field of photorefractive polymers is at the interface of the fields of organic photoconductors and of nonlinear optical polymers, and both fields are driven by strong technological interests. The progress in material development in these two areas is, therefore, expected to contribute strongly to the development of the field of photorefractive polymers.

At present, the most efficient materials that we presented are blends of a photoconductive polymer and nonlinear optical chromophores. These materials can have a glass transition temperature lower than room temperature, enabling efficient chromophore orientation by the internal space-charge field. As discussed in Section III.D.1, this effect results in a contribution of the anisotropy of the polarizability of the rodlike molecules (orientational birefringence) to the refractive index modulation, and there is clear evidence that this contribution dominates for the DMNPAA chromophore as discussed in Section V.C. Therefore, for photorefractive polymers with low glass transition temperature, a figure-of-merit is no longer the well-known $Q = N\beta\mu$ product as for electro-optic polymers, but can be described by the following figure-of-merit:

$$Q = N[(\alpha_{/\!/} - \alpha_{\perp})\mu^2 + \beta\mu] \tag{88}$$

where N is the density of chromophores, $(\alpha_{/\!/} - \alpha_{\perp})$ is the anisotropy of the polarizability of the molecule, and μ is its permanent dipole moment. Since the linear optical properties of poled materials are described by the $L_2(x)$ Langevin function in contrast to the second-order properties which are described by the higher order Langevin function $L_3(x)$, the change in the linear optical properties in the low poling field limit, scale as the square of the dipole moment as shown in Equation 88. This changes completely the selection criteria for nonlinear optical molecules and opens new choices and new design approaches for future chromophores optimized for photorefractivity. Note that this analysis is valid only in low glass transition temperature photorefractive polymers. So far, most of the low T_g polymers are blends and the density of chromophores and/or plasticizer is limited by segregation and crystallization problems. This can severely affect the aging of the materials and can be avoided by the development of fully functionalized polymers with a low glass transition temperature. Chromophore–chromophore cooperative effects can also possibly play a role in future material and enhance the orientation of the chromophores. An interesting development would be to study photorefractive liquid crystals. Promising preliminary results have already been demonstrated.[110]

Low T_g materials show high efficiency but have also drawbacks including mainly the existence of higher order gratings and the trade-off between orientational mobility vs. poling stability and long storage times. Therefore, the research effort in the field of photorefractive polymers should also be continued in developing high T_g photorefractive materials. Such materials can be prepoled and can show stable electro-optic properties[111] and consequently read-out of the hologram can possibly be achieved without applied electric field. Nevertheless, since the photogeneration is strongly field dependent in organics, an electric field is still needed during the recording of the hologram. At present, the values of the mobility for most of the amorphous photoconductors are small, but recent reports indicate that the mobility in random organic photoconductors can be as high as 10^{-2} cm^2/Vs[112] and seem very encouraging for the development of faster photorefractive polymers. The ability to read-out photorefractive gratings without partially erasing the information is an important issue that needs to be addressed in future materials. Preliminary encouraging results on nondestructive read-out have been obtained recently.[113] A better understanding of the structure–property relationship in photorefractive polymers will strongly contribute to their optimization. So far, the best efficiencies were measured in saturated polymers in which transport occurs through hopping between localized sites. Preliminary results have been obtained in conjugated polymers.[114] Owing to their versatility and their band-type transport mechanism, conjugated photorefractive polymers look promising and have to be further investigated. The electric field that is applied to the well-performing polymers is quite high at this stage, but can be further reduced by optimizing the efficiency of the chromophores. It is reasonable to believe that in the near future new chromophores with $\beta\mu$ product 10 times higher will be used, reducing the value of the electric field that has to be applied. But, the trade-off between low field values and high speed will remain during the recording step. An interesting future development would be to build a uniform and stable electric field in the material such as in electrets.[115] Besides the efficiency of the materials, other important parameters can become driving forces such as mechanical properties and long-term stability. Environmentally friendly materials are also preferred.

Progress can be achieved by developing new materials with better performance and simultaneously by optimizing the design of future devices. Photogeneration efficiency, for instance, remains a major limitation for high-speed recording at low writing powers, but it will probably be optimized by using bilayer systems as has been demonstrated successfully in the field of xerography.[116] In bilayer systems, the photogeneration and transport functions are separated in two different layers. In such devices, photosensitizers that are not soluble in the charge-transporting polymer can be used. For instance, thin layers can be deposited by other techniques such as vacuum evaporation. Furthermore, in bilayer systems, organic materials can be associated with inorganic photosensitizers which have usually a better photogeneration efficiency. Furthermore, if different sensitizers are used in different layers, one can stack several layers and record with different wavelengths in different layers.

Multi-layered systems can also be developed to increase the thickness and, consequently, the storage capacity without reducing the applied field that is needed for the buildup of the photorefractive effect. Preliminary results of stratified-volume systems have been demonstrated recently.[117] Furthermore, the high processing facilities of polymers opens the possibility to give them different shapes and fabricate rods and fibers.

C. OUTLOOK

Despite the high efficiencies of existing photorefractive polymers, a lot of effort is still needed to make faster materials and reduce the values of the electric field that need to be applied in order to make devices suitable for technological applications. However, owing to the progress that is accomplished in the synthesis of new molecules with better photoconducting and higher nonlinear optical properties, the future of photorefractive polymers looks bright. Photorefractive polymers can gain from the progress that is achieved simultaneously in the fields of organic photoconductors and electro-optic polymers. Besides material development, the understanding and the modeling of the physics of these new materials plays an important role as shown by the evidence of the anisotropy of the polarizability in the total diffraction efficiency of low T_g materials, for instance. There is a strong need to understand the structure–property relationship better in amorphous organic materials. At present, there exists no real model for photorefractivity in organic polymers and more fundamental studies will have to be carried out in order to lead to the development of a theoretical model of photorefractivity in organics. Furthermore, photorefractive polymers can be used as model materials for fundamental studies of more general physical processes. For instance, the superposition of a spatially modulated electric field and a uniform electric field opens new opportunities to study higher order effects with low-power laser sources. Their high efficiency provides good sensitivity for such fundamental investigations. Finally, the investigation of the photorefractive properties of polymers can offer a great opportunity to evidence new effects[118] such as photoisomerization gratings[119–121] that also give rise to reversible and efficient refractive index modulations.

ACKNOWLEDGMENTS

The authors would like to acknowledge Sandalphon and B. L. Volodin for valuable contributions to this work. They appreciated collaborations with Drs. H. K. Hall Jr., S. R. Lyon, and A. B. Padias who have been involved in this work since the early developments of photorefractive polymers. Fruitfull discussions with Dr. P. M. Allemand are also acknowledged. This work has been supported by the U.S. Air Force Office for Scientific Research and by the U.S. National Science Foundation. B. K. is also Chargé de Recherches at the CNRS (IPCMS, Strasbourg, France) and K. M. thanks the Deutsche Forschungsgemeinschaft for partial support for a postdoctoral fellowship.

REFERENCES

1. **Ashkin, A., Boyd, G. D., Dziedzic, J. M., Smith, R. G., Ballmann, A. A., Nassau, K.,** *Appl. Phys. Lett.*, 9, 72, 1966.
2. **Chen, F. S.,** *J. Appl. Phys.*, 38, 3418, 1967.
3. **Chen, F. S.,** *J. Appl. Phys.*, 40, 3389, 1969.

4. **Amodei, J. J.,** *Appl. Phys. Lett.,* 18, 22, 1971.
5. **Günter, P., Huignard, J.-P.,** *Photorefractive Materials and Their Applications,* Vols. I and II, Springer-Verlag, Berlin, 1988, 1989.
6. **Staebler, D. L. and Amodei, J. J.,** *J. Appl. Phys.,* 43, 1042, 1972.
7. **Marrakchi, A. and K., Rastani,** in *Photonic Switching and Interconnects,* Marrakchi, A., Ed., Marcel Dekker, New York, 1994.
8. **Moerner, W. E. and Silence, S. M.,** *Chem. Rev.,* 94, 127, 1994.
9. **Sutter, K., Hulliger, J., and Günter, P.,** *Solid State Commun.,* 74, 867, 1990.
10. **Sutter, K. and Günter, P.,** *J. Opt. Soc. Am. B,* 7, 2274, 1990.
11. **Ducharme, S., Scott, J.C., Twieg, R. J., and Moerner, W. E.,** Postdeadline paper OSA annual meeting, Boston, November 5–9, 1990.
12. **Ducharme, S., Scott, J. C., Twieg, R. J., and Moerner, W. E.,** *Phys. Rev. Lett.,* 66, 1846, 1991.
13. **Tamura, K., Padias, A. B., Hall, H. K., Jr., and Peyghambarian, N.,** *Appl. Phys. Lett.,* 60, 1803, 1992.
14. **Kippelen, B., Tamura, K., Peyghambarian, N., Padias, A. B., and Hall, H. K., Jr.,** *Phys. Rev. B,* 48, 10710, 1993.
15. **Kippelen, B., Tamura, K., Peyghambarian, N., Padias, A. B., and Hall, H. K., Jr.,** *J. Appl. Phys.,* 74, 3617, 1993.
16. **Yu, L., Chan, W., Bao, Z., Cao, S. X. F.,** *Macromolecules,* 26, 2216, 1993.
17. **Silence, S. M., Walsh, C. A., Scott, J. C., Matray, T. J., Twieg, R. J., Hache, F., Bjorklund, G. C., and Moerner, W. E.,** *Opt. Lett.,* 17, 1107, 1992.
18. **Silence, S. M., Walsh, C. A., Scott, J. C., and Moerner, W. E.,** *Appl. Phys. Lett.,* 61, 2967, 1992.
19. **Zhang, Y., Cui, Y., and Prasad, P. N.,** *Phys. Rev. B,* 46, 9900, 1992.
20. **Cui, Y., Zhang, Y., Prasad, P. N., Schildkraut, J. S., and Williams, D. J.,** *Appl. Phys. Lett.,* 61, 2132, 1992.
21. **Donckers, M. C. J. M., Silence, S. M., Walsh, C. A., Hache, F., Burland, D. M., Moerner, W. E., and Twieg, R. J.,** *Opt. Lett.,* 18, 1044, 1993.
22. **Kippelen, B., Sandalphon, Peyghambarian, N., Lyon, S. R., Padias, A. B., and Hall, H. K., Jr.,** *Electronics Lett.,* 29, 1873, 1993.
23. **Meerholz, K., Volodin, B., Sandalphon, Kippelen, B., and Peyghambarian, N.,** *Nature,* 371, 497, 1994.
24. **Law, K. Y.,** *Chem. Rev.,* 93, 449, 1993.
25. **Zyss, J.,** *Molecular Nonlinear Optics: Materials, Physics and Devices,* Academic Press, New York, 1994.
26. **Kuktharev, N. V.,** *Pis'ma Zh. Tekh. Fiz.,* 2, 114, 1976 (*Sov. Tech. Phys. Lett.,* 2, 438, 1976).
27. **Ye, P.,** *Introduction to Photorefractive Nonlinear Optics,* John Wiley & Sons, New York, 1993.
28. **Eichler, H. J., Günter, P., and Pohl, D. W.,** *Laser-Induced Dynamic Gratings,* Springer-Verlag, Berlin, 1986.
29. **Günter, P.,** *Phys. Rep.,* 93, 199, 1982.
30. **Roosen, G.,** *Int. J. Optoelectronics,* 4, 459, 1989.
31. **Glass, A. M., Nolte, D. D., Olson, D. H., Doran, G. E., Chemla, D. S., and Knox, W. H.,** *Opt. Lett.,* 15, 264, 1990.
32. **Nolte, D. D., Olson, D. H., Doran, G. E., Knox, W. H., and Glass, A. M.,** *J. Opt. Soc. Am. B,* 7, 2217, 1990.
33. **Nolte, D. D., Wang, Q., and Melloch, M. R.,** *Appl. Phys. Lett.,* 58, 2067, 1991.
34. **Wang, Q., Nolte, D. D., and Melloch, M. R.,** *Appl. Phys. Lett.,* 59, 256, 1991.
35. **Wang, Q., Brubaker, R. M., Nolte, D. D., and Melloch, N. R.,** *J. Opt. Soc. Am. B,* 9, 1626, 1992.
36. **Partovi, A., Glass, A. M., Olson, D. H., Zydik, G. J., Short, K. T., Feldman, R. D., and Austin, R. F.,** *Appl. Phys. Lett.,* 59, 1832, 1992.
37. **Partovi, A., Glass, A. M., Olson, D. H., Zydzik, G. J., O'Bryan, H. M., Chin, T. H., and Knox, W. H.,** *Appl. Phys. Lett.,* 62, 464, 1993.
38. **Kukhtarev, N. V., Markov, V. B., and Odulov, S. G.,** *Opt. Commun.,* 23, 338, 1977.
39. **Kukhtarev, N. V., Markov, V. B., Odulov, S. G., Soskin, M. S., and Vinetskii, V. L.,** *Ferroelectrics,* 22, 949, 1979.
40. **Vinetskii, V. L. and Kukharev, N. V.,** *Fiz. Tverd. Tela (Leningrad),* 16, 3714, 1975 (*Sov. Phys. Solid State,* 16, 2414, 1975).
41. **Vinetskii, V. L. and Kukharev, N. V.,** *Kvant. Elektron. (Moscow),* 5, 405, 1978 (*Sov. J. Quantum Electron.,* 8, 231, 1978).
42. **Meredith, G. R., Van Dusen, J. G., and Williams, D. J.,** *Macromolecules,* 15, 1385, 1982.
43. **Onsager, L.,** *Phys. Rev.,* 54, 554, 1938.
44. **Mozumder, A.,** *J. Chem. Phys.,* 60, 4300, 1974.
45. **Borsenberger, P. M. and Ateya, A. I.,** *J. Appl. Phys.,* 49, 4035, 1978.
46. **Mort, J.,** *Adv. Phys.,* 29, 367, 1980.
47. **Stolka, M.,** in *Encyclopedia of Polymer Science and Engineering,* Vol. 11, Mark, H. F. and Kroschwitz, J. I., Eds., John Wiley & Sons, New York, 1988, 154.
48. **Scher, H. and Montroll, E. W.,** *Phys. Rev. B,* 12, 2455, 1975.
49. **Schmidlin, S. W.,** *Phys. Rev. B,* 16, 2362, 1977.
50. **Haarer, D.,** in *Frontiers of Polymer Research,* Prasad, P. N. and Nigam, J. K., Eds., Plenum Press, New York, 1991, 297.
51. **Van der Auweraer, M., de Schryver, F. C., Borsenberger, P. M., and Bässler, H.,** *Adv. Mat.,* 6, 199, 1994.
52. **Gill, W. D.,** *J. Appl. Phys.,* 43, 5033, 1972.
53. **Fujino, M., Mikawa, H., and Yokoyama, M.,** *J. Non-Cryst. Solids,* 64, 163, 1984.

54. **Williams, D. J.,** in *Nonlinear Optical Properties of Organic Molecules and Crystals,* Chemla, D. S. and Zyss, J., Academic Press, New York, 1987.

55. **Singer, K. D., Kuzyk, M. G., and Sohn, J. E.,** *J. Opt. Soc. Am. B,* 4, 968, 1987.

56. **Burland, D. M., Miller, R. D., and Walsh, C. A.,** *Chem. Rev.,* 94, 31, 1994.

57. **Kuzyk, M. G., Moore, R. C., and King, L. A.,** *J. Opt. Soc. Am. B,* 7, 64, 1990.

58. **Boyd, G. T., Francis, C. V., Trend, J. E., and Ender, D. A.,** *J. Opt. Soc. Am. B,* 8, 887, 1991.

59. **Boyd, R. W.,** *Nonlinear Optics,* Academic Press, Boston, 1992.

60. **Wu, J. W.,** *J. Opt. Soc. Am. B,* 8, 142, 1991.

61. **Moerner, W. E., Silence, S. M., Hache, F., and Bjorklund, G. C.,** *J. Opt. Soc. Am. B,* 11, 320, 1994.

62. **Prasad, P. N. and Williams, D. J.,** *Introduction to Nonlinear Optical Effects in Molecules and Polymers,* John Wiley & Sons, New York, 1991.

63. **Nicoud, J. F. and Twieg, R. J.,** in *Nonlinear Optical Properties of Organic Molecules and Crystals,* Chemla, D. S. and Zyss, J., Eds., Academic Press, New York, 1987.

64. **Matsuzawa, N. and Dixon, D. A.,** *J. Phys. Chem.,* 96, 6222, 1992.

65. **Cheng, L. T., Tam, W., Stevenson, S. H., Meredith, G. R., Rikken, G., and Marder, S. R.,** *J. Phys. Chem.,* 95, 10631, 1991.

66. **Cheng, L. T., Tam, W., Marder, S. R., Stiegman, A. E., Rikken, G., and Spangler, C. W.,** *J. Phys. Chem.,* 95, 10643, 1991.

67. **Oudar, J. L. and Chemla, D. S.,** *J. Chem. Phys.* 66, 2664, 1977.

68. **Kogelnik, H.,** *Bell Syst. Tech. J.,* 48, 2909, 1969.

69. **Boyd, G. D. and Kleinmann, G. A.,** *J. Appl. Phys.,* 39, 3597, 1968.

70. **Feinberg, J., Heiman, D., Tanguay, A. R., Jr., and Hellwarth, R. W.,** *J. Appl. Phys.,* 5, 1297, 1980.

71. **Feinberg, J. and MacDonald, K. R.,** in *Photorefractive Materials and Their Applications,* Vol. II, Günter, P. and Huignard, J.-P., Eds., Springer-Verlag, Berlin, 1989.

72. **Arfken, G.,** *Mathematical Methods for Physicists,* Academic Press, Boston, 1985.

73. **Kuzyk, M. G., Sohn, J. E., and Dirk, C. W.,** *J. Opt. Soc. Am. B,* 7, 842, 1990.

74. **Vinetskii, V. L., and Kukharev, N. V., Odulov, S. G., and Soskin, M. S.,** *Usp. Fiz. Nauk,* 129, 113, 1979 (*Sov. Phys. Usp.,* 22, 742, 1979).

75. **Huignard, J.-P. and Marrakchi, A.,** *Opt. Commun.,* 38, 249, 1981.

76. **Valley, G.,** *J. Opt. Soc. Am. B,* 1, 868, 1984.

77. **Refregier, Ph., Solymar, L., Rajbenbach, H., and Huignard, J.-P.,** *J. Appl. Phys.,* 58, 45, 1985.

78. **Stepanov, S. I. and Petrov, M. P.,** *Opt. Commun.,* 53, 292, 1985.

79. **Schildkraut, J. S.,** *Appl. Phys. Lett.,* 58, 340, 1990.

80. **Schildkraut, J. S. and Cui, Y.,** *J. Appl. Phys.,* 72, 5055, 1992.

81. **Walsh, C. A. and Moerner, W. E.,** *J. Opt. Soc. Am. B,* 9, 1642, 1992.

82. **Ducharme, S., Jones, B., Takacs, J. M., and Zhang, L.,** *Opt. Lett.,* 18, 152, 1993.

83. **Kawakami, T. and Sonoda, N.,** *Appl. Phys. Lett.,* 62, 2167, 1993.

84. **Sansone, M. J., Teng, C. C., East, A. J., and Kwiatek, M. S.,** *Opt. Lett.,* 18, 1400, 1993.

85. **Sandalphon, Kippelen, B., Meerholz, K., Volodin, B., and Peyghambarian, N.,** Postdeadline paper, ILS-IX, Toronto, Canada, October 3–8, 1993.

86. **Volodin, B., Meerholz, K., Sandalphon, Kippelen, B., and Peyghambarian, N.,** *SPIE Proc.,* 2144, 1994.

87. **Meerholz, K., Sandalphon, Volodin, B., Kippelen, B., and Peyghambarian, N.,** *CLEO Proceedings,* Vol. 8, Anaheim, CA, 1994, 34.

88. **Roosen, G.,** private communication, 1994.

89. **Silence, S. M., Scott, J. C., Hache, F., Ginsburg, E. J., Jenkner, P. K., Miller, R. D., Twieg, R. J., and Moerner, W. E.,** *J. Opt. Soc. Am. B,* 10, 2306, 1993.

90. **Sutter, K., Hulliger, J., Schlesser, R., and Gunter, P.,** *Opt. Lett.,* 18, 778, 1993.

91. **Chemla, D. S. and Zyss, J.,** Eds., *Nonlinear Optical Properties of Organic Molecules and Crystals,* Academic Press, New York, 1987.

92. **Burland, D. M., Ed.,** *Optical Nonlinearities in Chemistry, Chem. Rev.,* 94, 1994.

93. **Ikeda, M.,** *J. Phys. Soc. Japan,* 60, 2031, 1991.

94. **Müller-Horsche, E., Haarer, D., and Scher, H.,** *Phys. Rev. B,* 35, 1273, 1987.

95. **Silence, S. M., Donckers, M. C. J. M., Walsh, C. A., Burland, D. M., Twieg, R. J., and Moerner, W. E.,** *Appl. Opt.,* 33, 2218, 1994.

96. **Scott, J. C., Pautmeir, Th., and Moerner, W. E.,** *J. Opt. Soc. Am. B,* 9, 2059, 1992.

97. **Kippelen, B., Sandalphon, Peyghambarian, N., Lyon, S. R., Padias, A. B., and Hall, H. K., Jr.,** in *Organic Thin Films for Photonic Applications Technical Digest 1993,* Vol. 17, Optical Society of America, Washington D. C., 1993, 228.

98. **Singer, K. D., Kuzyk, M. G., Holland, W. R., Sohn, J. E., Lalama, S. J., Comizolli, R. B., Katz, H. E., and Schilling, M. L.,** *Appl. Phys. Lett.,* 53, 1800, 1988.

99. **Teng, C. C. and Man, H. T.,** *Appl. Phys. Lett.,* 56, 1734, 1990.

100. **Schildkraut, J. S.,** *Appl. Opt.*, 29, 2839, 1990.
101. **Poga, C., Kuzyk, M. G., and Dirk, C. W.,** *J. Opt. Soc. Am. B*, 11, 80, 1994.
102. **Levy, Y., Dumont, M., Chastaing, E., Robin, P., Chollet, P. A., Gadret, G., and Kajzar, F.,** *Nonlinear Optics*, 4, 1, 1993.
103. **Yariv, A.,** *Quantum Electronics*, John Wiley & Sons, New York, 1989.
104. **Weiser, G.,** *Phys. Stat. Sol. A*, 18, 347, 1973.
105. **Halvorson, C., Kraabel, B., Heeger, A. J., Volodin, B. L., Meerholz, K., Sandalphon, and Peyghambarian, N.,** *Opt. Lett.*, 20, 76 1995.
106. **Gabor, D.,** *Nature*, 161, 777, 1948.
107. **Smith, H. M.,** *Holographic Recording Materials, Topics in Applied Physics*, Vol. 20, Springer-Verlag, New York, 1977.
108. **Manivannan, G. and Lessard, R. A.,** *Trends in Polymer Science*, 2, 282, 1994.
109. **Liphard, M., Goonesekera, A., Jones, B. E., Ducharme, S., Takacs, J. M., and Zhang, L.,** *Science*, 263, 367, 1994.
110. **Khoo, I. C., Li, H., and Liang, Y.,** *Opt. Lett.*, 19, 1723, 1994.
111. **Peng, Z., Bao, Z., and Yu, L.,** *J. Am. Chem. Soc.*, 116, 6003, 1994.
112. **Bässler, H.,** *Adv. Mat.*, 5, 662, 1993.
113. **Silence, S. M., Twieg, R. J., Bjorklund, G. C., and Moerner, W. E.,** *Phys. Rev. Lett.*, 73, 2047, 1994.
114. **Yu, L., Chen, Y., Chan, W. K., and Peng, Z.,** *Appl. Phys. Lett.*, 64, 2489, 1994.
115. **Sessler, G. M., Ed.,** *Electrets*, Springer-Verlag, Berlin, 1987.
116. **Umeda, M., Nimi, T., Hashimoto, M.,** *Jpn. J. Appl. Phys.*, 29, 2746, 1990.
117. **Stankus, J. J., Silence, S. M., Moerner, W. E., and Bjorklund, G. C.,** *Opt. Lett.*, 19, 1480, 1994.
118. **Silence, S. M., Donckers, M. C. J. M., Walsh, C. A., Burland, D. M., Moerner, W. E., and Twieg, R. J.,** *Appl. Phys. Lett.*, 64, 712, 1994.
119. **Natansohn, A., Xie, S., and Rochon, P.,** *Macromolecules*, 25, 5531, 1992.
120. **Sekkat, Z. and Dumont, M.,** *Appl. Phys. B*, 54, 486, 1992.
121. **Sandalphon, Kippelen, B., Peyghambarian, N., Lyon, S. R., Padias, A. B., and Hall, H. K., Jr.,** *Opt. Lett.*, 19, 68, 1994.
122. **Morita, R., Ogasawara, N., Umegaki, S., and Ito, R.,** *Jpn. J. Appl. Phys.*, 26, L1711, 1987.
123. **Allen, S., McLean, T. D., Gordon, P. F., Bothwell, B. D., Hursthouse, M. B., and Karaulov, S. A.,** *J. Appl. Phys.*, 64, 2583, 1988.

Chapter 9

Computational Evaluation of Third-Order Optical Nonlinearities

Hari Singh Nalwa

CONTENTS

I. INTRODUCTION

Computational methods are useful in establishing the relationship between organic structures and nonlinear optical properties. Many quantum chemistry approaches have been applied to calculate the polarizability and hyperpolarizabilities of organic molecules. Theoretical calculations can be used to estimate the dipole moment and derivatives of energy for individual structures. They can also provide first-hand information about the relationship between the magnitude of the polarizability and hyperpolarizabilities in accordance with molecular structures. In addition, theoretical results can be used as guidelines for the interpretation of experimental data. The magnitude of optical nonlinearity can be tailored by molecular engineering by analyzing the structure–property relationships. The third-order susceptibility values obtained by using different quantum chemistry approximations may differ significantly from each other. Although there have been reports of polarizability and hyperpolarizability values that were close to experimental values, their sufficient reliability remains a concern among theoreticians and experimentalists due to the possible error in molecular considerations. We have collected theoretical and experimental data on the polarizabilities and hyperpolarizabilities of a wide variety of organic materials. These materials were obtained from many different sources in order to understand structure–

property relationships. For example, one-dimensional π-conjugated systems have been investigated to theoretically estimate their third-order optical nonlinearities and to compare with experimental results. A variety of modifications to quantum chemistry approaches have been applied to evaluate precisely the magnitude of optical nonlinearities and to compare theoretical and experimental results. Such theoretical data will be presented to analyze the accuracy of techniques in correlating how the values of individual molecules fluctuate during molecular considerations and calculations.

1. Empirical methods
    a. Pariser-Parr-Pople (PPP) method
    b. Extended Huckel (EH) method
2. Semiempirical methods
    a. Complete neglect of differential overlap (CNDO) method
    b. Modified neglect of diatomic overlap (MNDO) method
    c. Intermediate neglect of differential overlap (INDO) method
3. *Ab initio* method

Table 1 lists the merits and demerits of quantum chemistry approaches that have been used in calculating polarizability and hyperpolarizabilities of organic molecules and polymers. It is worth mentioning that the theoretical formulations for molecules vary from one approach to another and lead to different values of polarizabilities. The polarizabilities of a single molecule may change by an order of magnitude or more depending upon the theoretical methods. A clear picture of molecular consideration can be obtained while viewing the parameters of empirical, semiempirical, and *ab initio* approaches. This chapter reviews the structure–property relationships between organic structure and their optical nonlinearities. The effects of atoms, bonds, π-conjugation chain length, electron donor–acceptor substituents, heteroatoms incorporated into conjugated systems, conformational changes, and dimensionality on polarizabilities and hyperpolarizabilities of organic molecules are discussed.

Throughout the text, we use different units of polarizability and hyperpolarizabilities to avoid confusion; therefore, the following conversion factors of atomic units (au) to electrostatic units (esu) and international system (SI) units should be considered.[1]

Dipole moment (μ) = 1 au = 2.542 Debye = 8.478×10^{-30} C m.
Polarizability (α) = 1 au = 0.148176×10^{-24} esu = 0.164867×10^{-40} $C^2m^2J^{-1}$
First hyperpolarizability (β) = 1 au = 0.863993×10^{-32} esu = 0.320662×10^{-52} $C^3m^3J^{-2}$
Second hyperpolarizability (γ) = 1 au = 0.503717×10^{-39} esu = 0.623597×10^{-64} $C^4m^4J^{-3}$

II. EMPIRICAL METHODS

A. PERTURBATION THEORY (PT-HUCKEL AND VPT-PARISER-PARR-POPLE METHODS)

McIntyre and Hameka[2] calculated second hyperpolarizabilities of different classes of organic materials by using the variation perturbation theory (VPT)-PPP, PT-Huckel, and a combination of PPP and

Table 1 A Comparison of Quantum Chemistry Approaches Used for Evaluating Polarizability and Hyperpolarizabilities

Parameters	Empirical (Huckel)	Semiempirical (CNDO)	Ab initio
1. Elements consideration	Molecules containing only C, N, O, H	Molecules containing C, N, O, H, S, P, Li, F, Cl, Fe, Ni, Pd	All elements possible
2. Computational time	Short-time CPU (easy)	Short-time CPU (easy)	Long-time CPU (difficult)
3. Electronic contribution	π electrons	σ + π electrons	All electrons
4. Precision	Moderate	Molecule-dependent	High precision
5. Size of calculations	~5,000 electrons (rough estimate)	~1,000 electrons (modified program)	~300–400 electrons
6. γ Calculations	Finite field (FF) method	Time-dependent perturbation (SOS)	FF, CPHF, time-dependent CPHF
7. Polymers	Suitable	Suitable	Most suitable

extended Huckel theory. In this method, all π electrons contributions are given by a PPP theory, whereas the σ and σ-π contributions are presented by the extended Huckel theory. The used perturbation approach for conjugated hydrocarbon is different since aromatic hydrocarbons have one or more degenerate Huckel orbitals. Therefore, the treatment is limited to perturbation theory of nondegenerate eigenstates. The perturbation treatment of molecular eigenstates by expanding in terms of other molecular eigenstates was considered. The following equation is valid only for the ground-state molecular energy E_0 in the nondegenerate.

$$E_0(X^4) = -\sum_{K\neq0}\sum_{L\neq0}\sum_{M\neq0} (E_r - E_0)^{-1}(E_L - E_0)^{-1}(E_M - E_0)^{-1}V_{0K}V_{KL}V_{LM}V_{M0}$$

$$+ \left[\sum_{K\neq0}(E_K - E_0)^{-1}V_{0K}V_{K0}\right]\left[\sum_{L\neq0}(E_L - E_0)^{-2}V_{0L}V_{L0}\right] \tag{1}$$

The molecular eigenstates K, L, and M and the energies E_K, E_L, and E_M are the unperturbed molecular energy eigenvalues. The matrix element V_{KL} is written by

$$V_{KL} = \left\langle K\left|\sum_i v(i)\right|L\right\rangle - \left\langle 0\left|\sum_i v(i)\right|0\right\rangle \tag{2}$$

because the perturbation is a sum of one-electron perturbation terms.

The $E_0(X^4)$ can be separated into three contributing parts

$$E_0(X^4) = -A(X^4) - B(X^4) + C(X^4) \tag{3}$$

where $A(X^4)$ is the contribution from the singly excited configurations L to the triple sum, $B(X^4)$ is the contribution from the doubly excited configuration, and $C(X^4)$ is the contribution that involves single excited configuration K and L only, which can be written as[2]

$$A(X^4) = 2\sum_i\sum_l\sum_j\sum_k (\epsilon_1 - \epsilon_i)^{-1}(\epsilon_1 - \epsilon_j)^{-1}(\epsilon_1 - \epsilon_k)^{-1}v_{i,l}v_{i,k}v_{k,j}v_{j,i}$$

$$+ 2\sum_i\sum_l\sum_m\sum_n (\epsilon_l - \epsilon_i)^{-1}(\epsilon_m - \epsilon_i)^{-1}(\epsilon_m - \epsilon_i)^{-1}v_{i,l}v_{l,m}v_{m,n}v_{n,i}$$

$$- 2\sum_i\sum_l\sum_j\sum_m (\epsilon_i - \epsilon_l)^{-1}(\epsilon_l - \epsilon_j)^{-1}(\epsilon_m - \epsilon_j)^{-1}v_{l,i}v_{l,m}v_{m,i}v_{j,i}$$

$$- 2\sum_i\sum_l\sum_j\sum_m (\epsilon_l - \epsilon_i k)^{-1}(\epsilon_m - \epsilon_i)^{-1}(\epsilon_n - \epsilon_j)^{-1}v_{i,l}v_{l,m}v_{m,j}v_{j,i}. \tag{4}$$

The summations over i, j, and k, and l, m, and n are taken over all molecular orbitals that are occupied in the molecular ground state and all other excited molecular orbitals, respectively. The excitation energies may be given as differences of molecular orbital energies $\epsilon_l - \epsilon_j$. The contribution from the doubly excited molecular configurations L to the triple sum is written by[2]

$$B(X^4) = 4\sum_i\sum_l\sum_j\sum_m (\epsilon_l - \epsilon_i)^{-1}(\epsilon_l - \epsilon_i)^{-1}(\epsilon_l + \epsilon_m - \epsilon_i - \epsilon_j)^{-1}v_{i,l}v_{l,i}v_{j,m}v_{m,j}$$

$$+ 4\sum_i\sum_l\sum_j\sum_m (\epsilon_l - \epsilon_i)^{-1}(\epsilon_m - \epsilon_j)^{-1}(\epsilon_l + \epsilon_m - \epsilon_l - \epsilon_j)^{-1}v_{i,l}v_{l,i}v_{j,m}v_{m,j}$$

$$- 2\sum_i\sum_l\sum_j\sum_m (\epsilon_l - \epsilon_i)^{-1}(\epsilon_i - \epsilon_j)^{-1}(\epsilon_l + \epsilon_m - \epsilon_i - \epsilon_j)^{-1}v_{i,l}v_{l,j}v_{j,m}v_{m,l}$$

$$- 2\sum_i\sum_l\sum_j\sum_m (\epsilon_l - \epsilon_i)^{-1}(\epsilon_m - \epsilon_i)^{-1}(\epsilon_l + \epsilon_m - \epsilon_i - \epsilon_j)^{-1}v_{i,l}v_{l,j}v_{j,m}v_{m,l} \tag{5}$$

and transformation of the remaining term give

$$C(X^4) = 4 \sum_i \sum_l \sum_j \sum_m (\epsilon_l - \epsilon_i)^{-1}(\epsilon_m - \epsilon_j)^{-2} v_{i,l} v_{l,i} v_{j,m} v_{m,j} \tag{6}$$

By substituting contributing terms, the fourth-order energy perturbation $E_0(X^4)$ due to a homogenous electric field can be obtained. Third-order nonlinear susceptibilities were obtained by

$$\chi^{(3)}_{xxxx} = -4E_0(X^4)(e^4\alpha^4/\beta^3) \tag{7}$$

$$\gamma_{xxxx} = -4E_0(x^4)(e^4\alpha^4/\beta^3) \tag{8}$$

where e is the electron charge, α is the carbon–carbon bond length, and the β is the Huckel resonance integral. The $E_0(X^4)$ and $E_0(x^4)$ are the perturbation energies. The sign of γ is the same as the sign of $E_0(x^4)$. The β is negative. Third-order nonlinear susceptibilities can be given by

$$\chi^{(3)}_{xxxx} = 4E_0(X^4)\gamma_0, \qquad \gamma = 4E_0(x^4)\gamma_0$$

$$\gamma_0 = (e^4\alpha^4/\beta^3) = 3.5 \times 10^{-36} \text{ esu} \tag{9}$$

where α and β were taken as 1.40 Å^3 and -0.192 au, respectively. Using these formulations, McIntyre and Hameka[2] calculated second hyperpolarizabilities of a variety of conjugated hydrocarbon molecules. Tables 2 and 3 list the hyperpolarizabilities of conjugated hydrocarbon molecules. The γ value increases as the length of π conjugation increases. In particular, pentacene and naphthoquinodimethane-2,6 show the largest γ values. Hyperpolarizabilities of conjugated hydrocarbon chains C_nH_{n+2} for various values of n were calculated. The γ values of conjugated hydrocarbon chains increase significantly with the increase of π conjugation length. After $n > 14$, the γ values become positive and increase as the number of conjugated bonds increases in the hydrocarbon chain. A γ value of 550×10^{-36} esu was predicted for $n = 22$, which compares with the experimental γ values of trans-β-carotene.

Zamani-Khamiri et al.[3] estimated the γ value of the benzene molecule as 0.8578×10^{-36} esu by using the extended Huckel method, and the electron contribution γ_π of 0.9782, 10^{-36} esu by means of the Pariser-Parr-Pople (PPP) method. The combination of various results leads to a theoretical γ value of 1.1332×10^{-36} esu. The same set of calculations was also applied to ethylene, trans-butadiene, and trans-hexatriene.[4] The total hyperpolarizabilities of these three molecules were obtained by using the conventional Rayleigh-Schrodinger perturbation theory in combination with the extended Huckel method. In the second case, the π electron perturbation energies of three molecules were estimated by using the PPP method. More precise theoretical predictions were obtained by combining the PPP results for the π electron contribution with the Huckel data for the σ and σ-π electron contributions. Table 4 lists the results of calculated hyperpolarizability for these three molecules. The hyperpolarizabilities of a group of 16 conjugated molecules with 8 to 14 carbon atoms were also calculated by using of the PPP method.[5] The γ values of naphthoquinodimethane-2,6, octatetraene, decapentaene, dodecahexaene, and tetradecaheptaene calculated by the PPP method were found to be 278.41×10^{-36}, 68.61×10^{-36}, 185.51×10^{-36}, 413.88×10^{-36}, and 775.55×10^{-36} esu, respectively. It was also observed that naphthalene, anthracene, and cis- and trans-stilbene have larger γ values for the PPP method than for the Huckel method by a factor of 5 showing the accuracy of two methods. Furthermore, the γ values were positive.

McIntyre and Hameka[6] also calculated the nonlinear susceptibilities of nitrogen-containing heterocycles and of nitriles by considering only the π electrons of the aromatic or conjugated systems and by approximating the ground-state molecular wave function using the Huckel SCF model. Some molecules with exceptionally large nonlinearities were found. Table 5 lists the susceptibilities of a group of nitrogen-containing heterocyclic molecules. The calculated hyperpolarizability values of a group of nitrile molecules are listed in Table 6. The introduction of a nitrogen atom into the ring, or the addition of a methyl group, shows little effect on the value of the π-electron susceptibility in heterocyclic molecules. The exceptionally large susceptibilities were obtained for 2,2,6,6-tetracyanonaphthoquinodimethane, as well as for TCNQ and 1,1,5,5-tetracyanonaphthoquinodimethane, indicating utility of these nitrile molecules for third-order nonlinear optics.

Table 2 Perturbation Energies and Hyperpolarizabilities of Selected Hydrocarbon Molecules

Number	Molecule	E_0 (X^4)	Hyperpolarizability (γ)
1	Benzene	0.0104	0.0222
2	Anthracene	1.9692	1.3404
3	Pentacene	11.5676	11.7112
4	Benz-3,4-phenanthrene	1.3807	2.4550
5	Triphenylene	0.9412	2.0079
6	Dibenz-1,2,5,6-anthracene	7.0423	7.3415
7	Perylene	1.0351	5.4723
8	Coronene	3.9446	8.4148
9	Biphenyl	0.0351	0.6723
10	*para*-Terphenyl	0.0681	3.9058
11	*meta*-Terphenyl	0.4381	1.4065
12	*ortho*-Terphenyl	0.6140	1.5234
13	*trans*-Stilbene	3.6447	2.8460
14	*cis*-Stilbene	0.5896	0.9479

The γ values are expressed in terms of $\gamma_0 = 3.5 \times 10^{-36}$ esu. From McIntyre, E. F. and Hameka, H. F., *J. Chem. Phys.*, 69, 4814, 1978. With permission.
The chemical structures of these molecules are:

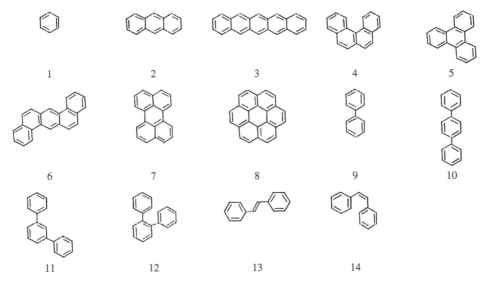

B. COUPLED SELF-CONSISTENT FIELD PERTURBATION THEORY (PPP)

DeMelo and Silbey[7] used a perturbation density matrix treatment to calculate γ values of linear conjugated chains C_nH_{2n+2} by describing a PPP Hamiltonian. The trend of γ values with increasing chain length, and the effect of different conformations such as solitons and polaron defects, were also examined. For regular polyenes and polaron chains, five independent components γ_{xxxx}, γ_{yyyy}, γ_{xxyy}, γ_{xyyy} and γ_{xxxy} exist, while for solitons only the first three components exist. For neutral solitons, the γ_{xxyy} and γ_{yyyy} components are positive and increase monotonically with the number of carbon sites (N). The γ_{xxxy} component of regular polyenes and neutral polarons is negative and increases as N increases while for large N it becomes positive. All other components γ_{xxxx}, γ_{yyyy}, γ_{xxyy}, and γ_{xyyy}, are negative and increase monotonically with increasing N. Table 7 lists the γ value of regular polyenes, neutral polarons, singly charged polarons, and bipolarons. For singly charged polarons, all five components are of negative sign and their magnitude increases monotonically with the increase of N. The γ values of neutral solitons and singly charged solitons for the largest chain are listed in Table 8.

The γ_{xxyy} component for charged solitons is positive and monotonically increases with N. For bipolarons, the γ_{xxyy} and γ_{yyyy} components are negative, but for all components the γ increases with N. The γ values strongly depend on the conjugation length. Figure 1 shows the variation of γ value for

Table 3 Perturbation Energies and Second Hyperpolarizabilities γ of Some Selected
Conjugated Hydrocarbon Molecules

Number	Molecule	E_0 (X^4)	Hyperpolarizability (γ)
1	Naphthalene	0.3895	0.2980
2	Phenanthrene	0.8032	1.2252
3	Naphthacene	5.1628	3.8970
4	Benz-1,2-anthracene	4.2812	4.1847
5	Chrysene	1.6915	2.8346
6	Pyrene	0.2242	1.2863
7	p-Benzoquinodimethane	0.0155	-16.4411
8	Naphthoquinodimethane-1,4	0.2318	-3.3613
9	o-Benzoquinodimethane	-0.6174	-0.6214
10	Naphthoquinodimethane-2,6	-282.6245	-331.4992
11	Dibenz-1,2,3,4-anthracene	4.2456	4.8531
12	Picene	4.0470	5.9224
13	Styrolene	0.0064	0.3052
14	1-Phenylbutadiene	-0.1435	1.5358
15	β,β-Naphthylethylene	2.2396	3.2028

The γ values are expressed in terms of $\gamma_0 = 3.5 \times 10^{-36}$ esu.
From McIntyre, E. F. and Hameka, H. F., *J. Chem. Phys.*, 69, 4814, 1978. With permission.
The chemical structures of the conjugated hydrocarbons mentioned above are:

Table 4 Theoretical Prediction of Hyperpolarizability in Terms of 10^{-36} esu

Method	Ethylene	*trans*-Butadiene	*trans*-Hexatriene
PPP π contribution	-0.0377	1.247	9.59
Huckel π contribution	-0.1091	1.026	-4.96
Huckel other	0.0094	0.098	0.32
Huckel total	-0.0997	1.124	-4.63
PPP + Huckel total	-0.0282	1.345	9.92

From Zamani-Khamiri, O., et al., *J. Chem. Phys.*, 72, 5906, 1980. With permission.

regular polyenes, neutral polarons, singly charged polarons, bipolarons, neutral solitons, and singly charged solitons. The results indicate that up to the $N \sim 20$ range, the onset of saturation with the chain length does not occur, either for the longitudinal γ_{xxxx} component or for the averaged polarizability γ^π. The second hyperpolarizability of conjugated chains was found to be extremely sensitive to the chain length along the chain direction.

Kavanaugh and Silbey[8] used a Morse oscillator, which has an excited state, to represent the monomer

Table 5 Perturbation Energies and Second Hyperpolarizabilities of a Group of Nitrogen-Containing Heterocyclic Molecules

Number	Molecule	E_0 (X^4)	Hyperpolarizability (γ)
1	Pyridine	0.01	0.02
2	Pyridazine	0.01	0.02
3	Pyrazine	0.00	0.02
4	Quinoline	0.40	0.31
5	Isoquinoline	0.39	0.30
6	2-Methyl quinoline	0.41	0.31
7	4-Methyl quinoline	0.41	0.32
8	6-Methyl quinoline	0.39	0.31
9	7-Methyl quinoline	0.40	0.31
10	8-Methyl quinoline	0.41	0.32
11	1,8-Naphthrydine	0.41	0.33
12	Phthalazine	0.40	0.32
13	Quninazoline	0.40	0.32
14	Quinoxaline	0.42	0.33
15	Acridine	2.02	1.46
16	Phenazine	2.22	1.60
17	Phenanthridine	0.81	1.26
18	3,4-Benzoacridine	4.37	4.38
19	9-Methyl-3,4-benzoacridine	4.51	4.52
20	5,7,9-Trimethyl-3,4-benzoacridine	4.53	4.59
21	1,2,5,6-Dibenzacridine	7.27	7.75
22	5,9-Dimethyl-1,2-benzoacridine	4.60	4.53
23	5,7,9-Trimethyl-1,2-benzoacridine	4.57	4.52
24	2,2′-Bipyridyl	0.90	0.72
25	2,3′-Bipyridyl	0.86	0.69
26	2,4′-Bipyridyl	0.90	0.72
27	3.3′-Bipyridyl	0.84	0.67
28	4,4′-Bipyridyl	0.82	0.67
29	4,4′-Biquinolyl	3.75	3.66
30	8,8′-Biquinolyl	3.83	3.56

The γ is in terms of $\gamma_0 = 3.5 \times 10^{-36}$ esu. From McIntyre, E. F., and Hameka, H. F., *J. Chem. Phys.*, 70, 2215, 1979. With permission.

The chemical structures of these heterocyclic molecules are shown below:

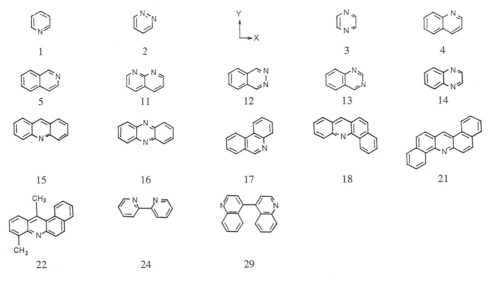

Table 6 Perturbation Energies and Second Hyperpolarizabilities of a Group of Nitrile Molecules

Number	Molecule	E_0 (X^4)	γ
1	Benzonitrile	0.02	0.42
2	Phthalonitrile	0.70	1.07
3	Pyromellitonitrile	4.71	7.50
4	4,4'-Dicyanobiphenyl	17.80	13.76
5	Tetracyanoquinodimethane (TCNQ)	−1,561.24	−1,150.35
6	9-Cyanoanthracene	−2.02	2.10
7	9-Cyanophenanthrene	0.88	2.98
8	9,9,10,10-Tetracyanoanthroquinodimethane	3.57	9.72
9	1,1,4,4-Tetracyanonaphthoquinodimethane	2.48	295.98
10	2,2,6,6-Tetracyanonaphthoquinodimethane	−10,569.65	−12,144.43
11	1,1,5,5-Tetracyanoanthroquinodimethane	35.63	−3,470.77

The γ is in terms of $\gamma_0 = 3.5 \times 10^{-36}$ esu. From McIntyre, E. F., and Hameka, H. F., *J. Chem. Phys.*, 70, 2215, 1979. With permission. The chemical structures and coordinates of the nitrile molecules are:

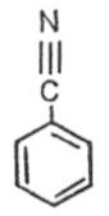

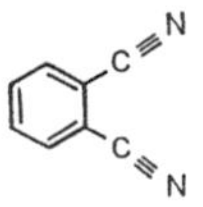

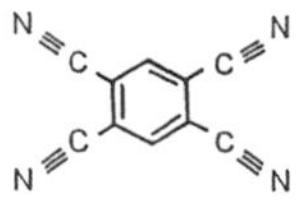

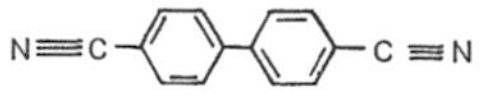

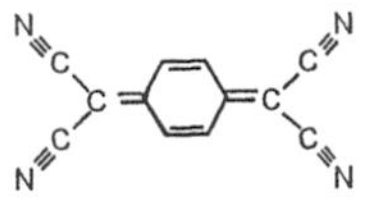

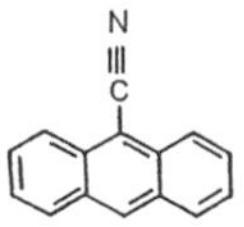

Table 7 The γ Value of Regular Polyenes, Neutral Polarons, Singly Charged Polarons, and Bipolarons for the Largest Chain ($N = 20$)

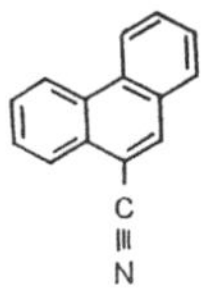

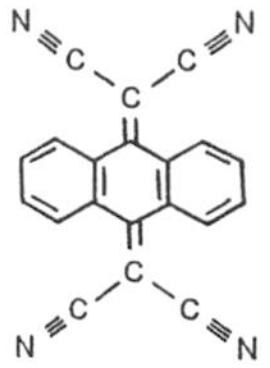

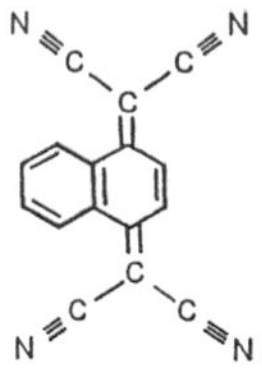

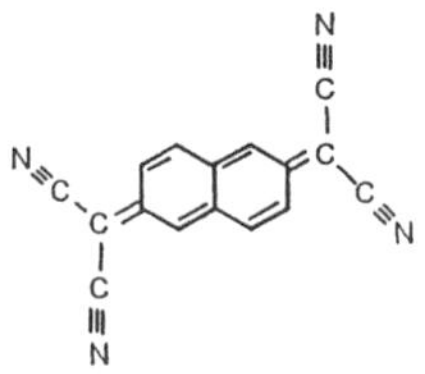

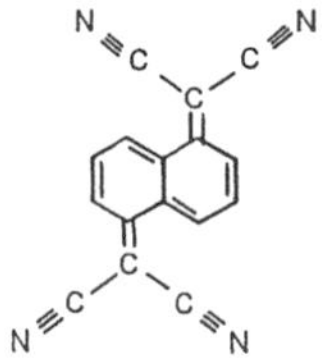

γ Component	Regular Polyens	Neutral Polarons	Singly Charged Polarons	Bipolarons
γ_{xxxx} (10^7 au)	1.379	3.745	−103.0	1.728
γ_{xxxy} (10^5 au)	3.514	6.0503	−172.2	83.26
γ_{xxyy} (10^5 au)	−1.627	−3.993	−239.7	−2.384
γ_{xyyy} (10^3 au)	−6.130	−12.80	−315.1	2.960
γ_{yyyy} (10^3 au)	−17.88	−32.38	−39.47	−4.278

From DeMelo, C. P. and Silbey, R., *J. Chem Phys.*, 88, 2567, 1988. With permission.

unit of the conjugated molecule, which decreases in separation as energy increases for the monomer. This model was applied to examine the chain-length (N) dependence of α, β, and γ. The model Hamiltonian was introduced, and the α, β, and γ for small N were described. The α, β, and γ were calculated for large N by using periodic boundary conditions and a perturbation technique. The Morse oscillator gives each molecule of the polymer a continuum of states at high energy. The Morse oscillator is not symmetric, hence gives non-zero values for dipole moment μ, as well as for α, β, and γ. To calculate the large N behavior of α, β, and γ, the model Hamiltonian is given by

Table 8 The γ Value of Neutral Solitons and Singly Charged Solitons for the Largest Chain $(N = 21)$

γ Component	Neutral Solitons	Singly Charged Solitons
γ_{xxxx} $(10^7$ au)	2.858	−5.791
γ_{xxyy} $(10^5$ au)	11.45	16.11
γ_{yyyy} $(10^3$ au)	15.01	0.9176

From DeMelo, C. P. and Silbey, R., *J. Chem Phys.*, 88, 2567, 1988. With permission.

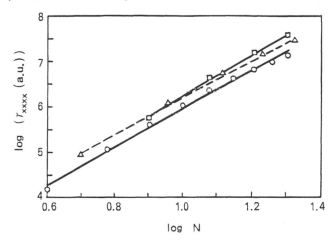

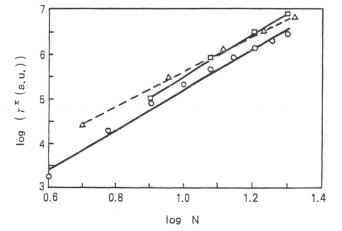

Figure 1(a) Variation of γ values for regular polyenes (○), neutral polarons (□), and neutral soliton (△), γ_{xxxx} = Longitudinal component, γ^π = Orientationally average. (From DeMelo, C. P. and Silbey, R., *J. Chem. Phys.*, **88**, 2567, 1988. With permission.)

$$H = \sum_n \left[\frac{1}{2} p_n^2 + D(1 - \exp(-ar_n))^2 \right] + K \sum_n r_n r_{n+1} + E \sum_n r_n \qquad (10)$$

where the first, second, and third terms refer to the sum of independent Morse oscillator Hamiltonians for each monomer units, coupling between monomer units and the linear coupling to the electric field, respectively. The periodic boundary conditions were used, and a perturbation method was applied, for the polarizabilities. The eigenvalue solution for a single Morse oscillator is written by

$$E_u = \left[\omega_0 \left(v + \frac{1}{2} \right) - \omega x \left(v + \frac{1}{2} \right)^2 \right] \qquad (11)$$

where ω_0 is the fundamental frequency of the oscillator, ωx is the anharmonicity, and D and α are related as

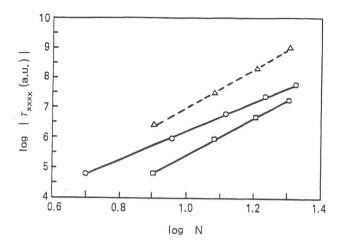

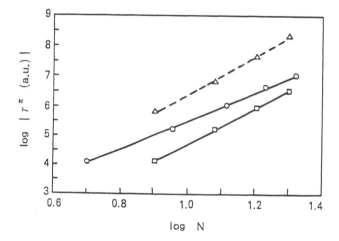

Figure 1 (b)   Variation of γ values for singly charged polarons (△), bipolarons (□), and charged solitons (○). γ_{xxxx} = Longitudinal component, γ^π = Orientationally average. (From DeMelo, C. P. and Silbey, R., *J. Chem. Phys.*, 88, 2567, 1988. With permission.)

$$\omega_0 = \alpha\sqrt{2D}, \qquad \omega x = \alpha^2/2 \tag{12}$$

A Taylor series as a function of energy E for the α, β, and γ at large N can be written by[8]

$$\frac{\alpha}{2} = \frac{N}{2\Omega_0^2} - \frac{7\omega x \omega_0^2}{4\Omega_0^4}\sum_k \frac{1}{\Omega_k} + \frac{9\omega x \omega_0^4}{8\Omega_0^4}\sum_k \frac{1}{\Omega_k^3} \tag{13}$$

$$-\frac{\beta}{6} = \frac{(2\omega x)^{1/2}}{2}\frac{\omega_0^2 N}{\omega_0^6} + \frac{(2\omega x)^{3/2}5\omega_0^2}{8\Omega_0^6}\sum_k \frac{1}{\Omega_k} - \frac{21(2\omega x)^{3/2}\omega_0^4}{16\Omega_0^6}\sum_k \frac{1}{\Omega_k^3} + \frac{27\omega_0^6(2\omega x)^{3/2}}{32\Omega_0^6}\sum_k \frac{1}{\Omega_k^3}, \tag{14}$$

$$\frac{\gamma}{24} = -\frac{7\omega x \omega_0^2 N}{12\Omega_0^8} + \frac{9\omega x N \omega_0^4}{4\Omega_0^{10}} - \frac{31\omega x^2 \omega_0^2}{24\Omega_0^8}\sum_k \frac{1}{\Omega_k} + \frac{109\omega x^2 \omega_0^4}{16\Omega_0^8}\sum_k \frac{1}{\Omega_k^3} - \frac{189\omega_0^6\omega x^2}{16\Omega_0^8}\sum_k \frac{1}{\Omega_k^5} \tag{15}$$

$$+ \frac{405\omega_0^8\omega x^2}{64\Omega_0^8}\sum_k \frac{1}{\Omega_k^7}.$$

where Ω_k is written by

$$\Omega_k^2 = 2Da^2 + 2K\cos(k)$$

$$k = \frac{2\pi}{N}\,m, \qquad m = 0, 1, \ldots, N - 1, \tag{16}$$

where K is the coupling between the polymer units. The α, β, and γ dependence at small N is given by[8]

$$\Delta\!\left(\frac{\alpha}{2N}\right) = \frac{K}{\Omega_0^2(\Omega_0^2 N - 2K)} \tag{17}$$

$$\Delta\!\left(\frac{-\beta}{6N}\right) = \frac{3\sqrt{2}KN^2(\omega x)^{1/2}\omega_0^2}{\Omega_0^2(N\Omega_0^2 - 2K)^3} \tag{18}$$

$$\Delta\!\left(\frac{\gamma}{24N}\right) = -\frac{N^4\omega_0^2\omega x K(28\Omega_0^2 - 135\omega_0^2)}{6\Omega_0^2(\Omega_0^2 N - 2K)^5} + \frac{N^3\omega_0^2\omega x K^2 20(7\Omega_0^2 - 27\omega_0^2)}{6\Omega_0^4(\Omega_0^2 N - 2K)^5}$$

$$- \frac{40N^2 K^3 \omega x \omega_0^2(7\Omega_0^2 - 27\omega_0^2)}{6\Omega_0^6(\Omega_0^2 N - 2K)^5} + \frac{N\omega_0^2\omega x K^4 40(7\Omega_0^2 - 27\omega_0^2)}{6\Omega_0^8(\Omega_0^2 N - 2K)^5}$$

$$- \frac{16K^5\omega x \omega_0^2(7\Omega_0^2 - 27\omega_0^2)}{6\Omega_0^6(\Omega_0^2 N - 2K)^5} \tag{19}$$

The analytical values for α, β, and γ for these equations reduce to the value for an N-independent oscillator system where K is zero. The above relationship demonstrates how the α, β, *and* γ depend on chain length N, the coupling term K, the energy Ω_k, and the anharmonicity ωx. Both the large N and small N relationships indicate a direct dependence on the anharmonicity. For large N, as N grows the α, β, and γ become linearly dependent on N. The small N indicates that the chain length at which α, β, and γ become linear with N may vary with the order of the polarizabilities, coupling constant K, and the anharmonicity ωx. The small N and large N behaviors are shown in Figures 2 and 3. This model can be used to simply calculate α, β, and γ and provides results similar to more complicated calculations and experiments.

Soos and Ramasesha[9] applied a diagrammatic valence bond (DVB) theory for calculating dynamic nonlinear susceptibilities of interacting π electrons in PPP and other quantum cell models. Static and dynamic NLO coefficients for *cis* and *trans*-polyenes up to $N = 12$ carbon were calculated. Table 9 lists $\gamma_{xxxx}(\omega,\omega,\omega)$ for PPP and Huckel models of N-site polyenes at $h\omega = 0.65$ and 0.30 eV up to $N = 12$.

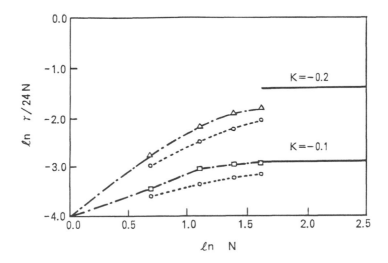

Figure 2 Dependence of hyperpolarizability on small chain length (N) using a Morse oscillator. (From Kavanaugh, T. C. and Silbey, R. J., *J. Chem. Phys.*, 95, 6924, 1991. With permission.)

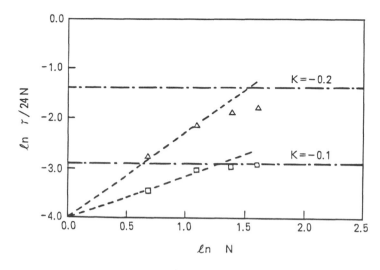

Figure 3 Dependence of hyperpolarizability on large chain length (*N*) using a Morse oscillator. (From Kavanaugh, T. C. and Silbey, R. J., *J. Chem. Phys.*, 95, 6924, 1991. With permission.)

Table 9 γ_{xxxx} (ω, ω, ω) (10^3 au) for PPP and Huckel Models of *N*-Site Polyenes at $\hbar\omega = 0.65$ or 0.30 eV

| | *trans*-Polyene | | | *cis*-Polyene |
| | PPP | | Huckel Model | PPP Model |
N	0.65 eV	0.30 eV	(0.30 eV)	(0.65 eV)
6	3.847	3.188	3.729	2.707
8	12.72	10.04	243.7	8.742
10	31.26	23.56	1049	21.61
12	63.25			44.42

From Soos, Z. G. and Ramasesha, R., *J. Chem. Phys.*, 90, 1067, 1989. With permission.

Since any perturbation may be taken into account, the DVB approach can be used for finite-dimensional models that conserve total spin.

III. SEMIEMPIRICAL METHODS

A. COUPLED SELF-CONSISTENT FIELD (SCF) PERTURBATION THEORY: COMPLETE NEGLECT OF DIFFERENTIAL OVERLAP/VERSION 2 (CNDO/2) METHOD

Papadopoulos, et al.[10,11] used the coupled SCF perturbation theory for calculating polarizabilities and hyperpolarizabilities of a number of polyene molecules. The energy of a closed-shell molecule described by a single determinant and perturbed by a uniform electric filed is written by

$$E(F) = E^{(01)} - \mu_\alpha F_\alpha - \frac{1}{2}\alpha_{\alpha\beta}F_\alpha F_\beta - \frac{1}{3!}\beta_{\alpha\beta\gamma}F_\alpha F_\beta F_\gamma - \frac{1}{4!}\gamma_{\alpha\beta\gamma\delta}F_\alpha F_\beta F_\gamma F_\delta - \cdots \quad (20)$$

where $E(F)$ and $E^{(0)}$ are the perturbed and unperturbed electronic energies, respectively. μ_α is the a component of the permanent dipole moment, $\alpha_{\alpha\beta}$ is the $\alpha\beta$ component of the polarizability, $\beta_{\alpha\beta\gamma}$ is the $\alpha\beta\gamma$ component of the first hyperpolarizability, and $\gamma_{\alpha\beta\gamma\delta}$ is the $\alpha\beta\gamma\delta$ component of the second hyperpolarizability. The α and γ components can be written as

$$\alpha_{\alpha\alpha} = -2E^{\alpha\alpha} \quad (21)$$

$$\gamma_{\alpha\alpha\alpha\alpha} = -24E^{\alpha\alpha\alpha\alpha} \quad (22)$$

$$\gamma_{\alpha\alpha\beta\beta} = -4E^{\alpha\alpha\beta\beta} \tag{23}$$

$E^{\alpha\alpha}$, $E^{\alpha\alpha\alpha\alpha}$, and $E^{\alpha\alpha\beta\beta}$ can be given in terms of x_α $y_{\alpha\alpha}$, and $y_{\alpha\beta}$ as

$$x_\alpha = \sum_{\substack{k(\text{occ}) \\ i\,(\text{usocc})}} \frac{C_k^*[f^\alpha + G(R^\alpha)]C_l}{\epsilon_k - \epsilon_l} C_k C_l^* \tag{24}$$

$$y_{\alpha\alpha} = \sum_{\substack{k(\text{occ}) \\ l(\text{usocc})}} \frac{C_k^*[h^{\alpha\alpha} + x_\alpha h^\alpha - h^\alpha x_\alpha]C_l}{\epsilon_k - \epsilon_l} C_k C_k^* \tag{25}$$

$$y_{\alpha\beta} = \sum_{\substack{k(\text{occ}) \\ l(\text{usocc})}} \frac{C_k^*(h^{\alpha\beta} + x_\alpha h^\beta + x_\beta h^\alpha - h^\alpha x_\beta)C_l}{\epsilon_k - \epsilon_l} C_k C_l^* \tag{26}$$

The C_k and ϵ_k are the kth unperturbed eigenvalue, f^α is the dipole moment matrix in the α direction, and h^α, $h^{\alpha\alpha}$, and $h^{\alpha\beta}$ are terms in the expanded h matrix. The x_α, $y_{\alpha\alpha}$, and $y_{\alpha\beta}$ are calculated iteratively until self-consistency is obtained. The average values for polarizability (α) and second hyperpolarizability (γ) are written as

$$\alpha = 1/3(\alpha_{xx} + \alpha_{yy} + \alpha_{zz}) \tag{27}$$

and

$$\gamma = 1/5(\gamma_{xxxx} + \gamma_{yyyy} + \gamma_{zzzz} + 2\gamma_{xxyy} + 2\gamma_{xxzz} + 2\gamma_{yyzz}) \tag{28}$$

The number of independent components to specify α and γ are related to the symmetry of the molecules. In the case of polyenes, all three components of α and all six components of γ were calculated. A CNDO/2 computational approach was used. The α and γ values of a variety of polyene molecules obtained by making use of this technique are listed in Table 10. Both α and γ values significantly increase as the π conjugation increases and the *trans* isomer shows larger nonlinearity than the *cis* form. Polarizabilities and hyperpolarizabilities of some aromatic molecules are shown in Table 11 and both α and γ increase with the increase in number of benzene rings. The chemical structures of these molecules have been shown earlier in the text. The different α and γ values of these aromatic molecules have been obtained in accordance with geometrical and other related considerations. Waite-and Papadopoulos[12] compared the nonlinearities of 1-nitronaphthalene and 2-nitronaphthalene which have α values of 164 and 167 a.u. and γ values of 122,000 and 129,000 au, respectively, indicating that the isomerism has little effect on the magnitude of γ values and an even smaller effect on α values. The α and γ values for some azabenzene molecules were calculated by using the PT-EB-CNDO method (Table 12). The hyperpolarizability values observed are

(i) γ(pyridazine) $>$ γ(pyrimidine) $>$ γ(pyrazine)
(ii) γ(triazine) $>$ γ(s-tetrazine) $>$ γ(s-triazine)

Hexazine indicates the effect of bonding on the magnitude of γ values. In these molecules, the part of γ is due to the interaction of nitrogen atoms which is equal to 6560 au, about twice that associated with noninteracting atoms.

Waite and Papadopoulos[13] used the CHF-PT-EB-CNDO method to study the intermolecular interactions of some formamide aggregates since amides form hydrogen bonding. Both α and γ showed an interesting effect of intermolecular interactions as the numer of molecules participating is increased. In addition, calculations were also extended to study the systematic effect of functional groups on nonlinearity. The normalized polarizabilities and hyperpolarizabilities of some multimers (dimers, trimers, and tetramers) of formamide are listed in Table 13 and the possible interactions are shown in respective figures.

Table 10 Polarizability and Hyperpolarizability Values of Some Polyene Molecules

Number	Molecule	α (au)	γ (au)
1	Ethylene	29.71	8,263
2	Cyclopropane	35.85	11,168
3	Allene	48.16	15,300
4	1,3-*cis*-butadiene	61.68	26,121
5	1,3-*trans*-butadiene	69.73	33,447
6	Butatriene	63.19	27,250
7	Cyclobutadiene	50.65	41,149
8	Cyclopentadiene	80.62	60,675
9	1,3,5-*cis*-hexatriene	120.21	65,749
10	1,3,5-*trans*-hexatriene	128.19	72,137
11	1-*cis*-octatetraene	188.77	117,299
12	3-*cis*-octatetraene	189.28	114,123
13	*cis-transoid*-octaterane	181.29	16,142
14	*trans-cisoid*-octatetraene	167.09	99,988
15	*trans*-octatetraene	203.81	124,313
16	Cyclo-octatetraene	129.67	101,494

From Papadopoulos, M. G., et al., *J. Chem. Phys.,* 77, 2527, 1982. With permission.
The α and γ values are given in atomic units. The chemical structures of polyene molecules are:

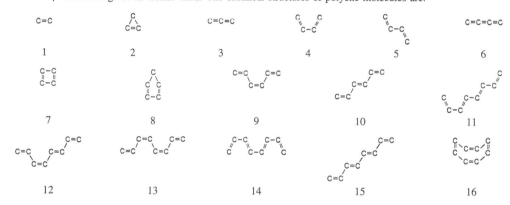

Table 11 Polarizabilities and Hyperpolarizabilities of Some Aromatic Molecules in Atomic Units

Number	Molecule	α (au)	γ (au)
1	Benzene	61.86	24,682
2	Napthalene	115.45	60,313
3	Biphenyl planar	145.12	72,835
4	Anthracene	183.72	111,226
5	Phenanthrene	173.24	101,421
6	Pyrene	205.73	134,400

From Waite, J., et al., *J. Chem. Phys.,* 77, 2536, 1982. With permission.

Table 12 Polarizabilities and Hyperpolarizabilities in Atomic Units of Azabenzene Molecules

Number	Molecule	Structure	α (au)	γ (au)
1	Pyridine		65.6	19,700
2	Pyridazine	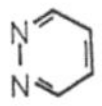	60.6	19,400
3	Pyrimidine	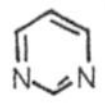	61.6	16,700
4	Pyrazine	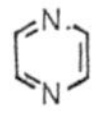	60.5	15,800
5	s-Triazine	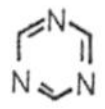	62.1	13,800
6	1,2,3-Triazine		66.0	26,300
7	s-Tetrazine		60.1	18,700
8	Hexazine		54.7	10,400

From White, J. and Papadopoulos, M. G., *J. Chem. Phys.*, 80, 3503, 1984. With permission.

Table 13 The Normalized Polarizabilities and Hyperpolarizabilities of Some Multimers

Multimer	Normalized α value	Normalized γ value
Dimer (FA)$_2$	1.0 (1.0)	1.0 (1.0)
Trimer (FA)$_3$	2.1 (1.5)	2.1 (1.5)
Tetramer (FA)$_4$	3.8 (2.0)	3.6 (2.0)

From Waite, J., and Papadopoulos, M. G., *J. Chem. Phys.*, 83, 4047, 1985. With permission.
The values shown in the parentheses are those if there was no interaction between the formamide (FA) molecules.[13]

The intermolecular interactions through hydrogen bonding among a dimer, trimer, and a tetramer are represented by:

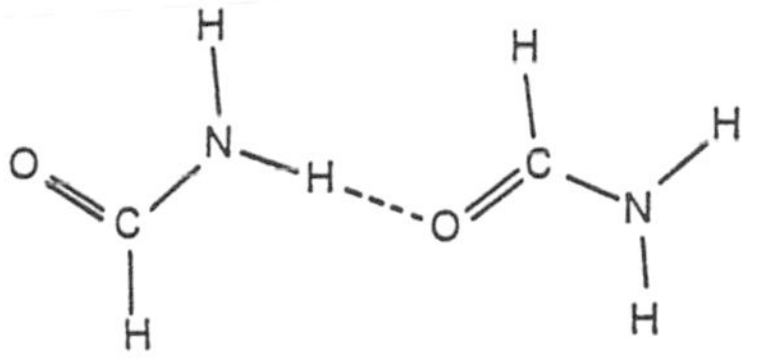

Dimer

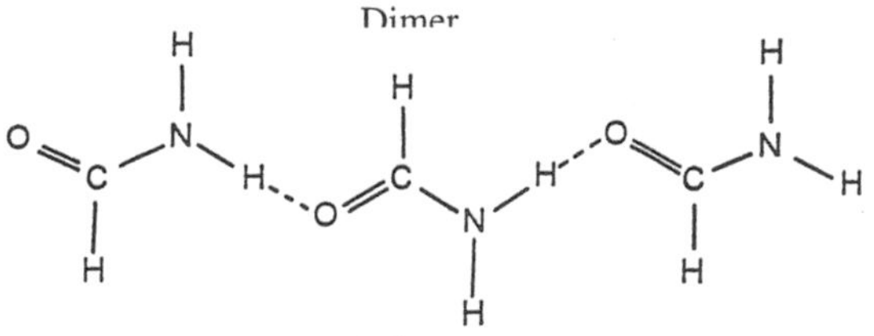

Trimer

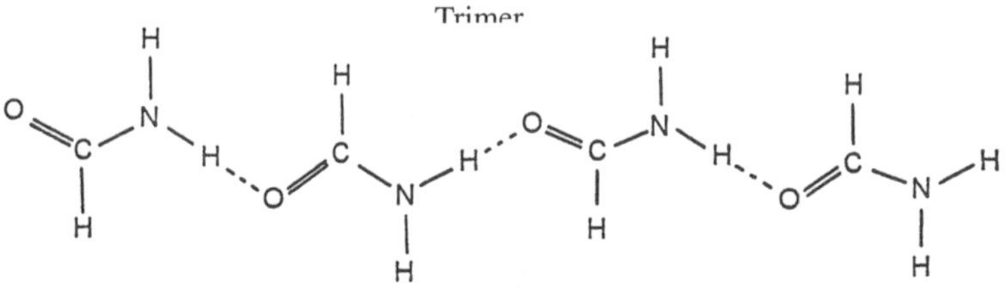

Tetramer

The study of functional groups showed the following sequences of α and γ values:

Polarizability (α): F < H < NH_2 < CH_3 < NH_2CO < CH_3NH < $(CH_3)_2N$ < $NHCH_3CO$ < $N(CH_3)_2CO$

Hyperpolarizability (γ): F < H < NH_2 < CH_3 < CH_3NH < $(CH_3)_2N$ < NH_2CO < $NHCH_3CO$ < $N(CH_3)_2CO$

In another study using the CHF-PT-EB-CNDO method, the effect of intermolecular processes (inversion and rotation) on the α and γ values in some aromatic amines has been demonstrated[1] (Table 14). Both α and γ show the effect of methylation, which was more dominant in the case of β values, since the β value increases by a factor of 30 as one goes from ammonia to trimethylammonia. A comparative study of aniline, N,N-dimethyl aniline, and 1-aminonaphthalene shows that reduction of charge-transfer leads to a reduction in β values. It was also found that the substitution of two hydrogen by two methyl groups leads to an increase in α value by 36.2% compared with a γ value increase of 25%. Waite and Papadopoulos[14] obtained polarizability of 146 au and second hyperpolarizability (γ) of 48,800 au for ferrocene, $Fe(C_5H_5)_2$, using the extended CHF-TP-EB-CNDO method.

(Ferrocene)

B. COMPLETE NEGLECT OF DIFFERENTIAL OVERLAP/SPECTROSCOPIC (CNDO/S) METHOD

Garito et al.[15] used an all-valenced electron self-consistent field (SCF) molecular orbital method in the standard, rigid lattice complete neglect of differential overlap/spectroscopic (CNDO/S) approximation. The hopping interaction between all pairs of atomic sites was included and the bond alternation was treated directly in the molecular coordinates. Garito et al.[15-18] estimated the values of γ_{ijkl} in a series of conjugated systems, applicable to conjugated polymers, and pointed out that the large $\gamma_{1111}(-\omega_4; \omega_1, \omega_2, \omega_3)$ value originates from three length-dependent factors:

1. The lowest optical excitation energy decreases as the length of conjugation increases.
2. The magnitude of the transitional dipole moments along the molecular chain axis increases with the increase of conjugation length; therefore, the larger transitional dipole moment would give rise to the larger γ values.
3. The various contributing terms of excited state help in increasing the γ values as the length of the chain increases.

Figure 4 shows the hyperpolarizability plot at a fundamental photon energy of 0.65 eV ($\lambda = 1.907$ μm) for a polyene system as a function of the number of the carbon atom sites (N) and the length for *cis*- and *trans*-polyenes.[15-18] The effect of the conformation using *trans* and *cis* forms of polyenes in which the number of the carbon atoms sites ranges from 6 to 16 was studied. The γ values of the *trans* conformation are larger than the *cis* conformation of equal conjugation length, and the percentage difference in γ values of the *trans* and *cis* conformation increases as the chain length increases. The power-law dependence yields an exponent of 4.7 for *cis* and 5.4 for *trans* conformations. The distance along the conjugation axis is shorter for *cis* conformation than for the *trans* conformation for an equal number of carbon atom sites. It was concluded that the γ values are more sensitive to the physical length of the conjugated chain than to the conformational effect. The extrapolation of power-law

Table 14 Polarizability and Hyperpolarizabilities of Ammonia and Its Methylated Derivatives in Atomic Units

Molecule	α	γ
NH_3	15.2	6,060
NH_2CH_3	25.6	5,950
$NH(CH_3)_2$	43.0	10,500
$N(CH_3)_3$	64.7	18,700

From Waite, J. and Papadopoulos, M. G., *J. Chem. Phys.*, 82, 1427, 1985. With permission.

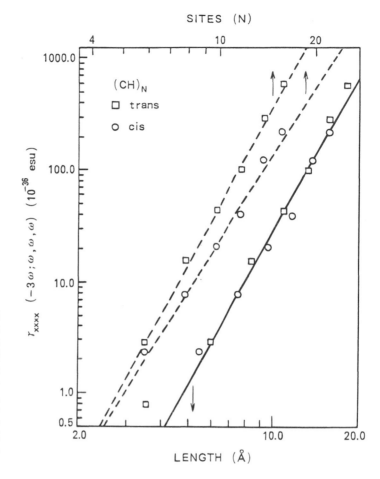

SITES (N)

Γ_{xxxx} $(-3\omega; \omega, \omega, \omega)$ $(10^{-36}$ esu$)$

$(CH)_N$
□ trans
○ cis

LENGTH (Å)

Figure 4 Hyperpolarizability plot at a fundamental photon energy of 0.65 eV ($\lambda = 1.907$ μm) for a polyene system as a function of the number of the carbon atom sites (N) and the length for a *cis*- and *trans*-polyenes. (From Helfin, J. R. and Garito, A. F. in *Electroresponsive Molecular and Polymeric Systems*, Vol. 2, Skotheim, T. A., Ed., Marcel Dekker, New York, 1991, 1–48. With permission.)

dependence of γ on chain length for the *trans*-polyenes indicates that the γ value starts saturating at 50 carbon atom sites, or at a length of about 60 Å. At 50 carbon atom sites, the γ values of the *trans* chain are about seven times larger than the *cis* conformation with a chain length of 60 Å. Authors stated that only a chain length of the order of 100 Å is sufficient for larger values of γ and any increases beyond this chain length, would have little effect.

Craig and Williams[19] calculated the hyperpolarizabilities of nickel dithiolene complexes having different substituent groups in the dithiolene ligand by the finite-field method in combination with the CNDO/S molecular orbital technique (Table 15). The substituent groups influence the polarizability and hyperpolarizabilities of nickel dithiolene complexes. Substitution of phenyl groups increases optical nonlinearity significantly and was found to be the most advantageous in molecules B and D. Furthermore, γ_{xxyy}, values of complex C are about 35,000 au larger than those of complex A because of the contribution from the benzene ring. In these complexes, large γ values arise from the interaction of the nickel 3d electrons with the conjugated π electron of the metal–ligand complex.

Table 15 Polarizabilities and Hyperpolarizabilities of Nickel Dithiolene Complexes in Atomic Units

Complex	α_{xx}	α_{yy}	α_{zz}	γ_{xxxx}	γ_{xxyy}
A	276.8	372.1	167.6	−37,865	86,305
B	559.7	658.1	246.3	12,978	330,191
C	282.1	525.0	143.0	3,334	120,769
D	439.8	700.0	158.9	51,695	239,233

From Craig, B. I. and Williams, G. R. J., *Adv. Mater., Opt. Electron.*, 1, 221, 1992. With permission.

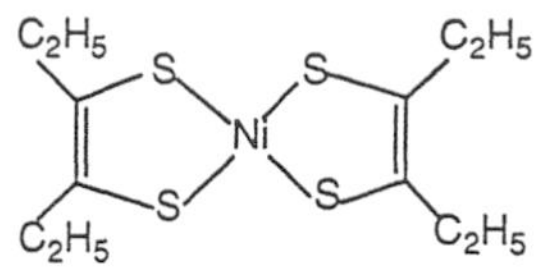

A

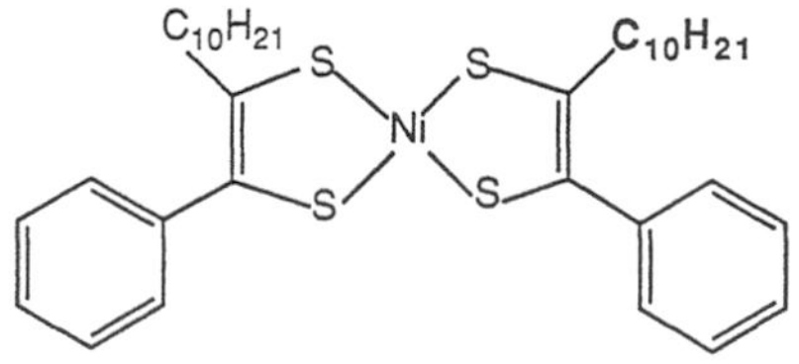

B

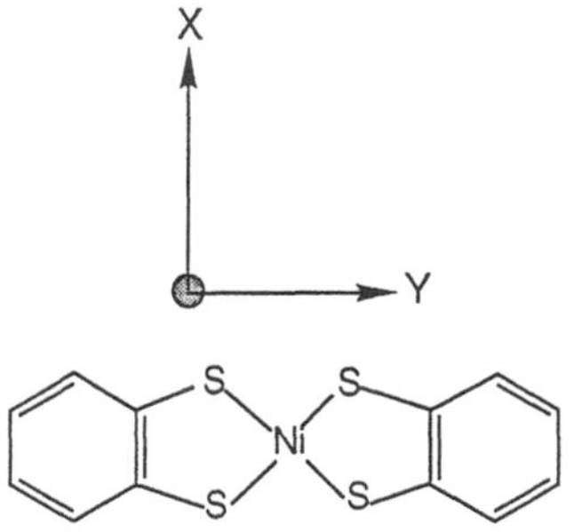

C

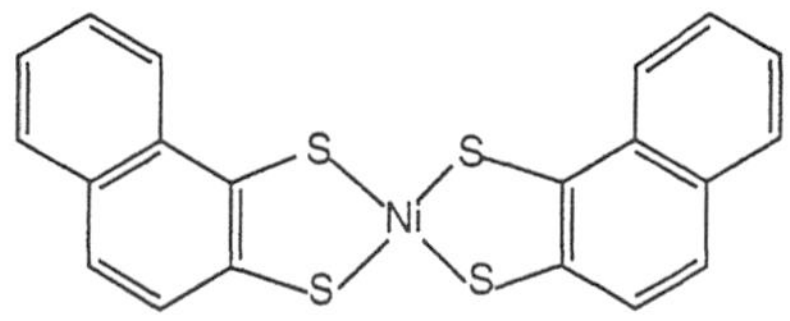

D

C. COMPLETE NEGLECT OF DIFFERENTIAL OVERLAP/SPECTROSCOPIC CONFIGURATION INTERACTION (CNDO/SCI) METHOD

Nakano et al.[20] used time-dependent perturbation theory for calculating second hyperpolarizabilities $\gamma_{ijkl}(-3\omega, \omega, \omega, \omega)$. The analytical expression is given by the following equation:

$$\gamma_{iiii}(-3\omega; \omega, \omega, \omega) = \sum_{n=1} (\mu_{n0}^i)^2 (\Delta\mu^i)^2 \frac{E_{n0}(E_{n0}^2 + (\hbar\omega)^2)}{(E_{n0}^2 - (3\hbar\omega)^2)(E_{n0}^2 - (2\hbar\omega)^2)(E_{n0}^2 - (\hbar\omega)^2)}$$

$$- \sum_{n=1} (\mu_{n0}^i)^4 \frac{E_{n0}}{(E_{n0}^2 - (3\hbar\omega)^2)(E_{n0}^2 - (\hbar\omega)^2)}$$

$$+ \sum_{\substack{m,n=1 \\ m \neq n}} (\mu_{n0}^i)^2 (\mu_{mn}^i)^2 \frac{E_{n0}^2 E_{m0} + 4E_{n0}(\hbar\omega)^2 - 3E_{m0}(\hbar\omega)^2}{(E_{n0}^2 - (3\hbar\omega)^2)(E_{n0}^2 - (\hbar\omega)^2)(E_{m0}^2 - (2\hbar\omega)^2)} \qquad (29)$$

For a static field, this equation is simplified to

$$\gamma_{iiii}(0; 0, 0, 0) = \sum_{n=1} \frac{(\mu_{n0}^i)^2 (\Delta\mu^i)^2}{E_{n0}^3} - \sum_{n=1} \frac{(\mu_{n0}^i)^4}{E_{n0}^3} + \sum_{\substack{m,n=1 \\ m \neq n}} \frac{(\mu_{n0}^i)^2 (\mu_{mn}^i)^2}{E_{n0}^2 E_{m0}}, \qquad (30)$$

where E_{n0} is the excitation energy, μ_{0n}, μ_{nm} are transition moments, and μ_{nn} refers to the difference of dipole moments between excited and ground states. The CNDO/S approximation includes the single-excitation configuration interaction (SCI), where only π electron orbitals are considered. Table 16 lists the γ_{zzzz} values of a variety of polyene molecules in a static electric field.

Among these disubstituted systems, polyenes (4) that have electron donor–acceptor groups at the opposite ends show the largest hyperpolarizability values. The γ values decrease in the order $4 > 3 > 2$. Furthermore, the γ values for C=N backbone are also larger than those of regular polyenes. The calculated γ values are positive for regular polyenes. The power law dependence on chain length shows an exponent of 4.14. The charged polyenes show negative signs for γ values and the exponent corresponds to 4.44. The exponent values were 4.10, 3.64, 2.37, 2.70, 4.53 for 1, 2, 3, 4, and 5, respectively. The disubstituted polyenes with donor–acceptor groups favor a large charge displacement in the main and hence provide a large second hyperpolarizability and seem favorable for large optical nonlinearities.

D. MODIFIED NEGLECT OF DIATOMIC OVERLAP (MNDO) METHOD

Williams[21] used the finite field technique in combination with the MNDO molecular orbital method for calculating second hyperpolarizability (γ) of a number of alternate and nonalternate hydrocarbon systems consisting of between 3 and 7 conjugated π bonds. This combination provides the simultaneous calculation of all appropriate tensor components of the polarizability (α) and hyperpolarizabilities for large organic molecules. These α, β, and γ tensors can be written as follows. The energy of a molecule in external field (E) is expressed by a power series in Equation 20 and can also be expressed as:

$$W(E) = W(0) - \mu_{i0}E_i - (2!)^{-1}\alpha_{ij}E_iE_j - (3!)^{-1}\beta_{ijk}E_iE_jE_k - (4!)^{-1}\gamma_{ijkl}E_iE_jE_kE_l - \quad (31)$$

where μ_{i0} is a component of a permanent dipole moment. The total dipole moment is given by the following expression:

Table 16 The γ_{zzzz} Values of a Variety of Polyene Molecules in a Static Electric Field, N is the Number of Carbon Atomic Sites

Number	Molecule	N	γ_{zzzz} (10^{-36} esu)
1		4	0.887
		6	5.60
		8	18.0
		10	43.9
		12	81.9
2	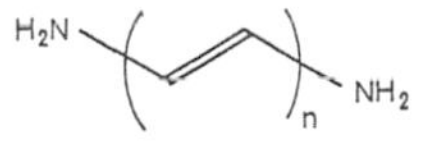 H_2N ... NH_2	4	1.86
		6	9.47
		8	27.4
		10	57.9
		12	99.3
3	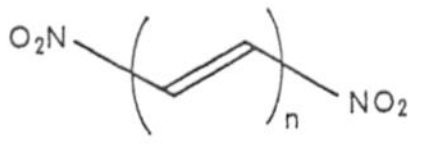 O_2N ... NO_2	4	12.2
		6	33.9
		8	65.0
		10	111
		12	166
4	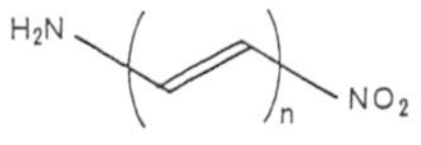 H_2N ... NO_2	4	17.0
		6	50.1
		8	110
		10	202
		12	327
5	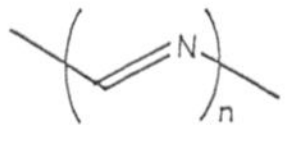	4	1.57
		6	11.1
		8	40.1
		10	97.7

From Nakano, M., et al., *Mol. Cryst. Liq. Cryst.*, 182A, 1, 1990. With permission.

$$\mu_i(E) = \mu_{i0} + \alpha_{ij}E_i + (2!)^{-1}\beta_{ijk}E_iE_k + (3!)^{-1}\gamma_{ijkl}E_jE_kE_l \ \ldots \ \ldots \tag{32}$$

The coefficients α_{ij}, β_{ijk}, and γ_{ijkl} may be obtained from using either the perturbation theory or a finite-field method. If choosing the finite field method, one may write the perturbation Hamiltonian for the molecule as

$$H^1 = H - \sum_i E \cdot r(i) \tag{33}$$

where E is a finite electric field and $r(i)$ the position operator for the ith electron. Following a variational Hartee-Fock procedure.

$$F_{\mu v} = F^0_{\mu v} + \sum_i E_i D^i_{\mu v} \tag{34}$$

where $F^0_{\mu v}$ is the unperturbed Hartee-Fock matrix element, E_i is the ith component of the finite field, and $D^i_{\mu v}$ is the ith component of the dipole moment matrix element. In this work, the F matrix elements have been calculated in the MNDO approximation.

The γ values calculated by this method for benzene, fulvene, heptafulvene, fulvalene, naphthalene, sesquifulvalene, biphenyl, and heptafulvalene were 0.73×10^{-36}, 1.23×10^{-36}, 4.10×10^{-36}, 6.34×10^{-36}, 8.12×10^{-36}, 11.45×10^{-36}, 27.20×10^{-36}, and 19.10×10^{-36}, esu, respectively. In another study, the hyperpolarizability of conjugated organic systems containing sulfur–nitrogen ring systems was studied.[21] The magnitude of the γ value strongly depends on the total number of π electrons in the conjugated system. Thiophene has a γ value of 1.1×10^{-36} esu while five-membered ring thiazole and isothiazole have γ values of 0.61×10^{-36} and 0.72×10^{-36} esu, respectively, close to the six-membered benzene ring. The trithiadiazepine shows a γ value of 8.7×10^{-36}, while the 14-π-electron system, benzotrithiadiazepine, has the largest value of -23.3×10^{-36} esu. The conjugated SN structure introduces an ionic character to the bonding in the ring and leads to large third-order nonlinearity. Second hyperpolarizabilities (γ) of conjugated polydiacetylene oligomers via the finite field MNDO method were also calculated. The influence of the increasing carbon chain length and substitution of different electron donor and electron acceptor groups on γ was investigated. The γ values of polyenyne oligomers increase with increasing conjugation.[22] The γ value changes by 2 orders of magnitude from a C_4H_4 chain (1.2×10^{-36} esu) to a $C_{14}H_{10}$ chain (595.4×10^{-36} esu). The γ values for C_6H_6, C_8H_6, $C_{10}H_8$, and $C_{12}H_8$, were 11.2×10^{-36}, 38.1×10^{-36}, 130.1×10^{-36}, and 347.3×10^{-36} esu, respectively. The γ values increase up to 1729.4×10^{-36} esu for the infinite polymer chain. Table 17 shows the effect of electron donating and electron accepting groups on γ value. A variety of symmetrical and asymmetrical substituted polyenyne oligomers were studied. The substitution of donor–acceptor groups shows a significant effect on γ value. The largest γ values were obtained for the SO_3H substituent due to the introduced charge asymmetry into the polyenyne backbone. A symmetrical substitution of an electron donor and acceptor group also leads to an increase in the γ values. For $R_1 = NH_2$, $R_2 = NO_2$ and $R_1 = NH_2$, $R_2 = SO_3H$, the magnitude of γ increases by a factor of 2 or more over that of unsubstituted polyenyne.

Williams[23] also used the finite-field MNDO method for a series of derivative polypyrrole and polythiophene oligomers. The influence of the chain length and substituent groups on the magnitude of γ was examined. The chemical structure, number of carbon atoms in the conjugated backbone chain, and γ values are shown in Table 18. The calculated results from the relationship $\gamma \sim N^x$, give $x \sim 5.1$ for the polypyrrole oligomers, $x \sim 5.4$ for the polythiophene oligomers, and $x \sim 5.5$ for the polyacetylene oligomers. The magnitude of γ increases with decreasing HOMO-LUMO energy gap. The magnitude of γ is larger for polyacetylene oligomers than that of an equivalent carbon chain length derivative of polypyrrole and polythiophene.

Matsuzawa and Dixon[24] calculated the polarizability (α) and the second hyperpolarizability (γ) of a number of aromatic hydrocarbons and fullerene (C_{60} and C_{70}) molecules using a finite-field approach with the PM-3 parametrization of the MNDO Hamiltonian. The calculated α and γ values were transformed to mean scalar values using the relationship

Table 17 Effect of Electron-Donating and Electron-Accepting Groups on γ

R$_1$	R$_2$	γ(10^{-36} esu)
H	H	11.2
CH$_3$	CH$_3$	12.1
OH	OH	12.7
NH$_2$	NH$_2$	13.3
NO$_2$	NO$_2$	13.6
CN	CN	15.1
SO$_3$H	SO$_3$H	32.2
H	CH$_3$	11.6
H	OH	12.1
H	COOH	12.1
H	NH$_2$	13.3
H	CN	13.4
H	NO$_2$	13.5
H	SO$_3$H	17.5
CH$_3$	NO$_2$	14.2
OH	NO$_2$	17.6
NH$_2$	CN	19.4
OH	SO$_3$H	19.5
NH$_2$	NO$_2$	22.8
NH$_2$	SO$_3$H	26.3

From Williams, G. R. J., *J. Mol. Struct. (Theochem.)*, 153, 191, 1987. With permission.

$$\alpha = \sum \alpha_{ii}/3 \tag{35}$$

$$\gamma = 6/4 \sum \gamma_{iijj}/5 \tag{36}$$

Table 19 lists the calculated α and γ values, HOMO and LUMO energies of the acene, pyrene, caronene, C$_{60}$ and C$_{70}$ molecules. The values of α and γ increase with increasing numbers of aromatic rings for acene. The α increases as $N^{1.3}$ while γ increases as $N^{3.7}$ for the acenes where N is the number of carbon atoms. Though second hyperpolarizability values increase significantly in the linear aromatic π-conjugated chain from benzene to hexacene, they decrease for pyrene and coronene which involve the addition of the aromatic rings in two dimensions. The two-dimensional addition of aromatic rings follows a γ dependence of $N^{2.7}$. For fullerenes, the extension of π conjugation by aromatic rings leads to three-dimensional systems. The C$_{60}$ molecule has the average of α per carbon atom of 0.107×10^{-23} cm^3 comparable to 0.113×10^{-23} cm^3 for each C–H units in benzene and the average γ per carbon atom of 0.40×10^{-36} less than anthracene. On the other hand, the C$_{70}$ molecule has the average α per carbon atom of 0.113×10^{-23} cm^3 and the average of γ per carbon atom of 0.65×10^{-36} esu, both higher than the C$_{60}$ molecule. This difference in α and γ values of fullerenes results from the bucky-ball structure of the C$_{60}$ molecule to the rugby-ball structure of the C$_{70}$ molecule, the latter providing, to some extent, planar units. The addition of ethylene bridges and the replacement of six-membered rings by five-membered rings resulted in an increase in γ values. The γ values of C$_{60}$ and C$_{70}$ molecules were found to be larger than expected from the calculated HOMO energy, presumably because of the balance between the number of carbon atoms and the degree of distortion. The γ values of the C$_{60}$ molecule were 17.2×10^{-36} esu from MNDO and 22.4×10^{-36} esu from the AM1 parametrizations.

E. INTERMEDIATE NEGLECT OF DIFFERENTIAL OVERLAP (INDO) METHOD

Li et al.[25] calculated the different components of hyperpolarizability of C$_{60}$ using both INDO/SCI and INDO/SDCI methods. The calculated $\langle\gamma\rangle(-\omega; \omega, \omega, -\omega) = 7.3 \times 10^{-34}$ at $\omega = 1.064$ μm and $\langle\gamma\rangle(-\omega;$

Table 18 Hyperpolarizabilities of Derivative Polypyrrole and Polythiophene Oligomers

Chemical Structure		N	$\gamma(10^{-36}$ esu)
	Polyacetylene	8	73.3
		10	244.6
		12	707.8
		14	1616.5
	Polypyrrole	8	20.8
		10	69.3
		14	358.9
	Polythiophene	8	28.8
		10	86.2
	R = H	12	331.3
	R = OH	12	314.0
	R = CN	12	307.6
		14	591.2

From Williams, G. R. J., *J. Mol. Struct. (Theochem.)*, 153, 185, 1987. With permission.
N refers to the number of carbon atoms in the conjugated backbone chain.

ω, ω, $-\omega$) = 6.9 × 10^{-34} at ω = 1.191 μm agreed with the experimental results (Table 20). These values are an order of magnitude larger than that reported by Matsuzawa and Dixon.[24]

Fichou et al.[26] first reported the nonlinear optical properties of a series of α-conjugated thiophene oligomers (α-nT) on the development of structure–property relationships. The α-conjugated thiophene oligomers (number of repeat thiophene units N = 3, 4, 5, 6, 8, and so on) have well-defined chemical structures and a precise and controlled π conjugation length. Second hyperpolarizability and third-order susceptibility of α-nT oligomers increase significantly up to N = 14, while the optical gap remained almost constant after N = 8. The calculated second hyperpolarizability of thiophene oliogomers increases by more than 2 orders of magnitude from the monomer (n = 1) to oligothiophene consisting of 14 units (N = 14) and shows no saturation, indicating that in order to optimize the optical nonlinearity in polythiophene, higher numbers of thiophene units are required and optical nonlinearity cannot be optimized in oligomers (N = 10) (an oligomer is defined by 10 monomer units in the backbone). This has also been demonstrated for polyene molecules.

Pierce[27,28] calculated second hyperpolarizabilities of ethylene, all-*trans*-1,3-butadiene, all-*trans*-1,3,5-hexatriene and all-*trans*-1,3,5,7-octatetraene using the sum-over-states method and the all-valence-electron, semiempirical INDO-SCF procedure combined with single- and double-excitation configuration interaction (SDCI). The INDO-SCI gave γ values of -0.208×10^{-36}, -0.667×10^{-36}, $-1.026 \times$

Table 19 Calculated α and γ Values, HOMO and LUMO Energies of the Acene, Pyrene, Caronene, C_{60} and C_{70} Molecules

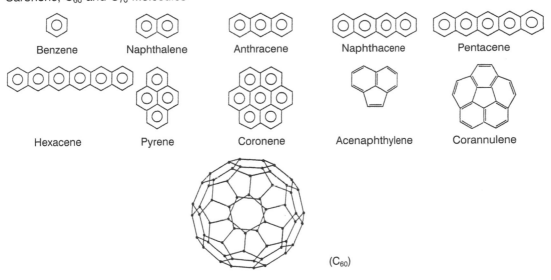

| Molecule | α
(10^{-23} cm^3) | γ (10^{-36} esu) HOMO | | |
		$\gamma\ (10^{-8})$	(eV)	LUMO (eV)
Benzene	0.68	0.5	−9.75	0.40
Naphthalene	1.24	3.4	−8.84	−0.41
Anthracene	1.95	11.2	−8.25	−0.97
Naphthacene	2.77	29.2	7.87	7.04
Pentacene	3.70	63.3	−7.61	−6.74
Hexacene	4.71	120	−7.43	−1.83
Pyrene	2.19	10.8	−8.25	−7.41
Coronene	3.37	25.3	−8.29	−1.06
Acenaphthylene	1.51	4.5	−9.06	−1.06
Camnulene	2.55	12.0	−8.96	−0.99
C_{60}	6.39	21.4	−9.48	−2.89
C_{70}	7.90	35.6	−9.01	−3.19

From Matsuzawa, N. and Dixon, P. A., *J. Phys. Chem.*, 96, 6241, 1992. With permission.

10^{-36}, and -1.750×10^{-36} esu and INDO-SDCI gave γ values of 0.034×10^{-36}, 0.901×10^{-36}, 4.527×10^{-36}, and 12.66×10^{-36} esu at ω = 0 for ethylene, butadiene, hexatriene, and octatetraene, respectively. The INDO-SDCI γ values were 0.051×10^{-36}, 1.564×10^{-36}, 9.312×10^{-36} and 29.82×10^{-36} esu at ω = 1.169 eV for ethylene, butadiene, hexatriene, and octatetraene, respectively. The INDO-SCI gave negative values of $\gamma(-3\omega;\ \omega,\ \omega,\ \omega)$, while the hyperpolarizabilities obtained from INDO-SDCI were positive values. The treatment of electron–electron interactions at the level of single and double-excitation CI is required to analyze the dynamic second hyperpolarizabilities of linear conjugated polyenes and has a significant impact on the calculations.

IV. *AB INITIO* METHOD: COUPLED-PERTURBED HARTEE-FOCK (CPHF) THEORY

For calculating the microscopic nonlinearities in the *ab initio* method, all σ and π electrons of the molecules are taken into account. Hurst et al.[29] performed a systematic study of a series of conjugated polyene systems via the *ab initio* coupled-perturbed Hartee-Focke theory. The second hyperpolarizabilities for polyenes: C_4H_6, C_6H_8, C_8H_{10}, $C_{10}H_{12}$, $C_{12}H_{14}$, $C_{14}H_{16}$, $C_{16}H_{18}$, $C_{20}H_{22}$, and $C_{22}H_{24}$ were calculated. The polyene geometries were optimized with 6-31G basis set, in C_{2h} symmetry. Among all the basis sets, the γ values obtained from the 6-31G+PD basis set were found to be close to the experimental

Table 20 Different Components of Hyperpolarizability of C_{60} Using Both INDO/SCI and INDO/SDCI Methods

Polarizability Tensor	INDO/SCI-SOS $\gamma(-\omega; \omega, \omega, -\omega)$ at $\omega = $ 1.064 µm	INDO/SDCI-SOS		
		$\gamma(-\omega; \omega, \omega, -\omega)$ $\omega = 1.064$ µm	$\gamma(-\omega; \omega, \omega, -\omega)$ $\omega = 1.91$ µm	$\gamma(-\omega; \omega, \omega, -\omega)$ $\omega = 0.532$ µm
γ_{xxxx}	8.83	7.30	6.90	3.11
γ_{xxyy}	3.02	2.48	2.32	−0.44
γ_{xxzz}	3.02	2.48	2.32	−0.43
γ_{yyyy}	8.85	7.30	6.90	3.11
γ_{yyzz}	3.03	2.49	2.32	−0.42
γ_{zzzz}	8.84	7.30	6.90	3.12
γ_{xyxy}	2.78	2.33	2.32	3.97
γ_{xzxz}	2.79	2.33	2.32	3.98
γ_{yxyx}	2.78	2.33	2.32	3.97
γ_{yzyz}	2.79	2.33	2.32	3.97
γ_{zxzx}	2.79	2.33	2.32	3.98
γ_{zyzy}	2.79	2.33	2.32	3.97
γ_{xyxy}	3.02	2.48	2.27	−0.44
γ_{xzzx}	3.02	2.48	2.27	−0.43
γ_{yxxy}	3.02	2.48	2.27	−0.44
γ_{yzzy}	3.03	2.49	2.27	−0.42
γ_{zxxz}	3.02	2.48	2.27	−0.43
γ_{zzyyz}	3.03	2.49	2.27	−0.42
$\langle \gamma \rangle$	8.84	7.30	6.90	3.11

The values are given in 10^{-34} esu. From Li et al., *Chem. Phys. Lett.*, 203, 560, 1993. With permission.

values. For example, the calculated γ values for C_4H_6 were 500, 1098, 1055, and 14,846 a.u. from STO-3G, 6-31G, 6-31G*, and 6-31G+PD basis sets, respectively. The STO-3G γ value of $C_{22}H_{24}$ was more than 3 orders of magnitude larger than C_4H_6, indicating a significant effect of increased π conjugation. The Cartesian axis x is the molecular plane; in the chain direction, the y axis is perpendicular to x and the z axis is the two-fold rotation axis. Therefore, the components in the polyene chain direction α_{xx} and γ_{xxxx} are the largest. The orientationally averaged polarizabilities and second hyperpolarizabilities for different polyene chain lengths in the STO-3G, 6-31G, 6-31G*, and 6-31G + PD basis sets are shown in Figure 5. The α and γ values are divided by the number of repeating units (N) to estimate the unit-cell properties of polyacetylene as N approaches infinity. Both α_{xx} and γ_{xxxx} vary with the number of polyene repeating units N. From the basis set 6-31G+PD, α_{xx} is proportional to $N^{1.51}$ between $N = 2$ and 3, and then increases to $N^{1.61}$ for N $\sim$ 6. The α_{xx} is proportional to $N^{1.40}$ for $N \sim 11$ with the 6-31G basis set. The N dependence of γ_{xxxx} increases from $N^{3.76}$ at $N \sim 2$ to $N^{3.98}$ for $N \sim 6$ and then decreases with greater N. The γ_{xxxx} is proportional to $N^{5.00}$ between $N = 2$ and 3, and then the γ_{xxxx} proportionally exponent decreases sharply with increasing N which reduces to $N^{3.04}$ for $N = 10$ and 11.

Bodart et al.[30] also utilized *ab initio* STO-3G calculations of polarizability for polycumulenes. The averaged polarizabilities $\langle \alpha \rangle$ were 6.5×10^{-24}, 6.8×10^{-24}, and 9.7×10^{-24} esu for hexatriene H-$(CH=CH)_3$-H, diallene H-$(CH=C=CH)_2$-H, and hexapentaene $CH_2=C=C=C=C=CH_2$, respectively. The total number of electrons n_e, the number of π electrons n_π and the length of carbon chain Lc, are $n_e = 44$, 42, and 40, $n_\pi = 6$, 8, and 10, and $Lc = 6.13$, 6.12, and 6.39 Å for hexatriene, diallene, and hexapetaene, respectively. A large polarizability results from the variation of the optical transitions. The $\langle \alpha \rangle$ for H-$(CH=C=C=C=CH)_x$-H were 34×10^{-24}, and 75×10^{-24} esu for $x = 2$ and 3 showing more than a two-fold increase as the length of π conjugation increase.

Nalwa et al., calculated second hyperpolarizabilities of a series of well-defined conjugation length polyazines by using the CPHF *ab initio* method.[31,32] Figures 6 and 7 show the optimized geometric structures in terms of bond lengths and bond angles for the monomer and the trimer azine derivatives respectively. The coordinate axes x and z are perpendicular to the π-conjugated polyazine backbone, while the y axis is along the periodicity direction. The important parameters that are influenced by the

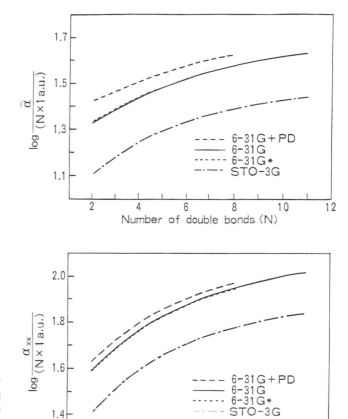

Figure 5(a) The plots of polarizability as a function of the number of repeating units for polyene chain. The α values were obtained from different basis sets. (From Hurst, G. J. B., Dupuis, M., and Clementi, E. J., *Chem. Phys.*, 89, 385, 1988. With permission.)

increased length of π-conjugated backbone are: (1) terminal nitrogen–nitrogen (N–N) bond length; (2) internal nitrogen–nitrogen (N–N) bond length; (3) internal nitrogen–carbon (C–N) bond length; (4) internal carbon–carbon (C–C) bond length; (5) bond angle between N–C–C; (6) bond angle between C–N–N; and (7) bond angle between N–N–C. For example, the C–C bond length in these oligomers was 1.48–1.49 Å, which is similar to that of the C–C bond length in polyenes (1.48 Å) and C–N was ca. 1.29 Å, slightly longer than the usual double bond C=N length, 1.27 Å. Because of the simplicity of synthesis, we have selected polyazine systems having electron-donating amino groups at both terminals, to compare with their model compounds. Table 21 lists the polarizability and second hyperpolarizability components of oligomer model compounds.

For second hyperpolarizability, the γ_{yyyy} component is by far the largest since for the polyazine linear chain, the y component of the transitional dipole moment is much larger than that of x and z components. There is a further increase in the γ_{yyyy} value with the increased chain length because of the larger number of contributing terms. The magnitude of the transitional moments along the chain axis increases steadily with chain length, since longer chains have a large transitional moments, therefore, larger γ_{yyyy} values are obtained. The oligomeric polyazine derivatives, having repeat units 1, 2, 3, 4, and 5, exhibit orientationally averaged $\langle\gamma\rangle$ values of 3.34×10^{-36}, 1.97×10^{-35}, 5.71×10^{-35}, 1.12×10^{-34}, and 1.75×10^{-34} esu, respectively, indicating an order of magnitude increase from the monomer to the trimer polyazine derivative. Figure 8 shows the log-log plot of γ_{yyyy} values vs. the number of nitrogen and carbon atomic sites for a well-defined conjugation length polyazine chain. The γ values increase much more rapidly as the π electron delocalization length of polyazine backbone increases; its value changes by a factor of 6 in going from the monomer to the dimer and by a factor of 17 in going from the monomer to the trimer. The γ value increases by a factor of more than 53 from the monomer to the pentamer polyazine derivative. Like other conjugated systems, a significant increase in γ values is observed for polyazines. The increase of γ_{yyyy} obeys $N^{3.2}$ power law for 3 repeat units ($N = 3$), then

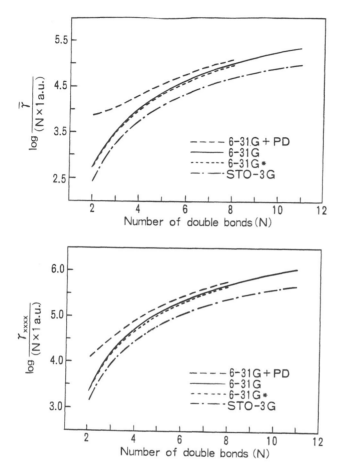

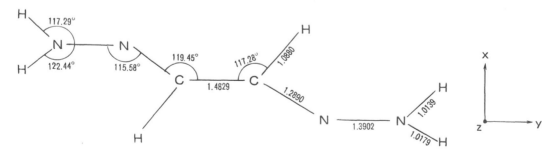

Figure 5(b) The plots of second hyper-polarizability as a function of the number of repeating units for polyene chain. The γ values were obtained from different basis sets. (From Hurst, G. J. B., Dupuis, M., and Clementi, E., *J. Chem. Phys.*, 89, 385, 1988. With permission.)

reduces to $N^{3.0}$ for $N = 4$. The orientationally averaged $\langle\gamma\rangle$ obeys $N^{3.35}$ simple power law for $N = 3$, then reduces to $N^{3.2}$ for $N = 4$ and $N^{3.07}$ for $N = 5$. The increase of α_{xx} obeys $N^{1.66}$ for $N = 5$, $N^{1.60}$ for $N = 4$, while the orientationally averaged $\langle\alpha\rangle$ obeys $N^{1.33}$ for $N = 5$, $N^{1.35}$ for $N = 4$, and $N^{2.14}$ for $N = 3$ power law. Theoretical studies predict the length dependence of α from 1.3 to 3.0 and γ from 3.2 to 5.2 and these results on well-defined chain length polyazine are in agreement. The same calculation technique was used by Hurst et al.[29] for polyenes; therefore, a comparison of polarizabilities of polyenes with polyazine is feasible. The averaged γ values of monomeric polyazine were found to be larger than those of regular polyenes having equivalent numbers of double bonds. Nakano et al.[20] calculated second hyperpolarizabilities for regular and donor–acceptor disubstituted polyenes, polydiacetylenes, and related species for low-lying excited states using the CNDO/S-CI method. These studies on monomeric polyazine

Figure 6 Optimized geometric structures in terms of bond lengths and bond angles for the monomer azine derivative. (From Nalwa, H. S., Hamada, T., Kakuta, A., and Mukoh, A., *Nonlinear Optics*, 7, 193, 1994. With permission.)

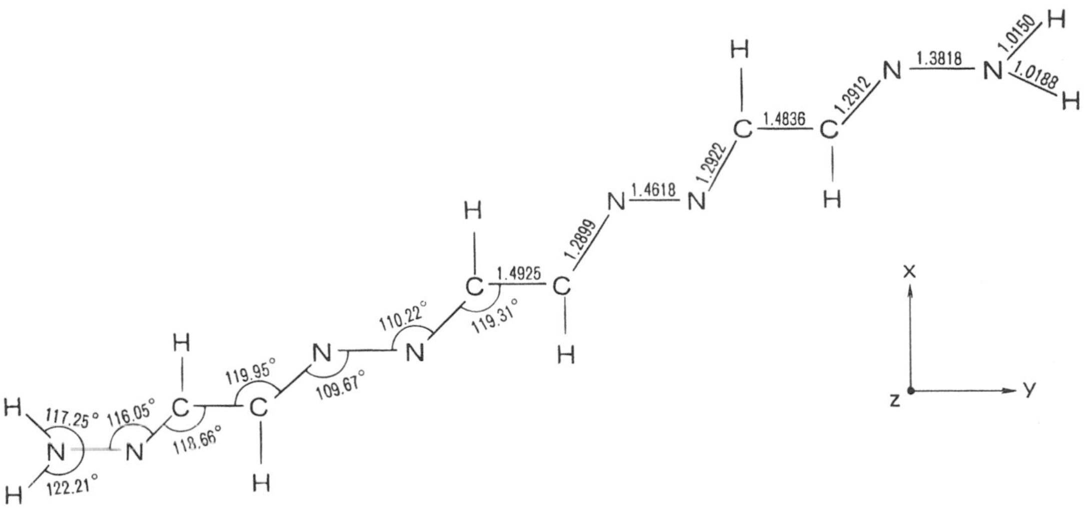

Figure 7 Optimized geometric structures in terms of bond lengths and bond angles for the trimer azine derivative. (From Nalwa, H. S., Hamada, T., Kakuta, A., and Mukoh, A., *Nonlinear Optics,* 7, 193, 1994. With permission.)

derivatives show the γ values at least 2 orders of magnitude larger than those of regular polyenes ($\gamma = 8.87 \times 10^{-37}$ esu) and 1 order of magnitude larger than that of diamino polyenes of the identical π electron delocalization length systems though it may change for longer chain lengths. The log-log plot of γ_{zzzz} vs. the number of carbon-atom sites yields an exponent of 3.64 for diamino polyenes which is close to results on diamino polyazine chains. As discussed earlier, the γ values of polyazine derivatives

Table 21 Dipole Polarizability Components of Polyazine Oligomers Obtained from 6-31G Basis Set[31,32]

Polarizability Tensor	Number of Repeating Unit in Polyazine Chain (N)				
	$N = 1$ $(C_2H_6N_4)$	$N = 2$ $(C_4H_8N_6)$	$N = 3$ $(C_6H_{10}N_8)$	$N = 4$ $(C_8H_{12}N_{10})$	$N = 5$ $(C_{10}H_{14}N_{12})$
α_{xx}	9.08	18.15	26.17	59.60	76.55
α_{yy}	12.48	26.74	44.90	38.92	49.93
α_{zz}	2.41	4.08	5.73	7.39	9.04
$\langle \alpha \rangle$	7.99	16.32	25.60	35.30	45.18
γ_{xxxx}	−683.78	−2,159.60	−2,668.91	4,158.13	12,408.28
γ_{xxxy}	−546.35	1,143.35	5,367.22	28,984.31	53,186.15
γ_{xxyy}	1,086.58	10,822.83	29,905.89	82,439.56	136,635.93
γ_{xxzz}	26.76	82.20	144.47	238.30	326.10
γ_{xyyy}	4,044.19	30,953.39	87,658.16	185,471.50	293,127.52
γ_{xyzz}	7.43	44.43	105.04	167.33	238.89
γ_{yyyy}	15,223.04	79,003.62	228,382.78	390,404.00	592,887.31
γ_{yyzz}	−16.83	6.75	62.30	106.97	168.91
γ_{zzzz}	12.73	20.47	27.87	35.34	42.49
$\langle \gamma \rangle$	3,349.00	19,737.6	57,193.41	112,033.43	175,919.99

The α is in Å^3 and γ is in 10^{-39} esu units. The orientationally averaged or mean values are $\langle \alpha \rangle = 1/3(\alpha_{xx} + \alpha_{yy} + \alpha_{zz})$ and $\langle \gamma \rangle = 1/5(\gamma_{xxxx} + \gamma_{yyyy} + \gamma_{zzzz} + 2\gamma_{xxyy} + 2\gamma_{xxzz} + 2\gamma_{yyzz})$. The polarizability axes x and z are perpendicular to the π-conjugated backbone while the y axis is along the periodicity direction.

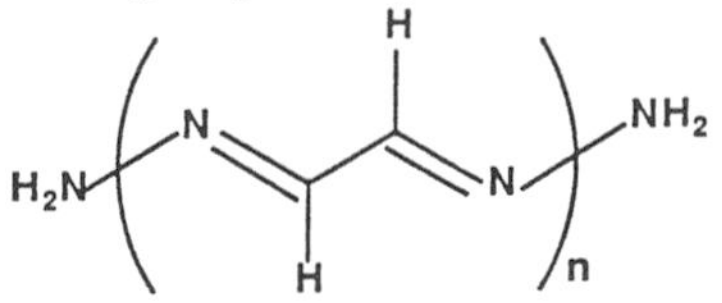

n = 1, 2, 3, 4, or 5

542

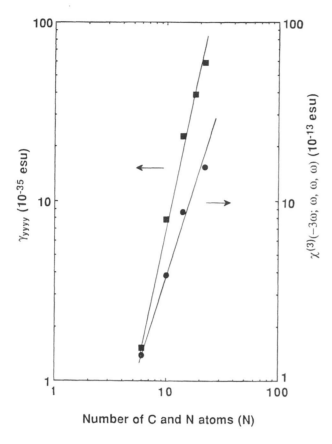

Figure 8 Log-log plot of γ_{yyyy} values vs. the number of nitrogen and carbon atomic sites for well-defined conjugation length polyazine chain. (From Nalwa, H. S., Hamada, T., Kakuta, A., and Mukoh, A., *Nonlinear Optics*, 7, 193, 1994. With permission.)

are much larger than those reported by Nakano et al.[20] for polyene derivatives; this noticeable difference may arise, from the CNDO/S-Cl calculations which are different from the *ab initio* technique.

Andre et al.[33,34] investigated the influence of the conjugated chain length on the dipole polarizability for *trans*-1,3,5-hexatriene (**2**), *trans*-1,3,5,7,9-decapentaene (**3**), 1,5-hexadiene-3-yne (**4**), and 1,5,9-decatriene-3,7-diyne (**5**). This was analyzed by *ab initio* CPHF using STO-3G basis set. For *trans*-1,3,5-hexatriene-3-yne (**2**), the α_{xx} and $\langle\alpha\rangle$ were 104.13 and 43.93 au, respectively. The α_{xx} were 217.25 au for *trans*-1,3,5,7,9-decapentaene (**3**), and 44.65 au for 1,5-hexadiene-3-yne (**4**), and $\langle\alpha\rangle$ was 85.28 au for 1,5,9-decatriene-3,7-diyne (**5**). The polarizability of polyenic backbone is larger because of the more efficient delocalization of the π electrons. The size dependence for α was $N^{1.399}$ and for α_{xx} was $N^{0.958}$.

(2)

(3)

(4)

(5)

Dory et al.[35] analyzed the STO-3G longitudinal polarizabilities for *all-trans*-1,3,5,7-octatetraene with various heteroatomic linkages (nitrogen or oxygen) grafted to and/or inserted in. The longitudinal polarizabilities with structural modifications were:

1. $H_2C=CH-CH=CH-CH=CH-CH=CH_2 = 140.44$ au
2. $HN=CH-CH=CH-CH=CH-CH=CH_2 = 134.86$ au
3. $H_2C=CH-CH=N-CH=CH-CH=CH_2 = 137.29$ au
4. $O=CH-CH=CH-CH=CH-CH=CH_2 = 119.32$ au
5. $HN=N-CH=CH-CH=CH-CH=CH_2 = 137.31$ au
6. $H_2C=CH-N=N-CH=CH-CH=CH_2 = 136.46$ au
7. $HN=CH-CH=CH-CH=CH-CH=NH = 140.44$ au
8. $O=N-CH=CH-CH=CH-CH=CH_2 = 121.53$ au
9. $O=CH-CH=CH-CH=CH-CH=O = 98.25$ au

Polarizability is not affected by the nitrogen substitution, but it leads to more stable compounds. Furthermore, the replacement of the methylene end group by oxygens leads to a decrease in polarizability. The STO-3G mean polarizabilities were 159.34 au for quaterphenyl, 125.23 au for quarterpyrrole, and 142.72 au for quaterthiophene. The 6-31G mean polarizabilities were 51.58 au for benzene, 75.59 au for styrene, 93.42 au for 3,6-dimethylene-1,4-cyclohexadiene, 37.98 au for pyrrole, 62.72 au for pyrrolethylene, 73.83 au for bismethylidene-2,5-thiacyclopentene, 74.29 au for pyrrole, 71.89 au for thienylethylene, and 82.56 au for bismethylidene-2,5-azacyclopentene. These calculations showed that polyenes are better than poly-ynes and polydiacetylenes. Doping with Na leads to a significant increase in linear polarizability for poly(*p*-phenylene), polypyrrole, and polythiophene. Furthermore, the quinoid structures leads to a polarizability increase of ~20% to counterparts in their aromatic vinylene structures. Bredas et al.[36] also calculated polarizabilities of isothianaphthene and thieno[3,4-c]thiophene oligomers using the *ab initio* STO-3G level. The fused ring derivatives of polythiopehne, i.e., poly(isothianaphthene) (PITN) and poly(thieno[3,4-c]thiophene) (PTT), are low bandgap conjugated polymers. For example, the bandgap of PITN (1 eV) is half that bandgap of polythiophene. The polarizabilities were 26.0, 61.3, and 101.7 au for the monomer, dimer, and trimer of thiophene, respectively. For isothianaphthene monomer and dimer, the calculated polarizabilities were 54.5 and 133.5 a.u., respectively. Interestingly, the polarizabilities of PTT oligomers were remarkably different and very large. The polarizabilities were 59.1, 211.8, and 645 au for the monomer, dimer, and trimer of thieno[3,4-c]thiophene, respectively. The polarizabilities of both polythiophene and PITN oligomers increase as the length of the chain increases. For PTT oligomers, the polarizability per repeat unit doubles from the dimer to the trimer and this polarizability is about 3 times larger than that of biphenyl analog. The calculated inter-ring bond length in the PTT dimer is ~1.42 Å, lower compared to 1.45 Å in both polythiophene and PITN dimers. In the case of a PTT trimer, the inter-ring bond length decreases further down to 1.38 Å, giving rise to a strong double-bond character. The quinoid-like structure leads to larger polarizabilities.

Table 22 *Ab initio* Longitudinal Electronic Polarizabilities (in atomic units) and Hyperpolarizabilities (in 10^3 a.u.) per SiH_2 (*n* is odd)

	STO-3G		6-31G		6-31G + P	
n	α	γ	α	γ	α	γ
1	27.3	4.87	48.5	29.4	49.5	34.9
3	30.6	13.50	55.2	71.2	55.9	7 6.3
5	31.5	20.58	58.8	114.7	59.5	119.0
7	32.1	24.79	60.7	148.8	61.4	152.6
9	32.4	26.95	61.8	172.0		
11	32.6	28.40	62.4	187.0		
13	32.6	28.95	62.8	193.1		

From Kirtman, B., and Hasan, M., *J. Chem. Phys.*, 96, 470, 1991. With permission.

(thieno [3,4-c]thiophene)

(isothianaphthene)

Kirtman and Hasan[37] calculated the longitudinal polarizability and hyperpolarizabilities for the C_6H_6 through $C_{26}H_{16}$ oligomers of PDA and PBT from the semiempirical INDO and *ab initio* SCF methods. For PDA, the leveling-off was noticed in the range C_{14}-C_{18} for polarizability and C_{22}-C_{26} for hyperpolarizability. In the case of PBT, there was no sign of saturation as the chain length increased. Kirtman and Hasan[38] reported the polarizabilities of the σ-conjugated *trans*-polysilane (*t*PSi) at the Hartee-Fock level by using the *ab initio* calculations on the oligomers $H_3Si-(SiH_2)_n-SiH_3$ with $n = 1, 3, 5, 7, 9, 11, 13$, and 15. Table 22 lists the second hyperpolarizabilities for polysilane oligomers. Chain length through $n = 13$ was sufficient for obtaining infinite chain polarizabilities. The calculated linear polarizabilities per SiH unit were of the same magnitude as that of prototype π-conjugated polymers polyacetylene and polydiacetylenes, while the second hyperpolarizabilities were found to be comparatively smaller. The γ values differ significantly between the 6-31G+P and 6-31 bases. The ratio of the 6-31G+P to the 6-31G values decreases regularly from 1.190 for $n = 1$ to 1.026 for $n = 7$. The STO-3G calculations give about 1/6 the hyperpolarizability. These theoretical results give third-order NLO susceptibility value of the order of 6.2×10^{-13} esu which were in agreement with the experimental data.

Karna and Dupuis[39] performed *ab initio* calculations of the frequency-dependent polarizabilities and hyperpolarizabilities of the haloform series CHX_3 where X is F, Cl, Br, and iodine. Table 23 lists the

Table 23 Static and Dynamic Polarizability and Hyperpolarizabilities for CHF_3, $CHCl_3$, $CHBr_3$, and CHI_3

	Molecules			
Parameters	CHF_3	$CHCl_3$	$CHBr_3$	CHI_3
Dipole moment (D)	1.918	0.511	1.053	0.786
Averaged $\langle\alpha\rangle$	2.35	7.72	10.43	15.00
Static $\langle\gamma\rangle$	65.5	524.2	1287.6	3523.2
OKE $\langle\gamma\rangle$	68.1	566.7	1418.7	4032.0
EFISH $\langle\gamma\rangle$	73.7	669.3	1754.6	5600.0
THG $\langle\gamma\rangle$	83.0	876.1	2630.9	6460.0
IDRI $\langle\gamma\rangle$	70.7	613.9	1575.4	4702.2
EFIOR $\langle\gamma\rangle$	68.1	567.3	1422.9	4053.3

The α is 10^{-24} and γ is 10^{-39} in esu. All dynamic quantities were calculated at 694.3 nm except CHI_3 where γ(THG) was obtained at 911.3 nm.
From Karna, S. P. and Dupuis, M., *Chem. Phys. Lett.*, 171, 201, 1990. With permission.

static and dynamic polarizability and hyperpolarizabilities for fluoroform CHF_3, chloroform $CHCl_3$, bromoform $CHBr_3$, and iodoform CHI_3. All dynamic quantities were calculated at the optical wavelength $\lambda = 694.3$ nm except CHI_3 where γ(THG) was obtained at 911.3 nm. The frequency-dependent second hyperpolarizabilities were calculated for optical Kerr effect (OKE) $\gamma(-\omega; 0, 0, \omega)$, DC-electric-field-induced second harmonic generation (EFISH) $\gamma(-2\omega; 0, \omega, \omega)$, third-harmonic generation (THG) $\gamma(-3\omega; \omega, \omega, \omega)$, intensity-dependent refractive index (IDRI) $\gamma(-\omega; \omega, \omega, -\omega)$ and DC-electric-field-induced optical rectification (EFIOR) $\gamma(-0; 0, \omega, -\omega)$. Both α and γ increase in going from fluoroform to iodoform; in particular, γ increases by 2 orders of magnitude. The dispersion of α and γ increases as the atomic number of the halogen atoms in the haloform increases and the dispersion of γ in CHI_3 increases with an increase in the applied optical frequency. The dispersion of γ(EFISH) in CHI_3 was larger than that in OKE or IDRI. The CHI_3 shows large optical nonlinearities because of polarizable lone-pair electrons of the iodine atoms.

Bratan et al.[40] calculated the second hyperpolarizability for polyenes and polyynes by using the tight-binding calculations. They demonstrated that the second hyperpolarizability for conjugated hydrocarbons with bond alternation increases rapidly up to <10 repeating units, and then more slowly beyond about 15 repeating units. Their finite-chain calculations indicated that molecular hyperpolarizability for $\geq$40 repeating units is of the same magnitude as that of infinite chains (Table 24). Therefore, these studies substantiate that the magnitude of γ increases with the increase of π-electron delocalization length up to an intermediate chain length, and that long conjugated structures may not be necessary for large optical nonlinearity, however more studies are needed to substantiate saturation effect.

Sekino and Barlett[41] calculated the frequency-dependent γ values of N_2 and *trans*-butadiene using the generalized time dependent Hartee-Fock theory for several frequencies of applied field. The calculated γ values of N_2 were 763.7 au from DC-SHG, 736.8 au from OKE, 752.3 au from IDRI, and 803.6 au from THG at 1.06 μm. The calculated γ values of *trans*-butadiene were 18,000 au from DC-SHG, 15,900 au from OKE, 716,800 au from IDRI, and 22,200 au from THG at 1.06 μm.

V. POLARIZABILITIES: A COMPARISON FROM DIFFERENT COMPUTATIONAL METHODS

Several approaches have been employed for the theoretical estimation of third-order optical nonlinearities of organic materials. As can be seen from various theoretical approaches, the magnitude of second hyperpolarizability significantly differs from each other because of the different quantum chemical consideration. Table 25 lists the second hyperpolarizabilities of benzene (C_6H_6), carbon tetrachloride (CCl_4), and urea (H_2N-CO-NH_2) molecules calculated from the *ab initio* CPHF theory from different basis sets. The second hyperpolarizability of benzene calculated from the *ab initio* CPHF method with different basis sets varies significantly.[42] The 6-31G basis set, augmented with diffuse p and d functions, estimates γ value of 16,659.4 au for benzene at the Hartree-Fock level. A γ value of 22,234 au for benzene was calculated from the Moller-Plesset 2 (MP2) method with 6-31G + pd basis set. In Table 26 we have summarized the calculated and experimental α and γ values of the benzene, naphthalene, anthracene and pyrene. The γ values obtained from 6-31G + pd for benzene are in agreement with the experimental results.[53-55] The γ values of 27,900 au for benzene and 27,400 au for CCl_4 have been experimentally determined.[55] The substitution of electron acceptor and donor groups on the benzene ring also remarkably influence the magnitude of third-order optical nonlinearity. For example, the 6-31G+pd basis estimates γ values of 20,977.6, 15,115.9, 27,414.2, 21,882.6, and 15,841.4 au for aniline, nitrobenzene, *p*-nitroaniline, *p*-diaminobenzene, and *p*-dinitrobenzene, respectively, at the Hartree-Fock level, showing the role of *para*-substituted acceptor and donor

Table 24 The γ_{xxxx} for Finite Chain Length Polyyne and Polyene

Repeating Units	γ_{xxxx}	
	Polyene	Polyyne
2	0.2592	0.4836
10	1.841×10^3	2.651×10^3
40	5.773×10^4	5.993×10^4

The units of γ_{xxxx} (the unnormalized hyperpolarizability) are $e^4\alpha^4/\beta_1^3$. From Bratan, D. N., et al., *J. Chem. Phys.*, 171, 201, 1990. With permission.

Table 25 Comparison of Second Hyperpolarizabilities γ of Benzene, Carbon Tetrachloride, and Urea Obtained from *ab initio* CPHF from Different Basis Sets (Atomic Units)

Basis Set	Benzene	CCl$_4$	Urea
STO-3G	183.4	−168.0	193.9
3-21G	713.2	−18.3	558.4
6-31G*	954.6	953.9	662.2
6-31G	1,013.6	460.9	728.0
6-31G + s	2,088.6	2,390.1	804.3
6-31G + p	7,979.2	3,621.6	1,973.8
6-31 + G	10,264.1	6,327.8	2,604.7
6-31G + d	11,297.8	9,950.7	3,799.0
6-31G + pd	16,659.4	10,242.2	5,210.7

From Ohta, K., The 3rd Symposium on Photonic Materials, Oct. 20–21, 1992, Tokyo, Japan. With permission.

groups in enhancing the magnitude of hyperpolarizability. On the other hand, the Moller-Plesset 2 theory with the 6-31G + pd basis estimates γ values of 31,704, 21,662, 50,256, 37,241, and 23,864 au for aniline, nitrobenzene, *p*-nitroaniline, *p*-diaminobenzene, and *p*-dinitrobenzene, respectively. The γ values calculated from the MP2 theory were found to be larger by a factor of 1.2 to 1.8. The γ values calculated from the MP2 theory with 6–31G + pd basis set for benzene and *p*-nitroaniline were 1.3 and 1.8 times those obtained by using the Hartree-Fock theory.

We have summarized the calculated and experimental α and γ values of the defined chain length polyenes obtained from various methods in Table 27. A useful comparison can be made not only with methods of calculations to show their reliability, but also to indicate the role of increased π-conjugation on both α and γ. Shelton[59] investigated the dispersion of the electronic part of the third-order nonlinear susceptibility of benzene vapor by measuring with EFISH technique (Figure 9). The hyperpolarizability values reported by Svendsen et al.[43] were about five times larger than experimental values. Other hyperpolarizabilities reported by McIntyre and Hameka[2] were five times smaller than the experimental results. There are discrepancies among γ values obtained from different methods. Therefore, the reliability of the values of second hyperpolarizability for organic molecules has to be considered. We have listed γ values for a wide variety of conjugated systems obtained by applying various quantum chemical approaches. Most of the calculations have been carried out with the semiempirical methods. The calculations may have an average error ranging between 10% and 50%. A number of studies performed on benzene indicate that quantum chemical approaches must be treated with caution since the magnitude of third-order optical nonlinearity varies significantly with different approaches.

VI. OPTICAL NONLINEARITIES OF SMALL MOLECULES: A COMPARISON BETWEEN THEORETICAL AND EXPERIMENTAL RESULTS

In this section we provide a comparison of computational data with experimental results on small molecules. Sekino and Bartlett[60] reported a systematic study of the polarizability and second hyperpolarizability of several small molecules from coupled-cluster (CC) and many-body perturbation theory (MBPT) methods (Table 28). The results on polar molecules HF, H_2O, NH_3, and H_2S were improved by adding lone-pair basis functions and were in agreement with experimental results, indicating the applicability of this level of theory to larger molecules. The HF has the largest discrepancy of about 20% among theory and experiments. In particular, polar molecules with lone pairs and hydrogens are found to be difficult to analyze. Guan et al.[61] also calculated polarizabilities and hyperpolarizabilities of CO, H_2, N_2, HF, H_2O, NH_3, and CH_4 by using coupled Hartee-Fock (HONDO-8) and local density approximations compared with experimental results. The α_{xx} values calculated from Hartree-Fock (HONDO-8) approximation and experiments (in parentheses) were found to be 11.27 (12.1), 4.57 (4.75), 9.83 (10.17), 4.48 (5.03), 9.16 (10.31), 12.72, (14.0), and 15.95 (17.5) Å^3 for CO, H_2, N_2, HF, H_2O, NH_3, and CH_4, respectively. On the other hand, the α_{zz} values calculated from Hartee-Fock (HONDO-8) approximation and experiments (in parentheses) were found to be 14.45 (15.7), 6.46 (6.79), 15.00 (14.87), 5.73 (6.51), 8.46 (9.91), and 13.27 (15.6) Å^3 for CO, H_2, N_2, HF, H_2O, and NH_3, respectively. The theoretical and experimental results on polarizabilities and those estimated by MBPT(2) and CCSD methods are in agreement.

Table 26 Summary of α and γ Values of Some Aromatic Molecules from Different Theoretical Approaches

Molecule	α	γ	Method	Ref.
Benzene	61.86	24,682	CNDO-SCF	11
	62.06	—	FPT-CNDO/S-CI	44
	51.90	—	FPT-CNDO/S-CI	45
	28.41	—	FPT-INDO	46
	37.07	—	PT-Huckel	3
	53.97	—	FPT-*ab initio* SCF	47
	48.72	—	FPT-MNDO	48
		926	PT-Huckel	2
		13,499	Ext.Huckel-PPP	3
		109	FPT-INDO	49
		73,258	VPT-CNDO/2	43
		1,449	MNDO	22
		258	PM-3 MNDO	50
	45.89	993	PM-3 MNDO	24
		2,581	*ab initio*	30
		16,659	*ab initio*	42
		1,469	INDO-SDCI	28
	51.58	—	*ab initio*	35
	66.80	—	Experimental	51
	69.51	—	Experimental	52
		24,538 ± 596	Experimental	53
		12,829 ± 5,641	Experimental	54
		18,582 ± 2,779	Experimental	55
Naphthalene	115.45	60,313	CNDO-SCF	11
	121.21	—	FPT-CNDO/S-CI	44
	93.81	—	FPT-CNDO/S-CI	45
	55.00	—	FPT-INDO	46
	81.00	—	PT-Huckel	3
	61.21	—	FPT-*ab initio* SCF	47
	111.8	—	Experimental	51
		90,292	VPT-PPP	5
		11,674	PT-Huckel	5
		12,508	MNDO	22
	83.68	6,750	PM-3-MNDO	24
Anthracene	183.72	111,226	CNDO-SCF	11
	225.61	—	FPT-CNDO/S-CI	44
	159.94	—	FPT-CNDO/S-CI	45
	89.01	—	FPT-INDO	46
	138.23	—	PT-Huckel	3
	99.25	—	FPT-*ab initio* SCF	47
	170.97	—	Experimental	51
		301,132	VPT-PPP	5
		55,867	PT-Huckel	5
		38,518	MNDO	22
	131.60	22,237	PM-3-MNDO	24
Pyrene	205.73	134,400	CNDO-SCF	11
	147.80	21,443	PM-3-MNDO	24
	190.09	—	Experimental	51

Both α and γ values are in atomic units.

VII. FACTORS AFFECTING THE MAGNITUDE OF OPTICAL NONLINEARITY

Third-order nonlinear optical responses are significantly affected by several factors. In particular, the π-conjugation length, donor–acceptor groups, conformation, and dimensionality play a major role in determining the magnitude of third-order optical nonlinearity. The effects of π-bonding sequence, charge-transfer complex formation and symmetry are also discussed.

Table 27 Summary of α and γ Values of Some Polyene Molecules from Different Theoretical Approaches

Molecule	α	γ	Method	Ref.
Ethylene	29.71	8263	CNDO-SCF	10
	17.49	131.7663	*ab initio*	31
	30.31	—	PT-CNDO/1	56
	22.94	—	FPT-MNDO/1	
		1,250	INDO-SDCI	28
		−336	PPP-Ext. Huckel	4
		5,956		53
		9,029 ± 202	Experimental	53
	28.48	—	Experimental	57
trans-Butadiene	69.73	33,447	CNDO-SCF	10
	25.15	—	*ab initio*	31
	20.9	1,759	PPP	7
		11,912		53
		1,788	INDO-SDCI	28
		16,021	PPP-Ext. Huckel	4
	32.08	—	VPT-PPP	4
	59.06	—	PT-Huckel	4
	61.91	—	SCF-PT	58
	53.27	27,400	*ab initio*	29
		14,900	SCF	41
		22,900	MBPT	41
		27,397 ± 1,549	Experimental	53
trans-Hexatriene	129.19	72,137	CNDO-SCF	10
	43.93	—	*ab initio*	31
	87.92	89,700	*ab initio*	29
	42.5	18,812	PPP	7
		19,059		53
		89,696 ± 8,338	Experimental	53
		8,990	INDO-SDCI	28
		118,166	PPP-Ext. Huckel	4
	72.99	—	VPT-PPP	4
	165.93	—	PT-Huckel	4
		3,847	PPP	9
		37,290	Huckel	9

Both α and γ values are in atomic units. MBPT = many-body perturbation theory; FPT = finite perturbation theory; VPT = variation perturbation theory.

A. EFFECT OF ATOMS AND BONDS

Meredith et al.[73] reported the average bond and group-plus-bond polarizabilities and second hyperpolarizabilities for monosubstituted benzene deduced by simple subtraction among aromatic compounds (Table 29) by the THG method. Scalar components of bond or bond-plus-group polarizabilities in monosubstituted benzenes were determined with the approximation that polarizabilities of ph-H bonds equal C–H bond polarizabilities. The averaged polarizabilities of pyridine were found to be smaller than that of benzene. To examine the bond effect, a comparison of halobenzenes can be made between *ab initio* calculations of the frequency-dependent polarizabilities and hyperpolarizabilities of the haloform series CHX_3, where X is F, Cl, Br, and I.[39]

Albert et al.[74] reported the frequency-dependent polarizabilities, static second hyperpolarizabilities, and THG coefficients of cumulenes and polyenynes calculated by the Pariser-Parr-Pople (PPP) model (Table 30). The polarizabilities and second hyperpolarizabilities were found to be the largest for cumulene, intermediate for polyenes, and smallest for polyenynes. The optical band gap of infinite cumulene was lowest at 0.75 eV whereas it was largest 4.37 eV for polyenynes. The exponents estimated from the free-electron calculations were 3 and 5 for α and γ values of cumulenes. For pentatetraene and hexapentaene, the exponent was 4.4 and slightly higher than polyenes. The γ values show some deviations

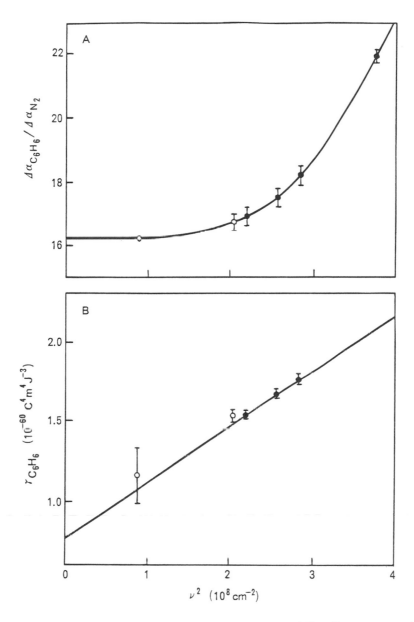

Figure 9 The dispersion of the electronic part of the second hyperpolarizability of benzene vapor by measuring with EFISH technique. (From Shelton, D. P., *J. Opt. Soc. Am. B,* 2, 1880, 1985. With permission.)

from the linear dependence in the log-log plot. The presence of heteroatoms leads to a decrease in both α and γ values.

B. EFFECT OF DELOCALIZATION LENGTH

Several examples have been provided earlier to show the effect of the increasing chain length on polarizability and hyperpolarizability. To sum up the theoretical data particularly on π-conjugated polymers, we are providing a few examples here. Shuai and Bredas[75] reported the static and dynamic second hyperpolarizability γ tensors of oligomers of polyacetylene, poly(*p*-phenylene vinylene) (PPV), and poly(thienylene vinylene) (PTV). The γ values were calculated from the valence effective Hamiltonian (VEH) method and were compared with SSH-SOS, MNDO finite-field, and Hartee-Fock *ab initio* calculations. Table 31 lists the averaged α and γ values for polyenes ranging in size from $N = 4$ to 30. The VEH-SOS results are comparable to the *ab initio* for $N = 4$ and 6. For large-sized N, the γ

Table 28 Theoretical and Experimental Polarizability and Hyperpolarizabilities of Some Small Molecules (in au)

Molecule		MBPT(2)					CCSD				DC-SHG
		$\langle\alpha\rangle$	α_{zz}	γ_{xxxx}	γ_{zzzz}	THG	$\langle\alpha\rangle$	α_{zz}	γ_{xxxx}	γ_{zzzz}	
H_2		5.23	6.56	260	800	670	5.22	6.56	590	810	710
	Exp.	5.43[62]				950[63]					801[63]
CO		13.09	15.70	1,380	1,740	2,200	12.92	12.56	1,360	1,740	1,800
	Exp.	13.08[64]									1,720[65]
N_2		11.45	14.41	780	1,220	1,200	11.61	14.82	810	1,200	1,100
	Exp.	11.8[66]				1,270[65]					1,030[65]
CO_2		17.85	27.84	1,010	1,120	1,500	17.44	27.06	930	900	1,300
	Exp.	17.50[67]				1,860[53]					1,332[65]
HF		5.72	6.40	600	370	700	5.59	6.32	600	370	580
	Exp.	5.52[68]	5.08[68]								830[69]
H_2O		9.87	8.81	2,800	1,540	2,700	9.61	9.54	2,700	1,390	2,000
	Exp.	9.81[67]									2,310[65]
NH_3		14.46	15.83	1,800	8,000	8,300	14.13	15.33	1,700	7,400	5,200
	Exp.	14.56[70]									6,090[65]
H_2S		24.74	24.45	14,200	5,100	20,200	24.88	24.18	13,500	5,100	11,000
	Exp.	26.6[71]									10,300[65]
C_2H_4		27.41	22.83	6,100	11,800	20,300	26.93	11.03	5,400	9,800	9,700
	Exp.	28.70[72]									9,030[53]

From Sekino, H. and Barlett, R. J., *J. Chem. Phys.*, 98, 3022, 1993. With permission.

Table 29 Scalar Components of Bond Polarizabilities and Second Hyperpolarizabilities Based on Strict Additivity and Group-Plus-Bond Polarizabilities for Monosubstituted Benzenes

Bond	α (10^{-24} esu)	γ (10^{-36} esu)
C-H	0.60	−0.0275
O-H	0.7871	0.5531
C-C	0.5996	0.6211
C-O	0.58	0.30
C-Cl	2.55	0.7725
C=C	1.47	0.61
C=O	1.41	0.99
C=S	4.01	2.2
C=N	1.89	0.33
Ph-F	0.44	−0.69
Ph-Cl	2.51	0.46
Ph-Br	3.57	1.52
Ph-I	5.40	4.31
Ph-CH$_3$	2.46	0.67
Ph-C	0.66	0.75
Ph-NH$_2$	3.06	1.83
Ph-NO$_2$	2.91	1.49
Ph-CN	2.63	0.23

Note: Ph refers to benzene ring.
From Meredith, G. B., et al., *J. Chem. Phys.*, 78, 1543, 1983. With permission.

values increase rapidly though their magnitude is within the same order as that of 6-31G *ab initio* values. For short and intermediate polyene chain lengths, the SSH-SOS values are close to the *ab initio* 6-31G+PD as well as comparable to MNDO-FF values for the long chains. The γ_{xxxx} value calculated for $N = 200$ polyene was found to be 3.75×10^{-31} esu. The $\chi^{(3)}$ value of 9.84×10^{-11} esu was calculated from the relationship

$$\chi^{(3)} = \gamma_{xxxx} F\sigma/(5Nc) \tag{37}$$

Table 30 Frequency-dependent Polarizabilities and Static Polarizabilities of Cumulenes, Polyenes, and Polyenynes

Molecule	α_{xx}(au) at $\hbar\omega$ (eV)				γ_{xxxx} (10^3 au) at $\hbar\omega$ (eV)			
	0.0	0.3	1.65	1.17	0.0	0.3	1.65	1.17
C = C = C	32.68	32.78	33.0	33.62	0.056	0.058	0.065	0.091
C = C = C = C	71.08	71.30	72.03	74.18	0.657	0.694	0.828	1.538
C = C = C = C = C	106.76	107.94	109.32	112.23	0.818	0.859	1.010	1.715
C = C = C = C = C = C	167.53	168.19	170.51	177.51	3.255	3.566	4.681	284.2
C = C–C = C	40.95				0.476	0.559		
C = C–C = C–C = C	83.94				3.009			
C ≡ C–C ≡ C	50.83	51.23	−52.08					
C = C–C ≡ C	40.28	40.35	40.68	41.69	0.326	0.339	0.386	0.608
C = C–C = C–C ≡ C	82.15	82.44	83.42	86.33	2.299	2.429	2.912	5.648
C = C–C ≡ C–C = C	84.68	85.57	86.60	89.68	2.093	2.218	2.680	5.679
Cyanoethylene		38.58	38.88	39.20		277.44	314.65	476.61
Cyanoimine		38.20	38.51	39.42		147.74	172.15	293.34
Cyanobutadiene		80.15	81.08	83.85		2239.81	2681.19	5073.16

Cyanoethylene: C = C–C ≡ N; cyanoimine: C = N–C ≡ N; cyanobutadiene: C = C–C = C–C ≡ N.
From Albert, I. D. L., et al., *J. Phys. Chem.*, 96, 10160, 1992. With permission.

Table 31 Averaged γ Values (10^{-36} esu) for Polyenes Ranging in Size from N = 4 to 30 and VEH Band Gap (E_g)

N	Band Gap (eV)	Shuai and Bredas[75]		Kurtz[76] MNDO-FF	Hurst et al.[29]		
		VEH	SSH-SOS		STO-3G	6-31G	6-31G + PD
4	5.41	0.51	1.36	2.14	0.25	0.55	7.48
6	4.13	6.42	14.20	15.18	2.74	4.97	17.69
8	3.49	36.70	59.34	52.06	11.33	20.54	41.41
10	3.08	127.50	163.42	125.45	31.12	57.73	89.87
12	2.78	333.55	349.16	243.30	66.69	127.85	174.13
14	2.59	721.20	628.90	408.15	120.90	239.94	303.98
16	2.43	1,358.62	1,003.40	618.27	197.47	407.40	491.71
18	2.32	2,305.96	1,464.36	869.25	290.29	619.66	
20	2.22	3,608.00	1,998.52	1,155.95	405.38	896.76	
22	2.14	5,295.17	2,590.92	1,472.84	526.39	1,198.92	
24	2.08	7,374.19	3,227.44	1,814.86			
26	2.03	9,838.30	3,896.22	2,178.10			
28	1.98	12,664.93	4,587.58	2,558.32			
30	1.93	15,826.55	5,294.38	2,952.42			

The average γ values are defined as $\gamma = 1/5 \; (\gamma_{xxxx} + \gamma_{yyyy} + 2\gamma_{xxyy})$.
From Shuai, Z. and Bredas, J. L., *Phys. Rev.*, B46, 4395, 1992. With permission.

where local-field correction F = 10, chain number density σ = 3.2×10^{14} cm^{-2}, one-dimensional lattice constant σ = 1.22 Å, and the factor 1/5 is for random orientation. This $\chi^{(3)}$ value corresponds to 1.85×10^{-10} esu for a well-oriented polyacetylene system where the factor is 3/8. The estimated $\chi^{(3)}$ value is in agreement with the experimental results. Figure 10 shows the VEH-SOS frequency-dependent spectra for polyene with N = 30. In the THG spectrum, three peaks correspond to $1B_u$, $2B_u$ and $2A_g$ resonance. The first two peaks have three-photon resonances while the third peak has a two-photon resonance. The DC Kerr spectrum consists of one broad peak since it is difficult to separate the various resonances which appear at about the same energies. The features of the VEH-SOS theoretical THG spectra of PPV for N = 1 and 4 oligomers were similar in characteristic to those of the N = 6 and 30 polyenes.

Table 32 lists the VEH-SOS α and γ values for phenyl-capped PPV and PTV oligomers. A comparison of the static second hyperpolarizabilities between PPV and PTV oligomers (N = 3) and linear polyenes $C_{30}H_{32}$ (N = 30) that contain 30 π electrons indicates that the γ values for the polyene chain is 16 and

Title: On the Electrodynamics of Moving Bodies
Title: On the Electrodynamics of Moving Bodies
Title: On the Electrodynamics of Moving Bodies

Title: On the Electrodynamics of Moving Bodies
Title: On the Electrodynamics of Moving Bodies

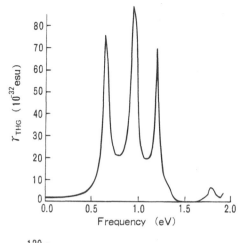

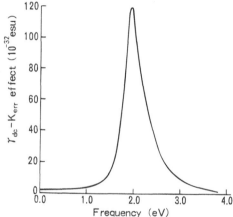

Figure 10 VEH-SOS frequency-dependent spectra for polyene with $N = 30$ for third harmonic generation (THG) and the DC Kerr effects. (From Shuai, Z. and Bredos, J. L., *Phys. Rev.*, B46, 4395, 1992. With permission.)

Table 32 VEH-SOS $\alpha(10^{-24}$ esu) and γ $(10^{-36}$ esu) Values and Band Gap (eV) for Phenyl-Capped PPV and PTV Oligomers

N	Number of π Electrons	PPV Oligomer				PTV Oligomer			
		E_g	α	γ	γ_{xxxx}	E_g	α	γ	γ_{xxxx}
1	14	3.81	22.57	42.63	184.67	3.00	27.07	67.90	304.51
2	22	3.11	44.86	326.06	1,520.03	2.26	58.66	682.56	3,419.38
3	30	2.81	66.32	961.57	4,886.22	1.96	97.11	2,868.90	14,375.08
4	38	2.65	89.53	1,923.41	9,777.14	1.80	138.94	7,527.03	37,689.30

From Shuai, Z. and Bredas, J. L., *Phys. Rev.*, B 46, 4345, 1992. With permission

5 times larger than that of the corresponding PPV and PTV oligomers. Furthermore, the magnitude of the static γ values is larger for PTV oligomers than for the PPV oligomers. This difference in γ values is related to the delocalization charge which is more facilitated in PTV than benzene outside the thiophene ring along the chain.

Ducasse et al.[77] calculated polarizabilities for finite and infinite chains of polyacetylene (PA), polythiophene (PT), and polypyrrole (PY) containing up to 400 atoms with the Huckel and PPP model by means of the finite-field (FF) formalism and by the perturbation method (SOS). Table 33 lists the polarizabilities per unit as a function of the total number of atoms N for PA, PT, and PY. The Huckel polarizabilities are larger than the PPP results. Furthermore the PPP-SOS α values are smaller than those obtained by the FF method. The difference between the SOS and the FF is not so large for small N compared with increasing N. The calculated polarizabilities vary in the order polyacetylene > polythiophene > polypyrrole. The application of the Genkin-Mednis formalism within the PPP provides

Table 33 Polarizabilities per Unit (in atomic units) as a Function of the Total Number of Atoms N for PA, PT, and PY

	Polyacetylene			Polypyrrole			Polythiophene		
N	Huckel	SOS	FF	Huckel	SOS	FF	Huckel	SOS	FF
20	439.0	96.5	136.0	157.3	50.6	70.5	228.5	65.3	94.6
40	677.1	115.2	179.2	209.4	57.8	86.2	333.7	79.4	125.6
60	776.5	121.5	198.3	228.1	60.2	91.9	374.4	84.2	136.8
80	827.3	124.7	209.5	237.7	61.5	95.2	394.7	86.6	143.2
100	857.9	126.5	216.4	243.1	62.2	97.3	407.0	88.1	147.4
200	918.9	130.3	230.5	254.3	63.6	101.4	431.6	90.9	156.4
300	939.3	131.6	253.3	258.0	64.0	102.8	439.9	91.9	159.4
400	949.5	132.2	237.6	259.9	64.3	103.5	443.9	92.4	160.9
infinite	980.1	143.1	244.7	265.6	64.9	105.5	456.2	93.8	165.4

From Ducasse, L., et al. *J. Chem. Phys.*, 97, 9389, 1992. With permission.

polarizabilities in excellent agreement with the asymptotic values obtained from the SOS method on oligomers.

From the γ values discussed, the following scaling law exponents have been reported for the various class of conjugated polymers such as polyenes, poly(*p*-phenylene vinylene (PPV), polyazomethine, polyazine, polyaromatics, and their donor or acceptor substituted derivatives (Table 34). Scaling law relates the α and γ, suggesting the scales with the length of conjugation. Theoretical studies predict a power law dependence of α with an exponent ranging from 1.3 to 3.0 and of γ with an exponent from 3.2 to 5.2 most of the results of π-conjugated polymers follow this.

Helfin and Garito[15] established the following expression for an isotropic distribution of chains

Table 34 Theoretical Values of the Scaling Law Exponent for α and γ for π-Conjugated Systems

	Exponents			
System	α	γ	Details and Methods	Ref.
Polyene oligomers	2.1	6	$N = 6$ (VEH method)	75
	1.2	3.3	$N = 30$	
		3.0	INDO method ($\omega = 0$)	28
	1.40		$N = 11$ $N = 2$ (*ab initio* 6-31G)	29
		3.67	$N = 2$ (*ab initio* 6-31G)	29
		3.98	$N = 6$, decrease $N > 6$	
		3.04	$N = 10$ to 11	
		3.8	CVB-PPP	9
		4.14	(CNDO/S-CI method)	20
		3.64	end-capped amino groups	20
		2.37	end-capped nitro groups	20
		2.70	end-capped amino and nitro groups	20
		3.85	neutral solitons	7
		6.32	charged polarons	7
		5.4	*trans*-polyene	16,17
		4.7	*cis*-polyene	16,17
		6.85	end-capped methyl groups	20
Polydiacetylene	1.399			31
Polyazomethine oligomers		4.53	(CNDO/S-CI method) end-capped amino groups	20
Polyazine oligomers	2.14	3.35	($N = 3$ (6-31G) end-capped amino group	31
	1.35	3.2	$N = 4$	
Polythiophene oligomers		5.5	derivatives (MNDO)	23
		4.5	N < 6 (CNDO)	26
Polypyrrole oligomers		5.1	derivatives (MNDO)	23

assumed as an independent source of nonlinear response with a single dominant tensor component γ_{xxxx} $(-3\omega, \omega, \omega, \omega)$:

$$\chi^{(3)}_{1111} = 1/5 \; N(f^\omega)^3 f^{3\omega} \gamma_{xxxx} \tag{38}$$

For isotropic media the Lorentz approximation is given by:

$$f(\omega) = (n^2 + 2)/3 \tag{39}$$

where n is the refractive index of the medium. Therefore, using typical values of $N = 10^{20}$ molecules/cm^3 and the refractive index of 1.8, one can roughly estimate the $\chi_{1111}{}^{(3)}$ values. For a chain of $N = 50$ carbon sites or a length of ~60 Å, the γ_{xxxx} should deviate from the power-law dependence and start to saturate. This suggests that the large values of $\chi^{(3)}$ may only require conjugated chains of intermediate length of the order of 100 Å and any increment beyond this limit may not be that significant. Theoretical and experimental data on a variety of sequentially built oligomers and their corresponding π-conjugated polymers suggest a saturation of third-order optical nonlinearity for larger chain lengths, depending upon the system.

C. EFFECT OF DONOR–ACCEPTOR GROUPS

Nakano et al.,[20] Williams,[21] and Nalwa et al. 3g have discussed several fundamental aspects of increasing the magnitude of third-order optical nonlinearity by substituting donor and acceptor functionalities on the π-conjugated chains such as polyenes, polydiacetylenes, polyazines, and polyazomethines. We are discussing several more examples to emphasize the role of different donor-acceptor groups on polarizability and hyperpolarizabilities. Meyers and Bredas[78] reported on the *ab initio* coupled Hartee-Fock calculations of polarizability and second hyperpolarizability of phenylene (Ph) and hexatriene (HT) systems end-capped by either two electron donor groups, two electron acceptor groups, or a donor and an acceptor group. The orientationally averaged α and γ values are given in Table 35. The donor group was amino while the acceptor groups were nitro and aldehyde. The α and γ are larger for the disubstituted hexatriene than for phenylene. Though donor and acceptor substituted phenylene and hexatriene have much larger α and γ values than those of benzene and hextriene molecules, for the donor and acceptor groups, the α and γ values of phenylene and hexatriene vary in the order NH$_2$/NO$_2$ > NO$_2$/NO$_2$ > NH$_2$/NH$_2$ > H/H. The α and γ values are the largest for the noncentrosymmetric push-pull type substitution, indicating that the symmetry reduction significantly enhances the magnitude of hyperpolarizability. The α and γ values of nitro and aldehyde group-substituted hexatriene are almost the same. The studies on phenylene, hexatriene, and NOT, COT, and OT suggest that symmetry reduction and dimensionality have a significant impact on third-order NLO properties.

Table 35 Orientationally Averaged α and γ Values of Donor–Acceptor Group-Substituted Phenylene and Hexatriene

X	Y	Phenylene		Hexatriene	
		$\langle\alpha\rangle$	$\langle\gamma\rangle$	$\langle\alpha\rangle$	$\langle\gamma\rangle$
H	H	7.1	0.4	10.2	4.3
NH$_2$	NH$_2$	9.1	1.3	13.0	10.6
NO$_2$	NO$_2$	10.9	3.1	15.5	21.7
NH$_2$	NO$_2$	10.5	5.6	15.7	29.8
CHO	CHO	—	—	15.8	21.7

The α is in 10^{-24} esu and γ is in 10^{-36} esu.
From Meyers, F. and Bredas, J. L., *Synth. Metals*, 49, 181, 1992. With permission.

Matsuzawa and Dixon[79] used the finite-field method with the PM-3 parametrization of the MNDO Hamiltonian to calculate the dipole moments, polarizability, and hyperpolarizabilities of benzenes (Table 36), biphenyls, fluorenes, styrene, stilbene, and tolans. To show the effect of the strength of donor and acceptor groups, examples of substituted benzenes, biphenyls, styrene, and tolans are provided here (see Tables 36 through 39). For *para*-disubstituted benzenes, thiomethyl, julolidine, $CHC(CN)_2$ and $C_2(CN)_3$ substituted benzenes, the μ, α, β, and γ values are larger compared with other substituents. Table 37 lists the results for substituted biphenyls. The cyano, sulfonyl, methoxy, nitro, and dimethylamine groups tend to increase the magnitude of μ, α, β, and γ values. The magnitude of linear and nonlinear optical parameters for donor–acceptor substituted styrenes also depends upon the strength of donor and acceptor functionalities and is the same as those of substituted biphenyls. In tolans, the benzene ring could be rotated with respect to each other as for biphenyls. The magnitude of optical nonlinearity also depends upon the torsion angle. In tolans also, the thiomethyl, nitro, dimethyamine, and cyano groups were found to be more effective than others. Beyond any doubt, the donor–acceptor substituted benzenes, biphenyls, styrenes, stilbenes, and tolans show larger optical nonlinearity than unsubstituted ones. The

Table 36 Polarizability and Hyperpolarizabilities of Donor–Acceptor-Substituted Benzenes

X	Y	$\mu(10^{-18}$ esu)	$\alpha(10^{-23}$ esu)	$\beta(10^{-30}$ esu)	$\gamma(10^{-36}$ esu)
SO_2Me	OH	5.6	1.2	0.0	2.5
$SO_2C_3F_1$	OMe	7.3	1.7	2.5	7.9
CN	Me	4.0	1.1	1.4	3.6
CN	Cl	2.8	1.1	4.6	7.0
CN	Br	2.5	1.1	1.5	3.6
CN	OPh	4.1	1.8	3.7	9.4
CN	OMe	4.2	1.2	3.0	4.7
CN	SMe	3.7	1.3	6.8	10
CN	NH_2	4.6	1.1	5.7	6.9
CN	NMe_2	4.9	1.4	7.2	10
COH	Me	3.0	1.0	1.6	3.7
COH	OPh	2.4	1.8	4.6	9.1
COH	OMe	2.6	1.1	3.7	5.0
COH	SMe	1.8	1.3	8.0	10
COH	NMe_2	3.8	1.3	6.8	11
$COCF_3$	OMe	4.2	1.2	4.9	6.5
NO	NMo_2	4.2	1.3	9.0	12
NO_2	Me	5.7	1.0	1.4	3.3
NO_2	Br	4.3	1.1	1.5	3.5
NO_2	OH	5.6	1.0	2.3	3.4
NO_2	OPh	6.2	1.8	4.5	7.9
NO_2	OMe	6.0	1.1	3.1	4.7
NO_2	SMe	5.4	1.3	8.1	11
NO_2	N_2H_3	5.8	1.2	8.0	11
NO_2	NH_2	6.6	1.1	6.3	6.8
NO_2	NMe_2	7.0	1.4	8.4	10
NO_2	CN	1.8	1.1	0.5	4.5
NO_2	COH	3.7	1.1	−0.4	3.8
$CHC(CN)_2$	OMe	4.7	1.7	9.0	21
$CHC(CN)_2$	NMe_2	6.2	2.0	16	40
$CHC(CN)_2$	Julolidine	6.5	2.4	17	42
$C_2(CN)_3$	NH_2	4.9	1.8	6.4	9.7
$C_2(CN)_3$	NMe_2	5.7	2.2	16	35
$C_2(CN)_3$	Julolidine	6.1	2.6	17	39

From Matsuzawa, N. and Dixon D. A., *J. Phys. Chem.*, 96, 6232, 1992. With permission.

Table 37 Polarizability and Hyperpolarizabilities of Donor–Acceptor Substituted Biphenyls

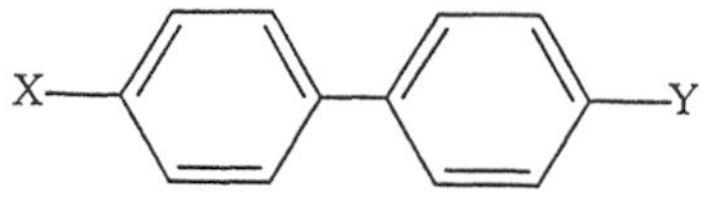

X	Y	μ $(10^{-18}$ esu)	α $(10^{-23}$ esu)	β $(10^{-30}$ esu)	γ $(10^{-36}$ esu)
H	H	0.0	1.5	0.0	6.4
CN	H	4.0	1.8	2.2	13
COCH$_3$	H	2.9	1.8	1.1	11
NO$_2$	H	5.9	1.7	3.7	13
SO$_2$C$_6$H$_{12}$OH	NH$_2$	7.0	3.1	9.8	34
CN	OH	4.2	2.0	8.5	29
COCH$_3$	OMe	4.0	2.1	3.9	19
NO$_2$	Br	4.6	2.0	4.9	17
NO$_2$	OH	6.0	1.8	7.9	20
NO$_2$	OMe	6.4	2.0	9.1	23
NO$_2$	NH$_2$	6.9	2.0	14	31
NO$_2$	NMe$_2$	7.2	2.2	17	39

From Matsuzawa, N. and Dixon D. A., *J. Phys. Chem.*, 96, 6232, 1992. With permission.

Table 38 Polarizability and Hyperpolarizabilities of Donor–Acceptor Substituted Stilbenes

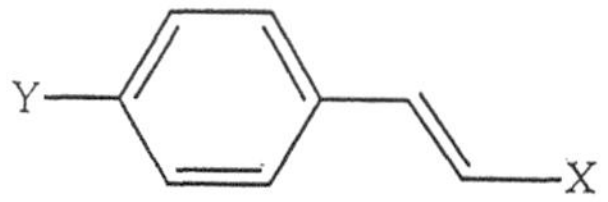

X	Y	μ $(10^{-18}$ esu)	$\alpha(10^{-23}$ esu)	$\beta(10^{-30}$ esu)	$\gamma(10^{-36}$ esu)
CN	OMe	4.6	1.6	7.5	20
CN	NMe$_2$	5.3	1.8	15	33
COH	Br	2.4	1.5	2.8	15
COH	OMe	3.0	1.5	8.3	19
COH	NMe$_2$	4.4	1.7	14	32
COMe	OMe	2.7	1.6	7.8	18
NO$_2$	H	6.1	1.3	4.3	10
NO$_2$	OH	6.2	1.4	9.5	16
NO$_2$	OMe	6.6	1.5	11	20
NO$_2$	NMe$_2$	7.7	1.8	21	37
NMe$_2$	NO$_2$	7.7	1.8	21	38

From Matsuzawa, N. and Dixon D. A., *J. Phys. Chem.*, 96, 6232, 1992. With permission.

effect of donor–acceptor groups is much larger for small molecules than stilbene and tolans. Daniel and Dupuis[80] illustrated the structure–property relationships by studying *ab initio* polarizabilities and hyperpolarizabilities of nitroaniline derivatives (Table 40).

Theoretical and experimental studies on a variety of π-conjugated short-chain molecules support the premise that the donor–acceptor functionalities can greatly boost the magnitude of third-order NLO properties.

D. EFFECT OF CONFORMATION

The effect of conformation on third-order optical nonlinearity has been described for polyenes that have *trans* and *cis* conformations. A systematic study of conformation-dependent optical nonlinearity is described here. The diagram of the geometry of *trans* and *cis*-polyacetylene is shown below.

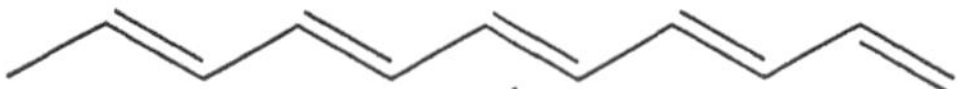

trans-polyene 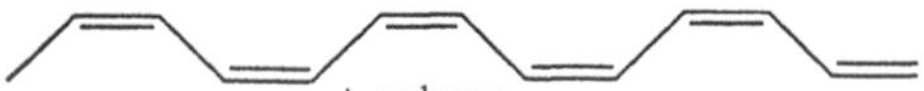   *cis*-polyene

Table 39 Polarizability and Hyperpolarizabilities of Donor–Acceptor Substituted Tolans

X	Y	μ (10^{-18} esu)	α (10^{-23} esu)	β (10^{-30} esu)	γ (10^{-36} esu)
SO$_2$Me	NH$_2$	6.3	2.8	18	66
CO$_2$Me	NH$_2$	2.8	2.7	17	60
COMe	SMe	2.4	2.8	8.5	65
COMe	NH$_2$	3.2	2.6	15	48
CN	SMe	4.0	2.8	17	79
CN	NH$_2$	5.3	2.6	19	71
CN	NHMc	5.4	2.7	20	76
CN	NMe$_2$	5.5	2.9	22	84
NO$_2$	Br	4.9	2.5	8.8	39
NO$_2$	OMe	6.9	2.5	15	48
NO$_2$	SMe	6.1	2.8	22	70
NO$_2$	NH$_2$	7.4	2.5	22	64
NO$_2$	NMe$_2$	7.7	2.8	27	82

From Matsuzawa, N. and Dixon, D. A., *J. Phys. Chem.*, 96, 6232, 1992. With permission.

Table 40 SCF Calculated Polarizabilities and Hyperpolarizabilities of Benzene Derivatives

Compound	X	Y	μ(D)	α(Å^3)	γ(10^{-36} esu)
Benzene			0.00	9.3	1.3
Aniline			1.5	11.3	1.3
nitrobenzene	H	H	5.6	12.3	1.0
p-Nitroaniline	NH$_2$	H	8.2	14.3	2.0
p-N-Methylaminonitrobenzene	NHCH$_3$	H	8.5	16.2	2.5
p-N,N-Dimethylaminonitrobenzene	N(CH$_3$)$_2$	H	8.8	18.2	3.2
2-Methyl-4-nitroaniline	NH$_2$	CH$_3$	8.2	16.1	2.2

From Daniel C. and Dupuis, M., *Chem. Phys. Lett.*, 171, 209, 1990. With permission.

Huang et al.[81] used the semiempirical Hartee-Fock method (CNDO/2) to calculate the polarizability and second hyperpolarizability of *trans*- and *cis*-polyene H–(CH=CH)$_n^{-H}$ (n = 3,4,5, 6,. . . . 18). Figure 11 shows the change of polarizability (α) and second hyperpolarizability (γ) as a function of chain length comparing the *trans* and *cis* isomers of polyacetylene. The molecular and π-system polarizabilities of *trans*-polyacetylene are larger than these of *cis* isomer. In particular, the molecular polarizability of polyenes originates from the π-electron system. The contribution of the π-electron system increases from 45% to 70% while going from n = 3 to n = 18. The contribution of the σ system in a –CH=CH– unit for various polyenes is independent of chain length. For a chain length of n = 18, the α^π and α values were 614.15 and 848.86 au, respectively, for *trans* isomer and 495.77 and 731.47 au, respectively, for the *cis* isomer. The larger polarizability of *trans* isomer can be explained by the length (l) of the projection of a unit cell –(CH=CH)– that includes both a C=C and a C–C bond. The l is 2.425 Å for *trans*-polyene and 2.125 Å for *cis*-polyene showing the projection of *trans* is larger than *cis* by 0.3 Å. As a result, the conjugation length of the *trans*-polyene π-system is longer than that of *cis*-polyene by a factor of n × 0.3 Å where n is the same. Likely, the γ values of of *trans* isomer are larger than those of the *cis* isomer. The γ shows a maximum at approximately n = 9 for *cis* isomer and n = 10–11 for

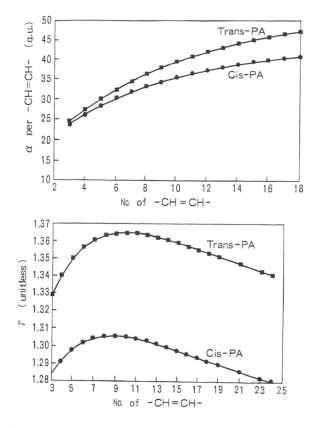

Figure 11 The change of polarizability (α) and second hyperpolarizability (γ) as a function of chain length comparing the *trans* and *cis* isomers of polyacetylene. (From Huang, Z., Jelski, D. A., Wang, R., Xie, D., Zhao, C., Xia, X., and George, T. F., *Can. J. Chem.*, 70, 372, 1992. With permission.)

the *trans* isomer. This indicates that the electrons delocalize over a larger length in the *trans* isomer than in the *cis* isomer. Therefore, the polarizability and γ values of *trans* and *cis* isomers are associated to the geometry and symmetry of the polyacetylene chain.

Papadopoulos and Waite[82] calculated the α and γ values of *cis* and *trans*-α–ω-dithienylopolyenes with the CHF-PT-EB-CNDO method to show the effect of geometry change. The α and γ values of both *cis* and *trans* isomers of α–ω-dithienylopolyenes for $n = 1, 2,$ and 3 were found to be the same, though they differ as the length of the conjugated polyene backbone increases for $n = 4$ and 5. The geometry differences between *cis* and *trans* isomers showed sensitivity by the anisotropy indices for ≥ 2.

(*cis*-α-ω-dithienylopolyenes)   (*trans*-α-ω-dithienylopolyenes)

Helfin and Garito[83] discussed the conformational dependence of optical nonlinearity for polyenes and their results are explained here. The effect of conformational changes on third-order optical nonlinearity was illustrated by comparing the *trans* and *cis* forms for polyene chains from $N = 6$ to 16 sites. The chains with an even number of short bonds ($N = 8, 12, 16, \ldots$) belong to C_{2h} group and have *trans* conformations (*all-trans*-polyene) with A_g and B_u states, while odd-number chains of short bonds ($N = 6, 10, 14, \ldots$) are noncentrosymmetric belonging to C_{2v} symmetry and possess *cis* conformation with A_1, A_2, B_1, and B_2 states. Therefore, the ground state is 1^1A_g and the next lowest energy state of the same symmetry is 2^1A_g. In the opposite symmetry, ground state is 1^1B_u where the superscript 1 refers to a single state of spin. The ^1B_u states are one-photon allowed and two-photon forbidden transitions from the ^1A_g ground states and hence called one-photon states. The excited ^1A_g states are two-photon allowed and one-photon forbidden and hence called two-photon states. The 1^1A_g and 1^1B_u states are significant due to their large transitional dipole moments. For $N = 4, 6,$ and 8, these are

Table 41 Calculated Symmetries, Excitation Energies, and Dipole Moments for the Low-Lying π-Electron States of *Trans*-Hexatriene ($n = 6$)

State	$\hbar\omega_{ng}$ (eV)	$\mu^x_{n,g}$ (Debye)	$\mu^x_{n,g}1\,{}^1B_u$ (Debye)
$2\,{}^1A_g$	4.59	0.00	2.42
$1\,{}^1B_u$	4.94	6.66	0.00
$2\,{}^1B_u$	5.22	0.17	0.00
$3\,{}^1A_g$	6.69	0.00	1.68
$4\,{}^1A_g$	6.80	0.00	0.63
$3\,{}^1B_u$	7.55	0.93	0.00
$5\,{}^1A_g$	7.97	0.00	11.40
$4\,{}^1B_u$	8.07	1.05	0.00
$6\,{}^1A_g$	8.62	0.00	0.87
$5\,{}^1B_u$	9.11	0.23	0.00

From Helfin, J. R. and Ganto, A. F., *Polymers for Lightwave and Integrated Optics*, Marcel Dekker, New York, 1992. With permission.

$4\,{}^1A_g$, $5\,{}^1A_g$, and $6\,{}^1A_g$ states at an energy of 9.18, 7.97, and 7.16 eV, respectively. The $2\,{}^1A_g$ and $1\,{}^1B_u$ states of the *cis* isomers slightly red-shift from the values of *trans* isomers by 0.02 and 0.10 eV, while the shift constantly increases with increasing chain length.

The origin of optical nonlinearity in polyenes was described by using hexatriene ($n = 6$) as an example. Tables 41 and 42 list the calculated symmetries, excitation energies, and key components of transition dipole moments for the low-lying π-electron states of *trans*- and *cis*-hexatriene ($n = 6$). For *trans*-hexatriene, the $2\,{}^1A_g$ two-photon state at 4.59 eV is the lowest-lying excited state and the $1\,{}^1B_u$ state at 4.94 eV is next to it. The $1\,{}^1B_u$ state leads to a large oscillator strength peak that appears in the near UV region of the optical absorption spectrum and its $\mu^x 1\,{}^1B_{u,g}$ is 6.66 debye. On the other hand, the $\mu^x 5\,{}^1A_g 1\,{}^1B_u$ has a very large dipole moment of 11.40 debye. For trans-octatetraene, the $\mu^x 1\,{}^1B_{u,g}$ is 7.81 debye and $\mu^x 6\,{}^1A_{g,1}\,{}^1B_u$ is 13.24 D. For *cis*-hexatriene, the $\mu^x 1\,{}^1B_{2,g}$ is 6.83 D and $\mu^x 5\,{}^1A_{g,1}\,{}^1B_2$ is 10.49 D. The $5\,{}^1A_g$ state at 7.97 eV is more important to second hyperpolarizability γ_{xxxx} (-2ω; ω, ω, 0) than the low-lying $2\,{}^1A_g$ two-photon state at 4.59 eV because of the dipole moment. These $2\,{}^1A_g$ and $5\,{}^1A_g$ states are approximately 60% composed of several double-excited configurations.

The dispersion curve for γ_{xxxx} (-2ω; ω, ω, 0) of *trans*-hexatriene from 0.5 eV to 3.0 eV shows the 2ω resonances of $2\,{}^1A_g$ state at 2.30 eV and $1\,{}^1B_u$ state at 2.47 eV. For DC-induced SHG, the calculated $\gamma_{ijkl}(-2\omega, \omega, \omega, 0)$ of *trans*-hexatriene at 0.65 eV was $\gamma_{xxxx} = 24.9 \times 10^{-36}$, $\gamma_{xyyx} = 2.0 \times 10^{-36}$, $\gamma_{xxyy} = 2.0 \times 10^{-36}$, $\gamma_{yxxy} = 2.0 \times 10^{-36}$, $\gamma_{yyxx} = 2.0 \times 10^{-36}$, and $\gamma_{yyyy} = 0.5 \times 10^{-36}$ esu. The calculated $\gamma_{ijkl}(-2\omega; \omega, \omega, 0)$ of *cis*-hexatriene at 0.65 eV was $\gamma_{xxxx} = 20.3 \times 10^{-36}$, $\gamma_{xyyx} = 1.8 \times 10^{-36}$, $\gamma_{xxyy} = 1.9 \times 10^{-36}$, $\gamma_{yxxy} = 1.8 \times 10^{-36}$, $\gamma_{yyxx} = 1.7 \times 10^{-36}$, and $\gamma_{yyyy} = 0.5 \times 10^{-36}$ esu. In the THG, the ω and 2ω resonances of ^1B_u state and 2ω resonance of ^1A_g state appear. For THG, the calculated γ_{ijkl} (-3ω; ω, ω, ω) of *trans*-hexatriene at 0.65 eV was $\gamma_{xxxx} = 4.7 \times 10^{-36}$, $\gamma_{xxyy} = 0.4 \times 10^{-36}$, γ_{yyxx}

Table 42 Calculated Symmetries, Excitation Energies, and Dipole Moments in Debye (D) for the Low-Lying π-Electron States of *cis*-Hexatriene ($n = 6$)

State	$\hbar\omega_{ng}$ (eV)	$\mu^x_{n,g}$ (D)	$\mu^y_{n,g}$ (D)	$\mu^x_n 1\,{}^1B_2$ (D)	$\mu^y_n 1\,{}^1B_2$ (D)
$2\,{}^1A_1$	4.55	0.00	0.08	1.72	0.00
$1\,{}^1B_2$	4.92	6.83	0.00	0.00	0.01
$2\,{}^1B_2$	5.24	0.14	0.00	0.00	1.67
$3\,{}^1A_1$	6.75	0.00	1.11	1.55	0.00
$4\,{}^1A_1$	6.80	0.00	0.62	0.24	0.00
$3\,{}^1B_2$	7.46	1.52	0.00	0.00	0.30
$4\,{}^1B_2$	7.72	1.84	0.00	0.00	0.09
$5\,{}^1A_1$	7.86	0.00	0.11	10.49	0.00
$6\,{}^1A_1$	8.78	0.00	1.73	1.12	0.00
$5\,{}^1B_2$	9.12	0.24	0.00	0.00	2.88

From Helfin, J. R. and Garito, A. F., *Polymers for Lightwave and Integrated Optics*, Marcel Dekker, New York, 1992. With permission.

$= 0.4 \times 10^{-36}$, and $\gamma_{yyyy} = 0.1 \times 10^{-36}$ esu. The optical nonlinearity at a fixed wavelength in all cases was higher for the *trans*-polyene chain than for the *cis* chain. The power law exponents in γ_{xxxx} (-3ω; ω, ω, ω) were 3.9 for *cis* and 4.0 for the *trans* chain.

For longer chain lengths, the large number of one-photon and two-photon states significantly contribute to $\gamma_{xxxx}(-\omega_4; \omega_{41}, \omega_2, \omega_3)$. The rapid increase of optical nonlinearity with increased chain length is associated with excitation energies and the magnitude of the transition dipole moments along the chain x direction. The excitation energies of the lowest-lying one photon 1^1B_u and two-photon 2^1A_g states of *all-trans*-polyenes decrease with increasing chain length. This decrease in excitation energies leads to an increase of the individual term's contribution to γ_{xxxx} ($-\omega_4$; ω_{41}, ω_2, ω_3). The transition dipole moment increases monotonically with increasing chain length. The correlated π-electron virtual transitions give rise to a charge redistribution over chain length and this leads to increased $\gamma_{xxxx}(-\omega_4; \omega_1, \omega_2, \omega_3)$. It was concluded that an intermdiate chain length of 100 Å should be sufficient to achieve large optical nonlinearities.

The effect of conformation on second hyperpolarizability has also been studied in polythiophene by Fichou et al.[26] The dependence of second hyperpolarizability as a function of θ, the angle between two successive thiophene monomer units for a helical polythiophene, was described.

E. EFFECT OF LOWERED SYMMETRY

The *trans*-octatetraene (*trans*-OT) has eight carbon atom sites. Figure 12 shows the calculatcd dispersion curve for $\gamma(-3\omega; \omega, \omega, \omega)$ of *trans*-OT as the function of the input photon energy.[15–18] The dotted curve represents the analogs calculated dispersion for *cis*-OT. The vertical dashed lines in the figure show the location of the first resonance at 1.47 eV ($\lambda = 0.84$ μm) due to 3ω resonance and second singularity at 2.08 eV ($\lambda = 0.6$ μm) due to the 2ω resonance of the 2^1A_g state. The 2^1A_g state will have only 2ω resonance while the ^1B_u state will have both 3ω and 2ω resonance in third harmonic generation resonance, selection rules for other third-order nonlinear processes are different, while the resonance characteristics of the *trans* and *cis* conformations are quite similar. Resonant γ values for the *trans* are higher than those for the *cis*-OT. In comparison,

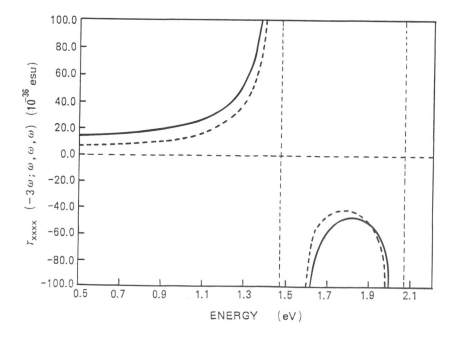

Figure 12 Calculated dispersion curve for $\gamma(-3\omega; \omega, \omega, \omega)$ of *trans*-OT (solid curve) and *cis*-OT (dotted curve) as a function of the input photon energy. (From Helfin, J. R. and Garito, A. F., in *Electroresponsive Molecular and Polymeric Systems*, Skotheim, T. A., ed., Marcel Dekker, New York, 1991, 1–48. With permission.)

the counter diagrams[15–18,82] of the *cis* and *trans* conformations of OT are shown in Figure 13, where ρ_n of the ground state and 6^1A_g states with 1^1B_u state were considered. The solid and dashed lines represent the increased and decreased charge density. The $g{\rightarrow}1^1B_u$ transition leads to modulated redistribution of charge with transition moment of 7.9 and 7.8 D for *cis* and *trans* conformations, respectively. The $1^1B_u{\rightarrow}6^1A_g$ transition results from the highly separated charge distribution; the transition moment of *cis* conformation is 12 D, whereas it is 13.2 D for the *trans* conformation. The variation in hyperpolarizability with the size of the chain is very similar in both cases. The schematic diagram of a noncentrosymmetric analog to OT is 1,1-dicyano, 8-*N,N*-dimethylamino-1,3,5,7-octatetraene (NOT) with electron-accepting cyano groups at one end and electron-donating dimethylamino groups at the other end. To understand the role of the lowered symmetry in third-order optical nonlinearity, examine the OT molecule. In this case, symmetry is lowered by introducing heteroatomic functionalities on the linear conjugated chain. Like OT, its third harmonic susceptibility also increases to the first resonance occurring at 1.0 eV due to 3ω resonance of the $2^1A'$ state. Unlike OT, NOT is a noncentrosymmetric molecule, therefore, the 3ω and 2ω resonance selection rules are not applicable in this case. As a result, every excited state allows 3ω, 2ω, and ω resonance, and the dispersion characteristics of $\gamma(-3\omega; \omega, \omega, \omega)$ show these resonances at frequencies beyond the first resonance. Compared to OT, the lowered symmetry of NOT generates a new type of virtual excitation process, where γ_{xxxx} dominates and increases the γ_{xxxx} values by an order of

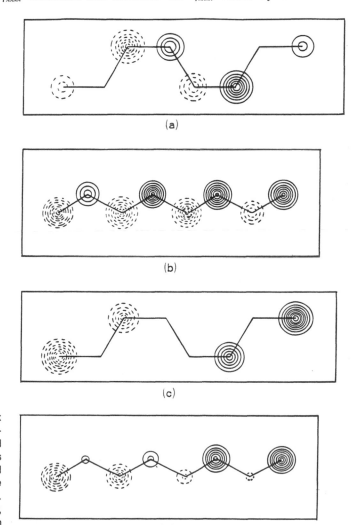

Figure 13 Transition density matrix diagrams for (a) *cis*-OT and (b) *trans*-OT ground states with 1^1B_u state and (c) *cis*-OT and (d) *trans*-OT 6^1Ag states with 1^1Bu state. (From Helfin, J. R. and Ganto, A. F., in *Electroresponsive Molecular and Polymeric Systems,* vol. 2., Skotheim, T. A., ed., Marcel Dekker, New York, 1991, 1–48. With permission.)

magnitude for NOT. The calculated nonresonant $\gamma_{xxxx}(-3\omega; \omega, \omega, \omega)$ values at 0.65 eV are 173 $\times 10^{-35}$ esu for NOT and 15.5×10^{-35} esu for OT.

Chemical structure of 1,1-dicyano-8-N,N'-dimethylamino-1,3,5,7-octatetraene (NOT)

F. EFFECT OF DIMENSIONALITY

Another effect on $\gamma_{ijkl}(-\omega_4; \omega_1, \omega_2, \omega_3)$ described by Garito et al.[15-19] is the dimensionality. We will present several examples of how third-order NLO properties of organic conjugated molecules are affected by the dimensionality. Phthalocyanines, porphyrins, and annulenes are examples of two-dimensional π-electron systems, while linear chain π-conjugated systems, such as polyenes, polyynes, polyenynes, polyazomethine, polyazines, and polystilbenes have one-dimensional structures. In the same series, OT and NOT are one-dimensional molecules. The cyclooctatetraene (COT) is the best example for describing microscopic nonlinearity in a two-dimensional conjugated cyclic structure. The calculated $\gamma_{ijkl}(-3\omega; \omega, \omega, \omega)$ values at the nonresonant fundamental photon energy of 0.65 eV are $\gamma_{xxxx} = 0.75 \times 10^{-36}$ esu and $\gamma_{xxyy} = 0.21 \times 10^{-36}$ esu for COT. The γ_{xxxx} component of OT is larger than that of any of the components of COT. The γ_g value for COT is 0.36×10^{-36} esu and has significant components in both the x and y directions, while for trans-OT the γ value is 3.4×10^{-36} esu. The cyclic structures of phthalocyanines and porphyrins are also two-dimensional charge-transfer molecules. In one-dimensional linear chains, the x component of the transition moment is larger than y and z components. Thus γ_{xxxx} is the one that affects the averaged susceptibility γ_g. In this case, the equation is reduced to

$$\gamma_g = 1/5\, \gamma_{ixxxx} \tag{40}$$

Now considering the cyclic structures, the γ_{xxxx} and γ_{yyyy} should be of the same magnitude. In addition, there should be a significant contribution from the γ_{xxyy} component. The corresponding equation can be written as:

$$\gamma_g = 1/5[\gamma_{xxxx} + \gamma_{yyyy} + 1/3(\gamma_{xxyy} + \gamma_{xyxy} + \gamma_{xyyx} + \gamma_{yyxx} + \gamma_{yxyx} + \gamma_{yxxy})] \tag{41}$$

Chemical structure of two-dimensional molecule cyclooctatetraene (COT).

The isotropically averaged third-order susceptibility is much less for the two-dimensional structure compared with the one-dimensional chain. The length for a two-dimensional cyclic structure is one-half of the circumference of the ring, compared to the full end-to-end chain length in a one-dimensional structure. The smaller nonresonant third-order susceptibility values are expected for two-dimensional conjugated cyclic structures but not for their one-dimensional analogs.

Ducuing[84] suggested the following relationship between the dimensionality of π-electron conjugation and second hyperpolarizability γ. The γ value should decrease as the square root of the increasing dimensionality number D

$$\gamma \propto D^{-1/2}$$

In this context, trans-polyenes, phthalocyanines, and fullerenes are the examples of one-,dimensional (1-D), two-dimensional (2-D) and three-dimensional (3-D) systems, respectively. Kajzar[85] pointed out that an experimental verification of this dependence is difficult due to the different resonance contributions to the γ values at measurement wavelengths. The γ values of trans-polyene as 1.5×10^{-31} esu, copper phthalocyanine (CuPc) as 2.2×10^{-31} esu and C_{60} as 3×10^{-33} esu were estimated without considering the resonance width. Almost similar γ values were obtained, except for CuPc, and this difference originates most likely from the packing. Experimental data also evidenced the effect of dimensionality on third-order optical nonlinearity, which follows an order: 1-D > 2-D > 3-D. Theoretical calculations by Matsuzawa and Dixon[24] also supported this order, showing the effect of dimensionality.

G. EFFECT OF CHARGE-TRANSFER COMPLEX FORMATION

Kakuta[86] reported hyperpolarizabilities of tetrathiofulvalene (TTF), tetracyanoquinodimethane (TCNQ), and TTF-TCNQ complex. The TTF-TCNQ is a well-known organic conductor and provides an example of super molecule calculations for charge-transfer complexes. The orientationally averaged $<\alpha>$ and $<\gamma>$ values for TTF-TCNQ complex calculated from the STO-3G set are several-fold larger than TTF and TCNQ molecules, indicating the role of charge-transfer formation in increasing the magnitude of optical nonlinearity. Table 43 lists the calculated hyperpolarizability of TTF, TCNQ, and TTF-TCNQ complex.

Nakano et al.[87] used the coupled Hartee-Fock (CHF) method based on the INDO approximation to calculate the γ values of alternate donor–acceptor stacks and of segregated charge-transfer clusters. The structure below shows the spatial arrangements of the alternate donor *trans*-diaminoethylene and acceptor *trans*-dicyanoethylene stacks having a fixed intermolecular distance of 3.0 Å. The donor and acceptor molecules were assumed to be placed in such a fashion that has maximum intermolecular interaction. The γ value increases linearly for the alternate stacks as the number of stack (N) increases from 2 to 8. The γ value for the alternate stack having 8 monomers was 8443 au, about 7 times larger than the dimer ($N = 2$) while the γ of the hexamer ($N = 6$) was 5 times larger that for the dimer. The neutral and ionic segregated stacks show relatively smaller γ values compared with CT complexes. These results were found to be in agreement with the mixed stack of tetracyanoquinodimethane (TCNQ) and perylene reported in the literature.[88,89] Therefore, charge-transfer complex formation seems effective for enlarging third-order optical nonlinearity.

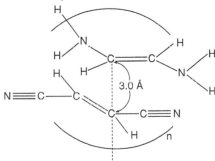

Table 43 Chemical Structures and Coordinate Systems and Calculated Polarizability (Å³) and Hyperpolarizability (10^{-39} esu) of TTF, TCNQ, and TTF-TCNQ Complex

Tensor	TTF	TCNQ	TTF-TCNQ Complex
α_{xx}	11.235	13.143	23.496
α_{yy}	20.633	36.557	62.953
α_{zz}	3.598	2.600	6.101
$<\alpha>$	11.822	17.433	30.85
γ_{xxxx}	1,247.699	2,183.746	3,012.216
γ_{xxyy}	1,762.759	12,686.199	15,841.380
γ_{xxzz}	207.211	83.920	271.222
γ_{yyyy}	4,300.524	10,570.857	31,380.498
γ_{yyzz}	1,175.176	−12.044	233.280
γ_{zzzz}	−385.586	−27.017	−381.171
$<\gamma>$	1,890.586	7,648.748	13,340.661

Courtesy of A. Kakuta, Hitachi, Ltd.

H. EFFECT OF π-BONDING SEQUENCE

Nalwa et al.[90] calculated polarizabilities (α) and second hyperpolarizabilities (γ) of well-defined π-conjugated systems: N=CH–CH=N, CH=N–N=CH, CH=N–CH=N, CH=CH–CH=CH, and C≡C–C≡C end-capped with electron-donating amino groups by *ab initio* coupled perturbed Hartee-Fock method to examine the effect of the nature of π-bonding sequence third-order optical nonlinearity. The chemical structures of five different π-π-conjugated systems end-capped with electron-donating amino groups are shown below. Table 44 lists the α and γ values of model compounds of Series I, II, and III. The $<\alpha>$ showed a different trend than the α_{yy} component from the same geometrical approximation because the α_{yy} component lies along the chain direction which is more sensitive to the geometry. Both dimeric and trimeric model molecules showed a different trend than that of Series I though they have exactly the same π-bonding sequence. The length of π-electron delocalization increases in Series II and III which may have an important contribution.

$$H_2N-\left(N=CH-CH=N\right)_n-NH_2$$

$$H_2N-\left(CH=N-N=CH\right)_n-NH_2$$

$$H_2N-\left(CH=N-CH=N\right)_n-NH_2$$

$$H_2N-\left(CH=CH-CH=CH\right)_n-NH_2$$

$$H_2N-\left(C\equiv C-C\equiv C\right)_n-NH_2$$

Series I (monomer) $n = 1$ for (1), (2), (3), (4), and (5)
Series II (dimer) $n = 2$ for (6), (7), (8), (9), and (10)
Series III (trimer) $n = 3$ for (11), (12), (13), (14), and (15)

From the *ab initio* calculations, the γ values of the azine (–N–CH=CH–N–) conjugated system were found to be the largest and those of CH–N–N=CH, the smallest for the Series I monomeric model compounds of the same chain length. On the other hand, the γ value varied in an order CH=CH–CH=CH > C≡C–C≡C > N–CH=CH–N > CH=N–CH=N > CH=N–N=CH for the Series II dimeric model compounds and the same order was followed for the Series III trimeric model compounds. In Series II and III, polyene (CH=CH) and polyyene (C≡C) bonds become more important than the polyazine (N=CH–CH=N) bond. The $<\gamma>$ and γ_{yyyy} of CH=CH bond in Series II were more than 1.72 and 1.83 times larger than that of N=CH–CH=N bond, respectively. The $<\gamma>$ and γ_{yyyy} of CH=CH bond in Series III are 2.56 and 2.74 times larger than that of N=CH–CH=N bond, respectively. The trend of the $<\gamma>$ and γ_{yyyy} variations in these three series is quite different with the exception of the CH=N–N=CH bond, which

Table 44 Polarizabilities in Å^3 and Second Hyperpolarizabilities in 10^{-39} esu of Series I, II, and III π-Conjugated Systems End-Capped with Amino Groups

Molecules	α_{yy}	$<\alpha>$	γ_{yyyy}	$<\gamma>$
(1)	12.37	7.84	14,677.46	3,232.52
(2)	12.00	7.42	4,720.50	1,029.72
(3)	12.84	7.63	9,557.36	2,016.50
(4)	13.13	8.51	11,818.31	2,370.13
(5)	15.43	6.90	10,310.00	2,384.90
(6)	26.54	16.07	77,003.16	19,264.77
(7)	27.25	15.49	58,627.10	12,956.63
(8)	30.36	15.80	83,911.54	17,425.71
(9)	32.08	18.81	141,048.05	33,155.04
(10)	39.07	15.59	114,950.11	23,450.98
(11)	42.76	25.26	207,564.91	56,011.50
(12)	45.19	24.62	199,785.61	46,476.15
(13)	52.10	24.92	280,522.87	56,998.07
(14)	55.88	31.16	568,879.74	143,678.37
(15)	68.37	26.16	445,537.66	89,636.07

From Nalwa, H. S., et al., *J. Phys. Chem.*, 99, 10766, 1995. With permission.

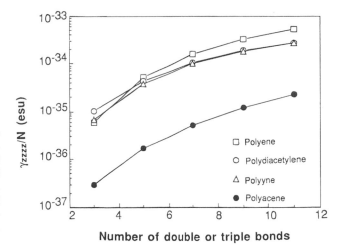

Figure 14 γ_{zzzz}/N vs. number of double/triple bonds for polyene, polyyne, polydiacetylene, and polyacenes showing the effect of π-bonding sequence on third-order optical nonlinearity. (From Kakuta, A., Imanishi, Y., Nalwa, H. S., Hamada, T., Ishihara, S., and Hattori, S., *Proc. Int. Symp. Nonlinear Photon. Mater.,* May 24–25 1994, Tokyo, p. 263. With permission.)

shows the smallest second hyperpolarizabilities in all cases. Furthermore, the presence of nitrogen atom in the conjugated backbone has a diminutive effect. The larger polarizabilities of a polyene chain (CH=CH) vs. a polyyne chain (C≡C) result from the more homogenous bonding sequence, the former being more polarizable. Even the polydiacetylene backbone with an acetylenic form is less polarizable than that of the butadiene form.[33,34] Pariser-Parr-Pople[74] and *ab initio*[91,92] calculations indicated that the largest γ values were for polyenes rather than polyynes and polydiacetylenes and significantly lower for polyacenes. The simple power law for $n = 3$ showed exponents for $<\gamma>$ as 2.59, 3.3, 3.73, 3.48, and 3.04 for N–CH=CH-N, CH=N–N=CH, CH=N–CH=N, CH=CH-CH=CH, and C≡C–C≡C π-conjugated systems, respectively. The effect of π-bonding sequence on the second hyperpolarizability is shown in Figures 14 and 15. The γ values of polyene, polyyne, polydiacetylene, and polyacene increase with the increase of π-conjugation length and follow an order polyene > polyyne > polydiacetylene > polyacene.[92] For example, the γ value of the polyene chain was 1.9 times that of polyyne and polydiacetylene chains and 22 times that of polyacene chains for double/triple bonds equal to 22 calculated from the CHFF/6-31G basis set.

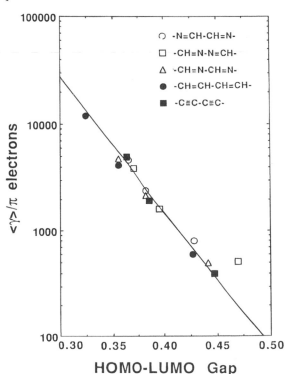

Figure 15 Plot of the mean γ/π-electrons as a function of HOMO-LUMO gap (au) for π-conjugated organic molecules of Series I, II and III. The HOMO-LUMO gap is in atomic unit where 1 au = 27.211608 eV. (From Nalwa, H. S., Mukai, J., and Kakuta, A., *J. Phys. Chem.,* 99, 10766, 1995.)

The γ values of polyene chains were more than 20 times larger that of polyacene at any given length; this indicates that one-dimensional π-conjugated systems are better candidates for third-order nonlinear optics. A similar observation was supported by the experimental results.[90] Figure 15 shows the plots of $<\gamma>/\pi$ electrons as a function of HOMO-LUMO gaps for the five different π-conjugated systems. A correlation between the γ values and the HOMO-LUMO gap exists. The HOMO-LUMO gap showed a decrease with the increase of the conjugation length. The polyacetylene system showed the smallest values among all π-bonding sequences which is also an indicator for the larger α and γ values for polyenes. The HOMO-LUMO gap of the π-conjugated systems substituted with N is significantly larger than the polyacetylene system and also showed lower α and γ values than polyenes, polyynes and polydiacteylenes. These *ab initio* calculations demonstrate that the nature of π-bonding sequence plays an important role for third-order optical nonlinearity, and that the π-bonding sequence should be taken into consideration while designing novel organic molecules for third-order nonlinear optics.

VIII. CONCLUSIONS

It is interesting to identify organic molecules and polymers with large third-order optical nonlinearities by applying quantum chemistry approaches.[93–96] These theoretical data provide at least the first-hand preliminary information on how third-order optical nonlinearity is affected by π-electron delocalization length, electron donor and acceptor substituents, lowered symmetry, heteroatoms incorporated into π-conjugated systems, conformational changes, dimensionality, and charge-transfer complex formation, etc. To some extent, a useful comparison may be made. A similar situation is generally encountered in experimental measurements of polarizabilities and third-order nonlinear susceptibilities.

It is clear that a good approximation of third-order optical nonlinearity of organic materials is possible through computational methods, but precise evaluations have not yet been reached to match experimental results. The following points become clear in evaluating organic materials for third-order nonlinear optics.

1. Besides C, N, H, S, and P, the CNDO method has been found useful for molecules containing elements like Li, F, Cl, Fe, Ni, Pd.[14,19,97–102]
2. Both α and γ values of organic molecules are affected by atomic and chemical bonds of molecules. In particular, C-Cl, C=S, and C=N, iodobenzene, aniline, nitrobenzene, and benzonitrile-type linkage were found to be more effective.
3. Both α and γ values of organic molecules increase with the increase of the delocalization length of molecules (e.g., polyenes, polydiacetylenes, polyazines, polyarylenes). Two-dimensional polycyclic aromatic hydrocarbons also yield large α and γ values shown by semiempirical calculations.[2,92,103].
4. The second hyperpolarizability is remarkably influenced by the π-bonding sequence and polyenes show larger polarizabilities than those of polyynes, polydiacetylenes, polyacenes, polyazine, polyazomethine, and other π-conjugated systems.
5. Both α and γ values are sensitive to the nature of the π-bonding sequence in linear conjugated molecules and polymers.
6. The nonresonant nonlinearities for noncentrosymmetric polyene linear chains are expected to be larger by at least an order of magnitude.
7. Large values of α and γ values are obtained for carbon linear chain molecules.
8. Electron donor and acceptor substituted molecules have larger values of α and γ than those of regular organic structures. The donor and acceptor groups substituted in a *para* position of a π-conjugated system generally give rise to larger polarizabilities than those substituted either by donor or acceptor groups. The donor–acceptor groups favor a large charge displacement in the main leading to the large second hyperpolarizability.
9. The two-dimensional π-conjugated cyclic systems are expected to have smaller optical nonlinearities than their one-dimensional analogs. Dimensionality affects the magnitude of α and γ values though this was difficult to verify from the currently available data.
10. The formation of a charge-transfer complex between an acceptor and a donor also leads to large third-order optical nonlinearities such as in the TTF-TCNQ complex.

ACKNOWLEDGMENTS

The author is especially grateful to Professor M. G. Papadopoulos of the National Hellenic Research Foundation in Athens, Greece, for critically reading the manuscript and for valuable comments and suggestions. The author also thanks Dr. A. Kakuta of Hitachi Research Laboratory for stimulating discussions and support throughout this work.

REFERENCES

1. **Waite, J. and Papadopoulos, M. G.,** The effect of intermolecular process on the polarizabilities and hyperpolarizabilities of some amines, *J. Chem. Phys.,* 82, 1427, 1985.
2. **McIntyre, E. F. and Hameka, H. F.,** Extended basis set calculations of nonlinear susceptibilities of conjugated hydrocarbonds, *J. Chem. Phys.,* 69, 4814, 1978.
3. **Zamani-Khamiri, O., McIntyre, E. F., and Hameka, H. F.,** Polarizability calculations with the SCF method. II. The benzene molecule, *J. Chem. Phys.,* 72, 1280, 1980.
4. **Zamani-Khamiri, O., McIntyre, E. F., and Hameka, H. F.,** Polarizability calculations with the SCF method. III. Ethyelne, butadiene and hexatriene, *J. Chem. Phys.,* 72, 5906, 1980.
5. **Zamani-Khamiri, O. and Hameka, H. F.,** Polarizability calculations with the SCF method. IV. Various conjugated hydrocarbons, *J. Chem. Phys.,* 73, 5693, 1980.
6. **McIntyre, E. F. and Hameka, H. F.,** Calculation of nonlinear susceptibilities of nitrogen containing heterocycles and of nitriles, *J. Chem. Phys.,* 70, 2215, 1979.
7. **DeMelo, C. P. and Silbey, R.,** Variational perturbation treatment for the polarizabilities of conjugated chains. II. Hyperpolarizabilities of polyenic chains, *J. Chem. Phys.,* 88, 2567, 1988.
8. **Kavanaugh, T. C. and Silbey, R. J.,** A simple model for polarizabilities of organic polymers, *J. Chem. Phys.,* 95, 6924, 1991.
9. **Soos, Z. G. and Ramasesha, R.,** Valence bond approach to exact nonlinear optical properties of conjugated systems, *J. Chem. Phys.,* 90, 1067, 1989.
10. **Papadopoulos, M. G., Waite, J., and Nicolaides, C. A.,** Calculations of induced moments in large molecules. II. Polarizabilities and second hyperpolarizabilities of some polyenes, *J. Chem. Phys.,* 77, 2527, 1982.
11. **Waite, J., Papadopoulos, M. G., and Nicolaides, C. A.,** Calculations of induced moments in large molecules. III. Polarizabilities and second hyperpolarizabilities in some aromatics, *J. Chem. Phys.,* 77, 2536, 1982.
12. **Waite, J. and Papadopoulos, M. G.,** The second hyperpolarizability of 1-nitronaphthalene, *J. Chem. Phys.,* 80, 3503, 1984.
13. **Waite, J. and Papadopoulos, M. G.,** Calculations of induced moments in large molecules. V. A study of intermolecular interactions and a functional group analysis of some amides, through the investigations of their polarizabilities and hyperpoarizabilities, a comparative study, *J. Chem. Phys.,* 83, 4047, 1985.
14. **Waite, J., and Papadopoulos, M. G.,** Calculation of polarizability and second hyperpolarizability of ferrocene, $Fe(C_5H_5)_2$, using the extended CHF-TP-EB-CNDO method, *Z. Naturforsch,* 42, 749, 1987.
15. **Garito, A. F., Heflin, J. R., Wong, K. Y., and Zamani-Khamiri, O.,** in *Nonlinear Optical Properties of Polymers, Materials Research Society Symposium Proceedings,* Heeger, A. J., Ulrich, D., and Orenstein, J., (Eds.) Vol. 109, 1988, 91.
16. **Heflin, J. R. and Garito, A. F.,** Electron correlation description of the nonlinear optical properties of lower dimensional conjugated structures, in *Electroresponsive Molecular and Polymeric Systems, Vol. 2.* Skotheim, T. A., Ed., Marcel Dekker, New York, 1991, 1.
17. **Grossman, C., Heflin, J. R., Wong, K. Y., Zamani-Khamiri, O., and Garito, A. F.,** in *Nonlinear Optical Effects in Organic Polymers,* Messier, J., Ed., NATO ASI Series E, Vol. 162, Kluwer Academic Press, Boston, 1989, 61.
18. **Heflin, J. R., Wong, K. Y., and Zamani-Khamiri, O and Garito, A. F.,** Nonlinear optical properties of linear chains and electron-correlation effects, *Phys. Rev.,* B38, 1573, 1988.
19. **Craig, B. I. and Williams, G. R. J.,** A theoretical study of the hyperpolarizabilities of nickel dithiolene molecules, *Adv. Mater., Opt. Electron.,* 1, 221, 1992.
20. **Nakano, M., Okumura, M., Yamaguchi, K., and Fueno, T.,** CNDO/S-CI calculations of hyperpolarizabilities. III. Regular polyenes, charged polyenes, disubstituted polyenes, polydiacetylene and related species, *Mol. Cryst. Liq. Cryst.,* 182A, 1, 1990.
21. **(a) Williams, G. R. J.,** Finite field calculations of molecular polarizability and hyperpolarizabilities for organic π electron systems, *J. Mol. Struct.,* (Theochem), 151, 215, 1987. **(b) Williams, G. R. J.,** Hyperpolarizability of conjugated organic systems. Part III. Organo sulfur-nitrogen ring systems, *J. Mol. Struct. (Theochem),* 153, 191, 1987.
22. **Williams, G. R. J.,** Hyperpolarizability of conjugated organic systems. Part II. Polydiacetylene (polyenyne) oligomers, *J. Mol. Struct. (Theochem),* 153, 185, 1987.
23. **Williams, G. R. J.,** Nonlinear optical properties of conjugated organic systems. Part IV. Small bandgap derivative polypyrrole and polythiophene oligomers, *J. Mol. Electron.,* 6, 99, 1990.

24. **Matsuzawa, N. and Dixon, D. A.,** Semiempirical calculations of the polarizability and second-order hyperpolarizability of C_{60}, C_{70}, and model aromatic compounds, *J. Phys. Chem.,* 96, 6241, 1992.

25. **Li, J., Feng, J., and Sun, J.,** Quantum chemical calculations on the spectra and nonlinear third-order optical susceptibility of C_{60}, *Chem. Phys. Lett.,* 203, 560, 1993.

26. **Fichou, D., Garnier, F., Charra, F., Kajzar, F., and Messier, J.,** Linear and nonlinear optical properties of thiophene oligomers, in *Organic Materials for Nonlinear Optics,* Hahn, R. A., and Bloor, D., Eds., Royal Society of Chemistry Special Publication No. 69, London, 1989, 176.

27. **Pierce, M. B.,** The importance of treating electron–electron interactions in the calculation of dynamic third-order polarizabilities for linear conjugated polyenes, in *Nonlinear Optical Properties of Polymers,* Materials Research Society Symposium Proceedings, Vol. 109, Heeger, A. J., Orenstein, J. and Ulrich, D. R., Eds. Materials Research Society, Pittsburgh, 1988, 109.

28. **Pierce, B. M.,** Theoretical analysis of third-order nonlinear optical properties of linear polyenes and benzene, *J. Chem. Phys.,* 91, 791, 1989.

29. **Hurst, G. J. B., Dupuis, M., and Clementi, E.,** *Ab initio* analytic polarizability, first and second hyperpolarizabilities of large conjugated organic molecules: Applications to polyenes C_4H_6 to $C_{22}H_{24}$, *J. Chem. Phys.,* 89, 385, 1988.

30. **Bodart, V. P., Delhalle, J., Dory, M., Fripita, J. G., and Andre, J. M.,** Finite hydrocarbon chains incorporating cumulenic structures: Perediction by *ab initio* calculations of their equilibrium geometry and electric polraizability, *J. Opt. Soc. Am. B,* 4, 1047, 1987.

31. **Nalwa, H. S., Hamada, T., Kakuta, A., and Mukoh, A.** Third-order nonlinear optical properties of polyazine and its oligomers, *Nonlinear Optics,* 6, 155, 1993.

32. **Nalwa, H. S., Hamada, T., Kakuta, A., and Mukoh, A.,** Polarizabilities and second hyperpolarizabilities of donor and acceptor substituted polyazines and polyacetylenes, *Nonlinear Optics,* 7, 193, 1994.

33. **Andre, J. Barbier, C. Bodart, V., and Delhalle J.** in *Nonlinear Optical Properties of Organic Molecules and Crystals,* Zyss J. and Chemla, D. S., Eds., Academic Press, Orlando, 1987, Vol. 2, p, 137.

34. **Andre, J. and Delhalle, J.,** Quantum chemistry and molecular engineering of oligomeric and polymeric materials for optoelectronics, *Chem. Rev.,* 91, 843, 1991.

35. **Dory, M., Bodart, V. P., Delhalle, Andre, J. M., and Bredas, J. L.,** Theoretical design of polymeric materials for nonlinear optics, in *Nonlinear Optical Properties of Polymers,* Materials Research Society Symposium Procedures, Vol. 109, Heeger, A. J., Orenstein, J., and Ulrich, D. R., Eds., Materials Research Society, Pittsburgh, 1988, 239.

36. **Bredas, J. L., Dory, M., Themans, B., Delhalle, J., and Andre, J. M.,** Electronic structure and nonlinear optical properties of aromatic polymers and their derivatives, *Synth. Metals,* 28, D533, 1989.

37. **Kirtman, B. and Hasan, M.,** Ab initio longitudinal polarizabilities and hyperpolarizabilities of polydacetylene and polybutatriene oligomers, *Chem. Phys. Lett.,* 157, 123, 1989.

38. **Kirtman, B. and Hasan, M.,** Linear and nonlinear polarizabilities of *trans*-polysilane from *ab initio* oligomer calculations, *J. Chem. Phys.,* 96, 470, 1991.

39. **Karna, S. P. and Dupuis, M.,** Frequency-dependent hyperpolarizabilities of haloforms from ab initio SCF calculations, *Chem. Phys. Lett.,* 171, 201, 1990.

40. **Bratan, D. N., Onuchic, J. N., and Perry, J. W.,** Nonlinear susceptibilities of finite conjugated organic polymers, *J. Chem. Phys.,* 91, 2696, 1987.

41. **Sekino, H. and Barlett, R. J.,** Hyperpolarizabilities of molecules with frequency dependence and electron correlation, *J. Chem. Phys.,* 94, 3665, 1991.

42. **Ohta, K.,** Ab initio molecular orbital calculation of third-order molecular hyperpolarizability, Extended Abstracts, Basic Technologies for Future Industries, The 3rd Symposium on Photonic Materials, October 20–21, 1992, Tokyo, Japan, p. 34 (In Japanese).

43. **Svendsen, E. N., Stroyer-Hansen, T., and Hameka, H. F.,** Calculation of the nonlinear electric susceptibility of benzene, *Chem. Phys. Lett.,* 54, 217, 1978.

44. **Marchese, F. T. and Jaffe, H. H.,** *Theor. Chim. Acta,* 45, 241, 1977.

45. **Mathies, R. and Albrecht, A. C.,** Experimental and theoretical studies on the excited state polarizabilities of benzene, naphthalene and anthracene, *J. Chem. Phys.,* 60, 2500, 1974.

46. **Bounds, P. J.,** *Chem. Phys. Lett.,* 70, 143, 1980.

47. **Chablo, A. and Hinchliffe, H.,** *Chem. Phys. Lett.,* 72, 149, 1980.

48. **Dewar, M. J. S., Yamaguchi, Y., and Suck, S. H.,** *Chem. Phys. Lett.,* 59, 541, 1978.

49. **Zyss, J.,** Hyperpolarizabilities of substituted conjugated molecules. I. Perturbed INDO approach to monosubstituted benzene, *J. Chem. Phys.,* 70, 3333, 1979.

50. **Burland, D. M., Walsh, C. A., Kajzar, F., and Sentein, C.,** Comparison of hyperpolarizabilities obtained with different experimental methods and theoretical techniques, *J. Opt. Soc. Am. B,* 8, 2269, 1991.

51. **LeFevre, R. J. W. and Radom, L.,** *J. Chem. Soc. B,* 1295, 1967.

52. **Bogaard, M. P., Buckingham, A. D., Corfield, M. G., Dummur, D. A., and White, A. H.,** *Chem. Phys. Lett.,* 12, 558, 1972.

53. **Ward, J. F. and Elliot, D. S.,** Measurements of molecular hyperpolarizabilities for ethylene, butadiene, hexatriene and benzene, *J. Chem. Phys.,* 69, 5438, 1978.

54. **Bogaard, M. P., Buckingham, A. D., and Ritche, G. L. D.,** *Mod. Phys.,* 18, 575, 1980.
55. **Levine, B. F. and Bethea, C. G.,** Second and third order hyperpolarizabilities of organic molecules, *J. Chem. Phys.,* 63, 2666, 1975.
56. **Shinoda, H. and Akutagawa, T.,** *Bull. Chem. Soc. Jpn.,* 48, 3431, 1975.
57. **Bridge, N. J. and Buckingham, A. D.,** *Proc. R. Soc. Lond. A,* 295, 334, 1966.
58. **Amos, A. T. and Hall, G. G.,** *Theor. Chim. Acta.,* 6, 159, 1966.
59. **Shelton, D. P.,** Dispersion of the nonlinear susceptibility measured for benzene, *J. Opt. Soc. Am. B,* 2, 1880, 1985.
60. **Sekino, H. and Bartlett, R. J.,** Molecular hyperpolarizabilities, *J. Chem. Phys.,* 98, 3022, 1993.
61. **Guan, J., Duffy, P., Carter, J. T., Chong, D. P., Casida, K. C., Casida, M. E., and Wrinn, M.,** Comparison of local field-density and Hartee-Fock calculations of molecular polarizabilities and hyperpolarizabilities, *J. Chem. Phys.,* 98, 4753, 1993.
62. **Newell, A. C. and Baird, R. C.,** *J. Appl. Phys.,* 36, 3751, 1965.
63. **Shelton, D. P.,** *Phys. Rev. A.,* 42, 2578, 1990.
64. **Werner, H. J. and Meyer, W.,** *Mol. Phys.,* 31, 855, 1976.
65. **Ward, J. F. and Miller., C. K.,** *Phys. Rev. A.,* 19, 826, 1979.
66. **Buckingham, A. B., Bogaard, M. P., Dunmur, D. A., Hobbs, C. P., and Orr, J.,** *Trans. Faraday Soc.,* 66, 1548, 1970.
67. **Spackman, M. A.,** *J. Phys. Chem.,* 93, 7594, 1989.
68. **Diercksen, G. H. F., and Sadlej, A. J.,** *J. Chem. Phys.,* 75, 1253, 1981.
69. **Bishop, D. M., and Maroulis, G.,** *J. Chem. Phys.,* 82, 2380, 1985.
70. **Zeiss, G. D. and Meath, W. J.,** *Mol. Phys.,* 33, 1155, 1977.
71. **Maryott, A. A. and Buckley, F.,** *Nat. Bur. Stand.* (US) Circ. 537, 1953.
72. **Bose, T. K. and Cole, R. H.,** *J. Chem. Phys.,* 54, 3829, 1971.
73. **Meredith, G. R., Buchalter, B., and Hanzlik, C.,** Third-order susceptibility determination by third harmonic generation. II. *J. Chem. Phys.,* 78, 1543, 1983.
74. **Albert, I. D. L., Pugh, D., Morley, J. O., and Ramasesha, S.,** Linear and nonlinear optical properties of cumulenes and polyenynes: a model exact study, *J. Phys. Chem.,* 96, 10160, 1992.
75. **Shuai, Z. and Bredas, J. L.,** Static and dynamic optical nonlinearities in conjugated polymers: Third-harmonic generation and the dc Kerr effect in polyacetylene, polyparaphenylene vinylene, and polythienylene vinylene, *Phys. Rev.,* B46, 4395, 1992.
76. **Kurtz, H. A.,** *Int. J. Quantum Chem. Symp.,* 24, 791, 1990.
77. **Ducasse, L., Villesuzanne, A., Hoarau, J., and Fritsch, A.,** π-electron polarizabilities of infinite organic polymers, *J. Chem. Phys.,* 97, 9389, 1992.
78. **Meyers, F. and Bredas, J. L.,** Geometric and electronic structure and nonlinear optical properties of push-push and pull-pull conjugated systems, *Synth. Metals,* 49, 181, 1992.
79. **Matsuzawa, N. and Dixon, D. A.,** Semiempirical calculations of hyperpolarizabilities for donor-acceptor molecules: Comparison to experiment, *J. Phys. Chem.,* 96, 6232, 1992.
80. **Daniel, C. and Dupuis, M.** Nonlinear optical properties of organic solids: *Ab initio* polarizability and hyperpolarizabilities of nitroaniline derivatives, *Chem. Phys. Lett.,* 171, 209, 1990.
81. **Huang, Z., Jelski, D. A., Wang, R., Xie, D., Zhao, C., Xia, X., and George, T. F.,** Polarizabilities of *trans* and *cis* polyacetylene and interactions among chains in crystalline polyacetylene, *Can. J. Chem.,* 70, 372, 1992.
82. **Papadopoulos, M. G. and Waite, J.,** Analysis of the polarizabilities and second hyperpolarizabilities of α,ω-dithienylopolyenes, *Nonlinear Optics,* 7, 65, 1994 and references therein.
83. **Helfin, J. R. and Garito, A. F.,** Third-order optical processes in linear chains: Electron correlation theory and experimental dispersion measurements, in *Polymers for Lightwave and Integrated Optics,* Hornak, L. A., Ed., Marcel Dekker, New York, 1992, 501.
84. **Ducuing, J.,** in *Nonlinear Optics,* Harper P. G. and Harper, B. S., Eds., Academic Press, London, 1977, p, 11.
85. **Kajzar, F.,** Impact of dimensionality, conjugation length, scaling laws and electronic structure on nonlinear optical properties of conjugated polymers, *Nonlinear Optics,* 5, 329, 1993.
86. **Kakuta, A.,** Organic nonlinear optical materials and their molecular design, in *Proceedings of the 5th International Kyoto Conference on New Aspects of Organic Chemistry.* Yoshida, Z., and Ohshira, Y., Eds., VCH, Weinheim, 1992, 379.
87. **Nakano, M., Yamaguchi, K., and Fueno, T.,** Coupled Hartee-Fock calculations of the third-order hyperpolarizabilities for mixed and segregated charge-transfer clusters, *Nonlinear Optics,* 1993, 6,289, 1994.
88. **T. Gotoh, Kondoh, T., and Egawa, K.,** Exceptionally large third-order optical nonlinearity of organic charge transfer complexes, *J. Opt. Soc. Am.,* B6, 703, 1990.
89. **Gong, Q., Xia, Z., Zou, Y. H., Meng, X., Wei, L., and Li, F.,** Large nonresonant third-order hyperpolarizabilities of organic charge-transfer complexes, *Appl. Phys. Lett.,* 59, 381, 1991.
90. **Nalwa, H. S., Mukai, J., and Kakuta, A.,** Effect of the π-bonding sequence on third-order optical nonlinearity evaluated by ab initio calculations, *J. Phys. Chem.,* 99, 10766, 1995.
91. **Hamada, T.,** 66th Fall Meeting of Chemical Society of Japan, Nishinomiya, Hyougo, Japan, September 27–30, 1993, p. 1G310.

92. **Kakuta, A., Imanishi, Y., Nalwa, H. S., Hamada, T., Ishihara, S., and Hattori, S.,** Organic superlattice structures for nonlinear optics, *Proc. Int. Symp. Nonlinear Photon. Mater.,* May 24–25, 1994, Tokyo, p. 263.

93. **Shelton, D. P. and Rice, J. E.,** Measurements and calculations of the hyperpolarizabilities of atoms and small molecules in gas phase. *Chem. Rev.,* 94, 3, 1994.

94. **Bredas, J. L., Adant, C., Tackx, C., Persoons, A., and Pierce, B. M.,** Third-order nonlinear optical response in organic materials: theoretical and experimental aspects. *Chem. Rev.,* 94, 243, 1994.

95. **Nalwa, H. S.,** Organic materials for third-order nonlinear optics, *Adv. Mater.,* 5, 341, 1993.

96. **Zyss, J.,** Ed, *Molecular Nonlinear Optics,* Academic Press, New York, 1994.

97. **Waite, J. and Papadopoulos, M. G.,** *Can. J. Chem.,* 66, 1140, 1988.

98. **Waite, J. and Papadopoulos, M. G.,** Calculations of induced moments in large molecules. 6. Scale of polarization for some functional groups. A comparative study. *J. Phys. Chem.,* 93, 43, 1989.

99. **Waite, J. and Papadopoulos, M. G.,** Polarization mechanisms and properties of substituted ferrocenes. A comparative study. *J. Phys. Chem.,* 95, 5426, 1991.

100. **Papadopoulos, M. G., Waite, J., Winter, C. S., and Oliver S. N.,** *Inorg. Chem.,* 32, 277, 1993.

101. **Waite, J., Papadopoulos, M. G., Oliver, S. N., and Winter, C. S.,** On the design of molecules with large nonlinearities. *Nonlinear Optics,* 6, 297, 1994.

102. **Papadopoulos, M. G. and Waite, J.,** *Opt. Mater.,* 3, 145, 1994.

103. **Lu, Y. J. and Lee, S. L.,** Semi-empirical calculations of the nonlinear optical properties of polycyclic aromatic compounds, *Chem. Phys.,* 179, 431, 1993.

Chapter 10

Measurement Techniques for Third-Order Optical Nonlinearities

Hari Singh Nalwa

CONTENTS

I. INTRODUCTION

In the previous chapter, we described various quantum chemistry approaches for calculating polarizability (α) and second hyperpolarizabilities (γ) of a wide variety of π-conjugated systems. Here, we will discuss various measurement techniques used to estimate third-order nonlinear optical susceptibility $\chi^{(3)}$ and second hyperpolarizabilities (γ). The relationship of third-order nonlinear optical interactions those are involved with the measurements of $\chi^{(3)}$ values is an important factor. As described in Chapter 9, the quoted values of second hyperpolarizability γ calculated by different quantum chemistry techniques differ significantly from each other even for a single material; so too are third-order nonlinear optical susceptibilities measured by different experimental methods. Different measurement techniques employed for $\chi^{(3)}$ evaluation not only involve different nonlinear interactions to the medium, but also different dispersion effects. Therefore, an understanding of different nonlinear optical interactions in materials is very important.

A. THIRD-ORDER NONLINEAR OPTICAL PROCESSES
The electromagnetic field of a laser beam impinging on a material generates an electrical polarization. The macroscopic polarization P, induced by the external field E, can be written by the following expression

$$P = \chi^{(1)}E + \chi^{(2)}EE + \chi^{(3)}EEE + \cdots\cdots + \chi^{(n)}E^n + \cdots\cdots \qquad (1)$$

0-8493-8923-2/97/$0.00+$.50

Here $\chi^{(1)}$, $\chi^{(2)}$, and $\chi^{(3)}$ are the first-, second-, and third-order nonlinear optical susceptibilities. An excellent description of various third-order nonlinear optical processes has been given by Flytzanis,[1] Shen,[2] and Reintjes.[3] Different third-order nonlinear optical processes are briefly mentioned here as given by Reintjes.[3] The numerous third-order nonlinear optical interactions at the optical frequencies ω_1, ω_2, and ω_3 can be expressed by the following equations where the P_i is the ith component of the nonlinear polarization and E is the electric field component at frequency ω in $i, j, \ldots k$ directions.

1. Third harmonic generation (THG) is given by:[3]

$$P_i\,(\omega_4 = 3\omega_1) = (1/4)\chi^{(3)}_{ijkl}\,(-\omega_4, \omega_1, \omega_1, \omega_1)E_j(\omega_1)E_k(\omega_1)E_l(\omega_1) \tag{2}$$

2. Sum and difference frequency mixing is written by:[3]

$$P_i(\omega_4 = 2\omega_1 + \omega_2) = (3/4)\chi^{(3)}_{ijkl}(-\omega_4; \omega_1, \omega_1, \omega_2)E_j(\omega_1)E_k(\omega_1)E_l(\omega_1) \tag{3}$$

$$P_i(\omega_4 = 2\omega_1 - \omega_2) = (3/4)\chi^{(3)}_{ijkl}(-\omega_4; \omega_1, \omega_1, -\omega_2)E_j(\omega_1)E_k(\omega_1)E_l(\omega_2) \tag{4}$$

$$P_i(\omega_4 = \omega_1 + \omega_2 + \omega_3) = (6/4)\chi^{(3)}_{ijkl}(-\omega_4; \omega_1, \omega_2, \omega_3)E_j(\omega_1)E_k(\omega_2)E_l(\omega_2) \tag{5}$$

$$P_i(\omega_4 = \omega_1 + \omega_2 - \omega_3) = (6/4)\chi^{(3)}_{ijkl}((-\omega_4, \omega_1\omega_2, -\omega_3)E_j(\omega_1)E_k(\omega_2)E_l(\omega_3) \tag{6}$$

$$P_i(\omega_4 = \omega_1 - \omega_2 - \omega_3) = (6/4)\chi^{(3)}_{ijkl}(-\omega_4, \omega_1, -\omega_2, -\omega_3)E_j(\omega_1)E_k(\omega_2)E_l(\omega_3) \tag{7}$$

3. Self-action effects are described as follows:[3]

   (a) Self-focusing or defocusing (nonlinear refractive index) is expressed by

$$P_i(\omega) = (3/4)\,\mathrm{Re}\,\chi^{(3)}_{iikk}(-\omega; \omega, \omega, -\omega)E_i(\omega)|E_k(\omega)|^2 \tag{8}$$

   (b) Two-photon absorption

$$P_i(\omega) = (3/4)\,\mathrm{Im}\,\chi^{(3)}_{iikk}(-\omega; \omega, \omega, -\omega)]E_i(\omega)|E_k(\omega)|^2 \tag{9}$$

   (c) Degenerate four-wave mixing (DFWM) can be expressed as

$$P_i(\omega) = (3/4)\,\chi^{(3)}_{ijkl}(-\omega; \omega, \omega, -\omega)E_j(\omega)E_k(\omega)E_l(\omega) \tag{10}$$

4. Raman effects are described as[3]

   (a) Raman-induced Kerr effect

$$P_i(\omega) = (3/4)\,\mathrm{Re}\,\chi^{(3)}_{iikk}(-\omega_1; \omega_1, \omega_2, -\omega_2)E_i(\omega_1)|E_k(\omega_2)|^2 \tag{11}$$

   (b) Stokes-Raman scattering

$$P_i(\omega_S) = (3/2)\,\mathrm{Im}\,\chi^{(3)}_{iikk}(-\omega_S; \omega_S, \omega_L, -\omega_L)E_i(\omega_S)|E_k(\omega_L)|^2 \tag{12}$$

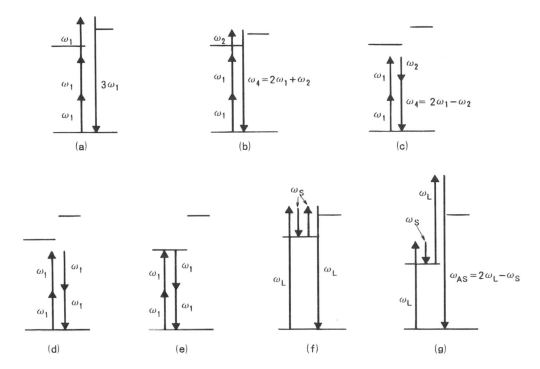

Figure 1. A schematic representation of various nonlinear optical interactions (a) third-harmonic generation, (b) four-wave sum-frequency mixing, (c) four-wave difference-frequency mixing, (d) nonlinear refractive index, (e) two-photon absorption, (f) stimulated Stokes-Raman scattering, and (g) anti-Stokes-Raman scattering. Photon energies are shown by arrows and the horizontal lines show the energy state of the medium. (From Reintjes, J. F., *Nonlinear Optical Parametric Process in Liquids and Gases*, Academic Press, New York, 1994. With permission.)

(c) Anti-Stokes-Raman scattering

$$P_l(\omega_{AS}) = (3/2)\,\text{Im}\,\chi^{(3)}_{ijkl}(-\omega_{AS};\,\omega_L,\,\omega_L,\,-\omega_S)E_j(\omega_L)E_k(\omega_L)E_l(\omega_S) \quad (13)$$

where Re and Im refer to the real and imaginary parts of $\chi^{(3)}$. The *S*, *AS*, and *L* refer to Stokes, anti-Stokes, and laser for the general expression for Raman interactions. Figure 1 shows a schematic representation of various nonlinear optical interactions referred in Equations 2 to 13. It can be seen that different nonlinear optical processes are involved with different measurement techniques. For example, the third-order optical nonlinearity $\chi^{(3)}(-3\omega;\,\omega,\,\omega,\,\omega)$, $\chi^{(3)}(-\omega;\,\omega,-\omega,\,\omega)$, $\chi^{(3)}(-2\omega;\,\omega,\,\omega,\,0)$, and $\chi^{(3)}(-\omega;\,0,\,0,\,\omega)$ are measured by the third harmonic generation (THG) technique, degenerate four-wave (DFWM) mixing, electric field-induced second harmonic generation (EFISH), and DC Kerr effect, respectively. In the preceding section, the experimental details of some of the most common techniques to measure third-order nonlinear optical susceptibility $\chi^{(3)}$ and second hyperpolarizability γ are discussed.

B. THIRD-ORDER NONLINEAR OPTICAL SUSCEPTIBILITY $\chi^{(3)}$ TENSORS

The third-order nonlinear optical susceptibility $\chi^{(3)}_{ijkl}$ has 81 separate tensor components, but the number of different tensor components reduces to 21 nonzero elements by symmetry[1] which can be written as

$$\chi^{(3)}_{xxxx} = \chi^{(3)}_{yyyy} = \chi^{(3)}_{zzzz}$$

$$\chi^{(3)}_{xyxy} = \chi^{(3)}_{xzxz} = \chi^{(3)}_{yxyx} = \chi^{(3)}_{yzyz} = \chi^{(3)}_{zyzy} = \chi^{(3)}_{zxzx}$$

$$\chi^{(3)}_{xxyy} = \chi^{(3)}_{xxzz} = \chi^{(3)}_{yyxx} = \chi^{(3)}_{yyzz} = \chi^{(3)}_{zzxx} = \chi^{(3)}_{zzyy}$$

574

Table 1 Theoretical Ratios of $\chi^{(3)}$ Tensor Components for Two-Beam amd Three-Beam Denegerate Four-Wave Mixing (DFWM) Experiments and Nondegenerate Pump-Probe of Absorption Variation in Isotropic Mixtures

Microscopic Effect	Two-Beam DFWM $\chi^{(3)}_{xyyx}/\chi^{(3)}_{xxxx}$	Three-Beam DFWM $\chi^{(3)}_{xyyx}/\chi^{(3)}_{xxxx}$	$\chi^{(3)}_{xxyy}/\chi^{(3)}_{xxxx}$	Nondegenerate Pump-Probe $\chi^{(3)}_{xyyx}/\chi^{(3)}_{xxxx}$
One-dimensional (Electronic)	1/3	1/3	1/3	1/3
Orientational (nuclear)	3/4	3/4	1/8	−1/2
Isotropic (thermal)	0	0	1/2	1

From Kajzar, F., et al., *Polymers for Lightwave and Integrated Optics,* Marcel Dekker New York, 1992. With permission.

$$\chi^{(3)}_{xyyx} = \chi^{(3)}_{xzzx} = \chi^{(3)}_{yxxy} = \chi^{(3)}_{yzzy} = \chi^{(3)}_{zxxz} = \chi^{(3)}_{zyzy}$$

$$\chi^{(3)}_{xxxx} = \chi^{(3)}_{xxyy} + \chi^{(3)}_{xyxy} + \chi^{(3)}_{xyyx} \tag{14}$$

Here the $\chi^{(3)}_{xxxx}$ is a diagonal component of $\chi^{(3)}$ in the x direction. The $\chi^{(3)}$ is a fourth-rank tensor that has only three independent components $\chi^{(3)}_{xxxx}$, $\chi^{(3)}_{xyxy}$, and $\chi^{(3)}_{xxyy}$. These components are related as $1/3\chi^{(3)}_{xxxx} = \chi^{(3)}_{xyxy} = \chi^{(3)}_{xxyy}$ in an isotropic medium far from the resonance for a purely electronic optical nonlinearity, and provide significant information about molecular interactions. Etchepare et al.[4] estimated theoretical ratios of $\chi^{(3)}$ tensor components as 1/3 for $\chi^{(3)}_{xyyx}/\chi^{(3)}_{xxxx}$ and $\chi^{(3)}_{xyxy}/\chi^{(3)}_{xxxx}$ and $\chi^{(3)}_{xxyy}/\chi^{(3)}_{xxxx}$ for electronic process. For the molecular process, the ratios were 3/4 both for $\chi^{(3)}_{xyyx}/\chi^{(3)}_{xxxx}$ and $\chi^{(3)}_{xyxy}/\chi^{(3)}_{xxxx}$ and 1/2 for $\chi^{(3)}_{xxyy}/\chi^{(3)}_{xxxx}$. The theoretical ratios of $\chi^{(3)}$ tensor components for two-beam and three-beam denegerate four-wave mixing (DFWM) experiments and nondegenerate pump and probe analysis in isotroic samples, which are most common in organic molecular and polymeric materials,[5] are given in Table 1. The x and y denote the parallel and perpendicular directions of polarizations. Various experimental techniques are available to determine the absolute, as well as some of the different tensor components of the real and imaginary parts of $\chi^{(3)}$ which are discussed in this chapter.

II. MEASUREMENT TECHNIQUES FOR THIRD-ORDER NONLINEAR OPTICAL SUSCEPTIBILITY

The $\chi^{(3)}$ measurements of organic molecular and polymeric materials can be performed in the forms of powder, solutions, single crystals, thin films, and liquid crystals. Organic materials offer tremendous opportunities in processing because of their architectural flexibility to be tailored into desired molecular structures required for fabrication. Thin films of organic materials for nonlinear optical (NLO) measurements can be obtained by a variety of fabrication techniques:

1. Sputtering (ion beam, magnetron, or electrical sputtering)
2. Chemical vapor deposition (CVD)
3. Vacuum evaporation [sublimation, ionized cluster beam (ICB), molecular beam epitaxy (MBE)]
4. Spin coating
5. Solution casting and dipping
6. Electrochemical polymerization
7. Langmuir-Blodgett (LB) technique

The magnitude of $\chi^{(3)}$ is significantly affected by the molecular orientation. Therefore, the thin-film deposition plays an important role. In addition, the phenomenon of nonlinear optical dichroism has also been observed in π-conjugated materials. The $\chi^{(3)}$ measurements on various forms of organic materials are discussed here and, furthermore, the details of material states such as solid phase, solutions, and gels used to evaluate third-order optical nonlinearity will be specified in Chapter 11. We have divided the measurement techniques for evaluating $\chi^{(3)}$ based on the third-order nonlinear optical processes.

A. THIRD HARMONIC GENERATION (THG) TECHNIQUE

In this section, we will describe various THG techniques that have traditionally been used for measuring third-order optical nonlinearity. In 1973, Hermann[6] reported the measurement of third-order nonlinear susceptibility in a quartz by comparing third harmonic generation and a two-step process: second harmonic generation and two-wave mixing. Here the following mechanism operates.

1. a direct process: $\omega + \omega + \omega \rightarrow 3\omega$
2. a two-step process involving second harmonic: $\omega + \omega \rightarrow 2\omega$; then mixing this second harmonic with the fundamental: $\omega + 2\omega \rightarrow 3\omega$. A laser beam of frequency ω reaches on a quartz plate Q_1, which generates a beam of frequency 2ω collinear with laser. By a two-wave mixing process operating at Q_2 plate, these two frequencies ω and 2ω give rise to frequency 3ω. There is no signal if the Q_2 plate is removed, hence mixing occurs in the Q_2 plate. By rotating Q_2, fringes are obtained by plotting $I_{3\omega}$/ $R_{3\omega}$ as a function of Q_2. By removing Q_1, the direct generation of the third harmonic ($\omega + \omega + \omega \rightarrow 3\omega$) in Q_2 is observed and fringes are recorded by rotating Q_2.

The following relationship holds for the output face of Q_1:

$$E_{3\omega}(r \mathinner{.\,.} t) = A_1 \chi^{(2)}(\omega, \omega) E_\omega(r \mathinner{.} t) E_{2\omega}(r \mathinner{.} t). \tag{15}$$

similarly at the output face of Q_2:

$$E_{3\omega}(r \mathinner{.\,.} t) = A_2 \chi^{(2)}(\omega, 2\omega) E_\omega(r \mathinner{.} t) E_{2\omega}(r \mathinner{.} t). \tag{16}$$

Therefore, combining these two equations

$$E_{3\omega}(r \mathinner{.\,.} t) = A_1 A_2 \chi^{(2)}(\omega, \omega) \chi^{(2)}(\omega, 2\omega) E_\omega^3(r \mathinner{.} t) \tag{17}$$

where A_1 and A_2 are constant quantities.

Q_1 and Q_2 also generate third-harmonic directly, if the harmonic coming out from Q_1 is filtered out, then Q_2 gives

$$E_{3\omega}'(r \mathinner{.\,.} t) = \chi^{(3)}(\omega, \omega, \omega) E_\omega^3(r \mathinner{.} t) \tag{18}$$

$E_{3\omega}$ and $E_{3\omega}'$ can be separately measured. Therefore a comparison of their intensities gives the value of the ratio

$$[\chi^{(3)}(\omega, \omega, \omega)/\chi^{(2)}(\omega, \omega)\chi^{(2)}(\omega, 2\omega)]^2 \tag{19}$$

hence $\chi^{(2)}(\omega, 2\omega)$ and $\chi^{(2)}(\omega, \omega)$ are almost equal, ω being far from the absorption band. The correct estimation of $\chi^{(3)}$ is strongly affected by the error in $\chi^{(2)}$ values. This method uses a direct measurement of the ratio; $\chi_{xxxx}^{(3)}/[\chi_{xxx}^{(2)}]^2$ without calibration. In this technique, choice of a proper sandwich filter placed between the Q_1 and Q_2 plates is very important. Fringes were obtained by plotting $S3\omega/R3\omega$ as a function of Q_2.

Hermann[6] later extended this method and measured the THG in cyanine solution as a function of the dye concentration. The method described below is applicable for liquids and was used first by Bey et al.[7] to measure the THG in a solution of fuchsin in hexafluoroacetone sesquihydrate. The TH intensity is expressed by:

$$I_{3\omega} \propto \frac{\chi^{(3)^2}[e^{-3\alpha_1 l} + e^{-\alpha_3 l} - 2e^{-1/2(3\alpha_1 + \alpha_3)l} \cos(\Delta k) l]}{(\Delta k)^2 + [1/2(\alpha_3 - 3\alpha_1)]^2} \tag{20}$$

Bloembergen[8] related it to the hyperpolarizability by the following equation:

$$\chi_{xxxx}^{(3)} = L^4(N_0 \langle \gamma_0 \rangle + N \langle \gamma \rangle) \tag{21}$$

where L is a Lorentz-type local field factor, N_0 and γ_0 are the concentration and the hyperpolarizability of the solvent molecules, and N and γ are related to the solute, respectively.

1. THG Measurements in Solutions

Meredith et al.[9,10] used a triple-wedge cell method for measuring $\chi^{(3)}$ and hyperpolarizabilities of liquids by THG. The method relies on use of a cell with wedge windows and a wedge liquid chamber placed in a vacuum chamber. The method has the restriction that the liquid should not absorb at the fundamental and harmonic wavelengths. The input and output windows are BK-7 glass with only the input window having an intentional wedge angle ($\alpha = 15$ min). There are five media, media 1 and 5 are vacua, 2 and 4 are silica windows, and 3 is the liquid under study and four interfaces. The exact phase mismatch between bound and free waves in windows and liquid must be known because THG is coherently detected. The considerations of atmospheric conditions are also important. Further details of the measurements, theoretical considerations, and their applications are given in references 9 and 10.

Kajzar and Messier[11] developed an original technique for THG measurements in liquids. This technique allows a high precision of third-order nonlinear susceptibility and refractive index dispersion measurements in air. This technique also facilitates measurements on liquids that absorb at fundamental and/or harmonic frequencies, as well as measurements of external parameters such as temperature, pressure, and electric or magnetic fields. Figure 2 shows third harmonic generation setup and the liquid cell design used in this THG procedure. The input and output windows of the cell are made of fused silica. Three distinct regions are taken into account: (1) before the focus with converging light, (2) at the focus where the beam is considered as a plane wave, and (3) at the diverging light region. The input and output liquid cell windows are assumed to be located in (1) and (2) regions, respectively, and region (3) is the liquid compartment.

For the input and output windows, in the case of harmonic generation in an infinite medium by a

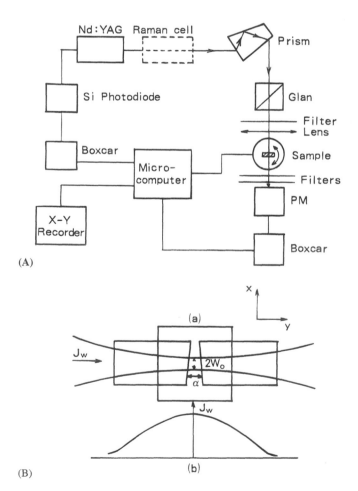

(A)

(B)

Figure 2. Experimental setup for third harmonic generation measurements (A) and the liquid cell design used in THG procedure (B). (From Kajzar, F. and Messier, J., *Rev. Sci. Instrum.*, 58, 2081, 1987. Kajzar, F., et al., *J. Appl. Phys.*, 60, 3040, 1986. With permission.)

focused laser beam, the following expressions are applied for the resultant harmonic field at a point (x, y, z) for light propagating in the y direction

$$E_{3\omega}(x, y, z) = \frac{4\pi P_{NL} I(y)}{\Delta \epsilon} \exp(ik_{3\omega} y)$$

$$\times \exp\left(\frac{-3(x^2 + z^2)}{W_0^2(1 + it)}\right) \tag{22}$$

where $\Delta \epsilon = \epsilon_\omega - \epsilon_{3\omega}$ is the dielectric constant dispersion between ω and 3ω frequency. The nonlinear polarization P_{NL} created in a medium by fundamental field E_ω is given by

$$P_{NL} = \frac{1}{4} \chi^{(3)}(-3\omega; \omega, \omega, \omega)E_\omega^3 \tag{23}$$

where $\chi^{(3)}(-3\omega; \omega,\omega,\omega,)$ is the third-order nonlinear optical susceptibility response for the THG process. The liquid between windows introduces a phase mismatch between fundamental and harmonic waves

$$E_{G2} = -E_{G1} e^{i\Delta\psi}(t_{GL}^\omega t_{LG}^\omega)^3 \tag{24}$$

where t_{GL}^ω and t_{LG}^ω refer to the transmission factor between glass–liquid (GL) and liquid–glass (LG) interface at ω frequency, respectively. The harmonic field generation in the liquid component can be written by

$$E_L(X) = \pi(\chi^{(3)}/\Delta\epsilon)_L (e^{i\Delta\psi} - 1)e^{i\psi_H}(E_\omega t_\omega)^3 \tag{25}$$

where $t\omega = 2n_\omega^G/(1 + n_\omega^G)$ is the transmission factor between air and input window. The harmonic field generated in the input window is given by

$$E_{G1} = \pi(\chi^{(3)}/\Delta\epsilon)_G (E_\omega t_\omega)^3 \tag{26}$$

and the resultant harmonic field at output of liquid cell is expressed by

$$E_R = \frac{\chi^{(3)}}{\Delta\epsilon} e^{i\psi_H}[t_{GL}^{3\omega} t_{LG}^{3\omega} - e^{i\Delta\psi}(t_{LG}^\omega t_{GL}^\omega)^3$$

$$+ \rho(e^{i\Delta\psi} - 1)(t_{GL}^\omega)^3]t_{3\omega}(t_\omega E_\omega)^3 \tag{27}$$

where

$$\rho = (\chi^{(3)}/\Delta\epsilon)_L/(\chi^{(3)}/\Delta\epsilon)_G, \tag{28}$$

and $t_{3\omega}$ is the transmission factor between glass and air at the harmonic frequency

$$t_{3\omega} = 2/(1 + n_{3\omega}^G) \tag{29}$$

The $\chi^{(3)}$ of the liquid sample (L) can be calibrated with a standard liquid (S) by the following expression

$$I_{3\omega}^L/I_{3\omega}^S = (E_R^L/E_R^S)^2 \tag{30}$$

where I is the third harmonic intensity. Using a standard sample, the $\chi^{(3)}$ values of organic liquids can be easily determined.

The precision of $\chi^{(3)}$ value depends on factors such as beam stability, focal length, operating wavelength, optical quality of windows, liquid cell position, and temperature stability. The Kajzar-Messier technique of THG allows the measurements of $\chi^{(3)}$ values with a very high precision.

The macroscopic third-order susceptibility for THG solution experiments is given by[12]

$$\chi^{(3)}(-3\omega; \omega, \omega, \omega) = N_p f_p(-3\omega; \omega, \omega, \omega)\gamma(-3\omega; \omega, \omega, \omega)_p$$
$$+ N_s f_s(-3\omega; \omega, \omega, \omega)\gamma(-3\omega; \omega, \omega, \omega)_s \quad (31)$$

where N_i is the number of solvent (s) or solute (p) molecules per unit volume and f_I is the product of a local-field factor:

$$F_I(-3\omega; \omega, \omega, \omega) = f(3\omega)f(\omega)^3 \quad (32)$$

Using local field factors for solvent and solute,

$$\chi^{(3)}(-3\omega; \omega, \omega, \omega) = f(N_p\gamma_p + N_s\gamma_s) \quad (33)$$

The second hyperpolarizability may be complex and is given by

$$\gamma_I = \gamma_I' + i\gamma_I'' \quad (34)$$

Therefore, one can determine the values of γ_s from THG measurements in pure liquid and values for γ_p from solutions if γ_s is already known. Kajzar[13] derived the following expressions for the concentration-dependent amplitude of the THG intensity, normalized to the intensity from the pure solvent:

$$I_N(C) = I_{3\omega}(C)/I_{3\omega}(0)$$
$$= \frac{(1/\beta^2)\{|[A + B(C)][(C\gamma_p'/M_p) + (\gamma_s/M_s)]|^2 + C^2|(\gamma_p''/M_p)B(C)|^2\}}{|A + B(C)(\gamma_s/M_s)|^2} \quad (35)$$

where $I_N(C)$ is the normalized THG intensity, and $I_{3\omega}(C)$ is the intensity for a solution of mass concentration C [where $C = m_p/m_s$ and $m_{p,s}$ is the weight fraction for the solute (p) or the solvent (s)]. The M_s and M_p refer to the molecular weights of the solvent and the solute, respectively. Other quantities shown in the above equation depend only on the refractive index, dielectric constant, and absorption coefficients. The solvent hyperpolarizability γ_s is considered real and is obtained separately. By a least-squares fit to the experimental measurements of the normalized TH intensity as a function of concentration, both the real and imaginary part of γ_p can be estimated.

2. THG Measurements of Thin Films

A THG method to determine third-order nonlinear optical susceptibilities of thin films described by Kajzar et al.[14] is presented here. The harmonic field generated in a plane parallel slab immersed between two linear media is given by

$$E_{3\omega} = A(e^{i\Delta} - 1) \quad (36)$$

where

$$A = (1/\pi)(\chi^{(3)}/\Delta\epsilon)e^{i\psi_{3\omega}}T(E_\omega)^3 \quad (37)$$

and

$$\Delta_- = \psi_\omega - \psi_{3\omega} = 3\omega(N_\omega - N_{3\omega})l/c, \quad (38)$$

where

$$N_{\omega(3\omega)} = n_{\omega(3\omega)} \cos \theta_{\omega(3\omega)} \tag{39}$$

The ω and 3ω are the fundamental and harmonic wave, respectively. The factor T arises from the transmission coefficient and boundary conditions, l is the slab thickness, θ is the propagation angle, $\Delta\epsilon$ is the dielectric constant dispersion which is equal to $\epsilon_\omega - \epsilon 3_\omega$, $n_{\omega(3\omega)}$ are the refractive indices, and E_ω is the incident electric field intensity. The resultant harmonic field for a polymer film deposited on a silica substrate is written by

$$E^R_{3\omega} = E^s_{3\omega} t^3_{3\omega} + E^p_{3\omega}(t^1_\omega)^3 \tag{40}$$

where s and p refer to substrate and (polymer) film, respectively, $t^3_{3\omega}$ is the transmission factor for the harmonic wave between polymer film and vacuum, and t^1_ω is the transmission factor of the fundamental wave between vacuum and substrate. Assuming that all transmission factors between substrate and polymer film are equal to 1, then for a transparent film at both fundamental and harmonic frequencies, the THG intensity is given by

$$I_{3\omega}(\theta) = \frac{256\pi^4}{c^2} T \left| \left(\frac{\chi^{(3)}}{\Delta\epsilon}\right)_s e^{-l\Delta^s_-/2} \sin\frac{\Delta^s_-}{2} \right.$$
$$\left. + \left(\frac{\chi^{(3)}}{\Delta\epsilon}\right)_p e^{l\Delta^p_-/2} \sin\frac{\Delta^p_-}{2} \right|^2 (I_\omega)^3, \tag{41}$$

again, s and p indicate the substrate and polymer, respectively. The phase mismatch $\Delta\phi$ can be written as $\Delta\phi = \phi_\omega - \phi_{3\omega} = 6\pi/\lambda \, (n_\omega\cos\theta_\omega - n_{3\omega}\cos\theta_{3\omega})l$, here θ_ω and $\theta_{3\omega}$ are the propagation angles at the ω and 3ω frequencies, respectively, and l is the nonlinear medium thickness. T is an overall transmission factor. In a general case for a substrate and polymer film, the expression can be given by

$$I_{3\omega}(\theta) = \frac{64\pi^4}{c^2} \left| e^{l(\psi^s_\omega + \psi^s_{3\omega})} \left(\frac{\chi^{(3)}}{\Delta\epsilon}\right)_s [T_1(1 - e^{-l\Delta^s_-}) \right.$$
$$\left. + \rho T_2 e^{i\phi}(e^{l\Delta^p_-} - 1)] \right|^2 (I_\omega)^3 \tag{42}$$

where

$$T_1 = A^s(t^{(1)}_\omega)^3 t^{(3)}_{3\omega} \tag{43}$$
$$T_2 = A^p(t^{(1)}_\omega t^{(2)}_\omega)^3 \tag{44}$$

ϕ is the phase of the polymer susceptibility in case of substrate equal to a zero and

$$\rho = \left| \frac{(\chi^{(3)}/\Delta\epsilon)_p}{(\chi^{(3)}/\Delta\epsilon)_s} \right| \tag{45}$$

$A^{s(p)}$ are the factors associated with boundary conditions and t_ω and $t_{3\omega}$ are the transmission factors of the fundamental and harmonic beams, respectively. A precise analysis of the $\chi^{(3)}$ phase can be obtained by using these relationships. If the $\chi^{(3)}$ and refractive indexes at ω and 3ω and thickness of the substrate are known, then the modulus and the phase of hyperpolarizability can also be determined. For a rapid screening, the following simple relation between minima I_m and maxima I_M harmonic intensity and the ratio ρ can be simplified to

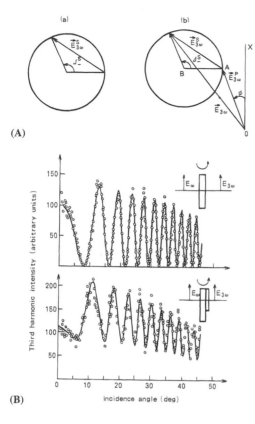

(A)

(B)

Figure 3. (A) the graphical diagram of the resulting electric field in a silica plate (a) and silica plate plus polymer film (b) in a vacuum. (B) Third harmonic intensity as a function of incidence angle for silica plate (a) and silica plate plus polymer film in vacuum (b). (From Kajzar, F. et al., *J. Appl. Phys.*, 60, 3040, 1986. With permission.)

$$\left(\frac{I_M^{1/2} + I_m^{1/2}}{I_M^{1/2} - I_m^{1/2}}\right)^2 = R^2 + 2R\sin\phi + 1 \tag{46}$$

where $R = \rho|\Delta^p_-|$ and for maxima and minima close to the normal incidence

$$R \simeq \pi|\chi_p^{(3)}\chi_s^{(3)}|l_p/l_c^s$$

where l_c^s is the coherence length in the substrate. Figure 3A shows the graphical diagram of resulting electric field in a silica plate and in a silica plate plus polymer film in a vacuum. Third harmonic intensity as a function of incidence angle for silica plate (a) and silica plate plus polymer film in a vacuum (b) are shown in Figure 3B. van Beek et al.[15] determined the magnitude of the thin-film nonlinear susceptibility from the THG Maker fringe measurements in which a dye β-carotene is dispersed into a host polystyrene matrix. The following expression was used to estimate the third-order nonlinear optical susceptibility:

$$|\chi_{\text{thin film}}^{(3)}| = |\chi_1^{\text{vol}}[\chi_{\beta\text{-carotene}}^{(3)\ \text{Res}} + \chi_{\beta\text{-carotene}}^{(3)\ \text{NonRes}}]$$
$$+ (1 - \chi_1^{\text{vol}})\chi_{\text{polystyrene}}^{(3)}| \tag{47}$$

where χ_1^{vol} is the β-carotene volume fraction in the thin film [$\chi_1^{\text{vol}} = m_1/d_1/(m_1/d_1 + m_2/d_2)$, where m and d are the masses and densities of components 1 and 2 in the thin films]. The nonlinear susceptibility of polystyrene is nonresonant, while that of β-carotene is separated into a complex resonant part. In the case of π-conjugated polymer thin films, the following factors are applied. The measured $\chi^{(3)}$ values are averaged over all π-conjugated polymer chain distribution. The tensor $\chi_{xxxx}^{(3)}$ along the π-conjuagted chain direction is given by:

$$\chi_{xxxx}^{(3)} = \langle\chi^{(3)}(-3\omega; \omega, \omega, \omega)\rangle/\langle\cos^4\theta\rangle \tag{48}$$

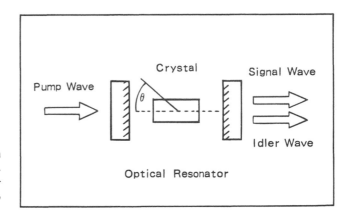

Figure 4. The basic component of an optical parametric oscillator (OPO). (From Fix, A., Schroder, T., and Wallenstein, R., *Laser Optoelectronic*, 23, 106, 1991. With permission.)

where θ is the angle between the π-conjugated chain and the incident light polarization direction. For a fully three-dimensional disorder chain, the $\langle\cos^4\theta\rangle = 1/5$. In the case the π-conjugated chains are parallel to the substrate, which means the disorder is two-dimensional and the $\langle\cos^4\theta\rangle = 3/8$.

3. THG Measurements by Optical Parametric Oscillator (OPO)

Gierulski et al.[16], at the BASF laboratory, used a frequency tunable optical parametric oscillator (OPO) with a tunable range of 0.420 to 3.0 μm wavelength. The OPO is based on a commercially available β-barium borate (β-BBO) crystal inside an optical resonator with a tunable range from the visible to the near infrared range. Fix et al.[17] described the full details of this measurement technique. The frequency selection is made by angle tuning the crystal, which is pumped by the third harmonic of an Nd-YAG laser. The NLO materials, β-barium borate (BBO) and lithium borate (LBO), were used for an OPO because of their high nonlinearity, wide transparency range, and high laser damage thresholds. The BBO and LBO-OPO offer the advantages of an all solid-state tunable source ranging in the ultraviolet to visible to near infrared, high peak and average power and high conversion efficiency. The basic components of an OPO are shown in Figure 4. A nonlinear crystal (BBO or LBO) is placed in an optical cavity which consists of two flat mirrors. The used BBO and LBO crystals were 11–15 mm long and cut at an angle of $\theta = 25$–40 degrees. The OPO resonator is 2.6–3.4 cm. The nonlinear interaction between the pump radiation—with frequency ω_p—and the NLO material generates radiation at the two frequencies, ω_s (signal wave) and ω_i (idler wave). The values of ω_s and ω_i can be determined and the calculated tuning range of signal λ_s and idler wave λ_i is shown in Table 2. The wavelength of the signal and the idler wave of the BBO-OPO as a function of the phase-matching angle is shown in Figure 5. The OPO is pumped by the second, third, fourth, or fifth harmonic of an Nd-YAG laser. For example, Figure 6 shows the wavelength of the signal and idler radiation measured at 355 and 532 nm. In case the OPO is pumped by the 355-nm third harmonic radiation, the measured and calculated wavelengths are in good agreement. In the case of 532-nm and 266-nm pump radiations, the measured tuning ranges were narrower than those of theoretical values. For fourth harmonic, a tuning range of 292–3000 nm was calculated but was limited to 302–2248 nm due to the decreasing reflectivity of the OPO resonator mirrors at $\lambda_{UV} < 310$ nm. The phase-matching angle θ changes only by 2 degrees for 532 nm and 10 degrees for 355 nm pump light. In the whole tuning range, only one crystal for each pump wavelength cut is required, for example, at 22 degrees (532 nm) and 30 degrees (355 nm). The change of phase-matching angle is larger with fourth and fifth harmonics, and corresponds to 21 degrees

Table 2 Calculated Tuning Range of Signal and Idler Wave

λ_p (nm)	θ (degree)	Theoretical Values		Experimental Values	
		λ_s (nm)	λ_i (nm)	λ_s (nm)	λ_i (nm)
532	20.7–22.8	647–1064	1064–3000	667–1064	1064–2628
355	23.1–33.1	403–710	710–3000	402–710	710–3036
266	26.7–47.6	292–532	532–3000	302–532	532–2248
213	31.3–72.6	229–426	426–3000	—	—

From Fix, A., et al., *Laser Optoelektronik*, 23;106, 1991. With permission.

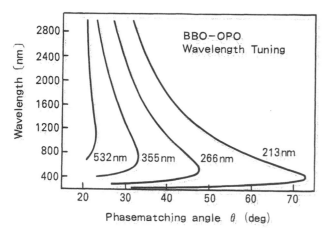

Figure 5. Wavelength of the signal and idler wave of the BBO-OPO as a function of the phase-matching angle. The OPO is pumped by the second, third, fourth, or fifth harmonic of a Nd-YAG laser. (From Fix, A., Schroder, T., and Wallenstein, R., *Laser Optoelectronic,* 23, 106, 1991. With permission.)

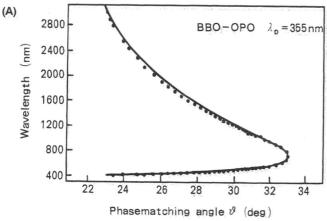

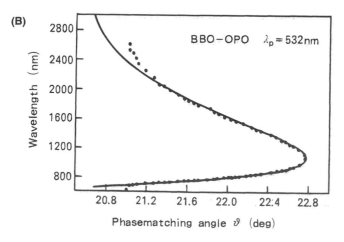

Figure 6. Wavelength of the signal and idler radiation measured at (A) 355 nm and (B) 532 nm. (From Fix, A., Schroder, T., and Wallenstein, R., *Laser Optoelectronic,* 23, 106, 1991. With permission.)

(266 nm) and 41 degrees (213 nm). The OPO system provides high power capability, high conversion efficiency, and very large tuning range.

Figure 7 shows the schematic diagram of the OPO THG experimental apparatus. The THG experiment is based on the Maker fringe technique. An Nd:YAG laser produces frequency tripled light at 355 nm that, after collimation, directly pumps the optical parametric oscillator. This OPO was developed by GWU, Lasertechnik, Germany. In this experiment, the sample is mounted on a goniometer within a vacuum cell to be protected from environmental effects. The laser beam can be split into a second reference arm to compensate for amplitude fluctuations of the OPO. From the detected THG signal, the $\chi^{(3)}$ is calculated against the reference measurement of quartz in the following equation[16]

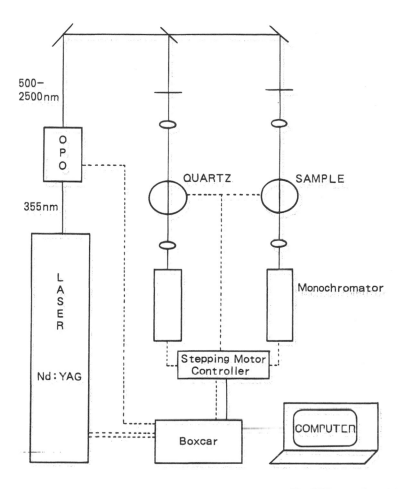

Figure 7. Schematic diagram of the OPO THG experimental apparatus. The THG experiment is based on the Maker fringe technique. A Nd:YAG laser produces frequency tripled light at 355 nm that after collimation directly pumps the optical parametric oscillator. The OPO system was developed by GWU, Lasertechnik, Germany. (Courtesy of BASF AG, Germany, from Gierulski, A., Naarman, H., Schrepp, W., Schrof, W., and Ticktin, A., Proc. 5th Toyota Conf. Nonlinear Optic. Mater., Nagoya, Japan, October 7–9, 1991. With permission.)

$$\chi^{(3)} = \frac{2\sqrt{P_{ms}}l_c}{\pi\sqrt{P_{mq}}l_s} \chi^{(3)}_{\text{quartz}} \tag{49}$$

$$P_{ms} = P_m \left[\frac{\alpha l_s/2}{1 - e^{-\alpha l_s/2}}\right]^2 \tag{50}$$

where P_m is the THG signal from the sample, P_{ms} the absorption corrected THG signal, αl_s the sample absorbance, and l_c and l_s the coherence length of quartz and sample thickness, respectively. The $\chi^{(3)}$ value of quartz is taken between 3.1 to 2.8 $\times$ 10^{-14} esu in the 1.0–2.1 μm wavelength range. This technique is simple and provides determination of $\chi^{(3)}$ in a wide range of wavelengths.

4. THG Measurements of Powder Materials

Kubodera[18] used a powder method for determing the $\chi^{(3)}$ values of organic solids in powder form. The THG intensity ($I_{3\omega}$) of a powder sample is given by

$$(I_{3\omega}) \alpha l_c^2 [\chi^{(3)}]^2 \tag{51}$$

Table 3 Powder THG Intensity and Solution $\chi^{(3)}$ Values of Some Known Materials

Material	Powder THG Relative Intensity	Solution THG $\chi^{(3)}$ (10^{-14} esu)
p-Nitroaniline (p-NA)	1	2.6
2-Methyl-4-nitroaniline (MNA)	15.2	2.9
DMSM	21.9	11.8
DANS	3.5	
DEANS	5.7	
PTS	10.7 (2.05 μm)	

From Kubodera, K., *Nonlinear Optics*, 1, 71, 1991. With permission.
The abbreviated names have the following chemical structures:

DMSM: 4-dimethylamino-N-methyl-4-stilbazolium methylsulfate

$(CH_3)_2N$—⟨ ⟩—$CH=CH$—⟨ ⟩—N^+-$CH_3CH_3SO_4^-$

DANS: 4-dimethylamino-4-nitrostilbene

$(CH_3)_2N$—⟨ ⟩—$CH=CH$—⟨ ⟩—NO_2

DEANS: 4-diethylamino-4-nitrostilbene

$(C_2H_5)_2N$—⟨ ⟩—$CH=CH$—⟨ ⟩—NO_2

PTS: poly-bis-[p-toluenesulfonate)-2,4-hexazine-1, 6-dole

$\begin{matrix} R \\ | \\ =\!\!=\!C\!-\!C\equiv C\!-\!C\!=\!\!= \end{matrix}_n$ R : CH_3OSO_2—⟨ ⟩—CH_3

where l_c is the THG coherence length. The l_c and $\chi^{(3)}$ are averaged over the crystal orientation of powder samples. The averaged l_c value of powder sample should be known for measuring $\chi^{(3)}$ values from THG intensities. Table 3 lists the measured THG intensity of a variety of powder samples.

MNA and DMSM are SHG-active materials and have THG intensities larger than those of SHG inactive materials. Therefore, $\chi^{(3)}$ values of SHG-active material was evaluated by solution THG. Figure 8 shows the powder THG of the above listed materials over several wavelengths. The measured THG intensities are caused by the cascading $\chi^{(2)}$ process being SHG and THG spectra identical. Solution measurements also supported this hypothesis. One can see that the major contribution to the TH intensity of MNA and DMSM are from the cascading process. The $\chi^{(3)}$ of p-NA and MNA is smaller than DMSM, DANS, and DEANS. Kubodera[18] also developed another method of measuring the $\chi^{(3)}$ of thin-film samples. The $\chi^{(3)}$ values of thin films were determined by comparing the TH intensity of reference using the following equations:

$$\chi^{(3)} = \chi_s^{(3)}(\sqrt{I_{3\omega}}/l_c)/(\sqrt{I_{3\omega,s}}/l_{c,s}) \text{ for } l >> l_c \tag{52}$$

$$\chi^{(3)} = (2/\pi)\chi_s^{(3)}(\sqrt{I_{3\omega}}/l_c)/(\sqrt{I_{3\omega,s}}/l_{c,s}) \text{ for } l << l_c \tag{53}$$

where $I_{3\omega}$ and $I_{3\omega,s}$ are TH peak intensities of thin films and reference sample, respectively. l and l_c are the film thickness and the coherence length of the sample, respectively. In this calculation, the refractive index of thin films are assumed to be the same as the reference.

5. THG Measurements in Nematic Liquid Crystals

Wong and Garito[19] developed a new method for the determination of the orientational order parameters based on the measurement of optical THG. A relationship was developed between the third-order macroscopic susceptibility and the microscopic molecular susceptibility in the nematic phase. THG was then measured with different polarizations leading to a determination of nonvanishing order parameters

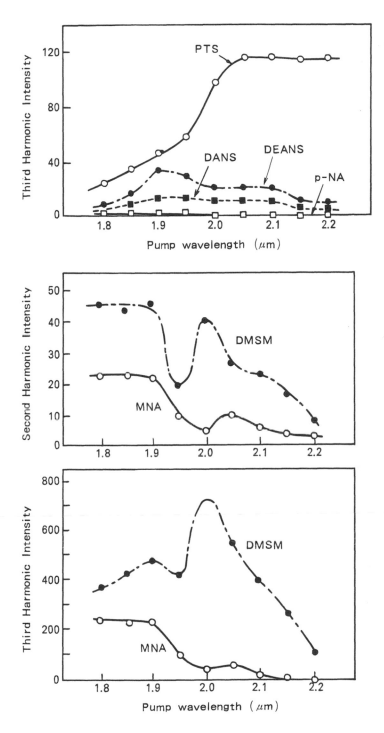

Figure 8. The powder THG of the materials listed in Table 3 over several wavelengths. (From Kubodera, K., *Nonlinear Optics,* 1, 71, 1991. With permission.)

$< P_2 >$ and $< P_4 >$ for a nematic-liquid-crystal N-(p-methoxybenzylidene)-p-butylaniline (MBBA). The wedge Maker fringe method was used to demonstrate the simultaneous determination of the order parameters as a function of polarization and temperature. The liquid crystal sample was formed into a wedge-shaped sample which allows for variations in thickness. The THG from the sample was measured as a function of the sample thickness and third-order susceptibility was estimated by calibrating from a material with known third harmonic susceptibility. The reference also provides information about the effect due to laser-power fluctuations. The liquid crystal was kept between two surface-treated glass plates spaced by a glass fiber, forming a wedge shape. The glass plate was coated with poly(vinylalcohol) following by a unidirectional rubbing for the desired uniaxial homogeneous alignment. The sample was mounted on a heating unit placed inside a vacuum chamber. The intensity of the THG signal originating from the liquid crystal can be written as

$$I_{3\omega}(l) = A(48\pi/c^2)^2\omega^2(\chi^{(3)})^2 l_c^2 I_\omega^3 \, \sin^2(\pi l/2l_c) \tag{54}$$

where

$$A = \left(\frac{2}{1+n_G^\omega}\right)^6 \left(\frac{2n_G^\omega}{n_S^\omega+n_G^\omega}\right)^6 \left(\frac{2n_G^{3\omega}}{1+n_G^{3\omega}}\right)^2 \left(\frac{1}{n_S^{3\omega}+n_S^\omega}\right)^2 \times \left(\frac{2n_S^{3\omega}}{n_S^{3\omega}+n_G^{3\omega}}\right) \left(\frac{n_S^\omega+n_S^{3\omega}}{n_S^{3\omega}+n_G^{3\omega}}\right) \left(\frac{n_S^\omega+n_G^{3\omega}}{n_S^{3\omega}+n_G^{3\omega}}\right) \tag{55}$$

where the subscript S and G refer to the liquid crystal (sample) and glass, respectively. The n is the refractive index, the I_ω is the fundamental beam intensity, l is the sample thickness, and l_c is the coherence length of the sample given by

$$l_c = \frac{\lambda}{6(n^{3\omega}-n^\omega)} \tag{56}$$

In this situation, the THG from the glass plate was shown to be negligible. The Maker fringe results are analyzed using a least-squares procedure for the fitting function

$$y = A_1 \, \sin^2\left(\frac{\pi l}{2A_3} + \frac{A_4}{2}\right) + A_2 \tag{57}$$

where $(A_1 + A_2)$ and A_2 are the maximum and minimum of each measured fringe, A_3 refers to the coherence length l_c of the sample, and A_4 is the phase offset.

The following equations express the relationship between the macroscopic and microscopic susceptibility for a nematic liquid crystal

$$\chi^{(3)}_{zzzz} = N(\delta + 2\xi\langle P_2\rangle + 8\eta\langle P_4\rangle) \tag{58}$$

$$\chi^{(3)}_{xxxx} = N(\delta - \xi\langle P_2\rangle + 3\eta\langle P_4\rangle) \tag{59}$$

where

$$\delta = \frac{1}{5}(\gamma_{z'z'z'z'} + \gamma_{x'x'x'x'} + \gamma_{y'y'y'y'} + \gamma_{z'z'x'x'} + \gamma_{x'z'x'x'} + \gamma_{z'z'y'y'} + \gamma_{y'z'z'y'} + \gamma_{x'x'y'y'} + \gamma_{y'x'x'y'}) \tag{60}$$

$$\xi = \frac{1}{14}(4\gamma_{z'z'z'z'} - 2\gamma_{x'x'x'x'} - 2\gamma_{y'y'y'y'} + \gamma_{z'z'x'x'} + \gamma_{x'z'z'x'} + \gamma_{z'z'y'y'} + \gamma_{y'z'z'y'} - 2\gamma_{x'x'y'y'}$$

$$- 2\gamma_{y'x'x'y'}) \tag{61}$$

$$\eta = \frac{1}{260}\,(8\gamma_{z'z'z'z'} + 3\gamma_{x'x'x'x'} + 3\gamma_{y'y'y'y'} - 12\gamma_{z'z'x'x'} - 12\gamma_{x'z'z'x'} - 12\gamma_{z'z'y'y'} - 12\gamma_{y'z'z'y'}$$

$$+ 3\gamma_{x'x'y'y'} + 3\gamma_{y'x'x'y'}) \qquad (62)$$

in the isotropic phase, both $< P_2 >$ and $<P_4 >$ vanish. By neglecting the small variation of the particle density through phase transition, the following relationship is obtained

$$\chi_{iso}^{(3)} = N\delta \qquad (63)$$

where $\chi^{(3)}{}_{iso}$ is the susceptibility for the isotropic phase where $< P_2 >$ and $< P_4 >$ vanish. The order parameters $< P_2 >$ and $< P_4 >$ can be estimated from the results of $\chi_{zzzz}^{(3)}$, $\chi_{xxxx}^{(3)}$, and $\chi_{iso}^{(3)}$, and if the parameters δ, ζ, and η are known. Using the equations noted above,

$$\chi_{zz}^{(3)}\,\frac{\chi_{zzzz}^{(3)} + 2\chi_{xxxx}^{(3)}}{\chi_{iso}^{(3)}} = 3 + K_4(P_4) \qquad (64)$$

where $K_4 = 14\,(\eta/\delta)$. Similarly

$$\frac{-3\chi_{zzzz}^{(3)} - 8\chi_{xxxx}^{(3)}}{\chi_{iso}^{(3)}} = K_2(P_2) - 5 \qquad (65)$$

where $K_2 = 14\,(\eta/\delta)$. These equations provide the estimation of the magnitude and temperature dependence of order parameters $< P_2 >$ and $< P_4 >$. The equations can also be written as

$$\gamma_{zzzz}^{(3)} = (\delta + 2\xi\langle P_2\rangle + 8\eta\langle P_4\rangle)M_\parallel^4 \qquad (66)$$

$$\chi_{xxxx}^{(3)} = (\delta - \xi\langle P_2\rangle + 3\eta\langle P_4\rangle)M_\perp^4$$

$$\chi_{iso}^{(3)} = \delta M_{iso}^4 \qquad (67)$$

where M is a tensor describing the dependence of the local field factor on the anisotropy of the bulk phase and diagonalized in the laboratory frame along the axes parallel and perpendicular to the nematic director. The corresponding quantity M refers to the isotropic phase and can be written

$$M_{iso} = (M_\parallel + 2M_\perp)/3 \qquad (68)$$

Equation 64 can be rewritten as

$$\frac{\chi_{zzzz}^{(3)}/M_\parallel^4 + 2\chi_{xxxx}^{(3)}/M_\perp^4}{\chi_{iso}^{(3)}/M_{iso}^4} = 3 + K_4\langle P_4\rangle \qquad (69)$$

The third-order nonlinear optical suscepbility can be measured as a function of temperature where z axis and x axis are parallel and perpendicular to the nematic director, respectively. These results were found to be in agreement with NMR and Raman scattering studies.

B. DEGENERATE FOUR-WAVE MIXING (DFWM) TECHNIQUE

DFWM refers to the nonlinear interactions of waves at a common frequency (all pump and probe beams are of the same frequency ω) but they are characteristically distinguishable by their direction of propagation or polarization. The DFWM technique has been widely used for measuring third-order nonlinear optical susceptibility; Gerritsen[20], for example, discussed DFWM for applications in real-time holography. After a decade, Yariv[21] proposed the use of DFWM in phase conjugation, oscillation, etc. There are a large number of research papers that have been published since the late 1970s on the aspects and analysis of the DFWM technique. Several excellent reviews and monographs discussing

588

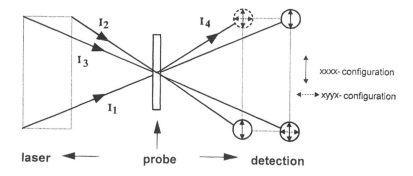

Figure 9. A folded boxcars geometry used in DFWM technique. (Courtesy of BASF AG, Germany.)

DFWM and its related nonlinear optical processes have been written by Yariv,[22] Hellwarth,[23] and Giuliano.[24] The concept of the polarization for third-order nonlinear optical interactions in DFWM has been written in Equation 10. Excellent texts by Reintjes,[25], Shen,[2] and others[26] are available on the DFWM technique. Shen[2] has described in details the four-wave mixing, phase conjugation by four-wave mixing, transient four-wave mixing, and four-wave mixing spectroscopy that includes coherent anti-Stokes Raman scattering (CARS), Raman-induced Kerr effect spectroscopy (RIKES), multiply resonant four-wave mixing, forced light scattering, and transient four-wave mixing spectroscopy and DFWM techniques. Moreover, Reintjes[25] described the DFWM geometries, theory (nonlinear phase shift, DFWM in absorbing media, and thermal effects) and experimental details as well as phase conjugation, holographic image reconstruction, and resonators with phase conjugate mirrors. Details of two-beam and three-beam DFWM experiments have been discussed by Kajzar et al.[5] Two-beam DFWM has a non-phase-matched geometry and the sample thickness should be smaller than coherence length. The selection of two independent beam polarizations provides information about dimensionality of microscopic effects. The pump-probe and optical Kerr shutter configurations are two-beam DFWM techniques. In the three-beam DFWM method, forward phase-matched DFWM has a three-dimensional geometry known as the folded boxcar geometry, while the backward phase-matched DFWM has a planar geometry known as the phase conjugation. Both folded boxcars and phase conjugation techniques yield the same information and are physically equivalent. Therefore folded boxcars, phase conjugation, and transient grating are examples of three-beam DFWM.

A folded boxcars geometry for the DFWM technique is shown in Figure 9. The DFWM experimental setup used to measure the third-order nonlinear optical susceptibility and response speed is shown in Figure 10. The theory of the nonlinear mechanism in understanding DFWM experiments in an absorbing medium was developed by Caro and Gower.[27] In a DFWM process, three imput beams $I_1(\omega),I_2(\omega)$, and $I_3(\omega)$ of equal frequency ω interact in a medium, where two of them, i.e., $I_1(\omega)$ and $I_2(\omega)$, counterpropagate; and the third $I_3(\omega)$, the probe beam, crosses at a small angle θ. The $I_4(\omega)$ beam of the same frequency is generated by the third-order nonlinear optical interaction $\chi^{(3)}$ that counterpropagates to $I_3(\omega)$. The phase conjugated $I_4(\omega)$ beam is separated by a beam splitter placed in the path of beam $I_3(\omega)$. The intensity of $I_4(\omega)$ is measured by a photodiode and boxcar combination. The intensities of the DFWM signal sample were compared to that of the reference sample under similar conditions. In DFWM, two common geometries—(1) retroreflection and (2) antiresonant ring geometries—are used. For a retroreflection geometry, the nonlinear process is written by

$$Q = 3\omega I_p\chi^{(3)}(-\omega; \omega, \omega, -\omega)\exp(-\alpha L')/4c^2n^2\epsilon_0 \qquad (70)$$

where

$$I_2(L) = I_1(O)\exp(-\alpha L') \qquad (71)$$

and for the antiresonant ring geometry, the expression changes to

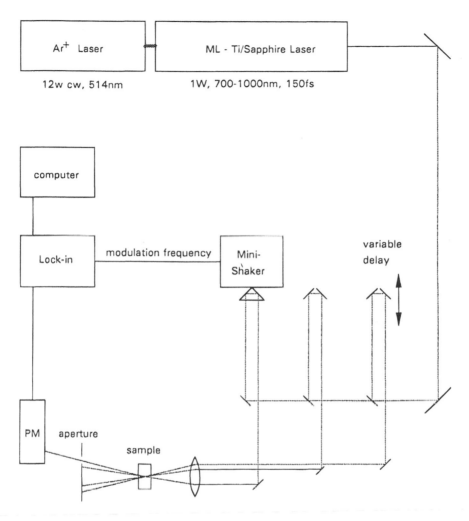

Figure 10. A DFWM experimental setup used to measure the third-order nonlinear optical susceptibility and its ultrafast dynamics. The details of this technique have been specified in Reference 3. (Courtesy of BASF AG, Germany.)

$$Q = 3\omega I_p \chi^{(3)}(-\omega; \omega, \omega, -\omega)\exp(-1/2\alpha L')/4c^2n^2\epsilon_0 \qquad (72)$$

where

$$I_2(L) = I_1(O) = I_p \qquad (73)$$

where Q is the coupling constant which is given by

$$Q = -3\omega/8cn\chi^{(3)}(-\omega; \omega, \omega, -\omega)E_1(O)E_2(L)\exp[-1/2\alpha L'] \qquad (74)$$

where α is the intensity absorption coefficient in the medium at frequency ω and n is the complex refractive index. The change in the refractive index in the nonlinear medium caused by heating by absorption of energy can be written as

$$Q = 3/4\chi^{(3)}(-\omega; \omega, \omega, -\omega)I/cn^2\epsilon_n \qquad (75)$$

Therefore, $\chi^{(3)}$ can be evaluated. Different mechanisms of thermally induced refractive index changes, the Kerr effect and saturable absorption describing DFWM process are possible. The following expression has been derived by Caro and Gower[27] for the predicted reflectivity in each case. The mechanism of optical nonlinearity in the medium may be caused by heating. The thermally induced refractive index in nonlinear medium is given by

$$\chi^{(3)} = \frac{dn}{dT} [4n^2 c\epsilon_0 \alpha\tau\Phi/3\rho C_p] \tag{76}$$

where T is the temperature in the medium, τ is the duration of electric fields, ρ and C_p are, respectively, the density and specifice heat at constant pressure, and ϕ is the friction of absorbed radiation. The very large $\chi^{(3)}$ values may be produced by thermal effects.

Another mechanism of DFWM in the ultraviolet may be the Kerr effect. The predicted reflectivity can be obtained from the experimentally determined Kerr susceptibility $\chi^{(3)}(-\omega; \omega, \omega, -\omega)$. The change in the refractive index in the medium is written by

$$|Q| = \left|\frac{\omega}{c} \delta n \exp(-\alpha L')\right| \tag{77}$$

The phase conjugate reflectivity (R) due to a Kerr mechanism is given by

$$R \simeq \left|\frac{\omega}{c\alpha} \delta n \exp(-\alpha L')[1 - \exp(-\alpha L)]\right|^2 \tag{78}$$

The third mechanism for phase conjugation using DFWM is of saturable absorption. A perturbation technique yields the following expression for the imaginary nonlinear susceptibility

$$|\chi^{(3)}| = \frac{4\alpha_0 c^2 n^2 \epsilon_0}{3\omega I_s(1 + 4I/I_s)^{3/2}} \tag{79}$$

α_0 is the signal intensity absorption coefficient. The value of $\chi^{(3)}$ obtained from saturable absorption can be very large. Of the three mechanisms, the thermal effects easily lead to high phase conjugation reflectivities.

The following relationship for absorbing media to calculate the $\chi^{(3)}$ value holds [27,28]

$$\chi^{(3)}_{sample} = \chi^{(3)}_{ref} \times \frac{\left(\frac{S_{sample}}{S_{ref}}\right)^{1/2}\left(\frac{n_{sample}}{n_{ref}}\right)^2\left(\frac{l_{ref}}{l_{sample}}\right)\alpha l_{sample}}{\exp\{(-\alpha l_{sample}/2)[1 - \exp(-\alpha l_{sample}]\}} \tag{80}$$

where ref shown in the equation is a CS$_2$ reference. The S is the coefficient of a cubic least-squares fit of phase conjugate signal against the laser intensity, n is the refractive index, l is the path length, and α is the linear absorption coefficient. Therefore, by measuring the values of n, one can estimate the $\chi^{(3)}$ value for the sample at the operating wavelength in the experiment. It is assumed that the reference CS$_2$ does not absorb at the measurement wavelength.

For nonabsorbing media, the equation is modified as follows[2]

$$\chi^{(3)}\text{sample} = \chi^{(3)}\text{ref}(n_{sample}/n_{ref})^2(l_{ref}/l_{sample})(I_{sample}/I_{ref})^{1/2} \tag{81}$$

The second hyperpolarizability (γ) can be calculated since it is related to $\chi^{(3)}$ by the following expression[28]

$$\chi^{(3)}(\text{solution}) = L^4 N\langle\gamma_{ijkl}\rangle \tag{82}$$

where L is the local field factor and N is the number of density of the sample. The $\chi^{(3)}$ of pure solute can be written as:

$$\chi^{(3)}(\text{pure}) = (N_{(\text{pure})}/N)\chi^{(3)}(\text{solution}) \tag{83}$$

where $N_{(\text{pure})}$ refers to the number density of the pure solid (solute).

A research team at Trinity College in Ireland[30] used forced light scattering from laser-induced grating, a technique that may be considered as a DFWM process in the forward direction. The relationship between the diffraction efficiency η in the first order may be written as:

$$\chi^{(3)} = \frac{4\epsilon_0 cn^2\lambda(\eta)^{1/2}}{3\pi\, dI_0} \tag{84}$$

where ϵ_0 is the permittivity of free space, c is the speed of light, n and d are the refractive index and thickness of the sample and I_0 is the input pulse intensity. This expression is valid for those materials that are transparent at the operating wavelength. Complimentary to the measurements of the transient decay, both the modulus and phase of $\chi^{(3)}(-\omega;\omega,\omega,-\omega)$ can be determined. For solutions, the $\chi_{\text{eff}}^{(3)}$ can be written by[30]

$$\chi_{\text{eff}}^{(3)} = [(\chi_{\text{solvent}}^{(3)} + \text{Re}\,\chi_{solute}^{(3)})^2 + (\text{Im}\,\chi_{solute}^{(3)})^2]^{1/2} \tag{85}$$

where $\text{Re}\chi^{(3)}$ and $\text{Im}\chi^{(3)}$ are the real and imaginary parts of optical susceptibility of the solute. By knowing the $\chi_{\text{solvent}}^{(3)}$, the components of $\chi_{\text{eff}}^{(3)}$ can be determined. The concentration dependence studies may yield the real and imaginary part of $\chi^{(3)}$. It is also possible to evaluate the change of refractive index by modifying the DFWM experiments. The DFWM technique has been used for a variety of organic materials such as charge-transfer complexes, fullerenes, metallophthalocyanines, metalloporphyrins, polydiacetylenes, polythiophenes, polyanilines, polysilane, polyacetylenes, etc. Such examples are given in chapter 11. Both resonant and nonresonant $\chi^{(3)}(-\omega; \omega, \omega, -\omega)$ values can be measured, depending upon the fundamental wavelength (ω) of the laser systems, comparing the optical absorption spectrum of sample with ω. In some cases, one- or two-photon resonant contributions may affect the magnitude of $\chi^{(3)}(-\omega; \omega, \omega, -\omega)$. For example, DFWM in saturable absorbers involves a single-photon resonance with large $\chi^{(3)}$ values in polydiacetylenes and π-conjugated polymers being reported.

C. Four-Wave Mixing (FWM) Technique

Like the DFWM method, four-wave mixing is useful for determining the magnitude of the third-order NLO susceptibility and optical response time. Shen[2] has described in detail the four-wave mixing technique, transient four-wave mixing, and four-wave mixing spectroscopy. In a four-wave mixing procedure, two incident laser beams with frequency ω_1 and ω_2 are focused in a nonlinear medium to generate a third beam of frequency $\omega_3 = 2\omega_1 - \omega_2$. The generated output beam of ω_3 is passed through a slit that blocks the ω_2 and ω_2 beams and through a double monochromator which blocks all frequencies away from ω_3. The intensity of ω_3 is measured by a photomultiplier and boxcar integrator. The intensity (I) of ω_3 is proportional to the total third-order nonlinear susceptibility which may include electronic susceptibility terms for the sample and the solvent and other terms arising when $\Delta\omega = \omega_1 - \omega_2$ is near Raman active molecular vibrational frequencies. When $\omega_1 - \omega_2$ is near such vibrations, the involved process is called coherent anti-Stokes Raman scattering (CARS).[31] Lynch et al.[32] described the following relationship for third-order nonlinear optical susceptibility:

$$\chi_T = \chi_s + \chi_p + \frac{A}{\omega_c - \Delta_\omega - i\Gamma_c}$$
$$+ \frac{NA'}{\omega_p - \Delta_\omega - i\Gamma_p} + \frac{NA''}{\omega_0 - 2\omega_1 - i\Gamma_0} \tag{86}$$

where N is the number density of the sample, A and Γ_C are the amplitude and widths, respectively of

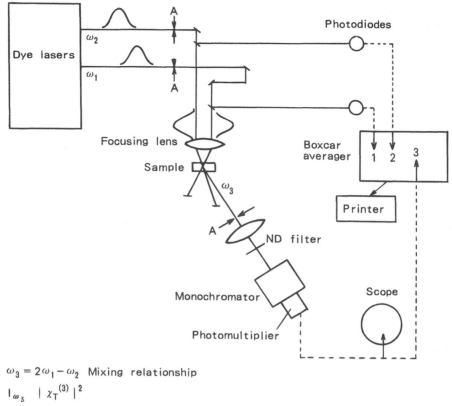

$$\omega_3 = 2\omega_1 - \omega_2 \quad \text{Mixing relationship}$$

$$I_{\omega_3} \quad | \chi_T^{(3)} |^2$$

Figure 11. Four-wave mixing experiment for determining hyperpolarizability. (From Frazier, C. C. et al., *Mater. Res. Soc. Symp. Proc.*, 109, 323, 1988. With permission.)

the solvent vibration (sample vibration). A'', ω_0 and Γ_0 are the amplitude, frequency, and width of an electronic state near $2\omega_1$ respectively. Shand and Chance[31] used this technique to measure the $\chi^{(3)}$ values of polydiacetylene in chloroform solutions (Figure 11). Frazier et al.[33] measured second hyperpolarizabilities of tetrahydrofuran solutions of metal polyynes. The molecular hyperpolarizability is given by:

$$\gamma = \chi^{(3)}/NL\omega_1 L\omega_1 L\omega_2 L\omega_3 \tag{87}$$

where N is the number density of the given species and each L is the local field correction factor [$L = (n^2 + 2)/3$] at the indicated frequency. The molecular hyperpolarizability of the sample at a given sample solution concentration can be estimated by using the known values of hyperpolarizability of the solvent and by measuring the ratio of $I\omega_3$ for the sample solution and pure solvent. The following equation describes the relationship:[33]

$$\frac{I\omega_3(\text{solvent} + \text{sample})}{I\omega_3(\text{solvent})} = \frac{|\chi_{\text{total}}^{(3)}|^2}{|\chi_{\text{solvent}}^{(3)}|^2} = \frac{(L\omega_1 L\omega_1 L\omega_2 L\omega_3)^2 (N_{\text{solvent}}\gamma_{\text{solvent}} + N_{\text{sample}}\gamma_{\text{sample}})^2}{(L\omega_1 L\omega_1 L\omega_2 L\omega_3)^2 (N_{\text{solvent}}\gamma_{\text{solvent}})^2} \tag{88}$$

Since for dilute solutions, the (solvent + sample) local field correction factors become nearly equivalent to the solvent local field correction factors, the relationship is modified to

$$\frac{I\omega_3(\text{solvent} + \text{sample})}{I\omega_3(\text{solvent})} = \frac{(N_{\text{solvent}}^2\gamma_{\text{solvent}}^2 + 2N_{\text{solvent}}\gamma_{\text{solvent}}N_{\text{sample}}\gamma_{\text{sample}} + N_{\text{sample}}^2\gamma_{\text{sample}}^2)}{N_{\text{pure solvent}}^2\gamma_{\text{solvent}}^2} \tag{89}$$

Frazier et al.[33] used the following expression to fit the experimental data, assuming dilute solutions

$$\frac{I\omega_3(\text{solvent} + \text{sample})}{I\omega_3(\text{solvent})} = \frac{|x_{\text{total}}^{(3)}|^2}{|x_{\text{solvent}}^{(3)}|^2} = 1 + \frac{2N_{\text{solvent}}N_{\text{sample}}\gamma'}{N_{\text{pure solvent}}^2\gamma_{\text{solvent}}}$$

$$+ \frac{N_{\text{sample}}^2\gamma'^2}{N_{\text{pure solvent}}^2\gamma_{\text{solvent}}^2} + \frac{N_{\text{sample}}^2\gamma''^2}{N_{\text{pure solvent}}^2\gamma_{\text{solvent}}^2} \qquad (90)$$

where $\gamma' = \gamma$ (nonresonant electronic hyperpolarizability of the compound) + $\text{Re}\gamma_{\text{TPA}}$ (dispersive part of the two-photon absorption) and $\gamma'' = \text{Im}\gamma_{\text{TPA}}$ (absorptive part of the two-photon absorption). Therefore, second hyperpolarizabilities can be obtained from the FWM technique. The γ values of poly(metalynes) were measured using the FWM technique. The third-order NLO susceptibility $\chi^{(3)}(-\omega_3; \omega_1, \omega_1, -\omega_2)$ can be determined.

D. OPTICAL KERR GATE (OKG) TECHNIQUE

Ho[34] reviewed the theoretical background, experimental techniques, and important parameters for direct measurement of ultrafast time-resolved processes by using the optical Kerr gate technique. This optical Kerr effect has been used for studying the electric dipole moment of liquids, determining the chemical structures, and for analyzing relaxation phenomena in physical, chemical, and biological processes at a picosecond scale. The history of the OKG effect dates back to 1956 when it was first reported by Buckingham.[35] In this chapter, we follow the approach given by Ho[34] and Rosen et al.[34b] in connection with $\chi^{(3)}$ effects and time-resolved spectroscopy. The optical Kerr gate technique is suitable in particular for the measurement of the $\chi^{(3)}$ of a liquid sample. The OKG method is applicable to a large dynamic range and is a low-cost and high-sensitivity technique.

Figure 12 shows the schematic diagram of an optical Kerr gate. In an optical Kerr gate setup, a Kerr active liquid such as carbon disulfide (CS_2) is placed between two polarizers. When a strong field is applied with the laser pulse, the birefringence is induced in the molecules of the Kerr active liquid. The light passing through the Kerr liquid is elliptically polarized and passes partially through the second polarizer. When the light coincides with the laser pulse, it then passes through the gate.

The time-dependent, field-induced nonlinear refractive index resulting from the difference between the parallel and perpendicular directions of the orienting laser is given by

$$\delta n = \delta n_{\parallel} - \delta n_{\perp}, \qquad (91)$$

and the time-dependent nonlinear refractive index is written as

$$\delta n(t) = \int_{-\infty}^{t}\int_{-\infty}^{t} f(t, t_2, t_3)E(t - t_2)E(t - t_3)\, dt_2\, dt_3, \qquad (92)$$

where the response of the refractive index to a pulse is written by

$$f(t, t_2, t_3) \equiv (2\pi/n_0)\chi^{(3)}(t, t_2, t_3) \qquad (93)$$

The net induced refractive index can be written by

$$\delta n(t) = n_2^f\langle E_1^2(t)\rangle + \sum_i \frac{n_{2i}^s}{\tau_i}\int_{-\infty}^{t}\langle E_1^2(t')\rangle\exp\left(-\frac{t - t'}{\tau_i}\right)dt', \qquad (94)$$

where $E_1(t)$ is the electric field of the orienting laser pulse with pulse duration time τ_L, τ_i is the relaxation time that induces the nonlinear refractive index in the isotropic Kerr medium, n_2^f and n_{2i}^s are the sum of all the nonlinear indexes of refraction and the refractive index from one particular mechanism, respectively, whose relaxation time τ_i is slower than the orienting laser pulse duration. The simplified form of the above equation can be written as

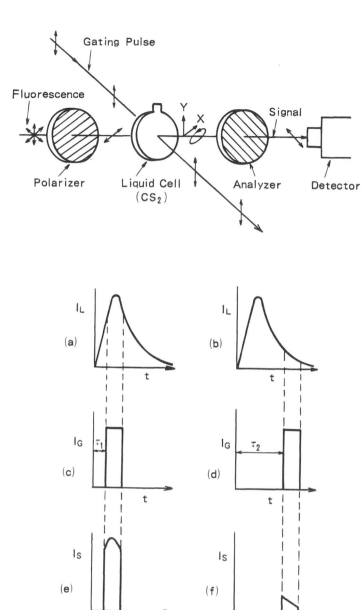

Figure 12. Schematic diagram of an optical Kerr gate for time-resolved luminescence spectroscopy. In optical Kerr gate setup, a Kerr active liquid such as carbon disulfide (CS_2) is placed between two polarizers. (From Rosen, D. L., et al., in *Semiconductor Probed by Ultrafast Laser Spectroscopy*, Vol. 2, Alfano, R. R., Ed., Academic Press, New York, 1984, 393. With permission.)

$$\delta n(t) = n_2^0 \langle E_1^2(t) \rangle + (n_2^0/\tau_0) \int_{-\infty}^{t} \langle E_1^2(t') \rangle \exp[-(t - t')/\tau_0] \, dt', \tag{95}$$

where τ_0 and n_2^0 are the molecular reorientation relaxation time of the Kerr medium and the nonlinear refractive index from the molecular reorientation motion. In case τ_0 is smaller than the laser pulse duration, then the above equation is simplified to

$$\delta n(t) = n_2 \langle E_1^2(t) \rangle \tag{96}$$

where n_2 is the sum of all the fast relaxation mechanisms and $\langle E_1^2(t) \rangle$ is the time average of the optical field over one period.

Table 4 Nonlinear Refractive Indexes of Kerr Media

Kerr Medium	n_2 (10^{-13} esu)	Relaxation Time (ps)
Carbon disulfide	200	1.8
Benzene	60	—
Chloroform	10.6	—
toluene	60	—
Nitrobenzene	250	32
Methanol	2.2	—
water	1.3	—
Hexadecane	15.5	210
Salol	460	550
p-ethoxy-benzylidene p-n-butylaniline (EBBA)	8000	500 (at 84°C)
p-methoxybenzylidene p-n-butylanilinc (MBBA)	2000	400 (at 56°C)

From Ho, P. P. and Alfano, R. R., *J. Chem. Phys.*, 67, 1004, 1977, Wong, C. K. L. and Shen, Y. R., *Phys. Rev.* A, 10, 1277, 1974. With permission.

The signal transmitted through the OKG when a signal beam passes through the medium with the induced birefringence of length L, and then phase retardation $\delta\phi(t)$ between the parallel and perpendicular component to the orienting field E is given by

$$\delta\phi(t) = (2\pi L/\lambda)n(t) \tag{97}$$

where λ is the wavelength of the signal beam. The state of the polarization of the signal beam is changed due to the phase retardation. The relative transmitted signal through a Kerr gate as a function of the delay time between the probe and the orienting pulses is written by

$$S_t(\tau_D) = \frac{1}{S_1} \int_b^a \langle E_2^2(t - \tau_D)\rangle \sin^2\left[\frac{\delta\phi(t)}{2}\right] \sin^2 2\theta \, dt, \tag{98}$$

where E_2 is the electric field of the signal beam, τ_D is the time difference between the signal laser pulse and the orienting laser pulse, θ is the polarization angle between electric field E_2 and E_1, and S_1 is the total normalization signal of the probe light while passing through Kerr gate. The polarization angle θ is generally 45° for the Kerr gate to achieve S_t and S_1 is written by

$$S_1 = \int_{-\infty}^{\infty} \langle E_2^2(t)\rangle \, dt \tag{99}$$

The time resolution of the OKG is affected by the incident angle to the Kerr cell between the orienting beam and the single beam. Usually the angular separation is about 3°–5°. The time resolution scale for the OGK is restricted by the duration of the laser pulse and the relaxation mechanism of the Kerr liquid. Several factors affect the measurements in the OKG technique. The phase retardation $\delta\phi(t)$ of the Kerr gate is linearly proportional to the cell length L. For $\delta\phi < 1$, the transmitted signal through Kerr gate is proportional to $(\delta\phi)^2$ which changes quadratically with the cell length. There is no significant difference in the time resolution Kerr cells up to 2 cm long but after this length, the time resolution is critically affected by the dispersion effect. The time resolution is also affected by the induced phase retardation $\delta\phi$ when it becomes larger.

The temperature dependence of n_2^0 in the OKG can be expressed by the following equation:

$$n_2^0(T) = (B/T([1 + A \exp(E/RT)) \tag{100}$$

where A and B are constant parameters, T is temperature in Kelvin, and E is the activation energy. The n_2^0 decreases as the temperature rises and reorientation relaxation time is an inverse function of the temperature. For example, the Kerr signal monotonically decreases as the temperature of the CS_2 gate increases. The Kerr signal is a function of the nonlinear refractive index and the inverse of the Kerr relaxation time. Table 4 lists the nonlinear refractive indexes of some Kerr media.[36,37] The OKE

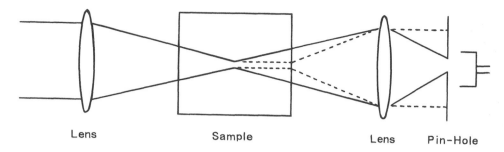

Figure 13. Schematic representation of the apparatus used in optical power limiting (OPL) method for measuring nonlinear refractive index (n_2). (From Soileau, M. J., Williams, W. E., and Van Stryland, E. E., *IEEE J. Quantum Electron.*, QE-19, 731, 1983. With permission.)

measurements have been performed on polydiacetylenes, polysilanes, and polythiophenes to determine their $\chi^{(3)}(-\omega; \omega, \omega, -\omega)$ values.

E. SELF-FOCUSING TECHNIQUES

Two main techniques have been widely used as self-focusing. The details of these methods are given below.

1. Optical Power Limiter (OPL) Technique

The optical power limiter (OPL) technique was developed by Soileau et al.[38] The schematic representation of the apparatus used in this technique is shown in Figure 13. The basic technique involves the tight focusing spatially Gaussian beam by lens 1 into the nonlinear cell (sample) and the measurement of the transmitted power by lens 2 through a pinhole system by a detector. The dotted lines represent the high-power situation. A time-dependent movement of the focus point occurs when the input power of the beam is increased more than the critical threshold, which causes defocusing of lens 2, with self-focusing taking place at the critical power of the beam. A detailed description of this technique is given by Winter et al.[39,40] For a tightly focused Gaussian beam, Marburger[41] established a relationship between the nonlinear refractive index n_2 and the critical power

$$Pc = 3.72c\lambda^2/32\pi^2 n_2 \text{ (esu)} \tag{101}$$

where Pc is the critical power, λ is the laser wavelength, and c is the speed of light. The n_2 is defined by

$$n = n_0 + n_2 \langle \zeta^2 \rangle \tag{102}$$

where $\langle \zeta^2 \rangle$ is the time-averaged optical electric field. Therefore, a direct relationship between the measured Pc for self-focusing and the n_2 exists. The n_2 is also related to the second hyperpolarizability (γ) by

$$n^2 = 4\pi\chi^{(3)}/n_0 \tag{103}$$

where

$$\gamma = \chi^{(3)}/L^4 N \tag{104}$$

where n_0 is the linear refractive index, $L = [(n^2 + 2)/3]$ is the local field correction, and N is the number density of the molecules. Therefore, using this method, third-order nonlinear optical susceptibility as well as hyperpolarizability can be measured.

The OPL technique offers the advantages of fast response time in picosecond scale, completely

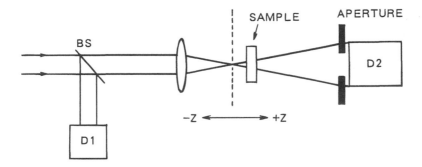

Figure 14. The Z-scan experiment apparatus for measuring nonlinear optics. The transmittance of a nonlinear medium is measured through a finite aperture located in the far field as a function of the sample position (z). (From Sheik-Bahae, M., Said, A. A., and Van Stryland, E. W., *Opt. Lett.*, 14, 955, 1989. With permission.)

passive operation, and a relatively low limiting power Pc (26 kW for carbon disulfide at 1.064 μm wavelength), and in this method Pc can be adjusted by varying nonlinear refractive index (n_2).

2. Z-Scan Method

Another self-focusing operated method called Z-scan was developed by Sheik-Bahae and co-workers.[42–48] The fundamentals, measurement details, analysis of experimental results, and usefulness of the Z-scan techniques are discussed here. This technique is a simple single-beam experimental method for the accurate measurements of the nonlinear refractive index (n_2) and the nonlinear absorption coefficient useful for a wide variety of materials. The sign and the magnitude of the nonlinear refractive index are determined from the transmittance curve (Z-scan). The n_2 is evaluated from a simple linear relationship between the observed transmittance changes and the induced phase distortion. The measurements of n_2 at different wavelengths have been performed in CS_2 and other materials. Furthermore, both thermo-optical and reorientational Kerr effects can be evaluated using nanosecond and picosecond pulses. Figure 14 shows the Z-scan experimental apparatus using a single Gaussian laser beam. The transmittance of a nonlinear medium is measured through a finite aperture located in the far field as a function of the sample position (z). When the sample is moved toward the focus (on the negative z side of the focus), the radiation increases, resulting in a negative lensing effect which tends to collimate the laser beam. As a consequence, the aperture transmittance is increased. To the contrary, when the sample is moved away from the focus (on the positive z side of the focus), the negative lensing effect tends to augment diffraction and the transmittance reaches to the original linear value for larger $+z$ side. This value is normalized to unity. The decreased transmittance occurs at $-z$ while increased transmittance occurs at the $+z$ side. Therefore, the sign of nonlinear refractive index (n_2) is measured from the Z-scan. The cubic optical nonlinearity expressed in terms of the nonlinear refractive index in esu units is given by[42]

$$n = n_0 + \frac{n_2}{2} |E|^2 = n_0 + \Delta n \tag{105}$$

where n_0 is the linear refractive index and E is the electric field. For a Gaussian beam in the $+z$ direction, the magnitude of E can be written by[43]

$$E(z, r, t) = E_0(t) \frac{w_0}{w(z)} \exp\left(-\frac{r^2}{w^2(z)} - \frac{ikr^2}{2R(z)}\right) e^{-i\phi(z,t)} \tag{106}$$

where $w^2(z) = w_0^2(1 + z^2/z_0^2)$ is the beam radius at z, $R(z) = z[1 + z_0^2/z^2]$ is the radius of the curvature of the wavefront at z, $kw_0^2/2$ is the diffraction length of the beam, $k = 2\pi/\lambda$ is the wave vector, and λ is the laser wavelength. $E_0(t)$ refers to the radiation electric field at the focus and the $e^{-i\phi(z,t)}$ contains all the radially uniform phase variations. The radial variation of the incident irradiance at a given position of the sample z can be given as

$$\Delta\phi(z, r, t) = \Delta\phi_0(z, r, t)\exp\left(-\frac{2r^2}{w^2(z)}\right) \tag{107}$$

with

$$\Delta\phi(z, t) = \frac{\Delta\phi_0(t)}{1 + z^2/z_0^2} \tag{108}$$

the on-axis phase shift at the focus is defined as

$$\Delta\phi_0(t) = k\Delta n_0(t)L_{\text{eff}} \tag{109}$$

and

$$L_{\text{eff}} = (1 - e^{-\alpha L})/\alpha$$

where L is the sample length, α is the linear absorption coefficient, $\Delta\phi$ is the phase at the exit surface of the sample, and $\Delta n_0(t)$ is the instantaneous on-axis nonlinear index change at the focus point ($z = 0$), which is given as

$$\langle \Delta n_0 \rangle \approx \frac{dn}{dT}\frac{F_0\alpha}{2\rho C_v} \tag{110}$$

where F_0 is the fluence, C_v is the specific heat, ρ is the density, and $1/2$ is the fluence averaging factor. Therefore, using these expressions, an index change Δn_0 can be calculated. The normalized instantaneous Z-scan power transmittance is obtained by[42]

$$T(z, t) = \frac{\int_0^{r_a} |E_a(\Delta\Phi_0, r, z, t)|^2 r dr}{S \int_0^{\infty} |E_a(0, r, z, t)|^2 r dr} \tag{111}$$

where r_a and S are the aperture radius and aperture transmittance in the linear regime, respectively. The plot of transmittance through an aperture in the far field gives a dispersion-shaped plot from which the value of n_2 can be deduced. The quantity ΔT_{p-v} is the difference between the normalized peak (maximum) and valley (minimum) transmittances. The ΔT_{p-v} shows linear dependence on $\Delta\phi_0$ when its variation was calculated as a function of $\Delta\phi_0$ from various aperture sizes. Both of these quantaties are related by the following expression[42]

$$\Delta T_{p-v} \approx p|\Delta\Phi_0| \text{ for } |\Delta\Phi_0| \leq \pi \tag{112}$$

The nonlinear refractive index n_2 is calculated from this expression which also shows the high sensitivity of the Z-scan method. The Z-scan of a 1-mm-thick carbon disulfide (CS_2) cell using 300-ns pulse at

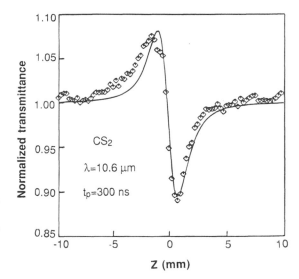

Figure 15. The Z-scan of a 1 mm thick carbon disulfide (CS$_2$) cell using a 300-ns pulse at 10.6 μm. The data show that a negative (self-defocusing) optical nonlinearity is indicated by the Z-scan peak-to-valley configuration. (From Sheik-Bahae, M., Said, A. A., and Van Stryland, E. W., *Opt. Lett.*, 14, 955, 1989. With permission.)

10.6 μm is shown in Figure 15. The negative (self-defocusing) optical nonlinearity is indicated by the Z-scan peak-to-valley configuration. The Z-scan of a 2.5 mm thick BaF$_2$ single crystal using a 27 ps pulse at 532 nm is shown in Figure 16. In this case, a positive (self-defocusing) optical nonlinearity can be observed. The Z-scan technique has been used to determine the optical nonlinearities of solutions and thin films.

Third-order nonlinear optical susceptibilities consist of real part and imaginary part $\chi^{(3)} = \chi_R^{(3)} + i\chi_I^{(3)}$. In two-photon absorption, the imaginary part is related to β through[43]

$$\chi_I^{(3)} = \frac{n_0^2 \epsilon_0 c^2}{\omega} \beta \tag{113}$$

while the real part is related to γ through

$$\chi_R^{(3)} = 2n_0^2 \epsilon_0 c \gamma \tag{114}$$

The intensity-dependent nonlinear refractive index n_2 and two-photon absorption coefficient β are related to the real and imaginary part of $\chi^{(3)}$ by the following equation[43]:

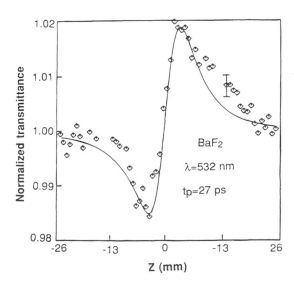

Figure 16. The Z-scan of a 2.5-mm-thick BaF$_2$ single crystal using 27-ps pulse at 532 nm. The data shows a positive optical nonlinearity. (From Sheik-Bahae, M., Said, A. A., and Van Stryland, E. W., *Opt. Lett.*, 14, 955, 1989. With permission.)

$$n_2(m^2/W) = \frac{3}{4\epsilon_0 c n_0^2} \, \text{Re} \, \chi^{(3)}(-\omega; -\omega, \omega, -\omega) \tag{115}$$

and

$$\beta(m/W) = \frac{3\omega}{2\epsilon_0 c^2 n_0^2} \, \text{Im} \, \chi^{(3)}(-\omega; -\omega, \omega, -\omega) \tag{116}$$

where n_0 and c are the linear refractive index and the speed of light in a vacuum, respectively. The conversion is written as n_2 (esu) $= (cn_0/40\pi)n_2$ (MKS). This technique has advantages compared with other methods because the optical configuration is simple and provides information on the sign, and the magnitude of the real and the imaginary part of the nonlinear response of the sample. This technique also offers the advantage of separately evaluating the nonlinear refractive index and nonlinear absorption via a second Z-scan with the aperture removed.

Since the introduction of the Z-scan in 1989, several modifications have been implemented to enhance its applicability. The Z-scan technique can also be employed for measuring the polarization dependence of the nonlinear refractive index n_2 and two-photon absorption coefficient (β) of single crystals such as KTP, GaAs, and BaF_2.[44] The anisotropy of n_2 and β in crystals is measured by incorporating a wave plate into the Z-scan apparatus. The n_2 and β are correlated to $\chi^{(3)}$.[45] The different $\chi^{(3)}$ tensors can be estimated by selecting specific crystal orientations and wave-vector propagation directions. The coefficient of anisotropy σ is given by

$$\sigma = \frac{\chi_{xxxx}^{(3)} - [\chi_{xxyy}^{(3)} + 2\chi_{xyyx}^{(3)}]}{\chi_{xxxx}^{(3)}} \tag{117}$$

where σ becomes zero for isotropic materials, i.e., $\chi_{xxxx}^{(3)} = \chi_{xxyy}^{(3)} = 2\chi_{xyyx}^{(3)}$.

This technique has been applied to investigate the nonlinear optical response of liquid crystals and π-conjugated polymers. The time-averaged index change $<\Delta n_0(t)>$ for transient effects induced by pulsed radiation can be expressed as[43]

$$\langle \Delta n_0(t) \rangle = \frac{\int_{-\infty}^{\infty} \Delta n_0(t) I_0(t) \, dt}{\int_{-\infty}^{\infty} I_0(t) \, dt} \tag{118}$$

where the time-averaged $<\Delta\Phi_0(t)>$ is related to the time-averaged index change $<\Delta n_0(t)>$ as shown above. The response and decay time of nonlinearity can be evaluated and for a temporally Gaussian pulse

$$\langle \Delta n_0(t) \rangle = \frac{\Delta n_0}{\sqrt{2}} \tag{119}$$

where Δn_0 is the peak-on-axis index change at the focus. The instantaneous index change for a cumulative nonlinearity with a decay time much faster than the pusle width is given by

$$\langle \Delta n_0(t) \rangle = A \int_{-\infty}^{t} I_0(t') \, dt' \tag{120}$$

where A is a constant dependent on the nature of the nonlinearity.

This technique has also been expanded to a simple dual-wavelength (two-color) Z-scan geometry for measuring third-order optical nonlinearities at one wavelength caused by a second.[46] A two-color Z-scan provides the sign and the magnitude of the nondegenerate nonlinear refractive index n_2 even in the presence of nondegenerate two-photon absorption coefficient (β). A time delay in one of the beams

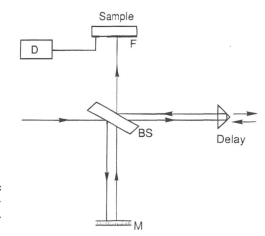

in a two-color Z-scan apparatus also permits the separate measurement of the sign and the magnitude of the temporal dependence of the nonlinear refraction and the nonlinear absorption.[47] The time-resolved Z-scan measurements allow determination of the bound electronic, nondegenerate nonlinear refractive index and nondegenerate two-photon absorption as well as the two-photon-generated free-carrier refraction and absorption as functions of time. The eclipsing Z-scan (EZ) measurement of $\lambda/10^4$ wave-front distortion is another simple modification that provides a high sensitivity for measuring small nonlinearly induced phase shifts, while differentiating between nonlinear refraction and nonlinear absorption and concurrently determining the sign of each of these effects.[48] This method is advantageous for measuring nonresonant optical nonlinearities of thin films without the need for waveguide coupling.

The Z-scan technique has become a versatile tool for measuring nondegenerate nonlinear refractive index (n_2) and nondegenerate two-photon absorption (β), $\chi^{(3)}$ tensors, anisotropy, response and decay time of nonlinearity. The Z-scan technique has been employed for studying third-order NLO effects in π-conjugated polymers such as poly(diacetylenes), polythiophenes, and organometallic compounds. The Z-scan technique is becoming a choice of many scientists for measuring third-order optical nonlinearities of a wide range of organic molecular and polymeric materials.

F. SATURATION ABSORPTION

Several reports have appeared on the direct measurements of the saturation absorption that provides degenerate one-photon resonant Kerr susceptibility from single crystal and thin films. Greene et al.[49] discussed the saturation absorption measurements on polydiacetylene *p*-toluene sulfonate (PDA-PTS) single crystal. Kajzar et al.[50] measured the saturation absorption by transmission on blue form of LB films of polydiacetylene from the photoacosutic technique. Figure 17 shows the schematic diagram of the experimental setup used to measure saturation absorption of PDA LB films. A high-energy pulse splits into two equal intensity pulses and then recombines at the sample position with a controllable time delay between them. The response time of the system can also be obtained by delaying the probe beam with respect to the pump beam. A piezoelectric transducer attached to the substrate detects the absorption of the picosecond pulses impinged on the sample. The detected signal is proportional to the integrated intensity of the energy deposited on the thin film. The measured signal $\delta\alpha/\alpha$ (where α is the linear absorption coefficient) is zero if the signals from two pulses are independent while, on the other hand, the $\delta\alpha/\alpha$ is nonzero if the absorption of one pulse is reduced due to bleaching (saturation absorption). The variation of the linear absorption coefficient is directly proportional to the relative change in the imaginary part of refractive index κ and can be given as

$$\delta\alpha/\alpha = (1/2)\delta\kappa/\kappa \tag{121}$$

where the factor 1/2 arises from the integration overtime, assuming a Gaussian shape of the laser pulse. The intensity-dependent refractive index can be written as

$$n = n_0 + n_2 I \tag{122}$$

where

$$n_2 = n_2^r + i\kappa_2 \tag{123}$$

and

$$\kappa_2 = \delta\kappa/I \tag{124}$$

The change of refractive index n_2 with light intensity is related to the imaginary part of Kerr susceptibility $\chi^{(3)}(-\omega; \omega, \omega, -\omega)$ as follows:

$$n_2 = 12\pi^2\chi^{(3)}(-\omega; \omega, \omega, -\omega)/n_0^2c\kappa_2 \tag{125}$$

The κ_2 measurements directly yield the imaginary part of $\chi^{(3)}(-\omega; \omega, \omega, -\omega)$ where n_0 is real at frequency ω. For a PDA LB film at 0.583 μm, the $\delta\alpha/\alpha = 0.063$ was obtained using a peak power density (I) of 37.5 MW/cm^2 by simultaneous arrival of both pulse and κ_2 of 2.3 $\times$ 10^{-10} cm^2/W which yields an imaginary part of $\chi^{(3)}(-\omega; \omega, \omega, -\omega) = 1.4 \times 10^{-8}$ esu. The ground-state recovery time was found to be less than the laser-pulse width of 3 ps. From the phase space-filling model for one-dimensional excitons, a value of ~50 Å for the length of the exciton along the polymer backbone was obtained. Greene et al.[49] measured the magnitude of the nonlinear refractive index as 3.0 $\times$ 10^{-8} cm^2/W at 1.97 eV for polydiacetylene p-toluene sulfonate (PDA-PTS) single crystal using saturation absorption. This saturation absorption gives rise to the imaginary part of $\chi^{(3)}(-\omega; \omega, \omega, -\omega) = 2.0 \times 10^{-5}$ esu, the largest third-order optical nonlinearity so far reported for a π-conjugated polymer. In the case of the PDA-PTS single crystal, an exciton length of 33 Å was deduced. The magnitude of $\chi^{(3)}(-\omega; \omega, \omega, -\omega)$ obtained from saturation absorption measurements was found to be 2–3 orders of magnitude larger than the $\chi^{(3)}$ values measured at the two- and three-photon resonance $\chi^{(3)}(-3\omega; \omega, \omega, \omega)$. The real and imaginary parts of $\chi^{(3)}$ are associated with the saturation absorption and were measured for polythiophene thin films by using the Z-scan technique.[51] In the case of the saturation absorption, the nonlinear refractive index has a real and an imaginary part. The real part of the $\chi^{(3)}$ is related to second hyperpolarizability γ and can be written as

$$\mathrm{Re}\ \chi^{(3)} = \gamma(cn_0^2/16\pi^2) \tag{126}$$

and the imaginary part of the $\chi^{(3)}$ is associated to the change of the absorption coefficient $\Delta\alpha$ and can be given as

$$\mathrm{Im}\ \chi^{(3)} = (cn_0^2/32\pi^2)(\Delta\alpha/\omega I_s) \tag{127}$$

where I_s is the saturation intensity. The Re $\chi^{(3)} = -6.1 \times 10^{-9}$ esu and Im $\chi^{(3)} = -1.9 \times 10^{-9}$ esu were measured for polythiophene thin films. The negative NLO susceptibility at 532 nm is related to the saturation of the absorption. Saturable absorption has also been discussed in section IIB dealing with DFWM method.

G. PHOTOINDUCED ABSORPTION

Heeger et al.[52] used photoinduced changes in absorption to observe nonlinear optical responses and their dynamics in polyacetylenes. Orenstein et al.[53] used transient photoinduced absorption on single crystal and solution-cast polycrystalline films of polydiacetylene PTS. The intensity-dependent exponential decay indicated that the photoinduced absorption was due to a photoexcitation. The photoinduced absorption spectrum showed a single sharp peak at 1.4 eV. Ruani et al.[54] used photoinduced absorption for studying the nonlinear response of polythienothiophene. Polyacetylene, polydiacetylene, and polythiophene indicated nonlinear optical process (photoinduced absorption, photoinduced bleaching, and photoluminescence) on the subpicosecond scale. The photoinduced absorption assists in evaluating the imaginary part of the optical nonlinearity.

H. ELECTRIC FIELD-INDUCED SECOND HARMONIC GENERATION (EFISH) METHOD

The most frequent techniques used for determining third-order optical nonlinearity are the electric field-induced second harmonic generation (EFISH) method, DFWM, THG, and Z-scan technique. The quadratic (β) and cubic hyperpolarizability (γ) tensors of organic molecules can be evaluated using solution-phase DC EFISH. The details of the EFISH measurements have already been described in our earlier monograph on second harmonic generation. Initially, Levine and Bethea[55] used DC EFISH technique for hyperpolarizability measurements in solution. Further details of DC EFISH can also be found in many other articles listed in the materials section. Like THG, the EFISH technique can also be applied to thin films and solutions. From the EFISH technique, both the modulus and phase of $\chi^{(3)}(-2\omega; \omega, \omega, 0)$ can be evaluated by performing measurements as a function of solute concentration. In centrosymmetric medium ($\beta = 0$), the nonlinear polarization at 2ω results from the term $\chi^{(3)}(-2\omega; \omega, \omega, 0)$. On the other hand, in the noncentrosymmetric medium, both first and second hyperpolarizabilities contribute to the $\chi^{(3)}(-2\omega; \omega, \omega, 0)$. Two approaches of the EFISH method have been expressed, the infinite-dilution method proposed by Singer and Garito[56] and the full-concentration-dependence least-squares method by Kajzar et al.[57] The γ and $\chi^{(3)}$ measurements by the EFISH method proposed by Kajzar et al.[12] are described here

$$\chi^{(3)}(-2\omega; \omega, \omega, 0) = N_p f_p(-2\omega; \omega, \omega, 0)\gamma_p(-2\omega; \omega, \omega, 0)$$
$$+ N_s f_s(-2\omega; \omega, \omega, 0)\gamma_s(-2\omega; \omega, \omega, 0) \tag{128}$$

where local-field factors are given as

$$f_i(-2\omega; \omega, \omega, 0) = f(2\omega)f(\omega)^2 f_0 \tag{129}$$

assuming $f_s = f_p$. The effective hyperpolarizability can be written as

$$\gamma_{Ieff} = \gamma_I(-2\omega; \omega, \omega, 0) + \mu_I\beta_I/5kT \tag{130}$$

where μ_I is the permanent dipole moment and β_I is the first hyperpolarizability. Equation 128 can be rewritten as:

$$\chi^{(3)}(-2\omega; \omega, \omega, 0) = \frac{N_A d(C)f(C)}{1 + C}\left[\frac{C(\gamma_p^{E'} + i\gamma_p^{E''})}{M_p} + \frac{\gamma_s^E}{M_s}\right] \tag{131}$$

where N_A is Avogadro's number, $f(C)$ is the concentration-dependent local-field factor, $d(C)$ is the concentration-dependent average solution density, and γ_s of the solvent is assumed to be real. In the EFISH measurement, the amplitude of the solution SH intensity generated at 2ω is compared with the harmonic from a standard such as a quartz single crystal and the normalized value of the intensity is given as[57]

$$I_N^{2\omega}(C) = \frac{I_{EFISH}^{2\omega}(C)}{I_Q^{2\omega}}$$
$$= \left(\frac{3}{2g}\right)^2\left[\frac{\chi_G^{(3)}/\Delta\epsilon_G}{\chi_Q^{(2)}/\Delta\epsilon_Q}(t_\omega^{OG}/t_\omega^{OQ})^2 t_{2\omega}^{GO}(T_L/T_Q)E_o\right]^2$$
$$\times \left\{\left[Y(C)\left(\frac{\gamma_s^{E'}}{M_s} + \frac{C\gamma_p^{E'}}{M_p}\right)(W + 1) - (V + 1)\right]^2\right.$$
$$\left.+ Y(C)^2 C^2\left(\frac{\gamma_p^{E''}}{M_p}\right)^2(W + 1)^2\right\} \tag{132}$$

and

$$I_{\text{EFISH}}^{2\omega}(x, C) = \frac{288\pi^3}{(cg)^2} \left(\frac{\chi_G^{(3)}}{\Delta\epsilon_G}\right)^2 [E_0 I_\omega (t_\omega^{GO})^2 t_{2\omega}^{GO} T_L]^2$$

$$\times \left\{ \left| Y\left(\frac{\gamma_s^{E'}}{M_s} + \frac{C\gamma_p^{E'}}{M_p}\right)[1 - W\exp(i\Delta\psi_E)] \right.\right.$$

$$\left.\left. -[1 - V\exp(i\Delta\psi_E)] \right|^2 + Y^2 C^2 \left(\frac{\gamma_p^{E''}}{M_p}\right) \times |1 - W\exp(i\Delta\psi_E)|^2 \right\} \qquad (133)$$

M_s and M_p are the weight fractions of the solute and the solvent, respectively. E_0 is the external applied DC electric field and other quantities are as follows:

$$g = \exp[2\omega K_{2\omega} l(x)/c] \qquad (134)$$

$$V = T_G (t_\omega^{GL} t_\omega^{LG})^2 T_1 t_{2\omega}^{LG} g \qquad (135)$$

$$W = T_L g / (T_2 t_{2\omega}^{LG}) \qquad (136)$$

$$Y = \frac{T_2}{T_1} \frac{\Delta\epsilon_G}{\Delta\epsilon_L} \frac{N_A f(C) d(C)}{(1 + C)\chi_G^{(3)}} \qquad (137)$$

$$T_1 = (n_\omega^G + n_\omega^G)/(n_{2\omega}^G + n_{2\omega}^L) \qquad (138)$$

$$T_2 = (n_{2\omega}^G + n_\omega^L)/(n_{2\omega}^G + n_{2\omega}^L) \qquad (139)$$

$$T_L = (n_\omega^L + n_{2\omega}^L)/(n_{2\omega}^L + n_{2\omega}^G) \qquad (140)$$

$$T_G = (n_{2\omega}^L + n_\omega^G)/(n_{2\omega}^L + n_{2\omega}^G) \qquad (141)$$

The $t_{\omega,2\omega}^{ij}$ are transmission factors and G is glass, L is liquid, and O is air. The phase mismatch between the harmonic and the fundamental is written as

$$\Delta\psi_E = \frac{4\pi l(x)}{\lambda_\omega} (n_\omega^L - n_{2\omega}^L), \qquad (142)$$

where $l(x) = 2x\tan\theta/2 + l_o$ where θ is the wedge angle and l_o is the thickness at the origin. Using concentration-dependent data and the normalized SH intensity, both the real and imaginary parts of the second hyperpolarizabilities can be determined from a least-squares fit to Equation 132. The EFISH measurements provide contributions from both β and γ values. For π-conjugated polymers, β is zero being centrosymmetric; therefore, the $\gamma(-2\omega; \omega, \omega, 0)$ can be determined. On the other hand, the separation of γ and β terms is rather difficult—though possible—and can be done by performing THG and EFISH measurements together. For long π-electron conjugated molecules having donor–acceptor groups at terminal ends, the $\gamma(-2\omega; \omega, \omega, 0)$ contribution is remarkably larger than the $\mu\beta/5kT$ term. Contrary to this, if the $\mu\beta/5kT$ term is much larger than the $\gamma(-2\omega; \omega, \omega, 0)$ contribution, then one may neglect this part. Both γ and $\chi^{(3)}$ of organic materials can be determined using the EFISH technique. Kajzar and co-workers[57–59] have used this technique for evaluation of third-order optical nonlinearities of polydiacetylenes thin films and solutions. The γ values of metallophthalocyanines have also been measured using EFISH method.

I. REFLECTION-TYPE POLARIZATION SPECTROSCOPY (RPS)

Kuwata[60] reported a reflection-type polarization spectroscopy (RPS) technique to evaluate $\chi^{(3)}$ value. Figure 18 shows the principle of RPS. In this method two beams—a strong pump beam and a weak probe beam—are used. The probe beam is linearly polarized along the x axis and illuminated to a sample surface along the z axis. The probe beam is very weak so that it does not induce any nonlinear

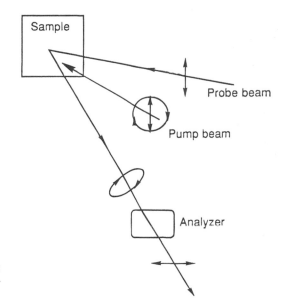

Figure 18. Principle of reflection type polarization spectroscopy (RPS). (From Kuwata, M., *J. Luminescence*, 38, 247, 1987. With permission.)

effects. On the other hand, the pump beam is responsible for inducing NLO effects. When the pump beam polarization is linear or circular and bisects the x and y axes, the nonlinear dielectric tensor ϵ_{nl} has off-diagonal components that change the polarization of the reflected probe beam. A cross-analyzer can detect this change as a positive signal. The nonlinear dielectric tensor ϵ_{nl} can be written in terms of $\chi^{(3)}$ and the intensity of the crossed component R as

$$R_I = \left| 6\pi \frac{(1 + r_i)^2}{(1 - r_i)} \left(\chi^{(3)}_{xxyy} \pm \chi^{(3)}_{xyxy} \right)(1 - |r_2|^2) \right|^2 \cdot |E_2|^4 \qquad (143)$$

where r_i is the linear reflection coefficient for the ith beam, $\pm$ refers to when the pump beam polarization is linear and circular. A photomultiplier detects the reflected probe beam and the photocurrent was measured with a boxcar integrator. Therefore, the measured values of R can yield $\chi^{(3)}$. Using this method, a $\chi^{(3)}$ value of 1 esu was measured for anthracene and on the order of 10^{-5} esu for CuCl and red-HgI$_2$ crystals. This method is very sensitive for measuring $\chi^{(3)}$ in highly absorptive solids and metallic materials.

J. MISCELLANEOUS MODIFIED MEASUREMENT TECHNIQUES

In this section, we describe the modified techniques from some of the most commonly used techniques. Toda and Verber[61] used a modified version of the Z-scan technique to examine the speed of nonlinear mechanism. The preliminary studies were performed on 2-methyl-4-nitroaniline (MNA) dissolved in chloroform. When a transparent sample of carbon disulfide was used, the tails of the two waveforms coalesced, while for MNA, the dominant index change had a slower rise time compared with carbon disulfide and a slower relaxation time than the duration of the exciting pulse. Chuang et al.[62] measured nonresonant $\chi^{(3)}_{IIII}$ by nearly degenerate four-wave mixing using chirped-pulse technology. This method is advantageous because it overcomes the intensity-related self-focusing and damage effects. It utilizes short pulses. The $\chi^{(3)}$ of nitrobenzene, chloronaphthalene, and 4'-octyloxy-1'-(4-decyloxybenzoate) were determined using this technique. Horan et al.[63] used a relatively simple and versatile DFWM setup for rapid testing of third-order NLO materials. Chamon et al.[64] reported a new femtosecond time division interferometry (TDI) technique for measuring nonlinear index changes from different components of the $\chi^{(3)}$ tensor. This technique allows direct measurements of n_2 in waveguide devices. The contribution to n_2 from different $\chi^{(3)}$ components can be obtained by measuring different permutations of pump and probe polarizations. Miyano et al.[65] reported a simple method that permits the determination of phase and magnitude of $\chi^{(3)}$ of thin films. This method is based on the Maker fringes generated by the interference between the third harmonic wave from the thin-film sample and from the substrate under a full 360-degree rotation. More details of these techniques can be found in respective references.

Excellent texts on sum or difference-frequency generation, harmonic generation, stimulated Raman scattering, stimulated light scattering, two-photon absorption, self-focusing, four-wave mixing and its spectroscopy, and multiphoton spectroscopy have been written by Reintjes,[25] Fisher,[26] and Shen.[2] Currently THG, DFWM, and EFISH are the most widely used techniques for measuring third-order optical nonlinearities of organic materials. Because it is difficult to compare the $\chi^{(3)}$ results obtained from various techniques due to the involvement of different nonlinear optical processes, these measurement techniques need to be standardized, keeping in mind the target of applications of organic materials in particular devices. Using the DFWM technique, one can measure simultaneously the ultrafast response and the magnitude of second hyperpolarizability and third-order optical nonlinearity. Certainly other techniques also have their merits in evaluating third-order nonlinear optical materials.

III. DYNAMICS OF THIRD-ORDER NONLINEAR OPTICAL PROCESSES

Recent research in nonlinear optics is aimed at substituting the electronic elements with photonics where the speed of light plays a role. Therefore, techniques are required to evaluate faster speed that varies from nanoseconds to femtoseconds depending upon the nonlinear optical processes and materials. Various third-order nonlinear optical processes are THG, DFWM, four-wave sum frequency mixing, four-wave difference frequency mixing, EFISH, degenerate two-photon absorption, nondegenerate photon absorption, absorption saturation, spontaneous Raman scattering, coherent anti-Stokes Raman scattering (CARS), coherent stoke Raman scattering (CSRS), Brillouin scattering, and hyper-Raman scattering, Optical Kerr effects, etc. Several techniques, such as DFWM, OKE, saturation absorption, and photoluminescence, can be used to evaluate the ultrafast optical nonlinearity in organic materials. In particular, DFWM techniques have been widely used to study the relaxation processes in organic materials. An excellent text on ultrafast spectroscopy has been written by Hermann and Wilhelmi[66] and several techniques can be summarized into two categories:

1. Fluorescent measurements (nanosecond, picosecond, and intensifying techniques);
2. Probe-pulse spectroscopy (single pulse pump and probe, repetitive pulse, pump and probe spectrometers, and pump and probe spectrometer with self-induced diffraction grating).

Literature on picosecond relaxations in solids[67] and time-resolved luminescence spectroscopy[34] are also available. In addition, femtosecond interferometry[68] and pump-probe[69,70] methods have also been utilized. The dynamics of nonlinear optical processes, electronic relaxation, and vibrational processes and selective excitations in organic molecules and polymers can be studied by using ultrafast spectroscopic techniques. Relaxation time constants of a wide variety of third-order NLO materials vary from picoseconds to femtoseconds.

Details of pump and probe spectroscopies have been described by Kajzar et al.[5] in a monograph. Pump and probe spectroscopy has been used for polydiacetylenes and π-conjugated polymers. It is particularly useful to determine the dynamics of the optical nonlinearity.

ACKNOWLEDGMENTS

The author is grateful to Drs. A. Ticktin, K. H. Haas, and A. Esser of BASF AG for providing illustrations of OPO-THG and DFWM techniques and for valuable comments. The author also wishes to thank Professor Eric W. van Stryland of the University of Central Florida for providing up-to-date information on optical limiting and Z-scan techniques.

REFERENCES

1. **Flytzanis, C.,** *Theory of Nonlinear Optical Susceptibilities, in Quantum Electronics: a Treatise,* Rabin, H., and Tang, C. L., Eds., Vol. 1, Nonlinear Optics, Part A, Academic Press, New York, 1975, 9–207.
2. **Shen, Y. R.,** *The Principles of Nonlinear Optics,* John Wiley, & Sons, New York, 1984.
3. **Reintjes, J. F.,** *Nonlinear Optical Parametric Process in Liquids and Gases,* Academic Press, New York, 1984.

4. **Etchepare, J., Grillon, G., Chambaret, J. P., Harmoniaux, G., and Orszag, A.,** Polarization selectivity in time-resolved transient phase grating, *Opt. Commun.,* 63, 329, 1987.
5. **Kajzar, F., Messier, J., Nunzi, J. M., and Raimond, P.,** Third-order materials: Processes and characterization, in *Polymers for Lightwave and Integrated Optics,* Hornak, L. A., Ed., Marcel Dekker, New York, 1992, 595.
6. **Hermann, J. P.,** Absolute measurements of third-order susceptibilities, *Opt. Commun.,* 9, 74, 1973.
7. **Bey, P. P., Giuliani, J. F., and Rabin, V.,** Phase-matched optical harmonic generation in liquid media employing anomalous dispersion, *IEEE, J. Quantum Electron.,* QE-4, 932, 1968.
8. **Bloembergen, N.,** *Nonlinear Optics,* Benjamin, New York, 1965.
9. **Meredith, G. R., Buchalter, B., and Hanzlik, C.,** Third-order optical susceptibility determination by third harmonic generation. I., *J. Chem. Phys.,* 78, 1533, 1983.
10. **Meredith, G. R., Buchalter, B., and Hanzlik, C.,** Third-order susceptibility determination by third harmonic generation. II., *J. Chem. Phys.,* 78, 1543, 1983.
11. **Kajzar, F. and Messier, J.,** Original technique for third-harmonic generation in liquids, *Rev. Sci. Instrum.* 58, 2081, 1987.
12. **Burland, D. M., Walsh, C. A., Kajzar, F., and Sentein, C.,** Comparison of hyperpolarizabilities obtained with different experimental methods and theoretical techniques, *J. Opt. Soc. Am. B,* 11, 2269, 1991.
13. **Kajzar, F.,** Cubic susceptibility of organic molecules in solution, in *Nonlinear Optical Effects in Organic Polymers,* Messier, J., Ed., Kluwer Academic Press, Boston, 1989, 225.
14. **Kajzar, F., Messier, J., and Rosilio, C.,** Nonlinear optical properties of thin films of polysilane, *J. Appl. Phys.,* 60, 3040, 1986.
15. **van Beek, J. B., Kajzar, F., and Albrecht, A. C.,** Resonant third-harmonic generation in all-*trans* β-carotene: the vibronic origins of the third-order nonlinear susceptibility in the visible region, *J. Chem. Phys.,* 95, 6400, 1991.
16. **Gierulski, A., Naarmann, H., Schrepp, W., Schrof, W., and Ticktin, A.,** Frequency tunable THG measurements of organic conjugated polymer films using an optical parametric oscillator (OPO), Proc. 5th Toyota Conference "Nonlinear Optical Materials," Nagoya, October 7–9, 1991, Japan.
17. **Fix, A., Schroder, T., and Wallenstein, R.,** New source of powerful tunable laser radiation in the ultraviolet, visible and near infrared, *Laser Optoelektronik,* 23, 106, 1991.
18. **Kubodera, K.,** Measurements of third-order nonlinear optical efficiencies, *Nonlinear Optics,* 1, 71, 1991.
19. **Wong, K. Y. and Garito, A. F.,** Third-harmonic generation study of orientational order in nematic liquid crystals, *Phys. Rev. A.,* 34, 5051, 1986.
20. **Gerritsen, H. J.,** Nonlinear effects in image formation, *Appl. Phys. Lett.,* 10, 239 1967.
21. **Yariv, A.,** On transmission and recovery of three-dimensional image information. *J. Opt. Soc. Am.,* 66, 310, 1976.
22. **Yariv, A.,** Phase conjugation optics and real-time holography, *IEEE J. Quantum Electron.,* QE-14, 650, 1978.
23. **Hellwarth, R. W.,** Generation of time-reversed wave fronts by nonlinear refraction, *J. Opt. Soc. Am.,* 67, 1, 1977.
24. **Guiliano, C. R.,** Applications of optical phase conjugation, *Phys. Today,* April, 1981, 27.
25. **Reintjes, J. F.,** Self action effects and degenerate four-wave mixing, in *Nonlinear Optical Parametric Process in Liquids and Gases,* Academic Press, New York, 1984, 327.
26. **Fisher, R. A.,** Ed., *Optical Phase Conjugation,* Academic Press, New York, 1984.
27. **Caro, R. G. and Gower, M. C.,** Phase conjugation by degenerate four-wave mixing in absorbing media, *IEEE J. Quantum Electron.,* QE-18, 1376, 1982.
28. **Shirk, J. S., Lindle, J. R., Bartoli, F. J., Hoffman, C. A., Kafafi, Z. H., and Snow, A. W.,** Off-resonance third-order optical nonlinearities of metal-substituted phthalocyanines, *Appl. Phys. Lett.,* 55, 1287, 1989.
29. **Carter, G. M., Thakur, M. K., Chen, Y. J., and Hryniewicz, J. V.,** Time and wavelength resolved nonlinear optical spectroscopy of a polydiacetylene in the solid state using picosecond dye laser pulses, *Appl. Phys. Lett.,* 47, 457, 1985.
30. **Byrne, H. J. and Blau, W.,** Multiphoton nonlinear interactions in conjugated organic polymers, *Synth. Metals,* 37, 231, 1990.
31. **Shand, M. L. and Chance, R. R.,** Third-order nonlinear mixing in polydiacetylene solutions, *J. Chem. Phys.,* 69, 4482, 1978.
32. **Lynch, R. T., Jr., Krmaer, S. D., Lotem, H., and Bloembergen, N.,** *Opt. Commun.,* 16, 372, 1976.
33. **Frazier, C. C., Chauchard, E. A., Cockerham, M. P., and Porter, P. L.,** Four-wave mixing in metal poly-ynes, *Mater. Res. Soc. Symp. Proc.,* 109, 323, 1988.
34. **(a) Ho. P. P.,** Picosecond Kerr gate, in *Semiconductor Probed by Ultrafast Laser Spectroscopy,* Vol. 2, Alfano, R. R., Ed., Academic Press, New York, 1984, 410. **(b) Rosen, D. L., Doukas, A. G., Katz, A., Budansky, Y., and Alfano, R. R.,** Techniques in time-resolved luminescence spectroscopy, in *Semiconductor Probed by Ultrafast Laser Spectroscopy,* vol. 2, Alfano, R. R., ed., Academic Press, New York, 1984, p. 393.
35. **Buckingham, A. D.,** *Proc. Phys. Soc. Lond. B,* 69, 344, 1956.
36. **Ho, P. P. and Alfano, R. R.,** Relaxation kinetics of salol in the supercooled liquid state investigated with the optical Kerr effect, *J. Chem. Phys.,* 67, 1004, 1977.

37. **Wong, C. K. L. and Shen, Y. R.,** Study of pretransitional behavior of laser-field-induced molecular alignment in isotropic nematic substances, *Phys. Rev. A,* 10, 1277, 1974.

38. **Soileau, M. J., Williams, W. E., and Van Stryland, E. E.,** Optical power limiter with picosecond response time, *IEEE J. Quantum Electron.,* QE-19, 731, 1983.

39. **Winter, C. S., Oliver, S. N., and Rash, J. D.,** Measurements of the nonlinear refractive index of some metallocenes by the optical power limiter technique, in *Nonlinear Optical Effects in Polymers,* Messier, J., et al. Eds., Kluwer Academic, Dordrecht, 1989, 247.

40. **Winter, C. S., Oliver, S. N., and Rash, J. D.** n_2 measurements of various forms of ferrocene, *Opt. Commun.,* 69, 45, 1988.

41. **Marburger, J. H.,** Process of Quantum Electronics, Sanders, J. H. and Stenholm, S., Eds, Pergamon Press, New York, 1977, 35.

42. **Sheik-Bahae, M., Said, A. A., and Van Stryland, E. W.,** High-sensitivity, single-beam n_2 measurements, *Opt. Lett.,* 14, 955, 1989.

43. **Sheik-Bahae, M., Said, A. A., Wei, T. H., Hagan, D. J., and Van Stryland, E. W.,** Sensitive measurement of optical nonlinearities using a single beam, *IEEE J. Quantum Electron.,* 26, 760, 1990.

44. **Desalvo, R., Sheik-Bahae, M., Said, A. A., Hagan, D. J., and Van Stryland, E. W.,** Z-scan measurements of the anisotropy of nonlinear refraction and absorption in crystals, *Opt. Lett.,* 18, 194, 1993.

45. **Van Stryland, E. W., Sheik-Bahae, M., Said, A. A., and Hagan, D. J.,** Characterization of nonlinear optical absorption and refraction, *Prog. Crystal Growth Charact.,* 27, 279, 1993.

46. **Sheik-Bahae, M., Wang, J., DeSalvo, R., Hagan, D. J., and Van Stryland, E. W.,** Measurement of nondegenerate nonlinearities using a two-color Z-scan, *Opt. Lett.,* 17, 258, 1992.

47. **Wang, J., Sheik-Bahae, M., Said, A. A., Hagan, D. J., and Van Stryland, E. W.,** Time-resolved Z-scan measurements of optical nonlinearities, *J. Opt. Soc. Am. B.,* 11, 1009, 1994.

48. **Xia, T., Hagan, D. J., Sheik-Bahae, M., and Van Stryland, E. W.,** Eclipsing Z-scan measurement of $\lambda/10^4$ wavefront distortion, *Opt. Lett.,* 19, 317, 1994.

49. **Greene, B. I., Orenstein, Millard, R. R., and Williams, L. R.,** Nonlinear optical response of excitons confined to one dimension, *Phys. Rev. Lett.,* 58, 2750, 1987.

50. **Kajzar, F., Rothberg, L., Etemad, S., Chollet, P. A., Grec, D., Boudet, A., and Jedju, T.,** Saturation absorption and Kerr susceptibility in polydiacetylene Langmuir-Blodgett films, *Opt. Commun.,* 66, 55, 1988.

51. **Yang, L., Dornsinville, R., Wang, Q. Z., Ye, P. X., Alfano, R. R., Zamboni, R., and Taliani, C.,** Excited-state nonlinearity in polythiophene thin films investigated by the Z-scan technique, *Opt. Lett.,* 17, 323, 1992.

52. **Heeger, A. J., Moses, D., and Sinclair, M.,** Semiconducting polymers: fast response nonlinear optical materials, *Synth. Metals,* 15, 95, 1986.

53. **Orenstein, J., Etemad, S., and Baker, G. L.,** Photoinduced absorption in polydiacetylene, *J. Phys. C. Solid State Phys.,* 17, L297, 1984.

54. **Ruani, G., A. J. Pal, Zamboni, R., Taliani, C., and Kajzar, F.,** Photoinduced absorption and nonlinear optical response in a polycondensed thiophene-based polymer PTT, in *Conjugated Polymeric Materials: Opportunities in Electronics Optoelectronics and Molecular Electronics,* vol. 182, Bredas, J. L. and Chance, R., R., Eds., Kluwer Press, Dordrecht, 1990, pp. 429–441.

55. **Levine, B. F. and Bethea, C. G.,** Molecular hyperpolarizabilities determined from conjugated and nonconjugated organic liquids, *Appl. Phys. Lett.,* 24, 445, 1974.

56. **Singer, K. D. and Garito, A.,** Measurement of molecular second order optical susceptibilities using dc induced second harmonic generation, *J. Chem. Phys.,* 75, 3572, 1981.

57. **Kajzar, F., Ledoux, I., and Zyss, J.,** Electric field induced optical second harmonic generation in polydiacetylene solutions, *Phys. Rev. A,* 36, 2210, 1987.

58. **Chollet, P. A., Kajzar, F., and Messier, J.,** Electric field induced optical second harmonic generation and polarization effects in polydiacetylene Langmuir-Blodgett multilayers, *Thin Solid Films,* 132, 1, 1985.

59. **Le Moigne, J., Thierry, A., Chollet, P. A., Kajzar, F. and Messier, J.,** Morphology, linear and nonlinear optical studies of poly(1,6-di(N-carbazolyl)-2,4-hexadiyne) thin films (pDCH), *J. Chem. Phys.,* 88, 6647, 1988.

60. **Kuwata, M.,** Optical nonlinearities in the exciton resonant region studied by polarization spectroscopy, *J. Luminescence,* 38, 247, 1987.

61. **Toda, H. and Verber, C. M.,** Simple technique to reveal a slow nonlinear mechanism in a z-scan like n_2 measurement, *Opt. Lett.,* 17, 1379, 1992.

62. **Chuang, Y. H., Li, Z. W., Meyerhofer, D. D., and Schmd, A.,** Nonresonant $\chi^{(3)}_{1111}$ obtained by nearly degenerate four-wave mixing using chirped-pulse technology, *Opt. Lett.,* 16, 7, 1991.

63. **Horan, P., Blau, W., Byrne, H., and Berglund, P.,** Simple setup for rapid testing of third-order nonlinear optical materials, *Appl. Opt.,* 29, 31, 1990.

64. **Chamon, C. de C, Sun, C. K., Haus, H. A., and Fujimoto, J. G.,** Femtosecond time division interferometry technique for measuring the tensor components $\chi^{(3)}$, *Appl. Phys. Lett.,* 60, 533, 1992.

65. **Miyano, K., Nishikawa, T., and Tomioka, A.,** A simple and accurate method to determine the third-order nonlinear susceptibility of thin films, *Opt. Commun.,* 91, 501, 1992.

66. **Hermann, J., and Wilhelmi, B.,** Laser fur Ultrakurze Lichtimpulse, Akademie-Verlag, Berlin, 1984.
67. **Takagahara, T.,** Picosecond relaxation in solids and nonlinear spectroscopy, in *Semiconductors Probed by Ultrafast Laser Spectroscopy,* Vol. II, Alfano, R. R., Ed., Academic Press, New York, 1984, 331.
68. **Tang, C. L. and Halbout, J. M.,** Femtosecond interferometry for nonlinear optics, *Chem. Phys. Lett.,* 40, 765, 1982.
69. **Cotter, D., Ironside, C. N., Ainsile, B. J., and Girldestone, H. P.,** Picosecond pump-probe interferometry of nonlinear-refractive materials, in *Ultrafast Phenomena VI,* Yamija, T., Ed., Springer-Verlag, Berlin, 1988, 369.
70. **Kobayashi, T.,** Third-order nonlinear optical processes and ultrafast dynamics in polymers, in *Polymers for Lightwave and Integrated Optics,* Hornak, L. A., Ed., Marcel Dekker, New York, 1992, p. 543.

Chapter 11

Organic Materials for Third-Order Nonlinear Optics

Hari Singh Nalwa

0-8493-8923-2/97/$0.00+$.50
© 1997 by CRC Press, Inc.

I. INTRODUCTION

A quantitative description of the molecular nonlinear optical response and its relationship to chemical structures is well established, although it is necessary to introduce briefly the fundamental aspects here again. A laser is a source of light and the light emitted by a laser has a very high degree of collimation and coherence. The light of a laser beam is highly monochromatic and can be focused to a minimum spot size on a material. Therefore, a focused laser beam can provide a high power per unit area. The laser light can stimulate the atomic or molecular systems to produce transition and release the stored energy also as light. This emission of light requires two energy levels that are separated by the photon energy of the light to be emitted. Therefore, what we have is a process of controlling the light by light in the atomic or molecular system. When the electromagnetic field of a laser beam is illuminated on an atom or a molecule, it induces electrical polarization that gives rise to many of the interesting properties that are optically nonlinear. The fundamental relationship between the polarization p induced in a molecule and the applied electric field E of incident electromagnetic wave at frequency ω can be expressed by:[1–10]

$$p_i(\omega_1) = \sum_j \alpha_{ij}(-\omega_1; \omega_2)E_j(\omega_2) + \sum_{jk} \beta_{ijk}(-\omega_1; \omega_2, \omega_3)E_j(\omega_2)E_k(\omega_3)$$

$$+ \sum_{jkl} \gamma_{ijkl}(-\omega_1; \omega_2, \omega_3, \omega_4)E_j(\omega_2)E_k(\omega_3)E_l(\omega_4) + \cdots \cdots \quad (1)$$

where $p_i(\omega_1)$ is the induced polarization in a microscopic medium at laser frequency ω_1 along the ith molecular axis, α the linear polarizability, β the first hyperpolarizability, and γ the second hyperpolarizability, E_j is the applied electric field component along the jth direction. The macroscopic polarization induced in the bulk media under high electromagnetic fields of a laser beam can be written in a power series as:

$$P_I(\omega_1) = \sum_J \chi_{IJ}^{(1)}(-\omega_1; \omega_2)E_J(\omega_2) + \sum_{JK} \chi_{IJK}^{(2)}(-\omega_1; \omega_2, \omega_3)E_J(\omega_2)E_K(\omega_3)$$

$$+ \sum_{JKL} \chi_{IJKL}^{(3)}(-\omega_1; \omega_2, \omega_3, \omega_4)E_J(\omega_2)E_K(\omega_3)E_L(\omega_4) + \cdots \cdots \quad (2)$$

Here $\chi^{(1)}$, $\chi^{(2)}$, $\chi^{(3)}$ are the first-, second-, and third-order nonlinear optical (NLO) susceptibilities, respectively. The even-order tensors vanish in a centrosymmetric medium, whereas odd-order tensors have no symmetry restrictions for their occurrence. Therefore third-order NLO effects can be observed in any media such as in solids, liquids, and gases. Various kinds of third-order NLO processes such as third harmonic generation (THG), quadratic electro-optic effect or Kerr effect, two-photon absorption, Raman, Brillouin, and Rayleigh scattering, etc., arise from $\chi^{(3)}$. The microscopic and macroscopic third-order optical nonlinearities are also related to each other by the following equation:

$$\chi_{IJKL}^{(3)} = N\gamma_{ijkl}f_{Ii}f_{Jj}f_{Kk}f_{Li} \quad (3)$$

where N is the molecular number density and f the local field factors due to intermolecular interactions. Therefore, it is feasible to estimate the general trend of second hyperpolarizability γ from third-order NLO susceptibility $\chi^{(3)}$, or vice versa, theoretically or experimentally. Third-order NLO susceptibility can be measured by techniques such as THG, degenerate four wave mixing (DFWM), electric-field induced second harmonic generation (EFISH), optical Kerr gate (OKG), two-photon absorption (TPA), Z-scan, etc. The individual tensor components for THG processes discussed here follow the work of Bloembergen[3] and Ward.[4] The quantum mechanical formula derived by time-dependent perturbation theory can be expressed as follows:

$$\gamma_{ijlk}(-3\omega; \omega, \omega, \omega) = k_{THG}\frac{P_{jkl}}{6}\left(\frac{e^4}{8h^3}\right)\chi\left[\sum_{n,n'm}\right.$$

$$r^i_{gn}r^j_{n'm}r^k_{mn}r^l_{ng}\left(\frac{1}{(\omega_{n'g}+3\omega)(\omega_{mg}+2\omega)(\omega_{ng}+\omega)}+\frac{1}{(\omega_{n'g}-3\omega)(\omega_{mg}-2\omega)(\omega_{ng}-\omega)}\right)$$

$$+ r^j_{gn}r^k_{n'm}r^l_{mn}r^i_{ng}\left(\frac{1}{(\omega_{n'g}-\omega)(\omega_{mg}-2\omega)(\omega_{ng}-3\omega)}+\frac{1}{(\omega_{n'g}+\omega)(\omega_{mg}+2\omega)(\omega_{ng}+3\omega)}\right)$$

$$+ r^j_{gn}r^i_{n'm}r^j_{mn}r^k_{ng}\left(\frac{1}{(\omega_{n'g}-\omega)(\omega_{mg}+2\omega)(\omega_{ng}+\omega)}+\frac{1}{(\omega_{n'g}+\omega)(\omega_{mg}-2\omega)(\omega_{ng}-\omega)}\right)$$

$$+ r^k_{gn}r^l_{n'm}r^i_{mn}r^j_{ng}\left(\frac{1}{(\omega_{n'g}-\omega)(\omega_{mg}-2\omega)(\omega_{ng}+\omega)}+\frac{1}{(\omega_{n'g}+\omega)(\omega_{mg}+2\omega)(\omega_{ng}-\omega)}\right)\right] \qquad (4)$$

where $k_{THG} = 4$ is the degeneracy factor for THG, P_{jkl} is the permutation operator, r^i_{gn} and $r^i_{m'n}$ are matrix elements of the ith components of the dipole operator for the molecule between the unperturbed ground and excited states and between two excited states, respectively, ω is the incident laser frequency, $\hbar\omega_{mg}$ corresponds to the excitation energy from the ground state ($|g\rangle$) to the excited doubly state ($|m\rangle$). Third-order optical nonlinearity of organic molecular and polymeric materials discussed in this monograph is in the terms of either third-order NLO susceptibility $\chi^{(3)}$, second hyperpolarizability (γ), or nonlinear refractive index (n_2) which are interrelated with each other and can be determined using different measurement techniques. The $\chi^{(3)}$ is a fourth-rank tensor and generally $\chi^{(3)}_{xxxx}$ component is reported where all the four waves are of the same polarization. The following relationship exists between three independent components for a purely electronic optical nonlinearity in an isotropic medium away from the resonance $1/3\chi^{(3)}_{xxxx} = \chi^{(3)}_{xxyy} = \chi^{(3)}_{xyxy}$. The components of the third-order NLO susceptibility $\chi^{(3)}$ are related to the third-order NLO coefficient (C) as follows:[11]

$$\chi^{(3)}_{jklm}(-3\omega; \omega, \omega, \omega) = 4C_{jklm}(-3\omega; \omega, \omega, \omega) \qquad (5)$$

Third-order NLO susceptibility is displayed by all crystal classes unlike second harmonic generation which is only exhibited by noncentrosymmetric materials. The $\chi^{(3)}$ is a frequency-dependent quantity; therefore, its wavelength dispersion provides valuable information about resonance process. The dispersion of $\chi^{(3)}$ in a three-level model can be expressed according to Orr and Ward:[5]

$$\chi^{(3)}(-3\omega; \omega, \omega, \omega) = (2\pi^3 Nf|\mu_{01}|^2)/h^3 \cdot |\mu_{12}|^2 F(\omega_1, \omega_2, \Gamma_1, \Gamma_2) - |\mu_{01}|^2 G(\omega_1, \omega_2, \Gamma_1, \Gamma_2) \qquad (6)$$

where N is the molecular density, f is the local field factor, ω_i are transition energies between fundamental and first dipole-allowed ($i = 1$) and forbidden ($i = 2$) levels, respectively, μ_{01} and μ_{12} are the dipole transition moments between the ground (0) and the first allowed excited state (1) and between states 1 and 2, respectively. Γ_i are the damping terms. The functions $F(\omega_1, \omega_2, \Gamma_1, \Gamma_2)$ and $G(\omega_1, \omega_2, \Gamma_1, \Gamma_2)$ refer to the dispersion of $\chi^{(3)}$. The $\chi^{(3)}$ measured by different techniques such as THG, DFWM, Kerr effect, TPA, EFISH, and Z-scan are not expected to be the same because they have fundamentally different origins of third-order optical nonlinearities. For example, the $\chi^{(3)}(-3\omega; \omega, \omega, \omega)$ by THG and $\chi^{(3)}(-\omega; \omega, -\omega, \omega)$ by optical Kerr effect may differ by several orders of magnitude on the same material because the latter corresponds to one-photon resonance. A complete expression of the $\chi^{(3)}$ value should be written as:

$$\chi^{(3)} = \chi^{(3)}_{Nonresonant} + \chi^{(3)}_{Resonant}$$

A. THIRD-ORDER OPTICAL NONLINEARITY CORRELATIONS

For many NLO devices, it is important to have a materials showing an intensity-dependent refractive index. In such cases, nonlinear refractive index n_2 is expressed as:[12]

$$n = n_0 + \frac{1}{2}n_2 c(\epsilon\epsilon_0)^{1/2}\langle E^2\rangle \qquad (6)$$

where n_0 is the linear refractive index, n is the refractive index of a medium, ϵ the dielectric constant of the material, ϵ_0 the permittivity of free space, c the speed of light and E the applied electric field. The relationship between n_2 and $\chi^{(3)}$ can be written as:

$$n_2 = \frac{1}{c}\left(\frac{1}{n_0\epsilon_0}\right)\chi^{(3)}_{xxxx} \tag{7}$$

where the diagonal component of the $\chi^{(3)}$ value with an electric field and polarization in the X direction is given. The intensity-dependent refractive index of the optical Kerr effect (OKE) is expressed as:

$$n = n_0 + n_2 I \tag{8}$$

where n_0 is the linear refractive index and n_2 is the nonlinear refractive index. The n_2 and $\chi^{(3)}$ values are related by[13]:

$$n_2 = \frac{12\pi^2\chi^{(3)}(-\omega;\ \omega,\ \omega,\ -\omega)}{cn_0} \tag{9}$$

when both are in esu units.

$$n_2 = \frac{5.26 \times 10^{-6}\chi^{(3)}}{n_0^2} \tag{10}$$

where n_2 is in MKS (m^2/W) units and $\chi^{(3)}$ is in esu units. The conversion of n_2 to $\chi^{(3)}$ can be done as follows:

$$n_2(m^2/V^2) = \frac{3}{8n}\chi^{(3)}(m^2/V^2) \tag{11}$$

and

$$\chi^{(3)}(m^2/V^2) = \frac{4\pi}{9 \times 10^8}\chi^{(3)}(esu) \tag{12}$$

The $\chi^{(3)}$ value should possess a sign, being the fundamental property of a material associated with the microscopic optical nonlinearity. The sign and the magnitude of the real and the imaginary parts of the excited state optical nonlinearities associated with the saturation of the single-photon absorption in polymer thin films have been studied using the Z-scan technique. In this technique, the transmittance with and without the aperture as a function of z from the polymer film at a given wavelength is recorded by two detectors in a single beam scan. The details of this technique can be found elsewhere.[14,15] In the case of the saturation absorption, n_2 has both a real and an imaginary part:

$$n_2 = \gamma + i\frac{c\Delta\alpha}{2\omega} = \gamma - i\frac{c\alpha_0}{2\omega I_s} \tag{13}$$

where α_0 is the linear absorption coefficient without absorption, I_s the saturation intenesity, and the absorption coefficient is given as:

$$\alpha = \frac{\alpha_0}{1 + I/I_s} \tag{14}$$

where I is the excitation intensity. The real part of $\chi^{(3)}$ is related to second hyperpolarizability γ as

$$\text{Re } \chi^{(3)} = \frac{cn_0^2}{16\pi^2} \gamma \tag{15}$$

and the imaginary part of $\chi^{(3)}$ is related to the change of the absorption $\Delta\alpha$ as follows:

$$\text{Im } \chi^{(3)} = -\frac{cn_0^2}{32\pi^2} \frac{c\alpha_0}{\omega I_s} \tag{16}$$

When the pump frequency reaches the single-photon transition frequency in the NLO material, then the resonant part of the $\chi^{(3)}(-\omega; \omega, -\omega, \omega)$ value is dominated by the following terms:[13]

$$\chi_R^{(3)} \propto i\left[\frac{1}{(\omega_{gi} - \omega + i\Gamma_{gi})^2} - \frac{1}{(\omega_{gi} - \omega)^2 + \Gamma_{gi}^2}\right] \tag{17}$$

This represents the two-level system with ^1Ag as the ground-state g and ^1Bu as the excited-state i, and Γ_{gi} is the damping term. The nonlinear refractive index n_2 can be converted to second hyperpolarizability γ by the following relationship:

$$n_2 = \gamma NL^4/n_0^3 \tag{18}$$

here n_0 is the linear refractive index, L is Lorentz field correction, and N is the molecular density. The second hyperpolarizability γ is related to $\chi^{(3)}$ by the following expression:

$$\langle\gamma_{xxxx}\rangle = \frac{\chi_{xxxx}^{(3)}}{L^4N} \tag{19}$$

where L is the local field factor $[(n^2 + 2)/3]$ and N is the number of density of the sample. The intensity-dependent nonlinear refractive index n_2 and two-photon absorption coefficient β are related to the real and imaginary part of $\chi^{(3)}$ by the following equation:[14,15]

$$n_2(\text{m}^2/\text{W}) = \frac{3}{4\epsilon_0 cn_0^2} \text{Re } \chi^{(3)}(-\omega; \omega, -\omega, \omega) \tag{20}$$

and

$$\beta(\text{m/W}) = \frac{3\omega}{2\epsilon_0 c^2 n_0^2} \text{Im } \chi^{(3)}(-\omega; \omega, -\omega, \omega) \tag{21}$$

where n_0 and c are the linear refractive index and the speed of light in a vacuum, respectively. The conversion is written as n_2 (esu) $= (cn_0/40\pi)n_2$ (MKS). The Z-scan technique has been used to investigate the nonlinear absorption. The discussion above demonstrated that the $\chi^{(3)}$, γ, n_2, and β are related to each other and can be deduced. This also supports the fact that complete information on third-order optical nonlinearity and its origin needs several types of measurements. The most frequently used techniques for measuring third-order NLO effects are THG, DFWM, Z-scan, EFISH, and optical Kerr effect.

The development of photonic technology during the past decade has intensified research activities on searching for new materials that display unusual and interesting NLO properties. New NLO materials are the key elements to future photonic technologies in which their functions can be integrated with other electrical, optical, and magnetic components that have become important in this era of optical telecommunications. The new photonic components have been made with NLO inorganic crystals and compound semiconductors although some of them are still at the laboratory level. Organic molecular

616

and polymeric materials are relatively newcomers in the field of nonlinear optics compared with inorganic materials. The most important advantages of organic materials are the possibility of unique diversity and the ability to be tailored with a variety of chromophores for the exogenous variables that stimulate photonic functions in a desirable manner; therefore, for example, different NLO processes can be adopted for applications. The applications of NLO materials are widespread in the field of solid state technology that includes harmonic generators, optical computing, telecommunications, laser lithography, image processing, sensors, diode lasers, and overall optical systems. This shows that the study of NLO properties is truly an interdisciplinary area of research because it needs expertise in various fields of science such as computation calculations for molecular modeling, chemical syntheses, molecular engineering, optical measurements, materials processing, and finally the fabrication of novel optoelectronic devices. Computer-aided calculations provide reasonable predictability at the molecular levels to design new materials of interest. The development of new materials can shed light on the theoretical understandings of the origins of NLO processes. Every year a wide variety of new NLO molecular and polymeric materials is discovered and the field rapidly expands. At the same time, the computational approaches of the structure–property relationships have been greatly broadened from molecular modeling to NLO phenomena. Therefore, the truly interdisciplinary nature of the field of nonlinear optics invites synthetic chemists, physicist, and technocrats to play a vital role while working together on new supermolecular structural materials specially designed for photonic applications. NLO polymers have many attractive features compared with oxygen-octahedra ferroelectrics such as lithium niobate, lithium tantalate, potassium niobate, barium titanate, and compound semiconductors such as GaAs, InP, CdSe, and ZnS in the field of integrated optics. Because the level of integration is increasing rapidly in photonic technologies, organic polymers seem more promising compared with inorganic counterparts because of their compatibility with a variety of materials used in fabrication technology. This article represents an attempt to give a systematic review of the field of nonlinear optics as the chemical physicist and polymer chemist see it. Here the emphasis will be on structure–property relationships that provide basic understanding on how these specialty molecules and polymers behave under different NLO processes and how to control their optical nonlinearities by using various molecular design and synthetic strategies.

Third-order NLO properties of inorganic materials such as CdS, As_2S_3, CdS_xSe_{1-x}, ZnS, PbI_4, AlGaAs, CuCl, CdS-CdSe, ZnSe, GaAs, CdS, InSb, Si, Ge, As_2S_3, Nb_2O_5-TiO_2-TeO_2, LiO_2-TiO_2-TeO_2, SiO_2, and $(C_{10}H_{21}NH_3)_2PbI_4$ have also been investigated. The third-order optical nonlinearities of inorganic materials are large but their response time is relatively slow.[16–20] Large third-order NLO susceptibilities have been measured in inorganic semiconductors using them as quantum wire and quantum dot materials; for example, the enhanced excitonic optical nonlinearities of typical quantum dot lattices of GaAs, CdS, and CuCl have been of considerable interest. Recently, research activities have been directed toward developing new organic molecular and polymeric materials. The superiority of organic materials has been realized because of their versatility and the possibility of tailoring material properties by molecularly engineering them for particular end-uses. In addition, organic materials also exhibit large nonlinear figures of merit, high damage thresholds, and ultrafast response time. With these views, recently the studies on third-order NLO properties of organic molecular and polymeric materials have been focused and progressing rapidly. The literature on third-order NLO materials is growing very rapidly and these is no single source that summarizes the chemical aspects of these scattered organic materials to develop structure–property relationships. The Society of Photo-optical Instrumentation Engineers (SPIE) annually publishes conference proceedings on NLO materials, processes, and devices. This chapter is the first in-depth presentation of recent information on organic materials. Attempts have been made to include all the published data on third-order NLO materials from many scientific journals and conferences, although some may have been excluded. A wide variety of organic molecular and polymeric materials that have been studied for third-order nonlinear optics are summarized here for the first time. We have divided third-order NLO organic molecular and polymeric systems based on their characteristics. The NLO properties of the following classes of organic molecular and polymeric materials are discussed in detail: (1) liquids; (2) molecular solids; (3) charge-transfer complexes; (4) conjugated polymers; (5) NLO dye-functionalized polymers; (6) organometallic compounds; (7) composites; (8) liquid crystals and (9) biomaterials.

Extensive efforts in synthesis and spectroscopic analysis of novel organic materials have been made. In past years, research activities on third-order NLO materials have expanded significantly, ranging from simple molecules to conjugated polymers to organometallic compounds to liquid crystals. As a result, this has lead to a fundamental understanding of the structure–property relationship for third-order NLO effects.

B. REFERENCE STANDARDS FOR $\chi^{(3)}$

There has been some debate on the magnitude of third-order nonlinear susceptibilities of standard reference materials. The third-order optical nonlinearity of a material depends on the relative $\chi^{(3)}$ value of the standard reference materials used. We summarize the $\chi^{(3)}$ values of fused silica, quartz, BK-7 glass and carbon disulfide on different wavelengths reported by various researchers (Table 1) because these have been used most frequently. Currently a wide range of wavelengths, for example, from 0.532 to 2.5 μm, are available with different laser systems for measuring third-order nonlinear susceptibility; therefore, resonant and nonresonant contributions to $\chi^{(3)}$ values can be evaluated. The $\chi^{(3)}$ value of the air as 7.0×10^{-18} esu at 1.064 μm and 1.8×10^{-17} esu at 1.907 μm has been measured by THG technique.[25]

Third-order optical nonlinearity can be expressed in terms of third-order nonlinear susceptibility $\chi^{(3)}$ or nonlinear refractive index (n_2). Therefore, it will be appropriate to provide conversion factors for third-order NLO susceptibility and hyperpolarizability from one unit to others.[34]

$$n_2 = 16\pi^2\chi^{(3)}/Cn^2$$

where C is the velocity of the light

$$n_2(\text{MKS}) = 5.26 \times 10^{-6}\chi^{(3)}/n^2 \text{ (esu)}$$

$$\chi^{(3)}(\text{m}^2\text{V}^{-2}) = 7.16 \times 10^7\chi^{(3)} \text{ (esu)}$$

Table 1 $\chi^{(3)}$ Values of the Standard Reference Materials Reported by Different Researchers

Reference Materials	$\chi^{(3)}$ (esu)	Wavelength (μm)	Techniques	Ref.
Fused silica glass	1.40×10^{-14}	1.907	THG	21,22
	1.47×10^{-14}	1.543	THG	23
	2.79×10^{-14}	1.907	THG	24
	2.79×10^{-14}	1.907	THG	25
	3.11×10^{-14}	1.064	THG	25
	1.03×10^{-14}	1.064	THG	26
	1.06×10^{-14}	1.910	THG	26
	0.99×10^{-14}	2.05	THG	26
BK-7 glass	4.67×10^{-14}	1.907	THG	24
	3.50×10^{-14}	1.064	DCSHG	21
	3.39×10^{-14}	1.543	DCSHG	21
	3.32×10^{-14}	1.907	DCSHG	21
	6.35×10^{-13}	1.064	THG	21
	6.00×10^{-13}	1.543	THG	21
	5.80×10^{-13}	1.907	THG	21
	5.68×10^{-13}	2.148	THG	21
Fused quartz	2.96×10^{-14}	1.89	THG	27
α-Quartz	3.81×10^{-14}	1.907	THG	21,22
Dynasil	2.82×10^{-14}	1.907	THG	21
Optosil	2.76×10^{-14}	1.907	THG	21
Carbon disulfide	4.0×10^{-13}	1.064	DFWM	28
	1.6×10^{-12}	1.907	THG	24
	2.73×10^{-13}		DFWM	29
	1.7×10^{-12}	0.532	DFWM	30
	6.8×10^{-13}	0.532	DFWM	31
	8.44×10^{-12}	1.064	OKE	32

The conversion factors from SI to esu for second hyperpolarizability (γ) are

$$\gamma \text{ (esu)} = 7.16 \times 10^{13} \times \gamma \text{ (SI)}$$

$$1 \text{ a.u.} = 0.503717 \times 10^{-39} \times \gamma \text{ (esu)}$$

$$1 \text{ a.u.} = 6.23597 \times 10^{-65} \times \gamma(C^4 m^4 J^{-3})$$

$$1 \text{ esu} = 7.4279 \times 10^{-25} \times \gamma(C^4 m^4 J^{-3})$$

The nonlinear dielectric constant ϵ_2 can be defined by $\epsilon = \epsilon_0 + 1/2 \; \epsilon_2 \; E^2$

$$\epsilon_2 \text{ (esu)} = 8\pi\chi^{(3)} \text{ (esu)} = 1.8 \times 10^9 \; \chi^{(3)} \text{ (SI)}$$

II. THIRD-ORDER NONLINEAR OPTICAL MATERIALS

A. ORGANIC LIQUIDS

The third-order NLO properties of a wide variety of organic liquids and solvents have been studied. These studies reveal how the third-order nonlinear parameters are influenced by the number of carbon atoms, substituents, donor–acceptor functionalities and π electron system. An important role of these factors can be seen from the following examples provided of organic liquids. In addition, some of these organic solvents are used for solubilizing organic solids and, as a result, knowledge of their third-order NLO susceptibilities and second hyperpolarizabilities is necessary to perform calculations in measurement techniques. The sign and magnitude of second hyperpolarizability of a solvent in such calculations is important because they are also used as reference samples; therefore, we have summarized various aspects of nonlinearities associated with liquids.

Oudar and Person[35] measured the third-order NLO susceptibility and second hyperpolarizabilities of benzene and stilbene derivatives from EFISH technique. 4-Dimethylamino-4'-nitrostilbene molecules possess the large second hyperpolarizability because of its system of conjugated double bonds is larger than that of benzene molecules. Furthermore, a strong electron acceptor nitro group and a strong donor dimethylamino group, which are attached at the opposite ends, provide a favorable charge-transfer interaction, which may also contribute to the large γ value, although such configurations are more favorable for obtaining the large first hyperpolarizability (β). Looking at the γ data, a conjecture may be made that the benzene molecules which possess either electron acceptor groups or electron donor groups exhibit slightly larger γ values. For example, nitrobenzene, benzonitrile, aniline, N,N-dimethylaniline, and p-nitroaniline molecules validate such assumption. 4-Dimethylamino-4'-nitrostilbene, with a strong donor $N(CH_3)_2$ group and a strong acceptor NO_2 group at the opposite ends of the molecules, shows molecular hyperpolarizability that is two orders of magnitude larger than that of nitrobenzene. The magnitude of the molecular second hyperpolarizability is strongly affected by the substituents and increases significantly with the size of the aromatic system.

Levine and Bethea[36] reported the third-order nonlinear properties of several conjugated and nonconjugated organic molecules. To examine the effect of the nonconjugated bond additivity on third-order nonlinear properties, a variety of simple substituted methane derivatives of longer and more complex structures either substituted or unsubstituted conjugated and nonconjugated chains and rings were investigated. The bond additivity concept tested for the unsubstituted alkanes C_nH_{2n+2}, methane ($n = 1$), hexane ($n = 6$), heptane ($n = 7$), and dodecane ($n = 12$) and on nitro-substituted hydrocarbons shows that it is a useful approximation for predicting the second hyperpolarizability. In addition, measured hyperpolarizability of two similar molecules, nitrobenzene (conjugated) and nitrocyclohexane (nonconjugated) indicates that in spite of the similar dipole moments, the hyperpolarizability of nitrobenzene was an order of magnitude larger than nitrocyclohexane, indicating the role of the π electrons.

Kajzar and Messier[25] measured third-order nonlinear susceptibilities of several organic solvents by THG at 1.064 and 1.907 μm wavelengths. All solvents are transparent at 1.064 μm and at harmonic frequencies, whereas some of them absorb slightly at 1.907 μm. The average molecular third-order susceptibility is a scalar quantity; therefore, the determination of bond hyperpolarizabilities within the bond additivity model is possible from listed γ values. For a series of three solvents, carbon tetrachloride (CCl_4), chloroform ($CHCl_3$), and dichloromethane (CH_2Cl_2), the variation of γ_{C-H} is influenced by a large C−H stretching vibration at 3.3 μm, whereas the saturated C−Cl bond shows the same value of γ at both wavelengths. Therefore the decrease in the hyperpolarizability at 1.907 μm with increasing number of C−H bonds for $CHCl_3$ and CH_2Cl_2 is affected by the frequency behavior of γ_{C-Cl} and γ_{C-H}. Other solvents such as benzene, dimethyl formamide (DMF), and acetone show the same dispersion in between 1.064 and 1.907 μm as that of fused silica.

In another study, Kajzar and Messier[37] measured the second hyperpolarizability of a family of alkanes and substituted alkanes with a general formula:

$$CH_{3-}(CH_2)_{n-2-}CH_2X \quad (X = H, Cl, Br, I; 6 < n < 16)$$

By using THG at 1.064 μm taking into account a systematic study of the bond-additivity model and the validity of local field factors (Table 2). The molecular hyperpolarizability γ was calculated by using the equation $\chi^{(3)} = \gamma \, NF^D$, where N is the number of the density of the molecule and F^D is a debye local field factor expressed by $F^D = (f_\omega^D)^3 f_{3\omega}^D$ and $F_{\omega,3\omega}^D \, D = (\epsilon_{\omega,3\omega} + 2)/3$. From this bond-additivity model, the molecular hyperpolarizability can be decomposed into a sum of hyperpolarizabilities of distinct $=CH_2(\gamma 2)$, $-CH_2X(\gamma X: X = Cl, Br, I)$ and $-CH_3(\gamma H)$ groups as follows:

$$\gamma = (n - 2)\gamma 2 + \gamma H + \gamma X \tag{20}$$

The linear dependence of molecular hyperpolarizability on the number of carbon atoms n can be written as:

$$\gamma = \chi^{(3)}/NF^D = (n - 2)\gamma 2 + \gamma H + \gamma X \tag{21}$$

Figure 1 represents the molecular hyperpolarizability dependence of alkanes and substituted alkanes on the number of carbon atoms calculated with an average local-field factor. The γ2 value depends strongly on the polar head which is largest for Cl-substituted and smallest for I-substituted alkane series. Figure 2 shows the calculated (solid lines) and measured $\chi^{(3)}$ values of alkanes and substituted alkanes as a function of chain length. At large chain lengths, the $\chi^{(3)}$ value tends to the same limiting values, whereas for $n = 0$, it is different depending on the polar head. The I−C bond is the most nonlinear bond, whereas the C−H bond is the smallest one, the hyperpolarizability depends strongly on the electron cloud size. Meredith et al.[38] measured third-order optical nonlinearities of several neat organic liquids by optical THG to examine the effects of chemical changes on third-order nonlinearities. Gonin et al[22] measured $\chi^{(3)}$ and γ values for a series of different solvents at 1.064 μm by EFISH and THG techniques. Table 3 summarizes the $\chi^{(3)}$ and γ values of some standard organic solvents obtained with

Table 2 Third-order NLO Susceptibilities of Alkanes and Substituted Alkanes Measured by THG at 1.064 μm at Room Temperature

Compound	n	$\chi^{(3)}$ (10^{-14} esu)
$CH_3(CH_2)_{n-2}CH_3$	6	4.67
	8	5.04
	10	5.35
	16	6.03
$CH_3(CH_2)_{n-2}CH_2Cl$	6	5.57
	10	6.10
	12	6.47
	14	6.66
$CH_3(CH_2)_{n-2}CH_2Br$	6	7.46
	10	6.87
	14	6.93
$CH_3(CH_2)_{n-2}CH_2I$	6	12.94
	8	11.47
	10	10.74
		7.11

From Kajzar, F. and Messier, F. *J. Opt. Soc. Am. B*, 4, 1040, 1987. With permission.

620

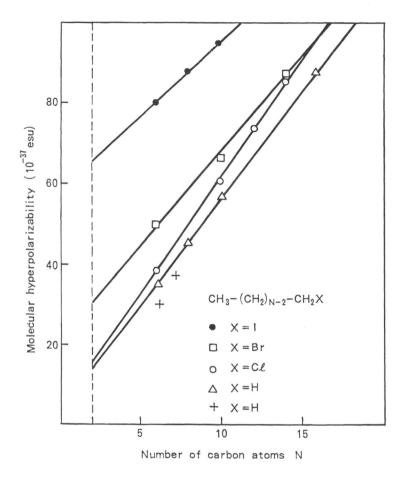

Figure 1 Molecular hyperpolarizability dependence on the number of carbon atoms for alkanes and substituted alkanes calculated with an average local-field factor. (From Kajzar, F. and Messier, F., *J. Opt. Soc. Am. B*, 4, 1040, 1987. With permission.)

various experimental techniques at different wavelengths. Third-order optical nonlinearities were affected by the atom, bond additivity, π-conjugation, and donor–acceptor substitution. Third-order optical nonlinearity increases noticeably for benzene derivatives having donor or acceptor groups. We have compiled theoretically obtained γ values in chapter 9; therefore, a comparison can be made with experimental values. Furthermore nonlinear refractive indexes (n_2) of various solvents measured by OKE are discussed in chapter 10. The nonlinear refractive index is related to third-order NLO susceptibility $\chi^{(3)}$ and second hyperpolarizability γ by $n_2 = 12\ \pi/n\ \chi^{(3)}\ (-\omega; \omega, \omega - \omega)$ and $n_2 = 12\pi\ NL/n < \gamma\ (-\omega; \omega, \omega - \omega)>$, where n is the linear refractive index.[13]

Meredith et al.[38] and Kajzar and Messier[37] determined the average bond polarizabilities and hyperpolarizabilities by bond-additivity model (Table 4). The γ values of C—Cl and C—H bonds were deduced from a series of three solvent: CCl_4, $CHCl_3$, and CH_2Cl_2. The C—H bond has a negative hyperpolarizability at 1.907 μm but the corresponding value at 1.064 μm was found to be small and positive. The γ values of 0.47×10^{-36} esu for C—CH_3, 0.03×10^{-36} esu for C—F, 0.57×10^{-36} esu for C—Cl, 0.6×10^{-36} esu for C—Br, 1.24×10^{-36} esu for C—I, 0.4×10^{-36} esu for C—NH_2, and 1.4×10^{-36} esu for C—NO_2 were estimated.[36] Obviously the α and γ parameters for benzene substituents will differ from those of nonaromatics. Levenson and Bloembergen[47] investigated the dispersion of the second hyperpolarizabilities of benzene, chlorobenzene, and nitrobenzene in the vicinity of Raman active vibrations. The dispersion of the third-order nonlinearity of benzene was described by a simple model applied only to crystals, whereas, in the case of nitrobenzene and chlorobenzene, an extension to a sum of resonant terms were considered. The hyperpolarizability of complex solute molecules can be determined by fitting of the Raman and three wave-mixing spectra from a well-characterized solvent such

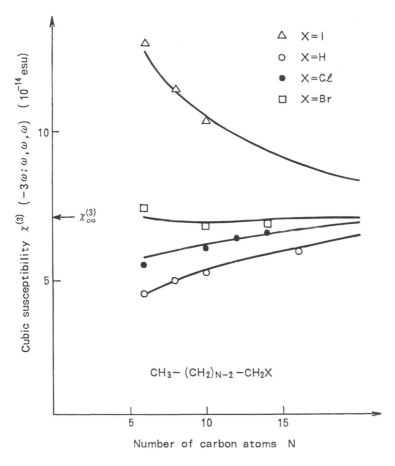

Figure 2 Calculated (solid lines) and measured $\chi^{(3)}$ values of alkanes and substituted alkanes as a function of chain length (From Kajzar, F. and Messier, F., *J. Opt. Soc. Am. 8,* 4, 1040, 1987. With permission.)

as benzene. The γ value of nitrobenzene was found to be an order of magnitude larger that of nitromethane, further confirming the fact that the delocalized π-electrons are effective in contributing more to γ than the localized nonconjugated electrons.[48] It is apparent from the analysis of the third-order nonlinear parameters of a variety of organic liquids that the $\chi^{(3)}$ values are strongly influenced by the number of carbon atoms, substituents, donor–acceptor functionalities, and π electron system incorporated in organic molecules and, in particular, the last two factors play a major role.

B. Molecular Solids
1. Dyes

Various measurements on liquids show that the $\chi^{(3)}$ and γ values are strongly affected by the presence of atoms, π-conjugation, and donor–acceptor groups. In the 1980s, various research teams performed measurements on benzene derivatives. The $\chi^{(3)}$ and γ values of substituted benzenes are discussed here. Oudar[43] also investigated the influence of donor–acceptor substitution on $\chi^{(3)}$ and γ values of large π-conjugated molecules such as stilbene and styrene derivatives using the EFISH technique. The γ value of *trans*-stilbene was found to be about 2 times larger than *cis*-stilbene, indicating the influence of conformation. Third-order optical nonlinearities of donor–acceptor-substituted stilbene and styrene were found to be affected by the extent of π-conjugation and strength of the substituents. The $\chi^{(3)}$ and γ values are anomalously large for *p*-nitroaniline (*p*NA) due to strong donor–acceptor charge-transfer interaction and the change occurs in an order *p*-NA > *o*-NA > *m*-NA.[45,46] Levine and Bethea[49] measured optical nonlinearities of *p-NA* and *trans*-dimethylamino-4′-nitrostilbene (DANS) using electric field-induced optical wave mixing (WM) ($\omega + 2\omega = 3\omega$) at 1.318 μm. The *p*-NA of concentration 12.1 wt% and DANS of concentration 5.94 wt% were dissolved in *trans*-stilbene and their crushed powders

622

Table 3 The $\chi^{(3)}$ and γ Values of Some Standard Organic Solvents Showing the Effect of Bond Additivity, π-conjugation, and Donor–Acceptor Substitution on Third-order Optical Nonlinearities

Compound	$\chi^{(3)}$ $(10^{-14}$ esu)	γ $(10^{-36}$ esu)	Wavelength (μm)	Technique	Ref.
Benzene	5.0	1.8	1.06	EFISH	35
	6.33	2.34	1.064	EFISH	36
	10.64	4.07	1.907	THG	37
	11.98	4.40	1.064	THG	37
	10.1	3.85	1.907	THG	38
		3.1	1.907	EFISH	41
		3.9	1.907	THG	41
		4.61	0.670	EFISH	42
	9.2	3.4	1.064	EFISH	43
	4.9		0.532	DFWM	31
Carbon tetrachloride	5.20	2.29	1.064	EFISH	36
	6.73	3.09	1.907	THG	38
	7.37	3.36	1.907	THG	37
	7.96	3.63	1.064	THG	37
	7.04	3.23	1.064	EFISH	39
		3.3	1.064	EFISH	40
	0.9		0.532	DFWM	31
Carbon disulfide	17.8	3.19	1.064	EFISH	36
	23	4.4	1.907	THG	38
	68		0.532	DFWM	31
n-Hexane	2.63	2.03	1.064	EFISH	36
		3.6	1.064	EFISH	40
	5.33	4.17	1.907	THG	37
	4.42	3.43	1.064	THG	37
	3.8	2.95	1.064	EFISH	44
Cyclohexane	6.47	3.32	1.064	THG	37
	4.91	2.58	1.064	EFISH	39
	5.60	3.02	1.907	THG	37
	0.9		0.532	DFWM	31
p-Xylene	5.94	3.08	1.064	EFISH	36
		4.3	1.907	EFISH	41
		5.2	1.907	THG	41
Toluene	6.80	2.99	1.064	EFISH	36
	9.76	4.49	1.064	EFISH	39
	9.81	4.55	1.907	THG	38
		4.6	1.907	THG	41
		4.1	1.907	EFISH	41
	3.8		0.532	DFWM	31
Pyridine	20.6	4.97	1.064	EFISH	36
	10.04	3.46	1.907	THG	37
	9.91	3.36	1.907	THG	38
Phenol	14.0	3.6	1.06	EFISH	35
		4.0	1.907	EFISH	41
		5.0	1.907	THG	41
Thiophenol		4.6	1.907	EFISH	41
		6.3	1.907	THG	41
Fluorobenzene	7.10	3.19	1.907	THG	38
	-6.0	-2.0	1.06	EFISH	35
		2.2	1.907	EFISH	41
		3.5	1.907	THG	41
Chlorobenzene	10.6	4.34	1.907	THG	38
	0 ± 3.0	0 ± 1.0	1.06	EFISH	35
		2.0	1.907	EFISH	41
		4.5	1.907	THG	41

Table 3 *Continued*

Compound	$\chi^{(3)}$ (10^{-14} esu)	γ (10^{-36} esu)	Wavelength (μm)	Technique	Ref.
Bromobenzene	14.0	5.40	1.907	THG	38
	6.0	1.8	1.06	EFISH	35
		4.4	1.907	EFISH	41
		5.4	1.907	THG	41
Iodobenzene	23.5	8.19	1.097	THG	38
	25.0	7.3	1.06	EFISH	35
		7.2	1.907	EFISH	41
		7.5	1.907	THG	41
Nitrobenzene	13.7	5.37	1.907	THG	38
	16.8	6.57	1.907	THG	37
	190	51	1.06	EFISH	35
		40.8	1.907	EFISH	41
		5.7	1.907	THG	41
	159	41.2	1.064	EFISH	44
	167	43.3	1.318	EFISH	45
	195	7.6	1.064	EFISH	46
	13		0.532	DFWM	31
Aniline	16.8	5.71	1.907	THG	38
		7.8	1.907	EFISH	41
		5.4	1.907	THG	41
	45	11	1.06	EFISH	35
	32.6	7.78	1.318	EFISH	45
	53	13	1.064	EFISH	46
N,N'-Dimethylaniline	40	15.2	1.06	EFISH	35
		14.1	1.907	EFISH	41
		8.1	1.907	THG	41
Acetone	4.10	1.88	1.907	THG	37
	4.69	1.80	1.064	THG	37
		2.10	1.064	EFISH	40
	8.3	2.7	1.064	EFISH	43
	1.0		0.532	DFWM	31
p-Dioxane	5.19	2.25	1.064	EFISH	39
		2.30	1.064	EFISH	40
		2.65	1.064	THG	40
		2.15	1.907	THG	40
		1.90	1.907	EFISH	41
		2.5	1.907	THG	41
	4.50	1.95	1.907	EFISH	21
	5.71	2.48	1.064	EFISH	21
	0.738	0.327	1.064	THG	21
	0.702	0.316	1.543	THG	21
	0.643	0.289	1.907	THG	21
	0.621	0.279	2.148	THG	21
Chloroform	5.78	2.29	1.907	THG	38
	5.72	2.27	1.907	THG	37
		2.71	1.064	THG	25
	8.13	2.63	1.064	EFISH	39
		2.4	1.064	EFISH	40
	7.0	2.71	1.064	THG	37
	8.8	2.8	1.064	EFISH	43
	1.9		0.532	DFWM	31
Benzonitrile	10.1	4.11	1.907	THG	38
	44	12.2	1.06	EFISH	35
		10.3	1.907	EFISH	41
		4.3	1.907	THG	41
Benzaldehyde		14.8	1.907	EFISH	41
		5.3	1.907	THG	41
Dimethyl formamide	5.57	2.20	1.064	THG	37
	5.05	2.03	1.907	THG	37
	15.40	4.20	1.064	EFISH	39

Table 3 *Continued*

Compound	$\chi^{(3)}$ (10^{-14} esu)	γ (10^{-36} esu)	Wavelength (μm)	Technique	Ref.
Dimethyl sulfoxide	−6.30	−1.35	1.064	EFISH	44
	13.84	3.00	1.064	EFISH	39
Methanol	2.86	0.80	1.907	THG	38
	3.12	0.87	1.907	THG	37
	2.98	0.83	1.064	THG	37
	4.60	0.85	1.064	EFISH	39
	5.3	1.0	1.064	EFISH	46
Ethanol	4.13	1.49	1.907	THG	37
	3.45	1.27	1.064	THG	37
	3.65	1.33	1.907	THG	38
	7.07	1.57	1.064	EFISH	39
	8.4	2.1	1.064	EFISH	43
Dichloromethane	6.13	1.94	1.064	THG	37
	4.77	1.55	1.907	THG	37
	10.33	2.41	1.064	EFISH	39
1,2-Dichloroethane	12.0	3.39		EFISH	36
Methane	−	0.42	1.064	EFISH	36
Heptane	2.88	2.40	1.064	EFISH	36
Dodecane	4.16	4.88	1.064	EFISH	36
Nitromethane	8.98	1.92	1.064	EFISH	36
1-Nitropropane	8.26	2.84	1.064	EFISH	36
2-Nitropropane	6.32	2.76	1.064	EFISH	36
Nitrocyclohexane	11.3	4.50	1.064	EFISH	36
Methylcyclohexane	5.93	3.97	1.907	THG	38
Methyl iodide	33.3	6.16	1.064	EFISH	36
Methylene iodide	48.3	7.32	1.064	EFISH	36
o-Diiodobenzene	24.2	6.12	1.064	EFISH	36
o-Fluoronitrobenzene	169.0	45.3	1.064	EFISH	36
m-Fluoronitrobenzene	107.0	31.4	1.064	EFISH	36
p-Fluoronitrobenzene	104.0	30.8	1.064	EFISH	36
α-Picoline	5.0		0.532	DFWM	31
Propanol		1.72	1.064	THG	39
	8.33	2.60	1.064	EFISH	39
2-Propanol	4.11	1.88	1.907	THG	38
2-Propanone	4.41	2.04	1.907	THG	38
Tetrahydrofuran	5.03	2.24	1.907	THG	38
Methylcyclohexane	5.93	3.97	1.907	THG	38
Acetonitrile	2.54	0.87	1.907	THG	38
Water	2.80	0.83	1.064	THG	37

Table 4 Average Bond Polarizabilities and Hyperpolarizabilities Calculated Within the Bond-Additivity Model

Bond	Meredith et al.[38] at λ = 1.907 μm		Kajzar and Messier[37] at λ = 1.064 μm
	α (10^{-24} esu)	γ (10^{-36} esu)	γ (10^{-36} esu)
C−H	0.60	−0.0275	0.05
C−Cl	2.55	0.7725	0.90
O−H	0.7871	0.5531	0.42
C−C	0.5996	0.6211	0.32
C=C	1.47	0.61	1.03
C−O	0.58	0.30	0.24
C=O	1.41	0.99	0.82
C=S	4.01	2.2	−
C=N	1.89	0.33	−

were heated in an oven to melt the mixture. The $\chi^{(3)}$ and γ values of some organic solids are summarized in Table 5. Kubodera[50] measured the $\chi^{(3)}$ values of *p*-NA, 2-methyl-4-nitroaniline (MNA), and *trans*-4′-dimethylamino-*N*-methyl-4-stilbazolium methylsulfate (DMSM) in 5 wt% DMF solutions. The powder samples of MNA, DMSM, and DANS showed THG intensities 15.2, 21.9, and 3.5 times larger, than that of *p*-NA respectively. The powder $\chi^{(3)}$ of 4-diethylamino-4′-nitrostilbene (DEANS) sample was 5.7 times larger than that of *p*-NA, and the crystal showed a $\chi^{(3)}$ value of 1.3×10^{-11} esu at 1.90 μm.[51] The $\chi^{(3)}$ values of MNA (2 wt%) and DANS (0.1 wt%) doped polymethyl methacrylate (PMMA) were 6.1×10^{-14} esu and 5.8×10^{-14} esu, respectively. On the other hand, DMSM (10 wt%) doped polyvinyl alcohol (PVA) was measured as 3.7×10^{-13} esu at 1.90 μm.[52] The charge-transfer compound, 4-(*N*,*N*-diethylamino-4′-nitrostyrene (DEANST) exhibited THG intensity about 720 times larger than *p*-NA and 20 times larger than DEANS at 1.064 μm.[53] The $\chi^{(3)}$ of DEANST crystal was measured as 7.4×10^{-12} esu for molecular longer axis. The DEANST doped PMMA films showed $\chi^{(3)}$ of 2×10^{-12} esu at 1.064 μm. Barzoukas et al.[54] measured second hyperpolarizabilities of *N*-(4-nitrophenyl)-L-prolinol (NPP), *N*-(4-nitrophenyl)-*N*-methylaminoacetonitrile (NPAN), and 2-(*N*-prolinol)-5-nitropyridine (PNP) by EFISH and THG techniques. The γ values of two charge-transfer molecules, 4,4′-octyloxycyanobiphenyl (OOCB) and 4,4′-methoxynitrostilbene (MONS), have been measured by EFISH and THG techniques.[22] The second hyperpolarizability in *p*-NA also depends on the solvent.[55] The γ value of MNA was also measured to examine the cascading contribution. These data showed that the $\chi^{(3)}$ of the same material differs depending on the material states and solvents.

In 1973, Hermann et al.[56] demonstrated that the third-order optical nonlinearity of conjugated organic molecules β-carotene, which has 11 double bonds is three orders of magnitude larger than that of benzene. Hermann and Ducuing[57] confirmed the relationship between the delocalized π-electrons and second hyperpolarizability. Figure 3 shows the variation of hyperpolarizability component with the number of double

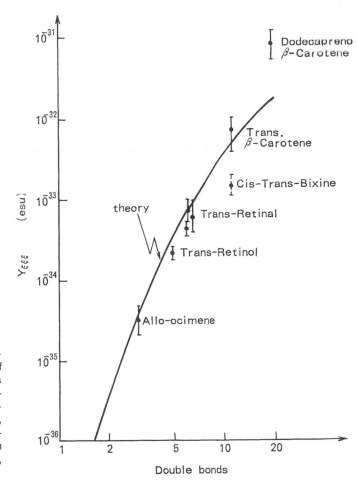

Figure 3 Variation of hyperpolarizability components with the number of double bonds. Conjugated systems of allo-ocimene, *trans*-retinol, *trans*-retinal, *trans*-β-carotene, and dodecapreno-β-carotene possess 3, 5, 6, 11, and 19 double bonds in their backbones, respectively. (From Hermann and Ducuing, *J. Appl. Phys.*, 45, 5100, 1974. With permission.)

626

Table 5 The $\chi^{(3)}$ Value in 10^{-14} esu and γ value in 10^{-36} esu of Some Organic Charge-Transfer Solids Showing the Effect of Donor–Acceptor Substitution

Compound	Solvent	$\chi^{(3)}$	γ	Wavelength (μm)	Technique	Ref.
o-Nitroaniline	Acetone	65.3	241	1.064	EFISH	46
		483	123	1.318	EFISH	45
	Methanol		9.6	1.90	THG	55
m-Nitroaniline	Acetone	18.1	155	1.064	EFISH	46
		332	85	1.318	EFISH	45
	Methanol		10.4	1.90	THG	55
p-Nitroaniline	Methanol	83	1,000	1.06	EFISH	35
	Acetone	85.5	1,100	1.064	EFISH	46
	Nitrobenzene	1970	496	1.318	EFISH	45
	trans-Stilbene	97	167	1.318	WM	49
	DMF	2.6	–	1.90	THG	50
			24	1.908	THG	55
	Methanol		27.8	1.908	THG	55
	DMF		33.7	1.908	THG	55
	Cyanobenzene		43	1.908	THG	55
	Nitrobenzene		38.9	1.908	THG	55
2-Methyl-4-nitroaniline	DMF	2.9	–	1.90	THG	50
	Chloroform	–	33	1.907	THG	40
	Dioxane	–	24	1.907	THG	40
2,4-Dinitroaniline	Acetone	53	596	1.064	EFISH	46
Dinitrophenylmethyl alaninate	Acetone	93	633	1.064	EFISH	46
trans-Stilbene	Neat	27.7	15.4	1.318	WM	49
4-Nitrostilbene	Benzene	28	750	1.06	EFISH	43
4-Aminostilbene	Benzene	16.2	200	1.064	EFISH	43
p-Dimethylamino-β-nitrostyrene	Chloroform	27.4	8,800	1.064	EFISH	43
p-Dimethylamino-l-phenyl-4-nitrobutadiene	Acetone	19.6	28,000	1.064	EFISH	43
DMSM	DMF	11.8		1.90	THG	50
DEANS	Neat	1300		1.90	THG	51
DEANST	Neat	740		1.064	THG	53
	PMMA	200		1.064	THG	53
NPAN	Chloroform		40	1.06	EFISH	54
			30	1.32	EFISH	54
			47	1.437	THG	54
			31	1.907	THG	54
NPP	Chloroform		80	1.06	EFISH	54
			50	1.32	EFISH	54
			103	1.437	THG	54
			48	1.907	THG	54
PNP	Chloroform		30	1.06	EFISH	54
			30	1.32	EFISH	54
			78	1.367	THG	54
			42	1.437	THG	54
			31	1.907	THG	54
MONS	Dioxane		60	1.064	THG	39
			428	1.064	EFISH	39
OOCB	Dioxane		79	1.064	THG	39
			203	1.064	EFISH	39

Table 5 *Continued*

Compound	Solvent	$\chi^{(3)}$	γ	Wavelength (μm)	Technique	Ref.
4-Dimethylaminostilbene	Benzene	30	470	1.064	EFISH	43
4-Chlorostilbene	Benzene	11.6	27	1.064	EFISH	43
4-Chloro-4'-nitrostilbene	Chloroform	21	690	1.064	EFISH	43
4-Chloro-4'-dimethylaminostilbene	Chloroform	33	1,000	1.064	EFISH	43
4-Nitro-4'-aminostilbene	Acetone	58	9,900	1.064	EFISH	43
DANS	Acetone	22	17,000	1.06	EFISH	35
	Acetone	24.3	17,000	1.064	EFISH	43
	trans-Stilbene	131	990		WM	49

p-Nitroaniline

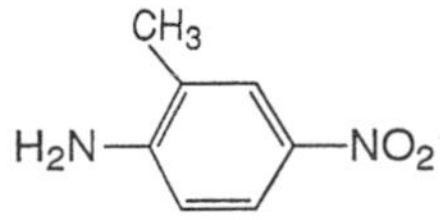

2-Methyl-4-nitroaniline

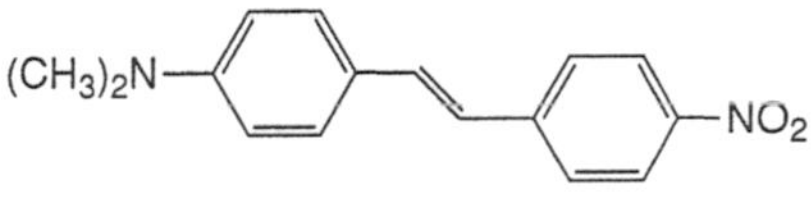

4-Dimethylamino-4'-nitrostilbene (DANS)

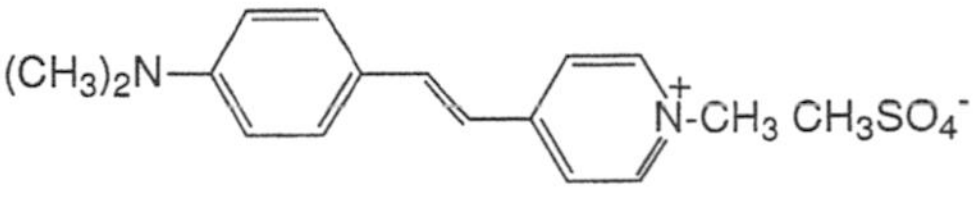

trans-4'-Dimethylamino-*N*-methyl-4-stilbazolium methylsulfate (DMSM)

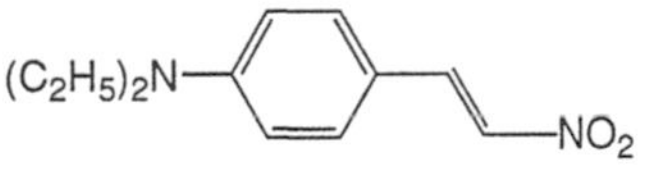

4-(*N,N*-Diethylamino-4'-nitrostyrene (DEANST)

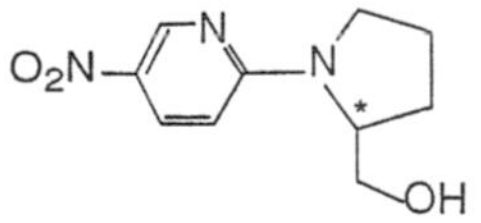

2-(*N*-Prolinol)-5-Nitropyridine (PNP)

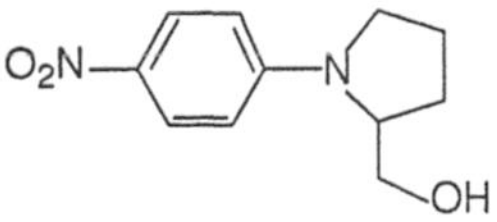

N-(4-nitrophenyl)-L-prolinol (NPP)

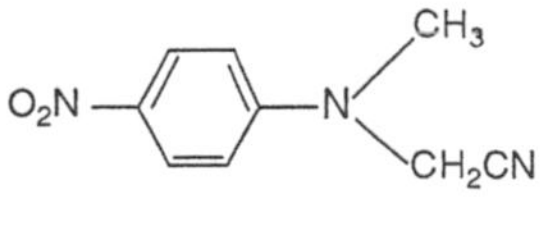

N-(4-Nitrophenyl)-4-methylamino-acetonitrile (NPAN)

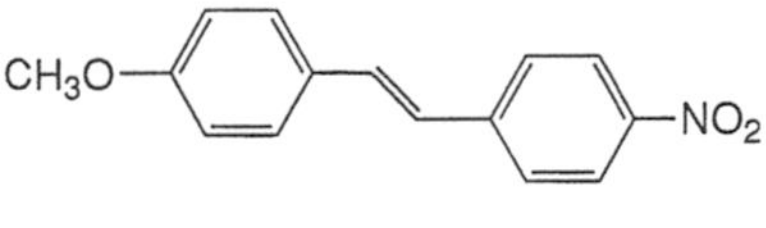

4,4'-Methoxynitrostilbene (MONS)

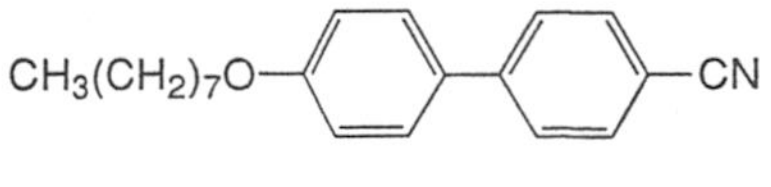

4,4'-Octyloxycyanobiphenyl (OOCB)

bonds. Conjugated systems of allo-ocimene, *trans*-retinol, *trans*-retinal, *trans*-β-carotene, and dodeca-preno-β-carotene possess 3, 5, 6, 11, and 19 double bonds in their backbone, respectively. The γ-values of allo-ocimene and dodecapreno-β-carotene are 9.7×10^{-36} esu and 1.7×10^{-32} esu, respectively, indicating a strong influence of delocalized π-electrons on third-order optical nonlinearity. The molten retinol and retinal exhibit electronic susceptibilities of 0.5×10^{-12} and 1.1×10^{-12} esu, respectively.

Sakai et al.[58] measured the second hyperpolarizability values of retinal derivatives by using the DFWM technique at 532 nm in dimethyl sulfoxide (DMSO). The chemical structures of three retinal derivatives are shown below together with their optical absorption maxima and optical nonlinearities. These organic molecules were synthesized from all-*trans*-retinal by the Knoevenagel reaction, except the first one which was obtained by the general method for Schiff base synthesis. The hyperpolarizabilities of these retinal derivatives were very large on resonance and there was a significant difference due to the nature of different substituent functionalities. The electron acceptor groups are more effective, the conjugated chain being an electron-rich system which acts as an electron donor. The hyperpolarizability increased by about two orders of magnitude as the dimethylamino groups were replaced by strong electron acceptor nitro groups. Depending on the chemical functionalities, the magnitude of resonant second hyperpolarizability was in the range of 10^{-31} to 10^{-29} esu. The large optical nonlinearity originated from electronic contribution. The relaxation time of these retinal derivatives in DMSO solution was shorter than 30 ps.

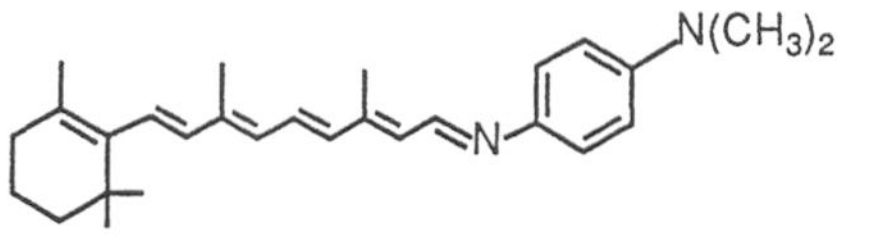

$\lambda_{max} = 440$ nm $\gamma = 0.65 \times 10^{-30}$ esu $\lambda_{max} = 480$ nm $\gamma = 15 \times 10^{-30}$ esu

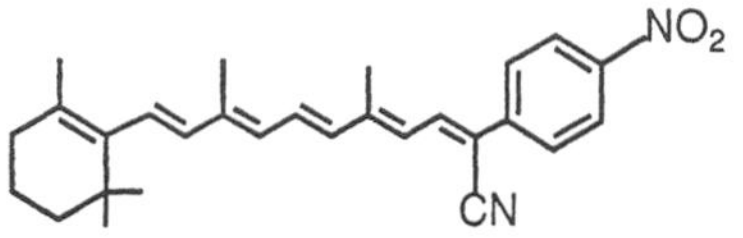

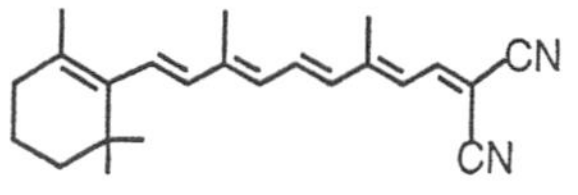

$\lambda_{max} = 470$ nm $\gamma = 1.8 \times 10^{-30}$ esu

Ikeda et al.[59] investigated the third-order optical nonlinearities of asymmetric carbocyanine dyes in solutions by using the DFWM technique. The samples were dissolved in DMSO at a concentration of about 0.001 mol/l and placed in a 1 mm thick quartz tube. The 1,1′,3,3,3′,3′-hexamethyl-2,2′-indotricarbocyanine iodide, 3,3′-diethyl-2,2′-oxotricarbocyanine iodide, and 3,3′-dimethyl-2,2′thiatricarbocyanine iodide show optical absorption maxima at 750, 700, and 770 nm, respectively. The benzothiazole derivative shows a bathochromic shift of 40 nm compared with corresponding benzoxazole derivative. These asymmetric dyes have very large hyperpolarizabilities and the converted γ value gives $\chi^{(3)}$ value of 3×10^{-8} esu for a 0.1 mol/l solution of benzothiazole derivative which is comparable with the organic metal $(BEDT-TTF)_2I_3$. The measurement wavelength of 532 nm is located at the edge of the large absorption band and, therefore, the resonance enhancement is less effective. The relatively small difference in third-order susceptibilities of these structurally similar dyes arises from the electron-donating behavior between oxygen and sulfur; the strong donating activity of oxygen probably forms a local charge transfer in the oxazole ring along the main conjugated chains.

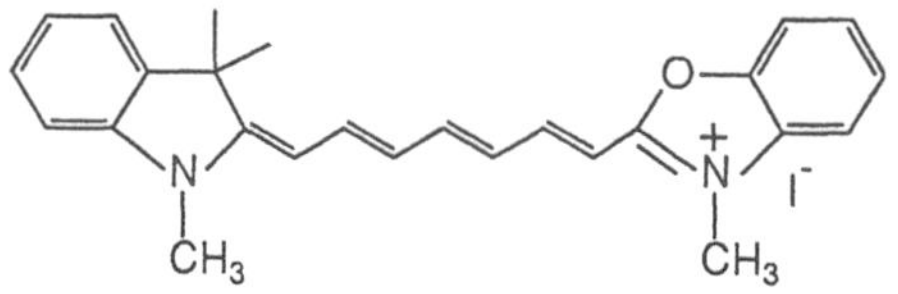

$\lambda_{max} = 720$ nm $\gamma = 3.0 \times 10^{-28}$ esu $\chi^{(3)} = 3.0 \times 10^{-8}$ esu

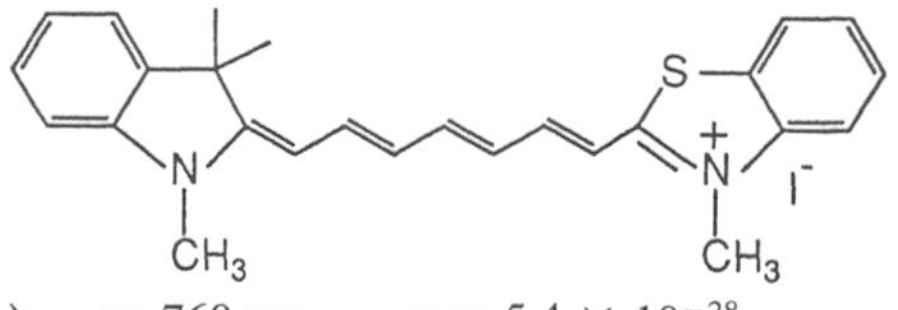

$\lambda_{max} = 760$ nm $\gamma = 5.4 \times 10^{-28}$ esu

Tomiyama et al.[60a] investigated third-order nonlinear properties of merocyanines to establish a relationship between the molecular aggregation and optical nonlinearity. Merocyanines investigated were 3-carboxymethyl-5-[2-(3-octadecyl-2-benzothiazolinylidene)ethylidene-rhodanine (referred to as MCS, where X = S) and 3-carboxymethyl-5-[2-(3-octadecyl-2-benzoselenazolinylidene)ethylidene-rhodanine (referred to as MCSe, where X = Se). The chemical structures of the merocyanines used are shown below. Thin films of MCS and MCSe were prepared by spin coating from chloroform solutions on fused silica glass slides. The color of these films changed from red to bluish purple when they were kept in an ammonia atmosphere at 60 °C for 12 h. The absorption band of ammonia-treated films was shifted to the lower energy side and, in addition, a new sharp absorption peak appeared at 612 nm called a J-band, which demonstrated the formation of J aggregates. Both MCS and MCSe show similar features. The vapor-deposited films can be converted completely to J aggregates, whereas, for spin-coated films, a complete conversion was difficult. The refractive indexes of J-aggregated films are larger than the nonaggregated films and becomes larger at the absorption edge. The nonlinear optical susceptibilities measured by the THG method at different wavelengths are listed in Table 6. The $\chi^{(3)}$ values for J-aggregated MCSe films are three times larger at the nonresonant region and ten times larger at the resonant region than the $\chi^{(3)}$ values of nonaggregated films. The maximum $\chi^{(3)}$ value of a J-aggregated MCSe film was 9×10^{-11} esu at 620 nm of third-harmonic wavelength. The MCS films also show similar enlargement. Therefore, intermolecular interaction (J aggregation) is an effective method for enhancing $\chi^{(3)}$ values. Tomiyama et al.[60b] reported $\chi^{(3)}$ values of 1.7×10^{-11}, 7×10^{-12}, and 4×10^{-12} esu at 1.5, 1.9, and 2.1 μm, respectively, for crystal violet.

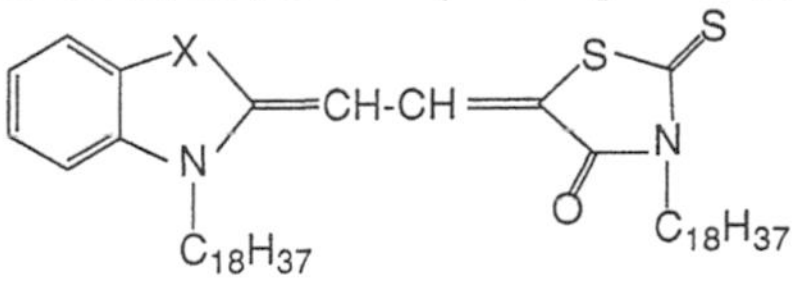

$$X = \text{S; MCS}$$
$$X = \text{Se; MCSe}$$

Kajzar et al.[61] measured the hyperpolarizability of Langmuir-Blodgett (LB) films of hemicyanine and Gedye merocyanine (protonated form). The LB films were deposited on both sides of the hydrophilic silica substrate. Figure 4 shows the hyperpolarizability values of the LB films relative to that of the silica substrate. The hyperpolarizability values measured at 1.064 μm are larger than those of 1.907 μm, resulting from the three-photon resonance. The hemicyanine LB film has a $\chi^{(3)}$ value of 1.4×10^{-12}, and 0.90×10^{-12} esu at 1.064 and 1.907 μm, respectively, which are about two orders of magnitude larger than silica and also larger that those of LB films of Gedye merocyanine.

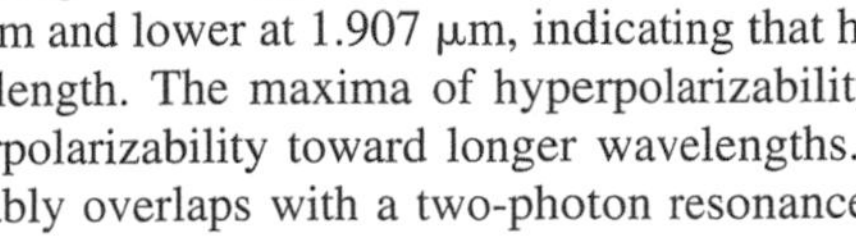

(Hemicyanine) 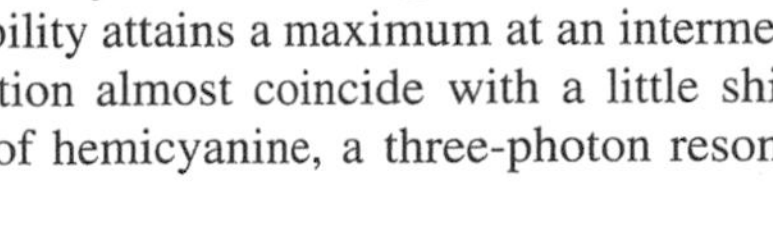   (Gedye merocyanine)

Figure 5 shows the wavelength dependence of hyperpolarizability of the hemicyanine LB films relative to the values of silica and the absorption spectrum of the dye layer plotted against the third-harmonic wavelength. The hyperpolarizability values increase systematically with wavelength and are larger near 1.4 μm and lower at 1.907 μm, indicating that hyperpolarizability attains a maximum at an intermediate wavelength. The maxima of hyperpolarizability and absorption almost coincide with a little shift of hyperpolarizability toward longer wavelengths. In the case of hemicyanine, a three-photon resonance probably overlaps with a two-photon resonance.

Table 6 Third-order Nonlinear Susceptibility of MCS and MCSe Films

Dye	Aggregation	$\chi^{(3)}$ (10^{-12} esu) at Wavelengths (μm)					
		1.5	1.65	1.68	1.8	1.86	2.10
MCS	Nonaggregation	0.9	2.6	–	1.6	1.5	1.0
	J aggregation	5.4	32.2	–	34.2	59.6	3.4
MCSe	Nonaggregation	1.9	–	7.3	3.2	3.0	1.7
	J aggregation	11.1	–	78.9	91.4	93.5	3.9

Courtesy of H. Tomiyama.

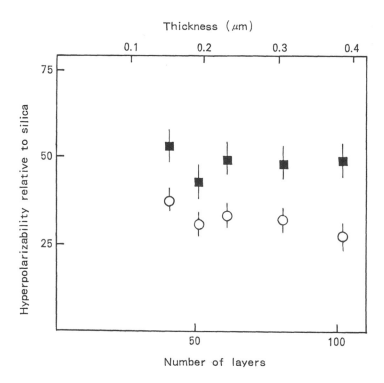

Figure 4 Hyperpolarizability values of the merocyanine LB films relative to the hyperpolarizability value of silica substrate as a function of film thickness in monolayers: (■) at 1.064 μm and (○) at 1.907 μm. (From Kajzar et al., *Thin Solid Films*, 160, 209, 1988. With permission.)

Stevenson et al.[62] reported polarizability and second hyperpolarizability of simple polyenes and cyanines measured by THG in solutions at 1.908 μm. Table 7 lists the THG optical nonlinearities of several conjugated compounds. Both α and γ values were smaller in tetramethylpolyenes than in the cyanines and an hormonicity was increased. The α values were slightly reduced on symmetric aza substitution of cyanines. The α and γ increase as the π-conjugation increases for polyenes 2,5-dimethyl-2,4-hexadiene (TMP-2), 2,7-dimethyl-2,4,6-octatriene (TMP-3), 2,9-dimethyl-2,4,6,8-decatetraene (TMP-4), and 2,3-dimethyl-1,3-butadiene (DMP-2).

Kobayashi and Sasaki[63] measured third-order NLO properties of pseudoisocyanine (PIC) J aggregates by absorption saturation and DFWM techniques. The DFWM $\chi^{(3)}$ values of J aggregates were 2.9 × 10^{-8} and 9.0 × 10^{-9} n_0^2 esu for PIC-Br and PIC-I, respectively. Blau[64] and Maloney et al.[65] measured third-order susceptibility and hyperpolarizabilities of a series of organic dyes. The abbreviated names, optical absorption maxima, and third-order optical nonlinearities are listed in Table 8. The chemical structures of these organic conjugated molecules are shown along with. For optical measurements, the DTTC and DNTPC were dissolved in methanol, A9860, IR5, and S501 in 1,2-dichloroethane, BDN in toluene, nigrosine in water, and β-carotene in ethanol. The optical nonlinearities are influenced by the change of the chemical structure and S501 shows the largest value. Penzkofer and Leupacher[66] measured optical nonlinearities of a variety of organic dye solutions by using the THG technique (Table 9).

Matsumoto et al.[67] reported the powder THG intensity of a series of phenyl azobenzene, stilbene, and benzylidene derivatives after esterification (Table 10). The THG intensity was determined by the powder technique with particle size of ~100 μm and for DMF solutions at a concentration of 2 wt%. The THG intensities of ester derivatives were found to be much larger than those of nonesterified compounds. Sulfonyl esterification yielded smaller enhancement in THG. The THG of methylbenzoate was 110 times that of *p*-NA. Esterified stilbene derivatives also showed larger THG values than those of nonesterified compounds. The THG enhancement was probably caused by the change in the crystal structure or molecular alignment induced by esterification. This study demonstrated that large THG can be achieved even in shorter π-electron-conjugated systems by molecular engineering.

The third-order NLO properties of the solutions of aromatic acetylenes, imdidazoles, azo dyes, hellicene, and other π-conjugated compounds measured by optical Kerr effect near the absorption edge have been reported by the research team at Ube industries.[68–72] Table 11 lists the $\chi^{(3)}$ values and optical properties of these compounds. Hellicenes show large $\chi^{(3)}$ values on the order of 10^{-10} esu. In particular, 5-nitrohexahellicene and tridecahellicene showed largest optical nonlinearity. The $\chi^{(3)}$ values increase with changing helicene structure, and the wavelength dependence of $\chi^{(3)}$ showed the small resonant

Harmonic wavelength (μm)

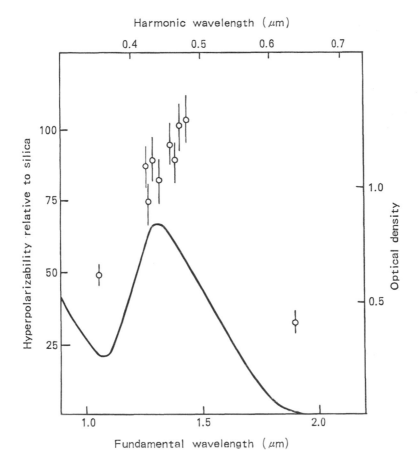

Figure 5 Wavelength dependence of hyperpolarizability of the hemicyanine LB films relative to that of silica and the solid line is absorption spectrum of the dye layer plotted against the third-harmonic wavelength. (From Kajzar et al., *Thin Solid Films*, 160, 209, 1988. With permission.)

enhancement. The 100% converted $\chi^{(3)}$ values were estimated in the range of 10^{-8} to 10^{-12} esu depending on the chemical structures and were larger than the nonresonant $\chi^{(3)}$ values. The response time was different for different optical processes associated with the induced refractive index.

Bollaert et al.[73] synthesized a trifunctional, highly symmetric probe, tris(p-isocyanatophenyl)amine (TIPA) and its butanol adduct TU. TIPA acts as a chemical sensor and as a cross-linking site; therefore, this chromophore can be incorporated into polyurethanes in large concentration. TIPA and TU show absorption maxima at 316 and 313 nm in acetonitrile, respectively. The γ values of 78×10^{-36} and 146×10^{-36} esu were measured using an optical phase conjugated interferometer at 532 nm for TIPA and TU in 1,2-dichloroethane solution, respectively.

Table 7 Polarizability and Second Hyperpolarizability of Simple Polyenes and Cyanines Measured by THG in Solutions at 1.908 μm

Compound	α (10^{-23} esu)	γ (10^{-36} esu)
$(CH_3)_2N$~~~$N(CH_3)_2^+$ Cl^-	2.0 ~ 2.1	0.53 ± 0.4
$(CH_3)_2N$~~~$N(CH_3)_2^+$ Cl^-	2.95	52 ± 1
$(CH_3)_2N$~~($)_n$~~$N(CH_3)_2^+$ Cl^- $n = 3,4$	3.95	−510 ± 50
(bis-thiazolidine cyanine, N^+ I^-, CH_3 CH_3)	3.15	16 ± 5
(bis-thiazolidine aza, N^+ I^-, CH_3 CH_3)	2.86	110 ± 80
(bis-thiazolidine trimethine, N^+ I^-, CH_3 CH_3)	4.3 ~ 4.6	−160/−50 ± 7
(bis-benzothiazole cyanine, N^+ I^-, CH_2-CH_3 CH_2-CH_3)	5.3	−30 ± 30
(bis-benzothiazole aza, N^+ I^-, CH_2-CH_3 CH_2-CH_3)	4.92	490 ± 40
(bis-benzothiazole trimethine, N^+ I^-, CH_2-CH_3 CH_2-CH_3)	10.8	−470 ± 300
DMP-2	1.13	4.9 ± 0.1
TMP-2	1.57	7.4 ± 0.1
1,3,5-Hexatriene	1.18	9.1 ± 0.4
TMP-3	2.55	38 ± 0.5
TMP-4	6.0	400 ± 50
β-Carotene	127	11,000 ± 2,500

From Stevenson et al. *Mater. Res. Soc. Symp. Proc.*, 1988. With permission.

Table 8 Chemical Structure, Absorption Maxima, and Optical Nonlinearities of Organic Dyes

Abbreviated Names	λ_{max} (μm)	$\chi^{(3)}$ (10^{-12} esu)	γ (10^{-30} esu)
β-Carotene	0.48	0.14	0.7
Nigrosine	0.56	1.8	5
DTTC	0.74	0.57	3.5
DNTPC	0.98	0.73	23
BDN	1.03	1.2	61
A9860	1.06	1.3	200
IR5	1.08	1.5	140
S501	1.47	9.0	90

From Blau, *Phys. Technol.*, 18, 250, 1987; Maloney et al., *Chem. Phys.*, 121, 21, 1988. With permission.

Nigrosine

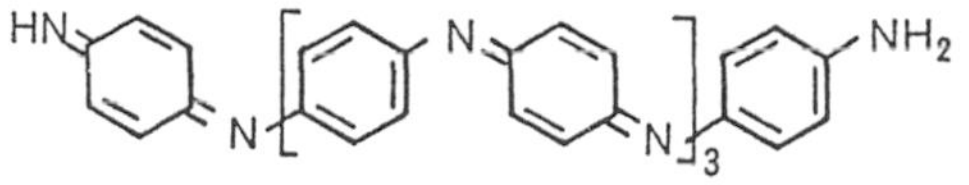

DTTC

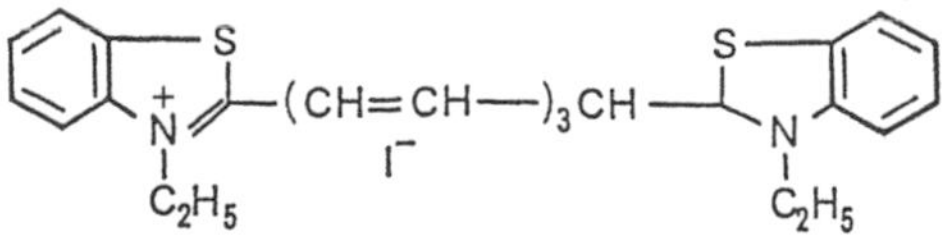

DNTPC

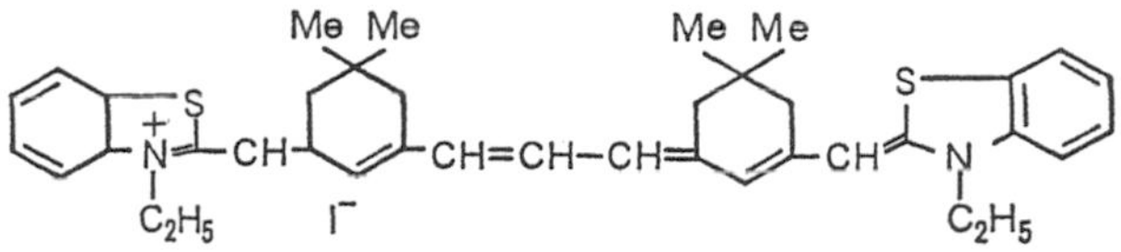

BDN

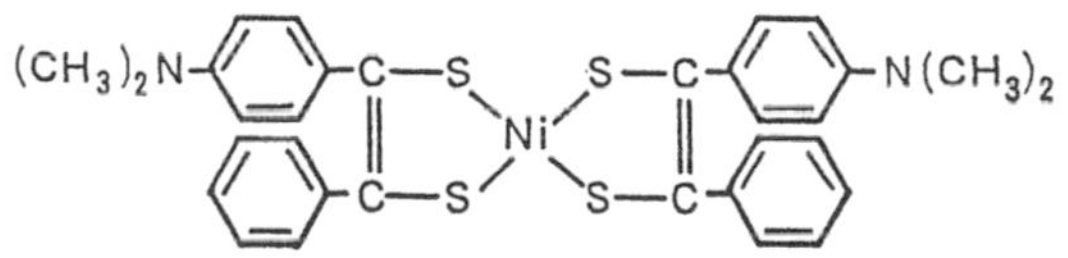

A9860

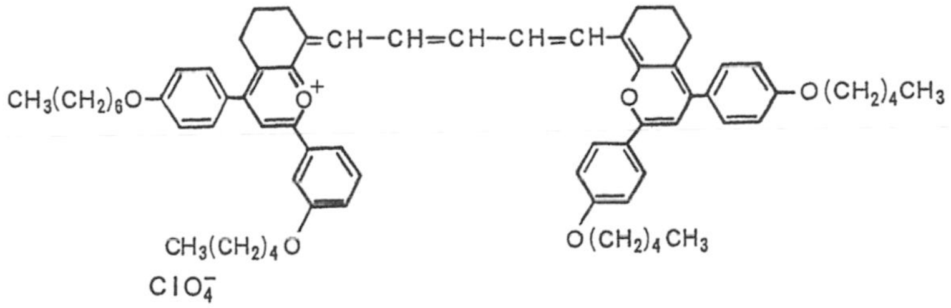

IR 5

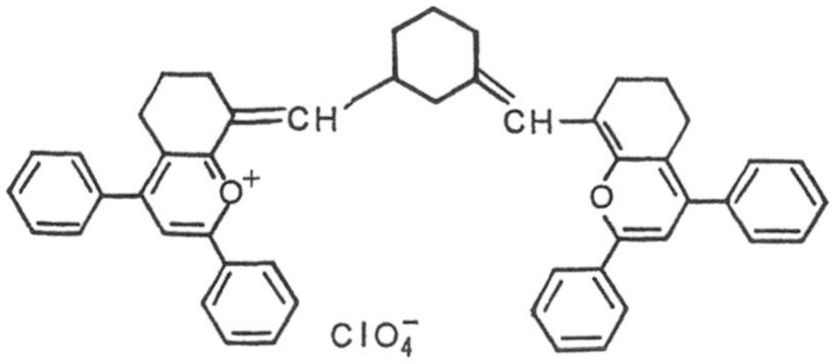

S 501

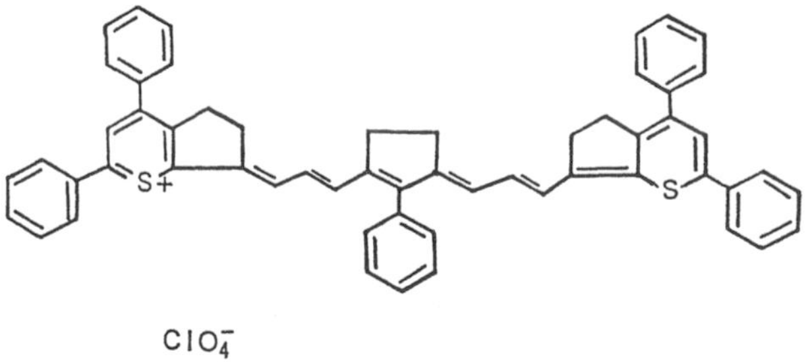

Table 9 Third-Order Nonlinear Susceptibility $\chi^{(3)}$ and Hyperpolarizability γ of Some Organic Dyes

Dye	Solvent	$\chi^{(3)}$ $(10^{-22}\ m^2V^{-2})^a$	γ $(10^{-60}\ cm^4V^{-3})^a$
Fuchsin	Methanol	5.1	12
Rhodamine 6G	Methanol	8.0	14
Methylene blue	Methanol	3.2	4.0
Safranine T	Hexafluoroisopropanol	1.7	3.3
PYC[b]	Hexafluoroisopropanol	2.0	17
HMICI[c]	Hexafluoroisopropanol	2.48	20

[a] Conversion factors: $\chi^{(3)}$ (esu) = $(9 \times 10^8/4\pi)\chi^{(3)}$ (SI); γ (esu) = $8.088 \times 10^{24}\ \gamma$ (SI).

[b] PYC = 1,3,1',3'-tetramethyl-2,2'-dioxopyrimido-6,6'-carbocyanine hydrogen sulfate;

[c] HMICI = 1,3,3,1',3',3'-hexamethylindocarbocyanine iodide.

From Penzkofer and Leupacher, *Organic Materials for Non-linear Optics*, Special Publ. No. 69, p. 203, 1989. With permission.

Table 10 THG Activity of Phenyl Azobenzene, Stilbene, and Benzylidene Derivatives

$$O_2N-\langle\bigcirc\rangle-A{=}B-\langle\bigcirc\rangle-N\begin{smallmatrix}CH_2\text{-}CH_3\\CH_2\text{-}CH_2\text{-}R\end{smallmatrix}$$

Compounds			Powder THG ($\times$ p-NA = 1) at Wavelength (μm)			THG in Solution (1.9 μm)
A	B	R	1.064	1.9	2.05	
N	N	OH	2	2	4.8	1.0
N	N	OCOCH$_3$	11	22	–	0.93
N	N	OCO$-$C$_6$H$_4$$-NO_2$	21	31	57	–
N	N	OCOCH$=$CH$_2$	17	–	60	0.93
N	N	OSO$_2$$-C_6H_4$$-CH_3$	–	4	–	–
N	N	OSO$_2$$-C_6H_4$$-NO_2$	–	3	–	–
CH	N	OH	25	17	–	1.0
CH	N	OCOCH$_3$	38	50	–	0.98
CH	N	OCO$-$C$_6$H$_4$$-NO_2$	43	57	44	0.92
CH	N	OCO$-$C$_6$H$_4$$-CH_3$	–	110	–	–
CH	N	OSO$_2$$-C_6H_4$$-CH_3$	–	7	–	–
CH	N	OSO$_2$$-C_6H_4$$-NO_2$	–	8	–	–
CH	CH	OH	14	8	–	2.1
CH	CH	OCOCH$_3$	33	20	–	1.8
CH	CH	OCO$-$C$_6$H$_4$$-NO_2$	18	27	23	2.0
CH	CH	OCOCH$=$CH$_2$	48	–	–	–

From Matsumoto et al. *Nonlinear Optics of Organics and Semiconductors*, T. Kobayashi, ed., Springer-Verlag, Berlin, 36, 236–239, 1989. With permission.

Hattori et al.[74] reported the third-order NLO properties of 1,4-dithio-keto-3,6-diphenyl-pyrrolo-[3,4-c]-pyrrole (DTPP) and 2,6-di(n-butylamino)-4,8-dihydroxy-1,5-naphthoquinone (d-BAHNQ). Both DTPP and d-BAHNQ formed hydrogen bonds in the crystal form. THG measurements on thin films yielded $\chi^{(3)}$ values of 4.8×10^{-11} esu for d-BAHNQ at 1.9 μm and 1.7×10^{-11} esu for DTPP at 2.1 μm. The $\chi^{(3)}$ of both d-BAHNQ and DTPP had a three-photon resonance enhancement because λ_{max} occured at 620 nm for d-BAHNQ and at 680 nm for DTPP thin films. Intermolecular hydrogen bonds were formed between O and NHC$_4$H$_9$ groups in d-BAHNQ; and, in the case of DTPP, sulfur and nitrogen formed intermolecular hydrogen bonds. Both of these molecules formed a molecular strip through intermolecular hydrogen bonds and the molecules aligned with their

Table 11   The $\chi^{(3)}$ Values of Aromatic Acetylenes, Imdidazoles, Azo Dyes, Hellicene, and Other π-Conjugated Compounds Measured by the Optical Kerr Effect

Material	Solvent	λ_{max} (nm)	Measurement Wavelength (nm)	$\chi^{(3)}$ (10^{-12} esu)	α (cm^{-1})
1	THF	435	644	30	20
2	THF	427	625	50	10
3	THF	369	555.7	40	20
4	THF	368	520.1	13	20
5	THF	466	690.5	No signal	20
6	THF	440	644	30	30
7	THF	447	560	50	30
8	THF	292	662.6	No signal	20
9	DMSO	398	555.7	30	10
10	THF	493	625	40	20
11	THF	–	490.5	60	30
12	THF	318	545	70	10
13	THF	251	486	20	10
14	THF	416	625	30	30
15	THF	462	625	10	20
16	DMSO	496	709.6	70	40
17	THF	446	555.7	30	10
18	THF	440	560	30	20
19	THF	–	662.6	50	20
20	THF	393	490.5	160	920
21	THF	415	550	220	50
22	THF	294	415.1	30	<2
23	ACN	447	690.5	70	40
24	THF	454	662.6	60	20
25	THF	402	510.1	80	70
26	THF	386	500	120	20
27	THF	443	650	50	50
28	THF	436	464.4	150	200
29	THF	458	513.3	160	100
30	THF	546	602.5	80	190
31	DMF	490	550	440	60
32	DMF	586	690.5	300	80
33	DMF	426	625	30	20
34	THF	396	458	20	120
35	THF	414	566.8	70	90
36	THF	440	560	30	20
37	THF	470	691.8	10	50
38	THF	498	700.1	3	220
39	THF	507	684	20	80
40	DMF	423	644	10	20
41	THF	385	490.5	15	20
42	THF	408	540.3	60	20
43	THF	414	435	20	10
44	THF	395	435	90	20
45	THF	415	435	50	23
46	THF		540	60	20
47	THF	442.5		1000	–
48			440	10000	20

From Morita et al., *J. Photopolym. Sci. Technol.*, 6, 229, 1993; Ashitaka, *2nd Symposium on Photonics Materials*, 1991; Ashitaka et al., *Sen-Gakkai Symp. Preprints*, 1992; Ashitaka et al., *Nonlinear Opt.*, 4, 281, 1993; Ashitaka and Sasabe, *Nonlinear Opt.*, 14, 81, 1995. With permission.

The chemical structures of the materials listed in Table 11 are as follows:

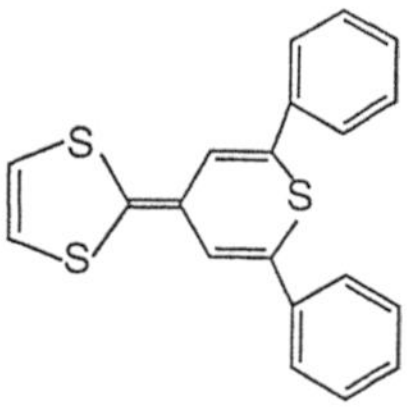

1

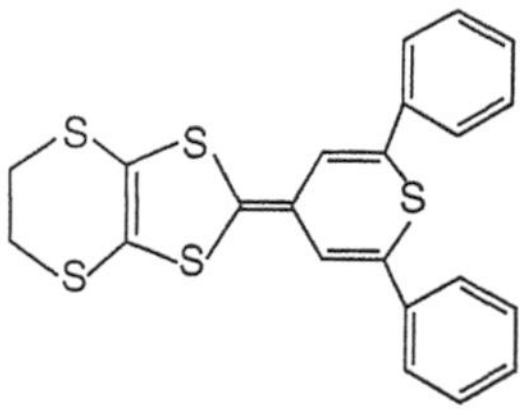

2

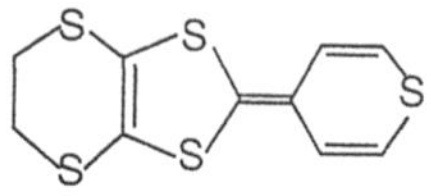

3

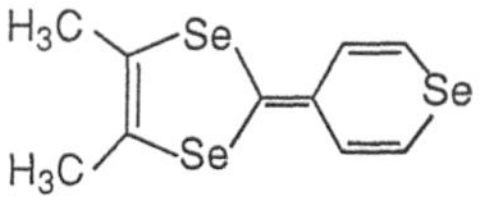

4

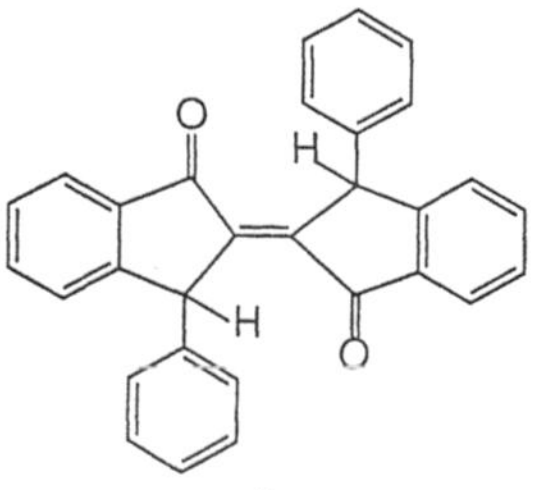

5

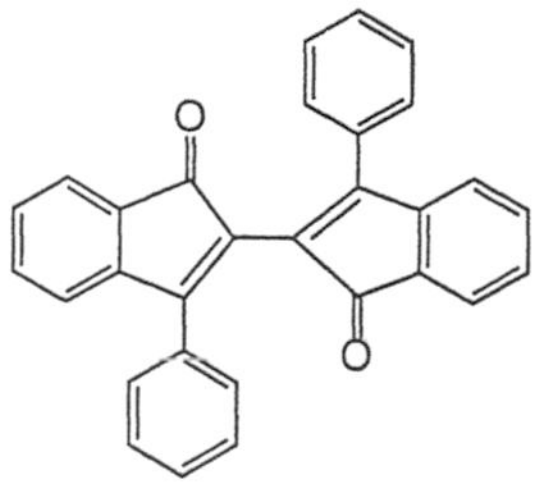

6

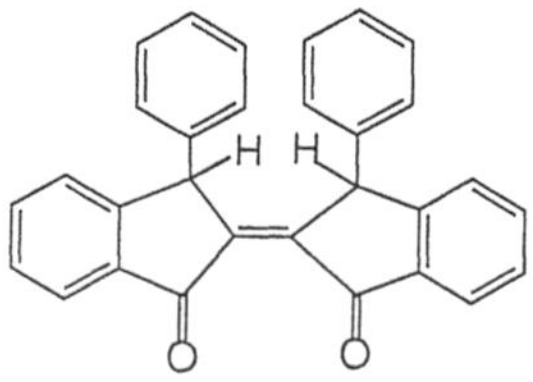

7

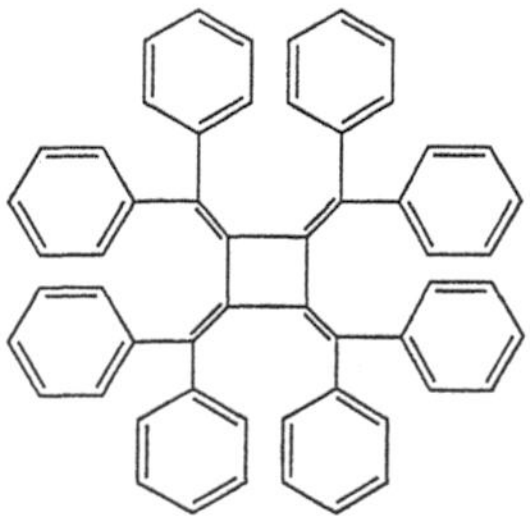

8

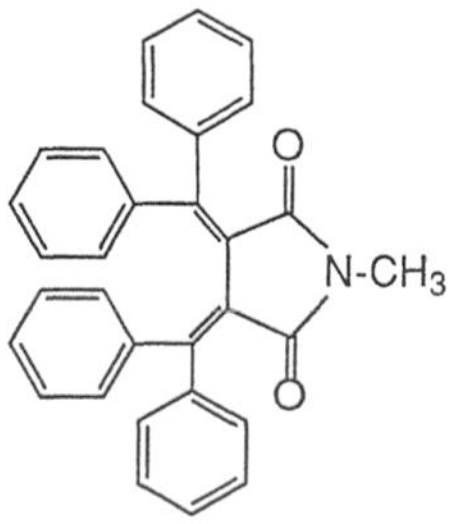

9

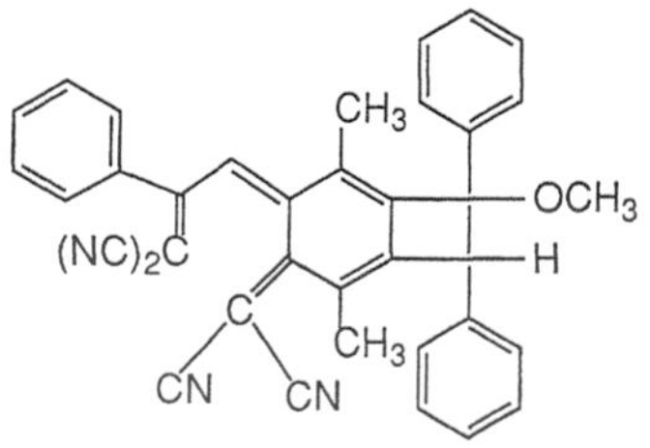

10

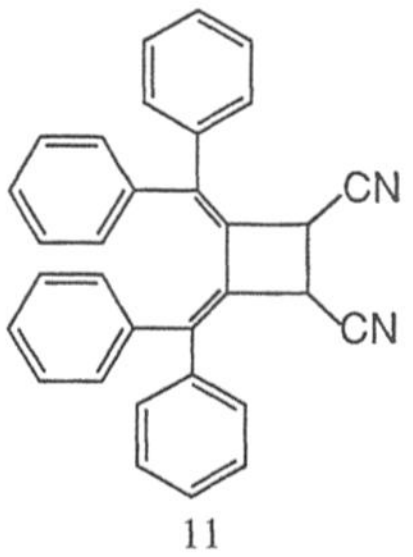

11

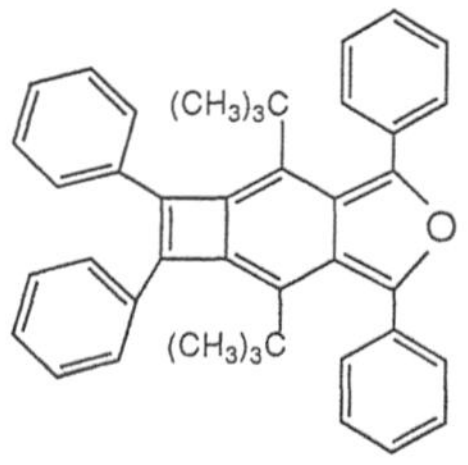

12

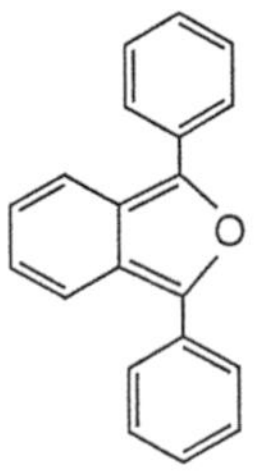

13

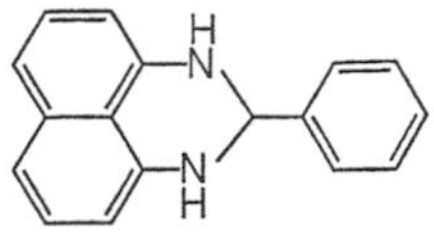

14

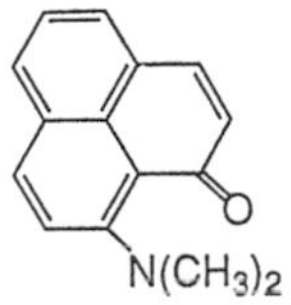

15

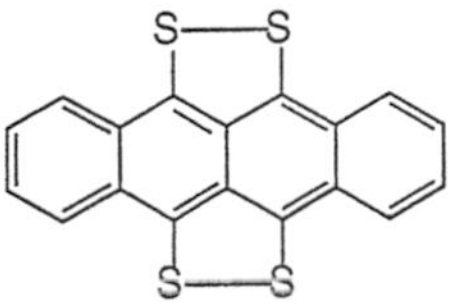

16

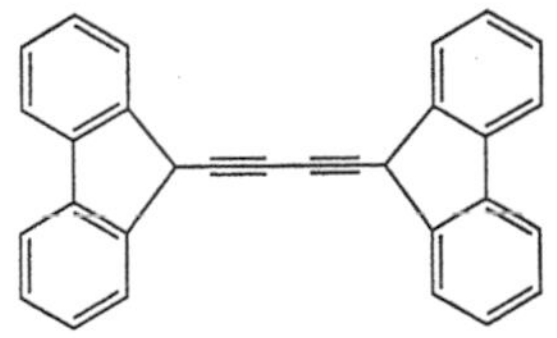

17

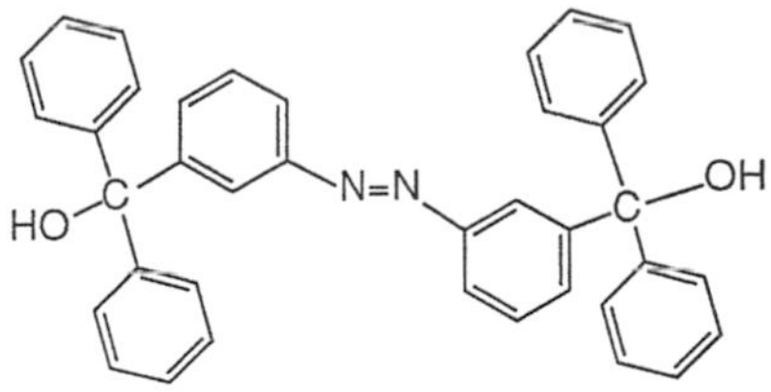

18

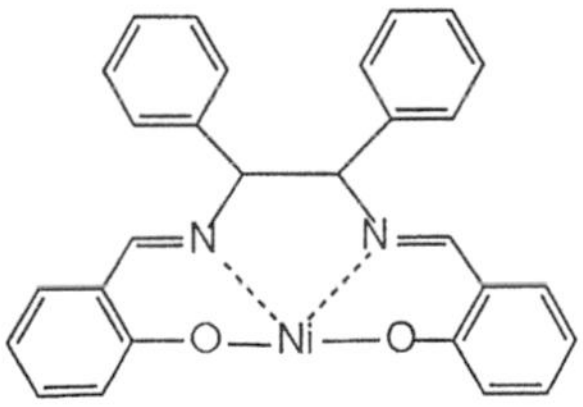

19

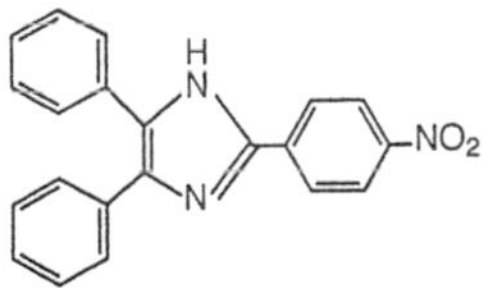

20

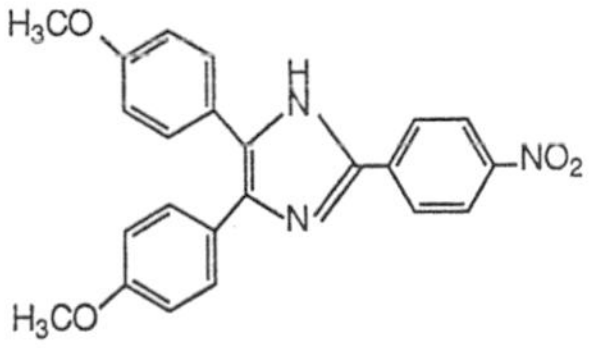

21

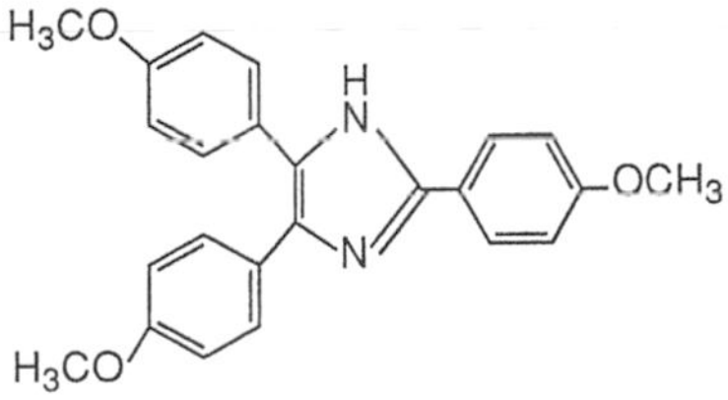

22

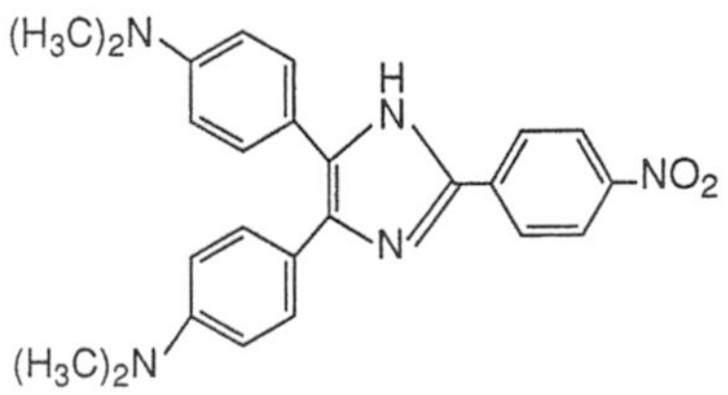

23

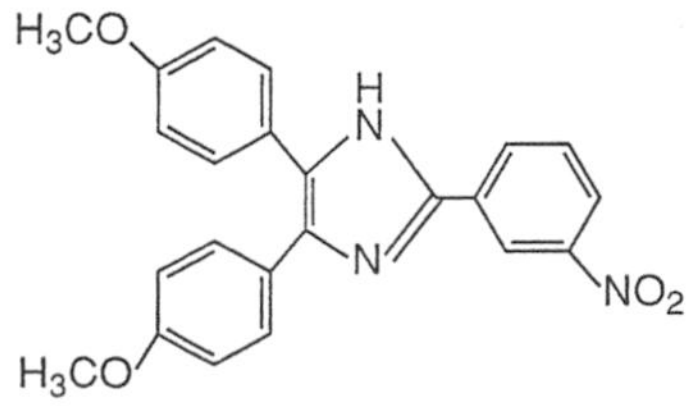

24

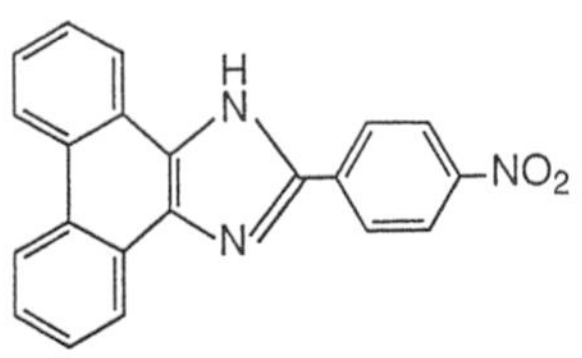

25

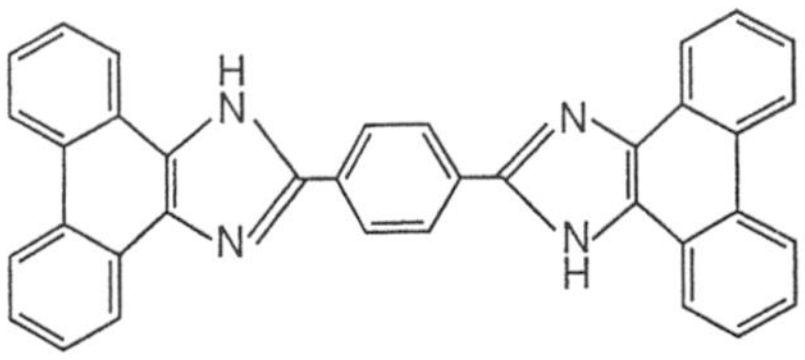

26

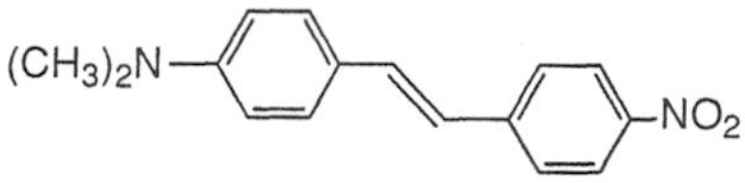

$(CH_3)_2N$—⟨benzene ring⟩—CH=CH—⟨benzene ring⟩—NO_2

27

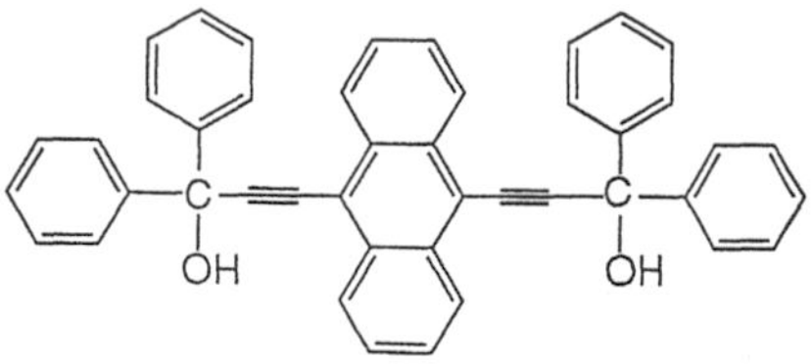

28

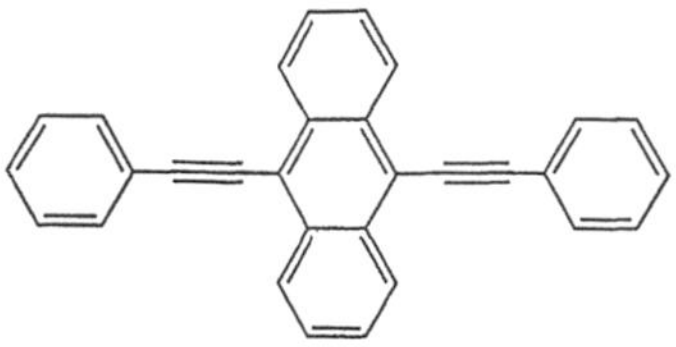

29

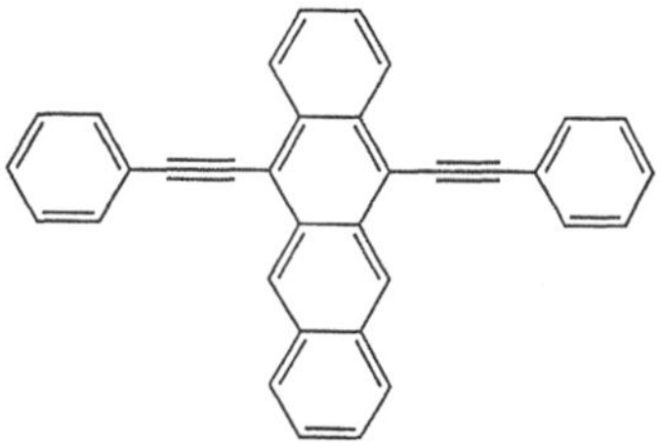

30

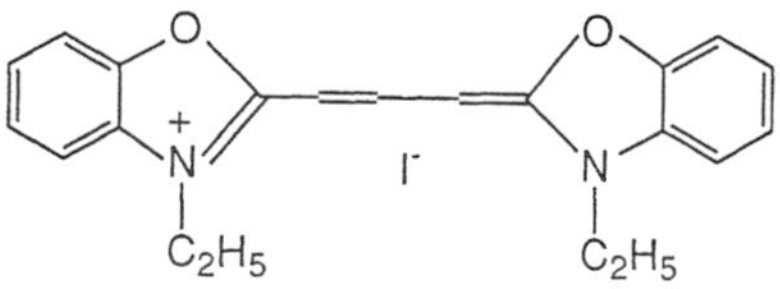

31

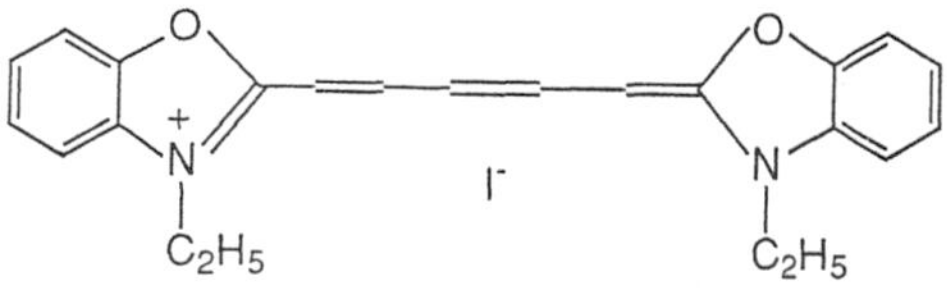

I⁻

32

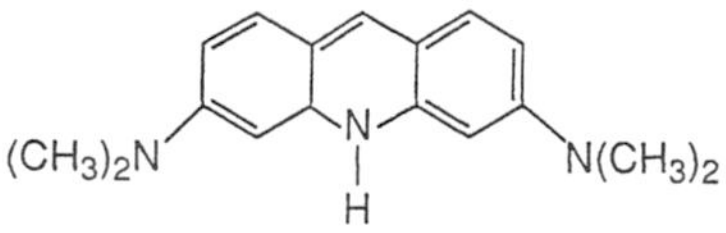

I⁻

33

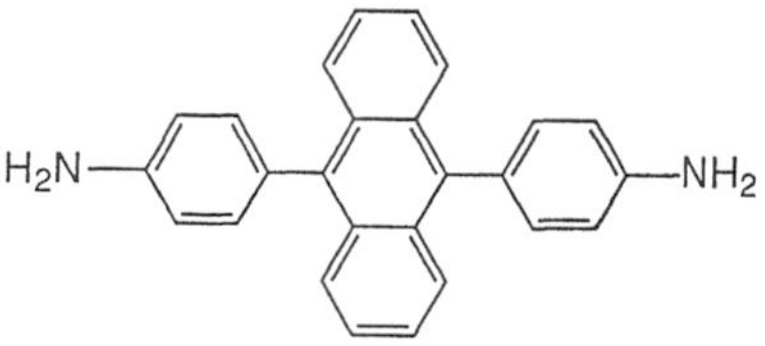

34

35

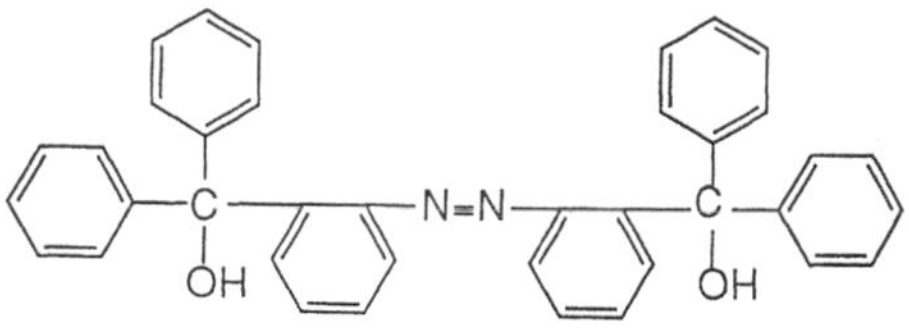

36

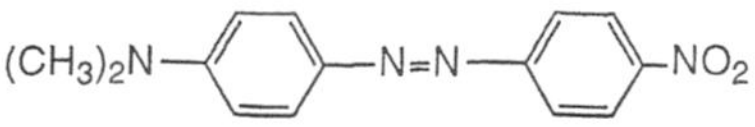

$(CH_3)_2N$—⟨benzene ring⟩—N=N—⟨benzene ring⟩—NO_2

37

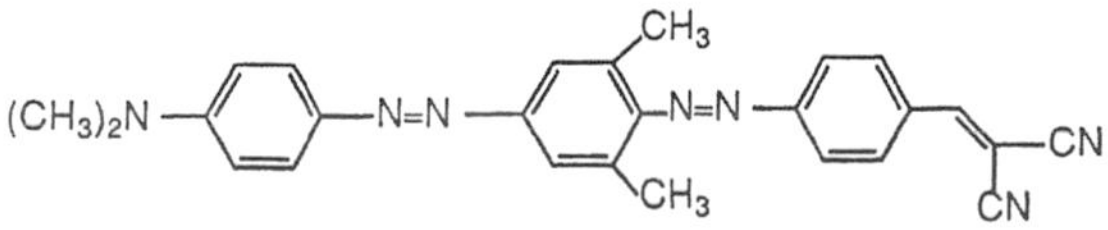

$(CH_3)_2N$—⟨benzene ring⟩—N=N—⟨ring with CH_3⟩—N=N—⟨benzene ring⟩—C=C(CN)(CN)

38

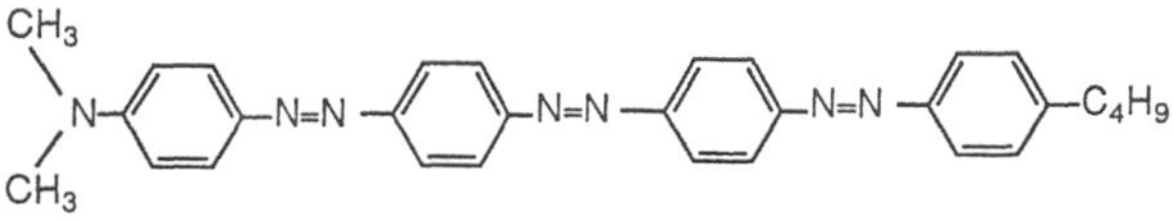

39

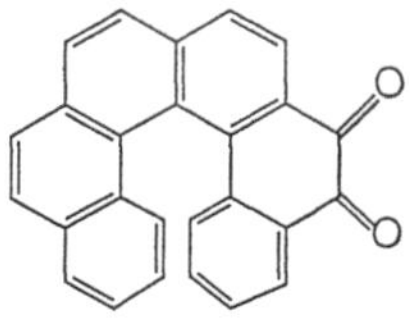

40

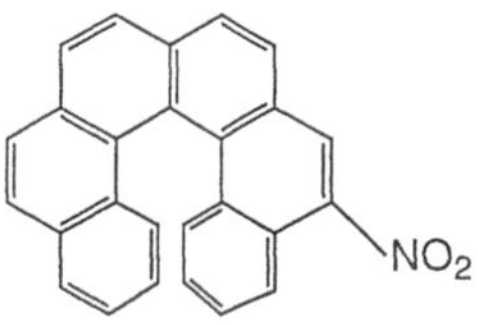

41

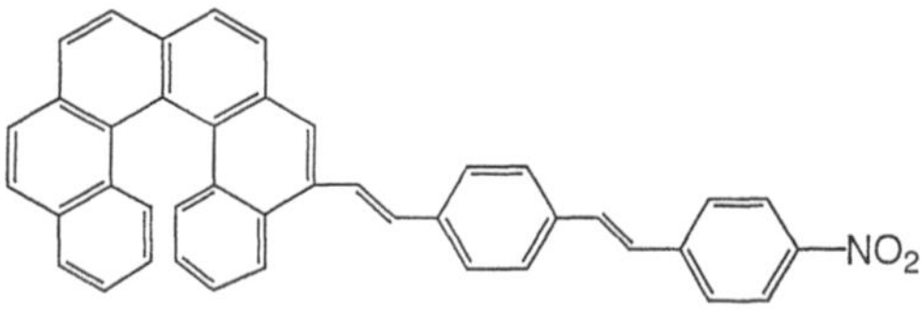

NO_2

42

NO_2

43

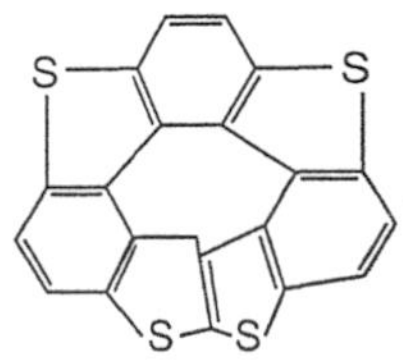

44

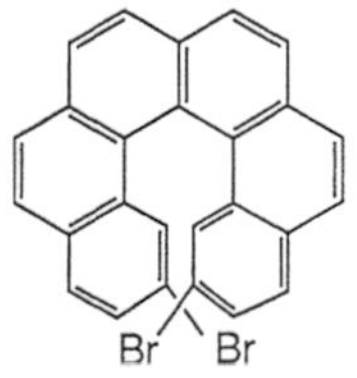

45

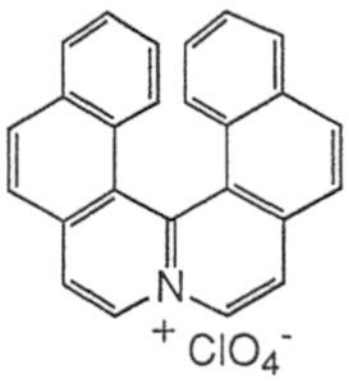

46

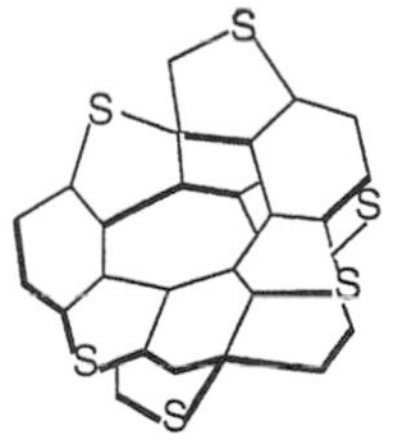

47

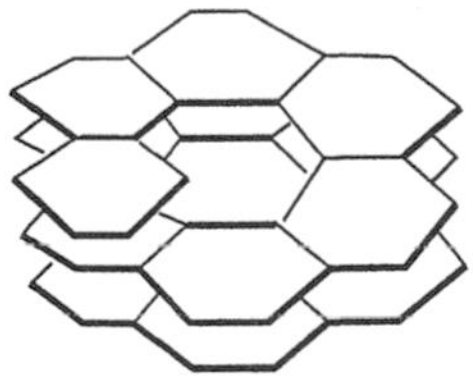

48

conjugated rings in a plane. The strong intermolecular interactions led to large third-order optical nonlinearity similar to π-conjugated polymers.

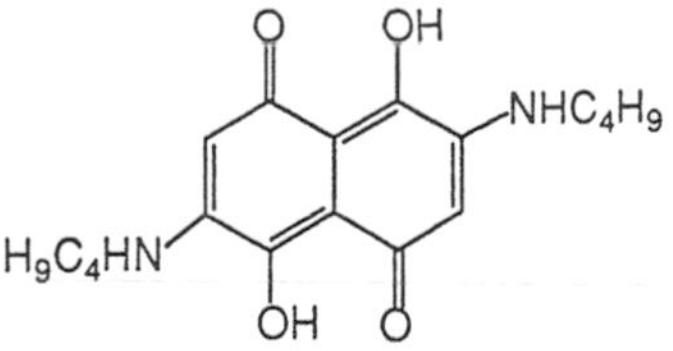

(d-BAHNQ)

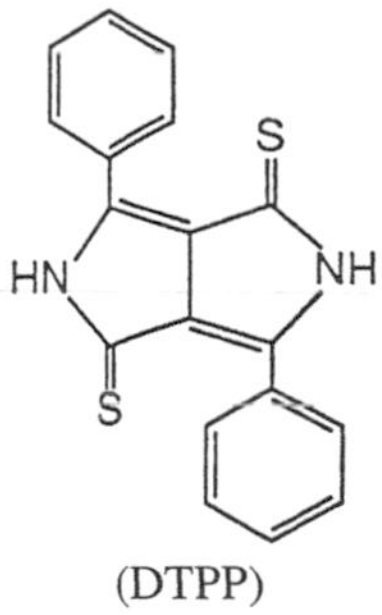

(DTPP)

2. Charge-Transfer Complexes

Besides organic molecules, electron donor and acceptor charge-transfer molecules are also good candidates for showing large third-order optical nonlinearities. In charge-transfer complexes, the optical nonlinearity originates from the supermolecular electronic polarization along the charge-transfer axis. The radical salts of bis(ethylenedithio)tetrathiofulvalene (BEDT-TTF) show two-dimensional electrical conductivity due to the nonplanar donor structure, and the β phase of (BEDT-TTF)$_2$I$_3$ is a superconductor at ambient pressure. The carrier density and delocalization are high in organic metals and encourage the study of third-order NLO properties. Huggard et al.[75] measured the third-order nonlinear susceptibility of one-dimensional α-[bis(ethylenedithio)tetrathiofulvalne] triiodide (α-(BEDT-TTF)$_2$I$_3$ complex by the DFWM technique, which showed a $\chi^{(3)}$ value of ~5 × 10^{-8} esu at 650 nm it strongly depended on the wavelength measurement. The wavelength dispersion of the optical nonlinearity points to a plasma effect because of the conduction electrons. The α-(BEDT-TTF)$_2$I$_3$ showed large $\chi^{(3)}$ values with four components: one was vibrational (Raman) and three other were electronic. The $\chi^{(3)}$ value was studied in the temperature range between 15 and 295 K.[76] The dependence of the decay times on temperature may be associated with the dynamics of the metal–insulator phase transition. The relaxation time at high-temperature phase

295 K was 307, 1705, 9418, and 3800 fs for electronic 1,2,3 and Raman processes, respectively. At low-temperature phase 15 K, the relaxation time was 313, 1,604, 19,157, and 25,000 for electronic 1, 2, 3 and Raman processes, respectively. (BEDT-TTF)$_2$I$_3$ complex is a two-dimensional charge-transfer complex and has a segregated stack structure in which donor (BEDT-TTF) and acceptor iodine form separate donor and acceptor stacks. Troung et al.[76] used the time-resolved, nondegenerate four-wave mixing technique to study the metal–insulator phase transitions of α-(BEDT-TTF)$_4$[Cu(SCN)$_2$]$_2$ and (BEDT-TTF)$_2$I$_3$ complexes. The brown plate crystals of the α-(BEDT-TTF)$_4$[Cu(SCN)$_2$]$_2$ were grown electrochemically by oxidation of BEDT-TTF in 1,1,2-trichloroethane or chlorobenzene. The χ$^{(3)}$ values of α-(BEDT-TTF)$_4$[Cu(SCN)$_2$]$_2$ complex in the high and low temperature (10 K) phases were measured by the DFWM technique with femtosecond time resolution and compared with the (BEDT-TTF)$_2$I$_3$ complex. The DFWM measurements indicated that the χ$^{(3)}$ signal of α-(BEDT-TTF)$_4$[Cu(SCN)$_2$]$_2$ complex was maximum parallel to the axis where the electronic transfer across the BEDT-TTF chain is favored. Both α-(BEDT-TTF)$_4$[Cu(SCN)$_2$]$_2$ and (BEDT-TTF)$_2$I$_3$ complexes showed one electronic component with the same lifetime around 8000 fs at 295 K; and below the metal–insulator phase transition, their lifetime increased to about 18,000 fs. The α-(BEDT-TTF)$_4$[Cu(SCN)$_2$]$_2$ showed two electronic contributions whereas (BEDT-TTF)$_2$I$_3$ complex showed three electronic components. The electronic component with the lifetime of 307 fs was considered to be related to the presence of I$_3^-$ in the (BEDT-TTF)$_2$I$_3$ complex. The second electronic component had lifetime of 950 and 1705 fs for α-(BEDT-TTF)$_4$[Cu(SCN)$_2$]$_2$ and (BEDT-TTF)$_2$I$_3$ complexes, respectively.

Bis(ethylenedithiolo)tetrathiafulvalene (BEDT-TTF)

Rangel-Rojo et al.[77] reported a nonlinear refractive index n_2 of 7.0 × 10^{-3} cm^2/GW for PMMA film doped with DEMI-3CNQ zwitterion using the z-scan technique with 15 ps duration pulses at 532 nm. The DEMI-3CNQ zwitterion was obtained by heat-induced reaction between the TCNQ and triethylammonium (TEA) ions.

Gotoh et al.[78] measured the third-order NLO properties of organic charge-transfer complexes possessing a mixed structure in which organic electron-donors and acceptors stack alternately plane to plane. The THG intensities of the charge-transfer complexes were measured with a power method. The particle size is smaller than 1 μm and poly[2,4-hexadyin-1,6-diol-bis(*p*-toluenesulfonate)] (PDA-PTS) was used as the reference whose THG was taken as unity. The acceptors were tetracyanoethylene (TCNE), 7,7,8,8-tetracyanoquinodimethane (TCNQ), *p*-benzoquinone (BQ), chloranil (CA), 9,10-phenanthrenequinone (PhQ), 2,3-dichloro-5,6-dicyano-*p*-benzoquinone (DDQ), pyromellitic anhydride (PMDA), and 2,4,7-trinitrofluorenone (TNF). The THG intensity of perylene/TCNE and perylene/TCNQ complexes are 4.1 and 2.4 times larger than those of PTS. No THG intensities were observed in perylene, TCNE, and TCNQ molecules. Therefore, supermolecular charge-transfer interactions lead to large third-order optical nonlinearities. A pyrene/PhQ complex showed THG intensity 1.3 times that of PTS. The THG intensity varied with the particle size as well as with the pump-light wavelength, and the THG intensity of complexes were larger in the wavelength region close to the charge-transfer absorption band. The THG intensity of perylene/TCNE complex was larger along the charge-transfer axis by a factor of 100 than the THG intensity perpendicular to the charge-transfer axis which indicated the anisotropy of the third-order optical nonlinearity. The perylene/TCNE and perylene/TCNQ complexes are one-dimensional CT molecules and have mixed-stack structures in which electron donor perylene and acceptors TCNE and TCNQ stack alternatively plane to plane. Gong et al.[79] measured second hyperpolarizability of perylene/TCNE, pyrene/TCNE, and naphthalene/TCNE complexes in tetrahydrofuran solutions by the DFWM method. The typical χ$^{(3)}$ values of 5.5 × 10^{-13} esu at a concentration of 5.2 × 10^{-3} *M* for perylene/TCNE, 5.3 × 10^{-13} esu at 2.4 × 10^{-2} *M* concentration for pyrene/TCNE and 6.0 × 10^{-13} esu at 4.8 × 10^{-2} *M* concentration for naphthalene/TCNE complexes were measured in THF, using carbon disulfide as a reference. The estimated χ$^{(3)}$ values were three to 4 orders of magnitude larger than the χ$^{(3)}$ value of CS$_2$. These charge-transfer complexes form an interesting class of NLO materials.

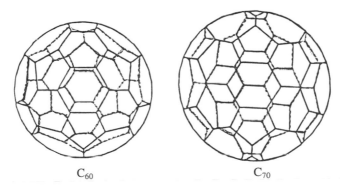

Tetracyanoethylene (TCNE)   7,7,8,8-Tetracyanoquinodimethane (TCNQ)

3. Fullerenes

Carbon clusters have attracted attention for more than two decades; they are important and fascinating materials because of their chemistry and unique physics coupled with material science.[80] Newly discovered highly stable icosahedral-cage molecule C_{60} fullerene is also a higher carbon clusters.[81] Solid C_{60} species is a polygon with 60 vertices and 32 faces, 20 of which are hexagonal and 12 of which are pentagonal, that has a soccer ball structure. Each C_{60} unit has electron-deficient pentagons and six electron-rich 6:6 bonds facing its 12 nearest neighbors. It is an aromatic molecule in which all valences are satisfied by two single bonds and one double bond and the outer and inner surfaces give rise to an ocean of π electrons. C_{60} has the symmetry of the icosahedral group, and the diameter of the C_{60} molecule is ~0.7 nm, which is capable of incorporating a variety of atoms. The laser vaporization of a variety of carbonaceous materials such as polyimides, coal, and polycyclic aromatic hydrocarbons yield C_{60}. The higher fullerenes such as C_{70}, C_{76}, C_{84}, C_{90}, C_{94}, and $C_{70}O$ can be isolated and characterized spectroscopically. The even-numbered carbon clusters of giant structures with as many as 600 carbon atoms can be formed by laser deposition of soot. Fullerenes are some of the most extensively investigated materials.[82–84]

C_{60} C_{70}

In C_{60}, all the atoms are connected by sp^2 bonds and the remaining 60 π electrons are distributed so that the C_{60} molecule attains a highly aromatic character. Therefore, C_{60} is an ideal candidate for third-order nonlinear optics. Blau et al.[85] reported for the first time the $\chi^{(3)}$ value and hyperpolarizability of C_{60} by the forward DFWM technique in benzene solutions using 50 ps pulses at 1.064 μm. Using the value of $L = 22$ Å and electron shell width $a = 2.5$ Å, the position of the first absorption maximum was calculated to be at 550 nm, the γ value to be 2×10^{-42} m⁵/V², and the $\chi^{(3)}$ value to be 1×10^{-10} esu in the solid state for C_{60} at 1.064 μm using the DFWM technique. The real part of the nonlinearity is positive and three times larger than the imaginary part. The NLO process considered in the model of a free electron in a spherical box demonstrates the complete delocalization of electrons.

Blau research team further confirmed the NLO susceptibilities of fullerenes using the Z-scan method and reported the same order of magnitude as that of the DFWM technique. Henri et al.[86] reported the third-order optical nonlinearity of solutions of C_{60} and C_{70} by Z-scan technique using a 5 ns pulse at different wavelengths. The solutions of concentration 10^{-5} M were prepared in benzene. The imaginary part of the third-order nonlinear susceptibility Im$\chi^{(3)}$ was estimated by:

$$\text{Im } \chi^{(3)} = \epsilon_0 c \eta_0 \lambda N_A C(\sigma_{ex} - 2\sigma_0)/16\pi I_{sat} \qquad (22)$$

where ϵ_o is the permittivity of free space, c is the velocity of light, η_o is the linear refractive index, and λ is the wavelength of the laser, N_A is Avagadro's constant, C is a molar concentration, and σ_{ex} and σ_o are excited-state and ground-state cross-sections. The $\chi^{(3)}$ values of 3.2×10^{-11} esu at 0.520

μm and 1.7×10^{-10} esu at 0.58 μm for C_{60} and $\chi^{(3)}$ values of 2.7×10^{-10} esu at 0.520 μm and 5.1×10^{-11} esu at 0.61 μm for C_{70} were reported.

Hoshi et al.[87] investigated the optical second- and third-order optical nonlinearity in C_{60} films. The 600 Å thick C_{60} film was deposited on a silica substrate by the molecular-beam epitaxy (MBE) system at 30 °C. Both second- and third-order optical nonlinearity depend on the polarization of light as well as on the angle of incidence considering the symmetry of the C_{60} film; the film shows weak second harmonic generation. The $\chi^{(3)}$ $(-3\omega; \omega, \omega, \omega)$ value of 2×10^{-10} esu has been reported at 1.064 μm from THG measurements by comparing the THG intensity with a standard fluoro-bridged aluminum phthalocyanine polymer. Apparently, the existense of a highly delocalized π-electron system around the periphery of the spherical C_{60} molecule causes high optical nonlinearity.

Kafafi et al.[88] conducted time-resolved DFWM experiments on pure C_{60} films vapor deposited onto optical substrate with a 35 ps Nd:YAG laser at 1.064 μm. The linear absorption coefficient α at the same wavelength was measured as 6 cm^{-1} from the NIR spectra of toluene solutions and the index of refraction n was 2.0. Figure 6 shows the phase conjugate signal of a C_{60} film as a function of the laser beam intensity. A cubic power dependence was observed below a power of 1 GW/cm^2, and a supercubic power dependence indicated a fifth-order component above 1 GW/cm^2. The observed signal intensity was fitted to a polynomial of the form $I_4 = a_3 I^3 + a_5 I^5$ and the $\chi^{(3)}_{xxxx}$ was determined from the following equation:

$$|\chi^{(3)\text{eff}}_{ijkl}| = |\chi^{(3)\text{ref}}_{ijkl}| \left(\frac{a_3}{a_{3\text{ref}}}\right)^{\frac{1}{2}} \left(\frac{n}{n_{\text{ref}}}\right)^2 \left(\frac{l_{\text{ref}}}{l}\right) \left(\frac{\alpha l}{e^{-\alpha l/2}(1 - e^{-\alpha l})}\right) \tag{23}$$

where is a_3 the coefficient of the cubic term in the least-squares fit of the phase conjugate signal, l is the path length, n is the refractive index, and α is linear absorption coefficient of the sample. The carbon disulfide was used as the reference sample. From this a $\chi^{(3)}$ value of 7×10^{-12} esu was measured and the $\chi^{(3)}/\alpha$ relationship gave a figure of merit of $\sim 1 \times 10^{-12}$ esu cm. A ratio $\chi^{(3)}_{xyyx}/\chi^{(3)}_{xxxx}$ of $1/7$ was determined from the tensor component. The temporal response of the phase-conjugated signal from a laser intensity of 1 GW/cm^2 was faster than 35 ps at a wavelength greater than any one-photon absorption bands. Knize and Partanent[89] pointed out that the second hyperpolarizability was 1.3×10^{-45} m^5/V^2 for C_{60}. Gong et al.[90] measured third-order optical nonlinearities of all C_{60} molecules (87% C_{60} and 13% C_{70}) by nanosecond DFWM experiments in

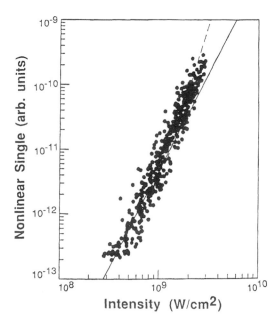

Figure 6 Diffraction beam intensity as a function of the laser beam intensity for a 21 μm thick C_{60} film. (From Kafafi, et al., *Chem. Phys. Lett.*, 188, 492, 1992. With permission.)

toluene solution. In the DFWM experiments, the intensity of the phase-conjugate beam generated was obtained from the following equation:

$$I_4 = (2\pi\omega L/cn)^2[\chi^{(3)}]^2 I_1 I_2 I_3 \tag{24}$$

where I_1, I_2, and I_3 are the intensities of the two pump beams and the probe beam, respectively, L is the interaction length, and the refractive index of 1.5 was used for the sample solution. The $\chi^{(3)}$ was calculated by using a CS_2 reference from the Equation 25:

$$\chi^{(3)} = \chi_{\text{ref}}^{(3)}(I_{4\text{sample}}/I_{4\text{ref}})^{1/2}(n_{\text{sample}}/n_{\text{ref}})^2 \tag{25}$$

where $\chi^{(3)}_{\text{ref}}$ is the NLO susceptibility of CS_2 taken as 4.0×10^{-13} esu. A $\chi^{(3)}$ of 2.6×10^{-12} esu was calculated for a C_{60} toluene solution of 7×10^{-3} M/l. A molecular hyperpolarizability γ of 1.6×10^{-31} esu was derived by $\gamma = \chi^{(3)}/NL^4$ relationship. The $\chi^{(3)}$ of 3.3×10^{-9} esu for the solid C_{60} compound was deduced from the corresponding C_{60} toluene solution using a density of 1.6 g/cm^3 and refractive index of 2.0.

Yang et al.[91] measured the third-order NLO susceptibility of C_{70} in a toluene solution with the DFWM technique using a 10-ns laser pulse at 1.06 μm. The $\chi_{1111}^{(3)}$ of 5.6×10^{-12} esu was estimated for the concentration of C_{70} in toluene at 0.474 g/l. The extracted $\chi_{1111}^{(3)}$ value was 2.5×10^{-8} esu for the solid compound of C_{70}, and the second hyperpolarizability of 1.2×10^{-30} esu was obtained from the susceptibility using the general relationship. Zhang et al.[92] reported third-order optical nonlinearities of both C_{60} and C_{70} in toluene solutions using a time-delayed DFWM technique with 30 ps pulses at 0.532 μm in both parallel and cross-polarization configurations. The purity of C_{60} was 99.9% whereas that of C_{70} was 90%. The cubic dependence of signal intensity on the reference intensity in a 7×10^{-5} M C_{60} in toluene reveals the origin of NLO susceptibility. The second hyperpolarizabilities γ_{1212} of 2.3×10^{-43} m^5/V^2 for C_{60} and of 8.6×10^{-41} m^5/V^2 for C_{70} were estimated. Vijaya et al.[93] determined $\chi^{(3)}$ of 2.8×10^{-18} m^2/V^2 for C_{60} solution of concentration 3.0×10^{-4} M from DFWM at 532 nm. The second hyperpolarizabilities γ of 1.6×10^{-41} m^5/V^2 for the solute.

Rosker et al.[94] measured $\chi^{(3)}$ values of thin films of C_{60} and C_{70} by time-resolved DFWM using 150 fs optical pulses at a wavelength of 0.633 μm. For both samples, the cubic dependence of the DFWM signal on the input beam intensity was observed. At the input fluence of 1 GW/cm^2, $\chi^{(3)}$ of 2.0×10^{-10} esu of C_{60} and 3.0×10^{-10} esu of C_{70} thin films were measured using the following relationship:

$$I_E/I_F = 4\pi^2\mu_0/n^4\lambda^2\epsilon_0^3(\chi^{(3)})^2 L^2 I_G I_H \tag{26}$$

where I_E/I_F is the phase–conjugate reflectivity, ϵ_o is the permittivity and μ_o is the permeability of vacuum, λ is the wavelength in vacuum, L is the interaction length, and n is the refractive index of the film. The refractive indexes of C_{60} and C_{70} films were estimated to be 2.0. These results agree with those reported by Kafafi et al.[88] The $\chi^{(3)}$ of C_{70} is larger than C_{60} because C_{70} has more quasi-free electrons that contribute to optical nonlinearity. For both C_{60} and C_{70}, a decay time constant of ~200 fs on the same scale as the optical pulse was measured and probably had an electronic origin. Additional components with time constants of a few hundred femtoseconds, a few picoseconds, and approximately 100 ps were observed. The time constants and their amplitude show a wavelength dependence. Aranda et al.[95] also measured real and imaginary parts of $\chi^{(3)}$ values of C_{60} and C_{70} using DFWM at 532 nm. The real component was found to be negative at this wavelength. The measured real and imaginary parts of second hyperpolarizability were in the range of 10^{-31} esu. At a concentration of 0.33 mg/ml, the figures-of-merit $\chi^{(3)}/\alpha$ of 2.9×10^{-13} esu cm for C_{60} and 2.0×10^{-13} esu cm for C_{70} were estimated.

Meth et al.[96] reported the dispersion of the third-order NLO susceptibility of polycrystalline C_{60} films over the fundamental energy range of 1.5 to 3.5 eV using the THG technique. The 0.32 μm thick C_{60} films were prepared on a fused silica substrate by the sublimation method. The C_{60} film had a refractive index of 2.031 at 1.064 μm determined from the m-line method. Figure 7 shows the dispersion of $\chi^{(3)}(-3\omega; \omega, \omega, \omega)$ of the C_{60} film together with the linear absorption spectrum where data are plotted at the third-harmonic energy (3ω). The $\chi^{(3)}(-3\omega; \omega, \omega, \omega)$ values follow the absorption spectrum, increasing with the increase of optical absorption. A peak of $\chi^{(3)}(-3\omega; \omega, \omega, \omega)$ at 2.8 eV corresponds to the absorption maximum of C_{60} film where the $\chi^{(3)}$ value reaches as high as 2.7×10^{-11} esu; therefore it has a three-photon resonant contribution. The $\chi^{(3)}(-3\omega; \omega, \omega, \omega)$ value at 0.52 eV is 4.1×10^{-12} esu, which is about an order of magnitude smaller because it is far from the resonance. The second hyperpolarizability $\langle\gamma\rangle$ of 4.3×10^{-34} esu were determined from the relationship $\chi^{(3)}(-3\omega; \omega, \omega, \omega) = N(f^{\omega})^3 f^{3\omega} \gamma(-3\omega; \omega, \omega, \omega)$, where N is the number density and f^{ω} and $f^{3\omega}$ are Lorenzt local field factors. From the EFISH technique, $\chi^{(3)}$ value of 9.6×10^{-12} esu and second hyperpolarizability $\langle\gamma\rangle$ of 7.5×10^{-34} esu at 0.68 eV (1.83 μm) were reported by Wang and Cheng.[97] The off-resonance $\chi^{(3)}$ values measured by THG were comparable to the DFWM experiments. The THG results on $\chi^{(3)}(-3\omega; \omega, \omega, \omega)$ and $\langle\gamma\rangle$ agreed with the EFISH and DFWM measurements although these techniques measure different optical parameters.

Wang and Cheng[97] reported the determination of hyperpolarizabilities of fullerenes and their charge-transfer complexes with electron donor N,N'diethylaniline (DEA) using solution phase dc electric field induced SHG (EFISH) measurements at a wavelength of 1.91 μm. The formation of charge-transfer complexes of C_{60}/DEA and C_{70}/DEA were confirmed by the increase in absorption strength by recording optical absorption spectra that show a new absorption band between the 560 and 750 nm region. The isotropic component of the second hyperpolarizability can be calculated at infinite dilution limit and EFISH γ can be determined from the following equation:

$$\text{EFISH}\langle\gamma\rangle_{1111}(-2\omega, \omega, \omega, 0) = \langle\gamma\rangle_{1111}(-2\omega, \omega, \omega, 0) + \mu\beta_{\mu}/5\,kT \tag{27}$$

where $\beta\mu$ is the vectorial projection of the first hyperpolarizability along the direction of molecular dipole μ. From EFISH measurements, the $\langle\gamma\rangle_{1111}(-2\omega; \omega,\omega,\omega,0)$ values of 7.5×10^{-34} esu for C_{60} and 1.3×10^{-33} esu for C_{70} were determined.

Wang et al.[98] reported SHG and THG properties of C_{60} films. Thin films of C_{60} were deposited on glass substrate by thermal sublimation at 350–400 °C under a pressure of 1.5×10^{-7} Torr. For SHG measurements, C_{60} films were subjected to corona poling at an electric field of 6 kV and the transmission SHG was measured while heating the films between room temperature to 170 °C. The temperature dependence of the SHG showed a peculiar behavior during heating and cooling. The SHG intensity began to increase at a temperature of 90 °C rising to a maximum at 140 °C and then decreased with increasing temperature.

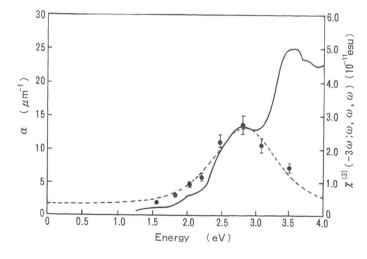

Figure 7 The dispersion of $\chi^{(3)}$ $(-3\omega;\omega,\omega,\omega)$ of C_{60} film along with the linear absorption spectrum (solid line). (From Meth et al., *Chem. Phys. Lett.*, 197, 26, 1992. With permission.)

The SHG intensity decreased during cooling. The comparative analysis of the data with a quartz plate revealed second-order NLO susceptibility $\chi^{(2)}_{zzz}$ of 2.1 $\times$ 10^{-9} esu for C_{60} films at room temperature. At the SHG intensity peak temperature of 140 °C, the $\chi^{(2)}_{zzz}$ reached to 20 $\times$ 10^{-9} esu for C_{60} films which is about an order of magnitude larger than its value at room temperature. On reheating, the films again showed SHG enhancement. Interestingly, such SHG enhancements were not observed either while poling was performed at room temperature or in the absence of the high electric field at 140 °C. Because C_{60} has a centrosymmetric crystal structure that restricts the origin of SHG from films, the origin of SHG from C_{60} films was determined from impurities involving either C_{60} isomers, higher fullerenes or $C_{60}O$. Another possible source of SHG activity was assumed to be the electric quadrupole or magnetic dipole contributions. The $\chi^{(3)}(-3\omega;\ \omega,\ \omega,\ \omega)$ measured by the THG method at 1.907 μm was 2 $\times$ 10^{-11} esu for C_{60} films. Kajzar et al.[99,100] reported optical SHG and THG measurements of C_{60} thin films. The $d_{11}(-2\omega;\ \omega,\ \omega)$ of 3.8 $\times$ 10^{-9} esu (1.6 pm/V) for C_{60} films was estimated to be about two times larger than the value reported by Wang and Cheng.[97] THG measurements showed two strong resonances in $\chi^{(3)}(-3\omega;\ \omega,\ \omega,\omega)$, a three-photon resonance with the T_{1u} state at 1.064 μm with a $\chi^{(3)}(-3\omega;\ \omega,\ \omega,\ \omega)$ of 6.1 $\times$ 10^{-11} esu and another three-photon resonance with vibronic components with total T_{1u} character made partially by vibronic mixing around 1.3 μm with a $\chi^{(3)}(-3\omega;\ \omega,\ \omega,\ \omega)$ value of 6.1 $\times$ 10^{-11} esu. The estimated second hyperpolarizability $\langle\gamma\rangle$ was 1.1 $\times$ 10^{-32} esu at 1.907 μm. Figure 8 shows the wavelength dispersion of $\chi^{(3)}(-3\omega;\ \omega,\ \omega,\ \omega)$ for a 530 nm thick C_{70} film at room temperature.[101] An intense and broad resonance of $\chi^{(3)}(-3\omega;\ \omega,\ \omega,\ \omega)$ with a maximum at 1.42 μm was observed, whereas $\chi^{(3)}(-3\omega;\ \omega,\ \omega,\ \omega)$ decreased significantly at higher energies. Neher et al.[102] measured second hyperpolarizabilities of C_{60} and C_{70} at 1.064, 1.5 and 2.0 μm by THG in a toluene solution. The C_{70} has larger γ values than C_{60}. The C_{70} has γ value of 5 $\times$ 10^{-32} esu in the three-photon resonant region (1.064 μm) and 4 $\times$ 10^{-33} esu in the nonresonant region having a negative real hyperpolarizability. For C_{70} the sign of real γ changes from positive to negative as the wavelength changes from 1.5 to 2.0 μm. In toluene, C_{60} shows a strong absorption peak at 334 nm with a weak broad absorption near 535 nm whereas C_{70} has a strong absorption peak at 472 nm and other absorption peaks in the 300 to 400 nm region. As a result, the optical nonlinearities of both C_{60} and C_{70} are affected by a three-photon resonance at 1.064 μm. Flom et al.[103] measured resonant optical response of thin films of C_{60} and C_{70} by the time-resolved DFWM technique using a picosecond tunable dye laser. The $\chi^{(3)}$ of C_{60} was 5 times larger at 597 nm than that obtained at 675 nm and 50 times larger than that measured at 1.064 μm. The C_{70} has a $\chi^{(3)}$ value of 2.1 $\times$ 10^{-9} esu, more than 5 times larger than C_{60} at 597 nm. The $\chi^{(3)}$ value of fullerenes in solid state was about 70 times larger than that $\chi^{(3)}$ values of solutions. The large third-order NLO susceptibilities at these wavelengths arise from the excited-state population. Fullerenes show interesting NLO properties in thin films[104] and in PMM films.[105] Flom et al.[106] measured the $\chi^{(3)}$ of thin films of

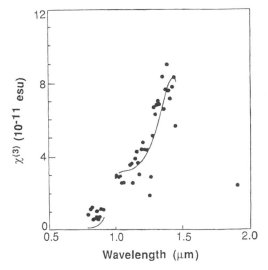

Figure 8 Wavelength dispersion of $\chi^{(3)}$ $(-3\omega;\omega,\omega,\omega)$ of 530 nm thick C_{70} film at room temperature. Solid line indicates the calculated $\chi^{(3)}(-3\omega;\ \omega,\omega,\omega)$ spectrum obtained from a three-level model. (From Kajzar et al., *Phys. Rev. Lett.*, 73, 1617, 1994; Kajzar, *SPIE Proc.*, 2025, 352, 1993. With permission.)

C_{60} and its composite with poly[2-methoxy-5-(2′-ethylhexyloxy)p-phenylenevinylene] (MEH-PPV) using DFWM and nonlinear transmission at 590.5 nm. The MEH-PPV and C_{60}/MEH-PPV exhibited a cubic intensity dependence and the $\chi^{(3)}$ value of the thin film composite was found to be larger than the sum of the $\chi^{(3)}$ values of the two components. Berrada et al.[107] synthesized a polymer of C_{60} while reacting difluorodiphenyl fulleroid with aromatic diamines. The $\chi^{(3)}$ of thin films was estimated 1.4×10^{-14} esu by the DFWM technique.

The $\chi^{(3)}$ and γ values of C_{60} and C_{70} have also been calculated using quantum chemistry approaches. The frequency dependence of third-order NLO susceptibilities of C_{60} calculated with a tight-binding model were found to be 1.22×10^{-12} esu near zero frequency and about 10^{-11} esu around the peak at 2.5 eV because three-photon resonance enhancement.[108] The magnitude of the $\chi^{(3)}$ value at $\omega = 0$ is similar to that reported by Meth et al.[96] using THG measurement, $\chi^{(3)} = 4 \times 10^{-12}$ esu at 1.6 eV, and by Kafafi et al.[88] using the DFWM technique, $\chi^{(3)} = 7 \times 10^{-12}$ esu at 3.6 eV. Using the INDO/SCl method, Li et al.[109] calculated $\langle\gamma\rangle\,(-\omega; \omega, -\omega, \omega) = 7.30 \times 10^{-34}$ esu at $\omega = 1.064$ µm and 3.11×10^{-34} esu at $\omega = 0.532$ µm and $\langle\gamma\rangle\,(-2\omega; \omega, \omega, 0) = 6.90 \times 10^{-34}$ esu at $\omega = 1.91$ µm. A $\langle\gamma\rangle\,(-\omega; \omega, -\omega, \omega) = 8.84 \times 10^{-34}$ esu at $\omega = 1.064$ µm was estimated by the INDO/SCI-SOS method. Shuai and Bredas[110] used the valence-effective-Hamiltonian-sum-over-states (VEH-SOS) approach to study the electronic structure and NLO properties of fullerenes. The averaged α values calculated were 154 Å^3 for C_{60} and 214 Å^3 for C_{70}. The average γ values calculated were 2.02×10^{-34} esu for C_{60} and 8.62×10^{-34} esu for C_{70}. The static γ value for C_{60} is very near that of 3×10^{-34} esu obtained from DFWM experiments.[88]

The results on third-order NLO properties of fullerenes have been summarized in reviews.[111,112] Experimentally and theoretically obtained $\chi^{(3)}$ and γ values of fullerenes are summarized in Table 12. NLO properties of fullerenes are comparable with other organic π-conjugated materials but controversies remain about the magnitude of $\chi^{(3)}$ and γ values. The γ value fluctuates between 10^{-30} and 10^{-34} esu depending on the material state and the measurement wavelengths and techniques of THG, DFWM, EFISH, and SA. Resonance effect may cause large variation in optical nonlinearity. Like other materials, off-resonance NLO properties of fullerenes are smaller than those of resonant ones. The sensitivity of fullerenes to atmosphere may cause such changes, although there are no reports available at this time. Third-order NLO susceptibilities and second hyperpolarizabilities of both C_{60} and C_{70} range over 3 orders of magnitude. C_{70} has more quasi-free electrons that gives rise to larger $\chi^{(3)}$ values for C_{70} than for C_{60}. Third-order optical nonlinearity measurements and theoretical calculations indicate that fullerenes are promising materials for third-order nonlinear optics.

C. CONJUGATED POLYMERS AND RELATED MODEL COMPOUNDS

In the 1970s, π-conjugated organic polymers emerged as a new class of electronic materials.[113,114] Like electrical conductivity, the π-electron-conjugated backbone holds the key to large third-order NLO effects in π-conjugated polymers; therefore, they attracted the attention of scientific community as novel materials for third-order nonlinear optics. Organic polymers are expected to possess the following desirable properties to show their potential for NLO devices:

1. Large third-order NLO susceptibility
2. Fast response time in subpicoseconds
3. High resistance to laser radiations (GW/cm^2)
4. Low dielectric constant
5. Wide bandwidth
6. Chemical tailoring of physical and optical properties
7. Ease of processing into ultrathin films, fibers, and even as liquid crystals
8. Mechanical strength and flexibility
9. Environmental stability
10. Room temperature operation
11. Compatibility with semiconductor-integrated structures
12. Nontoxicity

647

Table 12 Third-order NLO Susceptibilities and Hyperpolarizabilities of Fullerenes Measured by Different Research Groups

Fullerene (C_n)	Measurement Technique	Wavelength (μm)	$\chi^{(3)}$ (esu)	γ (esu)	Ref.
C_{60} (benzene)	DFWM	1.064	1×10^{-10}	1.1×10^{-30}	85,111
	SA	0.440	6.8×10^{-12}		86
		0.520	3.2×10^{-11}		86
		0.580	1.7×10^{-10}		86
C_{60} (toluene)	DFWM	0.532	2.0×10^{-10}	1.83×10^{-27}	93
	DFWM	0.532	10^{-13}	10^{-31}	92
	DFWM	0.532	1.4×10^{-13}	3.3×10^{-31}	95
C_{60} (benzene)	DFWM	0.532	1.2×10^{-13}	3.5×10^{-31}	95
C_{60} (styrene)	DFWM	0.532	1.2×10^{-13}	2.9×10^{-31}	95
C_{60} (xylene)	DFWM	0.532	0.8×10^{-13}	2.2×10^{-31}	95
C_{60} (film)	DFWM	1.064	7.0×10^{-12}	3.0×10^{-34}	88
		0.633	2.0×10^{-10}		94
	THG	1.064	2.0×10^{-10}		87
		1.32	3.0×10^{-11}		96
		1.064	1.4×10^{-11}		96
		1.91	9.0×10^{-12}		96
		2.37	4.0×10^{-12}		96
	THG	0.85	1.5×10^{-11}		99
		1.06	8.7×10^{-11}		99
		1.32	6.1×10^{-11}		99
		1.907	3.2×10^{-11}	1.1×10^{-32}	100
C_{60} (toluene)	THG	1.064	7.2×10^{-11}	1.3×10^{-33}	102
		1.50	3.0×10^{-11}	3.1×10^{-33}	102
		2.0	3.7×10^{-11}	1.6×10^{-33}	102
		2.37	4.0×10^{-12}		102
C_{60} (film)	DFWM	0.597	3.8×10^{-10}		103
		0.675	0.82×10^{-10}		103
C_{60} (toluene)	EFISH	1.91	1.6×10^{-11}	7.5×10^{-34}	97
C_{70} (toluene)	EFISH	1.91	4.4×10^{-11}	1.3×10^{-33}	97
		1.83	9.6×10^{-12}	7.5×10^{-34}	97
	THG	1.064	1.4×10^{-9}	5.0×10^{-32}	102
		1.5	5.4×10^{-10}	2.6×10^{-32}	102
		2.0	9.1×10^{-11}		102
$C_{60+}C_{70}$ (toluene)	DFWM	1.064	3.3×10^{-9}	1.6×10^{-31}	90
C_{70} (toluene)	DFWM	1.064	5.6×10^{-12}	1.2×10^{-30}	91
	DFWM	0.532	0.9×10^{-13}	7.3×10^{-31}	95
C_{70} (film)	DFWM	0.633	3.0×10^{-10}		94
	DFWM	0.597	2.1×10^{-9}		103
		0.675	6.4×10^{-10}		103
	THG	1.064	2.6×10^{-11}		104
		1.4	9.0×10^{-11}		100
		1.907	2.4×10^{-11}	1.1×10^{-32}	100
	DFWM	1.064	1.2×10^{-11}	5.0×10^{-34}	104
C_{70} (benzene)	SA	0.520	2.7×10^{-10}		86
		0.580	2.0×10^{-12}		86
		0.610	5.1×10^{-11}		86
C_{60}/DEA	THG	1.064		6.7×10^{-29}	96
$C_{60+}C_{70}$/PMMA	DFWM	0.608	1.0×10^{-10}	5.0×10^{-30}	105
C_{60}–polymer	DFWM	0.602	1.4×10^{-14}		107
C_{60}	TBC	$\omega = 0$	1.22×10^{-12}		108
C_{60}	INDO	$\omega = 0.532$		3.11×10^{-34}	109
		$\omega = 1.064$		7.3×10^{-34}	109
		$\omega = 1.91$		6.9×10^{-34}	109
C_{60}	VEH-SOS	$\omega = 0$	10^{-12}	2.02×10^{-34}	110
C_{70}	VEH-SOS	$\omega = 0$		8.62×10^{-34}	110

TBC = tight-binding calculations; DEA = N,N'-dethylaniline.

A combination of these physicochemical properties make organic π-conjugated polymers ideal candidates for third-order nonlinear optics. Certainly, polymeric materials may be subjected to some restrictions such as a limited temperature range of operation. Because organic polymers provide unlimited opportunities for chemical modifications, these hurdles could be overcome with the development of new materials and techniques. Most of the conjugated polymers can be processed into thin films which show resonant third-order NLO susceptibilities in the range of 10^{-5} to 10^{-12} esu. In particular, derivatives of polydiacetylenes, polyacetylenes, polythiophenes, polyaniline, heteroaromatic ladder polymers, etc. exhibit the largest third-order NLO susceptibilities. From application viewpoints, organic conjugated polymers possessing large third-order susceptibility at short wavelength are highly desirable for laser frequency conversion to fabricate high-power laser sources which can be used ultraviolet and near-infrared frequency ranges. One of the major difficulties in achieving this goal is that high-performance polymers having large optical nonlinearity coupled with small optical absorption.

1. Polydiacetylenes and Model Compounds

Polydiacetylenes (PDAs) are conjugated polymers prepared by the solid state polymerization of diacetylene monomers having a structural formula, $R_1 \equiv C = C \equiv C - R_2$, where R_1 and R_2 refer to the substituent side groups.[115-118] In the polymeric system, the conjugated carbon backbone holds the key for interesting electronic and NLO properties, whereas the side groups facilitate structure control and material processing. The long alkyl chain side groups attached to the conjugated backbone of the diacetylene monomers, such as in urethane $(-O-CONH-)$, have 2 fold advantages: first, they participate in the van der Waals interactions between monomers, and second the incorporated urethane groups lead to a planar structure on polymerization by forming hydrogen bonding between the adjacent side groups. PDAs possessing long alkyl chains $[-(CH_2)_n-]$ and urethane groups exhibit phase transitions observed with a color change. Polydiacetylene having the side group $R_1 = R_2 = (CH_2)_4 - O - CO - NH - CH_2 - COO - C_4H_9$ is called 4-BCMU, where BCMU stands for butoxycarbonylmethylurethane. In the case of $(CH_2)_n$, for example, if n is 3, 4, and 5, then the corresponding diacetylene will be termed as 3-BCMU, 4-BCMU, and 5-BCMU, respectively. A series of diacetylene monomers having the side chain $-CH_2 - O - CONH-$ (where in the alkyl chain, C_nH_{2n+1}, n is 1, 2, 3, 4, 5, 6, 7, 8, 9, 10, 12, 14, 16, and 18) have been prepared. In addition, to alkyl chain length, the acetylenic or cumulenic conjugated polymer backbone are formed when the corresponding diacetylene monomers are subjected to polymerization either by irradiation with ultraviolet, X-ray, Y-radiation or by annealing at elevated temperatures. The suitability of modifying chemical structures of substituted side groups enables PDAs to be prepared in the forms of single crystals, thin films, and solutions. Some of the representative examples of the diacetylene polymers are depicted below which are discussed in connection with third-order nonlinear optics:

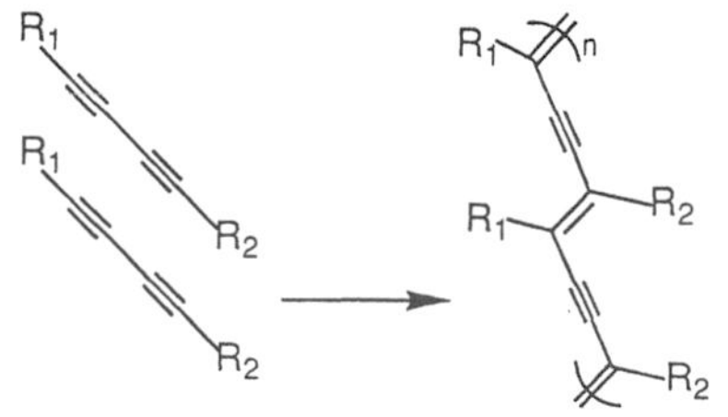

$$
\begin{aligned}
R_1 = R_2 &= -CH_2 - O - SO_2 - C_6H_4 - CH_3 & \text{(PDA-PTS)}\\
R_1 = R_2 &= -CH_2 - O - SO_2 - C_6H_4 - F & \text{(PDA-FBS)}\\
R_1 = R_2 &= -(CH_2)_4 - O - SO_2 - C_6H_4 - OCH_3 & \text{(PDA-PTS-12)}\\
R_2 = R_2 &= -CH_2 - O - SO_2 - C_6H_4 - OCH_3 & \text{(PDA-MBS)}\\
R_1 = R_2 &= -CH_2 - O - CO - NH - C_6H_5 & \text{(PDA-HDU)}\\
R_2 = R_2 &= -(CH_2)_4 - O - CO - NH - C_6H_5 & \text{(PDA-TCDU)}\\
R_1 = R_2 &= -(CH_2)_3 - O - CO - NH - CH_2 - COO - C_nH_{2n+1} & \text{(PDA-}n\text{BCMU)}
\end{aligned}
$$

$R_1 = R_2 = -(CH_2)_n - O - CO - NH - CH(CH_3) - C_6H_5$ (PDA-nSMBU)

$R_1 = R_2 = -(CH_2)_4 - O - CO - NH - C_6H_4 - Br$ (PDa-4-PBrPu)

$R_1 = R_2 = -CH_2 - O - C_6H_4 - NO_2$ (PDA-DNP)

$R_1 = R_2 = -CH_2 - NO_2$ (PDA-THD)

$R_1 = R_2 = CH_3 - (CH_2)_{15} - C - C \equiv C - C - (CH_2)_8 - COO-$ (PDA-AFA)

$R_1 = R_2 = CH_3 - (CH_2)_{13} - (C \equiv C)_3 - (CH_2)_8 - COOH$ (PDA-NTA)

$R_1 = R_2 = C_6H_4 - NH - CO - C_{17}H_{37}$ (PDA-$C_{17}H_{37}$)

$R_1 = R_2 = -(CH_2)_3 - OCO - NH - C_3H_6$ (PDA-C_3H_6)

$R_1 = R_2 = -CH_2 - OCO - NH - C_4H_9$ (PDA-C_4H_9)

$R_1 = R_2 = -(CH_2)_3 - OH$ (PDA-POL)

$R_1 = C_{12}H_{25}; R_2 = (CH_2)_8 - COOH$ (PDA-DOC)

$R_1 = R_2 = (CH_2)_4 - O - CO - NH - (CH_2)_{n-1} - CH_3,$ (PDA-C_4UC_n)

$R_1 = C_8H_{16} - (CONH)C_4H_8 - CH(NH_3^+) COO^-; R_2 = C_{16}H_{33}$ (PDA-NSL)

$R_1 = R_2$ $-CH_2-N$

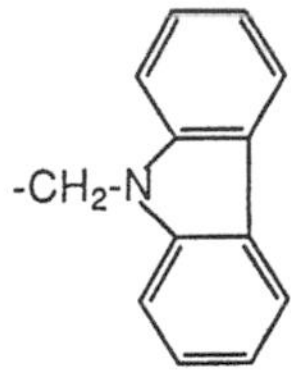

(PDA-DCHD)

$R_1 = R_2$

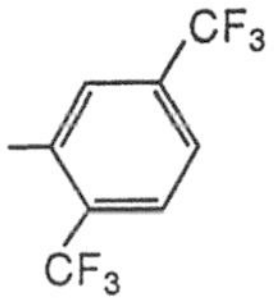

(PDA-DFMP)

$R_1 = R_2$

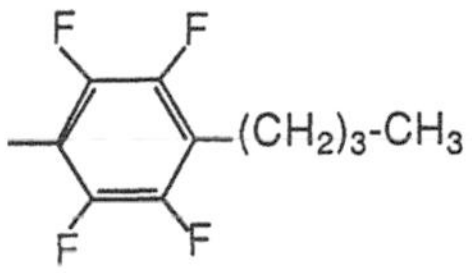

(PDA-BTFP)

$R_1 =$

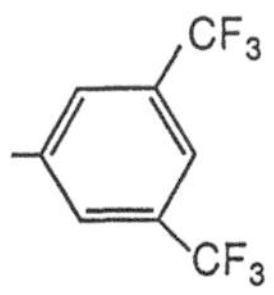

$R_2 =$

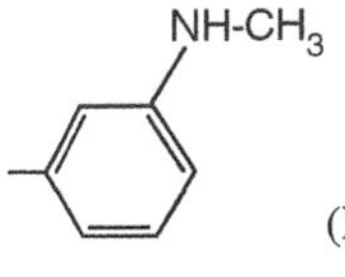

(PDA-MADF)

$R_1 =$

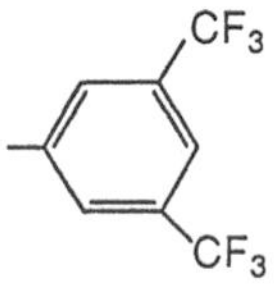

$R_2 = -CH_2-N$

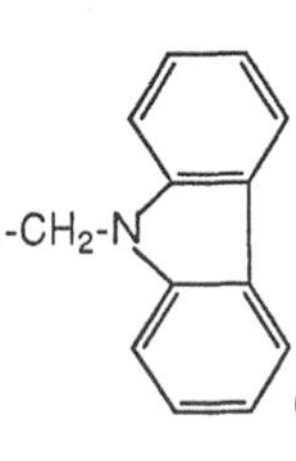

(PDA-DFCP)

The NLO properties of polydiacetylenes have been investigated by several research groups in the form of single crystals, solutions, thin films, and LB films. In 1976, Sauteret et al.[119] reported that single crystalline polydiacetylene having *p*-toluene sulfonate (PDA-PTS) side groups shows large nonresonant third-order NLO susceptibility. After this discovery, theoretical and experimental studies of cubic susceptibilities of polydiacetylenes were performed by a number of groups and similar results were reported. The magnitude of the third-order NLO susceptibility is about 10^{-11} esu, but varies slightly depending on the substituted side groups. These observations are for a model system of one-dimensional conjugated π-electron backbone of PDAs. Sauteret et al. report experimental $\chi^{(3)}$ values of 8.5×10^{-10} and 0.7×10^{-10} esu for poly-PTS and poly-TCDU respectively at 1.89 μm. Poly-PTS shows a damage threshold of about 50 GW/cm^2 in the picosecond regime. It has a fast response time of 10^{-14} seconds.

Carter et al.[120,121] investigated the time- and wavelength-resolved NLO spectroscopy of poly-PTS. Figure 9 shows the wavelength dependence of $\chi^{(3)}$ for poly-PTS. The $\chi^{(3)}$ value is very sensitive to the measurement wavelength and varies from 9×10^{-9} esu at 6515 Å to 5×10^{-10} esu at 7015 Å. DFWM experiments showed strong resonance enhancement when wavelength reached toward the $1B_u$ exciton energy. The response time is less than 6 ps. Degenerate four wave mixing measurements demonstrated the excited state lifetime of poly-PTS to be 1.8 ps. A four-wave mixing signal originating from a single crystal of poly-PTS platelets is shown in Figure 10. The response time of the nonresonant $\chi^{(3)}$ is shorter than 300 fs optical pulses.

Kajzar and Messier[122–126] investigated the third-order NLO susceptibility of PDAs Langmuir-Blodgett films at various different wavelengths. A blue form exhibits $\chi^{(3)}$ value of about 1.5×10^{-10} and 2.2

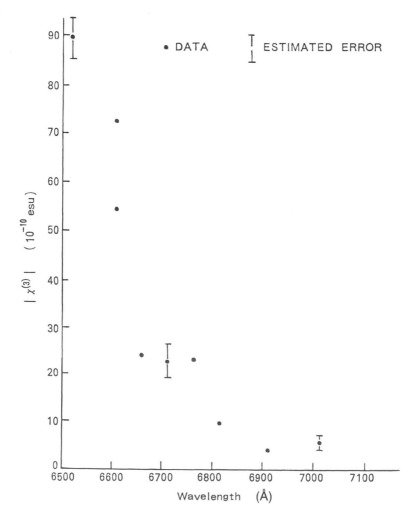

Figure 9 Wavelength dependence of $\chi^{(3)}$ for PDA-PTS. (From Carter et al., *Appl. Phys. Lett.*, 47, 457, 1985. With permission.)

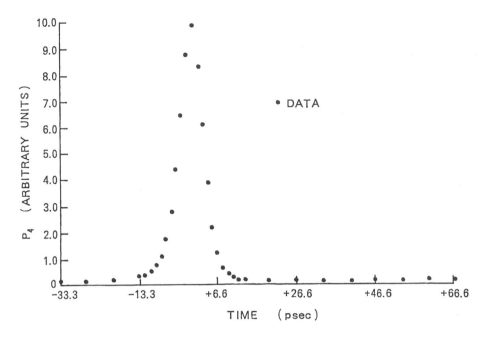

Figure 10 A four-wave mixing signal originating from a single crystal of PDA-PTS. (From Carter et al., *Appl. Phys. Lett.*, 47, 457, 1985. With permission.)

Table 13 Third-order Nonlinear Susceptibilities of Different Forms of PDAs

	$\chi^{(3)}$ (10^{-11} esu) at Wavelength (μm)		
Material	**1.064**	**1.35**	**1.907**
PDA LB film (blue form)	3.4	15.0	22.0
PDA LB film (red form)	0.9	1.9	3.4

Monomer $= CH_3-(CH_2)_{15}-C\equiv C-C\equiv C-(CH_2)_8-COOH$

Polymer $= CH_3-(CH_2)_{15}-C\equiv C-C\equiv C-(CH_2)_8-COOCd^{1/2}$ (PDA-AFA).

From Kajzar and Messier, *Thin Solid Films*, 132, 11, 1985. With permission.

$\times$ 10^{-10} esu at 1.35 and 1.907 μm, respectively. Table 13 lists the $\chi^{(3)}$ values of different forms of PDA LB films at three different wavelengths. The THG measurements on LB multilayers of PDAs show the presence of two resonances in $\chi^{(3)}$ value in the fundamental wavelength range of 0.8 to 1.9 μm. Figure 11 shows the third-order NLO susceptibility corrected for crystal disorder, as a function of the incident laser wavelength. Two-resonance enhancements: one at about 1.35 μm due to a two-photon resonance and the other at 1.907 μm due to a three-photon resonance were observed. The first resonance at the fundamental wavelength of 1.35 μm lies in the PDA transparency region. Chollet et al.[123] investigated the frequency dependence of wave dispersed THG and EFISH measurements of $\chi^{(3)}$ in PDA thin films. Two- and three-photon resonances, different for these two measurements, were observed in the wavelength range of 0.8 to 2.0 μm. Figures 12 and 13 show the measured $\chi^{(3)}$ value and the complex part of the refractive index for blue and red forms of LB films, respectively. The blue form of polymer film indicates that an enhancement at 1.35 μm and the large $\chi^{(3)}$ value at 1.907 μm are caused by a three-photon resonance. On the other hand, red form polymer film has a broad two-photon resonance at 1.25 μm. The EFISH measurements demonstrate a broader resonance enhancement in $\chi^{(3)}$ values than those of THG measurements. Temperature and frequency dependent $\chi^{(3)}$ in polydiacetylenes have also been measured.[123] The temperature-dependent studies show a negligible shift in two-photon resonance at low temperatures. The $\chi^{(3)}$ value does not change significantly between low temperature (7 K) and room temperature. The temperature-dependent $\chi^{(3)}$ data indicate an excitonic origin of resonant

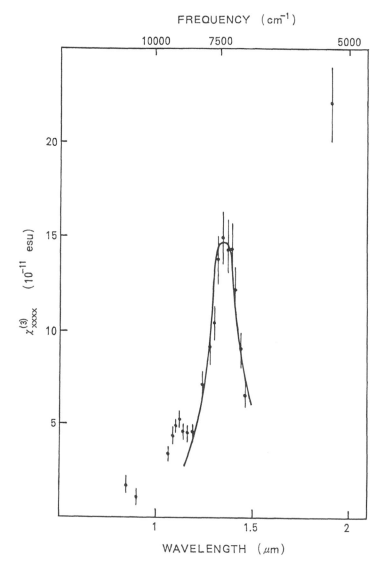

FREQUENCY (cm⁻¹)

Figure 11 Third-order nonlinear susceptibility $\chi^{(3)}$ value corrected for crystal disorder, as a function of the incident laser wavelength. There are two resonance enhancements: one at about 1.35 μm due to a two-photon resonance and the other at 1.907 μm due to a three-photon resonance. (From Kajzar and Messier, *Thin Solid Films*, 132, 11, 1985. With permission.)

nonlinearity and a strong bonding of excitons in polydiacetylenes. Optical absorption spectra of the blue form of LB films recorded at 7 K and at 260 °C show a small shift of excitonic maximum absorption wavelength toward larger wavelengths at low temperatures with a narrowing of the absorption peak, which indicates inhomogeneous broadening. Kajzar et al.[125] also measured the saturation absorption from LB films of a polydiacetylene in the picoseconds domain. The magnitude of the $\chi^{(3)}$ value of the imaginary part of Kerr susceptibility was found to be 1.4×10^{-8} esu and the associated ground-state recovery time was less than the laser pulse width of 3 ps. The $\chi^{(3)}$ value is about 2 orders of magnitude larger than the $\chi^{(3)}$ at the position of the two- and three-photon resonance reported above. Kajzar and Messier[122,126] conducted wave-dispersed THG measurements on LB films of polydiacetylene in the wavelength region 0.8 to 1.9 μm. Monomolecular films of the diacetylene molecule $[CH_3-(CH_2)_{15}-C\equiv C-C\equiv C-(CH_2)_8- COO]_2-Cd$ were transferred onto a silica substrate by the LB technique. The LB multilayers showed the existence of two-photon resonance in $\chi^{(3)}$: first a two-photon resonance around 1.35 μm and a three-photon resonance at 1.097 μm. The resonant $\chi^{(3)}$ values were 1.5×10^{-10} esu at 1.35 μm and 2.2×10^{-12} esu at 1.097 μm. Kajzar and Messier[122] also reported the solid state polymerization and NLO properties of diacetylene LB films of $CH_3-(CH_2)_{17}-C\equiv C-C\equiv C-COO= X^+$ where $X = $ H, NH_4, Ag, and Na, and $CH_3-(CH_2)_{17}-C\equiv C-C\equiv C-COO-]_2 M^{2+}$ where $M=$Cd, Cu, Hg, and Mn. The diacetylene monovalent salt of NH_4^+

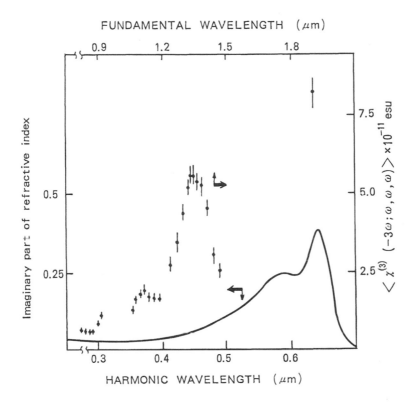

Figure 12 Third-order NLO susceptibility $\chi^{(3)}$ and imaginary part of the refractive index in blue form of LB film. (From Chollet et al., *Synth. Metals,* 18, 459, 1987. With permission.)

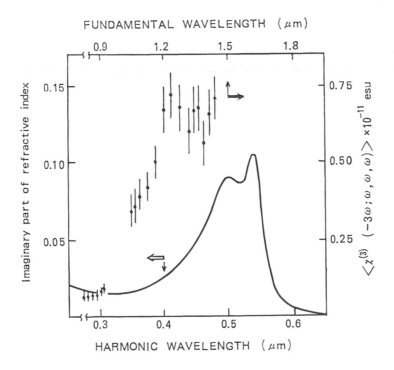

Figure 13 Third-order NLO susceptibility $\chi^{(3)}$ and imaginary part of the refractive index in red form of LB film (144 layer, polymer conversion rate ~30%). (From Chollet et al., *Synth. Metals,* 18, 459, 1987. With permission.)

and Ag⁺ and the acid form polymerize while Na⁺ and the divalent salts did not polymerize. The THG measurements conducted at the fundamental wavelength of 1.064 μm gave $\chi^{(3)}$ of 0.67 × 10⁻¹² esu for the polymer of the ammonium salt. The LB multilayers of the polymer with a diacetylene groups in the middle of the aliphatic chain [CH₃−(CH₂)₁₅−C≡C−C≡C(CH₂)₈−COO]₂−Cd showed a $\chi^{(3)}$ value of 0.69 × 10⁻¹² esu under similar conditions. The NLO susceptibility of the most stable ammonium salt polymer was almost the same as that of a Cd salt polymer. The EFISH measurements of [CH₃−(CH₂)₁₁−C≡C−C≡C−(CH₂)₈−COO]₂−Cd²⁺ LB multilayers with a fundamental wavelength ranging from 0.8 to 1.4 μm showed a $\chi^{(3)}$ maximum around 1.35 μm. Observation by the crossed polarizers indicated that the LB layers seemed to be made of single microcrystals oriented randomly around the normal to the substrate. Therefore the value of $\chi^{(3)}(-2\omega; \omega, \omega, 0)$ was an average for a single crystal over all orientations in the substrate plane. From the following equation

$$\chi^{(3)}_{xxxx}(-2\omega; \omega, \omega, 0) = 1/\langle\cos^4\theta\rangle\chi^{(3)}_{LB}(-2\omega; \omega, \omega, 0) \tag{28}$$

where $\langle\cos^4\theta\rangle = 3/8$, the value of $\chi^{(3)}$ at 1.15 μm was 1.3 × 10⁻¹² esu. This $\chi^{(3)}$ value was almost the same as the value of $\chi^{(3)}_{LB}(-3\omega; \omega,\omega,\omega) = 1.33 \times 10^{-12}$ esu estimated by THG measurements at 1.064 μm.[128] Polydiacetylene monovalent and divalent salts show interesting third-order NLO properties, the magnitude was within the range that of thin films.

Berkovic et al.[129] reported SHG and THG in Z-type LB films of an amphiphilic diacetylene derivative. The 10,12-nonacosadiynol-S-lysine (NSL) C₁₆H₃₃−C≡C−C≡C−C₈H₁₆−CONHC₄H₈−CH (NH₃⁺) COO⁻ was photopolymerized to yield a substituted PDA-NSL. The $\chi^{(3)}$ per unit volume was estimated as 1 × 10⁻¹² esu where the thickness of each polymer monolayer was 45 Å. The fused silica substrate was coated with an octadecyl trichlorosilane (OTS) monolayer and its $\chi^{(3)}$ was evaluated as 6 × 10⁻¹³ esu for 20 Å thickness. The $\chi^{(3)}$ of highly oriented PDA having the side group (CH₂)4−O− CO−NH−(CH₂)ₙ₋₁−CH₃ have also been reported.[130–132] The polymer abbreviated as PDA−C₄UCₙ where U is O−CO−NH and n = 3, 4, 5. The vacuum deposited films show strong dependence of orientation effect. Figure 14 represents the relative third harmonic intensities as a function of rotation angle. The THG intensity for parallel polarization is several folds larger than that of perpendicularly polarized showing anisotropy in $\chi^{(3)}$ values. The

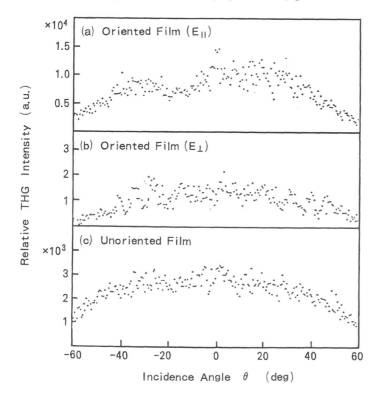

Figure 14 Relative third-harmonic intensities of PDA-C₄UCₙ as a function of rotation angle around a vertical axis at 1.90 μm pumping in the air at room temperature. (From Kanetake et al., *Appl. Phys. Lett.*, 54, 2287, 1989. With permission.)

$\chi^{(3)}$ values of 2.9×10^{-10} esu and 0.03×10^{-10} esu have been measured for parallel and perpendicular polarizations, respectively. A maximum value of 5×10^{-10} esu has been estimated for a pumping wavelength of 1.90 μm. The isotropic $\chi^{(3)}$ values of the polymers were 1.8×10^{-10}, 1.03×10^{-10}, and 0.94×10^{-10} esu for $n = 3, 4$, and 5, respectively. The THG intensities of unoriented films are independent of the polarization direction because the polymer chains are randomly oriented within the plane parallel to the substrate. The $\chi^{(3)}$ value is not remarkably affected by alkyl chain length and it ranges from 2.1 to 3.8×10^{-10} esu for oriented PDA films. The maximum $\chi^{(3)}$ value of 1.7×10^{-9} esu is achieved at the exciton absorption peak. The comparison of results indicates that the magnitude of $\chi^{(3)}$ for PDAs varies by two orders depending on the material and measurement wavelengths.

NLO susceptibilities of a variety of polydiacetylenes have been reported by Nakanishi et al.[133] The $\chi^{(3)}$ value for a poly-DCHD was 8×10^{-10} esu at 1.97 μm, which is larger than that of poly-PTS. The $\chi^{(3)}$ values of 10^{-6} esu on resonance (wavelength region 630–660 nm) and 10^{-7} esu near resonance (wavelength region 700–730 nm) for PDA-DCH thin single crystals for the polarization along the main chain direction has been reported using the DFWM technique.[134] The π-electron conjugation between the polymer main chain and aromatic substituents plays an important role in enhancing the $\chi^{(3)}$ value. The dihedral angles between polymer main chain and the phenyl substituents were 50° for poly-BTFP and 67° for poly-DFMP and were considered a measure of π-electron conjugation. The $\chi^{(3)}$ value of PDA-BTFP at nonresonant region of 2.1 μm was found to be 5 times larger than that of PDA-PTS. The directly bound aromatic substituents were more effective in enhancing the third-order optical nonlinearities of PDAs. Okada et al.[135,136] prepared PDAs containing π-electron conjugation between polymer backbone and side chains. In particular unsymmetrical diphenylbutadyne derivatives with trifluoromethyl substituents on one of the phenyl rings were synthesized. The PDA-MADF obtained by topochemical polymerization showed $\chi^{(3)}$ values larger than PDA-BTFP in the nonresonant region at the pumping wavelength of 2.16 μm. The large $\chi^{(3)}$ value probably arises because of the extended π-electron conjugation between polymer backbone and substituents. Ohsugi et al.[137] synthesized unsymmetrical diacetylene derivatives with aromatic ring on one side and carbazolylmethyl at the other side of the diacetylene moiety, to obtain polydiacetylenes with aromatic substituents directly bound to the polymer backbone. PDA-DFCP was obtained by topochemical polymerization. The $\chi^{(3)}$ value of poly-DFCP was larger than that of PDA-BTFP. The $\chi^{(3)}$ value results of PDA-DFCP is of the same magnitude to that of poly(diphenylacetylene) because of the similar degree of π-electron conjugation between the polymer backbone and phenyl ring.

Doi et al.[138] prepared poly-5-BCMU which has larger $\chi^{(3)}$ values than the well-known poly-4-BCMU. Table 14 lists the $\chi^{(3)}$ values of various poly-4BCMU and poly-5BCMU measured at different wavelengths using the THG technique. Philippart et al.[139] performed pump-probe femtosecond Kerr measurement in PDA-4-BCMU yellow in DMF solutions at 620 nm. A two-photon resonance was observed at 4 eV with $\beta \simeq 4 \times 10^{-3}$ cm/MW. The two-photon excited states have relaxation time of 12 ps. The $\chi^{(3)}$ value of 6×10^{-12} esu (8×10^{-20} m^2/V^2) was measured for pure polymer outside the Raman resonances and the molecular hyperpolarizability γ per repeat unit was calculated as 1.2×10^{-33} esu (1.7×10^{-47} m^5/V^2). Hsu et al.[140] reported that the $\chi^{(3)}$ values of red form single-crystalline PDA-4-BCMU were 5 times larger than those of spin-coated amorphous thin films due to the orientation of the polymer chains. The $\chi^{(3)}$ values of thin films at 1.319 μm were 2 times larger than the $\chi^{(3)}$ values at 1.064 μm because of either a three-photon or a two-photon resonance effect.

Third-order NLO susceptibilities at various wavelengths for the polydiacetylenes obtained by solid-state polymerization of tetrayne compounds have been reported.[141–143] The polydiacetylenes with n-BCMU groups are referred to as PDA-3BCMU-T ($n = 3$), PDA-4BCMU-T ($n = 4$), PDA-5BCMU-

Table 14 Third-order Nonlinear Susceptibilities of Spin-coated Poly(*n*-BCMU) Films

Polymer	λ_{max} (nm)	Thickness (μm)	$\chi^{(3)}$ (10^{-11} esu) at λ (μm)					
			1.50	1.62	1.74	1.86	1.98	2.10
Poly(4-BCMU)	700	0.06	5.4	5.0	6.1	10	9.2	6.8
		0.10	4.9	4.3	6.6	8.5	8.6	9.1
Poly(5-BCMU)	700	0.10	11	10	9.3	18	17	24
		0.26	5.8	5.4	9.6	12	11	15

From Doi et al., 52nd Meeting of the Japan Society of Applied Physics, 1991. With permission.

656

T ($n = 5$). The resonant $\chi^{(3)}(-3\omega; \omega, \omega, \omega)$ values of the spin-coated 0.10 μm-thick films of PDA-4BCMU-T, PDA-5BCMU-T, and PDA-PTS-T were 9.1×10^{-11}, 2.4×10^{-10}, and 9.9×10^{-11} esu at 2.1 μm wavelength, respectively. The $\chi^{(3)}(-3\omega; \omega, \omega, \omega)$ values at 1.50 μm were slightly smaller than those obtained at $1.62 \sim 2.10$ μm. The PDA-5ECMU-T polymerized by UV and γ irradiations showed different $\chi^{(3)}(-3\omega; \omega, \omega, \omega)$ values, slightly higher for those polymerized by γ irradiation because of intense deterioration of the polymer by UV radiation. The $\chi^{(3)}$ values of PDA-5BCMU-T-BTFP prepared from octateraynes were found to be lower than that of PDA-5BCMU-T polymer. PDA-5BCMU-T-BTFP polymer has two side groups: 5-BCMU-T on one side and BTFP on another side of the polymer backbone. The $\chi^{(3)}$ values of PDAs with arylurethane substituents of tetrayne compounds have also been reported.[144]. Shimada et al.[144] reported $\chi^{(3)}(-3\omega; \omega, \omega, \omega)$ values of PDA containing sulfur atoms directly bound to the main chain. The largest $\chi^{(3)}$ of 2.5×10^{-11} esu at 1.8 μm was obtained for thin films. In the THG measurements, the $\chi^{(3)}$ value of 1×10^{-14} esu for quartz was used as a reference over the pumping wavelength of 1.5 to 2.1 μm compared with 3.0×10^{-14} esu used in other measurements by this research group. The $\chi^{(3)}$ values of these PDAs with arylurethane groups vary with wavelength measurement and chemical structure.

Tetrayne monomer   →[γ-ray]→   Tetrayne PDA

$R = -(CH_2)_n-OCONHCH_2COO-(CH_2)_3-CH_3$
$n = 3$ (PDA-3BCMU-DT)
$n = 4$ (PDA-4BCMU-DT)
$n = 5$ (PDA-5BCMU-DT)

$R = -(CH_2)_5-OCONHCH_2COO-(CH_2)_2-CH_3$
(PDA-5ECMU-T)

$R = -(CH_2)_4-OSO_2-$⟨benzene⟩$-CH_3$
(PDA-PTS-T)

$R = -(CH_2)_5-OCONH-$⟨benzene⟩
(PDA-5PU-T)

$R = -(CH_2)_5-OCONH-$⟨benzene⟩$-Cl$
(PDA-5CPU-T)

$R = -(CH_2)_5-OCONH-$⟨benzene with Cl, Cl⟩
(PDA-5DCPU-T)

The PTS polymer shows the largest resonant $\chi^{(3)}$ value, 2×10^{-5} esu.[145] Bogler et al.[146] reported DFWM measured $\chi^{(3)}$ values of PTS between 2.0 and 1.68 eV. The magnitude of imaginary $\chi^{(3)}$ was $5 \pm 3 \times 10^{-9}$ esu in the two-photon region between 1.73 and 1.68 eV. In the region between 1.92 and 1.71 eV, the one-photon contribution varied $\chi^{(3)}$ between 3×10^{-7} and 2×10^{-9} esu. On resonance at ~ 2.0 eV the imaginary $\chi^{(3)}$ was 2×10^{-5} esu. Vogtmann et al.[147] measured the $\chi^{(3)}$ values of PDA-PTS, PDA-FBS, and PDA-4-BCMU single crystals with the DFWM technique. The $\chi^{(3)}/\alpha$ was 1.7×10^{-12} esu cm for the PDA-PTS single crystal.

Agrinskaya et al.[151] reported $\chi^{(3)}$ values of four symmetrical derivatives of diacetylene (1,6-bis-piperidine, 1,6-dimorpholine, 1,6-dimethylphenylamine, and 1,6-bis-diphenylamine) of 2,4-hexadiene measured by four-photon scattering under biharmonical pumping. These PDAs had an intermediate group CH_2 between the substituent and the diacetylene fragment and their $\chi^{(3)}$ values depended on the chemical structures of the substituent; with decreasing band gap the $\chi^{(3)}$ increased as a result of transition from donor substituent to an acceptor substituent (Table 15). The large $\chi^{(3)}$ values were observed on resonance and near resonance. Preparation of oriented films by vacuum evaporation gave $\chi^{(3)}$ values of 10^{-7} esu. The n_2/α values were calculated from: $n_2(cm^2/W) = \chi^{(3)}(esu)/19n_0^2$, where $n_0 = 1.8$ is the linear refraction coefficient for PDA. The relaxation time of these polymers were in picoseconds. Table 16 summarizes the $\chi^{(3)}$ values of various PDAs measured by different techniques at various wavelengths under different material conditions.[119-167] LeMoigne et al.[153-155] reported the effect of orientation on the $\chi^{(3)}$ value of PDA-p-DCH thin films obtained by epitaxial growth of the monomer on a single-crystal substrate. The $\chi^{(3)}$ of the mono-oriented (1.04×10^{-10} esu) films was an order of magnitude larger than unoriented films (1×10^{-11} esu) and the ratio of $\chi^{(3)}$ between the bi-oriented film and the three-dimensional isotropic film was 5. The $\chi^{(3)}_\parallel$ values of 4.0×10^{-8} esu and $\chi^{(3)}_\perp$ of 5.9×10^{-9} esu at 532 nm were measured by DFWM for a 19-layer LB films of poly{1-[(((butoxy-carbonyl)methyl)ami-doyl)oxy]-4-pyrimidyldiacetylene} PDA-BPOD by Cheong et al.[164] Molyneux et al.[165] measured the spectral dependence of the $\chi^{(3)}$ value of the red form of PDA-9BCMU in the near-resonant region and found that the imaginary component changed signs in the exciton absorption tail. The nonlinear refractive index n_2 was 1.9×10^{-10} cm^2/W at Im $\chi^{(3)} = 0$ position showing a high figure of merit. The n_2 values were 5.0×10^{-10}, 1.0×10^{-10}, and 3.2×10^{-11} cm^2/W at 2.08, 1.93, and 1.75 eV. The n_2 values were obtained from $\chi^{(3)}$ measurements through the relationship: $n_2(cm^2/W) = 0.0395$ Re $\chi^{(3)}$ (esu)$/n_0^2$. The figure of merit estimated from $n_2/\alpha\tau$ were 0.14 at 1.93 eV.

Byrne and Blau[168] measured optical nonlinearity of diacetylene-type polymers having anthracene and benzene rings in the backbone referred to as P28. The toluene solution of P28 shows a second hyperpolarizability $\gamma = 7.16 \times 10^{-28}$ esu (1×10^{-41} m^5/V^2). The optical nonlinearity of P28 is similar

Table 15 Polymer Content, α and $\chi^{(3)}$ at 0.57 µm, and Absorption Coefficients of PDAs with Different Subtituents

Substituents	Basicity	Polymer Contents (%)	$\chi^{(3)}$ (10^{-10} esu)	α(cm^{-1}) at 0.57 µm	n_2/α (cm^3/W)
1,6-bis-piperidine	11.1	0.5	0.001	50	10^{-16}
CH$_2$—N⬡					
1,6-dimorpholine	8.3	1	0.1 ~ 0.4	2×10^2	$1.2 \sim 5.0 \times 10^{-15}$
CH$_2$—N⬡O					
1,6-dimethylphenylamine	5.04	2	1.2	10^3	2.5×10^{-15}
CH$_2$—N(CH$_3$)⬡					
1,6-dimethylphenylamine^a	5.04	10	1000	10^5	2.5×10^{-14}
1,6-bis-diphenylamine	0.79	2	2.0	2×10^3	2.5×10^{-15}
CH$_2$—N(⬡)(⬡)					

From Agrinskaya et al., *Synth. Met.*, 54, 289, 1993. With permission.
^{a} Vacuum evaporated film.

Table 16 The Summary of $\chi^{(3)}$ Values of Polydiacetylenes

Polydiacetylenes	$\chi^{(3)}$ $(10^{-10}$ esu)	Wavelength (μm)	Experimental Technique	Ref.
PDA-PTS (single crystal)	8.5	1.89	THG	119
	1.6	2.62	THG	119
PDA-PTS (single crystal)	90	0.6515	DFWM	121
PDA-PTS (single crystal)	7.6	1.83	THG	133
	2.0	1.90	THG	133
	1.1	1.97	THG	133
	0.45	2.04	THG	133
	0.36	2.10	THG	133
PDA-PTS (single crystal)	2×10^5	1.97	SA	145
PDA-PTS (solution)	2×10^5	0.622	DFWM	146
PDA-PTS-12 (solution)	5.0	1.064	EFISH	162
	0.11	1.064	THG	163
	0.61	1.907	THG	163
PDA-FBS (single crystal)	1.0	0.720	DFWM	147
PDA-TCDU (single crystal)	0.7	1.89	THG	119
	0.37	2.62	THG	119
PDA-AFA (thin film)	4.3		EFISH	123,124
PDA-AFA (LB film-red)	0.20	1.22	THG	126
PDA-AFA (LB film-blue)	0.34	1.64	THG	122
PDA-AFA (LB film)	0.01	1.15	EFISH	161,124
PDA-AFA (thin film blue)	15	1.35	THG	124
PDA-AFA (thin film red)	0.19	1.35	THG	124
PDA-NTA (LB film)	0.12	1.90	THG	135
PDA-4-BCMU (thin film)	1.5		EFISH	161,124
(vacuum-deposited film)	2.9	1.90	THG	158–160
PDA-4-BCMU (red gel)	0.13	1.064	DFWM	152
(blue film)	0.60	1.064	DFWM	152
(yellow solution)	0.042	1.064	DFWM	152
PDA-4-BCMU (thin film)	1.0	1.86	THG	141
PDA-4-BCMU (amorphous film)	0.2	1.319	THG	140
	0.096	1.064	THG	140
PDA-4-BCMU (crystalline film-red)	0.49	1.064	THG	140
	1.0	1.319	THG	140
PDA-4-BCMU (thin film)	2.0	0.620	DFWM	148
PDA-4-BCMU (yellow-solution)	0.06	0.620	OKE	139
PDA-5-BCMU (thin film)	2.4	2.1	THG	138
PDA-3-BCMU (blue gel)	9.0	1.064	DFWM	152
PDA-p-DCH (thin film)	0.7		THG	153
(mono-oriented)	1.04	1.064	THG	154
(bi-oriented)	0.5		THG	154
PDA-DCH (oriented film)	6.0	1.907	THG	156
PDA-DCH	0.64	1.35	EFISH	135
PDA-DCHD (single crystal)	1.7	2.10	THG	133
	4.0	1.83	THG	133
	4.5	1.90	THG	133
	8.0	1.97	THG	133
	3.0	2.04	THG	133
PDA-DCH (single crystal)	2×10^4	0.637	DFWM	134
PDA-C$_4$UC$_3$ (oriented film)	3.8	1.90	THG	130
PDA-C$_4$UC$_4$ (oriented film)	2.9	1.90	THG	130
(red phase)	5.0	1.60	THG	131
(blue phase)	17	1.90	THG	131
PDA-C$_4$UC$_5$ (oriented film)	2.1	1.90	THG	130
PDA-BTFP (thin film)	7.7	1.83	THG	133
	4.3	1.94	THG	133
	1.8	2.10	THG	133
PDA-DFMP (thin film)	1.7	1.83	THG	133
	0.85	1.94	THG	133
	0.45	2.10	THG	133

Table 16 *Continued*

Polydiacetylenes	$\chi^{(3)}$ $(10^{-10}$ esu)	Wavelength (μm)	Experimental Technique	Ref.
PDA-DFCP (thin film)	1.1	1.90	THG	137
PDA-MADF (thin film)	8.0		THG	157
PDA-C_3H_6 (cast film)	0.052	1.90	THG	52,149
PDA-C_3H_7 (vacuum-deposited film)	3.8	1.90	THG	52,149
PDA-POL (vacuum-deposited film)	0.11	1.90	THG	52,158
PDA-$C_{17}H_{37}$ (cast film)	0.014	1.90	THG	52,158
PDA-C_4H_9 (vacuum deposited film)	0.14	1.90	THG	52,159
PDA-DOC (vacuum deposited film)	0.5	1.90	THG	52,160
PDA-3BCMU-T (thin film)	0.15	1.86	THG	142,143
	0.37	2.1	THG	142,143
PDA-4BCMU-T (thin film)	0.18	1.50	THG	142,143
	0.2	1.74	THG	142,143
	0.33	1.86	THG	142,143
	2.3	2.1	THG	142,143
PDA-5BCMU-T (thin film)	0.37	1.50	THG	142
	0.6	1.86	THG	142
	0.43	1.86	THG	143
	0.47	1.98	THG	143
	0.8	2.1	THG	142
	0.6	2.1	THG	143
	0.37	2.22	THG	143
	0.27	2.34	THG	143
	0.24	2.46	THG	143
PDA-5BCMU-T-BTFP (thin film)	0.23	1.50	THG	142
	0.31	1.74	THG	142
	0.37	2.1	THG	142
PDA-5ECMU-T (thin film)	0.11	1.62	THG	142
	0.17	1.98	THG	142
	0.23	2.1	THG	142
PDA-PTS-T (thin film)	0.99	2.1	THG	141
PDA-5PU-T (thin film)	0.044	1.5	THG	144
	0.014	1.62	THG	144
	0.048	1.74	THG	144
	0.043	2.10	THG	144
PDA-5-oCPU-T (thin film)	0.027	1.5	THG	144
	0.042	1.62	THG	144
	0.079	1.74	THG	144
	0.10	1.98	THG	144
	0.14	2.10	THG	144
PDA-5-mCPU-T (thin film)	0.013	1.5	THG	144
	0.04	1.74	THG	144
	0.051	1.86	THG	144
	0.045	2.10	THG	144
PDA-5-pCPU-T (thin film)	0.019	1.5	THG	144
	0.061	1.74	THG	144
	0.083	1.86	THG	144
	0.090	2.10	THG	144
PDA-5DCPU-T (thin film)	0.031	1.5	THG	144
	0.077	1.74	THG	144
	0.089	1.86	THG	144
	0.10	2.10	THG	144
PDA-BPP (thin film)	0.001	0.572	BHP	151
PDA-DML (thin film)	0.4	0.572	BHP	151
PDA-DMP (thin film)	1.2	0.572	BHP	151
PDA-DMP (vacuum-evap.^{a})	1000	0.572	BHP	151
PDA-BPA (yellow thin film)	2.0	0.572	BHP	151
PDA-NSL (LB film)	0.01	1.064	THG	129
PDA-BPOD (LF film)	400	0.532	DFWM	164

Table 16 *Continued*

Polydiacetylenes	$\chi^{(3)}$ $(10^{-10}$ esu)	Wavelength (μm)	Experimental Technique	Ref.
PDA-9-BCMU (red phase film)	330	0.595	PP	165
	120	0.641	PP	165
	20	0.708	PP	165
	0.1	1.907	THG	167
(blue phase film)	0.3	1.907	THG	167
PDA-3-SMBU	20	0.710	DFWM	167
PDA-9-SMBU	0.003	1.3	DFWM	167
PDA with S atoms (thin films)	0.23	1.53	THG	144b
	0.20	1.62	THG	144b
	0.25	1.80	THG	144b
	0.12	1.92	THG	144b
	0.16	2.10	THG	144b
	0.19	2.16	THG	144b

BHP = four-photon scattering under biharmonical pumping; PP = two-beam pump-probe.

[a] Vacuum-evaporated oriented PDA-DMP films.

to the optical nonlinearity of polydiacetylenes and shows no multiphoton enhancement. The $\chi^{(3)}$ of 2 $\times 10^{-12}$ esu was measured by THG from a thin, nonaligned film of rigid rod polyphenylenynylene having decyl-diether groups.[153]

$$(P28)$$

Oikawa et al.[169] reported third-order optical nonlinearities of poly(1,9-decadiyne) (PDD). The PDD was synthesized by oxidative polymerization of 1,9-decadiyne. The polymer can be spin coated. The films are opaque but become transparent after thermal treatment. The $\chi^{(3)}$ value of 1.2×10^{-12} esu was measured at 1.907 μm. Cross-linking is expected to increase the magnitude of the $\chi^{(3)}$ because of the expansion of the π-conjugation system. One-dimensional and two-dimensional π-conjugated systems can be obtained by cross-linking depending on the choice of the monomer compound.

$$(PDD)$$

Yu et al.[170] prepared a copolymer containing polyester and diacetylene segments by condensation of hexidyne-1,6-diol with terephthaloyl chloride. The copolymers are soluble in DMF and DMSO and can be processed into films from solutions. On heating the copolymer film to 150 °C for 30 min, the color changes from colorless to brown which indicates an increase in electron delocalization. The DFWM signal of the colorless film was hardly detected whereas the thermally treated sample showed a strong signal. The DFWM experiments performed at 532 nm showed an α of 3.9×10^3 cm^{-1} and a third-order susceptibility $\chi^{(3)}$ value of 3.2×10^{-10} esu comparable with polydiacetylene films.

Kondo et al.[171] reported the synthesis and $\chi^{(3)}$ value of soluble poly(phenylene-ethylene). The film cast from chloroform solution showed an absorption maximum at 335 nm, and the $\chi^{(3)}$ value measured by DFWM at 532 nm in a benzene solution was 1.8×10^{-8} esu.

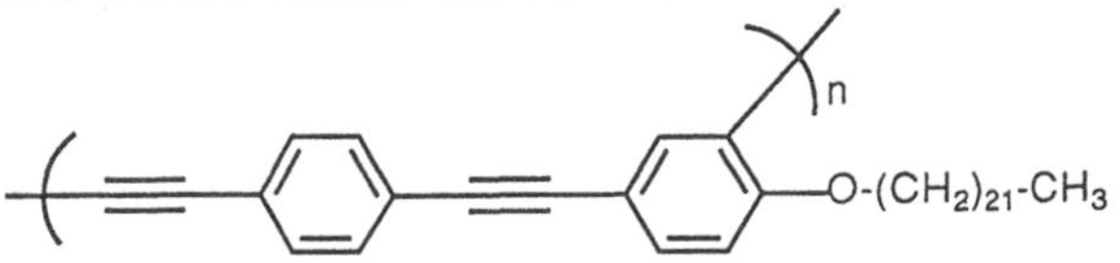

Abdelaven et al.[172] reported an $\chi^{(3)}$ value of 10^{-7} esu for a polydiacetylene derivative of 2-methyl-4-nitroaniline measured by DFWM at 532 nm. The response time of the derivatized PDA film was in picoseconds. Several waveguides of the PDA derivative were also fabricated.

a. Model Compounds of Polydiacetylenes

The study of model compounds of conjugated backbones can assist in establishing the structure–property relationships between optical nonlinearity and molecular structures. Byrne et al.[173] reported optical and third-order NLO properties of a series of enyne oligomers with increasing degrees of conjugation in n-hexane solution using the DFWM at 1.064 µm. Oligomeric compounds with well-defined diacetylene chain lengths were built up sequentially. The chemical structures of oligodiacetylene (ODA) are shown below where $n = 1$, 2, 3, and 5 and each compund has $4n + 2$ conjugated carbons; therefore, they are referred to as ODA-6, ODA-10, ODA-14, and ODA-22, respectively. The end-capped bulky *tert*-butyl groups provide high stability to the chain. The ODA-22 is less soluble in n-hexane, otherwise all other ODAs are readily soluble.

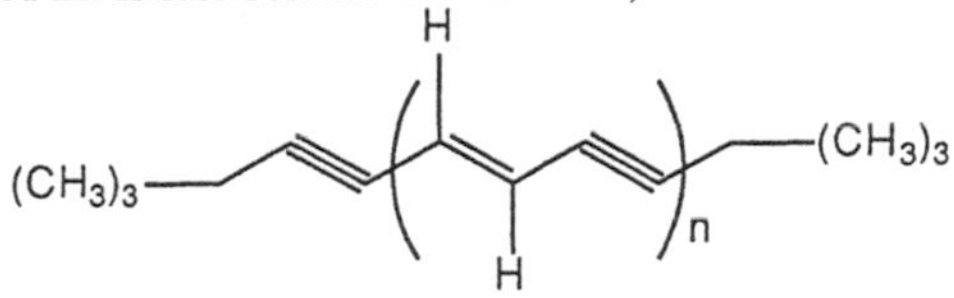

The absorption peaks were observed at 275 nm for ODA-6, at 332 nm for ODA-10, at 374 nm for ODA-14, and at 418 nm for ODA-22. The molar extinction coefficients of these ODAs ranged from 2.56×10^{-4} to 5.66×10^{-4} cm^2/mol. The $\chi^{(3)}$ measurements as a function of concentration indicated contributions from both real and imaginary components, and real components were found to be negative for longer chain lengths. All ODAs had a large contribution to the γ value from the imaginary component. The γ values of ODAs changed by two orders of magnitude as a function of chain length. The plot of γ values as a function of ODA chain length predicted a power-law dependence of about 4.0 ± 0.5.

Bosshard et al.[174a] reported third-order NLO properties of a series of donor–acceptor substituted tetraethynylethenes. The chemical structures, their λ_{max}, γ, and $\chi^{(3)}$ values are shown below. Third-order optical nonlinearities of chloroform solutions with concentrations between 0.25 and 0.36 wt% were measured by the THG at 1.907 µm. The $\chi^{(3)}$ values were extrapolated from measured γ values. First, the mono- and bis-aryl tetraethynylethenes having N,N'-dimethylamino ($N(CH_3)_2$) donor groups exhibit larger optical nonlinearities than those of nitro (NO_2) acceptor group compounds. Second, the compounds having stronger donor methoxy groups show 2 to 6-fold larger γ values than N,N'-dimethylamino donor groups. Third, acentric substitution in tetraethynylethenes such as those consisting of N,N'-dimethylamino and nitro groups leads to larger optical nonlinearities than those having the same groups, either NO_2 or $N(CH_3)_2$, due to the lowering of symmetry. The γ values of tetraethynylethenes increased with the increasing conjugation length as well as due to existence of two-dimensional conjugated paths.

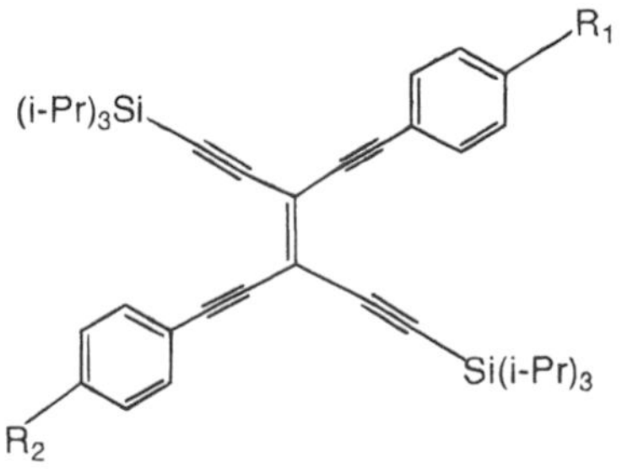

$R_1 = R_2 = NO_2$	$\gamma = 2.3 \times 10^{-34}$ esu	$\chi^{(3)} = 1.0 \times 10^{-12}$ esu
$R_1 = R_2 = N(CH_3)_2$	$\gamma = 4.2 \times 10^{-34}$ esu	$\chi^{(3)} = 1.93 \times 10^{-12}$ esu
$R_1 = N(CH_3)_2, R_2 = NO_2$	$\gamma = 1.0 \times 10^{-33}$ esu	$\chi^{(3)} = 4.73 \times 10^{-12}$ esu

$R_1 = R_2 = N(CH_3)_2 = R_3 = Si(i-Pr)_3$ $\gamma = 3.1 \times 10^{-34}$ esu $\chi^{(3)} = 1.43 \times 10^{-12}$ esu
$R_1 = R_2 = OCH_3, R_3 = Si(CH_3)_3$ $\gamma = 1.3 \times 10^{-34}$ esu $\chi^{(3)} = 7.01 \times 10^{-13}$ esu
$R_1 = N(CH_3)_2, R_2 = NO_2, R_3 = Si(i-Pr)_3$ $\gamma = 5.9 \times 10^{-35}$ esu $\chi^{(3)} = 2.80 \times 10^{-13}$ esu

$R_1 = R_2 = OCH_3, R_3 = NO_2$ $\gamma = 2.4 \times 10^{-34}$ esu $\chi^{(3)} = 1.22 \times 10^{-12}$ esu
$R_1 = R_2 = N(CH_3)_2, R_3 = NO_2$ $\gamma = 1.5 \times 10^{-33}$ esu $\chi^{(3)} = 7.16 \times 10^{-12}$ esu

$R_1 = R_2 = N(CH_3)_2, R_3 = R_4 = NO_2$ $\gamma = 3.2 \times 10^{-33}$ esu $\chi^{(3)} = 1.57 \times 10^{-11}$ esu

$\gamma = 1.1 \times 10^{-33}$ esu $\chi^{(3)} = 5.0 \times 10^{-12}$ esu

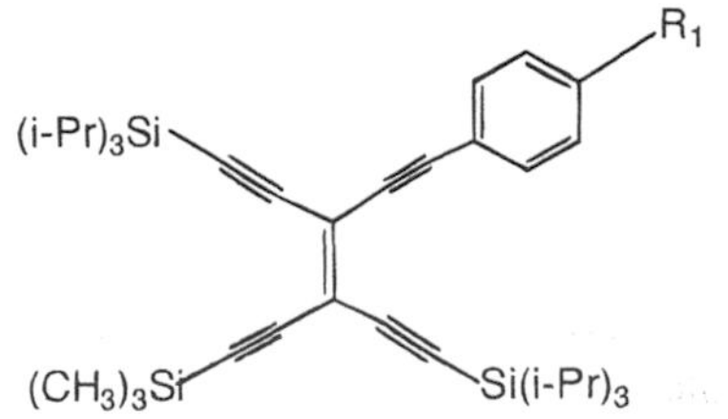

R_1 = NO$_2$ $\gamma = 9.5 \times 10^{-35}$ esu $\chi^{(3)} = 4.8 \times 10^{-13}$ esu

R_1 = N(CH$_3$)$_2$ $\gamma = 2.7 \times 10^{-34}$ esu $\chi^{(3)} = 1.36 \times 10^{-12}$ esu

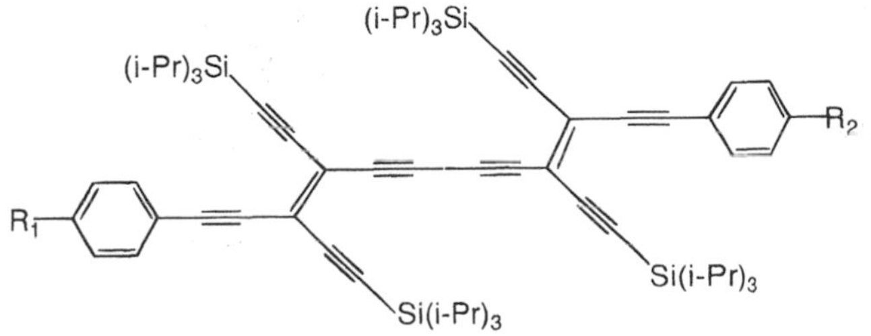

$R_1 = R_2$ = NO$_2$ $\gamma = 5.3 \times 10^{-34}$ esu $\chi^{(3)} = 1.43 \times 10^{-12}$ esu

$R_1 = R_2$ = N(CH$_3$)$_2$ $\gamma = 1.9 \times 10^{-33}$ esu $\chi^{(3)} = 5.37 \times 10^{-12}$ esu

R_1 = NO$_2$, R_2 = N(CH$_3$)$_2$ $\gamma = 3.6 \times 10^{-33}$ esu $\chi^{(3)} = 1.0 \times 10^{-11}$ esu

Wautelet et al.[174b] reported third-order NLO properties of low molecular weight oligomeric poly(phenylene-ethynylene) type compounds and their corresponding high molecular weight polymers. The $\chi^{(3)}$ values of thin films of conjugated oligomers and polymers having flexible alkyl chains were measured by the DFWM technique. The $\chi^{(3)}(-\omega; \omega, \omega, -\omega)$ values of 2.5×10^{-10} esu for PY-OCH$_3$, 1.6×10^{-10} esu for PY5 OC12, 4.6 $\times$ 10^{-10} esu for pPY-OC12, and also 4.6×10^{-10} esu for pPY-COOC12 were measured at 620 nm. Interestingly, the $\chi^{(3)}(-\omega; \omega, \omega, -\omega)$ value of the pentamer (PY5-OC12) is only three times lower than its corresponding polymer (pPY-OC12). The thin films of trimer PY-OCH$_3$ shows larger $\chi^{(3)}(-\omega; \omega, \omega, -\omega)$ value, only half the value of high molecular weight polymers.

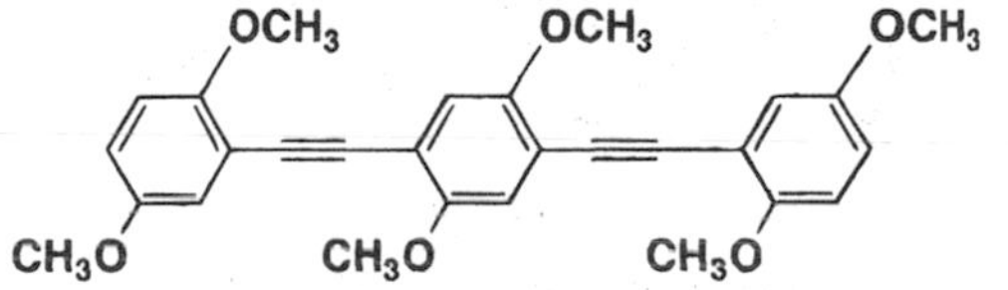

(PY3-OCH$_3$)

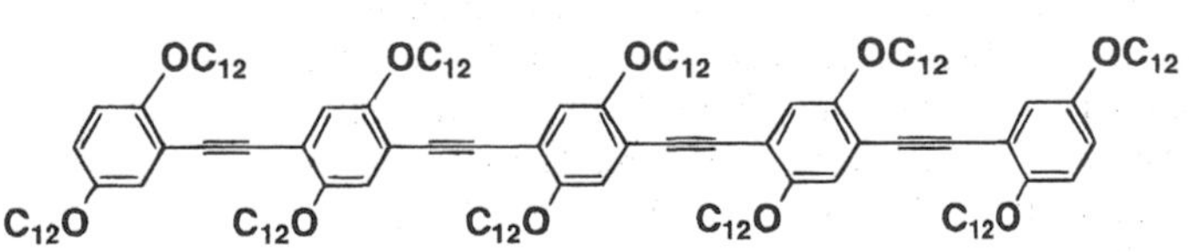

(PY5-OC12)

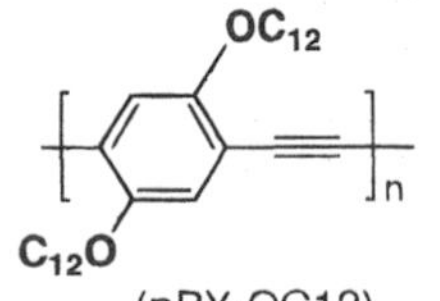

(pPY-OC12)

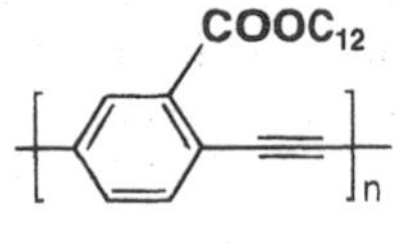

(pPY-COOC12)

Samuel et al.[174c] reported third-order NLO properties of oligo(3-ethylthiophene ethynylene)s with up to 16 repeat units. The chemical structure of oligo(3-ethylthiophene ethynylene)s (OETE) having repeat units $N = 2$ (OETE-2, π-electron = 16), $N = 4$ (OETE-4, π-electron = 32), $N = 8$ (OETE-8, π-electron = 64), and $N = 16$ (OETE-16, π-electron = 128) are shown below. The γ values of 0.47

$\times~10^{-34}$ esu for OETE-2, 2.77 $\times~10^{-34}$ esu for OETE-4, 12.6 $\times~10^{-34}$ esu for OETE-8, and 3.15 $\times~10^{-34}$ esu for OETE-16 in tetrahydrofuran solution were measured by THG at 1.907 µm. The scaling of second hyperpolarizability per repeat unit (γ/N) with chain length showed that the γ/N increases strongly up to $N = 8$ and then tends to level off for $N = 16$ indicating saturation effect. The power law fit yielded an exponent $\alpha = 2.4$, lower than the $\alpha = 2.8$ reported for the γ-THG of oligothiophenes with similar number of repeat units. Probably trimethylsilyl groups acting as electron donor groups boost the γ value of these molecules.

(OETE-2)

(OETE-4)

(OETE-8)

(OETE-16)

Perry et al.[175] and Yuan et al.[176] reported THG $\chi^{(3)}$ values for symmetric end-capped diacetylene and acetylene compounds. Table 17 lists the $\chi^{(3)}$ values for symmetric end-capped diacetylene compounds measured by the THG technique. The $\chi^{(3)}$ and hyperpolarizability values are larger for diacetylenes with longer π-conjugations. The bisferrocenyl compound shows the largest optical nonlinearity. Fleitz et al.[177] reported nonlinear absorption in diphenyl butadiene, diphenyl butadiyne, and a series of acetylene compounds using the Z-scan technique at 532 nm. The measured nonlinear absorption β coefficients were 0.8 $\times~10^{-11}$, 4.8 $\times~10^{-11}$, 2.7 $\times~10^{-11}$, 21 $\times~10^{-11}$, 20 $\times~10^{-11}$, and 27 $\times~10^{-11}$ cm/W at an intensity of 2.8 $\times~10^{10}$ W/cm^2 for diphenyl acetylene, diphenyl butadiene, diphenyl butadiyne, 1,4-bis(phenylethynyl) benzene, bis(4-biphenyl) acetylene, and 4,4'-bis(phenylethynyl)biphenyl, respectively. Very large two-photon absorption cross-sections in bis(4-biphenyl) acetylene and 4,4'-bis(phenylethynyl) biphenyl evidenced two-photon allowed states close to 37,590 cm^{-1} in these compounds.

Sylla et al.[178] reported $\chi^{(3)}$ of acetylenic analogues of tetrathisfulvalene (TTF) measured by DFWM at 532 nm. The $\chi^{(3)}_{xxxx}$ values were 7.68 $\times~10^{-13}$, 1.24 $\times~10^{-12}$, and 4.88 $\times~10^{-13}$ esu and the second hyperpolarizability γ_{xxxx} values were 1.72 $\times~10^{-31}$, 2.7 $\times~10^{-31}$, and 1.08 $\times~10^{-31}$ esu for (a), (b), and (c), respectively. The $\chi^{(3)}$ values follow the order (b) > (a) $\geqslant$ (c); this indicates that the polar electron-withdrawing methoxycarbonyl group (COOCH$_3$) contributes to large optical nonlinearity of the TTF derivatives. The γ_{xxxx} values of soluble acetylenic analogs of TTF are as high as 10^5 times that of carbon disulfide. The $\chi^{(3)}$ increases with solution concentration up to ~10 mg/ml for (b) and ~20 mg/ml for (a) and becomes independent as the concentration increases further. The response time of these TTF derivatives was shorter than 25 ps.

Table 17 $\chi^{(3)}$ and Hyperpolarizabilities of Symmetric End-capped Diacetylene and Acetylene Compounds Measured by the THG Technique at 1.064 and 1.907 μm

No.	Compound	$\chi^{(3)}$ (10^{-12} esu)		γ (10^{-36} esu)	
		1.064	1.907	1.064	1.907
1	$(CH_3)_3SiC{\equiv}CH$	—	0.07	—	5.4
2	$(CH_3)_3SiC{\equiv}CSi(CH_3)_3$	—	—	9.0	—
3	$(Et)_3Si{\cdot}C{\equiv}C{-}Si(CH_3)_3$	—	—	15	16
4	$(Et)_3Si{-}({\equiv})_n{-}Si(Et)_3$				
	$n = 2$	0.11	0.12	—	—
	$n = 4$	0.97	0.95	54	55
5	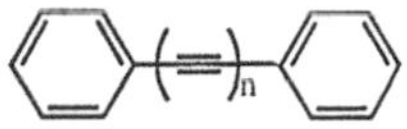				
	$n = 1$	0.67	0.46	27	18
	$n = 2$	1.80	0.92	68	38

From Perry et al., *Organic Materials for Nonlinear Optics*, Special Publication No. 69, 189, 1989; Perry, *Materials for Nonlinear Optics*, ACS Symposium Series 455, 67, 1991. With permission.

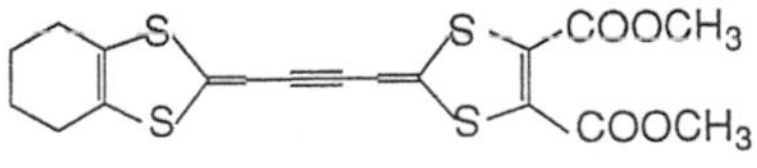

(a)

(b)

(c)

2. Polyacetylenes and Model Compounds

Polyacetylene, the simplest conjugated organic polymer, consists of an array of single and double bonds in a carbon chain.[113,114] Polyacetylene has two isomers: the *cis*- and *trans*-isomers. Both the *cis*- and *trans*-isomers can be obtained as free-standing flexible films. A nearly 100% *cis*-isomer can be obtained by performing polymerization at $-78\,°C$, whereas thermal isomerization at elevated temperatures (above 150 °C) under vacuum leads to the formation of a *trans*-isomer.

trans-polyacetylene

cis-polyacetylene

It is also possible to obtain the desired *cis/trans* ratio by performing isomerization at low temperatures. These two isomers have different physicochemical properties. For example, the pristine film of the *cis*-isomer is insulating, whereas the pristine film of the *trans*-isomer is semiconducting. The unique electrophysical properties of polyacetylene originate from the highly conjugated carbon backbone. Doped polyacetylenes show the largest electrical conductivity, as large as that of copper metal. The importance of polyacetylene system likely has also been realized for studying the third-order optical nonlinearity. There have been some major developments in preparing polyacetylene by different methods and they were named based on the procedure involved; for example, Shirakawa-PA, Naarmann-PA, Durham-PA, and Luttinger-PA. Several research groups, have made detailed studies on the third-order nonlinear properties.

Holvorson et al.[179] performed THG measurements over a wide range of wavelengths for both *trans*- and *cis*-isomers of polyacetylene. Figure 15 shows the third-order optical susceptibilities $\chi^{(3)}$ of *cis*- and *trans*-polyacetylene measured by the THG method over pump frequencies from 0.55 eV to 1.25 eV. The magnitude of $\chi^{(3)}(-3\omega; \omega, \omega, \omega)$ for the *trans*-isomer was an order of magnitude larger than that of the *cis*-isomer over the entire spectral range even at 3ω resonance maxima. Sinclaire et al.[180] investigated the $\chi^{(3)}$ value of *cis*-polyacetylene at -78 °C and, after isomerizing the same sample to *trans*-isomer, remeasured its THG. The magnitude of $\chi^{(3)}$ was 4.0×10^{-10} esu for *trans*-PA in which the component of the third-order susceptibility tensor with all indices parallels the polymer chain direction. Figure 16 shows the polarization dependence of THG from an aligned sample. The triangle indicates the measured power as a function of the angle between the chain direction and the polarization of the fundamental beam. The magnitude of the THG signal is about 30 times larger in the case where the fundamental field is polarized parallel to the chain direction compared to the magnitude of the THG signal in the case where the fundamental field polarization and chain directions are orthogonal. The dashed curve shows characteristic behavior anticipated for a perfectly aligned one-dimensional system. The comparison of $\chi^{(3)}$ values in two isomers indicates that the third-order susceptibility of *cis*-PA is more than an order of magnitude (15–20 times) smaller than that of *trans*-PA as measured on the same sample. The *trans*-PA has a 2-fold degenerate ground state where a solitonic processing involves in the fundamental nonlinear excitations. On the other hand, the ground-state degeneracy is lifted in *cis*-isomer; the important excitations are polarons and bipolarons. Lifting the ground-state degeneracy restrains the nonlinear response in *cis*-isomer. The difference in $\chi^{(3)}$ values for two isomers results from the fundamental change in polymer symmetry. The calculated and experimental results of $\chi^{(3)}$ values based on generation of solitons indicate the nonlinear ground-state fluctuations as the source of NLO properties of polyacetylene. The anisotropy of the THG for *trans*-isomer indicates that the nonlinear response is related to the nonlinear polarizability of the π-electrons in the conjugated polymer backbone. Sinclair et al.[180] concluded that in the conjugated polymers having a degenerate ground state, the nonlinear zero point fluctuations of the ground state lead to an important mechanism for NLO properties.

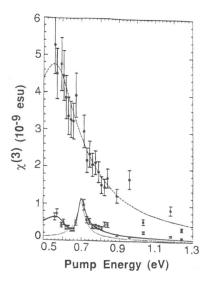

Figure 15 Third-order NLO susceptibility $\chi^{(3)}$ plots for *cis*-polyacetylene (open circle) and *trans*-polyacetylene (solid points) obtained by THG measurements. The solid line represents the data obtained from an effective medium theory analysis for both type of polyacetylenes. (From Holvorson et al., Paper presented at the 5th Toyota Conference at Aich, 1991; *Chem. Phys. Lett.*, 200, 364, 1992.)

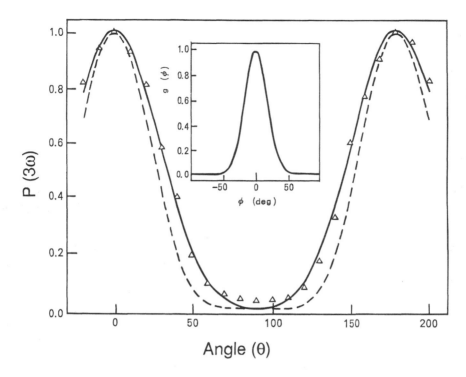

Figure 16 The polarization dependence of THG from an aligned polyacetylene sample. The triangle indicates the measured power as a function of the angle between the chain direction and the polarization of the fundamental beam. The dashed line shows the expected behavior for a perfectly aligned one-dimensional system. Characteristic behavior anticipated for a perfectly aligned one-dimensional system is shown in the inset. (From Sinlair et al., *Synth. Metals,* 28, D645, 1989. With permission.)

The resonant $\chi^{(3)}$ of 3×10^{-8} esu was also reported. A significantly large resonant $\chi^{(3)}$ value of 2×10^{-7} esu at 0.55 eV and an off-resonance $\chi^{(3)}$ value of about 10^{-8} esu were reported by the same group.[181] The THG $\chi^{(3)}$ dispersion showed two resonances, at approximately 0.55 eV caused by three-photon resonance to the $1B_u$ state and a two-photon resonance to a relaxed $2A_g$ state. Another resonance at about 0.80 eV corresponds to the two-photon resonance to the mA_g state. The two-photon absorption spectrum from 0.55 to 0.96 eV by photothermal deflection spectroscopy yielded imaginary $\chi^{(3)}$ values of 5.2×10^{-9} esu at 1.3 μm and 5.4×10^{-8} esu at 1.5 μm. The imaginary $\chi^{(3)}$ value at 0.55 eV was determined as 1.6×10^{-7} esu.[182]

Wintner et al.[183] measured $\chi^{(3)}$ values of oriented *trans*-$(CH)_n$ and reported a value of about 8.1×10^{-9} esu at 1.06 μm, which is an order of magnitude larger than the $\chi^{(3)}$ value of the nonoriented sample. The increment in $\chi^{(3)}$ value results from the orientation of the polymer chains and the high density attained by the material. Etemad and Baker[184] measured the THG $\chi^{(3)}$ spectrum of polyacetylene down to 0.38 eV (3.3 μm) using an infrared free electron laser (FEL). The $\chi^{(3)}$ value increases at 0.59 eV (2.1 μm), which was of the magnitude of 10^{-8} esu; this has been claimed to be the largest measured electronic $\chi^{(3)}$ value in the transparent gap region of any semiconductor. Wavelength dependence $\chi^{(3)}$ showed two peaks, at around 0.6 and 0.9 eV, which were interpreted as three-photon and two-photon resonance enhancements with the $1B_u$ state and the mA_g state, respectively. Wong and Vardeny[185] reported a $\chi^{(3)}(-\omega; \omega, \omega, -\omega)$ value of 5×10^{-8} esu at 620 nm for polyacetylene with a figure-of-merit $\chi^{(3)}/\alpha$ of 5×10^{-13} esu cm. The nonlinear refractive index of polyacetylene was calculated as 1.6×10^{-10} cm^2/W and n_2/α value of 1.6×10^{-15} cm^3/W. Krausz et al.[186] calculated the nonlinear refractive index n_2 of 7×10^{-6} M cm^2/W at 1.907 μm, a response time of $\sim$10 fs, and a laser damage threshold of 50 GW/cm^2. Durham polyacetylene was found to show larger optical nonlinearity than isotropic Shirakawa polyacetylene because of the chain orientation and highly dense morphology.[187]

BASF research team[188] reported THG and DFWM $\chi^{(3)}$ values of thin films of polyacetylene prepared by three different methods: Ziegler-Natta catalyst, Luttinger-Green process, and by thermal conversion of polyphenylvinylsulfoxide (PPVS). The Ziegler-Natta and Luttinger types and converted PPVS showed absorption maxima at 655, 570, and 480 nm and had conjugation lengths of 40, 30, and 13, respectively. Polyacetylene prepared by the Luttinger-Green process showed the largest $\chi^{(3)}$ value of 9.9×10^{-8} esu at 720 nm by DFWM technique. THG measured $\chi^{(3)}$ values were two orders of magnitude lower, although in the same range (10^{-10} esu), for all three samples. The response time measured by DFWM was on a scale of less than 10 ps. The wavelength dependence of $\chi^{(3)}$ values measured by THG showed two maxima at 550 and 650 nm for Luttinger-type polyacetylene, where the maximum at 650 nm is associated with the three-photon and two-photon resonances in *trans*-PA whereas the maximum at 550 nm is associated with the three-photon resonance in *cis*-PA. Two maxima indicate that significant *cis*-PA content remained even after the thermal treatment. The $\chi^{(3)}(-3\omega; \omega, \omega, \omega)$ of 1.7×10^{-10} esu was obtained for PA films prepared from poly(phenylvinylsulfoxide) PPVS precursor.

Beddard et al.[189] measured the $\chi^{(3)}$ values of polyacetylene and a norbornylene-polyacetylene copolymer synthesized by the Durham route using a variety of resonance four-wave mixing and transient absorption experiments. The $\chi^{(3)}_{xxxx}$ values at 625 nm were 1.1×10^{-7} esu for pure polyacetylene and 1.3×10^{-7} esu for the copolymer. At the same wavelength, the $\chi^{(3)}_{xyyx}$ components were 4.0×10^{-8} esu for both pure polyacetylene and the copolymer. The copolymer showed improved optical transparency and laser-induced damage threshold. The response time was on a scale of >10 ps. Table 18 compares the $\chi^{(3)}$ values of polyacetylene measured by various research groups.

The substitution of phenyl ring, i.e., poly(phenylacetylene) (PPA), exhibits $\chi^{(3)}$ values of about 1.2×10^{-12} esu, but further substitution at phenyl ring itself increases $\chi^{(3)}$ values significantly.[194] Strong changes in linear and NLO properties can be obtained by varying the substituents on the phenyl ring. The resonant $\chi^{(3)}(-3\omega; \omega, \omega, \omega)$ of 10^{-11} esu and $\chi^{(3)}(-\omega; \omega, -\omega, \omega)$ of higher than 10^{-9} esu were observed for substituted PPAs. The absorption coefficient α and $\chi^{(3)}(-\omega; \omega, -\omega, \omega)$ show a linear relationship.[195] The influence of substituents on the phenyl ring of PPA is represented in Figure 17. The $\chi^{(3)}$ values vary an order of magnitude with substituents. An *ortho*-trimethylsilyl-substituted PPA shows a maximum value of about 10^{-11} esu larger than that of a phenyl substituent (10^{-12} esu). The size of the substituents lowers the average $\chi^{(3)}$ by diminishing the density of polymer chain which is clearly shown for PPA containing methyl, ethyl, and octyl hydrocarbon chains. For PPAs, the phase of $\chi^{(3)}$ increases with absorption maximum and reaches 270° for a system containing $Si(CH_3)_3$ substituent. In this system, a two-photon resonance occurs with an electronic excitation state near 532 nm. The $\chi^{(3)}(-\omega; \omega, -\omega, \omega)$ values of PPAs are smaller than those of polyacetylene. Sasabe and Wada[196] reported the $\chi^{(3)}$ of PPA substituted with different groups at the *ortho*-position. PPA with *ortho*-substituted trimethylsilyl $Si(CH_3)_3$, trifluoromethyl CF_3, and $Ge(CH_3)_3$ groups showed λ_{max} at 536, 439, and 548

Table 18 Third-order NLO Susceptibility of Polyacetylenes

Polyacetylene	Form	$\chi^{(3)}$ (10^{-10} esu)	Wavelength (μm)	Experimental Technique	Ref.
PA (Shirakawa)	thin film	13	1.907	THG	190
		500	0.620	DFWM	148,185
		1800	2.25	THG	181
PA (Durham)	thin film	270	1.907	THG	187
		1100	0.625	DFWM	189
		1300	0.625	DFWM[a]	189
PA (Naarmann)	thin film	0.2	2.0	THG	191
		400	0.720	DFWM	188
PA (Luttinger)	thin film	1.7	0.550	THG	188
		990	0.720	DFWM	188
PA	solution	9.0	0.530	PC	192
PA	thin film	56	2.1	THG (FEL)	193
PA	single crystal	170	1.907	THG	186
		90	1.064	THG	186
PA	oriented film	81	1.064	THG	183

PC = phase conjugation.

[a] Copolymer.

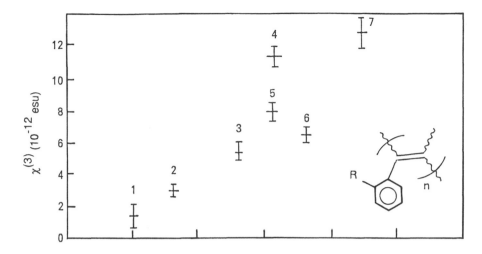

Figure 17 The influence of substituent groups on the phenyl ring on third-order NLO susceptibility of PPAs. (From Neher et al., *Chem. Phys. Lett.*, 163, 116, 1989. With permission.)

nm and $\chi^{(3)}(-3\omega; \omega, \omega, \omega)$ values of 1.7×10^{-11}, 3.0×10^{-12}, and 2.6×10^{-11} esu at 1.907 μm, respectively. The $\chi^{(3)}(-3\omega; \omega,\omega,\omega)$ values of PPA and poly(diphenylacetylene), one phenyl ring *ortho*-substituted with $Si(CH_3)_3$, were 5.4×10^{-13} and 8.8×10^{-13} esu at 1.907 μm, respectively. The studies on PPAs indicates that linear and NLO properties can be adjusted by molecular engineering. Unlike polyacetylene, the PPAs showed higher stability. Table 19 lists the linear and NLO properties of PPA. Other substituted PAs have also been found to show interesting NLO properties. For example, poly(ketene), in which hydrogens of the PA are replaced by hydroxy groups, has a third-order susceptibility value of 1.15×10^{-9} esu at 532 nm.[197]

(Polyketene)

A cholesteryl-4-pentynoate polyacetylene (CP5) shows the $\chi^{(3)}$ value of 2.1×10^{-21} m² V⁻² (9×10^{-12} esu) for the red film and 1.4×10^{-21} m² V⁻² for the yellow film.[198] Le Moigne et al.[199] investigated the NLO properties of a series of alkyl or aryl esters of poly(phenylacetylene) hereafter abbreviated as PPAC*n*. The THG measurements at 1.9 μm yielded $\chi^{(3)}(-3\omega; \omega, \omega, \omega)$ values of 6×10^{-13} esu for PPAC16 and 8×10^{-13} esu for PPAC1 thin films. The $\chi^{(3)}(-3\omega; \omega, \omega, \omega)$ values of PPAC*n* were found to be 7 or 8 times larger than the $\chi^{(3)}(-3\omega; \omega, \omega, \omega)$ values of cholesterylpolyacetylene (CP5) (9×10^{-12} esu). These results indicate that the improvement in $\chi^{(3)}$ of side chain PPA may be obtained by an increase of the content in the planar *trans* head-to-head configuration.

Table 19 Linear and NLO Properties of Poly(phenylacetylene) Derivatives

R	λ_{max} (nm)	α_{max} (10^4 cm^{-1})	$\chi^{(3)}$ (10^{-12} esu)	γ (10^{-34} esu)	Ref.
H	330	7.07	1.0	6.3	195
H	350	6.95	2.5 (3.0)	20	195
CH$_3$	414	6.78	4.9 (5.0)	50	195
CH$_3$	436	9.02	9.3 (9.2)	90	195
C$_2$H$_5$	429	6.24	5.4 (5.4)	110	195
C$_2$H$_5$	438	7.05	7.1 (6.8)	100	195
C$_8$H$_{17}$	462	8.70	6.4 (6.7)	50	195
Si(CH$_3$)$_3$	515	6.14	13.0 (12.0)	240	195
	536		17		196
Ge(CH$_3$)$_3$	548		26		196
CF$_3$	439		3.0		196

$R = (CH_2)_n$-H where $n = 1,2,4,8,16$; C$_6$H$_5$

(Cholesterylpolyacetylene)

Park et al.[200] synthesized soluble substituted PAs by the ring-forming metathesis polymerization of 4,4-disubstituted 1,6-heptadiynes to examine the effect of substituent chemical group (R) on the $\chi^{(3)}$ value of the conjugated poly(1,6-heptadiyne) main chain. The poly(4,4-disubstituted 1,6-heptadiyne) with substituents diethyl CH$_3$-CH$_2$, bis(1,1,1,3,3,3-hexafluoro-2-propyl) (CF$_3$)$_2$-CH, and bis(N-carbazole-n-hexyl) showed optical maximum at 540, 535, and 535 nm respectively; their $\chi^{(3)}$ values measured by the THG technique at 1.9 μm were approximately 10^{-11} esu (Table 20). The $\chi^{(3)}$ values may have had a three-photon resonance contribution because these polymers have slight absorption at the third-harmonic wave. For the poly(1,6-heptadienes) having silicon groups, the DFWM-measured $\chi^{(3)}$ values at 1.064 μm increased with the bulkiness of the substituents and were in accordance with the bathochromic shift of optical maximum.[201] The bulky group at 4-position introduced more planar conformation which increases the optical nonlinearity. The $\chi^{(3)}$ values of these conjugated polymers changed with the substituent groups.

The $\chi^{(3)}$ of conjugated poly(1,6-heptadiyne) having active NLO chromophores 4-(dimethylamino)-4'-(3-didropargyl acetoxy hexyl sulfone) stilbene measured by DFWM at 1.064 μm in 0.02 M DMF solution were found to be 2.6 × 10^{-11}, 3.0 × 10^{-11}, and 3.5 × 10^{-11} esu for 10, 20, and 60 mol%, respectively.[202]

Table 20 Third-order NLO Properties of Poly(41,6-heptadiyne) Derivatives

R_1	R_2	$\chi^{(3)}$ (10^{-11} esu)
$COOCH_2-CH_3$	$COOCH_2-CH_3$	6.5
$COOCH-(CF_3)_2$	$COOCH-(CF_3)_2$	4.1
H	$COO(CH_2)_2Si(CH_3)_3$	2.6
$COO(CH_2)_2Si(CH_3)_3$	$COO(CH_2)_2Si(CH_3)_3$	3.1
$OCOCF_3$	$OSi(CH_2)_2C(CH_3)_3$	3.6
$OCO(CF_2)_2CF_3$	$OSi(CH_2)_2C(CH_3)_3$	4.3
$OCO(CF_2)_6CF_3$	$OSi(CH_2)_2C(CH_3)_3$	4.1
$COO(CH_2)_6$-N (carbazole)	$COO(CH_2)_6$-N (carbazole)	2.6

From Park et al., *Appl. Phys. Lett.*, 65, 289, 1994; Kim et al., *Polym. Preprints*, 35, 244, 1994. With permission.

Marder et al.[203] prepared soluble polymers derived from the ring-opening metathesis copolymerization of cyclooctatetraene (COT) and 1,5-cyclooctadiene (COD). The $\chi^{(3)}$ and γ values of cyclooctatetraene (COT) and 1,5-cyclooctadine (COD) copolymers were evaluated by the wedged-cell THG technique at 1.907 μm. The NLO susceptibilities are based on the total mole fraction of monomer incorporated into polymer. The $\chi^{(3)}$ and γ values of the copolymer increased with the increase of COT fraction because of the increasing concentration of conjugated segments and the consequently increased conjugated backbone with a higher fraction of COT. The $\chi^{(3)}$ values were 3.3×10^{-13}, 6×10^{-13}, 1.3×10^{-12}, and 1.6×10^{-12} esu, and the γ values were 20×10^{-36}, 36×10^{-36}, 81×10^{-36}, and 1×10^{-34} esu for the copolymer containing 8, 15, 27, and 32% COT, respectively. The NLO activity of COD units in the copolymer was negligible when compared with COT units. These copolymers showed large third-order optical nonlinearities and low scattering losses. Poly(n-butylCOT) showed a $\chi^{(3)}$ value of 1×10^{-10} esu, the same as that of unoriented PA at the same wavelength. This study indicated that the ring-opening metathesis polymerization methodology is an effective way to control nonlinearity.

a. Model Compounds of Polyacetylene

The defined chain length polyenes have been the focus of theoretical and experimental studies. The polarizability and second hyperpolarizability of ethylene, butadiene, hexatriene, octateriene and other extended conjugated molecules have been calculated by various computational methods. Experimental studies have been conducted on third-order optical nonlinearities of oligomers of polyenes. In this section hyperpolarizabilities of polyene oligomers with an increasing number of double bonds and having push-pull substituents are discussed to establish a relationship between optical nonlinearities and electron delocalization length. Helfin et al.[21] measured second hyperpolarizability and $\chi^{(3)}$ values for hexatriene at several different wavelengths. The molecular structures of conformations for *all-trans* hexatriene and *cis-transoid* hexatriene are shown below. Table 21 lists the experimental and theoretical values for $\chi^{(3)}$ and γ for hexatriene measured from DC-SHG $(-2\omega; \omega, \omega, 0)$ and THG $(-3\omega; \omega, \omega, \omega)$ techniques.

all-trans hexatriene   *cis-transoid* hexatriene

The THG measurement at the fundamental wavelength $\lambda = 1.908$ mm gave $\gamma(-3\omega; \omega, \omega, \omega)$ of 9.1×10^{-36} esu and α of 1.18×10^{-23} esu for 1,3,5-hexatriene.[62] At the same wavelength, the $\gamma(-3\omega; \omega, \omega, \omega)$ of 7.4×10^{-36} esu and α of 1.57×10^{-23} esu for 2,5-dimethyl-2,4-hexadiene, $\gamma(-3\omega; \omega, \omega, \omega)$ of 4.9×10^{-36} esu and α of 1.13×10^{-23} esu for 2,5-dimethyl-1,3-butadiene, $\gamma(-3\omega; \omega, \omega, \omega)$ of 38×10^{-36} esu and α of 2.55×10^{-23} esu for 2,5-dimethyl-2,4,6-octatriene, and $\gamma(-3\omega; \omega, \omega, \omega)$ of $400 \pm 50 \times 10^{-36}$ esu and α of 6×10^{-23} esu for 2,9-dimethyl-2,4,6,8-decatetraene were obtained.

The excited state enhancement of the DFWM susceptibility of diphenylhexatriene (DPH) has also been studied.[203] An increase of a factor of 100 in nonresonant 1.064 μm DFWM signals was observed at the highest DPH concentrations when the 355 nm pump beams saturate the S_2 absorption than while the pump beam is turned off. The unpumped DFWM signal was independent of concentration showing that the ground-state hyperpolarizability for DPH is smaller than the experimental resolution of $\pm 50 \times 10^{-36}$ esu. On the other hand, pumped DFWM signal strongly depends on the DPH concentration. The $\chi^{(3)}$ value depended linearly on the concentration of the solution and yielded a hyperpolarizability $\gamma(-\omega; \omega, \omega, -\omega)$ value of $12,000 \pm 1,700 \times 10^{-36}$ esu. This study demonstrated that the value of $\chi^{(3)}$ can be enhanced by more than two orders of magnitude by optically exciting a linear molecule DPH at one wavelength into an electronic excited state without increasing optical absorption at the probe wavelength.

(DPH)

Perry et al.[175] reported THG $\chi^{(3)}$ values for phenyl-capped acetylenes. The $\chi^{(3)}$ values of 1.22×10^{-12} and 0.61×10^{-12} esu at 1.064 and 1.907 μm, respectively, were measured for stilbene. The $\chi^{(3)}$ value of 1.54×10^{-12} esu at 1.907 μm was measured for diphenyl-1,3-butadiene. The hyperpolarizability

Table 21 Experimental and Theoretical Values of $\chi^{(3)}$ and γ for Hexatriene Measured from DC-SHG $(-2\omega; \omega, \omega, 0)$, and THG $(-3\omega; \omega, \omega, \omega)$ Techniques

Wavelength (μm)	$\chi^{(3)}$ (10^{-14} esu)		γ (10^{-36} esu)			
			DCSHG		THG	
	DCSHG	THG	Expt.	Theory	Expt.	Theory
1.064	22.0	4.55	10.0 ± 1.5	8.9	1.94 ± 0.29	2.30
1.543	16.1	2.57	7.5 ± 1.1	7.5	1.18 ± 0.18	1.45
1.907	17.7	2.25	8.2 ± 1.2	7.1	1.04 ± 0.16	1.30

Adapted from Helfin et al., *J. Opt. Soc. Am. B*, 8, 2132, 1991. With permission.

values for stilbene were 55×10^{-36} and 29×10^{-36} esu at 1.064 and 1.907 μm, respectively. For diphenyl-1,3-butadiene, hyperpolarizability of 68×10^{-36} esu at 1.907 μm was measured. In another study, Persoon et al.[204] reported third-order NLO properties of solutions of α,ω-diphenylpolyenes ($n = 1 \sim 4$) measured by means of phase–conjugate interferometry at 532 nm and hyperpolarizabilities were compared with the values obtained from EFISH at 1.064 μm. The α,ω-diphenylpolyenes were *trans*-stilbene (DPE) ($n = 1$), 1,4-diphenyl-1,3-butadiene (DPB) ($n = 2$), 1,6-diphenyl-1,3,5-hexatriene (DPH) ($n = 3$), and 1,8-diphenyl-1,3,5,7-octatetraene (DPO) ($n = 4$). Table 22 lists the optical absorption peak and the γ and $\chi^{(3)}$ values of α,ω-diphenylpolyenes in dichloroethane solutions. Both γ and $\chi^{(3)}$ values increase as the length of polyene segment increases. The $\chi^{(3)}/N$ of DPB, DPH, and DPO were found to be 2.10, 3.07, and 184.2 relative to $\chi^{(3)}/N$ of DPE-dichloroethane, respectively. The hyperpolarizabilities measured from nonlinear interferometry were found to be much larger than those obtained from EFISH at 1.064 μm. Both THG and interferometry studies demonstrate that the $\chi^{(3)}$ values are larger for the longer π-conjugated phenyl-capped acetylenes.

$n = 1$ (DPE)
$n = 2$ (DPB)
$n = 3$ (DPH)

The real $\chi^{(3)}$ and imaginary $\chi^{(3)}$ values of a crystalline rhodanine derivative 5-(5-morphine)-2,4-pentaanilidineethylrodanine (MPER) were 10^{-6} esu and 10^{-7} esu, respectively.[205]

(MPER)

The molecular hyperpolarizability of a thiophene-substituted conjugated polyene molecule, 1,10-di(3-thienyl)-3,8-dimethyl-1,3,5,7,9-decapentaene (DTDMP), was measured at 1.064 and 1.907 μm in chloroform solution using a wedge-shaped liquid cell. The molecular hyperpolarizability $\gamma(-3\omega; \omega, \omega, \omega)$ at 1.064 μm was 2.0×10^{-33} esu, which is about 5 times larger than that measured at 1.907 μm fundamental wavelength ($\gamma = 0.4 \times 10^{-33}$ esu). The γ value at 1.064 μm has a three-photon resonance whereas the γ value at 1.907 μm is nonresonant. The normalized harmonic intensities of DTDMP at 1.907 μm show almost linear dependence as a function of concentration whereas a strong nonlinear dependence was observed at 1.064 μm.[206]

(DTDMP)

Table 22 Optical Absorption Peak, γ and $\chi^{(3)}$ Values of α,ω-Diphenylpolyenes in Dichloroethane Solutions

Polyene	λ_{max} (nm)	γEFISH (10^{-36} esu)	$\chi^{(3)}$ (10^{-14} esu) Interferometry	THG[a]
DPE	298	13.1 (391)[b]	2.43	61
DPB	331	35.8 (409)	2.62	154
DPH	358	52.6 (558)	3.55	

[a] THG values are at 1.907 μm from the reference.

[b] The γ values in parentheses are those obtained from interferometric measurements with CS_2 as a reference.

From Persoon et al., *SPIE Proc.*, 1409, 220, 1991. With permission.

Table 23 DFWM Measured $\chi^{(3)}$ Values of Bis-thienyl Polyenes

		$\chi^{(3)}$ ($-\omega; \omega, \omega, -\omega$) ($10^{-13}$ esu) at Wavelength (μm)			
		Neutral		Doped	
X	**n**	**0.532**	**1.064**	**0.532**	**1.064**
H	5	2.7	0.54	7.8	1.4
H	7	43	0.85	11	3.2
H	8	258	0.66	14	2.7

From Spangler et al., *Mater. Res. Soc. Symp. Proc.*, 328, 655, 1994. With permission.

Spangler et al.[207] reported $\chi^{(3)}(-\omega; \omega, \omega, -\omega)$ of end-capped bis-thienyl polyenes in neutral and oxidized doped states. The $\chi^{(3)}$ value of neutral samples increased more rapidly than doped ones as a function of conjugation length at 532 nm because of the one-photon resonance enhancement (Table 23). The trend of $\chi^{(3)}$ variation at 1.064 μm was reversed for neutral and doped species. Bis-thienyl polyenes end-capped with butylthio groups (X = BuS) showed $\chi^{(3)}$ values of 1.3×10^{-13}, 9.5×10^{-13}, 4.8×10^{-12}, 1.70×10^{-11}, 3.0×10^{-11}, and 10^{-10} esu at 532 nm for n = 3, 4, 5, 6, 7, and 8, respectively, showing an increase of three orders of magnitude as the π-conjugation length increased.

The π-conjugation length dependence of the third-order optical nonlinearity has also been investigated for polyene conjugated chains bearing various electron donor and electron acceptor end groups. Examining what combination of end groups would be advantageous while designing novel organic materials exhibiting large third-order optical nonlinearity would be of great interest. In this regard, only three combinations of electron donor and electron acceptor substituted conjugated systems are possible: (1) acceptor–acceptor end groups; (2) donor–donor end groups; and (3) donor and acceptor end groups.

Third-order NLO susceptibilities of a series of charge-transfer-conjugated polyene oligomers with an increasing number of double bonds (N = 1, 4, 6, and 8) were investigated with THG and EFISH techniques.[208] In these sulfur-containing asymmetric push-pull polyenes with variable π-conjugated chain lengths, the benzodithia group acts as a strong electron donor and an aldehyde group acts as a weak electron acceptor, whereas the methyl side groups assist in preserving the conformational stability. The push-pull polyene with N = 6 is different because it has a triple bond close to the middle of the chain and causes the detrimental effect on polarizability. Table 24 lists the optical linear absorption, dipole moment, and experimental γ values obtained from THG at 1.907 μm and a comparison of γ values from EFISH experiments with $\gamma(-2\omega; \omega, \omega, 0)$ values calculated from a THG experiment at ω = 1.34 μm. The γ_{EFISH} values are larger than THG values. The γ values increase rapidly with the length of conjugation and the γ value of N = 8 increases about two orders of magnitude than that of N = 1. The dependence of β and γ on the number N of double bonds follows an exponential law; β

Table 24 Optical Linear Absorption, Dipole Moment, and Second Hyperpolarizabilities of Push–Pull Polyenes

Polyene	λ_{max} (nm)	Dipole Moment (e$\mathring{A}$)^{2a}	$\gamma(10^{-33}$ esu)^b		
			γTHG (Expt.)	γEFISH (Expt.)	$\gamma(-2\omega; \omega, \omega, 0)$
N = 1	372	1.06	0.30	0.624	0.047
N = 4	456	8.31	4.11	28.8	3.9
N = 6	466	14.64	15.0	52.8	16
N = 8	500	22.9	56.6	168	95

^{a} The data on dipole moment are the squared transition dipole moment mean values where e is the electronic charge.

^{b} The experimental γ values were obtained from the THG at 1.907 μm and from the EFISH at 1.34 μm, and $\gamma(-2\omega; \omega, \omega, 0)$ values were deduced from the $\gamma(-3\omega; \omega, \omega, \omega)$ values; the data include the local field factor F.

From Messier et al., *Nonlinear Opt.*, 2, 53, 1992. With permission.

proportional to $N\alpha_1$ and γ proportional to $N\alpha_2$ with $\alpha_1 = 2.34$ and $\alpha_2 = 3.5$ without corrections for resonance and $\alpha_1 = 2.09$ and $\alpha_2 = 32.82$ when three- and two-photon resonance contributions were eliminated. The lower α_2 with the experimental coefficient indicates the asymmetrical nature of the push-pull polyenes. No saturation effect on γ chain length dependence was observed in the conjugated system.

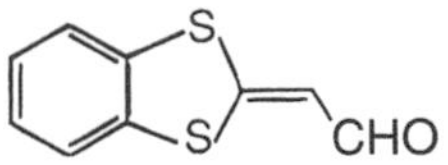

N = 1

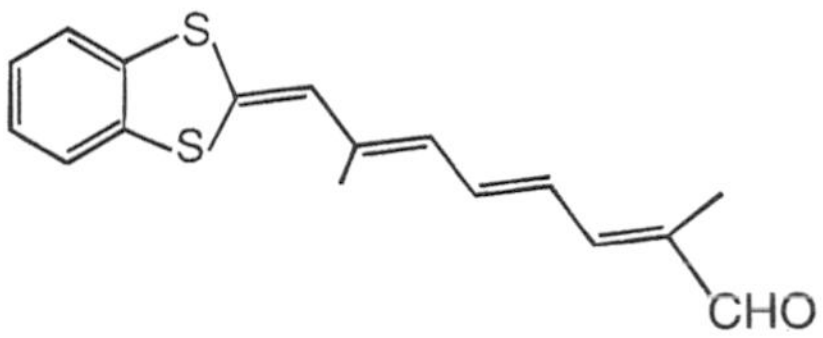

N = 4

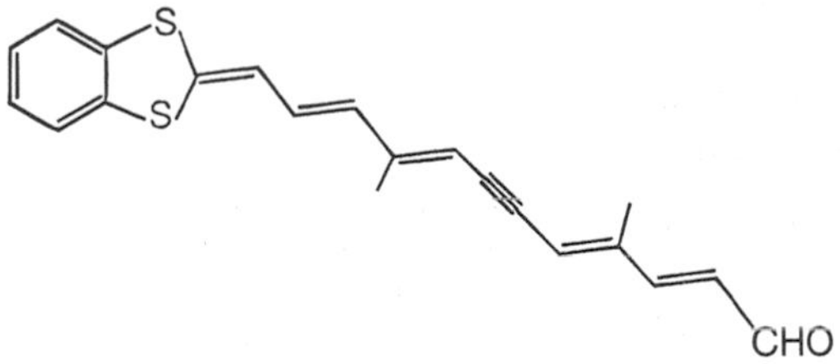

N = 6

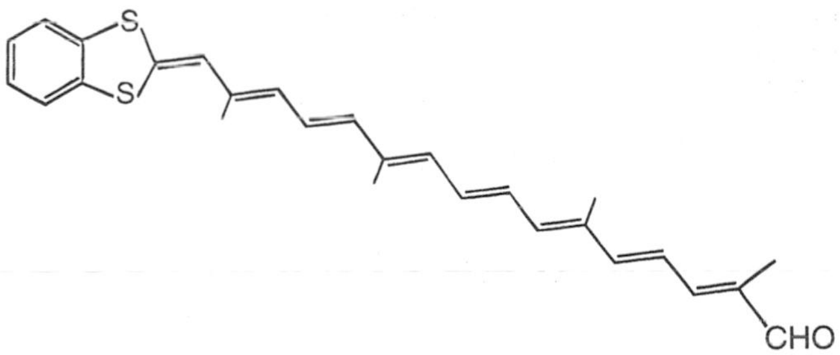

N = 8

A few examples of experimental and theoretical studies are provided here to develop a clear picture on how third-order optical nonlinearity can be increased. Puccetti et al.[209] conducted a systematic study on the various α,ω-disubstituted polyenes of increasing π-conjugation length and bearing electron-acceptor or electron-donor end groups and asymmetric, donor–acceptor polyenes for the same donor/acceptor couple to investigate the π-conjugation length dependence of their third-order optical nonlinearity. The γ values depend on increasing conjugation length for two series of symmetric polyenes: in Series I the electron-acceptor end groups are formyl moieties (CHO) and in Series II the electron-donor end groups are dimethylaniline [$(CH_3)_2N$] moieties. In Series III, the polyene chains have the same donor and acceptor end groups as of Series I and II. Figure 18 shows the plot of second hyperpolarizability as a function of the number of carbon-carbon double bonds (n) for the two series of symmetric polyenes: Series I (bis-acceptor compounds) and II (bis-donor compounds) measured at 1.34 μm by EFISH technique and an asymmetric polyenes Series III (donor–acceptor compounds) measured at 1.907 μm by the THG technique to avoid any second-order contribution. As expected, the increasing polyene conjugation length induced hyperchromic and bathochromic shifts and each series obeyed the linear dependence of λ_{max} with the square root of the number of carbon-carbon double bonds (n) as had been observed for polyenes. The logarithmic plots of the γ^{SHG} against n give an exponent of 2.3 for the bis-acceptor Series I and 3.0 for the bis-donor Series II. Exceptionally large γ values are observed for longest bis-donor polyenes and no saturation occurs up to 40 Å. The γ value was found to be 290 ×

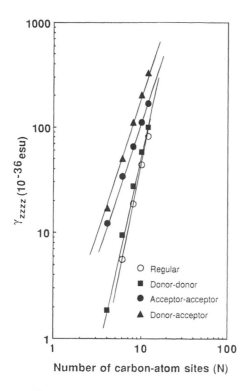

Figure 18 Plots of γ_{zzzz} values versus number of carbon-atom sites (N) for regular, donor–donor, acceptor–acceptor, and donor–acceptor polyenes. (From Puccetti et al., *J. Phys. Chem.,* 97, 9385, 1993. With permission.)

10^{-36} esu in a symmetric dialdehyde where the central double bond was replaced by a triple bond, indicating that the triple bond has a detrimental effect. The end groups also significantly affected the γ values; for example, the enhancement factor between the nitro and dimethylamino derivatives was 3.1. The γ value was 5.7 times larger for julolidino derivatives than for the nitro derivatives.

Series I

Series II

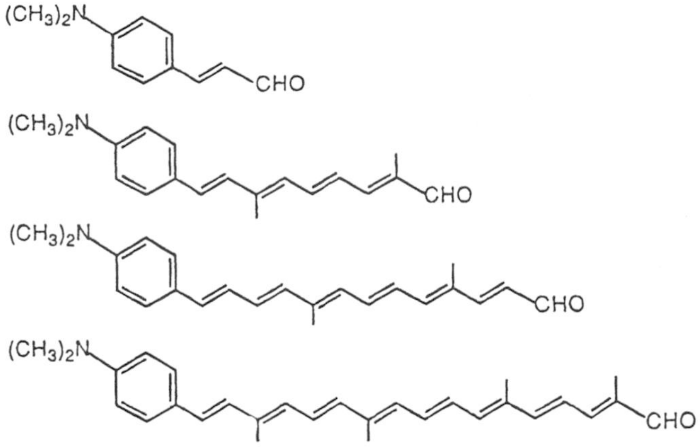

(CH$_3$)$_2$N—⟨⟩—CH=CH—CHO

(CH$_3$)$_2$N—⟨⟩—(polyene chain)—CHO

(CH$_3$)$_2$N—⟨⟩—(polyene chain)—CHO

(CH$_3$)$_2$N—⟨⟩—(polyene chain)—CHO

Series III

Marder et al.[210–213] pointed out that polarizabilities of polyene systems depend on the bond length alternation (BLA). For example, the optimization of third-order optical nonlinearities in polyenes possessing donor–acceptor groups need a specific degree of mixing between neutral and charge-separated resonance structures. The degree of mixing is associated with the strength of the donor–acceptor groups, ground-state polarization, and BLA. The difference between the average single and double carbon–carbon bond lengths in polymethine chain is defined as the BLA and this can be tailored by varying the strength of donor–acceptor groups. The tuning of BLA through chemical modification is shown below. A large positive BLA is observed for unsubstituted polyenes or with weak donor–acceptor groups (compound 1) because the ground-state structure resembles that of neutral resonance. On the other hand, the ground-state polarization increases and the BLA decreases with strong dicyanovinyl acceptor groups (compound 2) because of the increased mixing between the two limiting resonance forms. The BLA becomes zero with complete mixing of the limiting resonance forms such as in symmetrical cyanines (compound 3). A negative BLA is obtained as the strength of donor–acceptor groups increases (compound 4) further where the charge-separated resonance form mainly contributes to the ground state. Interestingly, the degree of BLA can be fine-tuned by using organic solvents of different polarity. The charge-separated structure can be stabilized to a larger extent by polar solvents rather than those of nonpolar solvents. An example how the BLA values change with donor–acceptor groups of varying strength and can be tailored by using organic solvents of different polarity is given here.

(1)   BLA = +0.06 Å

(2)   BLA = +0.01 Å

(3)   BLA = 0 Å

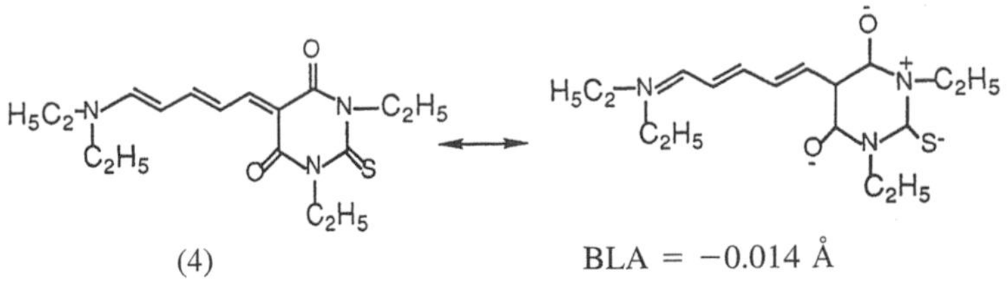

(4)   BLA = −0.014 Å

The compound (7) having dicyanovinyl groups has a BLA of +0.015 Å. The BLA decreased with barbituric and thiobarbituric acid acceptors. The γ values were measured by solvent-dependent THG experiments at 1.907 μm. Positive γ values were obtained for compounds 5 and 6. compound 1, which possessed a weak formyl group, showed a positive peak in γ value as the polarity of the solvent increased. The sign of γ values changed from positive to negative with polarity for compound 7 with dicyanovinyl groups. In compound 8, the γ value was negative and became larger with increasing solvent polarity whereas a clear peak with a negative γ value was observed for compound 9 in moderate polarity solvents such as chloroform and dichloroethane. The γ value of compound 10 increased in the most polar solvents. The γ values of molecules 11 through 18 which possessed zero BLA were found be to negative, which is in agreement with the relationship between polarizabilities and ground-state geometry. Table 25 lists the γ values of donor–acceptor polyenes.

(5)   (6)   (7)   (8)   (9)   (10)

Compounds 11, 12, and 13 have γ values of $+3 \times 10^{-36}$, -40×10^{-36}, and -370×10^{-36} esu, respectively. For compounds 14 and 15, the γ values were -45×10^{-36} and -500×10^{-36} esu, respectively. For compound 16, the γ value was positive 3×10^{-36} esu and changed to negative (-7×10^{-36} esu) as chain length increased.

$n = 1$ (11), $n = 2$ (12), $n = 3$ (13) $n = 1$ (14), $n = 2$ (15) $n = 1$ (116), $n = 2$ (17)

Table 25 Solvent-dependent γ Values of Donor–acceptor Polyenes

Solvents[a]	γ Values of Donor–acceptor Polyenes (10^{-36} esu)							
	(1)	(4)	(5)	(6)	(7)	(8)	(9)	(10)
CCl$_4$	–	–	–	–	40	–	–	–
Benzene	–	−85	34	60	15	−20	–	–
Dioxane	40	−170	25	70	−25	−100	−25	–
CHCl$_3$	95	−195	–	–	−42	−135	15	130
CH$_2$Cl$_2$	105	−175	27	72	−50	−145	30	167
CH$_3$CN	113	−130	–	–	−120	−205	79	228
CH$_3$NO$_2$	113	−125	–	–	−117	−220	73	227
CH$_3$OH	73	−10	–	–	−135	−166	150	291

[a] The polarity of the solvent increases from top to bottom.

From Bourhill et al., *SPIE Proc.*, 2143, 153, 1994. With permission.

Samuel et al.[214] performed a systematic study of the third-order optical nonlinearity of model polyene oligomers having double bonds from 28 to 240. The second hyperpolarizability measured by THG at 1.9 μm increased significantly as the number of double bonds increased. The polyene synthesized in living polymerization yielded soluble long chains with up to 240 double bonds. The γ/N in these model polyenes showed plateaus at about 120 double bonds where the γ value reached a maximum. Theory predicted the saturation of γ/N around typically 20 and 50 double bonds, therefore this is shown experimentally for the first time that third-order optical nonlinearity in polyenes requires up to 120 double bonds. Table 26 lists the γ values of model polyene oligomers.

All-trans β-carotene has 11 conjugated double bonds and represents a good model system of one-dimensional polyenes. *All-trans* β-carotene has been extensively investigated for third-order nonlinear optics and its $\chi^{(3)}$ and γ values have been measured over a wide wavelength range covering the entire absorption spectral region. The $\chi^{(3)}$ and γ values of *all-trans* β-carotene vary by several orders of magnitude depending on the measurement wavelength, material state, and techniques as summarized in Table 27. The $\chi^{(3)}$ value as high as 1.6×10^{-10} esu and γ value as high as 10^{-31} esu at the peak of the 1^1B_u resonance have been determined for *all-trans* β-carotene. The magnitude of $\chi^{(3)}$ and γ values is of the same order as that reported for polydiacetylenes. Ticktin[191] reported dispersion of the $\chi^{(3)}$ value of *all-trans* β-carotene-doped PMMA thin films in the region $1.2 \sim 2.0$ μm using OPO-THG measurements. The $\chi^{(3)}$ of *all-trans* β-carotene-doped PMMA thin films followed the absorption spectrum, and the measured resonant $\chi^{(3)}$ value was more than 2×10^{-11} esu at 1.4 μm, about 10 times larger than at 2.0 μm in the off-resonance region. van Beek et al.[215-217] studied the dispersion of the $\chi^{(3)}$ value of *all-trans* β-carotene-doped polystyrene thin films. The magnitude of the $\chi^{(3)}$ value of *all-trans* β-carotene-doped polystyrene thin films was 1.6×10^{-10} esu at the peak of the 1^1B_u resonance. The absorption peak at around 460 nm corresponds to a ^1B_u excited state. The $\chi^{(3)}$ value of an *all-trans* β-carotene in the 400 to 550 nm region can be explained by 1^1A_g ground state, the 1^1B_u excited state and an n^1A_g excited state. The dipole transition moment of 31 D was determined between the 1^1B_u excited state and the n^1A_g excited state around 22,200 cm^{-1}. The molecular excitations of *all-trans* β-carotene are similar to those in polyacetylenes and should exhibit similar effects.

Haas et al.[219] reported the $\chi^{(3)}$ THG measurements of four astaxanthenes of increasing conjugation length which structurally differ from β-carotene by having hydroxyl and ketone end groups. The

Table 26 Optical Absorption Maximum and Second Hyperpolarizabilities of Model Polyene Oligomers

N^a	λ_{max} (nm)	γ (10^{-34} esu) at 1.9 μm^b	γ_0 (10^{-34} esu)
28	466	81 ± 15	35 ± 7
39	486	141 ± 20	54 ± 8
50	516	247 ± 20	78 ± 10
68	530	553 ± 50	155 ± 16
88	538	1025 ± 150	267 ± 40
152	550	2731 ± 300	629 ± 70
240	552	3794 ± 500	854 ± 110

^{a} N is the average number of double bonds.

^{b} The experimental γ values were obtained from THG at 1.9 μm.

From Samuel et al., *Science*, 265, 1070, 1994. With permission.

Table 27 Third-Order Optical Nonlinearities of *All-trans* β-Carotene Measured in Solutions and Polymer Matrix

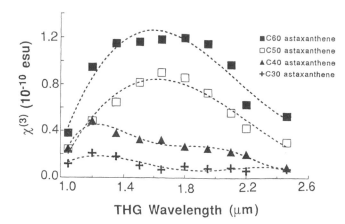

Matrix	$\chi^{(3)}$ (10^{-12} esu)	γ (10^{-36} esu)	Wavelength (μm)	Technique	Ref.
Benzene	2.0	8,000	1.89	THG	56
Dioxane	0.49	358	1.543	THG	21
Dioxane	3.09	1,580 ± 130	1.064	EFISH	21
Ethanol	0.14	70,000	1.064	DFWM	63
Benzene	38	–	1.221	THG	215
Solution	–	11,000 ± 2,500	1.908	THG	62
PMMA	20		1.40	THG	191
Polystyrene	160			THG	215–217
	1.25		0.979	THG	216,217
	1.08		1.008	THG	216,217
	1.9		1.90	THG	218

astaxanthene molecules, C_{30} ($n = 1$); C_{40} ($n = 2$); C_{50} ($n = 3$); and C_{60} ($n = 4$) were synthesized from the Witting reaction.

(Astaxanthene)

For THG measurements, thin films of four astaxanthenes having carbon numbers ranging from 30 to 60 were prepared by adding antioxidant stabilizer to the PMMA/astaxanthene/chloroform solution because pure astaxanthenes degrade in light and oxygen. Thin films were stored in the dark under vacuum before THG measurements. Figure 19 shows corresponding $\chi^{(3)}$ values of C_{30} ($n = 1$); C_{40} ($n = 2$); C_{50} ($n = 3$); and C_{60} ($n = 4$) astaxanthene molecules as a function of wavelength between 1.0

Figure 19 Wavelength dispersion of third-order NLO susceptibility $\chi^{(3)}$ values of astaxanthene molecules with increasing π-conjugation length. (From Haas et al., *J. Phys. Chem.*, 97, 8675, 1993. With permission.)

and 2.5 μm. Three-photon resonant transition dominates the dispersion spectra of $\chi^{(3)}$ with peaks occurring corresponding to optical absorption peaks. The magnitude of $\chi^{(3)}$ depends on the conjugation length of the astaxanthene. The highest optical nonlinearities were observed for C_{60} astaxanthene (17 double bonds in the main chain) for which the $\chi^{(3)}$ decreased slightly from a maximum of 1.2×10^{-10} esu to 5.4×10^{-11} esu at 2.46 μm where the absorption is less than 100 cm^{-1}. The $\chi^{(3)}$ values increased as a function of the number of double bonds, and a linear relationship between $\chi^{(3)}$ and the number of double bonds indicates that each double bond on the main chain contributes to 7×10^{-12} esu. Long-chain C_{60} astaxanthene exhibits the maximum $\chi^{(3)}$ value of 1.2×10^{-10} esu at the peak of the three-photon resonance with a figure of merit ($\chi^{(3)}/\alpha$) near 10^{-12} esu cm between 2.1 and 2.4 μm. Figure 20 shows the theoretically calculated and OPO-THG-measured $\chi^{(3)}$ values of astaxanthene molecules as a function of the number of double bonds. Molecular modeling calculations also supported the magnitude of $\chi^{(3)}$ values increasing linearly with the conjugation length.

Craig et al.[220] reported the dependence of hyperpolarizability on the conjugation length of a series of triblock copolymers containing model polyenes with conjugation lengths ranging from 4 to 16 double bonds. The typical chemical structure of the polynorbornene-polyene-polynorbornene triblock copolymer with 12-ene (12 double bonds) is depicted below. The hyperpolarizabilities of triblock copolymers were measured by EFISH at 1.34 μm. Polynorbornene homopolymer showed a γ value of 130×10^{-36} esu. The effective γ value changed from 6×10^{-34} esu for 6-ene to 4.7×10^{-33} esu for 13-ene for alternating copolymers. For isomerized copolymers, the effective γ value changed from 4.9×10^{-34} esu for 4-ene to 5.1×10^{-33} esu for 13-ene. The effective γ values of the copolymers varied with nominal conjugation length to the 2.9 and 3.0 powers for alternating and predominantly *all-trans*-isomers, respectively.

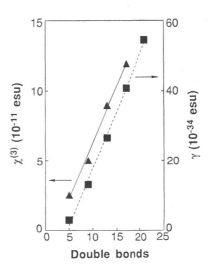

Craig et al.[221] further extended their studies to triblock copolymer containing discrete polyenes consisting of 9-, 10-, 11-, and 12-ene. All the copolymers have the same basic chemical structure: (olefin)$_{100}$-polyene-(olefin)$_{100}$. The chemical structure of a triblock copolymer with 9-ene is shown, discrete polyene with 10-, 11-, and 12-ene have an increasing number of π-bonds. These tricopolymers were orange to red. The hyperpolarizabilities of these discrete polyenes were measured by the EFISH at 1.34 μm in THF solution. The hyperpolarizabilities of the discrete polyenes were positive and increased with increasing π-conjugation lengths. The log-log plots of hyperpolarizability versus conjugation length yielded an exponent of 3.6 for $N = 9$, 10, and 11 similar to those reported earlier. The large $\gamma^{2\omega}$ values of discrete 9-, 10-, and 11-ene result from the greater *trans*-isomer contents in the discrete polyenes.

Figure 20 The π-conjugation chain-length-dependent THG-measured $\chi^{(3)}$ values and theoretically calculated γ values for astaxanthene molecules. (From Haas et al., *J. Phys. Chem.*, 97, 8675, 1993. With permission.)

Nickel et al.[222] reported $\chi^{(3)}$ of bis-anthracenyl polyenes substituted in the 10 and 10' positions with n-alkylthio groups using the DFWM technique. The $\chi^{(3)}$ values of the neutral bis-anthracenyl polyenes substituted with n-hexyl groups ($R = C_6H_{13}$, $n = 3$) in dichloromethane solution (0.14 wt%) were 2.3×10^{-14} and 7×10^{-14} esu at 0.532 and 1.064 μm, respectively. Upon oxidation with SbCl$_5$, the $\chi^{(3)}$ values increased to 3.1×10^{-13} and 1.3×10^{-12} esu at 0.532 and 1.064 μm, respectively. The absorption at 0.532 and 1.064 μm after doping was about the same, therefore the significant enhancement in the $\chi^{(3)}$ value at 1.064 μm was assumed to be a bipolaronic contribution rather than resonance enhancement. The neutral and doped bis-anthracenyl polyenes substituted with n-decylthio ($R = C_{10}H_{21}S$, $n = 3$) group showed $\chi^{(3)}$ values of 7×10^{-14} and 1.9×10^{-13} esu at 1.064 μm, respectively. The $\chi^{(3)}$ of the bis-anthracenyl polyenes having n-decylthio ($R = C_{10}H_{21}S$, $n = 4$) group was 7×10^{-14} esu in the neutral state, which increased by two orders of magnitude (6.7×10^{-12} esu) at 1.064 μm because of bipolaronic contributions. These $\chi^{(3)}$ values are large considering the low concentration (0.17 wt%); the exploration of $\chi^{(3)}$ to 100% concentration predicted a $\chi^{(3)}$ on the order of $\sim 3 \times 10^{-9}$ esu. These measurements indicated a significant enhancement of $\chi^{(3)}$ by bipolaron charge-state generation.

3. Polyphenylenes and Model Compounds

Poly(p-phenylene vinylene) consists of conjugated backbone like polyacetylene. McBranch et. al.[223] measured the $\chi^{(3)}$ of PPV and PDA-4BCMU using the THG method at a wavelength of 1.06 μm. Both polymers have the same value of $\chi^{(3)}$ of 10^{-11} esu. The index of refraction determined at 633 nm was 2 and 1.78 for PPV and PDA-4BCMU, respectively. The substituted poly(3,6-dimethoxy-p-phenylene vinylene) (MOPPV) has a $\chi^{(3)}$ value about equal to that of PPV whereas another substituted PPV having hexyloxy (-OC$_6$H$_{13}$) groups at 3- and 6-position structurally similar to MOPPV has a $\chi^{(3)}$ value one-third that of PPV. The THG measurements on stretched-oriented PPV suggests that main component of the $\chi^{(3)}$ tensor is parallel to the polymer chain direction and it is also sensitive to the degree of π-electron delocalization. The effect of increasing the π-electron conjugation on the $\chi^{(3)}$ values of PPV has been studied by Bradley and Mori[224] by thermal conversion of the sulfonium polyelectrolyte precursor polymer to fully conjugated PPV. The magnitude of $\chi^{(3)}$ values increases with increasing conversion from saturated precursor to fully conjugated polymer. For example, thermal conversion of the sulfonium polyelectrolyte precursor at 100 and 300 °C under vacuum for a period of 3 hr forms PPV polymers exhibiting $\chi^{(3)}$ values of 1.6×10^{-11} and 7.5×10^{-11} esu, respectively. The optical absorption studies and THG measurements performed at 1.064 μm suggest a resonant contribution for PPV. This systematic study demonstrates that the $\chi^{(3)}$ is sensitive to the extent of conjugation and that nonlinearity originates from the conjugated π-electron backbone.

The NTT group[225–228] reported THG in highly oriented poly(2,5-dimethoxy-p-phenylene vinylene) MO-PPV LB film prepared with an amphiphile. Figure 21 shows the dependence of TH intensity in MO-PPV LB films as a function of the number of monolayers and a quadratic dependence of TH intensity on the film thickness is demonstrated. The TH intensity in the direction parallel to the orientation axis is approximately 15 times larger than that in the direction perpendicular to the orientation axis in the MO-PPV LB films. The $\chi^{(3)}$ in the direction parallel and perpendicular to the orientation axis are 2.9×10^{-9} esu and 0.75×10^{-10} esu at 1.064 μm, respectively. The $\chi^{(3)}$ value is 0.32×10^{-10} esu for an unoriented film which is about 10 times smaller than that of an oriented MO-PPV film. The nonlinear response in MO-PPV-LB film is influenced by the orientation effect as well as by the π-

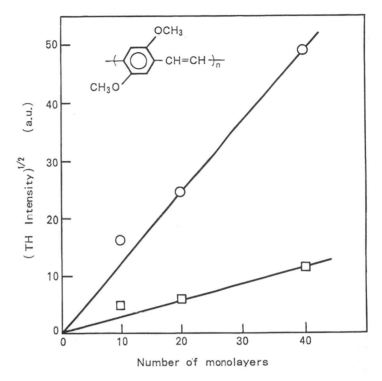

Figure 21 Dependence of third-harmonic (TH) intensity in MO-PPV LB films as a function of the number of monolayers. The circle shows the TH intensity parallel to the orientation axis and the square shows the TH intensity perpendicular to the orientation axis. (From Kamiyama et al., *Jpn. J. Appl. Phys.*, **29**, L840, 1990. With permission.)

electron conjugation length. Details of third-order nonlinearity for various conversion levels for MO-PPV and PPV are well discussed by Kurihara et al.[227] The synthetic route of MO-PPV is shown below.

Bubeck et al.[229] measured $\chi^{(3)}$ values of 1.2×10^{-10} esu from THG and 2×10^{-11} esu from DFWM for PPV. The $\chi^{(3)}$ value of poly(*p*-naphthalene vinylene) (PNV) was 5.0×10^{-12} esu from DFWM at 0.650 μm. Saitoh[230] synthesized PNV, poly(4,7-benzo[*b*]furandiyl-vinylene) (PBFV), and poly(4,7-benzo[b]thiophenediyl-vinylene) (PBTV). The optical band gaps of PNV, PBFV, and PBTV were 2.50, 2.40, and 2.27 eV, respectively. The $\chi^{(3)}$ value measured by THG method at 1.90 μm of PBTV was found to be more than two and four times that of PBFV and PNV, respectively. The $\chi^{(3)}$ values of oligomeric phenyl hexatrienylene derivatives by THG has also been evaluated.[231,232] This indicates that the largest $\chi^{(3)}$ values are observed for low band gap polymers. The $\chi^{(3)}$ values of both the thin films and of sol-gel-processed silica composites of PPV, poly(2-bromo-5-methoxy-1,4-phenylene vinylene) (BrMOPPV), and poly(2-butoxy-5-methoxy-1,4-phenylene vinylene) (BuOMOPPV) were reported at

Table 28 The $\chi^{(3)}$ Values of PPV and Analogous Polymers

Polymer	$\chi^{(3)}$ $(10^{-10}$ esu)	Wavelength (μm)	Experimental Technique	Ref.
PPV	2.0	1.064	THG	223
	16	1.064	THG	224
	0.75	1.064	THG	224
	1.3	1.064	THG	229
	2.0	0.620	DFWM	185
	0.2	0.65	DFWM	229
	0.072	1.90	THG	226,227
	1.2	1.55	THG	226,227
	0.54	1.85	THG	226,227
	1.4	1.475	THG	226,227
	0.047	1.86	THG	231
	4.0	0.602	DFWM	231
MO-PPV(LB Films)	2.9	1.064	THG	225
(thin-films)	0.32	1.064	THG	225
	0.54	1.85	THG	226
	1.60	1.60	THG	227
	0.30	2.10	THG	227
MEH-PPV	3.0	0.590	DFWM	106
HpPPV	0.01	1.86	THG	231
HpPPHT	0.035	1.86	THG	231
PNV	0.05	0.650	DFWM	229
	0.086	1.90	THG	230
PBFV	0.14	1.90	THG	230
PBTV	0.33	1.90	THG	230
BrMOPPV	9.0	0.602	DFWM	233
BuOMOPPV	20	0.602	DFWM	233
PCEMPV	0.83	0.602	DFWM	234a
PV-co-CEMPV	3.0	0.602	DFWM	234a
TMS-PPV	0.03	1.907	THG	234b
BTMS-PPV	0.04	1.907	THG	234b

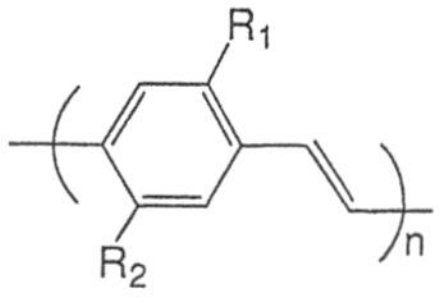

$R_1 = R_2 = H$, poly(p-phenylene vinylene) PPV
$R_1 = R_2 = OCH_3$, MOPPV
$R_1 = R_2 = C_7H_{15}$, HpPPV
$R_1 = Br, R_2 = OCH_3$, BrMOPPV
$R_1 = OC_4H_9, R_2 = OCH_3$, BuOMOPPV
$R_1 = Si(CH_3)_3, R_2 = H$, TMS-PPV
$R_1 = R_2 = Si(CH_3)_3$, BTMS-PPV

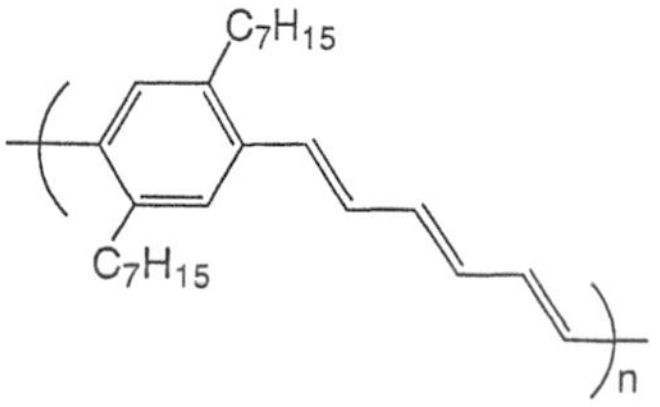

HpPPHT

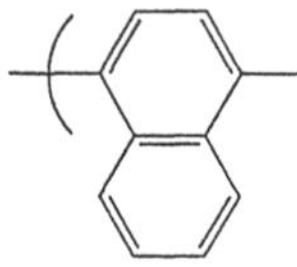

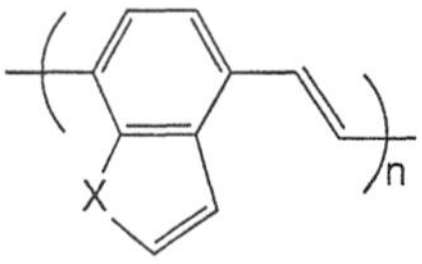

Poly(p-naphthalene vinylene) PNV $X = S$ PBTV
         $X = O$ PBFV

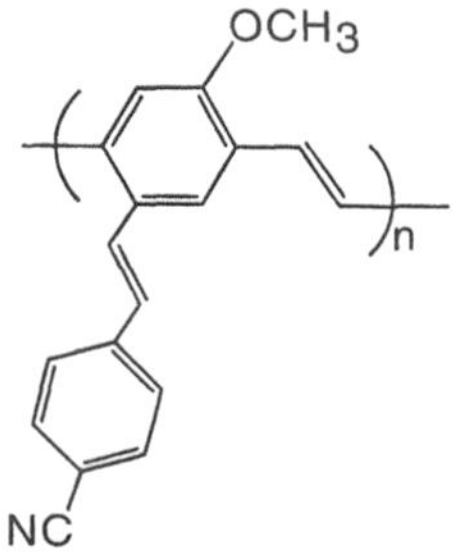

PCEMPV

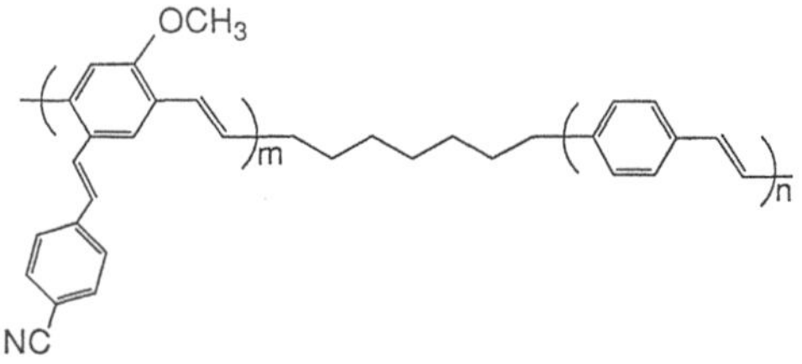

Poly(PV-co-CEMPV)

about 10^{-9} esu.[233] For BuOMOPPV, electron-donating methoxy and butoxy groups increased the π-electron density of the conjugated chain, this lead to larger $\chi^{(3)}$ values. On the other hand, relatively smaller $\chi^{(3)}$ values of BrMOPPV resulted from the reduced π-electron density of the conjugated chain by weakly electron-accepting bromine group. Hwang et al.[234] prepared poly[(2-2-(4-cyanophenyl)ethe-nyl)-5-methoxy-1,4-phenylene) vinylene] (PCEMPV) which has more extended π-conjugation length because of the substituted stilbene moiety. In this case, the electron-accepting nitrile group on stilbene moiety in PCEMPV also leads to smaller $\chi^{(3)}$ values. The copolymer of PCEMPV-containing PPV units showed larger $\chi^{(3)}$ values intermediate between PCEMPV and PPV homopolymers. The third-order NLO susceptibilities of PPV, derivatized PPV, and its copolymers measured by different techniques are listed in Table 28.

a. Model Compounds of Polyphenylenes

Zhao et al.[235] studied the influence of two-photon absorption on third-order NLO properties of soluble didecyloxy-substituted p-polyphenyl oligomer by using the DFWM technique. The soluble didecyloxy-substituted polyphenyls where $n = 0$, 1, and 2 are trimer, pentamer, and heptamer, respectively. The γ values were 2.2×10^{-35}, 6.9×10^{-35}, 3.5×10^{-34}, and 33×10^{-33} esu for the monomer, trimer, pentamer, and heptamer, respectively. This shows that the γ value increases by an order of magnitude from a pentamer to a heptamer with increasing π-conjugation length.

In another study, Marcy et al.[236] measured $\chi^{(3)}$ values of a series of solid-state poly(p-phenylene) oligomers by the DFWM at 650 nm with an optical pulse of 140 fs. The $\chi^{(3)}$ values were $< 1.6 \times 10^{-13}$ esu for both quaterphenyl ($n = 4$) and quinquephenyl ($n = 5$) and increased to 7.5×10^{-13}, 1.3×10^{-12}, and 1.7×10^{-12} esu for sexiphenyl, septiphenyl, and octiphenyl, respectively. The increase in $\chi^{(3)}$ value occurs for $n > 6$ and after that increases monotonically as the π-conjugation increases. The figure-of-merit $\chi^{(3)}/\alpha$ was 7.6×10^{-14} esu cm for $n = 4$ and almost the same for other oligomers. The $\chi^{(3)}$ values obtained in the solid state were found to be entirely different from those of solutions.

Perylene deriavtives are model compounds for the systematic study of planar polyconjugated materials. Schrader et al.[237] reported third-order NLO properties of a series of *tetra-tert*-butyl-perylene derivatives with different π-conjugation sizes from THG at 1.064 μm. These perylenes have two-dimensional features. The $\chi^{(3)}$ values of shorter rylenes, *tetra-tert*-butyl-biperylene (TTBPBP), *tetra-tert*-butyl-ter-rylene (TTBT), *tetra-tert*-butyl-quaterrylene (TTBQ) were 0.48×10^{-12}, 0.81×10^{-12}, and 1.57×10^{-12} esu, respectively. The three-photon resonance strongly enhanced the $\chi^{(3)}$ value of the *tetra-tert*-butyl-pentarylene (TTBP) ($\lambda_{max} = 785$ nm) which has a $\chi^{(3)}$ value of 8.0×10^{-11}, an order of magnitude larger than these derivatives. The $\chi^{(3)}$ values of perylene derivatives increased systematically with the interrelated quantities of the reciprocal of the S_0 S_1-excitation energy, the wavelength of the absorption maximum, and the π-conjugation length of the molecule.

(TTBT)   (TTBQ)

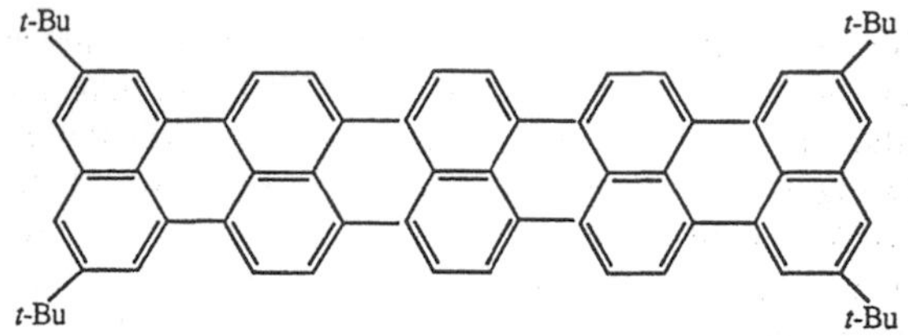

(TTBP) 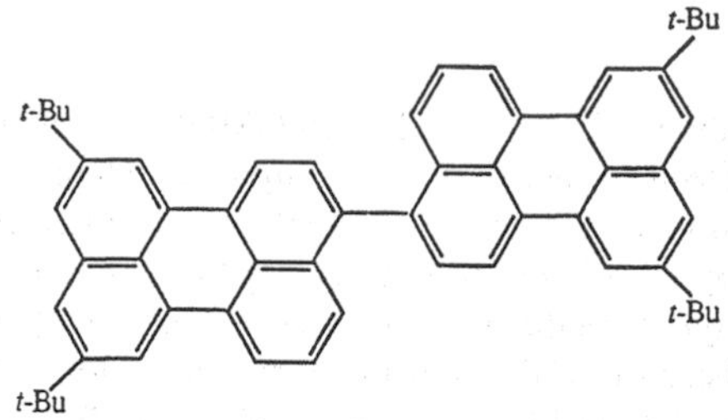　　　　　　　　　　　　(TTBBP)

Third-order NLO susceptibilities of anthracene derivatives have also been measured by optical Kerr shutter technique.[238,239] The 9,10-bis(4'-aminophenyl)anthracene (compound 1) (λ_{max} = 396 nm) and 9,10-bis(4'-(4''-amino-2,6-dimethylphenyl)-azophenyl) anthracene (compound 2) (λ_{max} = 414 nm) exhibit $\chi^{(3)}$ values at 458 nm of 20 times and at 566 nm 70 times that of the CS_2 reference material, respectively. The $\chi^{(3)}$ value of compound 2 that has extended conjugation and donor groups is 3.5 times larger than compound 1. The $\chi^{(3)}$ value of 4,4'-bis(diphenylhydromethyl)azobenzene (DPHM) (λ_{max} = 440 nm) was 30 times at 560 nm than that of the CS_2 reference.

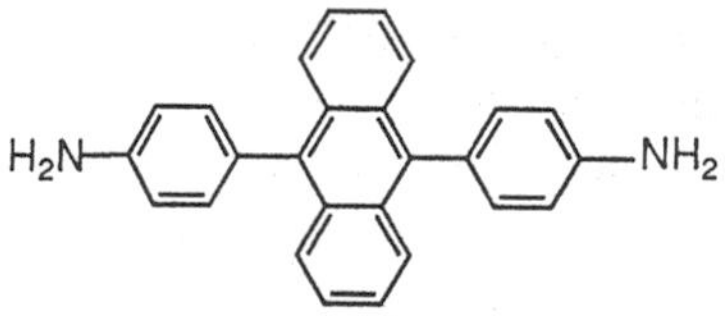

(1)

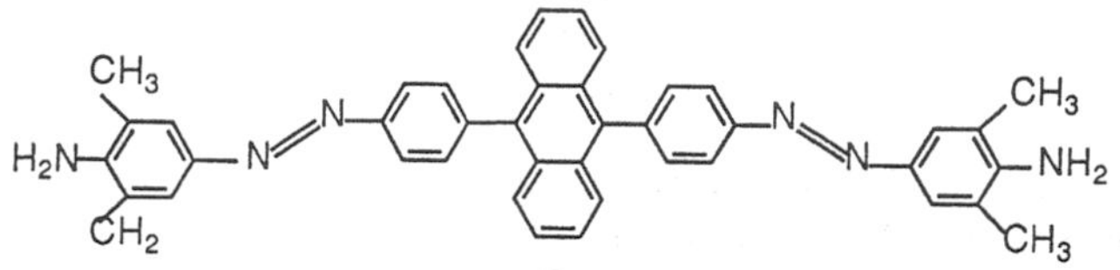

(2)

Matsuda et al.[231], Sonoda and Matsuda[232] reported the $\chi^{(3)}$ values of well-defined chain length oligomers having a phenylenevinylene-type structure. The PPV oligomers where n = 2 and 3 have 22 and 30 π-electrons. The $\chi^{(3)}$ measured by the THG at 1.86 μm were 4×10^{-14}, and 2×10^{-13} esu for n = 2 and n = 3, respectively. The band gap changed from 3.16 eV (n = 2) to 2.91 eV (n = 3) as the number of π-electrons increased.

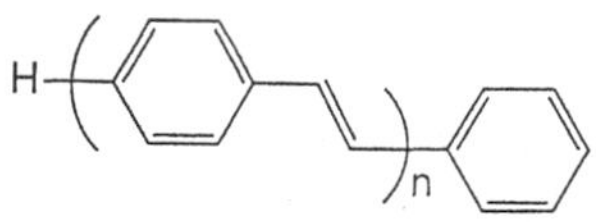

Table 29 $\chi^{(3)}$ Measured by THG at 1.86 μm for Well-defined Oligomers

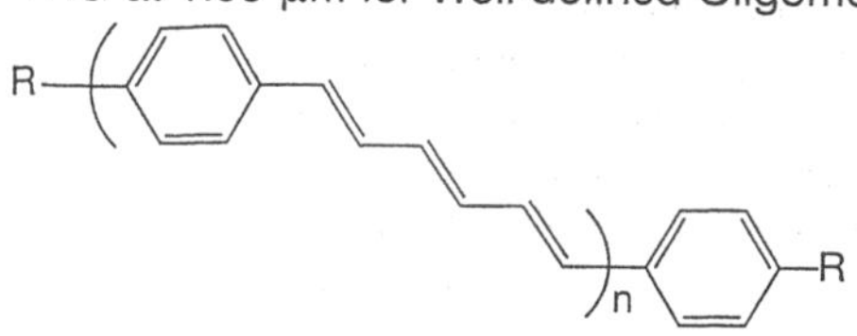

Oligomer	R	n	E_g (eV)	$\chi^{(3)}$ (10^{-13} esu)
DPH	H	1	3.17	0.35
PHT2	H	2	2.66	11
FPH	CHO	1	2.81	2.2
FPH2	CHO	2	2.53	7.8

From Matsuda et al., 55th Autumn Meeting of the Japan Society of Applied Physics, 1994. With permission.

Phenylenehexatrienylene derivatives showed larger $\chi^{(3)}$ values than PPV oligomers at the same wavelengths (Table 29). The $\chi^{(3)}$ value of phenylenehexatrienylene dimer (PHT2) was found to be more than 30 times the $\chi^{(3)}$ value of the monomer. The $\chi^{(3)}$ value of phenylenehexatrienylene having terminal aldehyde ($n = 1$, FPH) was about an order of magnitude larger than DPH. For FPH2, $\chi^{(3)}$ was 3.5 times larger than that of FPH although lower than PHT2.

4. Polythiophenes and Model Compounds

Polythiophene is another π-conjugated polymer which has created much interest because of its good environmental stability compared with polyacetylenes.[113,114] Polarons and bipolarons are associated with the geometry changes toward a quinoid structure. Yand et al.[240] investigated the third-order nonlinear properties of polycondensed thiophene-based polymers. Polythiophene, polythieno(3,2-b)-thiophene, and polydithieno(3,2,-b,2',3'-b)thiophene consist a thiophene ring as a common repeating unit in the polymer chain. Good optical quality thin films can be obtained by electrochemical polymerization. To introduce solubility in the parent polythiophene, an alkyl chain (C_nH_{2n+1}) is substituted at the 3-position of the thiophene ring. A number of poly(3-alkylthiophenes) have been studied in connection with this. The $\chi^{(3)}$ measurements on poly(3-alkylthiophene) where alkyl chain ranges from methyl (CH_3) to tetradecyl ($C_{14}H_{29}$) have also been made by various researchers. Usually, the values of third-order susceptibility is of the same magnitude except that relatively higher $\chi^{(3)}$ values are obtained for shorter alkyl chain and for measurements conducted in the resonant regions. For example, among these polymers, large $\chi^{(3)}$ values are observed at 532 and 602 nm because of the resonance enhancement because polythiophenes absorb strongly in this regime. Okawa et al.[241] investigated $\chi^{(3)}$ values of poly(3-alkylthiophene). The enhancement of the $\chi^{(3)}$ value by three-photon resonance was noticed at a fundamental wavelength of 1.54 μm. Figure 22 shows the chain length dependence of $\chi^{(3)}$ for alkyl chains ranging from hexyl to dodecyl. The difference in $\chi^{(3)}$ values as a function of the alkyl chain length occurs from the difference caused in the local field factors (refractive indices). These polythiophenes have optical maxima at about

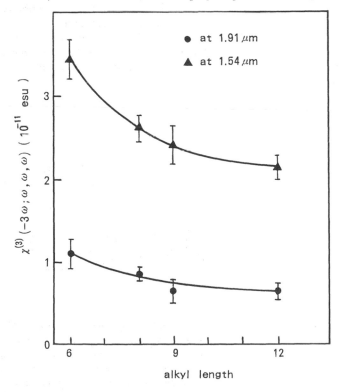

Figure 22 The chain-length dependence of $\chi^{(3)}$ values of poly(3-alkylthiophenes) at 1.91 and 1.54 μm. The enhancement of $\chi^{(3)}$ values by three-photon resonance was noticed at a fundamental wavelength of 1.54 μm. (From Okawa et al., *Mater. Res. Soc. Symp. Proc.*, 214, 23, 1991. With permission.)

541 nm, hence the $\chi^{(3)}$ values are several times larger at 1.54 μm than those at 1.907 μm. The side chains and the orientation of the main chain may also play a role. Sasabe et al.[242] reported a $\chi^{(3)}$ of 5×10^{-12} esu for $R = CH_2-OC_{12}H_{25}$ at 1.907 μm from THG measurements.

Callender et al.[243] reported $\chi^{(3)}$ values of a series of alkyl- and alkoxy-substituted polythiophenes measured by the THG technique. The $\chi^{(3)}$ values of poly(3-hexylthiophene), poly(3-tetradecylthiophene), poly(3-octyloxy,4-methylthiophene) (POMT), poly(3-dibutoxythiophene) (PDBuOT); and copolymers of 3-decylthiophene and 3-methylthiophenes [P(DT-3MeT)] thin films were found in the range 3.6×10^{-12} to 1.2×10^{-11} esu. The effects of substitution on the $\chi^{(3)}$ values were minor. The P(DT-3MeT) copolymer in a ratio of 50:50 shows a $\chi^{(3)}$ value of 6.6×10^{-12} esu, about twice that of 65:35 copolymer (3.6×10^{-12} esu).

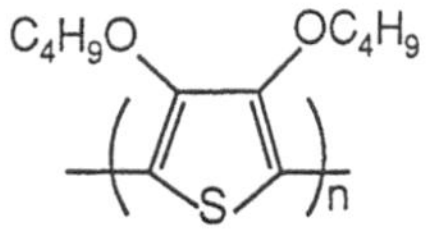

PDBT 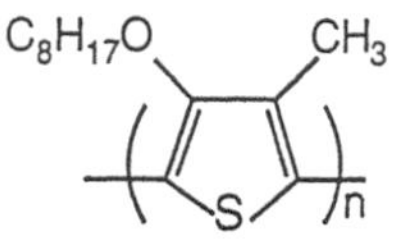   POMT 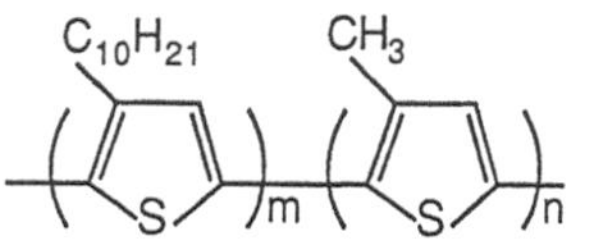   P(DT-3MeT)

Polythieno(3,2-b)-thiophene (PTT) and polydithieno(3,2-b,2',3'-b)thiophene (PDTT) consist of a thiophene ring as a common repeating unit in the polymer chain.[244] The polythiophene (PT) exhibits a $\chi^{(3)}$ value of about 3×10^{-11} esu at 1.06 μm which is about two orders of magnitude smaller than that of the values (10^{-9} esu) measured in the resonant region (532 nm). Above the band gap, the $\chi^{(3)}$ values are larger than 10^{-9} esu for PT, PTT, and PDTT and in particular the $\chi^{(3)}$ value reaches 10^{-8} esu for PDTT at 532 nm. Below the band gap, the $\chi^{(3)}$ values are about 10^{-11} esu for all three PT, PTT, and PDTT samples. Kajzar et al.[245] investigated third-order susceptibility of PTT poly[1,4-di(2-thienyl)benzene] (PDTB) by using the THG in the fundamental wavelength range of 1.064 μm to 1.907 μm. PDTB shows a $\chi^{(3)}$ value of 1.1×10^{-11} esu. The wavelength dependence shows a three-photon resonance enhancement in the 1.25~1.45 μm fundamental wavelength range. The possible differences in the optical nonlinearity of PTT and PDTB probably occurred due to the different conjugation lengths.

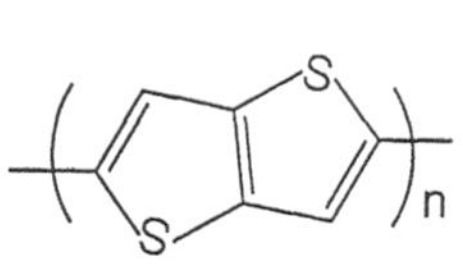

(PTT) 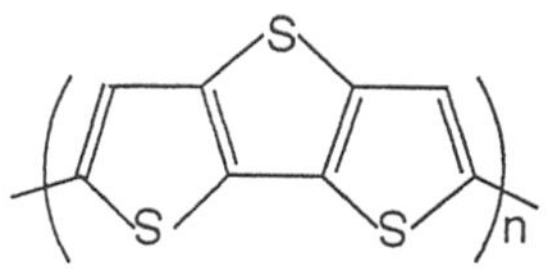   (PDTT)

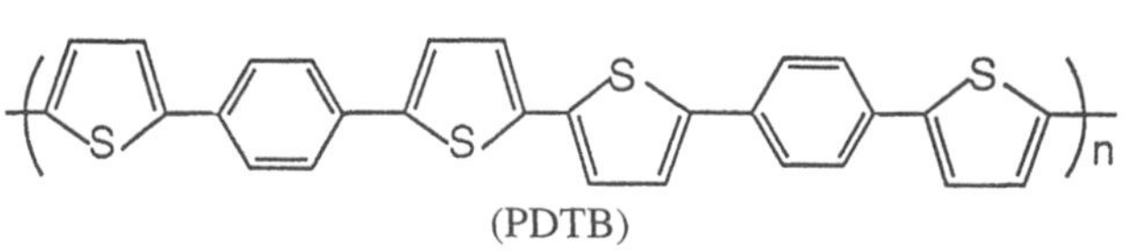

(PDTB)

Kaino et al.[246] report the nonlinear properties of poly(thiophene vinylene) (PTV) which consists of thiophene and acetylenic units. Figure 23 shows the variation of $\chi^{(3)}$ as a function of wavelength for both undoped and iodine-doped PTV samples. The $\chi^{(3)}$ value is almost the same for undoped as well as for iodine-doped PTV. The damage threshold of the iodine-doped PTV is almost half compared with the undoped films because the PTV films become darker after iodine doping which strongly enhances the absorption of laser beam irradiation. PTV exhibits a maximum $\chi^{(3)}$ value of 0.32×10^{-10} esu at 1.85 μm.

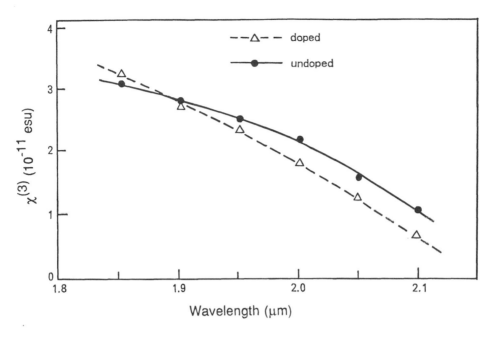

Figure 23 Variation of $\chi^{(3)}$ values as a function of wavelength for undoped and iodine-doped PTV samples. (From Kaino et al., *Appl. Phys. Lett.,* 53, 2002, 1988. With permission.)

Houlding et al.[247] report the third harmonic response of amorphous poly(3-methyl-4'-octyl-2,2'-bithiophene-5,5'-diyl) (P3MOT) as a function of wavelength and film thickness. The $\chi^{(3)}$ value decreases from 4.4×10^{-11} esu at 2.4 μm to 1.8×10^{-11} esu at 1.06 μm (Figure 24). The P3MOT prepared from a dimer of 3-methylthiophene and 3-octylthiophene contains a 1:1 ratio of corresponding units. Figure 25 shows the thickness dependence of $\chi^{(3)}$ for P3MOT. The thickness

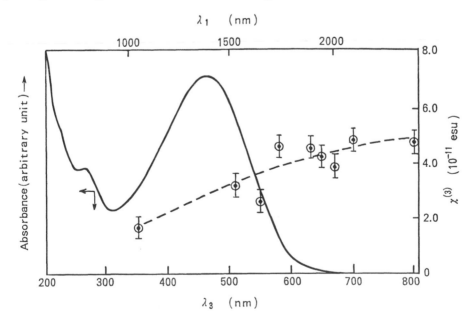

Figure 24 Variation of third-order NLO susceptibility of P3MOT as a function of wavelength for a 720 Å thick film. Top and bottom scales refer to the wavelength of fundamental beam and third-harmonic wavelength, respectively. (From Houlding et al., *Chem. Mater.,* 2, 169, 1990. With permission.)

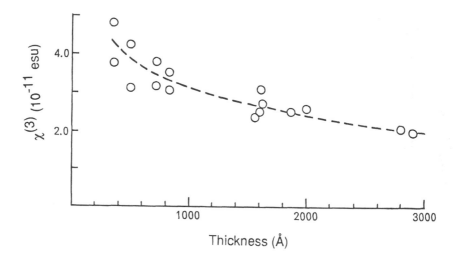

Figure 25 Variation of calculated third-order NLO susceptibility of P3MOT as a function of film thickness for fundamental wavelength of 1.740 μm. (From Houlding et al., *Chem. Mater.*, 2, 169, 1990. With permission.)

dependence is rather unusual and thinner films show large values of $\chi^{(3)}$. The P3MOT has a laser damage threshold as high as 30 MW/cm². Oxidation of a 650 Å thick P3MOT film by NOBF$_4$ leads to a significant decrease in third harmonic signal, and a doping level of more 15% causes damage on exposure to a laser source.

Jenekhe et al.[248,249] investigated the third-order susceptibility of new soluble, conjugated polythiophene derivatives containing benzenoid and quinoid moieties in a single strand. The $\chi^{(3)}$ value measured at 532 nm was 46×10^{-10} esu for a copolymer of polythiophene having a methyl group at the 3-position of the aromatic ring and two 3-methylthiophene units are linked together via a side group, $-C-C_6H_5$. Almost identical copolymers, poly(α-[5,5′-bithiophenediyl]benzylidene-block-α-[5,5′-bithiophenequinodi-methanediyl] (PBTBQ) and poly(α-[5,5′-bithiophenediyl]*p*-acetoxybenzylidene-block-α-[5,5′-bithiophenequinodi-methanediyl] (PBTABQ) show extremely large $\chi^{(3)}$ values of 2.7×10^{-7} and 4.5×10^{-8} esu, respectively, at 532 nm. The measured second hyperpolarizabilities γ were 1.2×10^{-31}, 1.6×10^{-29}, and 3.7×10^{-30} esu for PMTBQ, PBTBQ, and PBTABQ, respectively. The dispersion of $\chi^{(3)}(-3\omega; \omega, \omega, \omega)$ for several polythiophenes which possess mixed aromatic-quinonoid backbone structures was reported by Meth et al.[250] The precursor polymers poly(α-[5,5′-bithiophenediyl]*p*-benzylidene (PBTP), poly(α-[5,5′-bithiophenediyl]*p*-nitrobenzylidine (PBTNB), poly(α-[5,5′-bithiophenediyl]*p*-heptyloxybenzylidene (PBTHB), poly(α-[5,5′-terthiophenediyl]*p*-acetoxybenzylidene (PTTHB) and their derivatives are PBTBQ, PBTNBQ, PBTHBQ, and PTTHBQ. The dispersion of $\chi^{(3)}(-3\omega; \omega, \omega, \omega)$ in the range 0.5~1.4 eV was investigated. At 0.73 eV, the resonantly enhanced $\chi^{(3)}(-3\omega;\omega,\omega,\omega)$ values of P3HT, PTBQ, PBTBQ, PBTHBQ, PBTNBQ, and PTTHBQ were 46.0×10^{-12}, 5.32×10^{-12}, 7.76×10^{-12}, 4.20×10^{-12}, 10.3×10^{-12}, and 2.52×10^{-12} esu, respectively. On the other hand, at 1.38 eV, the $\chi^{(3)}(-3\omega; \omega, \omega, \omega)$ values of P3HT, PTBQ, PBTBQ, PBTHBQ, PBTNBQ, and PTTHBQ were 6.84×10^{-12}, 0.23×10^{-12}, 0.87×10^{-12}, 0.29×10^{-12}, 7.96×10^{-12}, and 0.82×10^{-12} esu, respectively. It seems that an aromatic and quinoid arrangement in both the ground and excited states is a condition favorable to the NLO process.

PTBQ: R_1 = H R_2 = H, R_3 = [benzene ring structure] $x = 1, y = 1$

PMTBQ: R_1 = H, R_2 = CH_3, R_3 = [benzene ring structure] $x = 1, y = 1$

PBTBQ: R_1 = H, R_2 = H, R_3 = [benzene ring structure] $x = 2, y = 2$

PBTABQ: R_1 = H, R_2 = H, R_3 = [benzene ring]$-O-\overset{\overset{\textstyle O}{\|}}{C}-CH_3$ $x = 2, y = 2$

PBTHBQ: R_1 = H, R_2 = H, R_3 = [benzene ring]$-OC_7H_{15}$ $x = 2, y = 2$

PTTHBQ: R_1 = H, R_2 = H, R_3 = [benzene ring]$-OC_7H_{15}$ $x = 3, y = 3$

PBTNBQ: R_1 = H, R_2 = H, R_3 = [benzene ring]$-NO_2$ $x = 2, y = 2$

Polythiophenes are intense colored materials; therefore, the third-order susceptibility has resonant contribution due to the high absorption. In an attempt to overcome this problem, Nalwa[251,252] prepared two random copolymers of 3-methylthiophene and methyl methacrylate; one is yellow and the other one is red. Both copolymers are solution processable and show good environmental stability. Thin films prepared by the solution-casting technique show different optical absorption characteristics.

(Co(3-MeTH/MMA))

The third-order NLO susceptibility $\chi^{(3)}$ value of thin films of 3-methylthiophene and methylmethacrylate (MMA) copolymers were measured using the DFWM and the THG techniques. The measured effective $\chi^{(3)}(-\omega; \omega, \omega - \omega)$ values were 7.0×10^{-12} esu for the yellow copolymer and 8.7×10^{-11} esu for the red copolymer at 602 nm.[251] The THG measured $\chi^{(3)}(-3\omega; \omega, \omega, \omega)$ values were 5.0×10^{-13} esu for a yellow copolymer and 1.3×10^{-11} esu for red copolymer at 1.860 μm.[252] The large $\chi^{(3)}$ of the red copolymer occured from the resonant contribution because the copolymer strongly absorbed at harmonic frequency (3ω). On the other hand, the yellow copolymer had no absorption at the harmonic frequency, therefore in this case the $\chi^{(3)}$ value was an off-resonance parameter. The large resonant optical nonlinearities of the red copolymer are attributed to the highly delocalized π-electron system of 3-methylthiophene segments, whereas the increased nonconjugated (MMA) segments significantly reduced the magnitude of optical nonlinearity in yellow copolymer. The absorption spectra clearly indicated that the number of 3-methylthiophene units was much lower than that of red copolymer and the increased components of MMA units induced a yellow color. The degree of polymerization probably was inhibited by the MMA segments resulting into a yellow copolymer composed of higher MMA segments and low 3-MeTH components. Therefore, the optical nonlinearities of copolymers were associated with the number of π-conjugated 3-MeTH segments. The NLO responses were somewhat restrained in the copolymer system because of the presence of nonconjugated MMA segments, but copolymerization led toward desired optical transparency.

Polythiophene and polyaniline homopolymers likely show high absorption. Polis et al.[253] prepared a copolymer of thiophene and aniline-reacting substituted p-phenylenediamine and 2,5-dibromothiophene in DMSO at 185 °C. This copolymer has $\chi^{(3)}(-\omega;\omega,\omega,-\omega)$ of about 4×10^{-10} esu. The $\chi^{(3)}$ value of poly(thiophene-2,5-diyl pyridine-2,5-diyl) (PTPY) and poly(thiophene-2,5-diyl 2,2'-bipyridine-5,5'-diyl) (PTBPY) were found to be larger than those of poly(p-phenylene pyridine-2,5-diyl) (PPPY) which has no electron-donating thiophene ring.[254] Third-order NLO susceptibilities measured by different experimental techniques for a variety of poly(3-alkylthiophenes) polythiophene and its copolymers are listed in Table 30.

Ogata[255] reported THG $\chi^{(3)}$ values of copolymers of poly(benzothiazole) PBT containing both thiophene and benzene units as a function of curing temperature. The $\chi^{(3)}$ value of prepared thin films of PTBT were 1.3×10^{-11} esu which increased to 1.6×10^{-11}, 1.9×10^{-11}, 2.6×10^{-11}, and 2.4×10^{-11} esu at curing temperature of 120, 180, 200, and 240 °C, respectively. The $\chi^{(3)}$ value of heat-treated films increased almost twice as much as prepared thin films. The $\chi^{(3)}$ value of copolymer increased with increasing contents of thiophene moiety which may be associated to the high electron-donating nature of thiophene moiety.

Yamamoto et al.[256,257] reported $\chi^{(3)}$ values of π-conjugated charge-transfer polymers consisting of electron-donating thiophene and electron-accepting quinoxaline. The quinoxaline rings have higher electron-accepting ability than the pyridine ring due to imine nitrogens. The $\chi^{(3)}$ of poly(2,3-diheptylquinoxaline-5,8-diyl) [PQx(diHep) and poly(2,3-diphenylquinoxaline-5,8-diyl) [PQx(diPh)] was measured at 5×10^{-11} esu at three-photon resonance region whereas the $\chi^{(3)}$ value at the nonresonant region was lower. Yamamoto et al.[258] reported $\chi^{(3)}$ values of poly(aryleneethynylene) (PAE)-type polymers $(-\,Ar-C{\equiv}C-Ar'-C{\equiv}C-)_n$ where Ar and/or Ar' contain the long alkyl chains and/or pyridine rings. The wavelength dependence of the $\chi^{(3)}$ value showed that PAE-1, PAE-2, and PAE-3 have $\chi^{(3)}$ values of $7 \sim 9 \times 10^{-12}$ esu at a near-resonance region and $\sim 5 \times 10^{-11}$ esu at a resonant region. The $\chi^{(3)}$ values of PAEs were found to be similar to PTV $\chi^{(3)}$ values and larger than those of poly(2,2'-bypyridine-5,5'-diyl) and poly(thiophene-2,5-diyl).

Table 30 Third-order NLO Susceptibilities of Polythiophenes

Polythiophenes	$\chi^{(3)}$ (10^{-10} esu)	Wavelength (μm)	Experimental Technique	Ref.
Poly(thiophene) R = H	10	0.532	DFWM	240
	6.0	0.620	DFWM	148
	3.52	1.907	THG	259
	66	0.532	DFWM	244
	30	0.605	DFWM	244
	0.3	1.064	DFWM	244
	0.53	1.50	THG	260
	0.16	1.90	THG	260
R = CH$_3$	0.49	1.50	THG	260
	0.22	1.90	THG	260
R = C$_4$H$_9$	0.20	1.30	THG	261
R = C$_6$H$_5$	0.81	1.50	THG	260
	0.68	1.90	THG	260
R = C$_6$H$_{11}$	0.044		THG	265
R = C$_6$H$_{13}$	0.12	1.053	THG	243
	7.0	0.620	DFWM	148
R = C$_{10}$H$_{21}$	0.1	1.064	DFWM	262
R = C$_{12}$H$_{25}$	7.0	0.590	DFWM	263
	5.0	0.602	DFWM	263
	4.0	0.705	DFWM	263

Table 30 *Continued*

Polythiophenes	$\chi^{(3)}$ $(10^{-10}$ esu)	Wavelength (μm)	Experimental Technique	Ref.
$R = C_{14}H_{29}$	0.056	1.053	THG	243
$R = CH_2-OC_{12}H_{25}$	0.05	1.907	THG	242
PMTBQ	46	0.532	DFWM	248
PBTABQ	450	0.532	DFWM	249
PBTBQ	2700	0.532	DFWM	249
POMT	0.06	1.053	THG	243
PDBuOT	0.075	1.053	THG	243
PTT	59	0.532	DFWM	244
	30	0.605	DFWM	244
PDTT	113	0.532	DFWM	244
	45	0.600	DFWM	244
	0.3	1.064	DFWM	244
PDTB	0.11		THG	245
PTPY	1.5	1.525	THG	254
P3MOT	0.44	2.4	THG	247
PTV	0.32	1.85	THG	246
	4.50	1.95	THG	227
Co(3-MeTH/MMA)-yellow	0.07	0.602	DFWM	252
	0.005	1.86	THG	252
Co(3-MeTH/MMA)-red	0.87	0.602	DFWM	252
	0.13	1.86	THG	252
PTBT	0.26	1.064	THG	255
PThQx(diHep)	0.5	1.7	THG	256
	0.1	2.2	THG	256
PTh(Qx(diPh)	0.016	1.6	THG	256
PAE-1	0.42	1.5	THG	257
PAE-2	0.47	1.5	THG	258
PAE-3	0.21	1.5	THG	258

The chemical structures of some polythiophenes are shown below.

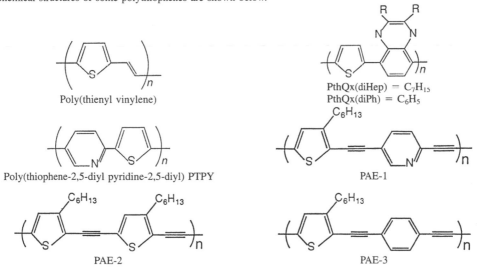

Poly(thienyl vinylene)

PthQx(diHep) = C_7H_{15}
PthQx(diPh) = C_6H_5

Poly(thiophene-2,5-diyl pyridine-2,5-diyl) PTPY

PAE-1

PAE-2

PAE-3

a. Model Compounds of Polythiophenes

Fichou et al. [266] first reported the third-order NLO properties of a series of α-conjugated thiophene oligomers (α-nT) having several repeat thiophene units N = 3,4,5,6,8, and so on. Figure 26 shows the variation of the second hyperpolarizability and third-order NLO susceptibility as a function of the number of the thiophene rings. Second hyperpolarizability and third-order susceptibility of α-nT oligomers increases significantly up to N = 14, whereas the optical gap remains almost constant after N = 8. The $\chi^{(3)}(-3\omega;\omega,\omega,\omega)$ values of 1.88×10^{-12} and 2.38×10^{-12} esu were recorded for the pentamer

694

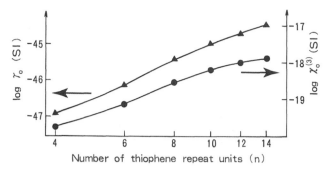

Figure 26 Variation of theoretical third-order NLO susceptibility and second hyperpolarizability as a function of the thiophene ring. (From Fichou et al., *Organic Materials for Nonlinear Optics*, Royal Society of Chemistry Special Publication No. 69, 1989. With permission.)

(α-5T) and hexamer (α-6T) of thiophene at 1.907 μm from the THG measurements respectively. The $\chi^{(3)}$ value of the trimer (α-3T) was found to be almost an order of magnitude smaller than the $\chi^{(3)}$ values of α-6T over the fundamental frequency range. Thienpont et al.[267] also measured second hyperpolarizability of a series of well-defined conjugated 3-alkyl-substituted oligothiophenes (N = 3,4,5,7,9, and 11). The conjugation length dependence of α and γ show the exponents of 2.4 and 4.6, respectively. The γ value of 3-alkyl-substituted oligothiophenes increased up to 11 repeat units, where γ values reached as high as 4.6×10^{-32} esu. The second hyperpolarizabilities of these well-defined oligothiophenes were measured by EFISH technique at the fundamental wavelength of 1.064 μm. Table 31 lists the NLO parameters for alkyl-substituted oligothiophenes. The γ values strongly depend on the conjugation chain length up to $n \geq 7$, and then it started slowing down as the π-electron system further increased. For example, the γ value changes by two orders of magnitude as the n = 3 increases to n = 7. The band gap also significantly decreases, which shows some preliminary saturation behavior at n = 9. The power law expressed by the extent of conjugation dependence on α and γ gave exponents of 2.4 and 4.6, respectively. Fichou et al.[266] point out that the optical nonlinearity becomes smaller if the thiophene ring rotates around the single bonds connecting thiophene units; and in a fully planar oligothiophene, the γ value may saturate at $n \geq 10$. Thienport et al.[267] conducted solvatochromic and thermochromic experiments on alkyl-substituted oligothiophenes and suggested that the molecules were not fully planar in a PMMA matrix at room temperature. A large γ value can be anticipated if the molecules can be planarized for long conjugated oligothiophenes. Hence, the conformation effect plays a great role in limiting the contribution of the conjugation length to optical nonlinearity.

Table 31 Third-order Nonlinear Hyperpolarizability, Polarizability, and Band Gap of a Series of Alkyl-substituted Oligothiophenes

Oligomer	Thiophene Units (n)	Hyperpolarizability γ (10^{-34} esu)[a]	Polarizability α (10^{-23} esu)[a]	Band Gap (eV)
1	3	9.9	1.1	3.67
2	4	22	1.4	3.13
3	5	107	2.6	3.01
4	7	360	6.6	2.81
5	9	370	–	2.78
6	11	460	7.0	2.77

From Thienpont et al., *Phys. Rev. Lett.*, 65, 2141, 1990. With permission.

[a] The hyperpolarizability and polarizability were measured by the EFISH method at 1.064 μm by dispersing a few weight percent of materials in a PMMA matrix at room temperature.

C_4H_9

3

C_4H_9 C_4H_9

4

$C_{12}H_{25}$

5

C_4H_9 C_4H_9 C_4H_9

6

Joshi et al.[268] measured $\chi^{(3)}$ values by laser self-trapping at 514.5 nm and by the DFWM at 1064 nm for chloroform and pyridine solutions of dicyanovinylthiophene (1), dicyanovinyldithienyl (2), dicyanovinylterthienyl (3), bis(dicyanovinyl)thiophene (4), bis(dicyanovinyl)dithienyl (5), and bis(dicyanovinyl)trithienyl (6). In these oligomers, thiophene, bithienyl, and terthienyl act as electron donors whereas the cyanovinylene groups act as electron acceptors, therefore they are donor–acceptor systems of defined conjugation length. The $\chi^{(3)}$ values measured by LST at 514.5 nm for chloroform solutions of oligomers 4, 5, and 6 were 2.9×10^{-7}, 1.1×10^{-4}, and 1.1×10^{-5} esu, respectively, and were affected by resonance and solution heating effects. The second hyperpolarizabilities of these donor–acceptor systems show a significant increase compared with oligothiophenes. Geisler et al.[269] measured $\chi^{(3)}$ and γ values of thin films of four oligomers of thienylene-ethynylenes and two oligomers of thienylenevinylenes in PMMA matrix by THG at 1.064 μm. The γ values ranged between 6×10^{-35} and 2.3×10^{-33} esu. THG $\chi^{(3)}$ value of thienylene-ethynylene (8) is larger than that of analogous vinylene compound (11), whereas compound 9 showed smaller value than that of compound 12; this is caused by the resonance contribution. Theoretical THG γ values were found larger for compound 9 than for compound 12 because of the resonance at 1.165 eV. Away from the resonance, the vinylenes show larger γ values than those of ethynylenes. The third-order optical nonlinearity was not increased significantly by adding carbon-carbon double or triple bonds between thiophene rings compared with thiophene oligomers. The $\chi^{(3)}$ and γ values measured by DFWM and THG techniques for three classes of thiophene oligomers are listed in Table 32.

5. Heteroaromatic Ladder Polymers

Heteroaromatic ladder polymers are an another class of conjugated polymers. Benzimidazobenzophenanthroline type BBL and BBB ladder polymers are known for their excellent environmental stability.[270,271] These polymers have similar chemical structures. In BBL the repeat units are linked together through two covalent bonds, whereas in BBB repeat units are linked by a single carbon-carbon bond. The difference in polymer chain linkage supports a complete ladder structure for BBL and a semiladder structure for BBB polymer. BBL and BBB can be obtained as free-standing thin films. Lindle et al.[272] measured the NLO properties of these polymers by the degenerate four-wave mixing technique at 1.064 μm. The nonresonant $\chi^{(3)}$ value (1.5×10^{-11} esu) of BBL polymer is three times larger than that of BBB polymer ($\chi^{(3)} = 5.5 \times 10^{-12}$ esu). The resonant $\chi^{(3)}$ value (2.0×10^{-9} esu) of BBL polymer measured at a wavelength of 0.532 μm is two orders of magnitude larger. The BBL polymer was electrochemically doped with tetrabutylammonium ions (BF$_4^-$) although it showed nine orders of magnitude difference in electrical conductivity, but $\chi^{(3)}$ was enhanced by about 30% which suggests that the mechanisms responsible for optical nonlinearity and electrical conductivity are different. The difference observed in optical nonlinearity of BBL and BBB polymer results from the coplanarity arising because of different structural features which in turn affect π-electron delocalization. BBL and BBB have a rigid backbone. Jenekhe et al.[273] measured values of ladder polymers by using THG technique.

Table 32 $\chi^{(3)}$ (10^{-12} esu) and γ Values (10^{-33} esu) of Oligothienyldicyanovinylenes, Thienylene-ethynylenes, and Thienylenevinylenes

Oligomers[a]	1	2	3	4	6	7	8	9	10	11	12
$\chi^{(3)}$	3.11	1.29	4.49	4.42	4.43	0.11	0.64	1.16	1.23	0.52	1.26
γ	8.7	2.2	370	280	–	0.063	0.55	1.5	2.3	0.5	1.65

[a] The $\chi^{(3)}$ and γ values of Compounds 1 through 6 were measured by DFWM at 1.064 μm,[268] whereas the $\chi^{(3)}$ and γ values for Compounds 7 through 12 were measured by THG at 1.064 μm for PMMA thin films.[269]

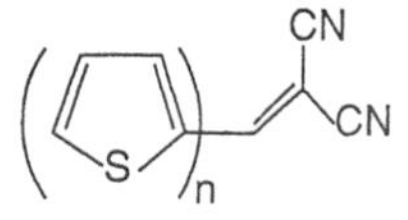

$n = 1$ (1), $n = 2$ (2), $n = 3$ (3)

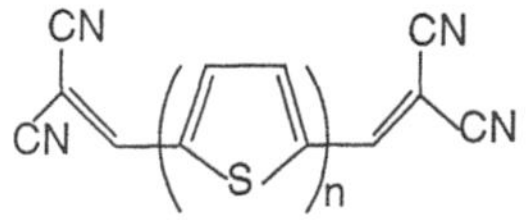

$n = 1$ (4), $n = 2$ (5), $n = 3$ (6)

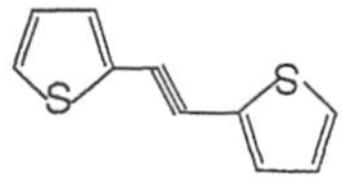

(7)

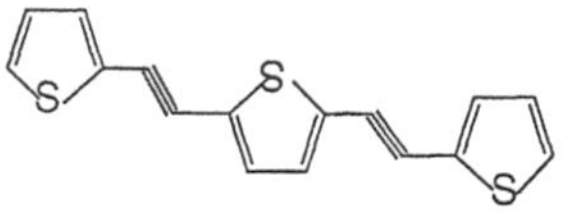

(8)

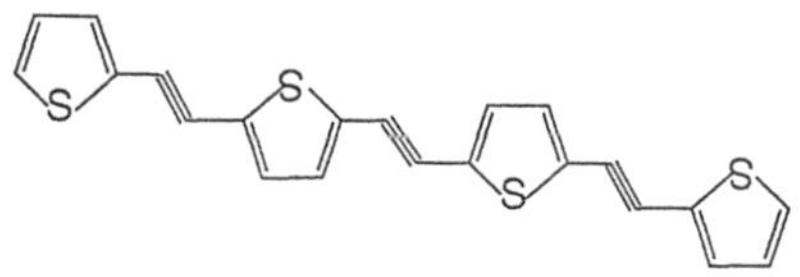

(9)

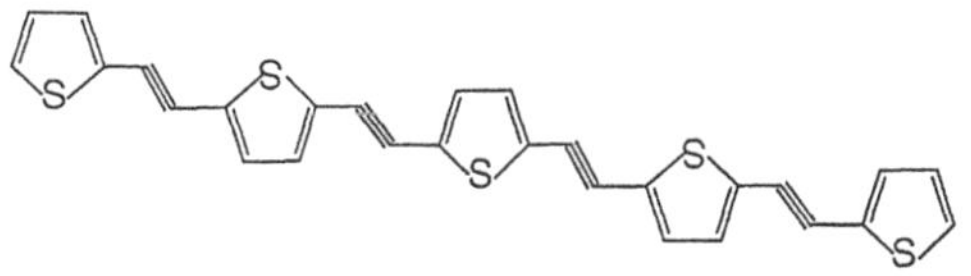

(10)

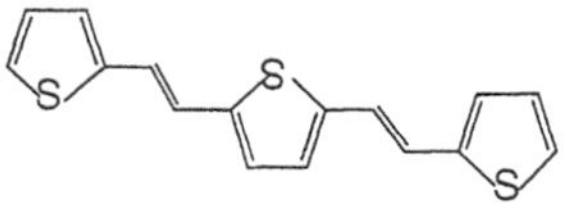

(11)

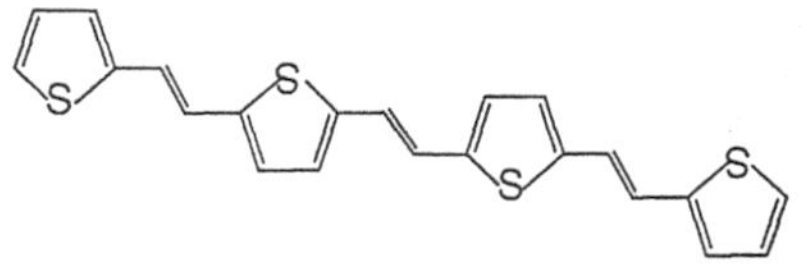

(12)

The magnitude of the $\chi^{(3)}$ was found to be enhanced significantly at three-photon resonance for BBL, BBB, and cis-BBB. The $\chi^{(3)}$ values of BBL, BBB, and cis-BBB were 2.90×10^{-11}, 0.96×10^{-11}, and 0.57×10^{-11} esu, respectively, at 1.05 μm. The $\chi^{(3)}$ values of BBL and BBB polymers are enhanced by factors of 50 and 17, respectively, compared with the model compound cis-BBB at 1.05 μm. Figure 27 shows the dispersion of $\chi^{(3)}$ values of BBL, BBB, and cis-BBB in wavelengths from 1.0 to 2.4 μm. The peak positions in the $\chi^{(3)}$ spectra of BBL, BBB, and cis-BBB at 1.695, 1.695, and 1.50 μm, respectively, show a three-photon resonance.[274] The large $\chi^{(3)}$ values of BBL relative to BBB polymer shows the effect of ladder structure and rod-like conformation.

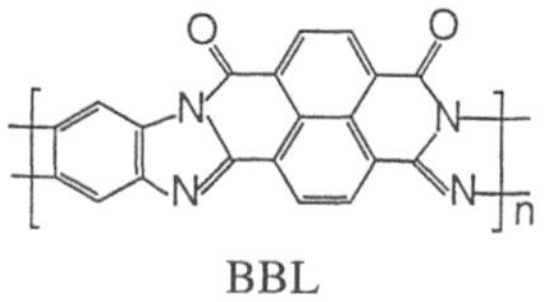

BBL 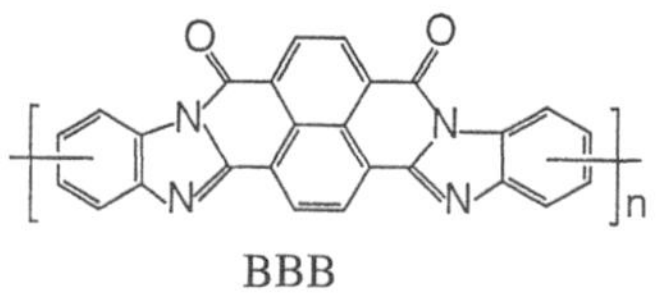   BBB 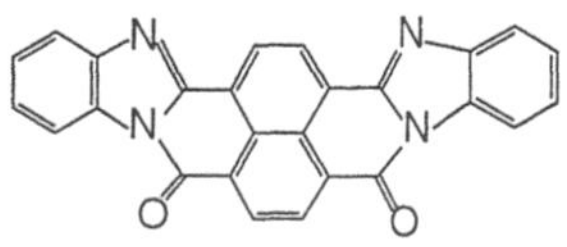   cis-BBB

Poly(p-phenylenebenzobisthiazole) (PBZT) is another heteroaromatic polymer that exhibits good environmental stability similar to BBL polymer, although it differs from a structural point of view. The polymers prepared by condensation reaction can be obtained as free-standing thin films as well as fibers which exhibit excellent mechanical properties because of their rigid-rod conformation. Benzazole units are the NLO components in PBZT backbone. Third-order NLO susceptibility of PBZT has been investigated by using a degenerate four-wave mixing technique.[275] The $\chi^{(3)}$ value of PBZT is about an order of magnitude higher than that of carbon disulfide (CS$_2$). The $\chi^{(3)}$ is anisotropic, and measurements as a function of angular orientation at two different sets of laser polarization show that the $\chi^{(3)}_{2222}$ (7.2 $\times 10^{-12}$ esu) component is larger than the $\chi^{(3)}_{1111}$ (5.4 $\times 10^{-12}$ esu) component. The maximum $\chi^{(3)}_{1111}$ is measured when the PBZT film is rotated 48° which corresponds to the mechanical draw direction.

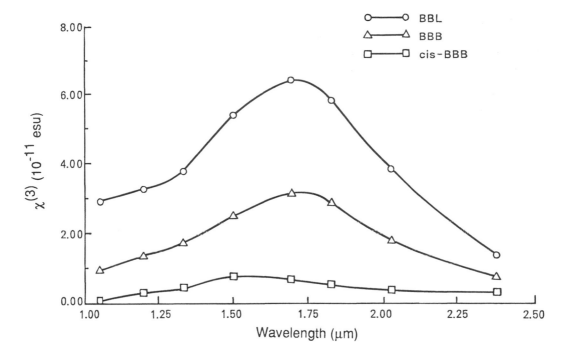

Figure 27 Dispersion of third-order NLO susceptibility of BBL, BBB, and *cis*-BBB in the wavelength range 1.0 to 2.4 μm. (From Jenekhe et al., *Mater. Res. Soc. Symp. Proc.*, 214, 55, 1991. With permission.)

Jenekhe et al.[276] reported the $\chi^{(3)}$ value of poly(*p*-phenylenebenzobisoxazole) (PBO) and compared it with PBZT. The $\chi^{(3)}$ values of both polymers were in the same range, showing no effect of the heteroatom. The $\chi^{(3)}$ values at resonance were one order of magnitude larger than those of the nonresonance region.

Poly(*p*-phenylenebenzobisoxazole) (PBO)　　　　Poly(*p*-phenylenebenzobisthiazole) (PBZT)

Polyquinoxaline-based aromatic conjugated ladder polymers have been investigated by Yu and co-workers.[277–282] Optical quality thin films of polyquinoxaline ladder polymers can be obtained by polycondensation from precursors having a large variety of substituents. The synthetic schemes of 2,5-aminovinyl-substituted 3,6-dichloroquinone are shown in Scheme 1.[277] Dichloroquinone monomers A through D react with tetraaminobenzene to prepare prepolymers and polymers, respectively.[277] These polymers show high thermal stabilities as well as high laser damage thresholds (<1 GW/cm^2). Table 33 lists the third-order nonlinearity, refractive indexes, and absorption coefficients of these ladder polymers. These ladder polymers show $\chi^{(3)}$ values in the range of 3.0×10^{-10} esu to 1.12×10^{-9} esu.

Table 33 The $\chi^{(3)}$, Refractive Indexes (*n*), and Absorption Coefficients (α) of Polyquinoxaline-based Ladder Polymers

Polymer	$\chi^{(3)}_{1111}$ (10^{-10} esu)	$\chi^{(3)}_{1221}$ (10^{-10} esu)	*n*	α (10^4 cm^{-1})	$\chi^{(3)}_{1111}/\alpha$ (10^{-14} esu cm)
A	9.78	1.8	1.96	2.32	3.23
B	8.90	1.9	1.92	2.54	3.50
C	7.40	1.6	1.92	1.88	3.93
D	15.5	3.0	1.92	4.08	3.80

From Yu et al., *Electroresponsive Organic Molecular and Polymeric Systems*, Skotheim, T. ed. Marcel Dekker, New York, 2, 113, 1991. With permission.

The different alkyl substituents on the amine have little influence on $\chi^{(3)}$ values. The ladder polymer D has the largest $\chi^{(3)}$ value of 15.5×10^{-10} esu. Figure 28 shows the DFWM signal–pulse energy as a function of the delay time for ladder polymer D. The decay time of nonlinearity is less than 6 ps. The effect of delocalization and oxidation is shown in Table 34. The lower $\chi^{(3)}$ value observed for prepolymer results from the incomplete condensation.

(Scheme 1)

Yu and Dalton[279] also synthesized triphenodioxazine rigid-rod and alophatic flexible-chain copolymers as depicted in the scheme 2. A prepolymer was obtained by reacting 1,5-bis(4-aminophenylene-1-oxy)pentane with chloronil. Polymer 1 was prepared by refluxing prepolymer in nitrobenzene with benzoyl chloride, whereas polymer 2 was prepared by reacting prepolymer with p-toluenesulfonyl chloride (TsCl) and 1-chloronaphthalene (C_8H_7Cl). Polymer 2 may have two possible structures; however, the lower one was found to be the best fit. Polymers 1 and 2 are thermally stable up to 480 °C and 360 °C, respectively. Polymer 1 showed a $\chi^{(3)}$ value of 4.5×10^{-9} esu as determined by the DFWM technique. Cao et al.[282] later reported the bipolaronic enhanced third-order optical nonlinearity of these ladder polymers. The ladder polymers are an important class of third-order NLO materials.

Spangler et al.[253] reported synthesis and incorporation of ladder polymer subunits in copolyamides, pendant polymers, and composites for enhanced nonlinear optical response. The chemical structures of

Table 34 Third-order Nonlinear Optical Susceptibilities of Heteroaromatic Ladder Polymers

Polymer	$\chi^{(3)}$ (10^{-11} esu)	Wavelength (μm)	Experimental Technique	Ref.
BBL	200	0.532	DFWM	279
	9.1	1.064	DFWM	279
	1.5	1.064	DFWM	272
(doped)	2.0	1.064	DFWM	272
	6.4	1.695	THG	273
	2.9	1.05	THG	273
BBB	0.55	1.064	DFWM	272
	3.15	1.695	THG	273
	0.96	1.05	THG	273
cis-BBB (monomer)	0.77	1.50	THG	273
	0.57	1.05	THG	273
PBZT	0.8	0.602	DFWM	275
	0.6	2.4	THG	276
	8.3	1.40	THG	276
PBO	0.81	2.4	THG	276
	7.0	1.28	THG	276
L-DCDEBQ-TAB	6	0.532	DFWM	280
OC-DCDEBQ-TAB	10	0.532	DFWM	280
L-DCDEBQ-TAB	30	0.532	DFWM	280
L-DCDPBQ-TAB	28	0.532	DFWM	280
L-DCPBQ-TAB	125	0.532	DFWM	280
L-DCPBQ-DABD	128	0.532	DFWM	280
L-DCHBQ-TAB	112	0.532	DFWM	280
20% POL 3-ring pendant	1.2	0.598	DFWM	283
20% POL pendant	2.6	0.598	DFWM	283
20% POL copolyamide	0.15	0.598	DFWM	283

DABQ = 3,3′diaminobenzidine; TAB = 1,2,4,5-tetraaminobenzene; DCDEBQ = 2,5-dichloro-3,6-bis(2-diethylaminovinyl)benzoquinone; DCDPBQ = 2,5-dichloro-3,6-bis(2-dipropylaminovinyl)benzoquinone; DCHBQ = 2,5-dichloro-3,6-bis(2-hexamethyleneiminylvinyl)-1,4-benzoquinone; OC = open-chain precursor polymer for DCDEBQ-TAB; DCPBQ = 2,5-dichloro-3,6-bis(2-piperidinovinyl)benzoquinone.

a polymer having POL ladder subunits as pendant groups to p-hydroxystyrene and a POL-copolyamide are shown below. DFWM studies showed that the 5-ring side-chain polymer with a NLO figure-of-merit of $\chi^{(3)}/\alpha$ of 10^{-14} esu cm was the most promising material. This polymer had a $\chi^{(3)}$ value twice as large as that of the 3-ring side-chain polymer. The $\chi^{(3)}$ value of main-chain copolyamide having POL subunits was an order of magnitude lower because of the dilution of π-conjugation from the nonactive segments of the polymer chain. The $\chi^{(3)}$ value of 2×10^{-12} esu was measured for the copolyamide with 10% PTV dimer. The copolyamide with incorporated polyenylic or PTV oligomers showed $\chi^{(3)}/\alpha$ values of 10^{-13} esu cm at 532 nm.

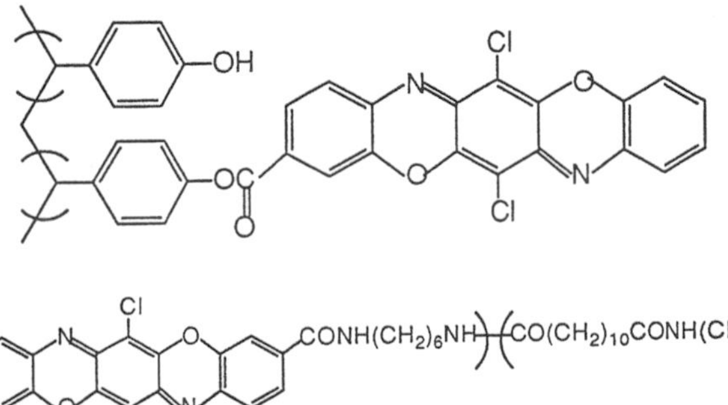

(Scheme 2)

Scheme labels: K₂CO₃ DMF; prepolymer; benzoyl chloride nitrobenzene; TsCl, C₈H₇Cl; polymer 1; OR; polymer 2; CH₃

6. Polyquinolines

Agrawal et al.[284] reported the third-order optical nonlinearities of a series of 12 systematically designed homopolymers and random copolymers of rigid-rod polyquinolines and polyanthrazolines. Table 35 lists the absorption maxima and the resonant and nonresonant $\chi^{(3)}$ values of rigid-rod polyquinolines and polyanthrazolines. The $\chi^{(3)}(-3\omega; \omega, \omega, \omega)$ dispersion in the 0.9 ~2.4 μm wavelength region indicates

Table 35 Absorption Maxima, Resonant and Nonresonant $\chi^{(3)}$ Values of Rigid-rod Polyquinolines and Polyanthrazolines

Polymer	λ_{max} (nm)	At 3λ (μm)	Resonant	Nonresonant (2.38 μm)
		$\chi^{(3)}(-3\omega; \omega, \omega, \omega)$ $(10^{-12}$ esu)		
PPQ	289, 410	1.2	11.2	2.3
PBPQ	311, 394	1.2	19.2	2.1
PBAPQ	305, 399	1.2	26.6	2.2
PSPQ	289, 372	1.2	8.1	2.2
PPPQ	299, 398	1.2	33.3	3.2
PBDA	338, 414	1.32	15.0	2.0
PBADA	336	1.05	31.5	
	426	1.32	19.3	2.8
PSDA	361	1.05	13.8	
	448	1.32	13.4	2.2
PPDA	338	1.05	17.7	
	443	1.32	10.9	1.8
PBPQ/PBAPQ	310, 397	1.2	26.4	2.0
PSPQ/PBAPQ	290, 390	1.2	17.8	2.2
PSPQ/PBPQ	290, 390	1.2	18.8	1.8

From Agrawal et al., *J. Phys. Chem.*, 96, 2837, 1992. With permission.

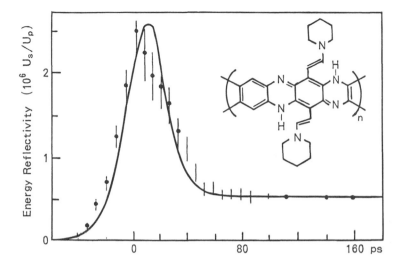

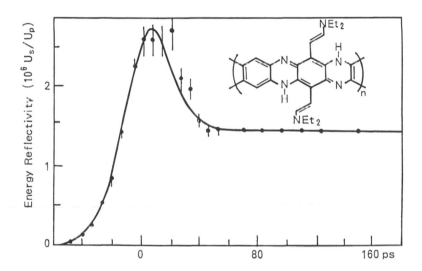

Figure 28 DFWM signal pulse energy versus delay time for ladder polymer A and C. Solid line indicates the theoretical calculations. (From Yu et al., *Electroresponsive Organic Molecular and Polymeric Systems*, Skotheim, T. A., ed., Marcel Dekker, New York, Vol. 2, 113, 1991. With permission.)

three-photon resonance enhancement where the $\chi^{(3)}$ ranges between 0.8×10^{-11} and 3.3×10^{-11} esu depending on the polymer backbone structures. Off-resonance at 2.38 μm, the $\chi^{(3)}(-3\omega; \omega, \omega, \omega)$ values of all the polymers were ~$(2.5 \pm 0.5) \times 10^{-12}$ esu. The order of increasing values of resonant $\chi^{(3)}(-3\omega; \omega, \omega, \omega)$ at 1.32 μm was PPDA < PSDA < PBDA < PBADA and at 1.0.5 μm was PBDA < PSDA < PPDA < PBADA. The resonant $\chi^{(3)}(-3\omega; \omega, \omega, \omega)$ values of the three copolymers were in the order PBAPQ/PBPQ > PSPQ/PBPQ ~ PSPQ/PBAPQ. The PBAPQ/PBPQ copolymer shows a 15% enhancement of resonant $\chi^{(3)}(-3\omega; \omega, \omega, \omega)$ compared with the molar average of 22.9×10^{-12} esu calculated from the $\chi^{(3)}$ values of homopolymers PBAPQ and PBPQ. PSPQ/PBPQ copolymer likely shows a 38% enhancement of $\chi^{(3)}$ compared with the calculated molar average of 13.7×10^{-12} esu. This indicates that the $\chi^{(3)}(-3\omega; \omega, \omega, \omega)$ of the copolymers are molar average of the $\chi^{(3)}$ values of the respective homopolymers and the $\chi^{(3)}$ values were not found to be affected noticeably by the disorder in the copolymer backbone.

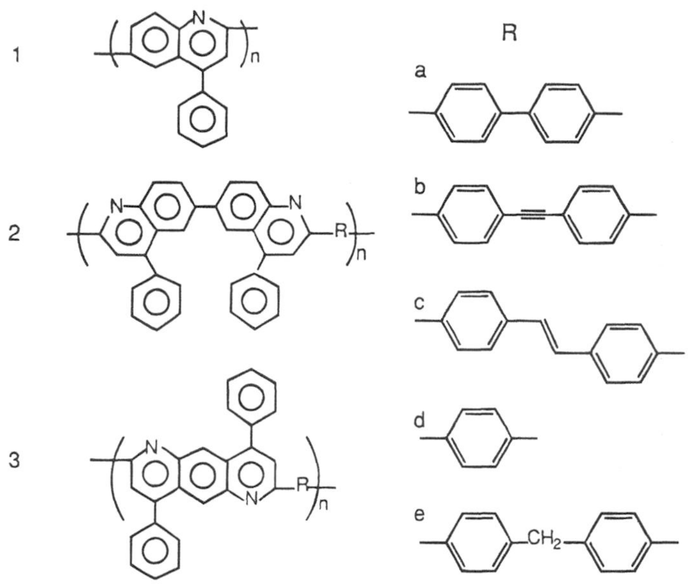

1

R

a

2

b

c

3

d

e

1 = PPQ

2a = PBPQ 3a = PBDA
2b = PBAPQ 3b = PBADA
2c = PSPQ 3c = PSDA
2d = PPPQ 3d = PPDA
2e = PDMPQ 3e = PDMDA

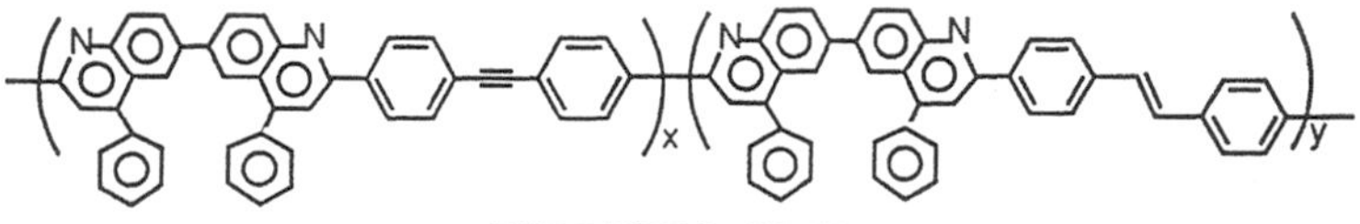

PSPQ/PBPQ (50:50)

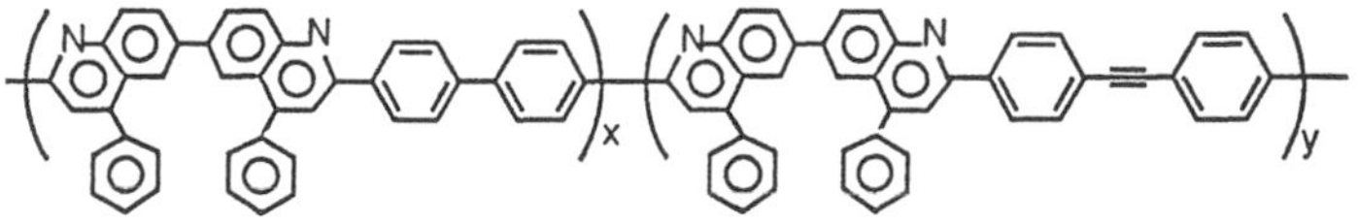

PSPQ/PBAPQ (50:50)

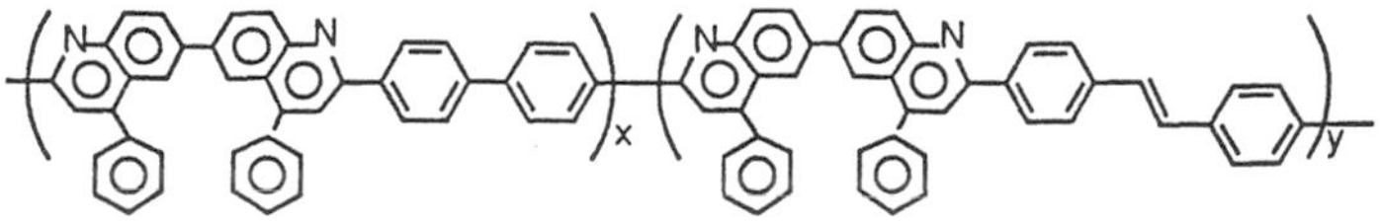

PSPQ/PBAPQ (50:50)

7. Polyanilines

Ginder et al.[255] reported the resonant $\chi^{(3)}$ value of polyaniline of 10^{-8} esu and emeraldine salt also showed the same values. Wong et al.[286] measured $\chi^{(3)}$ values of polyaniline by the DFWM technique at 620 nm. The emeraldine-based polyaniline showed a $\chi^{(3)}$ value of 8×10^{-10} esu which has the nonlinear figure-of-merit ($\chi^{(3)}/\alpha$) is 2×10^{-14} esu cm. The transient $\chi^{(3)}$ response for nonresonant excitation was < 100 fs, whereas resonant excitation has an ultrafast and a slow component. Halvorson et al.[287] measured $\chi^{(3)}$ values of the lecoemeraldine, emeraldine base, and pernigraniline isomers of polyanine from 0.55 eV to 1.25 eV using THG. Figure 29 shows the wavelength dependence of $\chi^{(3)}$ for lecoemeraldine, emeraldine base, and pernigraniline forms of polyanine. The peak $\chi^{(3)}$ value was 4.4×10^{-10} esu for emeraldine base and 1.3×10^{-10} esu for pernigraniline. The $\chi^{(3)}$ value of leucoemeraldine was ten times smaller than the $\chi^{(3)}$ value of either of these. The leucoemeraldine spectrum showed two resonances, one at 0.55 eV due to the presence of residual emeraldine base and at 1.18 eV due to three photon resonance. The resonant $\chi^{(3)}$ was 1.8×10^{-11} esu at 1.18 eV and the off-resonance $\chi^{(3)}$

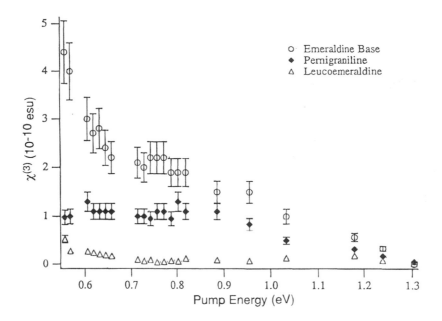

Figure 29 Wavelength dispersion of third-order NLO susceptibility $\chi^{(3)}$ values of different forms of polyaniline. (From Halvorson et al., *Synth. Metals*, 57, 3941, 1993. With permission.)

value of 4.5×10^{-12} esu at 0.75 eV for leucoemeraldine. The $\chi^{(3)}$ value of pernigraniline was smaller than in the nondegenerate ground-state emeraldine base.

In an other study, Mattes et al.[288] prepared high-quality silica glass incorporating polyaniline (PAN) and 2-ethyl-polyaniline (Et-PAN) both in emeraldine state into sol-gels to glass composites. The sol-gels were prepared by dissolving PAN and Et-PAN into a mutually compatible N-methylpyrrolidinone (NMP) solvent. The NMP solution of Et-PAN was added to a silica sol NMP mixture containing traces of acid. The NMP solutions of PAN were added to a silica sol-NMP mixture without the acid catalyst. The composition of the solutions were 7:1:2 volume ratio of silica sol:NMP:Et-PAN (or PAN)/NMP. The gelation took about 2 weeks. The film samples were dried at room temperature in the air for 3 weeks. The $\chi^{(3)}$ values were determined by the DFWM technique at 1.064 μm wavelength. The emeraldine base form of polyaniline dissolved in NMP showed a $\chi^{(3)}$ value of 1.6×10^{-11} esu which is about 10 times larger than that of the CS_2 reference solution. The ethylated polyaniline doped gel (sol-gel films containing 5 mg/ml of Et-PAN) exhibited $\chi^{(3)}$ values of 4.8×10^{-13} esu at 1.064 μm. Ando et al.[289] investigated a $\chi^{(3)}$ value of polyaniline over the wavelength range of 1.5 to 2.1 μm by the THG technique. Polyaniline showed the largest $\chi^{(3)}$ value of 9.0×10^{-11} esu at 1.907 μm because of the three-photon enhancement.

Osaheni et al.[290] measured $\chi^{(3)}$ values of polyanilines and derivatives by the picosecond THG technique in the wavelength range 0.9 ~ 2.4 μm. The $\chi^{(3)}$ values depended on the oxidation level and the dervatization of the p-phenyl rings. Table 36 lists the $\chi^{(3)}(-3\omega; \omega, \omega, \omega)$ values of polyanilines and derivatives measured by the picosecond THG technique. Polyemeraldine base with an oxidation level of ~50% had a larger $\chi^{(3)}(-3\omega; \omega, \omega, \omega)$ value than the fully oxidized form pernigraniline or poly(phenylaniline). Figure 30 shows wavelength-dependent $\chi^{(3)}$ values of PEMB, polypernigraniline (PPGN) and P4PAB. The methoxy substitution reduces the number density of the polymer repeat units and consequently leads to lower optical nonlinearity, besides this the $\chi^{(3)}(-3\omega; \omega, \omega, \omega)$ value of PMAB is larger than the basic polyemeraldine base and other derivatives due to the increased transition moment. The large $\chi^{(3)}(-3\omega; \omega, \omega, \omega)$ value of PEMB results from the higher number density. The magnitude of the $\chi^{(3)}(-3\omega; \omega, \omega\ \omega)$ in the 0.9 ~ 2.4 μm wavelength region was found to be in the decreasing order PMAB > PEMB > PDMAB > POTB. At the three-photon resonance peak, the $\chi^{(3)}(-3\omega; \omega, \omega, \omega)$ of PEMB is 2.64 times larger than that of PPGN. The poly(4-phenylaniline) P4PAB has an off-resonance $\chi^{(3)}(-3\omega; \omega, \omega, \omega)$ value of 10^{-12} esu and 1.8×10^{-12} esu at ~0.96 μm. The $\chi^{(3)}(-3\omega; \omega, \omega, \omega)$ of the fully reduced polyleukoemeraldine base (PLEMB) was estimated to be in the range 1.7×10^{-13} to 10^{-12} esu. The $\chi^{(3)}(-3\omega; \omega, \omega, \omega)$ of 50% oxidized polyanilines was more than an order of magnitude

Table 36 The $\chi^{(3)}(-3\omega; \omega, \omega, \omega)$ Values of Polyanilines and Derivatives

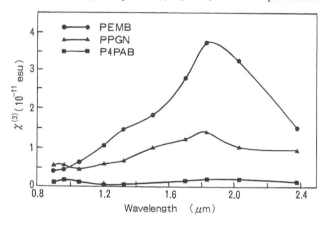

Polyaniline	λ_{max} (nm)	α (10^4 cm^{-1})	$\chi^{(3)}$ (10^{-11} esu) at 1.83 μm
$R_1 = R_2 = H$ (PEMB)	640	3.2	3.7
$R_1 = CH_3, R_2 = H$ (POTB)	605	2.3	1.8
$R_1 = OCH_3, R_2 = H$ (PMAB)	620	9.8	3.9
$R_1 = R_2 = OCH_3$ (PDMAB)	580	7.0	2.2
P4PAB	324	2.9	0.18
Polypernigraniline	560	2.8	1.4 (at 0.96 μm)

From Osaheni et al., *J. Phys. Chem.*, 96, 2830, 1992. With permission.

Figure 30 Wavelength-dependent $\chi^{(3)}$ values of polyanilines at different oxidation level. (From Osaheni et al., *J. Phys. Chem.*, 96, 2830, 1992. With permission.)

larger than that of the fully reduced form and the $\chi^{(3)}(-3\omega; \omega, \omega, \omega)$ of the intermediate oxidized state, PEMB, was 2.64 times that of the fully oxidized polypernigraniline.

P4PAB

8. Polyazines and Model Compounds

The third-order NLO properties of a polyazine and its well-defined chain length oligomers were reported by Nalwa et al.[291–295] The chemical structures of propylmethylpolyazine (PMPAZ) and sequentially built oligomers are shown below. The π-conjugation lengths of polyazines were controlled by sequentially building the oligomers. Polyazine is isolectronic with polyacetylene. Both polyacetylene and polyazine have a simple linear chain of atoms with alternating single and double bonds, but polyazine consists of pairs of nitrogen atoms substituted for pairs of carbon atoms in the polyacetylene chain. This arrangement of carbon and nitrogen atoms in the polyazine π-conjugated backbone offers environmental stability and leads to better optical transparency.

Oligoazines $n = 1, 2, 3,$ and 5

Propylmethylpolyazine

Table 37 Optical Absorption Maxima, THG Measured $\chi^{(3)}(-3\omega; \omega, \omega, \omega)$ Values at 1.5 μm, and *ab initio* Calculated γ Values of the Oligomers and the Polyazines

Repeating Units (*n*)	λ_{max} (nm)	γ (10^{-36} esu)	$\chi^{(3)}$ (10^{-12} esu)
1	263	3.349	0.139
2	265, 295	19.737	0.389
3	270, 300	57.193	0.867
4	–	112.033	–
5	270, 304	175.919	1.53
7	270, 307	–	2.48
PMPAZ	280, 320	–	8.0
MMPAZ	–	–	9.0

From Nalwa et al., *Jpn. J. Appl. Phys.*, 32, L193, 1993; Nalwa et al., *Synth. Met.*, 57, 3901, 1993.

The optical absorption spectra for a pentamer and PMPAZ were compared. The pentamer had two peaks; one at 270 nm and another at 304 nm; whereas PMPAZ exhibited absorption peaks at 280 nm and 320 nm. Table 37 lists the optical absorption maxima of various oligomers of polyazine. The monomer exhibits only one peak at 263 nm and an additional shoulder starts developing at 295 nm which grows into a separate absorption peak with further increase of the π-conjugation length. The optical absorption spectra indicate that the band gap of polymer repeat units in the oligomers decreases as the π-conjugation length of the polyazine backbone increases.[295]. The $\chi^{(3)}(-3\omega; \omega, \omega, \omega)$ values of oligomers increase substantially with increasing π-conjugation length. From the THG results, the $\chi^{(3)}$ values of the monomer and the dimer were found to be 1.39×10^{-13} and 3.85×10^{-13} esu, respectively. The $\chi^{(3)}$ value of the pentamer and the heptamer increased by an order of magnitude to 1.53×10^{-12} and 2.48×10^{-12} esu, respectively, at 1.5 μm, which are 11 and ~20 times larger than that of polyazine monomer. The oligometric azines used for THG measurement to determine $\chi^{(3)}$ values contained methyl groups at the carbon sites, whereas for theoretical calculations of static γ values only hydrogen atoms were considered instead of methyl groups. The γ values increased much more rapidly as the π-electron delocalization length of polyazine backbone increased; it changed by a factor of 5 from the monomer to the dimer and by a factor of 16 from the monomer to the trimer. The γ value increased by a factor of more than 61 in going from a monomeric to a pentameric polyazine derivative.[294] Therefore, the bulky methyl groups to some extent diluted the optical nonlinearity of polyazine derivatives. Both γ and $\chi^{(3)}$ ($-3\omega, \omega, \omega, \omega$) values increased substantially with the increasing conjugation length. The nonresonant $\chi^{(3)}$ values of polyazine repeat units in the oligomers were the same as the trend seen in the α-conjugated thiophene oligomers. It should be noted that the $\chi^{(3)}$ of the oligomers may be diluted to some extent because of the methyl pendant groups on the polyazine chain.

The PMPAZ shows a third-order NLO susceptibility of 8.0×10^{-12} esu at 1.5 μm. The optical absorption spectra of PMPAZ and oligomers did not exhibit any absorption around 500 nm, where a three-photon contribution may affect the third-order NLO susceptibility measured by the THG technique. Therefore a comparison of the optical spectrum with 3ω reveals the nonresonant origin of the $\chi^{(3)}(-3\omega, \omega, \omega, \omega)$ in PMPAZ and its oligomers. This indicates that the magnitude of the third-order NLO susceptibility $\chi^{(3)}$ value of PMPAZ and the pentamer in the off-resonance region is the same as that of other π-conjugated polymers. The $\chi^{(3)}$ value of permethylpolyazine (MMPAZ) that has only methyl side groups was about 9×10^{-12} esu at the same wavelength.[295] This was not surprising because different alkyl groups on carbon atoms should have a minor effect on chromophore density where a long alkyl chain dilutes optical nonlinearity. It was difficult to prepare good optical quality films of MMPAZ. Figure 31 shows the wavelength dependent $\chi^{(3)}$ value of a polyazine having a pyridine unit in the π-conjugated backbone. The magnitude of $\chi^{(3)}$ changes depended on the absorption features. Because of a three-photon resonance, $\chi^{(3)}$ reached as high as 7.8×10^{-12} esu at 1.5 μm. $\chi^{(3)}$ values of 10^{-11} esu and larger determined by the THG method have been obtained for modified polyazines. Although the $\chi^{(3)}$ values of polyazine are modest, these derivatives have better optical transparency than other conjugated polymers. Theoretical calculations as well as experimental results predicted the large third-order optical nonlinearities for polyazines, presumably originating from the multiple quantum-well structure of the main π-electron-conjugated backbone. Third-order optical nonlinearity of polyazine model compounds increased continuously up to heptamer without any saturation. For the heptamer, the

706

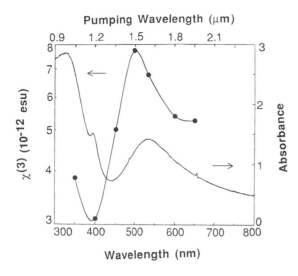

Figure 31 Wavelength-dependent $\chi^{(3)}$ values of a derivatized polyazine having a pyridine ring in the main π-conjugated backbone. The film thickness is 160 nm.

estimated chain length was about 35 Å. A comparison with polyazine indicated that the conjugation length of at least 20 or even more repeat units are required for optimizing optical nonlinearity. This agreed with thiophene oligomers where optical nonlinearity increased up to 14 repeat units whereas the band gap stayed constant after 8 repeat units.

The schematic diagram of the π-electron conjugated backbone structures of the polyacetylene and polyazine are depicted below. Polyacetylene is a simple linear chain of carbon atoms with alternating single and double bonds. Polyazine is also a simple linear chain of atoms with alternating single and double bonds, but with pairs of nitrogen atoms substituted for pairs of carbon atoms in polyacetylene chains. Therefore, polyazine can be viewed as isoelectronic to polyacetylene. The π-band structure of polyazine is different from that of polyacetylene because polyazine has four π bands: the lower two bands are filled and the upper two bands are empty, whereas polyacetylene has only two bands. Furthermore, the band gap of polyazine is 2.3 eV, about more than twice than that of polyacetylene (band gap = 1.1 eV). The significant difference between the NLO properties of polyacetylene and polyazine derivatives results from the presence of nitrogen atoms in the π-electron backbone. The charge density of carbon in polyacetylene is uniformly zero along the chain, whereas nitrogen in the polyazine chain has a charge of -0.39 ($+0.39$). Therefore the bonding pattern between the two conjugated polymers differs. The attractive charge separation between carbon and nitrogen atoms of polyazine is large, which causes shortening of bond distances of the carbon-nitrogen. In fact, the effect of nitrogen atoms is 2-fold. First, polyazine is environmentally stable, unlike polyacetylene. Second, the nitrogen atoms modify the topology of polyazine relative to polyacetylene. Polyacetylene has a degenerate ground state. The electronic structure of *trans*-polyacetylene opens the possibility of soliton midgap states. On the other hand, unlike *trans*-polyacetylene, the presence of nitrogen in polyazine creates a nondegenerate ground state which precludes the existence of solitons. Polyazine is the head–head polymer of methineimine $(CH=N)_n$. A one-bond shift of the π-bond along the chain would give a structure that has alternating azine and ethylene moieties. Therefore, polaron or bipolaron states have been proposed in polyazine. Polyacetylene structure supports soliton-like defects, whereas polyazine leads to polaron- or bipolaron-type deformation. In polyazine, defects are located at a carbon or at a nitrogen atom, whereas in polyacetylene defect centers are carbon atoms. Therefore, the possible NLO mechanism could be more complicated than that of polyacetylene. It is quite apparent that both conjugated polymers differ from chemical as well as electronic viewpoints. Not only the optical transparency of polyazine is a major advantage for third-order nonlinear optics, it also offers other advantages such as environmental stability, solution processability, and architectural flexibility. Experimental and theoretical results on third-order NLO susceptibility $\chi^{(3)}$ values and second hyperpolarizability values of oligomeric and polymeric azines predict that the large third-order optical nonlinearities for polyazines originate from the multiple quantum-well structure of the main π-electron-conjugated backbone.

Polyacetylene

Polyazine

9. Poly(aromatic amines) and Poly(aminothiazole)

Chandrasekhar et. al.[296-298] studied the third-order optical nonlinearity of three classes of polymers: poly(aromatic amines), poly(aminoquinolines), and poly(aminothiazoles) and other S-heterocycles. Poly(diphenylamine) (PDPA) and poly(4-amino biphenyl) (PABP), which are novel processible and stable conducting polymers that are analogs of polyaniline. These polymers are low molecular weight materials having on average about 12–17 monomer units per chain. In undoped state, PDPA and PABP polymers are light brown and very light green in solutions. PDPA shows small absorption at 532 and 1064 nm, whereas PABP absorbs significantly at 532 nm. The $\chi^{(3)}$ values for pure polymers were obtained by:

$$\chi^{(3)}(\text{pure}) = [N(\text{pure})/N(\text{solution})]\chi^{(3)}(\text{solution}) \qquad (27)$$

where N is the effective density. The measured densities for PDPA and PABP were 1.65 and 1.5 g/cm^3, respectively. The $\chi^{(3)}$ values of PDPA were 1.3×10^{-10}, 1.0×10^{-10}, and 7×10^{-11} esu; the $\chi^{(3)}$ of PABP values were 1.4×10^{-10}, 8×10^{-11}, and 6×10^{-11} esu at the concentrations of 3.3, 6.7, and 13 g/l, respectively. The $\chi^{(3)}$ values of three classes of polymers are shown in Table 38 for polymers with similar doping level and identical dopant (concentration approximately 3.3 g/l/DMF). The PDPA and PABP polymers have the largest $\chi^{(3)}$ values; the incorporation of heteroatom S or N into the aromatic backbone generally reduces the optical nonlinearity. For example, in poly(thiazoles) having N and S heteroatoms, the signals of all compounds were less than the CS$_2$ signal. The introduction of multiple fused aromatic rings in the polymer backbone significantly enhanced $\chi^{(3)}$ values, and the lower doping level gave rise to higher nonlinearities. The effect of various dopants on $\chi^{(3)}$ values was studied at a constant level of about 4.5%, identical concentration of 40 mM, and under identical electropolymerization conditions. For all polymers, the $\chi^{(3)}$ value was largest with BF$_4^-$ dopant, approximately the same for PF$_6^-$, and substantially reduced for tosylate and perchlorate dopans. The $\chi^{(3)}$ values of poly(diphenylamine) were 7.2×10^{-11} esu for BF$_4^-$ doping, 7.4×10^{-11} esu for PF$_6^-$, 3.7×10^{-11} esu for tosylate,

Table 38 The $\chi^{(3)}$ Values of Poly(aromatic amines), Poly(aminoquinolines), and Poly(s-heterocycles) Approximately 3–5% Doped with BF$_4^-$ Except Polyaniline

Polymer	$\chi^{(3)}$ (esu)
Poly(diphenyl amine)	1.3×10^{-10}
Poly(4-aminobiphenyl)	1.4×10^{-10}
Poly(N,N'-diphenyl benzidine)	9.0×10^{-11}
poly(1-aminopyrene)	2.7×10^{-10}
Polyaniline (emeraldine base)	1.0×10^{-11}
Poly(N-phenyl-2-naphthyl amine)	$>4.0 \times 10^{-12}$
Poly(3-aminoquinoline)	3.7×10^{-11}
Poly(5-aminoquinoline)	1.9×10^{-11}
Poly(6-aminoquinoline)	2.6×10^{-11}
Poly(8-aminoquinoline)	2.3×10^{-11}
Poly(3-aminopyridine)	2.2×10^{-11}
Poly(3-aminopyrazole)	3.6×10^{-11}
Poly(isothianaphthene)	5.0×10^{-10}
Poly(aminothiazole)	$<1.0 \times 10^{-12}$
Poly(aminobenzothiazole)	$<1.0 \times 10^{-12}$
Poly(amino-2,1,3-benzothiadizole)	$<1.0 \times 10^{-12}$

From Chandrasekhar et al., *Appl. Phys. Lett.*, 59, 1661, 1991; Chandrasekhar et al., *Synth. Metals*, 53, 175, 1993; Chandrasekhar et al., ACS Meeting, August 1992. With permission.

and 3.5×10^{-11} esu for perchlorate doping at a doping level of 4.5% and a concentration of 40 mM in DMF. The lower doping levels, approximately 1~2% gave much higher $\chi^{(3)}$ values than those for more doped (about 8%) polymers approximately 3 times larger, presumably because of the intrinsic lattice defects embodied in midgap states. Both the Z-scan and DFWM techniques indicated the purely off-resonance optical nonlinearities. The nonlinear response times were faster than 40 ps, which indicated the absence of long-lived thermal contributions.

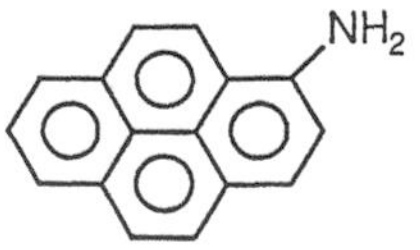

1-aminopyrene

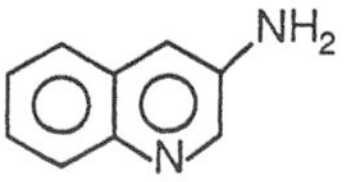

3-aminoquinoline

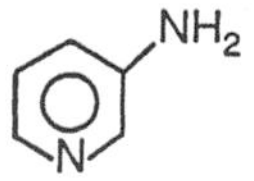

1-aminopyridine

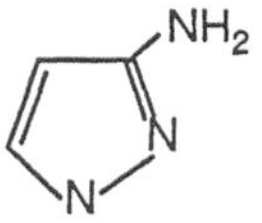

3-aminopyrazole

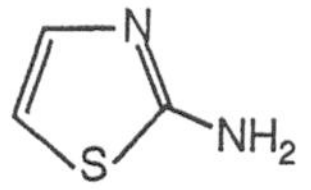

2-aminothiazole

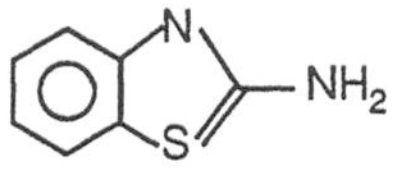

2-aminobenzothiazole

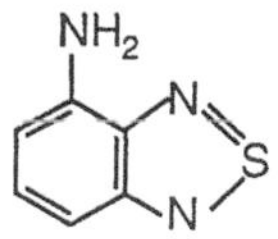

4-amino-2,1,3-benzothiadiazole

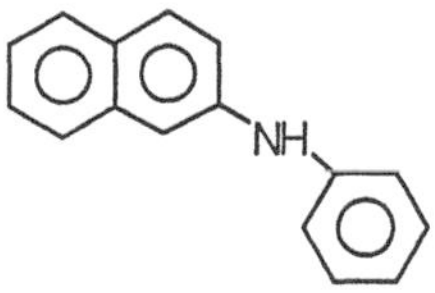

N-phenyl-2-naphthylamine

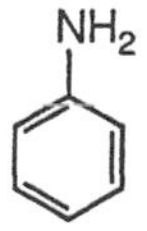

aniline

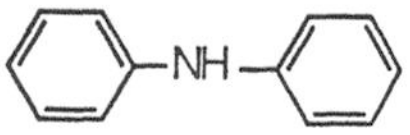

diphenylamine

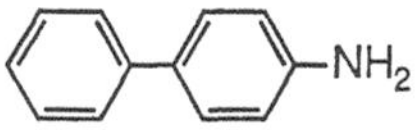

4-aminobiphenyl

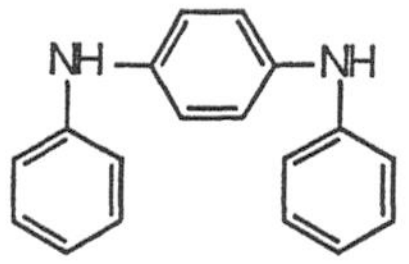

N,*N*'-diphenylbenzidine

isothianaphthene

10. Polycarbazoles

Poly(*N*-[1-butane 4-sulfonate benzyl triethylammonium]−3,6-carbazolyl) (PNBSC), its corresponding propane sulfonate derivatives, PNPSC and poly(*N*[2-pyrimido]3,6-carbazolyl) (PCZP), were studied by Wellinghoff et al.[299] Third-order optical nonlinearities of radical cations of these polymers, namely PNBCZ and PNPCZ, were measured by the DFWM technique at 568 nm. The third-order susceptibilities and hyperpolarizabilities of polycarbazole radical cations are shown in the Table 39. The electronic

Table 39 The $\chi^{(3)}$ and γ Values of Polycarbazole Radical Cations

Polymer	Solvent	$\chi^{(3)}$ (10^{-7} esu)	γ (10^{-30} esu)
PNPCZ	AlCl$_3$ in nitrobenzene	2.47	3.10
PNPCZ	AlCl$_3$ in nitromethane	2.27	2.84
PNPCZ	AlCl$_3$ in *n*-butylpyridinium chloride	1.34	1.12
PNBCZ	AlCl$_3$ in nitrobenzene	2.72	3.58
PNBCZ	AlCl$_3$ in nitromethane	2.70	3.54
PCZP	Nitrobenzene	1.30	1.09

From Wellinghoff et al., *Synth. Met.*, 41–43, 3203, 1991. With permission.

and reorientational contributions to the $\chi^{(3)}$ values are approximately 10^{-7} esu. The large optical nonlinearities are introduced by the delocalization of charge states into the polymer chains.

11. Schiff's Base Polymers

Schiff's base conjugated polymers are important due to their good thermal stability, mechanical strength and therotropic as well as lyotropic liquid crystallinity. Jenekhe et al.[300] reported the relationship between structures and third-order optical nonlinearity. The chemical structure of a series of aromatic Schiff-based polymers are shown below. The three-photon resonance enhanced $\chi^{(3)}(-3\omega,\omega,\omega,\omega)$ values of 1.6×10^{-11} esu for poly(1–4-phenylenemethylidynenitrilo-1,4-phenylene-nitrilomethylidyne) (PPI) at 1.2 μm and 5.4×10^{-11} esu for poly(1,4-phenylenemethylidynenitrilo—2,5-dimethoxy-1,4-phenylene-nitrilomethylidyne) (PMOPI) at 1.3 μm were measured by the THG method. The PMOPI had a $\chi^{(3)}$ value 5 times larger than that of PPI. The nonresonant $\chi^{(3)}$ values were 1.6×10^{-12} esu for PPI and 7.3×10^{-12} esu for PMOPI at 2.38 μm, respectively. The resonant $\chi^{(3)}(-3\omega;\omega,\omega,\omega)$ values were about an order of magnitude larger than the nonresonant values for these two aromatic poly(azomethines). PMOPI shows larger $\chi^{(3)}$ values than PHOPI and PPI over the energy range 0.5 to 1.5 eV and has two-photon and three-photon resonances. The PPI/PMPI copolymer shows a resonance $\chi^{(3)}(-3\omega,\omega,\omega,\omega)$ value of 16×10^{-12} esu whereas its nonresonant $\chi^{(3)}$ value at 2.38 μm was an order of magnitude smaller. Figure 32 represents the dispersion of the $\chi^{(3)}$ values of PPI, PMOPI, and PHOPI polymers. PHOPI shows a slight decrease in nonlinearity, probably because of the hydrogen bonding of the hydroxyl groups to the nitrogen atoms. In PPI derivatives, the methoxy donor group increased the nonlinearity, but the hydroxy donor group led to a decrease in nonlinearity. The dimethoxy substituted polymer, PMOPI, showed the largest $\chi^{(3)}$ value of 7.2×10^{-12} esu at off-resonance and $\chi^{(3)}$ of 5.2×10^{-11} esu on resonance. Interestingly, the random copolymer PPI/PMPI also exhibited larger $\chi^{(3)}$ values compared with those of constituent PPI and PMPI homopolymers. Polyazomethine (PPI) possesses a π-electron-conjugated backbone similar to the BBL polymer. The $\chi^{(3)}$ of BBL is about five times larger than that of PPI away from resonance, (Figure 33). The BBL molecule is held rigid by the two-amide bridges, whereas PPI is less constrained to rotate about single bond.

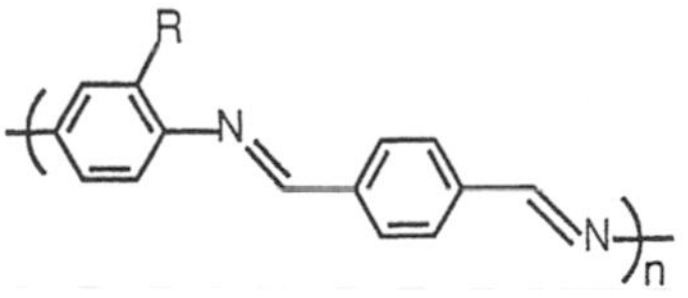

PPI: R = H; PMPI: R = CH$_3$

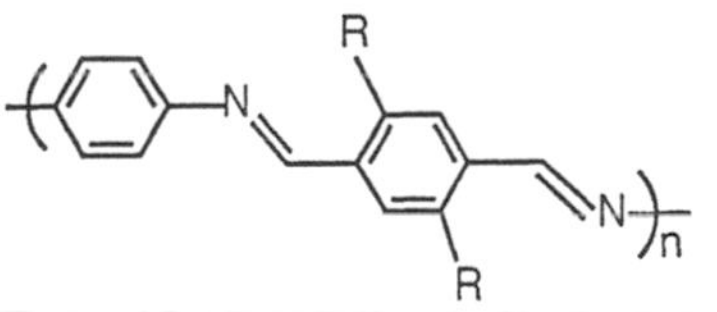

PHOPI: R = OH, PMOPI: R = OCH$_3$

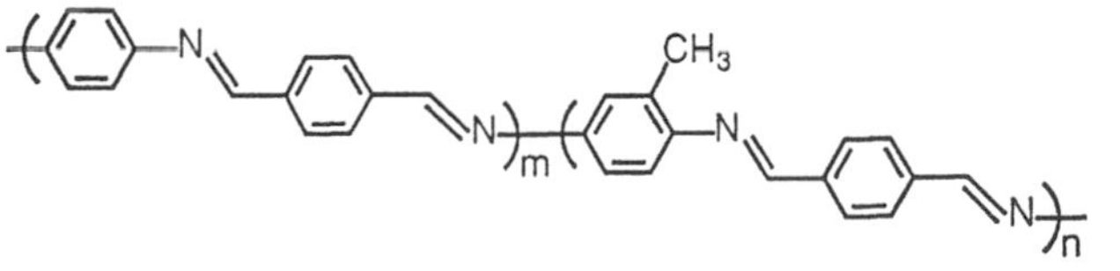

PPI/PMPI

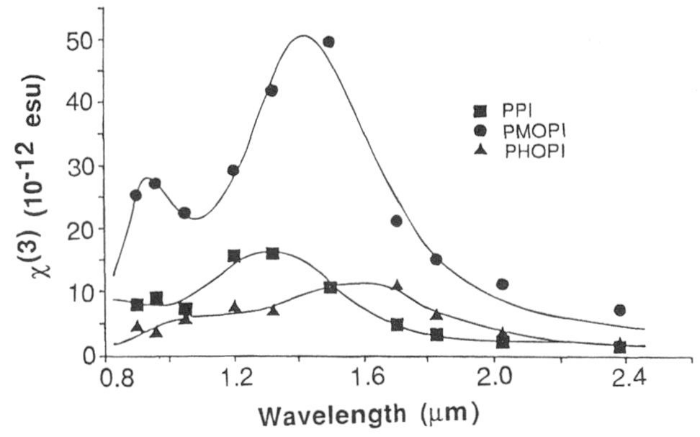

Figure 32 Wavelength-dependent THG measured $\chi^{(3)}$ values of polyazomethine polymers. (From Jenekhe et al., *Chem. Mater.,* 3, 985, 1991. With permission.)

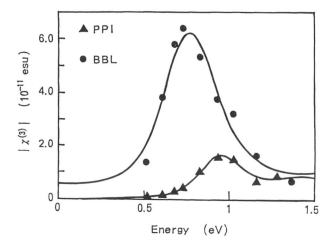

Figure 33 Comparison of the third-order NLO susceptibility $\chi^{(3)}$ value of a polyazomethine PPI with BBL polymer. (From Jenekhe et al., *Chem. Mater.*, 3, 985, 1991. With permission.)

Third-order NLO properties of wholly aromatic polyazomethines with flexible side chains were measured by Lee and Samoc.[301] The dialkoxy substituted aromatic polyazomethines were synthesized by the reaction of aromatic diamines and aromatic dialdehydes with flexible chains. The physical and NLO properties of these polymers are listed in Table 40. These polymers have dialkoxy chains where the alkyl group is either a dodecyl, or tetradecyl or hexadecyl substituted to a benzene ring. The $\chi^{(3)}$ values were determined by using the DFWM technique at 602 nm, with a 400 fs laser pulse. The lower $\chi^{(3)}$ values of these conjugated systems probably results from the noncoplanarity of azomethine linkages which reduces the π-electron conjugation along the polymer chains. The lower optical nonlinearity may also arise due to the low number density of π-electron conjugated polymer backbone units due to the side chains. These polyazomethine are a new class of NLO active conjugated polymers.

Table 40 The Physical and NLO Properties of Polyazomethines

PAM-I-14-A $X =$ (phenyl) $R = C_nH_{2n+1}$ ($n = 14$)

PAM-I-16-A $X =$ (biphenyl) $R = C_nH_{2n+1}$ ($n = 16$)

PAM-II-12-A $X =$ (biphenyl) $R = C_nH_{2n+1}$ ($n = 12$)

Polymer	Molecular Weight	λ_{max} (nm)	Refractive Index	$\chi^{(3)}$ (10^{-12} esu) at 602 nm
PAM-I-14-A	5900	446	1.67	2.4
PAM-I-16-A	4600	450	1.68	1.4
PAM-II-12-A	8400	422	1.66	2.4

From Lee and Samoc, *Polymer*, 32, 361, 1991. With permission.

12. Polyannulenes

Sasabe et al.[302] reported third-order susceptibility of tetrahydromethanoannulenes having well-defined π-conjugation lengths, which are two-dimensional π-electron systems. Annulenes were dissolved in PMMA and spin-coated on a quartz plate. THG experiments at 1.907 μm were conducted from thin films (0.25 μm) of annulenes. The $\chi^{(3)}$ values of annulene/PMMA films were 5.3×10^{-14} esu for $m = n = 1$, 9.3×10^{-14} esu for $m = 2$, $n = 1$, 1×10^{-13} esu for $m = n = 2$, and 7.9×10^{-14} esu for $m = 3$, $n = 2$. The $\chi^{(3)}$ values of annulenes increased with the increase of π-conjugation length. The $\chi^{(3)}/N$ values for annulenes are 4.4×10^{-34}, 9.5×10^{-34}, 13×10^{-34}, and 303×10^{-34} for 1, 2, 3, and 4 with PMMA, respectively.

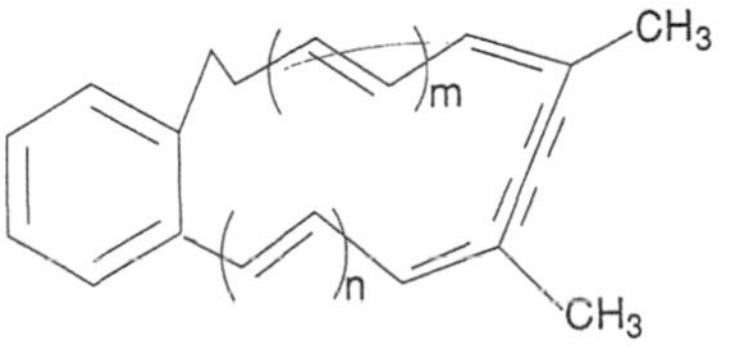

13. Polypyrrole

Ghoshal[303] reported third-order optical properties of chemically prepared polypyrrole by using the DFWM technique. The pristine polypyrrole film showed a $\chi^{(3)}$ value of 3.0×10^{-12} esu at 602 nm. The $\chi^{(3)}$ values of hydrochloric acid (HCl) and sulfuric acid (H_2SO_4) doped polypyrrole films were 3.5×10^{-12} esu whereas the $\chi^{(3)}$ value for a ferric chloride ($FeCl_3$) doped sample was 3.9×10^{-12} esu. The polypyrrole films treated with aqueous ammonia (NH_4OH) and iodine, chloranil, and 2,4,7-trinitrofluorenone (TNF) showed no DFWM signal despite of the several fold increase in power intensity. The response time of polypyrrole was of the order of 5 ps.

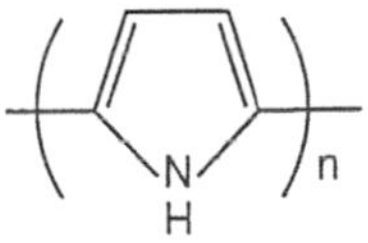

polypyrrole

14. Polyacenes

Flom et al.[304] studied three polyacene quinones doped in thin polymer films by using the DFWM technique. The structures and $\chi^{(3)}$ values of three polyacene quinones are shown below. Thin films of these compounds were prepared by dissolving a known amount in a volume of a 150 mg/ml polyvinyl chloride/THG solution. Although the largest optical nonlinearity is expected for compound 3 which has the longest conjugation length in the present study, the trend was quite different. The $\chi^{(3)}$ value of compound 1 is more than an order of magnitude larger than compound 2 and more than three orders of magnitude larger than compound 3. The origin of optical nonlinearity in compounds 1 and 2 has been described as an electronic phenomenon, their signals have two components, and their behavior was different from that observed for carbon disulfide. For compound 2, the slow response is about one third that of the fast response and the time constant is from 20 to 50 ps. A lifetime of about 600 ps has been reported for compound 1.

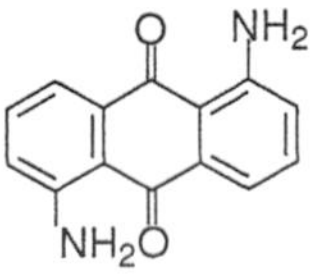

$\chi^{(3)}_{xxxx} = 2.1 \times 10^{-9}$ esu
$\chi^{(3)}_{xyyx} = 1.0 \times 10^{-9}$ esu

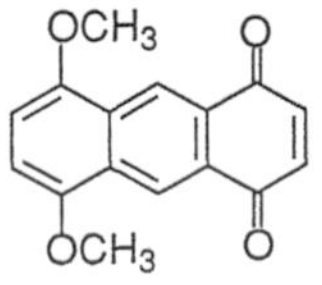

$\chi^{(3)}_{xxxx} = 8.5 \times 10^{-11}$ esu
$\chi^{(3)}_{xyyx} = 3.6 \times 10^{-11}$ esu

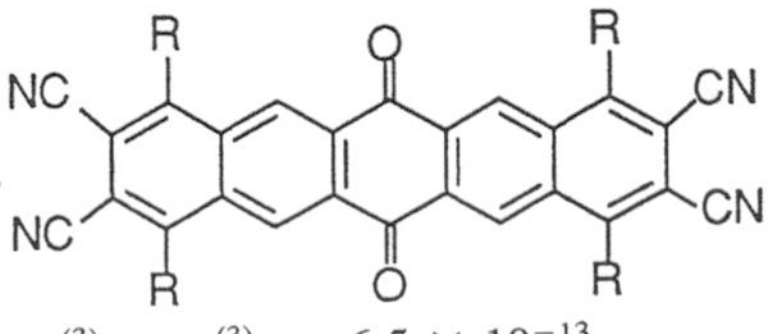

$\chi^{(3)}_{xxxx} = \chi^{(3)}_{xyyx} = 6.5 \times 10^{-13}$ esu
$R = p\text{-}t\text{-butylphenyl}$

15. Miscellaneous Organic Polymers

Kurihara et al.[305] reported main-chain azobenzene-type polymers for all-optical waveguide devices. The three-photon resonant $\chi^{(3)}(-3\omega; \omega, \omega, \omega)$ values were 14.5×10^{-11}, 8.5×10^{-11}, 7.5×10^{-11}, and 18

$\times$ 10^{-11} esu for PU-STAD, PE-SDAD, p-SBAC, and BCMU-STAD, respectively. The optical loss of 6 and 2 dB/cm and the refractive index of 1.725 and 1.617 at 1.3 μm were determined for PU-STAD and p-SBAC, respectively.

$R = OCONHCH_2(CH_2)_4CH_2NHCOO^-$ (PU-STAD)

$R = OCONHCH_2COOC_4H_9^-$ (BCMU-STAD)

$R = OCOCH_2(CH_2)_2CH_2COO^-$ (PE-SDAD)

$R = OCOCH_2(CH_2)_4CH_2{}^-COO^-$ (p-SBAC)

Kanbara et al.[306a] reported optical Kerr shutter measurements on symmetrically substituted benzylidene aniline derivative dyes of 2,5-dichloro-terephthal-bis-(4-N,N'-diethylaminoaniline doped in PMMA. The n_2 (Kerr) values were 6.0×10^{-15}, 9.1×10^{-15}, and 8.3×10^{-15}, and 1.3×10^{-14} cm^2/W for SBAC, BB-SBAC, AC-SBAC, and AC2-SBAC, respectively. The n_2(THG) values were 1.3×10^{-14}, 2.0×10^{-14}, 2.2×10^{-14}, and 3.3×10^{-14} cm^2/W for SBAC, BB-SBAC, AC-SBAC, and AC2-SBAC, respectively. The n_2 values for 30 wt% DEANST/nitrobenzene and 20 wt% DMSM/formamide were 1.9×10^{-13} and 1.6×10^{-13} cm^2/W, respectively. The response time of a solution of DEANST was measured as 4 ps. The picosecond switching operation was found to be feasible with the solution fiber waveguides.

$X = N(C_2H_5)_2$ SBAC

$X = N(C_4H_9)_2$ BB-SBAC

$X = N\text{-}C_2H_5$
|
$C_2H_4OCOCH_3$ AC-SBAC

$X = N\text{-}C_2H_5$
|
$C_2H_4OCOCH(CH_3)_2$ AC2-SBAC

Yamamoto et al.[307] synthesized a copolymer of pyridine-2,5-diyl and indigo-6,6'-diyl which forms good quality thin films and had absorption bands at 370 and 645 nm. The three-photon resonant $\chi^{(3)}$ values were about 10^{-11} esu at the absorption region of poly(pyridine-2,5-diyl) unit and more than 10^{-12} esu in the nonresonant region.

Kanbara et al.[306b] reported molecular length dependence of third-order NLO properties of the β-carotene series and benzene ring series by THG and optical Kerr shutter measurements. The nonresonant nonlinear refractive index n_2(THG) values were measured at 1.75 μm for the β-carotene series and at 2.1 μm for the benzene ring series.

β-ionone

Retinol

Retinal

The polyene compounds, β-ionone, retinol, retinal, and β-carotene, yielded nonresonant n_2(THG) values of 1.5×10^{-15}, 1.6×10^{-14}, 3.3×10^{-14}, and 4.4×10^{-12} cm^2/W for a solution concentration of 1 M, respectively. The 1 M n_2(THG) value for the carbon disulfide was estimated as 6.7×10^{-16} cm^2/W. On the other hand, the benzene ring series; nitrobenzene, MNA, 4-(N,N-diethylamino)-β-nitrostyrene (DEANST), 4-(N,N-diethylamino)-4'-nitrostilbene (DEANS), terephthalabis(4-(N,N-dihexylamino)aniline) (HH-SBA) and 4,4'-bis[p-N-ethyl-N((((butoxycarbonyl)-methyl)-carbamoyl)oxy)ethyl)amino]-o-(methylphenyl)azo]azobenzene (BCMU-STAD) yielded nonresonant n_2(THG) values of 5.2×10^{-16}, 1.2×10^{-15}, 1.2×10^{-14}, 9.3×10^{-14}, 2.5×10^{-13}, and 4.7×10^{-13} cm^2/W for a solution concentration of 1 M, respectively. The n_2(THG) values of the β-carotene series increased in proportion to the fifth power of the conjugated length. The same power dependence was found for the benzene ring series. The β-carotene series exhibited larger 1 M n_2(THG) values, whereas the benzene ring series yielded larger solution n_2(THG) values because of their high solubility. The β-carotene showed n_2(THG) values about 3000 times larger than those of β-ionone. Similarly, the n_2(THG) value of BCMU-STAD was found to be three orders of magnitude larger than that of nitrobenzene. The n_2(THG) value in the benzene ring series changed in a manner different from that in the β-carotene series. The solutions of MNA, DEANST showed a steeper n_2(THG) increase beyond the fifth-power law, probably an intramolecular charge-transfer formation besides the effect of increasing π-conjugation length, whereas the HH-SBA and BCMU-STAD solution exhibited saturation behavior. The n_2(THG) value was found to increase in proportion to the tenth power of the absorption edge from a plot between 1 M n_2(THG) values for solutions versus absorption edge. The n_2(THG) values of the β-carotene series was found to be 30 times larger than that of the benzene ring series having the same absorption edge. The β-carotene series exhibited larger third-order optical nonlinearities than the benzene ring series because of its molecular planarity. The contribution of the molecular orientation effect in all nonlinearities was estimated from n_2(Kerr)/n_2(THG). The n_2(Kerr)/n_2(THG) value decreased as the molecular length increased. Carbon disulfide or nitrobenzene with molecular orientation effect showed an n_2(Kerr)/n_2(THG) value of about 8. The longer chain molecules of β-carotene with electronic polarization effect showed n_2(Kerr)/n_2(THG) value of 3.2. The nonlinearities of the longer chain molecules have an electronic polarization contribution, whereas in medium-length molecules both molecular orientation effect and the electronic polarization effect may superimposed. The molecular orientation effect increased in proportion to the fourth power of the molecular length. The optical Kerr shutter measurements showed that the nonlinearity of the solution medium of the longer conjugation length molecules is dominated by the electronic polarization effect and reduces the contribution of the molecular orientation effect.

Yamashita et al.[308] reported third-order NLO properties of a variety of polyimides. The optical absorption and $\chi^{(3)}$ values of these polyimides vary with the chemical structure. A difference in $\chi^{(3)}(-3\omega; \omega, \omega, \omega)$ as large as one order of magnitude occurs and $\chi^{(3)}$ values are approximately 10^{-12} esu.

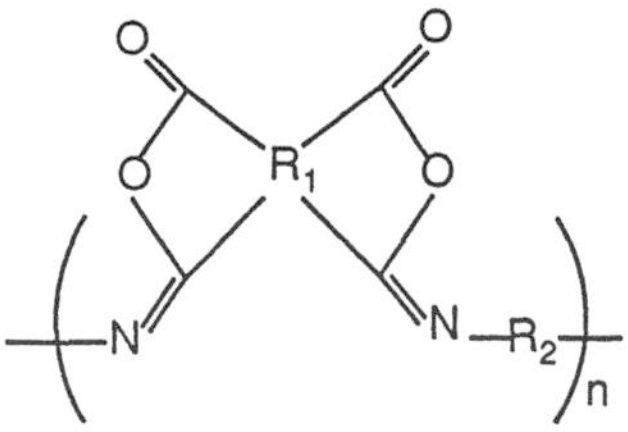

R_1	R_2	λ_{max}(nm)	$\chi^{(3)}(10^{-12}$ esu)
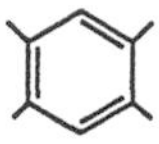	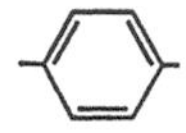	430	3.0
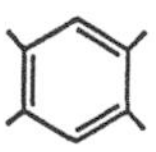	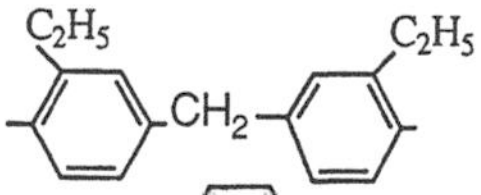	410	2.2
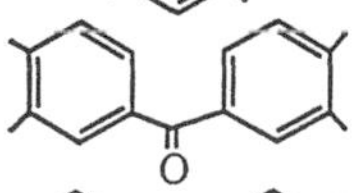	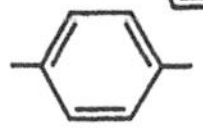	395	1.4
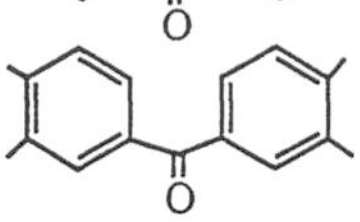	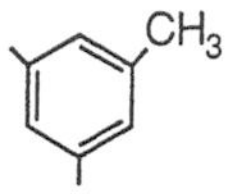	360	0.46

Hikita et al.[309] reported the fabrication of polyimide (PI) thin films by vapor deposition polymerization of PMDA and diamine monomers; decamethylenediamine (DMDA), 4,4'-diaminodiphenylether (DDE), and p-diaminobenzene (DAB). The structural change from precursor polyamic acid (PAA) into PI were studied spectroscopically. The chemical scheme of the vapor deposition polymerization of polyimide thin films are shown below;

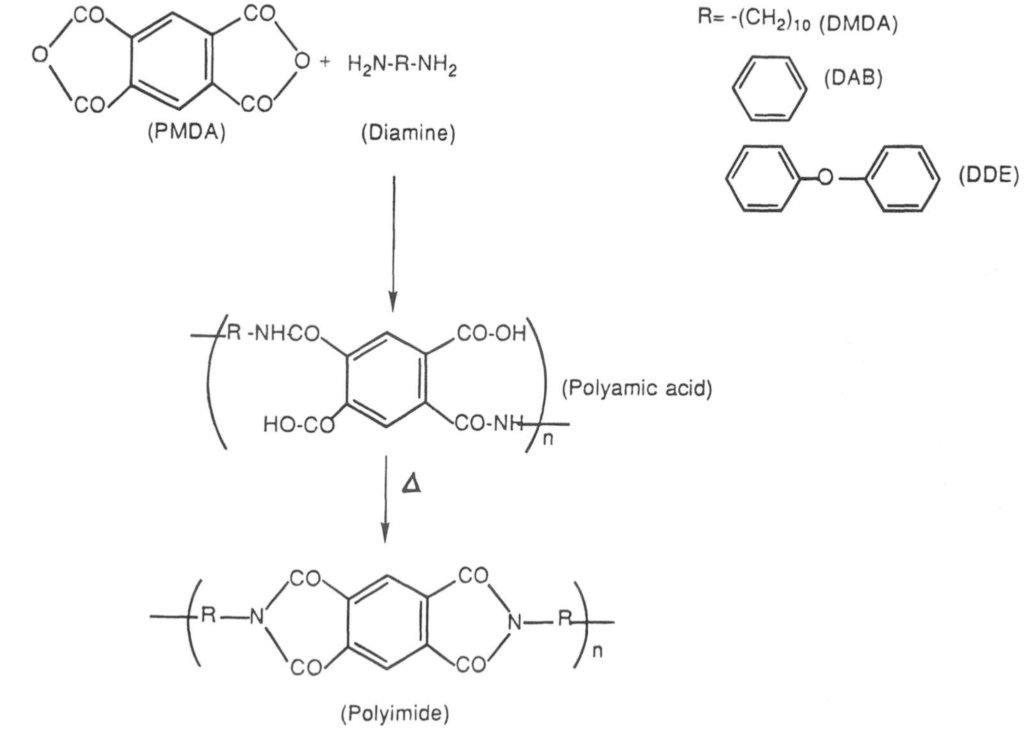

(PMDA) + H₂N-R-NH₂ (Diamine)

R= -(CH₂)₁₀ (DMDA)
(DAB)
(DDE)

(Polyamic acid)

Δ

(Polyimide)

Table 41 lists the $\chi^{(3)}$ values of precursor polyamic thin films and with different diamines and polyimide obtained by thermal annealing. The $\chi^{(3)}$ values of PI were found to be about twice those of PPA for all kinds of combination of monomers because of the extended π-conjugation through benzene ring of the diamine used either on both sides for PI(DAB) or on one side for PI(DDE) over the adjacent imide bond. The PI(DAB) thin films showed the largest $\chi^{(3)}$ value of 8.4×10^{-12} esu at 1.064 μm due to the extended conjugation via a benzene ring to the adjacent two imide groups. Thermal treatment of precursor leads to imidization for both PI(DAB) and PI(DDE) and hence larger delocalization giving rise to large optical nonlinearity. In PI(DMDA), the interruption of π-conjugation decreases $\chi^{(3)}$ values by an order of magnitude.

Table 41 $\chi^{(3)}$ Values of Precursor Polyamic Thin Films Obtained from Different Diamines and Polyimide Obtained by Thermal Annealing

Polymer	Thermal Treatment (°C)	$\chi^{(3)}(-3\omega; \omega, \omega, \omega)$ (10^{-12} esu)
PAA(DAB)		3.7
PI(DAB)	150	5.3
PI(DAB)	250	8.4
PAA(DDE)		4.2
PI(DDE)	150	4.4
PI(DDE)	250	5.8
PAA(DMDA)		0.33
PI(DMDA)	150	0.53
PI(DMDA)	250	0.49

From Hikita et al., *Nonlinear Opt.*, 2, 1, 1992. With permission.

Duray and Glatkowski[310] reported $\chi^{(3)}(-\omega; \omega, \omega, -\omega)$ of 1.2×10^{-12} esu for LARC-TPI-polyimide films from DFWM technique at 602 nm. The polyimides obtained from different anhydride and amine monomers show variation in optical nonlinearity and the $\chi^{(3)}$ value increases as high as 1.7×10^{-11} esu for another derivative which is more than 10 times larger than LARC-TPI.

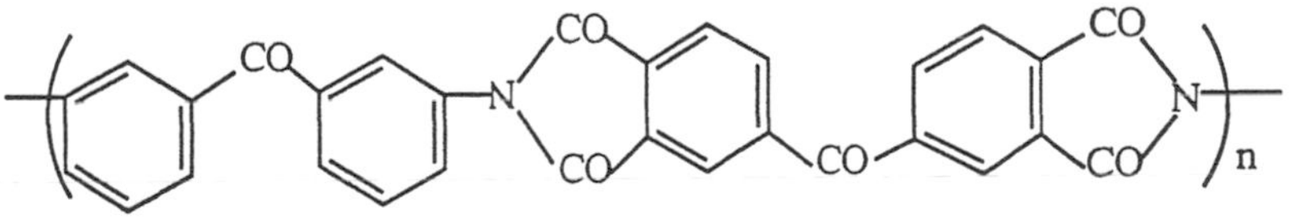

(LARC-TPI Polyimide)

Mates and Ober[311] synthesized thermotropic polymers with conjugated distyrylbenzene (DSB) and diphenylbutadiene (DPB) segments. The polyester containing DPB units have methyl and ethoxy substitutions (R). The $\chi^{(3)}$ values were determined by the DFWM method at 598 nm wavelength. The $\chi^{(3)}$ values are nonresonant because 598 nm is far from the absorption edge. The MBD5 is mesogen containing fewer double bonds than the other two polymers (8 double bonds vs. 11 bonds) and shows the lowest $\chi^{(3)}$ value as expected. The DSB-containing polyaramid shown below has a cutoff wavelength at 470 nm and shows the largest $\chi^{(3)}$ value because it has the highest volume concentration of chromophores. MBD5 with methyl group shows an absorption maximum at 351 nm and a $\chi^{(3)}$ value of 2.18×10^{-12} esu at 598 nm. EtO$_5$ having an ethoxy group shows λ_{max} at 370 nm and $\chi^{(3)}$ value of 2.62×10^{-12} esu at the same wavelength.

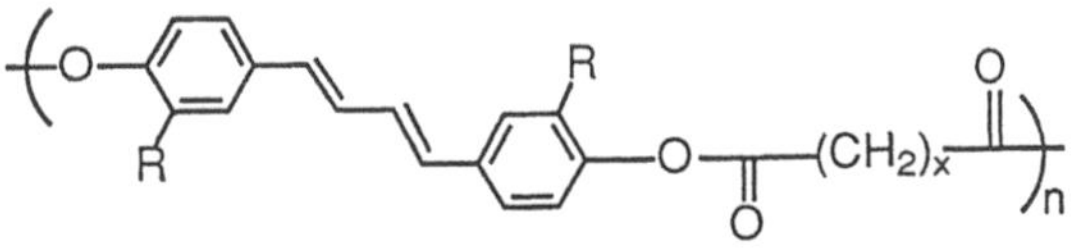

$R = H$ BD5   ($x = 5$)
$R = CH_3$ MBD5   ($x = 5$)

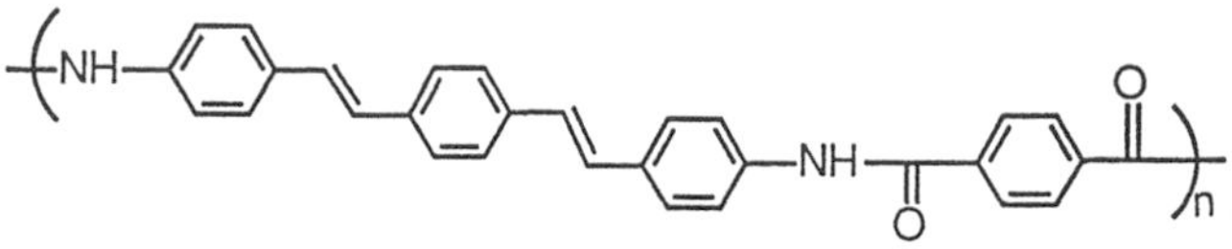

DSB-polyaramide

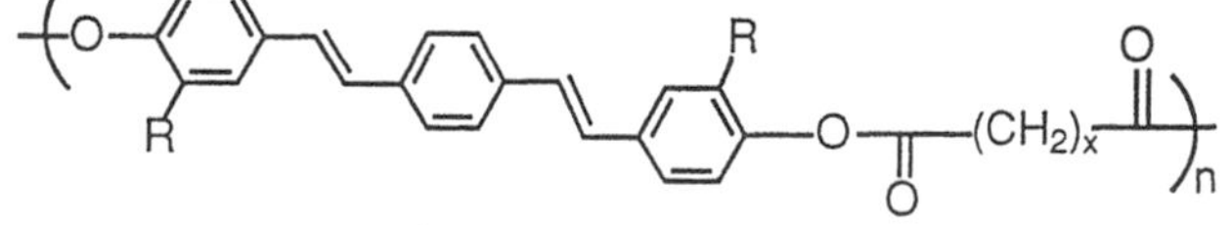

polyester containing DBS

D. NLO-CHROMOPHORE FUNCTIONALIZED POLYMERS

Organic conjugated polymers have been studied extensively for third-order nonlinear optics because they possess large optical nonlinearities and fast response times. In addition, their architectural flexibility provides easy modification of chemical structure according to the required functions for NLO properties and processibility into desired configurations such as ultrathin films, fibers, and liquid crystals. The π-electron-conjugated system gives rise to the large third-order optical nonlinearity of these polymeric materials. A very high concentration of NLO dyes can be incorporated into polymer chains via synthetic routes. This aspect encourages the study of NLO properties of other polymeric structures comprising either NLO functionalities in the side chains or copolymers having conjugated as well nonconjugated systems. Some examples are provided here. In addition to increased transparency, these polymeric materials also provide advantages similar to those of conjugated polymers. The details of NLO dye-functionalized polymers are discussed in the chapter 4 on second-order materials in which these systems have been widely used.

Third-order NLO properties of nonconjugated polymers have not been studied because a very low magnitude is anticipated from viewing their chemical structures. Using the THG technique, Amano et al.[312] measured a $\chi^{(3)}$ value of 4.0×10^{-14} esu for PMMA at 1.9 μm. Hadjichristov and Kircheva [313] studied nonconjugated polymers, PMMA and poly(vinyl toluene) (PVT), using optical four-wave mixing by the technique of multiplex coherent anto-Stoke Raman scattering (CARS). The $\chi^{(3)}$ values from the FWM signal intensity I were estimated from the relationship:

$$I(sample)/I(reference) = |\chi^{(3)}(sample)|^2/|\chi^{(3)}(reference)|^2$$

where the difference in the refractive index of two media was neglected. The electronic $\chi^{(3)}$ of 5.1×10^{-14} esu for PMMA and electronic plus Raman resonance $\chi^{(3)}$ of 3.4×10^{-14} esu for PVT were obtained. Pure electronic $\chi^{(3)}$ of PVT was difficult to estimate. Resonance Raman enhancement of $\chi^{(3)}$ was observed for PVT for vibrational frequencies from 900 to 1350 cm^{-1}. Raman resonance enhancement in PMMA was observed for vibrational frequency 1000 cm^{-1}.

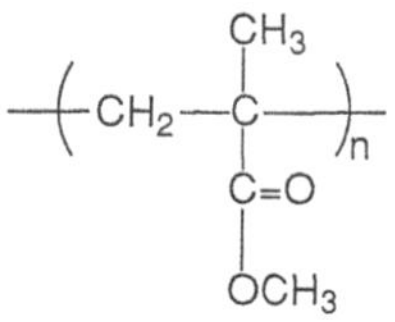

PMMA

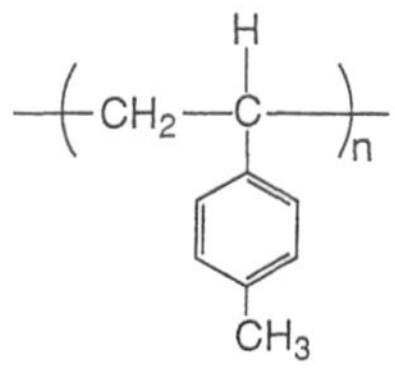

PVT

Novel NLO polymers can be designed and synthesized by covalently attaching dye moiety to conventional polymers such as PMMA, polystyrene, etc. In particular two types of NLO-chromophore functionalized systems have been investigated: NLO-dye functionalized polymers (side chain polymers) and NLO-dye introduced into the main chain (main chain polymers). Matsumoto et al.[314] studied the symmetrical cyanines having quinoline rings.

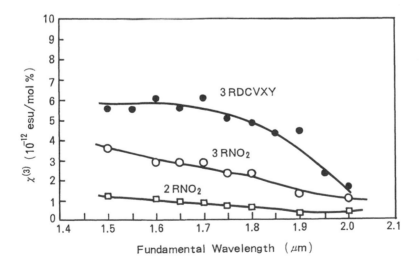

Dye 1: $R = H$
Dye 2: $R = Cl$

The ionic cyanine dyes and sodium polystyrene sulfonate were dissolved in an acetic acid-water solvent system. The polymers were doped as high as 50% by the dye, and transparent films of 0.2 to 0.3 μm were casted from the solution by spin coating. The $\chi^{(3)}$ values increased linearly as the dye concentration increased and showed some deviation at a concentration ~50%, probably because of the dye aggregation. The wavelength dependence studies show that a polymer film containing 50% cyanine dye 2 has a $\chi^{(3)}$ maximum around 1.9 μm at which the $\chi^{(3)}$ value reaches 2×10^{-11} esu. The $\chi^{(3)}$ of dye 1 was 7×10^{-12} esu at 1.9 μm. The ease of processing and large optical nonlinearity make these systems attractive for device fabrications.

Kaino et al.[315] measured third-order NLO susceptibilities of a variety of chromophore-functionalized polymeric materials; dye-attached side chain polymers and dye-introduced main chain polymers. PMMA was copolymerized with methyacrylate esters of the dicyanovinyl- or nitro-terminated diazo dye derivatives and Dispersed Red 1. The chemical structures of these azo dyes, referred to as 3RDCVXY, 3RNO$_2$, and 2RNO$_2$, are shown below. These azo dyes were synthesized by a diazo-coupling reaction. All these dyes had a diethylamino group as electron donor and dicyanovinyl and nitro groups as acceptors. Figure 34 shows the variation of $\chi^{(3)}$ values as a function of fundamental wavelengths, and the magnitude of $\chi^{(3)}$ increases with the decrease of fundamental wavelength. The $\chi^{(3)}$ value of 3RNO$_2$ was ~2.5×10^{-11} esu and of 2RNO$_2$ was ~0.8×10^{-11} esu for 7 mol% dye contents. The 3RDCVXY and 3RNO$_2$ dyes have three benzene rings connected with azo groups, hence their $\chi^{(3)}$ values are more than five and three times higher than 2RNO$_2$ dye due to the large conjugation length over the 1.5 to 2.0-μm range. The $\chi^{(3)}$ value of PMMA was 4×10^{-14} esu at 1.9 μm, therefore the large optical nonlinearities of these dye-attached polymers originated from dye moiety only. The $\chi^{(3)}$ values also increased linearly with increase of dye content up to 17 mol% for 3RNO$_2$ and 33 mol% for 2RNO$_2$. The $\chi^{(3)}$ value of 3RNO$_2$ was 4.8×10^{-11} esu at 1.5 μm and reached 2×10^{-10} esu for 50 mol% azo dye contents.

Figure 34 Variation of $\chi^{(3)}$ values as a function of fundamental wavelengths for azo-dye-attached polymers. (From Kaino et al., *Nonlinear Optics*, Elsevier, Amsterdam, 1992. With permission.)

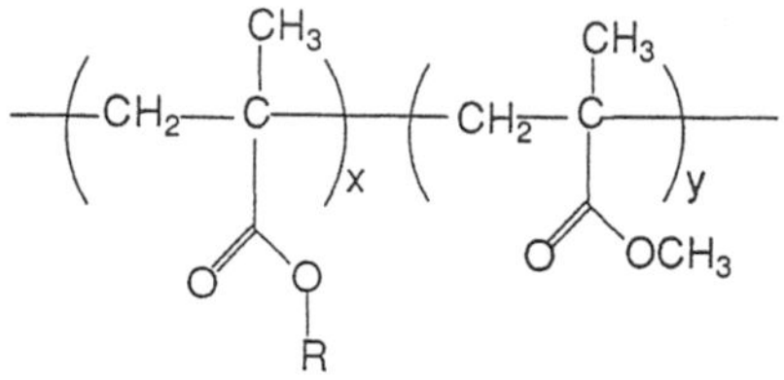

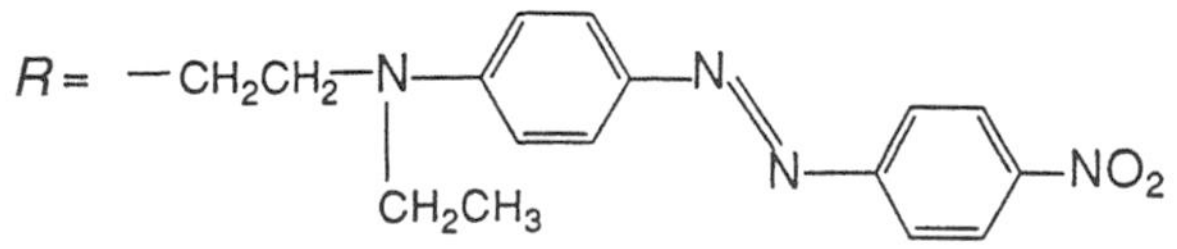

$$R = -CH_2CH_2-N(CH_2CH_3)-C_6H_4-N=N-C_6H_4-NO_2 \quad \text{2RNO}_2$$

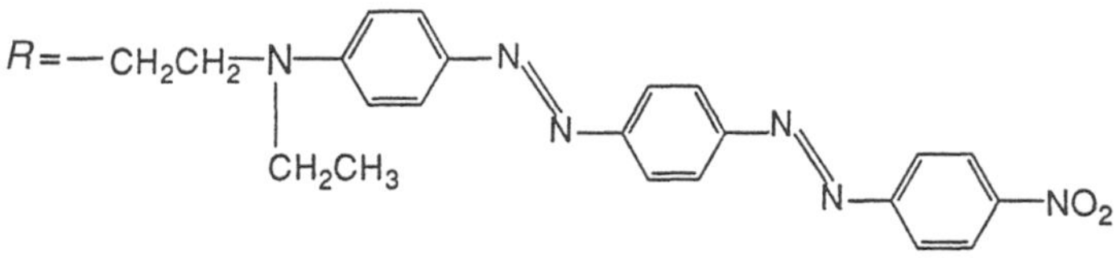

$$R = -CH_2CH_2-N(CH_2CH_3)-C_6H_4-N=N-C_6H_4-N=N-C_6H_4-NO_2 \quad \text{3RNO}_2$$

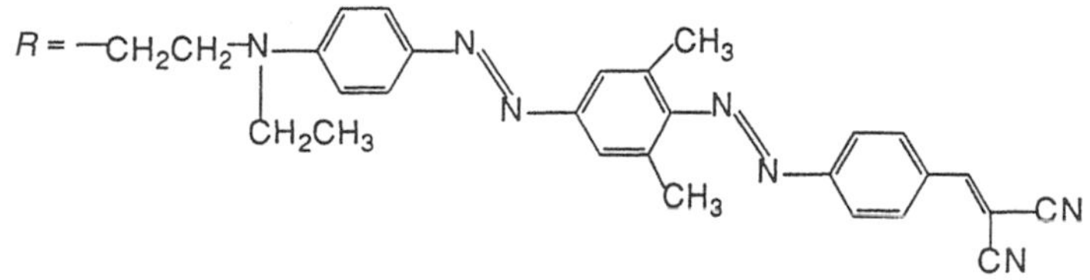

$$R = -CH_2CH_2-N(CH_2CH_3)-C_6H_4-N=N-\ldots \quad \text{3RDCVXY}$$

Table 42 lists the $\chi^{(3)}$ values of some main chain polymers of polyamic acids and polyimides.[316] The largest $\chi^{(3)}$ value is 1.2×10^{-11} esu for 4AS-PAA. The PAA with azo benzene (2A-PAA) shows larger $\chi^{(3)}$ values than the stilbene derivative of equivalent conjugation (2S-PAA). The $\chi^{(3)}$ values increased by one order of magnitude as the π-electron conjugation system containing two benzene rings and bonded either by N=N or C=C bonds increased to three benzene rings and by two orders of magnitude for four-benzene rings reaching a value of 10^{-11} esu. The $\chi^{(3)}$ dependence on the conjugation length gives an exponent of 4.5. The $\chi^{(3)}$ value was 10^{-11} esu for a four-benzene ring polyimide system. Therefore, these studies demonstrate that high $\chi^{(3)}$ values can be obtained in relatively short π-electron-conjugated chromophore-functionalized polymers by introducing a dye molecule into a polymer backbone and that processible polymeric materials with $\chi^{(3)}$ values higher than 10^{-10} esu can be obtained. Like π-electron-conjugated polymers, both chromophore-functionalized side chain and main chain polymers seem promising for third-order nonlinear optics.

The polymer ion-hemicyanine dye complexes were investigated by Nomura et al.[317] The polymer ion was poly(2-acrylamide-2-methylpropanesulfonic acid sodium salt) (AMPS). The hemicyanine dyes were (4-[2-(4'-dimethylaminophenyl)-vinyl]-1-methylpyridinium iodide named as HC-1 ($n = 1$) and (4-[4-4'-dimethylaminophenyl)-1,3-butadienyl]-1-methylpyridinium iodide named as HC-2 ($n = 2$). The structure of the polymer ion-dye system is shown below.

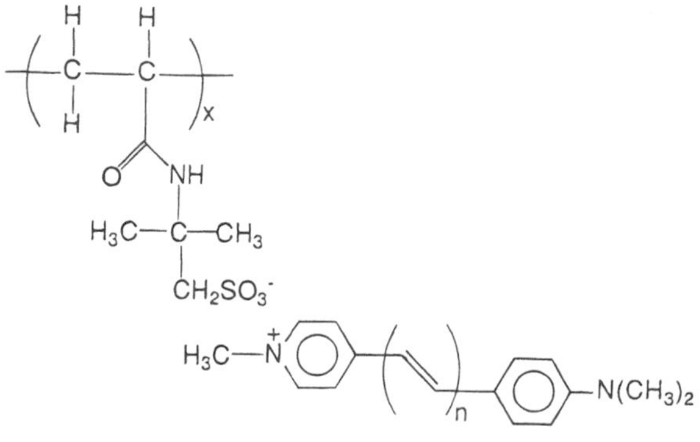

The hemicyanine dyes attached to the sulfonic groups of the polymer have a molar ratio of 0.56 and 0.62 for HC-1 and HC-2, respectively. The absorption spectra in chloroform showed peaks at 2.62 and

Table 42 The $\chi^{(3)}$ Values of Some Main Chain Polyamic Acids and Polyimides

Polymer Backbone	Polymer	$\chi^{(3)}$ (10^{-12} esu)		
		1.9 μm	2.0 μm	2.15 μm
Polyamic acid (PAA)	2S-PAA	0.54		
	2A-PAA	0.73		
	3A-PAA	2.7	3.1	2.8
	4AS-PAA	12.0		
Polyimide(PI)	3A-PI	2.8	2.5	1.8

From Kurihara et al., *J. Appl. Phys.*, 70, 17, 1991; Mori et al., *Jpn. J. Appl. Phys.*, 31, 896, 1992. With permission.

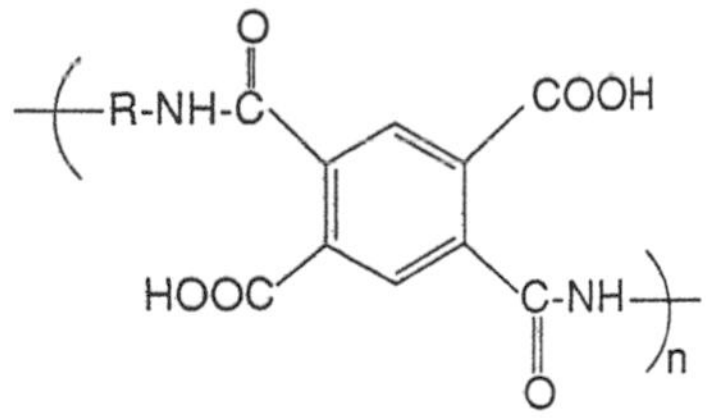

Polyamic acid (PAA)

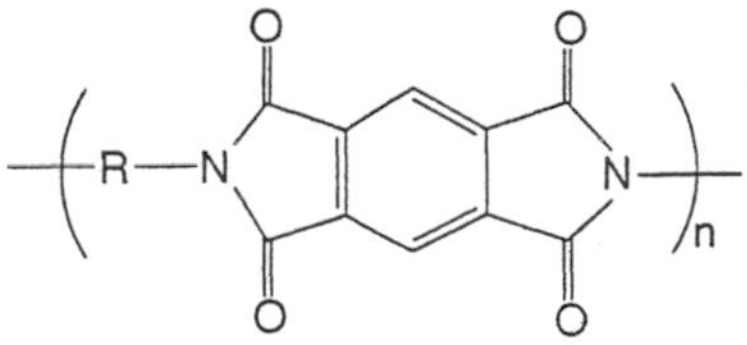

Polyimide (PI)

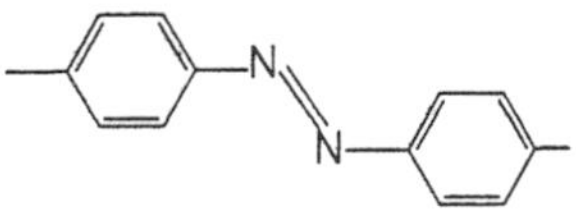

(2A-PAA)

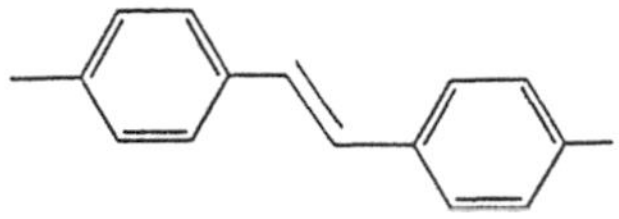

(2S-PAA)

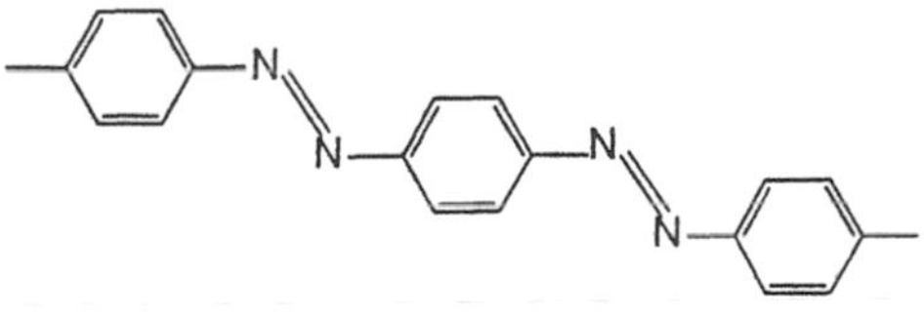

(3A-PAA, 3A-PI)

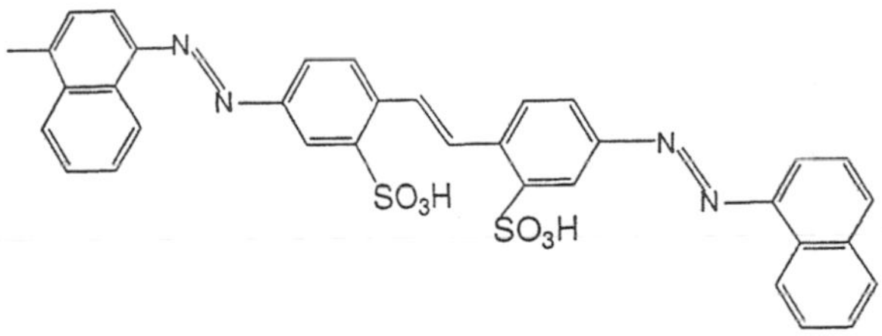

(4AS-PAA)

2.51 eV for HC-1 and HC-2, respectively. The HC-2 had longer π-electron-conjugated systems than HC-1, hence it showed larger $\chi^{(3)}$ values (3.6×10^{-10} esu at 2.05 eV) than those of HC-1 ($\chi^{(3)} = 2.1 \times 10^{-10}$ esu at 2.21 eV). The $\chi^{(3)}$ values were estimated by molar concentrations of the dye in the polymer films; $\chi^{(3)}/<M>$ were 5.5×10^{-11} esu dm³/mol for AMPS-HC-1 and 1.3×10^{-10} esu dm³/mol for AMPS-HC-2. The resonant nonlinear susceptibility $\chi^{(3)}$ THG/$<M>$ of AMPS-HC-2 with a little longer conjugation is about 1.5 times larger than that of AMPS-HC-1 system.

Meyer et al.[318] demonstrated that glasses can be made from polyelectrolyte (PEL) by adjusting the chemical structures of the ionic polymers. The NLO-dye containing PEL glasses were made by ion exchange by mixing an aqueous solution of the polymer (bromide), sodium tetrafluoroborate, and sodium salt of dye in desired molar ratio. The transparent film of polymer-dye salt was prepared from a DMF solution.

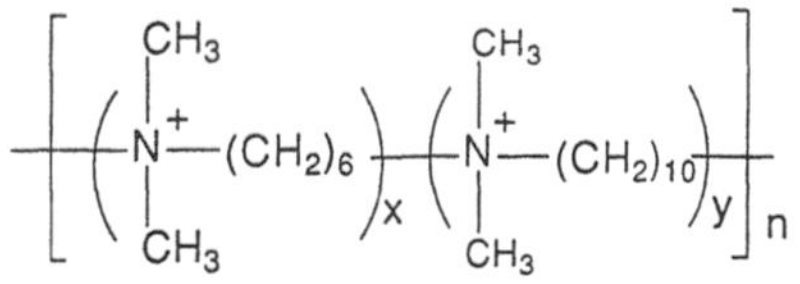

Polymer

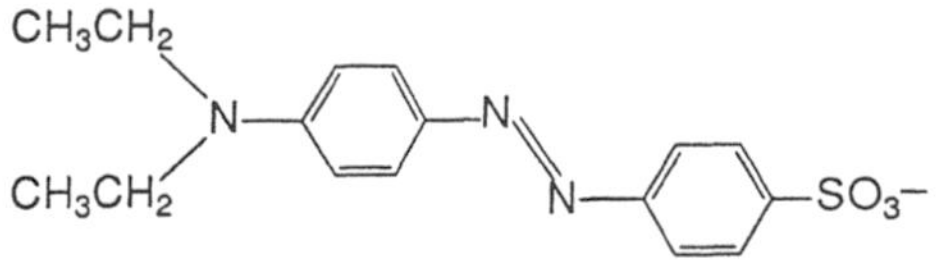

Dye

The $\chi^{(3)}$ values of glass PEL films are affected by dye concentration. The $\chi^{(3)}$ values measured by THG at 1.064 μm were 1.45×10^{-12} esu for 24% weight and 1.83×10^{-12} esu for dye contents of 31% by weight. The THG efficiency of PEL glasses increases as the dye content increases and reaches maximum with 62 wt% dye content (2.49×10^{-12} esu). The use of PEL glasses as a NLO medium has been accounted for due to the good solution processibility, incorporation of high dye concentration, optical quality, and structural variations by using various combinations of ionic polymers and appropriate counterions.

Torruellas et al.[319,320] investigated organic polymer functionalized by 4-dialkylaminonitrobenzene (1), 4-dialkylamino-4'-nitrostilbene (2), 4-dialkylamino-4'-nitrodiphenylbutadiene (3), and 4-dialkylamino-4'-nitrodiphenylhexatriene (4). The $\chi^{(3)}$ values were measured by the THG technique at different wavelengths. The $\chi^{(3)}(-3\omega; \omega, \omega, \omega)$ values of these side chain polymers range from 0.54×10^{-12} to 7.42×10^{-12} esu at 1.904 μm. The $\chi^{(3)}$ values at 1.904 μm and 1.579 μm change by a factor of 15 as the conjugation length increases. The $\chi^{(3)}$ value increases with the increase in the number of double bonds to the power 2.5 (Table 43). These polymers show no optical damage up to 1 GW/cm² of laser intensity. The dispersion of $\chi^{(3)}$ of side chain polymers containing 4-dialkylamino-4'-nitrostilbene and 4-alkoxy-4'-nitrostilbene side groups were measured in the range of 0.953 μm to 1.904 μm with THG. A scalar two-level model which includes microscopic local field cascading shows that it plays an important role in third-order susceptibility. In the near-infrared region, the phase demonstrates the presence of only one resonant transition near 3ω. Far from the three-photon resonance, the two-level model fails for the molecules with the smallest dipole moments.

P—N—⟨⟩—NO₂

(1)

P—N—⟨⟩—(⟨⟩)ₙ—⟨⟩—NO₂

(2) n = 1   (3) n = 2   (4) n = 3

here P denotes the polymer backbone as well as the spacer connecting the side group to the backbone. Structure 1 contains the 4-dialkylaminonitrobenzene. Structure 2 contains the 4-dialkylamino-4-nitrostilbene, Structure 3 contains the 4-dialkylamino-4'-nitrodiphenylbutadiene, and Structure 4 contains 4-dialkylamino-4'-nitrodiphenylhexatriene as the side groups.

Norwood et al.[321] also measured the $\chi^{(3)}$ values of side chain copolymers of 4-amino-4'-nitrostilbene (DANS) and 4-oxy-4'-nitrostilbene (ONS) chromophores with methylmethacrylate (MMA). The $\chi^{(3)}$ values of DANS/MMA (60/40) $\lambda_{max} = 438$ nm and ONS/MMA (50/50) $\lambda_{max} = 370$ nm that were measured by the picosecond DFWM technique at 598 nm were found to be 6.9 $\times 10^{-12}$ and 2×10^{-12} esu, respectively. The $\chi^{(3)}$ values obtained by the THG technique at 1.9 μm of DANS/MMA (60/40) and ONS/MMA (50/50) were found to be 5.0×10^{-12} and 2×10^{-12} esu, respectively.[320] The $\chi^{(3)}$ values of DANS/MMA were 3 times larger than those of ONS/MMA because of the dispersion effect. The magnitude of optical nonlinearity of these copolymers is the same as reported by Torruellas et al.[319,320]

The $\chi^{(3)}$ values of silicon-phthalocyanine (SiPc) copolymers and SiPc dispersed in PMMA were measured by the picosecond DFWM technique to examine the role of interactions between the

Table 43 The Density, Refractive Indexes, and $\chi^{(3)}$ Values of Side Chain Polymers

Polymer	Density (10^{27} m⁻³)	Refractive Index		$\chi^{(3)}$ (10^{-14} esu) at λ (μm)		
		0.632 μm	1.064 μm	1.064	1.579	1.904
1	1.25	1.599	1.568	210	56.7	54.0
2	1.1	1.690	1.623	2450	850.5	476
3	1.15	1.690	1.645	1395	1185	731
4	1.54	1.692	1.625	1535	1038	742

From Torruellas et al., *Chem. Phys. Lett.*, 175, 267, 1990; Torruellas et al., *J. Chem. Phys.*, 94, 6851, 1991. With permission.

SiPc monomers.[321] The $\chi^{(3)}_{1111}$ values of 9.5×10^{-11} and 9.5×10^{-11} esu and $\chi^{(3)}_{1221}$ values of 3.24×10^{-11} and 2.29×10^{-11} esu were obtained for 10% SiPc/MMA bis-methacrylate and 10% SiPc/ MMA mono-methacrylate (structure shown below) thin films, respectively. For a 10% by weight guest/host SiPc/MMA thin film, the $\chi^{(3)}_{1111}$ and $\chi^{(3)}_{1221}$ values were 2.57×10^{-11} and 1.04×10^{-11} esu. The size of $\chi^{(3)}_{1111}$ in the SiPc copolymers increases linearly for mole ratios less than 1% and saturates as the mole fraction becomes larger that 6 mol% (40% by weight) where the $\chi^{(3)}_{1111}$ reaches as high as 2.5×10^{-10} esu. The $\chi^{(3)}_{1111}$ decreases at 11.6 mol% SiPc monomer. The optical nonlinearity of copolymers is more than 3 times larger than the guest/host system. Furthermore, the ratio of the tensor components for copolymers is also larger than for the guest/host presumably due to the large instantaneous portion of the nonlinearity in the guest/host. The copolymers show a single exponential decay with small instantaneous and long time fractions while the guest/host system has a slow decay component. This comparison indicates the role of aggregation and Pc ring interaction in determining the magnitude and speed of the optical nonlinearity.

$Si(C_6H_{13})_3$

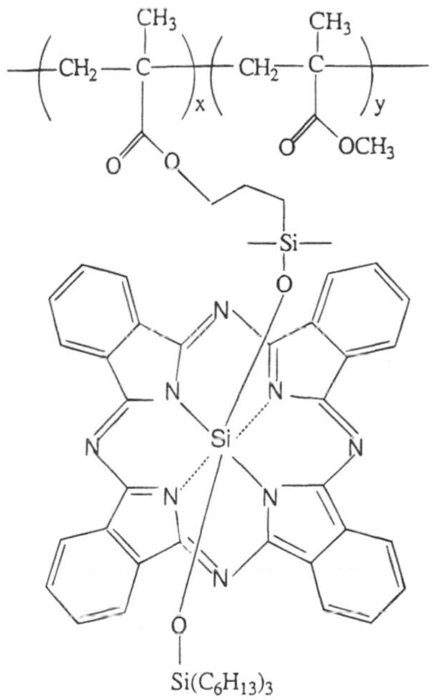

Kajzar and Zagorska[322] measured third-order NLO properties of functionalized polymers (structure shown below) by transverse THG before and after the corona poling to examine the cascading effects in $\chi^{(3)}(-3\omega; \omega, \omega, \omega)$. Three different polymers were PMMA-DR1 with 62 wt% of the chromophore concentration. The chromophore concentration was 28% for thermal cross-linking DIAM-Bis-A polymer and 10% for the photocross-linked 3DRCIN polymer. The PMMA-DR1 showed a $\chi^{(2)}$ value of 120 pm/V. The $\chi^{(3)}$ values were 1.4×10^{-11}, 3.8×10^{-12}, and 2.3×10^{-12} esu at 1.064 μm and 6.8×10^{-12}, 2.1×10^{-12}, and 6.1×10^{-12} esu at 1.097 μm for PMMA-DR1, DIAM-Bis-A, and 3DRCIN, respectively. PMMA-DR1 and DIAM-Bis-A showed $\chi^{(3)}$ values about twice as large at 1.064 μm as those at 1.907 μm because of a two-photon resonant contribution. Contrary to this, the measured $\chi^{(3)}$ at 1.907 μm was smaller for 3DRCIN polymer than at 1.064 μm, owing to the a larger three-photon resonant contribution to $\chi^{(3)}$ compared with PMMA-DR1 and DIAM-Bis-A. The tensor components $\chi^{(3)}_{pppp}$ were 6.8×10^{-12} esu before poling and 4.7×10^{-12} esu after poling and $\chi^{(3)}_{ssss}$ were 8.0×10^{-12} esu before poling and 7.5×10^{-12} esu after poling for PMMA-DR1 polymer showed no effect of poling. No cascading effect was observed in macroscopically ordered thin films due to the smaller thickness than the corresponding coherence length and for the transverse THG geometry which cancel the cascading effect.

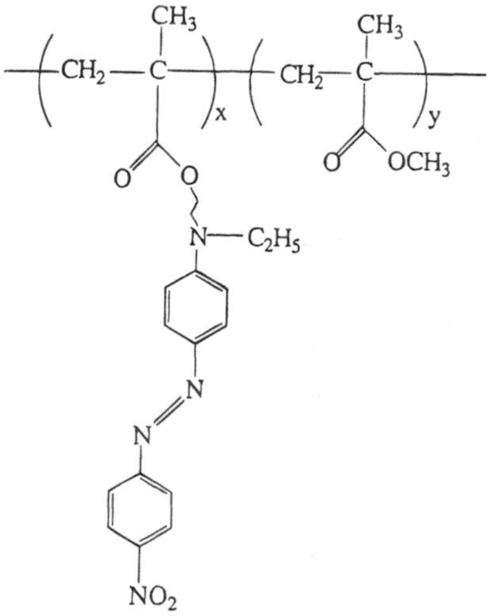

(PMMA-DR1)

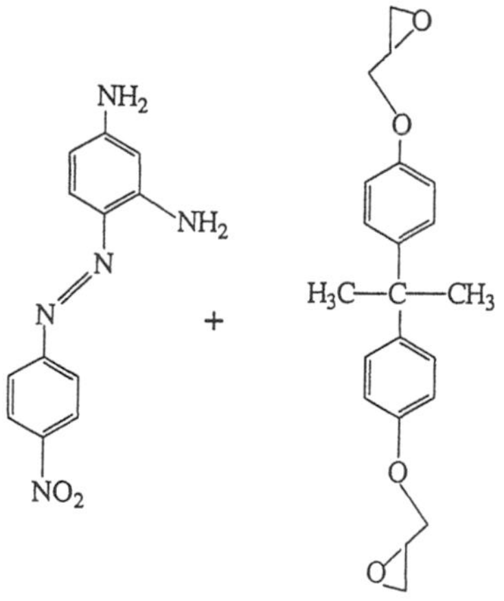

(DIAM-Bis-A)

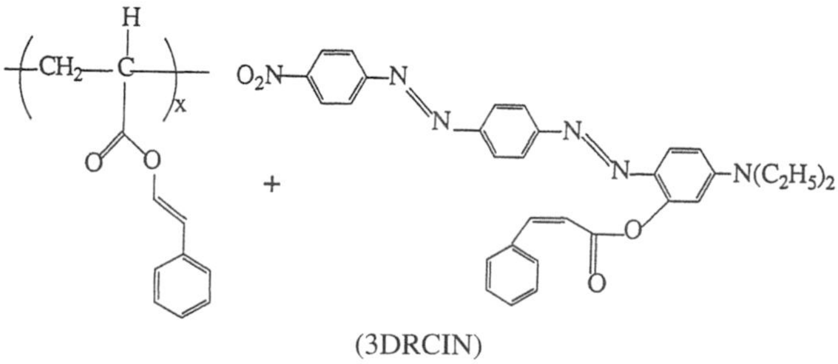

(3DRCIN)

E. ORGANOMETALLIC COMPOUNDS

After work with organic conjugated materials, the versatility of organometallic compounds has also been realized. Like organic materials, organometallic compounds also offer the advantages of architectural flexibility and ease of fabrication and tailoring. Organometallics have two types of charge-transfer transitions, i.e., metal-to-ligand and ligand-to-metal. The metal-ligand bonding displays large molecular hyperpolarizability caused by the transfer of electron density between the metal atom and the conjugated ligand systems. Furthermore, the overlapping of the electron orbitals of ligands with the metal ion-orbitals leads to larger intramolecular interactions in organometallic complexes. The diversity of central metal atoms, oxidation states, their size, and the nature of ligands helps in tailoring materials to optimize the charge-transfer interactions. Nalwa[323] reviewed progress made in the development and evaluation of the NLO properties of organometallic materials in 1991. Since then further progress has been made in this area and in-depth presentation of third-order NLO properties of a very wide variety of organometallic materials is discussed here.

1. Metallophthalocyanines

It is worth mentioning that metallophthalocyanines (MPcs) are not true organometallic compounds because there is no metal-to-carbon bonding in the macrocycle. Still in some sense, they may be regarded as organometallic materials. The MPcs possess exceptionally high thermal stability, ease of fabrication and processing, architectural flexibility, and environmental stability, which make them multipurpose materials for solid-state technology. Phthalocyanine is a two-dimensional π-conjugated molecule in which as many as 70 different metal atoms can be incorporated.[324,325] The chemical structure of the phthalocyanine molecule is shown below and the versatility of the macrocycle offers tremendous opportunities for tailoring of electronic and photonic functions through chemical modifications.

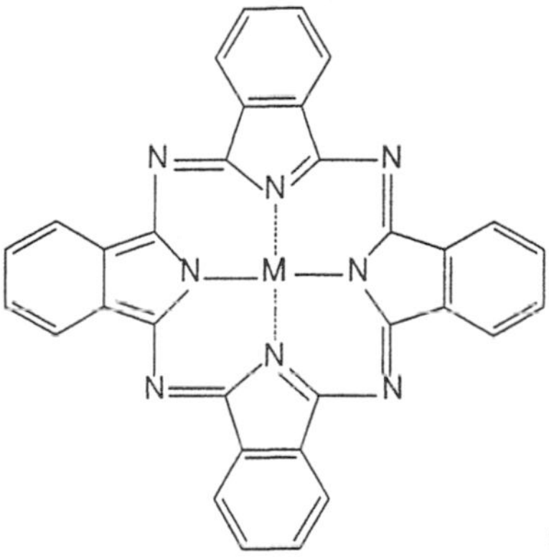

(MPc)

The first report on the third-order NLO properties of MPcs appeared in 1987. Ho et al.[326] reported THG in thin films of chlorogallium phthalocyanine (ClGa-Pc) and fluoroaluminum phthalocyanine (FAl-Pc). ClGa-Pc and FAl-Pc show $\chi^{(3)}$ values of 2.5×10^{-11} and 5.0×10^{-11} esu at 1.064 μm, respectively. Shirk et al.[327] measured third-order nonlinear susceptibility of H_2Pc, PbPc, and PtPc by using the DFWM technique. The nonresonant $\chi^{(3)}$ value of 2.0×10^{-10} esu for platinum phthalocyanine (PtPc) was 5 and 45 times larger than that of the PbPc and metal-free Pc, respectively. The $\chi^{(3)}$ and second hyperpolarizabilities of metal tetrakis(cumylphenoxy)phthalocyanines (MPcCP$_4$) in which the metal atom is Co, Ni, Cu, Zn, Pd, Pt, and Pb, were measured by the same team.[328,329] The $\chi^{(3)}$ values of MPcCP$_4$ are very large and vary between 7×10^{-12} esu and 2×10^{-10} esu with the metal substitution. The $\chi^{(3)}$ values of metal-substituted Pc (MPcCP$_4$) are about 2 to 45 times larger than the $\chi^{(3)}$ values of a metal-free phthalocyanine. The measured hyperpolarizability γ values of CoPcCP$_4$, NiPcCP$_4$, CuPcCP$_4$, ZnPcCP$_4$, PdPcCP$_4$, PtPcCP$_4$, PbPcCP$_4$, and H_2PcCP$_4$ were 5×10^{-32}, 4×10^{-32}, 3×10^{-32}, 5×10^{-33}, 1×10^{-32}, 1×10^{-31}, 1×10^{-32}, and 2×10^{-33} esu, respectively, at 1.064 μm and the nonlinear figure-of-merit $\chi^{(3)}/\alpha$ values were 1×10^{-13}, 1×10^{-13}, 5×10^{-13}, $> 7 \times 10^{-13}$, 3×10^{-13}, 2×10^{-13}, 2×10^{-12}, and $>4 \times 10^{-13}$ esu cm, respectively. The largest third-order optical nonlinearities were obesrved for CoPcCP$_4$, NiPcCP$_4$, and PtPcCP$_4$. In these MPcs, metal substitution introduces low-lying electronic states giving rise to large optical nonlinearities, Hosoda et al.[330,331] reported third-order optical nonlinearities in tetravalent metallophthalocyanine and soluble phthalocyanines with *tert*-butyl substituents. Thermal annealing increases the $\chi^{(3)}$ values of VOPc and TiOPc by a factor of 2.5. The $\chi^{(3)}$ values of the *tert*-butyl substituted were found to be smaller than that of unsubstituted phthalocyanines. A tetra-kis(*tert*-butyl) vanadylphthalocyanine (TBVOPc) shows $\chi^{(3)}$ values that are one order of magnitude smaller than those of VOPc. Hosoda et al.[332,333] also reported third-order optical nonlinearities of several MPcs. Third-order optical nonlinearity of several unsubstituted MPcs as well as tetra-alkylthio-substituted MPcs as a function of wavelength dependence were also reported.[334,335] The substituted MPcs materials have $\chi^{(3)}$ values in the range of 10^{-10} to 10^{-12}. Table 44 lists the $\chi^{(3)}$ values of a broad variety of MPcs compounds.

Wang et al.[336] measured resonant and nonresonant third-order optical properties of liquid solutions of silicon naphthalocyanine (SiNc) abbreviated as SINC. The THG measurements of SiNc (SINC) were performed in DMF solutions using the Maker fringe and showed isotropically averaged $\langle\gamma\rangle$ values of $(-410 \pm 30) \times 10^{-36}$ esu at 1.907 μm. The value of n_2 measured by saturable absorption at 770 nm for a SINC film was found to be 10^{-4} cm^2 kW^{-1}, which has been identified as the largest value for conjugated structures. A silicon naphthalocyanine with a 30% weight incorporated in PMMA shows $\chi^{(3)}_{xxxx}$ and $\chi^{(3)}_{xyyx}$ values of the order of 2.09×10^{-11} and 6.1×10^{-12} esu, respectively, from DFWM at 598 nm.[337] The SiNc shows a response time of $\sim$16 ps. On the other hand, a silicon phthalocyanine (SiPc) with a 10% weight incorporated in PMMA shows $\chi^{(3)}_{xxxx}$ and $\chi^{(3)}_{xyyx}$ values of 9.4×10^{-11} and 3.24×10^{-11} esu, respectively, at the same wavelength. The response time of 15 ps was evaluated by fitting coefficients for DFWM response.

The second hyperpolarizability and $\chi^{(3)}$ values of lanthanide bis(phthalocyanines), M(Pc)$_2$ (where M is Sc, Lu, Yb, Y, Gd, Eu, and Nd) and their anions, have been reported recently by Shirk et al.[338] These M(Pc)$_2$ show $\chi^{(3)}$ values in the range between 5×10^{-10} and 2×10^{-9} esu and γ values range between 15×10^{-32} and 48×10^{-32} esu at 1.064 μm depending on the metal atom. The γ values decrease in the order: ScPc$_2$ > YbPc$_2$ > LuPc$_2$ > YPc$_2$ > EuPc$_2 \geq$ GdPc$_2$ > NdPc$_2$. The measured figure-of-merit $\chi^{(3)}/\alpha$ values were 0.6×10^{-13} esu cm for ScPc$_2$, 1×10^{-13} esu cm for LuPc$_2$, YbPc$_2$, and NdPc$_2$, 2×10^{-13} esu cm for YPc$_2$, GdPc$_2$, EuPc$_2$, ScPc$_2^-$, LuPc$_2^-$, YPc$_2^-$, GdPc$_2^-$, and 3×10^{-13} esu cm for EuPc$_2^-$. Both the neutral species and the anions in EuPc$_2$ and GdPc$_2$ have about the same γ values whereas in ScPc$_2$ and LuPc$_2$ the neutral species have larger γ values than the anions. The larger γ values in ScPc$_2$, LuPc$_2$, and in YbPc$_2$ originate due to the proximity of the intervalence transition. The response time of the $\chi^{(3)}$ in LuPc$_2$ was found to be <35 ps.

The effect of donor-acceptor groups on the magnitude of third-order optical nonlinearity of metallophthalocyanines were investigated by measuring the $\chi^{(3)}(-3\omega;\omega,\omega,\omega)$ of VOPc(NH$_2$)$_4$, CuPc(NH$_2$)$_4$, NiPc(NH$_2$)$_4$, and FePc(COOH)$_4$.[339] The VOPc(NH$_2$)$_4$ showed the $\chi^{(3)}$, a factor of 4 and 3 larger than those of NiPc(NH$_2$)$_4$ and CuPc(NH$_2$)$_4$, because the axial oxygen group plays a major role in molecular arrangement. The $\chi^{(3)}$ values of tetra-substituted CuPc and NiPc were found to be slightly larger than the unsubstituted ones. A similar effect of peripheral substitution on $\chi^{(3)}$ values has been observed for NiPc derivatives.[329] The effect of the end group attached to phthalocyanine ring on $\chi^{(3)}$ values has also been investigated.[339] The $\chi^{(3)}$ values were found to decrease in an order cobalt[bis(3,4)phthalimidocarbonyl] phthalocyanine > cobalt[bis(3,4)dicarboxybenzoyl] phthalocyanine dianhydride >cobalt[bis(3,4)dicarboxybenzoyl] phthalocyanine > N,N'-diphenyl cobalt[bis(3,4)dicarboxybenzoyl] phthalocyanine diimide. The metallophthalocyanine possessing either donor or acceptor peripheral substituents showed $\chi^{(3)}$ values larger than that of the corresponding unsubstituted phthalocyanines with an exception of oxovanadium-phthalocyanine where the peripheral substituents required for processability or otherwise have an adverse effect on optical nonlinearity.

The effect of extended π-conjugation on $\chi^{(3)}(-3\omega; \omega, \omega, \omega)$ has been studied using MNcs.[340-343] The VONc, CuNc, ZnNc, PdNc, and NiNc possessing terminally *tetrakis-n*-pentoxy carbonyl (R=COO-C$_5$H$_{11}$) substituents showed a $\chi^{(3)}$ value of 10^{-12} esu at 2.1 μm. The $\chi^{(3)}(-3\omega; \omega, \omega, \omega)$ values of InClNc, IrNc, RuNc, and RhNc possessing terminally *tetrakis-tert*-butyl R=[C(CH$_3$)$_3$] substituents were also of the same order. In particular, the VONc(COO-C$_5$H$_{11}$)$_4$ showed $\chi^{(3)}$ an order of magnitude larger than those of other MNcs. The $\chi^{(3)}(-3\omega; \omega, \omega, \omega)$ values were affected by the nature of the central metal atoms although the effect was not so pronounced. The wavelength dependence of $\chi^{(3)}$ for VONc thin films is shown in Figure 35. The $\chi^{(3)}$ values at 2.1 μm are about 9 times larger than those measured at 1.74 μm. The VONc thin films when treated with dichloroethane vapors at room temperature showed the $\chi^{(3)}$ of 10^{-10} esu. The dichloroethane treatment gives rise to polymorphism in VOPc thin films leading to two phases: (a) a slipped-stacked arrangement on an adjacent molecule (Phase I) and (b) cofacially stacked arrangement aligned linearly along the metal-oxo bond (Phase II). The spin-coated films with 9.8% of (*t*-Bu)VOPc in PMMA showed $\chi^{(3)}$ values of 6.6×10^{-13} esu at 1.907 μm which increases to 4.4×10^{-12} esu on dichloroethane vapor treatment for 20 hr.[332] The vacuum-deposited thin films of VOPc show $\chi^{(3)}$ values of 9.3×10^{-11} esu whereas the thin films of pure *tert*-butyl-substituted VOPc and doped poly(methylmethacrylate) films show $\chi^{(3)}$ values of 5.0×10^{-12} esu for phase I and 6.0×10^{-12} esu for phase II.[23] In addition, the nature of central metal atoms and substituents

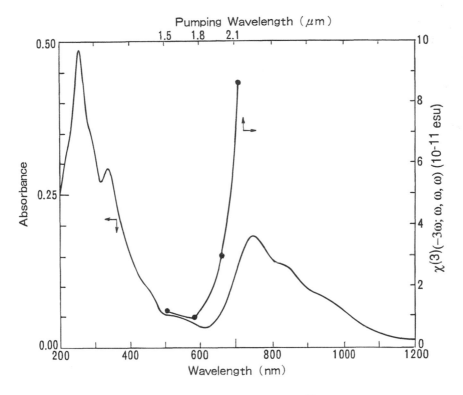

Figure 35 Optical absorption spectrum and wavelength-dependent $\chi^{(3)}(-3\omega;\omega,\omega,\omega)$ for thin films of oxovanadium naphthalocyanine derivative. (From Nalwa et al., *J. Phys. Chem.*, **97**, 1097, 1993. With permission.)

either attached axially or perpherally also play a very major role in determining the magnitude of the $\chi^{(3)}$ values of MNcs.

(Metallonaphthalocyanine)

Grund et al.[344] reported the third-order NLO properties of spin-cast films of soluble oligomeric bridged (phthalocyanato)ruthenium(II) complex [t-Bu$_4$PcRu(dib)]$_n$ which is axially linked by a p-diisocyanobenzene (dib) bridging ligands and a phthalocyanine ring being substituted by four $tert$-butyl groups. The $\chi^{(3)}(-\omega;\omega,\omega,-\omega)$ as high as $\sim1.5 \times 10^{-7}$ esu was measured at the absorption maximum by the DFWM technique. The THG measurements of a [t-Bu$_4$PcRu(dib)]$_n$ film yielded a $\chi^{(3)}(-3\omega;\omega,\omega,\omega)$ of 3.7×10^{-12} esu with a phase of $155 \pm 10°$. The $\chi^{(3)}(-\omega;\omega,\omega,-\omega)$ value was found to be more than 4 orders of magnitude larger than $\chi^{(3)}(-3\omega;\omega,\omega,\omega)$ because of the different optical processes. The relaxation time of 70 ps in [t-Bu$_4$PcRu(dib)]$_n$ thin films was determined at 600 nm probably having an electronic origin.

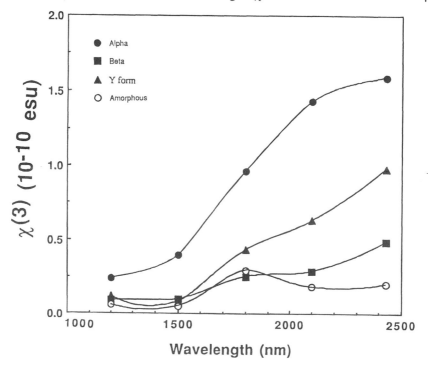

Nalwa et al.[345] established a relationship between the $\chi^{(3)}$ and crystal structures by studying different polymorphs of TiOPc. Figure 36 shows the wavelength dispersion of $\chi^{(3)}(-3\omega;\omega,\omega,\omega)$ for amorphous, α, β, and Y TiOPc vacuum-deposited thin films measured by the THG technique in the wavelength region of 1.2 to 2.43 μm. The α-TiOPc shows the largest $\chi^{(3)}$ of 1.59×10^{-10} esu at 2.43 μm which

Figure 36 Wavelength dependence of third-order NLO susceptibility $\chi^{(3)}$ values of different polymorphs of oxotitanium phthalocyanine. (From Nalwa et al., *J. Phys. Chem.*, 97, 10515, 1993. With permission.)

has a three-photon (3ω) resonance contribution. The large difference in $\chi^{(3)}$ values of TiOPc polymorphs occurs because of the dispersion of the $\chi^{(3)}$ in the resonant and nonresonant regions. In α-VOPc, the $\chi^{(3)}$ apparently changed with absorption intensity but it did not completely follow the absorption spectrum (Figure 37).The α-VOPc shows the largest $\chi^{(3)}(-3\omega; \omega, \omega, \omega)$ of 9.95×10^{-11} esu at 2.1 μm. Similar wavelength dependence of $\chi^{(3)}$ has been observed for the VONc derivative where the $\chi^{(3)}$ reaches as high as 8.6×10^{-11} esu at 2.1 μm because of the three-photon resonance. The $\chi^{(3)}$ values of Phase II of TiOPc and VOPc are quite large and similar to the π-conjugated organic polymers. The $\chi^{(3)}(-3\omega; \omega, \omega, \omega)$ of two different phases of TiOPc prepared by thermal treatment were reported by Hosoda et al.[330,331] A $\chi^{(3)}$ value of 5.3×10^{-11} esu at 2.1 μm by THG measurements has been reported for TiOPc. The $\chi^{(3)}$ values of 4.6×10^{-11} and 8.1×10^{-11} esu at 1.907 μm from THG measurements has been reported for α-VOPc and α-TiOPc, respectively. The $\chi^{(3)}$ of various MPcs have been reported by other groups.[346–351]

Diaz-Garcia et al.[352,353] reported real and imaginary parts of second hyperpolarizabilities γ of soluble metal-free and copper phthalocyanines determined from the concentration dependence of the THG intensity at 1.064 and 1.907 μm in chloroform. The measured γ values were analyzed with a four-level model; two excited one-photon allowed levels and one-photon forbidden level placed between both at about 500 nm. The negative real γ values at 1.064 which change the sign at 1.907 μm were measured. The imaginary part was positive at both wavelengths. The modulus γ values changed by an order of magnitude with the chemical structure (see Table 44).

$$R = R_1 = \text{(phenyl)}-O- = H_2Pc(O^-C_6H_5)_{16}$$

$$R = R_1 = -S-CH-(CH_3)_2 = H_2Pc[S^-CH(CH_3)_2]_{16}$$

$$R = -S-\text{(phenyl)}-CH_3 \quad R_1 = H = H_2Pc(O^-C_6H_5)_{16} = CuPc(C_{10}H_7S)_{15}$$

$$R = \text{(naphthyl)}-S- \quad R_1 = H = CuPc(CH_3 = C_6H_4S)_{15}$$

Diaz-Garcia et al.[354] reported second hyperpolarizabilities γ of soluble octasubstituted metal-free (H_2PcR_8) and metal-containing phthalocyanines [$CuPcR_8$, $NiPcR_8$, and $CoPcR_8$ where $R = OCH_2-CON(C_8H_{17})_2$] measured by THG at 1.34 μm and EFISH technique at 1.064 μm in chloroform solutions. The real (γ') and imaginary (γ'') components of molecular hyperpolarizability γ ($\gamma = \gamma + i\gamma''$) were determined from the concentration dependence of harmonic intensities with respect to that of pure solvent. The real γ values have negative sign. The $\gamma(-3\omega;\omega,\omega,\omega)$ of metal-free and metal-containing phthalocyanines were of the same magnitude. The $\gamma(-3\omega;\omega,\omega,\omega)$ value of H_2PcR_8 was an order of

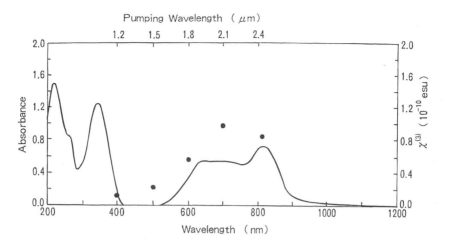

Figure 37 Wavelength dependence $\chi^{(3)}$ (-3ω; ω, ω, ω) for vacuum-deposited thin films of α-phase oxovanadium phthalocyanine (α-VOPc). (From Nalwa, 48th Okazaki Conference on Phthalocyanines, Jan. 26–28, 1994. With permission.)

magnitude larger than $\gamma(-2\omega;\omega,\omega,0)$ measured at 1.064 μm although no such difference was observed between the $\gamma(-3\omega;\omega,\omega,\omega)$ and $\gamma(-2\omega;\omega,\omega,0)$ values of $CuPcR_8$, $NiPcR_8$, and $CoPcR_8$. The γ values were affected by metal substitution and $CoPcR_8$ showed the largest hyperpolarizabilities. The enhancement of $\gamma(-2\omega;\omega,\omega,0)$ was considered to be associated with ω resonance with additional one-photon forbidden d-d level related with the metal in 1 to 1.3 μm range.

Diaz-Garcia et al.[355] measured real parts of second hyperpolarizabilities γ of soluble lipophilic-substituted metallotriazolehemiporphyrazines (MThp) from concentration dependence of the THG signal at 1.340 and 1907 μm in chloroform. The γ values at 1.904 μm for H_2Thp and ZnThp were smaller than 10^{-34} esu and the sign was uncertain. The CoThp showed the largest γ value of -6.1×10^{-33} esu at 1.340 μm and -11×10^{-34} esu at 1.904 μm. FeThp was found to the second effective compound with a γ value of -3.1×10^{-33} esu at 1.340 μm and -4.6×10^{-34} esu at 1.904 μm. The γ value varied with metal substitution as $H_2Thp <$ ZnThP < CuThp < NiThp < CoThp. The γ values for unfilled d-shell ions were several times higher than those of H_2Thp and ZnThp showing a remarkable effect of metal complexation with unfilled d-shell ions. At 1.340 μm, the γ values were an order of magnitude higher with negative sign due to a three-photon resonance enhancement approaching the allowed level at 360 nm or to two-photon lying above a forbidden π^* level.

H_2Thp M = H_2, MThp = M = Mn(II), Fe(II), Co(II), Ni(II), Cu(II), Zn(II)

Bubeck et al.[356] measured $\chi^{(3)}$ values of phthalocyaninatopolysiloxane LB films by THG at 1.064 μm. The ratio $\chi^{(3)}/n^{3/2}(\omega)\, n^{1/2}$ (3ω) was measured as 0.5×10^{-12} esu in the parallel direction and 5.0×10^{-12} esu in the perpendicular direction. The measured absorption coefficients $\alpha(\omega)$ and $\alpha(3\omega)$ were

1.5×10^3 and 3.8×10^4 cm^{-1} in the parallel direction and 4.0×10^3 and 8.5×10^4 cm^{-1} in the perpendicular direction, respectively. The $\chi^{(3)}$ of $1 \sim 3 \times 10^{-12}$ esu was estimated assuming the refractive indexes of 1.7 at ω and 2.3 at 3ω.

Maruno et al.[357] measured the $\chi^{(3)}$ value of TiOPc film by forming an ordered structure with dibenzo($b \cdot t$)phthalocyaninato-Zn(II) (ZnDBPc) as an underlying material with THG at 1.9 μm. Thin films were developed using organic molecular beam deposition (OMBD) because this enhances the ordering growth of the next layer of TiOPc. The thickness of double-layer film was about 10 nm for ZnDBPc and 50–80 nm for TiOPc. The $\chi^{(3)}$ value of 1.7×10^{-10} esu was measured for TiOPC film grown at 150 °C, which consists of α-phase crystals. The double-layer TiOPc ZnDBPc had a $\chi^{(3)}$ value of 3.0×10^{-10} esu because of the formation of highly ordered TiOPc thin films. Yamashita et al.[358] measured $\chi^{(3)}$ value of 2.1×10^{-10} esu for oxomolybdenum phthalocyanine (MoOPc) films developed by OMBD onto sapphire substrate. Imanishi et al.[359] reported multilayered ordered films of CuPc with naphthalene tetracarboxylic dianhydride (NTCDA) from an ultra-high vacuum OMBD method. The $\chi^{(3)}$ varies of 2.3×10^{-12} esu was measured for CuPc/NTCDA films at 1.907 μm. The total thickness of multilayers was about 600 Å, where CuPc and NTCDA each were 300 Å.

Reeves et al.[360] measured $\chi^{(3)}$ values of metal-free, cooper, platinum, and lead octakis(2-ethylhexylox-y)phthalocyanines (MPc-OEH)$_8$ in polystyrene films at 532 nm and in chloroform solutions at 1.064 μm by the picosecond DFWM technique. The resonant $\chi^{(3)}$ values at 532 nm were 5,600, 7,400, 3,300, and 10,000 times that of carbon disulfide in polystyrene matrix and the nonresonant $\chi^{(3)}$ values at 1.064 μm were 6, 28, 15, and 120 times that of carbon disulfide from 1 wt% chloroform solutions of (H$_2$Pc-OEH)$_8$, (CuPc-OEH)$_8$, (PbPc-OEH)$_8$, and (PtPc-OEH)$_8$, respectively. These $\chi^{(3)}$ values were for pure samples based on the exploration of the response from the measurements. The resonant $\chi^{(3)}$ values corresponded to 10^{-10} esu whereas the nonresonant $\chi^{(3)}$ values were approximately 10^{-12} esu.

Flom et al.[361] performed third-order NLO studies on lead-tetrakis(cumylphenoxy) phthalocyanine [PbPc(CP)$_4$], lead octa(α-butoxy)phthalocyanine [PbPc(α-BuO)$_8$] and lead-octa(α-phenoxy) phthalocyanine [PbPc(α-PO)$_8$] by performing nonlinear transmission (NLT), DFWM, and transient absorption experiments. The transient absorption spectra showed the spectral region of the reverse saturable absorption at about 175 nm for all of these PbPcs. They showed $\chi^{(3)}$ values of 10^{-12} esu. The real part of $\chi^{(3)}$ was 5.8×10^{-12}, 1.5×10^{-12}, and -1.5×10^{-12} esu for Pb(CP)$_4$, Pb(α-BuO)$_8$, and Pb(α-PO)$_8$, respectively. The nonlinear refractive indexes were 4.1×10^{-4}, 1.1×10^{-4}, and -9.1×10^{-5} cm^2/GW for Pb (CP$_4$), Pb(α-BuO)$_8$, and Pb(α-PO)$_8$, respectively. Flom et al.[362] conducted transient absorption and DFWM studies on thin films of Pb-phthalocyanine-urethane-butanediol copolymer. The DFWM experiments at 590 nm showed a $\chi^{(3)}_{xxxx}$ value of 1.8×10^{-11} esu and a $\chi^{(3)}_{xyyx}$ value of 2.3×10^{-12} esu. The excited state showed a rise time of <1 ps and a decay time of >10 ns. The nonlinear refractive index n_2 estimated from the measured $\chi^{(3)}$ value was 1.1×10^{-3} cm^2/GW. The copolymer also exhibited strong reverse saturable absorption in the visible region.

Wajiki et al.[363] reported the orientation dependence $\chi^{(3)}$ value of sublimed SnPc thin films as well as the effect of rubbing-processed substrate. The $\chi^{(3)}$ values of SnPc polycrystalline films with no-rubbing process were 30–50×10^{-12} esu. With rubbing-processed substrate, the $\chi^{(3)}$ values of SnPc films were found to be larger for perpendicular direction than those for the parallel direction. Furthermore, the $\chi^{(3)}$ values of SnPc films sublimed at 200 °C on rubbing-processed substrate were more than 4 times larger than those sublimed at room temperature. SnPc films sublimed at 200 °C on rubbing-processed substrate showed $\chi^{(3)}$ values of 8.4×10^{-11}, 1.8×10^{-10}, and 2.2×10^{-1} esu in perpendicular direction, whereas $\chi^{(3)}$ values were 3.0×10^{-11}, 4.3×10^{-11}, and 4.7×10^{-11} esu in parallel direction at 1.89, 2.10, and 2.19 μm, respectively.

The $\chi^{(3)}$ values of MPcs vary by several orders of magnitude depending on the chemical structures. Phthalocyanines exhibit two absorption bands; the Soret band in the near-UV region (300–400 nm) and the Q band in the visible region (600–800 nm) arising from the electronic transitions to the π–π^* states of E_u symmetry.[324,325] One of the reasons for the large difference noticed in $\chi^{(3)}$ values of MPcs is the dispersion of the $\chi^{(3)}$ values in resonant and nonresonant regions influenced by the optical wavelength at which measurements are performed. Furthermore, THG and DFWM techniques involve different optical processes; hence, the measurement technique is also important for evaluating the magnitude of optical nonlinearities. Furthermore, the nature of central metal atoms and substituents either attached axially or perpherally also play a very major role in determining the magnitude of the $\chi^{(3)}$ values of MPcs. The general trend shows that MPcs having axial ligands have significantly larger $\chi^{(3)}$ values than those MPcs having no axial ligand.

Table 44 Third-order NLO Susceptibility Data of MPcs

MPc	$\chi^{(3)}$ (10^{-12} esu)	Measurement Wavelength (μm)	Technique	Ref.
Cl-GaPc	25	1.064	THG	326
F-AlPc	50	1.064	THG	326
H_2PcCP_4(solution)	4.0	1.064	DFWM	328, 329
(thin film)	5000	1.064	DFWM	328, 329
$PbPcCP_4$	20	1.064	DFWM	328, 329
	5.8	0.5905	DFWM	361
$PbPc(\alpha\text{-BuO})_8$	1.5	0.5905	DFWM	361
$PbPc(\alpha\text{-PO})_8$	1.4	0.5905	DFWM	361
$PtPcCP_4$	200	1.064	DFWM	327, 328
$PdPcCP_4$	20	1.064	DFWM	327, 328
$ZnPcCP_4$	7	1.064	DFWM	328, 329
$CuPcCP_4$	40	1.064	DFWM	328, 329
$CoPcCP_4$	80	1.064	DFWM	328, 329
H_2Pc	3.0	1.907	THG	332
	4.0	1.064	DFWM	327
$H_2Pc(C_8H_{17}S)_4$	15	2.01	THG	335
$H_2Pc(O\text{-}C_6H_5)_{16}$	$\gamma = 3.42 \times 10^{-32}$	1.064	THG	353
	$\gamma = 3.60 \times 10^{-32}$	1.907	THG	353
$H_2Pc[S\text{-}CH(CH_3)_2]_{16}$	$\gamma = 3.81 \times 10^{-32}$	1.064	THG	353
	$\gamma = 0.06 \times 10^{-32}$	1.907	THG	353
TBH_2Pc	1.9	1.907	THG	331, 332
VOPc	8.6	1.50	THG	334
	4.1	1.543	THG	332
	30	1.90	THG	334
	38	1.907	THG	332
	40	2.10	THG	334
α-VOPc	9.0	1.543	THG	332
	81	1.907	THG	332
	93	1.907	THG	331
	40	2.10	THG	331
	93.3	2.10	THG	345
	130	1.90	THG	358
β-VOPc	38	1.907	THG	332
$VOPc(NH_2)_4$	6.3	2.10	THG	339
$VOPc(t\text{-Bu})_4$	7.0	1.543	THG	332
	2.6	1.907	THG	332
TBVOPc	6.0	1.907	THG	331, 332
$VOPc(C_6H_{13}S)_4$	3.3	1.50	THG	334
	9.8	1.90	THG	334
	14	2.10	THG	334
$VOPc(C_8H_{17}S)_4$	4.1	1.50	THG	334
	18	1.90	THG	334
	31	2.10	THG	334
CoPc	0.68	1.50	THG	334
	0.76	1.90	THG	334
	7.5	1.907	THG	332
	0.70	2.10	THG	334
NiPc	0.76	1.50	THG	334
	0.80	1.90	THG	334
	2.3	1.907	THG	332
	1.6	2.10	THG	334
$NiPc(NH_2)_4$	1.60	2.10	THG	339
$NiPc(t\text{-Bu})_4$	2.0	1.907	THG	331
$NiPcCP_4$	$\gamma = 4 \times 10^{-32}$	1.064	DFWM	329
$Ni(NP)_4$	$\gamma = 5 \times 10^{-32}$	1.064	DFWM	329
$Ni(OD)_4$	$\gamma = 1 \times 10^{-32}$	1.064	DFWM	329
PbPc	20	1.064	DFWM	329

Table 44 *Continued*

MPc	$\chi^{(3)}$ (10^{-12} esu)	Measurement		Ref.
		Wavelength (μm)	Technique	
PtPc	0.76	1.50	THG	334
	0.60	1.90	THG	334
	0.30	2.10	THG	334
	200	1.064	DFWM	327
SnPc	40	1.907	THG	332
	19	1.89	THG	363
	61	2.10	THG	363
	68	2.19	THG	363
Cl-AlPc	4.5	1.50	THG	334
	15	1.90	THG	334
	30	2.10	THG	334
Cl-InPc	13	1.50	THG	334
	130	1.90	THG	334
	94	2.10	THG	334
TiOPc	3.2	1.50	THG	334
	27	1.90	THG	334
	53	2.10	THG	334
	2.7	1.543	THG	332
	10	1.907	THG	332
α-TiOPc	159	2.43	THG	345
	170	1.90	THG	358
	14	1.543	THG	332
	46	1.907	THG	332
β-TiOPc	43	2.43	THG	345
Y-TiOPc	63	2.10	THG	345
α-MoOPc	210	1.90	THG	358
CuPc	1.3	1.50	THG	334
	1.5	1.90	THG	334
	1.1	2.10	THG	334
	3.4	1.097	THG	351
	4.06	1.064	THG	351
α-CuPc	0.1	1.90	THG	348
β-CuPc	1.6	1.90	THG	348
CuPc(NH$_2$)$_4$	2.0	2.10	THG	339
CuPc(C$_4$H$_9$S)$_4$	2.6	1.50	THG	334
	3.7	1.90	THG	334
	6.2	2.10	THG	334
CuPc(C$_6$H$_{13}$S)$_4$	2.5	1.50	THG	334
	20	1.90	THG	334
CuPc(C$_7$H$_{15}$S)$_4$	4.0	1.50	THG	334
	10	1.90	THG	334
	12	2.10	THG	334
CuPc(C$_8$H$_{17}$S)$_4$	3.4	1.50	THG	334
	23	1.90	THG	334
	50	2.10	THG	334
CuPc(C$_{10}$H$_{21}$S)$_4$	4.4	1.50	THG	334
	26	1.90	THG	334
	14	2.10	THG	334
CuPc(C$_{12}$H$_{25}$S)$_4$	2.0	1.50	THG	334
	8.7	1.90	THG	334
	13	2.10	THG	334
CuPc(C$_8$H$_{17}$S)$_8$	2.1	2.01	THG	335
CuPc(C$_{10}$H$_7$S)$_{15}$	$\gamma = 7.44 \times 10^{-32}$	1.064	THG	352
	$\gamma = 0.23 \times 10^{-32}$	1.907	THG	352
CuPc(CH$_3$-C$_6$H$_4$S)$_{15}$	$\gamma = 7.44 \times 10^{-32}$	1.064	THG	352
	$\gamma = 0.23 \times 10^{-32}$	1.907	THG	352

Table 44 *Continued*

MPc	$\chi^{(3)}$ $(10^{-12}$ esu$)$	Measurement Wavelength (μm)	Technique	Ref.
H_2PcR_8	$\gamma = 4.57 \times 10^{-32}$	1.340	THG	354
	$\gamma = 0.92 \times 10^{-32}$	1.064	EFISH	354
$CuPcR_8$	$\gamma = 2.45 \times 10^{-32}$	1.340	THG	354
	$\gamma = 3.94 \times 10^{-32}$	1.064	EFISH	354
$NiPcR_8$	$\gamma = 3.18 \times 10^{-32}$	1.340	THG	354
	$\gamma = 2.47 \times 10^{-32}$	1.064	EFISH	354
$CoPcR_8$	$\gamma = 4.48 \times 10^{-32}$	1.340	THG	354
	$\gamma = 4.39 \times 10^{-32}$	1.064	EFISH	354
$Fe(COOH)_4$	1.18	2.10	THG	339
MnPc	1.90	2.01	THG	349
$ScPc_2$	8.9	1.05	THG	349
$ScPc_2$	1700	1.064	DFWM	350
$ScPc_2$	$\gamma = 48 \times 10^{-32}$	1.064	DFWM	338
$LuPc_2$	$\gamma = 34 \times 10^{-32}$	1.064	DFWM	338
$YbPc_2$	$\gamma = 41 \times 10^{-32}$	1.064	DFWM	338
YPc_2	$\gamma = 26 \times 10^{-32}$	1.064	DFWM	338
$GdPc_2$	$\gamma = 22 \times 10^{-32}$	1.064	DFWM	338
$EuPc_2$	$\gamma - 22 \times 10^{-32}$	1.064	DFWM	338
$NdPc_2$	$\gamma = 15 \times 10^{-32}$	1.064	DFWM	338
$ScPc_2^-$	$\gamma = 20 \times 10^{-32}$	1.064	DFWM	338
$LuPc_2^-$	$\gamma = 22 \times 10^{-32}$	1.064	DFWM	338
YPc_2^-	$\gamma = 22 \times 10^{-32}$	1.064	DFWM	338
$GdPc_2^-$	$\gamma = 28 \times 10^{-32}$	1.064	DFWM	338
$EuPc_2^-$	$\gamma = 28 \times 10^{-32}$	1.064	DFWM	338
CuNc	3.0	1.06	THG	348
$CuNc(COO\text{-}C_5H_{11})_4$	2.0	2.10	THG	340
$VONc(COO\text{-}C_5H_{11})_4$	86	2.10	THG	340
$ZnNc(COO\text{-}C_5H_{11})_4$	1.88	2.10	THG	340
$PdNc(COO\text{-}C_5H_{11})_4$	1.28	2.10	THG	340
$NiNc(COO\text{-}C_5H_{11})_4$	1.59	2.10	THG	340
$InClNc[C(CH_3)_3]_4$	4.21	2.10	THG	340
$IrNc[C(CH_3)_3]_4$	1.05	2.10	THG	340
$RuNc[C(CH_3)_3]_4$	1.0	2.10	THG	340
$RhNc[C(CH_3)_3]_4$	1.16	2.10	THG	340
$SiNcX_2$	94	0.598	DFWM	337
	$\gamma = -3.14 \times 10^{-33}$	1.907	THG	336
	$\gamma = -5.0 \times 10^{-33}$	1.543	THG	336
TiOPc/ZnDBPc	300	1.90	THG	357
Pc-Polysiloxane	1.0	1.064	THG	356
CuPc/NTCA	2.3	1.90	THG	359
PbPc-copolymer	18	0.590	DFWM	362
$SiPc(EGY)_2$	2.2	1.064	DFWM	346
	288.0	0.532	DFWM	346
$SiPc(GLY)_2$	1.8	1.064	DFWM	346
$SiPc(TEG)_2$	0.3	1.064	DFWM	346
	14.0	0.532	DFWM	346
$SiPc(TEA)_2$	4.0	1.064	DFWM	346
	73.0	0.532	DFWM	346
$SiPc(PPz)_2$	158	1.064	DFWM	346
	419	0.532	DFWM	346
$SiPc(ISN)_2$	20.8	1.064	DFWM	346
	510	0.532	DFWM	346

The γ values are in esu unit.

CP = cumylphenoxy; NP = neopentoxy; OD = octadecyloxy; TB = *tert*-butyl; $R = OCH_2CON(C_8H_{17})_2$; EGY = $-OCH_2CH_2OH$; GLY = $-OCH_2CHOHCH_2OH$; TEG = $-OCH_2CH_2OCH_2CH_2OCH_2CH_2OH$; TEA = $-OCH_2CH_2N(CH_2\text{-}CH_2OH)_2$; PPz = $-OCH_2CH_2N(CH_2)_4NCH_2CH_2OH$; ISN = $-OCH_2CH_2N(CH_2CH_2OH)COC_5H_4N$.

2. Metalloporphyrins

Metalloporphyrins (M-PP) are another class of two-dimensional π-electron conjugated systems. Like metallophthalocyanines, they have a cyclic conjugated structure; a variety of central metal atoms can be incorporated into the ring structures and many modifications can be made at the peripheral sites of the ring. The structure–property relationships of M-PP in view of third-order optical nonlinearity are discussed here. In 1988, Meloney et al.[364] reported the third-order NLO properties of several tetraphenylporphyrin (TPP) compounds. The $\chi^{(3)}$ measurements on a metal-free (H$_2$TPP), zinc-containing (ZnTPP), and cobalt-containing (CoTPP) tetraphenylporphyrins were made in a toluene solution by using a DFWM technique at a wavelength of 532 nm. The effective third-order nonlinear susceptibilities of these materials were calculated considering the thermally induced refractive index changes; in these cases, the $\chi^{(3)}$ values were 2.07×10^{-11}, 2.0×10^{-11}, and 1.21×10^{-11} esu for H$_2$TPP, ZnTPP, and CoTPP, respectively. Sakaguchi et al.[365] measured the third-order optical nonlinearities of 5, 10, 15, 20-tetrakis(4-n-pentadecylphenyl) porphyrins containing central metal atoms of cobalt (Co), nickel (Ni), zinc (Zn), copper (Cu), and vanadium oxide (VO). The measurements were made in a benzene solution comprising concentrations of 0.02 to 0.05% by the DFWM technique, all the materials had $\chi^{(3)}$ values in the range of 10^{-11} esu, and these values were affected by the concentration. A C$_{15}$TPPCu with a 0.05% weight dissolved in benzene showed a $\chi^{(3)}$ value of approximately 1.8×10^{-11} esu. The C$_{15}$TPPNi showed the largest $\chi^{(3)}$ value, 10^{-10} esu at a concentration of 0.5 g/l. The effect of central metal on $\chi^{(3)}$ values was small and may be diluted because of bulky alkyl chains substituted on phenyl rings.

(1) H$_2$TPP: $R = C_6H_5$, M = H$_2$
(2) ZnTPP: $R = C_6H_5$, M = Zn,
(3) CoTPP: $R = C_6H_5$, M = Co
(4) C$_{15}$TPPH$_2$: $R = C_6H_4-CH_2-(CH_2)_{13}-CH_3$, M = H$_2$
(5) C$_{15}$TPPCo: $R = C_6H_4-CH_2-(CH_2)_{13}-CH_3$, M = Co
(6) C$_{15}$TPPNi: $R = C_6H_4-CH_2-(CH_2)_{13}-CH_3$, M = Ni
(7) C$_{15}$TPPCu: $R = C_6H_4-CH_2-(CH_2)_{13}-CH_3$, M = Cu
(8) C$_{15}$TPPZn: $R = C_6H_4-CH_2-(CH_2)_{13}-CH_3$, M = Zn
(9) C$_{15}$TPPVO: $R = C_6H_4-CH_2-(CH_2)_{13}-CH_3$, M = VO

Rao et al.[366] reported the $\chi^{(3)}$ measurements of several metal-free and metallotetrabenzoporphyrin derivatives in a tetrahydrofuran (THF) solution using a DFWM technique at a 532 nm wavelength. These benzoporphyrins were substituted at their *meso*-positions with a variety of aliphatic and aryl functional groups such m-fluorophenyl, p-methoxyphenyl, and p-methylphenyl. A zinc *meso*-tetra(p-dimethylaminophenyl) tetrabenzoporphyrin showed the largest $\chi^{(3)}$ value of 2.8×10^{-8} esu. Benzoporphyrins having electron-donating *meso*-phenyl groups showed large $\chi^{(3)}$ values. To point out the structural difference among these porphyrins, Compound 15, MgOMTBP is a special case that has eight methyl groups and eight hydrogens substituted on the phenyl rings. Guha et al.[367] analyzed imaginary and real parts of $\chi^{(3)}$ values of compounds 13 and 16 in THF by using picosecond and nanosecond laser pulses. The imaginary parts of the $\chi^{(3)}$ and γ values of these two compounds measured at 532 nm were 5 to 10 times larger than the real parts measured at 1.064 μm. The imaginary $\chi^{(3)}$ value of 9×10^{-13} esu and γ value of 8.0×10^{-31} esu for compound 13 and $\chi^{(3)}$ value of $\leq 1.8 \times 10^{-13}$ esu and γ value of $\leq 1.6 \times 10^{-31}$ esu for compound 16 were measured at a concentration of 0.46 g/l. At the same concentration, compounds 13 and 16 had a nonlinear refractive index n_2 of 7.8×10^{-10} and 9.1×10^{-10} cm^2/MW, respectively.

(10) ZnTMAPTBP: X = hydrogen; R = p-dimethylaminophenyl; M = Zn
(11) ZnTMTBP: X = hydrogen; R = methyl; M = Zn
(12) ZnTFPTBP: X = hydrogen; R = m-fluorophenyl; M = Zn
(13) ZnTMOPTBP: X = hydrogen; R = p-methoxyphenyl; M = Zn
(14) ZnTMPTBP: X = hydrogen; R = p-methylphenyl; M = Zn
(15) MgOMTBP: X = 8 hydrogens and 8 methyl groups; R = hydrogen; M = Mg
(16) ZnTPTBP: X = hydrogen; R = phenyl; M = Zn
(17) ZnHDFTBP: X = fluorine; R = hydrogen; M = Zn
(18) TBP: X = R = hydrogen; M = 2-hydrogens (tetrabenzoporphyrin)

To study the effect of increasing ring size and the number of π-electrons on third-order optical nonlinearity, Hosoda et al.[368] measured the $\chi^{(3)}$ values of spin-coated films of metal-free octaethyl porphyrin (H$_2$OEPP), MnCl-octaethyl[18]porphyrin (MnCl-OEPP), and N,N',N'',N'''-tetramethyl-octae-thyl[26]porphyrin-bistrifluoroacetate (TMOEPP) by THG at 1.907 μm. The refractive indexes $n\omega$ of 1.44, 1.48, and 1.46 and $n_{3\omega}$ of 1.78, 1.79, and 1.82 were measured by Kramers-Kronig analysis of reflectance data for H$_2$OEPP, MnCl-OEPP, and TMOEPP, respectively. The $\chi^{(3)}$ value of TMOEPP was about 5 times larger than H$_2$OEPP, which indicated the effect of the increased number of π-electrons in the cyclic conjugated chain on the $\chi^{(3)}$ value. The MnCl-OEPP also had slightly larger $\chi^{(3)}$ value than H$_2$OEPP.

(19) H$_2$OEPP: M = H$_2$
(20) MnCl-OEPP: M = MnCl

(21) TMOEPP

A magnesium octaphenyl tetrazaporphyrin (Mg-OPTAP) with a 5% weight incorporated in PMMA showed $\chi^{(3)}_{xxxx}$ and $\chi^{(3)}_{xyyx}$ values of about 1.17×10^{-11} and 3.03×10^{-12} esu, respectively, from DFWM at 598 nm.[337] The $\chi^{(3)}_{xxxx}/\chi^{(3)}_{xyyx} = 3$ indicated that the optical nonlinearity was predominantly electronic. The Mg-OPTAP showed a much faster response time ($\sim$44 ps). Anderson et al.[369] synthesized a soluble conjugated porphyrin polymer (23) by Glaser-Hay coupling of the $meso$-diethynyl zinc porphyrin. The real and imaginary components of the $\chi^{(3)}(-\omega;0,0,\omega)$ of the porphyrin polymer were calculated from the electroabsorption spectra. The $\chi^{(3)}(-\omega;0,0,\omega)$ of 7.3×10^{-8} esu was obtained at the peak resonance.

A comparison of the third-order optical nonlinearity on various metalloporphyrins can be made and the results are summarized in Table 45.

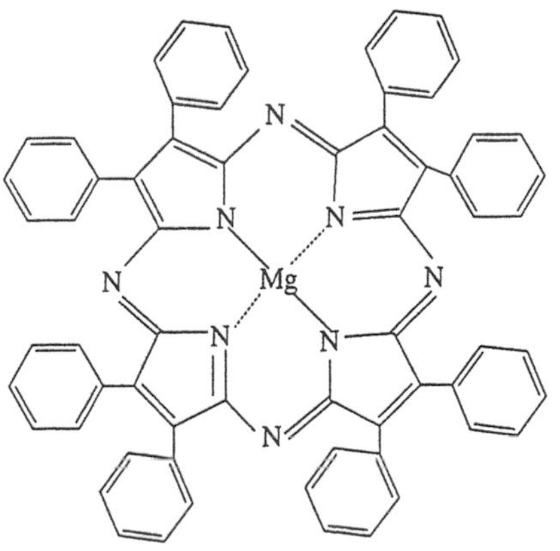

(22) 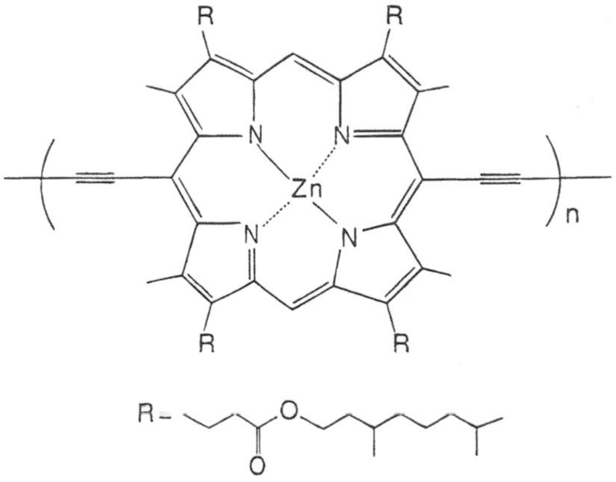   (23)

Bao and Yu[370] synthesized metal-free and Zn-containing new monomeric and polymeric porphyrins. These polymers not only showed interesting photoconductive and photovoltaic properties but also large optical nonlinearities. The $\chi^{(3)}$ values measured by DFWM at 0.532 μm were approximately 10^{-10} esu for thin films. Polymer solutions in chloroform (Compound 26) and in THF (Compound 27) showed $\chi^{(3)}$ values of 1×10^{-12} and 1.9×10^{-12} esu, respectively, which were more than two orders of magnitude smaller than those obtained for thin films. The $\chi^{(3)}$ value of Zn-porphyrin (25) was twice that of metal-free porphyrin (24) in THF solution. These $\chi^{(3)}$ values were enhanced because of the resonant contributions.

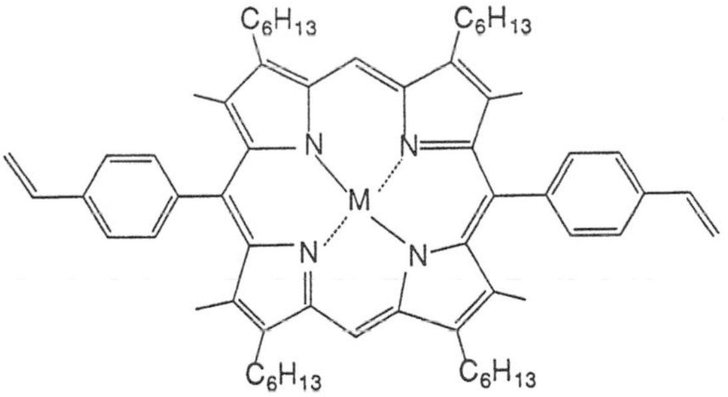

M=H$_2$ (24)
M=Zn (25)

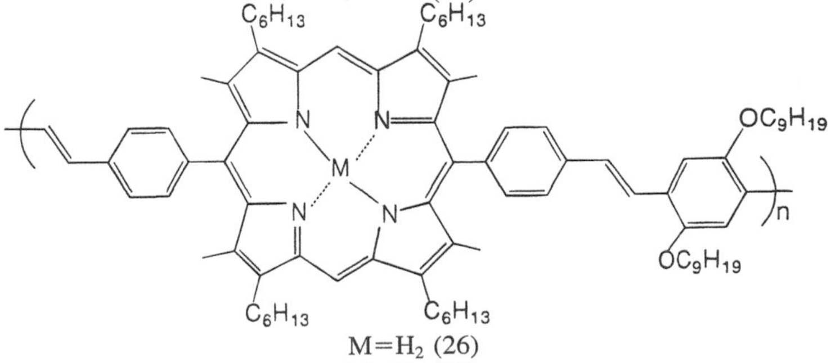

M=H$_2$ (26)
M=Zn (27)

Honeychuch[371] reported $\chi^{(3)}$ values of polymers formed from quinoline in the presence of trimethyl boron zinc chloride and protonic acids and films were formed by this method from 96% and 98% quinoline containing with poly(1,4-cyclohexylene carbonate, *meso*-porphyrin IX dimethyl ester, and Pd(II) *meso*-porphyrin IX dimethyl ester via heating and subsequent evaporation. The $\chi^{(3)}$ value of polycarbonate-prophyrin-quinoline polymer was 6×10^{-12} esu. The $\chi^{(3)}$ values of porphyrin and quinoline were 3×10^{-14} and 2×10^{-13} esu, respectively, showing that optical nonlinearity from 96% quinoline was significantly larger. The γ value of metal-free porphyrin was measured as $\sim 3 \times 10^{-34}$ esu.

Table 45 Third-order NLO Susceptibility Data of M-PPs Measured by Different Techniques

No.	M-PP	γ (10^{-34} esu)	$\chi^{(3)}$ (10^{-12} esu)	Wavelength (μm)	Measurement Technique	Ref.
(1)	H_2TPP		28.6	0.532	DFWM	364
(2)	ZnTPP		14.3	0.532	DFWM	364
(3)	CoTPP		7.16	0.532	DFWM	364
(4)	C_{15}TPPH$_2$		40.0	0.532	DFWM	365
(5)	C_{15}TPPCo		56.0	0.532	DFWM	365
(6)	C_{15}TPPNi		69.0	0.532	DFWM	365
(7)	C_{15}TPPCu		50.0	0.532	DFWM	365
(8)	C_{15}TPPZn		15.0	0.532	DFWM	365
(9)	C_{15}TPPVO		32.0	0.532	DFWM	365
(10)	ZnTMAPTBP	100,000	28,000	0.532	DFWM	366
(11)	ZnTMTBP	33,000	15,000	0.532	DFWM	366
(12)	ZnTFPTBP	38,000	13,000	0.532	DFWM	366
(13)	ZnTMOPTBP	48,000	14,000	0.532	DFWM	366
		10	0.078	1.064	OKE	367
(14)	ZnTMPTBP	40,000	12,000	0.532	DFWM	366
(15)	MgOMTBP	16,000	8,000	0.532	DFWM	366
(16)	ZnTPTBP	9,000	3,000	0.532	DFWM	366
		18	0.091	1.064	OKE	367
(17)	ZnHDFTBP	7,000	2,000	0.532	DFWM	366
(18)	TBP	5,000	3,000	0.532	DFWM	366
(19)	H_2OEPP	4.9	1.9	1.907	THG	368
(20)	MnCl-OEPP	6.0	2.6	1.907	THG	368
(21)	TMOEPP	24	10	1.907	THG	368
(22)	Mg-OPTAP		11.7	0.598	DFWM	337
(24)	H_2PP-Monomer		0.6	0.532	DFWM	370
(25)	ZnPP-Monomer		1.2	0.532	DFWM	370
(26)	H_2PP-Polymer		510	0.532	DFWM	370
(27)	ZnPP-Polymer		128	0.532	DFWM	370

3. Polysilanes

Because silicon is a metalloidal element, polysilanes are sometimes not treated as true organometallic compounds. Polysilanes display very interesting properties because of the Si-Si σ-electron delocalization. A typical σ-bonded polysilane backbone consists of two organic side groups attached to each silicon atom in the polymer chain. The uniqueness of polysilanes is their σ-bonded polymer backbone chain which contributes to the electronic delocalization that causes strong transitions for excitations. The optical absorption spectroscopic studies both in the solid state and in the solution demonstrated that the polysilanes strongly absorb in the UV region (300–400 nm). The intense optical absorption characteristics are either a $\sigma\sigma^*$ transition or a $\sigma3d\pi_{Si\text{-}Si}$ transition. Polysilanes are considered an important class of NLO materials because they offer several additional advantages from a practical viewpoint over conjugated polymers. Polysilanes are transparent, processible, and thermally and environmentally stable materials.[372,373]

Kajzar et al.[374] first reported third-order optical nonlinearities of thin films of poly(methylphenylsilane) (PMPS) using the THG technique. PMPS had an average $\chi^{(3)}$ value of 1.5×10^{-12} esu at 1.064 μm. Yang et al.[375] also measured the $\chi^{(3)}$ value of PMPS using the optical Kerr gate and DFMW techniques which show $\chi^{(3)}$ values of 2.0×10^{-12} and 1.6×10^{-12} esu, respectively. PMPS exhibited a nonlinear response faster than 3 ps which arose from electronic contribution. Baumert et al.[376] also performed film thickness as well as temperature and wavelength dependence measurements on PMPS using the THG technique. A 120 nm thick film of PMPS showed a $\chi^{(3)}$ value of 7.2×10^{-12} esu at 1.064 μm. Although there was no significant change as a function of temperature, the $\chi^{(3)}$ value showed noticeable change with film thickness. For example, when the film thickness of PMPS reduces from 1200 nm to 120 nm, the $\chi^{(3)}$ value increases from 1.9×10^{-12} to 4.2×10^{-12} esu, which is more than double.

Baumert et al.[376] also measured third-order nonlinear susceptibilities of poly(di-n-hexylsilane) (PDHS) and poly(di-n-hexylgermane) (PDHG). PDHS and PDHG showed $\chi^{(3)}$ values of 11.3×10^{-12} esu and 3.3×10^{-12} esu at 1.064 μm, respectively. Temperature, wavelength, and thickness dependence of the $\chi^{(3)}$ values had interesting features. The thinner films had the larger optical nonlinearities. McGraw et al.[377] reported a $\chi^{(3)}$ value of 2.9×10^{-12} esu for poly(n-octylmethylsilane) (POMS) at 0.532 μm. Shukla et al.[378] investigated the third-order nonlinear susceptibilities for a variety of polysilanes with different side groups and for composites of polysilanes and metals. The $\chi^{(3)}$ values changed with the change of the side groups, film thickness, and backbone conformation. The $\chi^{(3)}$ values of 120 nm thick films of poly(hexylphenylsilane) (PHPS), poly(hexylpentylsilane) (PHPnS), poly(ethylphenylsilane) (PEPS); poly(ter-butylphenylsilane) (PtBPS), and poly(diisohexylsilane) (PDIS) were approximately 10^{-12} esu. The composites of poly(di-n-hexylsilane) and silver particles showed interesting trends for $\chi^{(3)}$ values. Silver particles of 10-μm size showed no noticeable increase in $\chi^{(3)}$ value, whereas 1-μm-size particles with a 30% volume fraction showed significant increases in the $\chi^{(3)}$ value. The volume fractions of 30% and 50% had values of 2.25×10^{-11} and 2.35×10^{-11} esu. The value for pristine silver was about 2.5×10^{-9} esu. The remarkable increase in the increasing volume fraction of silver particles, which was smaller than that of the wavelength of the fundamental beam, resulted in enhancement of the Raman scattering cross-section which lead to large optical nonlinearity. Theoretically an increase as large as seven orders of magnitude in value has been predicted for such composites.

Wong et al.[379] measured the femtosecond dynamics of the nonlinear optical response in poly(diethynyl-silane) (PEDS) using the DFWM technique. PEDS is a solution-processible material that showed good environmental stability. The optical absorption spectrum of PEDS recorded in solid state showed an absorption maximum at 2.2 eV and an energy gap at 2.0 eV. The $\chi^{(3)}$ value at 602 nm (2 eV) was 3×10^{-9} esu, the largest ever reported for a polysilane. Using $\chi^{(3)} \sim \alpha$ relation, a $\chi^{(3)}$ value as high as 10^{-8} esu had been anticipated at 564 nm (2.2 eV). Likely, PEDS showed an ultrafast response time of 135 fs followed by a slower decay component of 750 fs. Hasegawa et al.[380] measured $\chi^{(3)}$ values of PDHS and poly(dibutylsilane) (PDBS) films by THG in the wavelength range from 1.90 to 2.15 eV. PDHS showed a clear $\chi^{(3)}$ maximum at 2.08 eV with a maximum value of 2.3×10^{-12} esu. On the other hand, PDBS showed no enhancement effect at the same fundamental photon energy region. The dipole-forbidden ^1A_g exciton detected by two-photon absorption and electroabsorption measurements in PDHS showed the involvement of the two-photon-resonant THG enhancement process.

Miller et al.[381] measured NLO susceptibility of a variety of polysilanes, and $\chi^{(3)}$ values ranged from 10^{-11} to 10^{-12} esu. The measured $\chi^{(3)}$ values depended on polymer orientation and to some extent on film thickness. The nonresonant $\chi^{(3)}$ values were insensitive to backbone conformation. Embs et al.[382] reported the preparation of LB films of $\sim$20 different polysilanes. Polysilane backbones having bis(p-butoxyphenyl) and bis(m-butoxyphenyl) substituents were subjected to third-order NLO effects. The $\chi^{(3)}$ value was higher with the polarization of the fundamental light parallel to the dipping direction than with perpendicular polarization. The angular dependence of the $\chi^{(3)}$ value of polysilane with bis(m-butoxyphenyl) group showed that the $\chi^{(3)}$ value was influenced by the oscillator strength of polysilane backbone rather than by the phenyl ring absorption. Annealing of the samples at elevated temperatures exhibited a strong increase in anisotropy and the largest absolute $\chi^{(3)}$ value of 4×10^{-12} esu was determined for an annealed sample. Polysiloxane having bis(m-butoxyphenyl) substituents showed the largest third-order susceptibility associated with the effective conjugated length along the polysilane backbone. Ho et al.[383] reported $\chi^{(3)}$ values of poly(methoxyphenylmethylsilane) (PMeOPMS) and poly(di-methylaminophenylmethylsilane) (PDMAPMS). The THG-measured $\chi^{(3)}$ values of PMeOPMS and PDMAPMS both having electron-donating groups were found to be three and two times larger than that of PPMS, respectively. This indicated that donor substituents enhance the third-order optical nonlinearity. The absorption maximum increased with the increasing donor strength of the substituent on the phenyl ring: PPMS (342 nm) < PMeOPMS (351 nm) < PDMAPMS (361 nm). The $\chi^{(3)}$ values measured at 1.064 μm for these polymers had a three-photon resonance effect. The $\chi^{(3)}$ value of PDMAPMS was smaller than the $\chi^{(3)}$ value of PMeOPMS, although the dimethylamino group was more electron donating than the methoxy group, which may be caused by the low molecular weight of PDMAPMS. Table 46 summarizes the chemical structures and third-order optical nonlinearities of various polysilanes where a useful comparison can be made.

Table 46 Third-order NLO Susceptibilities of Polysilanes and Polygermanes

Polymer	$\chi^{(3)}$ (10^{-12} esu)	Wavelength (μm)	Measurement Technique	Ref.
PPMS	1.5	1.064	THG	374
	7.2	1.064	THG	376
	6.8	1.064	THG	378
	4.2	1.907	THG	381
	2.1	1.064	THG	383
PDHS	11.3	1.064	THG	376
	11.0	1.064	THG	381
	1.3	1.907	THG	381
PDHG	6.5	1.076	THG	381
	1.4	1.907	THG	381
	2.1	1.064	THG	383b
POMS	2.9	0.532	THG	377
PHPS	6.2	1.064	THG	378
PHPnS	2.3	1.064	THG	378
PEPS	5.3	1.064	THG	378
PtBuPS	4.9	1.064	THG	378
PDIHS	1.8	1.064	THG	378
PDES	3000	0.620	DFWM	379
PDBS	0.5	0.595	THG	380
P4MPS	1.8	1.064	THG	381
PT4PhMS	4.9	1.064	THG	381
PT4PhMS	4.9	1.064	THG	381
PBN4PhS	5.2	1.064	THG	381
	0.9	1.907	THG	381
PDN4S	2.2	1.064	THG	381
	0.48	1.907	THG	381
PDN14S	4.5	1.064	THG	381
	0.23	1.907	THG	381
PB-p-BPS	0.9	1.064	THG	382
PB-m-BPS	4.0	1.064	THG	382
PMeOPMS	6.0	1.064	THG	383a
PDMAPMS	3.8	1.064	THG	383a
PDBG	5.2	1.064	THG	383b
PPMG	1.6	1.064	THG	383b
CP(DHG-DHS)	5.2	1.064	THG	383b

Abbreviations used in the Table 46 and the text are:

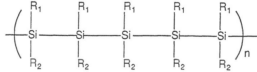

PPMS = poly(phenylmethylsilane): $R_1 = CH_3$, $R_2 = C_6H_5$
PMeOPMS = poly(p-methoxyphenylmethylsilane): $R_1 = CH_3$, $R_2 = C_6H_5-OCH_3$
PDMAPMS = poly(p-dimethylaminophenylmethylsilane): $R_1 = CH_3$, $R_2 = C_6H_5$; $-N(CH_3)_2$
PDHS = poly(di-n-hexylsilane): $R_1 = R_2 = C_6H_{13}$
PDHG = poly(di-n-hexylgermane): $R_1 = R_2 = C_6H_{13}$
(PDHG has Ge instead of Si in the chain)
POMS = poly(n-octylmethylsilane): $R_1 = CH_3$, $R_2 = C_8H_{17}$
PHPS = poly(n-hexylphenylsilane): $R_1 = C_6H_{13}$, $R_2 = C_6H_5$
PHPnS = poly(n-hexyl-n-pentylsilane): $R_1 = C_6H_{13}$, $R_2 = C_5H_{11}$
PEPS = poly(ethylpentylsilane): $R_1 = C_2H_5$, $R_2 = C_5H_{11}$
P4MPS = poly(4-methylpentylsilane): $R_1 = CH_3$, $R_2 = C_5H_{11}$
PBN4PhS = poly(bis-p-butylphenylsilane): $R_1 = R_2 = C_6H_4$-p-C_4H_9
PDN14S = poly(di-n-tetradecylsilane): $R_1 = R_2 = C_{14}H_{29}$
PEPhS = poly(ethylphenylsilane): $R_1 = C_2H_5$, $R_2 = C_6H_5$
PDBS = poly(di-n-butylsilane): $R_1 = R_2 = C_4H_9$
PB-p-BPS = poly-bis(p-butoxyphenyl)silane: $R_1 = R_2 = C_6H_4$-p-OC_4H_9
PB-m-BPS = poly-bis(m-butoxyphenyl)silane: $R_1 = R_2 = C_6H_4$-m-OC_4H_9
PDIHS = poly(diisohexylsilane),: $R_1 = R_2 = C_6H_{13}$

Table 46 *Continued*

PT4PhMS = poly(*tert*-butylphenylmethyl)silane: R_1 = $(CH_3)_3$; C_6H_4; R_2 = CH_3
PtBuPS = poly(*tert*-butylphenyl)silane): R_1 = $(CH_3)_3$; R_2 = C_6H_5
PDES = poly(diethynyl)silane: R_1 = $(C_4H_9)_2$, R_2 = C_4H_2
PDBG = poly(di-n-butylgermane); R_1 = R_2 = C_4H_9
PPMG = poly(phenylmethylgermane); R_1 = CH_3, R_2 = C_6H_5
CP(DHG−DHS) = Copolymer of poly(di-n-hexylgermane)-poly(di-n-hexylsilane) R_1 = R_2 = C_6H_{13}

Callender et al.[384] reported THG-measured $\chi^{(3)}$ values of a series of polysilane polymers and copolymers and polycarbosilanes with different alkyl and aryl side groups. The structures, film thicknesses, refractive indexes, and $\chi^{(3)}$ values of polysilanes and polycarbosilanes are given in Table 47. All alkyl- and aryl-substituted polysilanes showed $\chi^{(3)}$ values in the range of 3×10^{-11} and 4.5×10^{-12} esu. Poly(methylphenylsilane) showed the largest $\chi^{(3)}$ value of 4.4×10^{-12} esu, almost the same as reported by other groups. Polysilanes with pendant phenyl groups showed larger optical nonlinearities than unsymmetrical dialkyl polysilanes because the aryl group tended to increase the delocalization of the σ and σ^* states which are associated with the silicon backbone by mixing with π and π^* states of phenyl ring. The low $\chi^{(3)}$ values of polycarbosilanes resulted from the small effective conjugation lengths with short Si-Si sequences interrupted by nonconjugated carbon links.

4. Metal Polyynes

Third-order NLO properties of transition metal polyynes have been studied extensively by the research team at Martin Marietta Laboratories,[385–388] as early as 1987. To establish a relationship between polymer structure and optical nonlinearities, measurement techniques such as THG, four-wave mixing, optical Kerr gate, and self-focusing have been used. Very wide variations in the polymeric structures were made mainly on the benzene ring substituents of transition metal platinum- and palladium-containing polyynes and related oligomers, as well. Both the real and imaginary parts of the second hyperpolarizabilities for a series of platinum and palladium polyynes and related oligomers were measured and the variations of γ values with the extent of monomer repeat units were reported. The γ values of metallopolyynes are listed in Table 48. The γ' and γ'' refer to the real part of hyperpolarizability and the imaginary part of hyperpolarizability, respectively. The presence of large two-photon transitions in metal polyynes indicates that the imaginary part of the second hyperpolarizability mainly contributes to the third-order NLO susceptibility.

Table 47 The Structures, Film Thicknesses, Refractive Indexes, and $\chi^{(3)}$ Values of Polysilanes and Polycarbosilanes

Polymer	λ_{max} (nm)	Thickness (nm)	Refractive Indexes		$\chi^{(3)}$ (10^{-12} esu)
			1.064 μm	0.355 μm	
1	332	110	1.71	1.84	4.4
2	314	90	1.56	1.62	1.4
		596	1.56	1.63	0.75
3	322	227	1.74	1.84	1.2
4	322	87	1.75	1.88	3.4
		240	1.75	1.89	3.6
5	332	90	1.66	1.75	3.1
		462	1.66	1.77	1.6
6	295	227	1.61	1.63	0.34
7	291	198	1.67	1.69	0.37

From Callender et al., *J. Opt. Soc. Am.*, B9, 518, 1992. With permission.

The chemical units of polysilanes are: 1 = $(PhSiMe)_n$; 2 = $(c\text{-}HexylSiMe)_n$; 3 = $(n\text{-}propyl\ SiMe)_n$; 4 = $(PhSiMe)$ $(PhSiMe)_n$, random copolymer; 5 = $(PhSiMe)_m\ (n\text{-}hexylSiMe)_n$, block copolymer; 6 = $[(MeSiMe)_4(CH_2\text{-}Ph\text{-}CH_2)]_n$, random polycarbosilane; = $[(n\text{-}hexylSiMe)_8(CH_2\text{-}Ph\text{-}CH_2)_2]_n$, random polycarbosilane.

Table 48 Second Hyperpolarizabilities of Platinum and Palladium Polyynes

No.	Chemical Structure	No. of Repeat Units	γ' $(10^{-36}$ esu)	γ'' $(10^{-36}$ esu)
1a	PBu_3–Pt(PBu_3)(Cl)–C≡C–C₆H₄–C≡C–Pt(PBu_3)(Cl)–PBu_3 (cis)		11	224
1b	Cl–Pt(PBu_3)₂–C≡C–C₆H₄–C≡C–Pt(PBu_3)₂–Cl (trans)		19	827
2	Cl–Pt(PBu_3)₂–C≡C–C₆H₄–C≡C–Pt(PBu_3)₂–C≡C–C₆H₄–C≡C–Pt(PBu_3)₂–Cl		45	1196
3	Cl–Pt(PBu_3)₂–C≡C–C₆H₄–C≡C–C≡C–C₆H₄–C≡C–Pt(PBu_3)₂–Cl		88	2167
4	HC≡C–C₆H₄–C≡C–Pt(PBu_3)₂–C≡C–C₆H₄–C≡CH		53	759
5	HC≡C–C₆H₄–C≡C–Pt(PBu_3)₂–C≡C–C₆H₄–C≡C–Pt(PBu_3)₂–C≡C–C₆H₄–C≡CH		66	1328
6	SCN–Pt(PBu_3)₂–C≡C–C₆H₄–C≡C–Pt(PBu_3)₂–NCS		30	1134
7	–[Pt(PBu_3)₂–C≡C–C₆H₄–C≡C]ₙ–	112	37	1906
8	–[Pt(PBu_3)₂–C≡C–C₆H₂(CH_3)₂–C≡C]ₙ–	26	29	1200
9	–[Pt(PBu_3)₂–C≡C–C₆H₂(C_2H_5)₂–C≡C]ₙ–	15	43	956
10	–[Pt(PBu_3)₂–C≡C–C₆H₃(F)–C≡C]ₙ–	18	56	1260

Table 48 *Continued*

No.	Chemical Structure	No. of Repeat Units	γ' $(10^{-36}$ esu)	γ'' $(10^{-36}$ esu)
11		111 / 105 / 65	48 / 65 / 43	1724 / 1330 / 1586
12		oligomer	28	1324
13		76	18	1342
14		44	34	2148
15		62	19	2474
16		47.35	33	2263
17		4	19	1169
18		12	15	1753
19		67	22	2432
20		223 / 97	90 / 121	4558 / 4025

Table 48 *Continued*

No.	Chemical Structure	No. of Repeat Units	γ' (10^{-36} esu)	γ'' (10^{-36} esu)
21	[structure: $-[Pt(PBu_3)_2-C\equiv C-C_6H_2(CH_3)_2-C\equiv C-C\equiv C-C_6H_2(CH_3)_2-C\equiv C-]_n$]	52.38	116	2432
22	[structure: $-[Pt(PBu_3)_2-C\equiv C-C_6H_2(C_2H_5)_2-C\equiv C-C\equiv C-C_6H_2(C_2H_5)_2-C\equiv C-]_n$]	146	79	4933
23	[structure: $-[Pd(PBu_3)_2-C\equiv C-C_6H_4-C\equiv C-C\equiv C-C_6H_4-C\equiv C-]_n$]	oligomer (5,2)	66	2094
24	[structure: $-[Pd(PBu_3)_2-C\equiv C-C_6H_2(C_2H_5)_2-C\equiv C-C\equiv C-C_6H_2(C_2H_5)_2-C\equiv C-]_n$]	16	106	3490
25	[structure: $-[Pt(PBu_3)_2-C\equiv C-C_6H_4-C\equiv C-Pt(PBu_3)_2-C\equiv C-C_6H_4-C_6H_4-C\equiv C-]_n$]	66		4466

From Porter et al., *Polymer*, 32, 1756, 1991; Guha et al., *Opt. Lett.*, 14, 952, 1989; Frazier et al., *Polymer*, 28, 553, 1987; Frazier et al., *Mater. Res. Soc. Symp. Proc.*, 109, 323, 1988. With permission.

The following points were abstracted from the values of hyperpolarizabilities of a series of wide varieties of platinum and palladium compounds:

1. The *trans* arrangement of the phosphine ligands on the platinum metal center gives rise to large γ values.
2. The increase of conjugation length between the two metal centers results in a 4-fold increase in γ' and nearly a 3-fold increase in γ'' values.
3. The substitution of naphthalene for benzene shows a decrease in γ' value, whereas pyridine substitution showed no change in γ' values.
4. The γ' values of fluoro methoxy and ethyl-substituted polymers are slightly larger than the γ' values of unsubstituted polymer.
5. The γ' values of palladium polymers are lower than their platinum analogs.
6. The γ' values of palladium polymers are not significantly affected by the ring substitution changes.
7. The platinum and palladium polymers with two diethynylbenzenes in the repeat unit have larger hyperpolarizability values than analogous polymers.
8. The substitution of biphenyl for one of the diethynylbenzenes also shows larger γ'' values than the single diethynylbenzene repeat unit polymer. Higher values are expected with the further increase in the number of diethynylarenes in each repeat unit.

Blau et al.[384] reported third-order NLO properties of Group 10 metal alkynyls and their polymers. The hyperpolarizabilities of these materials were measured in solutions by the method of self-diffraction from laser-induced gratings at 1.064 μm. Table 49 lists the hyperpolarizabilities of metal—containing polyynes. These metal-containing polymers behave differently than polydiacetylenes. The real component of γ is negative although it was positive for polydiacetylene solutions. The optical nonlinearities of metal-containing monomers were found to be significantly larger than than those of short-chain enyne oligomers, which indicates the contribution of metal atoms to optical nonlinearity. Ni-containing polymer has larger hyperpolarizability than the Pt polyyne. Pt polymer has approximately 150 units in the backbone. This study demonstrated the strong role of the d-orbital electrons in the π-conjugated electron system of the backbone.

Page et al.[390] reported second hyperpolarizabilities of conjugated organometallic systems (structure shown below) by DFWM at 1.064 μm. Compounds 1, 2, and 3 showed that the magnitude of hyperpolarizability increases as the size of solubilizing groups increases, presumably because of the reduction of intermolecular interaction. A comparison between compounds 4 and 5 indicates that the presence of the bimetallic unit in the π-conjugated backbone leads to significantly larger γ values than that of the monometallic compounds. Therefore the large γ values of Compound 5 originate from electron delocalization in the excited state enhanced by the bimetalic palladium moiety.

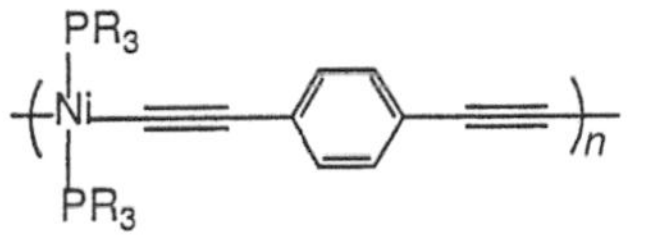

$R = n$-butyl (1) $R = n$-octyl (2) 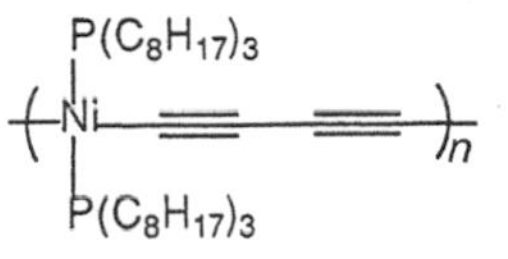 (3)

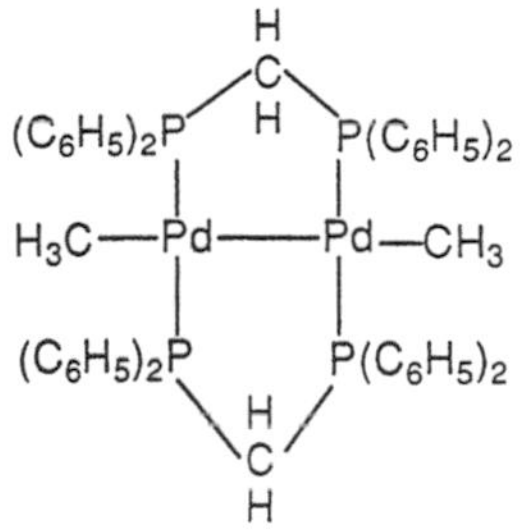

(4) 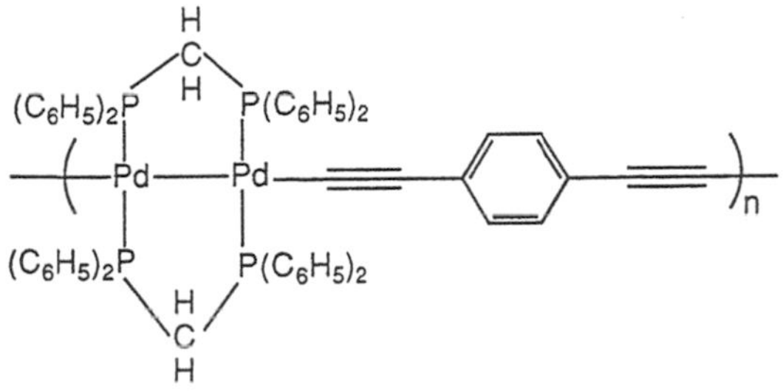 (5)

Agh-Atabay et al.[391] synthesized metal-containing polydiacetylenes reacting with $Co_2(CO)_8$ to introduce proportions of $Co_2(CO)_6$ groups attached as Co_2C_2 tetrahedrane units at the alkyne bonds of the conjugated eneyne backbone. The imaginary $\chi^{(3)}$ values obtained from picosecond transmission measurements at 532 nm in toluene solutions of pure polydiacetylene were $4.20 \times 10^{-22}/m^2/V^2$ and of derived metal-complexed polymer were 8.47×10^{-22} m^2/V^2 for 45% loading $Co_2(CO)_6$ groups. The largest two-photon absorption coefficients β_2 were 1.80×10^{-1} cm/GW for pure PDA

Table 49 Hyperpolarizabilities of Metal-containing Polyynes

Material	λ_{max} (nm)	γ Re (10^{-44} m^5V^{-2})	$\|\gamma$ Im$\|$ (10^{-43} m^5V^{-2})	$\|\gamma\|$ (10^{-43} m^5V^{-2})
$trans$-[Ni(C≡CPh)$_2$(PEt$_3$)$_2$]	370	−27.5	1.46	3.11
$trans$-[Pd(C≡CPh)$_2$Pd(PEt$_3$)$_2$]	370	−21.0	0.339	2.13
$trans$-[Pt(C≡CPh)$_2$(PEt$_3$)$_2$]	332	−11.2	0.215	1.14
$trans$-[Ni(C≡C−C≡CH)$_2$(PEt$_3$)$_2$]	336	−7.87	1.72	1.89
$trans$-[Pd(C≡C−C≡CH)$_2$(PEt$_3$)$_2$]	290	−3.85	0.09	0.396
$trans$-[Pt(C≡C−C≡CH)$_2$(PEt$_3$)$_2$]	318	−1.93	0.07	0.208
[N((PBu$_3$)$_2$−C≡C−C≡C]$_n$	412	−263	24.1	35.7
[P((PBu$_3$)$_2$−C≡C−C≡C]$_n$	364	−148	17.4	22.8
(1)		−0.1	0.02	0.02
(2)		−0.4	0.1	0.1
(3)		−0.4	0.03	0.05
(4)		−0.01	0.002	0.002
(5)		−0.2	0.02	0.02

From Blau et al., *J. Mater. Chem.*, 1, 245, 1991; Page et al., *Synth. Met.*, 63, 179, 1994. With permission.

and for the derived metal-complexed polymer were 1.84×10^{-1} cm/GW for 45% loading $Co_2(CO)_6$ groups.

5. Metallocenes

The third-order optical nonlinearity of a wide variety of metallocenes has been investigated at British Telecom Research Laboratories. Winter et al.[392,393] measured the nonlinear refractive index of four molten metallocenes, a liquid derivative of ferrocene, and a solution of ferrocene in ethanol by the optical power limiter technique (Table 50). The γ value of ferrocene is about 3 times larger than that of nitrobenzene for 10-ns pulses at 1.064 μm. The origin of the nonlinear refractive index may be electronic rather than rotational. The nonlinearity is also affected by the different metals and ring substitutents. Ruthenocene shows the largest optical nonlinearity which is more than twice the value of ferrocene and hafnocene. Whereas zirconocene is the lowest.

Kamata et al.[394] studied the effect of the metal–ligand charge-transfer transition on the $\chi^{(3)}$ values of metallocenes. The $\chi^{(3)}$ values of ferrocene (M = Fe), ruthenocene (M = Ru), and osmocene (M = Os) (Compound 1), 1-(4-nitrophenylhydrazono)methylferrocene (Compound 2), 1-(4-nitrophenylhydrazono)benzene (Compound 3), bis[1-(4-nitrophenylhydrazono)]ferrocene (Compound 4), and tricarbonylmangenese (Compound 5) measured by THG at 1.319 μm were 2.2×10^{-12}, 0.9×10^{-12}, 2.4×10^{-12}, 3.5×10^{-12}, 0.9×10^{-12}, 3.6×10^{-12}, and 1.4×10^{-12} esu, respectively. These results show that the $\chi^{(3)}$ values of these compounds mainly originate from the conjugated π-electron system of ligands and are also influenced by metal–ligand interactions.

Table 50 The Nonlinear Refractive Index n_2, Hyperpolarizability, and Critical Power of Metallocenes

M = Fe, Ru, Hf, and Zr

Metallocene	Concentration (molecules/cm³)	Critical Power (kW)	n_2 (10^{-11} esu)	γ (10^{-32} esu)
Ferrocene (solution)	5×10^{19}	1500	0.026	1.2
Ferrocene (melt)	3×10^{21}	28	1.48	3.9
BTMSF[a]	2×10^{21}	26	1.51	6.4
Ruthenocene		12	3.3	
Hafnocene		30	1.3	
Zirconocene		38	1.0	

From Winter et al., *Opt. Commun.*, 69, 45, 1988; Winter et al., *Nonlinear Optical Effects in Organic Polymers*, Messier, J., et al. Eds., Kluwer Academic Press, Dordrecht, 1989. p. 247. With permission.

[a] BTMSF = bis(trimethylsilyl)ferrocene.

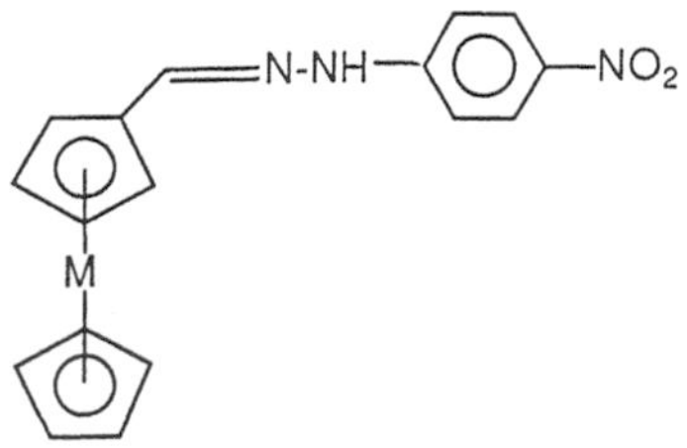

(1) 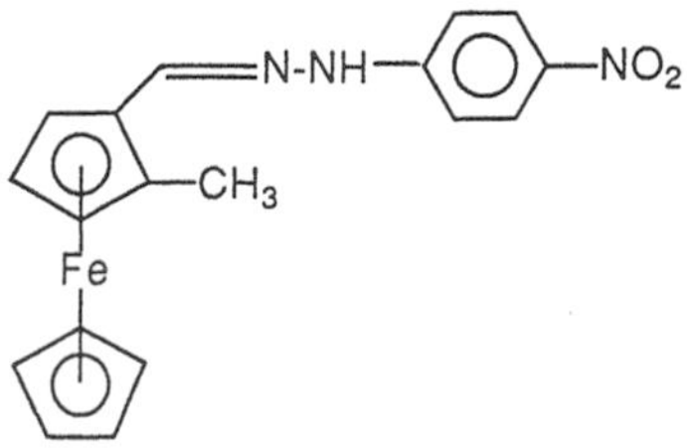　　　　　　　　　(2) 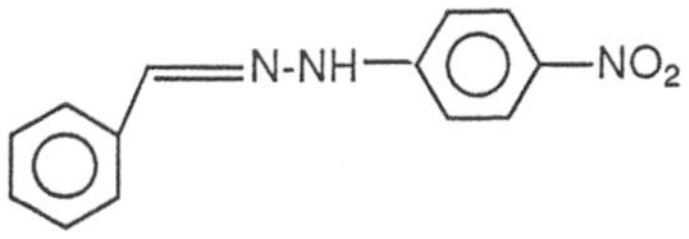　　　　　　　　　(3)

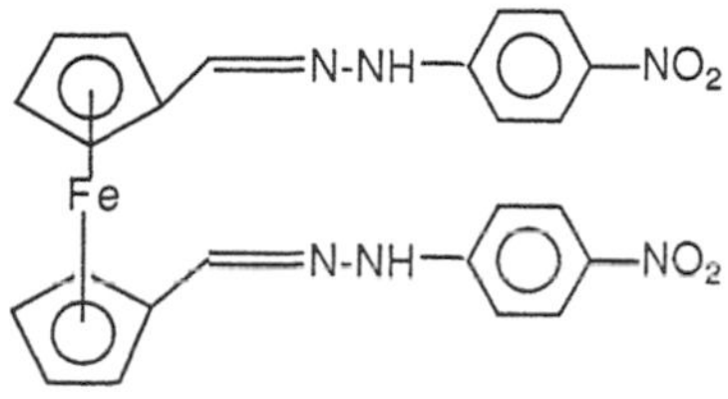

(4) 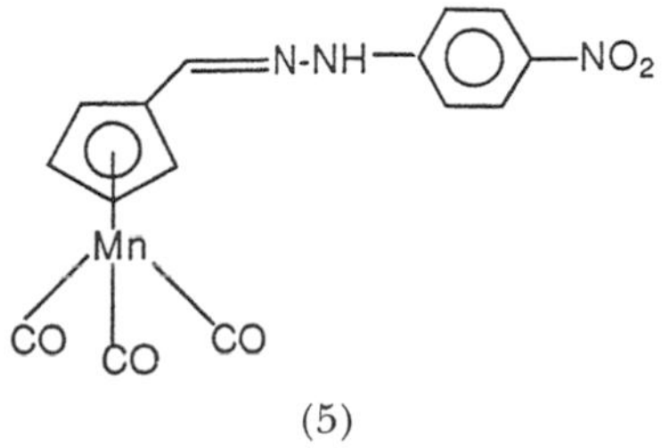　　　　　　　　　　　　　　　(5)

Oliver et al.[395] measured the $\chi^{(3)}$ values of solutions of tricyclopentadienylides of Dy, Er, Nd, Pr, and Yb abbreviated as Dy(cp)$_3$, Er(cp)$_3$, Nd(cp)$_3$, Pr(cp)$_3$, and Yb(cp)$_3$, respectively. The rare earth cyclopentadienyls (metallocenes) had metal ion electronic states with transitions in the range of 0.80 to 1.50 μm. The Yb(cp)$_3$ showed the largest $\chi^{(3)}$ coefficients of 1.3×10^{-20} m^2/V^2 with an α value of 0.58 cm^{-1}, which is about the value of carbon disulfide. Other metallocenes showed smaller values. From a self-focusing experiment, the measured n_2 value of 4.3×10^{-11} esu for Yb(cp)$_3$ converts to a $\chi^{(3)}$ coefficient of 7.3×10^{-20} m^2/V^2. Table 51 lists the $\chi^{(3)}$ coefficients of various metallocenes.

Myers et al.[396] measured third-order NLO properties of Group 4 metallocene halide and acetylide complexes, Cp$_2$MX$_2$ ($M = $ Ti, Zr, Hf; $X = $ F, Cl, Br, C $\equiv$ CC$_6$H$_5$), by the THG technique. The measured second hyperpolarizabilities of these complexes are listed in Table 52. The γ values increased significantly for metallocene-phenylacetylide complexes over the values for metallocene-dihalides. The Cp$_2$M units acted as an electron-accepting group. The metallocene-phenylacetylide complexes had the largest γ values ranging from 30×10^{-36} to 92×10^{-36} esu. The Cp$_2$Ti (C$\equiv$CC$_6$H$_5$)$_2$ had largest γ value compared with other metallocenes; the γ value decreased in the order Ti > Zr > Hf. This trend occurred because the Ti d-orbitals are closer in energy to the alkynyl π-orbitals than those of Zr or Hf d-orbitals.

The Du Pont group[397,398] measured hyperpolarizabilities of metallocene derivatives with the EFISH technique at 1.91 μm. The ferrocene derivative where $X = $ H and $Y = $ COCH$_3$ shows a μ value of 3.0×10^{-18}, an α value of 2.6×10^{-23}, a β value of 0.3×10^{-30}, and a γ value of 27×10^{-36} esu. The ruthenocene derivative where $X = $ CH$_3$ and $Y = $ NO$_2$ shows a μ value of 3.5×10^{-18}, an α value of 3.9×10^{-23}, and a β value of 0.6×10^{-30} esu. The π-conjugation of metallocenes was increased

Table 51 Optical Nonlinearities of Metallocenes for Solutions of 10^{19} molecules/cm^3 at 100 ps

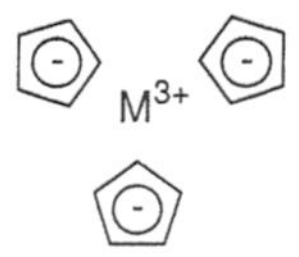

$M = $ Yb, Nb, Dy, Er, and Pr

Metallocene	λ_{max} (nm)	$\chi^{(3)}$ (10^{-20} m^2/V^2)	n_2 (10^{-11} esu)
Yb(cp)$_3$	1030	1.3	4.3
Nd(cp)$_3$	910	0.43	
Dy(cp)$_3$	1260	0.19	
Er(cp)$_3$	660	<0.1	
Pr(cp)$_3$	1395	<0.1	

From Oliver et al., *SPIE Proc.*, 1337, 81, 1990. With permission.

Table 52 Second Hyperpolarizabilities of Group 4 Metallocenes

Metallocene	λ_{max} (nm)	Solvent	γ (10^{-36} esu)
Cp_2TiF_2	324, 284	Chloroform	<3
Cp_2TiCl_2	394, 274	Chloroform	<5
Cp_2TiBr_2	428, 328, 276	Chloroform	<5
Cp_2ZrCl_2	338, 294	Chloroform	<5
$Cp_2Ti(C{\equiv}CC_6H_5)_2$	410	Tetrahydrofurna	92
$Cp_2Zr(C{\equiv}CC_6H_5)_2$	390	Tetrahydrofurna	58
$Cp_2Hf(C{\equiv}CC_6H_5)_2$	390	Tetrahydrofurna	51
$Cp_2Ti(C{\equiv}CC_6H_5)Cl$		Tetrahydrofurna	31
$(C_5H_4{-}n{-}Bu)_2Fe$		Neat	25

From Myers et al., *J. Am. Chem. Soc.*, 114, 7560, 1992. With permission.

further and Table 53 lists the optical nonlinearities of such metallocene derivatives. These compounds structurally resemble some nitrostilbenes. Pentamethyl, ethylenic cyano, and 2,4-dinitro substituents lead to large bathochromic shifts in the absorption bands; 4-nitrophenyl butadiene ferrocene is red shifted. In particular, the dipole moment of ruthenocenes is larger. The optical nonlinearities of the metallocene derivatives are influenced by metal atom, *cis-* and *trans*-conformations, π-electron conjugation, and terminal substituents. The optical nonlinearity of the *trans*-ferrocene derivative is larger than the *cis*-compound. Pentamethyl substitution also leads to large dipole moment and optical nonlinearity. The replacement of a cyclopentadienyl ring with a pentamethylcyclopentadienyl ring give rises to large optical nonlinearity. The β value of the (E)-ferrocene-4-(4-nitrophenyl)ethylene was 31×10^{-30} esu, larger than 4-methoxy-4'-nitrostilbene which has a β value of 29×10^{-30} esu. The replacement of Fe atom with Ru reduces β values because Ru has a higher ionization potential.

Table 53 NLO Parameters of Metallocene Derivatives Recorded in *p*-Dioxane Solvent

M	X	Y	μ (10^{-18} esu)	α (10^{-23} esu)	β (10^{-30} esu)	γ (10^{-36} esu)
Fe	H	H(*trans*)	4.5	3.9	31	–
Fe	H	H(*cis*)	4.0	3.8	13	
Fe	CH_3	H	4.4	5.3	40	
Fe	H	CN	5.3	4.2	21	
Fe	CH_3	CN	6.0	5.6	35	
Fe[a]	H	H	4.9	4.5	23	
Fe[b]	H	H	4.5	4.6	66	
Ru	H	H	5.3	4.2	12	114
Ru	CH_3	H	5.1	5.0	24	140
Ru	Ch_3	CN	5.9	5.5	24	105

From Cheng et al., *Mol. Cryst. Liq. Cryst.*, 198, 137, 1990; Calabrese et al., *J. Am. Chem. Soc.*, 113, 7227, 1991. With permission.

[a] Fe = 2,4-dinitro derivative.

[b] Fe = 4-nitrophenyl butadienyl.

Polin[399] from Universitat Innsbruck of Austria conducted THG measurements on ferrocenyl-substituted phenylacetylenes in THF solution at 1.064 μm. 1-Ferrocenylethynyl-4-nitrobenzene ($R = NO_2$) shows an absorption maximum at 340 nm and a γ value of 10.98×10^{-34} esu; two-photon resonance probably makes some contribution. 1-Acetyl-4-ferrocenylethynylbenzene ($R = COCH_3$) shows absorption maxima at 280 and 320 nm and a γ value of 2.6×10^{-34} esu. The DFWM-measured γ values were 1.1×10^{-33} ($R = NO_2$) and 2.6×10^{-34} ($R = COCH_3$) esu at 1.064 μm. Similar metallocenes with longer π-conjugation chain lengths and anthracene units in the chain were also synthesized, which are anticipated to have significantly larger third-order optical nonlinearity.

Perry et al.[175] measured $\chi^{(3)}$ values of symmetric ferroecene-capped diacetylene in toluene solution using the wedged cell THG technique. This bisferrocene compound had a $\chi^{(3)}$ value of 2.25×10^{-12} esu at 1.907 μm which may have been enhanced by three-photon resonance although the solution absorption spectrum indicates transparency around 636 nm. The $\chi^{(3)}$ value of this compound was enhanced more than twice the phenyl analog, which showed the role of ferroecene units in increasing optical nonlinearity. Yuan et al.[176] reported the γ value of 1.10×10^{-34} esu at 1.907 μm by the THG technique for bisferrocene having four triple bonds.

$R = NO_2, COCH_3$

Metallocenes are an interesting class of third-order NLO materials because they show large optical nonlinearity; however, their applications in devices may be limited because of the high two-photon absorption and low figure of merit.

6. Metal Dithiolenes

Oliver et al.[395] reported third-order optical nonlinearities of metal dithiolenes. Metal dithiolenes show a strong near-infrared absorption band which can be adjusted from 700 to 1400 nm by substituting the desired functional groups. Table 54 lists $\chi^{(3)}$ values of neutral nickel dithiolenes. The methyl- and phenyl-substituted nickel dithiolenes show nonlinear refractive indexes n_2 of 2.7×10^{-10} and 8.9×10^{-11} esu, respectively. The nickel complexes in which the ligand sulfur atoms are replaced by selenium (Compound 5) and the substituents are CF_3 show a nonlinear susceptibility of 1.4×10^{-13} esu. A platinum dithiolene complex (Compounds 5 through 9) having cyano (CN) substituents exhibits a $\chi^{(3)}$ value of 5.9×10^{-20} m^2/V^2 (4.2×10^{-12} esu). The $\chi^{(3)}$ values are affected by both the metal ion and substituents. A nickel complex with phenyl groups has $\chi^{(3)}$ values as large as 3.67×10^{-11} esu. A

Table 54 $\chi^{(3)}$ Values of Neutral Nickel Dithiolenes (M = Ni) for Solutions Corrected to 10^{18} molecules/cm^3 at 100 ps

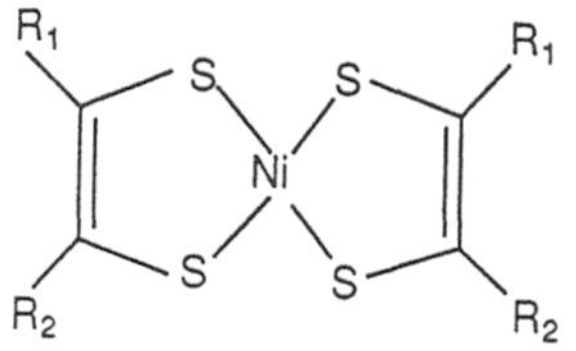

No.	R_1	R_2	λ_{max} (nm)	α (cm^{-1})	$\chi^{(3)}_{1111}$ (10^{-20} m^2/V^2)	$\chi^{(3)}_{1221}$ (10^{-20} m^2/V^2)
1	CH$_3$	CH$_3$	770	0.06	1.10	0.19
2	C$_6$H$_5$	CH$_3$	795	0.04	0.55	0.13
3	C$_6$H$_5$	C$_6$H$_5$	865	1.27	5.6	1.5
4	C$_6$H$_5$–OCH$_3$	C$_6$H$_5$–OCH$_3$	935	15.6	43.2	5.3
5	CF$_3$	CF$_3$	800	0.67	0.8	0.35
6	CN	CN	865	2.72	5.9	0.97
7	H	H	870	4.02	13.6	2.0
8	C$_6$H$_5$	CN	905	3.24	21.0	2.3
9	C$_6$H$_5$	C$_6$H$_5$	930	22.5	51.3	6.4

From Oliver et al., *SPIE Proc.*, 1337, 81, 1990. With permission.

bis[1,2-ethenedithiolato(2-)-S,S']platimun tetrabutylammonium shows a $\chi^{(3)}$ value of 7.16×10^{-14} esu. Because these metal complexes absorb between 700 and 1400 nm, these $\chi^{(3)}$ values are resonantly enhanced.[400]

Diaz-Garcia et al.[353] reported γ values of dithiolene and related coordination complexes. The γ values measued by THG were 5.5×10^{-33}, 4.0×10^{-33}, and 9.7×10^{-33} esu at 1.064 μm for Compounds 1, 2, 3, and 4, respectively. On the other hand, the γ values at 1.907 μm were measured as 8.8×10^{-33}, 3.2×10^{-34}, 3.6×10^{-33}, and 3.2×10^{-33} esu for Compounds 1, 2, 3, and 4, respectively. The THG measurements were conducted in chloroform for Compound 1 and in THF for Compounds 2, 3, and 4. Dithiolene compound 1 and aryldithiol compounds 2 and 3 showed larger hyperpolarizabilities at 1.907 μm, although naphthalene compound 4 showed the largest hyperpolarizability at 1.064 μm.

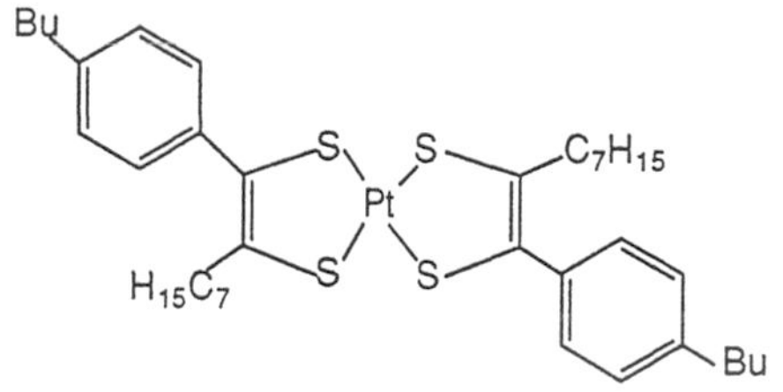

1

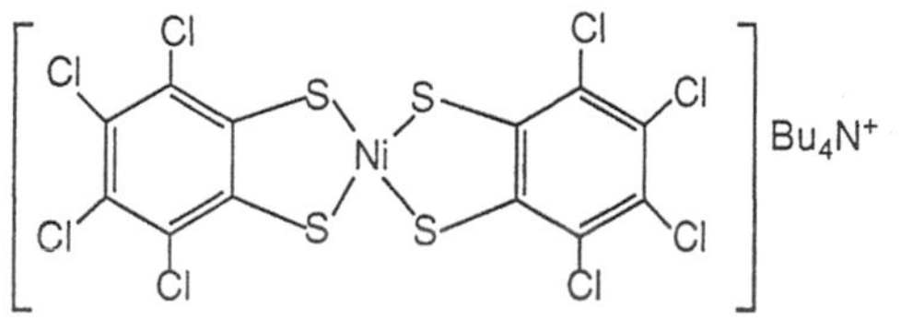

2

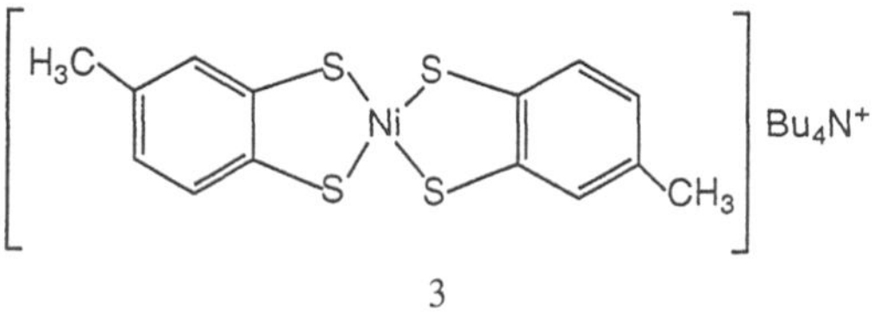

3

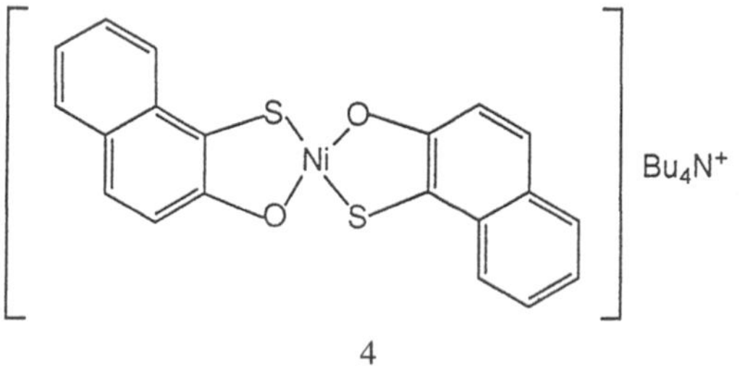

4

Fukaya et al.[401] reported $\chi^{(3)}$ values of Ni(dmbit)$_2$TBA as 3.6×10^{-11} esu at 1.319 μm from THG maker fringe and 6.5×10^{-9} esu at 0.532 μm by DFWM technique in DMF solutions.

Table 55 Dipole Moment, Polarizability, and Hyperpolarizabilities of Arene Metal Carbonyl Complexes, π-CTCB and σ-TPCB

Complex	X	Solvent	μ (10^{-18} esu)	α (10^{-23} esu)	β (10^{-30} esu)	γ (10^{-36} esu)
π-CTCB	H	Toluene	4.4	2.3	−0.8	2
π-CTCB	OCH$_3$	Toluene	4.7	2.7	−0.9	3
π-CTCB	NH$_2$	p-dioxane	5.5	2.7	−0.6	12
π-CTCB	N(CH$_3$)$_2$	Toluene	5.5	2.9	−0.4	10
π-CTCB	COOCH$_3$	p-dioxane	4.0	2.9	−0.7	6
σ-TPCB	H	Toluene	6.0	3.3	−4.4	8
σ-TPCB	C$_6$H$_5$	Chloroform	6.0	4.6	−4.5	12
σ-TPCB	C$_4$H$_9$	p-dioxane	7.3	4.2	−3.4	15
σ-TPCB	NH$_2$	DMSO	8	3.5	−2.1	15
σ-TPCB	COCH$_3$	Chloroform	4.5	4.0	−9.3	14
σ-TPCB	COH	Chloroform	4.6	3.7	−12	−

From Cheng et al., *Mol. Cryst. Liq. Cryst.*, 198, 137, 1990. With permission.

7. Arene Metal Carbonyl Complexes

Cheng et al.[397] measured hyperpolarizabilities of arene metal carbonyl complexes with the EFISH technique at 1.91 μm. Table 55 lists the dipole moment, polarizability, and hyperpolarizabilities of two series of compounds; chromium tricarbonyl benzene π-complexes (π-CTCB) and tungsten pentacarbonyl pyridine σ-complexes (σ-TPCB). The σ-TPCB complexes show larger optical nonlinearities which are affected significantly by the 4-pyridine substituents. The higher dipole moment of this series results from the pyridine fragment and the metal cluster contributions. The sensitivity at 4-position substituents is related to several factors such as additive effect, the strength of the pentacarbonyl tungsten acceptor, and d-π overlapping. The β values of metal carbonyl derivatives are negative, showing reversal of sign for charge transfer in the ground and excited states.

Cheng et al.[347] measured hyperpolarizabilities of square-planar metal benzenes. Table 56 lists the NLO properties of square-planar platinum and palladium benzene derivatives. These complexes are analogous to the donor–acceptor *para*-disubstituted benzenes. The optical nonlinearities of platinum compounds are higher than the palladium compounds because of the difference in inductive donating

Table 56 Dipole Moment, Polarizability, and Hyperpolarizabilities of Square-planar Metal Benzene Complexes

A	M	L	X	μ (10^{-18} esu)	α (10^{-23} esu)	β (10^{-30} esu)	γ (10^{-36} esu)
CHO	Pt	PEt$_3$	Br	2.5	5.8	2.1	37
NO$_2$	Pd	PEt$_3$	I	3.6	6.2	0.5	36
NO$_2$	Pd	PPh$_3$	I	5.5	1.05	1.5	50
NO$_2$	Pt	PEt$_3$	I	3.0	6.2	1.7	36
NO$_2$	Pt	PEt$_3$	Br	3.4	6.1	3.8	55

From Cheng et al., *Mol. Cryst. Liq. Cryst.*, 198, 137, 1990. With permission.

strength between the two metal atoms. The dipole moment, first and second hyperpolarizabilities increase with acceptor group strength and, as a result, the nitro derivatives had larger optical nonlinearities.

8. Vitamin B$_{12}$

Third-order NLO susceptibilities of various vitamins have also been studied; for example, retinol (vitamin A) and cyanocobalamin (vitamin B$_{12}$). Sakaguchi et al.[402] measured the $\chi^{(3)}$ value of vitamin B$_{12}$ by the DFWM technique in different solvents at 532 nm. In this novel technique, a steak camera was used to detect the time-dependent phase conjugation signal of the sample and the reference. The $\chi^{(3)}$ values of vitamin B$_{12}$ in various solvents were estimated relative to CS$_2$. A $\chi^{(3)}$ value of 7.82×10^{-12} esu was estimated for vitamin B$_{12}$ in butanol solution whereas an $\chi^{(3)}$ value of 7.48×10^{-12} esu was obtained both in ethanol and 2-hydroxyethylmethacrylate. The $\chi^{(3)}$ value of vitamin B$_{12}$ was 5 times larger in ethanol than the $\chi^{(3)}$ value in water. The $\chi^{(3)}$ values of vitamin B$_{12}$ in ethylene glycol, acetic acid, N-methyl-2-pyrrolidone, and N-vinyl-2-pyrrolidone were 3.91×10^{-12}, 6.63×10^{-12}, 6.29×10^{-12}, and 3.23×10^{-12} esu, respectively. The $\chi^{(3)}$ value of vitamin B$_{12}$ in DMSO and pyridine was 6.12×10^{-12} esu.

9. Miscellaneous Organometallic Complexes

Davey et al.[403] reported linear and nonlinear optical properties of *trans*-square-planar Group 10 metal thienyl complexes by forward DFWM at 1.064 μm. Table 57 lists the hyperpolarizabilities of Group 10 transition metal thienyl systems. The γ value increases with decreasing atomic number. The imaginary component of the γ value is large with no contribution from multiphoton absorption. The γ values of (PBu$_3$)$_2$(benzothiophene)$_2$Ni and (PEt$_3$)$_2$(thiophene)$_2$Ni are more than an order of magnitude larger than that of terthiophene, and Pd complex likely also has larger γ values. Polymeric analogs of the two nickel complexes were prepared by copolymerization. The [NiPBu$_3$)$_2$(thiophene)]$_n$ polymer showed a |γIm| value of 2.0×10^{-44} m^{5}V^{-2}) and a |γ| value of 2.6×10^{-44} m^{5}V^{-2}), whereas [NiPBu$_3$)$_2$(isothionaphthene)]$_n$ polymer showed a |γIm| value of 1.0×10^{-45} m^{5}V^{-2}) and a |γ| value of 3.0×10^{-45} m^{5}V^{-2}); this indicated that fused benzene ring does not enhance the hyperpolarizability of the polymer. The [NiPBu$_3$)$_2$(thiophene)]$_n$ polymer was found to be 25 times larger than poly(3-butylthiophene).

Table 57 Hyperpolarizabilities of Group 10 Transition Metal Thienyl Systems
$[(PX_3)_2(Y)_2M]$

X	Y	λ_{max} (nm)	M	γRe (10^{-47} m^{5}V^{-2})	$|\gamma Im|$ (10^{-46} m^{5}V^{-2})	$|\gamma|$ (10^{-46} m^{5}V^{-2})
C_2H_5	Thiophene	375	Ni	-51	4.6	6.8
C_2H_5	Thiophene	330	Pd	-6.9	1.3	1.5
C_2H_5	Thiophene	320	Pt	-2.5	0.6	0.65
n-C_4H_9	Benzothiophene	395	Ni	-20	10	10

From Davey et al., *Synth. Met.*, 58, 161, 1993. With permission.

Kafafi et al.[404] measured the $\chi^{(3)}$ values of solutions of Co(III) and Ni(II) complexes of o-aminobenzenethiol (ABT) by DFWM at 1.064 μm. The $\chi^{(3)}$ values of ABT increased significantly by metal substitution and both the cobalt and nickel complexes showed a γ value of 10^{-31} esu and a $\chi^{(3)}$ value of 10^{-9} esu. A nickel complex of ABT shows the largest $\chi^{(3)}$ value of 2.0×10^{-9} esu and a γ value of 2.0×10^{-31} esu. The molecular hyperpolarizability of the nickel complex was 2.5 times larger than that of the cobalt complex. For the nickel complex, an intense ligand-to-metal charge transfer (LMCT) band at 825 nm and a weak d-d transition were measured at 1.1 μm. The low-energy band may be responsible for its enhanced third-order optical nonlinearity. Lindle et al.[405] reported third-order optical nonlinearities of the first-row Co(d^7), Ni(d^8), and Cu(d^8) and the d^8 group Ni and Pt transition metal complexes of benzenedithiol (M(bdt)$_2$) using time-resolved DFWM at 1.064 μm. DFWM measurements performed on 5 mM DMF solution of M(bdt)$_2$ complexes yielded $\chi^{(3)}$ values of approximately 10^{-10} esu, and γ values of 10^{-33} esu. The figure-of-merit $\chi^{(3)}/\alpha$ were approximately 10^{-13} esu cm. The ratio $\chi^{(3)}_{xxxx}/\chi^{(3)}_{xyyx}$ was less than 1/3 indicating a pure electronic optical nonlinearity. The ABT showed the largest $\chi^{(3)}$ value at 1.064 μm, near its forbidden d-d absorption band. The NLO response of Co(bdt)$_2^{-1}$ was measured as <5 ps at 597 nm. Table 58 lists the components of $\chi^{(3)}$ and the figure of merit. The Pt(bdt)$_2^-$ showed the largest γ value and its $\chi^{(3)}$ value was an order of magnitude larger than that of Co(bdt)$_2^-$. The $\chi^{(3)}$ value of Ni(bdt)$_2$ was more than an order of magnitude larger than that of Ni(bdt)$_2^-$. The excited state showed a lifetime less than 35 ps contributing to optical nonlinearity. The effect of central metal was found to be more dominating at 597 nm. The $\chi^{(3)}$ value of the Co complex was two orders of magnitude larger than that of the Ni complex. Furthermore, o-aminobenzenethiol Pt complex showed a $\chi^{(3)}$ value more than two orders of magnitude larger than Pt(bdt)$_2^-$ complex. At 597 nm, the figure of merit also changed and was smaller than at 1.064 μm.

Table 58 The $\chi^{(3)}$ (10^{-11} esu), γ (10^{-33} esu), and $\chi^{(3)}/\alpha$ (10^{-14} esu cm) of Transition Metal Complexes of bdt and ABT Measured by the DFWM Technique

Complex[a]	At $\lambda = 1.064$ μm				At $\lambda = 0.597$ μm			
	γ_{xxxx}	$\chi^{(3)}_{xxxx}$	$\chi^{(3)}_{xyyx}$	$\frac{\chi^{(3)}_{xyyx}}{\alpha}$	γ_{xxxx}	$\chi^{(3)}_{xxxx}$	$\chi^{(3)}_{xyyx}$	$\frac{\chi^{(3)}_{xyyx}}{\alpha}$
Co(bdt)$_2^{-1}$	3.96	2.80	–	7.0	100	90	24	1.70
Ni(bdt)$_2^{-1}$	25.8	18.3	0.18	15.5	0.98	0.85	0.17	0.43
Cu(bdt)$_2^{-1}$	24.2	17.2	–	10.7	1.70	1.50	–	0.52
Pt(bdt)$_2^{-1}$	37.1	26.4	0.81	35.9	5.90	5.10	0.3	0.71
Ni(ABT)$_2^{-1}$	229	250	15.6	20.3	30	39	7.1	1.70
Pt(ABT)$_2^{-1}$	7.44	8.10	0.81	16.0	300	390	57	4.80

From Kafafi et al., *SPIE Proc.*, 1626, 440, 1992. With permission.

[a] Here ABT = (C$_6$H$_4$S$_2$) and dbt = (C$_6$H$_4$NHS).

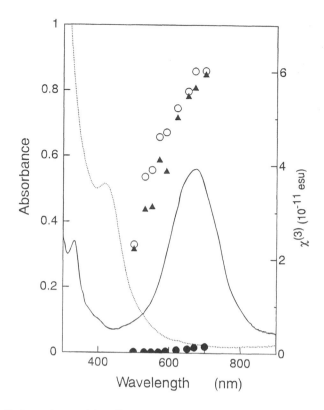

Figure 38 Wavelength dependence of the $\chi^{(3)}$ value for Pt(dmg)$_2$ (○) film before exposure to iodine, Pt(dmg)$_2$I$_2$ (●) film obtained after exposure to gaseous iodine and Pt(dmg)$_2$ (▲) after removing iodine from the doped film. Optical absorption spectra of Pt(dmg)$_2$ film is shown by the solid line and that of Pt(dmg)$_2$I$_2$ film is shown by the broken line. (From Kamata et al., *Appl. Phys. Lett.*, 65, 1343, 1994. With permission.)

Metal complexes of *o*-aminobenzenethiol (ABT)   Metal(II)-radical benzenedithiol (bdt) complexes

Fukaya and Kamata[407–409] reported THG-measured $\chi^{(3)}$ values of thin films of bis(dimethylglyoxima-to)nickel (II) (Ni(dmg)$_2$, bis(dimethylglyoximato)palladium (II) (Pd(dmg)$_2$), and bis(dimethylglyoxima-to)platinum (II) (Pt(dmg)$_2$). These compounds show strong bands at 550, 470, and 690 nm associated with d_z^2-p_z transition and at 410, 295, and 331 nm associated with metal-to-ligand charge-transfer transitions for Ni(dmg)$_2$, Pd(dmg)$_2$, and Pt(dmg)$_2$, respectively. The $\chi^{(3)}$ values of these complexes increase with the increase of absorption of the *d-p* transition because the three-photon resonance enhancement. The $\chi^{(3)}$ value of the Pt(dmg)$_2$ complex was an order of magnitude larger (~2 × 10^{-11} esu) at the absorption peak wavelength than the $\chi^{(3)}$ values of nickel and palladium complexes. Figure 38 shows the wavelength dependence of the $\chi^{(3)}$ value of Pt(dmg)$_2$ thin film before and after exposure to iodine and after removing iodine from the doped film. The Pt(dmg)$_2$ thin films show large $\chi^{(3)}$ values with three-photon resonance to the *d-p* transition band. Iodine exposure causes a decrease in the $\chi^{(3)}$ value which was approximately 10^{-13} esu, about two orders of magnitude lower than the value before iodine treatment. Iodine reaction probably circumvents metal–metal interaction as well as delocalization of electrons. After iodine removal, the initial $\chi^{(3)}$ value was again recovered at any wavelength. Figure 39 shows the $\chi^{(3)}$ versus α plots for third harmonic wavelength for nickel, palladium, and platinum complexes with several dionedioxime ligands.[408] The $\chi^{(3)}$

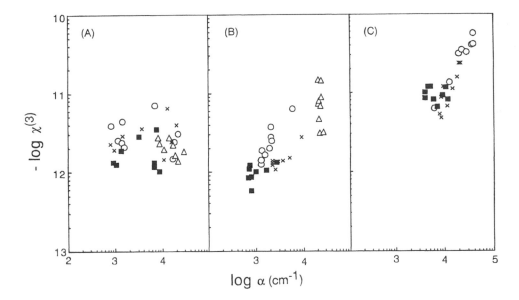

Figure 39 The $\chi^{(3)}$ versus α plots for corresponding third-harmonic wavelength for nickel (A), palladium (B), and platinum (C) complexes with several dionedioxime ligands. The ligands are (○) dmg, (■) dpg, (×) niox, and (△) bqd. (From Kamata et al., *Chem. Phys. Lett.*, 221, 194, 1994. With permission.)

value increases as the α increases; the Pd and Pt complexes showed significant dependence. The $\chi^{(3)}$ values of the Pt complexes were larger than the Pd complexes at the same α value. The $\chi^{(3)}$ values of the Ni complex were not as dependent on the α value. Ni complexes with dionedioxime ligands showed three-photon resonance behavior because of their strong absorption bands in the third-harmonic wavelength region. The three-photon resonance enhancement in Ni complexes was smaller than the three-photon resonance enhancements of Pd and Pt complexes. Therefore, the $\chi^{(3)}$ values of these complexes are strongly governed by the central metal ions. The $\chi^{(3)}$ values of 1.9×10^{-13} esu for Pd-salen, 2.4×10^{-13} esu for Ni-salphen, 2.2×10^{-13} esu for Pd-salphen and 1.3×10^{-13} esu for Pd-dpsalen were measured at 1.319 μm by THG method.[394]

$$M(dmg)_2 \qquad M(niox)_2 \qquad M(dpg)_2$$

$$M(bqd)_2 \qquad R = H\,M\text{-salen} \quad R = C_6H_5\,M\text{-dpsalpen} \qquad M\text{-salen}$$

Gong et al.[410] performed DFWM measurements in the dichloromethane solutions of 2,2,6,6-tetramethyl-pipridine-ol-N-oxyl p-dihydroxybenzaldehyde oxime benzoic ester (TPOHBOBE) and bis-(TPOHBOBE) cobalt (II). The measured $\chi^{(3)}$ values were 7.1×10^{-13} esu for TPOHBOBE at 5×10^{-3} M and 4.8×10^{-13} esu for bis-(TPOHBOBE) cobalt (II) at 2×10^{-4} M at 530 nm. The γ values deduced were found to be 4.6 × 10^{-32} esu for TPOHBOBE and 1.3 × 10^{-30} esu off-resonance for bis-(TPOHBOBE) cobalt (II). The corresponding $\chi^{(3)}$ values of solid samples were estimated as 5.6 × 10^{-10} esu for TPOHBOBE and 7.8 × 10^{-9} esu for bis-(TPOHBOBE) cobalt (II) using a density of 1.5 g/cm^3 and refractive index of 1.6 for both compounds. The conjugation length of bis-(TPOHBOBE) cobalt (II) is twice that of TPOHBOBE and the complex formation between cobalt and ligand gives rise to the charge-transfer process; as a result, the metal complex shows large third-order optical nonlinearity. The relation between γ and conjugation length l follows γ α^5 dependence.

(TPOHBOBE)

(bis-TPOHBOBE)

The black, plate-like 10 μm thick single crystals of organic metal (BEDT-TTF)$_4$Re$_6$Se$_5$Cl$_9$ showed a $\chi^{(3)}$ value of 1.6 × 10^{-8} esu using a refractive index $n = 1.9$ and a diffraction efficiency of 10^{-5} as reported by Sipp et al.[411] In this organic metal, the single sheets of BEDT-TTF donors alternate with layers of all-inorganic cluster anions. The high delocalization in the donor plane probably is favorable for large optical nonlinearity. This organic metal is monoclinic and belongs to space group P2$_1$/m. The (BEDT-TTF)$_8$SiW$_{12}$O$_{40}$ also has a similar sandwich structure and its single crystals show a $\chi^{(3)}$ value of approximately 10^{-8} esu.[412] The $\chi^{(3)}$ values of α-(BEDT-TTF)$_4$[Cu(NCS)$_2$]$_2$ and α-(BEDT-TTF)$_2$I$_3$ were measured by femtosecond, nondegenerate four-wave mixing.[77] The $\chi^{(3)}$ values of these complexes were also studied at 10 K and 295 K and the sensitivity of the decay times to the temperature can be related to the phase–transition dynamics. These two complexes have a life time of about 8000 fs at 295 K. The lifetime of the second component was 970 fs for α-(BEDT-TTF)$_4$[Cu(NCS)$_2$]$_2$ and 1705 fs for α-(BEDT-TTF)$_2$I$_3$.

Wada and Yamashita[413] calculated the $\chi^{(3)}(-3\omega;\omega,\omega,\omega)$ value of halogen-bridged mixed-valence Pt complexes: [Pt(en)$_2$][Pt(en)$_2$X$_2$](ClO$_4$), (X = Cl, Br, I) and en = ethylenediamine using a three-level model considering the ground state and odd and even charge-transfer excitons. These complexes have charge-transfer excitation from Pt^{2+q} to neighboring Pt^{4-q} state. The calculated $\chi^{(3)}$ values were in the range 10^{-8} to 10^{-12} esu and increased in the order Cl < Br < I because of the increase of the oscillator strengths and the decrease of the resonant energies. The THG-measured $\chi^{(3)}$ values were 1.2 × 10^{-12} esu at 1.9 μm and 4.1 × 10^{-12} esu at 1.06 μm for the X = Cl complex. On the other hand, the X = Br complex had $\chi^{(3)}$ values of 1.7 × 10^{-11} esu at 1.9 μm and 7.0 × 10^{-11} esu at 1.06 μm, an order of magnitude larger than X = Cl complex.[414,415] The incident laser was polarized parallel to the chain axis, and the parallel component of the $\chi^{(3)}$ value of the Pt-Cl complex was about an order of magnitude larger than the perpendicular one. The THG studies performed in the wavelength region from 0.6 to 1.0 μm showed a peak at 1.8 eV with an $\chi^{(3)}$ value of about 4 × 10^{-11} esu; this was assigned to the two-photon resonance peak of the optically forbidden charge-transfer exciton. These halogen-bridged mixed-valence metal complexes had linear chain structures with inversion centers. These were one-dimensional electron systems along the chain axis; their optical band gap can be varied over 1–3 eV by substitution of Cl or Br for I and of Ni or Pd for Pt. These halogen-bridged, mixed-valence *M-X* chain compounds have a one-dimensional structure that can be depicted by:

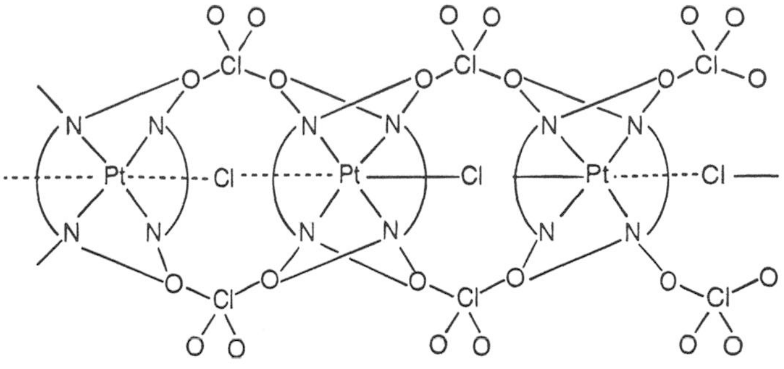

Yuan et al.[415] measured second hyperpolarizabilities of a series of organoboron compounds containing dimesitylboron moieties by THG technique at 1.907 μm. Symmetrical compounds having $B(mes)_2$ groups (mes = $2,4,6-(CH_3)_2C_6H_2$) at both ends for example; $(mes_2B–Y–B(mes)_2$ where Y = benezne, biphenyl or *trans-trans*-$CH=CH–C_6H_4)_n–CH=CH–(n = 1,2)$. In addition, push–pull compounds of the form $R-Y-B(mes)_2$ where R is electron donor group such as CH_3S, $(CH_3)_2N$, CH_2, etc. and Y = benzene, $C_6H_4–C≡C–$, $C_6H_4–CH=CH–$, $C_6H_4–CH=CH–C_6H_4$. The γ values were found to increase with increasing π-conjugation length for both symmetric and unsymmetric boron compounds. The γ value as high as 2.9×10^{-34} esu were measured for *trans*-$(mes)_2B–CH=CH–C_6H_4–CH=CH–B(mes)_2$. Also large γ value of 2.3×10^{-34} esu were measured for (E)–4,4′-$(CH_3)_2N–C_6H_4–CH=CH–C_6H_4-B(mes)_2$ and 2.25×10^{-34} esu were measured for (E)–4,4′–$(CH_3)_2N–C_6H_4–CH=CH–C_6H_4–NO_2$. The γ value of 8×10^{-36} esu were measured for (E)–Ph2P–$CH=CH–B(mes)_2$.

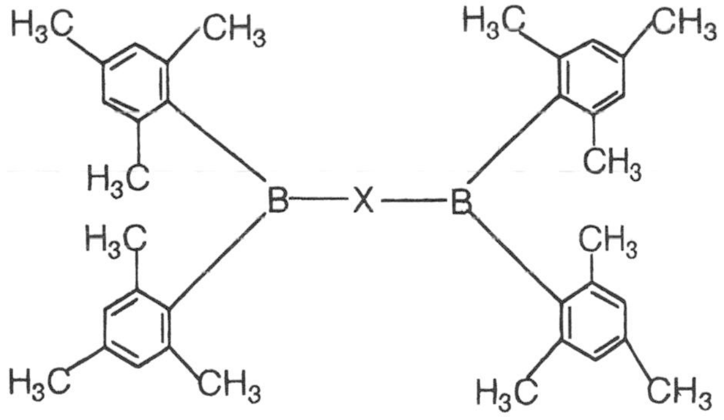

$X =$  $\gamma = 8.42 \times 10^{-35}$ esu

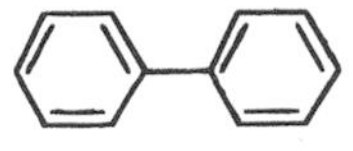

    $\gamma = 1.36 \times 10^{-34}$ esu

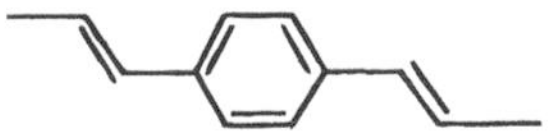

    $\gamma = 1.55 \times 10^{-34}$ esu

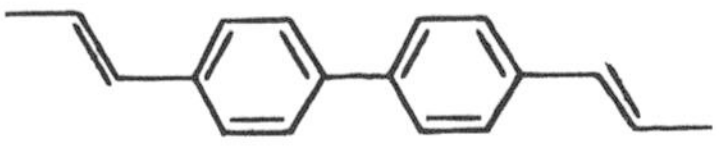

    $\gamma = 2.29 \times 10^{-34}$ esu

$R = (CH_3)_2N$ $\gamma = 2.4 \times 10^{-35}$ esu
$R = (CH_3)S$ $\gamma = 3.2 \times 10^{-35}$ esu
$R = OCH_3$ $\gamma = 2.3 \times 10^{-35}$ esu
$R = Br$ $\gamma = 2.0 \times 10^{-35}$ esu
$R = 1$ $\gamma = 2.7 \times 10^{-35}$ esu

$R = (CH_3)_2N$ $\gamma = 2.5 \times 10^{-35}$ esu
$R = CH_3S$ $\gamma = 7.5 \times 10^{-35}$ esu
$R = OCH_3$ $\gamma = 4.1 \times 10^{-35}$ esu
$R = H$ $\gamma = 2.7 \times 10^{-35}$ esu

$R = (CH_3)_2N$ $\gamma = 9.3 \times 10^{-35}$ esu
$R = NH_2$ $\gamma = 4.1 \times 10^{-35}$ esu
$R = CH_3S$ $\gamma = 8.1 \times 10^{-35}$ esu
$R = OCH_3$ $\gamma = 5.4 \times 10^{-35}$ esu
$R = H$ $\gamma = 3.7 \times 10^{-35}$ esu
$R = CN$ $\gamma = 3.0 \times 10^{-35}$ esu
$R = NO_2$ $\gamma = 3.2 \times 10^{-35}$ esu

Zhai et al.[416] reported second hyperpolarizabilities (γ) of molybdenum-based organometallic complexes in solution using independent DFWM and Z-scan techniques. The molybdenum complexes were dissolved in dry THF with concentrations ranging from 10^{-2} to 10^{-4} mol/l. Second hyperpolarizability of cis-Mo(CO)$_4$(PPh$_3$)$_2$, $trans$-Mo(CO)$_4$(PPh$_3$)$_2$, cis-Mo(CO)$_4$(PPh$_2$OMe)$_2$, cis-Mo(CO)$_4$(PPh$_2$Me)$_2$, Mo(CO)$_5$(PPh$_3$), and Mo(CO)$_5$(PPh$_2$NHMe) were found to be 6750, 2125, 375, 250, 120, and 73 times that of the carbon disulfide standard from the Z-scan measurements and 12593, 2444, 326, 296, 11, and 44 times that of carbon disulfide standard from the DFWM experiments, respectively. There was a good agreement between the DFWM and Z-scan γ values. These molybdenum complexes had donor phosphine ligands and acceptor carbonyl lignads which attained conjugation through the Mo metal. The magnitude of second hyperpolarizability is four times larger for cis-Mo(CO)$_4$(PPh$_3$)$_2$ than for $trans$-Mo(CO)$_4$(PPh$_3$)$_2$ because of the increased low-lying π-electron states arising from the noncentrosymmetric structure. The $\chi^{(3)}$ value for the cis-Mo(CO)$_4$(PPh$_3$)$_2$ was approximately 10^{-13} esu, which increased for higher concentrations. The γ value increased with the number of aromatic rings in the molybdenum complexes. The number of delocalized π-electrons was found to be the dominant factor affecting γ values rather than the charge-transfer mechanism.

Gale et al.[417] reported synthesis and γ values of a series of metal-free and nickel- and copper-containing salicyaldehyde-based organometallic complexes. Ligands 1 through 4 were prepared by refluxing two equivalents of salicyaldehyde with corresponding diamines in methanol. The Cu and Ni complexes were obtained by reacting ligands with metal acetates in hot methanol. The γ values measured by the DFWM method at 532 nm were 1.7×10^{-35}, 2.5×10^{-33}, 1.01×10^{-34}, and 2.8×10^{-35} esu for Ligands 1, 2, 3, and 4, respectively. Ligands 1 and 4 showed about the same γ values whereas Ligand 2 had the largest because of its extended π-conjugated system. The nickel complexes of ligands 1, 2, 3, and 4 showed γ values of 1.0×10^{-30}, 2.6×10^{-29}, 9.0×10^{-31}, and 6.7×10^{-32} esu, respectively. The copper complexes of Ligands 1 and 3 yielded γ values of 4.5×10^{-31} and 4.3×10^{-31} esu, respectively. The $\chi^{(3)}$ value of the nickel complex of Ligand 2 varied with the concentration of the complex in solution and reached as high as 5×10^{-12} esu. This Ni complex with ligand showed the largest γ value because of resonance enhancement.

1

2

3

4

F. Organic Composites
1. NLO Dye–Polymer Composites

The studies above have shown that organic materials can be investigated in the form of liquids, single crystals, liquid crystals, LB films, polymer thin films, and dye-functionalized polymers. Third-order NLO materials can be obtained by forming composites between guest dyes and host polymers. Kuzyk et al.[418–421] investigated doped polymeric materials as third-order optical materials. The third-order susceptibility of a doped polymer is given by:

$$\chi^{(3)}_{ijkl} = N\langle\gamma\rangle_{ijkl} \tag{28}$$

where N is the density of dopants and the bracket $<>$ indicates the orientational average of the molecular susceptibility tensor. The high-dopant concentration leads to the large optical nonlinearity. The optical nonlinearities of the dyes range between 10^{-12} and 10^{-14} esu. Table 59 lists the third-order optical nonlinearities of dye molecules. The third-order optical nonlinearities of substituted benzenes, m-dinitrobenzene and m-dicyanobenzene, dispersed in PMMA matrix were measured by electro-optic

Table 59 Third-order Optical Nonlinearities of Dyes Molecules Dispersed in PMMA Matrix

Dye Molecule	$\chi^{(3)}$ $(10^{-14}$ esu)	S_{1133} $(10^{-22}$ m^{2}V$^{-2})$	$\langle\gamma_{zzzz}\rangle$ $(10^{-34}$ esu)	N $(10^{20}$/cm$^3)$	λ_{max} (nm)
DNBA	3.1	2.0	2.3	1.37	451
DNTA	9.4	6.0	3.4	2.76	491
NFAI	15.7	11.7	6.6	2.40	483
DR1	17.0	12.9	7.0	2.44	491
NPCV	10.5	6.7	8.9	1.19	515
DCV	60	33.1	41	1.48	526
TCV	130	30.3	60	2.18	598
ISQ	12.9	10.7	68	0.19	658

From Kuzyk et al., *SPIE*, 1147, 198, 1989; Kuzyk et al., *J. Opt. Soc. Am. B*, 5, 84, 1990; Kuzyk et al., *Mater. Res. Soc. Symp. Proc.*, 214, 3, 1991; Dirk et al., *Mater. Res. Soc. Symp. Proc.*, 328, 643, 1994. With permission.

The chemical structures of the dye molecules are:

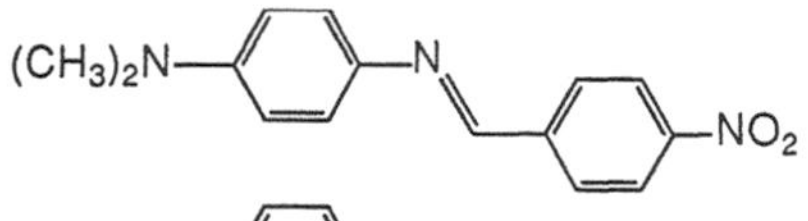

DNBA

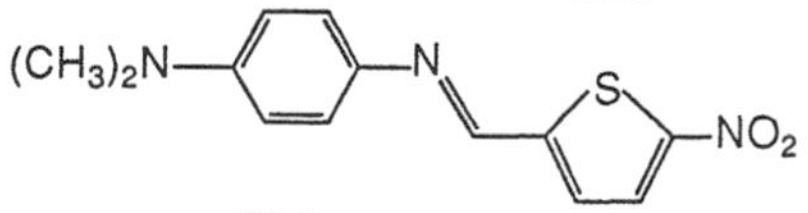

DNTA

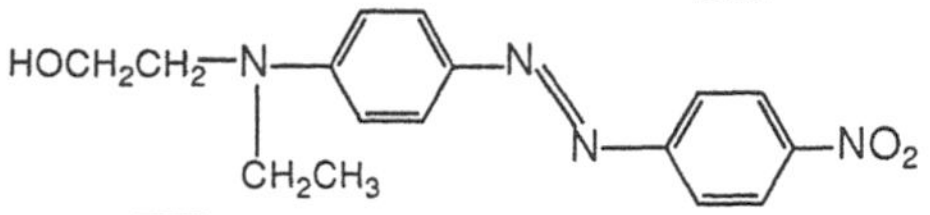

DR1

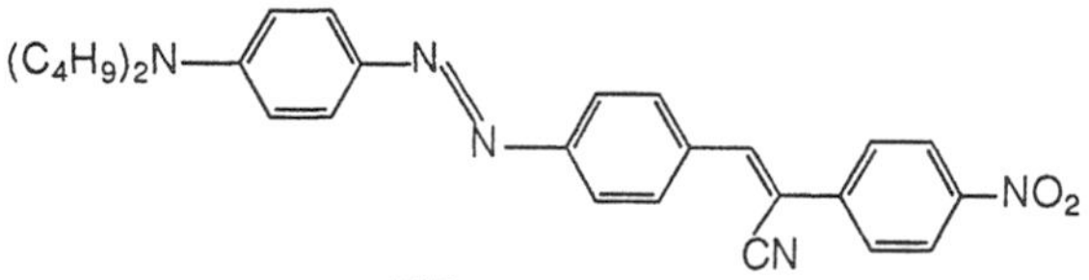

NPCV

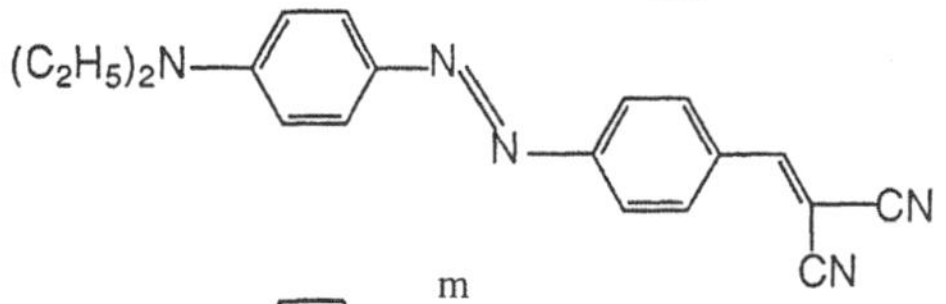

DCV

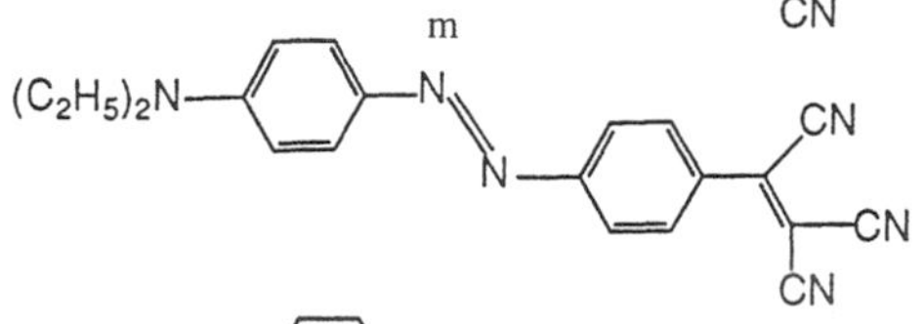

TCV

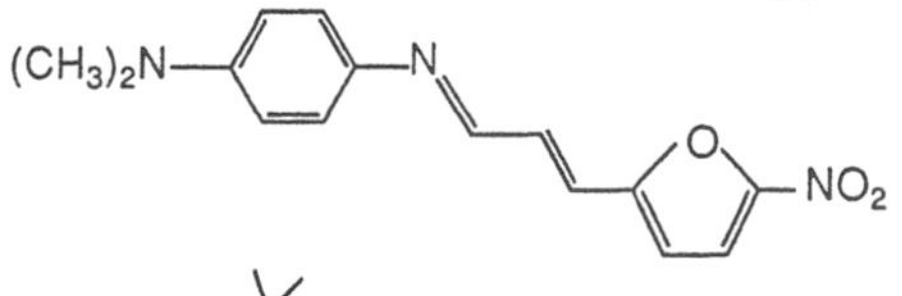

NFAI

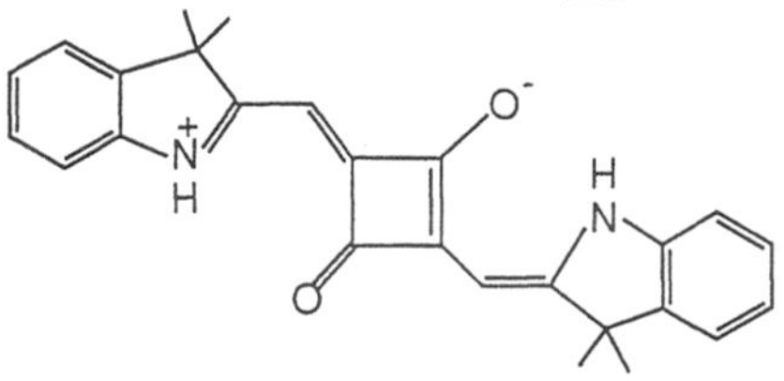

ISQ

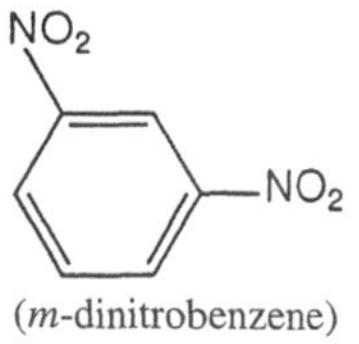

(*m*-dinitrobenzene)

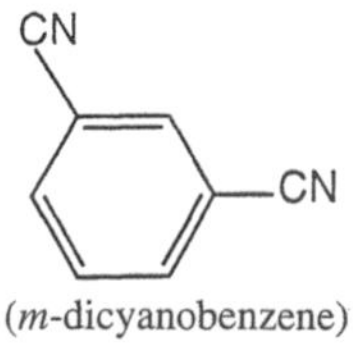

(*m*-dicyanobenzene)

technique.[123] The m-dinitrobenze with number density $N = 1.24 \times 10^{21}$ cm^{-3} has an orientationally averaged $<\gamma>$ value of 6.0×10^{-36} esu. The $<\gamma>$value of m-dicyanobenzene with number density $N = 1.24 \times 10^{21}$ cm^{-3} was measured as 8.0×10^{-36} esu. The calculated $<\gamma>$ values were 1.0×10^{-35} and 3.0×10^{-35} esu for m-dinitrobenzene and m-dicyanobenzene, respectively. These substituted benzenes have large optical nonlinearities because of the electron–acceptor groups.

As can be seen from the structure, typical squaraines have a highly polar electron-acceptor central squaraine moiety and two terminal electron-donor groups. According to the EFISH measurements, the γ values at 1.906 μm were -3.5×10^{-34} cm^7 esu^{-2} for ISQ and -7.5×10^{-34} cm^7 esu^{-2} for BSQ. Third-order optical nonlinearity showed a negative sign for both squarylium dyes. Dirk et al.[421] demonstrated that some squarylium dyes such as indole squarylium (ISQ) and anilnium squarylium (BSQ) exhibited negative second hyperpolarizability $\gamma(-2\omega;\omega,\omega,0)$ in case both ω and 2ω lie below the first electronic transition. Garito and co-workers[422–424] have performed many electron calculations on various squaraines. Second hyperpolarizability $\gamma_{xxxx}(-2\omega;\omega,\omega,0)$ values of bis(4-aminophenyl)squaraine (SQR), BSQ, and SQR1 calculated at 0.05 eV were -262×10^{-36}, -481×10^{-36}, and 191×10^{-36} esu, respectively. The sign of the γ value is positive for SQR1 where R is a hydrogen atom probably caused by the week donating nature of hydrogen atoms at both ends. The larger γ value of BSQ, which is twice that of SQR, originates from the strong dimethylamino (N(CH$_3$)$_2$) groups at both ends instead of amino (NH$_2$) groups in SQR. This indicates that both the sign and magnitude of the γ value depends on the strength of the donor group. The dispersion of $\gamma_{xxxx}(-2\omega;\omega,\omega,0)$ for BSQ in the nonresonant region indicated that both the real and imaginary parts have the onset of the two-photon absorption at 1.95 eV because of the $1^1 A_g$ to $1^1 B_{3u}$ transition.

Squaraines	X	R
SQR1	H	H
SQR	H	NH$_2$
BSQ	OH	N(CH$_3$)$_2$
	OH	N(C$_2$H$_5$)(C$_{18}$H$_{37}$)

The $\gamma_{xxxx}(-2\omega;\omega,\omega,0)$ value of -297×10^{-36} esu was obtained for BSQ at 0.65 eV. The femtosecond pump–probe experiments showed strong couplings between the first excited state and the high-lying two-photon states for bis[4-(N-ethyl-N-octadecylamino)-2-hydroxyphenyl] squaraine.

Park et al.[425] prepared the polymeric CT complexes of poly(vinylcarbazole) (PVK) and 2,4,7 trinitro-9-fluorenone (TNT). The molar ratio of TNF to the monomeric PVK unit ranged from 0.02 to 1.2. The $\chi^{(3)}$ values of the thin films charge-transfer complex increased linearly with the TNF contents and started saturating at 1.0×10^{-12} esu around 50 mol % of TNF. The saturated $\chi^{(3)}$ values of PVK/TNF and PVK/DDQ were 1.0×10^{-12} and 2.25×10^{-12} esu, respectively, by THG measurements. The lower CT excitation energy of PVK/DDQ may have been responsible for higher optical nonlinearity than PVK/TNF. The $\chi^{(3)}$ value of polymeric CT films originated from the CT complex formation.

Sakai et al.[426] prepared third-order NLO properties of hemicyanine, 3-methyl-2-(6-morpholino-1,3,5-hexatrienyl)-benzoazolium iodide (HC) doped MMA thin films. The $\chi^{(3)}$ values were measured by DFWM at 580 nm. The decay of DFWM had a fast component, 10 ps, and a slower component, $>$ 100 ps. The $\chi^{(3)}_{xxxx}$ value was as large at 2.8×10^{-9} esu and the effective γ_{xxxx} was estimated to be 1.2×10^{-28} esu. In the perpendicular configuration, $\chi^{(3)}_{xyyx}$ and γ_{xyyx} values were 7.0×10^{-10} esu and 2.9×10^{-29} esu, respectively. The ratio of $\chi^{(3)}_{xxxx}/\chi^{(3)}$ was about 4, which indicated the electronic origin.

Table 60 Nonlinear Refractive Indexes of Nickel Dithiolenes in PMMA films

No.	R_1	R_2	Concentration[a] (molecules/cm³) in PMMA	n_2(cm²/kW) DFWM	n_2(cm²/kW) Z-scan	Damage Threshold (GW/cm²)
1	C_6H_5	C_2H_5	10^{18}	4.7×10^{-11}	–	3.7
			10^{20}	3.1×10^{-9}	-2.4×10^{-11}	0.9
2	C_6H_5	C_4H_9	10^{18}	4.8×10^{-11}	-1.0×10^{-11}	3.8
			10^{20}	4.5×10^{-9}	-1.5×10^{-9}	1.0
3	C_6H_5	$C_{10}H_{21}$	10^{18}	6.2×10^{-11}	–	5.2
4	C_2H_5	C_2H_5	10^{18}	1.7×10^{-11}	–	5.0
			10^{20}		-0.5×10^{-9}	–

From Oliver et al., *SPIE*, 1775, 110, 1992. With permission.

[a] Samples with a concentration of 10^{18} molecules/cm³ were solutions, whereas 10^{20} molecules/cm³ were polymers.

Oliver et al.[427] measured the $\chi^{(3)}$ values and nonlinear refractive indexes n_2 of nickel dithiolenes having alkyl substituents using DFWM and Z-scan techniques at 1.064 μm. Compounds 1 through 3 could be incorporated to more than 40 wt% whereas Compound 4 tetraethyl Ni dithiolate at > 10 wt%. The nonlinear refractive index and damage thresholds are shown in Table 60. Tetraalkyl ($R_1 = R_2 =$ alkyl) and phenyl/alkyl ($R_1 =$ phenyl, $R_2 =$ alkyl) derivatives showed absorption maxima in the region 770 to 810 nm, and the ratio of $\chi^{(3)}_{1111}/\chi^{(3)}_{1221}$ was found to be 3/1. These materials had damage thresholds in the range from 9 to 15 GW/cm². Tetraphenyl, tetraaryl, and aryl/alkyl derivatives that had greater electron density in the dithiolene core, showed absorption maxima in the region 810 to 940 nm and had a ratio of $\chi^{(3)}_{1111}/\chi^{(3)}_{1221}$ between 4/1 and 10/1. The damage thresholds of these materials were about 1 GW/cm² or less. The phase-conjugation process had a time decay of >50 ps. The spin-coated films did not crystallize over many months. Compound 2 in PMMA showed an n_2 of 9×10^{-9} cm²/kW at a damage threshold of 1 GW/cm² for 2×10^{20} molecules/cm³ from DFWM. Bis[1-ethyl-2-phenylethene-1,2-dithiolato(2)-S,S]nickel showed n_2 as large as 10^{-8} cm²/kW. The n_2 values measured by Z-scan were lower than obtained from the DFWM technique. From the Z-scan technique, Re$\chi^{(3)}$/Im$\chi^{(3)}$ ratios were 3.9, 3.5, and 3.2 for compounds 1, 2, and 4 in PMMA at the concentration of 10^{20} molecules/cm³, respectively. These guest–host systems showed $\chi^{(3)}$ values of 7×10^{-10} esu, but showed increased absorption than that of solutions.

Lessard et al.[428] reported PVA-based systems for applications in holography and nonlinear optics. PVA was selected as host matrix because of its high solubility in water, ease of fabrication, and good environmental stability. The thin films on glass plates were prepared from composites of azo dyes chrysoidine, methyl orange, and mordant yellow 3R with PVA in aqueous solution with 6% by weight. From DFWM, the $\chi^{(3)}$ value of 1.9×10^{-8} m²/V² for 15-μm-thick film with a chrysoidine concentration of 2×10^{-3} M was obtained. The $\chi^{(3)}$ value of methyl orange with a chrysoidine concentration of 2×10^{-3} M for 30-μm-thick film was 2.4×10^{-8} m²/V² Mordant yellow 3R with a concentration of 2×10^{-2} M showed the largest $\chi^{(3)}$ value of 3.2×10^{-8} m²/V² for 30 μm thick film. The DFWM measurements were also carried out in rhodamine 6G/PVA films which showed high phase–conjugation reflectivities.

(Chrysoidin)

(Methyl orange)

NaO$_3$S—⬡—N≡N—⬡—OH

(Mordant yellow 3R)

Fujiwara et al.[429] reported $\chi^{(3)}$ values of mixed LB films of 4-docosylamino-4-nitroazobenze and tetracosanoic acid of 16 bilayers using the OPO-THG method. The $\chi^{(3)}$ value of 3.1×10^{-12} esu was determined at resonance, whereas 1.5×10^{-12} esu at off-resonance. Du et al.[430] reported a $\chi^{(3)}$ value of 5×10^{-8} esu, for LB films of cyanine blue dye and arachidic acid using the DFWM method at 532 nm and a γ value of 4.8×10^{-32} esu was estimated. Kodzasa et al.[431] reported $\chi^{(3)}$ values of the platinum-amido complex having a linear metal chain that has four Pt metals in one molecule bridged linearly by amino ligands. The composite having 34.8 wt% of the complex $[Pt_4(NH_3)_8(C_2H_{10}NO)_4]^{5+}$ in PVA showed $\chi^{(3)}$ values of 2.049×10^{-13}, 2.066×10^{-13}, 2.058×10^{-13}, and 0.995×10^{-13} esu at 1.8, 1.74, 1.62, and 1.5 μm, respectively.

2. Polymer–Polymer Composites

Vanherzeele et al.[432] reported the dispersion of the $\chi^{(3)}$ values of thin films of poly(p-phenylene benzobis-thiazole) and its molecular composites with nylon 66 and poly(trimethylhexamethylene) (PTMHT). The chemical structures of nylon 66 and PTMHT are shown below. The thin films from solutions of PBZT, PBZT/nylon 66, and PBZT/PTMHT in Lewis acid (GaCl$_3$ or AlCl$_3$)-nitromethane were prepared. Composition dependence of the $\chi^{(3)}$ value of thin films was measured at 1.9 μm (Figure 40). The PBZT/nylon 66 composite shows a linear behavior that is proportional to the mole fraction of the conjugated PBZT polymer. The PBZT/PTMHT composites versus PBZT composite deviates from the linear behavior. At 50 mol%, the $\chi^{(3)}$ value of BPZT/PTMHT composite was more than 50% larger than the corresponding PBZT/nylon 66 composite.

(Nylon 66)

(PTMHT)

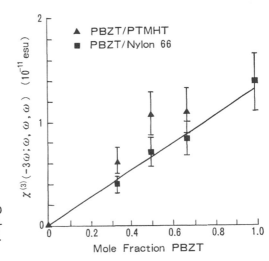

Figure 40 Composition dependence of third-order NLO susceptibility $\chi^{(3)}$ values for PBZT|nylon-66 and PBZT/pol(-trimethylhexamethylene). (From Vanherezeele et alk. *J. Opt. Soc. Am.*, 9, 524, 1992. With permission.)

The wavelength dependence of $\chi^{(3)}$ of pure PBZT, a 1:1 composite of PBZT/nylon 66 and a 1:1 composite of PBZT/PTMHT shows a resonant enhancement near 1 eV (1.3 μm) which is a three-photon resonance. The $\chi^{(3)}$ values of these composites were changing over the entire measurement range from 0.5 to 1.5 eV. The enhancement of $\chi^{(3)}$ values in molecular composites occurs because of the induced macroscopic ordering or orientational anisotropy of the PBZT. PBZT showed damage thresholds of ~50 GW/cm^2 for 30 to 50 ps pulses at 1.9 μm.

3. Semiconductor/Metal–Polymer Nanocomposites

It has been realized that the nanocrystals should have a great potential in the field of nonlinear optics. A model calculation of the electric-dipole third-order susceptibility of conduction electron in a metal sphere was published by Hacke et al.[433] From these studies, it was established that the third-order NLO susceptibility should vary roughly as the inverse third power of the radius for small spheres. In this connection, Stockman et al.[434] calculated the dispersion relations of excitations on fractals. Following these assumptions, third-order NLO properties of metal fractal cluster-doped polymers were reported by Kuzyk et al.[435] who suggested an enhanced third-order optical nonlinearity of fractal-metal-cluster doped poly(methyl methacrylate). Excitonic and surface-state enhancement of third-order optical nonlinearity in semiconductor microcrystallites have been reported for CdS, CdSe, CuCl, GaAs, Ag, and Au microcrystallites doped glasses and polymers.[436] The concept of quantum dot lattice for enhancing the excitonic optical nonlinearity is based on the enlarged exciton coherence length due to the cooperation of a large number of quantum dots. With this objective in mind, optical nonlinearities of quantum dot lattice composed of typical semiconductor materials have been estimated and as a consequence an enhancement of several orders of magnitude has been predicted. To obtain a large optical nonlinearity, the uniformity of the quantum dot size is important and has to be considered. This concept of reduction of dimensionality of material structures to one (quantum wire) and zero (quantum dot) have been exploited in developing novel third-order NLO materials. The enlarged third-order optical nonlinearity of quantized nanocrystallites of inorganic as well as organic materials dispersed in polymer matrices have been reported.

Ogawa et al.[437] reported $\chi^{(3)}$ values of 2×10^{-9} and 1×10^{-8} esu for polymer-films containing nanometer-sized particles of Ag and Au, respectively by DFWM. The $\chi^{(3)}/\alpha$ was on a scale of 10^{-12} esu.cm.

Birnboim and Ma[438] reported third-order optical nonlinearity of nanoparticle composites. The calculations for various multilayer spherical nanoparticle models in a host dielectric were (1) polymer core and metallic shell, (2) semoconductor core and metallic shell, (3) metallic core polymeric shell and metallic second shell, and (4) metallic core, semiconductor shell, and metallic second shell. The polymer, semiconductor, and metal were PDA, Si, and Ag, respectively. The host was water, GaAs$_2$ glass, or Si. Table 61 lists the optical nonlinearity of nanoparticle composites. The enhancement of four orders of magnitude was obtained in both $\chi^{(3)}$ values and the figures of merit without sacrificing the intrinsic

Table 61 Calculated $\chi^{(3)}$ Values of Nanoparticle Composites

Core	Shell	Shell	Host	Wavelength (nm)	$\chi^{(3)}$ (esu)
PDA	PDA	PDA	PDA	812	1×10^{-9}
PDA	PDA	PDA	PDA	673	1×10^{-7}
PDA	Ag	–	H$_2$O	812	2×10^{-8}
Ag	PDA	Ag	AsS$_2$	760	3×10^{-6}
Ag	PDA	Ag	AsS$_2$	620	1×10^{-5}
Ag	PDA	Ag	AsS$_2$	588	1×10^{-4}
Si	Si	Si	Si	800	1×10^{-7}
Si	Ag	–	Si	800	3×10^{-4}
Ag	Si	Ag	Si	800	5×10^{-4}
Ag	Si	Ag	Si	800	2×10^{-3}

From Birnboim and Ma, *Mater. Res. Soc. Symp. Proc.*, 164, 277, 1990. With permission.

speed of the nonlinearity from metal–polymer or metal–semiconductor composite. Nanoparticle composites offer tremendous flexibility in enhancement of third-order optical nonlinearity. The applications of these nanoparticles in phase–conjugate mirror (PCM) and optical bistability devices have been indicated.

LaPeruta et al.[439] reported composite NLO materials consisting of silver nanoparticles, ~11 nm in diameter in a polyurethane having a tricyanovinyl moiety. The polyurethane is dark red-violet, whereas the polyurethane/silver nanoparticle composite is light yellow. This color change depends on the volume fraction of silver; composites with low volume fractions of silver show optical absorptions identical with those of polyurethane. The DFWM measurememens were made on polymer/composite thin films having volume fractions of silver ~5 × 10⁻⁶. The $\chi^{(3)}$ values were 3.65×10^{-10}, 7.91×10^{-11}, and 5.86×10^{-11} esu at 532, 562, and 570 nm, respectively. At 532 nm, a sol solution containing an volume fraction of silver nanoparticles approximately equal to the composite showed a $\chi^{(3)}$ value of 9.63×10^{-11} esu.

Tokizaki et al.[440] reported $\chi^{(3)}$ value of 2×10^{-6} esu of CuCl quantum dots-doped glass using the DFWM technique. The figure-of-merit, $\chi^{(3)}/\alpha z$, was 20 esu.cm/sec for 5 nm radius at 80 k. Hayashi[441] reported $\chi^{(3)}$ values of CuCl or CuBr particles of diameter around 2 to 15 nm dispersed in PMMA polymer films. The Au and Ag ultrafine particles dispersed in poly(acrylonitrile-styrene) copolymer films showed $\chi^{(3)}$ values of approximately 2.1×10^{-8} esu because of the high filling fraction of particle in the films where the absorption coefficient (α) was 3430 cm⁻¹. The CdS/PHEM thin films showed $\chi^{(3)}$ values as large as 1.1×10^{-7} esu and absorption coefficients of 560 cm⁻¹. For partical applications in all optical switching where the figure-of-merit $\chi^{(3)}/\alpha$ is important, this study indicated that narrowing the particle size distribution of semiconductor particles dispersed in polymer matrices may lead to the improvement of the $\chi^{(3)}/\alpha$ ratio.

4. NLO Dye–Glass Composites

Kramer et al.[442] reported the third-order optical properties of fluorescein-doped boric acid glass. A mixture of boric acid and fluorescein dye (10^{-3} M concentration) were heated in a tube to a liquid state and the melt mixture was sandwiched between two glass slides at a temperature of ~160 °C. A 100-μm-thick film of boric acid glass containing fluorescein showed a large $\chi^{(3)}$ value of about 1 esu and a response time of 0.1 second. The phase–conjugation reflectivities as large as 0.6% were achieved by the DFWM technique. The composite had low saturation intensity of ~15 mW/cm². The NLO properties depended on the polarization state of the saturating field because the dye molecules were rigidly held and were not free to reorient during the relevant interaction time. These composites had low durability and poor homogeneity of the boric acid host, which limited the practicality of the composite system. Tompkin et al.[443] developed new composite materials based on lead-tin fluorophosphate glass containing acridine orange and acridine yellow dyes. The viscosity characteristics of lead-tin fluorophosphate glass facilitate in making melting and processing possible at a temperature compatible with many organic dyes. The tin content is the major factor which provides tailoring of physical and optical properties. The glasses containing tin ranging from 35 to 75 cation mol% form clear, transparent glasses. A lead content of 3 to 6 cation mol. % and a tin content of 40 to 60 cation mol. % provide exceptional durability. These glasses can be melted easily at about 500 °C and can be cooled to 300 °C. At these temperatures, the viscosity of the glass melt is low and organic guest molecules such as dyes can be dissolved completely without any thermal decomposition. The glasses can be polished because they are hard and durable. In a typical experiment, a low melting temperature base–glass composition (SnF_2 = 52.2%, SnO = 10.5%, PbF_2 = 5.1 %, and $1/2(P_2O_5)$ = 32.1%) was melted at 400 °C in air by using the carbon crucible and, after complete meltdown, was cooled to 300 °C. At this stage organic dye was added to the melt and the solution was mixed thoroughly. The lead-tin fluorophosphate glasses containing acridine yellow at a concnetration of 7.7×10^{17} molecules/cm³ and acridine orange at a concentration of 8.0×10^{17} molecules/cm³ show absorption peaks at 464 and 512 nm, respectively. Table 62 lists the NLO parameters for lead-tin fluorophosphate glass containing acridine yellow and acridine orange dyes. These composite have $\chi^{(3)}$ values of ~0.1 esu, saturation intensities of ~100 mW/cm² and response time of ~1 ms. The low melting temperature of lead-tin fluorophosphate glass allows doping with many organic dyes without any decomposition. Furthermore, the materials based on these glasses have good optical transparency and excellent environmental stability. These composites are useful in nonlinear devices where very large optical nonlinearities and millisecond response time are required.

(Acridine Yellow)   (Acridine Orange)

He et al.[444] studied the second- and third-order NLO properties of acridine yellow-doped lead-tin fluorophosphate glass by measurements on the surface. SHG was observed neither from the surface of pure lead-tin fluorophosphate (LTF) glass nor from the solution of acridine yellow dye in ethanol. The composite shows an estimated $\chi^{(2)}$ value of about 10^{-5} esu. The $\chi^{(3)}$ value was determined from the intensity of the THG by the following expression:

$$I_{(3\omega)} (\pi^2/4)L^2[\chi^{(3)}]^2I_{(\omega)}^3$$

where $L \sim \lambda = 1.064$ μm; $I_{(3\omega)}$ and $I_{(\omega)}$ are the intensities of THG and pump beam, respectively. From this relationship, the $\chi^{(3)}$ value of the composite was 5×10^{-3} esu. The laser-induced fluorescence studies indicate the three-photon excitation fluorescence. The advantages offered by lead-tin fluorophosphate glass are its low melting point, high optical quality, and durability. The molecules of acridine yellow dye are randomly oriented in the host matrix and held rigidly. In the case of THG, both bulk and surface enhancements contribute to large $\chi^{(3)}$ values.

Kumar et al.[445,446] measured the third-order optical nonlinearity from thin films of boric acid glass doped with Rhodomine 6G. Rossi et al.[447] reported the planar and nonlinear waveguiding in Rhodamine B-doped epoxy films. The 13-μm-thick films were prepared from a solution of one part epoxy resin, one part epoxy hardener, and one part 2-g/l Rhodamine B in ethanol solutions. The $\chi^{(3)}$ values of the film at the two wavelengths were $\sim 7.0 \times 10^{-20}$ m^2 V^{-2} at 616 nm and $\sim 5.0 \times 10^{-20}$ V^{-2} at 635 nm from the intensity dependence of the coupling angle. The molecular hyperpolarizabilities obtained from the bulk nonlinearity were 2.0×10^{-42} m^5 V^{-2}. Maloney and Blau[448] measured $\chi^{(3)}$ values of fluorescein-doped boric acid glass films using the DFWM technique. The $\chi^{(3)}$ value of the composite films with a fluorescein concentration of 8×10^{-3} M was 9×10^{-8} esu at 444 nm. This composite has a long-lived, lowest-lying triplet state on a time scale of 0.1 to 1.0 s; the transient NLO and fluorescence properties show significant dependence on the laser–pulse repetition rate in this time regime. Kramer et al.[442] reported the nonlinear susceptibility of fluorescein-doped boric acid glass as high as 1 esu. Organic dyes dispersed in inorganic glasses and other organic polymeric materials show large optical

Table 62 Third-order Nonlinear Susceptibility of Lead-tin Fluorophosphate Glass Containing Acridine Yellow and Acridine Orange Dyes

Composite	Concentration (molecule/cm³)	$\chi^{(3)}$ (esu)	Response Time (ms)	Measurement Wavelength (nm)
Acridine yellow	7.7×10^{17}	0.03	5.6	476
Acridine orange	8.0×10^{17}	0.04	1.9	514

From Tompkin et al., *J. Opt. Soc. Am. B*, 4, 1030, 1987. With permission.

nonlinearities. In the case of organic composites with glasses, although third-order optical nonlinearities are rather large, their drawbacks is a too slow decay time.

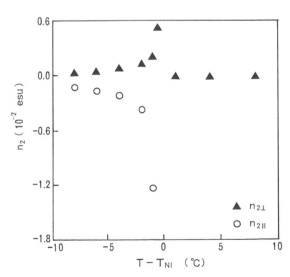

(fluorescein)

(Rhodomine 6G)

Speiser[449] and Beeson et al.[450] reported NLO properties of phenosafranin doped with sol-gel glasses or PVA by using pump–probe nonlinear absorption measurements. Phenosafranin sol-gel showed improved optical nonlinearities compared with eosin-Y-doped polymers. The triplet-triplet absorption band in the 600–850 nm spectral region indicated potential of phenosafranin for NLO devices.

(Phenosafranin)

G. LIQUID CRYSTALS

Liquid crystals are an interesting class of NLO materials because their physical properties can be controlled by achieving orientation from a modest field. The optical nonlinearities in liquids can arise from several mechanisms from collective reorientation to electronic. An excellent review on the NLO response of liquid crystal has been published by Palffy-Muhoray.[451] Li et al.[452] measured the nonlinear refractive index of a variety of liquid crystals by the Z-scan technique. The nonlinear refractive indexes for light polarized parallel to the director for 4-cyano-4'-n-pentylbisphenyl (5CB) (1), 4-cyano-4'-n-octylbisphenyl (8CB) (2), and a mixture of bisphenyl with terphenyls (E7) were reported to be -1.75×10^{-9}, -7.7×10^{-10}, and -1.15×10^{-9} esu, respectively, at 24 °C. For perpendicular polarization, the magnitude of the nonlinear refractive indexes were smaller with a positive sign. On the nanosecond timescale, the nonlinearities of these liquid crystals were two orders of magnitude larger than for Cs_2.

Figure 41 Temperature dependence of nonlinear refractive index for both parallel and perpendicular geometries for the nematic liquid crystal 5CB. (From Li et al., *Liq. Cryst.*, 16, 703, 1994. With permission.)

Figure 41 shows the temperature dependence of the nonlinear refractive indexes for both parallel and perpendicular geometries of the nematic liquid crystal 5CB. A strong nonlinear birefringence in the nematic phase is visible for parallel and perpendicular polarizations differing in magnitude as well as sign. These results were consistent with laser heating of the sample and the consequent reduction of orientational order. At the 10-ms timescale, n_2 values of 5CB measured for the parallel and perpendicular polarizations were -1.30×10^{-3} esu and $+0.26 \times 10^{-3}$ esu, respectively, at 514 nm.[453] Table 63 lists the n_2 values of a variety of liquid crystals at different wavelengths.[454] The d_{27} doped (<0.2%) 8CB exhibited n_2 values of -26×10^{-11} and $+3.7 \times 10^{-11}$ esu for parallel and perpendicular directions, respectively, at 22 °C using a 7-ns pulsewidth. The chemical structures of 4-(1-octynyl)-4'-cyanobiphenyl (OPL-7-1) (Structure 3), 4,4'-dipentylazoxybenzene (OPL-10-1) (Structure 4), 4,4'-dihexylazoxybenzene (OPL-10-2) (Structure 5), 4-cyano-4'n-alkyl-p-terphenyl (T15) (Structure 6), CB15 (Structure 7), cyclohexylcyclohexane (ZL1-1538) (Structure 8), and phenylpyrimidine (ZL1-2302) (Structure 9) are shown below and the optical nonlinearities are listed in Table 63.

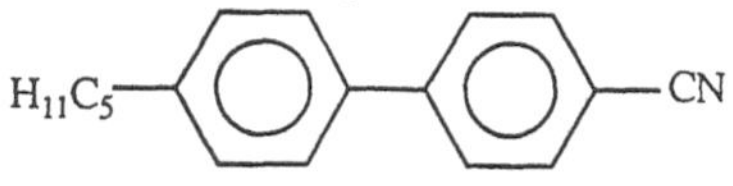

(1)   H$_{11}$C$_5$—⟨○⟩—⟨○⟩—CN

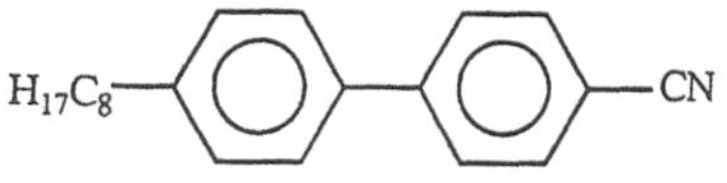

(2)   H$_{17}$C$_8$—⟨○⟩—⟨○⟩—CN

(3)   H$_{13}$C$_6$—≡—⟨○⟩—⟨○⟩—CN

(4)   H$_{11}$C$_5$—⟨○⟩—N=N(→O)—⟨○⟩—C$_5$H$_{11}$

(5)   H$_{13}$C$_6$—⟨○⟩—N=N(→O)—⟨○⟩—C$_6$H$_{13}$

(6)   H$_{11}$C$_5$—⟨○⟩—⟨○⟩—⟨○⟩—CN

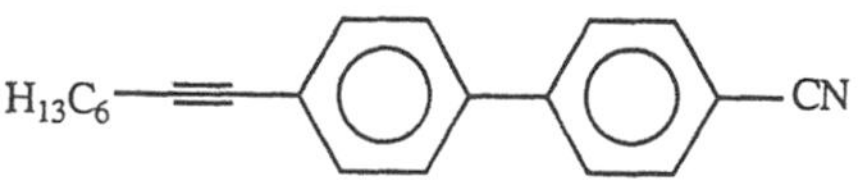

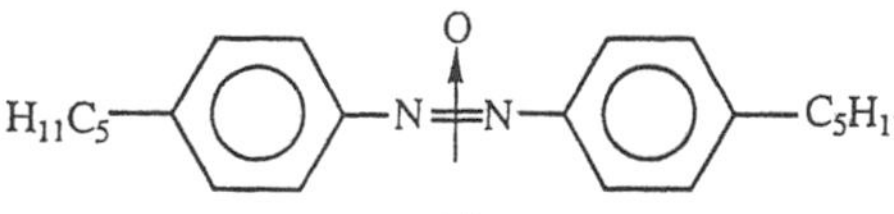

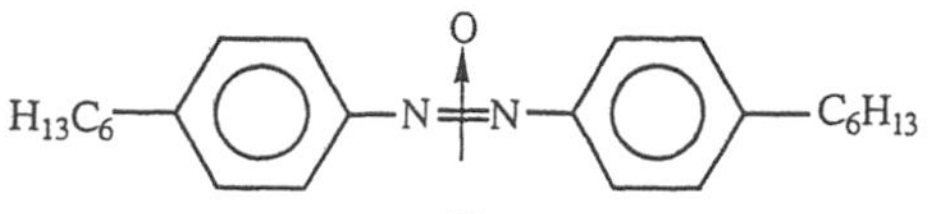

(7)   H$_5$C$_2$HCH$_2$C(CH$_3$)—⟨○⟩—⟨○⟩—CN

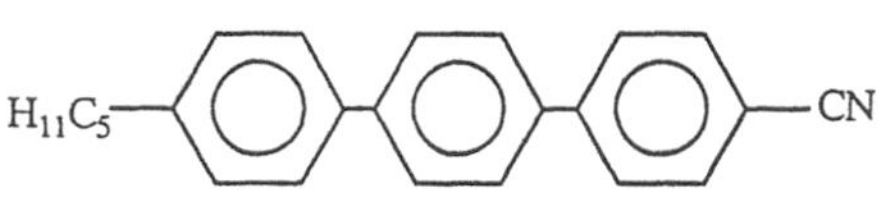

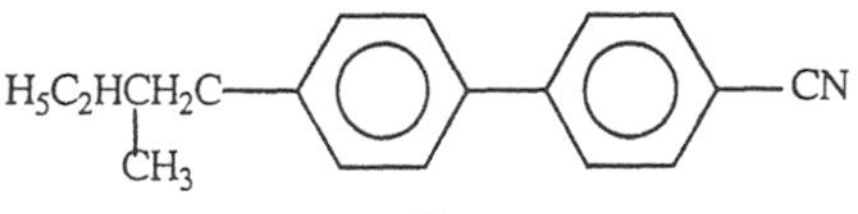

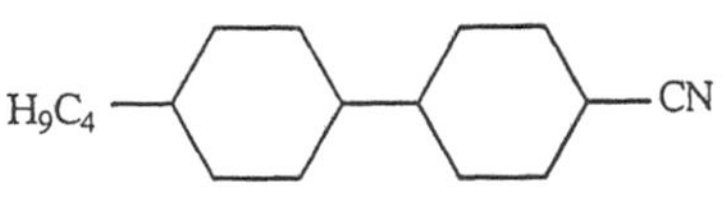

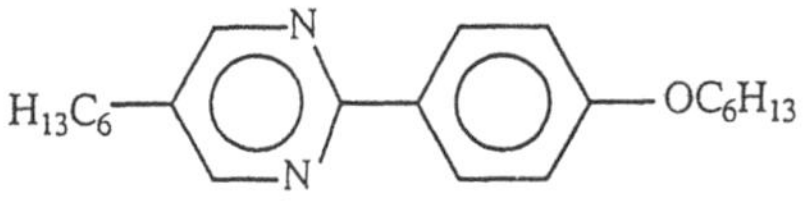

(8)   H$_9$C$_4$—⟨◯⟩—⟨◯⟩—CN

(9)   H$_{13}$C$_6$—⟨N○N⟩—⟨○⟩—OC$_6$H$_{13}$

Table 63 Nonlinear Refractive Indexes of Liquid Crystals

| | n_2 (esu) | | | | T (°C) |
| | λ = 532 nm (7 ps) | | λ = 514 nm (10 ms) | | |
Material	Parallel	Perpendicular	Parallel	Perpendicular	
(5CB)	-54×10^{-11}	$+8.3 \times 10^{-11}$	-10×10^{-4}	$+2.0 \times 10^{-4}$	24
(8CB)	-26×10^{-11}	$+3.7 \times 10^{-11}$			
	(22°C)	(24°C)			
(OPL-7-1)	-27×10^{-11}	$+7.0 \times 10^{-11}$			37
(OPL-10-1)	-9.9×10^{-11}	$+4.5 \times 10^{-11}$			40
(OPL-10-2)	-15×10^{-11}	$+2.0 \times 10^{-11}$			50
(T15)	-52×10^{-11}	$+21 \times 10^{-11}$	-0.68×10^{-4}	-0.092×10^{-4}	180
(ZL1-1538)	0	0	-0.43×10^{-4}	-0.13×10^{-4}	70
(ZL1-2303)	-20×10^{-11}	$+3.0 \times 10^{-11}$	-0.97×10^{-4}	$+0.082 \times 10^{-4}$	45
(CB15)	$+7.4 \times 10^{-11}$ (isotropic)		-0.25×10^{-4} (isotropic)		22
					25
(d_{27} doped 8CB)	-152×10^{-11}	$+10.5 \times 10^{-11}$			22
(E7)	-38×10^{-11}	$+3.3 \times 10^{-11}$			24

From Palffy-Muhoray, *Liquid Crystals: Applications and Uses*, Vol. 1, 493–545, 1991; Li et al. *Liq. Crystl.*, 16, 703, 1994; Palffy-Muhoray et al., *Mol. Crystl. Liq. Crystl.*, 207, 291, 1991; Li et al., *SPIE*, 1692, 107, 1992. With permission.

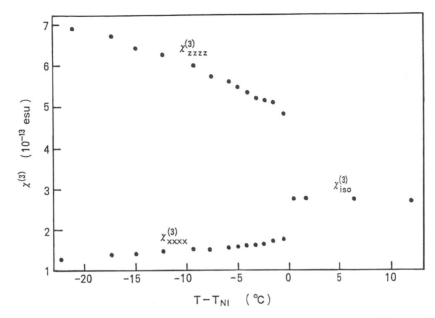

Figure 42 Measured third-order susceptibility of MMBA as a function of temperature where z-axis and x-axis are parallel and perpendicular to the nematic director, respectively. (From Wong and Garito, *Phys. Rev.,* A34, 5051, 1986. With permission.)

The temperature-dependent studies on 5CB on the nanosecond scale at 532 were carried out. The n_2 values in parallel and perpendicular polarizations in the nematic phase showed opposite signs, and an abrupt change in n_2 values across the nematic-isotropic transition was noticed. The nonlinear response for 5CB measured at the fundamental $\lambda = 1.064$ μm showed very weak nonlinearities. The 5CB showed the largest optical nonlinearity. The n_2 values were negative for both ZL1-1538 and T15. ZLI-1538 did not have any conjugation because it consisted of cyclohexane rather than benzene rings and the linear polarizability and n_2 values were small. The optical nonlinearities of these liquid crystals were relatively insensitive to structural change other than the presence of the benzene ring. The addition of the acetylene groups (OPL-7-10) did not significantly influence the optical nonlinearity. The azoxy group had little effect on the nanosecond response. The anthroquinone dye d_{27} doped 8CB showed significant enhancement in the optical nonlinearity without changing the sign of the n_2 values. The 5CB showed the largest optical nonlinearity among all the materials.

Wong and Garito[455] developed a relationship between the third-order macroscopic susceptibility and the microscopic molecular susceptibility in the nematic phase and measured THG with different polarization leading to a determination of nonvanishing order parameters $< P_2 >$ and $< P_4 >$ for a nematic liquid-crystal N-(p-methoxybenzylidene)-p-butylaniline (MBBA). The wedge-maker fringe method was used to demonstrate the simultaneous determination of the order parameters as a function of polarization and temperature. Third-order susceptibility of MBBA was measured as a function of temperature where z axis and x axis are parallel and perpendicular to the nematic director, respectively (Figure 42). These results were in agreement with NMR and Raman scattering studies. Ford et al.[456] measured $\chi^{(3)}$ values of a side-chain liquid-crystal polyacrylate (P-10) containing a $(CH_2)_{10}$ spacer chain and a 4-dimethylami-nostilbene-4'-carboxylic ester mesogen by the DFWM technique at 532 nm. The P-10 that contains a polarized stilbene side chain showed a resonant $\chi^{(3)}$ value of 1.8×10^{-12} esu, a factor 24 times larger than that of carbon disulfide reference. A 90-mm film containing 1.25% DANS in PMMA showed a $\chi^{(3)}$ value of 1.1×10^{-11} esu.

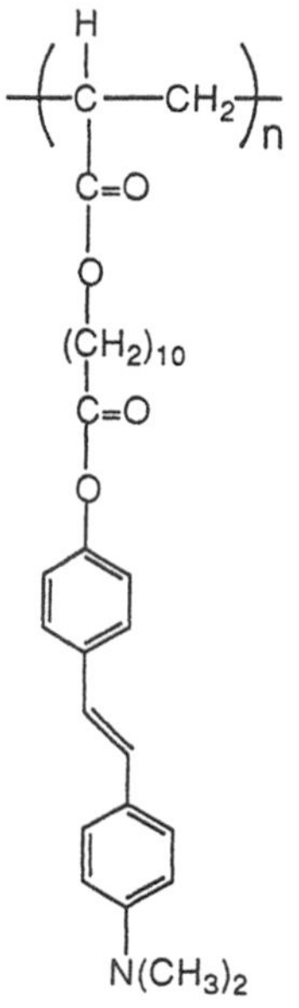

Smith and Coles[457] reported third-order NLO susceptibilities of liquid crystalline polymers possessing donor–acceptor-substituted side chains. The chemical structures of NLO chromophore-substituted liquid crystalline polymers (LCP) are shown below. The $\chi^{(3)}$ values from thin films were measured by THG at fundamental wavelengths of 1.064 and 1.579 μm. Table 64 lists the glass transition temperature, material phases, and third-order NLO susceptibilities of liquid crystalline polymers.

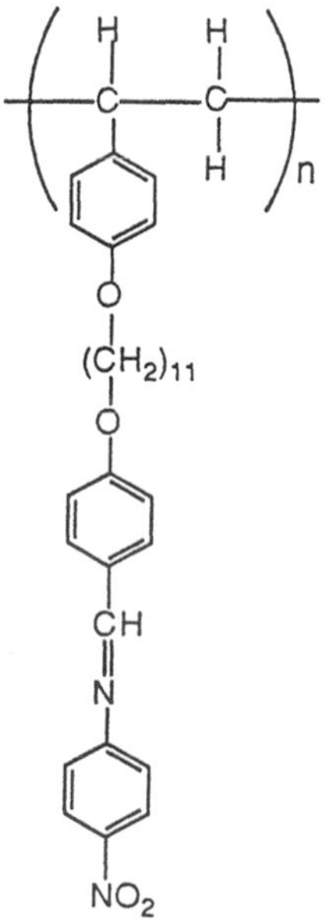

(LCP1)

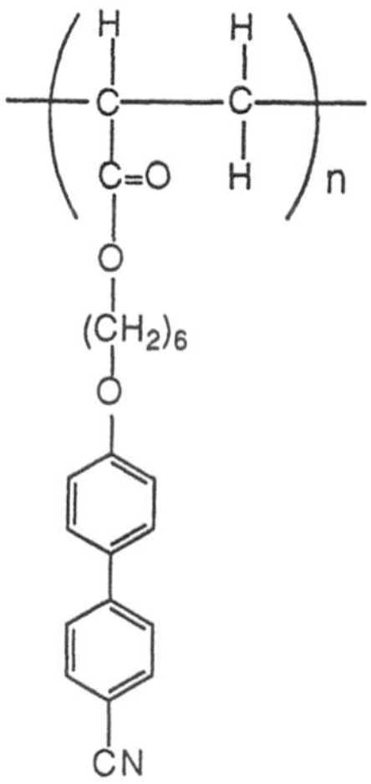

(LCP2)

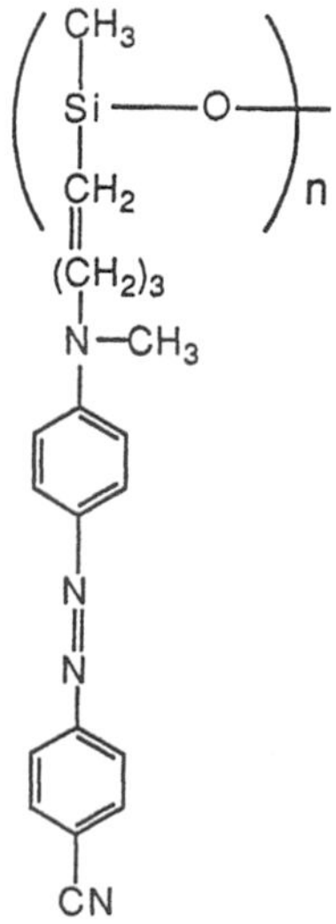

(LCP3)

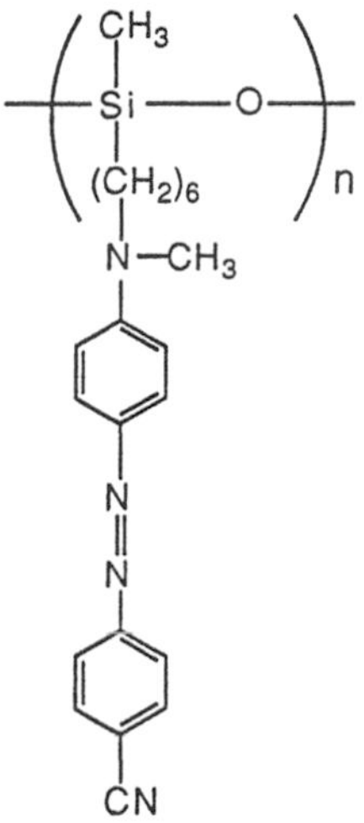

(LCP4)

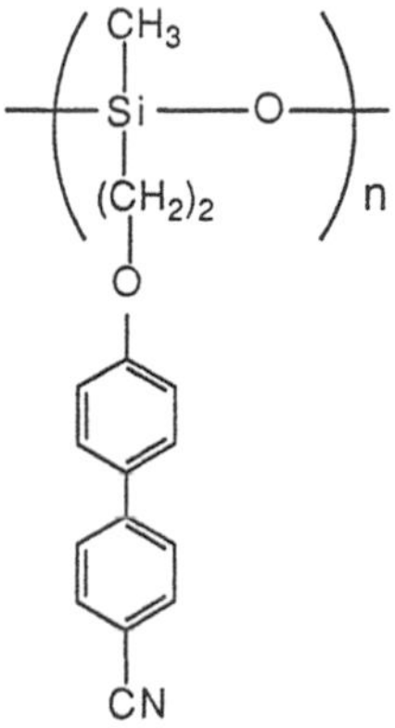

(LCP5)

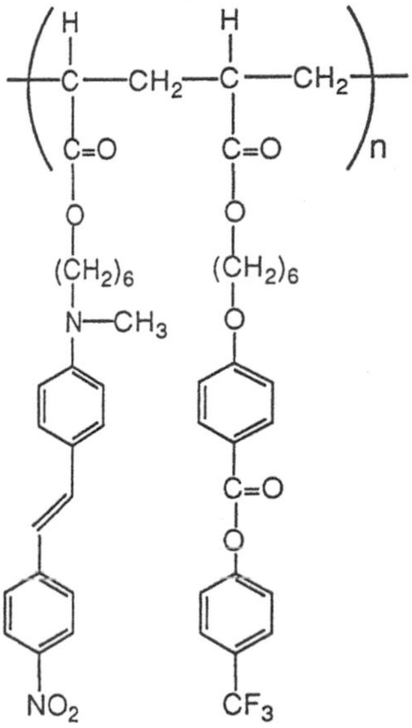

(LCP6)

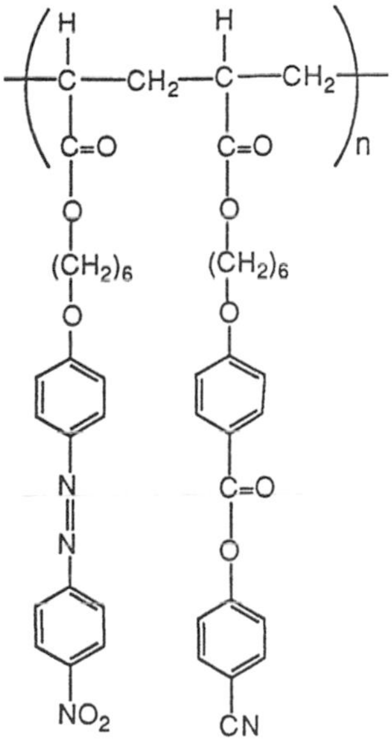

(LCP7)

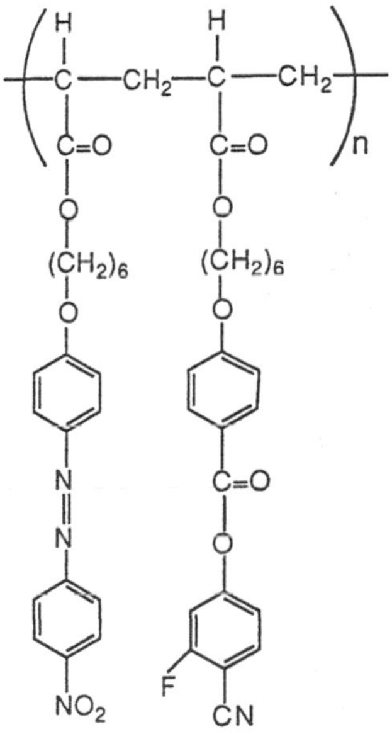

(LCP8)

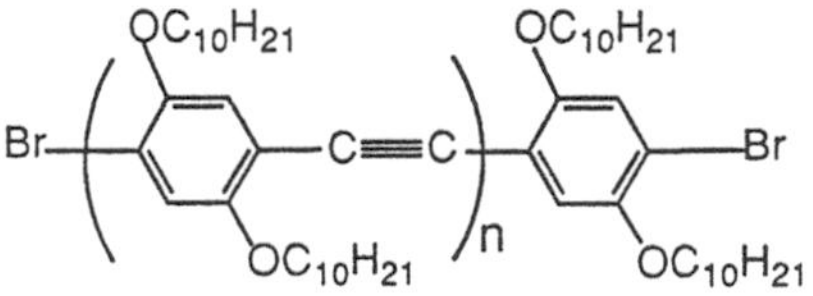

(LCP9)

The LCP3 showed the largest $\chi^{(3)}$ value of 5.8×10^{-12} esu at 1.064 μm. The largest $\chi^{(3)}$ value for all LCPs at 1.064 μm were due to three-photon resonance enhancement because the third harmonic at 355 nm lies within the absorption region. The LCPs having absorption at 526 nm, particularly with azo dye chromophores, show resonance enhancement at higher wavelengths compared to biphenyl side groups which have very low absorption in the visible region. The significant enhancement of $\chi^{(3)}$ values of the LCPs with azo and stilbene chromophores is caused by increased π-electron conjugation compared with biphenyl and benzoate groups. The polar benzoate side group in LCPs inhibit π-electron delocaliza-

Table 64 Liquid Crystal Polymers, Material Phases, and Third-order NLO Susceptibilities

Polymer	Glass Transition (°C)	Material Phases	$\chi^{(3)}(-3\omega; \omega, \omega, \omega)$ (10^{-13} esu)	
			1.064 μm	1.579 μm
LCP1	50	Meso 160°C	22	13.1
LCP2	32	Meso at 80°C	3.7	1.2
		Smectic at 124.5°C		
		Nematic at 132°C		
CLP3	40.5	Nematic at 104°C	58	5.6
LCP4	30	Smectic at 125°C	53.4	10
LCP5	30	Smectic-A at 145°C	4.6	0.6
LCP6	59	Smectic-A at 110°C	4	1.3
LCP7	14	Smectic-A at 90°C	16.1	3.2
		Nematic at 107.5°C		
LCP8	10	Meso at 90°C	3.5	2.1
LCP9	–	Smectic-A at 65°C	35	21
		Nematic at 100°C		

From Smith and Coles, *Liq. Cryst.*, 14, 937, 1993. With permission.

tion which causes a significant decrease in the $\chi^{(3)}$ values at both wavelengths. The $\chi^{(3)}$ values of these LCPs similar to those of typical π-conjugated polymers.

H. BIOMATERIALS

Third-order NLO properties of naturally occurring chemical species in Chinese tea, herbal medicine, black tea, and solutions of chloroform have been the suject of interesting studies. Zhang and co-workers[458] for the first time reported the nonlinear refractive index n_2 of Chinese green tea dissolved in ethyl alcohol. The nonlinear refractive index n_2 of 2.8×0^{-5} esu, as high as 10^6 times larger than that of carbon disulfide was measured by self-focusing and self-trapping modulation of a cw 632.8 nm He-Ne laser beam. The response time for both self-focusing and laser heating effects was 100 ms. Chlorophyll in green tea was considered responsible for observed large nonlinear Kerr effects. A similar type of nonlinear effect was also reported by Lin et al.[459] for a solution of chlorophyll in ethyl alcohol. They described the origin of nonlinear effect as thermal and negative intensity dependent and found that ink dissolved in alcohol behaved the same as chlorophyll. Lai and Diels[460] found a negative intensity-dependent index in the alcohol solution of green tea. Similar measurements performed on chlorophyll in water and alcohol identified chlorophyll as the responsible chemical species for the observed optical nonlinearity in green tea and its thermal origin. He et al.[461] observed multiple diffraction ring patterns induced by ultrafast laser at 532 nm from the solution of Chinese green tea. The optical nonlinearity was related to the reorientation and redistribution of the large and anisotropic biomolecules existing in the tea solution.

Cheung and Gayen[462] measured third-order optical nonlinearities of black-tea extract dissoved in water by z-scan and four-wave mixing techniques with 7 ns, 532 nm pulses. The z-scan measurements yielded real and imaginary $\chi^{(3)}$ values of -1.9×10^{-12} and -1.2×10^{-13} esu for the tea solution in water at 532 nm, respectively, and a linear absorption coefficient of 1.4 cm^{-1}. The thermally induced $\chi^{(3)}$ value was about 5.3×10^{-8} esu, 10^4 times larger than that of carbon disulfide. The nonlinear refraction and absorption coefficients were -5.7×10^{-18} m^2/W and -0.84×10^{-11} m^2/W, respectively. From the DFWM experiments, a $\chi^{(3)}$ value of 2.0×10^{-12} esu for the tea solution in water was measured, which was in good agreement with z-scan value. The biochemical species responsible for the large optical nonlinearity in the green tea solution in alcohol were identified as chlorophyll pigment, whereas oxidized polyphenolic compounds such as theaflavins and the arubigins were identified in the black tea, although other pigment may also have made some contributions. The main pigments in fresh tea solution are chlorophyll and carotenoids. Theaflavins showed a bright reddish color in solution whereas the arubigins were reddish brown pigments. Theaflavins and thearubigins were the oxidation products from the polymerization process in the tea fermentation.[463,464]

The nonlinear refractive index of a new kind of polymer doped with naturally occurring chlorophyll (structure shown below) were determined by self-focusing and saturable absorption.[465] The chromophores were extracted from Chinese tea in a chloroform solution. A new kind of polymer dissolved in chloroform was mixed with chlorophyll and 100 μm thin films were obtained from spin coating or solvent evaporation method. The solution extracted from Chinese tea was a mixture of several chromophores and the proportion of chlorophyll was rather small. From self-focusing experiments, a nonlinear refractive index n_2 of 1.0×10^{-7} cm^2/W was obtained which was 7 orders of magnitude larger than that of CS$_2$. The saturable absorption measurements yielded an n_2 value of 4.0×10^{-7} cm^2/W close to that of self-focusing method. The response time of this polymer-chlorophyll composite was less than 10 ns.

CH=CH$_2$ CH$_3$

H$_3$C

CH$_2$-CH$_3$

N   N

Mg

N   N

H$_3$C

CH$_3$

H$_3$C

Phytyl-O-CO-CH$_2$-CH$_2$

O

COOCH$_3$

Birge et al.[466,467] reported the third-order NLO properties of bacteriorhodopsin in the 0.0 to 1.2 eV optical region. Bacteriorhodopsin is a light transducing protein found in the purple membrane of *Halobacterium halobium*. This protein contains an all-*trans*-retinyl protonated Schiff base chromophore with six double bonds. The bacteriorhodopsin molecule consists of a 248-amino-acid polypeptide chain in which the retinylidene Schiff base chromophore is centered into the folded seven α-helices. On exposure to light, the bacteriorhodopsin undergoes a photocycle with a photo-induced *trans*- to *cis*-isomerization of the retinal. This photochromic effect indicates that bacteriorhodopsin could be an interesting biomaterial for photonic technologies.

For light-adapted bacteriorhodopsin, second hyperpolarizability $\gamma(-3\omega;\omega,\omega,\omega)$ values of $2,976 \times 10^{-36}$, $5,867 \times 10^{-36}$, $14,863 \times 10^{-36}$, $15,817 \times 10^{-36}$, and $10,755 \times 10^{-36}$ esu were estimated in D$_2$O at 0.25, 0.5, 0.66, 1.0, and 1.17 eV, respectively. The large third-order optical nonlinearities originate from the protein-bound chromophore of bacteriorhodopsin. The second hyperpolarizability was increased 5-fold by protonation of chromophore. Second hyperpolarizability of bacteriorhodopsin is similar to that of longer chain polyenes such as dodecapreno β-carotene that has 19 double bonds which exhibit second hyperpolarizability of $17,000 \times 10^{-36}$ esu at 0.66 eV. The protein-bound, positively charged chromophore in bacteriorhodopsin has a lowest lying, strongly allowed $^1B_u^+$ state which causes a large change in dipole moment upon excitation relative to the ground state ($\Delta\mu = 13.5$ D), this is responsible for a Type III enhancement of the second hyperpolarizability and yields at least 20-fold increases in γ values. Du and Liu[468] measured saturated absorption and the nonlinear refractive index of the LB films of bacteriorhodopsin found in the purple membrane of *H. halobium*. Bacteriorhodopsin consists of a 248-amino acid polypeptide chain folded into seven α-helices. The α-helices are arranged perpendicular to the plane of the purple membrane. Optical nonlinearity of bacteriorhodopsin was measured by the DFWM technique at 514.5 nm. The saturated absorption intensity of 0.42 cm^2 and the saturated nonlinear refractive index n_2 of 5×10^{-2} cm^2/W for the LB films were measured. The response time of the optical nonlinearity was a few milliseconds. Uji et al.[469] also reported the quantitative third-order susceptibility analysis of bacteriorhodopsin using picosecond resonant coherent anti-stoke Raman spectroscopy. Bacteriorhodopsin is promising material for optical switches for optical memory, holography recording, neural networks, and image processing.

III. MULTIPHOTON RESONANCE PHENOMENON

Kajzar and Messier[470] have provided an excellent description of various NLO phenomena, particularly as observed in π-conjugated organic polymers. The wavelength spectrum showing resonance effects and multiphoton resonances in $\chi^{(3)}$ values and nonlinear optical dichroism have been discussed in a variety of polymers. Two important features for determining the resonant and the nonresonant origins of the optical nonlinearity are the fundamental radiation of the laser system used to measure NLO susceptibility and the optical absorption characteristics of the individual material used. Resonance enhancement plays an important role in optical-frequency conversion. The $\chi^{(3)}$ value may have both a resonant and a nonresonant contribution:

$$\chi^{(3)} = \chi^{(3)}_{resonant} + \chi^{(3)}_{nonresonant}$$

Single-photon, two-photon, and three-photon resonance can be observed while measuring $\chi^{(3)}$ depending upon the wavelengths and technique. The origin of single-photon resonance lies in coinciding incident frequency with an allowed dipole transition from the ground state. Two-photon resonance results from the sum or difference of two input frequencies, whereas a three-photon resonance involves a combination of three incident frequencies. THG, DFWM, EFISH, and other measurement techniques can shed light on resonance effects. The wavelength dependence of $\chi^{(3)}$ on a variety of organic molecular and polymeric materials studied by various techniques are discussed here.

The magnitude of $\chi^{(3)}$ strongly depends on the measurement wavelength and its harmonic frequencies and the resonant $\chi^{(3)}$ values can be much larger than the nonresonant values as shown in many illustrations. Furthermore the $\chi^{(3)}(-3\omega;\omega,\omega,\omega)$, $\chi^{(3)}(-\omega;\omega,-\omega,\omega)$, $\chi^{(3)}(-2\omega;\omega,\omega,0)$, and $\chi^{(3)}(-\omega; 0, 0, \omega)$ values measured by the techniques of THG, DFWM, EFISH, and dc Kerr effect, respectively, are not expected to be the same because of the distinct NLO processes and resonance contributions associated with them. Moreover the experimental conditions, such as the measurement wavelength, pulse conditions, laser power, environment, and material states, can also influence the magnitude of optical nonlinearity. Consequently, all of these factors also make comparison of $\chi^{(3)}$ data difficult. An extensive survey of the literature on third-order NLO materials leads to the conclusion that the large difference in $\chi^{(3)}$ values can be reconciled from the viewpoints of optical process, or otherwise should be cautiously evaluated. A wide range of wavelength, for example, from 0.532 to 3.0 μm, is available with different laser systems for measuring third-order nonlinear susceptibility. We have taken the precaution, while mentioning $\chi^{(3)}$ values in the text, of describing the conditions of the measurement techniques and the measurement wavelength involved in different experimental procedures. The resonance enhancement of $\chi^{(3)}$ has already been discussed in a variety of materials such fullerenes, polyanilines, polyacetylenes, etc.

The wavelength dependence of the $\chi^{(3)}$ spectrum can shed light on multiphoton resonances occurring in organic molecular and polymeric materials. Comparing the fundamental radiation of a laser system used to measure $\chi^{(3)}$ values with that of optical absorption spectrum of an individual material used can determine the resonant and nonresonant contributions. The $\chi^{(3)}$ should have either a nonresonant value when there is no absorption in the medium at fundamental and harmonic wavelengths or a resonant value if the nonlinear medium is absorbing in the optical spectrum region. The $\chi^{(3)}$ value may fluctuate by several orders of magnitude between the resonant region and the nonresonant region. Single-photon (ω), two-photon (2ω), and three-photon (3ω) resonances can be detected while comparing wavelength-dependence $\chi^{(3)}$ value with the optical absorption spectrum. The single-photon (ω) resonance originates when coinciding incident frequency with an allowed dipole transition from the ground state. The origin of two-photon (2ω) resonance is the sum or difference of two input frequencies, whereas a three-photon (3ω) resonance results from a combination of three incident frequencies. Therefore, it becomes clear that one- two-, and three-photon resonances depend on the number of photons required to coincide with an excited level of an individual material. The value of the $\chi^{(3)}$ is increased remarkably by a one-photon resonance compared with the $\chi^{(3)}$ values of other resonance effects. The resonance contributions can be studied from THG, DFWM, EFISH, and other measurement techniques. The wavelength-dependent $\chi^{(3)}$ spectrum measured by THG yields information on two- and three-photon resonances. Multiphoton nonlinear interactions in charge-transfer complexes, fullerenes, π-conjugated polymers, and organometallic compounds have been investigated. Table 65 lists the existence of multiphoton resonances in organic molecular and polymeric materials measured by different experimental techniques.

Table 65 Multiphoton Resonances in Organic Molecules and Polymers

Materials	Measurement Technique	$\chi^{(3)}$ tensor	$\chi^{(3)}$ (10^{-12} esu)	Resonance	Ref.
C_{60} (thin film)	DFWM	$\chi^{(3)}(-\omega;\omega,\omega,-\omega)$	200	One photon (ω)	94
	THG	$\chi^{(3)}(-3\omega;\omega,\omega,\omega)$	61	Two photon (2ω)	99
	THG	$\chi^{(3)}(-3\omega;\omega,\omega,\omega)$	27	Three photon (3ω)	96
	THG	$\chi^{(3)}(-3\omega;\omega,\omega,\omega)$	87	Three photon (3ω)	100
C_{70} (thin film)	DFWM	$\chi^{(3)}(-\omega;\omega,\omega,-\omega)$	300	One photon (ω)	94
	THG	$\chi^{(3)}(-3\omega;\omega,\omega,\omega)$	26	Two photon (2ω)	101
	THG	$\chi^{(3)}(-3\omega;\omega,\omega,\omega)$	90	Three photon (3ω)	101
PDA-PTS (single crystal)	THG	$\chi^{(3)}(-3\omega;\omega,\omega,\omega)$	850	Three photon (3ω)	119
PDA-TCDU (single crystal)	THG	$\chi^{(3)}(-3\omega;\omega,\omega,\omega)$	160	Three photon (3ω)	119
PDA-4-BCMU (thin film)	EFISH	$\chi^{(3)}(-2\omega;\omega,\omega,0)$	150	Two photon (2ω)	123
PDA-AFA (LB flm)	EFISH	$\chi^{(3)}(-2\omega;\omega,\omega,0)$	8.0	Two photon (2ω)	125
PDA-C_4UC_n(thin film)	THG	$\chi^{(3)}(-3\omega;\omega,\omega,\omega)$	1,700	Three photon (3ω)	130
PDA-DCH (thin film)	THG	$\chi^{(3)}(-3\omega;\omega,\omega,\omega)$	70	Three photon (3ω)	153
trans-PA (thin film)	THG	$\chi^{(3)}(-3\omega;\omega,\omega,\omega)$	5,600	Three photon (3ω)	193
trans-PA (oriented film)	THG	$\chi^{(3)}(-3\omega;\omega,\omega,\omega)$	17,000	Three photon (3ω)	187
trans-PA (oriented film)	THG	$\chi^{(3)}(-3\omega;\omega,\omega,\omega)$	27,000	Three photon (3ω)	186
Astaxanthene (thin film)	THG	$\chi^{(3)}(-3\omega;\omega,\omega,\omega)$	120	Three photon (3ω)	219
PT (thin film)	DFWM	$\chi^{(3)}(-\omega;\omega,\omega,-\omega)$	6,300	One photon (ω)	240
PDTB(thin film)	THG	$\chi^{(3)}(-3\omega;\omega,\omega,\omega)$	11	Three photon (3ω)	245
PTT (thin film)	THG	$\chi^{(3)}(-3\omega;\omega,\omega,\omega)$	20	Three photon (3ω)	245
PTV (thin film)	THG	$\chi^{(3)}(-3\omega;\omega,\omega,\omega)$	450	Three photon (3ω)	246
PMTBQ (solution)	DFWM	$\chi^{(3)}(-\omega;\omega,\omega,-\omega)$	4,600	One photon (ω)	248
P3DDT (thin film)	DFWM	$\chi^{(3)}(-\omega;\omega,\omega,-\omega)$	4,500	One photon (ω)	263
PMMA-DR1 (thin film)	THG	$\chi^{(3)}(-3\omega;\omega,\omega,\omega)$	140	Two photon (2ω)	322
DIAM-Bis-A (thin film)	THG	$\chi^{(3)}(-3\omega;\omega,\omega,\omega)$	3.8	Two photon (2ω)	322
3DRCIN (thin film)	THG	$\chi^{(3)}(-3\omega;\omega,\omega,\omega)$	2.3	Two photon (2ω)	322
Polyaniline (thin film)	THG	$\chi^{(3)}(-3\omega;\omega,\omega,\omega)$	180	Three photon (3ω)	287
VONc derivative (thin film)	THG	$\chi^{(3)}(-3\omega;\omega,\omega,\omega)$	86	Three photon (3ω)	340
$[t\text{-}Bu_4PcRu(dib)]_n$ (thin film)	DFWM	$\chi^{(3)}(-\omega;\omega,\omega,-\omega)$	150,000	One photon (ω)	344
α-TiOPc (thin film)	THG	$\chi^{(3)}(-3\omega;\omega,\omega,\omega)$	159	Three photon (3ω)	345
α-VOPc (thin film)	THG	$\chi^{(3)}(-3\omega;\omega,\omega,\omega)$	99.5	Three photon (3ω)	345

IV. DYNAMICS OF ULTRAFAST OPTICAL NONLINEARITY

The recent advances in information processing have created the need for materials and technology that can provide ultrafast response time. As a result, recent research in nonlinear optics is aimed at substituting the electronics functions with photonics in which the speed of electricity is surpassed by the speed of light. This advent encourages us to do similar functions in processing technology with much faster speed. The response time may vary from millisecond to femtoseconds depending on the materials. In this case, highly π-conjugated systems seem very promising and, in particular, the π-conjugated polymeric materials have shown great potential. Conjugated polymers have been considered the most promising third-order NLO materials because they exhibit large optical nonlinearity coupled with ultrafast response speed on the scale of subpicoseconds. The dynamics of optical nonlinearity has been evaluated in organic third-order NLO materials by using techniques such as DFWM, OKE, pump and probe, saturation absorption, and photoluminescence. Kobayashi et al.[471–473] have studied the ultrafast NLO responses of polydiacetylenes and polythiophenes. The formation time and the decay time of self-trapped excitons are on the scale of femtoseconds and picoseconds, respectively. The formation time of the self-trapped excitons corresponds to the emission process of the strongly coupled phonon to the excitonic transitions. The formation times were 150 fs in PDA-3-BCMU and 140 fs in PDA-4-BCMU in the blue phase, 100 fs in red-oriented PDA-DFMP films, and approximately 70 fs in poly(3-methylthiophene) and 100 fs in poly(3-dodecylthiophene).[473] The difference in self-trapping time constants of excitons in PDA-3-BCMU, PDA-4-BCMU, and polythiophenes occurs because of the different chemical structures. Braunling and Becker[474] reported relaxation times in the range of 10 ps for poly(thienylmethylidenes) from saturable absorption and

Table 66 NLO Response Times of Various Organic $\chi^{(3)}$ Materials

Material	Response Speed	Ref.
Poly(3-methylthiophene)	0.62 ps (10 K)	471
Poly(3-dodecylthiophene)	0.45 ps (10 K)	471
P(3-MeTH/MMA)	2 ps	252
C_{60}	<35 ps	88
C_{60} (solution)	15 ~ 130 ps	475
C_{70}	200 fs	94
$[BEDT\text{-}TTF]_2I_3$	3800 fs	76
$[BEDT\text{-}TTF]_4Cu(SCN)_2]_2$	8000 fs	77
TTF derivative	25 ps	178
PDA-PTS	300 fs	120
PDA-PTS (single crystal)	1.8 ps	121
PDA-3BCMU	1.5 ps	471
PDA-DFMP	100 fs	471
PDA-4BCMU	140 fs	471
Polyacetylene	10 fs	186
	>10 ps	188
	>10 ps	189
trans-Polyacetylene	0.1 ps	476
cis-Polyacetylene	1.0 ps	476
trans-Polyacetylene	0.1 ps	477
Polyaniline	<100 fs	287
Polypyrrole	5 ps	303
Polyacenes	50 ps	304
SiNc	16 ps	337
$LuPc_2$	<35 ps	338
$[t\text{-}Bu_4PcRu(dib)]_n$	70 ps	344
Mg-OPTAP	44 ps	337
PEDS	135 fs	379
Acridine yellow doped glass	1 ms	443
TEA-TCNQ (film)	75 ps	478
Rhodamine doped epoxy (film)	2.05 ns	479
Nitrobenzene	32 ps	480
Hexadecane	210 ps	480
Salol	550 ps	480
p-Ethoxy-benzylidene *p-n*-butylanilne	500 ps (at 84 °C)	481
p-Methoxy-benzylidene *p-n*-butylanilne	400 ps (at 56 °C)	481
SBAC polymer	6 ps	482

between 5 and 60 ps for the DFWM technique. Table 66 summarizes the response speed of organic molecular and polymeric materials which range from picoseconds to femtoseconds. Some organic NLO materials show two components: fast and slower. Figure 43 shows the decay time of a copolymer of 3-methylthiophene and methyl methacrylate indicating a response time of 2 ps. The response times of PDA-4BCMU, polythiophene, polysilane, and polyaniline are shown in Figure 44. Most of the organic molecular and polymeric materials show ultrafast NLO responses on the scale of subpicosecond and a large optical damage threshold.

V. NONLINEAR OPTICAL DICHROISM

Kajzar[483] has discussed excellently the dependence of $\chi^{(3)}$ on molecular orientation in conjugated polymers that have one-dimensional π-electron delocalization; these details are mentioned here. The enhancement of $\chi^{(3)}$ in 1 D conjugated polymers is called NLO dichroism. In the case of mono-orientation, all polymer chains are aligned parallel to each other as well as parallel to a preferential direction on the substrate. Le Moigne et al.[154] demonstrated the formation of mono-oriented thin films of a DCH diacetylene monomer on a phthalic acid of a potassium single-crystal substrate. The $\chi^{(3)}$ tensor components can be given as:

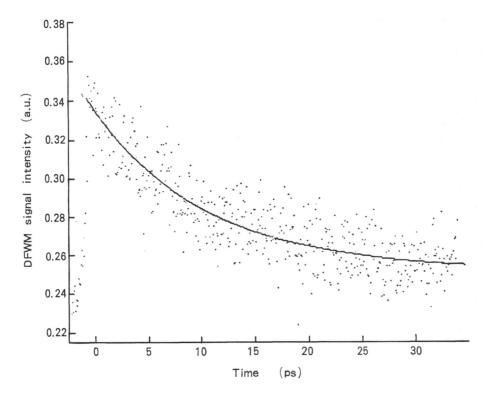

Figure 43 Decay time of a copolymer of 3-methylthiophene and methyl methacrylate. The NLO response seems restrained in the copolymer system because of the presence of nonconjugated MMA segments. (From Nalwa, *Thin Solid* Films, 225, 175, 1993. With permission.)

$$\chi^{(3)}_{xxxx} = NF\gamma_{xxxx} \cos^4 \Phi$$

$$\chi^{(3)}_{yyyy} = NF\gamma_{xxxx} \sin^4 \Phi$$

$$\chi^{(3)}_{xxyy} = \chi^{(3)}_{yyxx} = \chi^{(3)}_{xyxy} = 6NF\gamma_{xxxx} \sin^2 \Phi \cos^2 \Phi$$

where the *z*-axis is perpendicular to the substrate. This orientation could be examined by the THG technique because the orientation of polymer chain could be studied by rotating the sample between crossed polarizers, along an axis parallel to the beam propagation direction, and perpendicular to the thin film surface. For mono-oriented films, the dependence of harmonic intensity on the rotation angle θ is stronger and can be given as:

$$I_{3\omega}\Theta = AI^3_\omega \cos^8 \Theta$$

where *A* is a Θ-independent proportionality factor. Figure 45 shows THG intensity as a function of rotation angle θ from a mono-orientation *p*-DCH thin film on KAP single-crystal substrate.[156] THG restricts the three electric field polarizations to be parallel hence crossed tensor terms can not be measured, although DFWM is suitable in such cases. Le Moigne et al.[154] reported anisotropy $\chi^{(3)}_{xxxx}/\chi^{(3)}_{yyyy}$ >140 from THG experiments. Similarly, oriented films of PDA-C$_4$UC$_n$, where *n* = 3, 4, and 5, showed very high anisotropy in $\chi^{(3)}$ from the THG measurements.[130]

In bioriented thin films, the susceptibility tensors are written as:

$$\chi^{(3)}_{xxxx} = NF\gamma_{xxxx} (\cos^4 \Phi + \sin^4 \Phi)$$

$$\chi^{(3)}_{yyyy} = NF\gamma_{xxxx} (\sin^4 \Phi + \cos^4 \Phi)$$

$$\chi^{(3)}_{xxyy} = \chi^{(3)}_{yyxx} = \chi^{(3)}_{xyxy} = 6NF\gamma_{xxxx} \sin^2 \Phi \cos^2 \Phi$$

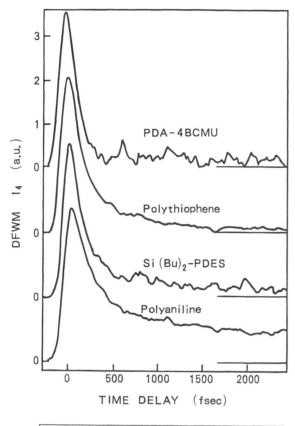

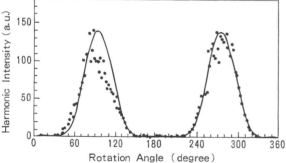

Figure 44 Decay time response for some of the π-electron-conjugated polymers: poly(4-BCMU), polythiophene, polyaniline, and a polysilane as a function of time decay in the DFWM experiment at 2 eV. (From Wong et al., *J. Appl. Phys.*, 70, 1896, 1991. With permission.)

Figure 45 Third-harmonic intensity as function of rotation angle θ from a mono-oriented p-DCH thin film on KAP single-crystal substrate. Points indicate the measured values and the solid line indicates the calculated values. (From LeMoigne et al., *Macromolecules*, 24, 2622, 1991. With permission.)

assuming polymer chains are aligned along the <100> or <010> direction of the (001) plane of cubic symmetry substrate. The dependence of harmonic intensity on the rotation angle can be given as:

$$I_{3\omega} = AI_\omega^3(\sin^4 \Theta + \cos^4 \Theta)^2$$

where A is again a Θ-independent proportionality factor. Figure 46 shows THG intensity as a function of rotation angle θ from a bioriented p-DCH 1300-Å-thick film grown on KBr substrate. Like mono-orientation, the crossed terms can not be seen from THG measurements.[154] The biorientation has been demonstrated in PDA thin films obtained by vacuum evaporation of a diacetylene monomer on a KBr thin crystal substrate and then applying thermal polymerization. The orthogonal biorientation can not be tested by the common linear dicroism technique because of the Pithogore law.

Orientation distribution depends on angle Φ. NLO dichroism has been discussed by Neher et al.[484] Embs et al.[382] reported that the third-order optical susceptibility $\chi^{(3)}$ of LB films of polysilanes are preferentially oriented into the dipping direction. The order parameter describing the degree of orientation with respect to the dipping direction was evaluated from the dichroic ratio of the band of the Si-Si

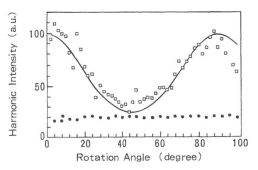

Figure 46 Third-harmonic intensity as function of rotation angle θ from a bioriented 1300 Å thick *p*-DCH film on KBr (open squares). Solid line indicates the least-square fit to experimental data. Closed circles show TH intensity from an unoriented film under the same experimental conditions. (From LeMoigne et al., *SPIE Proc.*, 1125, 9, 1990. With permission.)

chromophores at 397 nm for the *m*-alkoxyphenyl-substituted polysilane. The dichroic ratio of 2.9 for LB multilayers of poly(bis-*m*-butoxyphenylsilane) increases to more than 20 after annealing for 2 hr at 150°C and the spectra indicated a preferred orientation of polymer backbone along the dipping direction. All $\chi^{(3)}$ values were larger with the polarization of the fundamental light parallel to the dipping direction than that of perpendicular polarization showing the strong anisotropy.

The average for in-plane or in-space randomly distributed polymer chain is written as:

$$\langle \chi^{(3)} \rangle_{xxxx} = NF\gamma_{xxxx}\langle \cos^4 \Phi \rangle$$

where Φ is the angle between the polymer chain and the exciting optical fields assuming a priori as having the same direction[483,485]: 1 for a mono-orientation, i.e., all polymer chains parallel to a given direction; $<\cos^4 \Phi> = 3/8$ for a bidimensional disorder, i.e, all polymer chains parallel to a plane and randomly disoriented within this plane; 1/5 for a three-dimensional disorder. Therefore the $\chi^{(3)}$ value increases by a factor of 5 from a completely disordered system to a mono-oriented system. The angular dependence of $\chi^{(3)}$ has been shown for polydiacetylenes, polyacetylenes, and other organic polymers.

VI. PROGNOSIS

In summary, a very wide variety of organic materials ranging from liquids, dyes, π-conjugated polymers, dye-functionalized polymers, liquid crystals, organometallics, composites, and biomaterials have been used to study their third-order NLO properties. After almost a decade of evolution of third-order NLO materials, a new trend in fundamental understanding of the basic requirements for organic materials with large third-order NLO effects is emerging. One, two-, and three dimensional π-conjugated organic materials exhibit interesting NLO properties. It has become clear that the third-order optical nonlinearities can be significantly enhanced by applying molecular engineering strategies. Factors such as delocalization length, donor–acceptor substitution, conformation, chain orientation, dimensionality, charge-transfer complex formation, π-bonding sequence, and organometallic structures play important roles in determining the magnitude of $\chi^{(3)}$ values; these factors should be taken into consideration while designing novel organic structures possessing large $\chi^{(3)}$ values. The following criteria should be considered when designing an ideal third-order NLO material:

1. High π-electron density of NLO chromophores
2. Conjugated polymers with a degenerate ground state
3. Good optical transparency to reduce the optical loss
4. Molecular orientation for enhancing the $\chi^{(3)}$ value
5. Good processibility into thin films and optical quality
6. 1D π-conjugated materials are better than 2D and 3D π-conjugated materials
7. Donor and acceptor groups substituted to π-conjugated system are beneficial

The $\chi^{(3)}$ values of various classes of organic materials spans by more than 8 orders of magnitude between 10^{-5} and 10^{-13} esu at the resonance and off-resonance regions. The $\chi^{(3)}$ values as high as 1.4 esu have been measured for NLO dyes-glass composites although they have slow response speed on a millisecond scale. Organic π-conjugated materials such as polydiacetylenes, polyacetylenes, and

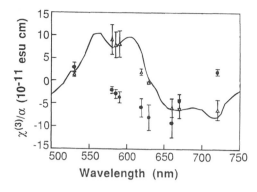

Figure 47 Wavelength dependence of the real part (solid circles) and imaginary part (open triangles) of the ratio of the $\chi^{(3)}$ values to linear absorption for a ladder copolymer. Solid line indicates the photo-induced absorption/bleaching response. (From Yu et al., *Organic Molecules for Nonlinear Optics and Photonics*, Kluwer Academic Publishers, Dordrecht, 1991. With permission.)

molecular aggregates (J-aggregates of pseudoisocyanines) show very large $\chi^{(3)}$ values and response times at a subpicosecond scale, although they suffer because of high optical losses. On the other hand, silica glass fiber show small $\chi^{(3)}$ values as well as low optical losses. Figure 47 shows the wavelength dependence of the real and imaginary parts of the ratio of the $\chi^{(3)}$ to linear absorption for a ladder copolymer.[486] Depending on the substituents and the nature of the heteroatom, the $\chi^{(3)}/\alpha$ values of 10^{-10} to 10^{-12} esu cm for thin films containing about 10% of electroactive five-ring ladders material have been reported. The $\chi^{(3)}/\alpha$ value on the order of 10^{-12} esu cm has been reported for Luttinger-type polyacetylene.[188] Most of the molecular and polymeric materials exhibit $\chi^{(3)}/\alpha$ values in the range of 10^{-15} to 10^{-13} esu cm. Figure 48 shows the NLO figure-of-merit ($\chi^{(3)}/\alpha$) versus response speed (τ) for a wide range of organic and inorganic materials currently available.[487] Organic molecular and polymeric materials currently fall behind inorganic nanocrystallites. A desired $\chi^{(3)}$ material should possess the following requisites to be useful for device applications in accordance with a target set for a Japanese National project:[69,191]

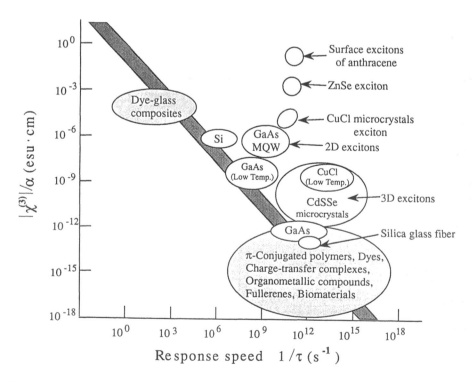

Figure 48 Third-order NLO figure-of-merit $\chi^{(3)}/\alpha$ versus response speed τ showing potential of various classes of inorganic and organic materials. (From Hanamura, *Quantum Optics*, Iwanami Press, Tokyo, 1992. With permission.)

1. High third-order optical nonlinearity ($\chi^{(3)} \geq 10^{-7}$ esu)
2. Low optical losses ($\alpha \leq 10^2$ cm^{-1})
3. Ultrafast response time (≤ 1 ps)

The most promising organic materials would be those possessing a combination of large NLO figure of merit (10^{-9} esu cm) coupled with ultrafast optical response. Therefore, the reduction of optical losses and enhancement of the $\chi^{(3)}$ are the key points to yield material that would be a good candidate for NLO devices. Currently, two combinations of third-order NLO materials are available; organic π-conjugated systems with large $\chi^{(3)}$ values and high optical losses, α of 10^{-1} cm^{-1} and silica glass fiber with small $\chi^{(3)}$ values and low optical losses, α of 10^{-6} cm^{-1}. Both of these classes have response speeds on a picosecond scale or higher. Organic materials with large third-order optical nonlinearity should have potential applications in photonic devices such as in optical fiber communication, data storage, optical computing, image processing, optical bistables, display, printers, dynamic holography, harmonic generators, frequency mixing, and optical switching. Several prototype devices have already been fabricated using organic third-order NLO materials. Two types of all-optical bistability devices were made from polydiacetylene LB films.[488] Thermal optical bistability at 514 nm using a Fabry-Perot cavity filled with a polydiacetylene/toluene solution or a polydiacetylene/polyvinyl chloride film has been reported.[489] The author suggested that thermally induced nonlinearities in organic materials may be an another way to reach $\chi^{(3)}$ values of 1 esu and larger, similar to the multi-quantum well structures and inorganic semiconductors. The optical bistable device is advantageous because of ultrafast switching and memory storage. A directional coupler was also fabricated from PDA-4-BCMU thin films.[490] A nonlinear Bragg mirror was fabricated from ultrathin layers of the saturably absorbing copolymer of siliconphthalocyanine and methylmethacrylate.[491] Optical limiting at an intensity of 5 to 10 MW/cm^2 for 6-ns pulses was demonstrated at a wavelength of 688 nm from a 23-layer stack. The all-optical phase modulation in waveguides from the single-crystal film of a PDA-PTS was demonstrated.[492] Two-photon absorption in polydiacetylene-based waveguides has also been reported.[483] The propagation losses of about 9 dB/cm and linear absorption coefficient α of 2 cm^{-1} for PDA-PTS and 2 to 22 dB/cm and α of 0.5 cm^{-1} for PDA-4-BCMU waveguides at 830 nm have been recorded.[494] Third-order NLO materials with $\chi^{(3)}$ values at shorter wavelengths are highly desirable for laser frequency conversion in the fabication of high-power laser sources that can be used in ultraviolet and near-infrared frequency regions. The applications of third-order NLO materials are attracting much attention from scientific community and various promising devices have been proposed and fabricated.[495–505] Both theoretical and experimental studies suggest that either π-conjugated polymers or organometallic complexes or inorganic nanostructured materials can fullfill the role of achieving large third-order optical nonlinearities. The demand for high-performance organic materials possessing large nonlinear optical figures of merit, high optical damage thresholds, ultrafast optical response, and good optical quality will continue and this can be achieved by designing novel molecules and polymers with the view of optimizing the optical nonlinearities and material performance by molecular engineering approaches. The development of NLO devices needs organic materials which must fit into the above-mentioned criterion. The availability of abundant synthetic approaches to amend organic materials into desired NLO devices for photonic industries should not be far from reach. Only time will tell how the scientific community will succeed in achieving the dreams of ultrafast technology.

ACKNOWLEDGMENTS

The author thanks Drs. A. Ticktin, K. H. Hass, A. Esser, and H. Saitoh of BASF Tsukuba, Drs. H. Matsuda and T. Kamata of the National Institute of Materials and Chemical Research, Dr. S. Okada of Tohoku University and Dr. J. Polin of Universität Innsbruk, Austria for providing their results on third-order NLO materials. The author is also grateful to Dr. Y. Imanishi for many valuable comments and assistance.

REFERENCES

1. **Armstrong, J. A., Bloembergen, N., Ducuing, P. S., and Pershan, P. S.,** Interactions between light waves in a nonlinear dielectric. *Phys. Rev.,* 127, 1918, 1962.
2. **Bloembergen, N. and Shen, Y. R.,** Quantum-theoretical comparison of nonlinear susceptibilities in parametric media, lasers, and Raman lasers. *Phys. Rev. A.;* 137, 133, 1964.
3. **Bloembergen, N.,** *Nonlinear Optics,* Benjamin, New York, 1965.

4. **Ward, J. F.,** Calculation of nonlinear optical susceptibilities using diagrammatic perturbation theory, *Phys. Rev.,* 37, 1, 1965.

5. **Orr, J. B. and Ward, J. F.,** Purterbation theory of the nonlinear optical polarization of an isolated system, *Mol. Phys.,* 20, 513, 1971.

6. **Flytzanis, C.,** Theory of nonlinear optical susceptibilities, in *Quantum Electronics,* Rabin, H., and Tang, C. C., Eds., Vol. 1, Academic Press, New York, 1975, 9.

7. **Fisher, R. A., Ed.,** *Optical Phase Conjugation,* Academic, New York, 1983.

8. **Chemla, D. S. and Zyss, J., Eds.,** *Nonlinear Optical Properties of Organic Molecules and Crystals,* Academic Press, New York, 1987.

9. **Zyss, J. Ed.,** *Molecular Nonlinear Optics,* Academic Press, New York, 1994.

10. (a) Thematic issue; Optical nonlinearities in chemistry, *Chem. Rev.,* 94, 3–278, 1994, and references therein. (b) **Li, D., Marks, T. J., and Ratner, M. A.,** Nonlinear optical phenomena in conjugated organic chromophores. Theoretical investigations via a π-electron formalison, *J. Phys. Chem.,* 96, 4325, 1992.

11. **Singh, S.,** in *Handbook of Laser Science and Technology,* Vol. III, Weber, M. J., Eds., CRC Press, Boca Raton, FL, 1986, 3.

12. **Yoon, H. N., Norwood, R. A., and Man, H. T.,** *Ullmann's Encyclopedia of Industrial Chemistry.* Vol. A17, VCH Publishers, Weinheim, 1991, 546.

13. **Shen, Y. R.,** *The Principle of Nonlinear Optics,* John Wiley & Sons, New York, 1984.

14. **Sheik-bahabe, M., Said, A. A., and Van Stryland, E. W.,** High-sensitivity, single-beam n_2 measurements, *Opt. Lett.,* 14, 955, 1989.

15. **Van Stryland, E. W., Sheik-bahabe, M., Said, A. A., and Hagan, D. J.,** Characterization of nonlinear optical absorption and refraction, *Prog. Crystl. Growth Character.,* 27, 279, 1993.

16. **Kobayashi, T.,** Introduction to nonlinear optical materials, *Nonlinear Opt.,* 1, 91, 1991.

17. **Miyata, S., Ed.,** *Proceedings of 5th Toyota Conference on Nonlinear Optics,* North-Holland, Amsterdam, 1992.

18. **Takagahara, T.,** Enhanced excitonic optical nonlinearity of quantum dot lattice, in *Nonlinear Optics,* Miyata, S., Ed., North-Holland Amsterdam, 1992, 85.

19. **Nasu, H., Uchigaki, T., Kamiya, K., Kanbara, H., and Kubodera, K.,** Nonresonant-type third-order nonlinearity of (PbO, Nb_2O_5)-TiO_2-TeO_2 glass measured by third-harmonic generation, *Jpn. J. Appl. Phys.,* 31, 3899, 1992.

20. **Kondo, T., Xu, C., and Ito, R.,** Linear and nonlinear optical properties of a natural quantum-well material ($C_1H_{21}NH_3)_2PbI_4$, in *Nonlinear Optics,* Miyata, S., Ed., North-Holland, Amsterdam, 1992, 323.

21. **Helfin, J. R., Cai, Y. M., and Garito, A. F.,** Dispersion measurements of electric-field-induced second-harmonic generation and third-harmonic generation in conjugated linear chains, *J. Opt. Soc. Am. B,* 8, 2132, 1991.

22. **Rodenberger, D. C., Helfin, J. R., and Garito, A. F.,** Excited-state enhancement of optical nonlinearities in linear conjugated molecules, *Nature,* 359, 309, 1992.

23. **Hosoda, M., Wada, T., Yamamoto, T., Kaneko, A., Garito, A. F., and Sasabe, H.,** Enhancement of third-order optical nonlinearities of soluble vanadyl phthalocyanines in doped polymer films, *Jpn. J. Appl. Phys.,* 31, 1071, 1992.

24. **Buchalter, B. and Meredith,** Third-order optical susceptibility of glasses determined by third harmonic generation, *Appl. Opt.,* 21, 3221, 1982.

25. **Kajzar, F. and Messier, J.,** Third-harmonic genertaion in liquids, *Phys. Rev. A,* 32, 2352, 1985.

26. **Mito, A.,** New reference materials for THG maker fringe measurement, in *Proceedings of the Second Symposium on Photonic Materials,* Tokyo, 1991, 126.

27. **Hermann, J. P.,** Absolute measurements of third order susceptibilities, *Opt. Commun.,* 9, 74, 1973.

28. **Moran, M. J., She, C. S., and Carman, R. L.,** Interferometric measurements of the nonlinear refractive index coefficient relative to CS_2 in laser-system related materials, *IEEE J. Quantum Electron.,* QE-11, 259, 1975.

29. **Pepper, D. M., Fakete, D., and Yariv, A.,** Observation of amplified phase-conjugate reflection and optical parametric oscillation by degenerate four-wave mixing in a transparent medium, *Appl. Phys. Lett.,* 33, 41, 1978.

30. **Sakaguchi, T., Shimizu, Y., Miya, M., Fukumi, T., Ohta, K., and Nagata, A.,** Third-order nonlinear optical susceptibilities of solutions of some mesogenic metallotetraphenylporphyrins by nanosecond degenerate four-wave mixing method, *Chem. Lett.,* 281, 1992.

31. **Xuan, N. P., Ferrier, J. L., Gazengel, J., and Rivoire, G.,** Picosecond measurements of the third-order susceptibility tensor in liquids, *Opt. Commun.,* 51, 433, 1984.

32. **Ho, P. P. and Alfano, R. R.,** *Phys. Rev.,* A20, 2170, 1979.

33. **Maker, P. D. and Terhune, R. W.,** *Phys. Rev.,* A137, 801, 1965.

34. **Palffy-Muhoray, P.,** The nonlinear optical response of liquid crystals, in *Liquid Crystals: Applications and Uses,* Vol. 1, Bahadur, B., Ed., World Scientific Publishing, Singapore, 1991, 493.

35. **Oudar, J. L. and Person, H. L.,** Second-order polarizabilities of some aromatic molecules, *Opt. Commun.,* 15, 258, 1975.

36. **Levine, B. F. and Bethea, C. G.,** Second and third-order hyperpolarizabilities of organic molecules, *J. Chem. Phys.,* 63, 2666, 1975.

37. **Kajzar, F. and Messier, F.,** Cubic hyperpolarizabilities and local electric field in alkanes and substituted alkanes, *J. Opt. Soc. Am. B,* 4, 1040, 1987.

38. **Meredith, G. R., Buchalter, B., and Hanzlik, C.,** Third-order susceptibility determination by third harmonic generation. II, *Chem. Phys.,* 78, 1543, 1983.

39. **Gonin, D., Noel, C., and Kajzar, F.,** EFISH and THG measurements on organic molecules in solution, *Nonlinear Opt.,* 8, 37, 1994.

40. **Burland, D. M., Walsh, C. A., Kajzar, F., and Sentein, C.,** Comparison of hyperpolarizabilities obtained with different experimental methods and theoretical techniques, *J. Opt. Soc. Am. B,* 8, 2269, 1991.

41. **Cheng, L., Tam, W., Meredith, G. R., Rikken, G., and Meijer, E.,** Nonresonant EFISH and THG studies of nonlinear optical property and molecular structure relations of benzene, stilbene and other arene derivatives, *SPIE Proc.,* 1147, 61, 1989.

42. **Shelton, D. P.,** Dispersion of the nonlinear susceptibility measured for benzene, *J. Opt. Soc. Am. B,* 2, 1880, 1985.

43. **Oudar, J. L.,** Optical nonlinearities of conjugated molecules. Stilbene derivatives and highly polar aromatic compounds, *J. Chem. Phys.,* 67, 446, 1977.

44. **Singer, K. D. and Garito, A. F.,** Measurements of molecular second order optical susceptibilities using dc induced second harmonic generation, *J. Chem. Phys.,* 75, 3572, 1981.

45. **Levine, B. F.,** Donor-acceptor charge transfer contributions to the second order hyperpolarizability, *Chem. Phys. Lett.,* 37, 516, 1976.

46. **Oudar, J. L. and Chemla, D. S.,** Hyperpolarizabilities of the nitroanilines and their relations to the excited state dipole moment, *J. Chem. Phys.,* 66, 2664, 1977.

47. **Levenson, M. D. and Bloembergen, N.,** Dispersion of the nonlinear optical susceptibilities of organic liquids and solutions, *J. Chem. Phys.,* 60, 1323, 1974.

48. **Levine, B. F. and Bethea, C. G.,** Molecular hyperpolarizabilities determined from conjugated and nonconjugated organic liquids, *Appl. Phys. Lett.,* 24, 445, 1974.

49. **Levine, B. F. and Bethea, C. G.,** Ultraviolet dispersion of the donor-acceptor charge transfer contribution to the second order hyperpolarizability, *J. Chem. Phys.,* 69, 5240, 1978.

50. **Kubodera, K.,** Measurements of third-order nonlinear optical efficiencies, *Nonlinear Opt.,* 1, 71, 1991.

51. **Kurihara, T, Kobayashi, H., Kubodera, K., and Kaino, T.,** $\chi^{(3)}$ evaluation for the molecular crystals of 4-(*N,N*-diethylamino)-4'-nitrostilbene (DEANS), *Chem. Phys. Lett.,* 165, 171, 1990.

52. **Kubodera, K. and Kaino, T.,** $\chi^{(3)}$ evaluation of organic nonlinear optical materials by THG measurements, in *Nonlinear Optics of Organics and Semiconductors,* Kobayashi, T., Ed., Springer-Verlag, Berlin, 1989, 163.

53. **Kurihara, T., Kobayash, H., Kubodera, K., and Kaino, T.,** Third-order nonlinear optical properties of DEANST: a new material for nonlinear optics, *Opt. Commun.,* 84, 149, 1991.

54. **Barzoukas, M., Fremaux, P., Josse D., Kajzar, F., and Zyss, J.,** Quadratic and cubic nonlinearities in solution of intramolecular charge transfer aromatic molecules: a quantum two-level approach, *Mater. Res. Soc. Symp. Proc.,* 109, 171, 1988.

55. **Meredith, G. R. and Buchalter, B.,** Solvent dependence of apparent third-order hyperpolarizability in para-nitroaniline, *J. Chem. Phys.,* 78, 1938, 1983.

56. **Hermann, J. P., Ricard, D., and Ducuing, J.,** Optical nonlinearities in conjugated systems: β-carotene, *Appl. Phys. Lett.,* 23, 178, 1973.

57. **Hermann, J. P. and Ducuing, J.,** Third-order polarizabilities of long chain molecules, *J. Appl. Phys.,* 45, 5100, 1974.

58. **Sakai, T., Kawabe, Y, Ikeda, H., and Kawasaki, K.,** Third-order nonlinear optical properties of retinal derivatives, *Appl. Phys. Lett.,* 56, 411, 1990.

59. **Ikeda, H., Sakai, T., and Kawasaki, K.,** Third-order optical nonlinearities of asymmetric carbocyanine dyes, *Chem. Lett.,* 1075, 1991.

60. **(a) Tomiyama, H., Matsuda, H., Okada, S., and Nakanishi, H.,** Third-order susceptibility of merocyanines enlarged by J-aggregation, in *Nonlinear Optics,* Miyata, S., Ed., Elsevier, Amsterdam, 1992, 305. (b) **Tomiyana, H., Okada, S., Matsuda, H., and Nakanishi, H.,** Novel third-order nonlinear optical materials composed of ionic polymers and chromophores. *SPIE Proc.,* 1337, 170, 1990.

61. **Kajzar, F., Girling, I. R., and Peterson I. R.** Third-order hyperpolarizability of centrosymmetric Langmuir-Blodgett films of stilbazolium dyes, *Thin Solid Films,* 160, 209, 1988.

62. **Stevenson, S. H., Donald, D. S., and Meredith, G. R.** Nonresonant third-order nonlinear polarizability in linear conjugated molecules, *Mater. Res. Soc. Symp. Proc.,* 109, 103, 1988.

63. **Kobayashi, S. and Sasaki, F.,** Enhancement of third-order optical nonlinearity in molecular aggregates: J-aggregates of pseudoisocyanine, *Nonlinear Opt.,* 4, 305, 1993.

64. **Blau, W.,** Organic materials for nonlinear optical devices, *Phys. Technol,* 18, 250, 1987.

65. **Maloney. C., Byrne, H., Dennis, W. M., Blau, W., and Kelly, J. M.,** Picosecond optical phase conjugation using conjugated organic molecules, *Chem. Phys.,* 121, 21, 1988.

66. **Penzkofer, A. and Leupacher, W.,** Third harmonic generation in organic dye solutions, in *Organic Materials for Non-linear Optics,* Special Publication No. 69, Hann, R. A. and Bloor, D., Eds., Royal Society of Chemistry, London, 1989, 203.

67. **Matsumoto, S., Kubodera, K., Kurihara, T., and Kaino, T.,** Effect of esterification on third harmonic generation

782

in several dyes, in *Nonlinear Optics of Organics and Semiconductors,* Vol. 36, Kobayashi, T., Ed., Springer-Verlag, Berlin, 1989, 236.

68. **Morita, K., Suehiro, T., Yokoh, Y., and Ashitaka, H.,** The development of organic third-order nonlinear optical materials, *J. Photopolym. Sci. Technol.,* 6, 229, 1993.

69. **Ashitaka, H.,** *Basic Technologies for Future Industries,* 2nd Symposium on Photonics Materials, extended abstracts, 1991, 174.

70. **Ashitaka, H., Yokoh, Y., Yokozawa, T., Morita, K., and Suehiro, T.,** *Sen-i-Gakkai Sympo.,* Preprints, 1992.

71. **Ashitaka, H., Yokoh, Y., Shimizu, R., Yokozawa, T., Morita, K., and Suehiro, T.,** *Nonlinear Opt.,* 4, 281, 1993.

72. **Ashitaka, H. and Sasabe, H.,** Chiral NLO materials; helicenes and other possible molecules. Nonlinear Opt., 14, 81, 1995.

73. **Bollaert, V., Schryver, F. C. D., Tackx, P., Persoons, A., Nusselder, J. J. H., and Put, J.,** A new fluorescent probe for network formation with interesting nonlinear optical properties, *Adv. Mater.,* 5, 268, 1993.

74. **Hattori, Y., Mizoguchi, A., Kubata, M., Uemiya, T., Tanaka, G., and Matsuoka, M.,** Third-order nonlinear optical properties of quninoid dyes, *SPIE Proc.,* 2285, 266, 1994.

75. **Huggard, P. G., Blau, W., and Schweitzer, D.,** Large third-order optical nonlinearity of the organic metal α-[bis(ethylenedithio)tetrathiofulvalne]triiodide), *Appl. Phys. Lett.,* 51, 2183, 1987.

76. (a) **Truong, K. D., Grenier, P., Houde, D., and Bandrauk, A. D.,** Stdies of third-order nonlinear optical susceptibility and the metal-insulator phase transition of α-(BEDT-TTF)$_2$I$_3$ via femtosecond four-wave mixing, *Chem. Phys. Lett.,* 196, 280, 1992. (b) **Truong, K. D., Grenier, P., Houde, D., and Bandrauk, A. D.,** Third-order optical nonlinearity of the α-phase of BEDT-TTF complexes and their phase transition by femtosecond nondegenerate four wave mixing, *Synth. Met.,* 57, 3968, 1993.

77. **Rangelo-Rojo, R., Kar, A-K., Wherrett, B. S., Carroll, M., Cross, G. H., and Bloor, D.,** Third-order optical nonlinearities of a polymeric film doped with a novel zwitterion, DEMI-3CNQ, Revista Mexicana de Fisica, 41, 832, 1995.

78. **Gotoh. T., Kondoh, T., Egawa, K., and Kubodera, K.,** Exceptionally large third-order optical nonlinearity of organic charge transfer complex, *J. Opt. Soc. Am. B,* 6, 703, 1989.

79. **Gong, Q., Xia, Z., Zou, Y. H., Meng, X., Wei., and Li, F.,** Large nonresonant third-order hyperpolarizabilities of organic charge transfer complexes, *Appl. Phys. Lett.,* 59, 381, 1991.

80. **Walker Jr., J. R.,** Ed., *Chemistry and Physics of Carbon,* Marcel Dekker, New York, 1967.

81. **Kroto, H. W., Health, J. R., O'Brien, S. C., Curl, R. F., and Smalley, R. E.,** C$_{60}$: Buckminsterfullerene, *Nature,* 318, 162, 1985.

82. **Kroto, H. W., Allaf, A. W., and Balm, S. P.,** C$_{60}$: Buckminsterfullerene, *Chem. Rev.,* 91, 1213, 1991.

83. **Taliani, C., Ruani, G., and Zamboni, R.,** Eds., *Fullerenes: Status and Perspectives,* World Scientific Publishing, Singapore, 1992.

84. **Billups, W. E. and Ciufolini, M. A.,** Eds., *Buckminsterfullerenes,* VCH Publishers, Weinheim, 1993.

85. **Blau, W. J., Byrne, H. J., Cardin, D. J., Dennis, T. J., Hare, J. P., Kroto, H. W., Taylor, R., and Walton, D. R. M.,** Large infrared nonlinear optical response of C$_{60}$, *Phys. Rev. Lett.,* 67, 1423, 1991.

86. **Henari, F., Callaghan, J., Stiel, H., Blau, W., and Cardin, D. J.,** Intensity-dependent absorption and resonant optical nonlinearity of C$_{60}$ and C$_{70}$ solutions, *Chem. Phys. Lett.,* 199, 144, 1992.

87. **Hoshi, H., Nakamura, N., Maruyama, Y., Nakagawa, T., Suzuki, S., Shiromaru, H., and Achiba, A.,** Optical second- and third-harmonic generation in C$_{60}$ film, *Jpn. J. Appl. Phys.,* 30, L1397, 1991.

88. **Kafafi, Z. H., Lindle, J. R., Pong, R. G. S., Bartoli, F. J., Lingg, L. J., and Milliken, J.,** Off-resonance nonlinear optical properties of C$_{60}$ studied by degenerate four-wave mixing, *Chem. Phys. Lett.,* 188, 492, 1992.

89. **Knize, R. J. and Partanen, J. P.,** Comments on large infrared nonlinear optical response of C$_{60}$, *Phys. Rev. Lett.,* 68, 2704, 1992.

90. **Gong, Q., Sun, Y., Xia, Z., Zou, Y. H., Gu, Z., Zhou, X., and Qing, D.,** Nonresonant third-order optical nonlinearity of all-carbon molecules C$_{60}$, *J. Appl. Phys.,* 71, 3025, 1992.

91. **Yang, S. C., Gong, Q., Xia, Z., Zou, Y. H., Wu, Y. Q., Qiang, D., Sun, Y. L., and Gu, Z. N.,** Large third-order nonlinear optical properties of fullerene in the infrared regime, *Appl. Phys.,* B55, 51, 1992.

92. **Zhang, Z., Wang, D., Ye, P., Li, Y., Wu, P., and Zhu, D.,** Studies of third-order optical nonlinearities in C$_{60}$-toluene and C$_{70}$-toluene solutions, *Opt. Lett.,* 17, 973, 1992.

93. **Vijaya, R., Murti, Y. V. G. S., Sundarajan, G., Mathews, C. K., and Rao, P. R. V.,** Degenerate four-wave mixing in the carbon cluster C$_{60}$, *Opt. Commun.,* 94, 353, 1992.

94. **Rosker, M. J., Marcy, H. O., Chang, T. Y., Khoury, J. T., Hansen, K., and Whetten, R. L.,** Time-resolved degenerate four-wave mixing in thin films of and using femtosecond optical pulses, *Chem. Phys. Lett.,* 196, 427, 1992.

95. **Aranda, F. J., Rao, D. V. G. L. N., Roach, J. F., and Tayebati, P.,** Third-order nonlinear optical interactions of C$_{60}$ and C$_{70}$, *J. Appl. Phys.,* 73, 7949, 1993.

96. **Meth, J. S., Vanherzeele, H., and Wang, Y.,** Dispersion of the third-order optical nonlinearity of C$_{60}$: a third-harmonic generation study, *Chem. Phys. Lett.,* 197, 26, 1992.

97. **Wang, Y. and Cheng, L. T.,** Nonlinear optical properties of fullerenes and charge-transfer complexes of fullerenes, *J. Phys. Chem.,* 96, 1530, 1992.

98. **Wang, X. K., Zhang, T. G., Lin, W. P., Liu, S. Z., Wong, G. K., Kappes, M. M., Chang, R. P. H., and Ketterson, J. B.,** Large second-harmonic response of C_{60} thin films, *Appl. Phys. Lett.,* 60, 810, 1992.

99. **Kajzar, F., Taliani, C., Zamboni, R., Rossini, S., and Danieli, R.,** Nonlinear optical properties of sublimed C_{60} thin films, *Synth. Met.,* 54, 21, 1993.

100. **Kajzar, F., Taliani, C., Danieli, R., Rossini, S., and Zamboni, R.,** Wave-dispersed third-order nonlinear optical properties of C_{60} thin films, *Chem. Phys. Lett.,* 217, 418, 1994.

101. (a) **Kajzar, F., Taliani, C., Danieli, R., Rossini, S., and Zamboni, R.,** Dispersion of third harmonic generation optical susceptibility in C_{70} thin films, *Phys. Rev. Lett.,* 73, 1617, 1994. (b) **Kajzar, F.,** Nonlinear optical properties of fullerenes, *SPIE Proc.,* 2025, 352, 1993.

102. **Neher, D., Stegeman, G. I., Tinker, F. A., and Peyghambarian, N.,** Nonlinear optical response of C_{60} and C_{70}, *Opt. Lett.,* 17, 1491, 1992.

103. **Flom, S. R., Pong, R. G. S., Bartoli, F. J., and Kafafi, Z. H.,** Resonant nonlinear optical response of the fullerenes C_{60} and C_{70}, *Phys. Rev. B,* 46, 15598, 1992.

104. **Lindle, J. R., Pong, G. S., Bartoli, F. J., and Kafafi, Z. H.,** Nonlinear optical properties of the fullerenes C_{60} and C_{70} at 1.064 μm, *Phys. Rev. B48,* 9447, 1993.

105. **Ji, W., Tang, S. H., Xu, G. Q., Chan, H. S. O., Ng, S. C., and Ng, W. W.,** Resonant optical nonlinearity of fullerenes in free-standing polymethyl methacrylate films, *J. Appl. Phys.,* 74, 3669, 1993.

106. **Flom, S. R., Sarkas, H. W., Pong, R. G. S., Bartoli, F. J., and Kafafi, Z. H.,** Nonlinear optical properties of a poly(phenylenevinylene)/C_{60} composite, *Polymer Preprints,* 35, 110, 1994.

107. **Berrada, M., Watanabe, T., and Miyata, S.,** Synthesis of C_{60} derivatives and their optical properties, *Sen-Gakkai Preprints,* 1994, G121, (in Japanese).

108. **Harigaya, K. and Abe, S.,** Dispersion of the third-order nonlinear optical susceptibility in C_{60} calculated with a tight-binding model, *Jpn. J. Appl. Phys.,* 31, L887, 1992.

109. **Li, J., Feng, J., and Sun, J.,** Quantum chemical calculations on the spectra and nonlinear third-order optical susceptibility of C_{60}, *Chem. Phys. Lett.,* 203, 560, 1993.

110. **Shuai, Z. and Bredas, J. L.,** Electronic structure and nonlinear optical properties of the fullerenes C_{60} and C_{70}: a valence-effective-Hamiltonian study, *Phys. Rev.,* B46, 16135, 1992.

111. **Blau, W. J. and Cardin, D. J.,** Nonlinear optical response of C_{60} and C_{70} fullerenes, *Mod. Phys. Lett. B,* 22, 1351, 1992.

112. **Nalwa, H. S.,** Organic materials for third-order nonlinear optics, *Adv. Mater.,* 5, 341, 1993.

113. **Skotheim, T. A.,** Ed., *Handbook of Conducting Polymers,* Vols. 1 and 2, Marcek Dekker, New York, 1986.

114. **Nalwa, H. S.,** ed., *Handbook of Organic Conductive Molecules and Polymers,* vols. 1–4, John Wiley & Sons, Chichester, 1997.

115. **Bloor, D. and Chance, R. R.,** Eds., *Polydiacetylenes: Synthesis, Structure, and Electronic Properties.* NATO ASI Series, Applied Science-no. 102, 1989.

116. **Bassler, H.,** Photopolymerization of diacetylenes, *Adv. Polym. Sci.,* 63, 1, 1984.

117. **Sixl, H.,** Spectroscopy of the intermediate states of the solid state polymerization reaction in diacetylene crystals, *Adv. Polym. Sci.,* 63, 49, 1984.

118. **Enkelmann, V.,** Structural aspects of the topochemical polymerization of diacetylenes, *Adv. Polym. Sci.,* 63, 91, 1984.

119. **Sauteret, C., Hermann, J. P., Frey, R., Pradere, F., Ducuing, J., Baugman, R. H., and Chance R. R.,** Optical nonlinearities in one-dimensional conjugated polymer crystals, *Phys. Rev. Lett.,* 36, 956, 1976.

120. **Carter, G. M., Takur, M. K., Chen, Y. J., and Hryniewicz, J. V.,** Time and wavelength resolved nonlinear optical spectroscopy of a polydiacetylene in the solid state using picosecond dye laser pulses, *Appl. Phys. Lett.,* 47, 457, 1985.

121. **Carter, G. M., Hryniewicz, J. V., Thakur, M. K., Chen, Y. J., and Mayler,** Nonlinear optical processes in polydiacetylene measured with femtosecond duration laser pulses, *Appl. Phys. Lett.,* 49, 998, 1986.

122. **Kajzar, F. and Messier, J.** Resonance enhancement in cubic susceptibility of Langmuir-Blodgett multilayers of polydiacetylene, *Thin Solid Films,* 132, 11, 1985.

123. **Chollet, P. A., Kajzar, F., and Messier, J.,** Nonlinear spectroscopy in polydiacetylenes, *Synth. Met.,* 18, 459, 1987.

124. (a) **Chollet, P. A., Kajzar, F., and Messier, J.,** in *Nonlinear Optics of Organics and Semiconductors,* Kobayashi, T., Ed., Springer, Berlin, 1988, 171. (b) **Kajzar, F., Messier, J., Nunzi, J. M., and Raimond, P.,** Third-order materials: processes and characterization, in *Polymers for Lightwave and Integrated Optics,* Hornak, L. A., ed., Marcel Dekker, New York, 1992, 595–645.

125. **Kajzar, F., Rothberg, L., Etemad, S., Chollet, P. A., Grec, D., Boudet, A., and Jedju, T.,** Saturation absorption and Kerr susceptibility in polydiacetylene Langmuir-Blodgett Film, *Opt. Commun.,* 66, 55, 1988.

126. **Kajzar, F. and Messier, J.,** Cubic nonlinear optical effects in conjugated polymers, *Polym. J.,* 19, 275, 1987.

127. **Kajzar, F. and Messier, J.,** *Thin Solid Films,* 99, 109, 1983.

128. **Kajzar, F., Messier, J., Zyss, J., and Ledoux, I.,** Nonlinear interferrometry in Langmuir-Blodgett multilayers of polydiacetylene, *Opt. Commun.,* 45, 133, 1983.

129. **Berkovic, G. and Popvitz-biro, R.,** Second- and third-harmonic generation by Z-type Langmuir-Blodgett multilayer films, *Mol. Cryst. Liq. Cryst.,* 211, 189, 1992.

130. **Kanetake, T., Ishikawa, K., Hasegawa, T., Koda, T., Takeda, K., Hasegawa, M., Kubodera, K., and Kobayashi, H.,** Nonlinear optical properties of highly oriented polydacetylene evaporated films, *Appl. Phys. Lett.,* 54, 2287, 1989.

131. **Hasegawa, T., Kanetake, T., Ishikawa, K., Koda, T., Takeda, K., Kobayashi, H., and Kubodera, K.,** Nonlinear optical properties of blue and red phase polydiacetylene films, *Mol. Cryst. Liq. Cryst.,* 183, 329, 1990.

132. **Koda, T., Kanetake, T., Ishikawa, K., Hasegawa, T., and Takeda, K.,** Growth and characterization of highly oriented polydiacetylene films prepared by vacuum deposition method, in *Internation Workshop on Crystal Growth of Organic Materials (CGOM'l) Proc.,* Miyata, S., Ed., 1989, 134.

133. **Nakanishi, H., Matsuda, H., Okada, S., and Kato, M.** Evaluation of nonlinear optical susceptibility of polydiacety-lenes by third harmonic generation, *Polym. Adv. Technol.,* 1, 75, 1990.

134. **Matsuda, H., Molyneux, S., Kar, A. K., Okada, S., Nakanishi, H., and Wherrett, B. S.,** Third-order nonlinear optical properties of PDA-DCH thin single crystals, paper presented at 3rd International Symposium on Organic Materials for Nonlinear Optics, OMNO'92, University of Oxford, August 19–21, 1992.

135. **Okada, S., Matsuda, H., Nakanishi, H., Kato, M., and Otsuka, M.,** New amphiphilic diacetylene compounds for nonlinear optics, *Thin Solid Films,* 179, 423, 1989.

136. **Okada, S., Ohsugi, M., Masaki, A., Matsuda, H., Takaragi, S., and Nakanishi, H.,** Preparation and nonlinear optical properties of polydiacetylenes from unsymetrical diphenylbutadiynes with trifluoromethyl substituents, *Mol. Cryst. Liq. Cryst.,* 183, 81, 1990.

137. **Ohsugi, M., Takaragi, S., Matsuda, H., Okada, S., Masaki, A., and Nakanishi, H.,** Polydiacetylene with directly bound aromatic substituents for the enlarged third order nonlinearity, *SPIE Proc.,* 1337, 162, 1990.

138. **Doi, T., Okada, S., Matsuda, H., Masaki, A., Minami, N., Nakanishi, H., and Hayamizu, K.,** Third harmonic generation of a polydiacetylene with diacetylenic groups directly bound to the main chain, paper presented at the 52nd Meeting of the Japan Society of Applied Physics, October 9–12, 1991, Okayama, Japan.

139. **Philippart, V., Dumont, M., Nunzi, J. M., and Charra, F.,** Femtosecond Kerr ellipsometry in polydiacetylne solutions: two-photon effects, *Appl. Phys.,* A56, 29, 1993.

140. **Hsu, C. C., Kawabe, Y., Ho, Z. Z., Peyghambarain, N., Polky, J. N., Krug, W., and Miao, E.,** Comparison of the $\chi^{(3)}$ values of crystalline and amorphous thin films of 4-butoxy-carbonyl-methyl-urethane polydiacetylene at 1.06 and 1.3 μm, *J. Appl. Phys.,* 67, 7199, 1990.

141. **Doi, T., Okada, S., Matsuda, H., Masaki, A., Minami, N., Hayamizu, K., Kikuchi, N., and Nakanishi, H.** Third-order nonlinear optical properties of the polymers obtained by solid-sate polymerization of tetrayne compounds, *Polym. Preprints, Jpn.,* 41, 2914, 1992 (in Japanese).

142. **Kawanami, H., Okada, S., Matsuda, Doi, T., Kabuto, C., Oikawa, H., and Nakanishi, H.,** Solid-state polymeriza-tion of tetraynes with urethane groups and nonlinear optical properties of their polymers, *Polym. Preprints, Jpn.* 42, 2890, 1993 (in Japanese).

143. **Okada, S., Doi, T., Mito, A., Hayamizu, K., Ticktin, A., Matsuda, H., Kikuchi, N., Masaki, A., Minami, N., Haas, K. H., and Nakanishi, H.,** Synthesis and third-order nonlinear optical properties of a polydiacetylene from an octatetrayne derivative with urethane groups, *Nonlinear Opt.,* 8, 121, 1994.

144. (a) **Kawanami, H., Okada, S., Matsuda, H., Doi, T., Kabuto, C., Oikawa, H., and Nakanishi, H.,** Synthesis of diacetylene-substituted polydiacetylenes with arylcarbonylurethnae substituted and their optical properties, *Nonlinear Opt.,* 14, 97, 1995. (b) **Shimada, S., Matsuda, H., Masaki, A. Okada, S., and Nakanishi, H.,** Synthesis and third-order nonlinear optical properties of the polydiacetylenes with sulfur atoms directly bound to the main chain, 13, 57, 1995.

145. **Greene, B. I., Orenstein, J., Millard, R. R., and Williams, L. R.,** Nonlinear optical response in excitons confined to one dimension, *Phys. Rev. Lett.,* 58, 2750, 1987.

146. **Bogler, J., Harvey, T. G., Ji, W., Kar, A. K., Molyneux, S., Wherrett, B. S., Bloor, D., and Norman, P.,** Near-resonant third-order optical nonlinearities in p-toluene sulfonate polydiacetylene, *J. Opt. Soc. Am.,* B9, 1552, 1992.

147. **Vogtmann, Th., Schmid, W., Fehn, Th., Bauer, F., Bauer, I., and Schwoerer, M.,** Nonlinear optics in polydiacetylene single crystal: anisotropy, spectral response and thermo effect, *Synth. Met.,* 57, 4018, 1993.

148. **Wong, K. S., Han, S. G., and Vardeny, Z. V.,** Femtosecond degenerate four wave mixing studies of third-order electronic nonlinearities in conjugated polymers, *Synth. Met.,* 41, 3209, 1991.

149. **Tomaru, S., Kobodera, K., Zembutsu, S., Takeda, K., and Hasegawa, M.,** Optical third-harmonic generation from polydiacetylene thin films deposited by vacuum evaporation, *Electron Lett.,* 23, 595, 1987.

150. **Kubodera, K., and Kaino, T.,** $\chi^{(3)}$ evaluation of organic nonlinear optical materials by THG measurements, in *Nonlinear Optics of Semiconductors and Organic Materials,* Kobayashi, T., Ed., Springer-Verlag, Berlin, 1989, 163.

151. **Agrinskaya, N. V., Guk, E. G., and Remizova, L. A.,** Nonlinear optical properties modification in some substituted polydiacetylene due to change of basicity constant of the side groups, *Synth. Met.,* 54, 289, 1993.

152. **Nunzi, J. M. and Grec, D.,** Picosecond phase conjugation in polydiacetylene gels, *J. Appl. Phys.,* 62, 2198, 1987.

153. **LeMoigne, J. Thierry, A., Chollet, P. A., Kajzar, F., and Messier, J.,** Morphology, linear and nonlinear optical studies of poly[1,6-di(N-carbazolyl)-2,4-hexadiyne] thin films (p-DCH), *J. Chem. Phys.,* 88, 6647, 1988.

154. **LeMoigne, Kajzar, F., and Messier, J.,** Single orientation in polydiacetylene films for nonlinear optics, molecular epitaxy of 1,6-bis(9-carbazolyl)-2,4-hexadiyne on organic crystal, *Macromolecules,* 24, 2622, 1991.

155. **LeMoigne, J., Moroni, M., Coles, H., Thierry, A., and Kajzar, F.,** Toward oriented polymeric structures for cubic nonlinear optics, *Mater. Res. Soc. Symp. Proc.,* 247, 65, 1992.

156. **LeMoigne, J., Theirry, A., and Kajzar, F.,** Highly oriented films of polydiacetylene for nonlinear optics: epitaxial growth of DCH monomer and optical properties of polymerized films, *SPIE Proc.,* 1125, 9, 1990.

157. **Matsuda, H., Okada, S., and Nakanishi, H.,** Molecular and crystal engineering of polydiacetylenes for enlarged nonlinear optical susceptibility, paper presented at Proc. 5th International Conference on Conventional Photoactive Solids, Okazaki, Japan, October 13–17, 1991, 264.

158. **Kurihara, T., Kubodera, K., Matsumoto, S., and Kaino, T.,** *Polym. Preprints Jpn.* 36, 1157, 1987 (in Japanese).

159. **Tomaru, S., Kobodera, K., Kurihara, T., and Zembutsu, S.,** *Jpn. J. Appl. Phys.,* 26, L1657, 1987.

160. **Kubodera, K. and Kobayashi, H.,** Determination of third-order nonlinear optical susceptibilities of organic materials by third-harmonic generation, *Mol. Cryst. Liq. Cryst.,* 182A, 103, 1990.

161. **Chollet, P. A., Kajzar, F., and Messier, J.,** Electric field induced optical second harmonic generation and polarization effects in polydiacetylene Langmuir-Blodgett multilayers, *Thin Solid films,* 132, 1, 1985.

162. **Kajzar, F., Ledoux, I., and Zyss, J.,** Electric field induced second harmonic generation in polydiacetylene solutions, *Phys. Rev.,* A36, 2210, 1987.

163. **Kajzar, F. and Messier, J.,** Third Harmonic generation and two photon absorption in polydiacetylene solutions, in *Polydiacetylenes: Synthesis, Structure, and Electronic Properties,* Bloor, D. and Chance, R. R., Eds., NATO ASI Series, Applied Science-No. 102, 1989, 325.

164. **Cheong, D. W., Kim, W. H., Samuelson, L. A., Kumar, J., and Tripathy, S. D.,** Oriented z-type Langmuir-Blodgett films from a soluble asymmetrically substituted polydiacetylene, *Macromolecules,* 29, 1416, 1996.

165. **Molyneux, S., Kar, A. K., Wherrett, B. S., Axon, T. L., and Bloor, D.,** Near-resonant refractive nonlinearity in polydiacetylene 9-BCMU thin films, *Opt. Lett.,* 18, 2093, 1993.

166. **Molyneux, S., Kar, A. K., Wherrett, B. S., Axon, T. L., and Bloor, D.,** Sub-picosecond investigation into the third-order nonlinear optical properties of polydiacetylenes, paper presented at the Organic Materials for Nonlinear Optics Meeting, 19–21 August, 1992, Oxford, UK.

167. **Axon, T. L., Bloor, D., Molyneux, S. Kar, A. K., and Wherrett, B. S.,** Characterization of the linear and nonlinear optical properties of polydiacetylenes, *SPIE Proc.,* 2025, 374, 1993.

168. **Byrne, H. J. and Blau, W.,** Multiphoton nonlinear interactions in conjugated organic polymers, *Synth. Met.,* 37, 231, 1990.

169. **Oikawa, H., Sekiya, J., Osawa, M., Wada, T., Yamada, A., Sasabe, H., and Uryu, T.,** A nonlinear optical waveguide of poly(1,9-decadiyne), *Polym. J.,* 23, 147, 1991.

170. **Yu, L., Chen, M., and Dalton, L. R.,** Synthesis and characterization of new polymers exhibiting large optical nonlinarity. II. Processible copolymers containing diacetylenic segments, *J. Polym. Sci. Polym. Chem. Ed.,* 29, 127, 1991.

171. **Kondo, K., Okuda, M., and Fujitani, T.,** Synthesis and nonlinear optics of soluble poly(phenylene-ethylene), *Macromolecules,* 26, 7382, 1993.

172. **Abdelaven, H., Frazier, D. O., Witherow, W. K., McManus, S. P., and Paley, M. S.,** Third-order nonlinear optical properties of polydiacetylenes, *Polym. Preprints,* 35, 249, 1994.

173. **Byrne, H. J., Blau, W., Giesa, R., and Schulz, R. C.,** Nonlinear optical studies of graded enyne oligomers, *Chem. Phys. Lett.,* 167, 484, 1990.

174. (a) **Bosshard, C., Spreiter, R., Gunter, P., Tykwinski, R. R., Schreiber, M., and Diederich, F.,** Structure-property relationships in nonlinear optical tetraethynylethenes, *Adv. Mater.,* 8, 231, 1996. (b) **Wautelet, P., Moroni, M., Oswald, L., Le Moigne, J., Pham, A., and Bigot, J.-Y.,** Rigid rod conjugated polymers for nonlinear optics. 2. Synthesis and characterization of phenylene-ethnylene oligomers, *Macromolecules,* 29, 446, 1996. (c) **Samuel, I. D. W., Ledoux, I., Delporte, C., Pearson, D. L., and Tour, J. M.,** Scaling of cubic polarizability with chain length in oligo(3-ethythiophene ethynylene)s, *Chem. Mater.,* 8, 819, 1996.

175. (a) **Perry, J. W., Stiegman, A. E., Marder, S. R., and Coulter, D. R.,** Second and third-order nonlinear optical properties of end-capped acetylenic oligomers, in *Organic Materials for Nonlinear Optics,* Hahn, R. A. and Bloor, D., Eds., Royal Society of Chemistry, London, Special Publication No. 69, 1989, 189. (b) **Perry, J. W.,** Nonlinear optical properties of molecules and materials, in *Materials for Nonlinear Optics: Chemical Perspectives,* Marder, S. R., Sohn, J. E., and Stucky, G. D., Eds., American Chemical Society Symposium Series 455, American Chemical Society, Washington, DC, 1991, 67.

176. **Yuan, Z., Taylor, N. J., Sun, Y., and Marder, T. B.,** Synthesis and second order nonlinear optical properties of three coordinate organboranes with diphenylphosphino and ferrocenyl groups as electron donor, *J. Organomet. Chem.,* 449, 27, 1993.

177. **Fleitz, P. A., McLean, D. G., and Sutherland, R. L.,** Nonlinear absorption in diphenyl butadiene, diphenyl butadiyne and a series of acetylene compounds using the Z-scan technique, *SPIE Proc.,* 2229, 33, 1994.

178. **Sylla, M., Zaremba, J., Chevalier, R., Rivoire, G., Khanous, A., and Gorgues, A.,** Third-order nonlinear optical susceptibility of acetylenic analogues of tetrathiafulvalene, *Synth. Met.,* 59, 111, 1993.

179. **Holvorson, C., Moses, D., Hagler, T. W., Cao, Y., and Heeger, A. J.** Frequency dependence of third harmonic generation in cis and trans-polyacetylene: importance of the degenerate ground state to nonlinear optical responses, *Nonlinear Optics,* Miyata, S., Ed., Elsevier, New York, 1992, 71.

180. **Sinlair, M., McBranch, D., Moses, D., and Heeger, A. J.,** Nonlinear optical conjugated polymers, *Synth. Met.,* 28, D645, 1989.

181. **Halvorson, C., Hagler, T. W., Moses, D., Cao, Y., and Heeger, A. J.,** Conjugated polymers with degenerate ground state: the route to high performance third-order nonlinear optical response, *Synth. Met.,* 57, 3961, 1993.

182. **Halvorson, C. and Heeger, A. J.,** Two-photon absorption spectrum of oriented *trans*-polyacetylene,*Chem. Phys. Lett.,* 216, 488, 1993.

183. **Wintner, E., Krausz, F., and Leising, G.,** Measurement of the third-order susceptibility of oriented trans-polyacetylene, *Synth. Met.,* 28, D155, 1988.

184. **Etemad, S. and Baker, G. L.,** Nonlinear optical response in polyacetylene and polydiacetylenes, *Synth. Met.,* 28, D159, 1988.

185. **Wong, K. S. and Vardeny, Z.,** Measurements of $\chi^{(3)}(\omega;\omega,\omega,-,\omega)$ in conducting polymers at λ = 620 nm, *Synth. Met.,* 49, 13, 1992.

186. **Krausz, F., Wintner, E., and Leising, G.,** Optical third-harmonic generation in polyacetylene, *Phys. Rev.,* B39, 3701, 1989.

187. **Drury, M. R.,** Observation third-harmonic generation in oriented Durham polyacetylene, *Solid State Commun.,* 68, 417, 1988.

188. (a) **Esser, A., Haas, K. H., Saitoh, H., Schrof, W., and Wuensch, J. R.,** Organic conjugated polymer films for nonlinear optics, Proceedings of International Symposium on Nonlinear Photonics Materials, May 24–25, 1994, Tokyo, Japan. (b) **Schrof, W., Wünsch, J. R., Esser, A., and Naarmann,** Nonlinear optics of conjugated polyacetylene thin films produced by different preparation techniques, *Nonlinear Opt.,* 10, 69, 1995.

189. **Beddard, G. S., McFadyen, G. G., Reid, G. D., and Thorne, J. R. G.,** Femtosecond optical processes in Durham polyacetylene and norbornylene-polyacetylene copolymer, *Chem. Phys.,* 172, 363, 1993.

190. **Kajzar, F., Etemad, S., Baker, G. L., and Messier, J.,** Frequency dependence of the large, electronic $\chi^{(3)}$ in polyacetylene, *Solid State Commun.,* 63, 1113, 1987.

191. **Ticktin, A.,** Organic conjugated polymer films for nonlinear optics, paper presented at the Proc. the 2nd Symposium on Photonic Materials, Basic Technologies for Future Industries, October 22–23, 1991, Tokyo, Japan.

192. **Dorsinville, R., Yang, L., Alfano, R. R., Tubino, R., and Destri, S.,** Picosecond nonlinear optical response using phase conjugation in soluble polyacetylene, *Solid State Commun.,* 68, 875, 1988.

193. **Fann, Fann, W. S., Benson, S., Madey, J. M. J., Etemad, S., Baker, G. L., and Kajzar, F.,** Spectrum of $\chi^{(3)}(-3\omega;\omega,\omega,\omega)$ in polyacetylene: an application of the free electron laser in nonlinear optical spectroscopy, *Phys. Rev. Lett.,* 62, 1492, 1989.

194. **Neher, D., Wolf, A., Bubeck, C., and Wegner, G.,** Third harmonic generation in polyphenylacetylene: exact determination of nonlinear optical susceptibility in ultrathin films, *Chem. Phys. Lett.,* 163, 116, 1989.

195. **Neher, D., Kalbeitzel, A., Wolf, A., Bubeck, C., and Wegner, G.,** Linear and nonlinear optical properties of substituted polyphenylacetylene thin films, *J. Phys. D: Appl. Phys.,* 24, 1193, 1991.

196. **Sasabe, H. and Wada, T.,** Third-order optical nonlinearities in π-conjugated systems, Proceedings of International Symposium on Nonlinear Photonics Materials, May 24–25, 1994, Tokyo, Japan.

197. **Olah, G. A., Zadok, E., Edler, R., Adamson, D. H., Kasha, W., and Surya Prakash, G. K.,** Friedel-Crafts dehydrohalogenative polymerization of acetyl and enolizable substituted acetyl halide (polyoxyacetylene), *J. Am. Chem. Soc.,* 111, 9123, 1989.

198. **Le Moigne, J., Hilberer, A., and Kajzar, F.,** Polyacetylenes with cholesteryl side groups: synthesis, linear and nonlinear optical properties of soluble conjugated polymers, *Makromol. Chem.,* 192, 515, 1992.

199. **Le Moigne, J., Hilberer, A., and Strazielle, C.,** Poly(phenylacetylene) derivatives for nonlinear optics, *Macromolecules,* 25, 6705,1992.

200. **Park, S. Y., Cho, H. N., Kim, N., Park, J. W., Jin, S. H., Choi, S. K., Wada, T., and Sasabe, H.,** Third-order optical nonlinearity of conjugated poly(4,4-disubstituted-1,6-heptadiyne)s, *Appl. Phys. Lett.,* 65, 289, 1994.

201. **Kim, S. H., Choi, S. K., Baek, S. H., Kim. S. C., Park, S. H., and Kim, H. K.,** Third-order nonlinear optical properties of poly(1,6-heptadiyne) derivatives, *Polym. Preprints,* 35, 244, 1994.

202. **Lee, H. J., Kang, S. J., Kim, H. K., Cho, H. N., Wu, J. W., and Choi, S. K.,** Synthesis and characterization of novel side chain NLO polymers based on poly(1,6-heptadiyne) derivatives, *Polym. Preprints,* 35, 142, 1994.

203. **Marder, S. R., Perry, J. W., Klavetter, F. L., and Grubbs, R. H.,** Third-order susceptibilities of soluble polymers derived from the ring-opening metathesis copolymerization of cyclooctatetraene and 1,5-cyclooctadiene, *Chem. Mater.,* 1, 171, 1989.

204. **Persoon, A., Wonterghem, B. V., and Tackx, P.,** Measurements of second hyperpolarizabilities of diphenylpolyenes by means of phase conjugate interferometry, *SPIE Proc.,* 1409, 220, 1991.

205. **Kawabe, Y., Sakai, T., Ikeda, H., Hasegawa, R., and Kawasaki, K.,** Reflection spectra and nonlinear optical response of a crystalline rhodanine derivative, in Extended abstracts of the 53rd Autumn Meeting, The Japan Society of Applied Physics, 1992, 1022 (in Japanese).

206. **Kajzar, F., Sentein, C., Zamboni, R., Ferracuti, P., Rossini, S., Danieli, R., Ruani, G., and Taliani, C.,** Third-order nonlinear optical response in thiophene substituted conjugated polyene: DTDMP, *Synth. Met.,* 1991, ISCM-90.

207. **Spangler, C. W., He, M. Q., Laquindanum, J., Dalton, L. R., Tang, N., Partanen, J., and Hellwarth,** Bipolaronic formation and nonlinear optical properties of bis-thienyl polyenes, *Mater. Res. Soc. Symp. Proc.,* 328, 655, 1994.

208. **Messier, J., Kajzar, F., Sentein, C., Borzoukas, M., Zyss, J., Blancharddesce, M., and Lehn, J. M.,** Chain-length dependence of the quadratic and cubic nonlinear optical susceptibilities of asymmetric push-pull polyenes, *Nonlinear Opt.,* 2, 53, 1992.

209. **Puccetti, G., Blanchard-Desce, M., Ledoux, I., Lehn, J. M., and Zyss, J.,** Chain-length dependence of the third-order polarizability of disubsituted polyenes. The effect of end groups and conjugation length, *J. Phys. Chem.,* 97, 9385, 1993.

210. **Marder, S. R., Beratan, D. N., and Cheng, L. T.,** *Science,* 252, 103, 1991.

211. **Marder, S. R., Perry, J. W., Tiemann, B. G., Gorman, C. B., Gilmour, S., Biddle, S., and Bourhill, G.,** Direct observation of reduced bond length alternation in donor/acceptor polyenes, *J. Am. Chem. Soc.,* 115, 2524, 1993.

212. **Marder, S. R., Perry, J. W., Bourhill, G., Gorman, C. B. Tiemann, B. G., and Mansour, K.,** *Science,* 261, 186, 1993.

213. **Bourhill, G., Cheng, L. T., Gorman, C. B., Lee, G., Marder, S. R., Perry, J. W., Perry, M. J., and Tiemann, B. G.,** Experimental demonstration of the relationship between the second- and third-order polarizabilities of conjugated donort-acceptor molecules, *SPIE Proc.,* 2143, 153, 1994.

214. **Samuel, I. D. F., Ledoux, I., Dhenout, C., Zyss, J., Fox, H. H., Schrock, R. R., and Silbey, R. J.,** Saturation of cubic optical nonlinearity in long chain polyene oligomers, *Science,* 265, 1070, 1994.

215. **van Beek, J. B., Kajzar, F., and Albrecht, A. C.,** Resonant third-harmonic generation in all-trans β-carotene: the vibronic origin of the third-order nonlinear susceptibility in the visible region, *J. Chem. Phys.,* 95, 6400, 1991.

216. **van Beek, J. B. and Albrecht, A. C.,** Third-harmonic generation from all-trans β-carotene in liquid solution, *Chem. Phys. Lett.,* 187, 269, 1991.

217. **van Beek, J. B., Kajzar, F., and Albrecht, A. C.,** Third-harmonic generation from all-trans β-carotene in polystyrene thin films: multiple reflection effects and the onset of a two-photon resonance, *Chem. Phys.,* 161, 299, 1992.

218. **Aramaki, S., Torruellas, W., Zanoni, R., and Stegeman, G. I.,** Tunable third-harmonic generation of trans β-carotene, *Opt. Commun.,* 85, 527, 1991.

219. **Haas, K. H., Ticktin, A., Esser, A., Fisch, H., Paust, J., and Schrof, J.,** $\chi^{(3)}$ dispersion measurements of the conjugated length dependence of acrotenoid derivatives, *J. Phys. Chem.,* 97, 8675, 1993.

220. **Craig, G. S. W., Cohen, R. E., Schrock, R. R., Silbey, R. J., Puccetti, G., Ledoux, I., and Zyss, J.,** Nonlinear optical analysis of a series of triblock copolymers containing model polyenes: the dependence of hyperpolarizability on conjugation length, *J. Am. Chem. Soc.,* 115, 860, 1993.

221. **Craig, G. S. W., Cohen, R. E., Schrock, R. R., Dhenaut, C., Ledoux, I., and Zyss, J.,** Synthesis and nonlinear optical analysis of triblock copolymers containing discrete polyenes, *Macromolecules,* 27, 1875, 1994.

222. **Nickel, E. G., Spangler, C. W., Tang, N., Hellwarth, R., and Dalton, L. R.,** Bipolaronic enhancement of $\chi^{(3)}$ in substituted bis-anthracenyl polyenes, *Nonlinear Opt.,* 6, 135, 1993.

223. **McBranch, D., Sinclaire, M., Heeger, A. J., Patil, A. O., Shi, S., Askari, S., and Wudl, F.,** Linear and nonlinear optical studies of poly(p-phenylene vinylene) derivatives and polydiacetylene-4BCMU, *Synth. Met.,* 29, E85, 1989.

224. **Bradley, D. D. C., and Mori, Y.,** Third harmonic generation in precursor route poly(p-phenylene vinylene), *Jpn. J. Appl. Phys.,* 28, 174, 1989.

225. **Kamiyama, K., Era, M., Tsutsui, T., and Saito, S.,** Third harmonic generation in highly oriented poly(2,5-dimethoxy-p-phenylene vinylene) films prepared by use of the Langmuir-Blodgett technique, *Jpn. J. Appl. Phys.,* 29, L840, 1990.

226. **Kaino, T., Kobayashi, H., Kubodera, K., Kurihara, T., Saito, S., Tsutsui, T., and Tokito, S.,** Optical third harmonic generation from poly(2,5-dimethox-*p*-phenylene vinylene) thin films, *Appl. Phys. Lett.,* 54, 1619, 1989.

227. **Kurihara, T., Mori, Y., Kaino, T., Murata, H., Takada, N., Tsutsui, T., and Saito, S.,** Spectra of $\chi^{(3)}(-\omega:\omega, \omega, \omega, \omega)$ in poly(2,5-dimethoxy *p*-phenylene vinylene) MO-PPV) for various conversion levels, *Chem. Phys. Lett.,* 183, 534, 1991. **Murata, H., Takeda, N., Tsutsui, T., Saito, S., Kurihara, T., and Kaino, T.,** Frequency dependence of third-order nonlinear susceptibilities in polyarylenevinylene thin films. *Synth. Met.,* 49–50, 131, 1992.

228. **Kaino, T. and Kurihara, T.,** Third-order nonlinear optical properties of polymeric materials, in *Polymers for Lightwave and Integrated Optics,* Hornak, L. A., Ed., Marcel Dekker, New York, 1992, 647.

229. **Bubeck, C., Kaltbeizel, A., Neher, D., Stenger-Smith, J. D., Wegner, G., and Wolf, A.,** Nonlinear optical properties of thin films of polymers with a one-dimensional conjugated π-electron system, in *Electronic Properties of Conjugated Polymers III,* Kuzmany, H., Mehring, M. and Roth, S., Eds., Springer-Verlag, Berlin, 1989, 214.

230. **Saitoh, H.,** Synthesis of full-conjugated polymers with benzofuran and benzothiophene ring and their electrical and nonlinear optical properties, Master's thesis, Mie University, Japan, 1992 (in Japanese).

231. **Matsuda, H., Sonoda, Y., Yase, K., Van Keuren, E., Suzuki, Y., Mito, M., and Takahashi, C.,** Nonlinear optical properties of linear π-conjugated oligomers: phenylenehexatrienylene derivatives, in 55th Autumn Meeting of the Japan Society of Applied Physics, 1994, Extended Abstract No. 20a-K-3, 1022 (in Japanese).

232. **Sonada, Y. and Matsuda, H.,** to be published.

233. **Wang, C. J., Lee, K. S., Prasad, P. N., Kim, J. C., Jin, J. I., and Kim, H. S.,** Study of third-order optical nonlinearity and electrical conductivity of sol-gel processed silica: poly(2-bromo-5-methoxy-p-phenylene vinylene) composite, *Polymer,* 33, 4145, 1992.

234. (a) **Hwang, D. H., Lee, J. I., Shim, H. K., Hwang, W. Y., Kim, J. J., and Jin, J. I.,** Synthesis of poly[(2-2-(4-cyanophenyl)ethynyl)-5-methoxy-1,4-phenylene) vinylene] and poly(1,4-phenylenevinylene) copolymers: electrical and the second and third-order nonlinear optical properties, *Macromolecules,* 27, 6000, 1994. (b) **Hwang, D. H., Lee, J. I., Shim, H. K., and Lee, G. J.,** Optical third-harmonic generation of poly(2-trimethylsilyl-1,4-phenylene vinylene and poly[2,5-bis(trimethylsilyl)1,4-phenylene vinylene], *Synth-Met.,* 71, 1721, 1995.

235. **Zhao, M., Cui, Y., Samoc, M., and Prasad, P. N.,** Influence of two-photon absorption on third-order nonlinear optical processes as studied by degenerate four-wave mixing: the study of soluble didecyloxy substituted polyphenyls, *J. Chem. Phys.,* 95, 3991, 1991.

236. **Marcy, H. O., Rosker, M. J., Warren, L. F., Reinhardt, B. A., Sinclair, M., and Seager, C. H.,** Time resolved degenerate four-wave mixing studies of solid-state poly(*p*-phenylene) oligomers, *J. Chem. Phys.,* 100, 3325, 1994.

237. **Schrader, S., Koch, K. H., Mathy, A., Bubeck, C., Mullen, K., and Wegner, G.,** Third harmonic generation in perylene derivatives, *Prog. Colloid Polym. Sci.,* 85, 143, 1991.

238. **Morita, K., Suehiro, T., Yokoo, T., Ashitaka, H., and Toda, F.,** Third-order nonlinear optics of aromatic acetylene derivatives, in Proceedings of 64th Japanese Chemical Society Meeting, No. 1, 1992, 195 (in Japanese).

239. **Suehiro, T., Morita, K., Yokoo, T., and Ashitaka, H.,** Synthesis of polyazo dyes and their third-order nonlinear optics, in Proceedings of 64th Japanese Chemical Society Meeting, No. 1, 1992, 195. (in Japanese).

240. **Yang, L., Dorsinville, R., Wang, Q. Z., Zou, W. K., Ho. P. P., Yang, N. L., Alfano, R. R., Zamboni, R., Danieli, R., Ruini, G., and Taliani, C.,** Third order optical nonlinearity in polycondensed thiophene-based polymers and polysilane polymers, *J. Opt. Soc. Am. B,* 6, 753, 1989.

241. **Okawa, H., Wada, T., Yamada, A., and Sasabe, H.,** Third-order optical nonlinearity of soluble polythiophenes, *Mater. Res. Soc. Symp. Proc.,* 214, 23, 1991.

242. **Sasabe, H., Wada, T., Sugiyama, T., Ohkawa, H., Yamada, A., and Garito, A. F.,** Third harmonic generation of polythiophene derivatives, in *Conjugated Polymeric Materials in Electronics, Optoelectronics and Molecular Electronics,* Bredas, J. L. and Chance, R. R., Eds., Kluwer Academic, London, 1990, 399.

243. **Callender, C. L., Karnas, S. J., Albert, J., Roux, C., and Leclerc, M.,** Third-harmonic generation measurements on thin films of novel substituted polythiophenes, *Opt. Mater.,* 1, 125, 1992.

244. **Dorsinville, R., Yang, L., Alfano, R. R., Zamboni, R., Danieli, G., and Taliani C.,** Nonlinear-optical response in polythiophene films using four-wave mixing technique, *Opt. Lett.,* 14, 1321, 1989.

245. **Kajzar, F., Ruani, G., Taliani, C., and Zamboni, R.,** Frequency variation of cubic susceptibility in the new conjugated polymers PTT and PDTB, *Synth. Met.,* 37, 223, 1990.

246. **Kaino, T., Kubodera, K., Kobayashi, T., Kurihara, S., Sato, T., Tsutsui, T., Takito, S., and Murata, H.,** Optical third-harmonic generation from poly(2,5-thienylene vinylene) thin films, *Appl. Phys. Lett.,* 53, 2002, 1988.

247. **Houlding, V. H., Nahata, A., Yardley, J. T., and Elsenbaumer, R. L.,** Optical third harmonic response of amorphous poly(3-methyl-4′-octyl-2,2′-bithiophene-5,5′-diyl) thin films, *Chem. Mater.,* 2, 169, 1990.

248. **Jenekhe, S. A., Lo, S. K., and Flom, S. R.,** Third-order nonlinear optical properties of asoluble conjugated polythiophene derivative, *Appl. Phys. Lett.,* 54, 2524, 1989.

249. **Jenekhe, S. A., Chen, W. C., Lo, S., and Flom, S. R.,** Large third-order optical nonlinearities in organic superlattices, *Appl. Phys. Lett.,* 57, 126, 1990.

250. **Meth, J. S., Vanherzeele, H., Chen, W., and Jenekhe, S. A.,** Interpreting the dispersion of $\chi^{(3)}(-3\omega;\omega,\omega,\omega)$ of polythiophenes, *Synth. Met.,* 49, 59, 1992.

251. **Nalwa, H. S.,** Third-order optical nonlinearity in a processable copolymer of 3-methylthiophene and methyl methacrylate, *J. Phys. D., Appl. Phys.,* 23, 745, 1990.

252. **Nalwa, H. S.,** Third-order nonlinear optical susceptibility measurements of thin films of 3-methylthiophene and methylmethacrylate copolymers, *Thin Solid Films,* 225, 175, 1993.

253. **Polis, D. W., Young, C. L., McLean, M. R., and Dalton, L. R.,** Synthesis and characterization of copolymers of aniline and thiophene, *Macromolecules,* 23, 3231, 1990.

254. **Kurihara, T., Kaino, T., Zhou, Z. H., Kanbara, T., and Yamamoto, T.,** New heteroaromatic polymers for third-order nonlinear optics, *Electron. Lett.,* 28, 681, 1992.

255. **Ogata, N.,** Intelligent polymers for opto-electronic applications, in Proc. of 4th Iketani Conference on Optically Nonlinear Organic Materials and Applications, Hawai, May 17–20, 1994, 15.

256. **Yamamoto, T., Kanbara, T., Ooba, N., and Tomaru, S.,** p-conjugated charge transfer polymer constituted of electron-donating thiophene and electron-withdrawing quinoxaline, *Chem. Lett.,* 1709, 1994.

257. **Yamada, W., Takagi, M., Kizu, K., Maruyama, T., and Yamamoto, T.,** Nonlinear optical properties of poly(arylen-eethynylene)s, in 55th Autumn Meeting of the Japan Society of Applied Physics, 1994, Extended Abstract No. 20a-K-3, 1023, (in Japanese).

258. **Yamamoto, T., Yamada, W., Takagi, M., Kizu, K., Maruyama, T., Ooba, N., Tomaru, S., Kurihara, T., Kaino, T., and Kubota, K.,** π-conjugated soluble poly(aryleneethynylene) type polymers. Preparation by palladium-catalyzed coupling reaction, nonlinear optical properties, doping, and chemical reactivity, *Macromolecules,* 27, 6620, 1994.

259. **Sugiyama, T., Wada, T., and Sasabe, H.,** Optical nonlinearity of conjugated polymers, *Synth. Met.,* 28, C323, 1989.

260. **Matsuda, H., Sato, M., Okada, S., Nakanishi, H., Kato, M., and Nishiyama, T.,** Nonlinear optical properties of thin films of polythienylene derivatives, *Polym. Preprints Jpn.,* 38, 1035, 1989.

261. **Kajzar, F., Messier, J., Sentein, C., Elsenbaumer, R. L., and Miller, G. G.,** Cubic susceptibility of polythiophene solutions and films, *SPIE Proc.,* 1147, 36, 1989.

262. **Neher, D., Wolf, A., Leclerc, M., Kaltbeitzel, A., Bubeck, C., and Wegner, G.,** Optical third harmonic generation in substituted poly(phenylacetylenes) and poly(3-decylthiophenes), *Synth. Met.* 37, 249, 1990.

263. **Singh, B. P., Samoc, M., Nalwa, H. S., and Prasad, P. N.,** Resonant third order nonlinear optical properties of poly(3-dodecylthiophene), *J. Chem. Phys.,* 92, 2756, 1990.

264. **Sasabe, H., Wada, T., Sugiyama, T., Ohkawa, T., Yamada, A., and Garito, A. F.,** Third harmonic generation of polythiophene derivatives, in *Conjugated Polymeric Materials in Electronics, Optoelectronics and Molecular Electronics,* Bredas, J. L., and Chance, R. R., Eds., Kluwer Academic, London, 1990, 399.

265. **Goedel, W. A., Somanathan, N. S., Enkelmann, V., and Wegner, G.,** Steric effects in 3-substituted polythiophenes: comparing band gap, nonlinear optical susceptibility and conductivity of poly(3-cyclohexylthiophene) and poly(3-hexylthiophene, *Makromol. Chem.,* 193, 1195, 1992.

266. **Fichou, D., Garnier, F., Charra, F., Kajzar, F., and Messier, J.,** Linear and nonlinear optical properties of thiophene oligomers, in *Organic Materials for Nonlinear Optics,* Hann, R. A., and Bloor, D., Eds., Royal Society of Chemistry, Special Publication No. 69, 1989, 176.

267. **Thienpont, H., Rikken, G. L. J. A., Meijer, E. W., ten Hoeve, W., and Wynberg, H.,** Saturation of the hyperpolarizability of oligothiophenes, *Phys. Rev. Lett.,* 65, 2141, 1990.

268. **Joshi, M. V., Cava, M. P., Lakshmikantham, M. V., Metzger, R. M., Abdeldayem, H., Henry, M., and Venkateswarlu, P.,** Third-order nonlinear optical susceptibilities by laser self-trapping and degenerate four-wave mixing in oligothienyldicyanovinylene solutions, *Synth. Met.,* 57, 3974, 1993.

269. **Geisler, T., Petersen, J. C., Bjornholm, T., Fischer, E., Larsen, J., Dehu, C., Bredas, J. L., Tormos, G. V., Nugara, P. N., Cava, M. P., and Metzger, R. M.,** Third-order nonlinear optical properties of oligomers of thienylene-ethynylenes and thienylenevinylenes, *J. Phys. Chem.,* 98, 10102, 1994.

270. **Nalwa, H. S.,** Optical and X-ray photoelectron spectroscopic studies of electrically conducting benzimidazobenzophenanthroline type ladder polymers, *Polymer,* 32, 802, 1991 (and references therein).

271. **Dalton, L. R., Thomson, J., and Nalwa, H. S.,** The role of extensively delocalized π-electrons in electrical conductivity, nonlinear optical properties and physical properties of polymers, *Polymer,* 28, 543, 1988.

272. **Lindle, J. R., Bartoli, F. J., Hoffman, C. A., Kim, O. K., Lee, Y. S., Shirk, J. S., and Kafafi, Z. H.,** Nonlinear optical properties benzimidazobenzophenanthroline type ladder polymers, *Appl. Phys. Lett.,* 56, 712, 1990.

273. **Jenekhe, S. A., Roberts, M., Agrawal, A. K., Meth, J. S., and Vanherzeele, H.,** Nonlinear optical properties of ladder polymers and their model compound, *Mater. Res. Soc. Symp. Proc.,* 214, 55, 1991.

274. **Meth, J. S., Vanherzeale, H., Jenekhe, S. A., Roberts, M. F., Agarwal, A. K., and Yang, C-J,** Dispersion of $\chi^{(3)}$ in fused aromatic ladder polymers and their precursors probed by the third harmonic generation, *SPIE,* 1560, 13, 1991.

275. **Rao, D. N., Swiaktiewicz, J., Chopra, P., Ghosal, S. K., and Prasad, P. N.,** Third-order nonlinear optical interactions in thin films of poly(p-phenylenebenzobisthiazole) polymer investigated by picosecond and subpicosecond degenerate four wave mixing, *Appl. Phys. Lett.,* 48, 1187, 1986.

276. **Jenekhe, S. A., Osaheni, J. A., Meth, J. S., and Vanherzeele, H.,** Nonlinear optical properties of poly(p-phenylene-benzobisoxazole), *Chem. Mater.,* 4, 683, 1992.

277. **Yu, L., Polis, D. W., McLean, M. R., and Dalton, L. R.,** Polymeric materials for third-order nonlinear optical susceptibility, in *Electroresponsive Organic* molecular and polymeric systems, Vol. 2, Skotheim, T. A., Ed., Marcel Dekker, Inc, New York, 1991, 113.

278. **Cao, X. T., Jiang, J. P., Bloch, D. P., Hellwarth, R. W., Yu, L., and Dalton, L. R.,** Picosecond nonlinear optical response of three rugged polyquinoxaline based aromatic conjugated ladder-polymer thin films, *J. Appl. Phys.,* 65, 5012, 1989.

279. **Yu, L., and Dalton, L. R.,** Synthesis and characterization of new electractive polymers, *Synth. Met.,* 29, E-463, 1989.

280. **Yu, L., Chen, M., and Dalton, L. R.,** Ladder Polymers: recent developments in synthesis, characterization and potential applications as electronic and optical materials, *Chem. Mater.,* 2, 649, 1990.

281. **Yu. L. and Dalton, L. R.,** Synthesis and characterization of new polymers exhibiting large optical nonlinearities. 3. Rigid-rod/flexible chain copolymers, *J. Am. Chem. Soc.,* 111, 8699, 1989.

282. **Cao, X. F., Jiang, J. P., Hellwarth, R. W., Yu, L., and Dalton, L. R.,** Bipolaronic enhanced third-order nonlinearity in organic ladder polymers, *SPIE Proc.,* 1337, 114, 1990.

283. **Spangler, C. W., Saindon, M. L., Nickel, E. G., Spochak, L. S., Polish, D. W., Dalton, L. R., and Norwood, R. A.,** Synthesis and incorporation of ladder polymer subunits in copolyamides, pendant polymers and composites for enhanced nonlinear optical response, *SPIE Proc.,* 1497, 408, 1991.

284. **Agrawal, A. K., Jenekhe, S. A., Vanherzeele, H., and Meth, J. S.,** Third-order nonlinear optical properties of conjugated rigid-rod polyquinolines, *J. Phys. Chem.,* 96, 2837, 1992.

285. **Ginder, J. M., Epstein, A. J., and McDiarmid, A. G.,** Electronic phenomena in polyaniline, *Synth. Met.,* 29, E395, 1989.

286. **Wong, K. S., Han, S. H., and Vardeny, Z. V.,** Studies of resonant and preresonant femtosecond degenerate four-wave mixing in oriented conducting polymers, *J. Appl. Phys.,* 70, 1896, 1991.

287. **Halvorson, C., Cao, Y., Moses, D., and Heeger, A. J.,** Third-order nonlinear optical susceptibility of polyaniline, *Synth. Met.,* 57, 3941, 1993.

288. **Mattes, B. R., Knobbe, E. T., Fuqua, P. D., Nishida, F., Chang, E. W., Pierce, B. M., Dunn, B., and Kaner, R. B.,** Polyaniline sol-gels and their third-order nonlinear optical effects, *Synth. Met.,* 41,3183, 1991.

289. **Ando, M., Matsuda, H., Okada, S., Nakanishi, H., Iyoda, T., and Shimidzu, T.,** Third-order nonlinear optical effect of polyaniline thin film, in Proc. Japanese Society of Applied Physics, 52nd Autumn Meeting, October 9–12, Okayama, Japan, 1991, 1128. (In Japanese).

290. **Osaheni, J. A., Jenekhe, A., Vanherzeele, H., Meth, J. S., Sun, Y., and MacDiarmid, A. G.,** Nonlinear optical properties of polyanilines and derivatives,*J. Phys. Chem.,* 96, 2830, 1992.

291. **Nalwa, H. S., Hamada, T., Kakuta, A., and Mukoh, A.,** Third-order nonlinear optical properties of polyazines and their well-defined π-conjuation chain length oligomers, *Nonlinear Opt.,* 6,155, 1993, and references therein.

292. **Nalwa, H. S., Hamada, A., Kakuta, A., and Mukoh, A.,** Studies of third-order nonlinear optical properties of polyazine and its oligomers, *Jpn. J. Appl. Phys.,* 32, L193, 1993.

293. **Nalwa, H. S. and Kakuta, A.,** Japanese Patent A-5-80373, 1993.

294. **Nalwa, H. S., Hamada, T., Kakuta, A., and Mukoh, A.,** Third-order nonlinear optical properties of polyazine derivatives, *Synth. Met.,* 57, 3091,1993.

295. **Nalwa, H. S., Hamada, T., Kakuta, A., and Mukoh, A.,** Polarizabilities and second hyperpolarizabilities of donor and acceptor substituted polyazines and polyacetylenes, *Nonlinear Opt.* 7, 193, 1994.

296. **Chandrasekhar, P., Thorne, J. R. G., and Hochstrasser, R. M.,** Third-order nonlinear optical properties of poly(diphenylamine) and poly(4-amino biphenyl), novel processible conducting polymers, *Appl. Phys. Lett.,* 59, 1661, 1991.

297. **Chandersekhar, P., Thorne, J. R. G., and Hochstrasser, R. M.,** Structure-property relationships in off-resonant third-order nonlinearities of conducting polymers 1. Poly(aromatic amines) and ring-S-heteroatom-containing polymers, *Synth. Met.,* 53,175, 1993.

298. **Chandrasekhar, P., Thorne, J. R. G., and Hochstrasser, R. M.,** Structure-property relationships in off-resonant third-order nonlinearities of electroactive polymers A: poly(aromatic amines) and poly(aminothiazoles), paper presented at the ACS Meeting, August 1992, Washington, 1992, 400.

299. **Wellinghoff, S. T., Hill, R. H., Naegeli, D. W., Lo, S. K., and Rogers, D.,** Synthesis and characterization of carbazole polymers exhibiting large nonlinear absorption and refractive index, *Synth. Met.,* 41,3203, 1991.

300. **Jenekhe, S. A., Yang, C. J., Vanherzeele, H., and Meth, J. S.,** Cubic nonlinear optics of polymer thin films. Effects of structure and dispersion on the nonlinear optical properties of aromatic Schiff base polymers, *Chem. Mater.,* 3, 985, 1991.

301. **Lee,K. S. and Samoc, M.,** Third-order nonlinear optical properties of wholly aromatic polyazomethines, *Polymer,* 32,361, 1991.

302. **Sasabe, H., Wada, T., Hosoda, M., Ohkawa, H., Hara, M., Yamada, A., and Garito, A. F.,** Photonics and organic nanostructures, *SPIE,* 1337,62, 1990.

303. **Goshal, S. K.,** Resonant third-order optical nonlinearity in polypyrrole, *Chem. Phys. Lett.,* 158, 65, 1989.

304. **Flom, S. R., Walker, G. C., Lynch, L. E., Miller, L. L., and Barbara, P. F.,** Degenerate four-wave mixing experiments on polyacene quinones,*Chem. Phys. Lett.,* 154, 193, 1989.

305. **Kurihara, T., Tomaru, S., Hikita, M., Shuto, Y., Kanbara, H., and Kaino, T.,** New main chain polymers for all-optical waveguide devices, *Polym. Preprints, Jpn.,* 41, 2935, 1992, (in Japanese).

306. (a) **Kanbara, H., Kobayashi, H., Kaino, T., Kurihara, T., Ooba, N., and Kubodera, K.,** Highly efficient ultrafast optical Kerr shutters with the use of organic nonlinear materials, *J. Opt. Soc. Am.,* 11,2216, 1994. (b) **Kanbara, H., Kobayashi, H., Kaino, T., Ooba, N. and Kurihara, T.,** Molecular length dependence of third-order nonlinear optical properties in conjugated organic materials, *J. Phys. Chem.,* 98, 12270, 1994.

307. **Yamamoto, T., Kizu, K., Maruyama, T., Ooba, N., Tomaru, S., and Kubota, K.,** Poly(pyridine-2,5-diyl) containing indigo unit and optical third harmonic generation from its film, *Chem. Lett.,* 913, 1994.

308. **Yamashita, S., Wada, T., and Sasabe, H.,** Third-order nonlinear optical properties of polyimides, in Proceedings of the 64th Japan Chemical Society Meeting, 1992, No. 1, 185. (In Japanese).

309. **Hikita, M., Yamada, S., and Mizutani, T.,** Relation between structure and third-order nonlinear optical effect in polyimide thin films prepared by vapour deposition polymerization using different diamine monomers, *Nonlinear Opt.,* 2, 1, 1992.

310. **Duray, M. A. and Glatkowski, P. J.,** Synthesis, characterization and processing of organic nonlinear optical polymers, *SPIE,* 1409, 215, 1991.

311. **Mates, T. and Ober, C. K.,** Thermotropic polymers with conjugated distyrylbenzene and diphenylbutadiene segments, *Polym. Preprints,* 204, 1990.

312. **Amano, M., Kaino, T., and Matsumoto, S.,** Third-order nonlinear optical properties of azo dye attached polymers, *Chem. Phys. Lett.,* 170, 515, 1990.

313. **Hadjichristov, G. B., and Kircheva, P. P.,** Optical four-wave mixing in bulk polymers, *Appl. Phys. B,* 55,373, 1992.

314. **Matsumoto, S., Kobodera, K., Kurihara, T., and Kaino, T.,** Third-order nonlinear optical properties of cyanine dyes and polymers films containing these dyes, *Opt. Commun.,* 76,147, 1990.

315. **Kaino, T., Amano, M., and Shuto, Y.,** Nonlinear optical properties of chromophore-functionalized polymeric materials, *Nonlinear Opt.,*Miyata, S., Ed., Elsevier, Amsterdam, 1992, 163.

316. **Kurihara, T., Oba, N., Mori, Y., Tomaru, S., and Kaino, T.,** New symmetrical o-conjugated molecules having large third-order optical nonlinearities, *J. Appl. Phys.,* 70, 17, 1991; Mori, Y., Kurihara, T., Kaino, T., and Tomaru, S., Molecular orbitals and third harmonic generation for symmetrically substituted benzylidene anilines, *Jpn. J. Appl. Phys.,* 31, 896, 1992. **Matsumoto, S., Kurihara, T., Kubodera, K., and Kaino, T.,** Third-order nonlinear optical properties of dye-attached polymers, *Mol. Cryst. Liq. Cryst.,* 182A, 115, 1990.

317. **Nomura, S., Kobayashi, T., Matsuda, H., Okada, S., Nakanishi, H., and Tomiyama, H.,** Electric-field dependence of absorption spetra of polymer ion-hemicyanine dye complexes, *Chem. Phys. Lett.,* 175, 389, 1990.

318. **Meyer, W. H., Pecherz, J., Mathy, A., and Wegner, G.,** Polyelectrolyte glasses with linear and nonlinear optical properties, *Adv. Mater.,* 3, 153, 1991.

319. **Torruellas, W. R., Zanoni, R., Marques, M. B., Stegeman, G. I., Mohlmann, G. R., Erdhuisen, E. W. P., and Horsthuis, W. H. G.,** Measurements of thid-order nonlinearities of side-chain-substituted polymers, *Chem. Phys. Lett.,* 175, 267, 1990.

320. **Torruellas, W. R., Zanoni, R., Stegeman, G. I., Mohlmann, G. R., Erdhuisen, E. W. P., and Horsthuis, W. H. G.,** The cubic susceptibility dispersion of alkoxy-nitro-stilbene (MONS) and di-alkyl-amino-nitro-stilbene (DANS) side chain substituted polymers: comparison with the two-level model, *J. Chem. Phys.,* 94, 6851, 1991.

321. **Norwood, R. A., Sounik, J. R., Popolo J., and Holcomb, D. R.,** Third-order nonlinear optical characterization of side-chain copolymers, *SPIE Proc.,* 1560, 54, 1991.

322. **Kajzar, F. and Zagorska, M.,** Third-order nonlinear optical properties of functionalized polymers, *Nonlinear Opt.,* 6, 181, 1993.

323. **Nalwa, H. S.** Organometallic materials for nonlinear optics, *Appl. Organomet. Chem.,* 5, 349, 1991.

324. **Leznoff, C. C. and Lever, A. B. P.,** *Phthalocyanines,* VCH Publishers, Inc., Weinheim, 1989.

325. **Thomas, A.L.,** *Phthalocyanine Research and Applications,* CRC Press, Boca Raton, Fl., 1990.

326. **Ho, Z. Z., Yu, C. Y., and Hetherington, W. M. III,** Third-harmonic generation in phthalocyanines, *J. Appl. Phys.,* 62, 716, 1987.

327. **Shirk, J. S., Lindle, J. R., Bartoli, F. J., Hoffman, C. A., Kafafi, Z. II., and Snow, A. W.,** Off-resonant third-order optical nonlinearities of metal substituted phthalocyanines, *Appl. Phys. Lett.,* 55,1287, 1989.

328. **Shirk, J. S., Lindle, J. R., Bartoli, F. J., Kafafi, Z. H., and Snow, A. W.,** Nonlinear optical properties of substituted phthalocyanines, in *Materials for Nonlinear Optics: Chemical Perspectives,* Marder,, S. R., Sohn, J. E. and Stucky, G. D., Eds., American Chemical Society Symposium Series 455, American Chemical Society, Washington, DC, 1991, 626.

329. **Shirk, J. S., Lindle, J. R., Bartoli, F. J., and Kafafi, Z. H.,** Third-order nonlinear optical properties of metallo-phthalocyanines, *Int. J. Nonlinear Opt.,* 1, 699, 1992.

330. **Hosoda, M., Wada, T., Yamada, A., Garito, A. F., and Sasabe, H.,** Phases and third-order optical nonlinearities in tetravalent metallophthalocyanine thin films, *Jpn. J. Appl. Phys.,* 30, L1486, 1991.

331. **Hosoda, M., Wada, T., Yamada, A., Garito, A. F., and Sasabe, H.,** Third-order nonlinear optical properties in soluble phthalocyanines with *tert*-butyl substituents, *Jpn. J. Appl. Phys.,* 30, 1715, 1991.

332. **Hosoda, M., Wada, T., Yamada, A., Garito, A. F., and Sasabe, H.,** Metallophthalocyanine thin films for nonlinear optics, *Nonlinear Opt.,* 3, 183, 1992.

333. **Wada, T., Hosoda, M., Garito, A. F., Sasabe, H., Terasaki, A., Kobayashi, T., Tada, H., and Koma, A.,** Third-order optical nonlinearities and femtosecond response in metallophthalocyanine thin films made by vacuum deposition, molecular beam epitaxy and spin coating, *SPIE Proc.,* 1560, 162, 1991 (and references therein).

334. **Matsuda, H., Okada, S., Masaki, A., Nakanishi, H., Suda, Y., Shigehara, K., and Yamada, A.,** Molecular structural view on the large third-order nonlinearity of phthalocyanine derivatives, *SPIE Proc.,* 1337, 105, 1990.

335. **Suda, Y., Shigehara, K., Yamada, A., Matsuda, H., Okada, S., Masaki, A., and Nakanishi, H.,** Reversible phase transition and third-order nonlinearity of phthalocyanine derivatives, *SPIE Proc.,* 1560, 75, 1991.

336. **Wang, N. Q., Cai, Y. M., Helfin, J. R., Wu, J. W., Rodenberger, D. C., and Garito, A. F.,** Resonant and non-resonant third-order optical properties of metallophthalocyanines, *Polymer,* 32, 1752, 1991.

337. **Norwood, R. A. and Sounik, J. R.,** Third-order nonlinear optical response in polymer thin films incorporating porphyrin derivatives, *Appl. Phys. Lett.,* 60, 295, 1992.

338. **Shirk, J. S., Lindle, J. R., Bartoli, F. J., and Boyle, M. E.,** Third-order optical nonlinearities of bis(phthalocyanines), *J. Chem. Phys.,* 96, 5847, 1992.

339. **Nalwa, H. S. and Kakuta, A.,** Third-order nonlinear optical properties of donor and acceptor substituted phthalocyanines, *Thin Solid Films,* 254, 218, 1995. Nalwa, H. S. and Kakuta, A., Japanese Patent, A-6-222409, 1994.

340. **Nalwa, H. S., Kobayashi, S., and Kakuta, A.,** Third-order optical nonlinearities properties of metallo-naphthalocyanine dyes, *Nonlinear Opt.,* 6, 169, 1993.

341. **Nalwa, H. S., Kakuta, A., and Mukoh, A.,** Third-order optical nonlinearities in *tetrakis-n*-pentoxy carbonyl metallo-naphthalocyanines, *Chem. Phys. Lett.,* 203, 109, 1993.

342. **Nalwa, H. S., Kakuta, A., and Mukoh, A.,** Third-order nonlinear optical properties of a vanadyl-naphthalocyanine derivative, *J. Phys. Chem.,* 97, 1097, 1993.

343. **Nalwa, H. S.,** Effect of polymorphism, donor-acceptor substitution and π-conjugation length on third-order nonlinear optical susceptibility of phthalocyanines, paper presented at 48th Okazaki Conference on "Phthalocyanines", Okazaki, Japan, January, 26–28, 1994.

344. **Grund, A., Kaltbeitzel, A., Mathy, A., Schwarz, R., Bubeck, C., Vermehren, P., and Hanack, M.,** Resonant nonlinear optical properties of spin-cast films of soluble oligomeric bridged (phthalocyanato)ruthenium(II) complexes, *J. Phys. Chem.,* 96, 7450, 1992.

345. (a) **Nalwa, H. S., Saito, T., Kakuta, A., and Iwayanagi, T.,** Third-order nonlinear optical properties of polymorphs of oxotitanium phthalocyanine (TiOPc), *J. Phys. Chem.,* 97, 10515, 1993.(b) **Nalwa, H. S., Saito, T., Kakuta, A., and Iwayanagi, T.,** Polymorphs of oxotitanium phthalocyanine exhibiting large third-order nonlinear optical susceptibility, Optical Society of America/American Chemical Society (OSA/ACS) *Organic Thin Films for Photonic Applications,* Royal York Hotel, Toronto, October 6–8, Technical Digest Series, 17, 144, 1993.

346. **Sinha, A. K., Bihari, B., Kamath, M., and Mandal, B. K.,** Thin film processing and nonlinear optical properties of novel axially modified phthalocyanine derivatives, *SPIE Proc.,* 2527, 18, 1995.

347. **Ishikawa, K., Kajita, M., Koda, T., Kobayashi, H., and Kubodera, K.,** Evidence of intermolecular interaction effect on nonlinear optical properties of copper phthalocyanine, *Mol. Crystl. Liq. Crystl.,* 218, 123, 1992.

348. **Kajita, M., Ishikawa, K., and Koda, T.,** Nonlinear optical properties of naphthalocyanines, in The Japanse Society of Applied Physics 51st Meeting, Abstract No. 26a-ZD-10, 1990, 1011. (in Japanese).

349. **Gieruski, A., Naarmann, H., Schrof, W., and Ticktin, A.,** Frequency tunable THG measurements of $\chi^{(3)}$ between 1–2.1 μm of organic conjugated polymer films using an optical parametric oscillator (OPO), *SPIE Proc.,* 1560, 172, 1991.

350. **Lindle, J. R., Shirk, J. S., Bartoli, F. J., Kafafi, Z. H., Snow, A. W., Boyle, M. E., and Kim, O.K.,** Large third-order optical nonlinearities of phthalocyanines, bisphthalocyanines and metal complexes of *o*-aminobenzenethiol, *Proc. Symp. New Mater. Nonlinear Opt.,* American Chemical Society, Symp. Proc., 20, 1991.

351. **Chollet, P. A., Kajzar, F., and Moigne, J. L.,** Structure and nonlinear optical properties of copper phthalocyanine thin films, *SPIE Proc.,* 1237, 87, 1990.

352. **Diaz-Garcia, M. A., Ledoux, I., Fernandez-Lazaro, F., Sastre, A., Torres, T., Angullo-Lopez, F., and Zyss, J.,** Third-order nonlinear optical properties of soluble metallotriazolehemiporphyrazines, *J. Phys. Chem.,* 98, 4495, 1994.

353. **Diaz-Garcia, M. A., Angullo-Lopez, F., Hutchings, M. G., Gordon, P. F., and Kajzar, F.,** Third-order hyperpolarizabilities of soluble organometallic compounds, *SPIE Proc.,* 2285, 227, 1994.

354. **Diaz-Garcia, M. A., Ledoux, I., Duro, J. A., Torres, T., Angullo-Lopez, F., and Zyss, J.,** Third-order nonlinear optical properties of soluble octasubstituted metallophthalocyanines, *J. Phys. Chem.,* 98, 8761, 1994.

355. **Diaz-Garcia, M. A., Angullo-Lopez, F., Hutchings, M. G., Gordon, P. F., and Kajzar, F.,** Third-order hyperpolarizabilities of soluble phthalocyanines, in Proc. of 4th Iketani Conference on Optically Nonlinear Organic Materials and Applications, Hawai, May 17–20, 1994, 138.

356. **Bubeck, C., Neher, D., Kaltbeitzel, A., Duda, G., Arndt, T., Sauer, T., and Wegner, G.,** Langmuir-Blodgett films of rigin rod polymers with controlled lateral orientation, in *Nonlinear Optical Effects in Organic Polymers,* Messier, J. et al., Eds., Kluwer Academic Publishers, London, 1989, 185.

357. **Murano, T., Yamashita, A., Hayashi, T., Kanbara, H., Konami, H., Hatano, M., Shiratori, Y., and Ohhira, K.,** Third-harmonic generation in highly ordered thin films of phthalocyaninato-oxytitanium (IV), *Nonlinear Opt.,* 7, 207, 1994.

358. **Yamashita, A., Matsumoto, S., Sakata, S., Kanbara, H., and Hayashi, T.,** Phase-controlled formation of triclinic metallophthalocyanine films by organic molecular beam deposition, *5th Symposium on Intelligent Materials,* 1996, Extended abstract No. A5-2, p. 86.

359. **Imanishi, Y., Hattori, S., Hamada, T., Ishihara, S., Nalwa, H. S., and Kakuta, A.,** New organic superlattice structures for nonlinear optics, *Nonlinear Opt.,* 73, 63, 1995.

360. **Reeves, R. J., Powell, R. C., Ford, W. T., Chang, Y. H., and Zu, W.,** Third-order nonlinear optical response of metallophthalocyanines in films and solutions, *Mater. Res. Soc. Symp. Proc.,* 247, 203, 1992.

361. **Flom, S. R., Shirk, J. S., Pong, R. G. S., Lindle, J. R., Batoli, F. J., Boyle, M. F., and Snow, A. W.,** Resonant third-order optical response in lead phthalocyanines, in Technical Digest of International Quantum Electronics Conference (IQEC), May 8–13, Anaheim, California, 9, QWC19, 95, 1994.

362. **Flom, S. R., Shirk, J. S., Pong, R. G. S., Lindle, J. R., Batoli, F. J., Boyle, M. F., Adkins, J. D., and Snow, A. W.,** Excited state absorption and dynamics in Pb-phthalocyanine copolymer, *SPIE Proc.,* 2143, 229, 1994.

363. **Wajiki, A., Yamashita, M., Lingbing, C., Mito, A., Morinaga, A., and Tako, T.,** Anisotropy of $\chi^{(3)}$ of sublimed tin-phthalocyanine film, in Proceedings of Korea-Japan Joint Forum, 1994, 53.

364. **Meloney, C., Byrne, H., Dennis, W. M., Blau, W., and Kelly, J. M.,** Picosecond optical phase conjugation using conjugated organic molecules, *Chem. Phys.,* 121, 21, 1988.

365. **Sakaguchi, T., Shimizu, Y., Miya, M., Fukumi, T., Ohta, K., and Nagata, A.,** Third-order nonlinear optical susceptibilities of solutions of some mesogenic metallotetraphenylporphyrins by nanosecond degenerate four-wave mixing method, *Chem. Lett.,* 281, 1992.

366. **Rao, D. V. G. L. N., Arando, F. J., Roach, J. F., and Remy, D. E.,** Third-order nonlinear optical interactions of some benzoporphyrins, *Appl. Phys. Lett.,* 58, 1241, 1991.

367. **Guha, S., Kang, K., Porter, P., Roach, J. E., Remy, D. E., Aranda, F. J., and Rao, D. V. G. L. N.,** Third-order optical nonlinearities of metallotetrabenzo-porphyrins and a platinum poly-yne, *Opt. Lett.,* 17, 264, 1992.

368. **Hosoda, M., Wada, T., Garito, A. F., and Sasabe, H.,** Third-order optical nonlinearities in porphyrins with extended π-electron systems, *Jpn. J. Appl. Phys.,* 31, L249, 1992.

369. **Anderson, H. L., Martin, S. J., and Bradley, D. D. C.,** Synthesis and third-order nonlinear optical properties of a conjugated porphyrin polymer, *Angew. Chem. Int. Ed. Engl.,* 33, 655, 1994.

370. **Bao, Z. and Yu, L.,** Novel photonic polymers containing porphyrin rings, *Proc. ACS Meeting Polym. Mater. Sci. Eng.* (PMSE), 71, 781, 1994.

371. **Honeychuck, R. V.,** Novel third-order NLO materials from 96% quinoline, *Polym. Preprints,* 32, 138, 1992.

372. **West, R.,** The polysilane high polymers, *J. Organomet. Chem.,* 300, 327, 1986.

373. **Miller, R. D. and Michl, J.,** Polysilane high polymers, *Chem. Rev.,* 89, 1359, 1989.

374. **Kajzar, F., Messier, J., and Rosilio, C.,** Nonlinear optical properties of thin films of polysilane, *J. Appl. Phys.,* 60, 3040, 1986.

375. **Yang, L., Wang, Q. Z., Ho, P. P., Dorsinville, R., Alfano, R. R., Zou, W. K., and Yang, N. L.,** Ultrafast time response of optical nonlinearity in polysilane polymers, *Appl. Phys. Lett.,* 53, 1245, 1988.

376. **Baumert, J. C., Bjorklund, G. C., Jundt, D. H., Jurich, M. C., Looser, H., Miller, R. D., Rabolt, J., Sooriyaku-maran, R., Swalen, J. D., and Tweig, R. J.,** *Appl. Phys. Lett.,* 53, 1147, 1988.

377. **McGraw, D. J., Siegman, A. E., Wallraff, G. M., and Miller, R. D.,** *Appl. Phys. Lett.,* 54, 1713, 1989.

378. **Shukla, P., Cotts, P. M., Miller, R. D., Ducharme, S., Asthana, R., and Zavislan, J.,** Nonlinear optical studies of polysilanes, *Mol. Cryst. Liq. Cryst.* 183, 241, 1990.

379. **Wong, K. S., Han, S. G., Vardeny, Z. V., Shinar, J., Pang, Y., Maghsoodi, I., Barton, T. J., Grigoras, S., and Parbhoo, B.,** Femtosecond dynamics of the nonlinear optical response in polydiethynylsilane, *Appl. Phys. Lett.,* 58, 1695, 1991.

380. **Hasegawa, T., Isawa, Y., Kishida, H., Koda, T., Tokura, Y., Tachibana, H., and Kawabata, Y.,** Two-photon resonant third-harmonic generation in polysilanes, *Phys. Rev. B,* 45, 6317, 1992.

381. **Miller, R. D., Schellenberg, F. M., Baumert, J. C., Looser, H., Shukla, P., Torruellas, W., Bjorklund, G. C., Kano, S., and Takahashi, Y.,** Nonlinear optical properties of substituted polysilanes and polygermanes, in *Materials for Nonlinear Optics: Chemical Perspectives,* Marder, S. R., Sohn, J. E., and Stucky, G. D., Eds., American Chemical Society Symposium Series 455, American Chemical Society, Washington, DC, 1991, 636.

382. **Embs, F. W., Wegner, G., Neher, D., Albouy, P., Miller, R. D., Willson, C. G., and Screpp, W.,** Preparation of oriented multilayers of poly(silane)s by the Langmuir-Blodgett tecgnique, *Macromolecules,* 24, 5068, 1991.

383. (a) **Ho, S., Maeda, N., Suzuki, T., and Sato, H.,** Preparation and characterization of polysilanes with electron donating substituent, *Polym. J.,* 24, 865, 1992. (b) **Kodaira, T., Watanabe, A., Ito, O., Matsuda, M., Tokura, S., Kira M., Nagano, S., and Mochida, T.,** Third-order nonlinear optical properties of thin films of organogermane homopolymers and organogormane-organosilane copolymers, *Adv. Mater.,* 7, 917, 1995.

384. **Callender, C. L., Carere, C. A., Albert, J., Zhou, L. L., and Worsfold, D. J.,** Determination of third-order nonlinear optical susceptibilities of polysilane thin films, *J. Opt. Soc. Am.,* 9, 518, 1992.

385. **Porter, P. L., Guha, S., Kang, K., and Frazier, C. C.,** Effects of structural variations of platinum and palladium poly-ynes on third-order non-linearity, *Polymer,* 32, 1756, 1991.

386. **Guha, S., Frazier, C. C., Porter, P. L., Kang, K., and Finberg, S. E.,** Measurement of the third-order hyperpolariza-bility of platinum poly-ynes, *Opt. Lett.,* 14, 952, 1989.

387. **Frazier, C. C., Guha, S., Chen, W. P., Cockerham, M. P., Porter, P. L., Chauchard, E. A., and Lee, C. H.,** Third-order optical nonlinearity in metal containing organic polymers, *Polymer,* 28, 553, 1987.

388. **Frazier, C. C., Chauchard, E. A., Cockerham, M. P., and Porter, P. L.,** Four-wave mixing in metal poly-ynes, *Mater. Res. Soc. Symp. Proc.,* 109, 323, 1988.

389. **Blau, W. J., Byrne, H. J., Cardin, D. J., and Davey, A. P.,** Nonlinear optical properties of group 10 metal alkynyls and their polymers, *J. Mater. Chem.,* 1, 245, 1991.

390. **Page, H., Blau, W., Davey, A. P., Lou, X., and Cardin, D. J.,** Nonlinear optical properties of some mono- and bimetallic organometallic systems, *Synth. Met.,* 63, 179, 1994.

391. **Agh-Atabay, N. M., Lindsell, W. E., Preston, P. N., Tomb, P. J., Lloyd, A. D., Tangel-Rojo, R., Spruce, G., and Wherrett, B. S.,** Synthesis, characterization and optical properties of metal-containing polydiacetylenes, *J. Mater. Chem.,* 2, 1241, 1992.

392. **Winter, C. S., Oliver, S. N., and Rush, J. D.,** n_2 measurements on various forms of ferrocene, *Opt. Commun.,* 69, 45, 1988.

393. **Winter, C. S., Oliver, S. N., and Rush, J. D.,** Measurement of the nonlinear refractive index of some metallocenes by the optical power limiter technique, in *Nonlinear Optical Effects in Organic Polymers,* Messier, J., et al., Eds., Kluwer Academic Press, Dordrecht, 1989, 247.

394. **Kamata, T., Fukaya, T., Kozasa, T., Matsuda, H., Mizukami, F., Tachiya, M., Ishikawa, R., Uchida, T., and Yamazaki, Y.,** Third harmonic generation measurements on evaporated thin films of coordination compounds, *Nonlinear Opt.,* 13, 31, 1995.

395. **Oliver, S. N., Winter, C. S., Rush, J. D., Underhill, A. E., and Hill, C.,** Organotransition metal and rare earth compounds with high resonant enhanced $\chi^{(3)}$ coefficients, *SPIE Proc.,* 1337, 81, 1990.

396. **Myers, L. K., Langhoff, C., and Thompson, M. E.,** Cubic nonlinear optical properties of group 4 metallocene halide and acetylide complexes, *J. Am. Chem. Soc.,* 114, 7560, 1992, **Thompson, M. E., Chiang, W., and Myers, L. K.,** Nonlinear optical properties of inorganic coordination polymers and organometallic complexes, *SPIE Proc.,* 1497, 423, 1991.

397. **Cheng, C. T., Tam, W., Meredith, G. R., and Marder, S. R.,** Quadratic hyperpolarizabilities of some organometallic compounds, *Mol. Crystl. Liq. Crystl.,* 198, 137, 1990.

398. **Calabrese, J. C., Cheng, L. T., Green, J. C., Marder, S. R., and Tam, W.,** Molecular second-order optical nonlinearities of metallocenes, *J. Am. Chem. Soc.,* 113, 7227, 1991.

399. **Polin, J.,** *Molecular Electronics and Molecular Electronic Devices,* Sienicki, K., Ed., CRC Press, Boca Raton, FL, 1994, Chap. 5; Bubeck, C. and Baumann, A., private communication, 1993.

400. **Winter, C. S., Oliver, S. N., Rush, J. D., Hill, C., and Underhill, A.,** Resonance enhanced third-order nonlinearities in metal dithiolenes, paper presented at Second International Symposium on Organic Materials for Nonlinear Optics (OMNO'90), University of Oxford, UK, September 4–6, 1990.

401. **Fukaya, T., Kamata, T., and Mizuno, M.,** Nonlinear optical properties of metal-dithiolene complex: Ni(dmbit)$_2$TBA, *Polym. Preprints Jpn.,* 41,2908, 1992 (in Japanese).

402. **Sakaguchi, T., Miya, M., Fukumi, T., Ohta, K., Shimizu, Y., and Nagata, A.,** Solvent effect of the third-order nonlinear optical susceptibility of vitamin B$_{12}$ revealed by degenerate four-wave mixing, *Chem. Express,* 7, 109, 1992.

403. **Davey, A. P., Byrne, H. J., Page, H., Blau, W., and Cardin, D. J.,** Nonlinear optical studies of group 10 transition-metal thienyl systems, *Synth. Met.,* 58, 161, 1993.

404. **Kafafi, Z. H., Lindle, J. R., Weisbecker, C. S., Bartoli, F. J., Shirk, J. S., Yoon, T. H., and Kim, O. K.,** Nonlinear optical studies on transition metal complexes of o-aminobenzenethiol, *Chem. Phys. Lett.,* 179, 79, 1991.

405. **Lindle, J. R., Weisbecker, C. S., Bartoli, F. J., Pong, R. G. S., and Kafafi, Z. H.,** Third- and fifth-order optical properties of transition metal complexes of benzenedithiol at 1.064 μm, *Mater. Res. Soc. Symp. Proc.,* 247, 277, 1992.

406. **Kafafi, Z. H., Lindle, J. R., Flom, S. R., Pong, R. G. S., Weisbecker, C. S., Claussen, R. S., and Bartoli, F. J.,** Third-order optical properties of square-planar transition metal complexes in the visible and near-infrared, *SPIE Proc.,* 1626, 440, 1992.

407. **Kamata, T., Fukaya, T., Matsuda, H., and Mizukami, F.,** Reversible control of the nonlinear optical activity of one-dimensional metal complexes, *Appl. Phys. Lett.,* 65, 1343, 1994.

408. **Kamata, T., Fukaya, T., Mizuno, M., Matsuda, H., and Mizukami, F.,** Third-order nonlinear optical properties of one-dimensional metal complexes, *Chem. Phys. Lett.,* 221, 194, 1994.

409. **Fukaya, T. and Kamata, T.,** Hetero atom compounds, paper presented at Proceedings of International Symposium on Nonlinear Photonics Materials, May 24–25, 1994, Tokyo, Japan.

410. **Gong, Q., Xia, Z., Zou, Y. H., Li, Y., Wu, P., and Zhu, D.,** Third-order nonlinear optical properties of some organic nitroxyls, *Appl. Phys. B,* 54, 181, 1992.

411. **Sipp, B., Klein, G., Lavoine, J. P., Penicaud, A., and Batail, P.,** Observation of large third-order optical susceptibility in (BEDT-TTF)$_4$Re$_6$Se$_5$Cl$_9$, *Chem. Phys.,* 144, 299, 1990.

412. **Davidson, A., Boubekeur, K., Penicaud, A., Auban, P., Lenoir, C., Batail, P., and Herve, G.,** *J. Chem. Soc. Chem. Commun.,* 1373, 1989.

413. **Wada, Y. and Yamashita, M.,** Third-order electric susceptibilities of halogen-bridged mixed-valence Pt complexes: [Pt(en)$_2$][Pt(en)$_2$X$_2$](ClO$_4$), X = Cl, Br, I, en=ethylenediamine, *Jpn. J. Appl. Phys.,* 29, 2744, 1990.

414. (a) **Iwasa, Y., Funatsu, E., Hasegawa, T., Koda, T., and Yamashita, M.,** Nonlinear optical study of quasi-one-dimensional platinum complexes: two-photon excitonic resonance effect, *Appl. Phys. Lett.,* 59, 2219, 1991.
     (b) **Funatsu, E., Iwasa, Y., Koda, T., Kobayashi, H., Kubodera, K., Yamashita, M., and Wada, Y.,** Nonlinear optical properties of 1-dimensional exciton in halogen-bridged platinum complexes, The Japanese Society of Applied Physics 51st Meeting, Abstract No. 26p-ZD-1, 1990, 1011, (*In Japanese*).

415. **Yuan, Z., Taylor, N. J., Ramachandran, R., and Marder, T. D.,** Third-order nonlinear optical properties of organoboron compounds: molecular structures and second hyperpolarizabilities, *Appl. Organometal. Chem.,* 10, 305, 1996.

416. **Zhai, T., Lawson, C. M., Burgess, G. E., Lewis, M. L., Gale, D. C., and Gray, G. M.,** Nonlinear optical studies of molybdenum metal organics, *Opt. Lett.,* 19, 871, 1994.

417. **Gale, D. C., Lawson, C. M., Zhai, T., and Gray, G. M.,** Four wave mixing measurements on metal organics, *SPIE Proc.,* 2229, 41, 1994.

418. **Kuzyk, M. G., Moore, R. C., Sohn, J. E., King, L. A., and Dirk, C. W.,** Quadratic electrooptic modulation of dye-doped polymers: measurement of third-order electronic nonlinear optical susceptibilities, *SPIE,* 1147, 198, 1989.

419. **Kuzyk, M. G., Dirk, C. W., and Sohn, J. E.,** Mechanisms of quadratic electro-optic modulation of dye-doped polymer systems, *J. Opt. Soc. Am. B,* 7, 842, 1990.

420. **Kuzyk, M. G., Paek, U. C., Dirk, C. W., and Andrews, M. P.,** Doped polymers as third-order nonlinear optical materials, *Mater. Res. Soc. Symp. Proc.,* 214, 3, 1991.

421. **Dirk, C. W., Cheng, L. T., and Kuzyk, M. G.,** A three-level model useful for exploring structure/property relationships for molecular third-order optical polarizabilities, *Mater. Res. Soc. Symp. Proc.,* 247, 73, 1992.

422. **Zhou, Q. L., Shi, R. F., Zamani-Khamari, O., and Garito, A. F.,** Negative third-order optical responses in squaraines, *Nonlinear Opt.,* 6, 145, 1993.

423. **Shi, R. F., Zhou, Q. L., and Garito, A. F.,** Negative third-order optical susceptibilities of squaraines, *Mater. Res. Soc. Symp. Proc.,* 328, 643, 1994.

424. **Yu, Y. Z., Shi, R. F., Garito, A. F., and Grossman, C. H.,** Origin of negative $\chi^{(3)}$ in suaraines: experimental observation of two-photon states, *Opt. Lett.,* 19, 786, 1994.

425. **Park, S. Y., Wada, T., and Sasabe, H.,** Third harmonic generation of polymeric charge-transfer complex films, *Mol. Crystl. Liq. Crystl.,* 227, 151, 1993.

426. **Sakai, T., Ikeda, H., Kawabe, Y., and Kawasaki, K.,** Resonant third-order nonlinear optical properties of hemicyanine doped PMMA thin film, *Mol. Crystl. Liq. Crystl.,* 227, 143, 1993.

427. **Oliver, S. N., Winter, C. S., Manning, R. J., Rush, J. D., Hill, C., and Underhill, A. E.,** Nickel dithiolene-PMMA guest-host polymers for all-optical signal switching, *SPIE,* 1775, 110, 1992.

428. **Lessard, R. A., Malouin, C., Changkakoto, R., and Manivannam, G.,** Dye-doped polyvinyl alcohol recording materials for holography and nonlinear optics, *Opt. Eng.,* 32, 665, 1993.

429. **Fujiwara, I., Ishibashi, T., and Asai, N.,** Third-order nonlinear optical properties in Langmuir-Blodgett films of 4-docosylamino-4-nitroazobenzene, *J. Appl. Phys.,* 75, 4759, 1994.

430. **Du, W., Zhang, X., Chen, K., and Liu, S.,** Third-order optical nonlinearity of absorbing dye Langmuir-Blodgett films, *Thin Solid Films,* 232, 237, 1993.

431. **Kodzasa, T., Kamata, T., Fukaya, T., Matsuda, H., Mizukami, F., and Matsumoto, K.,** Preparation and THG properties of transparent one-dimensional Pt complex dispersed alumina films, in Proceedings of Korea Japan Joint Forum, 1994, 60.

432. **Vanherzeele, H., Meth, J. S., Jenekhe, S. A., and Roberts, M. F.,** Dispersion of the third-order nonlinear-optical properties of poly(p-phenylene benzobisthiazole) and its molecular composites with polyamides, *J. Opt. Soc. Am.,* 9, 524, 1992.

433. **Hacke, F., Richard, D., and Flytzanis, C.,** Optical nonlinearities of small metal particles: surface-mediated resonance and quantum size effects, *J. Opt. Soc. Am. B,* 3, 1647, 1986.

434. **Stockman, M. I., George, T. F., and Shalaev, V. M.,** Field work and dispersion relations of excitations on fractals, *Phys. Rev. B,* 44, 115, 1991.

435. **Kuzyk, M. G., Ghebremichael, F., and Andrews, M. P.,** Nonlinear optical properties of metal fractal cluster-doed polymers, *Sen-i-Gakkai Symp. Preprints,* B114, 1992.

436. **Hanamura, E.,** Excitonic and surface-state enhancement of optical nonlinearity in semiconductor microcrystallites, in *Nonlinear Optics,* Miyata, S., Ed., North-Holland, Amsterdam, 1992, 3. **Kayanuma, Y.,** Resonant interaction of photons with randomly distributed quantum dots, in *Nonlinear Optics,* Miyata, S., Ed., North-Holland, Amsterdam, 1992, 79.

437. **Ogawa, S., Hayashi, Y., Kobayashi, N., Tokizaki, T., and Nakamura, A.,** Novel preparation method of metal particles dispersed in polymer films and their third-order optical nonlinearities, *Jpn. J. Appl. Phys.,* 33, L331, 1994.

438. **Birnboim, M. H. and Ma, W. P.,** Nonlinear optical properties of structured nanoparticle composites, *Mater. Res. Soc. Symp. Proc.,* 164, 277, 1990.

439. **LaPeruta, R., Van Wagenen, E. A., Roche, J. J., Kitipichai, P., Wnek, G. E., and Korenowski, G. M.,** Preparation and characterization of silver colloid/polymer composite nonlinear optical materials, *SPIE Proc.,* 1497, 357, 1991.

440. **Tokizaki, T., Kataoka, T., Nakamura, A., Sugimoto, N., and Manabe, T.,** Large enhancement of third-order optical susceptibility in CuCl quantum dots embedded in glass, *Jpn. J. Appl. Phys.,* 32, L 782, 1993.

441. **Hayashi, T.,** Ultrafine particle dispersed in organic polymers, in Extended Abstracts Basic Technologies for Future Industries, The 3rd Symposium on Photonic Materials, October 20–21, 1992, 96 (in Japanese).

442. **Kramer, M. A., Tompkin, W. R., and Boyd, R. W.,** Optical interactions in fluorescein-doped boric acid glass, *Phys. Rev. A,* 34, 2026, 1986.

443. **Tompkin, W. R., Boyd, R. W., Hall, D. W., and Tick, P. A.,** Nonlinear-optical properties of lead-tin fluorophosphate glass containing acridine dyes, *J. Opt. Soc. Am. B,* 4, 1030, 1987. **Tompkin, W. R., Malcuit, M. S., and Boyd, R. W.,** Enhancement of the nonlinear optical properties of fluorescin doped boric-acid glass through cooling, *Appl. Optics,* 29, 3921, 1990.

444. **He, K. X., Bryant, W., and Venkateswarlu, P.,** Nonlinear optical effects on the surface of acridine yellow-doped lead-tin fluorophosphate glass, *Appl. Phys. Lett.,* 59, 1935, 1991.

445. **Kumar, G. R., Singh, B. P., and Sharma, K. K.,** Optical phase conjugation in Rhodamine 6G doped boric acid glass, *Opt. Commun.,* 73, 81, 1989.

446. **Kumar, G. R., Singh, B. P., and Sharma, K. K.,** Continuous-wave self-diffraction in dye-doped glasses, *J. Opt. Soc. Am. B,* 8, 2119, 1991.

447. **Rossi, B., Byrne, H. J., Blau, W., Pratesi, G., and Sottini, S.,** Linear and nonlinear waveguiding in rhodamin-doped epoxy films, *J. Opt. Soc. Am. B,* 8, 2449, 1991.

448. **Maloney, C. and Blau, W.,** Transient nonlinear optical behavior of fluorescein-doped boric acid glass, *J. Opt. Soc. Am. B,* 9, 2225, 1992.

449. **Speiser, S.,** Nonlinear optical properties of phenosafranin doped substrates, *SPIE,* 1559, 238, 1991.

450. **Beeson, K. W., Yardley, J. T., and Speiser, S.,** Utilization of nonlinear optical absorption in eosin-Y for all optical switching, *Mol. Eng.,* 1, 1, 1991.

451. **Palffy-Muhoray, P.,** The nonlinear optical response of liquid crystals, in *Liquid Crystals: Applications and Uses,* Vol. 1, Bahadur, B., Ed., World Scientific, Singapore, 1991, 493.

452. **Li, L., Yuan, H. J., Hu, G., and Palffy-Muhoray, P.,** Third-order optical nonlinearities of nematic liquid crystals, *Liq. Crystl.,* 16, 703, 1994.

453. **Palffy-Muhoray, P., Yuan, H. J., Li, L., Lee, A., DeSalvo, J. R., Wei, T. H., Sheik-Bahae, M., Hagan, D. J., and Van Stryland, E. W.,** Measurements of the third-order optical nonlinearities of nematic liquid crystals, *Mol. Crystl. Liq. Crystl.,* 207, 291, 1991.

454. **Li, L., Hu, G., Palffy-Muhoray, P., Lee, M. A., and Yuan, H. J.,** Z-scan measurements of optical nonlinearities in nematic liquid crystals, *SPIE,* 1692, 107, 1992.

455. **Wong, K. Y. and Garito, A. F.,** Third harmonic generation study of orientational order in nematic liquid crystals, *Phys. Rev.,* A34, 5051, 1986.

456. **Ford, W. T., Bautista, M., Zhao, M., Reeves, R. J., and Powell, R. C.,** Nonlinear optical responses of a polarized stilbene side chain liquid crystalline polyacrylate, *Mol. Crystl. Liq. Crystl.,* 198, 351, 1991.

457. **Smith, D. A. and Coles, H. J.,** The second and third-order nonlinear optical properties of liquid crystaline polymers, *Liq. Crystl.,* 14, 937, 1993.

458. **Zhang, H. J., Dai, J. H., Wang, P. Y., and Wu, L. A.,** Self-focusing and self-trapping in new type of Kerr media with large nonlinearities, *Opt. Lett.,* 14, 695, 1989.

459. **Lin, H. H., Korpel, A., Mehrl, D., and Anderson, D. R.,** Nonlinear Chinese tea, *Opt. News,* 15, 55, 1989.

460. **Lai, M. and Diels, J. C.,** Thermal nonlinear effects in exotic media: application to the study of nonlinear interfaces, *Appl. Opt.,* 29, 3882, 1990.

461. **He, K. X., Abeledayem, H., Chandra Sekhar, P., Venkateswarlu, P., and George, M. C.,** Transient multiple diffraction rings induced by ultrafast laser from Chinese tea, *Opt. Commun.,* 81, 101, 1991.

462. **Cheung, Y. M. and Gayen, S. K.,** Optical nonlinearities of tea studied by Z-scan and four-wave mixing techniques, *J. Opt. Soc. Am. B,* 11, 636, 1994.

463. **Sanderson, G. W.,** The chemistry of tea and tea manufacturing, in *Structural and Fundamental Aspects of Phytochemistry,* Runeckle, V. C. and Tso, T. C., Eds., Academic, New York, 1978, 247.

464. **Roberts, E. A. H. and Williams, D. M.,** The phenolic substances of manufactured tea. III. Ultra-violet and visible absorption spectra, *J. Sci. Food Agric.,* 9, 217, 1958.

465. **Yang, S. C., Quin, Q. M., Zhang, L. P., Qui, P. H., and Wang, Z. J.,** Measurements for the nonlinear refractive index of a new kind of polymer material doped with chlorophyll using nanosecond laser pulses, *Opt. Lett.,* 16, 548, 1991.

466. **Birge, R. R., Masthay, M. B., Stuart, J. A., Tallent, J. R., and Zhang, C. F.,** Nonlinear optical properties of bacteriorhodopsin: assignment of the third-order polarizability based on two-photon absorption spectroscopy, *SPIE,* 1432, 129, 1991.

467. **Birge, R. R., Gross, R. B., Masthay, M. B., Stuart, J. A., Tallent, J. R., and Zhang, C. F.,** Nonlinear optical properties of bacteriorhodopsin and protein-based two-photon three-dimensional memories, *Nonlinear Opt.,* 3, 133, 1992.

468. **Du, W. and Liu, S.,** Saturated optical nonlinearity of bacteriorhodopsin Langmuir-Blodgett films derived from degenerate multiwave mixing, *Thin Solid Films,* 229, 122, 1993.

469. **Uji, L., Volodin, B. L., Popp, A., Delaney, J. K., and Atkinson, G. H.,** Picosecond resonant coherent anti-stoke Raman spectroscopy of bacteriorhodopsin: spectra and quantitative third-order susceptibility analysis of the light-adapted BR-570, *Chem. Phys.,* 182, 291, 1994.

470. **Kajzar, F. and Messier, J.,** Third-order nonlinear optical effects in conjugated polymers, in *Conjugated Polymers,* Bredas, J. L. and Silbey, R., Eds., Kluwer Academic Publishers, Dordrecht, 1991, 509.

471. **Kobayashi, T., Yoshizawa, M., Stamm, U., Taiji, M., and Hasegawa, M.,** *J. Opt. Soc. Am. B,* 7, 1558, 1990; *SPIE Proc.,* 1775, 85, 1992.

472. **Kobayashi, T.,** Ultrafast relaxation in conjugated polymers with large optical nonlinearity in *Nonlinear Optics,* Miyata, S., Ed., North-Holland, Amsterdam, 1992, 101.

473. **Kobayashi, T.,** Time resolved nonlinear spectroscopy of conjugated polymers, in *Molecular Nonlinear Optics,* Zyss, J., Ed., Academic Press, New York, 1994, 47.

474. **Braunling, H., Becker, R., and Blochl,** Poly(axene methylidenes)-low gap compounds with intriguing chemical and physical properties, *Synth. Met.,* 55, 833, 1993.

475. **Sension, R. J., Phillips, C. M., Szarka, A. Z., Romanow, W. J., McGhie, A. R., McCauley, J. P. Jr., Smith III, A. B., and Hochstrasser, R. M.,** Transient absorption studies in C_{60} in solutions, *J. Phys. Chem.,* 95, 6075, 1991.

476. **Shank, C. V., Yen, R., Fork, R. L., Orenstein, J., and Baker, G. L.,** Picosecond dynamics of photoexcited gap states in polyacetylene, *Phys. Rev. Lett.,* 49, 1660, 1982.

477. **Heeger, A. J., Moses, D., and Sinclair, M.,** Semiconducting polymers: fast response nonlinear optical materials, *Synth. Met.,* 15, 95, 1986.
478. **Cross, G. H., Carroll, M., Axon, T. L., Bloor, D., Rangel-Rojo, R., Kar, A. K., and Wherrett, B. S.,** Organic-semiconductor-doped polymer glasses as novel nonlinear media, *SPIE Proc.,* 1775, 54, 1992.
479. **Rossi, B., Byrne, H. J., and Blau, W.,** Degenerate four-wave mixing in rhodamine doped epoxy waveguides, *Appl. Phys. Lett.,* 58, 1712, 1991.
480. **Ho, P. P.,** Picosecond Kerr Gate, in *Semiconductor Probed by Ultrafast Laser Spectroscopy,* Vol. 2, Alfano, R. R., Ed., Academic Press, New York, 1984, 410.
481. **Ho, P. P. and Alfano, R. R.,** Relaxation kinetics of salol in the supercooled liquid state investigated with the optical Kerr effect, *J. Chem. Phys.,* 67, 1004, 1977.
482. **Ooba, N., Tomaru, S., Kurihara, T., Mori, Y., Shuto, Y., and Kaino, T.,** Resonant subpicosecond degenerate four-wave mixing for a mainchain polymer with a definite short π-conjugated, *Chem Phys. Lett.,* 207, 468, 1993.
483. **Kajzar, F.,** Impact of dimensionality, conjugation length, scaling laws and electronic structure on nonlinear optical properties of conjugated polymers, *Nonlinear Opt.,* 5, 329, 1993.
484. **Neher, D., Mittler-Neher, S., Cha, M., and Stegeman, G.,** *SPIE Proc.* 1560, 335, 1990.
485. **Kajzar, F.,** Organic molecules for guided wave quadratic and cubic optics, in *Guided Wave Nonlinear Optics,* Ostrowsky, D. B., and Reinisch, R., eds., Kluwer Academic Publishers, Dordrecht, 1992, pp. 87.
486. **Yu, L., Polis, D. W., Sapochak, L. S., and Dalton, L. R.,** Advances in the synthesis of electroactive materials and in the characterization of mechanisms of NLO activity, in *Organic Molecules for Nonlinear Optics and Photonics,* Messier, J., et al., Eds., Kluwer Academic Publishers, Dordrecht, 1991, 273.
487. **Hanamura, E.,** *Quantum Optics,* Iwanami Modern Physics Series, Vol. 8, Iwanami Press, Tokyo, 1992, 169.
488. **Sasaki, K., Fujii, K., Tomioka, T., and Kinoshita, T.,** All-optical bistabilities of polydiacetyelenes Langmuir-Blodgett film waveguides, *J. Opt. Soc. Am. B,* 5, 457, 1988.
489. **Blau, W.,** Low-power optical bistability and phase conjugation in polydiacetylene, *Opt. Commun.,* 64, 85, 1987.
490. **Townsend, P. D., Jackel, J. L., Baker, G. L., Shelburne, J. A., and Etemad, S.,** Observation of nonlinear optical transmission and switching phenomena in polydiacetylene-based directional couplers, *Appl. Phys. Lett.,* 55, 1829, 1989.
491. **Norwood, R. A., Sounik, J. R., Holcomb, D., Popolo, J., Swanson, D., Spitzer, R., and Hansen, G.,** Nonlinear Bragg mirror made from a silicon phthalocyanine/methyl methacrylate copolymer, *Opt. Lett.,* 17, 577, 1992.
492. **Takur, M.,** All-optical phase modulation in polydiacetylene waveguides. In: *Polymer for Lightwave and Integrated Optics,* Hornak, L. A., Ed., Marcel Dekker, Inc., New York, 1992, 667.
493. **Etemad, S., Townsend, P. D., Baker, G. L., and Soos, Z.,** Two-photon absorption in polydiacetylene-based waveguides: its origin and its consequences, in *Organic Molecules for Nonlinear Optics and Photonics,* Messier, J., et al., Eds., Kluwer Academic Publishers, Dordrecht, 1991, 489.
494. **Krug, W., Miao, E., Derstine, M., and Valera, J.,** Optical scattering and scattering losses of PTS and poly(4-BCMU) thin film waveguides in the near infrared, *J. Opt. Soc. Am. B,* 6, 726, 1989.
495. **Assanto, G.,** Third-order nonlinear integrated devices, in Guided-Wave Nonlinear Optics, Ostrowsky, D. B., and Reinisch, R., Eds., Kluwer Academic Publishers, Dordrecht, 1992, 257.
496. **Vitrant, G.,** Third-order nonlinear integrated optical resonators, in Guided-Wave Nonlinear Optics, Ostrowsky, D. B., and Reinisch, R., Eds., Kluwer Academic Publishers, Dordrecht, 1992, 285.
497. **Stolen, R. M.,** Overview of fiber nonlinear optics, in Guided-Wave Nonlinear Optics, Ostrowsky, D. B., and Reinisch, R., Eds., Kluwer Academic Publishers, Dordrecht, 1992, 371.
498. **Aramaki, S., Assanto, G., Stegeman, G. I., Horstius, W. H. G., and Mohlmann, G. R.,** Integrated Bragg reflectors in polymeric channels waveguides, *Opt. Commun.,* 94, 326, 1992.
499. **Stegeman, G. I.,** Materials requirements for nonlinear third-order phenomena in waveguides, in *Nonlinear Optics,* Miyata, S., Ed., North-Holland, Amsterdam, 1992, 337.
500. **Foote, P. D., Proudley, G. M., Beddard, G. S., McFadyen, G. G., Reid, G. D., Connors, L. M., Bell, M., Hall, T. J., and Bwell, K.,** Picosecond optical correlations using dynamic holography in polyacetylene, *Appl. Opt.,* 32, 174, 1993.
501. **Rochford, K. B., Zaroni, R., Gong, Q., and Stegeman, G. I.,** Fabrication of integrated optical structures in polydiacetylene films by irreversible photoinduced bleaching, *Appl. Phys. Lett.,* 55, 1161, 1989.
502. **Kanbara, H., Kobayashi, H., Kubodera, K., Kurihara, T., and Kaino, T.,** Optical Kerr shutter using organic nonlinear optical materials in capillary waveguides, *IEEE Photon. Technol. Lett.,* 3, 795, 1991.
503. **Kanbara, H., Asobe, M., Kubodera, K., Kaino, T., and Kurihara, T.,** All-optical picosecond switch using organic single-mode fiber waveguide, *Appl. Phys. Lett.,* 61, 2290, 1992.
504. **Neher, S. M., Otomo, A., Stegeman, G. I., Lee, C. Y., Mehta, R., Agrawal, A. K., and Jenekhe, S. A.,** Waveguiding in substate supported and freestanding films of insoluble conjugated polymers, Appl. Phys. Lett., 62, 115, 1993.
505. **Etemad, S., Baker, G. L., and Soos, Z. G.,** Third-order NLO process in polydicaetylenes: physics, materials, and devices, *Molecular Nonlinear Optics,* Zyss, J., Ed., Academic, New York, 1994, 433.

Chapter 12

Applications of Organic Materials in Third-Order Nonlinear Optics

George I. Stegeman

CONTENTS

I. INTRODUCTION

The use of organic materials for nonlinear optics dates back to the very earliest days of the field.[1] For example, liquids such as benzene and nitrobenzene were used to demonstrate self-focusing, stimulated Raman and similar phenomena. In that era the principal value of organics, especially in liquid form, was their availability. As nonlinear optics evolved, solid-state materials, especially semiconductors (because they had been highly developed for electronic applications) became progressively more important in nonlinear optics.[2] Although large nonlinear optical effects had been observed in conjugated polymers as early as the mid-1970s, systematic studies of the multiphoton nonlinear optics of these materials have been pursued for less than a decade and device demonstrations only in the last few years.[3]

In this chapter we review the devices reported to date in organic materials. To put in perspective the material properties needed for devices, in the next section we briefly summarize the multiphoton absorption and refraction phenomena that lead to a nonlinear optical response and ultimately applications. Next we deal with the device demonstrations that have been reported. Finally we summarize the status of the field, and comment on further work that needs to be done.

II. SUMMARY OF NONLINEAR OPTICS OF ORGANIC MATERIALS

The common terminology is that any process which leads to an intensity-dependent change in the refractive index or absorption coefficients belongs to nonlinear optics. Here we summarize the processes which have been used for nonlinear optical devices based on organic materials.

A. ELECTRONIC NONLINEARITIES

The electronic optical properties of materials can be understood in terms of their electronic energy levels and transitions between them. In semiconductors, there is a continuum of states for energies larger than a well-defined bandgap between the valence and conduction bands.[4] In contrast, in organic

799

(a)

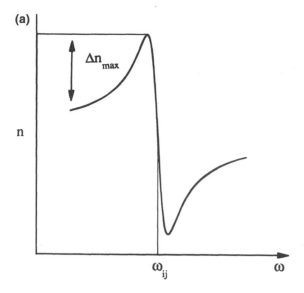

(b)

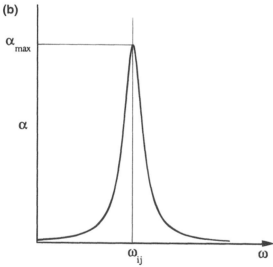

Figure 1 (a) Variation of n with photon frequency for a one-photon allowed transition. (b) Variation of α with photon frequency for a one-photon allowed transition.

systems there is a series of discrete, well-defined energy levels. At high enough energies, these discrete levels become sufficiently dense that they form a quasi-continuum, at least at room temperature.

Electric dipole-allowed transitions between the electronic energy levels E_i and E_j are responsible for the complex dielectric constant and hence the refractive index and absorption coefficient. For example, for input photon energies approximately equal to an allowed transition energy, $\hbar\omega \propto E_j - E_i = \hbar\omega_{ij}$, a peak occurs in the absorption spectrum, accompanied by dispersion in the refractive index (Figure 1). Normally the lower state for the transition is the ground state ($i \equiv g$). For electric-dipole transitions to be allowed, a change in the symmetry of the eigenfunctions between the lower and upper states is necessary. The larger the oscillator strength of the transition (transition electric dipole moment μ_{ij}), the larger the absorption maximum and the stronger the dispersion in refractive index. With reference to Figure 1, α_{max} and Δn_{max} are both proportional to $|\mu_{ij}|^2$. The wavelength dispersion in the linear optical properties is a superposition of absorption maxima and index dispersion due to all of the allowed one-photon transitions from the ground state.

The multiphoton optical response depends on dipole-allowed transitions involving many different electronic states (see, for example, reference 5). Here, transitions can also occur between multiple states, either simultaneously or sequentially. Both cases are dealt with in much more detail in Chapter 14. Here we will concentrate on the simultaneous case involving electric-dipole transitions and focus on

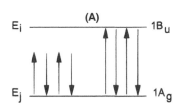 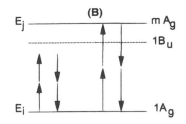

Figure 2 Schematic representation of various multiphoton processes which lead to n_2 and α_2. (A) Transition involving the one-photon state; (B) two-photon transitions. The state designations for a linear molecule are shown as $1A_g$, $1B_u$, and mA_g, respectively. Also, the nonresonant and resonant cases are indicated on the left- and right-hand sides in each case.

the interactions useful for signal-processing devices, hence limiting the discussion to the case where all of the inputs and outputs are at the same frequency. Within the scope of this review chapter it is not feasible to include all the contributions to the nonlinearities. For example, neglected here are transitions between vibronic sublevels which give a fine structure to the transitions between electronic levels, electric quadrapole, and magnetic dipole contributions.

The scenarios which govern the gross nonlinear response are shown schematically in Figure 2 for the interaction of two photons with an oversimplified molecular system. A and B are processes normally associated with the instantaneous response of third-order nonlinear optics. One photon changes the optical properties of the medium which can then be experienced by the second photon. The simplest case occurs for molecules with sufficient symmetry (inversion symmetry) to have only even and odd symmetry electronic states, usually labeled as A_g and B_u, respectively. The nonlinear optical (NLO) response depends on how many different even symmetry states are involved. Recalling that the ground state has A_g symmetry and therefore normally participates in all scenarios, the key question is whether an excited state of A_g symmetry (called a "two-photon" state) is involved or not.

The first case ("A", no excited two-photon state) is shown schematically in Figure 2A. The strength of the effect is proportional to $|\mu_{gj}|^4$. For photons far off resonance with the transition, the nonlinearity leads to the susceptibility $\chi^{(3)}(-\omega; \omega, -\omega, \omega)$ which is responsible for an instantaneous intensity-dependent refractive index n_2. This is the limit of the so-called "Kerr nonlinearity." For photon energies near resonance with the transition, the nonlinearity is greatly enhanced, but is also accompanied by absorption. For high photon fluxes, the excited state can become populated, resulting in saturation effects which appear as higher order effects for the nonlinear response. The variation in $n(\omega)$ and $\alpha(\omega)$ with intensity is sketched in Figure 3. As the intensity is increased, the population of excited molecules increases, decreasing the number of molecules with available (unoccupied) excited states and hence decreasing the macroscopic absorption coefficient. This is called "saturable" absorption. In parallel, the corresponding change in refractive index is "bleached" out. If one were to expand either the index or absorption in powers of intensity I, an approach valid for intensities approaching but still much less than the saturation intensity, $n = n_1 + n_2 I + n_3 I^2$ and $\alpha = \alpha_1 + \alpha_2 I + \alpha_3 I^2$ is usually adequate. For $\lambda_{max} < \lambda$, $n_2 < 0$ and $n_3 > 0$. The higher order absorptive terms given by α_3 arise due to "saturation"

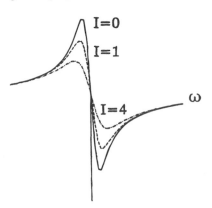

Figure 3 Spectral variation of the absorption and refractive index with increasing intensity for a one-photon allowed transition.

Table 1 Signs of Nonlinear Coefficients Associated with One- and Two-Photon Processes

	n_2	n_2	n_3	n_3	α_2	α_3
	$\omega > \omega_{ij}$	$\omega_{ij} > \omega$	$\omega > \omega_{ij}$	$\omega_{ij} > \omega$		
1 photon	+	−	−	+	−	+
2 photon	−	+	+	−	+	−

of the absorption. This leads to $\alpha_3 > 0$. The expected signs of the coefficients are listed in Table 1. Note that because the saturation of a state evolves in time (during a pulse of duration shorter than the excited state lifetime), the response of n_3 and α_3 is fluence dependent, evolving over the duration of the pulse.

The second process, labeled "B" in Figure 2, also leads to $n(I)$ and $\alpha(I)$, this time via the participation of an excited A_g state. The strength is proportional to $|\mu_{gj}\mu_{jg'}|^2$ where "g'" labels the excited, even symmetry state. The variation in $n(\omega)$ and $\alpha(\omega)$ with intensity is quite different from the preceding case (see Figure 4). For example, for frequencies less than the resonance frequency, Δn is positive. The higher order absorptive terms given by α_3 arise due to "saturation" of the absorption of the two-photon state. Shown in Table 1 are the signs of the coefficients for the different processes at different frequencies relative to the resonance frequency. The response of n_3 and α_3 is not instantaneous over an exciting pulse and hence they are fluence dependent. Note that the signs for α_2 are different for the one- and two-photon cases. This implies that at certain wavelengths they could cancel leading to an effective $\alpha_2 \to 0$ which has interesting repercussions for the figures-of-merit discussed later.[6]

The third-order response of a material usually refers to $n_2(\lambda)$ and $\alpha_2(\lambda)$ in the limit where $|n_2 I| \gg |n_3 I_2|$ and $|\alpha_2 I| \gg |\alpha_3 I^2|$. In fact, to the best of our knowledge, there is no molecule for which such a measurement has been made over its full spectrum. Based on the discussion above, the dispersion in the nonlinearity of a linear centrosymmetric molecule whose oscillator strength is dominated by a single one-photon and two-photon excited state is sketched in Figure 5. Standard nomenclature is that the nonlinear response within a few linewidths of the absorption maximum is called the "resonant" response. "Near-resonant" refers to the spectral region where there is a strong spectral dependence to the nonlinearity n_2 and "nonresonant" is defined by the long wavelength limit where n_2 becomes essentially independent

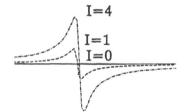

Figure 4 Spectral variation of the absorption and refractive index with increasing intensity for a two-photon allowed transition.

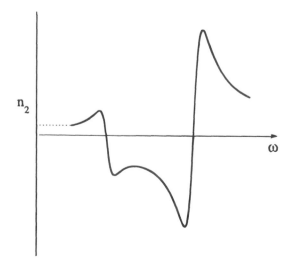

Figure 5 Wavelength variation of n_2 for a molecule with three states: the ground state, a one-photon allowed state, and a two-photon allowed state. The example is for a polyene.

of wavelength. In fact the case shown in Figure 5 corresponds to a polyene or conjugated polymer for which the dominant contribution to the nonresonant response is due to the two-photon state.[7] For a cyanine-like linear molecule, the long wavelength limit would show a negative value for n_2 because the one-photon contribution dominates in the nonresonant regime.[7] Note that for the wavelength region shown by the dashed lines, additional contributions to the nonlinearity due to vibrational transitions occur and further wavelength variation in the nonlinearity is actually expected.

If the molecule has insufficient symmetry to have well-defined symmetry states, all of the states have mixed one-photon (B_u) and two-photon (A_g) character and electric-dipole-allowed transitions occur between all of the energy levels. Therefore, both processes discussed above occur in parallel, leading to a complex nonlinear response. Whether a particular transition contributes to the nonlinearity or not depends on the product of the appropriate oscillator strengths.

Multiphoton processes involving the simultaneous interaction of three, four, etc. photons with a molecular system can also occur. In general they require very high intensities to interfere with the one- and two-photon contributions. The three-photon case leads to three-photon absorption and dispersion, i.e., to terms with $\Delta n \propto n_3 I^2 + n_4 I^3$, etc. and $\Delta\alpha \propto \alpha_3 I^2 + I^3\alpha_4$, etc. where terms in I^3 and beyond are due to saturation of the three-photon process. Far from any resonances the response is usually discussed in terms of the anharmonic response of electrons in their potential wells, i.e., $\chi^{(5)}(-\omega; \omega, -\omega, \omega, -\omega, \omega)$, and of course the details of the anharmonic response are rooted in the details of the electronic states. Note that the leading terms in this process are proportional to I^2 and interfere with the saturation terms due to two-photon processes, etc.

B. THERMAL NONLINEARITIES

Absorption can induce index changes via the thermo-optic effect. That is, the absorption of a photon leads to a local increase in the sample temperature T, which in turn leads to an index change via dn/dT. For illumination over a time period longer than the characteristic thermal relaxation time τ, the thermal nonlinearity n_{2t} is given approximately by[8]:

$$n_{2t} = \frac{\partial n}{\partial T}\frac{\alpha_1\tau}{\rho C_p} \tag{1}$$

where τ is the thermal relaxation time (which depends on the heat sinking conditions), ρ is the density, and C_p is the specific heat. Note, however, that if the pulse duration Δt is shorter than τ and the repetition rate is much less than $1/\tau$, the index change is fluence dependent. Furthermore, over the pulse envelope it varies as

$$\delta n(r, t) = \frac{n_{2t}}{\tau}\int_{-\infty}^{t} dt' I(r, t') \tag{2}$$

so that an effective nonlinearity in this case is $n_{2t,\text{eff}} \simeq n_{2t}\Delta t/\tau$. For the frequently encountered case of multiple pulses with repetition rate $> 1/\tau$ (for example, a cw mode-locked laser), it is the time-average intensity (over τ) that enters into estimating the thermo-optic contribution to the index change, independent of the duration or peak intensity of any single pulse in the pulse train. The thermo-optic effect can often be the dominant nonlinearity when mode-locked lasers are used.

It is very difficult to give an exact value for how large the effective thermal nonlinearity can be because the heat dissipation plays a key role and can be vastly different from case to case. For example, for an absorption coefficient of about 0.25 cm^{-1}, an effective nonlinearity of $n_2 \sim 4 \times 10^{-11}$ cm^2/W was estimated in a planar polymer waveguide based on a 1 μs thermal relaxation time.[9] This number will clearly vary with the absorption coefficient, but also by over an order of magnitude depending on the sample geometry and heat sinking. The key point is that thermo-optic effects are very large and can be best avoided by using single, very short pulses or materials with very small absorption coefficients.

C. MOLECULAR REORIENTATION

Intense optical fields can induce optical birefringence into a liquid composed of optically anisotropic molecules.[5] Preferential alignment of the molecules occurs along the polarization direction of a pump field via the difference in the molecular polarizabilities $\alpha_\parallel - \alpha_\perp$. The index increases along the direction of the pump electric field, and decreases in the plane orthogonal to E_{pump}. The induced birefringence is $\Delta n \propto [\alpha_\parallel - \alpha_\perp]^2 I$. Therefore a probe (signal) beam originally polarized at 45° to the pump beam experiences polarization rotation with propagation distance. In a Kerr shutter, a crossed polarizer (relative to the input) for the probe can be used to get a switched output.

III. APPLICATIONS OF NONLINEAR ORGANICS

The reported applications of organic materials fall into two broad categories: those that rely on multiphoton absorption and those that utilize a nonlinear refractive index change. For example, many optical limiting schemes for sensor protection rely on multiphoton absorption involving multiple one- and two-photon processes. These are discussed in detail in Chapter 14. Other devices, typically targeted for signal processing operations such as bistability, switching, etc. utilize intensity-induced refractive index changes.[10] This class of devices relies on producing a nonlinear index change.[11] A variety of materials have been used. In many cases it is not clear whether truly electronic nonlinearities were operative, or other effects were responsible for the observed device response.

There are two simple figures-of-merit (FOM) which can be used to assess whether a material is useful for all-optical devices at a specific wavelength.[10–12] They are based on two characteristics common to almost all devices. One is that a minimum, cumulative nonlinear phase shift is required for a device to function properly. The second is that for a device to be useful, it must have a reasonable signal throughput, i.e., the losses must be small. For a given material, these conditions can be quantified in terms of the maximum nonlinear phase shift possible in a sample of length L,

$$\Delta\phi^{NL} = \int_0^L n_2 \frac{2\pi}{\lambda_{vac}} I(z) \, dz \tag{3}$$

For example, if linear loss is dominant, $I(z) = I(0)\exp[-\alpha_1 z]$ which leads to an effective length $L_{eff} = [1 - \exp(-\alpha_1 L)]/\alpha_1$. Noting that $\alpha_1^{-1} > L_{eff}$, replacing L by α^{-1} means that $nI(0)k_{vac}/\alpha_1 > \Delta\phi^{NL}$. Based on detailed analysis of the nonlinear directional coupler (NLDC), the most versatile switching device reported to date, a minimum phase shift of 2π is necessary (reviewed in references 10 and 11). Therefore, a useful FOM for a NLDC is

$$W = \frac{\Delta n(\equiv n_2 I)}{\alpha_1 \lambda_{vac}} > \frac{\Delta\phi^{NL}}{2\pi} = 1 \tag{4}$$

That is, $W > 1$ is required for a NLDC with high throughput. Other devices require different minimum values of W varying from 0.50 to 2 (reviewed in references 10–12).

However, at high intensities two-photon absorption can dominate the loss. In this case, replacing L by $[\alpha_2 I]^{-1}$ leads to the FOM

$$T^{-1} = \frac{n_2}{\alpha_2 \lambda_{vac}} > 1 \tag{5}$$

Using similar arguments, α_3 leads to another FOM.[10] If these three FOM are not satisfied, then device operation is compromised and/or the throughput of the device is poor.

None of the materials that were used in the device studies discussed below were initially screened for these FOM. And that is the principal reason for the overall disappointing performance of organic materials in waveguide switching operations. Listed in Table 2 are the recently measured FOM for a

Table 2 Measured FOM for a Few Selected Organic Materials

Material	n_2 (cm²/W)	α (cm⁻¹)	W	T	λ (μm)
PTS (crystal)	2.2×10^{-12}	<0.8	>10	<0.1	1.60
poly-4BCMU	5×10^{-14}	<0.2	>2.5	<1.0	1.32
DANS (polymer)	8×10^{-14}	<0.2	>5.0	≈0.2	1.32

Note: An intensity of 1 GW/cm² is assumed.

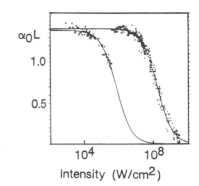

Figure 6 Saturation absorption measured in Si naphthalocyanine near λ_{max} for 10 ns (circles) and 30 ps (triangles) pulse widths. Agreement with theory (solid lines) is excellent. (From Wu, J. W., et al., *J. Opt. Soc. Am. B*, 4, 707, 1989. With permission.)

number of organic materials. It is clear that suitable materials do actually exist, and in fact single-crystal PTS looks very promising in the two communications windows at 1320 and 1550 nm.[13,14] In general, usually the nonresonant regime is needed for both FOM to be satisfied for waveguide devices which require high throughput. In addition, there are wavelength regions between the one- and two-photon absorption peaks, i.e., near-resonant wavelengths, where devices can operate with moderate losses. They are most suitable for parallel signal input devices.

A number of device-related experiments have been performed on the best materials available at the time of those experiments. These include absorption bleaching, optical bistability, nonlinear grating distributed feedback and coupling, and all-optical switching.

A. PARALLEL PROCESSING WITH BULK MEDIA

Bistability, initially projected as a key device for optical computing, has now been observed in a number of organic materials, dating back to the early work in liquid crystals. The recent work has dealt with materials chosen specifically for their large electronic nonlinearities, although frequently unwanted thermal effects were present, and in some cases dominant (for example, see references 15 and 16). The cleanest results were performed with a femtosecond laser and show a logic gate based on resonator bistability operating with a near-resonant nonlinearity.[17]

1. Etalon Devices: Optical Bistability and Logic Gates

There has been a sustained interest in bistability since the late 1970s for optical logic and computing.[2] A nonlinear optical material inside a high finesse cavity can lead to two possible output states (different transmission) at a given input power. They can correspond to a logical 1 (high transmission) or 0 (low transmission), the value depending on the detailed history of the prior illumination of the device.

The use of organic materials for this purpose started about a decade later.[15,16] Just as in other classes of materials, initial observations involved thermal nonlinearities, and unfortunately these are the easiest to observe unless the wavelength (relative to λ_{max}), pulse duration, and repetition rate are chosen carefully.[15] In fact, the first material to be studied thoroughly in the near-resonant regime where saturable absorption dominates the electronic response was the planar, symmetric molecule silicon naphthalocyanine.[16] The results shown in Figure 6 indicate a classical saturable absorber response. Armed with this information, the authors then chose the appropriate spectral regime 799 nm (slightly detuned from the absorption maximum at 807 nm) for absorptive (vs. thermal) bistability. An 80-nm-thick film was sufficient to observe bistability in a high-finesse Fabry-Perot etalon with an incident intensity of 160 kW/cm^2. In terms of the FOM, $W < 1$ when absorptive bistability dominates and hence the devices had low throughput.

A thin organic film was also used in a Fabry-Perot resonator structure to implement a logical NOR gate.[17] In an etalon geometry, a NOR gate is equivalent to using a control pulse to detune the resonator from maximum transmission for a probe beam. That is, an output signal (probe here) appears as long as no signal beam (control) is present. This detuning is shown in Figure 7 for the molecules 7-[4-[(4-hexyloxyphenyl)azo](naphthyl)azo-(2,3-dihydro-1,3-dimethyl-2-octyl)]perimidine (DNP) doped into PMMA. The control pulses at the peak of the absorption response (610 nm) were ~ 100 fs long and the probe signal was in the long wavelength tail of the absorption line at 770 nm. The ≈4 ps gate recovery time observed is real and attributable to the nonradiative lifetime of the DNP molecules. The required intensity unfortunately is high, of order GW/cm^2.

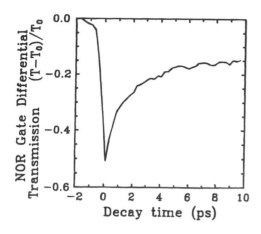

Figure 7 The differential transmission of a probe beam at 770 nm induced by the incidence of an intense pump beam at 610 nm for a cavity tuned to maximum transmission at the probe beam wavelength. (From Williams, V. S., et al., *Appl. Phys. Lett.*, 57, 2399, 1990. With permission.)

2. Nonlinear Bragg Reflector

A nonlinear Bragg reflector has been demonstrated using Si-phthalocyanine as the nonlinear material.[18] The sample was made by spinning multiple (23 total) layers of PMMA and Si-phthalocyanine (5%)/ MMA on top of each other, in an alternating fashion. The layer thicknesses were chosen so that the optical thickness of each corresponds to one quarter of the wavelength of the incident radiation in the respective media, i.e., $d = 0.25 \; \lambda/\cos\theta$ where θ is the angle of propagation in the material relative to the grating normal and λ is the wavelength in the material. As a result, the Bragg condition for retroreflection of the incident light is satisfied, i.e., $\Delta\mathbf{k} = \mathbf{k_s} - \mathbf{k_i} + \kappa \rightarrow 0$ where $\mathbf{k_s}$ and $\mathbf{k_i}$ are the scattered and incident wavevectors, respectively, and $\kappa = 4\pi/d$ is the grating wavevector. The transmission is reduced and part of the wave is reflected. The fraction of light reflected depends on the number of layers, and the refractive index discontinuity between alternate layers.

Increasing the intensity of the incident radiation changes the refractive index of the phthalocyanine layers. The nonlinearity leads to an intensity-dependent wavevector, i.e., $\mathbf{k_s}(I)$ and $\mathbf{k_i}(I)$. As a result $\Delta k \rightarrow \Delta\mathbf{k}(I)$ and it is possible to change the reflection response at a given wavelength. This detunes the structure away from the Bragg condition, i.e., a wavevector mismatch $\Delta k = 2k_{vac}n_2I$ is induced and the reflection coefficient is reduced. This result is shown in Figure 8 for $\lambda = 688$ nm, detuned to wavelengths longer than λ_{max} where there is a substantial refractive nonlinearity. Note the decrease in the reflection of the beam with increasing intensity. Operating in this near-resonant regime led to substantial total losses due to absorption. Such a device has potential applications for optical limiting, switching, etc.

3. Correlation Devices

Degenerate four-wave mixing is a nonlinear phenomenon usually associated with material characterization.[5] There are three input optical beams, two of them counter-propagating (pump beams, reference [ref] and object [ob]). The third input beam (read [re]) creates two interference gratings, one with each

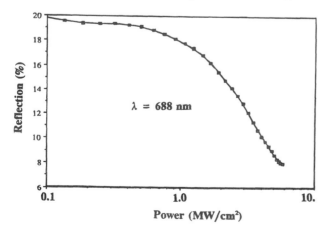

Figure 8 Intensity dependence of the reflectivity of a nonlinear Bragg mirror made of alternate layers containing phthalocyanine. (From Norwood, R. A., et al., *Opt. Lett.*, 17, 577, 1992. With permission.)

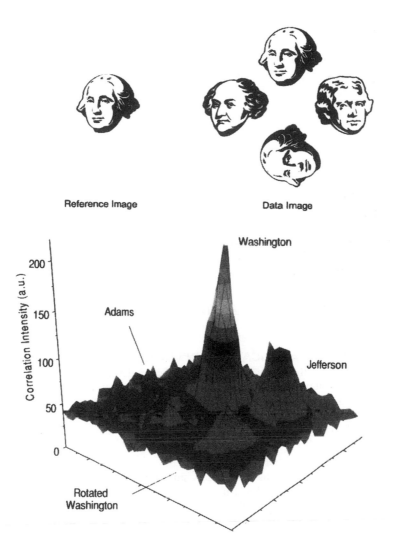

Figure 9 (A) Image of Washington. (B) Images of Washington (top, bottom rotated through 90°) and other former presidents. (C) Correlation of images in A and B. The largest correlation peak is due to the images of Washington with the same alignment. (From Halverson, C., et al., *Science*, 265, 1215, 1994. With permission.)

strong counter-propagating beam. Both of these gratings deflect the other counter-traveling beam back along the direction of the third (read) beam. The signal amplitude is given by $A_{ref}A_{re}A_{ob}^* + A_{ob}A_{re}A_{ref}^*$. When spatial phase and amplitude information is impressed upon one of the strong beams (for example, the object beam), the correlation of that information with anything impressed on the read beam is obtained through the first term. This process gives the correlation between the read and object beams.[19] Although this process can in principle lead to large signals, for films microns thick the signal strength is too weak to be used for any additional form of optical signal processing. This is not normally an impediment because the signal is detected and conversion efficiency is usually not an issue.

This approach has been used by Heeger and co-workers to demonstrate an optical image processor based on 160 fs pulses.[19] The spin-cast films were made from poly(1,6-heptadiester) (PHDE), a molecule which has a twofold degenerate ground state and which was soluble due to the addition of side groups. Shown in Figure 9 are the images impressed on two of the three input beams by spatial light modulators. The cross-correlation output is also shown in Figure 9C. The presence of a "Washington" input (Figure 9, A and B) on two of the beams is clearly "identified" by the strong correlation peak, and the correlation with the other images is small. The signal in this case is obtained by a CCD array and the small conversion efficiency (10^{-4}) is not an impediment. For this application, the FOM are not limitations,

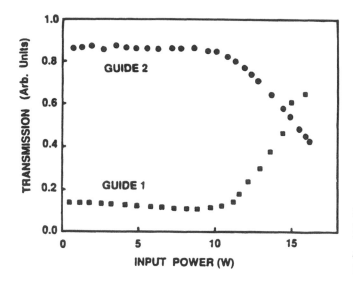

Figure 10 Variation in the channel outputs for long pulses (quasi-cw case) at 1064 nm. (From Townsend, P. D., et al., *Appl. Phys. Lett.*, 55, 1829, 1989. With permission.)

except perhaps for the slower decay time associated with absorptive grating relative to refractive index gratings.

B. SERIAL PROCESSING APPLICATIONS IN WAVEGUIDES

For these applications, signal throughput is critical because the device output is typically routed to other optical devices for either further processing, or transmission. Therefore the FOM play a critical role. A number of all-optical devices have been reported in waveguides. They include directional couplers, distributed grating devices, Kerr gates, and soliton interconnects. To the best of our knowledge, all of the results can be interpreted in terms of absorption mechanisms, either one- or two-photon, being dominant.

1. Nonlinear Directional Coupler

The nonlinear directional coupler is the most widely studied waveguide device for all-optical switching, logic, etc.[11] It consists of two waveguides, usually channels, which are parallel and located close enough so that the waveguide field of one channel overlaps the neighboring waveguide. When the waveguides are identical, this overlap leads to a periodic exchange between the waveguides with propagation distance. When the waveguides are sufficiently different, i.e., have different propagation wavevectors, the transfer between channels is inhibited. Such a wavevector detuning can be induced optically in one channel by injecting into it a high power beam, assuming of course that the channel contains a nonlinear material. That is, the index is changed locally by the high intensity via $\Delta n = n_2 I$ and hence the wavevector is changed. This leads to a power-dependent response of the outputs.

This device has been studied in polydiacetylene materials and organic semiconductors. The Bellcore group has worked with channel waveguides fabricated in the red form of poly-[5,7-dodecadiyn-1,12-diol-bis(n-butoxycarbonylmethylurethane)] (poly-4BCMU), investigated by lasers with different pulse widths.[20,21] The problem that they encountered is that at 1064 nm the nonlinear response was found to be dominated by one- and two-photon absorption. For long pulses, they obtained a thermo-optic based response (see Figure 10). For picosecond pulses, they were able to explain the observed response in terms of an increasing absorption with increasing intensity. The key result is the equalization of the output in the two channels at high input intensities, a characteristic feature of intensity-dependent absorption in a NLDC. In retrospect this case corresponds to $T > 1$, i.e., it violates the figures-of-merit discussed previously. This research is a classic example of the care that needs to be taken to interpret power-dependent responses in devices properly.

Sasaki's group at Keio University has investigated nonlinear directional couplers made from poly-4BCMU thin films deposited on glass waveguides. This particular device consisted of four layers instead of channels. Clear switching was observed at 1064 nm with 20 ns pulses.[22] The most recent results obtained by Sasaki et al. with a cw mode-locked Ti:sapphire laser operating at 885 nm with 100 fs pulses are shown in Figure 11. The switching results are different from those expected from a simple Kerr refractive nonlinearity. In that case the outputs should asymptotically approach either 0 or 1 at

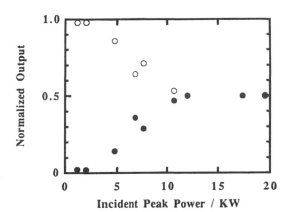

Figure 11 Switching in a planar polydiacetylene non-linear directional coupler implemented with 150-fs pulses from a mode-locked Ti:Sapphire laser at 825 nm. (From Sasaki, K., et al., *Materials Research Society Symposium Proceedings*, 247, 141, 1992. With permission.)

high input powers, depending on which channel is the input channel. The results are reminiscent of the Bellcore work which was dominated by two-photon absorption. Given the Bellcore results, and the large excited state absorption known to occur in these materials in the wavelength range 800–900 nm, the origin of the participating nonlinearities is not clear.[23] Further research is probably needed to resolve this uncertainty.

Experiments have been performed on directional couplers formed from 3,4,9,10-perylenetetracarboxylic dianhydride (PTCDA), an organic semiconductor.[24] The guided wave beams were at 1064 nm and the all-optical response was induced by illuminating the channels from above (externally) with a cw argon ion laser at 514.5 nm. Pump beam-induced switching of a few percent was observed in the 1064 nm input beams for an illumination intensity estimated to be 4 W/cm^2 when a cw pump beam at 515 nm was introduced. Because the pump beam falls near the exciton feature where strong absorption occurs, the switching was attributed to the index changes induced by the free carriers generated by the dissociated excitons. The effective nonlinearity is large, $\sim 5 \times 10^{-5}$ cm^2/W. In this case the probe beam was in a relatively transparent region of the spectrum and the pump beam was strongly absorbed. Typical nonlinearity relaxation times of 10 µs were measured, too long for ultrafast switching applications.

2. Nonlinear Bragg Reflector

As discussed previously, the transmission properties of a periodically stratified medium can change dramatically for wavelengths near the Bragg condition. In waveguides this is achieved by modulating the index periodically, or by corrugating one of the waveguide surfaces. This type of response has many applications, including switching, optical limiting, logic gates, etc.

There have been two reported implementations of nonlinear waveguide grating devices based on organic materials. In one case, the periodic structure was directly photobleached into a poly-4BCMU waveguide by illuminating it with an interference pattern near λ_{max}.[25] Light-induced breaking of bonds reduces locally the refractive index of the PDA. The nonlinear properties of the distributed feedback gratings were studied with a cw Nd:YAG laser operating at 1064 nm. Hysteresis loops were observed, but their power dependence does not seem to match that expected from a Kerr nonlinearity. However, the origin of the nonlinearity was not specified.

In the second case, the interface between a guiding film and a cover layer was corrugated by ion milling.[26] The nonlinear response of a distributed feedback grating was investigated in channel waveguides in the side-chain polymer 4-dialkylamino-4′-nitrostilbene (DANS) at 1550 nm. Near the Bragg wavelength the transmission of the grating was less than 1%. The power-dependent response for the transmission just off the Bragg condition is shown in Figure 12. It shows a power-dependent detuning from the Bragg wavelength and the consequent increase in transmission coefficient. A cw mode-locked color center laser operating with picosecond pulses was used. However, it appears that thermal effects played the dominant role. This was verified by reducing the pulse repetition rate while maintaining the peak pulse intensity and observing that the all-optical response disappeared. Note that reducing the repetition rate reduces the heat deposited in the sample and hence the nonlinear switching effect.

C. OTHER SWITCHING DEVICES

Another interesting approach to all-optical switching has been to use liquids in a hollow core fiber in a Kerr shutter geometry. Highly anisotropic molecules such as 4-(*N*,*N*-diethylamino)-4′-nitrostilbene

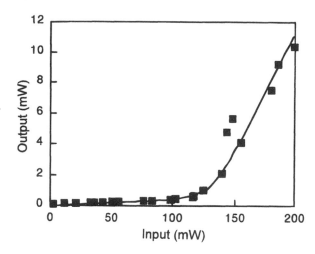

Figure 12 Transmitted vs. input power (average) for a nonlinear distributed feedback grating in DANS excited near its transmission minimum (Bragg condition). (From Aramaki, S., et al., *J. Lightwave Technol.*, 11, 1189, 1993. With permission.)

(DEANST) have been used because of their large nonlinearity due to molecular reorientation, ~6 × 10^{-14} cm^2/W, with a relaxation time of only a few picoseconds.[27,28] Switching powers of a few watts were achieved in the transparency window at 1064 nm by using fibers on the order of 1 m long. Their results are shown in Figure 13.

Bistability and switching have also been observed during the coupling process by which light is injected via gratings and prisms into planar waveguides.[9,25,29–34] In fact, this effect was the source of a lot of confusion in the early days of nonlinear integrated optics because many researchers found bistable responses in a variety of material systems. It is now widely accepted that bistability can only be obtained via diffusive nonlinearities (for example, the thermo-optic effect).[35] However, grating coupling with very fast single pulses has allowed the nonlinearity of a number of waveguiding materials to be measured.[9,32–34]

D. DARK SOLITON WAVEGUIDES

Thermal nonlinearities of organic molecules in solution have been used to demonstrate the existence and properties of dark spatial solitons.[36–40] Typically this nonlinearity on the long wavelength side of λ_{max} is negative and takes values of the order of -10^{-7} cm^2/W in the visible region of the spectrum. A dark soliton in a self-defocusing medium ($n_2 < 0$) consists of a specially shaped "hole" in a constant background illumination with a π phase change across its minimum. Because of the interplay between spatial diffraction and self-defocusing, this hole propagates unchanged with distance at certain intensities of the background light. Furthermore, because the hole region has a higher index than the surrounding medium, it forms a waveguide which can guide other (signal) beams. By changing the phase angle across the minimum, the "hole" can be steered into different directions, thus forming a reconfigurable interconnect for the signal.

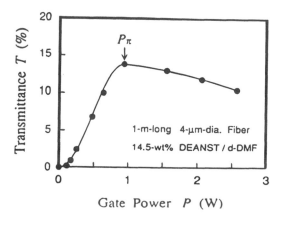

Figure 13 Probe transmission through cross-polaroids as a function of the gate (pump) power in a 1-meter-long, hollow core fiber filled with DEANST. (From Kanbara, H., et al., *J. Opt. Soc. Am. B.*, 11, 2216, 1994. With permission.)

IV. PROSPECTS

The progress with respect to using third-order nonlinearities in organics for signal processing has not been very good. Over the last few years many new materials with interesting nonlinearities have been reported. Unfortunately, rarely have the linear and two-photon absorption coefficient also been reported so that the nonlinear figures-of-merit for devices cannot be estimated. As a result, much of the work reported to date has been done with inappropriate materials.

One might ask whether more promising materials have been identified. Listed in Table 2 are a few materials for which the figures-of-merit are known. PTS, for example, has a very large electronic nonlinearity at the communications windows.[13,14] Furthermore, for bulk applications there are some interesting possibilities. For example, as discussed earlier, the sign of α_2 is positive for two-photon absorption and negative for saturating the one-photon nonlinearity. Therefore, in a limited wavelength region it should be possible to counteract two-photon absorption in this way. In fact, the initial experiments have been performed to verify that indeed the net α_2 does go to zero.[6] This region would be promising for bistability and logic gates since the nonlinearity is dominated by the one-photon peak and is still very large.

In semiconductors, the possibility of combining gain with nonlinearity has eliminated the loss problem and led to some interesting switching devices.[41-43] This can be expected in the polymer field, and in fact a polymer fiber amplifier has already been reported.[44]

REFERENCES

1. **Bloembergen, N.,** *Nonlinear Optics*, Benjamin Press, New York, 1964.
2. **Gibbs, H. M.,** *Optical Bistability: Controlling Light with Light*, Academic Press, Orlando, FL, 1985.
3. **Sauteret, C., Hermann, J. P., Frey, R., Praderc, F., Ducuing, J., Baughman, R. H., and Chance, R. R.,** Optical nonlinearities in one-dimensional conjugated polymer crystals, *Phys. Rev. Lett.*, 36, 956, 1976.
4. **Peyghambarian, N. and Koch, S. W.,** Semiconductor nonlinear materials, in *Nonlinear Photonics*, Gibbs, H. M., Khitrova, G., and Peyghambarian, N., Eds., Springer-Verlag, New York, 1990, p. 7.
5. **Boyd, R. W.,** *Nonlinear Optics*, Academic Press, Boston, MA, 1992.
6. **Molyneux, S., Kar, A. K., Wherrett, B. S., Axon, T. L., and Bloor, D.,** Near-resonant refractive nonlinearity in polydiacetylene 9-BCMU thin films, *Opt. Lett.*, 18, 2093, 1993.
7. **Pierce, B. M.,** A theoretical analysis of the third-order nonlinear optical properties of linear cyanines and polyenes, *SPIE Proceedings on Nonlinear Optical Organic Materials IV*, 1560, 148, 1991.
8. **Assanto, G., Fortenberry, R. M., Seaton, C. T., and Stegeman, G. I.,** Theory of pulsed excitation of nonlinear distributed prism couplers, *J. Opt. Soc. Am. B*, 5, 432, 1988.
9. **Burzynski, R., Singh, B. P., Prasad, P. N., Zanoni, R., and Stegeman, G. I.,** Nonlinear optical processes in a polymer waveguide: grating coupler measurements of electronic and thermal nonlinearities, *Appl. Phys. Lett.*, 53, 2011, 1988.
10. **Stegeman, G. I.,** Material figures of merit and implications to all-optical switching, *SPIE Proceedings on Nonlinear Optical Properties of Advanced Materials*, 1852, 75, 1993.
11. **Stegeman, G. I. and Miller, A.,** Physics of all-optical switching devices, in *Photonic Switching*, Vol I, Midwinter, J., Ed., Academic Press, Orlando, FL, 1993, p. 81.
12. **Stegeman, G. I. and Wright, E. M.,** All-optical waveguide switching, *J. Opt. Quant. Electron*, 22, 95, 1990.
13. **Lawrence, B., Cha, M. Kang, J. U., Torruellas, W. E., Stegeman, G. I., Baker, G., Meth, J., and Etemad, S.,** Large purely refractive nonlinear index of single crystal p-toluene sulfonate (PTS) at 1600 nm, *Electron. Lett.*, 30, 447, 1994.
14. **Kim, D. Y., Lawrence, B. L., Torruellas, W. E., Baker, G., and Meth, J.,** Assessment of single crystal PTS as an all-optical switching material at 1.3 µm, *Appl. Phys. Lett.*, 65, 1742, 1994.
15. **Blau, W.,** Low-power optical bistability and phase conjugation in polydiacetylene, *Opt. Commun.*, 64, 85, 1987.
16. **Wu, J. W., Helfin, J. R., Norwood, R. A., Wong, K. Y., Zamani-Kamiri, O., and Garito, A. F.,** Nonlinear-optical processes in lower-dimensional conjugated structures, *J. Opt. Soc. Am. B*, 4, 707, 1989.
17. **Williams, V. S., Ho, Z. Z., Peyghambarian, N., Gibbons, W. M., Grasso R. P., O'Brien, M. K., Shannon, P. J., and Sun, S. T.,** Picosecond all-optical logic gate in a nonlinear organic etalon, *Appl. Phys. Lett.*, 57, 2399, 1990.
18. **Norwood, R. A., Sounik, J. R., Holcomb, D., Popolo, J., Swanson, D., Spitzer, R., and Hansen, G.,** Nonlinear bragg mirror made from a silicon phthalocyanine/methyl methacrylate copolymer, *Opt. Lett.*, 17, 577, 1992.
19. **Halverson, C., Hays, A., Kraabel, B., Wu, R., Wudl, F., and Heeger, A. J.,** A 160-femtosecond optical image processor based on a conjugated polymer, *Science*, 265, 1215, 1994.
20. **Townsend, P. D., Jackel, J. L., Baker, G. L., Shelbourne III, J. A., and Etemad, S.,** Observation of nonlinear optical transmission and switching phenomena in polydiacetylene-based directional couplers, *Appl. Phys. Lett.*, 55, 1829, 1989.

21. **Schlotter, N. E., Jackel, J. L., Townsend, P. D., and Baker, G. L.,** Fabrication of channel waveguides in polydiacety-lenes: composite diffused glass/polymer structures, *Appl. Phys. Lett.*, 56, 13, 1990.

22. **Sasaki, K., Sasaki, S., and Furukawa, O.,** All-optical switches and all-optical bistability by nonlinear optical materials, *Materials Research Society Symposium Proceedings on Electrical, Optical and Magnetic Properties of Organic Solid State Materials*, Chiang, L. Y., Garito, A. F., and Sandman, D. J., Eds., 247, 141, 1992.

23. **Gass, P. A., Abram, I., Raj, R., and Schott, M.,** The nonlinear optical spectrum of polydiacetylene in the near infrared, *J. Chem. Phys.*, 100, 88, 1994.

24. **Zang, D. Y. and Forrest, S. R.,** Crystalline organic semiconductor optical directional couplers and switches using an index-matching layer. *IEEE Technol. Lett.* 4, 365, 1992.

25. **Sasaki, K., Fujii, K., Tomioka, T., and Kinoshita, T.,** All-optical bistabilities of polydiacetylene Langmuir-Blodgett film waveguides. *J. Opt. Soc. Am. B*, 5, 457, 1988.

26. **Aramaki S., Assanto, G., Stegeman, G. I., and Marciniak, M.,** Realization of integrated Bragg reflectors in DANS-polymer waveguides, *J. Lightwave Technol.*, 11, 1189, 1993.

27. **Kanbara, H., Asobe, M., Kubodera, K., Kaino, T., and Kurihara, T.,** All-optical picosecond switch using organic single-mode fiber waveguide, *Appl. Phys. Lett.*, 61, 2290, 1992.

28. **Kanbara, H., Kobayashi, H., Kaino, T., Kurihara, T., Oba, N., and Kubodera, K.,** Highly efficient ultrafast optical Kerr shutters with the use of organic nonlinear materials, *J. Opt. Soc. Am. B*, 11, 2216, 1994.

29. **Carter, G. M., Chen, Y. J., and Tripathy, S. K.,** Intensity-dependent index of refraction in multilayers of polydiacety-lene, *Appl. Phys. Lett.*, 43, 891, 1983.

30. **Valera, J. D., Seaton, C. T., Stegeman, G. I., Shoemaker, R. L., Xu Mai, and Liao, C.,** Demonstration of nonlinear prism coupling, *Appl. Phys. Lett.*, 45, 1013, 1884.

31. **Singh, B. P., and Prasad, P. N.,** Optical bistable behaviour of a planar quasi-waveguide interferometer made with a conjugated organic polymer film, *J. Opt. Soc. Am. B*, 5, 453, 1988.

32. **Sinclair, M., McBranch, D., Moses, D., and Heeger, A. J.,** Time-resolved waveguide modulation of a conjugated polymer, *Appl. Phys. Lett.*, 53, 2374, 1988.

33. **Marques, M. B., Assanto, G., Stegeman, G. I., Möhlmann, G. R., Erdhuisen, E. W. P., and Horsthuis, W. H. G.,** Large, nonresonant, intensity dependent refractive index of DANS side chain polymers in waveguides, *Appl. Phys. Lett.*, 58, 2613, 1991.

34. **McBranch D., Hays, A., and Heeger, A. J.,** Picosecond nondegenerate waveguide modulation in a conjugated polymer, *Opt. Commun.*, 81, 27, 1991.

35. **Vitrant, G., Reinisch, R., Paumier, J. Cl., Assanto, G., and Stegeman, G. I.,** Nonlinear prism coupler with non-locality, *Opt. Lett.*, 14, 898, 1989.

36. **Luther-Davies, B. and Xiaoping, Y.,** Waveguides and Y junctions formed in bulk media by using dark spatial solitons, *Opt. Lett.*, 17, 496, 1992.

37. **Luther-Davies, B. and Xiaoping, Y.,** Steerable optical waveguides formed in self-defocusing media by using dark spatial solitons, *Opt. Lett.*, 17, 1755, 1992.

38. **Swartzlander, G. A. and Law, C. T.,** Optical vortex solitons observed in Kerr nonlinear media, *Phys. Rev. Lett.*, 69, 2503, 1992.

39. **Bosshard, Ch., Mamyshev, P. V., and Stegeman, G. I.** All-optical steering of dark spatial soliton arrays and the beams guided by them, *Opt. Lett.*, 19, 90, 1994.

40. **Mamyshev, P. V., Bosshard, Ch., and Stegeman, G. I.,** Generation of a periodic array of dark spatial solitons in the regime of effective amplification, *J. Opt. Soc. Am. B*, 11, 1254, 1994.

41. **Davies, D. A. O., Fisher, M. A., Elton, D. J., Perrin, S. D., Adams, M. J., Kennedy, G. T., Grant, R. S., Roberts, P. D., and Sibbett, W.,** Nonlinear switching in InGaAsP laser amplifier directional couplers biased at transparency, *Electron. Lett.*, 29, 1710, 1993.

42. **Ellis, A. D. and Spirit, D. M.,** Compact 40 Gbits/s optical demultiplexer using a GaInAsP optical amplifier, *Electron. Lett.*, 29, 2115, 1993.

43. **Glesk, I., Sokoloff J. P., and Prucnal, P. R.,** Demonstration of all-optical demultiplexing of TDM data at 250 Gbits/s, *Electron. Lett.*, 30, 339, 1994.

44. **Tagaya, A., Koike, Y., Kinoshita, T., Nihei, E., Yamamoto, T., and Sasaki, K.,** Polymer optical fiber amplifier, *Appl. Phys. Lett.*, 63, 883, 1993.

Chapter 13

Organic and Metal-Containing Reverse Saturable Absorbers for Optical Limiters

Joseph W. Perry

CONTENTS

I. INTRODUCTION

Chromophores that exhibit a type of nonlinear absorption, termed reverse saturable absorption, are currently playing an important role as materials for optical limiting applications. Such materials may be used for the protection of eyes and sensors from intense light pulses, as well as in laser mode locking[33] and optical pulse shaping[3,39] and processing applications.[52] Reverse saturable absorber chromophores become more strongly absorbing as the incident optical intensity or energy is increased. This nonlinear optical response can be exhibited when chromophores have a weak ground-state absorption over some spectral range and strong excited-state absorption in the same wavelength range. Consequently, reverse saturable absorbers can be used in optical limiters which are nonlinear optical devices that are typically required to have a high transmittance at low intensities and have a low transmittance at high intensities.[15]

The protection of eyes and sensors is a very challenging problem and places extreme demands on the performance of optical limiters and reverse saturable absorber chromophores. As an illustration of this, consider the protection of eyes from laser pulses. For Q-switched laser pulses in the visible (say 10-ns, 532-nm pulses), the ED_{50} level (the exposure energy for which there is a 50% probability of producing a retinal lesion) is about 3 µJ,[51] whereas the ANSI standard maximum permissible exposure level is 0.1 µJ. In contrast, in the laboratory environment it is not uncommon for laser hazards to be present with pulse energies of 1–100 mJ, often in the form of an unanticipated surface reflection off an optical component. Even if one seeks only to keep pulse energies into the eye below 1 μJ, a very large attenuation of 10^3–10^5 is required. In addition, a high transmittance (say >50%) is needed to allow high visibility under ambient working conditions.

Although a variety of materials and mechanisms have been and are being explored for use in optical limiting,[53] there are many reasons for the recent interest in reverse saturable absorber chromophores for optical limiting, a few of which are listed here. First, for chromophores with a large ratio of excited-state to ground-state absorption cross sections ($\sigma_e/\sigma_g \gg 1$) there is potential for achieving large nonlinear attenuation and maintaining high linear transmittance. Second, since optical energy is absorbed and converted to heat as opposed to being spread, as in nonlinear scattering or refractive media, the limiting may be more reliable and may be applied in fast (highly convergent) optical systems. Third, chromophores with prompt singlet excited-state absorption and long-lived triplet-state absorption may be effectively applied to optical limiting of a wide range of pulse widths (picosecond to microsecond duration). Finally, our ability to modify systematically the photophysical properties of chromophores through rational changes in molecular structure enables molecular engineering approaches to the development of chromophores with enhanced optical limiting responses.

In the following sections of this chapter, an introduction to reverse saturable absorber chromophores and a review of recent developments in their application to optical limiting are presented. We begin with a review of the photophysical mechanism of intramolecular reverse saturable absorption. Rate equation models for the population dynamics of reverse saturable absorber chromophores under optical pumping are reviewed. Figures-of-merit for the application of reverse saturable absorbers to optical limiters are presented. We then turn to a review of currently known reverse saturable absorber materials and a brief summary of their properties. Recent progress made in using rational approaches to molecular structure modification for the enhancement and spectral tuning of the optical limiting responses of phthalocyanine complexes is then highlighted. We then examine the status of applications of reverse saturable absorber chromophores to high-performance optical limiting devices.

II. FUNDAMENTALS OF REVERSE SATURABLE ABSORPTION

A. BASIC CHARACTERISTICS

Chromophores can exhibit reverse saturable absorption when an excited state, which is populated by optical excitation, has an absorption cross section, σ_e, which is larger than the ground-state absorption cross section, σ_g, over a certain spectral range. If an optical pulse passes through a medium containing such chromophores, the transmittance decreases as the pulse intensity or fluence (energy per unit area) increases. A simplified version of the nonlinear absorptive response for reverse saturable absorber chromophores is illustrated in Figure 1. At low incident pulse energies, the transmittance is governed by the ground-state absorption and can be written as:

$$T_0 = \exp(-\sigma_g N_g L) \tag{1}$$

as would be measured in a spectrophotometer. As the initial ground-state population is transferred to the excited state upon excitation, the transmittance drops and approaches a value governed by the excited-state cross section:

$$T_{\text{sat}} = \exp(-\sigma_e N_g L) \tag{2}$$

Accordingly, in order to get a strong optical limiting response it is necessary to have a large value of σ_e/σ_g and to rapidly and efficiently transfer the ground-state population to the excited state. The rate constant for optical pumping of population from the ground state to an excited state is given by:

$$k_{ge} = \sigma_g I/h\nu = \sigma_g \phi \tag{3}$$

where I is the irradiance, ν is the photon frequency, and ϕ is the photon flux. Thus, to produce a large excited-state population σ_g must be sufficiently large to achieve saturation of population during the pulse. This leads to the perhaps counterintuitive conclusion that an adequate level of ground-state

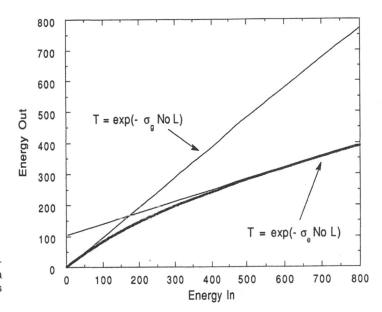

Figure 1 Illustration of the non-linear absorptive response of a reverse saturable absorber. Axes are in arbitrary units.

absorption is necessary for efficient optical limiting even though a large value of σ_e/σ_g must also be maintained.

An example of a chromophore with the spectral characteristics needed for reverse absorption is a tin phthalocyanine derivative (SnPc, Figure 2) whose visible ground-state electronic absorption is compared with its triplet–triplet absorption spectrum in Figure 2.[36] The ground-state spectrum shows a strong narrow absorption band at ~680 nm followed by a region of weak absorption from about 550 nm to about 450 nm. The triplet–triplet absorption shows a broad absorption band at with a maximum at 510 nm. Accordingly, this chromophore would be expected to show reverse saturable absorption over the 450–550 nm range, to the extent that the excited triplet state is produced during the laser pulse. As we will discuss below, SnPc indeed exhibits strong optical limiting of 8-ns duration pulses at 532 nm, with contributions from both excited triplet-state and excited singlet-state absorption. While SnPc shows a strong excited-state absorption band between ground-state bands, different relative positions of spectral bands may arise for other chromophores: the excited-state absorption band may be located on the long wavelength side of the lowest energy ground-state absorption band, or a strong excited-state absorption band may largely overlap a relatively weak ground-state absorption band.

B. KINETIC MODELS

Here we review the population dynamics of reverse saturable absorbers under optical pumping, as described by simple rate equation models.[17] This level of treatment, as opposed to more rigorous ones that include coherence effects, is sufficient for many systems investigated since dephasing is usually ultra-fast (on a subpicosecond time scale) under the conditions where large-molecule reverse saturable absorber chromophores are usually applied, i.e., liquid or solid solutions at room temperature. A generic energy level scheme for the lower electronic states of chromophores with a singlet ground state is shown in Figure 3. There are several general points to note: (1) optical pumping of excited-singlet population can lead to prompt sequential absorption to higher excited singlet states; (2) intersystem crossing from S_1 to T_1 is typically a slow process (10's to 100's of ns) but if k_{13}^{-1} is $\leq \tau_p$ (the pulse duration), appreciable excited triplet population can build up during the pulse leading to triplet–triplet absorption of the pulse; and (3) if vibrational relaxation is faster than the optical pumping, stimulated emission from the initially pumped vibronic level of S_1 to the ground state is negligible, thus the ground-state population may be essentially totally depleted by a sufficiently strong pulse.

Under the conditions of ultra-fast vibrational relaxation of vibronic levels, the rate equations describing the electronic state populations for the five-electronic-state model are:

$$\frac{dN_0}{dt} = -\sigma_{01}N_0\phi + k_{10}N_1 + k_{30}N_3 \tag{4}$$

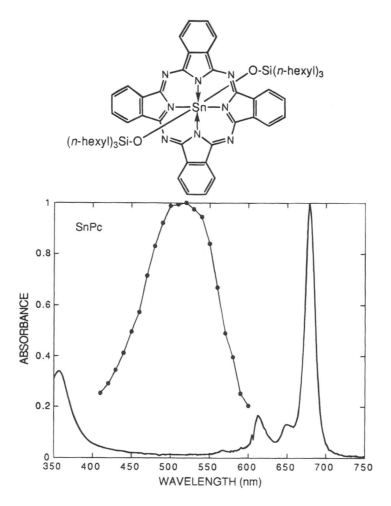

Figure 2 (a) Molecular structure of ((hexyl)$_3$SiO)$_2$SnPc. (b) Ground-state (——) and triplet–triplet ((•—•)) absorption spectra of ((hexyl)$_3$SiO)$_2$SnPc in toluene solution. (From Perry, J. W., Mansour, K., Marder, S. R., Perry, K. J., Alvarez, D., and Choong, I., *Opt. Lett.* 19, 625, 1994. With permission.)

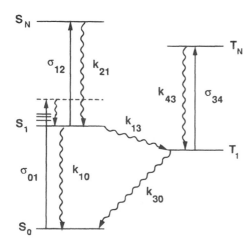

Figure 3 Schematic energy level diagram for a chromophore with a singlet ground state. Parameters are defined in the text.

$$\frac{dN_1}{dt} = \sigma_{01}N_0\phi - (\sigma_{12}\phi + k_{10} + k_{13})N_1 + k_{21}N_2 \tag{5}$$

$$\frac{dN_2}{dt} = \sigma_{12}\phi N_1 - k_{21}N_2 \tag{6}$$

$$\frac{dN_3}{dt} = -(\sigma_{34}\phi + k_{30})N_3 + k_{13}N_1 \tag{7}$$

$$\frac{dN_4}{dt} = \sigma_{34}\phi N_3 - k_{43}N_4 \tag{8}$$

subject to $N_0(0) = N_0(t) + N_1(t) + N_2(t) + N_3(t) + N_4(t)$ and the initial condition $N_1(0)$, $N_2(0)$, $N_3(0)$, and $N_4(0) = 0$, where N is the number density of the subscripted state ($S_0 = 0$, $S_1 \equiv 1$, $S_n = 2$, $T_1 \equiv 3$ and $T_n \equiv 4$), and σ and k are the absorption cross sections and spontaneous decay rates (sum of radiative and nonradiative terms) between subscripted states, respectively. The propagation equation for the optical pulse is:

$$\frac{d\phi}{dz} = -\sigma_{01}N_0\phi - \sigma_{12}N_1\phi - \sigma_{34}N_3\phi \tag{9}$$

With a specification of the material parameters ($N_0(0)$, the σ values, the various rate constants, the pathlength L) and the optical pulse characteristics (pulse energy, duration, temporal and transverse spatial profiles), these equations can be solved numerically to give the transmittance, after integration of the output pulse over time and transverse profile. On the other hand, for optically thin materials and constant intensity temporal and spatial profiles for the pulse, the rate equations can be solved analytically. We will not discuss these solutions here, but we will examine a simpler case that provides insight into the major features of the population dynamics, for nanosecond or longer pulses.

The electronic relaxation of upper excited states to the lowest excited state in the singlet or triplet manifold is ultra-fast, typically being on the order of picoseconds or shorter. Accordingly, for low to moderate irradiances, as for nanosecond duration pulses even of relatively high energy, the populations of the S_n and T_n states are negligible and one need consider only the populations of the S_0, S_1, and T_1 states. These populations determine the transmittance of the system. Under very high irradiance optical pumping, as may be obtained with picosecond or femtosecond pulses, the populations of the upper electronic states may not be negligible and the accumulation of population in these states can have dramatic effects on the transmittance, such as a turnaround of the reverse saturable absorption response at lower irradiance to a response of increasing transmittance (bleaching or saturable absorption) with increasing irradiance. This phenomenon has been observed[21] in picosecond pulse nonlinear transmittance measurements for a cyanine dye, HITCI, which has a relatively long upper singlet-state lifetime of ~ 10 ps. Nonetheless, as long as the upper singlet- (triplet-) state decay rate k_{21} (k_{43}) is much larger than the optical pumping rate $\sigma_{12}\phi$ ($\sigma_{34}\phi$), the upper state populations may be ignored. Additionally, the lifetime of the lowest triplet state is often microseconds or longer and its decay to the ground state can be ignored, at least for nanosecond time scale or shorter pulses. Thus, a simplified set of rate equations can be written, for a constant pumping intensity, as:

$$\frac{dN_0}{dt} = -\sigma_{01}N_0\phi + k_{10}N_1 \tag{10}$$

$$\frac{dN_1}{dt} = \sigma_{01}N_0\phi - (k_{10} + k_{13})N_1 \tag{11}$$

$$\frac{dN_3}{dt} = k_{13}N_1 - k_{30}N_3 \tag{12}$$

The solutions to this set of equations are:

$$N_0(t) = \frac{N_0(0)}{(\theta - \psi)} \{(\theta + \gamma)\exp(\theta t) - (\psi + \gamma)\exp(\psi t)\} \tag{13}$$

$$N_1(t) = \frac{\sigma_{01}N_0(0)}{(\theta - \psi)} \{\exp(\theta t) - \exp(\psi t)\} \tag{14}$$

$$N_3(t) = \frac{N_0(0)}{(\theta - \psi)} \{(\theta - \psi) + \psi \exp(\theta t) - \theta \exp(\psi t)\} \tag{15}$$

where $\theta = -(\Gamma - (\Gamma^2 - 4\Delta^2))/2$, $\psi = -(\Gamma + (\Gamma^2 - 4\Delta^2))/2$, $\gamma = k_{10} + k_{13}$, $\Gamma = \sigma_{01}\phi + \gamma$, and $\Delta = \sigma_{01}k_{13}$.

The time-dependent populations of S_0, S_1, and T_1 for optical pumping with varying intensities are shown in Figures 4a–c, for pulses of 8-ns duration and a relatively high intersystem crossing rate. In Figure 4a, the case of a moderately high intensity is shown. The S_0 population is depleted gradually over the duration of the pulse. The S_1 population reaches a quasi-steady state at about 3 ns but then decays as population is transferred to T_1, which builds up slowly during the pulse to about 0.5 of the initial S_0 population. Figure 4b shows the population dynamics for an intensity 2.5 times higher than in Figure 4a. The S_0 state is depleted to 0.1 of its initial value, S_1 reaches a quasi-steady state in 2 ns, and the triplet population builds up slowly to about 0.7 of the initial S_0 value. Figure 4c shows the populations for an intensity 25 times higher than for Figure 4a. The fluence corresponding to the intensity for this calculation is about 10 J/cm^2, which is above the damage level for many transparent materials and thus represents a limiting level of excitation for an 8-ns pulse. For this high intensity, the S_0 depletion is strong and prompt, the S_1 population reaches a maximum of 0.85 in 0.5 ns, but the T_1 population grows slowly, now with a time constant of about 4 ns, and reaches a value of 0.85 at the end of the pulse. Since T_1 population is fed by a nonradiative rate constant, k_{13}, its growth rate is limited even for extremely intense pulses.[7] Thus effective use of the triplet–triplet absorption for optical limiting can be made only if $k_{13}^{-1} < \tau_p$.

As a result of the time dependence of the populations of S_0, S_1, and T_1 during the pulse, the effective absorption cross section, σ_{eff}, for the system varies with time during the pulse. We can write σ_{eff} as a weighted average of the contributions of the various states:

$$\sigma_{eff}(t) = \frac{1}{N_0(0)} (\sigma_{01}N_0(t) + \sigma_{12}N_1(t) + \sigma_{34}N_3(t)). \tag{16}$$

The transmittance of the system is thus also a function of time during the pulse:

$$T(t) = \exp[-\sigma_{eff}N_0(0)L]. \tag{17}$$

Accordingly, the population *dynamics* during the pulse can be very important in determining the optical limiting response of a reverse saturable absorber.

C. STEADY-STATE BEHAVIOR: FAST ABSORBERS

The populations of the various states of a chromophore can attain values that become independent of time during the pulse when the time constant for return of population to the ground state is shorter than the pulse duration, which is the case of the so-called "fast absorbers."[18] This is the familiar steady-state condition for the populations and the corresponding solutions have long been used to describe the behavior of nonlinear absorbers. To illustrate the steady-state behavior, consider a system comprised involving S_0, S_1, and S_n states where, in addition to having ultra-fast relaxation from S_n to S_1, the relaxation time from S_1 to S_0 is shorter than the pulse duration. In accord with the requirement for fast overall return of excited molecules to the ground state, we have excluded the formation of triplets here since they would typically have a very slow relaxation to the ground state. At sufficiently high intensity the response time of the "fast" reverse saturable is essentially instantaneous. As a result of this fast response, there is minimal change in shape of the pulse after passing through the reverse saturable

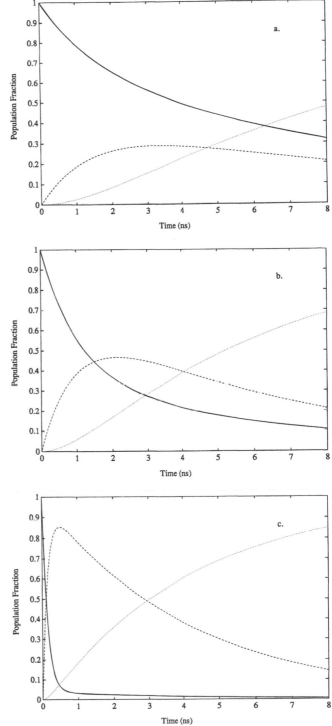

Figure 4 (a) Fractional populations of S_0 (——), S_1 (– – –), and T_1 (. . . .), levels during 8-ns pulses with different intensities as calculated using a simplified rate equation model. (a) $\phi = 1.34 \times 10^{26}$ photons/cm² s; (b) $\phi = 3.35 \times 10^{26}$ photons/cm² s; (c) $\phi = 3.35 \times 10^{27}$ photons/cm² s. Values of parameters used in the calculations: $k_{10} = k_{13} = 2.5 \times 10^8$ s⁻¹, $\sigma_{01} = 2 \times 10^{-18}$ cm².

absorber. For a "fast" three-level reverse saturable absorber, the steady-state solutions for the populations are, taking the upper excited-state lifetime to be negligible, as follows:

$$N_0 = N_0(0)\left(\frac{1}{1 + I/I_s}\right) \quad (18)$$

$$N_1 = N_0(0)\left(\frac{I/I_s}{1 + I/I_s}\right) \tag{19}$$

where I is the irradiance and I_s is the saturation irradiance, $I_s = h\nu/\sigma_{01}\tau_{10}$, and $\tau_{10} = 1/k_{10}$. In these solutions for "fast" absorbers, the distribution and saturation of population are dependent on the pulse irradiance. It follows that the effective cross section in steady state is given by:

$$\sigma_{\text{eff}} = \sigma_{01}\left(\frac{1}{1 + I/I_s}\right) + \sigma_{12}\left(\frac{I/I_s}{1 + I/I_s}\right). \tag{20}$$

When the input irradiance is low $\sigma \sim \sigma_{01}$ and when I is $\gg I_s$, σ_{eff} approaches σ_{12}. In the case of an optically thick three-level absorber, one can obtain, by using the steady-state populations, a transcendental equation for the transmittance[24]:

$$\ln(T/T_0) = (R^{-1} - 1)\ln[(R^{-1} + I_0/I_s)/(R^{-1} + TI_0/I_s)] \tag{21}$$

where R is σ_{12}/σ_{01}, I_0 is the incident irradiance, and T_0 is the low intensity transmittance. For $R > 1$, the response is that of a reverse saturable absorber, whereas for $R < 1$ the response is that of a normal saturable absorber. At very high irradiance, the transmittance approaches a value of T_0^R.

D. FLUENCE-DEPENDENT SATURATION: SLOW ABSORBERS

Another limit of the rate equation model results when the return of population to the ground state is slow compared to the pulse width, a case corresponding to a "slow" reverse saturable absorber.[18] In this case, the rate equations (Equations 10 and 11) are decoupled, since the terms involving $k_{10}N_1$ can be neglected, and can be solved analytically for a time-dependent pulse shape. The solution for the S_0 population for a pulse profile $\phi(t)$ is:

$$N_0(t) = N_0(0)\exp\left(-\sigma_{01}\int_0^t \phi(t')\, dt'\right) \tag{22}$$

and this equation can be recast as:

$$N_0(t) = N_0(0)\exp(-F/F_s) \tag{23}$$

where F is the pulse fluence and $F_s = h\nu/\sigma_{01}$ is the saturation fluence. Ignoring the T_1 population for the moment, the S_1 population follows as:

$$N_1(t) = N_0(0)(1 - \exp(-F/F_s)) \tag{24}$$

The effective absorption cross section for the reverse saturable absorber depends on the populations of S_0 and S_1 and therefore on the pulse fluence:

$$\sigma_{\text{eff}} = \frac{1}{N_0(0)}(\sigma_{01}N_0 + \sigma_{12}N_1) = \sigma_{01}\exp(-F/F_s) + \sigma_{12}(1 - \exp(-F/F_s)) \tag{25}$$

At very low fluence σ_{eff} starts at σ_{01} and as the fluence is increased σ_{eff} increases up to σ_{12}. It can be seen that to utilize fully the excited-state absorption of a chromophore for optical limiting, the fluence must be much higher than the saturation fluence, a point which has significant design implications as will be discussed below.

Accounting for the T_1 population in the fluence-dependent model leads to an explicit dependence of the effective absorption cross section on the $k_{isc}\tau_p$ product, where k_{isc} is the intersystem crossing rate, written as k_{13} above and in Figure 3. By using the following approximations: a constant input intensity during the pulse (which leads to $\sigma_{01}\phi t = F/F_s$), negligible relaxation of S_1 to S_0 during the pulse, and optically thin conditions, we can rewrite the solutions (Equations 13–15) to the three-state model above as:

$$N_0 = N_0(0)\exp[-F/F_s] \tag{26}$$

$$N_1 = c_1 N_0(0)\{\exp[-k_{isc}t] - \exp[-F/F_s]\} \tag{27}$$

$$N_3 = N_0(0)\{1 - c_1 \exp[-k_{isc}t] + c_2 \exp[-F/F_s]\} \tag{28}$$

where $c_1 = \sigma_{01}\phi/(\sigma_{01}\phi - k_{isc})$ and $c_2 = k_{isc}/(\sigma_{01}\phi - k_{isc})$. The form of the effective cross section is now dependent on the pulse fluence and the relative values of the pulse duration and time constant for intersystem crossing. In the strong pumping limit, it can be shown that:

$$\sigma_{eff}(t) = \sigma_{12}\exp[-k_{isc}t] + \sigma_{34}(1 - \exp[-k_{isc}t]). \tag{29}$$

The time behavior of the terms in this equation is consistent with the behavior calculated above for strong pumping as shown in Figure 4c. For an optically thin sample, one can show that the transmitted energy is:

$$E_{out} = E_{in}\left\{1 - N_0(0)L\frac{\sigma_{12}}{k_{isc}\tau_p}(1 - \exp[-k_{isc}\tau_p]) + \sigma_{34}\left(1 - \frac{1}{k_{isc}\tau_p}(1 - \exp[-k_{isc}\tau_p])\right)\right\}$$

$$= E_{in}\{1 - N_0(0)L(\sigma_{12}f_1 + \sigma_{34}f_3)\} \tag{30}$$

where f_i is the average fractional population of the ith state during the pulse. Accordingly, the $k_{isc}\tau_p$ product becomes an important figure-of-merit for chromophores with significant triplet–triplet absorption.[36]

E. SUMMARY OF CHROMOPHORE FIGURES-OF-MERIT

The important parameters characterizing the response of reverse saturable absorber chromophores are summarized in Table 1. As discussed above, the key parameters are the saturation irradiance or saturation fluence, which control the excited-state population, and the ratio of excited-state (or effective excited-state) to ground-state absorption cross section. When there is a triplet–triplet absorption contribution, the intersystem crossing rate is also important, and for strong pumping, the triplet population is controlled by the product $k_{isc}\tau_p$.

In many cases in the literature, nonlinear transmission or optical limiting data are reported without the analysis and photophysical measurements required for determination of the various cross sections and rate constants that characterize a reverse saturable absorber. When such data are discussed, we may consider the operational performance of the material in terms of the small signal transmittance, T_L, and the transmittance measured at a particular high input fluence or irradiance, T_{NL}. A simple comparative figure-of-merit can be taken as T_L/T_{NL}.[14] When T_{NL} approaches T_{sat} then the effective cross section ratio, can be simply estimated as $\sigma_e/\sigma_g = \ln(T_{NL})/\ln(T_L)$.

F. OPTICAL BANDWIDTHS FOR REVERSE SATURABLE ABSORBERS

The optical bandwidths associated with the linear transmittance and the excited-state absorption of reverse saturable absorbers are also important considerations in evaluating materials for optical limiting applications. While the importance of these bandwidths is commonly recognized, very little attention

Table 1 Figures-of-Merit for Reverse Saturable Absorbers

Excited State Transition(s)	Saturation Parameter(s)	Cross Section Ratio
S_1-S_n	$I_s = h\nu/\sigma_{01}\tau_{01}$ (fast absorber)	σ_{12}/σ_{01}
	$F_s = h\nu/\sigma_{01}$ (slow absorber)	
T_1-T_n	$F_s = h\nu/\sigma_{01}$ and $k_{isc}\tau_p$	σ_{34}/σ_{01}
S_1-S_n & T_1-T_n	$F_s = h\nu/\sigma_{01}$ and $k_{isc}\tau_p$	$(\sigma_{12}f_1 + \sigma_{34}f_3)/\sigma_{01}$

has been paid to quantifying the bandwidths in terms of specific performance requirements. Miles[31] has recently examined optical bandwidth issues for reverse saturable absorber-based optical limiters. He defined the bandwidth for linear transmittance, $\Delta\lambda_T$, by the condition that $\sigma_g \leq 2 \sigma_g^{min}$, where σ_g^{min} is the minimum value of σ_g. He defined the bandwidth for strong signal pulse suppression, $\Delta\lambda_S$ ($S = 1/T_{sat}$), by the condition that S be within a factor of 2 of its maximum value, i.e. $\sigma_e \geq \sigma_e^{max} - \ln2/N_0L$. By analyzing spectra of two phthalocyanine derivatives he showed that the bandwidth for strong signal suppression is significantly smaller than the commonly quoted full-width at half-maximum (FWHM) for the excited-state absorption band. The phthalocyanines typically have a $\Delta\lambda_T \sim 130$ nm centered at ~ 500 nm. The triplet–triplet absorption spectra of these molecules have a ~120–150 nm FWHM with a maximum around 510 nm, well overlapped with the transmission band in the ground-state absorption. On the other hand, the phthalocyanines examined gave values of $\Delta\lambda_S$ of 40–50 nm.

III. REVERSE SATURABLE ABSORBER CHROMOPHORES

A. EARLY OBSERVATIONS

In an early detailed study of the nonlinear transmission characteristics of saturable absorbers[12] wherein the inadequacy of a simple two-level model for the optical bleaching was demonstrated and the general presence of a typically weaker excited-state absorption was shown, solutions of the dyes sudanschwartz B and sulfonated indanthrone were observed to exhibit reverse saturable absorption at the ruby laser wavelength, 694.3 nm. No detailed study or analysis of the reverse saturable absorbers was made nor was there any deliberate attempt to utilize the reverse saturable absorption response for optical limiting or pulse-shaping applications. Nonetheless, the understanding of the response of saturable absorbers developed in this and other papers at that time and the observation of the unusual induced absorption response for the two dyes mentioned set the stage for the eventual growth of interest in and applications of reverse saturable absorbers.

Reverse saturable absorption has been observed for solutions of anthracene using 347-nm, 25-ns pulses.[10] The data were interpreted using a model involving three-step photoionization of anthracene and strong induced absorption by the radical ions thereby produced. The kinetic model was able to reproduce both the nonlinear transmission and the intensity dependence of the triplet yield. While the ionization efficiency of the upper excited state was inferred to be high, the magnitude of the nonlinear absorption was relatively small. This was probably due to fast recombination leading to a short ion lifetime. The effective cross section ratio for this system obtained at 347 nm was ~ 10 and the ground-state extinction coefficient, ϵ_g, was 2.75×10^3 M^{-1} cm^{-1}. A somewhat stronger reverse saturable absorption response was obtained with pyrene solutions. The mechanism of the UV reverse saturable absorption in these molecules is an intriguing one and such a process may be observable for other polycyclic aromatic hydrocarbons.

Considerable interest in reverse saturable absorbers subsequently developed in connection with the potential application of these materials to optical limiters, pulse shaping and mode locking of lasers, and spatial light modulation in the 1980s.

B. PORPHYRINS

The reverse saturable absorption characteristics of tetraphenylporphyrins, including free-base tetraphenylporphyrin (H_2TPP), ZnTPP, and CoTPP were examined, by using 80-ps, 532-nm pulses.[5] More recently, reverse saturable absorption has been reported in FeTPP.[9] The structure of a metallotetraphenylporphyrin is shown in Figure 5. The fluorescence lifetimes for H_2TPP and ZnTPP were 9 ns and 1.5 ns, respectively, whereas the CoTPP had a very low fluorescence quantum yield and the fluorescence lifetime was estimated to be <0.1 ns. The quenching of the fluorescence as a result of fast intersystem crossing in porphyrins containing paramagnetic transition metals is well known and presumably accounts

Table 2 Absorption Cross Sections for Tetraphenylporphyrins

Molecule	$\sigma_{01}(10^{-17}$ cm$^2)$	σ_{12}/σ_{01}	σ_{34}/σ_{01}
H_2TPP	1.6	3.8	
ZnTPP	2.7	2.6	
CoTPP	5		3.0

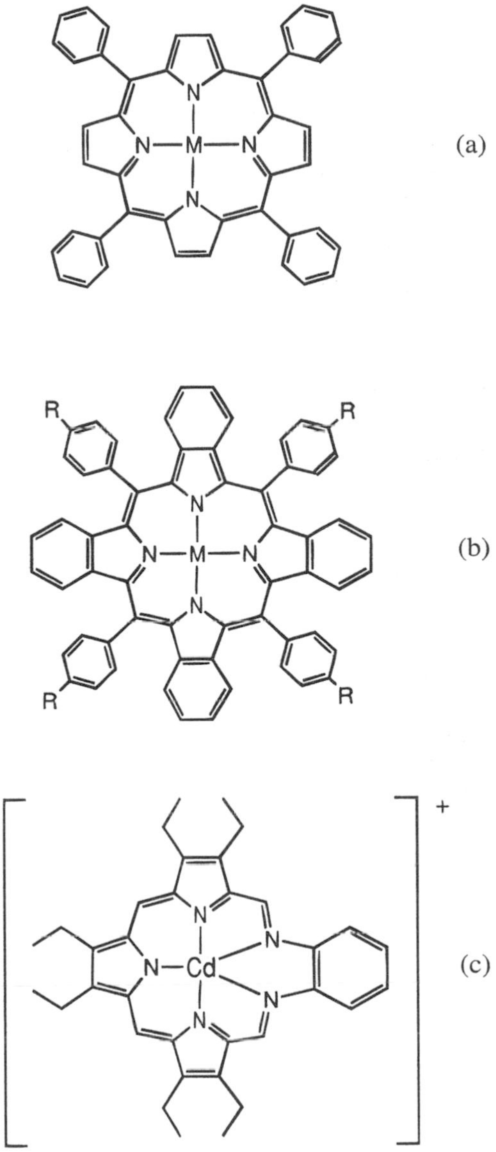

(a)

(b)

(c)

Figure 5 Molecular structures of metalloporphyrins examined for optical limiting: metallotetraphenylporphyrins (M = Zn or Co) (a), metallo(tetraphenyl) tetrabenzoporphyrins (M = Zn, R = H or methoxy) (b), and cadmium texaphrin cation (c).

for these results. For the 80-ps pulses, the porphyrin ring-based S_1-S_n absorption dominates the reverse saturable absorption in H_2TPP and ZnTPP and for CoTPP the T_1-T_n absorption is dominant. Nonlinear transmission data were analyzed using the five-state model to obtain estimates of the excited-state absorption cross sections. The estimated absorption cross section ratios are listed in Table 2. While these cross section ratios are relatively small compared to other more recently studied materials, this early work stimulated interest in the reverse saturable absorption properties of conjugated macrocyclic complexes.

Reverse saturable absorption has been reported for tetrabenzoporphyrins, which are porphyrin derivatives that have extended ring conjugation (see Figure 5).[13] Measurements made with 21-ps and 6-ns, 532-nm pulses showed that the nonlinear absorption response was dependent on the pulse fluence, as expected for a "slow" reverse saturable absorber. Independent experiments indicated an excited-state lifetime of > 10 ns, consistent with the finding of a fluence dependence. From the effective excited-state cross sections and linear absorption coefficients given, one finds $\sigma_{01} = 7.6 \times 10^{-18}$ cm^2 and 1.4×10^{-17} cm^2 and $\sigma_e/\sigma_{01} = 3.9$ and 2.1 for the (tetramethoxyphenyl)tetrabenzoporphyrin and (tetraphenyl)tetrabenzoporphyrin compounds, respectively. These cross section ratios are comparable to those of the tetraphenylporphyrins discussed above.

Nonlinear absorption has recently been reported[50] for cadmium texaphyrin chloride, whose structure is a type of expanded porphyrin and is shown in Figure 5. Measurements performed with 23-ps, 532-nm pulses show reverse saturable absorption at low fluence (below 0.03 J/cm^2) and induced bleaching at higher fluence, as was observed for HITCI. Only reverse saturable absorption was observed with 8-ns, 532-nm pulses. This compound has a high triplet quantum yield (0.9) and a short singlet lifetime (0.1 ns) has been inferred. Given the obviously fast intersystem crossing, the nanosecond response is dominated by the triplet absorption. The nonlinear transmission data were analyzed with a six-state rate equation model (the usual five-state model plus an extra upper excited state to allow for a second excited-singlet-state absorption). This analysis yielded values of $\sigma_{01} = 1.3 \times 10^{-17}$ cm^2, $\sigma_{12}/\sigma_{01} = 7.7$, and $\sigma_{34}/\sigma_{01} = 2.7$. The upper excited-state-absorption cross section was assessed to be $< 3 \times 10^{-18}$ cm^2. The ratio of triplet-state to ground-state cross sections is consistent with that of the tetraphenylporphyrins and the singlet ratio is a factor of two higher.

C. INDANTHRONE DERIVATIVES AND OLIGOMER

In a extension of the earlier observations on indanthrone; Hoffman et al.[19] examined the reverse saturable absorption of several indanthrone derivatives, (unsubstituted indanthrone, oxidized indanthrone, and chloroindanthrone) and an indanthrone oligomer with 10-ns pulses at 532 nm and 1064 nm. The structures of indanthrone and the oligomer are shown in Figure 6. Indanthrone exhibits a broad and relatively weak absorption in the red region of the spectrum, $\lambda_{max} = 641$ nm, $\epsilon_{max} \sim 2000$ M^{-1} cm^{-1}. Indanthrone was reported to have an $\epsilon_g = 1000$ M^{-1} cm^{-1} at 532 nm. All of the derivatives appeared to exhibit reverse saturable absorption at both wavelengths, although the detailed photophysical mechanism was not established. The data for the unsubstituted indanthrone and the oxidized indanthrone are consistent with a value of σ_e/σ_g of about 3, with the other derivatives being somewhat lower. The saturation fluence (or irradiance) for these compounds appeared to be rather high with the onset of nonlinear absorption being noticeable at values of about 1 to 2 J/cm^2. The oxidized indanthrone derivative seemed to be more resistant to a cumulative optical bleaching effect than the other compounds.

D. METAL CLUSTER COMPOUNDS

Transition metal cluster compounds exhibiting reverse saturable absorption were first reported for iron tricobalt metal clusters.[56] These clusters have a tetrahedral metal core bonded via metal–metal single

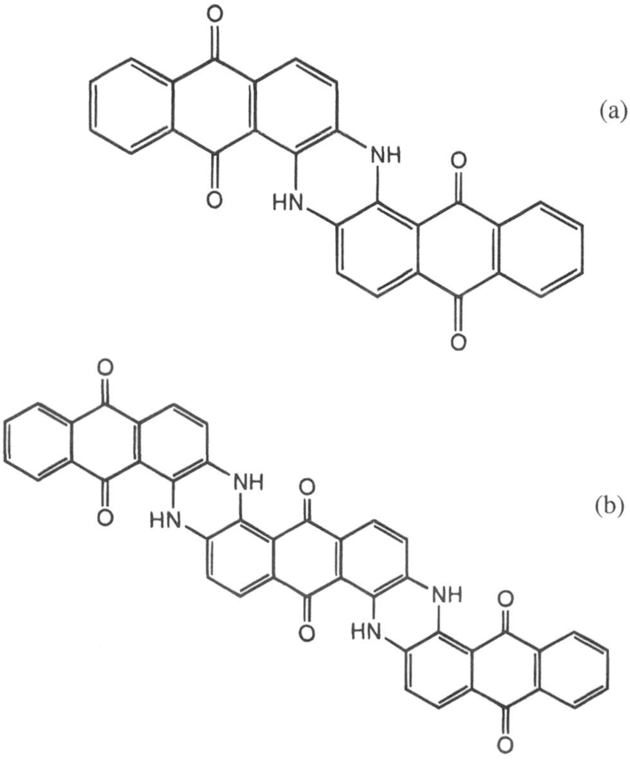

(a)

(b)

Figure 6 Molecular structures of indanthrone (a) and indanthrone oligomer (b).

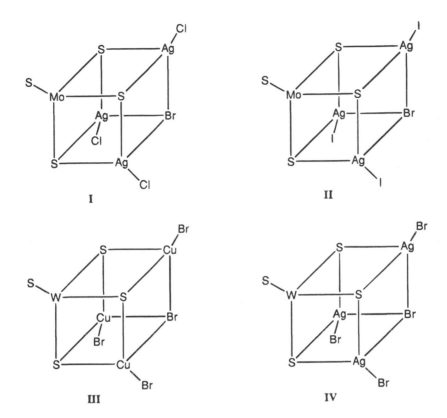

Figure 7 Structures of cubane-like metal cluster ions.

bonds. The initial clusters examined were: $HFeCo_3(CO)_{12}$, $[NEt_4]^+FeCo_3(CO)_{12}{}^-$, $HFeCo_3$-$(CO)_{10}(PMe_3)_2$, and $HFeCo_3(CO)_{10}(P(C_6H_5)_3)_2$. The clusters exhibit weak ground-state absorption ($\epsilon_g \sim 4000\ M^{-1}\ cm^{-1}$) in the visible ($\lambda_{max} \sim 500$–600 nm, depending on ligand and counterion) that has been assigned to metal–metal bonding to antibonding (σ–σ^*) transitions. A sizable effect on the optical limiting response was observed on substitution of two CO ligands on the parent compound $HFeCo_3(CO)_{12}$ with PMe_3 (see Figure 7) and $P(C_6H_5)_3$, with an enhancement observed for PMe_3 and a reduction in optical limiting response for $P(C_6H_5)_3$. The ground-state spectra for the two phosphine substituted compounds were virtually the same. These observations led the authors to conclude that there was a substantial increase in the excited-state absorption cross section for the PMe_3 substituted molecule and that the excited-state transition must have substantial ligand-to-metal or metal-to-ligand charge-transfer character. While these clusters are apparently photolytically unstable and air sensitive, the responses of the clusters stimulated interest in reverse saturable absorption properties of metal cluster compounds.

A more photochemically, thermally, and chemically stable tetrairon cluster compound, (Fe(CO)(cyclo-pentadienyl))$_4$ (King's complex, Figure 8), was also observed to exhibit optical limiting of 532-nm, 7-ns pulses.[55,57] A detailed study of the reverse saturable absorption in this cluster and two related clusters was conducted with fluence-dependent pump-probe experiments on the picosecond time scale.[1,2] Analysis of these measurements allowed the determination of σ_{12} and τ_{10}, and provided bounds on the upper excited singlet-state lifetime and the time constant for intersystem crossing. The derived parameters for (Fe(CO)(cyclopentadienyl))$_4$ are given in Table 3. It was found that King's complex exhibited only a

Table 3 Photophysical Parameters for (Fe(CO)Cp)$_4$

Parameter	Value
σ_{01} (532 nm)	$4.1 \times 10^{-18}\ cm^2$
σ_{12} (532 nm)	$8 \times 10^{-18}\ cm^2$
τ_{10}	120 ps

modest cross section ratio $\sigma_e/\sigma_g = 2$ and a relatively short singlet lifetime. In addition, longer time measurements showed essentially complete recovery, within experimental error, indicating a very low branching to the triplet state. These results showed that the nanosecond optical limiting response of this complex was not likely due to reverse saturable absorption. Further experiments showed the presence of a significant nonlinear scattering contribution to the nanosecond optical limiting in solutions of King's complex. This scattering was eliminated when the complex was incorporated into a polymethyl-methacrylate (PMMA) host and much weaker optical limiting was observed. It was concluded that the nanosecond pulse optical limiting response of King's complex was largely due to nonlinear scattering. Further investigations suggest that this scattering is thermal in origin.[6]

Optical limiting responses attributed to reverse saturable absorption have been reported for cubane-like mixed metal chalcogenide clusters.[45] The structures of some of the metal clusters examined are shown in Figure 8. These clusters show a very broad weak-to-moderate absorption across the whole visible spectrum. The strong signal nonlinear absorption for 532-nm, 7-ns pulses observed for I and II in acetonitrile was about twice as strong as that of the fullerene C_{60} in toluene.[45] Only a slight increase in nonlinear absorption was observed upon replacement of chloride ligands with iodide in going from I to II. The optical limiting data reported for I and II are consistent with a σ_e/σ_g of about 6–8. The effect of heavy-atom substitution in cubane-like mixed metal chalcogenide clusters was probed by replacing the Cu atoms in $WCu_3S_4Br_4$ (III) with Ag to give the cluster $WAg_3S_4Br_4$ (IV).[44] An increase in the nonlinear absorption was observed upon substitution with Ag and this was interpreted as a heavy-atom effect on the intersystem crossing rate. The reverse saturable absorption response of the cluster

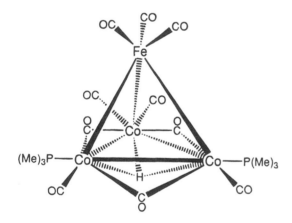

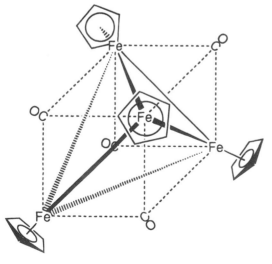

Figure 8 Structures of $P(Me)_3$ substituted $FeCo_3$ metal cluster (top) and Fe_4 metal cluster (King's complex) (bottom).

IV was stronger than that of C_{60} (at an input of 4 J/cm^2, T_L/T_{NL} = 6.4 for IV vs. 4 for C_{60}) but not quite as strong as that of I and II for which T_L/T_{NL} = 8 and 9, respectively, at the same input fluence.

Reverse saturable absorption has been observed for a very large mixed metal chalcogenide cluster $(n\text{-Bu}_4\text{N})_4[\text{Mo}_8\text{Cu}_{12}\text{O}_8\text{S}_{24}]$ as evidenced by Z-scan[42] measurements.[47] This cluster shows a moderately strong absorption band in the visible at 509 nm and a ground-state extinction coefficient of 7500 M^{-1} cm^{-1} at 532 nm. The focus of this paper was on the large resonant third-order optical nonlinearities of this cluster and the nonlinear absorption coefficient, α_2, was determined to be 0.23 cm/MW. However, insufficient data were provided on the excited-state lifetime for the excited-state cross section to be assessed. From the temporal profile of the transmitted pulses, it would appear that this cluster is a fast absorber, with a recovery time of $\ll$ 7 ns.

Shi and co-workers have reported on additional groups of metal chalcogenide clusters that have "nest"- and "butterfly"-shaped geometric structures. The "nest"-shaped clusters[20,46] were of the form $(n\text{-Bu}_4\text{N})_4[\text{MoOS}_3\text{Cu}_3(\text{NCS})\text{I}_3]$, $(n\text{-Bu}_4\text{N})_4[\text{Mo}_2\text{O}_2\text{S}_6\text{Cu}_6\text{Br}_2\text{I}_4]$, and $(n\text{-Bu}_4\text{N})_4[\text{Mo}_2\text{O}_2\text{S}_6\text{Cu}_6\text{Br}_2\text{Cl}_4]$, and the "butterfly"-shaped clusters[43] were of the form, $\text{WCu}_2\text{OS}_3(\text{PPh}_3)_4$ and $\text{MoCu}_2\text{OS}_3(\text{PPh}_3)_3$. These studies have found significant variations in nonlinear absorptive and refractive properties as the central core structure of the cluster is changed, with only a small dependence on the peripheral ligand substitution. For the cage structures, strong nonlinear absorption and weak self-focusing were observed. For the "nest" structures, strong nonlinear absorption and strong self-defocusing were observed. On going to the "butterfly" structures, weak nonlinear absorption and strong self-focusing were observed. This intriguing variation of properties with metal cluster structure has not yet been explained in terms of the cluster electronic states or photophysics, but the work shows that the properties of these clusters can be manipulated in an interesting way.

The metal cluster systems are quite interesting from the point of view of the very broad moderate-to-weak absorption across much of the visible and the potential for larger excited-state cross sections, via charge transfer transitions. However, there is clearly much to be done in characterizing the electronic states and the photophysics of these systems in order to develop an understanding of their reverse saturable absorption behavior. The photostability of these compounds, particularly at high fluence or irradiance, is an issue which requires due consideration for practical optical limiter applications.

E. FULLERENES

Because of their uniqueness as new forms of carbon, their high symmetry, and their three-dimensional conjugation, the fullerenes C_{60} and C_{70} attracted considerable interest regarding their optical properties. The weak absorption in the visible spectrum and the larger excited-singlet and triplet absorptions indicated that C_{60} was a candidate to exhibit reverse saturable absorption. C_{60} and C_{70} were first reported to exhibit optical limiting for 532-nm, 8-ns pulses by Tutt and Kost[54] and these authors attributed the response to reverse saturable absorption, with a possible contribution from nonlinear scattering. The C_{70} response was reported to be weaker than for C_{60}. For C_{60} solutions, the nonlinear absorption became noticeable at a fluence of ~0.1 J/cm^2 and at an input fluence of 1.5 J/cm^2 a solution with T_L = 0.63 showed a pulse suppression of 25 in a collimated beam optical geometry. Comparative optical limiting measurements indicated that the optical limiting in C_{60} was stronger than several FeCo$_3$ cluster complexes, indanthrone and chloroaluminum phthalocyanine (ClAlPc).

The role of reverse saturable absorption in the C_{60} response was probed in subsequent studies. Kost et al.[25] examined the optical limiting of C_{60} incorporated in a solid polymer host and this response was successfully modeled using the five-state model for reverse saturable absorption and literature values for the cross sections and rates. For a polymer sample with a T_L = 0.69, a pulse suppression of 3 was obtained. For C_{60}/toluene solutions, an additional contribution to the optical limiting was indicated. In comparative studies of several dyes, including C_{60}, in the polymer, it was shown that ClAlPc/polymer samples showed the strongest reverse saturable absorption. Henari et al.[16] reported reverse saturable absorption for C_{60} at different visible wavelengths and analyzed their results with a rate equation model. McLean et al.[30] performed measurements with 30-ps and 8-ns, 532-nm pulses and successfully modeled both sets of data using the five-state model. However a very large S_n decay rate was needed in the simulation to avoid bleaching at high fluence for the 30-ps calculations and the high fluence nanosecond data departed towards lower transmittance from the simulation, suggesting that other effects may be operative in both cases at fluences above about 1 J/cm^2. Joshi et al.[22] also studied C_{60} optical limiting but with 30-ns, 527-nm pulses. Their measurements and simulations also pointed to the action of a mechanism other than reverse saturable absorption at fluences above ~1 J/cm^2. While the details of

Table 4 Photophysical Parameters for C_{60}

Parameter	Value
σ_{01}	3.1×10^{-18} cm^2
σ_{12}	1.6×10^{-17} cm^2
σ_{34}	9.5×10^{-18} cm^2
k_{13}	8.5×10^{8} s^{-1}

the high fluence response are yet to be established, the optical limiting response of C_{60} at fluences below 1 J/cm^2 is well described by the reverse saturable absorption mechanism and the parameters in Table 4.

The excited-state absorption of C_{60} has been utilized to enhance the absorptive energy deposition in thermal defocusing optical limiters.[23]

F. CYANINES

While cyanine dyes are well known as saturable absorbers and laser dyes, they have received little attention as reverse saturable absorbers. Reverse saturable absorption has been reported for a cyanine laser dye, HITCI, whose structure is shown in Figure 9. An unusual effect was observed in the nonlinear transmission response of this compound as observed with 532-nm, 15-ps pulses.[21] The transmittance increased at high fluence, following a reverse saturable absorption response at low fluence. As discussed above, this response was accounted for and modeled as being due to the saturation of population in the upper excited singlet state (S_n), which results when the optical pumping rate exceeds the deactivation rate of this upper level. Apparently, in this compound the absorption from the S_n state to yet higher excited singlet states is quite small. Hence, despite the fact that the molecule has a large σ_{12}/σ_{01} value (see Table 5), when the population becomes saturated into the S_n level, the system becomes optically bleached. The implication of this result is a relatively long upper state lifetime and modeling of the data for HITCI provided an estimate of $\tau_{21} = 8$ ps. A rate equation treatment of the critical condition for reverse saturable absorption including a finite upper singlet-state lifetime and a finite intrastate vibrational relaxation time was presented and showed that having a $\sigma_e/\sigma_g > 1$ is not a sufficient general condition for reverse saturable absorption, but the condition becomes a complicated expression that itself depends on the irradiance. It was concluded, as was implicitly assumed in the treatments presented above, that to take advantage of a large σ_e/σ_g it is necessary that $k_{21}/k_{10} >> \sigma_e/\sigma_g$.

G. PHTHALOCYANINES

Phthalocyanine metallo-organic complexes have received considerable attention as reverse saturable absorbers for the Nd:YAG second harmonic at 532 nm. As shown above in Figure 2, metallophthalocyanines typically exhibit a strong sharp $\pi\pi^*$ (Q-band) absorption at ~680 nm, followed by a region of relatively weak absorption between 400 and 600 nm. In this same region, phthalocyanines exhibit strong excited-state S_1-S_n and T_1-T_n transitions. Reverse saturable absorption and optical limiting with

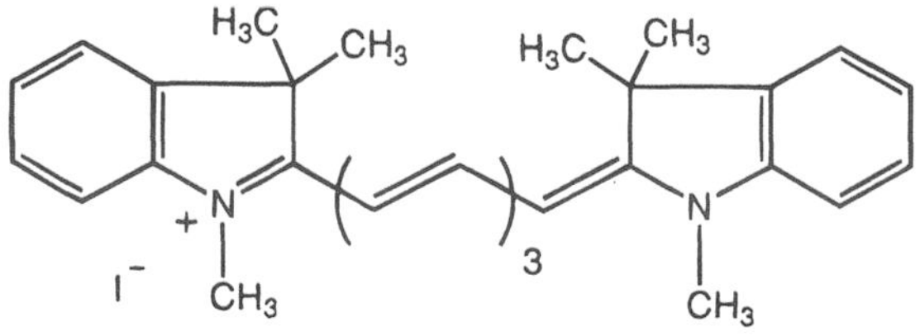

Figure 9 Structure of 1,3,3,1′,3′,3′-hexamethylin-dotricarbocyanine iodide (HITCI).

Table 5 Cross Sections at 532 nm and Time Constants for HITCI

Parameter	Value
σ_{01}	1.7×10^{-17} cm^2
σ_{12}	4.8×10^{-16} cm^2
τ_{10}	1.5 ns
τ_{21}	8 ps

phthalocyanines at 532 nm were first reported for chloroaluminum phthalocyanine (ClAlPc).[8] Reverse saturable absorption was observed for ClAlPc in methanol solution with pulses of 30-ps and ~5-ns duration. Interestingly, measurements performed with a mode-locked train of about 10 30-ps pulses, which were spaced by 10 ns, showed stronger optical limiting than what was calculated by assuming that each pulse in the train acted independently and was attenuated in accordance with the single pulse response. The enhanced limiting for the pulse train was interpreted as resulting from the buildup of triplet population during the roughly 100-ns pulse train envelope and a contribution to the limiting from the T_1-T_n absorption. These observations led to efforts to utilize the apparently stronger triplet–triplet absorption cross section at 532 nm for optical limiting of pulses of 1 to 10 ns in duration.

A relatively detailed picture of the photophysics of ClAlPc and other phthalocyanines has been developed. The values for the absorption cross sections for the S_0-S_1, S_1-S_n, and T_1-T_n transitions, the S_1 fluorescence lifetime, and the intersystem crossing rate have been established. The S_1-S_n absorption cross sections for ClAlPc and a silicon naphthalocyanine derivative, which is discussed below, were determined using a picosecond pulse Z-scan method.[58] In these experiments, the fluence-dependent nature of the reverse saturable absorption for short pulses (as described above) was confirmed by the Z-scan measurements performed with picosecond pulses of two different durations (29 and 61 ps) but the same pulse energy. The Z-scan measure of nonlinear absorption, for these pulses that differed in irradiance by a factor of 2 but had the same fluence, were identical. Examination of the photophysical properties of ClAlPc revealed that σ_{34}/σ_{01} (~30) is about 3 times larger than σ_{12}/σ_{01} (~10) but that τ_{isc} (i.e., k_{isc}^{-1}) has a value of 18 ns. On the time scale of a 1- to 10-ns laser pulse there would be little buildup of triplet population during the pulse and the reverse saturable absorption would be dominated by the excited singlet population and σ_{12}/σ_{01}.

Substantial enhancements in the reverse saturable absorption by phthalocyanines for 532-nm, 8-ns pulses have been realized by exploiting the heavy-atom effect.[27,34,36] In the heavy-atom effect, attachment of an atom of high atomic number to a chromophore leads to an increase in the intersystem crossing rate due to the increase in the spin-orbit coupling for π-electrons resulting from orbital mixing with the metal. The increase in the intersystem crossing rate provides a means for shortening the time scale for buildup of triplet population and making the more effective use of the triplet–triplet absorption for short pulse durations. It was shown that the use of central metal atoms of increased atomic number in phthalocyanine complexes leads to an increase in the effective excited-state absorption cross section and the optical limiting response, for nanosecond duration, 532-nm laser pulses. The nonlinear transmission response for phthalocyanines containing a series of metals and metalloids from group IIIA (Al, Ga, In) and IVA (Si, Ge, Sn and Pb) is shown in Figure 10. These results also showed that the phthalocyanines began to exhibit nonlinear transmission at very low fluences, < 0.01 J/cm^2, and possess saturation fluences (F_s) of about 0.2–0.5 J/cm^2. A comparative optical limiting response illustrating the magnitude of the enhancement provided by the heavy-atom approach is shown in Figure 11 for AlPc, SiNc, and PbPc derivatives, for an f/8 optical limiting device geometry.[36] Although the linear transmittance of the solution used was somewhat low (30%), the PbPc demonstrated a strong signal pulse suppression ($S = 1/T_{NL}$) of 260, and a transmittance ratio of $T_L/T_{NL} = 80$, which was the largest value reported for any reverse saturable absorber-based optical limiter.

Strong optical limiting via reverse saturable absorption has also been reported for another Pb-based heavy-atom phthalocyanine derivative, Pb tetrakis(cumylphenoxy)Pc or Pb(CP)$_4$Pc.[49] Solutions of highly concentrated samples (0.045 M) in a very thin optical cell (30 μm) were examined in an f/5 optical system. At input energies up to 2 μJ, where the response was shown to be dominated by nonlinear absorption and the limiting repeatable over many shots, the pulse suppression S was 40 and $T_L/T_{NL} = 27$. The limiter performed well on a single shot basis up to 20 μJ, giving $S = 87$ and $T_L/T_{NL} = 59$. It is not yet clear whether the strong optical limiting observed in these concentrated Pb(CP)$_4$Pc solutions is due to triplet–triplet absorption or some other mechanism.

The photophysical mechanism of reverse saturable absorption has been examined and the key photophysical parameters of the five-state model have been determined in a series of phthalocyanine derivatives.[27,34,36,58] Nonlinear transmission data obtained with 70-ps and 8-ns, 532-nm pulses (Figures 12 and 13) have been analyzed and fit to rate equation models involving three states (S_0, S_1, and S_n) for the 70-ps results and five states for the 8-ns results. Although the fit to the picosecond data is good for low-to-moderate pulse energies, it is interesting to note that the rate equation model predicts the onset of bleaching for the picosecond data at high pulse energies, as was observed and modeled in the case of HITCI discussed above, but this is not observed experimentally for the phthalocyanines studied.

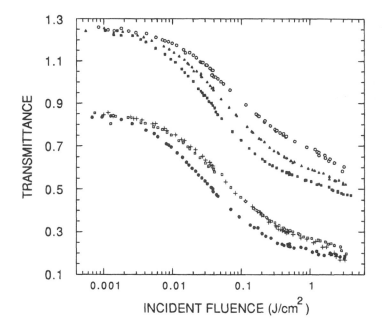

Figure 10 Nonlinear transmission data for 8-ns, 532-nm pulses for metallophthalocyanines. Upper set of data (displaced upwards by 0.4): ((hexyl)$_3$SiO)AlPc (○), ((hexyl)$_3$SiO)GaPc (▲), ((hexyl)$_3$SiO)InPc (■). Lower set of data: ((hexyl)$_3$SiO)$_2$SiPc (+), ((hexyl)$_3$SiO)$_2$GePc (□), ((hexyl)$_3$SiO)$_2$SnPc (●). (From Perry, J. W., Mansour, K., Marder, S. R., Perry, K. J., Alvarez, D., and Choong, I. *Opt. Lett.,* 19, 625, 1994. With permission.)

The implication here is that there is strong excited-state absorption from S_n to a yet higher excited singlet state in phthalocyanines at 532 nm; however, the lack of bleaching at high irradiance needs further investigation. The nanosecond data are typically fit very well by the rate equation model. A summary of the key photophysical parameters for various phthalocyanines is given in Table 6. These data show the reduction in fluorescence lifetime, the increase in triplet yield, and the enhancement in the effective cross section ratio due to the heavy-atom effect.

Reverse saturable absorption in paramagnetic phthalocyanine complexes has been reported. Enhanced nonlinear absorption was observed for VO(*t*-butyl)$_4$Pc and (Cu(SO$_3^-$)$_4$Pc)Na$^+_4$ relative to ClAlPc for 532-nm, 8-ns pulses.[34] Reverse saturable absorption, optical limiting, and optical bistability were also more recently reported for Cu(*t*-butyl)$_4$Pc for 532-nm, 15-ns laser pulses.[26] The paramagnetism due to unpaired spins on the metal facilitates intersystem crossing of the $\pi\pi^*$ excited state and this is believed to result in a strongly enhanced triplet–triplet contribution to the reverse saturable absorption. Li et al.

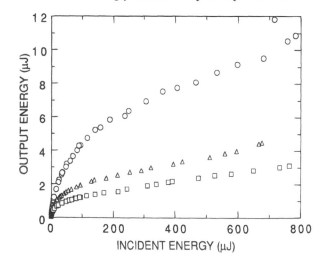

Figure 11 Optical limiting response for 8-ns, 532-nm pulses in an f/8 optical geometry for: ((hexyl)$_3$SiO)AlPc (○), ((hexyl)$_3$SiO)$_2$SiNc (△), Pb(*t*-butyl)$_4$Pc (□). (From Perry, J. W., Mansour, K., Marder, S. R., Perry, K. J., Alvarez, D., and Choong, I. *Opt. Lett.,* 19, 625, 1994. With permission.)

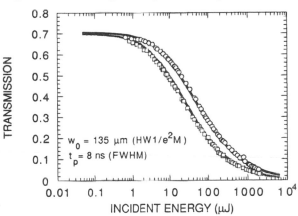

Figure 12 Picosecond pulse nonlinear transmission data for: ((hexyl)$_3$SiO)$_2$SnPc (□) and ((hexyl)$_3$SiO)$_2$SiPc (○). Curves are fits to the data using a numerical solution to the five-state rate equation model and propagation equation. w_0 is the beam radius at the focus. (From Mansour, K., Alvarez Jr., D., Perry, K. J., Choong, I., Marder, S. R., and Perry, J. W. *Proc. SPIE*, 1853, 132, 1993. With permission.)

showed that the full rate equation model provided a better fit of their data on Cu(t-butyl)$_4$Pc than did the steady-state model, consistent with the fact that the recovery time to the ground state was not much shorter than their pulse duration. They obtained a limit on the effective cross section ratio of σ_{eff}/σ_{01} > 10, which is consistent with the values obtained for other phthalocyanines.

H. NAPHTHALOCYANINES
Naphthalocyanines are phthalocyanine derivatives with ring conjugation that is symmetrically extended through benzo fusion to the phthalocyanine ring. A metalloid naphthalocyanine complex, bis(tri-*n*-

Figure 13 Nanosecond pulse nonlinear transmission data for: ((hexyl)$_3$SiO)$_2$SnPc (□) and ((hexyl)$_3$SiO)$_2$SiPc (○). Curves are fits to the data using a numerical solution to the five-state rate equation model and propagation equation. w_0 is the beam radius at the focus. (From Mansour, K., Alvarez Jr., D., Perry, K. J., Choong, I., Marder, S. R., and Perry, J. W. 1993. *Proc. SPIE*, 1853, 132, 1993. With permission.)

Table 6 Photophysical Parameters for Metallophthalocyanines

Molecule	τ_s (ns)	$\Phi_t{}^a$	$\sigma_{01}{}^b$	σ_{12}/σ_{01}	σ_{34}/σ_{01}	$\sigma_{eff}/\sigma_{01}{}^c$
AlPc	7.0	0.35	2.2	10	20	12
GaPc	3.3	0.51	2.2	—	—	16[d]
InPc	0.3	0.88	1.9	—	—	25[d]
SiPc	4.5	0.35	2.4	13	20	15
GePc	4.2	0.37	2.3	13	22	16
SnPc	2.0	0.62	2.1	11	32	24
PbPc	0.35	0.92	1.4	30	23	24
SiNc	3.15	0.20	2.8	14	32	18

Note: Phthalocyanines were metal-substituted mono- or bis-(*n*-hexyl)$_3$SiO- derivatives except for InPc and PbPc which were of the form: ClIn(t-butyl)$_4$Pc and Pb(t-butyl)$_4$Pc.

[a] Triplet quantum yield.

[b] Units of 10^{-18} cm^2.

[c] Calculated for an 8-ns pulse using $\sigma_{eff}/\sigma_0 = (\sigma_{12}f_1 + \sigma_{34}f_3)/\sigma_{01}$.

[d] Estimated from fits of nanosecond transmission data to equation 25.

hexylsiloxy) silicon naphthalocyanine or SiNc (Figure 14) was shown to exhibit enhanced reverse saturable absorption compared to ClAlPc, for both nanosecond and picosecond 532-nm pulses, as shown in Figure 11 for nanosecond pulses.[34,58] The increase in the conjugated ring size relative to phthalocyanines leads to an approximately 100-nm red shift of the Q-band absorption, $\lambda_{max} = 774$ nm in toluene, and a sizable increase in the σ_{12} and σ_{34} values at 532 nm (see entry in Table 6). The red shift of the ground-state absorption leads to a minimum absorption wavelength of about 560 nm for SiNc, and a transmission window better centered on the human eye spectral response compared to phthalocyanines.

A preliminary investigation of the utility of the heavy-atom approach to enhancement of reverse saturable absorption in naphthalocyanines has been reported.[35] A series of naphthalocyanine complexes was examined including SiNc, ((n-hexyl)$_3$SiO)$_2$SnNc, Pb(t-butyl)$_4$Nc, and ClIn(t-butyl)$_4$Nc. Initial comparative measurements on SiNc and ((n-hexyl)$_3$SiO)$_2$SnNc showed essentially no enhancement on going from Si to Sn, in contrast to observations on analogous phthalocyanines. This is likely due to the larger ring of the naphthalocyanine leading to a diminished effect of the central metal on the π-electron spin-orbit coupling, due to the decrease in orbital mixing with increasing distance. One would then expect to observe an enhancement only by going to yet heavier or more strongly coupled metals. On going from Si to In, a significant enhancement was observed as illustrated for ClIn(t-butyl)$_4$Nc in Figure 15. Since In is actually lighter than Sn, perhaps the enhancement for In is a result of it being more strongly coupled to the ring orbitals. On the other hand, the response of Pb(t-butyl)$_4$Nc was not appreciably stronger than for SiNc. A definitive interpretation of these results must await a more complete characterization of the photophysics of these compounds.

There has recently been increased interest in red-shifting the optical limiting response of phthalocyanine-based complexes. Alkoxy substitution at the α-carbon positions of phthalocyanines and naphthalocyanines results in substantial red shifts of the electronic absorption bands.[40,41] Reverse saturable absorption in the red region of the spectrum has recently been reported for some new octaalkoxynaphthalocyanines, including ClIn(O-n-butyl)$_8$Nc and F$_2$Sn(O-n-butyl)$_8$Nc, whose structures are shown in Figure 16.[35,37] The red shifts of the absorption bands upon donor substitution can be understood as a destabilization of the highest occupied π-bonding molecular orbital, due to mixing of this orbital with the lower energy oxygen p_z orbitals. This destabilization leads to a reduction of the $\pi\pi^*$ energy gap and to a red shift of the ground- and excited-state absorptions, thus to a red-shifted optical limiting band. The octaalkoxynaphthalocyanine molecules show strongly red-shifted electronic absorption and transmission windows compared to analogous unsubstituted naphthalocyanines. For example, the absorption minimum for ClIn(O-n-butyl)$_8$Nc and F$_2$Sn(O-n-butyl)$_8$Nc is at 635 nm, whereas for the nondonor-substituted SiNc the minimum is at about 560 nm. Shirk et al.[48] reported that for Pb(O-pentoxy)$_8$Nc there is a broad transient absorption with a $\lambda_{max} \sim 625$ nm that appears within 30 ps of excitation and shows little decay after 1.2 ns, which indicated a potential for reverse saturable absorption in the red with this class of chromophores.

Nonlinear transmittance responses for 630-nm, 5-ns pulses for ClIn(O-n-butyl)$_8$Nc, F$_2$Sn(O-n-butyl)$_8$Nc, and C$_{60}$ solutions are shown in Figure 17 and demonstrate that the octabutoxynaphthalocya-

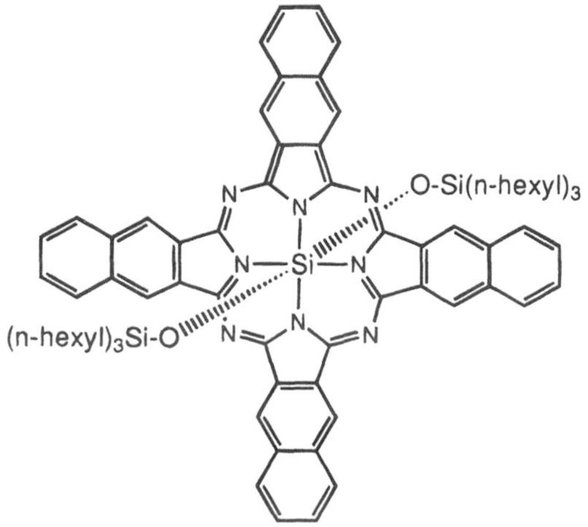

Figure 14 Molecular structure of ((hexyl)$_3$SiO)$_2$SiNc or SiNc.

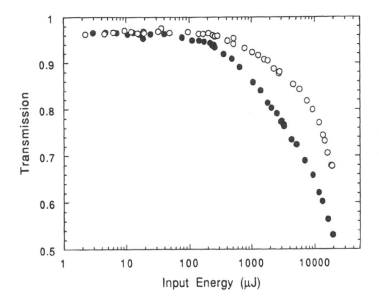

Figure 15 Nonlinear transmission responses for 8-ns, 532-nm pulses for: ((hexyl)$_3$SiO)$_2$SiNc (○) and ClIn (*t*-butyl)$_4$Nc (●). (From Perry, J. W., Mansour, K., Marder, S. R., Chen, C.-T., Miles, P., Kenney, M. E., and Kwag, G. In *Materials for Optical Limiting,* Vol. 374, Crane, R. L., Khoshnevisan, M., Lewis, K., and Van Stryland, E. W., Eds., Materials Research Society, Pittsburgh, 1995, p. 257. With permission.)

nines exhibit strong reverse saturable absorption at 630 nm.[35,37] A simple fluence-dependent analysis gives estimates of the ratio of effective excited-state to ground-state absorption cross sections of 22 and 17 for ClIn(O-*n*-butyl)$_8$Nc and F$_2$Sn(O-*n*-butyl)$_8$Nc, respectively. These values are comparable to those exhibited by previously studied phthalocyanines at 532 nm. However, the saturation fluence for C$_{60}$ is apparently an order of magnitude higher than for the octaalkoxynaphthalocyanines. This is due to the low value of σ_g for C$_{60}$ at 630 nm. The high saturation fluence precludes strongly saturating the excited-state population, despite the high triplet quantum yield for C$_{60}$. The low saturation fluences and high cross section ratios of the octaalkoxynaphthalocyanines suggest that high-suppression optical limiters for the red spectral region based on such reverse saturable absorbers are possible.

I. REVERSE SATURABLE ABSORBER DOPED POLYMER AND SOL-GEL MATERIALS

The development of solid-state reverse saturable absorber materials offers advantages for the fabrication of optical limiter devices. Organic polymers, inorganic sol-gels, and organically modified silica (ORMOSIL) sol-gels are attractive potential host materials from the point of view of processing and fabrication of bulk and thin film solid-state components. These host materials can be transparent in the

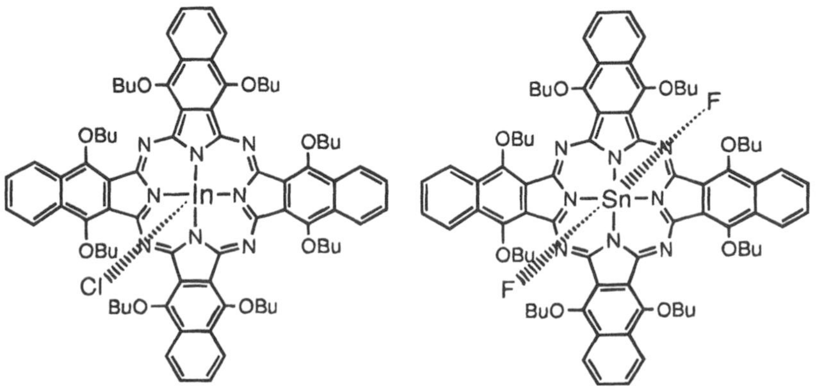

Figure 16 Molecular structures of metallo-octabutoxynaphthalocyanines.

visible and near-infrared spectral regions and can exhibit low optical scattering. Additionally, these materials can have high optical and thermomechanical damage resistance, in the range needed for optical limiting applications. The fact that polymer and sol-gel hosts can be found in which photophysical properties and reverse saturable absorption responses are not deleteriously affected, for a number of embedded chromophores that have been examined, is an indication of the potential for these materials.

The synthesis of phthalocyanine doped sol-gel based materials has been reported by Fuqua et al.[11] and Bentivegna et al.,[4] who also incorporated C_{60} into a sol-gel host. Fuqua et al. prepared silica xerogels containing $CuPC(SO_3^-)_4$, $((n\text{-hexyl})_3SiO)_2GePc$ and $((n\text{-hexyl})_3SiO)_2SnPc$. The Q-band absorption for $((n\text{-hexyl})_3SiO)_2SnPc$ showed little change on going from sol to xerogel, undergoing only a 14-nm red shift. An increase in photochemical stability for the SnPc in the xerogel as compared to aerated solution was observed. The $CuPC(SO_3^-)_4$ showed more complex spectral changes, indicating a breakup of dimers and the formation of a protonated species upon going from sol to xerogel. The $CuPc(SO_3^-)_4$ showed essentially identical reverse saturable absorption behavior in silica xerogel and in a pH 2 aqueous buffer. Quite strong reverse saturable absorption was observed for the SnPc, consistent with solution results. The response of the GePc doped silica xerogel appeared to be degraded relative to that for GePc in solution. Bentivegna et al. incorporated an AlPc into silica/zirconia xerogels and organically modified variants using methyltriethoxysilane (MTEOS) and vinyltriethoxysilane (VTEOS) and also incorporated C_{60} into VTEOS. Some variation in performance of the AlPc xerogels was seen with the best result obtained for VTEOS. C_{60} also exhibited reverse saturable absorption in VTEOS.

As discussed above, the optical limiting of C_{60} doped PMMA has been reported.[25] The results obtained provided evidence that reverse saturable absorption is dominant in the doped polymer and that C_{60} largely retains its solution photophysical properties in PMMA.

Mansour et al.[28] examined $((hexyl)_3SiO)_2SnPc$ and other phthalocyanine dyes in methyltrimethoxysilane (MTMOS) modified silica gel and in a PMMA host. Nonlinear transmission, triplet–triplet transient absorption spectra, and triplet quantum yields, as determined by picosecond pump/probe induced-bleaching recovery measurements were measured. These data showed only a small or no change in going from solution to PMMA or the ORMOSIL, showing that the photophysics of these molecules were largely unaffected by the hosts studied. As was observed with the silica xerogel, increased photochemical stability was observed for phthalocyanines in PMMA or MTMOS-ORMOSIL, relative to solutions. Several key figures-of-merit (σ_e/σ_g, the saturation fluence (F_s) and the optical damage fluence (F_D)) were determined for the phthalocyanine doped PMMA and MTMOS-ORMOSIL materials. Of particular interest were the results for materials containing $((hexyl)_3SiO)_2SnPc$ and $ClIn(t\text{-butyl})_4Pc$. These materials gave $\sigma_e/\sigma_g \sim 25$ (InPc/PMMA) and ~ 16 (SnPc/MTMOS-ORMOSIL), F_s values of

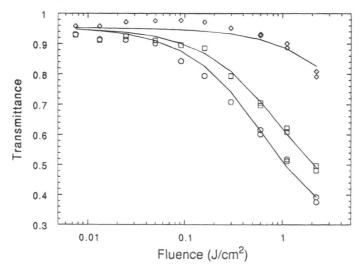

Figure 17 Nonlinear transmittance responses for 5-ns, 630-nm pulses for: C_{60} ($\diamond$), $F_2Sn(O\text{-}n\text{-butyl})_8Nc$ ($\square$), an $ClIn(O\text{-}n\text{-butyl})_8Nc$ ($\bigcirc$). (From Perry, J. W., Mansour, K., Marder, S. R., Chen, C.-T., Miles, P., Kenney, M. E., and Kwag, G. In *Materials for Optical Limiting*, Vol. 374, Crane, R. L., Khoshnevisan, M., Lewis, K., and Van Stryland, E. W., Eds., Materials Research Society, Pittsburgh, p. 257, 1995. With permission.)

~0.4 J/cm^2, and a damage threshold of 3 J/cm^2 (in f/5 optics for InPc/PMMA). The properties of these phthalocyanine doped solids are in the range needed for high-performance devices and initial efforts to utilize them in devices are discussed below.

IV. TOWARDS HIGH-PERFORMANCE DEVICES

While the performance of single homogeneous cells of phthalocyanines and other reverse saturable absorbers, as discussed above, has been investigated in a variety of optical limiting geometries, the results have fallen short of the desired performance levels of: strong signal suppression of 10^3 to 10^5, combined with a linear transmittance of >50%. Consideration of the absorption cross sections and excitation efficiencies shows that heavy-atom phthalocyanines, for example, possess the necessary characteristics to achieve this level of performance, so the deficiency is not associated with the intrinsic molecular properties. The difficulty of limiting total transmitted energy to some low prescribed level, say 1–3 μJ, and having the device be able to withstand tens of millijoules of input pulse energy without damage has led a several investigators to consider the use of multiple optical limiter cells in tandem or, more generally, the use of nonhomogeneous distributions of nonlinear material. Some of these initial efforts towards higher performance devices utilizing reverse saturable absorbers are summarized below.

A. TANDEM OPTICAL LIMITERS

In an effort to improve the dynamic range of optical limiter devices, Hagan et al.[14] investigated the performance of optical limiters comprised of tandem combinations of nonlinear optical materials. The idea behind this work was to have a "primary" limiter cell located at the focal plane of the optical system and to have a "protector" cell located upstream from the focus to provide attenuation of higher energy pulses and thereby increase the energy handling of the device. Hagan et al. studied several tandem limiters involving the following combinations: a naphthalocyanine (SiNc) solution followed by a ZnSe slab, a SiNc solution and a carbon black suspension, and a combination of two SiNc solutions, and the results are summarized in Table 7. These studies demonstrated that by using tandem arrangements of limiting materials that an improved energy handling ability could be obtained. Of particular interest are the results obtained with two SiNc cells in an f/10 optical system for 2.7-ns pulses. With an overall rather low linear transmittance of 0.21, a strong signal pulse suppression of 590 was obtained, but the device was able to handle input energy up to 8.7 mJ. These results are a sizable improvement in terms of energy handling over earlier attempts with single cells, which would typically be limited to about 1 mJ.

B. NONHOMOGENEOUS REVERSE SATURABLE ABSORBER MATERIALS

A significant conceptual step in the design of optical limiters utilizing reverse saturable absorbers was the realization that the problem of managing the fluence of a convergent optical beam in an absorbing material leads to a general requirement for a nonhomogeneous concentration profile of the absorber. Such a concept was the subject of a U.S. patent by McCahon and Tutt[29] and was discussed in a review by Tutt and Boggess.[53] McCahon and Tutt suggested a concentration profile of the form:

$$N(z) = N_0 \left[1 + \left(\frac{z}{z_0} \right)^{1.5} \right]^{-1} \tag{31}$$

and numerical calculations showed that this form should provide an increased dynamic range.

Table 7 Characteristics of Tandem Optical Limiters at 532 nm

Configuration	f/#	τ_p	T_L	T_{NL}	E_{max}[a]
CAP/ZnSe		30 ps	0.4		50 μJ
SiNc/ZnSe		30 ps	0.4		>80 μJ
SiNc/CBS	f/10	2.7 ns	0.3	0.002	10 mJ
CBS/SiNc	f/10	2.7 ns	0.3	0.0032	7 mJ
SiNc/SiNc	f/10	2.7 ns	0.21	0.0017	8.7 mJ

[a] Maximum energy before damage.

Miles[32] has recently considered both the management of fluence in the material and maximal utilization of the nonlinear response of the reverse saturable absorber, through an analytic formulation making use of a fluence-dependent saturation response for the reverse saturable absorber. In reverse saturable absorbers, maximal attenuation of the energy of the optical pulse requires strong saturation of the excited-state population. The key to establishing strong saturation along the optical path in the material is that the fluence along the path must be maintained above (say 2.5 times) the saturation fluence (F_s) and, at the same time, controlled so that it is below the limit imposed by optical damage. In a focusing optical system, it is necessary to achieve equal fractional rates of change of energy and beam cross-sectional area, in order to maintain a constant fluence. Miles showed that this condition can be achieved by using fully saturated absorbers with a nonhomogeneous concentration profile:

$$N(z) = \frac{2}{\sigma_e |z|}. \tag{32}$$

In a recent experiment, we set out to test this optical limiter design principle by using a practical approximation to the ideal hyperbolic concentration profile involving a geometrically expanding array of uniform concentration plates, as illustrated in Figure 18.[35,37,38] The individual plate thicknesses and locations in this design were chosen so that when the molecules are driven into saturation, the fluence is just below the maximum allowable value at the front face of each upstream plate and decreases to 2.5 F_s at each plate's exit face. To implement the limiter design, a series of ClIn(t-butyl)$_4$Pc doped PMMA plates were used, in order a avoid the significant thermal refractive beam spreading that occurs in solution. The results of an experiment with 8-ns, 532-nm pulses on a three-plate device in a highly convergent, f/5, optical system are shown in Figure 19 and confirm that an order of magnitude enhanced strong signal suppression can be obtained as compared to a homogeneous concentration, while keeping the linear transmittance constant. The strong signal suppression of 540 achieved with a linear transmittance of 70% in a such a fast optical system is the highest performance reported to date. This initial experiment shows that there is considerable promise for the use of hyperbolic concentration distributions of strong reverse saturable absorbers, such as phthalocyanines, in the development of high-performance optical limiter devices.

V. SUMMARY

Since the first report of reverse saturable absorption in dyes in 1967, reverse saturable absorption has been observed in a variety of classes of chromophore including: porphyrins, indanthrones, metal cluster compounds, fullerenes, phthalocyanines, and naphthalocyanines. Efforts to date to use such chromophores in optical limiting devices have resulted in performance that falls short of the desired levels (strong pulse attenuation of 10^3–10^5, and linear transmittance > 50%). Nonetheless, substantial progress towards the desired performance has been made recently. Chromophores that possess σ_e/σ_g ~25–30 in the visible spectrum and that can be efficiently pumped to the excited state have been synthesized. These

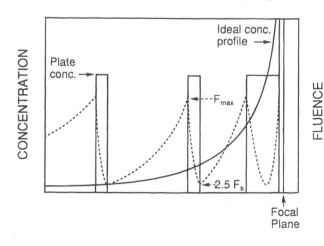

Figure 18 Schematic illustration of the ideal hyperbolic concentration profile for a reverse saturable absorber in a convergent optical beam and a simple approximation involving three geometrically spaced plates. Also shown is the axial profile of the fluence for the three-plate case. The horizontal axis is the position along the beam propagation direction.

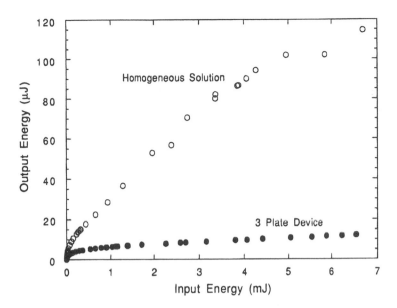

Figure 19 Optical limiting response of a homogeneous reverse saturable absorber cell and a geometrically spaced three-plate device. Data are for 8-ns, 532-nm pulses in an f/5 optical system with total output energy detection. The reverse saturable absorber is InPc(t-butyl)$_4$ in either toluene or in PMMA plates. Both samples had 70% linear transmittance.

σ_e/σ_g values are sufficiently large that nonlinear attenuations of the desired levels are feasible, in principle. Device design concepts, based on nonuniformly distributing the reverse saturable absorbers to utilize optimally their nonlinear absorptive properties, have recently been proposed. Initial experiments indicate that such approaches hold high potential for achieving the performance goals for optical limiters with available materials. There is now considerable interest in the field on device implementation of state-of-the-art materials in such novel limiter designs. There remain nonetheless significant scientific and technological challenges for the development of reverse saturable absorber materials including: (1) understanding of the optical limiting mechanism at high fluence and irradiance; (2) developing materials with large linear transmission and optical limiting bandwidths; (3) tailoring materials for optical limiting in specific spectral bands, including the near infrared; (4) enhancing σ_e/σ_g ratios to values considerably larger than 25; (5) enhancing the optical excitation efficiency while maintaining a high value of σ_e/σ_g; and (6) developing materials with improved optical damage thresholds.

VI. DEFINING TERMS

Reverse saturable absorption: A process of optically induced absorption. Results from excitation of states that have stronger absorption than the ground-state at the excitation wavelength.

Optical limiter: A device that attenuates strong optical signals to hold the output below a given level but maintains a high transmittance for low level signals.

Absorption cross section: The effective area for photon capture by a molecule. It is related to the absorption coefficient (α, as defined by $T = \exp(-\alpha L)$) by $\alpha = \sigma N$, and to the extinction coefficient (ϵ, as defined by $T = 10^{(-\epsilon CL)}$, where C is the molar concentration, by $\sigma = (1000 \ln 10 \; \epsilon)/N_A$, where N_A is Avogadro's number.

Z-scan: A method for measurement of nonlinear absorption and refraction involving the translation of a sample along a laser beam through a focus and monitoring either the total transmitted energy or the energy transmitted through a finite aperture.

Nonlinear attenuation (or suppression): The factor by which a strong laser pulse is reduced by a nonlinear material or device. The suppression is taken as $S = 1/T_{NL}$, where T_{NL} is the transmittance for a strong input pulse.

ACKNOWLEDGMENTS

This work was performed in part at the Jet Propulsion Laboratory, California Institute of Technology as part of its Center for Space Microelectronics Technology and was supported by the Ballistic Missile Defense Organization, Innovative Science and Technology Office and the U.S. Air Force Wright Laboratory, through an agreement with the National Aeronautics and Space Administration. Support at the Beckman Institute, California Institute of Technology from the Office of Naval Research, the Defense Advanced Research Projects Agency, the Office of Naval Research through the Center for Advanced Multifunctional Molecular Assemblies and Polymers, and the Air Force Office of Scientific Research is gratefully acknowledged.

REFERENCES

1. **Allan, G. R., Labergerie, D. R., Rychnovsky, S. J., Boggess, T. F., Smirl, A. L., and Tutt, L.** Picosecond reverse saturable absorption in King's complex $[(C_5H_5)Fe(CO)]_4$. *J. Phys. Chem.,* 96, 6313, 1992.
2. **Allan, G. R., Rychnovsky, S. J., Venzke, C. H., Boggess, T. F., and Tutt, L.** Picosecond investigations of the excited-state transition at 532 nm in King's complex $[(C_5H_5)Fe(CO)]_4$ and synthesized analogs. *J. Phys. Chem.,* 98, 216, 1994.
3. **Band, Y. B., Harter, D. J., and Bavli, R.** Optical pulse compressor composed of saturable and reverse saturable absorbers. *Chem. Phys. Lett.,* 126, 280, 1986.
4. **Bentivegna, F., Canva, M., Georges, P., Brun, A., Chaput, F., Malier, L., and Boilot, J. P.** Reverse saturable absorption in solid xerogel matrices. *Appl. Phys. Lett.,* 62, 1721, 1993.
5. **Blau, W., Byrne, H., Dennis, W. M., and Kelly, J. M.** Reverse saturable absorption in tetraphenylporphyrins. *Opt. Commun.,* 56, 25, 1985.
6. **Boggess, T. F., Allan, G. R., Rychnovsky, S. J., Labergerie, D. R., Venzke, C. H., Smirl, A. L., Tutt, L. W., Kost, A. R., McCahon, S. W., and Klein, M. B.** Picosecond investigations of optical limiting mechanisms in King's complex. *Opt. Eng.,* 32, 1063, 1993.
7. **Carmichael, I. and Hug, G. L.** A note on the total depletion method of measuring extinction coefficients of triplet-triplet transitions. *J. Phys. Chem.,* 89, 4036, 1985.
8. **Coulter, D. R., Miskowski, V. M., Perry, J. W., Wei, T. H., Van Stryland, E. W., and Hagan, D. J.** Optical limiting in solutions of metallo-phthalocyanines and naphthalocaynines. *Proc. SPIE,* 1105, 42, 1989.
9. **Fei, H. S., Han, L., Ai, X. C., Yin, R., and Shen, J. C.** Reverse saturable absorption and 3rd-order optical nonlinearity in porphyrin. *Chi. Sci. Bull.* 37, 298, 1992.
10. **Fisher, M. M., Veyret, B., and Weiss, K.** Nonlinear absorption and photoionization in the pulsed laser photolysis of anthracene. *Chem. Phys. Lett.,* 28, 60, 1974.
11. **Fuqua, P. D., Mansour, K., Alvarez Jr., D., Marder, S. R., Perry, J. W., and Dunn, B.** Synthesis and nonlinear optical properties of sol-gel materials containing phthalocyanines. *Proc. SPIE,* 1758, 499, 1992.
12. **Giuliano, C. R. and Hess, L. D.** Nonlinear absorption of light: optical saturation of electronic transitions in organic molecules with high intensity laser radiation. *IEEE J. Quantum Electronics,* QE-3, 858, 1967.
13. **Guha, S., Kang, K., Porter, P., Roach, J. F., Remy, D. E., Aranda, F. J., and Rao, D.** Third-order optical nonlinearities of metallotetrabenzoporphyrins and a platinum polyyne. *Opt. Lett.,* 17,264, 1992.
14. **Hagan, D. J., Xia, T., Said, A. A., Wei, T. H., and Van Stryland, E. W.** High dynamic range passive optical limiters. *Int. J. Nonlin. Opt. Phys.,* 2, 483, 1994.
15. **Harter, D. J., Shand, M. L., and Band, Y. B.** Power/energy limiter using reverse saturable absorption. *J. Appl. Phys.,* 56, 865, 1984.
16. **Henari, F., Callaghan, J., Stiel, H., Blau, W., and Cardin, D. J.** Intensity-dependent absorption and resonant optical nonlinearity of C_{60} and C_{70} solutions. *Chem. Phys. Lett.,* 199, 144, 1992.
17. **Hercher, M.** An analysis of saturable absorbers. *Appl. Optics,* 6, 947, 1967.
18. **Hercher, M., Chu, W., and Stockman, D. L.** An experimental study of saturable absorbers for ruby lasers. *IEEE J. Quantum Electronics,* QE-4, 954, 1968.
19. **Hoffman, R. C., Stetyick, K. A., Potember, R. S. and Mclean, D. G.** Reverse saturable absorbers—indanthrone and its derivatives. *J. Opt. Soc. Am. B Opt. Phys.,* 6, 772, 1989.
20. **Hou, H. W., Ye, X. G., Xin, X. Q., Liu, J., Chen, M. Q., and Shi, S.** Solid-state synthesis, crystal-structure, and nonlinear refractive and absorptive properties of the new cluster $(n\text{-}Bu_4N)_2[MoOS_3Cu_3BrCl_2]$. *Chem. Mater.,* 7, 472, 1995.
21. **Hughes, S., Spruce, G., Wherrett, B. S., Welford, K, R., and Lloyd, A. D.** The saturation limit to picosecond, induced absorption in dyes. *Opt. Commun.,* 100, 13, 1993.
22. **Joshi, M. P., Mishra, S. R., Rawat, H. S., Mehendale, S. C., and Rustagi, K. C.** Investigation of optical limiting in C_{60} solution. *Appl. Phys. Lett.,* 62, 1763, 1993.

23. **Justus, B. L., Kafafi, Z. H., and Huston, A. L.** Excited-state absorption-enhanced thermal optical limiting in C_{60}. *Opt. Lett.,* 18, 1603, 1993.

24. **Keyes, R. W.** Nonlinear absorbers of light. *IBM J. Res. Dev.,* 7, 354, 1963.

25. **Kost, A., Tutt, L., Klein, M. B., Dougherty, T. K., and Elias, W. E.** Optical limiting with C_{60} in polymethyl methacrylate. *Opt. Lett.,* 18, 334, 1993.

26. **Li, C. F., Zhang, L., Yang, M., Wang, H., and Wang, Y. X.** Dynamic and steady-state behaviors of reverse saturable absorption in metallophthalocyanine. *Phys. Rev. A.,* 49, 1149, 1994.

27. **Mansour, K., Alvarez Jr., D., Perry, K. J., Choong, I., Marder, S. R., and Perry, J. W.** Dynamics of optical limiting in heavy-atom substituted phthalocyanines. *Proc. SPIE,* 1853, 132, 1993.

28. **Mansour, K., Fuqua, P., Marder, S. R., Dunn, B., and Perry, J. W.** Solid-state optical limiting materials based on phthalocyanine containing polymers and organically-modified sol-gels. *Proc. SPIE,* 2143, 239, 1994.

29. **McCahon, S. W. and Tutt, L. W.** United States Patent 5,080,469, 1992.

30. **McLean, D G., Sutherland, R. L., Brant, M. C., Brandelik, D. M., Fleitz, P. A., and Pottenger, T.** Nonlinear absorption study of a C_{60}-toluene solution. *Opt. Lett.,* 18, 858, 1993.

31. **Miles, P.** Optical bandwidth determinants for excited-state absorptive limiters. In *Materials for Optical Limiting,* Crane, R. L., Khoshnevisan, M., Lewis, K., and Van Stryland, E. W., Eds., Materials Research Society Symposium Proceedings, Vol. 374, Materials Research Society, Pittsburgh, p. 51, 1995.

32. **Miles, P. A.** Bottleneck optical limiters—the optimal use of excited-state absorbers. *Appl. Optics,* 33, 6965, 1994.

33. **Penzkofer, A.** Passive Q-switching and mode-locking for the generation of nanosecond to femtosecond pulses. *Appl. Phys. B,* 46, 43, 1988.

34. **Perry, J. W., Khundkar, L. R., Coulter, D. R., Alvarez, D., Marder, S. R., Wei, T.-H., Sence, M. J., Van Stryland, E. W., and Hagan, D. J.** Excited state absorption and optical limiting in solutions of metallophthalocyanines. In *Organic Molecules for Nonlinear Optics and Photonics,* Messier, J., Kajzar, F., and Prasad, P., Eds., NATO ASI Series E: Applied Sciences, Vol. 194, Kluwer Academic Publishers, Dordrecht, 369, 1991.

35. **Perry, J. W., Mansour, K., Marder, S. R., Chen, C.-T., Miles, P., Kenney, M. E. and Kwag, G.** Approaches for optimizing and tuning the optical limiting response of phthalocyanine complexes. In *Materials for Optical Limiting,* Crane, R. L., Khoshnevisan, M., Lewis, K., and Van Stryland, E. W., Eds., Materials Research Society Symposium Proceedings, Vol. 374, Materials Research Society, Pittsburgh, p. 257, 1995.

36. **Perry, J. W., Mansour, K., Marder, S. R., Perry, K. J., Alvarez, D., and Choong, I.** Enhanced reverse saturable absorption and optical limiting in heavy-atom-substituted phthalocyanines. *Opt. Lett.,* 19, 625, 1994.

37. **Perry, J. W., Mansour, K., Miles, P., Chen, C. T., Marder, S. R., Kwag, G., and Kenney, M.** Phthalocyanine materials for optical limiting. In *Polymer Materials Science and Engineering,* ed. 72, American Chemical Society, Washington, D. C., 1995, p. 222.

38. **Perry, J. W., Mansour, K. M., Lee, I.-Y. S., Wu, X.-L., Bedworth, P. V., Chen, C.-T., Ng, D., Marder, S. R., Miles, P., Wada, T., Tian, M., and Sasabe, H.** Organic optical limiter with a strong nonlinear absorptive response. *Science,* in press, 1996.

39. **Reddy, K.** Applications of reverse saturable absorbers in laser science. *Curr. Sci.,* 61, 520, 1991.

40. **Rihter, B. D., Kenney, M. E., Ford, W. E., and Rodgers, M. A. J.** Synthesis and photoproperties of diamagnetic octabutoxyphthalocyanines with deep red optical absorbance. *J. Am. Chem. Soc.,* 112, 8064, 1990.

41. **Rihter, B. D., Kenney, M. E., Ford, W. E., and Rodgers, M. A. J.** Photochromic reactions involving palladium(II) octabutoxynaphthalocyanine and molecular oxygen. *J. Am. Chem. Soc.* 115, 8146, 1993.

42. **Sheik-bahae, M., Said, A. A., and Van Stryland, E. W.** High-sensitivity, single-beam n_2 measurements. *Opt. Lett.,* 14, 955, 1989.

43. **Shi, S., Hou, H. W., and Xin, X. Q.** Solid-state synthesis and self-focusing and nonlinear absorptive properties of 2 butterfly-shaped clusters $WCu_2OS_3(PPh_3)_4$ and $MoCu_2OS_3(PPh_3)_3$. *J. Phys. Chem.,* 99, 4050, 1995.

44. **Shi, S., Ji, W., Lang, J. P., and Xin, X. Q.** New nonlinear-optical chromophore—synthesis, structures, and optical limiting effect of transition-metal clusters $(n-Bu_4N)_4[WM_3Br_4S_4]$ (M=Cu and Ag). *J. Phys. Chem.,* 98, 3570, 1994.

45. **Shi, S., Ji, W., Tang, S. H., Lang, J. P., and Xin, X. Q.** Synthesis and optical limiting capability of cubane-like mixed-metal clusters $(n-Bu_4N)_3[MoAg_3BrX_3S_4]$ (X = Cl and I). *J. Am. Chem. Soc.,* 116, 3615, 1994.

46. **Shi, S., Ji, W., Xie, W., Chong, T. C., Zeng, H. C., Lang, J. P., and Xin, X. Q.** The mixed-metal cluster $(n-Bu_4N)_2[MoCu_3OS_3(NCS)_3]$—the first example of a nest-shaped compound with large third-order polarizability and optical limiting effect. *Materials Chem. Phys.,* 39, 298, 1995.

47. **Shi, S., Ji, W., and Xin, X. Q.** Synthesis and superior 3rd-order nonlinear-optical properties of the cluster $(n-Bu_4N)_4[Mo_8Cu_{12}O_8S_{24}]$. *J. Phys. Chem.,* 99, 894, 1995.

48. **Shirk, J. S., Flom, S. R., Lindle, J. R., Bartoli, F. J., Snow, A. W., and Boyle, M. E.** Nonlinear absorption in phthalocyaninnes and naphthalocyanines. In *Materials Research Society Symposium Proceedings,* Vol. 328. Materials Research Society, Pittsburgh, 1994, 661.

49. **Shirk, J. S., Pong, R. G. S., Bartoli, F. J., and Snow, A. W.** Optical limiter using a lead phthalocyanine. *Appl. Phys. Lett.,* 63, 1880, 1993.

50. **Si, J. H., Yang, M., Wang, Y. X., Zhang, L., Li, C. F., Wang, D. Y., Dong, S. M., and Sun, W. F.** Nonlinear excited-state absorption in cadmium texaphyrin solution. *Appl. Phys. Lett.,* 64, 3083, 1994.

51. **Sliney, D. H.** In *Laser Induced Damage in Optical Materials,* Bennett, H. E., Gunther, A. E., Milan, D., and Newman, B. E., Eds., NBS Special Publication 669, 1984, 355.

52. **Speiser, S. and Orenstein, M.** Spatial light modulation via optically induced absorption changes in molecules. *Appl. Optics,* 27, 2944, 1985.

53. **Tutt, L. W. and Boggess, T. F.** A review of optical limiting mechanisms and devices using organics, fullerenes, semiconductors and other materials. *Prog. Quantum Electronics,* 17, 299, 1993.

54. **Tutt, L. W. and Kost, A.** Optical limiting performance of C_{60} and C_{70} solutions. *Nature,* 356, 225, 1992.

55. **Tutt, L. W., McCahon, S., Kost, A., Klein, M., Boggess, T. F., Allan, G. R., Rychnovsky, S. R., Labergerie, D. R., and Smirl, A. L.** Organometallics for optical limiting devices, *1st International Conference on Intelligent Materials,* Oiso, Japan. T. Tagaki, T., Ed., Technomic, Lancaster, PA, 1992, 165.

56. **Tutt, L. W. and McCahon, S. W.** Reverse saturable absorption in metal cluster compounds. *Opt. Lett.,* 15, 700, 1990.

57. **Tutt, L. W. and McCahon, S. W.** International Patent WO 91/14964, 1991.

58. **Wei, T. H., Hagan, D. J., Sence, M. J., Var Stryland, E. W., Perry, J. W., and Coulter, D. R.** Direct measurements of nonlinear absorption and refraction in solutions of phthalocyanines. *Appl. Phys. B.,* 54, 46, 1992.

Chapter 14

Application of Nonlinear Optics to Passive Optical Limiting

Eric W. Van Stryland, D. J. Hagan, T. Xia, and A. A. Said

CONTENTS

I. INTRODUCTION

The nonlinear optical (NLO) properties of materials can be used to reduce significantly the transmittance of a device for high inputs. Such a device is called an optical limiter (sometimes referred to as an optical power limiter, OPL). For many applications, including protection of optical sensors from laser-induced damage (LID), it is desirable for an optical limiter to have high linear transmittance. The output of what is often referred to as an ideal optical limiter is shown in Figure 1. The limiter has a high linear transmittance, a variable and potentially low limiting threshold (the input corresponding to the breakpoint in the curve), a fast reponse (e.g., picoseconds or faster), a broadband response (e.g., the entire visible spectrum), and a large dynamic range defined as the ratio of the linear transmittance to the transmittance at the highest possible input (often 10^4 or greater is desired for sensor protection). In most cases, the limiting does not occur with a sharp threshold as indicated by Figure 1, but changes from high to low transmittance gradually. Some applications need the material to be in thin film form. The best known optical limiting materials are photochromics used in sunglasses which darken in sunlight. This nonlinearity is often due to a chemical reaction induced by light (sometimes in organic dyes) which recovers in the dark due to thermal effects. These materials satisfy many of the criteria mentioned above except they are too slow in response and recovery time for most applications. We concentrate here on nonlinear properties useful for limiting of picosecond to microsecond pulses.

Very little progress has been made at developing such optical limiters in thin film form to limit nanosecond or shorter pulses without an optical system to concentrate the light into the nonlinear material. However, significant progress has been made for applications where the nonlinear optical material can be placed at or near a focal plane. The irradiance or fluence (energy per unit area) at the focal plane can be high enough to give sufficient nonlinear absorption (NLA), nonlinear refraction (NLR), nonlinear scattering (NLS), or combinations of these nonlinearities to limit the output. These are the passive NLO responses used for optical limiting. In this context, passive means that the input light is the energy source for lowering the transmittance. For pulsed sources it is usually fluence, rather than total integrated energy, that determines whether or not a device damages. Thus fluence-dependent nonlinearities are of special interest.

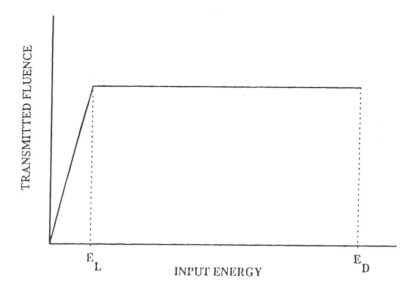

Figure 1 The output energy, power, or fluence of an ideal optical limiting device as a function of the input energy, power, or fluence. E_L is the energy at which limiting begins and E_D is the energy at which the limiting device damages.

In this chapter we discuss the increasing use of NLO organic materials and carbon derivatives for such sensor protection applications. We begin in Section II with a brief discussion of previous optical limiting research as well as a general overview of the useful nonlinear mechanisms. In Section III we describe some of the physical processes available for limiting with organics. In Section IV we look at the use of materials displaying the ultrafast bound-electronic nonlinearities of two-photon absorption (2PA) of coefficient β, and nonlinear refraction of coefficient n_2. We then discuss in Section V reverse saturable absorbing (RSA) materials which are currently under intensive experimental investigation. These materials are discussed in detail by Perry in Chapter 13. Limiting in liquid crystals is briefly discussed in Section VI and pure carbon compounds, carbon black suspensions, and fullerenes are overviewed in Section VII. We also briefly examine the optical geometry of optical limiting devices used to optimize the dynamic range. Section VIII looks at thick limiters and Section IX shows how tandem limiters can extend the dynamic range of devices. Overall conclusions are given in Section X.

II. BACKGROUND

We will not give a complete history of optical limiting research, but mention a few experiments, devices, or materials that exemplify the uses and potential uses of organic materials for optical limiting. The first limiter was based on thermal lensing in nitrobenzene. Leite et al.[1] demonstrated a device that used thermal blooming in this liquid which after spatial filtering regulated the output power of a cw Ar+ laser. The principle of this limiting process is shown in Figure 2. In 1983, CS$_2$ was tested as a limiting

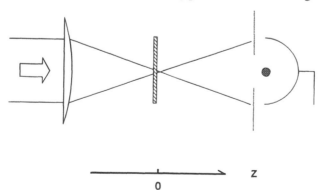

Figure 2 A schematic of a typical optical limiter experiment. Moving the sample along the z direction and monitoring the transmittance of the aperture also results in a Z-scan. Removal of the aperture in front of the detector monitors only nonlinear absorption.

medium for nanosecond pulses.[2] These experiments, using a focused geometry similar to that of Leite et al., showed that the mechanisms that limit the transmission of this device are catastrophic self-focusing, along with absorption and scattering associated with the resulting laser-induced breakdown plasma. The response time of the reorientational nonlinearity responsible for the self-focusing is extremely broadband and has a 2 ps relaxation time.[3] CS_2 meets most of the requirements for a useful optical limiting material; however, the critical power (≈ 8 kW in the visible) is too high for most applications. Liquid crystals can have refractive nonlinearities orders of magnitude higher than that of CS_2, but the response times are correspondingly slower, usually nanoseconds or longer.[4] Some of the largest nonlinearities exhibited to date are in semiconductors.[5] Unfortunately, these extremely large nonlinearities are associated with near band-gap resonance and are maximized in a region of relatively high linear absorption. In addition, solids undergo irreversible optical damage. Even so, effective limiting has been demonstrated in semiconductors using other mechanisms. In 1969 Geusic et al.[6] reported limiting in Si that was attributed to stepwise nonlinear absorption with 1.06 μm radiation. More recently, two-step absorption processes in GaP have been used to produce low threshold limiters.[7] Boggess et al.[8] demonstrated fluence limiting in Si that was due to a combination of two-step nonlinear absorption with a refractive contribution induced by the photoexcitation of free carriers. Power-limiting experiments were conducted by Ralston and Chang[9] in a series of semiconductors such as CdS, GaAs, and CdSe. This was the first reported use of two-photon absorption (2PA), an $Im\{\chi^{(3)}\}$ process, for optical limiting. $\chi^{(3)}$ is the third-order nonlinear optical susceptibility which in this chapter we only use to describe ultrafast responses. Boggess et al.[10] were the first to use the combined effects of 2PA and carrier defocusing to obtain optical limiting. The geometry used was to focus picosecond 1.06-μm pulses onto the surface of a thin sample of GaAs and monitor the fluence by measuring the transmittance of an aperture in an arrangement similar to that shown in Figure 2 except a second collecting lens was inserted prior to the aperture. Such experiments showed that limiting depended on the position of the sample with respect to focus as shown by Miller et al.[1] which eventually led to the development of the Z-scan by Sheik-Bahae et al.[12,13]. Since for thin samples the damage-prone surfaces are subjected to the maximum fluence of the input pulses, the range over which these devices function without incurring damage is low. In 1988, Van Stryland et al.[14,15] demonstrated that if samples of optical thickness much greater than the Rayleigh range are used, the large nonlinearities of the semiconductor can actually be used to self-protect the limiting device from damage. For example, at high inputs, the 2PA-induced carrier self-defocusing occurring prior to focus keeps the fluence below the damage threshold. All these different nonlinearities observed in semiconductors and used for optical limiting have now been observed in organics.[16,17]

In particular very large 2PA coefficients have been reported in polydiacetylenes. For example, Lawrence et al.[1] measured the 2PA spectrum of polydiacetylene paratoluene-sulfonate (PTS) using Z-scan and found a peak near $2\hbar\omega = 2.7$ eV of $\beta \simeq 700$ cm/GW. The potential of such materials for sensor protection applications relies on improving their linear transmissive properties, i.e., low linear absorption, low scattering loss, and good index homogeneity. Besides large 2PA which blocks the energy transmission, self-lensing nonlinearities can effectively reduce the fluence transmission in an image plane. If the lensing is due to bound-electronic effects, the response is ultrafast, described by $Re\{\chi^{(3)}\}$, and the index change characterized by n_2. At a fixed wavelength in a material it is often the case that both 2PA and n_2 effects are present. In fact these nonlinearities are related by causality in a manner analogous to the Kramers-Kronig relations relating the linear index to linear absorption.[19-22] The combination of these nonlinearities can be an effective approach to both reduce the energy transmittance and spread the beam of short pulses at an image plane.[14,15] In addition, 2PA often leads to excited states that can themselves linearly absorb light.[16]. This is also usually accompanied by refractive index changes.[17] These nonlinearities are further discussed in Section IV on ultrafast nonlinearities.

Such excited states can also be created by linear absorption which *requires* that the optical limiter have linear loss; however, this type of nonlinearity in organics is one of the most promising for applications in sensor protection. This so-called reverse saturable absorption (RSA) is discussed in Section V.

In principle, 2PA-based limiters and or ultrafast nonlinear refraction (n_2)-based limiters offer the advantage that linear loss is not required to initiate limiting. In addition the response time is essentially instantaneous. However, both processes are strictly irradiance dependent so that the energy of long pulses will be relatively large before limiting occurs.

The advantage of excited-state absorption, (ESA) initiated by linear absorption (i.e., RSA) is that the NLA can become significant for quite low inputs, and the NLA is fluence dependent as is the detector damage. Thus the nonlinearity is cumulative, getting larger as the pulse width increases for a fixed irradiance, i.e., longer pulses are effectively limited. These RSA materials are discussed in detail in Chapter 13.

In the next section we discuss these physical processes and their characterization.

III. PHYSICAL PROCESSES

Researchers often define a change in absorption as $\alpha_2 I$ (or βI) and a change in index as $n_2 I$ and then quote experimental values of α_2 and n_2. The numbers quoted can be misleading unless a clear explanation of the underlying physical mechanisms leading to these changes is given. As examples we further examine two-photon absorption (2PA), excited-state absorption (ESA), and thermal refraction, and we discuss how to describe these nonlinearities.

As defined in most nonlinear optics texts the total material polarization, P, that drives the wave equation for the electric field, E, is (ignoring nonlocality)

$$P(t) = \epsilon_0 \int_{-\infty}^{\infty} \chi^{(1)}(t - t_1)E(t_1) \, dt_1 + \int_{-\infty}^{\infty} \int_{-\infty}^{\infty} \chi^{(2)}(t - t_1, t - t_2)E(t_1)E(t_2) \, dt_1 \, dt_2$$

$$+ \int_{-\infty}^{\infty} \int_{-\infty}^{\infty} \int_{-\infty}^{\infty} \chi^{(3)}(t - t_1, t - t_2, t - t_3)E(t_1)E(t_2)E(t_3) \, dt_1 \, dt_2 \, dt_3 + \ldots \ , \tag{1}$$

where $\chi^{(n)}$ is defined as the nth order time-dependent response function or time-dependent susceptibility. Thus the nonlinear polarization is electric field dependent. As an example, for harmonic generation (second harmonic from $\chi^{(2)}$ of third from $\chi^{(3)}$ the nonlinearity by necessity follows the field. The only material response capable of this is the bound-electronic response, i.e., the so-called "instantaneous" response.

The 2PA coefficient β arises from the imaginary part of the ultrafast third-order nonlinear susceptibility, $\chi^{(3)}$. Similarly the bound-electronic n_2 comes from the real part of this susceptibility. We note here that in the literature n_2 is used to discuss everything from thermal and reorientational (e.g., for CS_2), to changes in index from saturation of absorption to ultrafast $\chi^{(3)}$ nonlinearities. Here we use n_2 only to describe the ultrafast index change.

For the bound-electronic response, two optical pulses with the same electric field but different pulsewidths result in the same polarization. This statement is also approximately true if the pulse widths of both pulses are much longer than the material response time contained in the susceptibilities in Equation 1. However, the integrals allow for memory. Thus slower responding materials can have their nonlinear response build up with time. For example, in CS_2, the molecules can be reoriented due to the difference in polarizability along the long and short axes of the molecule which changes the macroscopic index as seen by the incoming light. For pulses shorter than the $\simeq$ 2-ps reorientational response time this buildup can be measured. Or, for a thermal nonlinearity, the light heats up the material which changes the linear index.

Fourier transformation of Equation 2 results in the usually quoted frequency-dependent susceptibility $\chi^{(n)}(\omega_1, \omega_2 \ldots \omega_n)$. Memory, which was previously explicitly included in the response function, is lost in the dispersion. Thus, irradiance, I, and fluence, F, dependences are treated equally, which can lead to confusion.

Consider thermal nonlinearities as an example. It is more instructive to account properly for the physics under investigation by quoting the index change (Δn) with temperature and the temperature change with input power for cw inputs, or fluence, F, for pulsed inputs than to quote a $\chi^{(3)}$. The overall index change for a thermal nonlinearity assuming quasi-steady state is

$$\Delta n = \frac{dn}{dT} \frac{\alpha F}{\rho c_v}. \tag{2}$$

where dn/dT is the thermo-optic coefficient, ρ the density, and c_v the specific heat. However, this equation is valid only for pulses long compared to the time for the acoustic waves initially generated by absorption to have traversed the laser beam and short enough that thermal diffusion can be ignored. For other pulse widths transient effects must be accounted for and the index change will not be proportional to F. For example, for tightly focused beams and nanosecond pulses the thermal nonlinearity is transient. Defining a $\chi^{(3)}$ for a process such as thermal refraction can be misleading since the value will depend on the pulse width as well as the focusing geometry.

There exist orders of magnitude differences appearing in the literature for the third-order susceptibility of the same material.[23,24] Many of these discrepancies are due to measurements of different nonlinear processes occurring in the same material under different experimental conditions, e.g., input pulse width.

Such processes as reorientational, electrostrictive, thermal, saturation, and excited-state nonlinearities can also be thought of as two-step processes, or cascaded $\chi^{(1)}{:}\chi^{(1)}$ nonlinearities. For example, for ESA, light first induces a transition to the excited state (an $\mathrm{Im}\{\chi^{(1)}\}$ process) and then the excited state absorbs (a second $\mathrm{Im}\{\chi^{(1)}\}$ process), i.e., two *linear* absorption processes. This cascading description is useful when the pulse width is shorter than or comparable to the material response time.

For these types of slow cumulative nonlinearities the irradiance (or field) may no longer be the important input parameter. For ESA nonlinearities, 1 GW/cm² for a picosecond pulse could give the same Δn as 1 kW/cm² for a microsecond pulse. In any case, quoting α_2 or n_2 without a clear explanation of the physical mechanisms can be misleading. This is particularly true for excited-state nonlinearities.

Both 2PA and ESA are third order in nature, but only 2PA is irradiance dependent. As the intermediate level in the 2PA process becomes resonant, the intermediate level (i.e., first excited state) can have a *real* population as opposed to virtual. This population has a finite lifetime while the nonresonant (virtual) *lifetime* is determined by the uncertainty principle and the energy mismatch between its energy and the input photon energy. Thus, for input pulses shorter than the lifetime of this intermediate state, the nonlinearity will be fluence dependent, i.e., the absorption grows over the duration of the pulse as the excited-state population increases. While this ESA can be considered a resonant 2PA process, a description of the process by $\chi^{(3)}$ is no longer as useful near resonance because of this change to a dependence on the time-integrated irradiance, i.e., the fluence. The ESA terminology and analysis becomes more useful. Therefore, quoting an α_2 gives a number that is pulse width specific. A longer pulse would have given a larger α_2. The nonlinear absorption is now governed by strictly linear quantities, the ground and ESA cross sections (σ_g and σ_e). The nonlinear response is better described by a sequential $\mathrm{Im}\{\chi^{(1)}\}{:}\mathrm{Im}\{\chi^{(1)}\}$ process. In particular, for low inputs (where the excited-state population remains much less than the ground-state population) the nonlinear response can be described by [25,26]:

$$\frac{dI}{dz} = -\sigma_g N_g I - \sigma_e N_e I \tag{3}$$

where $N_g + N_e = N$ is the number density of absorbers. The excited-state density is created by linear absorption from the ground state, the generation rate being

$$\frac{dN_e}{dt} = \frac{\sigma_g I}{\hbar\omega} \tag{4}$$

These equations can be combined for $N_g \gg N_e$ to give,

$$\frac{dF}{dz} = -\sigma_g N_g F - \frac{\sigma_e \sigma_g N_g}{2\hbar\omega} F^2 \tag{5}$$

where $\hbar\omega$ is the photon energy. Note the similarity in form to the 2PA equation,

$$\frac{dI}{dz} = -\alpha I - \beta I^2 \tag{6}$$

where $\alpha = \sigma_g N_g$.

However, for high inputs, for the resonant process, saturation of linear absorption (e.g., depletion of the ground state) must be taken into account.[27,28] This considerably complicates the fluence loss equation as discussed in Section V, but the ultimate loss for very high inputs is easily determined. It is simply given by the absorption coefficient at saturation ($N_e = N_g$), $\alpha' = (\sigma_g + \sigma_e)N/2$. In practice, the excited electronic state of organics rapidly relaxes to a thermally equilibrated distribution of the rotational-vibrational continuum. This actually simplifies the analysis (except for femtosecond pulses), and, in principle, the absorbers can all be excited, giving a maximum absorption coefficient of $\sigma_e N$ (again assuming $\sigma_e > \sigma_g$). We briefly discuss these processes in Section V and they are discussed in considerable detail in the chapter by Perry.[1]

For optical limiting applications, knowing whether a quoted σ_2 is due to 2PA or ESA is extremely important. Let's look at an example where, for simplicity, no spatial or temporal integrations are performed. We are given an "α_2" in the literature of $\alpha_2 = 100$ cm/GW and a linear absorption coefficient of 0.1 cm^{-1}. The measurements were performed with 10 ns laser pulses and the authors quote the energy and irradiance so that we can calculate the fluence and the length. If this is a 2PA coefficient we expect a transmittance of $T_{NL} = 0.83$ at $I = 1$ MW/cm^2 where the linear transmittance is $T_L = 0.90$. If on the other hand, the loss is from ESA, with a fluence of 10^{-2} J/cm^2 we find $\sigma_e = 2.0 \times 10^{-17}$ cm^2 assuming $\sigma_g = 1.0 \times 10^{-18}$ cm^2, $N_g = 10^{17}$ cm^{-3}, and $\hbar\omega = 10^{-19}$ J (i.e., for the above values the losses from 2PA or from ESA are assumed the same). However, if the experiment is repeated for 1-μs pulses with the same irradiance, we would see the same loss for 2PA but low transmittance (≈ 0.14) for ESA since the fluence is now 1 J/cm^2. Similarly, if 100-ps pulses are used with the same irradiance, the ESA loss becomes negligible ($F = 10^{-4}$ J/cm^2). Simply stated, 2PA is irradiance dependent and ESA is fluence dependent.

From these arguments we see that if we want to perform optical limiting of picosecond pulses, 2PA may be more useful (assuming normal values of β from 1–100 cm/GW). If we need to limit longer pulses from nanoseconds to microseconds, ESA may be more useful.

An important note on ESA is that in the above example the largest loss possible is α changing from 0.1 cm^{-1} to 20 times larger or 2 cm^{-1} when all the absorbers are promoted to the excited state. This "saturation" of ESA is much less likely for 2PA (the population of the 2PA excited state would have to become comparable to that of the ground state which would require very large β's and high irradiance).

IV. ULTRAFAST NONLINEARITIES IN ORGANICS

In the last decade, organic materials have emerged as potentially viable media for optical switching and limiting. Many researchers have been involved in the investigation of the nonlinear optical properties of various types of organics.[27-32] It is thought that the number of conjugated bonds (alternating double-single bonds between carbon atoms which leads to large linear oscillator strengths) will lead to large nonlinearities. Because the bonding π-electrons are loosely bound to the individual nuclear sites, their orbitals extend over long distances. Additionally, the joint density of states is raised in going from three to one dimension as in polymers.[33] This is similar to what happens in going from bulk semiconductors to quantum wires. In linear polymeric molecules such as polydiacetylenes, bound-electronic n_2's of 5×10^{-3} cm^2/GW have been reported by Lawrence et al.[17] and 2PA coefficients of 700 cm/GW have been reported by Lawrence et al.[18]

While for all-optical-switching applications it is desirable to have large n_2 but small β (see Chapter 12 by Stegeman), for optical limiting it is desirable for both n_2 and β to be large. The 2PA absorbs energy and, thus, lowers the transmitted energy, while the NLR will spread the beam lowering the transmitted fluence. This has been shown to work well in semiconductors.[14,15,34]

Recently there has been considerable progress in the understanding of the relationship between electronic nonlinearities and the structure of polyene chains with terminal electron donor and acceptor groups. Quantum chemical calculations[35] and experimental measurements of first and second hyperpolarizabilities[36,37] for such systems suggest that the hyperpolariabilities are dependent on the bond length alternation in the conjugated chain, which varies from about 0.11 Å for a neutral polyene to -0.11 Å for a fully polarized (zwitterionic) polyene. Cyanine or cyanine-like molecules are those limiting structures. The structure property trends that emerge from these studies indicate that the second hyperpolarizability, γ, which is directly related to the ultrafast electronic n_2, exhibits three extrema as the structure changes from the polyene to zwitterion type. Two of these extrema correspond to maximal positive γ and occur when the bond length alternation is about 0.07 and -0.07 Å. For conjugated chains with zero bond length alternation, γ has a maximal negative value and the magnitude is several times that at positive extrema. This picture thus provides structural guidelines for designing molecules which have optimized positive or negative γ, for their given length. Molecules with very large negative γ may give rise to large negative n_2 materials which would exhibit strong ultrafast self-defocusing.

While the nonlinearities optical properties of organics have been studied for some time, there has not been a great deal of work performed using the ultrafast nonlinearities for optical limiting. However, for optical limiting of short pulses, these ultrafast nonlinearities may offer considerable potential. The potential for short pulse protection lies in the claims of (1) large nonlinearities; (2) mechanical, chemical, thermal, and photostabilities; and (3) high optical damage thresholds. In fact, the questions that remain for their application for optical limiting are just those that they are touted for; namely, large nonlinearities, stability, in particular photostability, and high damage threshold. For example, polymethylmethacrylate (PMMA) has a reported damage threshold to 10-ns pulses of ≈ 2–3 J/cm^2.[38] In addition the linear optical quality of many of these materials [e.g., polydiacetylene (PTS)] in solid form exhibits nonuniformities and inhomogeneities that lead to significant scattering. These problems may be extrinsic but need to be solved before organic materials will be widely utilized. Additionally, for optical limiting where broadband operation is needed, the magnitude of the combined nonlinearities of NLA and NLR need to be large over the entire spectral region of concern. These nonlinearities have not in general been measured over sufficiently large spectral ranges to know their applicability. A noted exception is the nonlinear spectrum of PTS as measured by Lawrence et al.[18]

The magnitude of the largest reported 2PA coefficient in organic materials, $\beta \approx 700$ cm/GW, is in principle large enough to limit the transmittance to levels of 1 μJ output in a 10-ns pulse if used in a "thick" limiter geometry (described in Section VIII). However, this large value is at the peak in the 2PA spectrum and the material, PTS, is one of very few available in millimeter size single crystals. Also it suffers from considerable scattering as well as only linearly transmitting at wavelengths longer than ≈ 0.7 μm. On the other hand, this large nonlinearity in the near IR shows the potential of organic materials if the several problems mentioned can be overcome. In addition, the large nonlinear refractive index of $\approx 5 \times 10^{-3}$ cm^2/GW ($\approx 2 \times 10^{-9}$ esu)[17] is capable of producing a significant phase shift for 1-μJ, 10-ns pulses which helps to spread the energy in a focal plane. In fact this magnitude of n_2, if broadband, is approximately what is needed to sufficiently limit the fluence of nanosecond pulses acting alone.

However, some organic materials displaying large 2PA also show higher order nonlinearities. These may be associated with excited-state nonlinearities from the 2PA generated excited states. Such higher order nonlinearities were found to be useful for optical limiting in semiconductors[14,15] and hold the potential in organics to give effective optical limiting for nanosecond as well as shorter pulses. We give examples here of two organic materials studied at 532 nm. These nonlinearities were studied in organics as early as 1974 by Kleinschmidt et al.[34] and later by several others.[16,17,40,41]

The materials discussed here are a bisbenzethiozole-substituted thiophene compound (BBTDOT)[40] and a didecyloxy-substitute polyphenyl compound (DDOS).[40] Measurements were performed on samples of BBTDOT and DDOS.[10] The linear absorption for the 2-mm path length spectroscopic cells was too low to measure ($< 2\%$ loss) for both solutions. This is consistent with the maxima in the linear absorption occurring at ≈ 400 nm for BBTDOT and ≈ 300 nm for DDOS. Linear spectra for these materials are given in Zhao et al.[40,42] Experiments were performed with 32-ps (FWHM) pulses at 532 nm. These materials have been characterized with degenerate four-wave mixing at 602 nm[40,42] as well as by Z-scan at 532 nm.[10] The nonlinearities observed are analogous to those previously studied in semiconductors.[43,44] As in semiconductors, two-photon absorption (2PA), ultrafast nonlinear refraction (NLR) and excited-state nonlinearities induced by 2PA are observed. The molecular structures of these compounds are shown in Figure 3.

Making measurements of the nonlinear absorption at several input irradiances and defining an effective 2PA coefficient, β_{eff}, by $dI/dz = -\beta_{eff}I^2$, the plot for BBTDOT of Figure 5a gives a straight line of positive slope. Thus, a higher order effect, such as ESA accessed via 2PA, is contributing to the NLA. The behavior of this material is similar to semiconductors irradiated in the 2PA regime, and hence an analysis similar to that of Said et al.[44] is appropriate. In the case of semiconductors, the NLA is caused by 2PA and by the absorption due to free carriers generated via this 2PA. The propagation equation for the irradiance within the sample is given by

$$\frac{dI}{dz} = -\beta I^2 - \sigma_a N_e I, \tag{7}$$

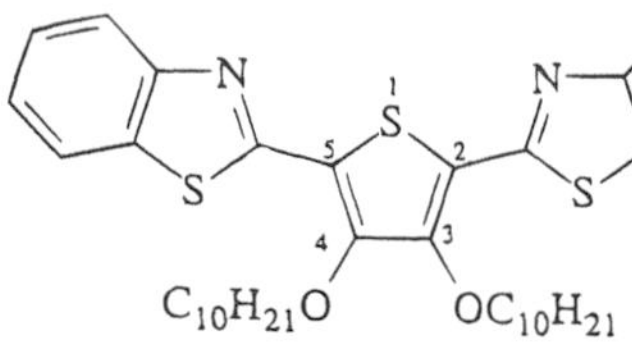

2,5-bis(2-benzothiazoyl)-3,4-didecyloxy-thiophene

(BBTDOT)

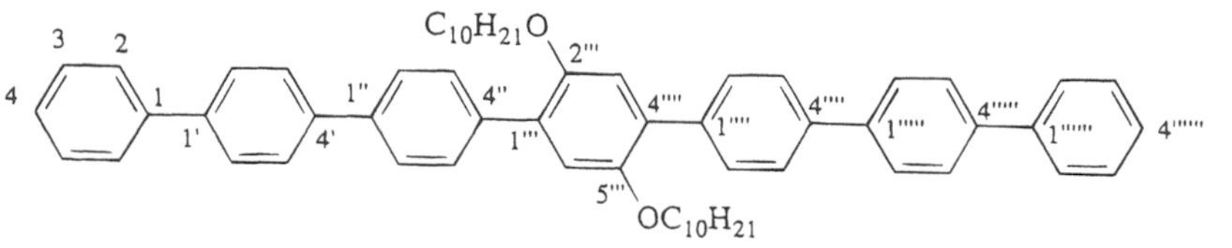

2''',5'''-Didecyloxy-1,1':4',1'':4'',1''':4''',1'''':4'''',1''''':4''''',1''''''-septiphenyl

(DDOS)

Figure 3 The chemical structure of BBTDOT and DDOS.

where β is the actual 2PA coefficient, σ_a is the ESA cross section, and N_e is the density of excited states created by 2PA. The excited-state generation rate is[44,45]

$$\frac{dN_e}{dt} = \frac{\beta I^2}{2\hbar\omega}. \tag{8}$$

where $\hbar\omega$ is the incident photon energy. The data of Figure 4a are described well by Equations 7 and 8, giving a value for β and σ_a. The ESA from the 2PA generated excited states acts as a fifth-order nonlinearity described by a sequential $\mathrm{Im}\{\chi^{(3)}\}{:}\mathrm{Im}\{\chi^{(1)}\}$ (2PA followed by linear absorption).[46]

Upon examining the NLR data of BBTDOT at several input irradiance levels, a trend similar to that for the NLA data is seen. The sample exhibits both positive third-order and positive fifth-order contributions to the NLR. The third-order contribution is due to the bound-electronic NLR of coefficient n_2, and the fifth-order effect can be described by an index change caused by the production of excited states via 2PA. Hence, the change in the refractive index is

$$\Delta n = (n_2 I + \sigma_r\, N/k), \tag{9}$$

where $k = 2\pi/\lambda$ and σ_r is the excited-state refractive index cross section. Using the same procedure as in the nonlinear absorption case, we plot Δn divided by the input irradiance as a function of I_0 in Figure 4b.[16,43] Also shown is the total index change before subtraction of the NLR from the cell and solvent.

The sample DDOS also shows increasing NLA with increasing I_0 as seen in BBTDOT. From the closed aperture Z-scan data for DDOS we determine a negative intercept for a plot of $\Delta n/I_0$ vs. I_0 leading to a negative n_2 and γ_R. The slope of this plot is negative, indicating that σ_r is also negative for DDOS.

The combination of the third- and fifth-order nonlinearities has the potential to provide effective limiting to considerably longer pulses than just the ultrafast third-order effects can provide. This comes about since the fifth-order effect can accumulate with time as the number of excited states created by 2PA increases.

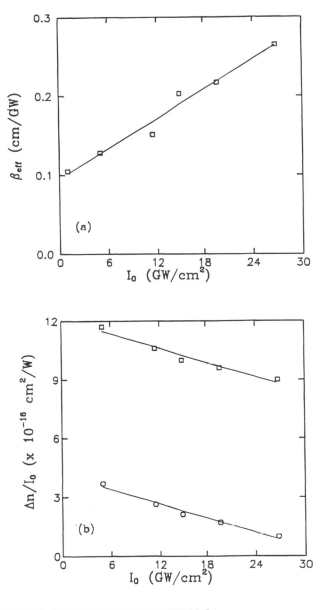

Figure 4 (a) $\Delta\alpha/I_0 = \beta_{\text{eff}}$ vs. I_0 for the BBTDOT sample with least-squares fit (solid line). (b) $\Delta n/I_0$ vs. I_0 with solvent (squares) and with solvent n_2 subtracted (circles) along with fits (solid lines).

V. REVERSE SATURABLE ABSORBING MATERIALS

Reverse saturable absorbing (RSA) molecules are one of the most promising candidates for use in passive optical limiters.[25–28,38,47] The excited-state absorbtion (ESA) cross section in such materials is larger than that of the ground state, hence the absorption increases with increasing input fluence, F (energy per unit area); thus the name reverse saturable absorber.[48]

The photophysics of these molecules is discussed in detail in Chapter 13. Here we use an oversimplified quasi-three-level model, shown in Figure 5a, to describe the basic physical phenomena. Figure 5b also shows a more realistic five-level model with both singlet and triplet manifolds. We use a ground-state absorption cross section, σ_g, and an effective excited-state absorption cross section, σ_e with no decay to the ground state but extremely fast decay ($r_{32} \simeq 0$) from the uppermost level back to the first excited state. This allows a single excited molecule to make many transitions within a pulse. If a single pulse width is to be modeled, this three-level system can give good agreement with data.[25–28,30,49,50] If the pulse width is subsequently changed, the effective three-level ESA cross section can be changed to fit, but in general there will still be a fluence dependence of this effective cross section that is not included in the three-level model.

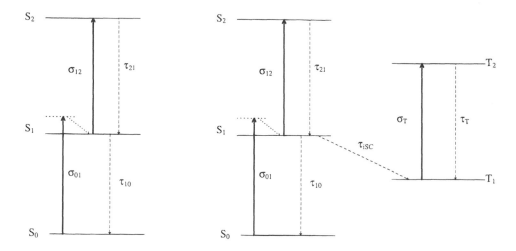

Figure 5 (a) Simplified three-level structure for RSA molecules along with the more realistic five-level model shown in (b). For the three-level system calculation the lifetimes are taken to be $\tau_{21} = 0$ and $\tau_{10} = \infty$.

As discussed in Section III, for low inputs, the loss from ESA can be described by Equation 5, which is analogous to 2PA except that the loss is to the fluence rather than irradiance. This immediately shows the interest in ESA for optical limiting of long pulses. The nonlinearity grows with fluence and builds up in time. Thus later portions of a pulse are more effectively limited than early portions. Here we give a somewhat more realistic three-level model including ground-state depletion following Miles.[28]

For longer pulses in many systems the first excited state can undergo intersystem crossing to the triplet manifold where the lifetime is long (see Fig. 5b), thus facilitating ground-state depletion. Fortunately the ESA within the triplet manifold can also be larger than the ground-state absorption and in several phthalocyanines has been found to be even larger than the ESA within the singlet manifold. Thus efforts have been undertaken to increase the intersystem crossing rate by heavy atom substitution in the phthalocyanine ring (See Chapter 13).

In this three-level model, the depth, z, dependence of the irradiance, I, is given by Equation 3, while the rate of change of ground-state population density, n_g, is

$$\frac{dn_g}{dt} = -\sigma_g n_g \frac{I}{\hbar\omega} \tag{10}$$

We now require overall conservation of population density, N_0, assuming the uppermost level relaxes instantaneously to the first excited state, i.e., no saturation of the ESA is allowed, and the first excited state does not decay within the time of the input pulse width. This requires

$$N_0 = n_g + n_e \tag{11}$$

where n_e is the excited-state population density. Defining $F_t = \int_{\infty}^{t} I(t')dt'$, we have

$$n_g = N_0 \exp(-\sigma_g F_t/\hbar\omega) = N_0 e^{-F_t/F_S} \tag{12}$$

$$n_e = N_0(1 - e^{-F_t/F_S}) \tag{13}$$

where F_s is defined by Equation 12. Plugging Equations 12 and 13 into Equation 3 yields

$$\frac{\partial I}{\partial z} = -N_0 I \{\sigma_g e^{-F_t/F_S} + \sigma_e (1 - e^{-F_t/F_S})\} \tag{14}$$

The term in curly brackets, { }, is the effective absorption cross section as defined by Miles.[28] However, integrating Equation 14 over time from $-\infty$ to ∞ to give the fluence change with z yields

$$\frac{\partial F}{\partial z} = -N_0 \left\{ \frac{F_S}{F} (\sigma_e - \sigma_g)[e^{-F/F_S} - 1] + \sigma_e \right\} F = -N_0 \sigma_{eff} F \tag{15}$$

where the last equality defines the effective absorption cross section, σ_{eff}. This equation must in general be solved numerically. However, Miles[28] introduced a way to obtain the maximum possible dynamic range for limiting devices based on RSA by keeping the peak fluence constant within a thick nonlinear material at the highest input allowed by the damage threshold of the material. This optimization is obtained by allowing the density of molecules to vary along the propagation direction which requires a solid host material. This allows an analytic solution for the density distribution as well as for the limiting energy.[30,49,50] Allowing the density to vary was originally suggested by McCahon and Tutt.[51]

The effective absorption cross section, σ_{eff}, defined in Equation 15, grows as a function of the input fluence as shown in Figure 6. Here we take $\sigma_g = 1$, $\sigma_e = 20$ and $F_S = 1$. For low fluence $\sigma_{eff} = 1$, the ground-state cross section, while at very high input fluence $\sigma_{eff} = 20$, the excited-state cross section. Clearly to obtain the most effective limiting (largest overall loss) F must be very much larger than F_S ($F_S = 1$ in Fig. 6). The physical meaning of F_S is that the ground state becomes depleted (i.e., saturated) for $F >> F_S$. Thus the loss tends to $\partial F/\partial z = -N_0 \sigma_e F$, the maximum possible loss when all the molecules are in the excited state. Therefore, as pointed out by Miles,[28] it is desirable to work in the saturated regime which in turn requires F_S to be small. This also implies that the ground-state absorption cross section should be large, i.e., $F_S = \hbar\omega/\sigma_g$, but still much smaller than that of the excited state. Limitations of the analytic model for optimized RSA limiters can be determined by numerical solutions to the five-level model coupled to the wave equation, as discussed by Hagan et al.[49] and Xia et al.[50]

It is expected that thermal nonlinearities for from a few nanoseconds to longer pulses will play an important role in these and other limiters where absorption is used. For example, for liquid-based

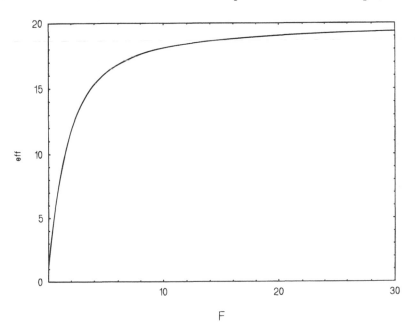

Figure 6 The effective absorption cross section σ_{eff}, defined in Equation 15, as a function of the input fluence to a three-level RSA material taking $\sigma_g = 1$, $\sigma_e = 20$, and $F_S = 1$.

limiters, the index change with increasing temperature results in self-defocusing which spreads the beam. If the thickness of the nonlinear material is long enough for the phase distortion induced by the thermal index change to propagate to change the irradiance distribution within the sample, the fluence limiting process of RSA will be reduced. On the other hand, the fluence may be reduced directly by the self-defocusing at an image plane. Clearly careful design of the limiter will be needed to utilize these effects optimally. This will require computer codes that take into account the full nonlinear propagation as determined by the wave equation as well as accounting for the nonlocal behavior of the thermal as well as excited-state nonlinearities which can diffuse and/or decay in time.[50]

We should also point out that the use of liquid-based vs. solid-state materials for optical limiting leads to considerably different designs. Liquid-based limiters can self-heal after breakdown allowing the absorption and scattering by the breakdown plasma or bubble formation to be utilized in the design of the limiting device. Damage to solid materials is irreversible, and therefore undesirable, so that fluences within the device must be kept below this damage threshold. Thick limiters[14,15,28,51] where the nonlinear material is much thicker than the depth of focus of the input beam, and "tandem" optical limiters[27,28,38,52] with multiple nonlinear elements may allow "self-protection" of the limiter element or elements.

VI. LIQUID CRYSTALLINE MATERIALS

Liquid crystals are best known for their use in active devices where an electric field aligns the molecules to generate standard liquid crystal displays. Potentially, active limiting devices could be devised using this technology, but the response time would be relatively slow (perhaps as fast as microseconds).

For passive optical limiiting several researchers have shown that liquid crystals can give effective limiting.[53–57] However, many liquid crystalline phases may not be suitable for limiting in anything but thin film form due to significant linear scattering. As these material offer the potential for active as well as passive devices they may, in principle, be used for limiting from picoseconds to cw.

Nematic liquid crystals have received the most investigation for passive optical limiting and show the largest nonlinear birefringence.[58–60] Khoo et al.[61] have investigated such liquid crystal response using pulses from picosecond to millisecond duration. They observed a fast component for the NLR which builds up in about 100 ns and a slower component which rises to a maximum in about 10 ms. These effects are attributed to temperature and density changes as described in Khoo and Normandin[62,63]. In addition a very fast component of the NLR was observed due to the bound-electronic response as discussed above in Section IV. As with many other organic materials the large anisotropic molecules have conjugated p-orbitals along their length to support charge separation.[25] They can also be reoriented by the applied optical field. This can change the index leading to self-lensing. Liquid crystals can also be used for switching at an interface using total internal reflection geometries.[64]

Soileau et al.[65] investigated the NLO response of liquid crystals in their isotropic phase and found that they can have significant 2PA and used both the 2PA and bound-electronic n_2 effects for optical limiting of 30-ps, 532-nm light at energies as low as 0.15 μJ.

Also reports by Yuan et al.[66] show nonlinearities in the liquid crystal CB-15 that change considerably as the input pulse width is varied from cw to picoseconds. One possible interpretation consistent with the current data is that the nonlinear response is a combination of 2PA and excited-state nonlinearities as observed in BBTDOT and DDOS discussed in Section IV.

VII. PURE CARBON COMPOUNDS: NONLINEAR SCATTERING

A. CARBON BLACK SUSPENSIONS

Carbon black suspensions (CBS) as used for optical limiting are dilute colloidal suspensions of carbon black in a liquid using a surfactant, e.g., ethanol or water plus surfactant. This is one of the most extensively studied materials for optical limiting and one of the "best" found to date.[67–70] The material can be most easily prepared by simply diluting commercial india ink to make it more transparent. The mixture should then be filtered with a ≈0.02-μm filter which normally leaves particles of ≈35 nm in size with possible agglomerates up to 0.2 μm. Solutions with greater than 90 linear transmittance can provide substantial limiting. Other procedures and characteristics are described in Mansour et al.[68]

The mechanism for limiting is a combination of processes including avalanche ionization and bubble formation, both of which lead to enhanced optical scattering of the input beam. The observed nonlinear scattering results in limiting the transmitted beam fluence. The mechanism using water as the dispersant goes as follows.[68] The tiny suspended carbon particles are rapidly heated by strong linear absorption giving rise to thermionic emission, which in the presence of the strong electric field leads to avalanche ionization. The resulting microplasmas then rapidly expand into the surrounding liquid and strongly scatter the incident light for the duration of the plasma ($\approx 10^2$ ns). Subsequently at incident energies above the threshold for obtaining a clamped output, the heating leads to bubble formation and further scattering lasting for microseconds. These conclusions imply that the limiting is broadband and fluence rather than irradiance dependent, both of which agree with observations. In addition, since the carbon particles are ionized during the process (and vaporized), the suspension must be replenished after each laser exposure, again as observed. The mechanism in other solvents having a smaller latent heat of fusion may be somewhat different. Fein et al.[70] report bubble formation prior to the avalanche ionization when solvents such as toluene or ethyl-ether are used. In either case, however, the material is vaporized after irradiation and must be replenished. This is considered a drawback for applications of this material.

Another idea for utilizing bubble formation in the suspensions is to use the index change at an interface between glass and CBS.[71] For the interface tilted at an angle such that the vapor glass interface results in total internal reflection, the transmittance switches to zero.

Experiments performed to confirm the NLO mechanisms include optical limiting as a function of pulse width (from 30 ps to 30 ns), focal spot size, limiting experiments of carbon deposited on glass substrates, simultaneous monitoring of transmittance, absorption and scattering (nonlinear scattering dominates for inputs near threshold), several nonlinear refraction experiments (showing that it is not a significant contributor to the limiting), time-resolved transmittance measurements using both a single beam and a pulse-probe technique, time-resolved emission spectra measurements where ionized carbon lines were observed, and angular distribution measurements of the side scattered light for different fluences which show rapid growth of scattering centers with increasing fluence.[68] These experiments confirmed the fluence dependence and physical description of the process.

The power needed to initiate limiting in CBS with ≈ 10 ns pulses of $\approx 10^2$ W peak power (energy of ≈ 1 μJ) is compared to ≈ 8 kW for CS_2 in the visible. The onset of limiting is nearly independent of the concentration of carbon black particles. However, samples with a higher concentration of particles, and in turn lower transmittance for low input light levels, block the output light more effectively at higher incident fluences than samples with a low concentration of carbon particles.

The process involved in CBS is what is commonly referred to as laser-induced damage, and CBS has a very low damage threshold. This understanding of the limiting mechanisms also tells us that we will not be able to lower significantly (i.e., 10×) the limiting threshold or increase the bandwidth. The reasons are that the carbon is "black" and therefore broadband and highly absorbing. To lower the threshold we must either increase the absorption (we cannot significantly do this) or lower the ionization threshold. Flooding the suspension with ionizing radiation may lower this ionization threshold. At the moment a carbon black suspension is one of the most effective optical limiting materials available for nanosecond laser pulses.

It has been argued that limiters based on CBS will be very inefficient for wide field of view systems since the Mie scattering for the large induced scatterers is concentrated in the forward direction and the detector system will collect this light. However, experiments have shown good optical limiting even with relatively small f-number systems.[68] Damage to the detector element is fluence dependent and even though the light is scattered in the forward direction and still may be detected, it may be spread over a larger area (similar to the operation of refractive limiters). This is because CBS is usually used in a "thick" sample geometry (see Section VIII) and when limiting initiates, the plasmas or bubbles are formed near the focal point; however, well above threshold these scattering centers are produced prior to the focal plane so that the forward scattering is not imaged onto the detector.

B. FULLERENES, C_{60}

Other forms of carbon have also been studied for their optical limiting properties including the Buckminsterfullerenes, e.g., C_{60} and C_{70}.[72–79] The limiting in these materials is due to the larger excited-state absorption than ground-state absorption, i.e., reverse saturable absorption as discussed in Section V. A comparison of C_{60} solutions and C_{60} in PMMA was performed by Kost et al.[77] and it was found that in solutions, nonlinear scattering was also a significant factor in limiting the transmittance. Measurements by Brandelik et al.[72] using Z-scan also indicated that nonlinear refraction was present in toluene solutions probably due to thermal lensing in the tight focusing geometry. While the RSA process in the visible is not as effective as several of the organometalics discussed in Section V, this material is still being studied for its limiting properties in the near IR (around 760 nm) where the peak of the excited-state absorption occurs. Unfortunately, this is a region of very small linear absorption making it difficult to saturate the transition and thereby fully utilize the RSA.

VIII. THICK SAMPLE LIMITING

The experiments performed to characterize CBS showed the fluence dependence of its NLO response. However, limiting experiments do not always follow such a fluence dependence for a variety of reasons. Characterization experiments are usually performed on samples whose thickness is both less than the depth of focus (or Rayleigh range) and less than the distance required for nonlinearly induced phase distortions to propagate to produce irradiance variations. Under such circumstances the sample is considered "thin" and the beam's spatial profile can be easily determined throughout the thickness of the sample. In particular the NLO wave equation can be separated into an equation describing the magnitude of the irradiance (field envelope) and an equation describing the phase as given in Section IV (see Equations 7 and 9). Samples that do not satisfy the thin sample criteria are considered "thick" and the wave equation must usually be numerically integrated to find the output irradiance and phase. However, if the nonlinearity is dominated by losses, either absorbing or scattering, we can determine the expected dependence of output on increasing input.

We find that the fundamental fluence dependence of limiting in CBS or RSA dyes manifests itself quite differently in the thick sample geometry. This is analogous to the irradiance-dependent two-photon absorption (2PA) appearing as a power dependence in tight focusing geometries as described below.[68]

In the case of 2PA the irradiance change dI with depth in the sample dz is

$$\frac{dI}{I} = -\beta I dz \qquad (16)$$

where β is the 2PA coefficient, giving the change in transmittance as approximately

$$\Delta T = \frac{\Delta I}{I} = -\beta I L_{eff} \qquad (17)$$

where L_{eff} is the effective interaction length. For a "thin" cell, $L_{eff} = L$, the sample thickness (assuming negligible linear loss). For a thick cell, the depth of focus (Z_0) of the lens determines the approximate range over which the nonlinear interaction occurs and is given by

$$Z_0 = \frac{\pi \omega_0^2}{\lambda} \qquad (18)$$

where ω_0 is the beam radius at the focus of the lens ($HW1/e^2M$). The area of the beam is given by

$$A = \frac{\pi \omega_0^2}{2} \qquad (19)$$

From Equation 16 the transmission change is proportional to the product of the irradiance times the effective interaction length, $L_{eff} \simeq Z_0$ giving

$$\Delta T \simeq -\beta I Z_0 = -\frac{\text{Power}}{A} \beta Z_0 = -\text{Power} \frac{2\beta}{\pi \omega_0^2} \frac{\pi \omega_0^2}{\lambda} = -\text{Power} \frac{2\beta}{\lambda} \qquad (20)$$

Now the nonlinear change in transmittance appears power dependent and is independent of the beam size at focus. That is, ΔT is independent of the focal length of the input lens of f/number. For refractive nonlinearities a similar argument shows that the overall phase shift obtained from $d\phi/dz = kn_2I$ is $\Delta\phi \propto kn_2$Power/λ.

For fluence-dependent nonlinearities the argument follows the same lines and results in an energy-dependent nonlinearity independent of focusing for a thick sample. The above analysis is approximate, and combinations of nonlinear mechanisms can be responsible for limiting in different NLO materials leading to more complicated dependencies. Also at very high inputs the approximation breaks down.

The above analysis gives an idea of how thick sample geometries can give results considerably different from the usual thin sample limits that are normally discussed. As previously stated there are often multiple nonlinear mechanisms present in a given material and predicting the outcome of an experiment, or interpreting experimental results in a thick sample geometry can become difficult. With the rapid increase in computer speed, numerical calculations will become more helpful in the future. An approximate method for characterizing simple nonlinearities in thick media is given in Sheik-Bahae et al.[80]

IX. TANDEM DEVICES

In addition to a low threshold and high linear transmittance, a practical limiting device must itself maintain a high resistance to irreversible laser damage. This increases the problems in making a practical device, because tightly focusing onto the nonlinear material to reduce the limiting threshold may cause a corresponding decrease in the damage threshold. Nonlinear materials in liquid form have the ability to self-heal, so the damage threshold is set by the windows of the cell containing the liquid. For collimated beams, even liquid limiter cells suffer from permanent optical damage at input fluences about an order of magnitude larger than the limiting threshold. Practical devices require a damage threshold several orders of magnitude larger.

While "thick" sample limiters, discussed in Section VIII, have been shown to self-protect,[14,15] they can suffer from reduced linear transmittance due to their considerable thickness. A similar, but more versatile implementation of self-protecting devices is to place two or more limiting elements in tandem in the optical path.[27,28,49,50,52] The limit of the tandem device for a large number of elements gives the distributed limiter.[28,49,56] A geometry with just two elements is shown schematically in Figure 7, where element 1, the "primary limiter," is placed at or near focus, while element 2, the "protector," is placed in front of the primary at a region of lower irradiance.[27] The basic concept is that the primary limiter provides a low limiting threshold while the protector prevents optical damage to the primary limiter. The damage energy threshold of the system is thus determined by the damage threshold of the protector which occurs at a much higher input energy than the primary by itself. This is because it is positioned far from the focus, where the fluence is lower. The dynamic range of a tandem device of this type, based solely on reverse saturable absorbing materials, can be approximately given by the product of the dynamic ranges of the individual elements.[27] As pointed out by Miles,[28] allowing the primary limiter to be positioned away from the waist allows the multiple element tandem device to be designed for a given maximum input energy. A tandem device is more versatile than the thick limiter, as it gives more freedom to optimize the geometry and allows the use of different materials to play the role of limiting and protecting elements. However, such versatility creates a vast parameter space that must be explored in order to optimize devices of this type. In addition to the choice of materials combinations, one must

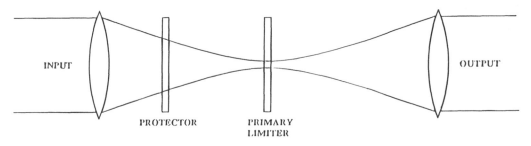

Figure 7 A schematic of a two-element tandem limiting device with the primary limiter located at focus.

also consider geometric variables such as the number of elements, their relative positions, their thicknesses and, in the case of solutions or suspensions, their concentrations. Nevertheless, using a few simple rules for RSA dyes, researchers have done preliminary optimization which has resulted in the current state-of-the-art in nanosecond limiters.[27,38,52] For other nonlinear processes where nonlinear refraction is present the design of tandem devices becomes quite complicated as propagation between elements can drastically alter the focusing conditions.[49,50]

X. CONCLUSIONS

Organic materials and carbon compounds have considerable promise as the nonlinear element or elements in practical optical limiting devices where an intermediate focal plane is present. Limiting of pulses from femtoseconds to cw is possible, but different nonlinear materials and processes may be needed for different pulse width regimes. For a recent review of nonlinear optical responses of organic materials the reader is referred to Bredas et al.[81] The reader is referred to Tutt and Boggess[82] and Hermann[83] for recent review articles on limiting mechanisms and materials. Carbon black suspensions are currently the best studied materials for nanosecond inputs and under certain criteria give the "best" limiting performance, e.g., broadest wavelength range, neutral density in appearance (no color distortion). However, the limited output is not low enough for some applications, and the fact that the suspension must be replenished is considered to be a drawback. For the important application of protecting sensors from damage by pulses of $\simeq 1$ ns to 1 μs duration, reverse saturable absorbing organic dyes appear to have the greatest potential, although recent reports of limiting in some inorganic clusters appear interesting.[84,85] In RSA materials there is a very severe trade-off between the performance of the limiter and the allowable linear transmittance. However, the compromise between these should allow useful devices to operate with tolerable linear transmittance (i.e., $\simeq 70$). The remaining questions for these materials, as well as other types of materials, are the spectral width over which they will protect sensors, the spectral window over which they can have high linear transmittance, and their environmental stability. It may be possible to expand the spectral range of optical limiting by using a combination of different RSA dyes. Also for these and other materials the role of thermal nonlinear refraction is not fully understood. For example, this nonlinearity may reduce the effectiveness of RSA or, perhaps with proper design, help the overall limiting of the transmitted fluence (i.e., spread the beam in the focal plane). This is still a very active area of research both in terms of materials development as well as optimization of use and spectroscopic characterization.

Organic materials, in various forms, are the least well understood nonlinear optical materials. On the other hand, they also offer great potential since organic chemistry has a long history of being able to engineer a material for a specific application. As the materials are not yet well characterized or understood, this research is still in its infancy. A close working relation between materials synthesis, characterization, and modeling will be essential for advances in this field.

Work at utilizing multiple nonlinear responses in combination such as 2PA plus ESA plus thermal lensing may be fruitful. This may allow covering a range of pulse widths. The combination of nonlinearities in thick media or tandem limiters is difficult to model and can require supercomputer codes. The output of these codes along with further experimentation is needed to determine the ultimate usefullness of organics for sensor protection applications.

A nonlinear mechanism used for optical limiting not discussed in this chapter where organics may play a role in the future is photorefraction. Development of organic photorefractive materials is currently a very active research area.[86,87] The potential for optical limiting is to use these new materials in, for example, beam fanning limiters[88] or photorefractive excisors.[89–91]

A final point regarding limiters based on nonlinear refraction or using nonlinear refraction for part of the limiting effect is that the apertures used in experiments performed to study these materials are not a neccesary part of an optical limiting device. The aperture is an experimental means of measuring the transmitted fluence rather than energy. As the damage threshold of typical sensor devices is fluence dependent, the reduction in fluence due to beam spreading is helpful in protecting the device and is present at the detector with or without the aperture present. This is consistently a point of confusion within the community.

ACKNOWLEDGMENTS

We gratefully acknowledge the support of the Naval Air Warfare Center Joint Service Agile Program Contract N66269-C-93-0256. The authors also thank Joe Perry of Jet Propulsion Laboratories for several useful discussions.

REFERENCES

1. **Leite, R. C. C., Porto, S. P. S., and Damen, P. C.,** The thermal lens effect as a power-limiting device. *Appl. Phys. Lett.*, 10, 100, 1967.
2. **Soileau, M. J., Williams, W. E., and Van Stryland, E. W.,** Optical power limiter with picosecond response time, *IEEE J. Quantum Electron.*, QE-19, 731, 1983.
3. **Ippen, E. P. and Shank, C. V.,** Picosecond response of a high-repetition rate CS_2 optical Kerr gate, *Appl. Phys. Lett.*, 26, 92, 1975.
4. **Narashimha Rao, D. V. G. L. and Jayaraman, S.,** Self-focusing of laser light in the isotropic phase of a nematic liquid crystal, *Appl. Phys. Lett.*, 23, 539, 1973.
5. **Miller, D. A. B., Seaton, C. T., Prise, M. E., and Smith, S. D.,** Band-gap resonant nonlinear refraction in III-V semiconductors, *Phys. Rev. Lett.*, 47, 197, 1981.
6. **Geusic, J. E., Singh, S., Tipping, D. W., and Rich, T. C.,** Three-photon stepwise optical limiting in silicon, *Phys. Rev. Lett.*, 19, 1126, 1969.
7. **Rychnovsky, S. J., Allen, G. R., Venzke, C. H., Smirl, A. L., and Boggess, T. F.,** Optical nonlinearities and optical limiting in GaP at 532 nm, *Proc. SPIE*, 1692, 191, 1992.
8. **Boggess, T. F., Moss, S. C., Boyd, I. W., and Smirl, A. L.,** Nonlinear-optical energy regulation by nonlinear refraction and absorption in silicon, *Opt. Lett.*, 9, 291, 1984.
9. **Ralston, J. M. and Chang, K. R.,** Optical limiting in semiconductors, *Appl. Phys. Lett.*, 15, 164, 1969.
10. **Boggess, T. F., Smirl, A. L., Moss, S. C., Boyd, I. W., and Van Stryland, E. W.,** Optical limiting by two-photon absorption and nonlinear refraction in GaAs, *IEEE J. Quantum Electron.*, QE-21, 488, 1985.
11. **Hill, J. R., Parry, G., and Miller, A.,** Nonlinear refractive index changes in CdHgTe at 175 K with 10.6 μm radiation, *Opt. Commun.*, 43, 151, 1982.
12. **Sheik-Bahae, M., Said, A. A., and Van Stryland, E. W.,** High sensitivity, single beam n_2 measurements, *Opt. Lett.*, 14, 1989.
13. **Sheik-Bahae, M., Said, A. A., Wei, T. H., Hagan, D. J., and Van Stryland, E. W.,** Sensitive measurement of optical nonlinearities using a single beam, *IEEE J. Quantum Electron.*, 26, 760, 1990.
14. **Hagan, D. J., Van Stryland, E. W., Soileau, M. J., and Wu, Y. Y.,** Self-protecting semiconductor optical limiters, *Opt. Lett.*, 13, 315, 1988.
15. **Van Stryland, E. W., Wu, Y. Y., Hagan, D. J., Soileau, M. J., and Mansour, K.,** Optical limiting with semiconductors, *J. Opt. Soc. Am. B*, 5, 1981, 1988.
16. **Said, A. A., Wamsley, C., Hagan, D. J., Van Stryland, E. W., Reinherdt, B. A., Roderer, P., and Dillard, A. G.,** Third and fifth order optical nonlinearities in organic materials, *Chem. Phys. Lett.*, 228, 646, 1994.
17. **Lawrence, B. L., Cha, M., Torruellas, W. E., Stegeman, G., Etemad, S., and Baker, G.,** Z-scan measurements of third and fifth order nonlineariites in single crystal PTS at 1064 nm, *Nonlinear Opt.*, 10, 193, 1995.
18. **Lawrence, B., Torruellas, W. E., Cha, M., Sundheimer, M. L., Stegeman, G. I., Meth, J., Etemad, S., and Baker, G.,** Identification and role of two-photon excited states in a π conjugated polymer, *Phys. Rev. Lett.*, 73, 597, 1994.
19. **Sheik-Bahae, M., Hagan, D. J., and Van Stryland, E. W.,** Dispersion and band-gap scaling of the electronic kerr effect in solids associated with two-photon absorption, *Phys. Rev. Lett.*, 65, 96, 1989.
20. **Sheik-Bahae, M., Hutchings, D. C., Hagan, D. J., and Van Stryland, E. W.,** Dispersion of bound electronic nonlinear refraction in solids, *IEEE J. Quantum Electron.*, QE-27, 1296, 1991.
21. **Hutchings, D. C., Sheik-Bahae, M., Hagan, D. J., and Van Stryland, E. W.,** Kramers-Kronig relations in nonlinear optics, *Opt. Quantum Electron.*, 24, 1, 1992.
22. **Bassani, F. and Scandolo, S.,** Dispersion relations in nonlinear optics, *Phys. Rev.* B44, 8446, 1991.
23. **Chase, L. and Van Stryland, E.,** Nonlinear Refractive Index: inorganic materials, in *Handbook of Laser Science and Technology: suppl., 2. Optical Materials,* sect. 8, Weber, M., Ed., CRC Press, Boca Raton, FL, 1994, 29. Also see Van Stryland, E. and Chase, L., Two photon absorption: inorganic materials, in *Handbook of Laser Science and Technology: suppl. 2. Optical Materials,* sect. 8, Weber, M., Ed., CRC Press, Boca Raton, FL, 1994, 299.
24. **Garito, A. F. and Kuzyk, M. G.,** Nonlinear Refractive Index: organic materials, in *Handbook of Laser Science and Technology; suppl. 2. Optical Materials,* sect. 8, Weber, M., Ed., CRC Press, Boca Raton, FL, 1994, 289. Also see Garito, A. F. and Kuzyk, M. G., Two photon absorption: organic materials, in *Handbook of Laser Science and Technology; suppl. 2. Optical Materials,* sect. 8, Weber, M., Ed., CRC Press, Boca Raton, FL, 1994, 329.

25. **Coulter, D. R., Miskowski, V. M., Perry, J. W., Wei, T. H., Van Stryland, E. W., and Hagan, D. J.,** Optical limiting in solutions of metallo-phthalocyanines and naphthalocyanines, *Proc. SPIE,* 1105, 42, 1989. See also Perry, J. W., Khundkar, L. R., Coulter, D. L., Alvarez, Jr., D., Marder, S. R., Wei, T. H., Sence, M. J., Van Stryland, E. W., and Hagan, D. J., *Organic Molecules for Nonlinear Optics and Photonics,* Messier, J., Kajzar, F., and Prasad, P., Eds., NATO ASI Series E, Vol. 194, Kluwer Academic Publishers, Dordrecht, 1991, 369.

26. **Wei, T. H., Hagan, D. J., Sence, M. J., Van Stryland, E. W., Perry, J. W., and Coulter, D. R.,** Direct measurements of nonlinear absorption and refraction in solutions of phthalocyanines, *Appl. Phys.,* B54, 46, 1992.

27. **Hagan, D. J., Xia, T., Said, A. A., Wei, T. H., and Van Stryland, E. W.,** High dynamic range passive optical limiters, *Int. J. Nonlinear Opt. Phys.,* 2, 483, 1993.

28. **Miles, P. A.,** Bottleneck optical limiters: the optimal use of excited-state absorbers, *Appl. Opt.,* 33, 6965, 1994.

29. **Khoo, I. C. and Wu, S. T.,** *Optics and Nonlinear Optics of Liquid Crystals,* World Scientific, Singapore, 1993.

30. **Prasad, P. N. and Williams, D. J.,** *Introduction to Nonlinear Optical Effects in Molecules and Polymers,* Wiley, New York, 1991.

31. **Zyss, J.,** *J. Mol. Electron.* 1, 15, 1985.

32. **Prasad, P. N. and Ulrich, D. R., Eds.,** *Nonlinear Optical and Electroactive Polymers,* Plenum, New York, 1989.

33. **Flytzanis, C.,** Dimensionality effects and scaling laws in nonlinear optical susceptibilities, in *Nonlinear Optical Properties of Organic Molecules and Crystals,* vol. 2, chap. III-4, Chemla, D. S. and Zyss, J., Eds., Academic Press, Orlando, FL, 1987.

34. **Van Stryland, E. W., Vanherzeele, H., Woodall, M. A., Soileau, M. J., Smirl, A. L., Guha, S., and Boggess, T. F.,** Two-photon absorption, nonlinear refraction and optical limiting in semiconductors, *Opt. Eng.,* 24, 613, 1985.

35. **Meyers, F., Marder, S. R., Pierce, B. M., and Bredas, J. L.,** Electric field modulated nonlinear optical properties of donor acceptor polyenes: sum over excited states investigation of the relationship between molecular polarizabilities (α, β and γ) and bond length alternation, *J. Am. Chem. Soc.,* 116, 10703, 1994.

36. **Bourhill, G., Bredas, J.-L., Cheng, L. T., Marder, S. R., Meyers, F., Perry, J. W., and Tiemann, B. G.,** Experimental demonstration of the dependence of the first hyperpolarizability od donor-acceptor substituted polyenes on the ground state polarization and bond length alternation, *J. Am. Chem. Soc.,* 116, 2619, 1994.

37. **Marder, S. R., Perry, J. W., Bourhill, G., Gorman, C., Tiemann, B., and Mansour, K.,** Relation between bond-length alternation and second electronic hyperpolarizability of conjugated organic molecules, *Science,* 261, 186, 1993. See also Marder, S. R., Gorman, C. B., Meyers, F., Perry, J. W., Bourhill, G., Bredas, J.-L., and Pierce, B. M., A unified description of linear and nonlinear polarization in organic polymethine dyes, *Science,* 265, 632, 1994.

38. **Mansour, K., Fuqua, P., Marder, S. R., Dunn, B., and Perry, J. W.,** Solid state optical limiting materials based on phthalocyanine containing polymers and qqorganically modified sol-gels, *Proc. SPIE,* 2143, 239, 1994.

39. **Kleinschmidt, J., Rentsch, S., Tottleben, W., and Wilhelmi, B.,** Measurement of strong nonlinear absorption in stilbene-chloroform solutions, explained by the superposition of two-photon absorption and one-photon absorption from the excited state, *Chem. Phys. Lett.,* 24, 133, 1974.

40. **Zhao, M., Cui, Y., Samoc, M., Prasad, P. N., Unroe, M. R., and Reinhardt, B. A.,** Influence of two-photon absorption on third-order, nonlinear optical processes as studied by degenerate four-wave mixing: the study of soluble didecyloxy substituted polyphenyls, *J. Chem, Phys.,* 95, 3991, 1991.

41. **Lindle, J. R., Weisbecker, C., Bartoli, F., Pong, R., and Kafafi, Z.,** Third and fifth order optical properties of transition metal complexes of benzenedithiol at 1.064 μm, *Mat. Res. Soc. Symp. Proc.,* 247, 277, 1992.

42. **Zhao, M., Samoc, M., Prasad, P. N., Reinhardt, B. A., Unroe, M. R., Prazak, M., Evers, R. C., Kane, J. J., Jariwala, C., and Sinsky, M.,** Studies of third-order optical nonlinearities of model compounds containing benzothiazole, benzimidazole, and benzoxazole units, *Chem. Mater.,* 2, 670, 1990.

43. **Canto-Said, E. J., Hagan, D. J., Young, J., and Van Stryland, E. W.,** Degenerate four-wave mixing measurements of high order nonlinearities in semiconductors, *IEEE J. Quantum Electron.,* QE-27, 2274, 1991.

44. **Said, A. A., Sheik-Bahae, M., Hagan, D. J. Wei, T. H., Wang, J., Young, J., and Van Stryland, E. W.,** Determination of Bound and free-carrier nonlinearities in ZnSe, GaAs, CdTe, and ZnTe, *J. Opt. Soc. Am.,* B9, 405, 1992.

45. **Bechtel, J. H. and Smith, W. L.,** Two-photon absorption in semiconductors with picosecond pulses, *Phys. Rev.,* B13, 3515, 1976.

46. **Van Stryland, E. W., Sheik-Bahae, M., Said, A. A., and Hagan, D. J.,** Characterization of nonlinear optical absorption and refraction, *J. Prog. Crystal Growth Characterization,* 27, 279, 1993.

47. **Shirk, J., Pong, R., Bartoli, F., and Snow, A.,** Optical limiter using a lead phthalocyanine, *Appl. Phys. Lett.,* 63, 1880, 1993.

48. **Giuliano C. R. and Hess, L. D.,** Nonlinear absorption of light optical saturation of electronic transitions in organic molecules with high intensity laser radiation, *IEEE J. Quantum Electron.,* QE-3, 338, 1967.

49. **Hagan, D. J., Xia, T., Dogariu, A., Said, A. A., and Van Stryland, E.,** Optimization of reverse saturable absorbers limiters: material requirements and design considerations, *Mat. Res. Soc. Symp. Proc.,* in press.

50. **Xia, T., Dogariu, A., Hagan, D. J., and Van Stryland, E. W.,** Optimization of optical limiting devices based on excited state absorption, *Appl. Opt.,* in press.

51. **McCahon, S. W. and Tutt, L. W.,** U.S. patent 5,080,469, Jan. 14, 1992.

52. **Said, A. A., Wei, T. H., DeSalvo, J. R., Sheik-Bahae, M., Hagan, D. J., and Van Stryland, E. W.,** Self-protecting optical limiters using cascading geometries, *Proc. SPIE,* 1692, 37, 1992.

53. **DeMartino, R. N., Khanarian, G., Leslie, T., Sansone, M., Stamatoff, J., Yoon, H., and Mitchell, R.,** Organic and polymeric materials for nonlinear devices, *Proc. SPIE,* 1105, 2, 1989.

54. **Ozaki, M., Tagawa, A., Hatai, T., Sadohara, Y., Ohmori, Y., and Yoshino, K.,** Nonlinear optical response in ferroelectric liquid crystal, *Mol. Cryst. Liq. Cryst.,* 199, 213, 1991.

55. **Khoo, I. C., Zhou, P., Michael, R., Lindquist, R., and Mansfield, R.,** Optical switching by a dielectric-cladded nematic film, *IEEE J. Quantum Elect.,* 25, 1755, 1989.

56. **Khoo, I. C., Michael, R., and Finn, G.,** Self-phase modulation and optical limiting of a low-power CO_2 laser with a nematic liquid crystal film, *Appl. Phys. Lett.,* 52, 2108, 1988.

57. **Yuan, H., Li, L. and Palffy-Muhoray, P.,** Nonlinear birefringence of nematic liquid crystals, *Proc. SPIE,* 1307, 363, 1990; Palffy-Muhoray, P., Yuan, H., Li, L., Lee, M., DeSalvo, J., Wei, T., Sheik-Bahae, M., Hagan, D., and Van Stryland, E., Measurement of third-order optical nonlineairities of nematic liquid crystals, *Mol. Cryst. Liq. Cryst.,* 207, 291, 1991.

58. **Khoo, I. C. and Shen, Y. R.,** Liquid crystals: nonlinear optical properties and processes, *Opt. Eng.,* 24, 579, 1985.

59. **Khoo, I. C.,** Nonlinear optics of liquid crystals, in *Progress in Optics,* Vol. XXVI, Wolf, E., Ed., North Holland, Amsterdam, 1988.

60. **Tabiryan, N. V., Sukhov, A. V., and Ya. Zel'dovich, B.,** The orientational optical nonlinearity of liquid crystals, *Mol. Cryst. Liq. Cryst.,* 136, 1, 1986.

61. **Khoo, I. C., Lindquist, R. G., Michael, R. R., Mansfield, R. J., Zhou, P., and Lopresti, P.,** Picosecond-millisecond optical nonlinearities of liquid crystals for limiting, switching and modulation applications, *Proc. SPIE,* 1307, 336, 1990. See also Khoo, I. C., Lindquist, R. G., Michael, R. R., Mansfield, R. J., and Lopresti, P., Dynamics of picosecond laser induced density, temperature, and flow-reorientation effects in the mesophases of liquid crystals, *J. Appl. Phys.,* 69, 3853, 1991.

62. **Khoo, I. C. and Normandin, R.,** Nanosecond-laser-induced optical wave mixing and ultrasonic wave generation in the nematic phase of liquid crystals, *Opt. Lett.,* 9, 285, 1984.

63. **Khoo, I. C. and Normandin, R.,** Optically-induced molecular reorientation in nematic liquid crystals and nonlinear optical processes in the nanosecond regime, *IEEE J. Quant. Electron.,* 23, 267, 1987.

64. **Khoo, I. C., Michael, R., Mansfield, R., Lindquist, R., Zhou, P., Cipparrone, G., and Simoni, F.,** Experimantal studies of the dynamics and parametric dependences of switching from total internal reflection to transmission and limiting effects, *J. Opt. Soc. Am.,* B8, 1464, 1991.

65. **Soileau, M. J., Van Stryland, E. W., Guha, S., Sharp, E., Wood, G., and Pohlmann, J.,** Nonlinear optical Properties of liquid crystals in the isotropic Phase, *Mol. Cryst. Liq. Cryst.,* 143, 139, 1987.

66. **Yuan, H. J., Li, L., and Palffy-Muhoray, P.,** Z-scan measurements of liquid crystals using top-hat beams, *Proc. SPIE,* 2229, 131, 1994.

67. **Nashold, K. M., Brown, R., Walter, D., and Honey, R. C.,** Temporal and spatial characterization of optical breakdown in a suspension of small absorbing particles, *Proc. SPIE,* 1105, 114, 1989.

68. **Mansour, K., Soileau, M. J., and Van Stryland, E. W.,** Nonlinear optical properties of carbon black suspensions, *J. Opt. Soc. Am.,* B9, 110, 1992.

69. **Mansour, K., Soileau, M. J., and Van Stryland, E. W.,** Optical limiting in media with absorbing microparticles, *Proc. SPIE,* 1105, 91, 1989.

70. **Fein, A., Kotler, Z., Bar-Sagi, J., Jackel, J., Shaier, P., and Zinger, B.,** Nonlinear transmission characteristics of carbon-black suspensions, *Nonlinear Opt.,* 1994.

71. **Lawson, C. M., Euliss, G., and Michael, R.,** Nanosecond laser-induced cavitation in carbon micropartical suspensions: applications in nonlinear interface switching, *Appl. Phys. Lett.,* 58, 2195, 1991.

72. **Brandelik, D., McLean, D., Schmitt, M., Epling, B., Colclasure, C., Tondiglia, V., Pachter, R., Obermeier, K., and Crane, R.,** *Proc. Mat. Res. Soc.,* 247, 361, 1991.

73. **Tutt, L. W. and Kost, A.,** Optical limiting performance of C_{60} and C_{70} solutions, *Nature,* 356, 225, 1992.

74. **Henari, F., Callaghan, J., Stiel, H., Blau, W., and Cardin, D.,** Intensity-dependent absorption and resonant optical nonlinearty of C_{60} and C_{70} solutions, *Chem. Phys. Lett.,* 199, 144, 1992.

75. **McLean, D., Sutherland, R., Brant, M., and Brandelik, D.,** Nonlinear absorption study of a C_{60}-toluene solution, *Opt. Lett.,* 18, 858, 1993.

76. **Wray, J., Liu, K., Chen, C., Garrett, W., Payne, M., Goedert, R., and Templeton, D.,** Optical power limiting of fullerenes, *Appl. Phys. Lett.,* 64, 2785, 1994.

77. **Kost, A., Tutt, L., Klein, M. B., Dougherty, T. K., and Elias, W. E.,** Optical limiting with C_{60} in polymethyl methacrylate, *Opt. Lett.,* 18, 334, 1993.

78. **Justus, B. L., Kafafi, Z., and Huston, A.,** Excited-state absorption-enhanced thermal optical limiting in C_{60}, *Opt. Lett.,* 18, 1603, 1993.

79. **Lindle, J. R., Pong, R., Bartoli, F., and Kafafi, Z.,** Nonlinear optical properties of the fullerenes C_{60} and C_{70} at 1.064 μm *Phys. Rev.,* B48, 9447, 1993.

80. **Sheik-Bahae, M., Said, A. A., Hagan, D. J., Soileau, M. J., and Van Stryland, E. W.,** Nonlinear refraction and optical limiting in "Thick" Media, *Opt. Eng.,* 30, 1228, 1990.

81. **Bredas, J. L., Adant, P., Tackx, P., and Persoons, A.,** Third-order nonlinear optical response in organic materials: theoretical and experimental aspects, *Chem. Rev.,* 94, 243, 1994.

82. **Tutt, L. W., and Boggess, T. F.,** A review of optical limiting mechanisms and devices using organics, fullerenes, semiconductors and other materials, *Prog. Quant. Electron.,* 17, 299, 1993.

83. **Hermann, J. A. and Staromlynska, J.,** Trends in optical switches, limitrers and discriminators, *Int. J. Nonlinear Opt. Phys.,* 2, 271, 1993.

84. **Shi, S., Ji, W., Lang, J., and Xin, X.,** New nonlinear optical chromophore: synthesis, structures, and optical limiting effect of transition-metal clusters (n-Bu$_4$N)$_3$[WM$_3$Br$_4$S$_4$ (M=Cu and Ag)], *J. Phys. Chem.,* 98, 3570, 1994.

85. **Shi, S., Ji, W., Tang., H., Lang, J., and Xin, X.,** Synthesis and optical limiting capability of cubane-like mixed metal clusters (n-Bu$_4$N)$_3$[MoAg$_3$BrX$_3$S$_4$ (X=Cl and I)], *J. Am. Chem. Soc.,* 116, 3615, 1994.

86. **Ducharme, S., Scott, J. C., Twieg, R., and Moerner, W.,** Observation of the photorefractive effect in a polymer, *Phys. Rev. Lett.,* 66, 1846, 1991.

87. **Meerholz, K., Volodin, B. L., Sandalphon, Kippelen, B., and Peyghambarian, N.,** A photorefractive polymer with high optical gain and diffraction efficiency near 100%, *Nature,* 371, 497, 1994.

88. **Cronin Golumb, M. and Yariv, A.,** Optical limiters using photorefractive nonlinearities, *J. Appl. Phys.* 57, 4906, 1985.

89. **McCahon, W. W. and Klein, M. B.,** Coherent beam excisors using the photorefractive effect in BaTiO$_3$, *Proc. SPIE,* 1105, 119, 1989.

90. **Klein, M. B. and Dunning, G. J.,** Performance optimization of photorefractive excisors, *Proc. SPIE,* 1692, 73, 1992.

91. **Shultz, J. L., Salamo, G., Sharp, E., Wood, G., Andrson, R., and Neurgaonkar, R. R.,** Enhancing the response time of photorefractive beam fanning, *Proc. SPIE,* 1692, 78, 1992.

INDEX

Electron migration, 491
Electron orbitals, 20
Electrons, 89
Electro-optical polymer modulator, 324
Electro-optic chromophores, 466
Electro-optic coefficient, 310, 320, 327, 484
  temporal decay, 275
Electro-optic coefficients, 95, 249, 304
  acoustic phonons, 393
  clamped, 394
  electronic part, 393
  frequency dependence, 393
  optic phonons, 393
  resonance effects, 393
  upper limits, 399–400
Electro-optic deflectors, 430
Electro-optic effect
  electronic contributions, 399
  measurement
    amplitude modulation technique, 405
    attenuated total reflection technique, 406
    interferometric technique, 402
    pulsed electro-optic measurements, 408
    reflection technique, 406
    second harmonic generation, 409
  poling, 444
Electro-optic materials
  stability, 401
Electro-optic modulation
  bandwidth, 431
  figure-of-merit dispersion, 397
  power requirements, 431
Electro-optic photorefractive grating, 480
Electro-optic poled polymers, 466
Electro-optic polymers, 319, 325, 326
Electro-optic stability, 297
Electro-optic tensor elements, 249, 470
Electrostatic interaction, 359
Electrostatic units, 516
Electrostriction constant, 246
Ellipsoid, 353
Ellipsometric measurements, 504
Ellipsometry, 63–65, 374
ELO, See Epitaxial lift-off
Emeraldine base, 702
eoo interaction, 354
Epitaxial lift-off (ELO), 326
Epoxy resin, 300
Ester groups, 256
Esterification, 630
Etalon devices, 805
Ethanol, 381, 624
Ethenyl, 115
Ethylene, 518
Ethylene derivatives, 188
Ethylene glycol chain (PEG), 376

Excimer lasers, 90
Excitation energy, 517, 532, 559
Excitation intensity, 615
Excited-state absorption, 843
Excited-state properties, 20
Extended Huckel method, 516
Extended Huckel theory, 517
Extinction coefficients, 69
Extraordinary beam, 353

F

Fabrication tolerance, 377
Fabry-Perot etalon, 805
Fabry-Perot interference, 71
Fast absorbers, 818
F-DEANST, 466
Femtosecond interferometry, 606
Ferrocene, 744
Ferrocenyl, 112
Ferrocenyl derivatives, 191–192
Ferroelectric copolymer, 253
Ferroelectric liquid crystalline polymers, 312
Ferroelectric liquid crystalline thiadiazole derivatives, 190
Ferroelectric liquid crystals (FLCs), 236
Ferroelectric polymers, 130, 259, 309, 312
  SHG efficiency, 314
Ferroelectric properties, 246, 309
Ferroelectrics, 465, 468, 616
Fiberoptic systems, 316
Figure-of-merit, 393, 432, 466, 492, 616, 804
Figure-of-merit dispersion, 397–398
Film deposition, 230
Finite field formalism, 552
Finite field technique, 18, 33
First hyperpolarizability, 91, 105, 516
First-order component, 469
First-order Langevin functions, 95
Five-layer waveguide, 369
  conversion efficiency, 376
  efficient doubling, 369
  fabrication tolerance, 369
  normalized field distribution, 370
  overlap integral, 369, 370
  propagation constants, 370
FLC, See Ferroelectric liquid crystals
Fluence, 598, 844
Fluence dependence, 802
Fluence-dependent saturation, 820
Fluorene, 115
Fluorescein-doped boric acid glass, 763
Fluorescent measurements, 606
Fluorinated polydialkyl-fumalate, 376
Fluorinated stilbenes, 40
Fluorinated sulfone, 314